SEPARATION PROCESS PRINCIPLES

SEPARATION PROCESS PRINCIPLES

J. D. Seader

Department of Chemical and Fuels Engineering
University of Utah

Ernest J. Henley

Department of Chemical Engineering
University of Houston

John Wiley & Sons, Inc.

New York/Chichester/Weinheim
Brisbane/Singapore/Toronto

Acquisitions Editor   Wayne Anderson
Marketing Manager   Katherine Hepburn
Production Editor   Tony VenGraitis
Cover Designer   Kenny Beck
Illustration Coordinator   Jaime Perea
Outside Production Coordination   Spectrum Publisher Services, Inc.

This book was set in Times Ten by Bi-Comp, Inc. and printed and bound by Hamilton Printing
The cover was printed by Phoenix Color

This book is printed on acid-free paper. ∞

The paper in this book was manufactured by a mill whose forest management programs include
sustained yield harvesting of its timberlands. Sustained yield harvesting principles ensure that the
numbers of trees cut each year does not exceed the amount of new growth.

ISBN 0-471-58626-9

Printed in the United States of America

10 9 8 7 6

About the Authors

J. D. Seader is Professor of Chemical Engineering at the University of Utah. He received B.S. and M.S. degrees from the University of California at Berkeley and a Ph.D. from the University of Wisconsin. From 1952 to 1959, Seader designed processes for Chevron Research, and from 1959 to 1965, he conducted rocket engine reseach for Rocketdyne. Before joining the faculty at the University of Utah, he was a professor at the University of Idaho. Combined, he has authored or coauthored 102 technical articles, six books, and four patents, and also coauthored the section on distillation in the 6th and 7th editions of *Perry's Chemical Engineers' Handbook.* Seader has been a trustee of CACHE for 26 years, serving as an executive officer from 1980 to 1984. For 20 years, he directed the use and distribution of Monsanto's FLOWTRAN process simulation computer program for various universities. Seader also served as a director of AIChE from 1983 to 1985. In 1983, he presented the 35th Annual Institute Lecture of AIChE, and in 1988, received the Computing in Chemical Engineering Award of the CAST Division of AIChE.

Ernest J. Henley is Professor of Chemical Engineering at the University of Houston. He received his B.S. degree from the University of Delaware and his Dr. Eng. Sci. from Columbia University, where he served as a professor from 1953 to 1959. Henley also has held professorships at the Stevens Institute of Technology, the University of Brazil, Stanford University, Cambridge University, and the City University of New York. He has authored or coauthored 72 technical articles and 12 books, the most recent one being *Probabilistic Risk Management for Scientists and Engineers.* For 17 years, he was a trustee of CACHE, serving as President from 1975 to 1976 and directing the efforts that produced the seven-volume set of "Computer Programs for Chemical Engineering Education" and the five-volume set, "AIChE Modular Instruction." An active consultant, Henley holds nine patents and serves on the Board of Directors of Maxxim Medical, Inc., Procedyne, Inc., Lasermedics, Inc., and Nanodyne, Inc.

Preface

This textbook is intended for use in undergraduate chemical engineering curriculums. The material is suitable for courses in equilibrium-stage processes, stagewise separation processes, mass transfer operations, separation processes, and rate-controlled separations. Those schools that teach a two-semester sequence in equilibrium stages and mass transfer may find that this textbook satisfies all the needs of that sequence. Some schools may find some of the material suitable for graduate courses in separations.

In 1963, E. J. Henley and H. K. Staffin authored a book, entitled *Stagewise Process Design,* that introduced chemical engineering students to nondiffusional aspects of material and energy balances under phase equilibria constraints, using mainly graphical methods. Most of that book was incorporated, in 1981, into a greatly expanded textbook, *Equilibrium-Stage Separation Operations in Chemical Engineering,* by E. J. Henley and J. D. Seader. The objective of this expanded book was to enhance the 1963 book by adding material on the mathematics and science associated with staged calculations as implemented in commercial, steady-state process simulation computer programs, which were becoming widely available and relied on numeric, rather than graphical, methods of solution. Today, the use of simulation programs is taught to undergraduate students in virtually every chemical engineering department. These programs are easy to use, but, to avoid convergence problems and impossible specifications, the user must have a firm understanding of the fundamentals of chemical engineering. Hopefully, the 1981 textbook provided that understanding.

Since publication of our 1981 textbook, interest in the design and simulation of separation operations using mass transfer (rate-based) principles has increased considerably. This resulted from the availability of improved packings for packed columns used in absorption, distillation, and stripping, and also from the development of theory and applications for the less mature separation operations of adsorption, crystallization, and membrane separations. At the same time, batch distillation, for which rigorous, computer-based calculation methods have been developed, has found wider application. Also, greatly improved procedures for the development of separation processes using enhanced distillation (azeotropic, extractive, pressure-swing, and reactive) have been published, and new applications of supercritical-fluid extraction and chromatography have been commercialized. Our 1981 textbook contained little material on these topics; however, this new textbook contains substantial material on these important topics, as well as those covered in the 1981 textbook. Both equilibrium-based and rate-based methods are covered extensively.

This textbook is organized and divided into four parts. Part one, which consists of five chapters, presents introductory concepts. Chapter 1 describes the many ways in which chemical mixtures are separated industrially. Chapter 2 reviews solution thermodynamics, for both equilibrium-based and rate-based approaches to separation operations. This chapter can be omitted and used only for reference if students have completed or are taking concurrently a course in chemical engineering thermodynamics. Chapter 3 covers the basic principles of diffusion and mass transfer required for the rate-based approach to separation operations. The use of phase equilibrium and material-balance equations to solve a wide range of single equilibrium-stage separations is covered in Chapter 4, while Chapter 5 introduces the student to cascades of equilibrium stages.

The remaining three parts of the textbook are organized according to the method of separation. In Part two, separations achieved by phase creation or addition are presented. Chapters 6 through 8 cover absorption and stripping of dilute solutions, binary distillation, and ternary liquid–liquid extraction. Chapters 9 through 11 detail computer-based methods used in simulation programs for vapor–liquid and liquid–liquid separations. Chapter 12 presents new rate-based methods for multicomponent, multistage separations, while Chapter 13 focuses on batch distillation.

Separations by barriers and solid agents are presented in Part three, with membrane separations in Chapter 14 and adsorption, ion exchange, and chromatography in Chapter 15. This first edition of *Separation Process Principles* does not include Part four, which consists of three chapters on separations that involve a solid phase: leaching and washing; crystallization, desublimation, and evaporation; and drying. These chapters may be obtained from the senior author. If there is sufficient demand for these three chapters, they will be included in the next edition of this textbook.

Almost every topic in this textbook is illustrated by a detailed example and is accompanied by at least three homework exercises. There are a total of 157 examples and 538 homework exercises, and solutions to most homework exercises are included in an instructor's manual. In addition, the authors plan to add examples and exercises to a web site located at www.wiley.com/college/seader.

An attempt has been made to present the development of industrial equipment and the accompanying theory for each separation operation, together with pertinent references to the literature, in an historical context. To assist students in gaining a suitable understanding of this descriptive material, the authors have prepared extensive sets of questions for each chapter (available on the Web site).

The authors wish to acknowledge Professors Vincent Van Brunt of the University of South Carolina, William L. Conger of Virginia Polytechnic Institute and State University, William A. Heenan of Texas A&M University–Kingsville, James H. McMicking of Wayne State University, and Ross Taylor of Clarkson University, who provided advice and detailed reviews for many of the chapters. The draft of the manuscript was typed by Anna Zoe Simmons and Christie J. Perry of the University of Utah, with the effort facilitated by Vickie S. Jones. Chemical engineering students at the University of Utah and the University of Houston, who used early drafts of many of the chapters in their undergraduate courses in equilibrium-stage calculations, mass transfer, and separations, also provided valuable suggestions that improved the presentations of this material. Finally, we are indebted to A. Wayne Anderson of John Wiley & Sons, who provided valuable guidance from a publisher's perspective.

J. D. Seader
Ernest J. Henley

Contents

Chapter 5   **Cascades   232**

Chapter 6   **Absorption and Stripping of Dilute Mixtures   270**

Nomenclature

Latin Capital and Lowercase Letters

A constant in equations of state; constant in Margules equation; area for mass transfer; area for heat transfer; area; coefficient in Freundlich equation; absorption factor=L/KV; total area of a tray; frequency factor

A_a active area of a sieve tray

A_b active bubbling area of a tray

A_d downcomer cross-sectional area of a tray

A_{da} area for liquid flow under downcomer

A_h hole area of a sieve tray

A_{ij} binary interaction parameter in van Laar equation

$\overline{A}_{ij}$ binary interaction parameter in Margules two-constant equation

A_j, B_j, C_j, D_j material-balance parameters defined by (10-7) to (10-11)

A_M membrane surface area

A_p pre-exponential (frequency) factor

A_w specific surface area of a particle

a activity; constants in the ideal-gas heat capacity equation; constant in equations of state; interfacial area per unit volume; surface area; characteristic dimension of a solid particle; equivalents exchanged in ion exchange; interfacial area per stage

$\overline{a}$ interfacial area per unit volume of equivalent clear liquid on a tray

a_h specific hydraulic area of packing

a_{mk} group interaction parameter in UNIFAC method

a_v surface area per unit volume

B constant in equations of state, bottoms flow rate; number of binary azeotropes

B^0 rate of nucleation per unit volume of solution

b molar availability function $= h - T_0 s$; constant in equations of state; component flow rate in bottoms; surface perimeter

C general composition variable such as concentration, mass fraction, mole fraction, or volume fraction; number of components; constant; capacity parameter in (6-40); constant in tray liquid holdup expression given by (6-50); rate of production of crystals

C_1 constant in (6-115)

C_2 constant in (6-116)

C_D drag coefficient

C_G constant in (6-122) and Table 6.8

C_L constant in (6-121) and Table 6.8

C_F entrainment flooding factor in Fig. 6.24 and (6-42)

C_h packing constant in Table 6.8

C_o orifice coefficient

C_P specific heat at constant pressure; packing constant in Table 6.8

$C_{P_V}^o$ ideal gas heat capacity at constant pressure

c concentration; constant in the BET equation; speed of light

c^* liquid concentration in equilibrium with gas at its bulk partial pressure

c' concentration in liquid adjacent to a membrane surface

c_m metastable limiting solubility of crystals

c_s humid heat; normal solubility of crystals

c_t total molar concentration

Δc_{limit} limiting supersaturation

D diffusivity; distillate flow rate; amount of distillate; desorbent (purge) flow rate; discrepancy functions in inside-out method of Chapter 10.

D_B bubble diameter

D_E eddy diffusion coefficient in (6-36)

D_e, D_{eff} effective diffusivity

D_H diameter of perforation for a sieve tray

D_i impeller diameter

D_{ij} mutual diffusion coefficient of i in j

D_K Knudsen diffusivity

D_L longitudinal eddy diffusivity

$\overline{D}_N$ arithmetic-mean diameter

D_0 diffusion constant in (3-57)

D_P, D_p effective packing diameter; particle diameter

$\overline{D}_P$ average of apertures of two successive screen sizes

D_s surface diffusivity

$\overline{D}_s$ surface (Sauter) mean diameter

D_T tower or vessel diameter

$\overline{D_v}$ volume-mean diameter

$\overline{D_w}$ mass-mean diameter

d component flow rate in distillate

d_e equivalent drop diameter; pore diameter

d_H hydraulic diameter $= 4r_H$

d_m molecule diameter

d_p droplet or particle diameter; pore diameter

d_{vs} Sauter mean diameter defined by (8-35)

E activation energy; dimensionless concentration change defined in (3-80); extraction factor defined in (4-24); amount or flow rate of extract; turbulent diffusion coefficient; voltage; wave energy; evaporation rate

E^0 standard electrical potential

E_b radiant energy emitted by a black body

E_D activation energy of diffusion in a polymer

$E_{i,j}$ residual of equilibrium equation (10-2)

E_{MD} fractional Murphree dispersed-phase efficiency

E_{MV} fractional Murphree vapor efficiency

E_{OV} fractional Murphree vapor point efficiency

E_o fractional overall stage (tray) efficiency

E_p activation energy

$E_{\lambda,b}$ radiant energy of a given wavelength emitted by a black body

$E\{t\}dt$ fraction of effluent with a residence time between t and $t + dt$

$\mathscr{E}$ number of independent equations in Gibbs phase rule

ΔE^{vap} molar internal energy of vaporization

e entrainment rate; heat transfer rate across a phase boundary

F Faraday's constant $= 96{,}500$ coulomb/g-equivalent; feed flow rate; force; F-factor defined below (6-63)

F_b buoyancy force

F_d drag force

F_F foaming factor in (6-42)

F_g gravitational force

F_{HA} hole-area factor in (6-42)

F_{LV}, F_{LG} kinetic-energy ratio defined in Fig. 6.24

F_P Packing factor in Table 6.8

F_{ST} surface tension factor in (6-42)

F_V solids volumetric velocity in volume per unit cross-sectional area per unit time

$F\{t\}$ fraction of eddies with a contact time less than t

$\mathscr{F}$ number of degrees of freedom

f pure-component fugacity; Fanning friction factor; function; component flow rate in feed; residual

f_f fraction of flooding velocity

f_i fugacity of component i in a mixture

f_v volume shape factor

$\overline{f}$ partial fugacity

f_ω function of the acentric factor in the S-R-K and P-R equations

G Gibbs free energy; amount or flow rate of gas; mass velocity; volumetric holdup on a tray; rate of growth of crystal size

G' solute-free gas molar flow rate

G_{ij} binary interaction parameter in NRTL equation

g molar Gibbs free energy; acceleration due to gravity

g_c universal constant $= 32.16$ lbm $\cdot$ ft/lbf $\cdot$ s^2

g_{ij} energy of interaction in NRTL equation

H Henry's law coefficient defined in Table 2.3; Henry's law constant defined in (3-56); height or length of vessel; molar enthalpy

$\overline{H}$ partial molar enthalpy

H' Henry's law coefficient defined by (6-110)

H_j residual of energy balance equation (10-5)

ΔH_{ads} heat of adsorption

ΔH_{cond} heat of condensation

ΔH_{crys} heat of crystallization

ΔH_{dil} heat of dilution

ΔH_{sol}^{sat} integral heat of solution at saturation

ΔH_{sol}^{∞} heat of solution at infinite dilution

ΔH^{vap} molar enthalpy of vaporization

H_G height of a transfer unit for the gas phase $= l_T/N_G$

H_i distance of impeller above tank bottom

H_L height of a transfer unit for the liquid phase $= l_T/N_L$

H_{OG} height of an overall transfer unit based on the gas phase $= l_T/N_{OG}$

H_{OL} height of an overall transfer unit based on the liquid phase $= l_T/N_{OL}$

$\mathscr{H}$ humidity

$\mathscr{H}_m$ molar humidity

$\mathscr{H}_P$ percentage humidity

$\mathscr{H}_R$ relative humidity

$\mathscr{H}_s$ saturation humidity

$\mathscr{H}_w$ saturation humidity at temperature T_w

HETP   height equivalent to a theoretical plate

HETS   height equivalent to a theoretical stage (same as HETP)

HTU   height of a transfer unit

h	molar enthalpy; heat-transfer coefficient; specific enthalpy; liquid molar enthalpy; height of a channel; height; Planck's constant = 6.626 $\times$ 10^{-34} J $\cdot$ s/molecule
h_d	dry tray pressure drop as head of liquid
h_{da}	head loss for liquid flow under downcomer
h_{dc}	clear liquid head in downcomer
h_{df}	height of froth in downcomer
h_f	height of froth on tray
h_l	equivalent head of clear liquid on tray
h_L	specific liquid holdup in a packed column
h_t	total tray pressure drop as head of liquid
h_w	weir height
h_σ	pressure drop due to surface tension as head of liquid
I	electrical current
i	current density
J_i	molar flux of i by ordinary molecular diffusion relative to the molar average velocity of the mixture
j_D	Chilton–Colburn j-factor for mass transfer $\equiv N_{St_M}(N_{Sc})^{2/3}$
j_H	Chilton–Colburn j-factor for heat transfer $\equiv N_{St}(N_{Pr})^{2/3}$
j_M	Chilton–Colburn j-factor for momentum transfer $\equiv f/2$
K	equilibrium ratio for vapor–liquid equilibria; equilibrium partition coefficient in (3-59) and for a component distributed between a fluid and a membrane; overall mass-transfer coefficient; adsorption equilibrium constant
K'	overall mass-transfer coefficient for UM diffusion
K_a	chemical equilibrium constant based on activities
K_c	solubility product; overall mass-transfer coefficient for crystallization
K_D	equilibrium ratio for liquid–liquid equilibria
K'_D	equilibrium ratio in mole- or mass-ratio compositions for liquid–liquid equilibria
K_G	overall mass-transfer coefficient based on the gas phase with a partial pressure driving force
K_{ij}	molar selectivity coefficient in ion exchange
K_L	overall mass-transfer coefficient based on the liquid phase with a concentration driving force
K_s	capacity parameter defined by (6-49)
K_W	wall factor given by (6-102)
K_X	overall mass-transfer coefficient based on the liquid phase with a mole ratio driving force
K_x	overall mass-transfer coefficient based on the liquid phase with a mole-fraction driving force

K_Y	overall mass-transfer coefficient based on the gas phase with a mole ratio driving force
K_y	overall mass-transfer coefficient based on the gas phase with a mole-fraction driving force
K_r	restrictive factor for diffusion in a pore
k	thermal conductivity; mass-transfer coefficient in the absence of the bulk-flow effect
k'	mass-transfer coefficient that takes into account the bulk-flow effect as in (3-218) and (3-219)
k_c	mass-transfer coefficient based on a concentration, c, driving force; thermal conductivity of crystal layer
k_g	mass-transfer coefficient for the gas phase based on a partial pressure, p, driving force
k_{ij}	binary interaction parameter
k_i	mass-transfer coefficient for integration into crystal lattice
k_N	constant
k'_N	constant
k_x	mass-transfer coefficient for the liquid phase based on a mole-fraction driving force
k_y	mass-transfer coefficient for the gas phase based on a mole-fraction driving force
$\overline{L}$	liquid molar flow rate in stripping section
L	liquid; length; height; liquid flow rate; underflow flow rate; crystal size
L'	solute-free liquid molar flow rate; liquid molar flow rate in an intermediate section of a column
L_B	length of adsorption bed
L_e	entry length
L_{pd}	predominant crystal size
L_S	liquid molar flow rate of sidestream
LES	length of equilibrium (spent) section of adsorption bed
LUB	length of unused bed in adsorption
L_w	weir length
l	constant in UNIQUAC and UNIFAC equations; component flow rate in liquid; length
l_{ij}	binary interaction parameter
l_M	membrane thickness
l_T	packed height
M	molecular weight; mixing-point amount or flow rate, molar liquid holdup
M_i	moles of i in batch still
$M_{i,j}$	residual of component material-balance equation (10-1)
M_T	mass of crystals per unit volume of magma
M_t	total mass
m	slope of equilibrium curve; mass flow rate; mass

m_c mass of crystals per unit volume of mother liquor

$\overline{m}_i$ molality of i in solution

m_p mass of adsorbent or particle

m_s mass of solid on a dry basis; solids flow rate

m_v mass evaporated; rate of evaporation

m_x tangent to the vapor–liquid equilibrium line in the region of liquid-film mole fractions as in Fig. 3.22

m_y tangent to the vapor–liquid equilibrium line in the region of gas-film mole fractions as in Fig. 3.22

MTZ length of mass-transfer zone in adsorption bed

N number of phases; number of moles; molar flux $= n/A$; number of equilibrium (theoretical, perfect) stages; rate of rotation; number of transfer units; cumulative number of crystals of size, L, and smaller; number of stable nodes; molar flow rate

N_A number of additional variables; Avogadro's number $= 6.023 \times 10^{23}$ molecules/mol

N_a number of actual trays

N_{Bi} Biot number for heat transfer

N_{Bi_M} Biot number for mass transfer

N_D number of degrees of freedom

N_E number of independent equations

N_{Eo} Eotvos number defined by (8-49)

N_{Fo} Fourier number for heat transfer $= \alpha t/a^2 =$ dimensionless time

N_{Fo_M} Fourier number for mass transfer $= Dt/a^2 =$ dimensionless time

N_{Fr} Froude number $=$ inertial force/gravitational force

N_G number of gas-phase transfer units defined in Table 6.7

N_L number of liquid-phase transfer units defined in Table 6.7

N_{Le} Lewis number $= N_{Sc}/N_{Pr}$

N_{Lu} Luikov number

N_{min} mininum number of stages for specified split

N_{Nu} Nusselt number $= dh/k =$ temperature gradient at wall or interface/temperature gradient across fluid ($d =$ characteristic length)

N_{OG} number of overall gas-phase transfer units defined in Table 6.7

N_{OL} number of overall liquid-phase transfer units defined in Table 6.7

N_{Pe} Peclet number for heat transfer $= N_{Re} N_{Pr} =$ convective transport to molecular transfer

N_{Pe_M} Peclet number for mass transfer $= N_{Re} N_{Sc} =$ convective transport to molecular transfer

N_{po} Power number defined in (8-21)

N_{Pr} Prandtl number $= C_P \mu/k =$ momentum diffusivity/thermal diffusivity

N_R number of redundant equations

N_{Re} Reynolds number $= du\rho/\mu =$ inertial force/ viscous force ($d =$ characteristic length)

NRX number of reactions

N_{Sc} Schmidt number $= \mu/\rho D =$ momentum diffusivity/mass diffusivity

N_{Sh} Sherwood number $= dk_c/D =$ concentration gradient at wall or interface/concentration gradient across fluid ($d =$ characteristic length)

N_{St} Stanton number for heat transfer $= h/GC_P$

N_{St_M} Stanton number for mass transfer $= k_c \rho/G$

NTU number of transfer units

N_T total number of crystals per unit volume of mother liquor; number of transfer units for heat transfer

N_t number of equilibrium (theoretical) stages

N_V number of variables

N_{We} Weber number defined by (8-37)

$\mathcal{N}$ number of moles

n molar flow rate; moles; constant in Freundlich equation; number of pores per cross-sectional area of membrane; number of crystals per unit size per unit volume

n_c number of crystals per unit volume of mother liquor

n^0 initial value for number of crystals per unit size per unit volume

n_+, n_- valences of cation and anion, respectively

P pressure; power; electrical power

P', P'' difference points

$\mathscr{P}$ parachor; number of phases in Gibbs phase rule

P_c critical pressure

P_M permeability

$\overline{P}_M$ permeance

P_r reduced pressure, P/P_c

P^s vapor pressure

P_P^s vapor pressure in a pore

P_0 adsorbate vapor pressure at test conditions

p partial pressure

p^* partial pressure in equilibrium with liquid at its bulk concentration

p_j, q_j material-balance parameters for Thomas algorithm in Chapter 10

Q rate of heat transfer across a boundary; volume of liquid; volumetric flow rate

Q_C rate of heat transfer from condenser

Q_L volumetric liquid flow rate

Q_{ML} volumetric flow rate of mother liquor

Q_R rate of heat transfer to reboiler

Q_k area parameter for functional group k in UNIFAC method

q relative surface area of a molecule in UNIQUAC and UNIFAC equations; heat flux; loading or concentration of adsorbate on adsorbent; feed condition in distillation defined as the ratio of increase in liquid molar flow rate across feed stage to molar feed rate

$\bar{q}$ volume-average adsorbate loading defined for a spherical particle by (15-103)

q^e surface excess in liquid adsorption

q_L liquid flow rate across a tray

R universal gas constant:
1.987 cal/mol · K or Btu/lbmol · °F
8314 J/kgmol · K or Pa · m³/kmol · K
82.06 atm · cm³/mol · K
0.7302 atm · ft³/lbmol · °R
10.73 psia · ft³/lbmol · °R;
molecule radius; amount or flow rate of raffinate; ratio of solvent to insoluble solids; reflux ratio; drying-rate flux; inverted binary mass-transfer coefficients defined by (12-31) and (12-32)

R' drying-rate per unit mass of bone-dry solid

R_c drying-rate flux in the constant-rate period

R_f drying-rate flux in the falling-rate period

R_k volume parameter for functional group k in UNIFAC method

R_L liquid-phase withdrawal factor in (10-80)

$R_{\min}$ minimum reflux ratio for specified split

R_p particle radius

R_V vapor-rate withdrawal factor in (10-81)

r relative number of segments per molecule in UNIQUAC and UNIFAC equations; radius; ratio of permeate to feed pressure for a membrane; distance in direction of diffusion; reaction rate; fraction of a stream exiting a stage that is removed as a sidestream; molar rate of mass transfer per unit volume of packed bed

r_c radius at reaction interface

r_H hydraulic radius = flow cross section/wetted perimeter

r_p pore radius

r_s radius at surface of particle

r_w radius at tube wall

S solid; rate of entropy; total entropy; solubility equal to H in (3-56); cross-sectional area for flow; solvent flow rate; mass of adsorbent; stripping factor = KV/L; surface area; inert solid flow rate; flow rate of crystals; supersaturation; belt speed; number of saddles

S_{ij} separation factor in ion exchange

S_g surface area per unit volume of a porous particle

S_x residual of liquid-phase mole-fraction summation equation (10-3)

S_y residual of vapor-phase mole-fraction summation equation (10-4)

s molar entropy; fractional rate of surface renewal; relative supersaturation

s_p particle external surface area

SF   split fraction defined by (1-2)

SP   separation power or relative split ratio defined by (1-4); salt passage defined by (14-70)

SR   split ratio defined by (1-3)

T temperature

T_c critical temperature

T_g glass-transition temperature for a polymer

T_{ij} binary interaction parameter in UNIQUAC and UNIFAC equations

T_m melting temperature for a polymer

T_0 datum temperature for enthalpy; reference temperature; infinite source or sink temperature

T_r reduced temperature = T/T_c

T_s source or sink temperature

T_v moisture evaporation temperature

t time; residence time

$\bar{t}$ average residence time

t_b time to breakthrough in adsorption

t_c contact time in the penetration theory

t_E elution time in chromatography

t_F feed pulse time in chromatography

t_L contact time of liquid in penetration theory; residence time of crystals to reach size L

t_{res} residence time

U superficial velocity; overall heat-transfer coefficient; liquid sidestream molar flow rate; reciprocal of extraction factor

U_a superficial vapor velocity based on tray active bubbling area

U_f flooding velocity

u velocity; interstitial velocity

$\bar{u}$ bulk-average velocity; flow-average velocity

$\overline{u_r}$ relative or slip velocity

u_{all} allowable velocity

u_c velocity of concentration wave in adsorption

u_G gas velocity

u_{ij} energy of interaction in UNIQUAC equation

u_L superficial liquid velocity

u_{mf} minimum fluidization velocity

u_o	hole velocity for sieve tray; superficial gas velocity in a packed column
u_s	superficial velocity
u_0	characteristic rise velocity of a droplet
V	vapor; volume; vapor flow rate; overflow flow rate
V'	vapor molar flow rate in an intermediate section of a column
V_B	boilup ratio
V_H	holdup as a fraction of dryer volume
V_{LH}	volumetric liquid holdup
V_{ML}	volume of mother liquor in magma
V_p	pore volume per unit mass of particle
V_V	volume of a vessel
$\overline{V}$	vapor molar flow rate in stripping section
$\mathcal{V}$	number of variables in Gibbs phase rule
v	molar volume; velocity; component flow rate in vapor; volume of gas adsorbed
$\overline{v}$	average molecule velocity
v_i	species velocity relative to stationary coordinates
v_{i_D}	species diffusion velocity relative to the molar average velocity of the mixture
v_c	critical molar volume
v_H	humid volume
v_M	molar average velocity of a mixture
v_p	particle volume
v_r	reduced molar volume, v/v_c
v_s	molar volume of crystals
v_0	superficial velocity
Σ_v	summation of atomic and structural diffusion volumes in (3-36)
W	rate of work; width of film; bottoms flow rate; amount of adsorbate; washing factor in leaching $= S/RF_A$; baffle width; moles of liquid in a batch still; moisture content on a wet basis; vapor sidestream molar flow rate; weir length
$W_{\min}$	minimum work of separation
WES	weight of equilibrium (spent) section of adsorption bed
WUB	weight of unused adsorption bed
W_s	rate of shaft work
w	mass fraction; width of a channel; weighting function in (10-90)
X	mole or mass ratio; mass ratio of soluble material to solvent in underflow; moisture content on a dry basis; general variable; parameter in (9-34)
X^*	equilibrium moisture on a dry basis
X_B	bound moisture content on a dry basis
X_c	critical free moisture content on a dry basis
X_T	total moisture content on a dry basis
X_i	mass of solute per volume of solid
X_m	mole fraction of functional group m in UNIFAC method
x	mole fraction in liquid phase; mole fraction in any phase; distance; mass fraction in raffinate; mass fraction in underflow; mass fraction of particles
x'	normalized mole fraction $= x_i/\sum_{j=1}^{C} x_j$
$\mathbf{x}$	vector of mole fractions in liquid phase
x_n	fraction of crystals of size smaller than L
Y	mole or mass ratio; mass ratio of soluble material to solvent in overflow; pressure-drop factor for packed columns defined by (6-98); concentration of solute in solvent; parameter in (9-34)
y	mole fraction in vapor phase; distance; mass fraction in extract; mass fraction in overflow
$\mathbf{y}$	vector of mole fractions in vapor phase
Z	compressibility factor $= Pv/RT$; total mass; height
Z_f	froth height on a tray
Z_L	length of liquid flow path across a tray
$\overline{Z}$	lattice coordination number in UNIQUAC and UNIFAC equations
z	mole fraction in any phase; overall mole fraction in combined phases; distance; overall mole fraction in feed; dimensionless crystal size; length of liquid flow path across tray
$\mathbf{z}$	vector of mole fractions in overall mixture

Greek Letters

α	thermal diffusivity, $k/\rho C_P$; relative volatility; surface area per adsorbed molecule
α^*	ideal separation factor for a membrane
α_{ij}	relative volatility of component i with respect to component j for vapor-liquid equilibria; parameter in NRTL equation
$\alpha_j, \beta_j \gamma_j$	energy-balance parameters defined by (10-23) to (10-26)
β_{ij}	relative selectivity of component i with respect to component j for liquid–liquid equilibria
Γ	film flow rate/unit width of film; thermodynamic function defined by (12-37)

Γ_κ residual activity coefficient of functional group k in UNIFAC equation

γ specific heat ratio; activity coefficient

Δ change (final−initial)

δ solubility parameter; film thickness; velocity boundary layer thickness; thickness of the laminar sublayer in the Prandtl analogy

δ_c concentration boundary layer thickness

δ_{ij} Kronecker delta

ϵ exponent parameter in (3-40); fractional porosity; allowable error; tolerance in (10-31)

ϵ_b bed porosity (external void fraction)

ϵ_D eddy diffusivity for diffusion (mass transfer)

ϵ_H eddy diffusivity for heat transfer

ϵ_M eddy diffusivity for momentum transfer

ϵ_p particle porosity (internal void fraction)

η Murphree vapor-phase plate efficiency in (10-73)

θ area fraction in UNIQUAC and UNIFAC equations; dimensionless concentration change defined in (3-80); correction factor in Edmister group method; cut equal to permeate flow rate to feed flow rate for a membrane; contact angle; fractional coverage in Langmuir equation; solids residence time in a dryer; root of the Underwood equation, (9-28)

θ_L average liquid residence time on a tray

κ Maxwell-Stefan mass-transfer coefficient in a binary mixture

Λ_{ij} binary interaction parameter in Wilson equation

λ mV/L; radiation wavelength

λ_+, λ_- limiting ionic conductances of cation and anion, respectively

λ_{ij} energy of interaction in Wilson equation

μ chemical potential or partial molar Gibbs free energy; viscosity

ν momentum diffusivity (kinematic viscosity), μ/ρ; wave frequency; stoichiometric coefficient

$\nu_k^{(i)}$ number of functional groups of kind k in molecule i in UNIFAC method

ξ fractional current efficiency; dimensionless distance in adsorption defined by (15-115); dimensionless warped time in (11-2)

π osmotic pressure; product of ionic concentrations

ρ density

ρ_b bulk density

ρ_M crystal density

ρ_p particle density

ρ_s true (crystalline) solid density

σ surface tension; interfacial tension; Stefan-Boltzmann constant $= 5.670 \times 10^{-8}\,\text{W/m}^2 \cdot \text{K}^4$

σ_I interfacial tension

$\sigma_{s,L}$ interfacial tension between crystal and solution

τ tortuosity; shear stress; dimensionless time in adsorption defined by (15-116); retention time of mother liquor in crystallizer; convergence criterion in (10-32)

τ_{ij} binary interaction parameter in NRTL equation

τ_w shear stress at wall

υ number of ions per molecule

Φ, Φ' volume fraction; parameter in Underwood equations (9-24) and (9-25)

$\overline{\Phi}$ local volume fraction in the Wilson equation

$\phi\{t\}$ probability function in the surface renewal theory

ϕ pure-species fugacity coefficient; association factor in the Wilke-Chang equation; recovery factor in absorption and stripping; volume fraction; concentration ratio defined by (15-125)

$\overline{\phi}$ partial fugacity coefficient

ϕ_{df} froth density

ϕ_e effective relative density of froth defined by (6-48)

ϕ_s particle sphericity

Ψ segment fraction in UNIQUAC equation; V/F in flash calculations; E/F in liquid–liquid equilibria calculations for single-stage extraction; sphericity defined before Example 15.7

Ψ_o dry-packing resistance coefficient given by (6-104)

ψ fractional entrainment; loading ratio defined by (15-126); sphericity

ω acentric factor defined by (2-45); segment fraction in UNIFAC method

Subscripts

A solute

a,ads adsorption

avg average

B bottoms

b bulk conditions; buoyancy

bubble bubble-point condition

C	condenser; carrier; continuous phase
c	critical; convection; constant-rate period
cum	cumulative
D	distillate, dispersed phase; displacement
d	drag; desorption
d,db	dry bulb
des	desorption
dew	dew-point condition
ds	dry solid
E	enriching (absorption) section
e	effective; element
eff	effective
F	feed
f	flooding; feed; falling-rate period
G	gas phase
GM	geometric mean of two values, A and B = square root of A times B
g	gravity
gi	gas in
go	gas out
H,h	heat transfer
I, I	interface condition
i	particular species or component
in	entering
irr	irreversible
j	stage number
k	particular separator; key component
L	liquid phase; leaching stage
LM	log mean of two values, A and B = (A − B)/ ln(A/B)

LP	low pressure
M	mass transfer; mixing-point condition; mixture
m	mixture; maximum
max	maximum
min	minimum
N	stage
n	stage
O	overall
$o,0$	reference condition; initial condition
out	leaving
OV	overhead vapor
P	permeate
R	reboiler; rectification section; retentate
r	reduced; reference component; radiation
res	residence time
S	solid; stripping section; sidestream; solvent; stage; salt
s	source or sink; surface condition; solute; saturation
T	total
t	turbulent contribution
V	vapor
W	batch still
w	wet solid-gas interface
w,wb	wet bulb
ws	wet solid
X	exhausting (stripping) section
x,y,z	directions
δ	at the edge of the laminar sublayer
0	surroundings; initial
∞	infinite dilution; pinch-point zone

Superscripts

E	excess; extract phase
F	feed
ID	ideal mixture
(k)	iteration index
LF	liquid feed
o	pure species; standard state; reference condition
p	particular phase

R	raffinate phase
s	saturation condition
VF	vapor feed
$^{-}$	partial quantity; average value
∞	infinite dilution
$(1), (2)$	denotes which liquid phase
I, II	denotes which liquid phase
$*$	at equilibrium

Abbreviations

Angstrom	1×10^{-10} m
ARD	asymmetric rotating disk contactor
atm	atmosphere
avg	average
BET	Brunauer-Emmett-Teller
BP	bubble-point method
B-W-R	Benedict-Webb-Rubin equation of state
bar	0.9869 atmosphere or 100 kPa
barrer	membrane permeability unit, 1 barrer = 10^{-10} cm³ (STP) · cm/(cm² · s · cm Hg)
bbl	barrel
Btu	British thermal unit
C_i	paraffin with i carbon atoms
$C_i^=$	olefin with i carbon atoms
C-S	Chao-Seader equation
°C	degrees Celsius, K-273.2
cal	calorie
cfh	cubic feet per hour
cfm	cubic feet per minute
cfs	cubic feet per second
cm	centimeter
cmHg	pressure in centimeters head of mercury
cP	centipoise
cw	cooling water
EMD	equimolar counter diffusion
EOS	equation of state
ESA	energy separating agent
ESS	error sum of squares
eq	equivalents
°F	degrees Fahrenheit, °R-459.7
FUG	Fenske-Underwood-Gilliland
ft	feet
GLC-EOS	group-contribution equation of state
GP	gas permeation
g	gram
gmol	gram-mole
gpd	gallons per day
gph	gallons per hour
gpm	gallons per minute
gps	gallons per second
H	high boiler
HHK	heavier than heavy key component
HK	heavy-key component
Hp	horsepower

h	hour
I	intermediate boiler
in.	inch
J	joule
K	degrees Kelvin
kg	kilogram
kmol	kilogram-mole
L	liter; low boiler
LHS	left-hand side of an equation
LK	light-key component
LLK	lighter than light key component
L-K-P	Lee-Kessler-Plocker equation of state
LM	log mean
LW	lost work
lb	pound
lbf	pound-force
lbm	pound-mass
lbmol	pound-mole
M	molar
MSMPR	mixed-suspension, mixed-product removal
MSC	molecular-sieve carbon
MSA	mass separating agent
MW	megawatts
m	meter
meq	milliequivalents
mg	milligram
min	minute
mm	millimeter
mmHg	pressure in mm head of mercury
mmol	millimole (0.001 mole)
mol	gram-mole
mole	gram-mole
N	newton; normal
NLE	nonlinear equation
NRTL	nonrandom, two-liquid theory
nbp	normal boiling point
ODE	ordinary differential equation
PDE	partial differential equation
POD	Podbielniak extractor
P-R	Peng–Robinson equation of state
ppm	parts per million (usually by weight)
PSA	pressure-swing adsorption
psi	pounds force per square inch
psia	pounds force per square inch absolute

PV   pervaporation

RDC   rotating-disk contactor

RHS   right-hand side of an equation

R-K   Redlich-Kwong equation of state

R-K-S   Redlich-Kwong-Soave equation of state (same as S-R-K)

RO   reverse osmosis

RTL   raining-bucket contactor

°R   degrees Rankine

SC   simultaneous-correction method

SG   silica gel

S.G.   specific gravity

SR   stiffness ratio; sum-rates method

S-R-K   Soave-Redlich-Kwong equation of state

STP   standard conditions of temperature and pressure (usually 1 atm and either 0°C or 60°F)

s   second

scf   standard cubic feet

scfd   standard cubic feet per day

scfh   standard cubic feet per hour

scfm   standard cubic feet per minute

stm   steam

TSA   temperature-swing adsorption

UMD   unimolecular diffusion

UNIFAC UNIQUAC functional group activity coefficients

UNIQUAC universal quasi-chemical theory

VOC   volatile organic compound

VPE   vibrating-plate extractor

vs   versus

VSA   vacuum-swing adsorption

wt   weight

y   year

yr   year

μm   micron = micrometer

Mathematical symbols

d differential

e   exponential function

erf(x)   error function of $x - \dfrac{1}{\sqrt{\pi}} \int_0^x \exp(-\eta^2)d\eta$

erfc(x) complementary error function of $x = 1 - \text{erf}(x)$

exp   exponential function

f function

i   imaginary part of a complex value

ln   natural logarithm

log   logarithm to the base 10

∂ partial differential

{ }   delimiters for a function

‖   delimiters for absolute value

Σ sum

π product; pi

Chapter 1

Separation Processes

The separation of chemical mixtures into their constituents has been practiced, as an art, for millennia. Early civilizations developed techniques to (1) extract metals from ores, perfumes from flowers, dyes from plants, and potash from the ashes of burnt plants, (2) evaporate sea water to obtain salt, (3) refine rock asphalt, and (4) distill liquors. The human body could not function for long if it had no kidney, a membrane that selectively removes water and waste products of metabolism from blood.

Separations, including enrichment, concentration, purification, refining, and isolation, are important to chemists and chemical engineers. The former use *analytical separation methods,* such as chromatography, to determine compositions of complex mixtures quantitatively. Chemists also use small-scale *preparative separation techniques,* often similar to analytical separation methods, to recover and purify chemicals. Chemical engineers are more concerned with the manufacture of chemicals using economical, large-scale separation methods, which may differ considerably from laboratory techniques. For example, in a laboratory, chemists separate and analyze light-hydrocarbon mixtures by gas-liquid chromatography, while in a large manufacturing plant a chemical engineer uses distillation to separate the same hydrocarbon mixtures.

This book presents the principles of large-scale component separation operations, with emphasis on methods applied by chemical engineers to produce useful chemical products economically. Included are treatments of classical separation methods, such as distillation, absorption, and liquid-liquid extraction, as well as newer methods, such as adsorption and membrane separation. Separation operations for all three phases (gas, liquid, and solid) are covered. Knowing the principles of separation operations, chemical engineers can successfully develop, design, and operate industrial processes.

1.1 INDUSTRIAL CHEMICAL PROCESSES

The chemical industry manufactures products that differ in chemical content from process feeds, which can be (1) naturally occurring raw materials, (2) plant or animal matter, (3) chemical intermediates, (4) chemicals of commerce, or (5) wastes products. Especially common are oil refineries [1], which, as indicated in Figure 1.1, produce a variety of useful products. The relative amounts of these products produced from, say, 150,000 bbl/day of crude oil depend on the constituents of the crude oil and the types of refinery processes. Typical processes are distillation to separate the crude oil into various boiling-point fractions

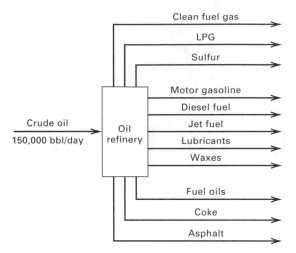

Figure 1.1 Refinery for converting crude oil into a variety of marketable products.

or cuts, alkylation to combine small hydrocarbon molecules into larger molecules, catalytic reforming to change the structure of medium-size hydrocarbon molecules, fluid catalytic cracking to break apart large hydrocarbon molecules, hydrocracking to break apart even larger molecules, and other processes to convert the crude-oil residue to coke and lighter fractions.

A chemical process involves different modes of operation, conducted in either a *batchwise, continuous,* or *semicontinuous* manner. The operations may be classified as either *key operations,* which are unique to chemical engineering because they involve changes in chemical composition, or *auxiliary operations,* which are necessary to the success of the key operations but may be designed by mechanical engineers as well because the auxiliary operations do not involve changes in chemical composition. The key operations are (1) chemical reaction and (2) separation of a mixture of chemicals. The auxiliary operations include separation of phases, heat addition or removal (to change temperature or phase condition), shaft-work addition or removal (to change pressure), mixing or dividing of streams or batches of material, solids agglomeration, size reduction of solids, and separation of solids by size.

The key operations for the separation of chemical mixtures into new mixtures and/or essentially pure components are of central importance in the chemical industry. Most of the equipment in the average chemical plant is there to purify raw materials, intermediates, and products by the separation techniques described qualitatively in this chapter and discussed in detail in subsequent chapters.

Block flow diagrams are used to represent chemical processes. These diagrams indicate, by square or rectangular blocks, the two key processing steps of chemical reaction and separation and, by connecting lines, the major process streams that flow from one processing step to another. Considerably more detail is shown in *process flow diagrams,* which also include auxiliary operations and utilize symbols that depict more realistically the type of equipment employed. The block flow diagram of a continuous process for manufacturing hydrogen chloride gas from evaporated chlorine and electrolytic hydrogen [2] is shown in Figure 1.2. The heart of the process is a chemical reactor, where the high-temperature gas-phase combustion reaction, $H_2 + Cl_2 \rightarrow 2HCl$, occurs. The only auxiliary equipment required consists of pumps and compressors to deliver feeds to the reactor and product to storage, and a heat exchanger to cool the product. For this process, no separation operations are necessary because complete conversion of chlorine occurs in the reactor.

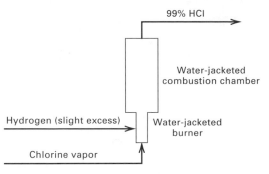

Figure 1.2 Synthetic process for anhydrous HCl production.

A slight excess of hydrogen is used and the product, consisting of 99% HCl and small amounts of H_2, N_2, H_2O, CO, and CO_2, requires no purification. Such simple commercial processes that require no separation of chemical species are very rare.

Some industrial chemical processes involve no chemical reactions, but only operations for separating chemicals and phases, together with auxiliary equipment. A block flow diagram for such a process is shown in Figure 1.3, where wet natural gas is continuously separated into six light-paraffin hydrocarbon components and mixtures by a train of separators [3]. A train or sequence of separators is used because it is rarely economical and often impossible to produce more than two products with a single piece of separation equipment.

Many industrial chemical processes involve at least one chemical reactor accompanied by one or more separation trains. An example is the continuous direct hydration of ethylene to ethyl alcohol [4]. The heart of the process is a reactor packed with solid catalyst particles, operating at 572 K and 6.72 MPa (570°F and 975 psia), in which the hydration reaction, $C_2H_4 + H_2O \rightarrow C_2H_5OH$, takes place. Because of thermodynamic equilibrium limitations,

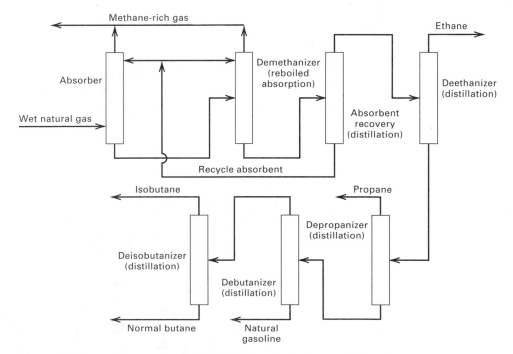

Figure 1.3 Process for recovery of light hydrocarbons from casinghead gas.

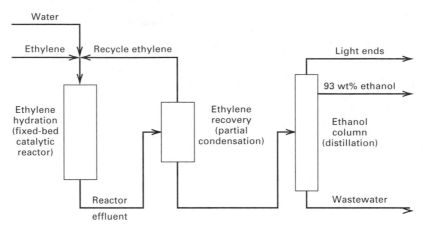

Figure 1.4 Hypothetical process for hydration of ethylene to ethanol.

the conversion of ethylene is only 5% per pass through the reactor. The unreacted ethylene is recovered in a separation step and recycled back to the reactor. By this recycle technique, which is common to many industrial processes, essentially complete conversion of the ethylene fed to the process is achieved. If pure ethylene were available as a feedstock and no side reactions occurred, the relatively simple process in Figure 1.4 could be constructed, in which two by-products (light ends and waste water) are also produced. This process uses a reactor, a partial condenser for ethylene recovery, and distillation to produce aqueous ethyl alcohol of near-azeotropic composition (93 wt%). Unfortunately, a number of factors frequently combine to increase greatly the complexity of the process, particularly with respect to separation equipment requirements. These factors include impurities in the ethylene feed, and side reactions involving both ethylene and feed impurities such as propylene. Consequently, the separation system must also deal with diethyl ether, isopropyl alcohol, acetaldehyde, and other chemicals. The resulting industrial process, shown in Figure 1.5, is much more complicated. After the hydration reaction, a partial condenser and high-pressure water absorber recover unreacted ethylene for recycling. The pressure of the liquid from the bottom of the absorber is reduced, causing partial vaporization. The vapor is separated from the remaining liquid in the low-pressure flash drum. Vapor from the low-pressure flash is scrubbed with water in an absorber to remove alcohol and prevent its loss to the vent gas. Crude concentrated ethanol containing diethyl ether and acetaldehyde is distilled overhead in the crude distillation (stripper) column and then catalytically hydrogenated in the vapor phase to convert acetaldehyde to ethanol. Diethyl ether is removed by distillation in the light-ends tower and scrubbed with water in an absorption tower. The final product is prepared by distillation in the final purification tower, where 93 wt% aqueous ethanol product is withdrawn several trays below the top tray, light ends are concentrated in the so-called pasteurization tray section above the product-withdrawal tray and recycled to the catalytic hydrogenation reactor, and wastewater is removed from the bottom of the tower. Besides the separation equipment shown, additional separation steps may be necessary to concentrate the ethylene feed to the process and remove impurities that poison the catalysts in the reactors. In the development of a new process from the laboratory stage through the pilot-plant stage, experience shows that more separation steps than originally anticipated are usually needed.

The above examples serve to illustrate the importance of separation operations in the majority of industrial chemical processes. Such operations are employed not only to separate a feed mixture into other mixtures and relatively pure components, to recover solvents for recycle, and to remove wastes, but also, when used in conjunction with chemical

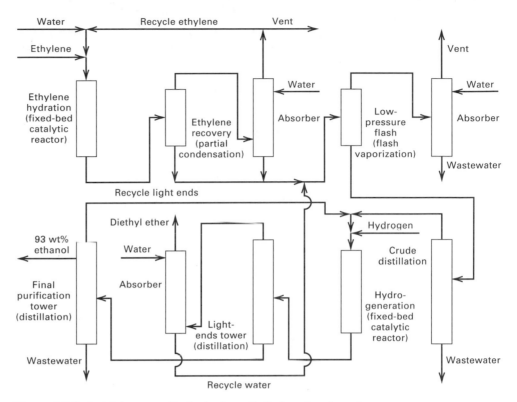

Figure 1.5 Industrial process for hydration of ethylene to ethanol.

reactors, to purify reactor feeds, recover reactants from reactor effluents for recycle, recover by-products, and recover and purify products to meet required specifications. Sometimes a separation operation, such as absorption of SO_2 into limestone slurry, may be accompanied by a chemical reaction that serves to facilitate the separation. In this book, emphasis is on separation operations that do not rely on concurrent chemical reactions; that topic is best covered in chemical reaction engineering, but is given some attention here in Chapter 11.

1.2 MECHANISM OF SEPARATION

The mixing of chemicals to form a mixture is a spontaneous, natural process that is accompanied by an increase in entropy or randomness. The inverse process, the separation of that mixture into its constituent chemical species, is not a spontaneous process; it requires an expenditure of energy. A mixture to be separated usually originates as a single, homogenous phase (solid, liquid, or gas). If it exists as two or more immiscible phases, it is often best to first use some mechanical means based on gravity, centrifugal force, pressure reduction, or an electric and/or magnetic field to separate the phases. Then, appropriate separation techniques are applied to each phase.

A schematic diagram of a general separation process is shown in Figure 1.6. The feed mixture can be vapor, liquid, or solid, while the two or more products may differ in composition from each other and the feed and may differ in phase state from each other and/or from the feed. The separation is accomplished by forcing the different chemical species (components) in the feed into different spatial locations by any of five general separation techniques, or combinations thereof, as shown in Figure 1.7. The most common industrial technique, Figure 1.7a, involves the creation of a second phase (vapor, liquid,

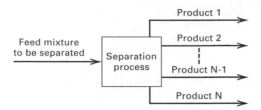

Figure 1.6 General separation process.

or solid) that is immiscible with the feed phase. The creation is accomplished by energy (heat and/or shaft-work) transfer to or from the process or by pressure reduction. A second technique, Figure 1.7b, is to introduce the second phase into the system in the form of a solvent that selectively dissolves some of the species in the feed mixture. Less common, but of growing importance, is the use of a barrier, Figure 1.7c, which restricts and/or enhances the movement of certain chemical species with respect to other species. Also of growing importance are techniques that involve the addition of solid particles, Figure 1.7d, which act directly or as inert carriers for other substances so as to cause separation. Finally, external fields, Figure 1.7e, of various types are sometimes applied for specialized separations.

For all of the general techniques of Figure 1.7, the separations are achieved by enhancing the rate of mass transfer by diffusion of certain species relative to mass transfer of all species by bulk movement within a particular phase. The driving force and direction of mass transfer by diffusion is governed by thermodynamics, with the usual limitations of equilibrium. Thus, both transport and thermodynamic considerations are crucial in separa-

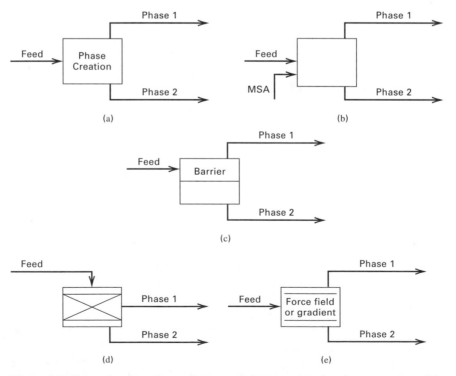

Figure 1.7 General separation techniques: (a) separation by phase creation; (b) separation by phase addition; (c) separation by barrier; (d) separation by solid agent; (e) separation by force field or gradient.

tion operations. The rate of separation is governed by *mass transfer,* while the extent of separation is limited by *thermodynamic equilibrium.* These two topics are treated in Chapters 2 and 3. Fluid mechanics also plays an important role, and applicable principles are included in appropriate chapters, particularly with respect to phase separation, pressure drop, and entrainment.

The extent of separation achieved between or among the product phases for each of the chemical species present in the feed depends on the exploitation of differences in molecular, thermodynamic, and transport properties of the species in the different phases present. Some properties of importance are:

1. Molecular properties

Molecular weight	Polarizability
van der Waals volume	Dielectric constant
van der Waals area	Electric charge
Molecular shape (acentric factor)	Radius of gyration
Dipole moment	

2. Thermodynamic and transport properties

   | Vapor pressure | Adsorptivity |
   | Solubility | Diffusivity |

Values of these properties for many substances are available in handbooks, specialized reference books, and journals. Some of these properties can also be estimated using computer-aided process simulation programs. When they are not available, these properties must be estimated or determined experimentally if a successful application of the appropriate separation operation(s) is to be achieved.

1.3 SEPARATION BY PHASE ADDITION OR CREATION

If the feed mixture is a homogeneous, single-phase solution (gas, liquid, or solid), a second immiscible phase must often be developed or added before separation of chemical species can be achieved. This second phase is created by an *energy-separating agent* (ESA) and/or added as a *mass-separating agent* (MSA). Application of an ESA involves heat transfer and/or transfer of shaft work to or from the mixture to be separated. Alternatively, vapor may be created from a liquid phase by reducing the pressure. An MSA may be partially immiscible with one or more of the species in the mixture. In this case, the MSA frequently remains the constituent of highest concentration in the added phase. Alternatively, the MSA may be completely miscible with a liquid mixture to be separated, but may selectively alter the partitioning of species between liquid and vapor phases. This facilitates a more complete separation when used in conjunction with an ESA, as in extractive distillation.

When two immiscible fluid phases are contacted, intimate mixing of the two phases is important in enhancing mass transfer rates so that the thermodynamic maximum degree of partitioning of species can be approached more rapidly. After phase contact, the separation operation is completed by employing gravity and/or an enhanced technique, such as centrifugal force, to disengage the two phases. Table 1.1 is a compilation of the more common industrial separation operations based on interphase mass transfer between two phases, one of which is created by an ESA or added as an MSA. Graphic symbols that are suitable for process flow diagrams are included in the table. Vapor and liquid and/or solid phases are designated by *V, L,* and *S,* respectively. Design procedures have become fairly routine for the operations prefixed by an asterisk (*) in the first column of Table 1.1. Such procedures have been incorporated as mathematical models into widely used commercial computer-aided chemical process simulation and design (CAPD) programs for continuous, steady-state operations and are treated in considerable detail in subsequent chapters of this book. Batchwise modes of these operations are also treated in this book when appropriate.

Table 1.1 Separation Operations Based on Phase Creation or Addition

Separation Operation	Symbol[a]	Initial or Feed Phase	Created or Added Phase	Separating Agent(s)	Industrial Example[b]
Partial condensation or vaporization* (1)		Vapor and/or liquid	Liquid or vapor	Heat transfer (ESA)	Recovery of H_2 and N_2 from ammonia by partial condensation and high-pressure phase separation (Vol. 2, pp. 494–496)
Flash vaporization* (2)		Liquid	Vapor	Pressure reduction	Recovery of water from seawater (Vol. 24, pp. 343–348)
Distillation* (3)		Vapor and/or liquid	Vapor and liquid	Heat transfer (ESA) and sometimes work transfer	Purification of styrene (Vol. 21, pp. 785–786)
Extractive distillation* (4)		Vapor and/or liquid	Vapor and liquid	Liquid solvent (MSA) and heat transfer (ESA)	Separation of acetone and methanol (Suppl. Vol., pp. 153–155)
Reboiled absorption* (5)		Vapor and/or liquid	Vapor and liquid	Liquid absorbent (MSA) and heat transfer (ESA)	Removal of ethane and lower molecular weight hydrocarbons for LPG production (Vol. 14, pp. 384–385)
Absorption* (6)		Vapor	Liquid	Liquid absorbent (MSA)	Separation of carbon dioxide from combustion products by absorption with aqueous solutions of an ethanolamine (Vol. 4, pp. 730–735)
Stripping* (7)		Liquid	Vapor	Stripping vapor (MSA)	Stream stripping of naphtha, kerosene, and gas oil side cuts from crude distillation unit to remove light ends (Vol. 17, pp. 199–201)

Table 1.1 (*Continued*)

Separation Operation	Symbol[a]	Initial or Feed Phase	Created or Added Phase	Separating Agent(s)	Industrial Example[b]
Refluxed stripping (steam distillation)* (8)		Vapor and/or liquid	Vapor and liquid	Stripping vapor (MSA) and heat transfer (ESA)	Separation of products from delayed coking (Vol. 17, pp. 210–215)
Reboiled stripping* (9)		Liquid	Vapor	Heat transfer (ESA)	Recovery of amine absorbent (Vol. 17, pp. 229–232)
Azeotropic distillation* (10)		Vapor and/or liquid	Vapor and liquid	Liquid entrainer (MSA) and heat transfer (ESA)	Separation of acetic acid from water using *n*-butyl acetate as an entrainer to form an azeotrope with water (Vol. 3, pp. 365–368)
Liquid–liquid extraction* (11)		Liquid	Liquid	Liquid solvent (MSA)	Recovery of aromatics (Vol. 9, pp. 707–709)
Liquid–liquid extraction (two-solvent)* (12)		Liquid	Liquid	Two liquid solvents (MSA_1 and MSA_2)	Use of propane and cresylic acid as solvents to separate paraffins from aromatics and naphthenes (Vol. 17, pp. 223–224)
Drying (13)		Liquid and often solid	Vapor	Gas (MSA) and/or heat transfer (ESA)	Removal of water from polyvinylchloride with hot air in a fluid-bed dryer (Vol. 23, pp. 901–904)

(continues)

9

Table 1.1 (*Continued*)

Separation Operation	Symbol[a]	Initial or Feed Phase	Created or Added Phase	Separating Agent(s)	Industrial Example[b]
Evaporation (14)		Liquid	Vapor	Heat transfer (ESA)	Evaporation of water from a solution of urea and water (Vol. 23, pp. 555–558)
Crystallization (15)		Liquid	Solid (and vapor)	Heat transfer (ESA)	Crystallization of *p*-xylene from a mixture with *m*-xylene (Vol. 24, pp. 718–723)
Desublimation (16)		Vapor	Solid	Heat transfer (ESA)	Recovery of phthalic anhydride from noncondensible gas (Vol. 17, pp. 741–742)
Leaching (liquid-solid extraction) (17)		Solid	Liquid	Liquid solvent	Extraction of sucrose from sugar beets with hot water (Vol. 21, pp. 907–908)
Foam Fractionation (18)		Liquid	Gas	Gas bubbles (MSA)	Recovery of detergents from waste solutions (Vol. 10, pp. 544–545)

* Design procedures are fairly well accepted.

[a] Trays are shown for columns, but alternatively packing can be used. Multiple feeds and side streams are often used and may be added to the symbol.

[b] Citations refer to volume and page(s) of *Kirk-Othmer Encyclopedia of Chemical Technology*, 3rd ed., John Wiley, and Sons, New York (1978–1984).

When the feed mixture includes species that differ widely in their tendency to vaporize and condense, *partial condensation* or *partial vaporization,* Separation Operation (1) in Table 1.1, may be adequate to achieve the desired separation or recovery of a particular component. A vapor feed is partially condensed by removing heat, and a liquid feed is partially vaporized by adding heat. Alternatively, partial vaporization can be caused by *flash vaporization,* Operation (2), by reducing the pressure of the feed with a valve. In both of these operations, after partitioning of species by interphase mass transfer has occurred, the resulting vapor phase is enriched with respect to the species that are most volatile (most easily vaporized), while the liquid phase is enriched with respect to the least volatile species. After this single contact, the two phases, which, except near the critical region, are of considerably different density, are separated, generally by gravity.

Often, the degree of species separation achieved by a single partial vaporization or partial condensation step is inadequate because the volatility differences among species in the feed mixture are not sufficiently large. In that case, it may still be possible to achieve a desired separation of the species in the feed mixture, without introducing an MSA, by employing *distillation,* Operation (3) in Table 1.1, the most widely utilized industrial separation method. Distillation involves multiple contacts between countercurrently flowing liquid and vapor phases. Each contact consists of mixing the two phases to promote rapid partitioning of species by mass transfer, followed by phase separation. The contacts are often made on horizontal trays (referred to as *stages*) arranged in a vertical column as shown in the symbol for distillation in Table 1.1. Vapor, while flowing up the column, is increasingly enriched with respect to the more volatile species. Correspondingly, liquid flowing down the column is increasingly enriched with respect to the less volatile species. Feed to the distillation column enters on a tray somewhere between the top and bottom trays, and often near the middle of the column. The portion of the column above the feed entry is called the *enriching* or *rectification section,* and that below is the *stripping section.* Feed vapor starts up the column; feed liquid starts down. Liquid is required for making contacts with vapor above the feed tray, and vapor is required for making contacts with liquid below the feed tray. Often, vapor from the top of the column is condensed in a condenser by cooling water or a refrigerant to provide contacting liquid, called *reflux.* Similarly, liquid at the bottom of the column passes through a reboiler, where it is heated by condensing steam or some other heating medium to provide contacting vapor, called *boilup.*

When volatility differences between species to be separated are so small as to necessitate more than about 100 trays in a distillation operation, *extractive distillation,* Operation (4), is often considered. Here, an MSA, acting as a solvent, is used to increase volatility differences between selected species of the feed, thereby reducing the number of required trays to a reasonable value. Generally, the MSA, which must be completely miscible with the liquid phase throughout the column, is less volatile than any of the species in the feed mixture and is introduced to a stage near the top of the column. Reflux to the top tray is utilized to minimize MSA content in the top product. A subsequent separation operation, usually distillation, is used to recover the MSA for recycling back to the extractive distillation column.

If condensation of vapor leaving the top of a distillation column is not easily accomplished by heat transfer to cooling water or a refrigerant, a liquid MSA called an *absorbent* may be introduced to the top tray in place of reflux. The resulting separation operation is called *reboiled absorption,* (5). If the feed is all vapor and the stripping section of the column is not needed to achieve the desired separation, the operation is referred to as *absorption,* (6). This operation may not require an ESA and is frequently conducted at ambient temperature and high pressure. Constituents of the vapor feed dissolve in the absorbent to varying extents depending on their solubilities. Vaporization of a small fraction of the absorbent also generally occurs.

The inverse of absorption is *stripping,* Operation (7) in Table 1.1. Here, a liquid mixture is separated, generally at elevated temperature and ambient pressure, by contacting liquid feed with a stripping agent. This MSA eliminates the need to reboil the liquid at the bottom of the column, which may be important if the liquid is not thermally stable. If contacting trays are also needed above the feed tray in order to achieve the desired separation, a *refluxed stripper,* (8), may be employed. If the bottoms product from a stripper is thermally stable, it may be reboiled without using an MSA. In that case, the column is called a *reboiled stripper,* (9). Additional separation operations may be required to recover, for recycling, MSAs used in absorption and stripping operations.

The formation of minimum-boiling azeotropic mixtures makes *azeotropic distillation,* (10), another useful tool in those cases where separation by distillation is not feasible. In the example cited in Table 1.1, the MSA, *n*-butyl acetate, which forms a heterogeneous (i.e., two liquid phases present), minimum-boiling azeotrope with water, is used as an entrainer to facilitate the separation of acetic acid from water. The azeotrope is taken overhead and then condensed and separated into acetate and water layers. The MSA is recirculated and the distillate water layer and bottoms acetic acid are removed as products.

Liquid-liquid extraction, (11) and (12), using one or two solvents, respectively, is widely used when distillation is impractical, especially when the mixture to be separated is temperature-sensitive and/or more than about 100 distillation stages would be required. When one solvent is used, it selectively dissolves only one or a fraction of the components in the feed mixture. In a two-solvent extraction system, each solvent has its own specific selectivity for dissolving the components of the feed mixture. Thus, if a feed mixture consists of species A and B, solvent C might preferentially dissolve A, while solvent D dissolves B. As with extractive distillation, additional separation operations are generally required to recover, for recycling, solvent from streams leaving the extraction operation.

A variation of liquid-liquid extraction is *supercritical-fluid extraction,* where the extraction temperature and pressure are slightly above the critical point of the solvent. In this region, solute solubility in the supercritical fluid changes drastically with small changes in temperature and pressure. Following extraction, the pressure of the extract can be reduced to release the solvent, which is then recycled. For the processing of foodstuffs, the supercritical fluid is an inert substance such as CO_2, which will not contaminate the product.

Since many chemicals are processed wet but sold in dry, solid forms, one of the more common manufacturing steps is a *drying* operation, (13), which involves removal of a liquid from a solid by vaporization of the liquid. Although the only basic requirement in drying is that the vapor pressure of the liquid to be evaporated be higher than its partial pressure in the gas stream, the design and operation of dryers represents a complex problem in heat transfer, fluid mechanics, and mass transfer. In addition to the effect of such external conditions as temperature, humidity, air flow rate, and degree of solid subdivision on drying rate; the effect of internal conditions of liquid and vapor diffusion, capillary flow, equilibrium moisture content, and heat sensitivity in the solid must be considered. Although drying is a multiphase mass transfer process, equipment design procedures differ from those of any of the other processes discussed in this chapter because the thermodynamic concepts of equilibrium are difficult to apply to typical drying situations, where the concentration of vapor in the gas is so far from saturation, and concentration gradients in the solid are such that mass transfer-driving forces are undefined. Also, heat transfer rather than mass transfer may well be the limiting rate process. Therefore, the typical dryer design procedure is for the process engineer to send a few tons of representative, wet sample material for pilot-plant tests by one or two reliable dryer manufacturers and to purchase the equipment that produces a satisfactorily dried product at the lowest cost. The different types of commercial dryers are discussed in detail in *Perry's Chemical Engineers' Handbook* [5].

Evaporation, Operation (14) in Table 1.1, is generally defined as the transfer of volatile components of a liquid into a gas by volatilization caused by heat transfer. Humidification and evaporation are synonymous in the scientific sense; however, *humidification* or *dehumidification* implies that one is intentionally adding vapor to or removing vapor from a gas. Major applications of evaporation are humidification, conditioning of air, cooling of water, and the concentration of aqueous solutions.

Crystallization, (15), is carried out in many organic, and almost all inorganic, chemical manufacturing plants where the desired product is a finely divided solid. Since crystallization is essentially a purification step, the conditions in the crystallizer must be such that impurities do not precipitate with the desired product. In *solution crystallization,* the mixture, which includes a solvent, is cooled and/or the solvent is evaporated to cause crystallization. In *melt crystallization,* two or more soluble species, in the absence of a solvent, are separated by partial freezing. A particularly versatile melt crystallization technique is *zone melting* or *refining,* which relies on selective distribution of impurity solutes between a liquid and a solid phase to achieve a separation. Many metals are refined by this technique, which, in its simplest form, involves moving a molten zone slowly through an ingot by moving the heater or drawing the ingot past the heater.

Sublimation is the transfer of a substance from the solid to the gaseous state without formation of an intermediate liquid phase, usually at a relatively high vacuum. Major applications have been in the removal of a volatile component from an essentially nonvolatile one. Examples are separation of sulfur from impurities, purification of benzoic acid, and freeze-drying of foods. The reverse process, *desublimation,* (16), is also practiced, for example, in the recovery of phthalic anhydride from gaseous reactor effluent. The most common application of sublimation in everyday life occurs in the use of Dry Ice as a refrigerant for storing ice cream, vegetables, and other perishables. The sublimed gas, unlike water, does not puddle and spoil the frozen materials.

Liquid-solid extraction, often referred to as *leaching,* (17), is widely used in the metallurgical, natural product, and food industries under batch, semicontinuous, or continuous operating conditions. The major problem in leaching is to promote diffusion of the solute out of the solid and into the liquid solvent. The most effective way of doing this is to reduce the dimensions of the solid to the smallest feasible particle size. For large-scale applications, in the metallurgical industries in particular, large, open tanks are used in countercurrent operation. The major difference between solid-liquid and liquid-liquid systems centers about the difficulty of transporting the solid, or the solid slurry, from stage to stage. For this reason, the solid may be left in the same tank, with only the liquid transferred from tank to tank. In the pharmaceutical, food, and natural-product industries, countercurrent solid transport is often provided by fairly complicated mechanical devices. A supercritical fluid is sometimes used as the solvent in leaching.

In adsorptive bubble separation methods, surface-active material collects at solution interfaces, establishing a concentration gradient between a solute in the bulk and in the surface layer. If the (very thin) surface layer can be collected, partial solute removal from the solution will have been achieved. The major application of this phenomenon is in ore flotation processes, where solid particles migrate to and attach themselves to rising gas bubbles and literally float out of the solution. This is essentially a three-phase system. *Foam fractionation,* (18), a two-phase adsorptive bubble separation method, is a process where natural or chelate-induced surface activity causes a solute to migrate to rising bubbles and is thus removed as a foam. This method is not covered in this book.

Each equipment symbol shown in Table 1.1 corresponds to the simplest configuration for the operation represented. More complex versions are possible and frequently desirable. For example, a more complex version of the reboiled absorber, Separation Operation (5) in Table 1.1, is shown in Figure 1.8. This reboiled absorber has two feeds, an intercooler, a side stream, and both an interreboiler and a bottoms reboiler. Acceptable design procedures must handle such complex situations. It is also possible to conduct chemical reactions simultaneously with separation operations in a single column. Siirola [6] describes the

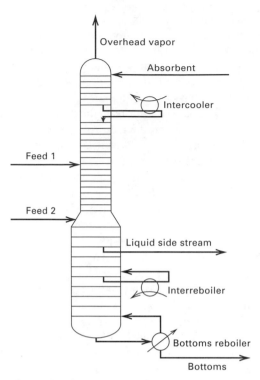

Figure 1.8 Complex reboiled absorber.

evolution of an advanced commercial process for producing methyl acetate by the esterification of methanol and acetic acid. The process is conducted in a single column in an integrated process that involves three reaction zones and three separation zones.

1.4 SEPARATION BY BARRIER

The use of microporous and nonporous membranes as semipermeable barriers for application to difficult and highly selective separations is rapidly gaining adherents in industrial separation processes. Membranes are usually fabricated from natural fibers, synthetic polymers, ceramics, or metals, but they may also consist of liquid films. Solid membranes are fabricated into flat sheets, tubes, hollow fibers, or spiral-wound sheets. For the microporous membranes, separation is effected by differing rates of diffusion through the pores; while for nonporous membranes, separation occurs because of differences in both solubility in the membrane and the rate of diffusion through the membrane.

Table 1.2 lists the more widely used membrane separation operations. *Osmosis,* Operation (1) in Table 1.2, involves the transfer, by a concentration gradient, of a solvent through a membrane into a mixture of solute and solvent. The membrane is almost nonpermeable to the solute. In *reverse osmosis,* (2), transport of solvent in the opposite direction is effected by imposing a pressure, higher than the osmotic pressure, on the feed side. Using a nonporous membrane, reverse osmosis successfully desalts water. *Dialysis,* (3), is the transport, by a concentration gradient, of small solute molecules, sometimes called crystalloids, through a porous membrane. The molecules unable to pass through the membrane are small, insoluble, nondiffusible particles, sometimes referred to as colloids.

Microporous membranes can be used in a manner similar to reverse osmosis to selectively allow small solute molecules and/or solvents to pass through the membrane and to prevent large dissolved molecules and suspended solids from passing through. *Microfiltration,* (4),

Table 1.2 Separation Operations Based on a Barrier

Separation Operation	Symbol[a]	Initial or Feed Phase	Separating Agent	Industrial Example[b]
Osmosis (1)		Liquid	Nonporous membrane	—
Reverse osmosis* (2)		Liquid	Nonporous membrane with pressure gradient	Desalinization of sea water (Vol. 24, pp. 349–353)
Dialysis (3)		Liquid	Porous membrane with pressure gradient	Recovery of caustic from hemicellulose (Vol. 7, p. 572)
Microfiltration (4)		Liquid	Microporous membrane with pressure gradient	Removal of bacteria from drinking water (Vol. 15, p. 115)
Ultrafiltration (5)		Liquid	Microporous membrane with pressure gradient	Separation of whey from cheese (Vol. 15, pp. 562–564)
Pervaporation* (6)		Liquid	Nonporous membrane with pressure gradient	Separation of azeotropic mixtures (Vol. 15, pp. 116–117)
Gas permeation* (7)		Vapor	Nonporous membrane with pressure gradient	Hydrogen enrichment (Vol. 20, pp. 709–710)
Liquid membrane (8)		Vapor and/or liquid	Liquid membrane with pressure gradient	Removal of hydrogen sulfide (Vol. 15, p. 119)

* Design procedures are fairly well accepted.
[a] Single units are shown. Multiple units can be cascaded.
[b] Citations refer to volume and page(s) of *Kirk-Othmer Encyclopedia of Chemical Technology*, 3rd ed., John Wiley and Sons, New York (1978–1984).

refers to the retention of molecules typically in the size range from 0.02 to 10 μm. *Ultrafiltration,* (5), refers to the range from 1 to 20 nm. To retain even smaller molecules, reverse osmosis, sometimes called *hyperfiltration,* can be used down to 0.1 nm.

Although reverse osmosis can be used to separate organic and aqueous-organic liquid mixtures, high pressures are required. Alternatively, *pervaporation,* (6), in which the species being absorbed by and transported through the nonporous membrane are evaporated, can be used. This method, which uses much lower pressures than reverse osmosis, but where the heat of vaporization must be supplied, is used to separate azeotropic mixtures.

The separation of gas mixtures by selective *gas permeation,* (7), through membranes using pressure as the driving force is a relatively simple process, first used in the 1940s with porous fluorocarbon barriers to separate $^{235}UF_6$ and $^{238}UF_6$ at great expense because it required enormous amounts of electric power. More recently, nonporous polymer membranes are being used commercially to enrich gas mixtures containing hydrogen, recover hydrocarbons from gas streams, and produce nitrogen-enriched and oxygen-enriched air.

Liquid membranes, (8), of only a few molecules in thickness can be formed from surfactant-containing mixtures that locate at the interface between two fluid phases. With such a membrane, aromatic hydrocarbons can be separated from paraffinic hydrocarbons. Alternatively, the membrane can be formed by imbibing the micropores with liquids that are doped with additives to facilitate transport of certain solutes, such as CO_2 and H_2S.

1.5 SEPARATION BY SOLID AGENT

Separation operations that use solid mass-separating agents are listed in Table 1.3. The solid, usually in the form of a granular material or packing, acts as an inert support for a thin layer of absorbent or enters directly into the separation operation by selective adsorption of or chemical reaction with certain species in the feed mixture. Adsorption is confined to the surface of the solid adsorbent, unlike absorption, which occurs throughout the bulk of the absorbent. In all cases, the active separating agent eventually becomes saturated with solute and must be regenerated or replaced periodically. Such separations are often conducted batchwise or semicontinuously. However, equipment is available to simulate continuous operation.

Adsorption, Separation Operation (1) in Table 1.3, is used to remove components present in low concentrations in nonadsorbing solvents or gases and to separate the components in gas or liquid mixtures by selective adsorption on solids, followed by desorption to regenerate the adsorbents, which include activated carbon, aluminum oxide, silica gel, and synthetic sodium or calcium aluminosilicate zeolite adsorbents (molecular sieves). The sieves differ from the other adsorbents in that they are crystalline and have pore openings of fixed dimensions, making them very selective. A simple adsorption device consists of a cylindrical vessel packed with a bed of solid adsorbent particles through which the gas or liquid flows. Regeneration of the adsorbent is conducted periodically, so two or more vessels are used, one vessel desorbing while the other(s) adsorb(s). If the vessel is arranged vertically, it is usually advantageous to employ downward flow. With upward flow, jiggling of the bed can cause particle attrition and a resulting increase in pressure drop and loss of material. However, for liquid flow, better distribution is achieved by upward flow. Regeneration is accomplished by one of four methods: (1) vaporizing the adsorbate with a hot purge gas (*thermal-swing adsorption*), (2) reducing the pressure to vaporize the adsorbate (*pressure-swing adsorption*), (3) inert purge stripping without change in temperature or pressure, and (4) displacement desorption by a fluid containing a more strongly adsorbed species.

Chromatography, Separation Operation (2) in Table 1.3, is a method for separating the components of a feed gas or liquid mixture by passing the feed through a bed of packing. The feed may be volatilized into a carrier gas, and the bed may be a solid adsorbent (gas-

Table 1.3 Separation Operations Based on a Solid Agent

Separation Operation	Symbol[a]	Initial or Feed Phase	Separating Agent	Industrial Example[b]
Adsorption* (1)		Vapor or liquid	Solid adsorbent	Purification of *p*-xylene (Vol. 24, pp. 723–725)
Chromatography* (2)		Vapor or liquid	Solid adsorbent or liquid adsorbent on a solid support	Separation of xylene isomers and ethylbenzene (Vol. 24, pp. 726–727)
Ion exchange* (3)		Liquid	Resin with ion-active sites	Demineralization of water (Vol. 13, pp. 700–701)

* Design procedures are fairly well accepted.

[a] Single units are shown. Multiple units can be cascaded.

[b] Citations refer to volume and page(s) of *Kirk-Othmer Encyclopedia of Chemical Technology*, 3rd ed., John Wiley and Sons, New York (1978–1984).

solid chromatography) or a solid inert support that is coated with a very viscous liquid that acts as an absorbent (gas-liquid chromatography). Because of selective adsorption on the solid adsorbent surface or absorption into the liquid absorbent, followed by desorption, the different components of the feed mixture move through the bed at different rates, thus effecting the separation. In *affinity chromatography,* a macromolecule (called a ligate) is selectively adsorbed by a ligand (e.g., an ammonia molecule in a coordination compound) that is covalently bonded to a solid support particle. Ligand–ligate pairs include inhibitors–enzymes, antigens–antibodies, and antibodies–proteins. Chromatography in its various forms is finding use in bioseparations.

Ion exchange, (3), resembles adsorption in that solid particles are used and regeneration is necessary. However, a chemical reaction is involved. In water softening, a typical ion-exchange application, an organic or inorganic polymer in its sodium form removes calcium ions by exchanging calcium for sodium. After prolonged use, the (spent) polymer, which becomes saturated with calcium, is regenerated by contact with a concentrated salt solution.

1.6 SEPARATION BY EXTERNAL FIELD OR GRADIENT

External fields can be used to take advantage of the differing degrees of response of molecules and ions to forces and gradients. Table 1.4 lists the most common techniques, with combinations of these techniques with each other and with previously described separation methods also being possible.

Centrifugation, Operation (1) in Table 1.4, establishes a pressure field that separates fluid mixtures according to molecular weight. This technique is used to separate $^{235}UF_6$ from $^{238}UF_6$, and large polymer molecules according to molecular weight.

If a rather large temperature gradient is applied to a homogeneous solution, concentration gradients can be established and *thermal diffusion,* (2), is induced. It has been used to enhance the separation of uranium isotopes in gas permeation processes.

Natural water contains 0.000149 atom fraction of deuterium. When water is decomposed by *electrolysis,* (3), into hydrogen at the cathode and oxygen at the anode, the deuterium concentration in the hydrogen produced is lower than that in the water. Until 1953, this process was the only commercial source of heavy water (D_2O). In *electrodialysis,* (4), cation- and anion-permeable membranes carry a fixed charge, preventing the migration of species of like charge. This operation can be used to desalinize (remove salts from) sea water. A somewhat related process is *electrophoresis,* (5), which exploits the different migration velocities of charged colloidal or suspended species in an electric field. Positively

Table 1.4 Separation Operations by Applied Field or Gradient

Separation Operation	Initial or Feed Phase	Force Field or Gradient	Industrial Example[a]
Centrifugation (1)	Vapor	Centrifugal force field	Separation of uranium isotopes (Vol. 23, pp. 531–532)
Thermal diffusion (2)	Vapor or liquid	Thermal gradient	Separation of chlorine isotopes (Vol. 7, p. 684)
Electrolysis (3)	Liquid	Electrical force field	Concentration of heavy water (Vol. 7, p. 550)
Electrodialysis (4)	Liquid	Electrical force field and membrane	Desalinization of sea water (Vol. 24, pp. 353–359)
Electrophoresis (5)	Liquid	Electrical force field	Recovery of hemicelluloses (Vol. 4, p. 551)
Field-flow fractionation (6)	Liquid	Laminar flow in force field	

[a] Citations refer to volume and page(s) of *Kirk-Othmer Encyclopedia of Chemical Technology,* 3rd ed., John Wiley and Sons, New York (1978–1984).

charged species, such as dyes, hydroxide sols, and colloids, migrate to the cathode; while most small, suspended, negatively charged particles are attracted to the anode. By changing the solvent from an acidic to a basic condition, migration direction can sometimes be changed, particularly for proteins. Electrophoresis is a highly versatile method for separating biochemicals.

Another separation technique for biochemicals and difficult-to-separate heterogeneous mixtures of micromolecular and colloidal materials is *field flow fractionation*, (6). For the mixture to be separated, an electrical field, magnetic field, or thermal gradient is established in a direction perpendicular to a laminar-flow field. Components of the mixture are driven to different locations in the stream; thus, they travel in the flow direction at different velocities, so a separation is achieved.

1.7 COMPONENT RECOVERIES AND PRODUCT PURITIES

Separation operations are subject to the conservation of mass. Accordingly, if no chemical reactions occur and the process operates in a continuous, steady-state fashion, then for each component, i, in a mixture of C components, the molar (or mass) flow rate in the feed, $n_i^{(F)}$, is equal to the sum of the product molar (or mass) flow rates, $n_i^{(p)}$, for that component in the N product phases, p. Thus, referring to Figure 1.6,

$$n_i^{(F)} = \sum_{p=1}^{N} n_i^{(p)} = n_i^{(1)} + n_i^{(2)} + \cdots + n_i^{(N-1)} + n_i^{(N)} \tag{1-1}$$

To solve (1-1) for values of $n_i^{(p)}$, from specified values of $n_i^{(F)}$, we need an additional $N-1$ independent expressions involving $n_i^{(p)}$. This gives a total of NC equations in NC unknowns. For example, if a feed mixture containing C components is separated into N product phases, $C(N-1)$ additional expressions are needed. General forms of these expressions, which deal with the extent of separation, are considered in this and the next section. If more than one stream is fed to the separation process, $n_i^{(F)}$ is the summation for all feeds.

Equipment for separating components of a mixture is designed and operated to effect desired *component recoveries* and/or *product purities*. In Figure 1.9, the block flow diagram

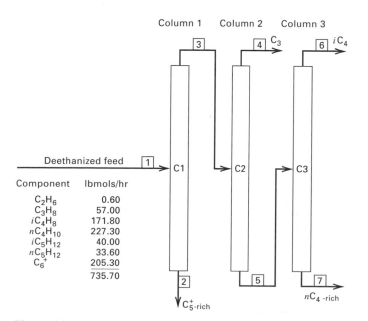

Figure 1.9 Hydrocarbon recovery process.

for a hydrocarbon separation system, the feed is the bottoms product from a reboiled absorber used to deethanize (i.e., remove ethane and components of smaller molecular weight) a mixture of refinery gases and liquids. The separation process of choice in this example is a sequence of three distillation operations, each conducted in a multistage column. The composition of the feed to the process is included in Figure 1.9, where components are rank-listed by decreasing volatility, and hydrocarbons heavier (i.e., of greater molecular weight) than normal pentane and in the hexane (C_6)-to-undecane (C_{11}) range are lumped together in a so-called C_6^+ fraction. The three distillation columns of Figure 1.9 separate the deethanized feed into four products: a C_5^+-rich bottoms, a C_3-rich distillate, an iC_4-rich distillate, and a nC_4-rich bottoms. For each column, each component in the feed is partitioned between the overhead and the bottoms, according to a unique split fraction or split ratio that depends on (1) the component thermodynamic and transport properties in the vapor and liquid phases, (2) the number of contacting stages, and (3) the relative vapor and liquid flows through the column. The *split fraction,* SF, for component i in separator k is

$$SF_{i,k} = \frac{n_{i,k}^{(1)}}{n_{i,k}^{(F)}} \qquad (1\text{-}2)$$

where $n^{(1)}$ and $n^{(F)}$ refer to component molar flow rates in the first product and the feed, respectively. Alternatively, a *split ratio,* SR, may be defined as

$$SR_{i,k} = \frac{n_{i,k}^{(1)}}{n_{i,k}^{(2)}} = \frac{SF_{i,k}}{(1 - SF_{i,k})} \qquad (1\text{-}3)$$

where $n^{(2)}$ refers to a component molar flow rate in the second product. Alternatively, SF and SR can be defined in terms of component mass flow rates.

If the process shown in Figure 1.9 is part of an operating plant with the measured material balance of Table 1.5, the split fractions and split ratios in Table 1.6 are determined from (1-2) and (1-3). In Table 1.5, it is seen that only two of the four products are relatively pure: C_3 overhead from the second column and iC_4 overhead from the third column. The molar purity of C_3 in the C_3 overhead is (54.80/56.00) or 97.86%, while the iC_4 overhead purity is (162.50/175.50) or 92.59% iC_4. The nC_4-rich bottoms from column 3 has a nC_4 purity of only (215.80/270.00) or 79.93%. Each of the three columns is designed to make a split between two adjacent components (called the *key components*) in the list of components ordered in decreasing volatility. As seen in Table 1.5, the three key splits are nC_4H_{10}/iC_5H_{12}, C_3H_8/iC_4H_{10}, and iC_4H_{10}/nC_4H_{10} for Columns 1, 2, and 3, respectively. From the split fractions listed in Table 1.6, it is seen that all splits are relatively sharp (SF >

Table 1.5 Operating Material Balance for Hydrocarbon Recovery Process

	lbmol/h in Stream						
Component	1 Feed to C1	2 C_5^+-rich	3 Feed to C2	4 C_3	5 Feed to C3	6 iC_4	7 nC_4-rich
C_2H_6	0.60	0.00	0.60	0.60	0.00	0.00	0.00
C_3H_8	57.00	0.00	57.00	54.80	2.20	2.20	0.00
iC_4H_{10}	171.80	0.10	171.70	0.60	171.10	162.50	8.60
nC_4H_{10}	227.30	0.70	226.60	0.00	226.60	10.80	215.80
iC_5H_{12}	40.00	11.90	28.10	0.00	28.10	0.00	28.10
nC_5H_{12}	33.60	16.10	17.50	0.00	17.50	0.00	17.50
C_6^+	205.30	205.30	0.00	0.00	0.00	0.00	0.00
Total	735.60	234.10	501.50	56.00	445.50	175.50	270.00

Table 1.6 Split Fractions and Split Ratios for Hydrocarbon Recovery Process

	Column 1		Column 2		Column 3		Overall Percent Recovery
Component	SF	SR	SF	SR	SF	SR	
C_2H_6	1.00	Large	1.00	Large	—	—	100
C_3H_8	1.00	Large	0.9614	24.91	1.00	Large	96.14
iC_4H_{10}	0.9994	1,717	0.0035	0.0035	0.9497	18.90	94.59
nC_4H_{10}	0.9969	323.7	0.00	0.00	0.0477	0.0501	94.94
iC_5H_{12}	0.7025	2.361	0.00	0.00	0.00	0.00	29.75
nC_5H_{12}	0.5208	1.087	0.00	0.00	0.00	0.00	47.92
C_6^+	0.00	Small	—	—	—	—	100

0.95 for the light key and SF < 0.05 for the heavy key), except for column 1, where the split ratio for the heavy key (iC_5H_{12}) is not sharp at all, and ultimately causes the nC_4 rich bottoms to be relatively impure in nC_4, even though the split between the two key components in the third column is relatively sharp.

In Table 1.6, for each column, values of SF and SR decrease in the order of the ranked component list. It is also noted in Table 1.6 that SF may be a better quantitative measure of degree of separation than SR because SF is bounded between 0 and 1, while SR can range from 0 to a very large value.

Two other common measures of extent of separation can be applied to each column, or to the separation system as a whole. One measure is the *percent recovery* in a designated system product of each component in the feed to the system. These values, as computed from the data of Table 1.5, are listed in the last column of Table 1.6. As shown, the component recoveries are all relatively high (>95%) except for the two pentane isomers. The other measure of extent of separation is *product purity*. These purities for the main component were computed for all except the C_5^+-rich product, which is [(11.90 + 16.10 + 205.30)/234.10] or 99.66% pure with respect to the pentanes and heavier. Such a product is a *multicomponent product*. One of the most common multicomponent products is gasoline. Product impurity levels and a designation of the impurities are included in product specifications for chemicals of commerce. The product purity with respect to each component in each of the three final products for the hydrocarbon recovery process, as computed from the process operating data of Table 1.5, is given in Table 1.7, where the values are also extremely important because maximum allowable percentages of impurities are compared to the product specifications. The C_5^+ fraction is not included because it is an intermediate that is sent to an isomerization process. From the comparison in Table 1.7,

Table 1.7 Comparison of Measured Product Purities with Specifications

	mol% in Product					
	Propane		Isobutane		Normal Butane	
Component	Data	Spec	Data	Spec	Data	Spec
C_2H_6	1.07	15 max	0	—	0	—
C_3H_8	97.86	93 min	1.25	3 max	0	1 max
iC_4H_{10}	1.07	2 min	92.60	92 min	83.11	80 min
nC_4H_{10}	0	—	6.15	7 max		
C_5^+	0	—	0	—	16.89	20 max
Total	100.00		100.00		100.00	

it is seen that two products easily meet their specifications, while the iC_4 product just meets its specification. If the process is equipped with effective controllers, it might be possible to reduce the energy input to the process and still meet C_3 and nC_4-rich product specifications.

1.8 SEPARATION POWER

Some of the separation operations in Table 1.1 are often inadequate for making a sharp split between two key components of a feed mixture, and can only effect the desired recovery of a single key component. Examples are Operations 1, 2, 6, 7, 8, 9, 11, 13, 14, 15, 16, and 17 in Table 1.1. For these, either a single separation stage is utilized as in Operations 1, 2, 13, 14, 15, 16, and 17, or the feed enters at one end (not near the middle) of a multistage separator as in Separation Operations 6, 7, 8, 9, and 11. The split ratio, SR, split fraction, SF, recovery, or purity that can be achieved for the key component depends on a number of factors. For the simplest case of a single separation stage, the factors that influence SR and SF values include: (1) the relative molar amounts of the two phases leaving the separator and (2) thermodynamic, mass transport, and other properties of the key components in the mixture. For multistage separators, an additional factor must be added, namely, (3) the number of stages and their configuration. The quantitative relationships involving these factors are unique to each type of separator. Therefore, detailed presentation and discussion of these relationships is deferred to subsequent chapters of this book, where individual separation operations are discussed in some detail. A general but brief discussion of some of the important property factors is given in the next section.

When multistage separators are utilized and the feed mixture enters somewhere near the middle of the separator, such that the separator consists of two sections of stages, one on either side of the feed stage, it is often possible to achieve a relatively sharp separation between two key components. One section acts to remove one key component, while the other section acts to remove the other key component. Examples are Separation Operations 3, 4, 5, 10, and 12 in Table 1.1. For these operations, a convenient measure of the relative degree of separation between two components, i and j, is the *separation power* (also referred to as the *relative split ratio*), SP, of the separation equipment, defined in terms of the component splits achieved, as measured by the compositions of the two products, (1) and (2):

$$\mathrm{SP}_{i,j} = \frac{C_i^{(1)}/C_i^{(2)}}{C_j^{(1)}/C_j^{(2)}} \tag{1-4}$$

where C is some measure of composition such as mole fraction, mass fraction, or concentration in moles or mass per unit volume. Most commonly, mole fractions or concentrations are used, but in any case, the separation power is readily converted to the following forms in terms of split fractions or split ratios:

$$\mathrm{SP}_{i,j} = \frac{\mathrm{SR}_i}{\mathrm{SR}_j} \tag{1-5}$$

$$\mathrm{SP}_{i,j} = \frac{\mathrm{SF}_i/\mathrm{SF}_j}{(1 - \mathrm{SF}_i)/(1 - \mathrm{SF}_j)} \tag{1-6}$$

Achievable values of SP depend on the number of stages and the relative thermodynamic and mass transport properties of components i and j. In general, when applied to the two key components, components i and j and products 1 and 2 are selected so that $\mathrm{SP}_{i,j} > 1.0$. Then, a large value corresponds to a relatively high degree of separation or high separation power; a small value larger than but close to 1.0 corresponds to a low degree of separation power. For example, if $\mathrm{SP} = 10{,}000$ and $\mathrm{SR}_i = 1/\mathrm{SR}_j$, then, from (1-5),

Table 1.8 Main Separation Factors for Hydrocarbon
Recovery Process

Key-Component Split	Column	Separation Factor, SP
nC_4H_{10}/iC_5H_{12}	C1	137.1
C_3H_8/iC_4H_{10}	C2	7103
iC_4H_{10}/nC_4H_{10}	C3	377.6

$SR_i = 100$ and $SR_j = 0.01$, corresponding to a sharp separation. However, if $SP = 9$ and $SR_i = 1/SR_j$, then $SR_i = 3$ and $SR_j = \frac{1}{3}$, corresponding to a nonsharp separation.

For the hydrocarbon recovery process of Figure 1.9, the values of SP in Table 1.8 are computed from the data in Tables 1.5 or 1.6 for the main split in each of the three separators. The separation factor in distillation column C1 is relatively small because the split for the heavy key, iC_5H_{12}, is not sharp. The largest separation factor occurs in column C2, where the separation is relatively easy because of the fairly large volatility difference between the two keys. Much more difficult is the butane-isomer split in column C3, where only a moderately sharp split is achieved.

1.9 SELECTION OF FEASIBLE SEPARATION PROCESSES

The selection of a best separation process must frequently be made from among a number of feasible candidates. When the feed mixture is to be separated into more than two products, a combination of two or more operations may be best. Even when only two products are to be produced, a hybrid process of two or more operations may be most economical.

The important factors in the selection of feasible separation operations are listed in Table 1.9. These factors have to do with feed and product conditions, property differences

Table 1.9 Factors That Influence the Selection of Feasible
Separation Operations

A. Feed conditions
    1. Composition, particularly concentration of species to
        be recovered or separated
    2. Flow rate
    3. Temperature
    4. Pressure
    5. Phase state (solid, liquid, and/or gas)
B. Product conditions
    1. Required purities of products
    2. Temperatures
    3. Pressures
    4. Phase states
C. Property differences that may be exploited
    1. Molecular
    2. Thermodynamic
    3. Transport
D. Characteristics of separation operation
    1. Ease of scale-up
    2. Ease of staging
    3. Temperature, pressure, and phase-state requirements
    4. Physical size limitations
    5. Energy requirements

that can be exploited, and certain characteristics of the candidate separation operations. The most important feed conditions are composition and flow rate, because the other conditions (temperature, pressure, and phase condition) can be altered by pumps, compressors, and heat exchangers to fit the required conditions of a particular candidate separation operation. In general, however, the vaporization of a liquid feed that has a high heat of vaporization, the condensation of a vapor feed with a refrigerant, and/or the compression of a vapor feed can add significantly to the cost. Some separation operations, such as those based on the use of barriers or solid agents, perform best on feeds that are dilute in the species to be recovered. The most important product conditions are the required purities because, again, the other conditions listed can be altered by energy transfer after the separation is achieved.

In general, as demonstrated by Sherwood, Pigford, and Wilke [7] and updated recently by Keller [8], the cost of recovering and purifying a chemical contained in a mixture can depend strongly on the concentration of that chemical in the mixture. Keller's correlation is given in Figure 1.10, where it is seen that the more dilute the chemical is in the mixture, the higher its sales price is.

When very pure products are required, either large differences in certain properties must exist or significant numbers of stages must be provided. It is important to consider both molecular and bulk thermodynamic and transport properties, some of which are listed near the end of Section 1.2. Data and estimation methods for many bulk properties are given by Reid, Prausnitz, and Poling [9] and for both molecular and bulk properties by Daubert and Danner [10].

Some separation operations are well understood and can be readily designed from a mathematical model and/or scaled up to a commercial size from laboratory data. The results of a survey by Keller [8], shown in Figure 1.11, show that the degree to which a separation operation is technologically mature correlates well with its commercial use. Operations based on a barrier are more expensive to stage than those based on the use

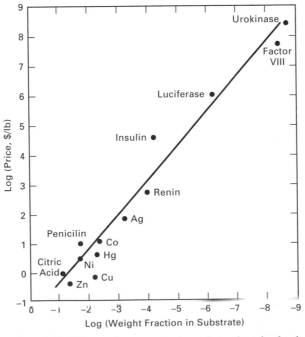

Figure 1.10 Effect of concentration of product in feed material on price [8].

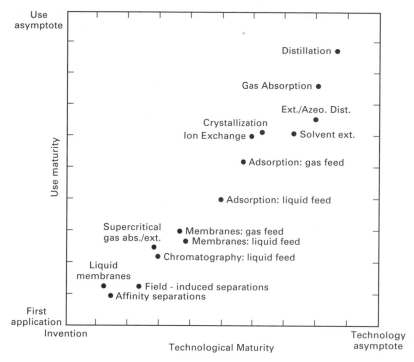

Figure 1.11 Technological and use maturities of separation processes [8].

of a solid agent or the creation or addition of a second phase. Some operations are limited to a maximum size. For capacities requiring a larger size, parallel units must be provided. The choice of single or parallel units must be given careful consideration. Except for size constraints or fabrication problems, the capacity of a single unit can be doubled for an additional investment cost of only about 50%. If two parallel units are installed, the additional investment is 100%. Table 1.10 is a list of the more common separation operations ranked according to ease of scale-up. Those operations ranked near the top are frequently designed without the need for any laboratory data or pilot-plant tests. Operations near the middle usually require laboratory data, while operations near the bottom require pilot-plant tests on actual feed mixtures. Also included in the table is an indication of the ease of providing multiple stages and to what extent parallel units may be required to handle high capacities. A detailed discussion of the selection of alternative techniques for the

Table 1.10 Ease of Scale-up of the Most Common Separation Operations

Operation in Decreasing Ease of Scale-up	Ease of Staging	Need for Parallel Units
Distillation	Easy	No need
Absorption	Easy	No need
Extractive and azeotropic distillation	Easy	No need
Liquid-liquid extraction	Easy	Sometimes
Membranes	Repressurization required between stages	Almost always
Adsorption	Easy	Only for regeneration cycle
Crystallization	Not easy	Sometimes
Drying	Not convenient	Sometimes

separation of components from both homogeneous and heterogeneous phases, with many examples, is given by Woods [11]. Ultimately, the process having the lowest operating, maintenance, and capital costs is selected.

EXAMPLE 1.1

Propylene and propane are among the light hydrocarbons produced by thermal and catalytic cracking of heavy petroleum fractions. Propane is valuable as a fuel by itself and in liquefied natural gas (LPG), and as a feedstock for producing propylene and ethylene. Propylene is used to make acrylonitrile monomer for synthetic rubber, isopropyl alcohol, cumene, propylene oxide, and polypropylene. Although propylene and propane have close boiling points, they are traditionally separated by distillation. Representative conditions are shown in Figure 1.12, where it is seen that a large number of stages is needed and the reflux and boilup flow rates compared to the feed flow rate are also large. Accordingly, considerable attention has been given to the possible replacement of distillation with a more economical and less energy-intensive operation. Based on the factors in Table 1.9, the characteristics in Table 1.10, and the list of species properties that might be exploited, given at the end of Section 1.2, propose some feasible alternatives to distillation to produce the feed and products in Figure 1.12.

SOLUTION

First, note that the component feed and product flow rates in Figure 1.12 satisfy (1-1), the conservation of mass. Table 1.11 compares properties of the two species, taken mainly from Daubert and Danner [10], where it is seen that the only listed property that might be exploited is the dipole moment. Because of the asymmetric location of the double bond in propylene, its dipole moment is significantly greater than that of propane, making propylene a polar compound,

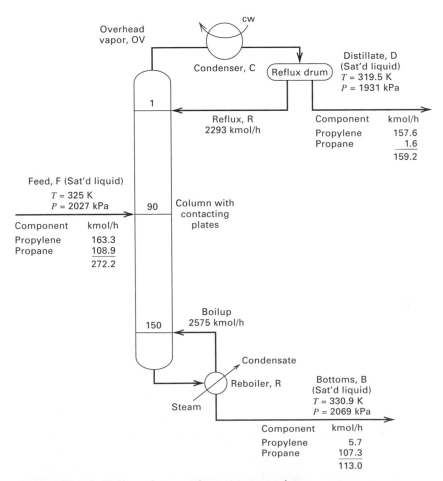

Figure 1.12 Distillation of a propylene–propane mixture.

Table 1.11 Comparison of Properties for Example 1.1

Property	Propylene	Propane
Molecular weight	42.081	44.096
van der Waals volume, $m^3/kmol$	0.03408	0.03757
van der Waals area, $m^2/kmol \times 10^{-8}$	5.060	5.590
Acentric factor	0.142	0.152
Dipole moment, debyes	0.4	0.0
Radius of gyration, $m \times 10^{10}$	2.254	2.431
Normal melting point, K	87.9	85.5
Normal boiling point, K	225.4	231.1
Critical temperature, K	364.8	369.8
Critical pressure, MPa	4.61	4.25

although weakly so (some define a polar compound as one with a dipole moment greater than 1 debye). Separation operations that can exploit this difference are

1. Extractive distillation with a polar solvent such as furfural or an aliphatic nitrile that will reduce the volatility of propylene (Ref.: U.S. Patent 2,588,056, March 4, 1952).
2. Adsorption with silica gel or a zeolite that will selectively adsorb propylene [Ref.: *J. Am. Chem. Soc.,* **72,** 1153–1157 (1950)].
3. Facilitated transport membranes using impregnated silver nitrate to carry propylene selectively through the membrane [Ref.: *Recent Developments in Separation Science,* Vol. IX, 173–195 (1986)]. ∎

SUMMARY

1. Almost all industrial chemical processes include equipment for separating chemicals contained in the process feed(s) and/or produced in reactors within the process.
2. More than 25 different separation operations are commercially important.
3. The extent of separation achievable by a particular separation operation depends on exploitation of the differences in certain properties of the species.
4. The more widely used separation operations involve the transfer of species between two phases, one of which is created by energy transfer or the reduction of pressure, or by introduction as a MSA.
5. Less commonly used separation operations are based on the use of a barrier, a solid agent, or a force field to cause species being separated to diffuse at different rates and/or to be selectively absorbed or adsorbed.
6. Separation operations are subject to the conservation of mass. The degree of separation of a component in a separator is indicated by a split fraction, SF, given by (1-2), and/or by a split ratio, SR, given by (1-3).
7. For a sequence, system, or train of separators, overall component recoveries and product purities are of prime importance and are related by material balances to the individual SF and/or SR values for the separators in the system.
8. Some separation operations, such as absorption, are capable of only a specified degree of separation for a single species. Other separation operations, such as distillation, can effect a sharp split between two so-called key components.
9. The degree of separation between two key components by a particular separation operation can be indicated by a separation power (separation factor), SP, given by (1-4) and related to SF and SR values by (1-5) and (1-6).
10. For given feed(s) and product specifications, the best separation process must frequently be selected from among a number of feasible candidates. The choice may depend on factors listed in Table 1.9. The cost of recovering and purifying a chemical depends on its concentration in the feed mixture. The extent of industrial use of a separation operation depends on the technological maturity of the operation.

REFERENCES

1. *Kirk-Othmer Encyclopedia of Chemical Technology,* 3rd ed., John Wiley and Sons, New York, Vol. 17, pp. 183–256 (1982).

2. Maude, A.H., *Trans. AIChE,* **38,** 865–882 (1942).

3. Considine, D.M., Ed., *Chemical and Process Technology Encyclopedia,* McGraw-Hill, New York, pp. 760–763 (1974).

4. Carle, T.C., and D.M. Stewart, *Chem. Ind.* (London), May 12, 1962, 830–839.

5. Perry, R.H., and C.H. Chilton, Eds., *Perry's Chemical Engineers' Handbook,* 6th ed., McGraw-Hill, New York (1984).

6. Siirola, J.J., *AIChE Symp. Ser.,* **91**(304), 222–233 (1995).

7. Sherwood, T.K., R.L. Pigford, and C.R. Wilke, *Mass Transfer,* McGraw-Hill, New York (1975).

8. Keller, G.E., II, *AIChE Monogr. Ser.,* **83**(17) (1987).

9. Reid, R.C., J.M. Prausnitz, and B.E. Poling, *The Properties of Gases and Liquids,* 4th ed., McGraw-Hill, New York (1987).

10. Daubert, T.E., and R.P. Danner, *Physical and Thermodynamic Properties of Pure Chemicals—Data Compilation,* DIPPR, AIChE, Hemisphere, New York (1989).

11. Woods, D.R., *Process Design and Engineering Practice,* Prentice-Hall, Englewood Cliffs, NJ (1995).

EXERCISES

Section 1.1

1.1 The book, *Chemical Process Industries,* 4th edition, by R. Norris Shreve and J. A. Brink, Jr. (McGraw-Hill, New York, 1984), contains process descriptions, process flow diagrams, and technical data for processes used commercially in 38 chemical industries. For each of the following processes, draw a block flow diagram of just the reaction and separation steps and describe the process in terms of just those steps, giving careful attention to the particular chemicals being formed in the reactor and separated in each of the separation operations:
(a) Coal chemicals, pp. 72–74
(b) Natural gas purification, pp. 84–86
(c) Acetylene, pp. 115–117
(d) Magnesium compounds, pp. 174–177
(e) Chlorine and caustic soda, pp. 214–219
(f) Potassium chloride, pp. 269–270
(g) Ammonia, pp. 278–282
(h) Sulfuric acid, pp. 299–310
 (i) Fluorocarbons, pp. 321–323
 (j) Uranium, pp. 338–340
(k) Titanium dioxide, pp. 388–390
 (l) Cottonseed oil, pp. 468–471
(m) Glycerin, pp. 502–503
(n) Industrial alcohol, pp. 530–534
(o) Polyethylene, pp. 587–588
(p) Formaldehyde, pp. 596–598
(q) Styrene, pp. 630–635
 (r) Natural-gas liquids, pp. 660–661

Section 1.2

1.2 Explain in detail, using thermodynamic principles, why the mixing of pure chemicals to form a homogeneous mixture is a so-called spontaneous process, while the separation of that mixture into its pure (or nearly pure) species is not.

1.3 Explain in detail, using thermodynamic principles, why the separation of a mixture into essentially pure species or other mixtures of differing compositions requires the transfer of energy to the mixture or a degradation of its energy.

Section 1.3

1.4 Compare the advantages and disadvantages of making separations using an ESA versus using an MSA.

1.5 Every other year, the magazine, *Hydrocarbon Processing,* publishes a petroleum refining handbook, which gives process flow diagrams and data for more than 75 commercial processes. For each of the following processes in the November 1990 handbook, list the separation operations of the type given in Table 1.1 and indicate what chemical(s) is(are) being separated:
(a) Hydrotreating (Chevron), p. 114
(b) Ethers (Phillips), p. 128
(c) Alkylation (Exxon), p. 130
(d) Treating of BTX cut (GKT), p. 136

1.6 Every other year, the magazine, *Hydrocarbon Processing,* publishes a petrochemical handbook, which gives process flow diagrams and data for more than 50 commercial processes. For each of the following processes in the March 1991 handbook, list the separation operations of the type given in Table 1.1 and indicate what chemical(s) is(are) being separated:
(a) Linear alkylbenzene (UOP), p. 130
(b) Methyl amines (Acid-Amine Technologies), p. 133
(c) Butene-2 (Phillips), p. 144
(d) Caprolactam (SNIA), p. 150
(e) Ethylene glycols (Scientific Design), p. 156
(f) Styrene (Monsanto), p. 188

Section 1.4

1.7 Explain why osmosis is not used as a separation operation.

1.8 The osmotic pressure, π, of sea water is given approximately by the expression $\pi = RTc/M$, where c is the concentration of the dissolved salts (solutes), in g/cm^3 and M is the average molecular weight of the solutes as ions. If pure water is to be recovered from sea water at 298 K and containing 0.035 g of salts/cm^3 of sea water and $M = 31.5$, what is the minimum required pressure difference across the membrane in kPa?

1.9 It has been shown that a liquid membrane of aqueous ferrous ethylenediaminetetraacetic acid, maintained between two sets of microporous, hydrophobic hollow fibers that are packed in a permeator cell, can selectively and continuously remove sulfur dioxide and nitrogen oxides from the flue gas of power-generating plants. Prepare a detailed drawing of a possible device to carry out such a separation. Show all locations of inlet and outlet streams, the arrangement of the hollow fibers, and a method for handling the membrane liquid. Should the membrane liquid be left in the cell or circulated? Would a sweep fluid be needed to remove the oxides?

Section 1.5

1.10 Explain the differences, if any, between adsorption and gas-solid chromatography.

1.11 In gas-liquid chromatography, is it essential that the gas flow through the packed tube in plug flow? Discuss in detail.

Section 1.6

1.12 In electrophoresis, explain why most small suspended particles are negatively charged.

1.13 In field-flow fractionation, could a turbulent-flow field be used? Why or why not?

Section 1.7

1.14 The feed to column C3 of the distillation sequence in Figure 1.9 is as given in Table 1.5. However, the separation is to be altered so as to produce a distillate that is 95 mol% pure isobutane with a recovery of isobutane in the distillate (SF) of 96%.
(a) Compute the flow rates in lbmol/h of each component in each of the two products leaving column C3.
(b) What is the percent purity of the normal butane in the bottoms product?
(c) If the purity of the isobutane in the distillate is fixed at 95%, what percent recovery of isobutane in the distillate will maximize the percent purity of normal butane in the bottoms product?

1.15 Five hundred kmol/h of a liquid mixture of light alcohols containing, by moles, 40% methanol (M), 35% ethanol (E), 15% isopropanol (IP), and 10% normal propanol (NP) is distilled in a sequence of two distillation columns. The distillate from the first column is 98% pure M with a 96% recovery of M. The distillate from the second column is 92% pure E with a 95% recovery of E from the process feed.
(a) By material balances, assuming negligible propanols in the distillate from the first column, compute the flow rates in kmol/h of each component in each feed, distillate, and bottoms. Draw a labeled block flow diagram like Figure 1.9. Include the results of the material balances in a table like

Table 1.5 and place the table below your block flow diagram.
(b) Compute the mole-percent purity of the propanol mixture leaving as bottoms from the second column in the sequence.
(c) If the recovery of ethanol is fixed at 95%, what is the maximum purity that can be achieved for the ethanol in the distillate from the second column?
(d) If instead, the purity of the ethanol is fixed at 92%, what is the maximum recovery of ethanol (based on the process feed) that can be achieved?

1.16 A mixture of ethanol and benzene is separated in a network of distillation and membrane separation steps. In one intermediate step, a near-azeotropic liquid mixture of 8,000 kg/h of 23 wt% ethanol in benzene is fed to a pervaporation membrane consisting of a thin ionomeric film of perfluorosulfonic acid polymer cast on a porous Teflon support. The membrane is selective for ethanol such that the vapor permeate contains 60 wt% ethanol, while the nonpermeate liquid contains 90 wt% benzene.
(a) Draw a flow diagram of the pervaporation step using the appropriate symbol from Table 1.2 and include on the diagram all of the given information.
(b) Compute the component flow rates in kg/h in the feed stream and in the two product streams and enter these results on the diagram.
(c) What separation operation could be used to purify the vapor permeate further?

Section 1.8

1.17 The Prism gas permeation process developed by the Monsanto Company is highly selective for hydrogen when using hollow-fiber membranes of materials such as silicone-coated polysulphone. In a typical application, a gas at 16.7 MPa and 40°C, containing the following components in kmol/h: 42.4 H_2, 7.0 CH_4, and 0.5 N_2, is separated into a nonpermeate gas at 16.2 MPa and a permeate gas at 4.56 MPa.
(a) If the membrane is nonpermeable to nitrogen, the Prism membrane separation index (SP) for hydrogen relative to methane is 34.13, and the split fraction (SF) for hydrogen to the permeate gas is 0.6038, calculate the kmol/h of each component and the total flow in kmol/h of both the nonpermeate gas and the permeate gas.
(b) Compute the percent purity of the hydrogen in the permeate gas.
(c) Using an average heat capacity ratio, γ, of 1.4, estimate the outlet temperatures of the two exiting gas streams by assuming the ideal gas law and reversible expansions for each gas and no heat transfer between the two exiting gas streams.
(d) Draw a process flow diagram of the membrane process and indicate on the diagram for each stream the pressure, temperature, and component flow rates.

1.18 Nitrogen gas can be injected into oil wells to increase the recovery of crude oil (enhanced oil recovery). Usually, natural gas is produced with the oil and it is desirable to recover the nitrogen from the gas for reinjection into the well. Furthermore, the natural gas must not contain more than 3 mol% nitrogen if the natural gas is to be put into a pipeline. A total of 170,000 SCFH (based on 60°F and 14.7 psia) of natural gas containing 18% N_2, 75% CH_4, and 7% C_2H_6 at 100°F and 800 psia is to be processed for N_2 removal. A two-step separation process has been proposed consisting of (1) membrane separation with a nonporous glassy polyimide membrane, followed by (2) pressure-swing adsorption using molecular sieves to which the permeate gas is fed. The membrane separator is highly selective for N_2 (SP_{N_2,CH_4} = 16) and completely impermeable to ethane. The pressure-swing adsorption step selectively adsorbs methane, giving 97% pure methane product in the adsorbate, with an 85% recovery of CH_4 fed to the adsorber. The nonpermeate (retentate) gas from the membrane step and adsorbate from the pressure-swing adsorption step are combined to give a methane stream that contains 3.0% N_2. The pressure drop across the membrane is 760 psia. The permeate at 20°F is compressed in two stages to 275 psia and cooled to 100°F before entering the adsorption step. The adsorbate gas, which exits the adsorber during regeneration at 100°F and 15 psia, is compressed in three stages to 800 psia and cooled to 100°F before being combined with nonpermeate gas to give the final pipeline natural gas.

(a) Draw a process flow diagram of the separation process using appropriate symbols from Tables 1.2 and 1.3. Include the gas compressors and heat exchangers. Label the diagram with all of the data given above, and number all process streams.

(b) Compute by material balances, using the data above, the component flow rates of N_2, CH_4, and C_2H_6 in lbmol/h for all process streams entering and exiting the two separation operations. Place the results in a material balance table similar to Table 1.5.

Section 1.9

1.19 A mixture of ethylbenzene (EB) and the three isomers (ortho, meta, and para) of xylene are widely available in petroleum refineries.

(a) Based on the differences in normal boiling points, verify that the separation between *meta*-xylene (MX) and *para*-xylene (PX) by distillation is far more difficult than the separations between EB and PX, and MX and *ortho*-xylene (OX).

(b) Prepare a list of properties for MX and PX similar to Table 1.11. From that list, which property differences might be the best ones to exploit to separate a mixture of these two xylenes.

(c) Explain why melt crystallization and adsorption are used commercially to separate MX and PX.

1.20 When a mixture of ethanol and water is distilled at ambient pressure, the products are a distillate of ethanol and water of near-azeotrope composition (89.4 mol% ethanol) and a bottoms product of nearly pure water. Based on differences in certain properties of ethanol and water, explain how the following separation operations might be able to recover almost pure ethanol from the distillate:

(a) Extractive distillation

(b) Azeotropic distillation

(c) Liquid-liquid extraction

(d) Crystallization

(e) Pervaporation membrane

(f) Adsorption

1.21 A stream containing 7,000 kmol/h of water and 3,000 parts per million (ppm) by weight of ammonia at 350 K and 1 bar is to be processed to remove 90% of the ammonia. What type of separation operation would you use? If it involves a mass-separating agent, propose one.

Chapter 2

Thermodynamics of Separation Operations

Thermodynamic properties and equations play a major role in separation operations, particularly with respect to energy requirements, phase equilibria, and sizing equipment. This chapter discusses applied thermodynamics for separation processes. Equations for energy balances, entropy and availability balances, and for determining phase densities and phase compositions at equilibrium are developed. These involve thermodynamic properties, including specific volume or density, enthalpy, entropy, availability, and fugacities and activities together with their coefficients, all as functions of temperature, pressure, and phase composition. Methods for estimating properties for ideal and non-ideal mixtures are summarized.

2.1 ENERGY, ENTROPY, AND AVAILABILITY BALANCES

Most commercial separation operations utilize large quantities of energy in the form of heat and/or shaft work. A study by Mix et al. [1] reports that two quads (1 quad = 10^{15} Btu) of energy were consumed by distillation separations in petroleum, chemical, and natural-gas processing plants in the United States in 1976. This amount of energy was 2.7% of the total U.S. energy consumption of 74.5 quads and is equivalent to the energy obtained from approximately 1 million bbl of crude oil per day over a one-year period. This amount of oil can be compared to 13 million bbl/day, the average amount of crude oil processed by petroleum refineries in the United States in early 1991. At a crude oil price of approximately \$20/bbl, the energy consumption by distillation in the United States is approximately \$10 trillion per year. Thus, it is of considerable interest to know the extent of energy consumption in a separation process, and to what degree energy requirements might be reduced. Such energy estimates can be made by applying the first and second laws of thermodynamics.

Consider the continuous, steady-state flow system for a general separation process in Figure 2.1. One or more feed streams flowing into the system are separated into two or more product streams that flow out of the system. For all these streams, we denote the molar flow rates by n, the component mole fractions by z_i, the temperature by T, the pressure by P, the molar enthalpies by h, the molar entropies by s, and the molar availabilities by b. If chemical reactions occur in the process, enthalpies and entropies are referred to the elements, as discussed by Felder and Rousseau [2]; otherwise they can be referred to the compounds. Heat flows in or out of the system are denoted by Q, and shaft work crossing the boundary of the system is denoted by W_s. At steady state, if kinetic, potential,

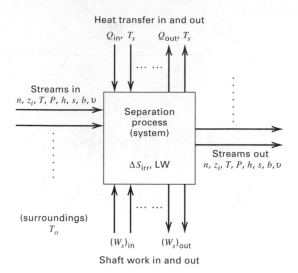

Figure 2.1 General separation system.

and surface energy changes are neglected, the first law of thermodynamics (also referred to as the conservation of energy or the energy balance), states that the sum of all forms of energy flowing into the system equals the sum of the energy flows leaving the system:

(stream enthalpy flows + heat transfer + shaft work)$_{\text{leaving system}}$
$$- \text{(stream enthalpy flows + heat transfer + shaft work)}_{\text{entering system}} = 0$$

In terms of symbols, the energy balance is given by Eq. (1) in Table 2.1, where all flow rate, heat transfer, and shaft-work terms are positive. Molar enthalpies may be positive or negative depending on the reference state.

All separation processes must satisfy the energy balance. Inefficient separation processes require large transfers of heat and/or shaft work both into and out of the process; efficient processes require smaller levels of heat transfer and/or shaft work. The first law of thermodynamics provides no information on energy efficiency, but the second law of thermodynamics (also referred to as the entropy balance), given by Eq. (2) in Table 2.1, does. In words, the entropy balance is

(Stream entropy flows + entropy flows by heat transfer)$_{\text{leaving system}}$
$$- \text{(stream entropy flows + entropy flows by heat transfer)}_{\text{entering system}}$$
$$= \text{production of entropy by the process}$$

In the entropy balance equation, the heat sources and sinks in Figure 2.1 are at absolute temperatures T_s. For example, if condensing steam at 150°C supplies heat, Q, to the reboiler of a distillation column, $T_s = 150 + 273 = 423$ K. If cooling water at an average temperature of 30°C removes heat, Q, in a condenser, $T_s = 30 + 273 = 303$ K. Unlike the energy balance, which states that energy is conserved, the entropy balance predicts the production of entropy, ΔS_{irr}, which is the irreversible increase in the entropy of the universe. This term, which must be a positive quantity, is a quantitative measure of the thermodynamic inefficiency of a process. In the limit, as a reversible process is approached, ΔS_{irr} tends to zero. Note that the entropy balance contains no terms related to shaft work.

Although ΔS_{irr} is a measure of energy inefficiency, it is difficult to relate to this measure because it does not have the units of energy/time (power). A more useful measure or process inefficiency can be derived by combining Eqs. (1) and (2) in Table 2.1 to obtain a combined statement of the first and second laws of thermodynamics, which is given as

Table 2.1 Universal Thermodynamic Laws for a Continuous, Steady-State Flow System

Energy balance:

$$(1) \quad \sum_{\substack{\text{out of} \\ \text{system}}} (nh + Q + W_s) - \sum_{\substack{\text{in to} \\ \text{system}}} (nh + Q + W_s) = 0$$

Entropy balance:

$$(2) \quad \sum_{\substack{\text{out of} \\ \text{system}}} \left(ns + \frac{Q}{T_s}\right) - \sum_{\substack{\text{in to} \\ \text{system}}} \left(ns + \frac{Q}{T_s}\right) = \Delta S_{\text{irr}}$$

Availability balance:

$$(3) \quad \sum_{\substack{\text{in to} \\ \text{system}}} \left[nb + Q\left(1 - \frac{T_0}{T_s}\right) + W_s\right] - \sum_{\substack{\text{out of} \\ \text{system}}} \left[nb + Q\left(1 - \frac{T_0}{T_s}\right) + W_s\right] = \text{LW}$$

Minimum work of separation:

$$(4) \quad W_{\text{min}} = \sum_{\substack{\text{out of} \\ \text{system}}} nb - \sum_{\substack{\text{in to} \\ \text{system}}} nb$$

Second-law efficiency:

$$(5) \quad \eta = \frac{W_{\text{min}}}{\text{LW} + W_{\text{min}}}$$

where $b = h - T_0 s$ – availability function

$\text{LW} = T_0 \Delta S_{\text{irr}}$ = lost work

Eq. (3) in Table 2.1. To perform this derivation, it is first necessary to define an infinite source of or sink for heat transfer at the absolute temperature, $T_s = T_0$, of the surroundings. This temperature is typically about 300 K and represents the largest source of coolant associated with the processing plant being analyzed. This might be the average temperature of cooling water, air, or a nearby river, lake, or ocean. Heat transfer associated with this surrounding coolant and transferred from (or to) the process is termed Q_0. Thus, in both Eqs. (1) and (2) in Table 2.1, the Q and Q/T_s terms include contributions from Q_0 and Q_0/T_0, respectively.

The derivation of Eq. (3) in Table 2.1 can be made, as shown by de Nevers and Seader [3], by combining Eqs. (1) and (2) to eliminate Q_0. The resulting equation is referred to as an *availability* (or *exergy*) balance, where the term availability means "available for complete conversion to shaft work." The stream availability function, b, as defined by

$$b = h - T_0 s \qquad \text{(2-1)}$$

is a measure of the maximum amount of stream energy that can be converted into shaft work if the stream is taken to the reference state. It is similar to Gibbs free energy, $g = h - Ts$, but differs in that the infinite surroundings temperature, T_0, appears in the equation instead of the stream temperature, T. Terms in Eq. (3) containing Q are multiplied by $(1 - T_0/T_s)$, which, as shown in Figure 2.2, is the reversible Carnot heat-engine cycle efficiency, representing the maximum amount of shaft work that can be produced from Q at T_s, where the residual amount of energy $(Q - W_s)$ is transferred as heat to a sink at T_0. Shaft work, W_s, remains at its full value in Eq. (3). Thus, although Q and W_s have the same thermodynamic worth in Eq. (1) of Table 2.1, heat transfer has less worth in Eq. (3). This is because shaft work can be converted completely to heat (by friction), but

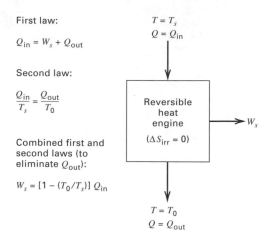

First law:

$$Q_{in} = W_s + Q_{out}$$

Second law:

$$\frac{Q_{in}}{T_s} = \frac{Q_{out}}{T_0}$$

Combined first and second laws (to eliminate Q_{out}):

$$W_s = [1 - (T_0/T_s)]\, Q_{in}$$

(Inside box)
$T = T_s$
$Q = Q_{in}$

Reversible heat engine
$(\Delta S_{irr} = 0)$

$\longrightarrow W_s$

$T = T_0$
$Q = Q_{out}$

Figure 2.2 Carnot heat engine cycle for converting heat to shaft work.

heat cannot be converted completely to shaft work unless the heat is available at an infinite temperature.

Availability, like entropy, is not conserved in a real, irreversible process. The total availability (i.e., ability to produce shaft work) passing into a system is always greater than the total availability leaving a process. Thus Eq. (3) in Table 2.1 is written with the "into system" terms first. The difference is the *lost work*, LW, which is also called the loss of availability (or exergy), and is defined by

$$LW = T_0\, \Delta S_{irr} \qquad\qquad (2\text{-}2)$$

Lost work is always a positive quantity. The greater its value, the greater is the energy inefficiency. In the lower limit, as a reversible process is approached, lost work tends to zero. The lost work has the same units as energy, thus making it easy to attach significance to its numerical value. In words, the availability balance is

(Stream availability flows + availability of heat + shaft work)$_{\text{entering system}}$
$-$ (system availability flows + availability of heat
$+$ shaft work)$_{\text{leaving system}}$ = loss of availability (lost work)

For any separation process, lost work can be computed from Eq. (3) in Table 2.1. Its magnitude depends on the extent of process irreversibilities, which include fluid friction, heat transfer due to finite temperature-driving forces, mass transfer due to finite concentration or activity-driving forces, chemical reactions proceeding at finite displacements from chemical equilibrium, mixing of streams at differing conditions of temperature, pressure, and/or composition, and so on. Thus, to reduce the lost work, driving forces for momentum transfer, heat transfer, mass transfer, and chemical reaction must be reduced. Practical limits to this reduction exist because, as driving forces are decreased, equipment sizes increase, tending to infinitely large sizes as driving forces approach zero.

For a separation process that occurs without chemical reaction, the summation of the stream availability functions leaving the process is usually greater than the same summation for streams entering the process. In the limit for a reversible process (LW = 0), Eq. (3) of Table 2.1 reduces to Eq. (4), where W_{min} is the minimum work required to conduct the separation and is equivalent to the difference in the heat transfer and shaft work terms in Eq. (3). This minimum work is independent of the nature (or path) of the separation process. The work of separation for an actual irreversible process is always greater than the minimum value computed from Eq. (4).

From Eq. (3) of Table 2.1, it is seen that as a separation process becomes more irreversible, and thus more energy inefficient, the increasing LW causes the required equivalent work of separation to increase by the same amount. Thus, the equivalent work of separation for an irreversible process is given by the sum of lost work and minimum work of separation. The *second-law efficiency* can be defined as

$$(\text{fractional second-law efficiency}) = \left(\frac{\text{minimum work of separation}}{\text{equivalent actual work of separation}} \right)$$

In terms of symbols, the efficiency is given by Eq. (5) in Table 2.1.

EXAMPLE 2.1

For the propylene–propane separation of Figure 1.12, using the following thermodynamic properties for certain streams, as estimated from the Soave–Redlich–Kwong equation of state discussed in Section 2.5, and the relations given in Table 2.1, compute in SI units:

(a) The condenser duty, Q_C
(b) The reboiler duty, Q_R
(c) The irreversible entropy production, assuming 303 K for the condenser cooling-water sink and 378 K for the reboiler steam source
(d) The lost work, assuming $T_0 = 303$ K
(e) The minimum work of separation
(f) The second-law efficiency

Stream	Phase Condition	Enthalpy (h), kJ/kmol	Entropy (s), kJ/kmol-K
Feed (F)	Liquid	13,338	−4.1683
Overhead vapor (OV)	Vapor	24,400	24.2609
Distillate (D) and reflux (R)	Liquid	12,243	−13.8068
Bottoms (B)	Liquid	14,687	−2.3886

SOLUTION

Place the condenser (C) cooling water and the reboiler (R) steam outside the distillation system. Thus, Q_C and Q_R cross the boundary of the system. The following calculations are made with SI units.

(a) Compute condenser duty from an energy balance around the condenser. From Eq. (1), Table 2.1, noting that the overhead-vapor flow rate is given by $n_{OV} = n_R + n_D$ and $h_R = h_D$, the condenser duty is:

$$Q_C = n_{OV}(h_{OV} - h_R)$$
$$= (2{,}293 + 159.2)(24{,}400 - 12{,}243) = 29{,}811{,}000 \text{ kJ/h}$$

(b) Compute reboiler duty from an energy balance around the entire distillation operation. (An energy balance around the reboiler cannot be made because data are not given for the boilup rate.) From Eq. (1), Table 2.1,

$$Q_R = n_D h_D + n_B h_B + Q_C - n_F h_F$$
$$= 159.2(12{,}243) + 113(14{,}687) + 29{,}811{,}000 - 272.2(13{,}338)$$
$$= 29{,}789{,}000 \text{ kJ/h}$$

(c) Compute the production of entropy from an entropy balance around the entire distillation system. From Eq. (2), Table 2.1,

$$\Delta S_{\text{irr}} = n_D s_D + n_B s_B + Q_C/T_C - n_F s_F - Q_R/T_R$$
$$= 159.2(-13.8068) + 113(-2.3886) + 29{,}811{,}000/303$$
$$- 272.2(-4.1683) - 29{,}789{,}000/378$$
$$= 18{,}246 \text{ kJ/h-K}$$

(d) Compute lost work from its definition at the bottom of Table 2.1:

$$LW = T_0 \, \Delta S_{irr}$$
$$= 303(18,246) = 5,529,000 \text{ kJ/h}$$

Alternatively, compute lost work from an availability balance around the entire distillation system. From Eq. (3), Table 2.1, where the availability function, b, is defined near the bottom of Table 2.1,

$$LW = n_F b_F + Q_R(1 - T_0/T_R) - n_D b_D - n_B b_B - Q_C(1 - T_0/T_C)$$
$$= 272.2[13,338 - (303)(-4.1683)]$$
$$+ 29,789,000(1 - 303/378)$$
$$- 159.2[12,243 - (303)(-13.8068)] - 113[14,687 - (303)(-2.3886)]$$
$$- 29,811,000 \, (1 - 303/303)$$
$$= 5,529,000 \text{ kJ/h} \qquad \text{(same result)}$$

(e) Compute the minimum work of separation for the entire distillation system. From Eq. (4), Table 2.1,

$$W_{min} = n_D b_D + n_B b_B - n_F b_F$$
$$= 159.2[12,243 - (303)(-13.8068)] + 113[14,687 - (303)(-2.3886)]$$
$$- 272.2[13,338 - (303)(-4.1683)]$$
$$= 382,100 \text{ kJ/h}$$

(f) Compute the second-law efficiency for the entire distillation system. From Eq. (5), Table 2.1,

$$\eta = \frac{W_{min}}{LW + W_{min}}$$
$$= \frac{382,100}{5,529,000 + 382,100} = 0.0646 \quad \text{or} \quad 6.46\%$$

This low second-law efficiency is typical of a difficult distillation separation, which in this case requires 150 theoretical stages with a reflux ratio of almost 15 times the distillate rate.

■

2.2 PHASE EQUILIBRIA

Analysis of separations equipment frequently involves the assumption of phase equilibria as expressed in terms of Gibbs free energy, chemical potentials, fugacities, or activities. For each phase in a multiphase, multicomponent system, the total Gibbs free energy is

$$G = G(T, P, N_1, N_2, \ldots, N_C)$$

where N_i = moles of species i. At equilibrium, the total G for all phases is a minimum, and methods for determining this minimum are referred to as *free-energy minimization techniques*. Gibbs free energy is also the starting point for the derivation of commonly used equations for expressing phase equilibria. From classical thermodynamics, the total differential of G is given by

$$dG = -S \, dT + V \, dP + \sum_{i=1}^{C} \mu_i \, dN_i \qquad \textbf{(2-3)}$$

where μ_i is the chemical potential or partial molar Gibbs free energy of species i. When (2-3) is applied to a closed system consisting of two or more phases in equilibrium at uniform temperature and pressure, where each phase is an open system capable of mass transfer with another phase, then

$$dG_{system} = \sum_{p=1}^{N} \left[\sum_{i=1}^{C} \mu_i^{(p)} \, dN_i^{(p)} \right]_{P,T} \qquad \textbf{(2-4)}$$

where the superscript (p) refers to each of N phases in equilibrium. Conservation of moles of each species requires that

$$dN_i^{(1)} = -\sum_{p=2}^{N} dN_i^{(p)} \tag{2-5}$$

which, upon substitution into (2-4), gives

$$\sum_{p=2}^{N}\left[\sum_{i=1}^{C}(\mu_i^{(p)} - \mu_i^{(1)})\, dN_i^{(p)}\right] = 0 \tag{2-6}$$

With $dN_i^{(1)}$ eliminated in (2-6), each $dN_i^{(p)}$ term can be varied independently of any other $dN_i^{(p)}$ term. But this requires that each coefficient of $dN_i^{(p)}$ in (2-6) be zero. Therefore,

$$\mu_i^{(1)} = \mu_i^{(?)} = \mu_i^{(3)} = \ldots = \mu_i^{(N)} \tag{2-7}$$

Thus, the chemical potential of a particular species in a multicomponent system is identical in all phases at physical equilibrium.

Fugacities and Activity Coefficients

Chemical potential cannot be expressed as an absolute quantity, and the numerical values of chemical potential are difficult to relate to more easily understood physical quantities. Furthermore, the chemical potential approaches an infinite negative value as pressure approaches zero. For these reasons, the chemical potential is not favored for phase equilibria calculations. Instead, fugacity, invented by G. N. Lewis in 1901, is employed as a surrogate.

The partial fugacity of species i in a mixture is like a pseudo-pressure, defined in terms of the chemical potential by

$$\bar{f}_i = C \exp\left(\frac{\mu_i}{RT}\right) \tag{2-8}$$

where C is a temperature-dependent constant. Regardless of the value of C, it is shown by Prausnitz, Lichtenthaler, and Azevedo [4] that (2-7) can be replaced with

$$\bar{f}_i^{(1)} = \bar{f}_i^{(2)} = \bar{f}_i^{(3)} = \cdots = \bar{f}_i^{(N)} \tag{2-9}$$

Thus, at equilibrium, a given species has the same partial fugacity in each existing phase. This equality, together with equality of phase temperatures and pressures,

$$T^{(1)} = T^{(2)} = T^{(3)} = \cdots = T^{(N)} \tag{2-10}$$

and

$$P^{(1)} = P^{(2)} = P^{(3)} = \cdots = P^{(N)} \tag{2-11}$$

constitutes the required conditions for phase equilibria. For a pure component, the partial fugacity, $\bar{f}_i$, becomes the pure-component fugacity, f_i. For a pure ideal gas, fugacity is equal to the pressure, and for a component in an ideal gas mixture, the partial fugacity is equal to its partial pressure, $p_i = y_i P$. Because of the close relationship between fugacity and pressure, it is convenient to define their ratio for a pure substance as

$$\phi_i = \frac{f_i}{P} \tag{2-12}$$

where ϕ_i is the pure-species fugacity coefficient, which has a value of 1.0 for an ideal gas.

For a mixture, partial fugacity coefficients are defined by

$$\bar{\phi}_{iV} \equiv \frac{\bar{f}_{iV}}{y_i P} \tag{2-13}$$

$$\bar{\phi}_{iL} \equiv \frac{\bar{f}_{iL}}{x_i P} \tag{2-14}$$

such that as ideal gas behavior is approached, $\bar{\phi}_{iV} \rightarrow 1.0$ and $\bar{\phi}_{iL} \rightarrow P_i^s/P$, where $P_i^s =$ vapor (saturation) pressure.

At a given temperature, the ratio of the partial fugacity of a component to its fugacity in some defined standard state is termed the *activity*. If the standard state is selected as the pure species at the same pressure and phase condition as the mixture, then

$$a_i \equiv \frac{\bar{f}_i}{f_i^o} \tag{2-15}$$

Since at phase equilibrium, the value of f_i^o is the same for each phase, substitution of (2-15) into (2-9) gives another alternative condition for phase equilibria,

$$a_i^{(1)} = a_i^{(2)} = a_i^{(3)} = \cdots = a_i^{(N)} \tag{2-16}$$

For an ideal solution, $a_{iV} = y_i$ and $a_{iL} = x_i$.

To represent departure of activities from mole fractions when solutions are nonideal, *activity coefficients* based on concentrations in mole fractions are defined by

$$\gamma_{iV} \equiv \frac{a_{iV}}{y_i} \tag{2-17}$$

$$\gamma_{iL} \equiv \frac{a_{iL}}{x_i} \tag{2-18}$$

For ideal solutions, $\gamma_{iV} = 1.0$ and $\gamma_{iL} = 1.0$.

For convenient reference, thermodynamic quantities that are useful in phase equilibria calculations are summarized in Table 2.2.

K-Values

An *equilibrium ratio* is the ratio of mole fractions of a species present in two phases at equilibrium. For the vapor–liquid case, the constant is referred to as the *K-value* or vapor–liquid equilibrium ratio:

$$K_i \equiv \frac{y_i}{x_i} \tag{2-19}$$

For the liquid–liquid case, the ratio is referred to as the distribution coefficient or liquid–liquid equilibrium ratio:

$$K_{D_i} \equiv \frac{x_i^{(1)}}{x_i^{(2)}} \tag{2-20}$$

For equilibrium-stage calculations involving the separation of two or more components, separation factors are defined by forming ratios of equilibrium ratios. For the vapor–liquid case, *relative volatility* is defined by

$$\alpha_{ij} \equiv \frac{K_i}{K_j} \tag{2-21}$$

Table 2.2 Thermodynamic Quantities for Phase Equilibria

Thermodynamic Quantity	Definition	Physical Significance	Limiting Value for Ideal Gas and Ideal Solution
Chemical potential	$\mu_i \equiv \left(\dfrac{\partial G}{\partial N_i}\right)_{P,T,N_j}$	Partial molar free energy, $\bar{g}_i$	$\mu_i = \bar{g}_i$
Partial fugacity	$\bar{f}_i \equiv C \exp\left(\dfrac{\mu_i}{RT}\right)$	Thermodynamic pressure	$\bar{f}_{iV} = y_i P$ $\bar{f}_{iL} = x_i P_i^s$
Fugacity coefficient of a pure species	$\phi_i \equiv \dfrac{f_i}{P}$	Deviation to fugacity due to pressure	$\phi_{iV} = 1.0$ $\phi_{iL} = \dfrac{P_i^s}{P}$
Partial fugacity coefficient of a species in a mixture	$\bar{\phi}_{iV} \equiv \dfrac{\bar{f}_{iV}}{y_i P}$ $\bar{\phi}_{iL} \equiv \dfrac{\bar{f}_{iL}}{x_i P}$	Deviations to fugacity due to pressure and composition	$\bar{\phi}_{iV} = 1.0$ $\bar{\phi}_{iL} = \dfrac{P_i^s}{P}$
Activity	$a_i \equiv \dfrac{\bar{f}_i}{f_i^0}$	Relative thermodynamic pressure	$a_{iV} = y_i$ $a_{iL} = x_i$
Activity coefficient	$\gamma_{iV} \equiv \dfrac{a_{iV}}{y_i}$ $\gamma_{iL} \equiv \dfrac{a_{iL}}{x_i}$	Deviation to fugacity due to composition	$\gamma_{iV} = 1.0$ $\gamma_{iL} = 1.0$

For the liquid–liquid case, the *relative selectivity* is

$$\beta_{ij} \equiv \frac{K_{D_i}}{K_{D_j}} \tag{2-22}$$

Equilibrium ratios can be expressed by the quantities in Table 2.2 in a variety of rigorous formulations. However, the only ones of practical interest are developed as follows. For vapor–liquid equilibrium, (2-9) becomes

$$\bar{f}_{iV} = \bar{f}_{iL}$$

To form an equilibrium ratio, these partial fugacities are commonly replaced by expressions involving mole fractions as derived from the definitions in Table 2.2:

$$\bar{f}_{iL} = \gamma_{iL} x_i f_{iL}^0 \tag{2-23}$$

or

$$\bar{f}_{iL} = \bar{\phi}_{iL} x_i P \tag{2-24}$$

and

$$\bar{f}_{iV} = \bar{\phi}_{iV} y_i P \tag{2-25}$$

If (2-24) and (2-25) are used with (2-19), a so-called *equation-of-state form* of the K-value is obtained:

$$K_i = \frac{\bar{\phi}_{iL}}{\bar{\phi}_{iV}} \tag{2-26}$$

This expression has received considerable attention, with applications of importance being the Starling modification of the Benedict, Webb, and Rubin (B–W–R–S) equation of state

Table 2.3 Useful Expressions for Estimating K-Values for Vapor–Liquid Equilibria ($K_i \equiv y_i/x_i$)

	Equation	Recommended Application
Rigorous forms:		
(1) Equation-of-state	$K_i = \dfrac{\bar{\phi}_{iL}}{\bar{\phi}_{iV}}$	Hydrocarbon and light gas mixtures from cryogenic temperatures to the critical region
(2) Activity coefficient	$K_i = \dfrac{\gamma_{iL}\,\phi_{iL}}{\bar{\phi}_{iV}}$	All mixtures from ambient to near-critical temperature
Approximate forms:		
(3) Raoult's law (ideal)	$K_i = \dfrac{P_i^s}{P}$	Ideal solutions at near-ambient pressure
(4) Modified Raoult's law	$K_i = \dfrac{\gamma_{iL} P_i^s}{P}$	Nonideal liquid solutions at near-ambient pressure
(5) Poynting correction	$K_i = \gamma_{iL}\,\phi_{iV}^s \left(\dfrac{P_i^s}{P}\right) \exp\left(\dfrac{1}{RT}\displaystyle\int_{P_i^s}^{P} v_{iL}\,dP\right)$	Nonideal liquid solutions at moderate pressure and below the critical temperature
(6) Henry's law	$K_i = \dfrac{H_i}{P}$	Low-to-moderate pressures for species at supercritical temperature

[5], the Soave modification of the Redlich–Kwong (S–R–K or R–K–S) equation of state [6], the Peng–Robinson (P–R) equation of state [7], and the Plöcker et al. modification of the Lee–Kesler (L–K–P) equation of state [8].

If (2-23) and (2-25) are used, a so-called *activity coefficient form* of the K-value is obtained:

$$K_i = \frac{\gamma_{iL} f_{iL}^o}{\bar{\phi}_{iV} P} = \frac{\gamma_{iL}\,\phi_{iL}}{\bar{\phi}_{iV}} \qquad \textbf{(2-27)}$$

Since 1960, (2-27) has received considerable attention with applications to important industrial systems presented by Chao and Seader (C–S) [9], with a modification by Grayson and Streed [10].

Table 2.3 is a summary of useful formulations for estimating K-values for vapor–liquid equilibrium. Included are the two rigorous expressions given by (2-26) and (2-27), from which the other approximate formulations are derived. The so-called Raoult's law or ideal K-value is obtained from (2-27) by substituting from Table 2.2, for an ideal gas and ideal gas and liquid solutions, $\gamma_{iL} = 1.0$, $\phi_{iL} = P_i^s/P$, and $\bar{\phi}_{iV} = 1.0$. The modified Raoult's law relaxes the assumption of an ideal liquid solution by including the liquid-phase activity coefficient. The Poynting correction form for moderate pressures is obtained by approximating the pure-component liquid fugacity coefficient in (2-27) by the expression

$$\phi_{iL} = \phi_{iV}^s \frac{P_i^s}{P} \exp\left(\frac{1}{RT}\int_{P_i^s}^{P} v_{iL}\,dP\right) \qquad \textbf{(2-28)}$$

where the exponential term is the Poynting factor or correction. If the liquid molar volume is reasonably constant over the pressure range, the integral in (2-28) becomes $v_{iL}(P - P_i^s)$. For a light gas species, whose critical temperature is less than the system temperature, the Henry's law form for the K-value is convenient provided that a value of H_i, the empirical Henry's law coefficient, is available. This constant for a particular species, i, depends on liquid-phase composition, temperature, and pressure. Included in Table 2.3 are recommendations for the application of each of the vapor–liquid K-value expressions.

Regardless of which thermodynamic formulation is used for estimating K-values, the accuracy depends on the particular correlations used for the thermodynamic properties required (i.e., vapor pressure, activity coefficient, and fugacity coefficients). For practical applications, the choice of K-value formulation is a compromise among considerations of accuracy, complexity, convenience, and past experience.

For liquid–liquid equilibria, (2-9) becomes

$$\bar{f}_{iL}^{(1)} = \bar{f}_{iL}^{(2)} \tag{2-29}$$

where superscripts (1) and (2) refer to the two immiscible liquid phases. A rigorous formulation for the distribution coefficient is obtained by combining (2-23) with (2-20) to obtain

$$K_{D_i} = \frac{x_i^{(1)}}{x_i^{(2)}} = \frac{\gamma_{iL}^{(2)} f_{iL}^{o(2)}}{\gamma_{iL}^{(1)} f_{iL}^{o(1)}} = \frac{\gamma_{iL}^{(2)}}{\gamma_{iL}^{(1)}} \tag{2-30}$$

For vapor–solid equilibria, a useful formulation can be derived if the solid phase consists of just one of the components of the vapor phase. In that case, the combination of (2-9) and (2-25) gives

$$f_{iS} = \bar{\phi}_{iV} y_i P \tag{2-31}$$

At low pressure, $\bar{\phi}_{iV} = 1.0$ and the solid fugacity can be approximated by the vapor pressure of the solid to give for the vapor-phase mole fraction of the component forming the solid phase:

$$y_i = \frac{(P_i^s)_{\text{solid}}}{P} \tag{2-32}$$

For liquid–solid equilibria, a similar useful formulation can be derived if again the solid phase is a pure component. Then the combination of (2-9) and (2-23) gives

$$f_{iS} = \gamma_{iL} x_i f_{iL}^o \tag{2-33}$$

At low pressure, the solid fugacity can be approximated by vapor pressure to give, for the component in the solid phase,

$$x_i = \frac{(P_i^s)_{\text{solid}}}{\gamma_{iL}(P_i^s)_{\text{liquid}}} \tag{2-34}$$

Example 2.2

Estimate the K-values of a vapor-liquid mixture of water and methane at 2 atm total pressure for temperatures of 20 and 80°C.

SOLUTION

At the conditions of temperature and pressure, water will exist mainly in the liquid phase and will follow Raoult's law, as given in Table 2.3. Because methane has a critical temperature of -82.5°C, well below the temperatures of interest, it will exist mainly in the vapor phase and follow Henry's law, as given in Table 2.3. From *Perry's Chemical Engineers' Handbook,* 6th ed., pp. 3-237 and 3-103, the following vapor pressure data for water and Henry's law coefficients for CH_4 are obtained:

T, °C	P^s for H_2O, atm	H for CH_4, atm
20	0.02307	3.76×10^4
80	0.4673	6.82×10^4

K-values for water and methane are estimated from Eqs. (3) and (6), respectively, in Table 2.3, using $P = 2$ atm, with the following results:

T, °C	K_{H_2O}	K_{CH_4}
20	0.01154	18,800
80	0.2337	34,100

The above K-values confirm the assumptions of the phase distribution of the two species. The K-values for H_2O are low, but increase rapidly with increasing temperature. The K-values for methane are extremely high and do not change rapidly with temperature for this example. ■

2.3   IDEAL GAS, IDEAL LIQUID SOLUTION MODEL

Design procedures for separation equipment require numerical values for phase enthalpies, entropies, densities, and phase equilibrium ratios. Classical thermodynamics provides a means for obtaining these quantities in a consistent manner from $P–v–T$ relationships, which are usually referred to as *equation-of-state models*. The simplest model applies when both liquid and vapor phases are ideal solutions (all activity coefficients equal 1.0) and the vapor is an ideal gas. Then the thermodynamic properties can be computed from unary constants for each of the species in the mixture in a relatively straightforward manner using the equations given in Table 2.4. In general, these ideal equations apply only at

Table 2.4 Thermodynamic Properties for Ideal Mixtures

Ideal gas and ideal gas solution:

$$(1)\ \ v_V = \frac{V}{\sum\limits_{i=1}^{C} N_i} = \frac{M}{\rho_V} = \frac{RT}{P} \qquad M = \sum_{i=1}^{C} y_i M_i$$

$$(2)\ \ h_V = \sum_{i=1}^{C} y_i \int_{T_0}^{T} (C_P^o)_{iV}\, dT = \sum_{i=1}^{C} y_i h_{iV}^o$$

$$(3)\ \ s_V = \sum_{i=1}^{C} y_i \int_{T_0}^{T} \frac{(C_P^o)_{iV}}{T}\, dT - R\ln\left(\frac{P}{P_0}\right) - R\sum_{i=1}^{C} y_i \ln y_i \qquad \text{where the first term is } s_V^o$$

Ideal liquid solution:

$$(4)\ \ v_L = \frac{V}{\sum\limits_{i=1}^{C} N_i} = \frac{M}{\rho_L} = \sum_{i=1}^{C} x_i v_{iL} \qquad M = \sum_{i=1}^{C} x_i M_i$$

$$(5)\ \ h_L = \sum_{i=1}^{C} x_i (h_{iV}^o - \Delta H_i^{\text{vap}})$$

$$(6)\ \ s_L = \sum_{i=1}^{C} x_i \left[\int_{T_0}^{T} \frac{(C_P^o)_{iV}}{T}\, dT - \frac{\Delta H_i^{\text{vap}}}{T} \right] - R\ln\left(\frac{P}{P_0}\right) - R\sum_{i=1}^{C} x_i \ln x_i$$

Vapor–liquid equilibria:

$$(7)\ \ K_i = \frac{P_i^s}{P}$$

Reference conditions (datum): h, ideal gas at T_0 and zero pressure; s, ideal gas at T_0 and 1 atm pressure.
Refer to elements if chemical reactions occur; otherwise refer to components.

near-ambient pressure, up to about 50 psia (345 kPa), for mixtures of isomers or components of similar molecular structure.

For the vapor, the molar volume and density are computed from Eq. (1), the ideal gas law, in Table 2.4, which involves the molecular weight, M, of the mixture and the universal gas constant, R. For a mixture, the ideal gas law assumes that both Dalton's law of additive partial pressures and Amagat's law of additive pure-species volumes apply.

The molar vapor enthalpy is computed from Eq. (2) by integrating, for each species, an equation in temperature for the zero-pressure heat capacity at constant pressure, $C^o_{P_V}$, starting from a reference (datum) temperature, T_0, to the temperature of interest, and then summing the resulting species vapor enthalpies on a mole-fraction basis. Typically, T_0 is taken as 0 K or 25°C. Although the reference pressure is zero, pressure has no effect on the enthalpy of an ideal gas. A common empirical representation of the effect of temperature on the zero-pressure vapor heat capacity of a pure component is the following third-degree polynomial equation:

$$C^o_{P_V} = a_1 + a_2 T + a_3 T^2 + a_4 T^3 \tag{2-35}$$

where the constants a_k depend on the species. Values of the constants for hundreds of compounds are tabulated by Reid, Prausnitz, and Poling [11]. Because $C_P = dh/dT$, (2-35) can be integrated for each species to give the ideal gas species molar enthalpy:

$$h^o_V = \int_{T_0}^{T} C^o_{P_V}\, dT = \sum_{k=1}^{4} \frac{a_k(T^k - T^k_0)}{k} \tag{2-36}$$

The vapor molar entropy is computed from Eq. (3) in Table 2.4 by integrating $C^o_{P_V}/T$ from T_0 to T for each species, summing on a mole-fraction basis, adding a term for the effect of pressure referenced to a datum pressure, P_0, which is generally taken to be 1 atm (101.3 kPa), and adding a term for the entropy change of mixing. Unlike the ideal vapor enthalpy, the ideal vapor entropy includes terms for the effects of pressure and mixing. The reference pressure is not taken to be zero, because the entropy is infinity at zero pressure. If (2-35) is used for the heat capacity,

$$\int_{T_0}^{T} \left(\frac{C^o_{P_V}}{T}\right) dT = a_1 \ln\left(\frac{T}{T_0}\right) + \sum_{k=1}^{3} \frac{a_{k+1}(T^k - T^k_0)}{k} \tag{2-37}$$

The liquid molar volume and density are computed from the pure species molar volumes using Eq. (4) in Table 2.4 and the assumption of additive volumes (not densities). The effect of temperature on pure-component liquid density from the freezing point to the critical region at saturation pressure is correlated well by the empirical two-constant equation of Rackett [12]:

$$\rho_L = AB^{-(1-T/T_c)^{2/7}} \tag{2-38}$$

where values of the constants A, B, and the critical temperature, T_c, are tabulated for approximately 700 organic compounds by Yaws et al. [13].

The vapor pressure of a pure liquid species is well represented over a wide range of temperature from below the normal boiling point to the critical region by an empirical extended Antoine equation:

$$\ln P^s = k_1 + k_2/(k_3 + T) + k_4 T + k_5 \ln T + k_6 T^{k_7} \tag{2-39}$$

where the constants k_k depend on the species. Values of the constants for hundreds of compounds are built into the physical-property libraries of all computer-aided process simulation and design programs. Constants for other empirical vapor pressure equations are tabulated for hundreds of compounds by Reid et al. [11]. At low pressures, the enthalpy of vaporization is given in terms of vapor pressure by classical thermodynamics:

$$\Delta H^{\text{vap}} = RT^2 \left(\frac{d \ln P^s}{dT} \right) \qquad (2\text{-}40)$$

If (2-39) is used for the vapor pressure, (2-40) becomes

$$\Delta H^{\text{vap}} = RT^2 \left[-\frac{k_2}{(k_3 + T)^2} + k_4 + \frac{k_5}{T} + k_7 k_6 T^{k_7 - 1} \right] \qquad (2\text{-}41)$$

The enthalpy of an ideal liquid mixture is obtained by subtracting the molar enthalpy of vaporization from the ideal vapor molar enthalpy for each species, as given by (2-36), and summing these, as shown by Eq. (5) in Table 2.4. The entropy of the ideal liquid mixture, given by Eq. (6), is obtained in a similar manner from the ideal gas entropy by subtracting the molar entropy of vaporization given by $\Delta H^{\text{vap}}/T$.

The final equation in Table 2.4 gives the expression for the ideal K-value, previously included in Table 2.3. Although it is usually referred to as the Raoult's law K-value, where Raoult's law is given by

$$p_i = x_i P_i^s \qquad (2\text{-}42)$$

the assumption of Dalton's law is also required:

$$p_i = y_i P \qquad (2\text{-}43)$$

Combination of (2-42) and (2-43) gives the Raoult's law K-value:

$$K_i \equiv \frac{y_i}{x_i} = \frac{P_i^s}{P} \qquad (2\text{-}44)$$

The extended Antoine equation, (2-39) (or some other suitable expression) can be used to estimate vapor pressure. Note that the ideal K-value is independent of phase compositions, but is exponentially dependent on temperature, because of the vapor pressure, and inversely proportional to pressure. From (2-21), the relative volatility using (2-44) is independent of pressure.

EXAMPLE 2.3

Styrene is manufactured by catalytic dehydrogenation of ethylbenzene, followed by vacuum distillation to separate styrene from unreacted ethylbenzene [14]. Typical conditions for the feed to a commercial distillation unit are 77.5°C (350.6 K) and 100 torr (13.33 kPa) with the following vapor and liquid flows at equilibrium:

Component	n, kmol/h	
	Vapor	**Liquid**
Ethylbenzene (EB)	76.51	27.31
Styrene (S)	61.12	29.03

Based on the property constants given below, and assuming that the ideal gas, ideal liquid solution model of Table 2.4 is suitable at this low pressure, estimate values of v_V, ρ_V, h_V, s_V, v_L, ρ_L, h_L, and s_L in SI units, and the K-values and relative volatility.

Property Constants for Eqs. (2-35), (2-38), (2-39)
(In all cases, T is in K)

	Ethylbenzene	**Styrene**
M, kg/kmol	106.168	104.152
$C_{P_V}^o$, J/kmol-K:		
a_1	−43,098.9	−28,248.3
a_2	707.151	615.878
a_3	−0.481063	−0.40231
a_4	1.30084×10^{-4}	9.93528×10^{-5}
P^s, Pa:		
k_1	86.5008	130.542
k_2	−7,440.61	−9,141.07
k_3	0	0
k_4	0.00623121	0.0143369
k_5	−9.87052	−17.0918
k_6	4.13065×10^{-18}	1.8375×10^{-18}
k_7	6	6
ρ_L, kg/m^3:		
A	289.8	299.2
B	0.268	0.264
T_c, K	617.9	617.1

$R = 8.314$ kJ/kmol-K or kPa-m^3/kmol-K $= 8{,}314$ J/kmol-K

SOLUTION

Phase mole-fraction compositions and average molecular weights: From $y_i = (n_{iV})/n_V$, $x_i = (n_{iL})/n_L$,

	Ethylbenzene	**Styrene**
y	0.5559	0.4441
x	0.4848	0.5152

From Eq. (1), Table 2.4,

$$M_V = (0.5559)(106.168) + (0.4441)(104.152) = 105.27$$
$$M_L = (0.4848)(106.168) + (0.5152)(104.152) = 105.13$$

Vapor molar volume and density: From Eq. (1), Table 2.4,

$$v_V = \frac{RT}{P} = \frac{(8.314)(350.65)}{(13.332)} = 219.2 \text{ m}^3/\text{kmol}$$

$$\rho_V = \frac{M_V}{v_V} = \frac{105.27}{219.2} = 0.4802 \text{ kg/m}^3$$

Vapor molar enthalpy (datum = ideal gas at 298.15 K and 0 kPa): From (2-36) for ethylbenzene,

$$h_{\text{EB}_V}^o = -43098.9(350.65 - 298.15) + \left(\frac{707.151}{2}\right)(350.65^2 - 298.15^2)$$

$$- \left(\frac{0.481063}{3}\right)(350.65^3 - 298.15^3) + \left(\frac{1.30084 \times 10^{-4}}{4}\right)(350.65^4 - 298.15^4)$$

$$= 7{,}351{,}900 \text{ J/kmol}$$

Similarly,

$$h_{\text{S}_V}^o = 6{,}957{,}700 \text{ J/kmol}$$

From Eq. (2), Table 2.4, for the mixture,

$$h_V = \sum y_i h^o_{iV} = (0.5559)(7,351,900) + (0.4441)(6,957,100) = 7,176,800 \text{ J/kmol}$$

Vapor molar entropy (datum = pure components as vapor at 298.15 K, 101.3 kPa): From (2-37), for each component,

$$\int_{T_0}^{T} \left(\frac{C^o_{P_V}}{T}\right) dT = a_1 \ln\left(\frac{T}{T_0}\right) + a_2(T - T_0) + \frac{a_3}{2}(T^2 - T_0^2) + \frac{a_4}{3}(T^3 - T_0^3)$$

$$= 22,662 \text{ J/kmol-K} \qquad \text{for ethylbenzene}$$
$$= 21,450 \text{ J/kmol-K} \qquad \text{for styrene}$$

From Eq. (3), Table 2.4, for the mixture,

$$s_V = [(0.5559)(22,662.4) + (0.4441)(21,450.3)] - 8,314 \ln\left(\frac{13.332}{101.3}\right)$$

$$- 8,314[(0.5559) \ln(0.5559) + (0.4441) \ln(0.4441)]$$

$$= 44,695 \text{ J/kmol-K}$$

Note that the terms for the pressure effect and the mixing effect are significant for this problem.
Liquid molar volume and density: From (2-38), for ethylbenzene,

$$\rho_{EB_L} = (289.8)(0.268)^{-(1-350.65/617.9)^{2/7}} = 816.9 \text{ kg/m}^3$$

$$v_{EB_L} = \frac{M_{EB}}{\rho_{EB_L}} = 0.1300 \text{ m}^3/\text{kmol}$$

Similarly,

$$\rho_{S_L} = 853.0 \text{ kg/m}^3$$
$$v_{S_L} = 0.1221 \text{ m}^3/\text{kmol}$$

From Eq. (4), Table 2.4, for the mixture,

$$v_L = (0.4848)(0.1300) + (0.5152)(0.1221) = 0.1259 \text{ m}^3/\text{kmol}$$
$$\rho_L = \frac{M_L}{v_L} = \frac{105.13}{0.1259} = 835.0 \text{ kg/m}^3$$

Liquid molar enthalpy (datum = ideal gas at 298.15 K): Use Eq. (5) in Table 2.4 for the mixture. For the enthalpy of vaporization of ethylbenzene, from (2-41),

$$\Delta H^{vap}_{EB} = 8,314(350.65)^2 \left[\frac{-(-7,440.61)}{(0 + 350.65)^2} + 0.00623121 + \frac{-(9.87052)}{(350.65)} + 6(4.13065 \times 10^{-18})(350.65)^5\right]$$

$$= 39,589,800 \text{ J/kmol}$$

Similarly,

$$\Delta H^{vap}_S = 40,886,700 \text{ J/kmol}$$

Then, applying Eq.(5), Table 2.4, using $h^o_{EB_V}$ and $h^o_{S_V}$ from above,

$$h_L = [(0.4848)(7,351,900 - 39,589,800) + (0.5152)(6,957,700 - 40,886,700)]$$
$$= -33,109,000 \text{ J/kmol}$$

Liquid molar entropy (datum = pure components as vapor at 298.15 K and 101.3 kPa): From Eq. (6), Table 2.4 for the mixture, using values for $\int_{T_0}^{T} (C_{P_V}^o/T) \, dT$ and ΔH^{vap} of EB and S from above,

$$s_L = (0.4848) \left(22{,}662 - \frac{39{,}589{,}800}{350.65} \right) + (0.5152) \left(21{,}450 - \frac{40{,}886{,}700}{350.65} \right)$$

$$- 8.314 \ln \left(\frac{13.332}{101.3} \right) - 8{,}314[0.4848 \ln(0.4848) + 0.5152 \ln(0.5152)]$$

$$= -70{,}150 \text{ J/kmol-K}$$

K-values: Because Eq. (7), Table 2.4 will be used to compute the K-values, first estimate the vapor pressures using (2-39). For ethylbenzene,

$$\ln P_{EB}^s = 86.5008 + \left(\frac{-7{,}440.61}{(0 + 350.65)} \right)$$

$$+ 0.00623121(350.65) + (-9.87052) \ln(350.65) + 4.13065 \times 10^{-18}(350.65)^6$$

$$= 9.63481$$

$$P_{EB}^s = \exp(9.63481) = 15{,}288 \text{ Pa} = 15.288 \text{ kPa}$$

Similarly,

$$P_S^s = 11.492 \text{ kPa}$$

From Eq. (7), Table 2.4,

$$K_{EB} = \frac{15.288}{13.332} = 1.147$$

$$K_S = \frac{11.492}{13.332} = 0.862$$

Relative volatility: From (2-21),

$$\alpha_{EB,S} = \frac{K_{EB}}{K_S} = \frac{1.147}{0.862} = 1.331$$

2.4 GRAPHICAL CORRELATIONS OF THERMODYNAMIC PROPERTIES

For manual calculations, graphs of thermodynamic properties of pure compounds are very useful. This section presents representative plots for some common chemicals to illustrate effects of temperature and pressure. Such plots can be readily generated by computer-aided process design and simulation programs.

Saturated liquid densities as a function of temperature are plotted for some hydrocarbons in Figure 2.3. The density decreases rapidly as the critical temperature is approached until it becomes equal to the density of the vapor phase at the critical point. The liquid density curves are well correlated by the modified Rackett equation (2-38).

Figure 2.4 is a plot of liquid-state vapor pressures for some common chemicals, covering a wide range of temperature from below the normal boiling point to the critical temperature, where the vapor pressure terminates at the critical pressure. In general, the curves are found to fit the extended Antoine equation (2-39) reasonably well. This plot is particularly useful for determining the phase state (liquid or vapor) of a pure substance and for computing Raoult's law K-values from (2-44) [or Eq. (3) in Table 2.3].

Curves of ideal gas, zero-pressure enthalpy over a wide range of temperature are given in Figure 2.5 for light paraffin hydrocarbons. The datum is the liquid phase at 0°C, at

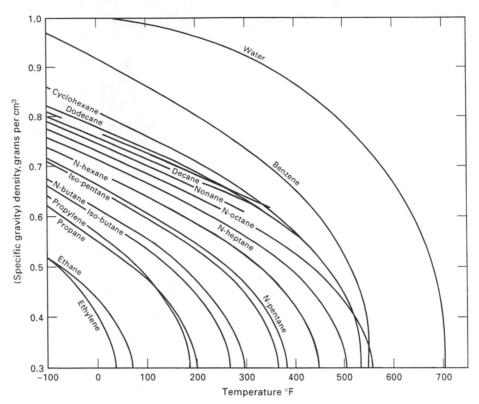

Figure 2.3 Hydrocarbon fluid densities. [Adapted from G.G. Brown, D.L. Katz, G.G. Oberfell, and R.C. Alden, *Natural Gasoline and the Volatile Hydrocarbons,* Nat'l Gas Assoc. Amer., Tulsa, OK (1948)]

which the enthalpy is zero. The derivatives of these curves fit the third-degree polynomial (2-35) for the ideal gas heat capacity reasonably well. Curves of ideal-gas entropy of several light gases, over a wide range of temperature, are given in Figure 2.6.

Enthalpies (heats) of vaporization are plotted as a function of saturation temperature in Figure 2.6 for light paraffin hydrocarbons. These values are independent of pressure and decrease to zero at the critical point, where vapor and liquid phases become indistinguishable. These plots, together with the ideal-gas enthalpy of Figure 2.5, can be used with Eq. (5) of Table 2.4 to compute liquid enthalpy up to temperatures corresponding to a vapor pressure of about 350 kPa.

Nomographs for estimating K-values of hydrocarbons and light gases are presented in Figures 2.8 and 2.9, which are taken from Hadden and Grayson [15]. In both charts all K-values collapse to 1.0 at a pressure of 5,000 psia (34.5 MPa). This pressure, called the *convergence pressure,* depends on the boiling range of the components in the mixture. For example, in Figure 2.10 the components of the mixture (N_2 to nC_{10}) cover a very wide boiling-point range, resulting in a convergence pressure of close to 2,500 psia. For narrow-boiling mixtures, such as a mixture of ethane and propane, the convergence pressure is generally less than 1,000 psia. The K-value charts of Figures 2.8 and 2.9 apply strictly to a convergence pressure of 5,000 psia, but can be used without correction to make preliminary estimates. A detailed procedure for correcting for the convergence pressure is given by Hadden and Grayson [15].

No simple charts are available for estimating liquid–liquid equilibrium constants (distribution coefficients) because of the pronounced effect of composition. However, for ternary

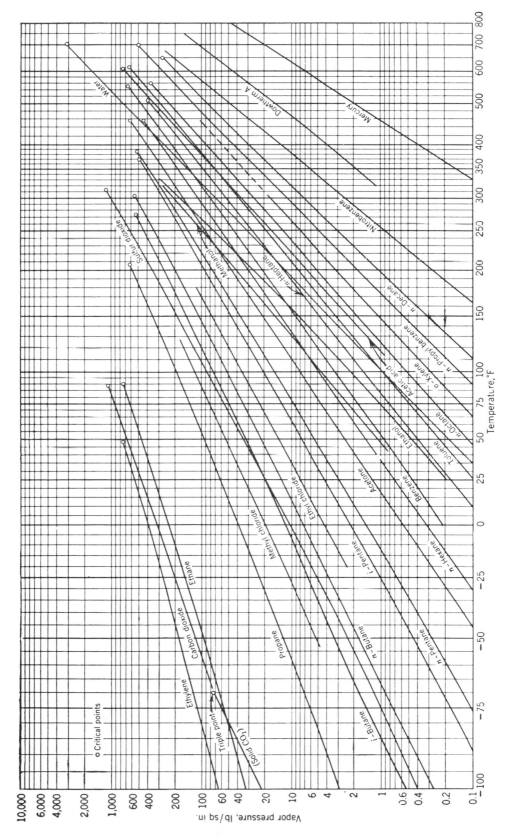

Figure 2.4 Vapor pressure as a function of temperature. [Adapted from A.S. Faust, L.A. Wenzel, C.W. Clump, L. Maus, and L.B. Andersen, *Principles of Unit Operations*, John Wiley and Sons, New York (1960).]

49

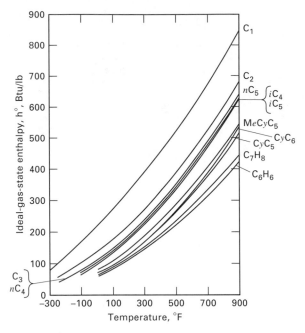

Figure 2.5 Ideal-gas-state enthalpy of pure components. [Adapted from *Engineering Data Book*, 9th ed., Gas Processors Suppliers Association, Tulsa (1972).]

systems that are dilute in the solute and involve almost immiscible solvents, an extensive tabulation of distribution coefficients for the solute is given by Robbins [16].

EXAMPLE 2.4

Petroleum refining begins with the distillation, at near-atmospheric pressure, of crude oil into fractions of different boiling ranges. The fraction boiling from 0 to 100°C, the light naphtha, is a blending stock for gasoline. The fraction boiling from 100 to 200°C, the heavy naphtha, undergoes subsequent chemical processing into more useful products. One such process is steam cracking to produce a gas containing ethylene, propylene, and a number of other compounds,

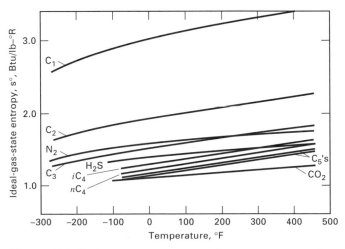

Figure 2.6 Ideal-gas-state entropy of pure components. [Adapted from *Engineering Data Book*, 9th ed., Gas Processors Suppliers Association, Tulsa (1972).]

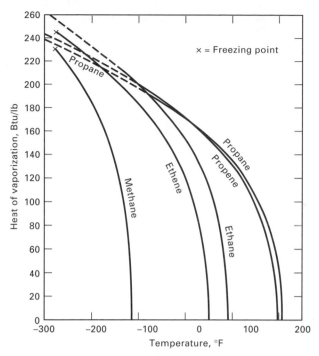

Figure 2.7 Heat of vaporization of light olefins and paraffins. [Adapted from American Petroleum Institute, Technical Data Book, Washington, DC (Aug. 1963).]

including benzene and toluene. This gas is then sent to a distillation train to separate the mixture into a dozen or more products. In the first column, hydrogen and methane are removed by cryogenic distillation at 3.2 MPa (464 psia). At a tray in the distillation column where the temperature is 40°F, use the appropriate K-value nomograph to estimate K-values for H_2, CH_4, C_2H_4, and C_3H_6.

SOLUTION

At 40°F, Figure 2.8 applies. The K-value of hydrogen depends on the other compounds in the mixture. Because appreciable amounts of benzene and toluene are present, locate a point (call it A) midway between the points for "H_2 in benzene" and "H_2 in toluene." Next, locate a point (call it B) at 40°F and 464 psia on the T–P grid. Connect points A and B with a straight line and read a value of $K = 100$ where the line intersects the K scale.

In a similar way, with the same location for point B, read $K = 11$ for methane. For ethylene (ethene) and propylene (propene), the point A is located on the normal boiling-point scale and the same point is used for B. Resulting K-values are 1.5 and 0.32, respectively. ∎

2.5 NONIDEAL THERMODYNAMIC PROPERTY MODELS

Unlike the equations of Table 2.1, which are universally applicable to all pure substances and mixtures, whether ideal or nonideal, no universal equations are available for computing, for nonideal mixtures, values of thermodynamic properties such as density, enthalpy, entropy, fugacities, and activity coefficients as functions of temperature, pressure, and phase composition. Instead, two types of models are used: (1) P–v–T equation-of-state models and (2) activity coefficient or free-energy models. These are based on *constitutive equations* because they depend on the constitution or nature of the components in the mixture.

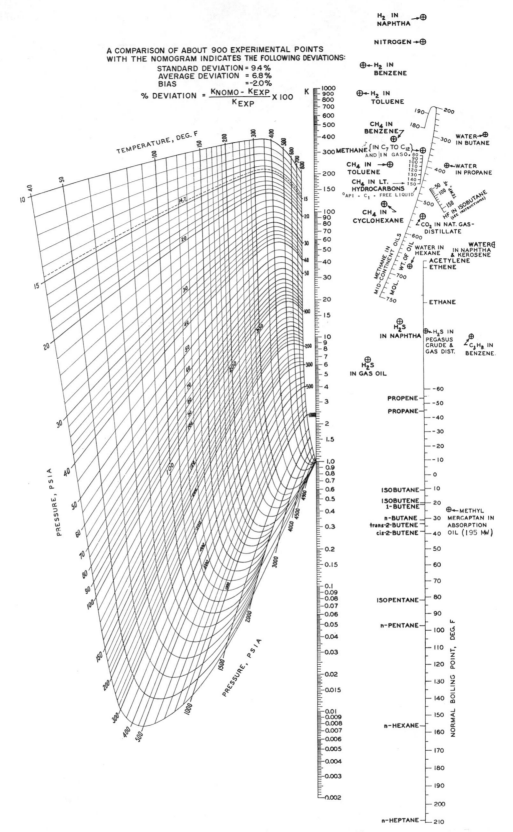

Figure 2.8 Vapor–liquid equilibria, 40 to 800°F. [From S.T. Hadden and H.G. Grayson, *Hydrocarbon Proc. and Petrol. Refiner,* **40,** 207 (Sept. 1961), with permission.]

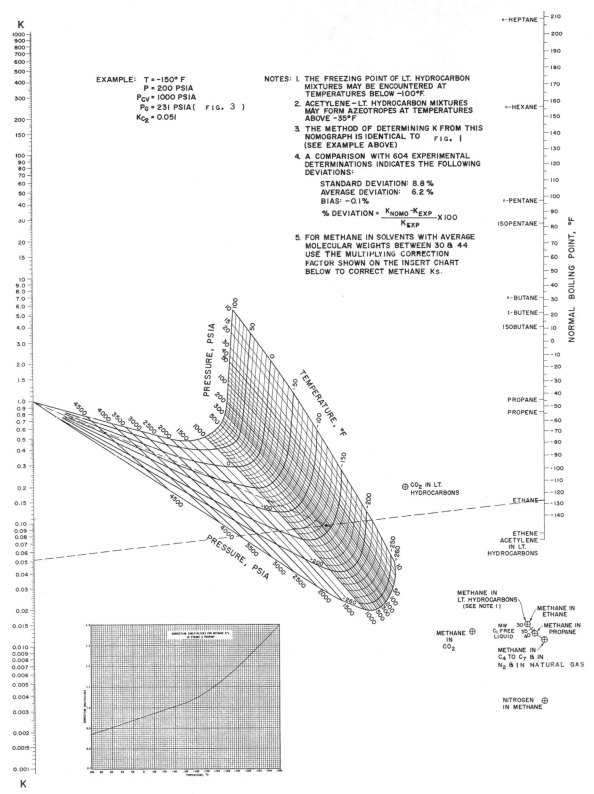

Figure 2.9 Vapor–liquid equilibria, −260 to 100°F. [From S.T. Hadden and H.G. Grayson, *Hydrocarbon Proc. and Petrol. Refiner,* **40,** 207 (Sept. 1961), with permission.]

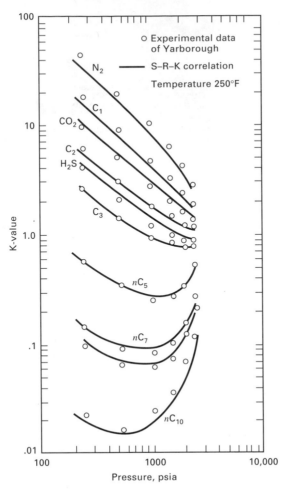

Figure 2.10 Comparison of experimental K-value data and S–R–K correlation.

$P–v–T$ Equation-of-State Models

The first type of model is a relationship between molar volume (or density), temperature, and pressure, usually referred to as a $P–v–T$ equation of state. A large number of such equations has been proposed, mostly for the vapor phase. The simplest is the ideal gas law, which applies only at low pressures or high temperatures because it neglects the volume occupied by the molecules and intermolecular forces among the molecules. All other equations of state attempt to correct for these two deficiencies. The equations of state that are most widely used by chemical engineers are listed in Table 2.5.

Not included in Table 2.5 is the van der Waals equation, $P = RT/(v - b) - a/v^2$, where a and b are species-dependent constants that can be estimated from the critical temperature and pressure. The van der Waals equation was the first successful approach to the formulation of an equation of state for a nonideal gas. It is rarely used by chemical engineers because its range of application is too narrow. However, its development did suggest that all species might have equal reduced molar volumes, $v_r = v/v_c$, at the same reduced temperature, $T_r = T/T_c$, and reduced pressure, $P_r = P/P_c$. This finding, referred to as the *law* (*principle* or *theorem*) *of corresponding states*, was utilized to develop the generalized equation of state given as Eq. (2) in Table 2.5. That equation defines the *compressibility*

Table 2.5 Useful Equations of State

Name	Equation	Equation Constants and Functions
(1) Ideal gas law	$P = \dfrac{RT}{v}$	None
(2) Generalized	$P = \dfrac{ZRT}{v}$	$Z = Z\{P_r, T_r, Z_c \text{ or } \omega\}$ as derived from data
(3) Redlich–Kwong (R–K)	$P = \dfrac{RT}{v-b} - \dfrac{a}{v^2 + bv}$	$b = 0.08664RT_c/P_c$ $a = 0.42748R^2T_c^{2.5}/P_cT^{0.5}$
(4) Soave–Redlich–Kwong (S–R–K or R–K–S)	$P = \dfrac{RT}{v-b} - \dfrac{a}{v^2 + bv}$	$b = 0.08664RT_c/P_c$ $a = 0.42748R^2T_c^2[1 + f_\omega(1 - T_r^{0.5})]^2/P_c$ $f_\omega = 0.48 + 1.574\omega - 0.176\omega^2$
(5) Peng–Robinson (P–R)	$P = \dfrac{RT}{v-b} - \dfrac{a}{v^2 + 2bv - b^2}$	$b = 0.07780RT_c/P_c$ $a = 0.45724R^2T_c^2[1 + f_\omega(1 - T_r^{0.5})]^2/P_c$ $f_\omega = 0.37464 + 1.54226\omega - 0.26992\omega^2$

factor, Z, which is a function of P_r, T_r, and the critical compressibility factor, Z_c, or the *acentric factor*, ω, which is determined from experimental P–v–T data. The acentric factor, introduced by Pitzer et al. [17], accounts for differences in molecular shape and is determined from the vapor pressure curve:

$$\omega = \left[-\log\left(\frac{P^s}{P_c}\right)_{T_r=0.7} \right] - 1.000 \tag{2-45}$$

This definition results in a value for ω of zero for symmetric molecules. Some typical values of ω are 0, 0.263, 0.489, and 0.644 for methane, toluene, n-decane, and ethyl alcohol, respectively, as taken from the extensive tabulation of Reid et al. [11].

In 1949, Redlich and Kwong [18] published an equation of state that, like the van der Waals equation, contains only two constants, both of which can be determined from T_c and P_c, by applying the critical conditions

$$\left(\frac{\partial P}{\partial v}\right)_{T_c} = 0 \quad \text{and} \quad \left(\frac{\partial^2 P}{\partial v^2}\right)_{T_c} = 0$$

However, the R–K equation, given as Eq. (3) in Table 2.5, is a considerable improvement over the van der Waals equation. A study by Shah and Thodos [19] showed that the simple R–K equation, when applied to nonpolar compounds, has an accuracy that compares quite favorably with equations containing many more constants. Furthermore, the R–K equation can approximate the liquid-phase region.

If the R–K equation is expanded to obtain a common denominator, a cubic equation in v results. Alternatively, Eqs. (2) and (3) in Table 2.5 can be combined to eliminate v to give the compressibility factor, Z, form of the R–K equation:

$$Z^3 - Z^2 + (A - B - B^2)Z - AB = 0 \tag{2-46}$$

where

$$A = \frac{aP}{R^2T^2} \tag{2-47}$$

$$B = \frac{bP}{RT} \tag{2-48}$$

Equation (2-46), which is cubic in Z, can be solved analytically for three roots (e.g., see *Perry's Handbook,* 6th ed., p. 2-15). In general, at supercritical temperatures, where only one phase can exist, one real root and a complex conjugate pair of roots are obtained.

Below the critical temperature, where vapor and/or liquid phases can exist, three real roots are obtained, with the largest value of Z (largest v) corresponding to the vapor phase—that is, Z_V—and the smallest Z (smallest v) corresponding to the liquid phase—that is, Z_L. The intermediate value of Z is of no practical use.

To apply the R–K equation to mixtures, *mixing rules* are used to average the constants a and b for each component in the mixture. The recommended rules for vapor mixtures of C components are

$$a = \sum_{i=1}^{C} \left[\sum_{j=1}^{C} y_i y_j (a_i a_j)^{0.5} \right] \qquad \textbf{(2-49)}$$

$$b = \sum_{i=1}^{C} y_i b_i \qquad \textbf{(2-50)}$$

EXAMPLE 2.5

Glanville, Sage, and Lacey [20] measured specific volumes of vapor and liquid mixtures of propane and benzene over wide ranges of temperature and pressure. Use the R–K equation to estimate specific volume of a vapor mixture containing 26.92 wt% propane at 400°F (477.6 K) and a saturation pressure of 410.3 psia (2,829 kPa). Compare the estimated and experimental values.

SOLUTION

Let propane be denoted by P and benzene by B. The mole fractions are

$$y_P = \frac{0.2692/44.097}{(0.2692/44.097) + (0.7308/78.114)} = 0.3949$$

$$y_B = 1 - 0.3949 = 0.6051$$

The critical constants for propane and benzene are given by Reid et al. [11]:

	Propane	**Benzene**
T_c, K	369.8	562.2
P_c, kPa	4,250	4,890

From the equations for the constants b and a in Table 2.5 for the R–K equation, using SI units,

$$b_P = \frac{0.08664(8.3144)(369.8)}{4,250} = 0.06268 \text{ m}^3/\text{kmol}$$

$$a_P = \frac{0.42748 \, (8.3144)^2 (369.8)^{2.5}}{(4,250)(477.59)^{0.5}}$$
$$= 836.7 \text{ kPa-m}^6/\text{kmol}^2$$

Similarly, $b_B = 0.08263 \text{ m}^3/\text{kmol}$

$$a_B = 2,072 \text{ kPa-m}^6/\text{kmol}^2$$

From (2-50), $b = (0.3949)(0.06268) + (0.6051)(0.08263) = 0.07475 \text{ m}^3/\text{kmol}$

From (2-49), $a = y_P^2 a_P + 2 y_P y_B (a_P a_B)^{0.5} + y_B^2 a_B$
$$= (0.3949)^2 (836.7) + 2(0.3949)(0.6051)[(836.7)(2,072)]^{0.5}$$
$$+ (0.6051)^2 (2,072) = 1,518 \text{ kPa-m}^6/\text{kmol}^2$$

From (2-47) and (2-48) using SI units,

$$A = \frac{(1,518)(2,829)}{(8.314)^2 (477.59)^2} = 0.2724$$

$$B = \frac{(0.07475)(2,829)}{(8.314)(477.59)} = 0.05326$$

From (2-46), we obtain the cubic Z form of the R–K equation:

$$Z^3 - Z^2 + 0.2163Z - 0.01451 = 0$$

Solving this equation gives one real root and a conjugate pair of complex roots:

$$Z = 0.7314, \qquad 0.1314 + 0.04243\,i, \qquad 0.1314 - 0.04243i$$

The one real root is assumed to be that for the vapor phase.

From Eq. (2) of Table 2.5, the molar volume is

$$v = \frac{ZRT}{P} = \frac{(0.7314)(8.314)(477.59)}{2,829} = 1.027 \text{ m}^3/\text{kmol}$$

The average molecular weight of the mixture is computed to 64.68 kg/kmol. The specific volume is

$$\frac{v}{M} = \frac{1.027}{64.68} = 0.01588 \text{ m}^3/\text{kg} = 0.2543 \text{ ft}^3/\text{lb}$$

Glanville et al. report experimental values of $Z = 0.7128$ and $v/M = 0.2478$ ft^3/lb, which are within 3% of the above estimated values. ∎

Following the success of earlier work by Wilson [21], Soave [6] added a third parameter, the acentric factor, ω, defined by (2-45), to the R–K equation. The resulting, so-called Soave–Redlich–Kwong (S–R–K) or Redlich–Kwong–Soave (R–K–S) equation of state, given as Eq. (4) in Table 2.5, was immediately accepted for application to mixtures containing hydrocarbons and/or light gases because of its simplicity and accuracy. The main improvement was to make the parameter a a function of the acentric factor and temperature so as to achieve a good fit to vapor pressure data of hydrocarbons and thereby greatly improve the ability of the equation to predict properties of the liquid phase.

Four years after the introduction of the S–R–K equation, Peng and Robinson [7] presented a further modification of the R–K and S–R–K equations in an attempt to achieve improved agreement with experimental data in the critical region and for liquid molar volume. The Peng–Robinson (P–R) equation of state is listed as Eq. (5) in Table 2.5. The S–R–K and P–R equations of state are widely applied in process calculations, particularly for saturated vapors and liquids. When applied to mixtures of hydrocarbons and/or light gases, the mixing rules are given by (2-49) and (2-50), except that (2-49) is often modified to include a binary interaction coefficient, k_{ij}:

$$a = \sum_{i=1}^{C} \left[\sum_{j=1}^{C} y_i y_j (a_i a_j)^{0.5} (1 - k_{ij}) \right] \tag{2-51}$$

Values of k_{ij}, back-calculated from experimental data, have been published for both the S–R–K and P–R equations. Knapp et al. [22] present an extensive tabulation. Generally, k_{ij} is taken as zero for hydrocarbons paired with hydrogen or other hydrocarbons.

Although the S–R–K and P–R equations were not intended to be applied to mixtures containing polar organic compounds, they are finding increasing use in such applications by employing large values of k_{ij}, in the vicinity of 0.5, as back-calculated from experimental data. However, a preferred procedure for mixtures containing polar organic compounds is to use the recent theoretically based mixing rule of Wong and Sandler, which is discussed in detail in Chapter 11 and which bridges the gap between a cubic equation of state and an activity coefficient equation.

Another theoretical basis for polar and nonpolar substances is the virial equation of state due to Thiesen [23] and Onnes [24]. A common representation of the virial equation, which can be derived from the statistical mechanics of the forces between the molecules, is a power series in $1/v$ for Z:

$$Z = 1 + \frac{B}{v} + \frac{C}{v^2} + \dots \tag{2-52}$$

An empirical modification of the virial equation is the Starling form [5] of the Benedict–Webb–Rubin (B–W–R) equation of state for hydrocarbons and light gases in both the gas and liquid phases. Walas [25] presents an extensive discussion of B–W–R-type equations, which because of the large number of terms and species constants (at least 8), is not widely used except for pure substances at cryogenic temperatures. A more useful modification of the B–W–R equation is a generalized corresponding-states form developed by Lee and Kesler [26] with an important extension to mixtures by Plocker et al. [8]. All of the constants in the L–K–P equation are given in terms of the acentric factor and reduced temperature and pressure, as developed from P–v–T data for three simple fluids ($\omega = 0$), methane, argon, and krypton, and a reference fluid ($\omega = 0.398$), n-octane. The equations, constants, and mixing rules in terms of pseudo-critical properties are given by Walas [25]. The Lee–Kesler–Plocker (L–K–P) equation of state describes vapor and liquid mixtures of hydrocarbons and/or light gases over wide ranges of temperature and pressure.

Derived Thermodynamic Properties from P–v–T Models

In the previous subsection, several useful P–v–T equations of state for the estimation of the molar volume (or density) or pure substances and mixtures in either the vapor or liquid phase were presented. If a temperature-dependent ideal-gas heat capacity or enthalpy equation, such as (2-35) or (2-36), is also available, all other vapor- and liquid-phase properties can be derived in a consistent manner by applying the classical integral equations of thermodynamics given in Table 2.6. These equations, in the form of departure (from the ideal gas) equations of Table 2.4, and often referred to as residuals, are applicable to vapor or liquid phases.

When the ideal gas law, $P = RT/v$, is substituted into Eqs. (1) to (4) of Table 2.6, the results for the vapor, as expected, are

$$(h - h_V^o) = 0 \qquad \phi = 1$$
$$(s - s_V^o) = 0 \qquad \overline{\phi} = 1$$

Table 2.6 Classical Integral Departure Equations of Thermodynamics

At a given temperature and composition, the following equations give the effect of pressure above that for an ideal gas

Mixture enthalpy:

$$(1) \ (h - h_V^o) = Pv - RT - \int_\infty^v \left[P - T\left(\frac{\partial P}{\partial T}\right)_v \right] dv$$

Mixture entropy:

$$(2) \ (s - s_V^o) = \int_\infty^v \left(\frac{\partial P}{\partial T}\right)_v dv - \int_\infty^v \frac{R}{v} dv$$

Pure-component fugacity coefficient:

$$(3) \ \phi_{iV} = \exp\left[\frac{1}{RT}\int_0^P \left(v - \frac{RT}{P}\right) dP\right] = \exp\left[\frac{1}{RT}\int_v^\infty \left(P - \frac{RT}{v}\right) dv - \ln Z + (Z - 1)\right]$$

Partial fugacity coefficient:

$$(4) \ \overline{\phi}_{iV} = \exp\left\{\frac{1}{RT}\int_V^\infty \left[\left(\frac{\partial P}{\partial N_i}\right)_{T,V,N_j} - \frac{RT}{V}\right] dV - \ln Z\right\}$$

$$\text{where } V = v \sum_{i=1}^C N_i$$

However, when the R–K equation is substituted into the equations of Table 2.6, the following results for the vapor phase are obtained after a rather tedious exercise in calculus:

$$h_V = \sum_{i=1}^{C} (y_i h_{iV}^o) + RT \left[Z_V - 1 - \frac{3A}{2B} \ln\left(\frac{B}{Z_V}\right) \right] \tag{2-53}$$

$$s_V = \sum_{i=1}^{C} (y_i s_{iV}^o) - R \ln\left(\frac{P}{P^o}\right) - R \sum_{i=1}^{C} (y_i \ln y_i) + R \ln(Z_V - B) \tag{2-54}$$

$$\phi_V = \exp\left[Z_V - 1 - \ln(Z_V - B) - \frac{A}{B} \ln\left(1 + \frac{B}{Z_V}\right) \right] \tag{2-55}$$

$$\overline{\phi}_{iV} = \exp\left[(Z_V - 1)\frac{B_i}{B} - \ln(Z_V - B) - \frac{A}{B}\left(2\sqrt{\frac{A_i}{A}} - \frac{B_i}{B}\right) \ln\left(1 + \frac{B}{Z_V}\right) \right] \tag{2-56}$$

The results for the liquid phase are identical if y_i and Z_V (but not h_{iV}^o) are replaced by x_i and Z_L, respectively. It may be surprising that the liquid-phase forms of (2-53) and (2-54) account for the enthalpy and entropy of vaporization, respectively. This is because the R–K equation of state, as well as the S–R–K and P–R equations, are continuous functions in passing between the vapor and liquid regions, as shown for enthalpy in Figure 2.11. Thus, the liquid enthalpy is determined by accounting for the following 4 effects for a pure species at a temperature below the critical. From Eq. (1), Table 2.6, the four contributions to enthalpy in Figure 2.11 are as follows:

$$h_L = h_V^o + Pv - RT - \int_{\infty}^{v} \left[P - T\left(\frac{\partial P}{\partial T}\right)_v \right] dv$$

$$= \underbrace{h_V^o}_{\substack{\text{(1) Vapor } \textit{at} \text{ zero pressure}}} + \underbrace{(Pv)_{V_s} - RT - \int_{\infty}^{v_{V_s}} \left[P - T\left(\frac{\partial P}{\partial T}\right)_v \right] dv}_{\text{(2) Pressure correction for vapor to saturation pressure}}$$

$$\underbrace{- T\left(\frac{\partial P}{\partial T}\right)_s (v_{V_s} - v_{L_s})}_{\text{(3) Latent heat of vaporization}} + \underbrace{[(Pv)_L - (Pv)_{L_s}] - \int_{v_{L_s}}^{v_L} \left[P - T\left(\frac{\partial P}{\partial T}\right)_v \right] dv}_{\substack{\text{(4) Correction to liquid for pressure in excess} \\ \text{of saturation pressure}}} \tag{2-57}$$

where the subscript s refers to the saturation pressure.

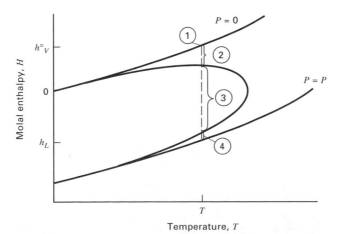

Figure 2.11 Contributions to enthalpy.

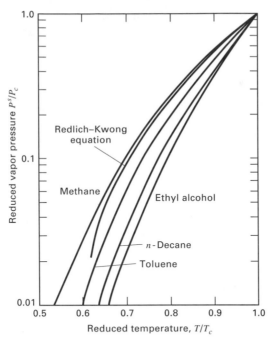

Figure 2.12 Reduced vapor pressure.

The fugacity coefficient, ϕ, of a pure species at temperature T and pressure P from the R–K equation, as given by (2-55), applies to the vapor for $P < P_i^s$. For $P > P_i^s$, ϕ is the fugacity coefficient of the liquid. Saturation pressure corresponds to the condition of $\phi_V = \phi_L$. Thus, at a temperature $T < T_c$, the saturation pressure (vapor pressure), P^s, can be estimated from the R–K equation of state by setting (2-55) for the vapor equal to (2-55) for the liquid and solving for P, which then equals P^s, by an iterative procedure. The results, as given by Edmister [27] are plotted in reduced form in Figure 2.12. The R–K vapor pressure curve does not satisfactorily represent data for a wide range of molecular shapes, as witnessed by the experimental curves for methane, toluene, n-decane, and ethyl alcohol on the same plot. This failure represents one of the major shortcomings of the R–K equation and is the main reason why Soave [6] modified the R–K equation by introducing the acentric factor in such a way as to greatly improve agreement with experimental vapor pressure data. Thus, while the critical constants, T_c and P_c alone are insufficient to generalize thermodynamic behavior, a substantial improvement is made by incorporating into the P–v–T equation a third parameter that represents the generic differences in the reduced vapor pressure curves.

As seen in (2-56), partial fugacity coefficients depend on pure species properties, A_i and B_i, and mixture properties, A and B. Once $\overline{\phi}_{iV}$ and $\overline{\phi}_{iL}$ are computed from (2-56), a K-value can be estimated from (2-26).

The most widely used P–v–T equations of state for separation calculations involving vapor and liquid phases are the S–R–K, P–R, and L–K–P relations. These equations are combined with the integral departure equations of Table 2.6 to obtain useful equations for estimating the enthalpy, entropy, fugacity coefficients, partial fugacity coefficients of vapor and liquid phases, and K-values. The results of the integrations are even more complex than (2-53) to (2-56) and are unsuitable for manual calculations. However, computer programs for making calculations with these equations are rapid, accurate, and readily available. Such programs are incorporated into widely used steady-state computer-aided process design and simulation programs.

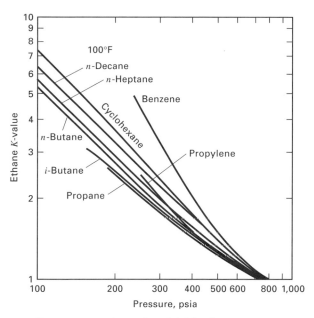

Figure 2.13 K-values of ethane in binary hydrocarbon mixtures at 100°F.

Ideal K-values as determined from Eq. (7) in Table 2.4, depend only on temperature and pressure, and not on composition. Most frequently, ideal K-values are applied to mixtures of nonpolar compounds, particularly hydrocarbons such as paraffins and olefins. Figure 2.13 shows experimental K-value curves for a light hydrocarbon, ethane, in various binary mixtures with other, less volatile hydrocarbons at 100°F (310.93 K) for pressures from 100 psia (689.5 kPa) to *convergence pressures* between 720 and 780 psia (4.964 MPa to 5.378 MPa). At the convergence pressure, separation by operations involving vapor–liquid equilibrium becomes impossible because all K-values become 1.0. The temperature of 100°F is close to the critical temperature of 550°R (305.6 K) for ethane. Figure 2.13 shows that ethane does not form ideal solutions at 100°F with all the other components because the K-values depend on the other component, even for paraffin homologs. For example, at 300 psia, the K-value of ethane in benzene is 80% higher than in propane.

The ability of equations of state, such as S–R–K, P–R, and L–K–P equations, to predict the effect of composition as well as the effect of temperature and pressure on K-values of multicomponent mixtures of hydrocarbons and light gases is shown in Figure 2.10. The mixture contains 10 species ranging in volatility from nitrogen to n-decane. The experimental data points, covering almost a 10-fold range of pressure at 250°F, are those of Yarborough [28]. Agreement with the S–R–K equation is very good.

EXAMPLE 2.6

In the high-pressure, high-temperature thermal hydrodealkylation of toluene to benzene ($C_7H_8 + H_2 \rightarrow C_6H_6 + CH_4$), excess hydrogen is used to minimize cracking of aromatics to light gases. In practice, conversion of toluene per pass through the reactor is only 70%. To separate and recycle hydrogen, hot reactor effluent vapor of 5,597 kmol/h at 500 psia (3,448 kPa) and 275°F (408.2 K) is partially condensed to 120°F (322 K), with product phases separated in a flash drum. If the composition of the reactor effluent is as follows, and the flash drum pressure is 485 psia (3,344 kPa), calculate equilibrium compositions and flow rates of vapor and liquid leaving the flash drum and the amount of heat that must be transferred using a computer-aided, steady-state simulation program with each of the equation-of-state models discussed above. Compare the results, including flash drum K-values and enthalpy and entropy changes.

Component	Mole Fraction
Hydrogen (H)	0.3177
Methane (M)	0.5894
Benzene (B)	0.0715
Toluene (T)	0.0214
	1.0000

SOLUTION

The computations were made with the DESIGN II program of the ChemShare Corporation, using the S–R–K, P–R, and L–K–P equations of state. The results at 120°F and 485 psia are as follows:

	Equation of State		
	S–R–K	**P–R**	**L–K–P**
Vapor flows, kmol/h:			
Hydrogen	1,777.1	1,774.9	1,777.8
Methane	3,271.0	3,278.5	3,281.4
Benzene	55.1	61.9	56.0
Toluene	6.4	7.4	7.0
Total	5,109.6	5,122.7	5,122.2
Liquid flows, kmol/h:			
Hydrogen	1.0	3.3	0.4
Methane	27.9	20.4	17.5
Benzene	345.1	338.2	344.1
Toluene	113.4	112.4	112.8
Total	487.4	474.3	474.8
K-values:			
Hydrogen	164.95	50.50	466.45
Methane	11.19	14.88	17.40
Benzene	0.01524	0.01695	0.01507
Toluene	0.00537	0.00610	0.00575
Enthalpy change, GJ/h	35.267	34.592	35.173
Entropy change, MJ/h-K	−95.2559	−93.4262	−95.0287
Percent of benzene and toluene condensed	88.2	86.7	87.9

Because the reactor effluent is mostly hydrogen and methane, the effluent at 275°F and 500 psia, and the equilibrium vapor at 120°F and 485 psia are nearly ideal gases ($0.98 < Z < 1.00$), despite the moderately high pressures. Thus, the enthalpy and entropy changes are dominated by vapor heat capacity and latent heat effects, which are largely independent of which equation of state is used. Consequently, the enthalpy and entropy changes among the three equations of state differ by less than 2%.

Significant differences exist for the K-values of H_2 and CH_4. However, because the values are in all cases large, the effect on the amount of equilibrium vapor is very small. Reasonable K-values for H_2 and CH_4, based on experimental data, are 100 and 13, respectively. K-values for benzene and toluene differ among the three equations of state by as much as 11% and 14%, respectively, which, however, causes less than a 2% difference in the percentage of benzene and toluene condensed. Raoult's law K-values for benzene and toluene, based on vapor pressure data, are 0.01032 and 0.00350, which are considerably lower than the values computed from each of the three equations of state because deviations to fugacities due to pressure are important in the liquid phase and, particularly, in the vapor phase.

Note that the material balances are precisely satisfied for each equation of state. However, the user of a computer-aided design and simulation program should never take this as an indication that the results are correct. ■

2.6 ACTIVITY COEFFICIENT MODELS FOR THE LIQUID PHASE

In Sections 2.3 and 2.5, methods based on equations of state are presented for predicting thermodynamic properties of vapor and liquid mixtures. In this section, predictions of liquid properties based on *Gibbs free-energy models* for predicting liquid-phase activity coefficients and other excess functions such as volume and enthalpy of mixing are developed. Regular solution theory, which can be applied to mixtures of nonpolar compounds using only constants for the pure components, is the first model presented. This is followed by a discussion of several models that can be applied to mixtures containing polar compounds, provided that experimental data are available to determine the *binary interaction parameters* in these models. If not, group-contribution methods, which have been extensively developed, can be used to make estimates. All the models discussed can be applied to predict vapor–liquid phase equilibria; and some can estimate liquid–liquid equilibria, or even solid–liquid and polymer–liquid equilibria.

Except at high pressures, dependency of K-values on composition is due primarily to nonideal solution behavior in the liquid phase. Prausnitz, Edmister, and Chao [29] showed that the relatively simple *regular solution theory* of Scatchard and Hildebrand [30] can be used to estimate deviations due to nonideal behavior for hydrocarbon–liquid mixtures. They expressed K-values in terms of (2-27), $K_i = \gamma_{iL} \phi_{iL} / \overline{\phi}_{iV}$. Chao and Seader [9] simplified and extended application of this equation to a general correlation for hydrocarbons and some light gases in the form of a compact set of equations especially suitable for use with a digital computer.

Simple models for the liquid-phase activity coefficient, γ_{iL}, based only on properties of pure species are not generally accurate. However, for hydrocarbon mixtures, regular solution theory is convenient and widely applied. The theory is based on the premise that nonideality is due to differences in van der Waals forces of attraction among the different molecules present. Regular solutions have an endothermic heat of mixing, and all activity coefficients are greater than 1. These solutions are regular in the sense that molecules are assumed to be randomly dispersed. Unequal attractive forces between like and unlike molecule pairs tend to cause segregation of molecules. However, for regular solutions the species concentrations on a molecular level are identical to overall solution concentrations. Therefore, excess entropy due to segregation is zero and entropy of regular solutions is identical to that of ideal solutions, in which the molecules are randomly dispersed.

Activity Coefficients from Gibbs Free Energy

Activity coefficient equations often have their basis in Gibbs free-energy models. For a nonideal solution, the molar Gibbs free energy, g, is the sum of the molar free energy for an ideal solution and an excess molar free energy g^E for nonideal effects. For a liquid solution,

$$g = \sum_{i=1}^{C} x_i g_i + RT \sum_{i=1}^{C} x_i \ln x_i + g^E = \sum_{i=1}^{C} x_i (g_i + RT \ln x_i + \overline{g}_i^E) \qquad \textbf{(2-58)}$$

where $g \equiv h - Ts$ and excess molar free energy is the sum of the partial excess molar free energies. The partial excess molar free energy is related by classical thermodynamics to the liquid-phase activity coefficient by

$$\frac{\overline{g}_i^E}{RT} = \ln \gamma_i = \left[\frac{\partial(N_i g^E/RT)}{\partial N_i}\right]_{P,T,N_j} = \frac{g^E}{RT} - \sum_k x_k \left[\frac{\partial(g^E/RT)}{\partial x_k}\right]_{P,T,x_r} \qquad (2\text{-}59)$$

where $j \neq i$, $r \neq k$, $k \neq i$, and $r \neq i$.

The relationship between excess molar free energy and excess molar enthalpy and entropy is

$$g^E = h^E - Ts^E = \sum_{i=1}^{C} x_i(\overline{h}_i^E - T\overline{s}_i^E) \qquad (2\text{-}60)$$

Regular Solution Model

For a multicomponent, regular liquid solution, the excess molar free energy is based on nonideality due to differences in molecular size and intermolecular forces. The former are expressed in terms of liquid molar volume and the latter in terms of the enthalpy of vaporization. The resulting model is

$$g^E = \sum_{i=1}^{C} (x_i v_{iL}) \left[\frac{1}{2}\sum_{i=1}^{C}\sum_{j=1}^{C} \Phi_i \Phi_j (\delta_i - \delta_j)^2\right] \qquad (2\text{-}61)$$

where Φ is the volume fraction assuming additive molar volumes, as given by

$$\Phi_i = \frac{x_i v_{iL}}{\sum_{j=1}^{C} x_j v_{jL}} = \frac{x_i v_{iL}}{v_L} \qquad (2\text{-}62)$$

and δ is the solubility parameter, which is defined in terms of the volumetric internal energy of vaporization as

$$\delta_i = \left(\frac{\Delta E_i^{\text{vap}}}{v_{iL}}\right)^{1/2} \qquad (2\text{-}63)$$

Applying (2-59) to (2-61) gives an expression for the activity coefficient in a regular solution:

$$\ln \gamma_{iL} = \frac{v_{iL}\left(\delta_i - \sum_{j=1}^{C} \Phi_j \delta_j\right)^2}{RT} \qquad (2\text{-}64)$$

Because $\ln \gamma_{iL}$ varies almost inversely with absolute temperature, v_{iL} and δ_j are frequently taken as constants at some convenient reference temperature, such as 25°C. Thus, the estimation of γ_L by regular solution theory requires only the pure-species constants v_L and δ. The latter parameter is often treated as an empirical constant determined by back-calculation from experimental data. For species with a critical temperature below 25°C, v_L and δ at 25°C are hypothetical. However, they can be evaluated by back-calculation from phase equilibria data.

When molecular size differences, as reflected by liquid molar volumes, are appreciable, the following Flory–Huggins size correction can be added to the regular solution free-energy contribution:

$$g^E = RT \sum_{i=1}^{C} x_i \ln\left(\frac{\Phi_i}{x_i}\right) \qquad (2\text{-}65)$$

Substitution of (2-65) into (2-59) gives

$$\ln \gamma_{iL} = \ln \left(\frac{v_{iL}}{v_L}\right) + 1 - \left(\frac{v_{iL}}{v_L}\right) \tag{2-66}$$

The complete expression for the activity coefficient of a species in a regular solution, including the Flory–Huggins correction, is

$$\gamma_{iL} = \exp \left[\frac{v_{iL}\left(\delta_i - \sum_{j=1}^{C} \Phi_j \delta_j\right)^2}{RT} + \ln \left(\frac{v_{iL}}{v_L}\right) + 1 - \frac{v_{iL}}{v_L} \right] \tag{2-67}$$

The Flory–Huggins correction, which was not included in the treatment by Chao and Seader [9] reduces the magnitude of the activity coefficient, and its use is recommended.

EXAMPLE 2.7

Yerazunis, Plowright, and Smola [31] measured liquid-phase activity coefficients for the *n*-heptane/toluene system over the entire concentration range at 1 atm (101.3 kPa). Estimate activity coefficients for the range of conditions using regular solution theory both with and without the Flory–Huggins correction. Compare estimated values with experimental data.

SOLUTION

Experimental liquid-phase compositions and temperatures for 7 of 19 points are as follows, where H denotes heptane and T denotes toluene:

T, °C	x_H	x_T
98.41	1.0000	0.0000
98.70	0.9154	0.0846
99.58	0.7479	0.2521
101.47	0.5096	0.4904
104.52	0.2681	0.7319
107.57	0.1087	0.8913
110.60	0.0000	1.0000

At 25°C, liquid molar volumes are $v_{H_L} = 147.5$ cm^3/mol and $v_{T_L} = 106.8$ cm^3/mol. Solubility parameters are 7.43 and 8.914 (cal/cm^3)$^{1/2}$, respectively, for H and T. As an example, consider mole fractions in the above table for 104.52°C. From (2-62), volume fractions are

$$\Phi_H = \frac{0.2681(147.5)}{0.2681(147.5) + 0.7319(106.8)} = 0.3359$$

$$\Phi_T = 1 - \Phi_H = 1 - 0.3359 = 0.6641$$

Substitution of these values, together with the solubility parameters, into (2-64) gives

$$\gamma_H = \exp \left\{ \frac{147.5[7.430 - 0.3359(7.430) - 0.6641(8.914)]^2}{1.987(377.67)} \right\} = 1.212$$

Values of γ_H and γ_T computed in this manner for all seven liquid-phase conditions are plotted in Figure 2.14.

Applying (2-67), with the Flory–Huggins correction, to the same data point gives

$$\gamma_H = \exp \left[0.1923 + \ln \left(\frac{147.5}{117.73}\right) + 1 - \left(\frac{147.5}{117.73}\right) \right] = 1.179$$

Values of γ_H and γ_T computed in this manner are included in Figure 2.14. Deviations from experiment are not greater than 12% for regular solution theory and not greater than 6% when the Flory–Huggins correction is included. Unfortunately, such good agreement is not always

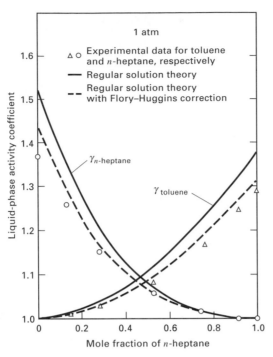

Figure 2.14 Liquid-phase activity coefficients for *n*-heptane/
toluene system at 1 atm.

obtained with nonpolar hydrocarbon solutions, as shown, for example, by Hermsen and Prausnitz
[32], who studied the cyclopentane/benzene system.   ∎

Chao–Seader Correlation

In the Chao–Seader (C–S) *K*-value correlation, the R-K equation of state, Eq. (3) in Table
2.5, is used to compute $\overline{\phi}_{iV}$, which is close to unity at low pressures. As pressure increases,
$\overline{\phi}_{iV}$ remains close to 1 for very volatile components in the mixture. For components of
low volatility, $\overline{\phi}_{iV}$ is much less than 1 as pressure approaches the convergence pressure of
the mixture.

Chao and Seader developed an empirical expression for ϕ_{iL} in terms of T_r, P_r, and ω
using the generalized correlation of Pitzer et al. [17], which is based on the equation of
state given as Eq. (2) in Table 2.5. For hypothetical liquid conditions ($P < P^s$ or $T > T_c$),
which occur very frequently in mixtures of components having widely different boiling
points or components that are noncondensable at mixture temperature, the correlation
was achieved by back-calculating ϕ_{iL} from vapor–liquid equilibrium data. The C–S equation
for ϕ_{iL} is

$$\log \phi_{iL} = \log \phi_{iL}^{(0)} + \omega_i \log \phi_{iL}^{(1)} \tag{2-68}$$

where
$$\log \phi_{iL}^{(0)} = A_0 + \frac{A_1}{T_{r_i}} + A_2 T_{r_i} + A_3 T_{r_i}^2 + A_4 T_{r_i}^3$$
$$+ (A_5 + A_6 T_{r_i} + A_7 T_{r_i}^2)P_{r_i} + (A_8 + A_9 T_{r_i})P_{r_i}^2 - \log P_{r_i} \tag{2-69}$$

and
$$\log \phi_{iL}^{(1)} = A_{10} + A_{11}T_{r_i} + \frac{A_{12}}{T_{r_i}} + A_{13}T_{r_i}^3 + A_{14}(P_{r_i} - 0.6) \tag{2-70}$$

The constants for (2-70) are

$$A_{10} = -4.23893 \qquad A_{12} = -1.22060$$
$$A_{11} = 8.65808 \qquad A_{13} = -3.15224 \qquad A_{14} = -0.025$$

Grayson and Streed [10] presented revised constants for A_0 through A_9 as follows:

	Simple Fluid, $\omega = 0$	Methane	Hydrogen
A_0	2.05135	1.36822	1.50709
A_1	−2.10899	−1.54831	2.74283
A_2	0	0	−0.02110
A_3	−0.19396	0.02889	0.00011
A_4	0.02282	−0.01076	0
A_5	0.08852	0.10486	0.008585
A_6	0	0.02529	0
A_7	−0.00872	0	0
A_8	−0.00353	0	0
A_9	−0.00203	0	0

Use of these revised constants, rather than the original constants of Chao and Seader, permits application of the C–S correlation to higher temperatures and pressures and gives improved predictions for hydrogen. An alternative improvement to the correlation of ϕ_{iL} for hydrogen, for temperatures to 730 K, is given by Jin, Greenkorn, and Chao [59].

As shown in Figures 2.15 and 2.16, the C–S correlation can be unreliable at low temperatures and generally is not recommended at temperatures below about 0°F. The empirical equations for ϕ_{iL} are applicable at reduced temperatures from 0.5 to 1.3. When the vapor is an ideal gas solution obeying the ideal gas law and the liquid solution is ideal, ϕ_{iL} is the ideal K-value.

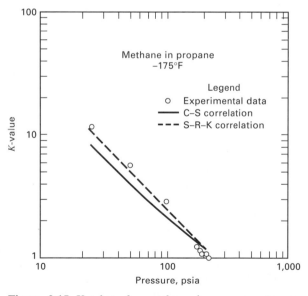

Figure 2.15 K-values for methane in propane at cryogenic conditions. (Data from R.H. Cavett, "Monsanto Physical Data System," paper presented at AIChE meeting, 1972; and E.W. West and J.H. Erbar, "An Evaluation of Four Methods of Predicting Thermodynamic Properties of Light Hydrocarbon Systems," paper presented at NGPA meeting, 1973.)

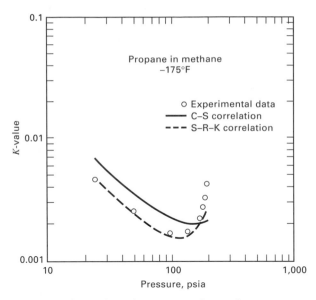

Figure 2.16 K-values for propane in methane at cryogenic conditions. (Data from R.H. Cavett, "Monsanto Physical Data System," paper presented at AIChE meeting, 1972; and E.W. West and J.H. Erbar, "An Evaluation of Four Methods of Predicting Thermodynamic Properties of Light Hydrocarbon Systems," paper presented at NGPA meeting, 1973.)

Nonideal Liquid Solutions

When liquids contain dissimilar polar species, particularly those that can form or break hydrogen bonds, the ideal liquid solution assumption is almost always invalid and the regular solution theory is not applicable. Ewell, Harrison, and Berg [33] provide a very useful classification of molecules based on the potential for association or solvation due to hydrogen bond formation. If a molecule contains a hydrogen atom attached to a donor atom (O, N, F, and in certain cases C), the active hydrogen atom can form a bond with another molecule containing a donor atom. The classification in Table 2.7 permits qualitative estimates of deviations from Raoult's law for binary pairs when used in conjunction with Table 2.8. Positive deviations correspond to values of $\gamma_{iL} > 1$. Nonideality results in a variety of variations of γ_{iL} with composition, as shown in Figure 2.17 for several binary systems, where the Roman numerals refer to classification groups in Tables 2.7 and 2.8. Starting with Figure 2.17a and taking the other plots in order, we offer the following explanations for the nonidealities. Normal heptane (V) breaks ethanol (II) hydrogen bonds, causing strong positive deviations. In Figure 2.17b, similar but less positive deviations occur when acetone (III) is added to formamide (I). Hydrogen bonds are broken and formed with chloroform (IV) and methanol (II) in Figure 2.17c, resulting in an unusual positive deviation curve for chloroform that passes through a maximum. In Figure 2.17d, chloroform (IV) provides active hydrogen atoms that can form hydrogen bonds with oxygen atoms of acetone (III), thus causing negative deviations. For water (I) and n-butanol (II) in Figure 2.17e, hydrogen bonds of both molecules are broken, and nonideality is sufficiently strong to cause formation of two immiscible liquid phases (*phase splitting*) over a wide region of overall composition.

Nonideal solution effects can be incorporated into K-value formulations in two different ways. We have already described the use of $\bar{\phi}_i$, the partial fugacity coefficient, in conjunction with an equation of state and adequate mixing rules. This is the method most frequently

Table 2.7 Classification of Molecules Based on Potential for Forming Hydrogen Bonds

Class	Description	Example
I	Molecules capable of forming three-dimensional networks of strong H-bonds	Water, glycols, glycerol, amino alcohols, hydroxylamines, hydroxyacids, polyphenols, and amides
II	Other molecules containing both active hydrogen atoms and donor atoms (O, N, and F)	Alcohols, acids, phenols, primary and secondary amines, oximes, nitro and nitrile compounds with α-hydrogen atoms, ammonia, hydrazine, hydrogen fluoride, and hydrogen cyanide
III	Molecules containing donor atoms but no active hydrogen atoms	Ethers, ketones, aldehydes, esters, tertiary amines (including pyridine type), and nitro and nitrile compounds without α-hydrogen atoms
IV	Molecules containing active hydrogen atoms but no donor atoms that have two or three chlorine atoms on the same carbon atom as a hydrogen or one chlorine on the carbon atom and one or more chlorine atoms on adjacent carbon atoms	$CHCl_3$, CH_2Cl_2, CH_3CHCl_2, CH_2ClCH_2Cl, $CH_2ClCHClCH_2Cl$, and $CH_2ClCHCl_2$
V	All other molecules having neither active hydrogen atoms nor donor atoms	Hydrocarbons, carbon disulfide, sulfides, mercaptans, and halohydrocarbons not in class IV

Table 2.8 Molecule Interactions Causing Deviations from Raoult's Law

Type of Deviation	Classes	Effect on Hydrogen Bonding
Always negative	III + IV	H-bonds formed only
Quasi-ideal; always positive or ideal	III + III III + V IV + IV IV + V V + V	No H-bonds involved
Usually positive, but some negative	I + I I + II I + III II + II II + III	H-bonds broken and formed
Always positive	I + IV (frequently limited solubility) II + IV	H-bonds broken and formed, but dissociation of class I or II is more important effect
Always positive	I + V II + V	H-bonds broken only

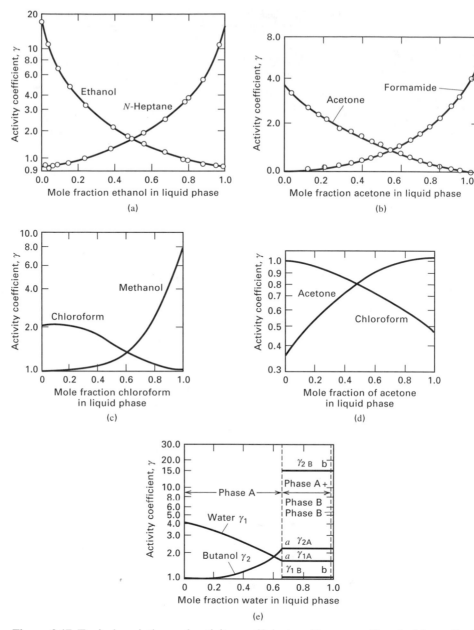

Figure 2.17 Typical variations of activity coefficients with composition in binary liquid systems: (a) ethanol(II)/n-heptane(V); (b) acetone(III)/Formamide(I); (c) chloroform(IV)/methanol(II); (d) acetone(III)/chloroform(IV); (e) water(I)/n-butanol(II).

used for handling nonidealities in the vapor phase. However, $\overline{\phi}_{iV}$ reflects the combined effects of a nonideal gas and a nonideal gas solution. At low pressures, both effects are negligible. At moderate pressures, a vapor solution may still be ideal even though the gas mixture does not follow the ideal gas law. Nonidealities in the liquid phase, however, can be severe even at low pressures. Earlier in this section, $\overline{\phi}_{iL}$ was used to express liquid-phase nonidealities for nonpolar species. When polar species are present, mixing rules can be modified to include binary interaction parameters, k_{ij}, as in (2-51).

The other technique for handling solution nonidealities is to retain in the K-value formulation $\overline{\phi}_{iV}$ but replace $\overline{\phi}_{iL}$ by the product of γ_{iL} and ϕ_{iL}, where the former quantity accounts for deviations from nonideal solutions. Equation (2-26) then becomes

$$K_i = \frac{\gamma_{iL}\phi_{iL}}{\overline{\phi}_{iV}} \tag{2-71}$$

which was derived previously as (2-27). At low pressures, from Table 2.2, $\phi_{iL} = P_i^s/P$ and $\overline{\phi}_{iV} = 1.0$, so (2-71) reduces to a modified Raoult's law K-value, which differs from (2-44) only in the γ_{iL} term:

$$K_i = \frac{\gamma_{iL}P_i^s}{P} \tag{2-72}$$

At moderate pressures Eq. (5) of Table 2.3 is preferred over (2-72).

Regular solution theory is useful only for estimating values of γ_{iL} for mixtures of nonpolar species. However, many empirical and semitheoretical equations exist for estimating activity coefficients of binary mixtures containing polar and/or nonpolar species. These equations contain binary interaction parameters, which are back-calculated from experimental data. Some of the more useful equations are listed in Table 2.9 in binary-pair form. For a given activity coefficient correlation, the equations of Table 2.10 can be used to determine excess volume, excess enthalpy, and excess entropy. However, unless the dependency on pressure of the parameters and properties used in the equations for activity coefficient is known, excess liquid volumes cannot be determined directly from Eq. (1) of Table 2.10. Fortunately, the contribution of excess volume to total mixture volume is generally small for solutions of nonelectrolytes. For example, consider a 50 mol% solution of ethanol in n-heptane at 25°C. From Figure 2.17a, this is a highly nonideal, but miscible, liquid mixture. From the data of Van Ness, Soczek, and Kochar [34], excess volume is only 0.465 cm³/mol, compared

Table 2.9 Empirical and Semitheoretical Equations for Correlating Liquid-Phase Activity Coefficients of Binary Pairs

Name	Equation for Species 1	Equation for Species 2
(1) Margules	$\log \gamma_1 = Ax_2^2$	$\log \gamma_2 = Ax_1^2$
(2) Margules (two-constant)	$\log \gamma_1 = x_2^2[\overline{A}_{12} + 2x_1(\overline{A}_{21} - \overline{A}_{12})]$	$\log \gamma_2 = x_1^2[\overline{A}_{21} + 2x_2(\overline{A}_{12} - \overline{A}_{21})]$
(3) van Laar (two-constant)	$\ln \gamma_1 = \dfrac{A_{12}}{[1 + (x_1 A_{12})/(x_2 A_{21})]^2}$	$\ln \gamma_2 = \dfrac{A_{21}}{[1 + (x_2 A_{21})/(x_1 A_{12})]^2}$
(4) Wilson (two-constant)	$\ln \gamma_1 = -\ln(x_1 + \Lambda_{12}x_2)$ $+ x_2 \left(\dfrac{\Lambda_{12}}{x_1 + \Lambda_{12}x_2} - \dfrac{\Lambda_{21}}{x_2 + \Lambda_{21}x_1} \right)$	$\ln \gamma_2 = -\ln(x_2 + \Lambda_{21}x_1)$ $- x_1 \left(\dfrac{\Lambda_{12}}{x_1 + \Lambda_{12}x_2} - \dfrac{\Lambda_{21}}{x_2 + \Lambda_{21}x_1} \right)$
(5) NRTL (three-constant)	$\ln \gamma_1 = \dfrac{x_2^2 \tau_{21} G_{21}^2}{(x_1 + x_2 G_{21})^2} + \dfrac{x_1^2 \tau_{12} G_{12}}{(x_2 + x_1 G_{12})^2}$ $G_{ij} = \exp(-\alpha_{ij}\tau_{ij})$	$\ln \gamma_2 = \dfrac{x_1^2 \tau_{12} G_{12}^2}{(x_2 + x_1 G_{12})^2} + \dfrac{x_2^2 \tau_{21} G_{21}}{(x_1 + x_2 G_{21})^2}$ $G_{ij} = \exp(-\alpha_{ij}\tau_{ij})$
(6) UNIQUAC (two-constant)	$\ln \gamma_1 = \ln \dfrac{\Psi_1}{x_1} + \dfrac{Z}{2} q_1 \ln \dfrac{\theta_1}{\Psi_1}$ $+ \Psi_2 \left(l_1 - \dfrac{r_1}{r_2} l_2 \right) - q_1 \ln(\theta_1 + \theta_2 T_{21})$ $+ \theta_2 q_1 \left(\dfrac{T_{21}}{\theta_1 + \theta_2 T_{21}} - \dfrac{T_{12}}{\theta_2 + \theta_1 T_{12}} \right)$	$\ln \gamma_2 = \ln \dfrac{\Psi_2}{x_2} + \dfrac{Z}{2} q_2 \ln \dfrac{\theta_2}{\Psi_2}$ $+ \Psi_1 \left(l_2 - \dfrac{r_2}{r_1} l_1 \right) - q_2 \ln(\theta_2 + \theta_1 T_{12})$ $+ \theta_1 q_2 \left(\dfrac{T_{12}}{\theta_2 + \theta_1 T_{12}} - \dfrac{T_{21}}{\theta_1 + \theta_2 T_{21}} \right)$

Table 2.10 Classical Partial Molar Excess
Functions of Thermodynamics

Excess volume:

(1) $(\bar{v}_{iL} - \bar{v}_{iL}^{ID}) \equiv \bar{v}_{iL}^{E} = RT \left(\dfrac{\partial \ln \gamma_{iL}}{\partial P} \right)_{T,x}$

Excess enthalpy:

(2) $(\bar{h}_{iL} - \bar{h}_{iL}^{ID}) \equiv \bar{h}_{iL}^{E} = -RT^2 \left(\dfrac{\partial \ln \gamma_{iL}}{\partial T} \right)_{P,x}$

Excess entropy:

(3) $(\bar{s}_{iL} - \bar{s}_{iL}^{ID}) \equiv \bar{s}_{iL}^{E} = -R \left[T \left(\dfrac{\partial \ln \gamma_{iL}}{\partial T} \right)_{P,x} + \ln \gamma_{iL} \right]$

ID = ideal mixture; E = excess because of nonideality.

to an estimated ideal solution molar volume of 106.3 cm³/mol. Once the partial molar excess functions are estimated for each species, the excess functions are computed from the mole fraction sums.

Margules Equations

The Margules equations (1) and (2) in Table 2.9 date back to 1895, and the two-constant form is still in common use because of its simplicity. These equations result from power-series expansions in mole fractions for $\bar{g}_i^E$ and conversion to activity coefficients by means of (2-59). The one-constant form is equivalent to symmetrical activity coefficient curves, which are rarely observed experimentally.

van Laar Equation

Because of its flexibility, simplicity, and ability to fit many systems well, the van Laar equation is widely used. It was derived from the van der Waals equation of state, but the constants, shown as A_{12} and A_{21} in Eq. (3) of Table 2.9, are best back-calculated from experimental data. These constants are, in theory, constant only for a particular binary pair at a given temperature. In practice, they are frequently computed from isobaric data covering a range of temperature. The van Laar theory expresses the temperature dependence of A_{ij} as

$$A_{ij} = \frac{A_{ij}'}{RT} \tag{2-73}$$

Regular solution theory and the van Laar equation are equivalent for a binary solution if

$$A_{ij} = \frac{v_{iL}}{RT} (\delta_i - \delta_j)^2 \tag{2-74}$$

The van Laar equation can fit activity coefficient–composition curves corresponding to both positive and negative deviations from Raoult's law, but cannot fit curves that exhibit minima or maxima such as those in Figure 2.17c.

When data are isothermal, or isobaric over only a narrow range of temperature, determination of van Laar constants is conducted in a straightforward manner. The most accurate procedure is a nonlinear regression to obtain the best fit to the data over the entire range of binary composition, subject to minimization of some objective function. A less accurate, but extremely rapid, manual calculation procedure can be used when experimental data can be extrapolated to infinite-dilution conditions. Modern experimental techniques are available for accurately and rapidly determining activity coefficients at infinite dilution. Applying Eq. (3) of Table 2.9 to the conditions $x_i = 0$ and then $x_j = 0$, we have

$$A_{ij} = \ln \gamma_i^\infty, \qquad x_i = 0$$

and
$$A_{ji} = \ln \gamma_j^\infty, \qquad x_j = 0 \tag{2-75}$$

For practical applications, it is important that the van Laar equation predicts azeotrope formation correctly, where $x_i = y_i$ and $K_i = 1.0$. If activity coefficients are known or can be computed at the azeotropic composition—say, from (2-72), ($\gamma_{iL} = P/P_i^s$, since $K_i = 1.0$)—these coefficients can be used to determine the van Laar constants directly from the following equations obtained by solving simultaneously for A_{12} and A_{21}:

$$A_{12} = \ln \gamma_1 \left(1 + \frac{x_2 \ln \gamma_2}{x_1 \ln \gamma_1} \right)^2 \tag{2-76}$$

$$A_{21} = \ln \gamma_2 \left(1 + \frac{x_1 \ln \gamma_1}{x_2 \ln \gamma_2} \right)^2 \tag{2-77}$$

These equations are applicable to activity coefficient data obtained at any single composition.

Mixtures of self-associated polar molecules (class II in Table 2.7) with nonpolar molecules such as hydrocarbons (class V) can exhibit the strong nonideality of the positive-deviation type shown in Figure 2.17a. Figure 2.18 shows experimental data of Sinor and Weber [35]

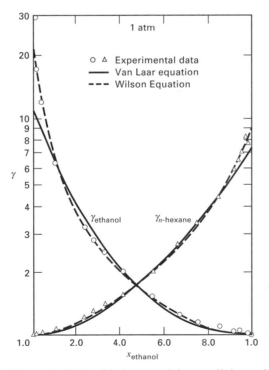

Figure 2.18 Liquid-phase activity coefficients for ethanol/ *n*-hexane system. [Data from J.E. Sinor and J.H. Weber, *J. Chem. Eng. Data*, **5**, 243–247 (1960).]

for ethanol (1)/n-hexane (2), a system of this type, at 101.3 kPa. These data were correlated with the van Laar equation by Orye and Prausnitz [36] to give $A_{12} = 2.409$ and $A_{21} = 1.970$. From $x_1 = 0.1$ to 0.9, the fit of the data to the van Laar equation is reasonably good; in the dilute regions, however, deviations are quite severe and the predicted activity coefficients for ethanol are low. An even more serious problem with these highly nonideal mixtures is that the van Laar equation may erroneously predict formation of two liquid phases (phase splitting) when values of activity coefficients exceed approximately 7.

Local Composition Concept and Wilson Equation

Since its introduction in 1964, the Wilson equation [37], shown in binary form in Table 2.9 as Eq. (4), has received wide attention because of its ability to fit strongly nonideal, but miscible, systems. As shown in Figure 2.18, the Wilson equation, with the binary interaction parameters of $\Lambda_{12} = 0.0952$ and $\Lambda_{21} = 0.2713$ determined by Orye and Prausnitz [36], fits experimental data well even in dilute regions where the variation of γ_1 becomes exponential. Corresponding infinite-dilution activity coefficients computed from the Wilson equation are $\gamma_1^\infty = 21.72$ and $\gamma_2^\infty = 9.104$.

In the Wilson equation, the effects of differences both in molecular size and intermolecular forces are incorporated by an extension of the Flory–Huggins relation (2-65). Overall solution volume fractions ($\Phi_i = x_i v_{iL}/v_L$) are replaced by local volume fractions, $\overline{\Phi}_i$, which are related to local molecule segregations caused by differing energies of interaction between pairs of molecules. The concept of local compositions that differ from overall compositions is shown schematically for an overall equimolar binary solution in Figure 2.19, which is taken from Cukor and Prausnitz [38]. About a central molecule of type 1, the local mole fraction of molecules of type 2 is shown as $\frac{5}{8}$, while the overall composition is $\frac{1}{2}$.

For local volume fraction, Wilson proposed:

$$\overline{\Phi}_i = \frac{v_{iL} x_i \exp(-\lambda_{ii}/RT)}{\sum\limits_{j=1}^{C} v_{jL} x_j \exp(-\lambda_{ij}/RT)} \tag{2-78}$$

where energies of interaction $\lambda_{ij} = \lambda_{ji}$, but $\lambda_{ii} \neq \lambda_{jj}$. Following the treatment by Orye and Prausnitz [36], substitution of the binary form of (2-78) into (2-65) and defining the binary interaction parameters as

$$\Lambda_{12} = \frac{v_{2L}}{v_{1L}} \exp\left[-\frac{(\lambda_{12} - \lambda_{11})}{RT} \right] \tag{2-79}$$

$$\Lambda_{21} = \frac{v_{1L}}{v_{2L}} \exp\left[-\frac{(\lambda_{12} - \lambda_{22})}{RT} \right] \tag{2-80}$$

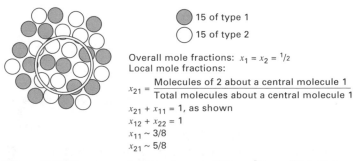

Figure 2.19 The concept of local compositions. [From P.M. Cukor and J.M. Prausnitz, *Int. Chem. Eng. Symp. Ser. No. 32*, **3**, 88 (1969).]

leads to the following equation for a binary system:

$$\frac{g^E}{RT} = -x_1 \ln(x_1 + \Lambda_{12}x_2) - x_2 \ln(x_2 + \Lambda_{21}x_1) \tag{2-81}$$

The Wilson equation is very effective for dilute compositions where entropy effects dominate over enthalpy effects. The Orye–Prausnitz form of the Wilson equation for the activity coefficient, as given in Table 2.9, follows from combining (2-59) with (2-81). Values of $\Lambda_{ij} < 1$ correspond to positive deviations from Raoult's law, while values > 1 correspond to negative deviations. Ideal solutions result from $\Lambda_{ij} = 1$. Studies indicate that λ_{ii} and λ_{ij} are temperature-dependent. Values of v_{iL}/v_{jL} depend on temperature also, but the variation may be small compared to temperature effects on the exponential terms in (2-79) and (2-80).

The Wilson equation is readily extended to multicomponent mixtures by neglecting ternary and higher molecular interactions and assuming a pseudo-binary mixture. The following multicomponent Wilson equation involves only binary interaction constants:

$$\ln \gamma_k = 1 - \ln \left(\sum_{j=1}^{C} x_j \Lambda_{kj} \right) - \sum_{i=1}^{C} \left(\frac{x_i \Lambda_{ik}}{\sum\limits_{j=1}^{C} x_j \Lambda_{ij}} \right) \tag{2-82}$$

where $\Lambda_{ii} = \Lambda_{jj} = \Lambda_{kk} = 1$.

As mixtures become highly nonideal, but still miscible, the Wilson equation becomes markedly superior to the Margules and van Laar equations. The Wilson equation is consistently superior for multicomponent solutions. Values of the constants in the Wilson equation for many binary systems are tabulated in the DECHEMA collection of Gmehling and Onken [39]. Two limitations of the Wilson equation are its inability to predict immiscibility, as in Figure 2.17e, and maxima and minima in the activity coefficient–mole fraction relationships, as shown in Figure 2.17c.

When insufficient experimental data are available to determine binary Wilson parameters from a best fit of activity coefficients over the entire range of composition, infinite-dilution or single-point values can be used. At infinite dilution, the Wilson equation in Table 2.9 becomes

$$\ln \gamma_1^{\infty} = 1 - \ln \Lambda_{12} - \Lambda_{21} \tag{2-83}$$

$$\ln \gamma_2^{\infty} = 1 - \ln \Lambda_{21} - \Lambda_{12} \tag{2-84}$$

An iterative procedure is required to obtain Λ_{12} and Λ_{21} from these nonlinear equations. If temperatures corresponding to γ_1^{∞} and γ_2^{∞} are not close or equal, (2-79) and (2-80) should be substituted into (2-83) and (2-84) with values of $(\lambda_{12} - \lambda_{11})$ and $(\lambda_{12} - \lambda_{22})$ determined from estimates of pure-component liquid molar volumes.

When the experimental data of Sinor and Weber [35] for *n*-hexane/ethanol, shown in Figure 2.18, are plotted as a *y–x* diagram in ethanol (Figure 2.20), the equilibrium curve crosses the 45° line at an ethanol mole fraction of $x = 0.332$. The measured temperature corresponding to this composition is 58°C. Ethanol has a normal boiling point of 78.33°C, which is higher than the normal boiling point of 68.75°C for *n*-hexane. Nevertheless, ethanol is more volatile than *n*-hexane up to an ethanol mole fraction of $x = 0.322$, the minimum-boiling azeotrope. This occurs because of the relatively close boiling points of the two species and the high activity coefficients for ethanol at low concentrations. At the azeotropic composition, $y_i = x_i$; therefore, $K_i = 1.0$. Applying (2-72) to both species,

$$\gamma_1 P_1^s = \gamma_2 P_2^s \tag{2-85}$$

If species 2 is more volatile in the pure state $(P_2^s > P_1^s)$, the criteria for formation of a minimum-boiling azeotrope are

$$\gamma_1 \geq 1 \tag{2-86}$$

$$\gamma_2 \geq 1 \tag{2-87}$$

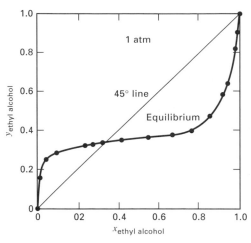

Figure 2.20 Equilibrium curve for *n*-hexane/ethanol system.

and

$$\frac{\gamma_1}{\gamma_2} < \frac{P_2^s}{P_1^s} \qquad (2\text{-}88)$$

for x_1 less than the azeotropic composition. These critieria are most readily applied at $x_1 = 0$. For example, for the *n*-hexane (2)/ethanol (1) system at 1 atm (101.3 kPa), when the liquid-phase mole fraction of ethanol approaches zero, the temperature approaches 68.75°C (155.75°F), the boiling point of pure *n*-hexane. At this temperature, $P_1^s = 10$ psia (68.9 kPa) and $P_2^s = 14.7$ psia (101.3 kPa). Also from Figure 2.18, $\gamma_1^\infty = 21.72$ when $\gamma_2 = 1.0$. Thus, $\gamma_1^\infty/\gamma_2 = 21.72$, but $= P_2^s/P_1^s = 1.47$. Therefore, a minimum-boiling azeotrope will occur.

Maximum-boiling azeotropes are less common. They occur for relatively close-boiling mixtures when negative deviations from Raoult's law arise such that $\gamma_i < 1.0$. Criteria for their formation are derived in a manner similar to that for minimum-boiling azeotropes. At $x_1 = 1$, where species 2 is more volatile,

$$\gamma_1 = 1.0 \qquad (2\text{-}89)$$

$$\gamma_2^\infty < 1.0 \qquad (2\text{-}90)$$

and

$$\frac{\gamma_2^\infty}{\gamma_1} < \frac{P_1^s}{P_2^s} \qquad (2\text{-}91)$$

For an azeotropic binary system, the two binary interaction parameters Λ_{12} and Λ_{21} can be determined by solving Eq. (4) of Table 2.9 at the azeotropic composition, as shown in the following example.

EXAMPLE 2.8

From measurements by Sinor and Weber [35] of the azeotropic condition for the ethanol/*n*-hexane system at 1 atm (101.3 kPa, 14.696 psia), calculate Λ_{12} and Λ_{21}.

SOLUTION

Let E denote ethanol and H denote *n*-hexane. The azeotrope occurs at $x_E = 0.332$, $x_H = 0.668$, and $T = 58°C$ (331.15 K). At 1 atm, (2-72) can be used to approximate K-values. Thus, at azeotropic conditions, $\gamma_i = P/P_i^s$. The vapor pressures at 58°C are $P_E^s = 6.26$ psia and $P_H^s = 10.28$ psia. Therefore,

$$\gamma_E = \frac{14.696}{6.26} = 2.348$$

$$\gamma_H = \frac{14.696}{10.28} = 1.430$$

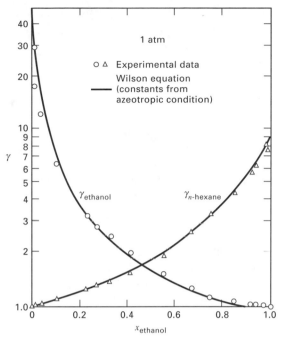

Figure 2.21 Liquid-phase activity coefficients for ethanol/ *n*-hexane system.

Substituting these values together with the above corresponding values of x_i into the binary form of the Wilson equation in Table 2.9 gives

$$\ln 2.348 = -\ln(0.332 + 0.668\Lambda_{\text{EH}}) + 0.668 \left(\frac{\Lambda_{\text{EH}}}{0.332 + 0.668\Lambda_{\text{EH}}} - \frac{\Lambda_{\text{HE}}}{0.332\Lambda_{\text{HE}} + 0.668} \right)$$

$$\ln 1.430 = -\ln(0.668 + 0.332\Lambda_{\text{HE}}) - 0.332 \left(\frac{\Lambda_{\text{EH}}}{0.332 + 0.668\Lambda_{\text{EH}}} - \frac{\Lambda_{\text{HE}}}{0.332\Lambda_{\text{HE}} + 0.668} \right)$$

Solving these two nonlinear equations simultaneously by an iterative procedure, we obtain $\Lambda_{\text{EH}} = 0.041$ and $\Lambda_{\text{HE}} = 0.281$. From these constants, the activity coefficient curves can be predicted if the temperature variations of Λ_{EH} and Λ_{HE} are ignored. The results are plotted in Figure 2.21. The fit of experimental data is good except, perhaps, for near-infinite-dilution conditions, where $\gamma_{\text{E}}^{\infty} = 49.82$ and $\gamma_{\text{H}}^{\infty} = 9.28$. The former value is considerably greater than the value of 21.72 obtained by Orye and Prausnitz [36] from a fit of all experimental data points. However, if Figures 2.18 and 2.21 are compared, it is seen that widely differing $\gamma_{\text{E}}^{\infty}$ values have little effect on γ in the composition region $x_{\text{E}} = 0.15$ to 1.00, where the two sets of Wilson curves are almost identical. For accuracy over the entire composition range, commensurate with the ability of the Wilson equation, data for at least three well-spaced liquid compositions per binary are preferred. ■

The Wilson equation can be extended to liquid–liquid or vapor–liquid–liquid systems by multiplying the right-hand side of (2-81) by a third binary-pair constant evaluated from experimental data [37]. However, for multicomponent systems of three or more species, the third binary-pair constants must be the same for all constituent binary pairs. Furthermore, as shown by Hiranuma [40], representation of ternary systems involving only one partially miscible binary pair can be extremely sensitive to the third binary-pair Wilson constant. For these reasons, application of the Wilson equation to liquid–liquid systems has not been widespread. Rather, the success of the Wilson equation for prediction of activity coefficients for miscible liquid systems has greatly stimulated further development of the local-composition concept of Wilson in an effort to obtain more universal expressions for liquid-phase activity coefficients.

NRTL Equation

The nonrandom, two-liquid (NRTL) equation developed by Renon and Prausnitz [41,42] as listed in Table 2.9, represents an accepted extension of Wilson's concept. The NRTL equation is applicable to multicomponent vapor–liquid, liquid–liquid, and vapor–liquid–liquid systems. For multicomponent vapor–liquid systems, only binary-pair constants from the corresponding binary-pair experimental data are required. For a multicomponent system, the NRTL expression for the activity coefficient is

$$\ln \gamma_i = \frac{\sum\limits_{j=1}^{C} \tau_{ji} G_{ji} x_j}{\sum\limits_{k=1}^{C} G_{ki} x_k} + \sum\limits_{j=1}^{C} \left[\frac{x_j G_{ij}}{\sum\limits_{k=1}^{C} G_{kj} x_k} \left(\tau_{ij} - \frac{\sum\limits_{k=1}^{C} x_k \tau_{kj} G_{kj}}{\sum\limits_{k=1}^{C} G_{kj} x_k} \right) \right] \tag{2-92}$$

where

$$G_{ji} = \exp(-\alpha_{ji} \tau_{ji}) \tag{2-93}$$

The coefficients τ are given by

$$\tau_{ij} = \frac{g_{ij} - g_{jj}}{RT} \tag{2-94}$$

$$\tau_{ji} = \frac{g_{ji} - g_{ii}}{RT} \tag{2-95}$$

where g_{ij}, g_{jj}, and so on, are energies of interaction between molecule pairs. In the above equations, $G_{ji} \neq G_{ij}$, $\tau_{ij} \neq \tau_{ji}$, $G_{ii} = G_{jj} = 1$, and $\tau_{ii} = \tau_{jj} = 0$. Often $(g_{ij} - g_{jj})$ and other constants are linear in temperature. For ideal solutions, $\tau_{ji} = 0$.

The parameter α_{ji} characterizes the tendency of species j and species i to be distributed in a nonrandom fashion. When $\alpha_{ji} = 0$, local mole fractions are equal to overall solution mole fractions. Generally α_{ji} is independent of temperature and depends on molecule properties in a manner similar to the classifications in Tables 2.7 and 2.8. Values of α_{ji} usually lie between 0.2 and 0.47. When $\alpha_{ji} < 0.426$, phase immiscibility is predicted. Although α_{ji} can be treated as an adjustable parameter, to be determined from experimental binary-pair data, more commonly α_{ji} is set according to the following rules, which are occasionally ambiguous:

1. $\alpha_{ji} = 0.20$ for mixtures of saturated hydrocarbons and polar nonassociated species (e.g., n-heptane/acetone).
2. $\alpha_{ji} = 0.30$ for mixtures of nonpolar compounds (e.g., benzene/n-heptane), except fluorocarbons and paraffins; mixtures of nonpolar and polar nonassociated species (e.g., benzene/acetone); mixtures of polar species that exhibit negative deviations from Raoult's law (e.g., acetone/chloroform) and moderate positive deviations (e.g., ethanol/water); mixtures of water and polar nonassociated species (e.g., water/acetone).
3. $\alpha_{ji} = 0.40$ for mixtures of saturated hydrocarbons and homolog perfluorocarbons (e.g., n-hexane/perfluoro-n-hexane).
4. $\alpha_{ji} = 0.47$ for mixtures of an alcohol or other strongly self-associated species with nonpolar species (e.g., ethanol/benzene); mixtures of carbon tetrachloride with either acetonitrile or nitromethane; mixtures of water with either butyl glycol or pyridine.

UNIQUAC Equation

In an attempt to place calculations of liquid-phase activity coefficients on a simple, yet more theoretical basis, Abrams and Prausnitz [43–45] used statistical mechanics to derive an expression for excess free energy. Their model, called UNIQUAC (universal quasi-chemical), generalizes a previous analysis by Guggenheim and extends it to mixtures of molecules that differ appreciably in size and shape. As in the Wilson and NRTL equations, local concentrations are used. However, rather than local volume fractions or local mole fractions, UNIQUAC uses the local area fraction θ_{ij} as the primary concentration variable.

The local area fraction is determined by representing a molecule by a set of bonded segments. Each molecule is characterized by two structural parameters that are determined relative to a standard segment taken as an equivalent sphere of a mer unit of a linear, infinite-length polymethylene molecule. The two structural parameters are the relative number of segments per molecule, r (volume parameter), and the relative surface area of the molecule, q (surface parameter). Values of these parameters computed from bond angles and bond distances are given by Abrams and Prausnitz [43] and Gmehling and Onken [39] for a number of species. For other compounds, values can be estimated by the group-contribution method of Fredenslund et al. [46]

For a multicomponent liquid mixture, the UNIQUAC model gives the excess free energy as

$$\frac{g^E}{RT} = \sum_{i=1}^{C} x_i \ln\left(\frac{\Psi_i}{x_i}\right) + \frac{\overline{Z}}{2}\sum_{i=1}^{C} q_i x_i \ln\left(\frac{\theta_i}{\Psi_i}\right) - \sum_{i=1}^{C} q_i x_i \ln\left(\sum_{j=1}^{C} \theta_i T_{ji}\right) \qquad (2\text{-}96)$$

The first two terms on the right-hand side account for *combinatorial* effects due to differences in molecule size and shape; the last term provides a *residual* contribution due to differences in intermolecular forces, where

$$\Psi_i = \frac{x_i r_i}{\sum\limits_{i=1}^{C} x_l r_l} = \text{segment fraction} \qquad (2\text{-}97)$$

$$\theta = \frac{x_i q_i}{\sum\limits_{i=1}^{C} x_i q_i} = \text{area fraction} \qquad (2\text{-}98)$$

where $\overline{Z}$ = lattice coordination number set equal to 10, and

$$T_{ji} = \exp\left(\frac{u_{ji} - u_{ii}}{RT}\right) \qquad (2\text{-}99)$$

Equation (2-96) contains only two adjustable parameters for each binary pair, $(u_{ji} - u_{ii})$ and $u_{ij} - u_{jj})$. Abrams and Prausnitz show that $u_{ji} = u_{ij}$ and $T_{ii} = T_{jj} = 1$. In general, $(u_{ji} - u_{ii})$ and $u_{ij} - u_{jj})$ are linear functions of temperature.

If (2-59) is combined with (2-96), an equation for the liquid-phase activity coefficient for a species in a multicomponent mixture is obtained:

$$\ln\gamma_i = \ln\gamma_i^C + \ln\gamma_i^R = \underbrace{\ln(\Psi_i/x_i) + (\overline{Z}/2\, q_i \ln(\theta_i/\Psi_i) + l_i - (\Psi_i/x_i)\sum_{j=1}^{C} x_j l_j}_{C,\ \text{combinatorial}}$$

$$\underbrace{+\, q_i\left[1 - \ln\left(\sum_{j=1}^{C}\theta_j T_{ji}\right) - \sum_{j=1}^{C}\left(\frac{\theta_j T_{ij}}{\sum\limits_{k=1}^{C}\theta_k T_{kj}}\right)\right]}_{R,\ \text{residual}} \qquad (2\text{-}100)$$

where

$$l_j = \left(\frac{\overline{Z}}{2}\right)(r_j - q_j) - (r_j - 1) \tag{2-101}$$

For a binary mixture of species 1 and 2, (2-100) reduces to Eq. (6) in Table 2.9 for $\overline{Z} = 10$.

UNIFAC Equation

Liquid-phase activity coefficients must be estimated for nonideal mixtures even when experimental phase equilibria data are not available and when the assumption of regular solutions is not valid because polar compounds are present. For such predictions, Wilson and Deal [47] and then Derr and Deal [48], in the 1960s, presented methods based on treating a solution as a mixture of functional groups instead of molecules. For example, in a solution of toluene and acetone, the contributions might be 5 aromatic CH groups, 1 aromatic C group, and 1 CH_3 group from toluene; and 2 CH_3 groups plus 1 CO carbonyl group from acetone. Alternatively, larger groups might be employed to give 5 aromatic CH groups and 1 CCH_3 group from toluene; and 1 CH_3 group and 1 CH_3CO group from acetone. As larger and larger functional groups are used, the accuracy of molecular representation increases, but the advantage of the group-contribution method decreases because a larger number of groups is required. In practice, about 50 functional groups are used to represent literally thousands of multicomponent liquid mixtures.

To estimate the partial molar excess free energies, $\overline{g}_i^E$, and then the activity coefficients, size parameters for each functional group and binary interaction parameters for each pair of functional groups are required. Size parameters can be calculated from theory. Interaction parameters are back-calculated from existing phase equilibria data and then used with the size parameters to predict phase equilibrium properties of mixtures for which no data are available.

The UNIFAC (UNIQUAC Functional-group Activity Coefficients) group-contribution method, first presented by Fredenslund, Jones, and Prausnitz [49] and further developed for use in practice by Fredenslund, Gmehling, and Rasmussen [50], has several advantages over other group-contribution methods: (1) It is theoretically based on the UNIQUAC method; (2) the parameters are essentially independent of temperature; (3) size and binary interaction parameters are available for a wide range of types of functional groups; (4) predictions can be made over a temperature range of 275 to 425 K and for pressures up to a few atmospheres; and (5) extensive comparisons with experimental data are available. All components in the mixture must be condensable.

The UNIFAC method for predicting liquid-phase activity coefficients is based on the UNIQUAC equation (2-100), wherein the molecular volume and area parameters in the combinatorial terms are replaced by:

$$r_i = \sum_k v_k^{(i)} R_k \tag{2-102}$$

$$q_i = \sum_k v_k^{(i)} Q_k \tag{2-103}$$

where $v_k^{(i)}$ is the number of functional groups of type k in molecule i, and R_k and Q_k are the volume and area parameters, respectively, for the type-k functional group.

The residual term in (2-100), which is represented by $\ln \gamma_i^R$, is replaced by the expression

$$\ln \gamma_i^R = \underbrace{\sum_k v_k^{(i)} \left(\ln \Gamma_k - \ln \Gamma_k^{(i)}\right)}_{\text{all functional groups in mixture}} \tag{2-104}$$

where Γ_k is the residual activity coefficient of the functional group k in the actual mixture, and $\Gamma_k^{(i)}$ is the same quantity but in a reference mixture that contains only molecules of type i. The latter quantity is required so that $\gamma_i^R \rightarrow 1.0$ as $x_i \rightarrow 1.0$. Both Γ_k and $\Gamma_k^{(i)}$ have the same form as the residual term in (2-100). Thus,

$$\ln \Gamma_k = Q_k \left[1 - \ln \left(\sum_m \theta_m T_{mk} \right) - \sum_m \frac{\theta_m T_{mk}}{\sum_n \theta_n T_{nm}} \right] \tag{2-105}$$

where θ_m is the area fraction of group m, given by an equation similar to (2-98),

$$\theta_m = \frac{X_m Q_m}{\sum_n X_n Q_m} \tag{2-106}$$

where X_m is the mole fraction of group m in the solution,

$$X_m = \frac{\sum_j v_m^{(j)} x_j}{\sum_j \sum_n (v_n^{(j)} x_j)} \tag{2-107}$$

and T_{mk} is a group interaction parameter given by an equation similar to (2-99),

$$T_{mk} = \exp \left(-\frac{a_{mk}}{T} \right) \tag{2-108}$$

where $a_{mk} \neq a_{km}$. When $m = k$, then $a_{mk} = 0$ and $T_{mk} = 1.0$. For $\Gamma_k^{(i)}$, (2-105) also applies, where θ terms correspond to the pure component i. Although values of R_k and Q_k are different for each functional group, values of a_{mk} are equal for all subgroups within a main group. For example, main group CH_2 consists of subgroups CH_3, CH_2, CH, and C. Accordingly,

$$a_{CH_3,CHO} = a_{CH_2,CHO} = a_{CH,CHO} = a_{C,CHO}$$

Thus, the amount of experimental data required to obtain values of a_{mk} and a_{km} and the size of the corresponding bank of data for these parameters are not as large as might be expected. Tables of recommended values of UNIFAC parameters R, Q, a_{mk}, and a_{km} are updated periodically. Values for 44 common groups are given by Gmehling, Rasmussen, and Fredenslund, [51], Macedo et al. [52], and Tiege et al. [53]. A detailed example of the estimation of activity coefficients by the UNIFAC method is given by Reid et al. [11].

To better account for the effect of temperature and improve predictions of excess enthalpy, a modified UNIFAC method was developed by Larsen, Rasmussen, and Fredenslund [54]. They use three coefficients to describe the temperature dependence of the binary interaction parameters in the following expression:

$$a_{mk} = a_{mk,1} + a_{mk,2}(T - T_0) + a_{mk,3}[T \ln(T_0/T) + (T - T_0)] \tag{2-109}$$

where T_0 is taken as 298.15 K. They also replace the combinatorial terms in (2-100) with a Flory–Huggins-type expression:

$$\ln \gamma_i^C = \ln \left(\frac{\omega_i}{x_i} \right) + 1 - \left(\frac{\omega_i}{x_i} \right) \tag{2-110}$$

where

$$\omega_i = \frac{x_i r_i^{2/3}}{\sum_i x_i r_i^{2/3}} \tag{2-111}$$

Parameters for the modified UNIFAC method are given for 21 groups by Larsen et al. [54].

Liquid–Liquid Equilibria

When species are notably dissimilar and activity coefficients are large, two and even more liquid phases may coexist at equilibrium. For example, consider the binary system of methanol (1) and cyclohexane (2) at 25°C. From measurements of Takeuchi, Nitta, and Katayama [55], van Laar constants are $A_{12} = 2.61$ and $A_{21} = 2.34$, corresponding, respectively, to infinite-dilution activity coefficients of 13.6 and 10.4 obtained using (2-75). These values of A_{12} and A_{21} can be used to construct an equilibrium plot of y_1 against x_1 assuming an isothermal condition. By combining (2-72), where $K_i = y_i/x_i$, with

$$P = \sum_{i=1}^{C} x_i \gamma_{iL} P_i^s \qquad (2\text{-}112)$$

one obtains the following relation for computing y_i from x_i:

$$y_1 = \frac{x_1 \gamma_1 P_1^s}{x_1 \gamma_1 P_1^s + x_2 \gamma_2 P_2^s} \qquad (2\text{-}113)$$

Vapor pressures at 25°C are $P_1^s = 2.452$ psia (16.9 kPa) and $P_2^s = 1.886$ psia (13.0 kPa). Activity coefficients can be computed from the van Laar equation in Table 2.9. The resulting equilibrium plot is shown in Figure 2.22, where it is observed that over much of the liquid-phase region, three values of y_1 exist. This indicates phase instability. Experimentally, single liquid phases can exist only for cyclohexane-rich mixtures of $x_1 = 0.8248$ to 1.0 and for methanol-rich mixtures of $x_1 = 0.0$ to 0.1291. Because a coexisting vapor phase exhibits only a single composition, two coexisting liquid phases prevail at opposite ends of the dashed line in Figure 2.22. The liquid phases represent solubility limits of methanol in cyclohexane and cyclohexane in methanol.

For two coexisting equilibrium liquid phases, the relation $\gamma_{iL}^{(1)} x_i^{(1)} = \gamma_{iL}^{(2)} x_i^{(2)}$ must hold. This permits determination of the two-phase region in Figure 2.22 from the van Laar or other suitable activity coefficient equation for which the constants are known. Also shown in Figure 2.22 is an equilibrium curve for the same binary system at 55°C based on data of Strubl et al. [56]. At this higher temperature, methanol and cyclohexane are completely

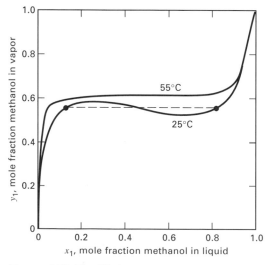

Figure 2.22 Equilibrium curves for methanol/cyclohexane systems. [Data from K. Strubl, V. Svoboda, R. Holub, and J. Pick, *Collect. Czech. Chem. Commun.*, **35**, 3004–3019 (1970).]

miscible. The data of Kiser, Johnson, and Shetlar [57] show that phase instability ceases to exist at 45.75°C, the critical solution temperature. Rigorous thermodynamic methods for determining phase instability and, thus, existence of two equilibrium liquid phases are generally based on free-energy calculations, as discussed by Prausnitz et al. [4]. Most of the empirical and semitheoretical equations for the liquid-phase activity coefficient listed in Table 2.9 apply to liquid–liquid systems. The Wilson equation is a notable exception. When using the UNIFAC method to predict liquid-phase activity coefficients for use with liquid–liquid equilibria (LLE), the parameters given in [51], [52], and [53] should not be used. A set of parameters back-calculated from LLE experimental data, given by Magnussen [58], is preferred, especially for a temperature of 25°C. However, if the modified UNIFAC method [54] is used, the accompanying parameters can be applied, but should not be expected to give accurate estimates because the parameters were not back-calculated from experimental LLE data.

SUMMARY

1. Separation processes are often energy-intensive. Energy requirements are determined by applying the first law of thermodynamics. Estimates of minimum energy needs can be made by applying the second law of thermodynamics with an entropy balance or an availability balance.

2. Phase equilibrium is expressed in terms of vapor–liquid and liquid–liquid K-values, which are formulated in terms of fugacity and activity coefficients.

3. For separation systems involving an ideal gas mixture and an ideal liquid solution, all necessary thermodynamic properties can be estimated from just the ideal gas law, a vapor heat capacity equation, a vapor pressure equation, and an equation for the liquid density as a function of temperature.

4. Graphical correlations of pure-component thermodynamic properties are widely available and useful for making rapid, manual calculations at near-ambient pressure for an ideal solution.

5. For nonideal vapor and liquid mixtures containing nonpolar components, certain P–v–T equation-of-state models such as S–R–K, P–R, and L–K–P can be used to estimate density, enthalpy, entropy, fugacity coefficients, and K-values.

6. For nonideal liquid solutions containing nonpolar and/or polar components, certain free-energy models such as Margules, van Laar, Wilson, NRTL, UNIQUAC, and UNIFAC can be used to estimate activity coefficients, volume and enthalpy of mixing, excess entropy of mixing, and K-values.

REFERENCES

1. Mix, T.W., J.S. Dweck, M. Weinberg, and R.C. Armstrong, *AIChE Symp. Ser., No. 192,* Vol. 76, 15–23 (1980).

2. Felder, R.M., and R.W. Rousseau, *Elementary Principles of Chemical Processes,* 2nd ed., John Wiley & Sons, New York (1986).

3. de Nevers, N., and J.D. Seader, *Latin Am. J. Heat and Mass Transfer,* **8,** 77–105 (1984).

4. Prausnitz, J.M., R.N. Lichtenthaler, and E.G. de Azevedo, *Molecular Thermodynamics of Fluid-Phase Equilibria,* 2nd ed., Prentice-Hall, Englewood Cliffs, NJ, pp. 18–20 (1986).

5. Starling, K.E., *Fluid Thermodynamic Properties for Light Petroleum Systems,* Gulf Publishing, Houston, TX (1973).

6. Soave, G., *Chem. Eng. Sci.,* **27,** 1197–1203 (1972).

7. Peng, D.Y., and D.B. Robinson, *Ind. Eng. Chem. Fundam.,* **15,** 59–64 (1976).

8. Plöcker, U., H. Knapp, and J.M. Prausnitz, *Ind. Eng. Chem. Process Des. Dev.,* **17,** 324–332 (1978).

9. Chao, K.C., and J.D. Seader, *AIChE J.,* **7,** 598–605 (1961).

10. Grayson, H.G., and C.W. Streed, Paper 20-P07, Sixth World Petroleum Conference, Frankfurt, June 1963.

11. Reid, R.C., J.M. Prausnitz, and B.E. Poling, *The Properties of Gases and Liquids,* 4th ed., McGraw-Hill, New York (1986).

12. Rackett, H.G., *J. Chem. Eng. Data,* **15,** 514–517 (1970).

13. Yaws, C.L., H.-C. Yang, J.R. Hopper, and W.A. Cawley, *Hydrocarbon Processing,* **71** (1), 103–106 (1991).

14. Frank, J.C., G.R. Geyer, and H. Kehde, *Chem. Eng. Prog.,* **65** (2), 79–86 (1969).

15. Hadden, S.T., and H.G. Grayson, *Hydrocarbon Process., Petrol. Refiner,* **40** (9), 207–218 (1961).

16. Robbins, L.A., Section 15, "Liquid-Liquid Extraction," in R.H. Perry, D. Green, and J.O. Maloney, Eds., *Perry's Chemical Engineers' Handbook,* 6th ed., McGraw-Hill, New York (1984).

17. Pitzer, K.S., D.Z. Lippman, R.F. Curl, Jr., C.M. Huggins, and

D.E. Petersen, *J. Am. Chem. Soc.*, **77**, 3433–3440 (1955).

18. Redlich, O., and J.N.S. Kwong, *Chem. Rev.*, **44**, 233–244 (1949).

19. Shah, K.K., and G. Thodos, *Ind. Eng. Chem.*, **57** (3), 30–37 (1965).

20. Glanville, J.W., B.H. Sage, and W.N. Lacey, *Ind. Eng. Chem.*, **42**, 508–513 (1950).

21. Wilson, G.M., *Adv. Cryogenic Eng.*, **11**, 392–400 (1966).

22. Knapp, H., R. Doring, L. Oellrich, U. Plöcker, and J.M. Prausnitz, *Vapor-Liquid Equilbria for Mixtures of Low Boiling Substances,* Chem. Data. Ser., Vol. VI, DECHEMA (1982).

23. Thiesen, M., *Ann. Phys.*, **24**, 467–492 (1885).

24. Onnes, K., *Konink. Akad. Wetens,* p. 633 (1912).

25. Walas, S.M., *Phase Equilbria in Chemical Engineering,* Butterworth, Boston (1985).

26. Lee, B.I., and M.G. Kessler, *AIChE J.*, **21**, 510–527 (1975).

27. Edmister, W.C., *Hydrocarbon Processing,* **47** (9), 239–244 (1968).

28. Yarborough, L., *J. Chem. Eng. Data*, **17**, 129–133 (1972).

29. Prausnitz, J.M., W.C. Edmister, and K.C. Chao, *AIChE J.*, **6**, 214–219 (1960).

30. Hildebrand, J.H., J.M. Prausnitz, and R.L. Scott, *Regular and Related Solutions,* Van Nostrand Reinhold, New York (1970).

31. Yerazunis, S., J.D. Plowright, and F.M. Smola, *AIChE J.*, **10**, 660–665 (1964).

32. Hermsen, R.W., and J.M. Prausnitz, *Chem. Eng. Sci.*, **18**, 485–494 (1963).

33. Ewell, R.H., J.M. Harrison, and L. Berg, *Ind. Eng. Chem.*, **36**, 871–875 (1944).

34. Van Ness, H.C., C.A. Soczek, and N.K. Kochar, *J. Chem. Eng. Data*, **12**, 346–351 (1967).

35. Sinor, J.E., and J.H. Weber, *J. Chem. Eng. Data*, **5**, 243–247 (1960).

36. Orye, R.V., and J.M. Prausnitz, *Ind. Eng. Chem.*, **57** (5), 18–26 (1965).

37. Wilson, G.M., *J. Am. Chem. Soc.*, **86**, 127–130 (1964).

38. Cukor, P.M., and J.M. Prausnitz, *Inst. Chem. Eng. Symp. Ser. No. 32*, **3**, 88 (1969).

39. Gmehling, J., and U. Onken, *Vapor-Liquid Equilibrium Data Collection,* DECHEMA Chem. Data Ser., 1-8, (1977–1984).

40. Hiranuma, M., *J. Chem. Eng. Japan*, **8**, 162–163 (1957).

41. Renon, H., and J.M. Prausnitz, *AIChE J.*, **14**, 135–144 (1968).

42. Renon, H., and J.M. Prausnitz, *Ind. Eng. Chem. Process Des. Dev.*, **8**, 413–419 (1969).

43. Abrams, D.S., and J.M. Prausnitz, *AIChE J.*, **21**, 116–128 (1975).

44. Abrams, D.S., Ph.D. thesis in chemical engineering, University of California, Berkeley, 1974.

45. Prausnitz, J.M., T.F. Anderson, E.A. Grens, C.A. Eckert, R. Hsieh, and J.P. O'Connell, *Computer Calculations for Multicomponent Vapor-Liquid and Liquid-Liquid Equilibria,* Prentice-Hall, Englewood Cliffs, NJ (1980).

46. Fredenslund, A., J. Gmehling, M.L. Michelsen, P. Rasmussen, and J.M. Prausnitz, *Ind. Eng. Chem. Process Des. Dev.*, **16**, 450–462 (1977).

47. Wilson, G.M., and C.H. Deal, *Ind. Eng. Chem. Fundam.*, **1**, 20–23 (1962).

48. Derr, E.L., and C.H. Deal, *Inst. Chem. Eng. Symp. Ser. No. 32*, **3**, 40–51 (1969).

49. Fredenslund, A., R.L. Jones, and J.M. Prausnitz, *AIChE J.*, **21**, 1086–1099 (1975).

50. Fredenslund, A., J. Gmehling, and P. Rasmussen, *Vapor-Liquid Equilibria Using UNIFAC, A Group Contribution Method,* Elsevier, Amsterdam (1977).

51. Gmehling, J., P. Rasmussen, and A. Fredenslund, *Ind. Eng. Chem. Process Des. Dev.*, **21**, 118–127 (1982).

52. Macedo, E.A., U. Weidlich, J. Gmehling, and P. Rasmussen, *Ind. Eng. Chem. Process Des. Dev.*, **22**, 676–678 (1983).

53. Tiegs, D., J. Gmehling, P. Rasmussen, and A. Fredenslund, *Ind. Eng. Chem. Res.*, **26**, 159–161 (1987).

54. Larsen, B.L., P. Rasmussen, and A. Fredenslund, *Ind. Eng. Chem. Res.*, **26**, 2274–2286 (1987).

55. Takeuchi, S., T. Nitta, and T. Katayama, *J. Chem. Eng. Japan*, **8**, 248–250 (1975).

56. Strubl, K., V. Svoboda, R. Holub, and J. Pick, *Collect. Czech. Chem. Commun.*, **35**, 3004–3019 (1970).

57. Kiser, R.W., G.D. Johnson, and M.D. Shetlar, *J. Chem. Eng. Data*, **6**, 338–341 (1961).

58. Magnussen, T., *Ind. Eng. Chem. Process Des. Dev.*, **20**, 331–339 (1981).

59. Jin, Z.-L., R.A. Greenkorn, and K.C. Chao, *AIChE J.*, **41**, 1602–1604 (1995).

EXERCISES

Section 2.1

2.1 A hydrocarbon stream in a petroleum refinery is to be separated at 1,500 kPa into two products under the conditions shown below. Using the data given, compute the minimum work of separation, W_{min}, in kJ/h for $T_0 = 298.15$ K.

	Feed	Product 1	Product 2
Phase condition	Liquid	Vapor	Liquid
Temperature, K	364	313	394
Enthalpy, kJ/kmol	19,480	25,040	25,640
Entropy, kJ/kmol-K	36.64	33.13	54.84

	kmol/h	
Component	**Feed**	**Product 1**
Ethane	30	30
Propane	200	192
n-Butane	370	4
n-Pentane	350	0
n-Hexane	50	0

2.2 In petroleum refineries, a mixture of paraffins and cycloparaffins is commonly reformed in a fixed-bed catalytic reactor to produce blending stocks for gasoline and aromatic precursors for making petrochemicals. A typical multicomponent product from catalytic reforming is a mixture of ethylbenzene with the three xylene isomers. If this mixture is separated, these four chemicals can then be subsequently processed to make styrene, phthalic anhydride, isophthalic acid, and terephthalic acid. Compute, using the following data, the minimum work of separation in Btu/h for $T_0 = 560°R$ if the mixture below is separated at 20 psia into three products.

		Split Fraction (SF)		
Component	**Feed, lbmol/h**	**Product 1**	**Product 2**	**Product 3**
Ethylbenzene	150	0.96	0.04	0.000
p-Xylene	190	0.005	0.99	0.005
m-Xylene	430	0.004	0.99	0.006
o-Xylene	230	0.00	0.015	0.985

	Feed	**Product 1**	**Product 2**	**Product 3**
Phase condition	Liquid	Liquid	Liquid	Liquid
Temperature, °F	305	299	304	314
Enthalpy, Btu/lbmol	29,290	29,750	29,550	28,320
Entropy, Btu/lbmol-°R	15.32	12.47	13.60	14.68

2.3 Distillation column C3 in Figure 1.9 separates stream 5 into streams 6 and 7, according to the material balance in Table 1.5. A suitable column for the separation, if carried out at 700 kPa, contains 70 plates with a condenser duty of 27,300,000 kJ/h. Using the following data and an infinite surroundings temperature, T_0, of 298.15 K, compute:
(a) The duty of the reboiler in kJ/h.
(b) The irreversible production of entropy in kJ/h-K, assuming the use of cooling water at a nominal temperature of 25°C for the condenser and saturated steam at 100°C for the reboiler
(c) The lost work in kJ/h.
(d) The minimum work of separation in kJ/h.
(e) The second-law efficiency

	Feed (Stream 5)	**Distillate (Stream 6)**	**Bottoms (Stream 7)**
Phase condition	Liquid	Liquid	Liquid
Temperature, K	348	323	343
Pressure, kPa	1,950	700	730
Enthalpy, kJ/kmol	17,000	13,420	15,840
Entropy, kJ/kmol-K	25.05	5.87	21.22

2.4 A spiral-wound, nonporous cellulose acetate membrane separator is to be used to separate a gas containing H_2, CH_4, and C_2H_6. The permeate will be 95 mol% pure H_2 and will contain no ethane. The relative split ratio (separation power), SP, for H_2 relative to methane will be 47. Using the

following data and an infinite surroundings temperature of 80°F, compute:
(a) The irreversible production of entropy in Btu/h-R
(b) The lost work in Btu/h
(c) The minimum work of separation in Btu/h
(d) The second-law efficiency
What other method(s) might be used to make the separation?

	Feed flow rates, lbmol/h
H_2	3,000
CH_4	884
C_2H_6	120

Stream properties:

	Feed	**Permeate**	**Retentate**
Phase condition	Vapor	Vapor	Vapor
Temperature, °F	80	80	80
Pressure, psia	365	50	365
Enthalpy, Btu/lbmol	8,550	8,380	8,890
Entropy, Btu/lbmol-R	1.520	4.222	2.742

Section 2.2

2.5 Which of the following K-value expressions, if any, is (are) rigorous? For those expressions that are not rigorous, cite the assumptions involved.
(a) $K_i = \overline{\phi}_{iL}/\overline{\phi}_{iV}$; (b) $K_i = \phi_{iL}/\phi_{iL}$; (c) $K_i = \phi_{iL}$
(d) $K_i = \gamma_{iL}\phi_{iL}/\overline{\phi}_{iV}$; (e) $K_i = P_i^s/P$; (f) $K_i = \gamma_{iL}\phi_{iL}/\gamma_{iV}\phi_{iV}$
(g) $K_i = \gamma_{iL}P_i^s/P$

2.6 Experimental measurements of Vaughan and Collins [*Ind. Eng. Chem.,* **34,** 885 (1942)] for the propane–isopentane system at 167°F and 147 psia show for propane a liquid-phase mole fraction of 0.2900 in equilibrium with a vapor-phase mole fraction of 0.6650. Calculate:
(a) The K-values for C_3 and iC_5 from the experimental data.
(b) Estimates of the K-values of C_3 and iC_5 from Raoult's law assuming vapor pressures at 167°F of 409.6 and 58.6 psia, respectively.

Compare the results of (a) and (b). Assuming the experimental values are correct, how could better estimates of the K-values be achieved? To respond to this question, compare the rigorous expression $K_i = \gamma_{iL}\phi_{iL}/\overline{\phi}_{iV}$ to the Raoult's law expression $K_i = P_i^s/P$.

2.7 Mutual solubility data for the isooctane (1)/furfural (2) system at 25°C are [*Chem. Eng. Sci.,* **6,** 116 (1957)]

	Liquid Phase I	**Liquid Phase II**
x_1	0.0431	0.9461

Compute:
(a) The distribution coefficients for isooctane and furfural
(b) The relative selectivity for isooctane relative to furfural
(c) The activity coefficient of isooctane in liquid phase 1 and the activity coefficient of furfural in liquid phase 2 assuming $\gamma_2^{(1)} = 1.0$ and $\gamma_1^{(2)} = 1.0$.

2.8 In petroleum refineries, streams rich in alkylbenzenes and alkylnaphthalenes result from catalytic cracking operations. Such streams can be hydrodealkylated to more valuable products such as benzene and naphthalene. At 25°C, solid naphthalene (normal melting point = 80.3°C) has the following solubilities in various liquid solvents [*Naphthalene*, API Publication 707, Washington, DC (Oct. 1978)], including benzene:

Solvent	Mole Fraction Naphthalene
Benzene	0.2946
Cyclohexane	0.1487
Carbon tetrachloride	0.2591
n-Hexane	0.1168
Water	0.18×10^{-5}

For each solvent, compute the activity coefficient of naphthalene in the liquid solvent phase using the following equations for the vapor pressure in torr of solid and liquid naphthalene:

$$\ln P_{\text{solid}}^s = 26.708 - 8{,}712/T$$

$$\ln P_{\text{liquid}}^s = 16.1426 - 3992.01/(T - 71.29)$$

where T is in K.

Section 2.3

2.9 A binary ideal gas mixture of A and B undergoes an isothermal separation at T_0, the infinite surroundings temperature. Starting with Eq. (4), Table 2.1, derive an equation for the minimum work of separation, $W_{\min}$, in terms of mole fractions of the feed and the two products. Use your equation to prepare a plot of the dimensionless group, $W_{\min}/RT_0 n_F$, as a function of mole fraction of A in the feed for:
(a) A perfect separation
(b) A separation with $SF_A = 0.98$, $SF_B = 0.02$
(c) A separation with $SR_A = 9.0$ and $SR_B = \frac{1}{9}$
(d) A separation with $SF = 0.95$ for A and $SP_{A,B} = 361$

How sensitive is $W_{\min}$ to product purities? Does $W_{\min}$ depend on the particular separation operation used?

Prove, by calculus, that the largest value of $W_{\min}$ occurs for a feed with equimolar quantities of A and B.

2.10 The separation of isopentane from n-pentane by distillation is difficult (approximately 100 trays are required), but is commonly practiced in industry. Using the extended Antoine vapor pressure equation, (2-39), with the constants

below and in conjunction with Raoult's law, calculate relative volatilities for the isopentane/n-pentane system and compare the values on a plot with the following smoothed experimental values [*J. Chem. Eng. Data*, **8**, 504 (1963)]:

Temperature, °F	α_{iC_5, nC_5}
125	1.26
150	1.23
175	1.21
200	1.18
225	1.16
250	1.14

What do you conclude about the applicability of Raoult's law in this temperature range for this binary system?
Vapor pressure constants for (2-39) with vapor pressure in kPa and T in K

	iC_5	nC_5
k_1	13.6106	13.9778
k_2	−2,345.09	−2,554.60
k_3	−40.2128	−36.2529
k_4, k_5, k_6	0	0

2.11 Operating conditions at the top of a vacuum distillation column for the separation of ethylbenzene from styrene are given below, where the overhead vapor is condensed in an air-cooled condenser to give subcooled reflux and distillate. Using the property constants in Example 2.3, estimate the heat transfer rate (duty) for the condenser in kJ/h.

	Overhead Vapor	Reflux	Distillate
Phase condition	Vapor	Liquid	Liquid
Temperature, K	331	325	325
Pressure, kPa	6.69	6.40	6.40
Component flow rates, kg/h:			
Ethylbenzene	77,500	66,960	10,540
Styrene	2,500	2,160	340

2.12 Toluene can be hydrodealkylated to benzene, but the conversion per pass through the reactor is only about 70%. Consequently, the toluene must be recovered and recycled. Typical conditions for the feed to a commercial distillation unit are 100°F, 20 psia, 415 lbmol/h of benzene and 131 lbmol/h of toluene. Based on the property constants below, and assuming that the ideal gas, ideal liquid solution model of Table 2.4 applies at this low pressure, estimate values of v_L, ρ_L, h_L, and s_L in American engineering units.
Property constants for (2-35), (2-38) and (2-39), where in all cases, T is in K, are

	Benzene	Toluene
M, kg/kmol	78.114	92.141
$C^0_{P_V}$, J/kmol-K:		
a_1	−33,920	−24,350
a_2	473.9	512.5
a_3	−0.3017	−0.2765
a_4	7.13×10^{-5}	4.911×10^{-5}
P^s, torr:		
k_1	15.9008	16.0137
k_2	−2,788.51	−3,096.52
k_3	−52.36	−53.67
k_4, k_5, k_6	0	0
ρ_L, kg/m³:		
A	304.1	290.6
B	0.269	0.265
T_c	562.0	593.1

Section 2.4

2.13 Measured conditions for the bottoms from a depropanizer distillation unit in a small refinery are given below. Using the data in Figure 2.3 and assuming an ideal liquid solution (volume of mixing = 0), compute the liquid density in lb/ft³, lb/gal, lb/bbl (42 gal), and kg/m³.

Phase Condition	Liquid
Temperature, °F	229
Pressure, psia	282
Flow rates, lbmol/h:	
C_3	2.2
iC_4	171.1
nC_4	226.6
iC_5	28.1
nC_5	17.5

2.14 Isopropanol, containing 13 wt% water, can be dehydrated to obtain almost pure isopropanol at a 90% recovery by azeotropic distillation with benzene. When condensed, the overhead vapor from the column splits into two immiscible liquid phases. Use the relations in Table 2.4 with data in Perry's Handbook and the operating conditions below to compute the rate of heat transfer in Btu/h and kJ/h for the condenser.

	Overhead	Water-Rich Phase	Organic-Rich Phase
Phase	Vapor	Liquid	Liquid
Temperature, °C	76	40	40
Pressure, bar	1.4	1.4	1.4
Flow rate, kg/h:			
Isopropanol	6,800	5,870	930
Water	2,350	1,790	560
Benzene	24,600	30	24,570

2.15 A hydrocarbon vapor–liquid mixture at 250°F and 500 psia contains N_2, H_2S, CO_2, and all the normal paraffins from methane to heptane. Use Figure 2.8 to estimate the K-value of each component in the mixture. Which components will have a tendency to be present to a greater extent in the equilibrium vapor?

2.16 Acetone, a valuable solvent, can be recovered from air by absorption in water or by adsorption on activated carbon. If absorption is used, the conditions for the streams entering and leaving are as listed below. If the absorber operates adiabatically, estimate the temperature of the exiting liquid phase.

	Feed Gas	Absorbent	Gas Out	Liquid Out
Flow rate, lbmol/h:				
Air	687	0	687	0
Acetone	15	0	0.1	14.9
Water	0	1,733	22	1,711
Temperature, °F	78	90	80	—
Pressure, psia	15	15	14	15
Phase	Vapor	Liquid	Vapor	Liquid

Some concern has been expressed about the possible explosion hazard associated with the feed gas. The lower and upper flammability limits for acetone in air are 2.5 and 13 mol%, respectively. Is the mixture within the explosive range? If so, what can be done to remedy the situation?

Section 2.5

2.17 Subquality natural gas contains an intolerable amount of nitrogen impurity. Separation processes that can be used to remove nitrogen include cryogenic distillation, membrane separation, and pressure-swing adsorption. For the latter process, a set of typical feed and product conditions is given below. Assume a 90% removal of N_2 and a 97% methane natural-gas product. Using the R–K equation of state with the constants listed below, compute the flow rate in thousands of actual cubic feet per hour for each of the three streams.

	N_2	CH_4
Feed flow rate, lbmol/h:	176	704
T_c, K	126.2	190.4
P_c, bar	33.9	46.0

Stream conditions are

	Feed (Subquality Natural Gas)	Product (Natural Gas)	Waste Gas
Temperature, °F	70	100	70
Pressure, psia	800	790	280

2.18 Use the R–K equation of state to estimate the partial fugacity coefficients of propane and benzene in the vapor mixture of Example 2.5.

2.19 Use a computer-aided, steady-state simulation program to estimate the K-values, using the P–R or S–R–K equation of state, of an equimolar mixture of the two butane isomers and the four butene isomers at 220°F and 276.5 psia. Compare these values with the following experimental results [*J. Chem. Eng. Data*, **7**, 331 (1962)]:

Component	K-value
Isobutane	1.067
Isobutene	1.024
n-Butane	0.922
1-Butene	1.024
trans-2-Butene	0.952
cis-2-Butene	0.876

2.20 The disproportionation of toluene to benzene and xylenes is carried out in a catalytic reactor at 500 psia and 950°F. The reactor effluent is cooled in a series of heat exchangers for heat recovery until a temperature of 235°F is reached at a pressure of 490 psia. The effluent is then further cooled and partially condensed by the transfer of heat to cooling water in a final exchanger. The resulting two-phase equilibrium mixture at 100°F and 485 psia is then separated in a flash drum. For the reactor effluent composition given below, use a computer-aided, steady-state simulation program with the S–R–K and P–R equations of state to compute the component flow rates in lbmol/h in both the resulting vapor and liquid streams, the component K-values for the equilibrium mixture, and the rate of heat transfer to the cooling water. Compare the results.

Component	Reactor Effluent, lbmol/h
H_2	1,900
CH_4	215
C_2H_6	17
Benzene	577
Toluene	1,349
p-Xylene	508

Section 2.6

2.21 For an ambient separation process where the feed and products are all nonideal liquid solutions at the infinite surroundings temperature, T_0, Eq. (4) of Table 2.1 for the minimum work of separation reduces to

$$\frac{W_{min}}{RT_0} = \sum_{out} n \left[\sum_i x_i \ln(\gamma_i x_i) \right] - \sum_{in} n \left[\sum_i x_i \ln(\gamma_i x_i) \right]$$

For the liquid-phase separation at ambient conditions (298 K, 101.3 kPa) of a 35 mol% mixture of acetone (1) in water (2) into 99 mol% acetone and 98 mol% water products, calculate the minimum work in kJ/kmol of feed. Liquid-phase activity coefficients at ambient conditions are correlated reasonably well by the van Laar equations with $A_{12} = 2.0$ and $A_{21} = 1.7$. What would the minimum rate of work be if acetone and water formed an ideal liquid solution?

2.22 The sharp separation of benzene and cyclohexane by distillation at ambient pressure is impossible because of the formation of an azeotrope at 77.6°C. K.C. Chao [Ph.D. thesis, University of Wisconsin (1956)] obtained the following vapor–liquid equilibrium data for the benzene (B)/cyclohexane (CH) system at 1 atm:

T, °C	x_B	y_B	γ_B	γ_{CH}
79.7	0.088	0.113	1.300	1.003
79.1	0.156	0.190	1.256	1.008
78.5	0.231	0.268	1.219	1.019
78.0	0.308	0.343	1.189	1.032
77.7	0.400	0.422	1.136	1.056
77.6	0.470	0.482	1.108	1.075
77.6	0.545	0.544	1.079	1.102
77.6	0.625	0.612	1.058	1.138
77.8	0.701	0.678	1.039	1.178
78.0	0.757	0.727	1.025	1.221
78.3	0.822	0.791	1.018	1.263
78.9	0.891	0.863	1.005	1.328
79.5	0.953	0.938	1.003	1.369

Vapor pressure is given by (2-39), where constants for benzene are in Exercise 2.12 and constants for cyclohexane are $k_1 = 15.7527$, $k_2 = -2766.63$, and $k_3 = -50.50$.

(a) Use the data to calculate and plot the relative volatility of benzene with respect to cyclohexane versus benzene composition in the liquid phase. What happens to the relative volatility in the vicinity of the azeotrope?

(b) From the azeotropic composition for the benzene/cyclohexane system, calculate the constants in the van Laar equation. With these constants, use the van Laar equation to compute the activity coefficients over the entire range of composition and compare them, in a plot like Figure 2.18, with the above experimental data. How well does the van Laar equation predict the activity coefficients?

2.23 Benzene can be used to break the ethanol/water azeotrope so as to produce nearly pure ethanol. The Wilson constants for the ethanol(1)/benzene(2) system at 45°C are $\Lambda_{12} = 0.124$ and $\Lambda_{21} = 0.523$. Use these constants with the Wilson equation to predict the liquid-phase activity coefficients for this system over the entire range of composition and compare them, in a plot like Figure 2.18, with the following experimental results [*Austral. J. Chem.*, **7**, 264 (1954)]:

x_1	$\ln \gamma_1$	$\ln \gamma_2$
0.0374	2.0937	0.0220
0.0972	1.6153	0.0519
0.3141	0.7090	0.2599
0.5199	0.3136	0.5392
0.7087	0.1079	0.8645
0.9193	0.0002	1.3177
0.9591	−0.0077	1.3999

2.24 For the binary system ethanol(1)/isooctane(2) at 50°C, the infinite-dilution, liquid-phase activity coefficients are $\gamma_1^\infty = 21.17$ and $\gamma_2^\infty = 9.84$.

(a) Calculate the constants A_{12} and A_{21} in the van Laar equations.

(b) Calculate the constants Λ_{12} and Λ_{21} in the Wilson equations.

(c) Using the constants from (a) and (b), calculate γ_1 and γ_2 over the entire composition range and plot the calculated points as log γ versus x_1.

(d) How well do the van Laar and Wilson predictions agree with the azeotropic point where $x_1 = 0.5941$, $\gamma_1 = 1.44$, and $\gamma_2 = 2.18$?

(e) Show that the van Laar equation erroneously predicts separation into two liquid phases over a portion of the composition range by calculating and plotting a y–x diagram like Figure 2.22.

Chapter 3

Mass Transfer and Diffusion

Mass transfer is the net movement of a component in a mixture from one location to another location where the component exists at a different concentration. Often, the transfer takes place between two phases across an interface. Thus, the absorption by a liquid of a solute from a gas involves mass transfer of the solute through the gas to the gas–liquid interface, across the interface, and into the liquid. Mass transfer models are used to describe processes such as the passage of a species through a gas to the outer surface of a porous adsorbent particle and into the pores of the adsorbent, where the species is adsorbed on the porous surface. Mass transfer is also the selective permeation through a nonporous polymeric material of a component of a gas mixture. Mass transfer is not the flow of a fluid through a pipe. However, mass transfer might be superimposed on that flow. Mass transfer is not the flow of solids on a conveyor belt.

Mass transfer occurs by two basic mechanisms: (1) *molecular diffusion* by random and spontaneous microscopic movement of individual molecules in a gas, liquid, or solid as a result of thermal motion; and (2) *eddy* (turbulent) *diffusion* by random macroscopic fluid motion. Molecular and/or eddy diffusion frequently involves the movement of different species in opposing directions. When a net flow occurs in one of these directions, the total rate of mass transfer of individual species is increased or decreased by this bulk flow or *convection* effect, which is a third mechanism of mass transfer. As will be shown later, molecular diffusion is extremely slow, whereas eddy diffusion, when it occurs, is orders of magnitude more rapid. Therefore, if large-scale separation processes are to be conducted in equipment of a reasonable size, fluids must be agitated, interfacial areas maximized, and distances in the direction of diffusion minimized.

In a binary mixture, molecular diffusion occurs because of one or more different potentials or driving forces, including differences (gradients) of concentration (ordinary diffusion), pressure (pressure diffusion), temperature (thermal diffusion), and external force fields (forced diffusion) that act unequally on the different chemical species present. Pressure diffusion requires a large pressure gradient, which is achieved for gas mixtures with a centrifuge. Thermal diffusion columns or cascades can be employed to separate liquid and gas mixtures by establishing a temperature gradient across the mixture. More widely applied is forced diffusion in an electrical field, to cause ions of different charges to move in different directions at different speeds.

In this chapter, only molecular diffusion caused by concentration gradients is considered, because this is the most common type of molecular diffusion in commercial separation processes. Furthermore, emphasis is on binary systems, for which molecular diffusion theory is simple and applications are relatively straightforward. Multicomponent molecular diffusion, which is important in many applications, is considered briefly in Chapter 11. Diffusion in multicomponent systems is much more complex than diffusion in binary systems, and is a more appropriate topic for advanced study using a specialized text such as Taylor and Krishna [1].

Molecular diffusion occurs in solids and in fluids that are stagnant or in laminar or turbulent motion. Eddy diffusion occurs in fluids in turbulent motion. When both molecular diffusion and eddy diffusion occur, they take place in parallel and are additive. Furthermore, they take place because of the same concentration difference (gradient). When mass transfer occurs under turbulent flow conditions, but across an interface or to a solid surface, conditions may be laminar or nearly stagnant near the interface or solid surface. Thus, even though eddy diffusion may be the dominant mechanism in the bulk of the fluid, the overall rate of mass transfer is controlled by molecular diffusion because the eddy diffusion mechanism is damped or even eliminated as the interface or solid surface is approached.

Mass transfer of one or more species results in a total net rate of bulk flow or flux in one direction relative to a fixed plane or stationary coordinate system. When a net flux occurs, it carries all species present. Thus, the molar flux of an individual species is the sum of all three mechanisms. If N_i is the molar flux of species i with mole fraction x_i, and N is the total molar flux, with both fluxes in moles per unit time per unit area in a direction perpendicular to a stationary plane across which mass transfer occurs, then

$$N_i = x_i N + \text{molecular diffusion flux of } i + \text{eddy diffusion flux of } i \qquad \text{(3-1)}$$

where $x_i N$ is the bulk-flow flux. Each term in (3-1) is positive or negative depending on the direction of the flux relative to the direction selected as positive. When the molecular and eddy diffusion fluxes are in one direction and N is in the opposite direction, even though a concentration difference or gradient of i exists, the net mass transfer flux, N_i, of i can be zero.

3.1 STEADY-STATE ORDINARY MOLECULAR DIFFUSION

Suppose a cylindrical glass vessel is partly filled with water containing a soluble red dye. Clear water is carefully added on top so that the dyed solution on the bottom is undisturbed. At first, a sharp boundary exists between the two layers, but after a time the upper layer becomes colored, while the layer below becomes less colored. The upper layer is more colored near the original interface between the two layers and less colored in the region near the top of the upper layer. During this color change, the motion of each dye molecule is random, undergoing collisions mainly with water molecules and sometimes with other dye molecules, moving first in one direction and then in another, with no one direction preferred. This type of motion is sometimes referred to as a *random-walk process,* which yields a mean-square distance of travel for a given interval of time, but not a direction of travel. Thus, at a given horizontal plane through the solution in the cylinder, it is not possible to determine whether, in a given time interval, a given molecule will cross the plane or not. However, on the average, a fraction of the molecules in the solution below the plane will cross over into the region above and the same fraction will cross over in the opposite direction. Therefore, if the concentration of dye molecules in the lower region is greater than in the upper region, a net rate of mass transfer of dye molecules will take place from the lower to the upper region. After a long period of time, the concentration of dye will be uniform throughout the solution. Based on these observations, it is clear that:

1. Mass transfer by ordinary molecular diffusion occurs because of a concentration. difference or gradient; that is, a species diffuses in the direction of decreasing concentration.
2. The mass transfer rate is proportional to the area normal to the direction of mass transfer and not to the volume of the mixture. Thus, the rate can be expressed as a flux.
3. Mass transfer stops when the concentration is uniform.

Fick's Law of Diffusion

The above observations were quantified by Fick in 1855, who proposed an extension of Fourier's 1822 heat conduction theory. Fourier's first law of heat conduction is

$$q_z = -k \frac{dT}{dz} \tag{3-2}$$

where q_z is the heat flux by conduction in the positive z direction, k is the thermal conductivity of the medium, and dT/dz is the temperature gradient, which is negative in the direction of heat conduction. Fick's first law of molecular diffusion is also a proportionality between a flux and a gradient. For a binary mixture of A and B,

$$J_{A_z} = -D_{AB} \frac{dc_A}{dz} \tag{3-3}$$

and

$$J_{B_z} = -D_{BA} \frac{dc_B}{dz} \tag{3-4}$$

where, in (3-3), J_{A_z} is the molar flux of A by ordinary molecular diffusion relative to the molar average velocity of the mixture in the positive z direction, D_{AB} is the mutual diffusion coefficient of A in B, and dc_A/dz is the concentration gradient of A, which is negative in the direction of ordinary molecular diffusion. Similar definitions apply to (3-4). The molar fluxes of A and B are in opposite directions. If the gas, liquid, or solid mixture through which diffusion occurs is isotropic, then values of k and D_{AB} are independent of direction. Nonisotropic (anisotropic) materials include fibrous and laminated solids as well as single, noncubic crystals. The diffusion coefficient is also referred to as the diffusivity and the mass diffusivity (to distinguish it from thermal and momentum diffusivities).

Many alternative forms of (3-3) and (3-4) are used, depending on the choice of driving force or potential in the gradient. For example, we can express (3-3) as

$$J_A = -cD_{AB} \frac{dx_A}{dz} \tag{3-5}$$

where, for convenience, the z subscript on J has been dropped, c = total molar concentration, and x_i = mole fraction of species i.

Velocities in Mass Transfer

It is also useful to formulate expressions for velocities of the various chemical species in the mixture. These velocities are based on the molar flux, N, and the diffusion flux, J. The molar average velocity of the mixture, v_M, relative to stationary coordinates is given for a binary mixture as

$$v_M = \frac{N}{c} = \frac{N_A + N_B}{c} \tag{3-6}$$

Similarly, the velocity of species i, defined in terms of N_i, is relative to stationary coordinates:

$$v_i = \frac{N_i}{c_i} \tag{3-7}$$

Combining (3-6) and (3-7) with $x_i = c_i/c$ gives

$$v_M = x_A v_A + x_B v_B \tag{3-8}$$

Alternatively, species diffusion velocities, v_{i_D}, defined in terms of J_i, are relative to the molar average velocity and are defined as the difference between the species velocity and the molar average velocity for the mixture:

$$v_{i_D} = \frac{J_i}{c_i} = v_i - v_M \tag{3-9}$$

When solving mass transfer problems involving net movement of the mixture, it is not convenient to use fluxes and flow rates based on v_M as the frame of reference. Rather, it is preferred to use mass transfer fluxes referred to stationary coordinates with the observer fixed in space. Thus, from (3-9), the total species velocity is

$$v_i = v_M + v_{i_D} \tag{3-10}$$

Combining (3-7) and (3-10),

$$N_i = c_i v_M + c_i v_{i_D} \tag{3-11}$$

Combining (3-11) with (3-5), (3-6), and (3-7),

$$N_A = \frac{n_A}{A} = x_A N - c D_{AB} \left(\frac{dx_A}{dz} \right) \tag{3-12}$$

and

$$N_B = \frac{n_B}{A} = x_B N - c D_{BA} \left(\frac{dx_B}{dz} \right) \tag{3-13}$$

where in (3-12) and (3-13), n_i is the molar flow rate in moles per unit time, A is the mass transfer area, the first terms on the right-hand sides are the fluxes resulting from bulk flow, and the second terms on the right-hand sides are the ordinary molecular diffusion fluxes. Two limiting cases are important:

1. Equimolar counterdiffusion (EMD) and
2. Unimolecular diffusion (UMD)

Equimolar Counterdiffusion

In equimolar counterdiffusion (EMD), the molar fluxes of A and B are equal, but opposite in direction; thus,

$$N = N_A + N_B = 0 \tag{3-14}$$

Thus, from (3-12) and (3-13), the diffusion fluxes are also equal, but opposite in direction:

$$J_A = -J_B \tag{3-15}$$

This idealization is closely approached in distillation. From (3-12) and (3-13), we see that in the absence of fluxes other than molecular diffusion,

$$N_A = J_A = -c D_{AB} \left(\frac{dx_A}{dz} \right) \tag{3-16}$$

and

$$N_B = J_B = -c D_{BA} \left(\frac{dx_B}{dz} \right) \tag{3-17}$$

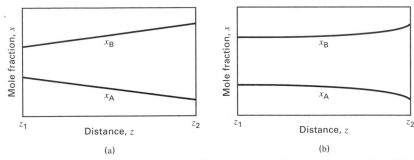

Figure 3.1 Concentration profiles for limiting cases of ordinary molecular diffusion in binary mixtures across a stagnant film: (a) equimolar counterdiffusion (EMD); (b) unimolecular diffusion (UMD).

If the total concentration, pressure, and temperature are constant and the mole fractions are maintained constant (but different) at two sides of a stagnant film between z_1 and z_2, then (3-16) and (3-17) can be integrated from z_1 to any z between z_1 and z_2 to give

$$J_A = \frac{cD_{AB}}{z - z_1}(x_{A_1} - x_A) \tag{3-18}$$

and

$$J_B = \frac{cD_{BA}}{z - z_1}(x_{B_1} - x_B) \tag{3-19}$$

Thus, in the steady state, the mole fractions are linear in distance, as shown in Figure 3.1a. Furthermore, because c is constant through the film, where

$$c = c_A + c_B \tag{3-20}$$

by differentiation,

$$dc = 0 = dc_A + dc_B \tag{3-21}$$

Thus,

$$dc_A = -dc_B \tag{3-22}$$

From (3-3), (3-4), (3-15), and (3-22),

$$\frac{D_{AB}}{dz} = \frac{D_{BA}}{dz} \tag{3-23}$$

Therefore, $D_{AB} = D_{BA}$.

This equality of diffusion coefficients is always true in a binary system of constant molar density.

EXAMPLE 3.1

Two bulbs are connected by a straight tube, 0.001 m in diameter and 0.15 m in length. Initially the bulb at end 1 contains N_2 and the bulb at end 2 contains H_2. The pressure and temperature are maintained constant at 25°C and 1 atm. At a certain time after allowing diffusion to occur between the two bulbs, the nitrogen content of the gas at end 1 of the tube is 80 mol% and at end 2 is 25 mol%. If the binary diffusion coefficient is 0.784 cm²/s, determine:

(a) The rates and directions of mass transfer of hydrogen and nitrogen in mol/s
(b) The species velocities relative to stationary coordinates, in cm/s

SOLUTION

(a) Because the gas system is closed and at constant pressure and temperature, mass transfer in the connecting tube is equimolar counterdiffusion by molecular diffusion.

The area for mass transfer through the tube, in cm^2, is $A = 3.14(0.1)^2/4 = 7.85 \times 10^{-3}$ cm^2. The total gas concentration is $c = \dfrac{P}{RT} = \dfrac{1}{(82.06)(298)} = 4.09 \times 10^{-5}$ mol/cm^3. Take the reference plane at end 1 of the connecting tube. Applying (3-18) to N$_2$ over the length of the tube,

$$n_{N_2} = \frac{cD_{N_2,H_2}}{z_2 - z_1}[(x_{N_2})_1 - (x_{N_2})_2]A$$

$$= \frac{(4.09 \times 10^{-5})(0.784)(0.80 - 0.25)}{15}(7.85 \times 10^{-3})$$

$$= 9.23 \times 10^{-9} \text{ mol/s} \qquad \text{in the positive } z \text{ direction}$$

$$n_{H_2} = 9.23 \times 10^{-9} \text{ mol/s} \qquad \text{in the negative } z \text{ direction}$$

(b) For equimolar counter diffusion, the molar average velocity of the mixture, v_M, is 0. Therefore, from (3-9), species velocities are equal to species diffusion velocities. Thus,

$$v_{N_2} = (v_{N_2})_D = \frac{J_{N_2}}{c_{N_2}} = \frac{n_{N_2}}{Acx_{N_2}} = \frac{9.23 \times 10^{-9}}{[(7.85 \times 10^{-3})(4.09 \times 10^{-5})x_{N_2}]}$$

$$= \frac{0.0287}{x_{N_2}} \qquad \text{in the positive } z \text{ direction}$$

Similarly,

$$v_{H_2} = \frac{0.0287}{x_{H_2}} \qquad \text{in the negative } z \text{ direction}$$

Thus, species velocities depend on species mole fractions, as follows:

z, cm	x_{N_2}	x_{H_2}	v_{N_2}, cm/s	v_{H_2}, cm/s
0 (end 1)	0.800	0.200	0.0351	−0.1435
5	0.617	0.383	0.0465	−0.0749
10	0.433	0.567	0.0663	−0.0506
15 (end 2)	0.250	0.750	0.1148	−0.0383

Note that species velocities vary across the length of the connecting tube, but at any location, z, $v_M = 0$. For example, at $z = 10$ cm, from (3-8),

$$v_M = (0.433)(0.0663) + (0.567)(-0.0506) = 0 \qquad \blacksquare$$

Unimolecular Diffusion

In unimolecular diffusion (UMD), mass transfer of component A occurs through stagnant component B. Thus,

$$N_B = 0 \tag{3-24}$$

and

$$N = N_A \tag{3-25}$$

Therefore, from (3-12),

$$N_A = x_A N_A - cD_{AB}\frac{dx_A}{dz} \tag{3-26}$$

which can be rearranged to a Fick's law form,

$$N_A = -\frac{cD_{AB}}{(1 - x_A)}\frac{dx_A}{dz} \tag{3-27}$$

The factor $(1 - x_A)$ accounts for the bulk flow effect. For a mixture dilute in A, the bulk flow effect is negligible or small. In mixtures more concentrated in A, the bulk flow effect can be appreciable. For example, in an equimolar mixture of A and B, $(1 - x_A) = 0.5$ and the molar mass transfer flux of A is twice the ordinary molecular diffusion flux.

For the stagnant component, B, (3-13) becomes

$$0 = x_B N_A - c D_{BA} \frac{dx_B}{dz} \tag{3-28}$$

or

$$x_B N_A = c D_{BA} \frac{dx_B}{dz} \tag{3-29}$$

Thus, the bulk flow flux of B is equal but opposite to its diffusion flux.

At quasi-steady-state conditions, that is, with no accumulation, and with constant molar density, (3-27) becomes in integral form:

$$\int_{z_1}^{z} dz = -\frac{c D_{AB}}{N_A} \int_{x_{A_1}}^{x_A} \frac{dx_A}{1 - x_A} \tag{3-30}$$

which upon integration yields

$$N_A = \frac{c D_{AB}}{z - z_1} \ln \left(\frac{1 - x_A}{1 - x_{A_1}} \right) \tag{3-31}$$

Rearrangement to give the mole-fraction variation as a function of z yields

$$x_A = 1 - (1 - x_{A_1}) \exp \left[\frac{N_A(z - z_1)}{c D_{AB}} \right] \tag{3-32}$$

Thus, as shown in Figure 3.1b, the mole fractions are nonlinear in distance.

An alternative and more useful form of (3-31) can be derived from the definition of the log mean. When $z = z_2$, (3-31) becomes

$$N_A = \frac{c D_{AB}}{z_2 - z_1} \ln \left(\frac{1 - x_{A_2}}{1 - x_{A_1}} \right) \tag{3-33}$$

The log mean (LM) of $(1 - x_A)$ at the two ends of the stagnant layer is

$$(1 - x_A)_{LM} = \frac{(1 - x_{A_2}) - (1 - x_{A_1})}{\ln[(1 - x_{A_2})/(1 - x_{A_1})]} = \frac{x_{A_1} - x_{A_2}}{\ln[(1 - x_{A_2})/(1 - x_{A_1})]} \tag{3-34}$$

Combining (3-33) with (3-34) gives

$$N_A = \frac{c D_{AB}}{z_2 - z_1} \frac{(x_{A_1} - x_{A_2})}{(1 - x_A)_{LM}} = \frac{c D_{AB}}{(1 - x_A)_{LM}} \frac{(-\Delta x_A)}{\Delta z} \tag{3-35}$$

EXAMPLE 3.2

As shown in Figure 3.2, an open beaker, 6 cm in height, is filled with liquid benzene at 25°C to within 0.5 cm of the top. A gentle breeze of dry air at 25°C and 1 atm is blown by a fan across the mouth of the beaker so that evaporated benzene is carried away by convection after it transfers through a stagnant air layer in the beaker. The vapor pressure of benzene at 25°C is 0.131 atm. The mutual diffusion coefficient for benzene in air at 25°C and 1 atm is 0.0905 cm²/s. Compute:

(a) The initial rate of evaporation of benzene as a molar flux in mol/cm²-s
(b) The initial mole fraction profiles in the stagnant layer
(c) The initial fractions of the mass transfer fluxes due to molecular diffusion
(d) The initial diffusion velocities, and the species velocities (relative to stationary coordinates) in the stagnant layer

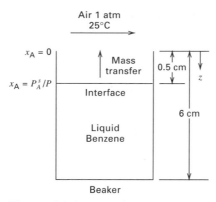

Figure 3.2 Evaporation of benzene from a beaker—Example 3.2.

(e) The time in hours for the benzene level in the beaker to drop 2 cm from the initial level, if the specific gravity of liquid benzene is 0.874. Neglect the accumulation of benzene and air in the stagnant layer as it increases in height.

SOLUTION

Let A = benzene, B = air.

$$c = \frac{P}{RT} = \frac{1}{(82.06)(298)} = 4.09 \times 10^{-5} \, \text{mol/cm}^3$$

(a) Take $z_1 = 0$. Then $z_2 - z_1 = \Delta z = 0.5$ cm. From Dalton's law, assuming equilibrium at the liquid benzene–air interface,

$$x_{A_1} = \frac{p_{A_1}}{P} = \frac{0.131}{1} = 0.131 \qquad x_{A_2} = 0$$

$$(1 - x_A)_{LM} = \frac{0.131}{\ln[(1-0)/(1-0.131)]} = 0.933$$

From (3-35),

$$N_A = \frac{(4.09 \times 10^{-5})(0.0905)}{0.5}\left(\frac{0.131}{0.933}\right) = 1.04 \times 10^{-6} \, \text{mol/cm}^2\text{-s}$$

(b)

$$\frac{N_A(z - z_1)}{cD_{AB}} = \frac{(1.04 \times 10^{-6})(z - 0)}{(4.09 \times 10^{-5})(0.0905)} = 0.281\, z$$

From (3-32), $x_A = 1 - 0.869 \exp(0.281\, z)$ **(1)**

Using (1), the following results are obtained:

z, cm	x_A	x_B
0.0	0.1310	0.8690
0.1	0.1060	0.8940
0.2	0.0808	0.9192
0.3	0.0546	0.9454
0.4	0.0276	0.9724
0.5	0.0000	1.0000

These profiles are only slightly curved.

(c) From (3-27) and (3-29), we can compute the bulk flow terms, $x_A N_A$ and $x_B N_A$, from which the molecular diffusion terms are obtained.

	$x_i N$ Bulk Flow Flux, mol/cm²-s × 10⁶		J_i Molecular Diffusion Flux, mol/cm²-s × 10⁶	
z, cm	A	B	A	B
0.0	0.1360	0.9040	0.9040	−0.9040
0.1	0.1100	0.9300	0.9300	−0.9300
0.2	0.0840	0.9560	0.9560	−0.9560
0.3	0.0568	0.9832	0.9832	−0.9832
0.4	0.0287	1.0113	1.0113	−1.0113
0.5	0.0000	1.0400	1.0400	−1.0400

Note that the molecular diffusion fluxes are equal but opposite, and the bulk flow flux of B is equal but opposite to its molecular diffusion flux, so that its molar flux, N_B, is zero.

(d) From (3-6),

$$v_M = \frac{N}{c} = \frac{N_A}{c} = \frac{1.04 \times 10^{-6}}{4.09 \times 10^{-5}} = 0.0254 \text{ cm/s} \qquad (2)$$

From (3-9), the diffusion velocities are given by

$$v_{i_d} = \frac{J_i}{c_i} = \frac{J_i}{x_i c} \qquad (3)$$

From (3-10), the species velocities relative to stationary coordinates are

$$v_i = v_{i_d} + v_M \qquad (4)$$

Using (2) to (4), we obtain

	v_{i_d} Molecular Diffusion Velocity, cm/s		J_i Species Velocity, cm/s	
z, cm	A	B	A	B
0.0	0.1687	−0.0254	0.1941	0
0.1	0.2145	−0.0254	0.2171	0
0.2	0.2893	−0.0254	0.3147	0
0.3	0.4403	−0.0254	0.4657	0
0.4	0.8959	−0.0254	0.9213	0
0.5	∞	−0.0254	∞	0

Note that v_B is zero everywhere, because its molecular diffusion velocity is negated by the molar mean velocity.

(e) The mass transfer flux for benzene evaporation can be equated to the rate of change of liquid benzene. Letting z = distance down from the mouth of the beaker and using (3-35) with $\Delta z = z$,

$$N_A = \frac{c D_{AB}}{z} \frac{(-\Delta x_A)}{(1 - x_A)_{LM}} = \frac{\rho_L}{M_L} \frac{dz}{dt} \qquad (5)$$

Separating variables and integrating,

$$\int_0^t dt = t = \frac{\rho_L (1 - x_A)_{LM}}{M_L c D_{AB}(-\Delta x_A)} \int_{z_1}^{z_2} z \, dz \qquad (6)$$

The coefficient of the integral on the right-hand side of (6) is constant at

$$\frac{0.874(0.933)}{78.11(4.09 \times 10^{-5})(0.0905)(0.131)} = 21.530 \text{ s/cm}^2$$

$$\int_{z_1}^{z_2} z \, dz = \int_{0.5}^{2.5} z \, dz = 3 \text{ cm}^2$$

From (6), $t = 21,530(3) = 64,590$ s   or   17.94 h ∎

3.2 DIFFUSION COEFFICIENTS

Diffusivities, and molecular diffusivities, are defined for a binary mixture by (3-3) and (3-4). Measurement of diffusion coefficients must involve a correction for any bulk flow using (3-12) and (3-13) with the reference plane being such that there is no net molar bulk flow.

The binary diffusivities, D_{AB} and D_{BA}, are mutual or binary diffusion coefficients. Other coefficients include D_{i_M}, the diffusivity of i in a multicomponent mixture; D_{ii}, the self-diffusion coefficient; and the tracer or interdiffusion coefficient. In this chapter, and throughout this book, the focus is on the mutual diffusion coefficient.

Diffusivity in Gas Mixtures

As discussed by Reid, Prausnitz, and Poling [2], a number of theoretical and empirical equations are available for estimating the value of $D_{AB} = D_{BA}$ in gases at low to moderate pressures. The theoretical equations, based on Boltzmann's kinetic theory of gases, the theorem of corresponding states, and a suitable intermolecular energy potential function, as developed by Chapman and Enskog, predict D_{AB} to be inversely proportional to pressure and almost independent of composition, with a significant increase for increasing temperature. Of greater accuracy and ease of use is the following empirical equation of Fuller, Schettler, and Giddings [3], which retains the form of the Chapman–Enskog theory but utilizes empirical constants derived from experimental data:

$$D_{AB} = D_{BA} = \frac{0.00143 T^{1.75}}{P M_{AB}^{1/2} [(\Sigma_V)_A^{1/3} + (\Sigma_V)_B^{1/3}]^2} \tag{3-36}$$

where D_{AB} is in cm^2/s, P is in atm, T is in K,

$$M_{AB} = \frac{2}{(1/M_A) + (1/M_B)} \tag{3-37}$$

and Σ_V = summation of atomic and structural diffusion volumes from Table 3.1, which includes diffusion volumes of some simple molecules.

Experimental values of binary gas diffusivity at 1 atm and near-ambient temperature range from about 0.10 to 10.0 cm^2/s. Reid et al. [2] compared (3-36) to experimental data for 51 different binary gas mixtures at 1 atm over a temperature range of 195–1,068 K. The average deviation was only 5.4%, with a maximum deviation of 22%. Only 7 of 69 estimated values deviated from experimental values by more than 10%. When an experimental diffusivity is available at values T and P that are different from the desired conditions, (3-36) indicates that D_{AB} is proportional to $T^{1.75}/P$, which can be used to obtain the desired value.

Table 3.1 Diffusion Volumes from Fuller, Ensley, and Giddings [*J. Phys. Chem,* **73,** 3679–3685 (1969)] for Estimating Binary Gas Diffusivity by the Method of Fuller et al. [3]

Atomic Diffusion Volumes Atomic and Structural Diffusion Volume Increments			
C	15.9	F	14.7
H	2.31	Cl	21.0
O	6.11	Br	21.9
N	4.54	I	29.8
Aromatic ring	−18.3	S	22.9
Heterocyclic ring	−18.3		
Diffusion Volumes of Simple Molecules			
He	2.67	CO	18.0
Ne	5.98	CO_2	26.7
Ar	16.2	N_2O	35.9
Kr	24.5	NH_3	20.7
Xe	32.7	H_2O	13.1
H_2	6.12	SF_6	71.3
D_2	6.84	Cl_2	38.4
N_2	18.5	Br_2	69.0
O_2	16.3	SO_2	41.8
Air	19.7		

EXAMPLE 3.3

Estimate the diffusion coefficient for the system oxygen (A)/benzene (B) at 38°C and 2 atm using the method of Fuller et al.

SOLUTION

From (3-37), $M_{AB} = \dfrac{2}{(1/32) + (1/78.11)} = 45.4$

From Table 3.1, $(\Sigma_V)_A = 16.3$ and $(\Sigma_V)_B = 6(15.9) + 6(2.31) - 18.3 = 90.96$

From (3-36), at 2 atm and 311 K,

$$D_{AB} = D_{BA} = \frac{0.00143(311)^{1.75}}{(2)(45.4)^{1/2}[16.3^{1/3} + 90.96^{1/3}]^2} = 0.0494 \text{ cm}^2/\text{s}$$

At 1 atm, the predicted diffusivity is 0.0988 cm^2/s, which is about 3% below the experimental value of 0.102 cm^2/s. The experimental value for 38°C can be extrapolated by the temperature dependency of (3-36) to give the following prediction at 200°C:

$$D_{AB} \text{ at 200°C and 1 atm} = 0.102 \left(\frac{200 + 273}{38 + 273} \right)^{1.75} = 0.212 \text{ cm}^2/\text{s} \qquad \blacksquare$$

For binary mixtures of light gases, at pressures to about 10 atm, the pressure dependence on diffusivity is adequately predicted by the simple inverse relation (3-36), that is, $PD_{AB} =$ a constant for a given temperature and gas mixture. At higher pressures, deviations from this relation are handled in a manner somewhat similar to the modification of the ideal gas law by the compressibility factor based on the theorem of corresponding states. Although few reliable experimental data are available at high pressure, Takahasi [4] has published a tentative corresponding-states correlation, shown in Figure 3.3, patterned after an earlier correlation for self-diffusivities by Slattery [5]. In the Takahashi plot, $D_{AB}P/(D_{AB}P)_{LP}$ is given as a function of reduced temperature and pressure, where $(D_{AB}P)_{LP}$ is

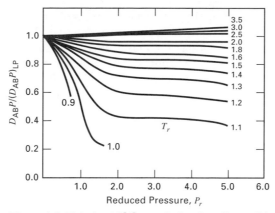

Figure 3.3 Takahashi [4] correlation for effect of high pressure on binary gas diffusivity.

at low pressure where (3-36) applies. Mixture critical temperature and pressure are molar-average values. Thus, a finite effect of composition is predicted at high pressure. The effect of high pressure on diffusivity is important in supercritical extraction, discussed in Chapter 11.

EXAMPLE 3.4

Estimate the diffusion coefficient for a 25/75 molar mixture of argon and xenon at 200 atm and 378 K. At this temperature and 1 atm, the diffusion coefficient is 0.180 cm²/s. Critical constants are

	T_c, **K**	P_c, **atm**
Argon	151.0	48.0
Xenon	289.8	58.0

SOLUTION

Calculate reduced conditions:

$$T_c = 0.25(151) + 0.75(289.8) = 255.1 \text{ K}; \quad T_r = T/T_c = 378/255.1 = 1.48$$

$$P_c = 0.25(48) + 0.75(58) = 55.5; \quad P_r = P/P_c = 200/55.5 = 3.6$$

From Figure 3.3, $\dfrac{D_{AB}P}{(D_{AB}P)_{LP}} = 0.82$

$$D_{AB} = \frac{(D_{AB}P)_{LP}}{P}\left[\frac{D_{AB}P}{(D_{AB}P)_{LP}}\right] = \frac{(0.180)(1)}{200}(0.82) = 7.38 \times 10^{-4} \text{ cm/s} \quad ■$$

Diffusivity in Liquid Mixtures

Diffusion coefficients in binary liquid mixtures are difficult to estimate because of the lack of a rigorous model for the liquid state. An exception is the case of a dilute solute (A) of very large, rigid, spherical molecules diffusing through a stationary solvent (B) of small molecules with no slip of the solvent at the surface of the solute molecules. The resulting relation, based on the hydrodynamics of creeping flow to describe drag, is the Stokes–Einstein equation:

$$(D_{AB})_\infty = \frac{RT}{6\pi\mu_B R_A} \tag{3-38}$$

where R_A is the radius of the solute molecule. Although (3-38) is very limited in its application to liquid mixtures, it has long served as a starting point for more widely applicable empirical correlations for the diffusivity of solute (A) in solvent (B), where both A and B are of the same approximate molecular size. Unfortunately, unlike the situation in binary gas mixtures, $D_{AB} \neq D_{BA}$ in liquid mixtures because molar density varies with composition. Also, both diffusivities can vary greatly with composition. Because the Stokes–Einstein equation does not provide a basis for extending dilute conditions to more concentrated conditions, the extensions of (3-38) have been restricted to binary liquid mixtures dilute in A, up to 5 and perhaps 10 mol%.

One such extension, which gives reasonably good predictions for small solute molecules when water is the solvent, is the empirical Wilke–Chang [6] equation:

$$D_{AB} = \frac{7.4 \times 10^{-8}(\phi_B M_B)^{1/2} T}{\mu_B v_A^{0.6}} \qquad (3\text{-}39)$$

where the units are cm^2/s for D_{AB}; cP for the solvent viscosity, μ_B; K for T; and cm^3/mol for v_A, the liquid molar volume of the solute at its normal boiling point. The parameter ϕ_B is an association factor for the solvent, which for water is 2.6. Note that the effects of temperature and viscosity are identical to the prediction of the Stokes–Einstein equation, while the effect of the radius of the solute molecule is replaced by v_A, which can be estimated by summing the atomic contributions in Table 3.2, which also lists values of v_A for dissolved light gases.

Table 3.2 Molecular Volumes of Dissolved Light Gases and Atomic Contributions for Other Molecules at the Normal Boiling Point

	Atomic Volume (m³/kmol) × 10³		Atomic Volume (m³/kmol) × 10³
C	14.8	Ring	
H	3.7	Three-membered, as in	−6
O (except as below)	7.4	ethylene oxide	
Doubly bonded as carbonyl	7.4	Four-membered	−8.5
Coupled to two other elements:		Five-membered	−11.5
In aldehydes, ketones	7.4	Six-membered	−15
In methyl esters	9.1	Naphthalene ring	−30
In methyl ethers	9.9	Anthracene ring	−47.5
In ethyl esters	9.9		
In ethyl ethers	9.9		Molecular Volume (m³/kmol) × 10³
In higher esters	11.0		
In higher ethers	11.0	Air	29.9
In acids (−OH)	12.0	O_2	25.6
Joined to S, P. N	8.3	N_2	31.2
N		Br_2	53.2
Doubly bonded	15.6	Cl_2	48.4
In primary amines	10.5	CO	30.7
In secondary amines	12.0	CO_2	34.0
Br	27.0	H_2	14.3
Cl in RCHClR′	24.6	H_2O	18.8
Cl in RCl (terminal)	21.6	H_2S	32.9
F	8.7	NH_3	25.8
I	37.0	NO	23.6
S	25.6	N_2O	36.4
P	27.0	SO_2	44.8

Source: G. Le Bas, *The Molecular Volumes of Liquid Chemical Compounds,* David McKay, New York (1915).

EXAMPLE 3.5

Use the Wilke–Chang equation to estimate the diffusivity of aniline (A) in a 0.5 mol% aqueous solution at 20°C. At this temperature, the solubility of aniline in water is about 4 g/100 g of water or 0.77 mol% aniline. The experimental diffusivity value for an infinitely dilute mixture is 0.92×10^{-5} cm²/s.

SOLUTION

$$\mu_B = \mu_{H_2O} = 1.01 \text{ cP at } 20°C$$

v_A = liquid molar volume of aniline at its normal boiling point of 457.6 K = 107 cm³/mol

$\phi_B = 2.6$ for water $\qquad M_B = 18$ for water $\qquad T = 293$ K

From (3-39),

$$D_{AB} = \frac{(7.4 \times 10^{-8})[2.6(18)]^{0.5}(293)}{1.01(107)^{0.6}} = 0.89 \times 10^{-5} \text{ cm}^2/\text{s}$$

This value is about 3% less than the experimental value for an infinitely dilute solution of aniline in water. ■

Although Wilke and Chang determined values of ϕ_B for methanol (1.9), ethanol (1.5), and unassociated solvents (1.0), more recent correlations due to Hayduk and Minhas [7] give better agreement with experimental values. For a dilute solution of one normal paraffin (C_5 to C_{32}) in another (C_5 to C_{16}),

$$(D_{AB})_\infty = 13.3 \times 10^{-8} \frac{T^{1.47} \mu_B^{\epsilon}}{v_A^{0.71}} \qquad \textbf{(3-40)}$$

where

$$\epsilon = \frac{10.2}{v_A} - 0.791 \qquad \textbf{(3-41)}$$

and the other variables have the same units as in (3-39).

For general nonaqueous solutions,

$$(D_{AB})_\infty = 1.55 \times 10^{-8} \frac{T^{1.29}(\mathscr{P}_B^{0.5}/\mathscr{P}_A^{0.42})}{\mu_B^{0.92} v_B^{0.23}} \qquad \textbf{(3-42)}$$

where $\mathscr{P}$ is the parachor, which is defined as

$$\mathscr{P} = v\sigma^{1/4} \qquad \textbf{(3-43)}$$

When the units of the liquid molar volume, v, are cm³/mol and the surface tension, σ, are g/s² (dynes/cm), then the units of the parachor are cm³-g$^{1/4}$/s$^{1/2}$-mol. Normally, at near-ambient conditions, $\mathscr{P}$ is treated as a constant, for which an extensive tabulation is available from Quayle [8], who also provides a group-contribution method for estimating parachors for compounds not listed. Table 3.3 gives values of parachors for a number of compounds, while Table 3.4 contains structural contributions for predicting the parachor in the absence of data.

The following restrictions apply to (3-42):

1. Solvent viscosity should not exceed 30 cP.
2. For organic acid solutes and solvents other than water, methanol, and butanols, the acid should be treated as a dimer by doubling the values of $\mathscr{P}_A$ and v_A.
3. For a nonpolar solute in monohydroxy alcohols, values of v_B and $\mathscr{P}_B$ should be multiplied by $8\mu_B$, where the viscosity is in centipoise.

Liquid diffusion coefficients for a solute in a dilute binary system range from about 10^{-6} to 10^{-4} cm²/s for solutes of molecular weight up to about 200 and solvents with viscosity up to about 10 cP. Thus, liquid diffusivities are five orders of magnitude less than diffusivities for binary gas mixtures at 1 atm. However, diffusion rates in liquids are not necessarily

Table 3.3 Parachors for
Representative Compounds

	Parachor, cm^3-$g^{1/4}/s^{1/2}$-mol
Acetic acid	131.2
Acetone	161.5
Acetonitrile	122
Acetylene	88.6
Aniline	234.4
Benzene	205.3
Benzonitrile	258
n-Butyric acid	209.1
Carbon disulfide	143.6
Cyclohexane	239.3
Chlorobenzene	244.5
Diphenyl	380.0
Ethane	110.8
Ethylene	99.5
Ethyl butyrate	295.1
Ethyl ether	211.7
Ethyl mercaptan	162.9
Formic acid	93.7
Isobutyl benzene	365.4
Methanol	88.8
Methyl amine	95.9
Methyl formate	138.6
Naphthalene	312.5
n-Octane	350.3
1-Pentene	218.2
1-Pentyne	207.0
Phenol	221.3
n-Propanol	165.4
Toluene	245.5
Triethyl amine	297.8

Source: Meissner, *Chem. Eng. Prog.,* **45,** 149–153 (1949).

five orders of magnitude lower than in gases because, as seen in (3-5), the product of the concentration (molar density) and the diffusivity determines the rate of diffusion for a given concentration gradient in mole fraction. At 1 atm, the molar density of a liquid is three times that of a gas and, thus, the diffusion rate in liquids is only two orders of magnitude lower than in gases at 1 atm.

EXAMPLE 3.6

Estimate the diffusivity of formic acid (A) in benzene (B) at 25°C and infinite dilution, using the appropriate correlation of Hayduk and Minhas. The experimental value is 2.28×10^{-5} cm^2/s.

SOLUTION

Equation (3-42) applies, with $T = 298$ K

$$\mathscr{P}_A = 93.7 \ cm^3\text{-}g^{1/4}/s^{1/2}\text{-mol} \qquad \mathscr{P}_B = 205.3 \ cm^3\text{-}g^{1/4}/s^{1/2}\text{-mol}$$

$$\mu_B = 0.6 \ cP \ at \ 25°C \qquad\qquad v_B = 96 \ cm^3/mol \ at \ 80°C$$

However, because formic acid is an organic acid, $\mathscr{P}_A$ is doubled to 187.4.

Table 3.4 Structural Contributions for Estimating the Parachor

Carbon–hydrogen:		R−[−CO−]−R′(ketone)	
C	9.0	R + R′ = 2	51.3
H	15.5	R + R′ = 3	49.0
CH₃	55.5	R + R′ = 4	47.5
CH₂ in −(CH₂)ₙ		R + R′ = 5	46.3
$n < 12$	40.0	R + R′ = 6	45.3
$n > 12$	40.3	R + R′ = 7	44.1
		−CHO	66
Alkyl groups			
1-Methylethyl	133.3	O (not noted above)	20
1-Methylpropyl	171.9	N (not noted above)	17.5
1-Methylbutyl	211.7	S	49.1
2-Methylpropyl	173.3	P	40.5
1-Ethylpropyl	209.5	F	26.1
1,1-Dimethylethyl	170.4	Cl	55.2
1,1-Dimethylpropyl	207.5	Br	68.0
1,2-Dimethylpropyl	207.9	I	90.3
1,1,2-Trimethylpropyl	243.5	Ethylenic bonds:	
C₆H₅	189.6	Terminal	19.1
		2,3-position	17.7
Special groups:		3,4-position	16.3
−COO−	63.8		
−COOH	73.8	Triple bond	40.6
−OH	29.8		
−NH₂	42.5	Ring closure:	
−O−	20.0	Three-membered	12
−NO₂	74	Four-membered	6.0
−NO₃ (nitrate)	93	Five-membered	3.0
−CO(NH₂)	91.7	Six-membered	0.8

Source: Quale [8].

From (3-42),

$$(D_{AB})_\infty = 1.55 \times 10^{-8} \left[\frac{298^{1.29}(205.3^{0.5}/187.4^{0.42})}{0.6^{0.92}96^{0.23}} \right]$$

$$= 2.15 \times 10^{-5} \, \text{cm}^2/\text{s}$$

which is within 6% of the the experimental value.   ■

    The Stokes–Einstein and Wilke–Chang equations predict an inverse dependence of viscosity with liquid diffusivity. The Hayduk–Minhas equations predict a somewhat smaller dependence of viscosity. From data covering several orders of magnitude variation of viscosity, the liquid diffusivity is found to vary inversely with the viscosity raised to an exponent closer to 0.5 than to 1.0. The Stokes–Einstein and Wilke–Chang equations also predict that $D_{AB}\mu_B/T$ is a constant over a narrow temperature range. Because μ_B decreases exponentially with temperature, D_{AB} is predicted to increase exponentially with temperature. For example, for a dilute solution of water in ethanol, the diffusivity of water increases by a factor of almost 20 when the absolute temperature is increased 50%. Over a wide temperature range, it is preferable to express the effect of temperature on D_{AB} by an

Arrhenius-type expression,

$$(D_{AB})_\infty = A \exp\left(\frac{-E}{RT}\right) \tag{3-44}$$

where, typically the activation energy for liquid diffusion, E, is no greater than 6,000 cal/mol.

Fick's first law for ordinary molecular diffusion uses a concentration or mole-fraction driving force. While this is adequate for gases, except at very high pressures and for ideal liquid solutions, experimental evidence sheds doubt on its validity for nonideal solutions. For that case, it is argued that for thermodynamic consistency a chemical potential driving force should be applied, at least for conditions close to equilibrium. The resulting modification of (3-5) is

$$J_A = -\frac{c}{RT} D_{AB} x_A \left[\left(\frac{\partial \mu_A}{\partial x_A}\right)_{T,P} \frac{dx_A}{dz}\right] \tag{3-45}$$

where μ = chemical potential. From thermodynamics,

$$\frac{x_A}{RT}\left(\frac{\partial \mu_A}{\partial x_A}\right)_{T,P} = \left(\frac{\partial \ln a_A}{\partial \ln x_A}\right)_{T,P}$$

Therefore, (3-45) becomes

$$J_A = -cD_{AB}\left(\frac{\partial \ln a_A}{\partial \ln x_A}\right)_{T,P} \frac{dx_A}{dz} \tag{3-46}$$

However, $$a_A = \gamma_A x_A \tag{3-47}$$

or $$\ln a_A = \ln \gamma_A + \ln x_A \tag{3-48}$$

Therefore,

$$\left(\frac{\partial \ln a_A}{\partial \ln x_A}\right)_{T,P} = \left(1 + \frac{\partial \ln \gamma_A}{\partial \ln x_A}\right)_{T,P} \tag{3-49}$$

and (3-46) becomes

$$J_A = -cD_{AB}\left(1 + \frac{\partial \ln \gamma_A}{\partial \ln x_A}\right)_{T,P}\frac{dx_A}{dz} \tag{3-50}$$

Based on this nonideal form of Fick's law, Vignes [9] has shown that, except for strongly associated binary mixtures such as chloroform/acetone, which exhibit a rare negative deviation from Raoult's law, infinite-dilution binary diffusivities, $(D)_\infty$, can be combined with mixture activity-coefficient data or correlations to predict liquid binary diffusion coefficients D_{AB} and D_{BA} over the entire composition range. The Vignes equations are:

$$D_{AB} = (D_{AB})_\infty^{x_B}(D_{BA})_\infty^{x_A}\left(1 + \frac{\partial \ln \gamma_A}{\partial \ln x_A}\right)_{T,P} \tag{3-51}$$

$$D_{BA} = (D_{BA})_\infty^{x_A}(D_{AB})_\infty^{x_B}\left(1 + \frac{\partial \ln \gamma_B}{\partial \ln x_B}\right)_{T,P} \tag{3-52}$$

where it should be noted that by the Gibbs–Duhem equation for thermodynamic consistency, the last (derivative) term in (3-51) is equal to the last term in (3-52). When using activity coefficient data with these equations, this equality is true only if the data meet the consistency test.

EXAMPLE 3.7

At 298 K and 1 atm, infinite-dilution diffusion coefficients for the methanol (A)/water (B) system are 1.5×10^{-5} cm^2/s and 1.75×10^{-5} cm^2/s for AB and BA, respectively.

Activity coefficient data for the same conditions as estimated from the UNIFAC method are as follows:

x_A	γ_A	x_B	γ_B
0.0	2.245	1.0	1.000
0.1	1.748	0.9	1.013
0.2	1.470	0.8	1.044
0.3	1.300	0.7	1.087
0.4	1.189	0.6	1.140
0.5	1.116	0.5	1.201
0.6	1.066	0.4	1.269
0.7	1.034	0.3	1.343
0.8	1.014	0.2	1.424
0.9	1.003	0.1	1.511
1.0	1.000	0.0	1.605

Use the Vignes equations to estimate diffusion coefficients over the entire composition range.

SOLUTION

Using a spreadsheet to compute the derivatives in (3-51) and (3-52), which are found to be essentially equal at any composition, and the diffusivities from the same equations, the following results are obtained:

x_A	D_{AB}, cm^2/s	D_{BA}, cm^2/s
0.20	1.11×10^{-5}	1.10×10^{-5}
0.30	1.08×10^{-5}	1.08×10^{-5}
0.40	1.12×10^{-5}	1.12×10^{-5}
0.50	1.18×10^{-5}	1.19×10^{-5}
0.60	1.28×10^{-5}	1.28×10^{-5}
0.70	1.38×10^{-5}	1.38×10^{-5}
0.80	1.50×10^{-5}	1.50×10^{-5}

If the diffusivity is assumed linear with mole fraction, the value at $x_A = 0.50$ is 1.625×10^{-5}, which is almost 40% higher than the predicted value of 1.18×10^{-5}. ■

In an electrolyte solute, the diffusion coefficient of the dissolved salt, acid, or base depends on the ions, since they are the diffusing entities. However, in the absence of an electric potential, only the molecular diffusion of the electrolyte molecule is of interest. The infinite-dilution diffusivity of a single salt in an aqueous solution in cm^2/s can be estimated from the Nernst–Haskell equation:

$$(D_{AB})_\infty = \frac{RT[(1/n_+) + (1/n_-)]}{F^2[(1/\lambda_+) + (1/\lambda_-)]} \tag{3-53}$$

where n_+ and n_- = valences of the cation and anion, respectively
λ_+ and λ_- = limiting ionic conductances in (A/cm^2)(V/cm)(g-equiv/cm^3)
F = Faraday's constant = 96,500 C/g-equiv
T = temperature, K
R = gas constant = 8.314 J/mol-K

Values of λ_+ and λ_- at 25°C are listed in Table 3.5. At other temperatures, these values are multiplied by $T/334\mu_B$, where T and μ_B are in kelvins and centipoise, respectively. As the concentration of the electrolyte increases, the diffusivity at first decreases rapidly by

Table 3.5 Limiting Ionic Conductances in Water at 25°C, in $(A/cm^2)(V/cm)(g\text{-}equiv/cm^3)$

Anion	λ_-	Cation	λ_+
OH^-	197.6	H^+	349.8
Cl^-	76.3	Li^+	38.7
Br^-	78.3	Na^+	50.1
I^-	76.8	K^+	73.5
NO_3^-	71.4	NH_4^+	73.4
ClO_4^-	68.0	Ag^+	61.9
HCO_3^-	44.5	Tl^+	74.7
HCO_2^-	54.6	$(\frac{1}{2})Mg^{2+}$	53.1
$CH_3CO_2^-$	40.9	$(\frac{1}{2})Ca^{2+}$	59.5
$ClCH_2CO_2^-$	39.8	$(\frac{1}{2})Sr^{2+}$	50.5
$CNCH_2CO_2^-$	41.8	$(\frac{1}{2})Ba^{2+}$	63.6
$CH_3CH_2CO_2^-$	35.8	$(\frac{1}{2})Cu^{2+}$	54
$CH_3(CH_2)_2CO_2^-$	32.6	$(\frac{1}{2})Zn^{2+}$	53
$C_6H_5CO_2^-$	32.3	$(\frac{1}{3})La^{3+}$	69.5
$HC_2O_4^-$	40.2	$(\frac{1}{3})Co(NH_3)_6^{3+}$	102
$(\frac{1}{2})C_2O_4^{2-}$	74.2		
$(\frac{1}{2})SO_4^{2-}$	80		
$(\frac{1}{3})Fe(CN)_6^{3-}$	101		
$(\frac{1}{4})Fe(CN)_6^{4-}$	111		

Source: Reid, Prausnitz, and Poling [2].

about 10% to 20% and then rises to values at a concentration of 2 normal that approximate the infinite-dilution value.

EXAMPLE 3.8

Estimate the diffusivity of KCl in a dilute solution of water at 18.5°C. The experimental value is 1.7×10^{-5} cm²/s. At concentrations up to 2N, this value varies only from 1.5×10^{-5} to 1.75×10^{-5} cm²/s.

SOLUTION

At 18.5°C, $T/334\mu = 291.7/[(334)(1.05)] = 0.832$. Using Table 3.5, at 25°C, the corrected limiting ionic conductances are

$$\lambda_+ = 73.5(0.832) = 61.2 \text{ and } \lambda_- = 76.3(0.832) = 63.5$$

From (3-53), $D_\infty = \dfrac{(8.314)(291.7)[(1/1) + ((1/1)]}{96,500^2[(1/61.2) + (1/63.5)]} = 1.62 \times 10^{-5}$ cm²/s

which is close to the experimental value. ∎

For dilute, aqueous, nonelectrolyte solutions, the Wilke–Chang equation (3-39) can be used for small solute molecules of liquid molar volumes up to 500 cm³/mol, which corresponds to molecular weights to almost 600. In biological applications, diffusivities of protein macromolecules having molecular weights greater than 1,000 are of interest. In general, molecules with molecular weights to 500,000 have diffusivities at 25°C that range from 1×10^{-7} to 8×10^{-7} cm²/s, which is two orders of magnitude smaller than values of diffusivity for molecules with molecular weights less than 1,000. Data for many globular and fibrous protein macromolecules are tabulated by Sorber [10]. In the absence of data, the following semiempirical equation given by Geankoplis [11] and patterned after the Stokes–Einstein equation can be used:

$$D_{AB} = \frac{9.4 \times 10^{-15} T}{\mu_B (M_A)^{1/3}}$$

(3-54)

where the units are those of (3-39).

Also of interest in biological applications are diffusivities of small, nonelectrolyte molecules in aqueous gels containing up to 10 wt% of macromolecules such as protein mixtures (gelatin) and certain polysaccharides (agar), which have a great tendency to swell. Diffusivities are given by Friedman and Kraemer [12]. In general, the diffusivities of small solute molecules in gels are not less than 50% of the values for the diffusivity of the solute in water, with values decreasing with increasing weight percent of gel.

Diffusivity in Solids

Diffusion in solids takes place by different mechanisms depending on the diffusing atom, molecule, or ion; the nature of the solid structure, whether it be porous or nonporous, crystalline, or amorphous; and the type of solid material, whether it be metallic, ceramic, polymeric, biological, or cellular. Crystalline materials may be further classified according to the type of bonding, as molecular, covalent, ionic, or metallic, with most inorganic solids being ionic. However, ceramic materials can be ionic, covalent, or most often a combination of the two. Molecular solids, such as the noble gases, have relatively weak forces of attraction among the atoms or molecules. In covalent solids, such as quartz silica, two atoms share two or more electrons equally. In ionic solids, such as inorganic salts, one atom loses one or more of its electrons by transfer to one or more other atoms, thus forming ions. In metals, positively charged ions are bonded through a field of electrons that are free to move. Unlike diffusion coefficients in gases and low-molecular-weight liquids, which cover a range of only one or two orders of magnitude, diffusion coefficients in solids cover a range of many orders of magnitude. Despite the great complexity of diffusion in solids, Fick's first law can be used to describe diffusion if a measured diffusivity is available. However, when the diffusing solute is a gas, its solubility in the solid must also be known. If the gas dissociates upon dissolution in the solid, the concentration of the dissociated species must be used in Fick's law. In this section, many of the mechanisms of diffusion in solids are mentioned, but because they are exceedingly complex to quantify, the mechanisms are considered only qualitatively. Examples of diffusion in solids are considered, together with measured diffusion coefficients that can be used with Fick's first law. Emphasis is on diffusion of gas and liquid solutes through or into the solid, but movement of the atoms, molecules, or ions of the solid through itself is also considered.

Porous Solids

When solids are porous, predictions of the diffusivity of gaseous and liquid solute species in the pores can be made. These methods are considered only briefly here, with details deferred to Chapters 14, 15, and 16, where applications are made to membrane separations, adsorption, and leaching. This type of diffusion is also of great importance in the analysis and design of reactors using porous solid catalysts. It is sufficient to mention here that any of the following four mass transfer mechanisms or combinations thereof may take place:

1. Ordinary molecular diffusion through pores, which present tortuous paths and hinder the movement of large molecules when their diameter is more than 10% of the pore diameter

2. Knudsen diffusion, which involves collisions of diffusing gaseous molecules with the pore walls when the pore diameter and pressure are such that the molecular mean free path is large compared to the pore diameter
3. Surface diffusion involving the jumping of molecules, adsorbed on the pore walls, from one adsorption site to another based on a surface concentration-driving force
4. Bulk flow through or into the pores

When treating diffusion of solutes in porous materials where diffusion is considered to occur only in the fluid in the pores, it is common to refer to an effective diffusivity, D_{eff}, which is based on (1) the total cross-sectional area of the porous solid rather than the cross-sectional area of the pore and (2) on a straight path, rather than the pore path, which may be tortuous. If pore diffusion occurs only by ordinary molecular diffusion, the effective diffusivity can be expressed in terms of the ordinary diffusion coefficient, D, by

$$D_{eff} = \frac{D\epsilon}{\tau} \tag{3-55}$$

where ϵ is the fractional porosity (typically 0.5) of the solid and τ is the pore-path tortuosity (typically 2 to 3), which is the ratio of the pore length to the length if the pore were straight in the direction of diffusion. The effective diffusivity is either determined experimentally without knowledge of the porosity or tortuosity or predicted from (3-55) based on measurement of the porosity and tortuosity and use of the predictive methods for ordinary molecular diffusivity. As an example of the former, Boucher, Brier, and Osburn [13] measured effective diffusivities for the leaching of processed soybean oil (viscosity = 20.1 cP at 120°F) from 1/16-in.-thick porous clay plates with liquid tetrachloroethylene solvent. The rate of extraction was controlled by the rate of diffusion of the soybean oil from the clay plates. The measured value of D_{eff} was 1.0×10^{-6} cm^2/s. As might be expected from the effects of porosity and tortuosity, the effective value is about one order of magnitude less than the expected ordinary molecular diffusivity, D, of oil in the solvent.

Crystalline Solids

Diffusion through nonporous crystalline solids depends markedly on the crystal lattice structure and the diffusing entity. As discussed in Chapter 17 on crystallization, only seven different lattice structures are possible. For the cubic lattice (simple, body-centered, and face-centered), the diffusivity is the same in all directions (isotropic). In the six other lattice structures (including hexagonal and tetragonal), the diffusivity can be different in different directions (anisotropic). Many metals, including Ag, Al, Au, Cu, Ni, Pb, and Pt, crystallize into the face-centered cubic lattice structure. Others, including Be, Mg, Ti, and Zn, form anisotropic hexagonal structures. The mechanisms of diffusion in crystalline solids include:

1. Direct exchange of lattice position by two atoms or ions, probably by a ring rotation involving three or more atoms or ions
2. Migration by small solutes through interlattice spaces called interstitial sites
3. Migration to a vacant site in the lattice
4. Migration along lattice imperfections (dislocations), or grain boundaries (crystal interfaces)

Diffusion coefficients associated with the first three mechanisms can vary widely and are almost always at least one order of magnitude smaller than diffusion coefficients in low-viscosity liquids. As might be expected, diffusion by the fourth mechanism can be faster than by the other three mechanisms. Typical experimental diffusivity values, taken mainly from Barrer [14], are given in Table 3.6. The diffusivities cover gaseous, ionic, and

Table 3.6 Diffusivities of Solutes in Crystalline Metals and Salts

Metal	Solute	T, °C	D, cm²/s
Ag	Au	760	3.6×10^{-10}
	Sb	20	3.5×10^{-21}
	Sb	760	1.4×10^{-9}
Al	Fe	359	6.2×10^{-14}
	Zn	500	2×10^{-9}
	Ag	50	1.2×10^{-9}
Cu	Al	20	1.3×10^{-30}
	Al	850	2.2×10^{-9}
	Au	750	2.1×10^{-11}
Fe	H_2	10	1.66×10^{-9}
	H_2	100	1.24×10^{-7}
	C	800	1.5×10^{-8}
Ni	H_2	85	1.16×10^{-8}
	H_2	165	1.05×10^{-7}
	CO	950	4×10^{-8}
W	U	1727	1.3×10^{-11}
AgCl	Ag^+	150	2.5×10^{-14}
	Ag^+	350	7.1×10^{-8}
	Cl^-	350	3.2×10^{-16}
KBr	H_2	600	5.5×10^{-4}
	Br_2	600	2.64×10^{-4}

metallic solutes. The values cover an enormous 26-fold range. Temperature effects can be extremely large.

Metals

Important practical applications exist for diffusion of light gases through metals. To diffuse through a metal, a gas must first dissolve in the metal. As discussed by Barrer [14], all light gases do not dissolve in all metals. For example, hydrogen dissolves in such metals as Cu, Al, Ti, Ta, Cr, W, Fe, Ni, Pt, and Pd, but not in Au, Zn, Sb, and Rh. Nitrogen dissolves in Zr, but not in Cu, Ag, or Au. The noble gases do not dissolve in any of the common metals. When H_2, N_2, and O_2 dissolve in metals, they dissociate and may react to form hydrides, nitrides, and oxides, respectively. More complex molecules such as ammonia, carbon dioxide, carbon monoxide, and sulfur dioxide also dissociate. The following example illustrates how pressurized hydrogen gas can slowly leak through the wall of a small, thin pressure vessel.

EXAMPLE 3.9

Gaseous hydrogen at 200 psia and 300°C is stored in a small, 10-cm-diameter steel pressure vessel having a wall thickness of 0.125 in. The solubility of hydrogen in steel, which is proportional to the square root of the hydrogen partial pressure in the gas, is equal to 3.8×10^{-6} mol/cm³ at 14.7 psia and 300°C. The diffusivity of hydrogen in steel at 300°C is 5×10^{-6} cm²/s. If the inner surface of the vessel wall remains saturated at the existing hydrogen partial pressure and the hydrogen partial pressure at the outer surface is zero, estimate the time, in hours, for the pressure in the vessel to decrease to 100 psia because of hydrogen loss by dissolving in and diffusing through the metal wall.

SOLUTION

Integrating Fick's first law, (3-3), where A is H_2 and B is the metal, assuming a linear concentration gradient, and equating the flux to the loss of hydrogen in the vessel,

$$-\frac{dn_A}{dt} = \frac{D_A A \, \Delta c_A}{\Delta z} \tag{1}$$

Because $p_A = 0$ outside the vessel, $\Delta c_A = c_A$ = solubility of A at the inside wall surface in mol/cm³ and $c_A = 3.8 \times 10^{-6} \left(\frac{p_A}{14.7}\right)^{0.5}$, where p_A is the pressure of A in psia inside the vessel. Let p_{A_o} and n_{A_o} be the initial pressure and moles of A, respectively, in the vessel. Assuming the ideal gas law and isothermal conditions,

$$n_A = n_{A_o} p_A / p_{A_o} \tag{2}$$

Differentiating (2) with respect to time,

$$\frac{dn_A}{dt} = \frac{n_{A_o}}{p_{A_o}} \frac{dp_A}{dt} \tag{3}$$

Combining (1) and (3),

$$\frac{dp_A}{dt} = -\frac{D_A A (3.8 \times 10^{-6}) p_A^{0.5} p_{A_o}}{n_{A_o} \Delta z (14.7)^{0.5}} \tag{4}$$

Integrating and solving for t, $t = \dfrac{2 n_{A_o} \Delta z (14.7)^{0.5}}{3.8 \times 10^{-6} D_A A p_{A_o}} (p_{A_o}^{0.5} - p_A^{0.5})$

Assuming the ideal gas law, $n_{A_o} = \dfrac{(200/14.7)[(3.14 \times 10^3)/6)]}{82.05(300 + 273)} = 0.1515$ mol

The mean-spherical shell area for mass transfer, A, is

$$A = \frac{3.14}{2}[(10)^2 + (10.635)^2] = 336 \text{ cm}^2$$

The time for the pressure to drop to 100 psia is

$$t = \frac{2(0.1515)(0.125 \times 2.54)(14.7)^{0.5}}{3.8 \times 10^{-6}(5 \times 10^{-6})(336)(200)}(200^{0.5} - 100^{0.5}) = 1.2 \times 10^6 \text{ s} = 332 \text{ h} \qquad \blacksquare$$

Silica and Glass

Another area of great interest is the diffusion of light gases through various forms of silica, whose two elements, Si and O, make up about 60% of the earth's crust. Solid silica can exist in three principal crystalline forms (quartz, tridymite, and cristobalite) and in various stable amorphous forms, including vitreous silica (a noncrystalline silicate glass or fused quartz). Table 3.7 includes diffusivities, D, and solubilities as Henry's law constants, H, at 1 atm for helium and hydrogen in fused quartz as calculated from correlations of experimental data by Swets, Lee, and Frank [15] and Lee [16], respectively. The product of the diffusivity and the solubility is called the permeability, P_M. Thus,

$$P_M = DH \tag{3-56}$$

Unlike metals, where hydrogen usually diffuses as the atom, hydrogen apparently diffuses as a molecule in glass. For both hydrogen and helium, diffusivities increase rapidly with

Table 3.7 Diffusivities and Solubilities of Gases in Amorphous Silica at 1 atm

Gas	Temp, C	Diffusivity cm²/s	Solubility mol/cm³-atm
He	24	2.39×10^{-8}	1.04×10^{-7}
	100	1.64×10^{-7}	1.32×10^{-7}
	300	2.26×10^{-6}	1.82×10^{-7}
	500	9.99×10^{-6}	9.9×10^{-8}
	1,000	5.42×10^{-5}	1.34×10^{-7}
H_2	300	6.11×10^{-8}	3.2×10^{-14}
	500	6.49×10^{-7}	2.48×10^{-13}
	1,000	9.26×10^{-6}	2.49×10^{-12}
O_2	1,000	6.25×10^{-9}	

increasing temperature. At ambient temperature the diffusivities are three orders of magnitude lower than in liquids. At elevated temperatures the diffusivities approach those observed in liquids. Solubilities vary only slowly with temperature. Hydrogen is orders of magnitude less soluble in glass than helium. For hydrogen, the diffusivity is somewhat lower than in metals. Diffusivities for oxygen are also included in Table 3.7 from studies by Williams [17] and Sucov [18]. At 1000°C, the two values differ widely because, as discussed by Kingery, Bowen, and Uhlmann [19], in the former case, transport occurs by molecular diffusion; while in the latter case, transport is by slower network diffusion as oxygen jumps from one position in the silicate network to another. The activation energy for the latter is much larger than for the former (71,000 cal/mol versus 27,000 cal/mol). The choice of glass can be very critical in high-vacuum operations because of the wide range of diffusivity.

Ceramics

Diffusion rates of light gases and elements in crystalline ceramics are very important because diffusion must precede chemical reactions and causes changes in the microstructure. Therefore, diffusion in ceramics has been the subject of numerous studies, many of which are summarized in Figure 3.4, taken from Kingery et al. [19], where diffusivity is plotted as a function of the inverse of temperature in the high-temperature range. In this form, the slopes of the curves are proportional to the activation energy for diffusion, E, where

$$D = D_o \exp\left(-\frac{E}{RT}\right) \tag{3-57}$$

An insert at the middle-right region of Figure 3.4 relates the slopes of the curves to activation energy. The diffusivity curves cover a ninefold range from 10^{-6} to 10^{-15} cm²/s, with the largest values corresponding to the diffusion of potassium in β-Al_2O_3 and one of the smallest values for carbon in graphite. In general, the lower the diffusivity, the higher is the activation energy. As discussed in detail by Kingery et al. [19], diffusion in crystalline oxides depends not only on temperature but also on whether the oxide is stoichiometric or not (e.g., FeO and $Fe_{0.95}O$) and on impurities. Diffusion through vacant sites of nonstoichiometric oxides is often classified as metal-deficient or oxygen-deficient. Impurities can hinder diffusion by filling vacant lattice or interstitial sites.

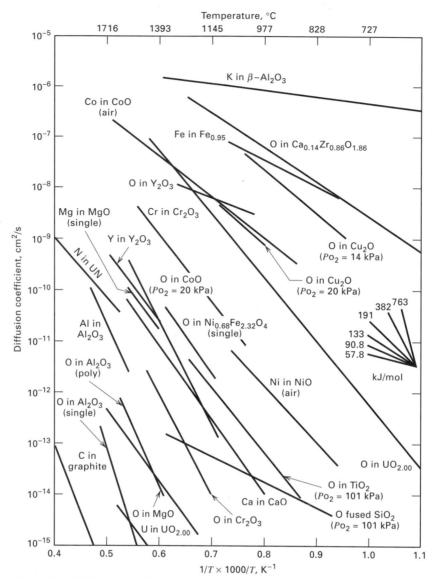

Figure 3.4 Diffusion coefficients for single- and poly-crystalline ceramics. [From W.D. Kingery, H.K. Bowen, and D.R. Uhlmann, *Introduction to Ceramics,* 2nd ed., Wiley Interscience, New York (1976), with permission.]

Polymers

Thin, dense, nonporous polymer membranes are widely used to separate gas and liquid mixtures. As discussed in detail in Chapter 14, diffusion of gas and liquid species through polymers is highly dependent on the type of polymer, whether it be crystalline or amorphous and, if the latter, glassy or rubbery. Commercial crystalline polymers are about 20% amorphous. It is mainly through the amorphous regions that diffusion occurs. As with the transport of gases through metals, transport of gaseous species through polymer membranes is usually characterized by the solution-diffusion mechanism of (3-56). Fick's first law, in the following integrated forms, is then applied to compute the mass transfer flux.

Gas species:

$$N_i = \frac{H_i D_i}{z_2 - z_1}(p_{i_1} - p_{i_2}) = \frac{P_{M_i}}{z_2 - z_1}(p_{i_1} - p_{i_2}) \tag{3-58}$$

where p_i is the partial pressure of the gas species at a polymer surface.

Liquid species:

$$N_i = \frac{K_i D_i}{z_2 - z_1}(c_{i_1} - c_{i_2}) \tag{3-59}$$

where K_i, the equilibrium partition coefficient, is equal to the ratio of the concentration in the polymer to the concentration, c_i, in the liquid adjacent to the polymer surface. The product $K_i D_i$ is the liquid permeability.

Values of diffusivity for light gases in four polymers, given in Table 14.6, range from 1.3×10^{-9} to 1.6×10^{-6} cm²/s, which is orders of magnitude less than for diffusion of the same species in a gas.

Diffusivities of liquids in rubbery polymers have been studied extensively as a means of determining viscoelastic parameters. In Table 3.8, taken from Ferry [20], diffusivities are given for different solutes in seven different rubber polymers at near-ambient conditions. The values cover a sixfold range, with the largest diffusivity being that for *n*-hexadecane in polydimethylsiloxane. The smallest diffusivities correspond to the case where the temperature is approaching the glass transition temperature, where the polymer becomes glassy in structure. This more rigid structure hinders diffusion. In general, as would be expected, smaller molecules have higher diffusivities. A more detailed study of the diffusivity of *n*-hexadecane in random styrene/butadiene copolymers at 25°C by Rhee and Ferry [21] shows a large effect on diffusivity of fractional free volume in the polymer.

Diffusion and permeability in crystalline polymers depend on the degree of crystallinity. Polymers that are 100% crystalline permit little or no diffusion of gases and liquids. For example, the diffusivity of methane at 25°C in polyoxyethylene oxyisophthaloyl decreases from 0.30×10^{-9} to 0.13×10^{-9} cm²/s when the degree of crystallinity increases from 0

Table 3.8 Diffusivities of Solutes in Rubbery Polymers

Polymer	Solute	Temperature, K	Diffusivity, cm²/s
Polyisobutylene	*n*-Butane	298	1.19×10^{-9}
	i-Butane	298	5.3×10^{-10}
	n-Pentane	298	1.08×10^{-9}
	n-Hexadecane	298	6.08×10^{-10}
Hevea rubber	*n*-Butane	303	2.3×10^{-7}
	i-Butane	303	1.52×10^{-7}
	n-Pentane	303	2.3×10^{-7}
	n-Hexadecane	298	7.66×10^{-8}
Polymethylacrylate	Ethyl alcohol	323	2.18×10^{-10}
Polyvinylacetate	*n*-Propyl alcohol	313	1.11×10^{-12}
	n-Propyl chloride	313	1.34×10^{-12}
	Ethyl chloride	343	2.01×10^{-9}
	Ethyl bromide	343	1.11×10^{-9}
Polydimethylsiloxane	*n*-Hexadecane	298	1.6×10^{-6}
1,4-Polybutadiene	*n*-Hexadecane	298	2.21×10^{-7}
Styrene-butadiene rubber	*n*-Hexadecane	298	2.66×10^{-8}

(totally amorphous) to 40% [22]. A measure of crystallinity is the polymer density. The diffusivity of methane at 25°C in polyethylene decreases from 0.193×10^{-6} to 0.057×10^{-6} cm^2/s when the specific gravity increases from 0.914 (low density) to 0.964 (high density) [22]. A plasticizer can cause the diffusivity to increase. For example, when polyvinylchloride is plasticized with 40% tricresyl triphosphate, the diffusivity of CO at 27°C increases from 0.23×10^{-8} to 2.9×10^{-8} cm^2/s [22].

EXAMPLE 3.10

Hydrogen diffuses through a nonporous polyvinyltrimethylsilane membrane at 25°C. The pressures on the sides of the membrane are 3.5 MPa and 200 kPa. Diffusivity and solubility data are given in Table 14.9. If the hydrogen flux is to be 0.64 kmol/m^2-h, how thick in micrometers should the membrane be?

SOLUTION

Equation (3-58) applies. From Table 14.9,

$$D = 160 \times 10^{-11}\,\text{m}^2/\text{s} \qquad H = S = 0.54 \times 10^{-4}\,\text{mol/m}^3\text{-Pa}$$

From (3-56), $P_M = DH = (160 \times 10^{-11})(0.54 \times 10^{-4}) = 86.4 \times 10^{-15}\,\text{mol/m-s-Pa}$

$$p_1 = 3.5 \times 10^6\,\text{Pa} \qquad p_2 = 0.2 \times 10^6\,\text{Pa}$$

Membrane thickness $= z_2 - z_1 = \Delta z = P_M(p_1 - p_2)/N$

$$\Delta z = \frac{86.4 \times 10^{-15}(3.5 \times 10^6 - 0.2 \times 10^6)}{[0.64(1000)/3600]} = 1.6 \times 10^{-6}\,\text{m} = 1.6\,\mu\text{m}$$

As discussed in Chapter 14, polymer membranes must be very thin to achieve reasonable gas permeation rates. ∎

Cellular Solids and Wood

As discussed by Gibson and Ashby [23], cellular solids consist of solid struts or plates that form edges and faces of cells, which are compartments or enclosed spaces. Cellular solids such as wood, cork, sponge, and coral exist in nature. Synthetic cellular structures include honeycombs, and foams (some with open cells) made from polymers, metals, ceramics, and glass. The word *cellulose* means "full of little cells."

A widely used cellular solid is wood, whose annual world production of the order of 10^{12} kg is comparable to the production of iron and steel. Chemically, wood consists of lignin, cellulose, hemicellulose, and minor amounts of organic chemicals and elements. The latter are extractable, and the former three, which are all polymers, give wood its structure. Green wood also contains up to 25 wt% moisture in the cell walls and cell cavities. Adsorption or desorption of moisture in wood causes anisotropic swelling and shrinkage. The structure of wood, which often consists of (1) highly elongated hexagonal or rectangular cells, called tracheids in softwood (coniferous species, e.g., spruce, pine, and fir) and fibers in hardwood (deciduous or broad-leaf species, e.g., oak, birch, and walnut); (2) radial arrays of rectangular-like cells, called rays, which are narrow and short in softwoods but wide and long in hardwoods; and (3) enlarged cells with large pore spaces and thin walls, called sap channels because they conduct fluids up the tree. The sap channels are less than 3 vol% of softwood, but as much as 55 vol% of hardwood.

Because the structure of wood is directional, many of its properties are anisotropic. For example, stiffness and strength are 2 to 20 times greater in the axial direction of the tracheids or fibers than in the radial and tangential directions of the trunk from which the wood is cut. This anisotropy extends to permeability and diffusivity of wood penetrants, such as moisture and preservatives. According to Stamm [24], the permeability of wood to liquids in the axial direction can be up to 10 times greater than in the transverse direction.

Movement of liquids and gases through wood and wood products takes time during drying and treatment with preservatives, fire retardants, and other chemicals. This movement takes place by capillarity, pressure permeability, and diffusion. Nevertheless, wood is not highly permeable because the cell voids are largely discrete and lack direct interconnections. Instead, communication among cells is through circular openings spanned by thin membranes with submicrometer-sized pores, called pits, and to a smaller extent, across the cell walls. Rays give wood some permeability in the radial direction. Sap channels do not contribute to permeability. All three mechanisms of movement of gases and liquids in wood are considered by Stamm [24]. Only diffusion is discussed here.

The simplest form of diffusion is that of a water-soluble solute through wood saturated with water, such that no dimensional changes occur. For the diffusion of urea, glycerine, and lactic acid into hardwood, Stamm [24] lists diffusivities in the axial direction that are about 50% of ordinary liquid diffusivities. In the radial direction, diffusivities are about 10% of the values in the axial direction. For example, at 26.7°C the diffusivity of zinc sulfate in water is 5×10^{-6} cm^2/s. If loblolly pine sapwood is impregnated with zinc sulfate in the radial direction, the diffusivity is found to be 0.18×10^{-6} cm^2/s [24].

The diffusion of water in wood is much more complex. Moisture content determines the degree of swelling or shrinkage. Water is held in the wood in different ways: It may be physically adsorbed on cell walls in monomolecular layers, condensed in preexisting or transient cell capillaries, or absorbed in cell walls to form a solid solution.

Because of the practical importance of the lumber drying rates, most diffusion coefficients are measured under drying conditions in the radial direction across the fibers. The results depend on temperature and swollen-volume specific gravity. Typical results are given by Sherwood [25] and Stamm [24]. For example, for beech with a swollen specific gravity of 0.4, the diffusivity increases from a value of about 1×10^{-6} cm^2/s at 10°C to 10×10^{-6} cm^2/s at 60°C.

3.3. ONE-DIMENSIONAL STEADY-STATE AND UNSTEADY-STATE MOLECULAR DIFFUSION

Fick's first law of ordinary molecular diffusion relates the molar flow flux through species B of a diffusing species, A, across a given plane of area A, to the diffusivity and the concentration gradient at the plane. In terms of the mass transfer rate by diffusion only across the plane, (3-3) becomes

$$n_A = -D_{AB}A\left(\frac{dc_A}{dz}\right) \tag{3-60}$$

Steady State

For steady-state, one-dimensional diffusion, with constant D_{AB}, (3-60) can be integrated for various geometries, the most common results being analogous to heat conduction.

1. Plane wall with a thickness, $z_2 - z_1$:

$$n_A = D_{AB}A\left(\frac{c_{A_1} - c_{A_2}}{z_2 - z_1}\right) \tag{3-61}$$

2. Hollow cylinder of inner radius r_1 and outer radius r_2, with diffusion in the radial direction outward:

$$n_A = 2\pi L\frac{D_{AB}(c_{A_1} - c_{A_2})}{\ln(r_2/r_1)} \tag{3-62}$$

or

$$n_A = D_{AB}A_{LM}\left(\frac{c_{A_1} - c_{A_2}}{r_2 - r_1}\right) \tag{3-63}$$

where A_{LM} = log mean of the areas $2\pi rL$ at r_1 and r_2
L = length of the hollow cylinder

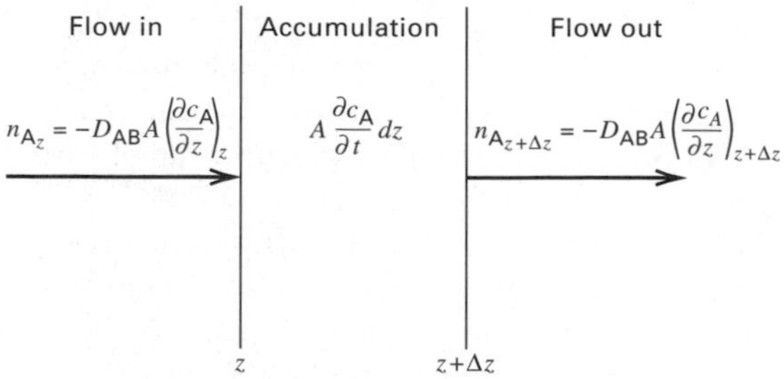

Flow in	Accumulation	Flow out

$$n_{A_z} = -D_{AB}A\left(\frac{\partial c_A}{\partial z}\right)_z \qquad A\frac{\partial c_A}{\partial t}dz \qquad n_{A_{z+\Delta z}} = -D_{AB}A\left(\frac{\partial c_A}{\partial z}\right)_{z+\Delta z}$$

z $\qquad$ $z+\Delta z$

Figure 3.5 Unsteady-state diffusion through a differential volume $A\ dz$.

3. Spherical shell of inner radius r_1 and outer radius r_2, with diffusion in the radial direction outward:

$$n_A = \frac{4\pi r_1 r_2 D_{AB}(c_{A_1} - c_{A_2})}{r_2 - r_1} \qquad (3\text{-}64)$$

or

$$n_A = D_{AB}A_{GM}\left(\frac{c_{A_1} - c_{A_2}}{r_2 - r_1}\right) \qquad (3\text{-}65)$$

where A_{GM} = geometric mean of the areas $4\pi r^2$ at r_1 and r_2.

When $r_1/r_2 < 2$, the arithmetic mean area is no more that 4% greater than the log mean area. When $r_1/r_2 < 1.33$, the arithmetic mean area is no more than 4% greater than the geometric mean area.

Unsteady State

Equation (3-60) is applied to unsteady-state molecular diffusion by considering the accumulation or depletion of a species with time in a unit volume through which the species is diffusing. Consider the one-dimensional diffusion through species B of species A through a differential control volume with diffusion in the z direction only, as shown in Figure 3.5. Assume constant total concentration, $c = c_A + c_B$, constant diffusivity, and negligible bulk flow. The molar flow rate of species A by diffusion at the plane $z = z$ is given by (3-60)

$$n_{A_z} = -D_{AB}A\left(\frac{\partial c_A}{\partial z}\right)_z \qquad (3\text{-}66)$$

At the plane, $z = z + \Delta z$, the diffusion rate is

$$n_{A_{z+\Delta z}} = -D_{AB}A\left(\frac{\partial c_A}{\partial z}\right)_{z+\Delta z} \qquad (3\text{-}67)$$

The accumulation of species A in the control volume is

$$A\frac{\partial c_A}{\partial t}\Delta z \qquad (3\text{-}68)$$

Since rate in − rate out = accumulation,

$$-D_{AB}A\left(\frac{\partial c_A}{\partial z}\right)_z + D_{AB}A\left(\frac{\partial c_A}{\partial z}\right)_{z+\Delta z} = A\left(\frac{\partial c_A}{\partial t}\right)\Delta z \qquad (3\text{-}69)$$

Rearranging and simplifying,

$$D_{AB} \left[\frac{(\partial c_A/\partial z)_{z+\Delta z} - (\partial c_A/\partial z)_z}{\Delta z} \right] = \frac{\partial c_A}{\partial t} \tag{3-70}$$

In the limit, as $\Delta z \rightarrow 0$,

$$\frac{\partial c_A}{\partial t} = D_{AB} \frac{\partial^2 c_A}{\partial z^2} \tag{3-71}$$

Equation (3-71) is Fick's second law for one-dimensional diffusion. The more general form, for three-dimensional rectangular coordinates, is

$$\frac{\partial c_A}{\partial t} = D_{AB} \left(\frac{\partial^2 c_A}{\partial x^2} + \frac{\partial^2 c_A}{\partial y^2} + \frac{\partial^2 c_A}{\partial z^2} \right) \tag{3-72}$$

For one-dimensional diffusion in the radial direction only, for cylindrical and spherical coordinates, Fick's second law becomes, respectively,

$$\frac{\partial c_A}{\partial t} = \frac{D_{AB}}{r} \frac{\partial}{\partial r} \left(r \frac{\partial c_A}{\partial r} \right) \tag{3-73}$$

and

$$\frac{\partial c_A}{\partial t} = \frac{D_{AB}}{r^2} \frac{\partial}{\partial r} \left(r^2 \frac{\partial c_A}{\partial r} \right) \tag{3-74}$$

Equations (3-71) to (3-74) are analogous to Fourier's second law of heat conduction where c_A is replaced by temperature, T, and diffusivity, D_{AB}, is replaced by thermal diffusivity, $\alpha = k/\rho C_P$. Analytical solutions to these partial differential equations in either Fick's law or Fourier's law form are available for a variety of boundary conditions. Many of these solutions are derived and discussed by Carslaw and Jaeger [26] and Crank [27]. Only a few of the more useful solutions are presented here.

Semiinfinite Medium

Consider the semiinfinite medium shown in Figure 3.6, which extends in the z direction from $z = 0$ to $z = \infty$. The x and y coordinates extend from $-\infty$ to $+\infty$, but are not of interest because diffusion takes place only in the z direction. Thus, (3-71) applies to the region $z \geq 0$. At time $t \leq 0$, the concentration is c_{A_o} for $z \geq 0$. At $t = 0$, the surface of the semiinfinite medium at $z = 0$ is instantaneously brought to the concentration $c_{A_s} > c_{A_o}$ and held there for $t > 0$. Therefore, diffusion into the medium occurs. However, because the medium is infinite in the z direction, diffusion cannot extend to $z = \infty$ and, therefore, as $z \rightarrow \infty$, $c_A = c_{A_o}$ for all $t \geq 0$. Because the partial differential equation (3-71) and its one boundary (initial) condition in time and two boundary conditions in distance are linear in the dependent variable, c_A, an exact solution can be obtained. Either the

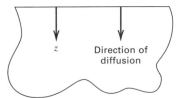

Figure 3.6 One-dimensional diffusion into a semiinfinite medium.

method of combination of variables [28] or the Laplace transform method [29] is applicable. The result, in terms of the fractional accomplished concentration change, is

$$\theta = \frac{c_A - c_{A_o}}{c_{A_s} - c_{A_o}} = \text{erfc}\left(\frac{z}{2\sqrt{D_{AB}t}}\right) \tag{3-75}$$

where the complementary error function, erfc, is related to the error function, erf, by

$$\text{erfc}(x) = 1 - \text{erf}(x) = 1 - \frac{2}{\sqrt{\pi}}\int_0^x e^{-\eta^2}\,d\eta \tag{3-76}$$

The error function is included in most spreadsheet programs and handbooks, such as *Handbook of Mathematical Functions* [30]. The variation of $\text{erf}(x)$ and $\text{erfc}(x)$ is as follows:

x	$\text{erf}(x)$	$\text{erfc}(x)$
0	0.0000	1.0000
0.5	0.5205	0.4795
1.0	0.8427	0.1573
1.5	0.9661	0.0339
2.0	0.9953	0.0047
∞	1.0000	0.0000

Equation (3-75) is used to compute the concentration in the semiinfinite medium, as a function of time and distance from the surface, assuming no bulk flow. Thus, it applies most rigorously to diffusion in solids, and also to stagnant liquids and gases when the medium is dilute in the diffusing solute.

In (3-75), when $(z/2\sqrt{D_{AB}t}) = 2$, the complementary error function is only 0.0047, which represents less than a 1% change in the ratio of the concentration change at $z = z$ to the change at $z = 0$. Thus, it is common to refer to $z = 4\sqrt{D_{AB}t}$ as the penetration depth and to apply (3-75) to media of finite thickness as long as the thickness is greater than the penetration depth.

The instantaneous rate of mass transfer across the surface of the medium at $z = 0$ can be obtained by taking the derivative of (3-75) with respect to distance and substituting it into Fick's first law applied at the surface of the medium. Thus, using the Leibnitz rule for differentiating the integral of (3-76), with $x = z/2\sqrt{D_{AB}t}$,

$$n_A = -D_{AB}A\left(\frac{\partial c_A}{\partial z}\right)_{z=0} = D_{AB}A\left(\frac{c_{A_s} - c_{A_o}}{\sqrt{\pi D_{AB}t}}\right)\exp\left(-\frac{z^2}{4D_{AB}t}\right)\bigg|_{z=0} \tag{3-77}$$

Thus,

$$n_A|_{z=0} = \sqrt{\frac{D_{AB}}{\pi t}}A(c_{A_s} - c_{A_o}) \tag{3-78}$$

We can also determine the total number of moles of solute, $\mathcal{N}_A$, transferred into the semiinfinite medium by integrating (3-78) with respect to time:

$$\mathcal{N}_A = \int_o^t n_A|_{z=0}\,dt = \sqrt{\frac{D_{AB}}{\pi}}A(c_{A_s} - c_{A_o})\int_o^t \frac{dt}{\sqrt{t}} = 2A(c_{A_s} - c_{A_o})\sqrt{\frac{D_{AB}t}{\pi}} \tag{3-79}$$

EXAMPLE 3.11

Determine how long it will take for the dimensionless concentration change, $\theta = (c_A - c_{A_o})/(c_{A_s} - c_{A_o})$, to reach 0.01 at a depth $z = 1$ m in a semiinfinite medium, which is initially at a solute concentration c_{A_o}, after the surface concentration at $z = 0$ increases to c_{A_s}, for diffusivities representative of a solute diffusing through a gas, a liquid, and a solid.

SOLUTION

For a gas, assume $D_{AB} = 0.1$ cm^2/s. We know that $z = 1$ m $= 100$ cm.

From (3-75) and (3-76), $\theta = 0.01 = 1 - \mathrm{erf}\left(\dfrac{z}{2\sqrt{D_{AB}t}}\right)$

Therefore, $\mathrm{erf}\left(\dfrac{z}{2\sqrt{D_{AB}t}}\right) = 0.99$

From tables of the error function,

$$\left(\frac{z}{2\sqrt{D_{AB}t}}\right) = 1.8214$$

Solving,

$$t = \left[\frac{100}{1.8214(2)}\right]^2 \frac{1}{0.10} = 7{,}540\ \mathrm{s} = 2.09\ \mathrm{h}$$

In a similar manner, the times for typical gas, liquid, and solid media are:

Semiinfinite Medium	D_{AB}, cm^2/s	Time for $\theta = 0.01$ at 1 m
Gas	0.10	2.09 h
Liquid	1×10^{-5}	2.39 year
Solid	1×10^{-9}	239 centuries

These results show that molecular diffusion is very slow, especially in liquids and solids. In liquids and gases, the rate of mass transfer can be greatly increased by agitation to induce turbulent motion. For solids, it is best to reduce the diffusion path to as small a dimension as possible. ∎

Medium of Finite Thickness with Sealed Edges

Consider a rectangular parallelepiped medium of finite thickness $2a$ in the z direction, and either infinitely long dimensions in the y and x directions or finite lengths of $2b$ and $2c$, respectively, in those directions. Assume that in Figure 3.7a the edges parallel to the z direction are scaled, so diffusion occurs only in the z direction and initially the concentration of the solute in the medium is uniform at c_{A_o}. At time $t = 0$, the two unsealed surfaces of the medium at $z = \pm a$ are brought to and held at concentration $c_{A_s} > c_{A_o}$. Because of symmetry, $\partial c_A/\partial z = 0$ at $z = 0$. Assume constant D_{AB}. Again (3-71) applies, and an exact solution can be obtained because both (3-71) and the boundary conditions are linear in c_A. By the method of separation of variables [28] or the Laplace transform method [29], the result from Carslaw and Jaeger [26], in terms of the fractional unaccomplished concentration change, E, is

$$E = 1 - \theta = \frac{c_{A_s} - c_A}{c_{A_s} - c_{A_o}} = \frac{4}{\pi} \sum_{n=0}^{\infty} \frac{(-1)^n}{(2n+1)} \exp[-D_{AB}(2n+1)^2 \pi^2 t/4a^2] \cos\frac{(2n+1)\pi z}{2a} \quad \text{(3-80)}$$

or, in terms of the complementary error function,

$$E = 1 - \theta = \frac{c_{A_s} - c_A}{c_{A_s} - c_{A_o}} = \sum_{n=0}^{\infty} (-1)^n \left[\mathrm{erfc}\frac{(2n+1)a - z}{2\sqrt{D_{AB}t}} + \mathrm{erfc}\frac{(2n+1)a + z}{2\sqrt{D_{AB}t}}\right] \quad \text{(3-81)}$$

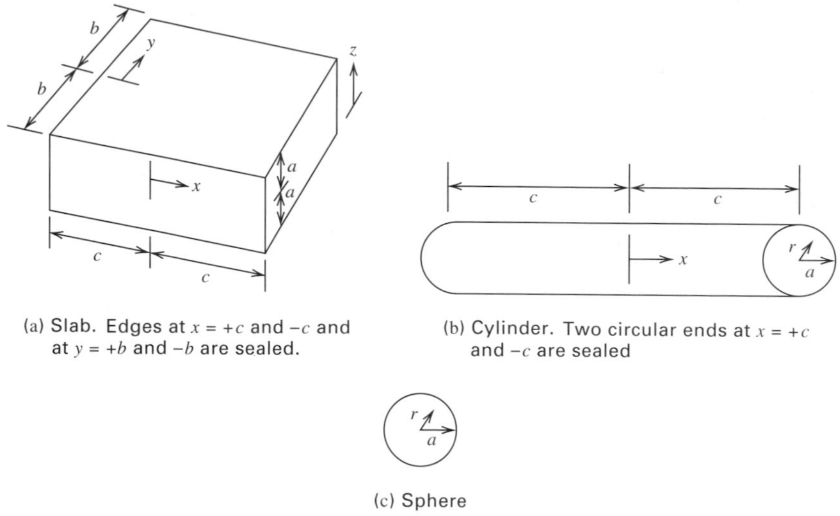

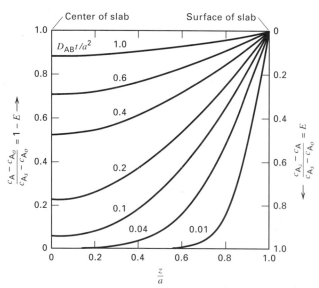

(a) Slab. Edges at $x = +c$ and $-c$ and
at $y = +b$ and $-b$ are sealed.

(b) Cylinder. Two circular ends at $x = +c$
and $-c$ are sealed

(c) Sphere

Figure 3.7 Unsteady-state diffusion in media of finite dimensions.

For large values of $D_{AB}t/a^2$, the Fourier number for mass transfer, the infinite series solutions of (3-80) and (3-81) converge rapidly, but for small values (e.g., short times), they do not. However, in the latter case, the solution for the semiinfinite medium applies for $D_{AB}t/a^2 < \frac{1}{16}$. A convenient plot of the exact solution is given in Figure 3.8.

The instantaneous rate of mass transfer across the surface of either unsealed face of the medium (i.e., at $z = \pm a$), is obtained by differentiating (3-80) with respect to z, evaluating the result at $z = a$, followed by substitution into Fick's first law to give

$$n_A|_{z=a} = \frac{2D_{AB}(c_{A_s} - c_{A_o})A}{a} \sum_{n=0}^{\infty} \exp\left[-\frac{D_{AB}(2n+1)^2\pi^2 t}{4a^2}\right] \tag{3-82}$$

Figure 3.8 Concentration profiles for unsteady-state diffusion in a slab. [Adapted from H.S. Carslaw and J.C. Jaeger, *Conduction of Heat in Solids,* 2nd ed., Oxford University Press, London (1959).]

We can also determine the total number of moles transferred across either unsealed face by integrating (3-82) with respect to time. Thus,

$$\mathcal{N}_A = \int_o^t n_A|_{z=a}\, dt = \frac{8(c_{A_s} - c_{A_o})Aa}{\pi^2} \sum_{n=0}^{\infty} \frac{1}{(2n+1)^2} \left\{ 1 - \exp\left[-\frac{D_{AB}(2n+1)^2\pi^2 t}{4a^2} \right] \right\} \quad \textbf{(3-83)}$$

In addition, the average concentration of the solute in the medium, $c_{A_{avg}}$, as a function of time, can be obtained in the case of a slab from:

$$\frac{c_{A_s} - c_{A_{avg}}}{c_{A_s} - c_{A_o}} = \frac{\int_o^a (1 - \theta)\, dz}{a} \quad \textbf{(3-84)}$$

Substitution of (3-80) into (3-84) followed by integration gives

$$E_{avg_{slab}} = (1 - \theta_{ave})_{slab} = \frac{c_{A_s} - c_{A_{avg}}}{c_{A_s} - c_{A_o}} = \frac{8}{\pi^2} \sum_{n=0}^{\infty} \frac{1}{(2n+1)^2} \exp\left[-\frac{D_{AB}(2n+1)^2\pi^2 t}{4a^2} \right] \quad \textbf{(3-85)}$$

This equation is plotted in Figure 3.9. Concentrations are in mass solute per mass dry solid or mass solute/volume.

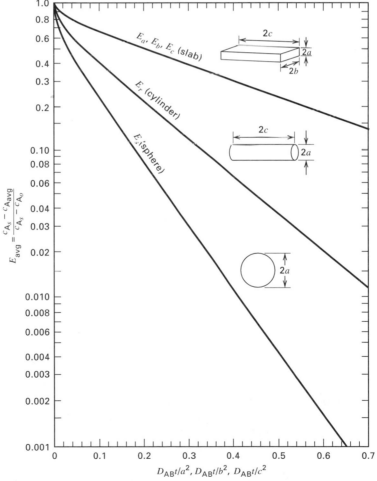

Figure 3.9 Average concentrations for unsteady-state diffusion. [Adapted from R.E. Treybal, *Mass-Transfer Operations*, 3rd ed., McGraw-Hill, New York (1980).]

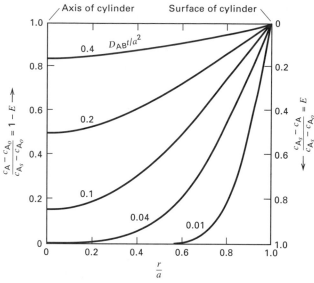

Figure 3.10 Concentration profiles for unsteady-state diffusion in a cylinder. [Adapted from H.S. Carslaw and J.C. Jaeger, *Conduction of Heat in Solids,* 2nd ed., Oxford University Press, London (1959).]

When the edges of the slab in Figure 3.7a are not sealed, the method of Newman [31] can be used with (3-72) to determine concentration changes within the slab. In this method, E or E_{avg} is given in terms of the E values from the solution of (3-71) for each of the coordinate directions by

$$E = E_x E_y E_z \tag{3-86}$$

Corresponding solutions for infinitely long circular cylinders and spheres are available in Carslaw and Jaeger [26] and are plotted in Figures 3.9, 3.10, and 3.11. For a short cylinder, where the ends are not sealed, E or E_{ave} is given by the method of Newman as

$$E = E_r E_x \tag{3-87}$$

Some materials, such as crystals and wood, have thermal conductivities and diffusivities that vary markedly with direction. For these anisotropic materials, Fick's second law in the form of (3-72) does not hold. Although the general anisotropic case is exceedingly complex, as shown in the following example, the mathematical treatment is relatively simple when the principal axes of diffusivity coincide with the coordinate system.

EXAMPLE 3.12

A piece of lumber, measuring $5 \times 10 \times 20$ cm, initially contains 20 wt% moisture. At time 0, all six faces are brought to an equilibrium moisture content of 2 wt%. Diffusivities for moisture at 25°C are 2×10^{-5} cm^2/s in the axial (z) direction along the fibers and 4×10^{-6} cm^2/s in the two directions perpendicular to the fibers. Calculate the time in hours for the average moisture content to drop to 5 wt% at 25°C. At that time, determine the moisture content at the center of the piece of lumber. All moisture contents are on a dry basis.

SOLUTION

In this case, the solid is anisotropic, with $D_x = D_y = 4 \times 10^{-6}$ cm^2/s and $D_z = 2 \times 10^{-5}$ cm^2/s, where dimensions $2c$, $2b$, and $2a$ in the x, y, and z directions are 5, 10, and 20 cm, respectively. Fick's second law for an isotropic medium, (3-72), must be rewritten for this anisotropic mate-

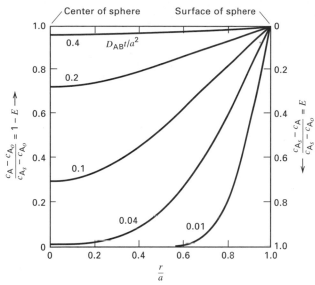

Figure 3.11 Concentration profiles for unsteady-state diffusion in a sphere. [Adapted from H.S. Carslaw and J.C. Jaeger, *Conduction of Heat in Solids,* 2nd ed., Oxford University Press, London (1959).]

rial as

$$\frac{\partial c_A}{\partial t} = D_x \left[\frac{\partial^2 c_A}{\partial x^2} + \frac{\partial^2 c_A}{\partial y^2} \right] + D_z \frac{\partial^2 c_A}{\partial z^2} \tag{1}$$

as discussed by Carslaw and Jaeger [26].

To transform (1) into the form of (3-72), let

$$x_1 = x \sqrt{\frac{D}{D_x}} \qquad y_1 = y \sqrt{\frac{D}{D_x}} \qquad z_1 = z \sqrt{\frac{D}{D_z}} \tag{2}$$

where D is chosen arbitrarily. With these changes in variables, (1) becomes

$$\frac{\partial c_A}{\partial t} = D \left(\frac{\partial^2 c_A}{\partial x_1^2} + \frac{\partial^2 c_A}{\partial y_1^2} + \frac{\partial^2 c_A}{\partial z_1^2} \right) \tag{3}$$

Since this is the same form as (3-72) and since the boundary conditions do not involve diffusivities, we can apply Newman's method, using Figure 3.9, where concentration, c_A, is replaced by weight-percent moisture.

From (3-86) and (3-85),

$$E_{ave_{slab}} = E_{avg_x} E_{avg_y} E_{avg_z} = \frac{c_{A_{ave}} - c_{A_s}}{c_{A_o} - c_{A_s}} = \frac{5 - 2}{20 - 2} = 0.167$$

Let $D = 1 \times 10^{-5}$ cm^2/s.

z_1 Direction (axial):

$$a_1 = a \left(\frac{D}{D_z} \right)^{1/2} = \frac{20}{2} \left(\frac{1 \times 10^{-5}}{2 \times 10^{-5}} \right)^{1/2} = 7.07 \text{ cm}$$

$$\frac{Dt}{a_1^2} = \frac{1 \times 10^{-5} t}{7.07^2} = 2.0 \times 10^{-7} t, \text{ s}$$

y_1 Direction:

$$b_1 = b\left(\frac{D}{D_y}\right)^{1/2} = \frac{10}{2}\left(\frac{1\times10^{-5}}{4\times10^{-6}}\right)^{1/2} = 7.906 \text{ cm}$$

$$\frac{Dt}{b_1^2} = \frac{1\times10^{-5}t}{7.906^2} = 1.6\times10^{-7}t, \text{ s}$$

x_1-Direction:

$$c_1 = c\left(\frac{D}{D_x}\right)^{1/2} = \frac{5}{2}\left(\frac{1\times10^{-5}}{4\times10^{-6}}\right)^{1/2} = 3.953 \text{ cm}$$

$$\frac{Dt}{c_1^2} = \frac{1\times10^{-5}t}{3.953^2} = 3.953^2 = 6.4\times10^{-7}t, \text{ s}$$

Use Figure 3.9 iteratively with assumed values of time in seconds to obtain values of E_{avg} for each of the three coordinates until (3-86) equals 0.167.

t, h	t, s	$E_{avg_{z_1}}$	$E_{avg_{y_1}}$	$E_{avg_{x_1}}$	E_{avg}
100	360,000	0.70	0.73	0.46	0.235
120	432,000	0.67	0.70	0.41	0.193
135	486,000	0.65	0.68	0.37	0.164

Therefore, it takes approximately 136 h.

For 136 h = 490,000 s, the Fourier numbers for mass transfer are

$$\frac{Dt}{a_1^2} = \frac{(1\times10^{-5})(490,000)}{7.07^2} = 0.0980$$

$$\frac{Dt}{b_1^2} = \frac{(1\times10^{-15})(490,000)}{7.906^2} = 0.0784$$

$$\frac{Dt}{c_1^2} = \frac{(1\times10^{-5})(490,000)}{3.953^2} = 0.3136$$

From Figure 3.8, at the center of the slab,

$$E_{center} = E_{z_1}E_{y_1}E_{x_1} = (0.945)(0.956)(0.605) = 0.547$$

$$= \frac{c_{A_s} - c_{A_{center}}}{c_{A_s} - c_{A_o}} = \frac{2 - c_{A_{center}}}{2 - 20} = 0.547$$

Solving,

$$c_A \text{ at the center} = 11.8 \text{ wt\% moisture} \qquad\blacksquare$$

3.4   MOLECULAR DIFFUSION IN LAMINAR FLOW

Many mass transfer operations involve diffusion in fluids in laminar flow. The fluid may be a film flowing slowly down a vertical or inclined surface, a laminar boundary layer that forms as the fluid flows slowly past a thin plate, or the fluid may flow through a small tube or slowly through a large pipe or duct. Mass transfer may occur between a gas and a liquid film, between a solid surface and a fluid, or between a fluid and a membrane surface.

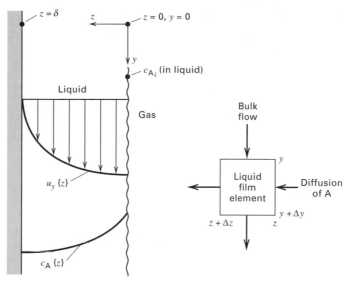

Figure 3.12 Mass transfer from a gas into a falling laminar liquid film.

Falling Liquid Film

Consider a thin liquid film, of a mixture of A and nonvolatile B, falling in laminar flow at steady state down one side of a vertical surface and exposed to pure gas, A, on the other side of the film, as shown in Figure 3.12. The surface is infinitely wide in the x direction. In the absence of mass transfer of A into the liquid film, the liquid velocity in the z direction, u_z, is zero. In the absence of end effects, the equation of motion for the liquid film in fully developed laminar flow in the downward y direction is

$$\mu \frac{d^2 u_y}{dz^2} + \rho g = 0 \tag{3-88}$$

Usually, fully developed flow, where u_y is independent of the distance y, is established quickly. If δ is the thickness of the film and the boundary conditions are $u_y = 0$ at $z = \delta$ (no-slip condition at the solid surface) and $du_y/dz = 0$ at $z = 0$ (no drag at the liquid–gas interface), (3-88) is readily integrated to give a parabolic velocity profile:

$$u_y = \frac{\rho g \delta^2}{2\mu} \left[1 - \left(\frac{z}{\delta} \right)^2 \right] \tag{3-89}$$

Thus, the maximum liquid velocity, which occurs at $z = 0$, is

$$(u_y)_{\max} = \frac{\rho g \delta^2}{2\mu} \tag{3-90}$$

The bulk average velocity in the liquid film is

$$\bar{u}_y = \frac{\int_0^\delta u_y \, dz}{\delta} = \frac{\rho g \delta^2}{3\mu} \tag{3-91}$$

Thus, the film thickness for fully developed flow is independent of location y and is

$$\delta = \left(\frac{3\overline{u}_y \mu}{\rho g} \right)^{1/2} = \left(\frac{3\mu \Gamma}{\rho^2 g} \right)^{1/3} \tag{3-92}$$

where Γ = liquid film flow rate per unit width of film, W.

For film flow, the Reynolds number, which is the ratio of the inertial force to the viscous force, is

$$N_{\text{Re}} = \frac{4 r_H \overline{u}_y \rho}{\mu} = \frac{4\delta \overline{u}_y \rho}{\mu} = \frac{4\Gamma}{\mu} \tag{3-93}$$

where r_H = hydraulic radius = (flow cross section)/(wetted perimeter) = $(W\delta)/W = \delta$ and, by the equation of continuity, $\Gamma = \overline{u}_y \rho \delta$.

As reported by Grimley [32], for $N_{\text{Re}} < 8$ to 25, depending on the surface tension and viscosity, the flow in the film is laminar and the interface between the liquid film and the gas is flat. The value of 25 is obtained with water. For 8 to $25 < N_{\text{Re}} < 1,200$, the flow is still laminar, but ripples and waves may appear at the interface unless suppressed by the addition of wetting agents to the liquid.

For a flat liquid–gas interface and a small rate of mass transfer of A into the liquid film, (3-88) to (3-93) hold and the film velocity profile is given by (3-89). Now consider a mole balance on A for an incremental volume of liquid film of constant density, as shown in Figure 3.12. Neglect bulk flow in the z direction and axial diffusion in the y direction. Then, at steady state, neglecting accumulation or depletion of A in the incremental volume,

$$-D_{\text{AB}}(\Delta y)(\Delta x)\left(\frac{\partial c_{\text{A}}}{\partial z} \right)_z + u_y c_{\text{A}}|_y (\Delta z)(\Delta x)$$
$$= -D_{\text{AB}}(\Delta y)(\Delta x)\left(\frac{\partial c_{\text{A}}}{\partial z} \right)_{z+\Delta z} + u_y c_{\text{A}}|_{y+\Delta y}(\Delta z)(\Delta x) \tag{3-94}$$

Rearranging and simplifying (3-94),

$$\left[\frac{u_y c_{\text{A}}|_{y+\Delta y} - u_y c_{\text{A}}|_y}{\Delta y} \right] = D_{\text{AB}}\left[\frac{(\partial c_{\text{A}}/\partial z)_{z+\Delta z} - (\partial c_{\text{A}}/\partial z)_z}{\Delta z} \right] \tag{3-95}$$

In the limit, as $\Delta z \to 0$ and $\Delta y \to 0$,

$$u_y \frac{\partial c_{\text{A}}}{\partial y} = D_{\text{AB}} \frac{\partial^2 c_{\text{A}}}{\partial z^2} \tag{3-96}$$

Substituting (3-89) into (3-96),

$$\frac{\rho g \delta^2}{2\mu}\left[1 - \left(\frac{z}{\delta} \right)^2 \right] \frac{\partial c_{\text{A}}}{\partial y} = D_{\text{AB}} \frac{\partial^2 c_{\text{A}}}{\partial z^2} \tag{3-97}$$

This equation was solved by Johnstone and Pigford [33] and later by Olbrich and Wild [34], for the following boundary conditions:

$$c_{\text{A}} = c_{\text{A}_i} \text{ at } z = 0 \text{ for } y > 0$$
$$c_{\text{A}} = c_{\text{A}_0} \text{ at } y = 0 \text{ for } 0 < z < \delta$$
$$\partial c_{\text{A}}/\partial z = 0 \text{ at } z = \delta \text{ for } 0 < y < L$$

where L = height of the vertical surface. The solution of Olbrich and Wild is in the form of an infinite series, giving c_{A} as a function of z and y. However, of more interest is the average concentration at $y = L$, which, by integration, is

$$\bar{c}_{A_y} = \frac{1}{\bar{u}_y \delta} \int_0^\delta u_y c_{A_y} \, dz \tag{3-98}$$

For the condition $y = L$, the result is

$$\frac{c_{A_i} - \bar{c}_{A_L}}{c_{A_i} - c_{A_0}} = 0.7857 e^{-5.1213\eta} + 0.09726 e^{-39.661\eta} + 0.036093^{-106.25\eta} + \ldots \tag{3-99}$$

where

$$\eta = \frac{2 D_{AB} L}{3 \delta^2 \bar{u}_y} = \frac{8/3}{N_{Re} N_{Sc} (\delta/L)} = \frac{8/3}{(\delta/L) N_{Pe_M}} \tag{3-100}$$

$$N_{Sc} = \text{Schmidt number} = \frac{\mu}{\rho D_{AB}} = \frac{\text{momentum diffusivity, } \mu/\rho}{\text{mass diffusivity, } D_{AB}} \tag{3-101}$$

$$N_{Pe_M} = N_{Re} N_{Sc} = \text{Peclet number for mass transfer} = \frac{4 \delta \bar{u}_y}{D_{AB}} \tag{3-102}$$

The Schmidt number is analogous to the Prandtl number, used in heat transfer:

$$N_{Pr} = \frac{C_P \mu}{k} = \frac{(\mu/\rho)}{(k/\rho C_P)} = \frac{\text{momentum diffusivity}}{\text{thermal diffusivity}}$$

The Peclet number for mass transfer is analogous to the Peclet number for heat transfer:

$$N_{Pe_H} = N_{Re} N_{Pr} = \frac{4 \delta \bar{u}_y C_P \rho}{k}$$

Both Peclet numbers are ratios of convective transport to molecular transport.

The total rate of absorption of A from the gas into the liquid film for height L and width W is

$$n_A = \bar{u}_y \delta W (\bar{c}_{A_L} - c_{A_0}) \tag{3-103}$$

Mass Transfer Coefficients

Mass transfer problems involving fluids are most often solved using mass transfer coefficients, analogous to heat transfer coefficients. For the latter, Newton's law of cooling defines a heat transfer coefficient, h:

$$Q = hA \, \Delta T \tag{3-104}$$

where Q = rate of heat transfer

A = area for heat transfer (normal to the direction of heat transfer)

ΔT = temperature-driving force for heat transfer

For mass transfer, a concentration driving force replaces ΔT. As discussed later in this chapter, because concentration can be expressed in a number of ways, different mass transfer coefficients are defined. If we select Δc_A as the driving force for mass transfer, we can write

$$n_A = k_c A \, \Delta c_A \tag{3-105}$$

which defines a mass transfer coefficient, k_c, in mol/time-area-driving force, for a concentration driving force. For the falling laminar film, we take $\Delta c_A = c_{A_i} - \bar{c}_A$, which varies with vertical location, y, because even though c_{A_i} is independent of y, the average film concentration, $\overline{c}_A$, increases with y. To derive an expression for k_c, we equate (3-105) to Fick's first law at the gas–liquid interface:

$$k_c A(c_{A_i} - \overline{c}_A) = -D_{AB} A \left(\frac{\partial c_A}{\partial z} \right)_{z=0} \qquad (3\text{-}106)$$

Although this is the most widely used approach for defining a mass transfer coefficient, in this case of a falling film it fails because $(\partial c_A / \partial z)$ at $z = 0$ is not defined. Therefore, for this case we use another approach as follows. For an incremental height, we can write for film width W,

$$n_A = \overline{u}_y \delta W \, d\overline{c}_A = k_c(c_{A_i} - \overline{c}_A) W \, dy \qquad (3\text{-}107)$$

This defines a local value of k_c, which varies with distance y because $\overline{c}_A$ varies with y. An average value of k_c, over a height L, can be defined by separating variables and integrating (3-107):

$$k_{c_{avg}} = \frac{\int_0^L k_c \, dy}{L} = \frac{\overline{u}_y \delta \int_{c_{A_0}}^{c_{A_L}} [d\overline{c}_A / (c_{A_i} - \overline{c}_A)]}{L} = \frac{\overline{u}_y \delta}{L} \ln \frac{c_{A_i} - c_{A_0}}{c_{A_i} - \overline{c}_{A_L}} \qquad (3\text{-}108)$$

In general, the argument of the natural logarithm in (3-108) is obtained from the reciprocal of (3-99). For values of η in (3-100) greater than 0.1, only the first term in (3-99) is significant (error is less than 0.5%). In that case,

$$k_{c_{avg}} = \frac{\overline{u}_y \delta}{L} \ln \frac{e^{5.1213\eta}}{0.7857} \qquad (3\text{-}109)$$

Since $\ln e^x = x$,

$$k_{c_{avg}} = \frac{\overline{u}_y \delta}{L} (0.241 + 5.1213\eta) \qquad (3\text{-}110)$$

In the limit, for large η, using (3-100) and (3-102), (3-110) becomes

$$k_{c_{avg}} = 3.414 \frac{D_{AB}}{\delta} \qquad (3\text{-}111)$$

In a manner suggested by the Nusselt number, $N_{Nu} = h\delta/k$ for heat transfer, where $\delta = $ a characteristic length, we define a Sherwood number for mass transfer, which for a falling film of characteristic length δ is

$$N_{Sh_{avg}} = \frac{k_{c_{avg}} \delta}{D_{AB}} \qquad (3\text{-}112)$$

From (3-111), $N_{Sh_{avg}} = 3.414$, which is the smallest value that the Sherwood number can have for a falling liquid film.

The average mass transfer flux of A is given by

$$N_{A_{avg}} = \frac{n_{A_{avg}}}{A} = k_{c_{avg}} (c_{A_i} - \overline{c}_A)_{mean} \qquad (3\text{-}113)$$

For values of $\eta < 0.001$ in (3-100), when the liquid film flow regime is still laminar without ripples, the time of contact of the gas with the liquid is short and mass transfer is confined to the vicinity of the gas–liquid interface. Thus, the film acts as if it were infinite in thickness. In this limiting case, the downward velocity of the liquid film in the region of mass transfer is just $u_{y_{\max}}$, and (3-96) becomes

$$u_{y_{\max}} \frac{\partial c_A}{\partial y} = D_{AB} \frac{\partial^2 c_A}{\partial z^2} \tag{3-114}$$

Since from (3-90) and (3-91), $u_{y_{\max}} = 3\bar{u}_y/2$, (3-114) can be rewritten as

$$\frac{\partial c_A}{\partial y} = \left(\frac{2D_{AB}}{3\bar{u}_y}\right) \frac{\partial^2 c_A}{\partial z^2} \tag{3-115}$$

where the boundary conditions are

$$c_A = c_{A_0} \text{ for } z > 0 \text{ and } y > 0$$
$$c_A = c_{A_i} \text{ for } z = 0 \text{ and } y > 0$$
$$c_A = c_{A_0} \text{ for large } z \text{ and } y > 0$$

Equation (3-115) and the boundary conditions are equivalent to the case of the semiinfinite medium, as developed above. Thus, by analogy to (3-71), (3-75) and (3-76) the solution is

$$E = 1 - \theta = \frac{c_{A_i} - c_A}{c_{A_i} - c_{A_0}} = \text{erf}\left(\frac{z}{2\sqrt{2D_{AB}y/3\bar{u}_y}}\right) \tag{3-116}$$

Assuming that the driving force for mass transfer in the film is $c_{A_i} - c_{A_0}$, we can use Fick's first law at the gas–liquid interface to define a mass transfer coefficient:

$$N_A = -D_{AB} \left.\frac{\partial c_A}{\partial z}\right|_{z=0} = k_c(c_{A_i} - c_{A_0}) \tag{3-117}$$

The error function is defined as

$$\text{erf } z = \frac{2}{\sqrt{\pi}} \int_0^z e^{-t^2} \, dt \tag{3-118}$$

Using the Leibnitz rule with (3-116) to differentiate this integral function:

$$\left.\frac{\partial c_A}{\partial z}\right|_{z=0} = -(c_{A_i} - c_{A_0}) \sqrt{\frac{3\bar{u}_y}{2\pi D_{AB}y}} \tag{3-119}$$

Substituting (3-119) into (3-117) and introducing the Peclet number for mass transfer from (3-102), we obtain an expression for the local mass transfer coefficient as a function of distance down from the top of the wall:

$$k_c = \sqrt{\frac{3D_{AB}^2 N_{Pe_M}}{8\pi y\delta}} = \sqrt{\frac{3D_{AB}\Gamma}{2\pi y\delta\rho}} \tag{3-120}$$

The average value of k_c over the height of the film, L, is obtained by integrating (3-120) with respect to y, giving

$$k_{c_{\text{avg}}} = \sqrt{\frac{6D_{AB}\Gamma}{\pi\delta\rho L}} = \sqrt{\frac{3D_{AB}^2}{2\pi\delta L} N_{Pe_M}} \tag{3-121}$$

Combining (3-121) with (3-112) and (3-102),

$$N_{Sh_{\text{avg}}} = \sqrt{\frac{3\delta}{2\pi L} N_{Pe_M}} = \sqrt{\frac{4}{\pi\eta}} \tag{3-122}$$

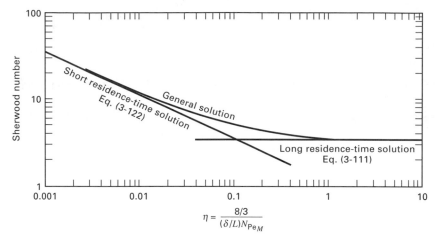

Figure 3.13 Limiting and general solutions for mass transfer to a falling laminar liquid film.

where, by (3-108), the proper mean to use with $k_{c_{avg}}$ is the log mean. Thus,

$$(c_{A_i} - \overline{c}_A)_{mean} = (c_{A_i} - \overline{c}_A)_{LM} = \frac{(c_{A_i} - c_{A_0}) - (c_{A_i} - \overline{c}_{A_L})}{\ln\left[(c_{A_i} - c_{A_0})/(c_{A_i} - \overline{c}_{A_L})\right]} \qquad \textbf{(3-123)}$$

When ripples are present, values of $k_{c_{avg}}$ and $N_{Sh_{avg}}$ can be considerably larger than predicted by these equations.

In the above development, asymptotic, closed-form solutions are obtained with relative ease for large and small values of η, defined by (3-100). These limits, in terms of the average Sherwood number, are shown in Figure 3.13. The general solution for intermediate values of η is not available in closed form. Similar limiting solutions for large and small values of appropriate parameters, usually dimensionless groups, have been obtained for a large variety of transport and kinetic phenomena, as discussed by Churchill [35]. Often the two limiting cases can be patched together to provide a reasonable estimate of the intermediate solution, if a single intermediate value is available from experiment or the general numerical solution. The procedure is discussed by Churchill and Usagi [36]. The general solution of Emmert and Pigford [37] to the falling laminar liquid film problem is included in Figure 3.13.

EXAMPLE 3.13

Water (B) at 25°C, in contact with pure CO_2 (A) at 1 atm, flows as a film down a vertical wall 1 m wide and 3 m high at a Reynolds number of 25. Using the following properties, estimate the rate of adsorption of CO_2 into water in kmol/s:

$$D_{AB} = 1.96 \times 10^{-5} \text{ cm}^2/\text{s}; \rho = 1.0 \text{ g/cm}^3; \mu_L = 0.89 \text{ cP}$$

Solubility of CO_2 in water at 1 atm and 25°C = 3.4×10^{-5} mol/cm³.

SOLUTION

From (3-93), $\Gamma = \dfrac{25(0.89)(0.001)}{4} = 0.00556 \dfrac{\text{kg}}{\text{m} - \text{s}}$

From (3-101), $N_{Sc} = \dfrac{\mu}{\rho D_{AB}} = \dfrac{(0.89)(0.001)}{(1.0)(1,000)(1.96 \times 10^{-5})(10^{-4})} = 454$

From (3-92), $\delta = \left[\dfrac{3(0.89)(0.001)(0.00556)}{1.0^2(1,000)^2(9.807)}\right]^{1/3} = 1.15 \times 10^{-4} \text{ m}$

From (3-90), and (3-91), $\bar{u}_y = (2/3)u_{y_{max}}$. Therefore,

$$\bar{u}_y = \frac{2}{3}\left[\frac{(1.0)(1,000)(9.807)(1.15 \times 10^{-4})^2}{2(0.89)(0.001)}\right] = 0.0486 \text{ m/s}$$

From (3-100), $\eta = \dfrac{8/3}{(25)(454)[(1.15 \times 10^{-4})/3]} = 6.13$

Therefore, (3-111) applies, giving $k_{c_{avg}} = \dfrac{3.41(1.96 \times 10^{-5})(10^{-4})}{1.15 \times 10^{-4}} = 5.81 \times 10^{-5} \text{ m/s}$

To determine the rate of absorption, $\bar{c}_{A_L}$ must be determined. From (3-103) and (3-113), for a unit width of 1 m,

$$n_A = \bar{u}_y\delta(\bar{c}_{A_L} - c_{A_0}) = k_{c_{avg}}A\frac{(\bar{c}_{A_L} - c_{A_0})}{\ln[(c_{A_i} - c_{A_0})/(c_{A_i} - \bar{c}_{A_L})]}$$

Thus,

$$\ln\left(\frac{c_{A_i} - c_{A_0}}{c_{A_i} - \bar{c}_{A_L}}\right) = \frac{k_{c_{avg}}A}{\bar{u}_y\delta}$$

Solving for $\bar{c}_{A_L}$, $\bar{c}_{A_L} = c_{A_i} - (c_{A_i} - c_{A_0})\exp\left(-\frac{k_{c_{avg}}A}{\bar{u}_y\delta}\right)$

$A = (1)(3) = 3 \text{ m}^2 \qquad c_{A_0} = 0; c_{A_i} = 3.4 \times 10^{-5} \text{ mol/cm}^3 = 3.4 \times 10^{-2} \text{ kmol/m}^3$

$$\bar{c}_{A_L} = 3.4 \times 10^{-2}\left\{1 - \exp\left[-\frac{(5.81 \times 10^{-5})(3)}{(0.0486)(1.15 \times 10^{-4})}\right]\right\} = 3.4 \times 10^{-2} \text{ kmol/m}^3$$

Thus, the exiting liquid film is saturated with CO_2, which implies equilibrium at the gas–liquid interface. From (3-103),

$$n_A = 0.0486(1.15 \times 10^{-4})(3.4 \times 10^{-2}) = 1.9 \times 10^{-7} \text{ kmol/s} \qquad \blacksquare$$

Boundary-Layer Flow on a Flat Plate

Consider the flow of a fluid (B) over a thin, flat plate parallel with the direction of flow of the fluid upstream of the plate, as shown in Figure 3.14. A number of possibilities for mass transfer of another species, A, into B exist: (1) The plate might consist of material A, which is slightly soluble in B. (2) Component A might be held in the pores of an inert solid plate, from which it evaporates or dissolves into B. (3) The plate might be an inert, dense polymeric membrane, through which species A can pass into fluid B. Let the fluid velocity profile upstream of the plate be uniform at a free-system velocity of u_0. As the

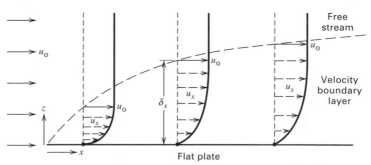

Figure 3.14 Laminar boundary-layer development for flow across a flat plate.

fluid passes over the plate, the velocity u_x in the direction of flow is reduced to zero at the wall, which establishes a velocity profile due to drag. At a certain distance z, normal to and out from the solid surface, the fluid velocity is 99% of u_o. This distance, which increases with increasing distance x from the leading edge of the plate, is arbitrarily defined as the velocity boundary-layer thickness, δ. Essentially all flow retardation occurs in the boundary layer, as first suggested by Prandtl [38]. The buildup of this layer, the velocity profile in the layer, and the drag force can be determined for laminar flow by solving the equations of continuity and motion (Navier–Stokes equations), for the x direction. For a Newtonian fluid of constant density and viscosity, in the absence of pressure gradients in the x and y (normal to the x–z plane) directions, these equations for the region of the boundary layer are

$$\frac{\partial u_x}{\partial x} + \frac{\partial u_z}{\partial z} = 0 \tag{3-124}$$

$$u_x \frac{\partial u_x}{\partial x} + u_z \frac{\partial u_x}{\partial z} = \frac{\mu}{\rho}\left(\frac{\partial^2 u_x}{\partial z^2}\right) \tag{3-125}$$

The boundary conditions are

$$u_x = u_o \text{ at } x = 0 \text{ for } z > 0 \qquad u_x = 0 \text{ at } z = 0 \text{ for } x > 0$$
$$u_x = u_o \text{ at } z = \infty \text{ for } x > 0 \qquad u_z = u_0 \text{ at } z = 0 \text{ for } x > 0$$

The solution of (3-124) and (3-125) in the absence of heat and mass transfer, subject to these boundary conditions, was first obtained by Blasius [39] and is described in detail by Schlicting [40]. The result in terms of a local friction factor, f_x, a local shear stress at the wall, τ_{w_x}, and a local drag coefficient at the wall, C_{D_x}, is

$$\frac{C_{D_x}}{2} = \frac{f_x}{2} = \frac{\tau_{w_x}}{\rho u_o^2} = \frac{0.322}{N_{\mathrm{Re}_x}^{0.5}} \tag{3-126}$$

where

$$N_{\mathrm{Re}_x} = \frac{x u_o \rho}{\mu} \tag{3-127}$$

Thus, the drag is greatest at the leading edge of the plate, where the Reynolds number is smallest. Average values of the drag coefficient are obtained by integrating (3-126) from $x = 0$ to L, giving

$$\frac{C_{D_{\mathrm{avg}}}}{2} = \frac{f_{\mathrm{ave}}}{2} = \frac{0.664}{(N_{\mathrm{Re}_L})^{0.5}} \tag{3-128}$$

The thickness of the velocity boundary layer increases with distance along the plate:

$$\frac{\delta}{x} = \frac{4.96}{N_{\mathrm{Re}_x}^{0.5}} \tag{3-129}$$

A reasonably accurate expression for the velocity profile was obtained by Pohlhausen [41], who assumed the empirical form $u_x = C_1 z + C_2 z^3$.

If the boundary conditions,

$$u_x = 0 \text{ at } z = 0 \qquad u_x = u_o \text{ at } z = \delta \qquad \partial u_x/\partial z = 0 \text{ at } z = \delta$$

are applied to evaluate C_1 and C_2, the result is

$$\frac{u_x}{u_o} = 1.5\left(\frac{z}{\delta}\right) - 0.5\left(\frac{z}{\delta}\right)^3 \tag{3-130}$$

This solution is valid only for a laminar boundary layer that by experiment persists to $N_{Re_x} = 5 \times 10^5$.

When mass transfer of A into the boundary layer occurs, the following species continuity equation applies at constant diffusivity:

$$u_x \frac{\partial c_A}{\partial x} + u_z \frac{\partial c_A}{\partial z} = D_{AB}\left(\frac{\partial^2 c_A}{\partial x^2}\right) \tag{3-131}$$

If mass transfer begins at the leading edge of the plate and if the concentration in the fluid at the solid–fluid interface is constant, the additional boundary conditions are

$c_A = c_{A_o}$ at $x = 0$ for $z > 0$, $c_A = c_{A_i}$ at $z = 0$ for $x > 0$,
   and $c_A = c_{A_o}$ at $z = \infty$ for $x > 0$

If the rate of mass transfer is low, the velocity profiles are undisturbed. The solution to the analogous problem in heat transfer was first obtained by Pohlhausen [42] for $N_{Pr} > 0.5$, as described in detail by Schlichting [40]. The results for mass transfer are

$$\frac{N_{Sh_x}}{N_{Re_x} N_{Sc}^{1/3}} = \frac{0.332}{N_{Re_x}^{0.5}} \tag{3-132}$$

where

$$N_{Sh_x} = \frac{x k_{c_x}}{D_{AB}} \tag{3-133}$$

and the driving force for mass transfer is $c_{A_i} - c_{A_o}$.

The concentration boundary layer, where essentially all of the resistance to mass transfer resides, is defined by

$$\frac{c_{A_i} - c_A}{c_{A_i} - c_{A_o}} = 0.99 \tag{3-134}$$

and the ratio of the concentration boundary-layer thickness, δ_c, to the velocity boundary thickness, δ, is

$$\delta_c/\delta = 1/N_{Sc}^{1/3} \tag{3-135}$$

Thus, for a liquid boundary layer, where $N_{Sc} > 1$, the concentration boundary layer builds up more slowly than the velocity boundary layer. For a gas boundary layer, where $N_{Sc} \approx 1$, the two boundary layers build up at about the same rate. By analogy to (3-130), the concentration profile is given by

$$\frac{c_{A_i} - c_A}{c_{A_i} - c_{A_o}} = 1.5\left(\frac{z}{\delta_c}\right) - 0.5\left(\frac{z}{\delta_c}\right)^3 \tag{3-136}$$

Equation (3-132) gives the local Sherwood number. If this expression is integrated over the length of the plate, L, the average Sherwood number is found to be

$$N_{Sh_{avg}} = 0.664 \, N_{Re_L}^{1/2} \, N_{Sc}^{1/3} \tag{3-137}$$

where

$$N_{Sh_{avg}} = \frac{L k_{c_{avg}}}{D_{AB}} \tag{3-138}$$

EXAMPLE 3.14

Air at 100°C, 1 atm, and a free-stream velocity of 5 m/s flows over a 3-m-long, thin flat plate of napthalene, causing it to sublime.

(a) Determine the length over which a laminar boundary layer persists.
(b) For that length, determine the rate of mass transfer of napthalene into air.
(c) At the point of transition of the boundary layer to turbulent flow, determine the thicknesses of the velocity and concentration boundary layers.

Assume the following values for physical properties:

Vapor pressure of napthalene = 10 torr   Viscosity of air = 0.0215 cP
Molar density of air = 0.0327 kmol/m³   Diffusivity of napthalene in air = 0.94×10^{-5} m²/s

SOLUTION

(a) $N_{Re_x} = 5 \times 10^5$ for transition. From (3-127),

$$x = L = \frac{\mu N_{Re_x}}{u_o \rho} = \frac{(0.0215)(0.001)(5 \times 10^5)}{(5)(0.0327)(29)} = 2.27 \text{ m}$$

at which transition to turbulent flow begins.

(b) $c_{A_o} = 0$ $c_{A_i} = \frac{10(0.0327)}{760} = 4.3 \times 10^{-4} \text{ kmol/m}^3$

From (3-101), $N_{Sc} = \frac{\mu}{\rho D_{AB}} = \frac{(0.0215)(0.001)}{(0.0327)(29)(0.94 \times 10^{-5})} = 2.41$

From (3-137), $N_{Sh_{avg}} = 0.664(5 \times 10^5)^{1/2}(2.41)^{1/3} = 630$

From (3-138), $k_{c_{avg}} = \frac{630(0.94 \times 10^{-5})}{2.27} = 2.61 \times 10^{-3} \text{ m/s}$

For a unit width,

$A = 2.27 \text{ m}^2$

$n_A = k_{c_{avg}} A(c_{A_i} - c_{A_o}) = 2.61 \times 10^{-3}(2.27)(4.3 \times 10^{-4}) = 2.55 \times 10^{-6} \text{ kmol/s}$

(c) From (3-129), at $x = L = 2.27$ m, $\delta = \frac{3.46(2.27)}{(5 \times 10^5)^{0.5}} = 0.0111$ m

From (3-135), $\delta_c = \frac{0.0111}{(2.41)^{1/3}} = 0.0083$ m ∎

Fully Developed Flow in a Straight, Circular Tube

Figure 3.15 shows the formation and buildup of a laminar velocity boundary layer when a fluid flows from a vessel into a straight, circular tube. At the entrance, plane a, the

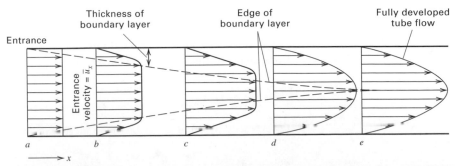

Figure 3.15 Buildup of a laminar-velocity boundary layer for flow in a straight, circular tube.

velocity profile is flat. A velocity boundary layer then begins to build up as shown at planes *b, c,* and *d.* In this region, the central core outside the boundary layer has a flat velocity profile where the flow is accelerated over the entrance velocity. Finally, at plane *e,* the boundary layer fills the tube. From here the velocity profile is fixed and the flow is said to be fully developed. The distance from the plane *a* to plane *e* is the entry region.

For fully developed laminar flow in a straight, circular tube, by experiment, the Reynolds number, $N_{Re} = D\bar{u}_x \rho / \mu$, where $\bar{u}_x$ is the flow-average velocity in the axial direction, *x,* and *D* is the inside diameter of the tube, must be less than 2,100. For this condition, the equation of motion in the axial direction for horizontal flow and constant properties is

$$\frac{\mu}{r}\frac{\partial}{\partial r}\left(r\frac{\partial u_x}{\partial r}\right) - \frac{dP}{dx} = 0 \tag{3-139}$$

where the boundary conditions are

$$r = 0 \text{ (axis of the tube)}, \partial u_x / \partial r = 0 \text{ and}$$
$$r = r_w \text{ (tube wall)}, u_x = 0$$

Equation (3-139) was integrated by Hagen in 1839 and Poiseuille in 1841. The resulting equation for the velocity profile, expressed in terms of the flow-average velocity, is

$$u_x = 2\bar{u}_x\left[1 - \left(\frac{r}{r_w}\right)^2\right] \tag{3-140}$$

or, in terms of the maximum velocity at the tube axis,

$$u_x = u_{x_{max}}\left[1 - \left(\frac{r}{r_w}\right)^2\right] \tag{3-141}$$

From the form of (3-141), the velocity profile is parabolic in nature.

The shear stress, pressure drop, and Fanning friction factor are obtained from solutions to (3-139):

$$\tau_w = -\mu\left(\frac{\partial u_x}{\partial r}\right)\bigg|_{r=r_w} = \frac{4\mu\bar{u}_x}{r_w} \tag{3-142}$$

$$-\frac{dP}{dx} = \frac{32\mu\bar{u}_x}{D^2} = \frac{2f\rho\bar{u}_x^2}{D} \tag{3-143}$$

with

$$f = \frac{16}{N_{Re}} \tag{3-144}$$

The entry length to achieve fully developed flow is defined as the axial distance, L_e, from the entrance to the point at which the centerline velocity is 99% of the fully developed flow value. From the analysis of Langhaar [43] for the entry region,

$$\frac{L_e}{D} = 0.0575 N_{Re} \tag{3-145}$$

Thus, at the upper limit of laminar flow, $N_{Re} = 2,100$, $L_e/D = 121$, a rather large ratio. For $N_{Re} = 100$, the ratio is only 5.75. In the entry region, Langhaar's analysis shows the friction factor is considerably higher than the fully developed flow value given by (3-144). At $x = 0$, *f* is infinity, but then decreases exponentially with *x,* approaching the fully developed flow value at L_e. For example, for $N_{Re} = 1,000$, (3-144) gives $f = 0.016$, with $L_e/D = 57.5$. In the region from $x = 0$ to $x/D = 5.35$, the average friction factor from Langhaar is 0.0487, which is about three times higher than the fully developed value.

In 1885, Graetz [44] obtained a theoretical solution to the problem of convective heat transfer between the wall of a circular tube, held at a constant temperature, and a fluid flowing through the tube in fully developed laminar flow. Assuming constant properties and negligible conduction in the axial direction, the energy equation, after substituting (3-140) for u_x, is

$$2\overline{u}_x \left[1 - \left(\frac{r}{r_w} \right)^2 \right] \frac{\partial T}{\partial x} = \frac{k}{\rho C_P} \left[\frac{1}{r} \frac{\partial}{\partial r} \left(r \frac{\partial T}{\partial r} \right) \right] \qquad \textbf{(3-146)}$$

The boundary conditions are

$x = 0$ (where heat transfer begins), $T = T_0$, for all r

$x > 0, r = r_w, T = T_i \qquad x > 0, r = 0, \partial T/\partial r = 0$

The analogous species continuity equation for mass transfer, neglecting bulk flow in the radial direction and diffusion in the axial direction, is

$$2\overline{u}_x \left[1 - \left(\frac{r}{r_w} \right)^2 \right] \frac{\partial c_A}{\partial x} = D_{AB} \left[\frac{1}{r} \frac{\partial}{\partial r} \left(r \frac{\partial c_A}{\partial r} \right) \right] \qquad \textbf{(3-147)}$$

with analogous boundary conditions.

The Graetz solution of (3-147) for the temperature profile or the concentration profile is in the form of an infinite series, and can be obtained from (3-146) by the method of separation of variables using the method of Frobenius. A detailed solution is given by Sellars, Tribus, and Klein [45]. From the concentration profile, expressions for the mass transfer coefficient and the Sherwood number are obtained. When x is large, the concentration profile is fully developed and the local Sherwood number, N_{Sh_x}, approaches a limiting value of 3.656. At the other extreme, when x is small such that the concentration boundary layer is very thin and confined to a region where the fully developed velocity profile is linear, the local Sherwood number is obtained from the classic Leveque [46] solution, presented by Knudsen and Katz [47]:

$$N_{Sh_x} = \frac{k_{c_x} D}{D_{AB}} = 1.077 \left[\frac{N_{Pe_M}}{(x/D)} \right]^{1/3} \qquad \textbf{(3-148)}$$

where, $\qquad N_{Pe_M} = \frac{D\overline{u}_x}{D_{AB}} \qquad \textbf{(3-149)}$

The limiting solutions, together with the general Graetz solution, are shown in Figure 3.16, where it is seen that $N_{Sh_x} = 3.656$ is valid for $N_{Pe_M}/(x/D) < 4$ and (3-148) is valid for

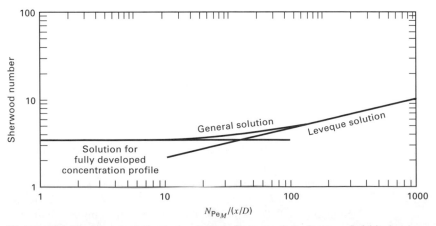

Figure 3.16 Limiting and general solutions for mass transfer to a fluid in laminar flow in a straight, circular tube.

$N_{Pe_M}/(x/D) > 100$. The two limiting solutions can be patched together if one point of the general solution is available where the two solutions intersect.

Over a length of tube where mass transfer occurs, an average Sherwood number can be derived by integrating the general expression for the local Sherwood number. An empirical representation for that average, proposed by Hansen [48], is

$$N_{Sh_{avg}} = 3.66 + \frac{0.0668[N_{Pe_M}/(x/D)]}{1 + 0.04[N_{Pe_M}/(x/D)]^{2/3}} \qquad \textbf{(3-150)}$$

which is based on a log-mean concentration driving force.

EXAMPLE 3.15

Linton and Sherwood [49] conducted experiments on the dissolution of cast tubes of benzoic acid (A) into water (B) flowing through the tubes in laminar flow. They obtained good agreement with predictions based on the Graetz and Leveque equations. Consider a 5.23-cm-inside-diameter by 32-cm-long tube of benzoic acid, preceded by 400 cm of straight metal pipe of the same inside diameter where a fully developed velocity profile is established. Pure water enters the system at 25°C at a velocity corresponding to a Reynolds number of 100. Based on the following property data at 25°C, estimate the average concentration of benzoic acid in the water leaving the cast tube before a significant increase in the inside diameter of the benzoic acid tube occurs because of dissolution.

Solubility of benzoic acid in water $= 0.0034$ g/cm^3
Viscosity of water $= 0.89$ cP $= 0.0089$ g/cm-s
Diffusivity of benzoic acid in water at infinite dilution $= 9.18 \times 10^{-6}$ cm^2/s

SOLUTION

$$N_{Sc} = \frac{0.0089}{(1.0)(9.18 \times 10^{-6})} = 970$$

$$N_{Re} = \frac{D\bar{u}_x\rho}{\mu} = 100$$

from which,

$$\bar{u}_x = \frac{(100)(0.0089)}{(5.23)(1.0)} = 0.170 \text{ cm/s}$$

From (3-149),

$$N_{Pe_M} = \frac{(5.23)(0.170)}{9.18 \times 10^{-6}} = 9.69 \times 10^4$$

$$\frac{x}{D} = \frac{32}{5.23} = 6.12$$

$$\frac{N_{Pe_M}}{(x/D)} = \frac{9.69 \times 10^4}{6.12} = 1.58 \times 10^{-4}$$

From (3-150),

$$N_{Sh_{avg}} = 3.66 + \frac{0.0668(1.58 \times 10^4)}{1 + 0.04(1.58 \times 10^4)^{2/3}} = 44$$

$$k_{c_{avg}} = N_{Sh_{avg}} \left(\frac{D_{AB}}{D}\right) = 44 \frac{(9.18 \times 10^{-6})}{5.23} = 7.7 \times 10^{-5} \text{ cm/s}$$

Using a log-mean driving force,

$$n_A = \bar{u}_x S(\bar{c}_{A_x} - c_{A_0}) = k_{c_{avg}} A \frac{[(c_{A_i} - c_{A_0}) - (c_{A_i} - \bar{c}_{A_x})]}{\ln[(c_{A_i} - c_{A_0})/(c_{A_i} - \bar{c}_{A_x})]}$$

where S is the cross-sectional area for flow. Simplifying,

$$\ln\left(\frac{c_{A_i} - c_{A_0}}{c_{A_i} - \overline{c}_{A_x}}\right) = \frac{k_{c_{\text{avg}}} A}{\overline{u}_x S}$$

$$c_{A_0} = 0 \text{ and } c_{A_i} = 0.0034 \text{ g/cm}^3$$

$$S = \frac{\pi D^2}{4} = \frac{(3.14)(5.23)^2}{4} = 21.5 \text{ cm}^2 \text{ and } A = \pi Dx = (3.14)(5.23)(32) = 526 \text{ cm}^2$$

$$\ln\left(\frac{0.0034}{0.0034 - \overline{c}_{A_x}}\right) = \frac{(7.7 \times 10^{-5})(526)}{(0.170)(21.5)} = 0.0111$$

$$\overline{c}_{A_x} = 0.0034 - \frac{0.0034}{e^{0.0111}} = 0.000038 \text{ g/cm}^3$$

Thus, the concentration of benzoic acid in the water leaving the cast tube is far from saturation. ∎

3.5 MASS TRANSFER IN TURBULENT FLOW

In the two previous sections, diffusion in stagnant media and in laminar flow were considered. For both cases, Fick's law can be applied to obtain rates of mass transfer. A more common occurrence in engineering is turbulent flow, which is accompanied by much higher transport rates, but for which theory is still under development and the estimation of mass transfer rates relies heavily on empirical correlations of experimental data and analogies with heat and momentum transfer.

As shown by the famous dye experiment of Osborne Reynolds [50] in 1883, a fluid in laminar flow moves parallel to the solid boundaries in streamline patterns. Every particle of fluid moves with the same velocity along a streamline and there are no fluid velocity components normal to these streamlines. For a Newtonian fluid, the momentum transfer, heat transfer, and mass transfer are by molecular transport, governed by Newton's law of viscosity, Fourier's law of heat conduction, and Fick's law of molecular diffusion, respectively.

In turbulent flow, the rates of momentum, heat, and mass transfer are orders of magnitude greater than for molecular transport. This occurs because streamlines no longer exist and particles or eddies of fluid, which are large compared to the mean free path of the molecules in the fluid, mix with each other by moving from one region to another in fluctuating motion. This eddy mixing by velocity fluctuations occurs not only in the direction of flow but also in the direction normal to flow, with the latter being of more interest. Momentum, heat, and mass transfer now occur by two parallel mechanisms: (1) molecular motion, which is slow; and (2) turbulent or eddy motion, which is rapid except near a solid surface, where the flow velocity accompanying turbulence decreases to zero.

In 1877, Boussinesq [51] modified Newton's law of viscosity to account for eddy motion. Analogous expressions were subsequently developed for turbulent-flow heat and mass transfer. For flow in the x direction and transport in the z direction normal to flow, these expressions are written in the following forms in the absence of bulk flow in the z direction:

$$\tau_{zx} = -(\mu + \mu_t)\frac{du_x}{dz} \tag{3-151}$$

$$q_z = -(k + k_t)\frac{dT}{dz} \tag{3-152}$$

$$N_{A_z} = -(D_{AB} + D_t)\frac{dc_A}{dz} \tag{3-153}$$

where the double subscript, zx, on the shear stress, τ, stands for x momentum in the z direction. The molecular contributions, μ, k, and D_{AB}, are properties of the fluid and depend on chemical composition, temperature, and pressure; the turbulent contributions, μ_t, k_t, and D_t, depend on the mean fluid velocity in the direction of flow and on position in the fluid with respect to the solid boundaries.

In 1925, in an attempt to quantify turbulent transport, Prandtl [52] developed an expression for μ_t in terms of an eddy mixing length, which is a function of position. By analogy, the same mixing length is valid for turbulent-flow heat transfer and mass transfer. To use this analogy, (3-151) to (3-153) are rewritten in diffusivity form:

$$\frac{\tau_{zx}}{\rho} = -(\nu + \epsilon_M)\frac{du_z}{dz} \tag{3-154}$$

$$\frac{q_z}{C_P\rho} = -(\alpha + \epsilon_H)\frac{dT}{dz} \tag{3-155}$$

$$N_{A_z} = -(D_{AB} + \epsilon_D)\frac{dc_A}{dz} \tag{3-156}$$

where ϵ_M, ϵ_H, are ϵ_D are momentum, heat, and mass eddy diffusivities, respectively; ν is the momentum diffusivity (kinematic viscosity), μ/ρ; and α is the thermal diffusivity, $k/\rho C_P$. As a first approximation, the three eddy diffusivities may be assumed equal. This assumption is reasonably valid for ϵ_H and ϵ_D, but experimental data indicate that $\epsilon_M/\epsilon_H = \epsilon_M/\epsilon_D$ is sometimes less than 1.0 and as low as 0.5 for turbulence in a free jet.

Reynolds Analogy

If (3-154) to (3-156) are applied at a solid boundary, they can be used to determine transport fluxes based on transport coefficients, with driving forces from the wall, i, to the bulk fluid, designated with an overbar, $\overline{}$:

$$\frac{\tau_{zx}}{\overline{u}_x} = -(\nu + \epsilon_M)\frac{d(\rho u_x/\overline{u}_x)}{dz}\bigg|_{z=0} = \frac{f\rho}{2}\overline{u}_x \tag{3-157}$$

$$q_z = -(\alpha + \epsilon_H)\frac{d(\rho C_P T)}{dz}\bigg|_{z=0} = h(T_i - \overline{T}) \tag{3-158}$$

$$N_{A_z} = -(D_{AB} + \epsilon_D)\frac{dc_A}{dz}\bigg|_{z=0} = k_c(c_{A_i} - \overline{c}_A) \tag{3-159}$$

We define dimensionless velocity, temperature, and solute concentration by

$$\theta = \frac{u_x}{\overline{u}_x} = \frac{T_i - T}{T_i - \overline{T}} = \frac{c_{A_i} - c_A}{c_{A_i} - \overline{c}_A} \tag{3-160}$$

If (3-160) is substituted into (3-157) to (3-159),

$$\frac{\partial\theta}{\partial z}\bigg|_{z=0} = \frac{f\overline{u}_x}{2(\nu + \epsilon_M)} = \frac{h}{\rho C_P(\alpha + \epsilon_H)} = \frac{k_c}{(D_{AB} + \epsilon_D)} \tag{3-161}$$

This equation defines the analogies among momentum, heat, and mass transfer. Assuming that the three eddy diffusivities are equal and that the molecular diffusivities are either everywhere negligible or equal,

$$\frac{f}{2} = \frac{h}{\rho C_P \overline{u}_x} = \frac{k_c}{\overline{u}_x} \tag{3-162}$$

Equation (3-162) defines the Stanton number for heat transfer,

$$N_{St} = \frac{h}{\rho C_P \bar{u}_x} = \frac{h}{G C_P} \tag{3-163}$$

and the Stanton number for mass transfer,

$$N_{St_M} = \frac{k_c}{\bar{u}_x} = \frac{k_c \rho}{G} \tag{3-164}$$

Equation (3-162) is referred to as the Reynolds analogy. It can be used to estimate values of heat and mass transfer coefficients from experimental measurements of the Fanning friction factor for turbulent flow, but only when $N_{Pr} = N_{Sc} = 1$. Reynolds postulated the existence of the analogy in 1874 [53] and derived it in 1883 [50].

Chilton–Colburn Analogy

A widely used extension of the Reynolds analogy to Prandtl and Schmidt numbers other than 1 was presented by Colburn [54] for heat transfer and by Chilton and Colburn [55] for mass transfer. They showed that the Reynolds analogy for turbulent flow could be corrected for differences in velocity, temperature, and concentration distributions by incorporating N_{Pr} and N_{Sc} into (3-162) to define the following three j-factors:

$$j_M \equiv \frac{f}{2} = j_H \equiv \frac{h}{G C_P} (N_{Pr})^{2/3} = j_D \equiv \frac{k_c \rho}{G} (N_{Sc})^{2/3} \tag{3-165}$$

Equation (3-165) is the Chilton–Colburn analogy or the Colburn analogy for estimating average transport coefficients for turbulent flow.

In general, j-factors are uniquely determined by the geometric configuration and the Reynolds number. Based on the analysis, over many years, of experimental data on momentum, heat, and mass transfer, the following representative correlations have been developed for turbulent transport to or from smooth surfaces. Other correlations are presented in other chapters.

1. Flow through a straight, circular tube of inside diameter D:

$$j_M = j_H = j_D = 0.023(N_{Re})^{-0.2}$$
$$\text{for } 10,000 < N_{Re} = DG/\mu < 1,000,000 \tag{3-166}$$

2. Average transport coefficients for flow across a flat plate of length L:

$$j_M = j_H = j_D = 0.037(N_{Re})^{-0.2}$$
$$\text{for } 5 \times 10^5 < N_{Re} = L u_o \rho / \mu < 5 \times 10^8 \tag{3-167}$$

3. Average transport coefficients for flow normal to a long circular cylinder of diameter D, where the drag coefficient includes both form drag and skin friction, but only the skin friction contribution applies to the analogy:

$$(j_M)_{\text{skin friction}} = j_H = j_D = 0.193(N_{Re})^{-0.382}$$
$$\text{for } 4,000 < N_{Re} < 40,000 \tag{3-168}$$

$$(j_M)_{\text{skin friction}} = j_H = j_D = 0.0266(N_{Re})^{-0.195}$$
$$\text{for } 40,000 < N_{Re} < 250,000 \tag{3-169}$$

$$\text{with } N_{Re} = \frac{DG}{\mu}$$

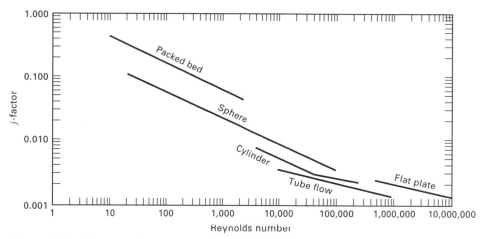

Figure 3.17 Chilton–Colburn j-factor correlations.

4. Average transport coefficients for flow past a single sphere of diameter D:

$$(j_M)_{\text{skin friction}} = j_H = j_D = 0.37(N_{\text{Re}})^{-0.4}$$
$$\text{for } 20 < N_{\text{Re}} = \frac{DG}{\mu} < 100{,}000$$

(3-170)

5. Average transport coefficients for flow through beds packed with spherical particles of uniform size D_P:

$$j_H = j_D = 1.17(N_{\text{Re}})^{-0.415}$$
$$\text{for } 10 < N_{\text{Re}} = \frac{D_P G}{\mu} < 2{,}500$$

(3-171)

The above correlations are plotted in Figure 3.17, where the curves do not coincide because of the differing definitions of the Reynolds number. However, the curves are not widely separated. When using the correlations in the presence of appreciable temperature and/or composition differences, Chilton and Colburn recommend that N_{Pr} and N_{Sc} be evaluated at the average conditions from the surface to the bulk stream.

Prandtl Analogy

Other improvements to and extensions of the Reynolds analogy have been based on more theory and less empiricism. The first major improvement was by Prandtl [56] in 1910. He divided the flow into two regions: (1) a thin laminar sublayer of thickness δ next to the wall boundary, where only molecular transport occurs; and (2) a turbulent region dominated by eddy transport, with $\epsilon_M = \epsilon_H = \epsilon_D$. Integrating (3-154) to (3-156) over the laminar sublayer region and then the turbulent region, and combining the results to eliminate the temperature and concentration at the interface between the two regions, the following expressions are obtained:

$$N_{\text{St}} = \frac{f/2}{1 + (u_\delta/\overline{u})(N_{\text{Pr}} - 1)}$$

(3-172)

and

$$N_{\text{St}_M} = \frac{f/2}{1 + (u_\delta/\overline{u})(N_{\text{Sc}} - 1)}$$

(3-173)

But from the universal velocity profile for turbulent flow [47],

$$\frac{u_\delta}{\bar{u}} = 5\sqrt{\frac{f}{2}}$$

Therefore, the final equations for the Prandtl analogy for turbulent flow in a straight, circular tube become

$$N_{St} = \frac{f/2}{1 + 5\sqrt{f/2}(N_{Pr} - 1)} \qquad \textbf{(3-174)}$$

$$N_{St_M} = \frac{f/2}{1 + 5\sqrt{f/2}(N_{Sc} - 1)} \qquad \textbf{(3-175)}$$

Equations (3-174) and (3-175) simplify to the Reynolds analogy when $N_{Pr} = N_{Sc} = 1$.

Further important theoretical improvements to the Reynolds analogy were made by von Karman, Martinelli, and Deissler, as discussed in detail by Knudsen and Katz [47]. The first two investigators inserted a buffer zone between the laminar sublayer and turbulent core. Deissler gradually reduced the eddy diffusivities as the wall was approached. However, except for very low N_{Pr} (< 0.5), the empirical Chilton–Colburn analogy is preferred over these other analogies for practical applications because it is applicable to a wide range of configurations and is fitted to experimental data to an accuracy acceptable for engineering calculations.

EXAMPLE 3.16

Linton and Sherwood [49] conducted experiments on the solution of cast tubes of cinnamic acid (A) into water (B) flowing through the tubes in turbulent flow. In one run, with a 1.90-cm-i.d. tube, $N_{Re} = 62,000$, and $N_{Sc} = 3,000$, they measured a Stanton number for mass transfer, N_{St_M}, of 0.0000136. Compare this experimental value with predictions by the Reynolds, Chilton–Colburn, and Prandtl analogies.

SOLUTION

From (3-165) and (3-166), $\dfrac{f}{2} = 0.023(62,000)^{-0.2} = 0.00253$

Reynolds analogy:

From (3-162), $N_{St_M} = \dfrac{f}{2} = 0.00253$

Chilton–Colburn analogy:

From (3-165), $N_{St_M} = \dfrac{f/2}{N_{Sc}^{2/3}} = \dfrac{0.00253}{(3,000)^{2/3}} = 0.0000122$

Prandtl analogy:

From (3-175), $N_{St_M} = \dfrac{0.00253}{1 + 5\sqrt{0.00253}(3,000 - 1)^{2/3}} = 0.0000335$

Of the three analogies, only the Chilton–Colburn analogy gives a reasonably accurate prediction for the large Schmidt number of this example. The correlations of Friend and Metzner [57] are also valid for large Schmidt and Prandtl numbers. ∎

3.6 MODELS FOR MASS TRANSFER AT A FLUID–FLUID INTERFACE

In the three previous sections, diffusion and mass transfer within solids and fluids were considered, where the interface was a smooth solid surface. Of greater interest in separation processes is mass transfer across an interface between a gas and a liquid or between two liquid phases. Such interfaces exist in absorption, distillation, extraction, and stripping. At

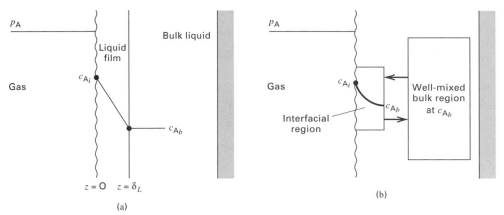

Figure 3.18 Theories for mass transfer from a fluid–fluid interface into a liquid: (a) film theory; (b) penetration and surface-renewal theories.

fluid–fluid interfaces, turbulence may persist to the interface. The following theoretical models have been developed to describe mass transfer from a fluid to such an interface.

Film Theory

A simple theoretical model for turbulent mass transfer to or from a fluid-phase boundary was suggested in 1904 by Nernst [58], who postulated that the entire resistance to mass transfer in a given turbulent phase is in a thin, stagnant region of that phase at the interface, called a film. This film is similar to the laminar sublayer that forms when a fluid flows in the turbulent regime parallel to a flat plate. This is shown schematically in Figure 3.18a for the case of a gas–liquid interface, where the gas is pure component A, which diffuses into nonvolatile liquid B. Thus, a process of absorption of A into liquid B takes place, without desorption of B into gaseous A. Because the gas is pure A at total pressure $P = p_A$, there is no resistance to mass transfer in the gas phase. At the gas–liquid interface, equilibrium is assumed so the concentration of A, c_{A_i}, is related to the partial pressure of A, p_A, by some form of Henry's law, for example, $c_{A_i} = H_A p_A$. In the thin, stagnant liquid film of thickness δ, molecular diffusion only occurs with a driving force of $c_{A_i} - c_{A_b}$. Since the film is assumed to be very thin, all of the diffusing A passes through the film and into the bulk liquid. If, in addition, bulk flow of A is neglected, the concentration gradient is linear as in Figure 3.18a. Accordingly, Fick's first law, (3-3), for the diffusion flux integrates to

$$J_A = \frac{D_{AB}}{\delta}(c_{A_i} - c_{A_b}) = \frac{cD_{AB}}{\delta}(x_{A_i} - x_{A_b}) \tag{3-176}$$

If the liquid phase is dilute in A, the bulk-flow effect can be neglected and (3-176) applies to the total flux:

$$N_A = \frac{D_{AB}}{\delta}(c_{A_i} - c_{A_b}) = \frac{cD_{AB}}{\delta}(x_{A_i} - x_{A_b}) \tag{3-177}$$

If the bulk-flow effect is not negligible, then, from (3-31),

$$N_A = \frac{cD_{AB}}{\delta}\ln\left[\frac{1 - x_{A_b}}{1 - x_{A_i}}\right] = \frac{cD_{AB}}{\delta(1 - x_A)_{LM}}(x_{A_i} - x_{A_b}) \tag{3-178}$$

where

$$(1 - x_A)_{LM} = \frac{x_{A_i} - x_{A_b}}{\ln[(1 - x_{A_b})/(1 - x_{A_i})]} = (x_B)_{LM} \qquad (3\text{-}179)$$

In practice, the ratios D_{AB}/δ in (3-177) and $D_{AB}/\delta(1 - x_A)_{LM}$ in (3-178) are replaced by mass transfer coefficients k_c and k_c', respectively, because the film thickness, δ, which depends on the flow conditions, is not known.

The film theory, which is easy to understand and apply, is often criticized because it appears to predict that the rate of mass transfer is directly proportional to the molecular diffusivity. This dependency is at odds with experimental data, which indicate a dependency of D^n, where n ranges from about 0.5 to 0.75. However, if D_{AB}/δ is replaced with k_c, which is then estimated from the Chilton–Colburn analogy, Eq. (3-165), we obtain k_c proportional to $D_{AB}^{2/3}$, which is in better agreement with experimental data. In effect, δ depends on D_{AB} (or N_{Sc}). Regardless of whether the criticism of the film theory is valid, the theory has been and continues to be widely used in the design of mass transfer separation equipment.

EXAMPLE 3.17

Sulfur dioxide is absorbed from air into water in a packed absorption tower. At a certain location in the tower, the mass transfer flux is 0.0270 kmol SO_2/m^2-h and the liquid-phase mole fractions are 0.0025 and 0.0003, respectively at the two-phase interface and in the bulk liquid. If the diffusivity of SO_2 in water is 1.7×10^{-5} cm²/s, determine the mass transfer coefficient, k_c, and the film thickness.

SOLUTION

$$N_{SO_2} = \frac{0.027(1,000)}{(3,600)(100)^2} = 7.5 \times 10^{-7} \frac{\text{mol}}{\text{cm}^2\text{-s}}$$

For dilute conditions, the concentration of water is $c = \dfrac{1}{18.02} = 5.55 \times 10^{-2}$ mol/cm³

From (3-177), $k_c = \dfrac{D_{AB}}{\delta} = \dfrac{N_A}{c(x_{A_i} - x_{A_b})} = \dfrac{7.5 \times 10^{-7}}{5.55 \times 10^{-2}(0.0025 - 0.0003)} = 6.14 \times 10^{-3}$ cm/s

Therefore, $\delta = \dfrac{D_{AB}}{k_c} = \dfrac{1.7 \times 10^{-5}}{6.14 \times 10^{-3}} = 0.0028$ cm

which is very small and typical of turbulent-flow mass transfer processes.   ■

Penetration Theory

A more realistic physical model of mass transfer from a fluid–fluid interface into a bulk liquid stream is provided by the penetration theory of Higbie [59], shown schematically in Figure 3.18b. The stagnant-film concept is replaced by Boussinesq eddies that, during a cycle, (1) move from the bulk to the interface; (2) stay at the interface for a short, fixed period of time during which they remain static so that molecular diffusion takes place in a direction normal to the interface; and (3) leave the interface to mix with the bulk stream. When an eddy moves to the interface, it replaces another static eddy. Thus, the eddies are intermittently static and moving. Turbulence extends to the interface.

In the penetration theory, unsteady-state diffusion takes place at the interface during the time the eddy is static. This process is governed by Fick's second law, (3-71), with boundary conditions

$c_A = c_{A_b}$ at $t = 0$ for $0 \le z \le \infty$; $c_A = c_{A_i}$ at $z = 0$ for t > 0; and $c_A = c_{A_b}$ at $z = \infty$ for $t > 0$

These are the same boundary conditions as in unsteady-state diffusion in a semiinfinite medium. Thus, the solution can be written by a rearrangement of (3-75):

$$\frac{c_{A_i} - c_A}{c_{A_i} - c_{A_b}} = \text{erf}\left(\frac{z}{2\sqrt{D_{AB}t_c}}\right) \tag{3-180}$$

where t_c = "contact time" of the static eddy at the interface during one cycle. The corresponding average mass transfer flux of A in the absence of bulk flow is given by the following form of (3-79):

$$N_A = 2\sqrt{\frac{D_{AB}}{\pi t_c}}(c_{A_i} - c_{A_b}) \tag{3-181}$$

or

$$N_A = k_c(c_{A_i} - c_{A_b}) \tag{3-182}$$

Thus, the penetration theory gives

$$k_c = 2\sqrt{\frac{D_{AB}}{\pi t_c}} \tag{3-183}$$

which predicts that k_c is proportional to the square root of the molecular diffusivity, which is at the lower limit of experimental data.

The penetration theory is most useful when mass transfer involves bubbles or droplets, or flow over random packing. For bubbles, the contact time, t_c, of the liquid surrounding the bubble is taken as the ratio of bubble diameter to bubble rise velocity. For example, an air bubble of 0.4 cm diameter rises through water at a velocity of about 20 cm/s. Thus, the estimated contact time, t_c, is 0.4/20 = 0.02 s. For a liquid spray, where no circulation of liquid occurs inside the droplets, the contact time is the total time for the droplets to fall through the gas. For a packed tower, where the liquid flows as a film over particles of random packing, mixing can be assumed to occur each time the liquid film passes from one piece of packing to another. Resulting contact times are of the order of about 1 s. In the absence of any method of estimating the contact time, the liquid-phase mass transfer coefficient is sometimes correlated by an empirical expression consistent with the 0.5 exponent on D_{AB}, given by (3-183) with the contact time replaced by a function of geometry and the liquid velocity, density, and viscosity.

EXAMPLE 3.18

For the conditions of Example 3.17, estimate the contact time for Higbie's penetration theory.

SOLUTION

From Example 3.17, $k_c = 6.14 \times 10^{-3}$ cm/s and $D_{AB} = 1.7 \times 10^{-5}$ cm²/s. From a rearrangement of (3-183), $t_c = \dfrac{4D_{AB}}{\pi k_c^2} = \dfrac{4(1.7 \times 10^{-5})}{3.14(6.14 \times 10^{-3})^2} = 0.57$ s ∎

Surface Renewal Theory

The penetration theory is not satisfying because the assumption of a constant contact time for all eddies that temporarily reside at the surface is not reasonable, especially for stirred tanks, contactors with random packings, and bubble and spray columns where the bubbles and droplets cover a wide range of sizes. In 1951, Danckwerts [60] suggested an improvement to the penetration theory that involves the replacement of the constant eddy contact time with the assumption of a residence-time distribution, wherein the probability of an eddy at the surface being replaced by a fresh eddy is independent of the age of the surface eddy.

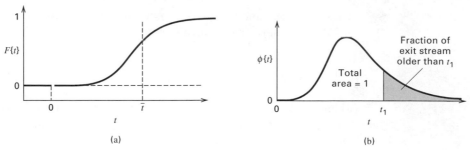

Figure 3.19 Residence-time distribution plots: (a) typical F curve; (b) typical age distribution. [Adapted from O. Levenspiel, *Chemical Reaction Engineering,* 2nd ed., John Wiley and Sons, New York (1972).]

Following the Levenspiel [61] treatment of residence-time distribution, let $F(t)$ be the fraction of eddies with a contact time of less than t. For $t = 0$, $F\{t\} = 0$, and $F\{t\}$ approaches 1 as t goes to infinity. A plot of $F\{t\}$ versus t, as shown in Figure 3.19, is referred to as a residence time or age distribution. If $F\{t\}$ is differentiated with respect to t, we obtain another function:

$$\phi\{t\} = dF\{t\}/dt$$

where $\phi\{t\}\, dt$ = the probability that a given surface eddy will have a residence time t. The sum of probabilities is

$$\int_0^\infty \phi\{t\}\, dt = 1 \tag{3-184}$$

Typical plots of $F\{t\}$ and $\phi\{t\}$ are shown in Figure 3.19, where it is seen that $\phi\{t\}$ is similar to a normal probability curve.

For steady-state flow in and out of a well-mixed vessel, Levenspiel shows that

$$F\{t\} = 1 - e^{-t/\bar{t}} \tag{3-185}$$

where $\bar{t}$ is the average residence time. This function forms the basis, in reaction engineering, of the ideal model of a continuous stirred-tank reactor (CSTR). Danckwerts selected the same model for his surface renewal theory, using the corresponding $\phi\{t\}$ function:

$$\phi\{t\} = s e^{-st} \tag{3-186}$$

where $s = 1/\bar{t}$ = fractional rate of surface renewal. As shown in Example 3.19 below, plots of (3-185) and (3-186) are much different from those in Figure 3.19.

The instantaneous mass transfer rate for an eddy with an age t is given by (3-181) for the penetration theory in flux form as

$$N_{A_t} = \sqrt{\frac{D_{AB}}{\pi t}}\,(c_{A_i} - c_{A_b}) \tag{3-187}$$

The integrated average rate is

$$(N_A)_{\text{avg}} = \int_0^\infty \phi\{t\} N_{A_t}\, dt \tag{3-188}$$

Combining (3-186), (3-187), and (3-188), and integrating:

$$(N_A)_{\text{avg}} = \sqrt{D_{AB}s}\,(c_{A_i} - c_{A_b}) \tag{3-189}$$

Thus,
$$k_c = \sqrt{D_{AB}s} \tag{3-190}$$

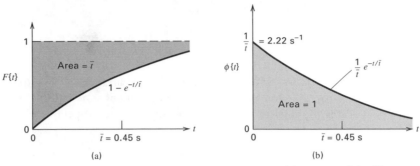

Figure 3.20 Age distribution curves for Example 3.19: (a) F curve; (b) $\phi\{t\}$ curve.

The more reasonable surface renewal theory predicts the same dependency on molecular diffusivity as the penetration theory. Unfortunately, s, the fractional rate of surface renewal, is as elusive a parameter as the constant contact time, t_c.

EXAMPLE 3.19

For the conditions of Example 3.17, estimate the fractional rate of surface renewal, s, for Danckwert's theory and determine the residence time and probability distributions.

SOLUTION

From Example 3.17, $k_c = 6.14 \times 10^{-3}$ cm/s and $D_{AB} = 1.7 \times 10^{-5}$ cm²/s

From (3-190), $s = \dfrac{k_c^2}{D_{AB}} = \dfrac{(6.14 \times 10^{-3})^2}{1.7 \times 10^{-5}} = 2.22 \text{ s}^{-1}$

Thus, the average residence time of an eddy at the surface is $1/2.22 = 0.45$ s.

From (3-186), $\phi\{t\} = 2.22 e^{-2.22t}$ \hfill **(1)**

From (3-185), the residence-time distribution is given by $F\{t\} = 1 - e^{-t/0.45}$, \hfill **(2)**

where t is in seconds. Equations (1) and (2) are plotted in Figure 3.20. These curves are much different from the curves of Figure 3.19. ∎

Film-Penetration Theory

Toor and Marchello [62], in 1958, combined features of the film, penetration, and surface renewal theories to develop a film-penetration model, which predicts a dependency of the mass transfer coefficient k_c, on the diffusivity, that varies from $\sqrt{D_{AB}}$ to D_{AB}. Their theory assumes that the entire resistance to mass transfer resides in a film of fixed thickness δ. Eddies move to and from the bulk fluid and this film. Age distributions for time spent in the film are of the Higbie or Danckwerts type.

Fick's second law, (3-71), still applies, but the boundary conditions are now

$$c_A = c_{A_b} \text{ at } t = 0 \text{ for } 0 \le z \le \infty,$$

$$c_A = c_{A_i} \text{ at } z = 0 \text{ for } t > 0; \text{ and } c_A = c_{A_b} \text{ at } z = \delta \text{ for } t > 0$$

Infinite-series solutions are obtained by the method of Laplace transforms. The rate of mass transfer is then obtained in the usual manner by applying Fick's first law (3-117) at the fluid–fluid interface. For small t, the solution, given as

$$N_{A_t} = (c_{A_i} - c_{A_b}) \left(\frac{D_{AB}}{\pi t} \right)^{1/2} \left[1 + 2 \sum_{n=1}^{\infty} \exp\left(-\frac{n^2 \delta^2}{D_{AB} t} \right) \right]$$ \hfill **(3-191)**

converges rapidly. For large t,

$$N_{A_t} = (c_{A_i} - c_{A_b}) \left(\frac{D_{AB}}{\delta} \right) \left[1 + 2 \sum_{n=1}^{\infty} \exp \left(-n^2 \pi^2 \frac{D_{AB} t}{\delta^2} \right) \right] \qquad \text{(3-192)}$$

Equation (3-188) with $\phi\{t\}$ from (3-186) can then be used to obtain average rates of mass transfer. Again, we can write two equivalent series solutions, which converge at different rates. Equations (3-191) and (3-192) become, respectively,

$$N_{A_{avg}} = k_c(c_{A_i} - c_{A_b}) = (c_{A_i} - c_{A_b})(s D_{AB})^{1/2} \left[1 + 2 \sum_{n=1}^{\infty} \exp \left(-2n\delta \sqrt{\frac{s}{D_{AB}}} \right) \right] \qquad \text{(3-193)}$$

$$N_{A_{avg}} = k_c(c_{A_i} - c_{A_b}) = (c_{A_i} - c_{A_b}) \left(\frac{D_{AB}}{\delta} \right) \left[1 + 2 \sum_{n=1}^{\infty} \frac{1}{1 + n^2 \pi^2 \frac{D_{AB}}{s\delta^2}} \right] \qquad \text{(3-194)}$$

In the limit, for a high rate of surface renewal, $s\delta^2/D_{AB}$, (3-193) reduces to the surface renewal theory, (3-189). For low rates of renewal, (3-194) reduces to the film theory, (3-177). At conditions in between, k_c is proportional to D_{AB}^n, where n is in the range of 0.5 to 1.0. The application of the film-penetration theory is difficult because of lack of data on δ and s, but the predicted effect of the molecular diffusivity brackets experimental data.

3.7 TWO-FILM THEORY AND OVERALL MASS TRANSFER COEFFICIENTS

Separation processes that involve contacting two fluid phases generally require consideration of mass transfer resistances in both phases. In 1923, Whitman [63] suggested an extension of the film theory to two fluid films in series. Each film presents a resistance to mass transfer, but concentrations in the two fluids at the interface are in equilibrium. That is, there is no additional interfacial resistance to mass transfer. This concept has found extensive application in modeling of steady-state gas–liquid and liquid–liquid separation processes, when the fluid phases are in laminar or turbulent flow. The assumption of equilibrium at the interface is satisfactory unless mass transfer rates are very high or surfactants accumulate at the interface.

Gas–Liquid Case

Consider the steady-state mass transfer of A from a gas phase, across an interface, into a liquid phase. It could be postulated, as shown in Figure 3.21a, that a thin gas film exists on one side of the interface and a thin liquid film exists on the other side with the controlling factors being molecular diffusion through each of the films. However, this postulation is not necessary, because instead of writing

$$N_A = \frac{(D_{AB})_G}{\delta_G}(c_{A_b} - c_{A_i})_G = \frac{(D_{AB})_L}{\delta_L}(c_{A_i} - c_{A_b})_L \qquad \text{(3-195)}$$

we can express the rate of mass transfer in terms of mass transfer coefficients that can be determined from any suitable theory, with the concentration gradients visualized more realistically as in Figure 3.21b. In addition, we can use any number of different mass transfer coefficients, depending on the selection of the driving force for mass transfer. For the gas phase, under dilute or equimolar counter diffusion (EMD) conditions, we write the mass transfer rate in terms of partial pressures:

$$N_A = k_g(p_{A_b} - p_{A_i}) \qquad \text{(3-196)}$$

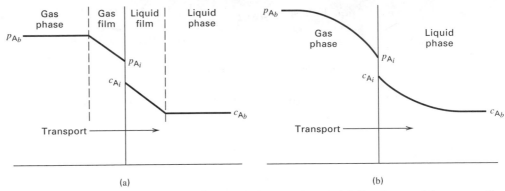

Figure 3.21 Concentration gradients for two-resistance theory: (a) film theory; (b) more realistic gradients.

For the liquid phase, we might use molar concentrations:

$$N_A = k_c(c_{A_i} - c_{A_b}) \tag{3-197}$$

At the phase interface, c_{A_i} and p_{A_i} are in equilibrium. Applying a version of Henry's law different from that in Table 2.3,

$$c_{A_i} = H_A p_{A_i} \tag{3-198}$$

Equations (3-196) to (3-198) are a commonly used combination for vapor–liquid mass transfer. Computations of mass transfer rates are generally made from a knowledge of bulk concentrations, which in this case are p_{A_b} and c_{A_b}. To obtain an expression for N_A in terms of an overall driving force for mass transfer, (3-196) to (3-198) are combined in the following manner to eliminate the interfacial concentrations, c_{A_i} and p_{A_i}. Solve (3-196) for p_{A_i}:

$$p_{A_i} = p_{A_b} - \frac{N_A}{k_g} \tag{3-199}$$

Solve (3-197) for c_{A_i}:
$$c_{A_i} = c_{A_b} + \frac{N_A}{k_c} \tag{3-200}$$

Combine (3-200) with (3-198) to eliminate c_{A_i} and combine the result with (3-199) to eliminate p_{A_i} to give

$$N_A = \frac{p_{A_b} H_A - c_{A_b}}{(H_A/k_g) + (1/k_c)} \tag{3-201}$$

It is customary to define: (1) a fictitious liquid-phase concentration $c_A^* = p_{A_b} H$, which is the concentration that would be in equilibrium with the partial pressure in the bulk gas; and (2) an overall mass transfer coefficient, K_L. Thus, if $H = H_A$, (3-201) is rewritten as

$$N_A = K_L(c_A^* - c_{A_b}) = \frac{(c_A^* - c_{A_b})}{(H_A/k_g) + (1/k_c)} \tag{3-202}$$

where
$$\frac{1}{K_L} = \frac{H_A}{k_g} + \frac{1}{k_c} \tag{3-203}$$

in which K_L is the overall mass transfer coefficient based on the liquid phase. The quantities H_A/k_g and $1/k_c$ are measures of the mass-transfer resistances of the gas phase and the liquid phase, respectively. When $1/k_c \gg H_A/k_g$, (3-203) becomes

$$N_A = k_c(c_A^* - c_{A_b}) \qquad \text{(3-204)}$$

Since resistance in the gas phase is then negligible, the gas-phase driving force is $p_{A_b} - p_{Ai} \approx 0$ and $p_{A_b} \approx p_{A_i}$.

Alternatively, (3-196) to (3-198) can be combined to define an overall mass transfer coefficient based on the gas phase. The result is

$$N_A = \frac{p_{A_b} - c_{A_b}/H_A}{(1/k_g) + (1/H_A k_c)} \qquad \text{(3-205)}$$

In this case, it is customary to define: (1) a fictitious gas phase partial pressure $p_A^* = c_{A_b}/H_A$, which is the partial pressure that would be in equilibrium with the bulk liquid; and (2) an overall mass transfer coefficient for the gas phase, K_G, based on a partial-pressure driving force. Thus, (3-205) can be rewritten as

$$N_A = K_G(p_{A_b} - p_A^*) = \frac{(p_{A_b} - p_A^*)}{(1/k_g) + (1/H_A k_c)} \qquad \text{(3-206)}$$

where

$$\frac{1}{K_G} = \frac{1}{k_g} + \frac{1}{H_A k_c} \qquad \text{(3-207)}$$

In this, the resistances are $1/k_g$ and $1/(H_A k_c)$. When $1/k_g \gg 1/H_A k_c$,

$$N_A = k_g(p_{A_b} \quad p_A^*) \qquad \text{(3-208)}$$

Since the resistance in the liquid phase is then negligible, the liquid-phase driving force is $c_{A_i} - c_{A_b} \approx 0$ and $c_{A_i} \approx c_{A_b}$.

The choice between using (3-202) or (3-206) is arbitrary, but is usually made on the basis of which phase has the largest relative mass transfer resistance; if the liquid, use (3-202); if the gas, use (3-206). Another common combination for vapor–liquid mass transfer uses mole fraction-driving forces, which define another set of mass transfer coefficients:

$$N_A = k_y(y_{A_b} - y_{A_i}) = k_x(x_{A_i} - x_{A_b}) \qquad \text{(3-209)}$$

In this case, phase equilibrium at the interface can be expressed in terms of the K-value for vapor–liquid equilibrium. Thus,

$$K_A = y_{A_i}/x_{A_i} \qquad \text{(3-210)}$$

Combining (3-209) and (3-210) to eliminate y_{A_i} and x_{A_i},

$$N_A = \frac{y_{A_b} - x_{A_b}}{(1/K_A k_y) + (1/k_x)} \qquad \text{(3-211)}$$

This time we define fictitious concentration quantities and overall mass transfer coefficients for mole fraction-driving forces. Thus, $x_A^* = y_{A_b}/K_A$ and $y_A^* = K_A x_{A_b}$. If the two values of K_A are equal, we obtain

$$N_A = K_x(x_A^* - x_{A_b}) = \frac{x_A^* - x_{A_b}}{(1/K_A k_y) + (1/k_x)} \qquad \text{(3-212)}$$

$$\text{and } N_A = K_y(y_{A_b} - y_A^*) = \frac{y_{A_b} - y_A^*}{(1/k_y) + (K_A/k_x)} \qquad \text{(3-213)}$$

where
$$\frac{1}{K_x} = \frac{1}{K_A k_y} + \frac{1}{k_x} \tag{3-214}$$

and
$$\frac{1}{K_y} = \frac{1}{k_y} + \frac{K_A}{k_x} \tag{3-215}$$

When using correlations to estimate mass transfer coefficients for use in the above equations, it is important to determine which coefficient (k_g, k_c, k_y, or k_x) is correlated. This can usually be done by checking the units or the form of the Sherwood or Stanton numbers. Coefficients correlated by the Chilton–Colburn analogy are k_c for either the liquid or gas phase. The different coefficients are related by the following expressions:

Liquid phase:

$$k_x = k_c c = k_c \left(\frac{\rho_L}{M}\right) \tag{3-216}$$

Ideal gas phase:

$$k_y = k_g P = (k_c)_g \frac{P}{RT} = (k_c)_g c \tag{3-217}$$

Typical units are

	SI	American Engineering
k_c	m/s	ft/h
k_g	kmol/s-m²-kPa	lbmol/h-ft²-atm
k_y, k_x	kmol/s-m²	lbmol/h-ft²

When unimolecular diffusion (UMD) occurs under nondilute conditions, the effect of bulk-flow must be included in the above equations. For binary mixtures, one method for doing this is to define modified mass transfer coefficients, designated with a prime, as follows.

For the liquid phase, using k_c or k_x,

$$k' = \frac{k}{(1 - x_A)_{LM}} = \frac{k}{(x_B)_{LM}} \tag{3-218}$$

For the gas phase, using k_g, k_y, or k_c,

$$k' = \frac{k}{(1 - y_A)_{LM}} = \frac{k}{(y_B)_{LM}} \tag{3-219}$$

The expressions for k' are most readily used when the mass transfer rate is controlled mainly by one of the two resistances. Experimental mass transfer coefficient data reported in the literature are generally correlated in terms of k rather than k'.

Liquid–Liquid Case

For mass transfer across two liquid phases, equilibrium is again assumed at the interface. Denoting the two phases by $L^{(1)}$ and $L^{(2)}$, (3-212) and (3-213) can be rewritten as

$$N_A = K_{x^{(2)}}(x_A^{(2)*} - x_{A_b}^{(2)}) = \frac{x_A^{(2)*} - x_{A_b}^{(2)}}{(1/K_{D_A} k_x^{(1)}) + (1/k_x^{(2)})} \tag{3-220}$$

and
$$N_A = K_x^{(1)}(x_{A_b}^{(1)} - x_A^{(1)*}) = \frac{x_{A_b}^{(1)} - x_A^{(1)*}}{(1/k_x^{(1)}) + (K_{D_A}/k_x^{(2)})} \tag{3-221}$$

where
$$K_{D_A} = \frac{x_{A_i}^{(1)}}{x_{A_i}^{(2)}} \tag{3-222}$$

Case of Large Driving Forces for Mass Transfer

When large driving forces exist for mass transfer, phase equilibria ratios such as H_A, K_A, and K_{D_A} may not be constant across the two phases. This occurs particularly when one or both phases are not dilute with respect to the diffusing solute, A. In that case, expressions for the mass transfer flux must be revised.

For example, if mole fraction-driving forces are used, we write, from (3-209) and (3-213),

$$N_A = k_y(y_{A_b} - y_{A_i}) = K_y(y_{A_b} - y_A^*) \tag{3-223}$$

Thus,

$$\frac{1}{K_y} = \frac{y_{A_b} - y_A^*}{k_y(y_{A_b} - y_{A_i})} \tag{3-224}$$

or
$$\frac{1}{K_y} = \frac{(y_{A_b} - y_{A_i}) + (y_{A_i} - y_A^*)}{k_y(y_{A_b} - y_{A_i})} = \frac{1}{k_y} + \frac{1}{k_y}\left(\frac{y_{A_i} - y_A^*}{y_{A_b} - y_{A_i}}\right) \tag{3-225}$$

From (3-209),

$$\frac{k_x}{k_y} = \frac{(y_{A_b} - y_{A_i})}{(x_{A_i} - x_{A_b})} \tag{3-226}$$

Combining (3-225) and (3-226),

$$\frac{1}{K_y} = \frac{1}{k_y} + \frac{1}{k_x}\left(\frac{y_{A_i} - y_A^*}{x_{A_i} - x_{A_b}}\right) \tag{3-227}$$

In a similar manner,

$$\frac{1}{K_x} = \frac{1}{k_x} + \frac{1}{k_y}\left(\frac{x_A^* - x_{A_i}}{y_{A_b} - y_{A_i}}\right) \tag{3-228}$$

A typical curved equilibrium line is shown in Figure 3.22 with representative values of $y_{A_b}, y_{A_i}, y_A^*, x_A^*, x_{A_i}$, and x_{A_b} indicated. Because the line is curved, the vapor–liquid equilibrium ratio, $K_A = y_A/x_A$ is not constant across the two phases. As shown, K_A decreases with increasing concentration of A. Denote two slopes of the equilibrium line by

$$m_x = \left(\frac{y_{A_i} - y_A^*}{x_{A_i} - x_{A_b}}\right) \tag{3-229}$$

and
$$m_y = \left(\frac{y_{A_b} - y_{A_i}}{x_A^* - x_{A_i}}\right) \tag{3-230}$$

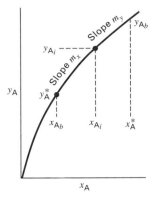

Figure 3.22 Curved equilibrium line.

Substituting (3-229) and (3-230) into (3-227) and (3-228), respectively, gives

$$\frac{1}{K_y} = \frac{1}{k_y} + \frac{m_x}{k_x} \qquad \text{(3-231)}$$

and

$$\frac{1}{K_x} = \frac{1}{k_x} + \frac{1}{m_y k_y} \qquad \text{(3-232)}$$

EXAMPLE 3.20

Sulfur dioxide (A) is absorbed into water in a packed column. At a certain location, the bulk conditions are 50°C, 2 atm, $y_{A_b} = 0.085$, and $x_{A_b} = 0.001$. Equilibrium data for SO_2 between air and water at 50°C are

p_{SO_2}, **atm**	c_{SO_2}, **lbmol/ft³**
0.0382	0.00193
0.0606	0.00290
0.1092	0.00483
0.1700	0.00676

Experimental values of the mass transfer coefficients are as follows.

Liquid phase: $k_c = 0.18$ m/h

Gas phase: $k_g = 0.040 \dfrac{\text{kmol}}{\text{h-m}^2\text{-kPa}}$

Using mole fraction-driving forces, compute the mass transfer flux by:

(a) Assuming an average Henry's law constant and a negligible bulk-flow effect
(b) Utilizing the actual curved equilibrium line and assuming a negligible bulk-flow effect
(c) Utilizing the actual curved equilibrium line and taking into account the bulk-flow effect

In addition,

(d) Determine the relative magnitude of the two resistances and the values of the mole fractions at the interface from the results of part (c).

SOLUTION

The equilibrium data are converted to mole fractions by assuming Dalton's law, $y_A = p_A/P$, for the gas and using $x_A = c_A/c$ for the liquid. The concentration of the liquid is close to that of pure water or 3.43 lbmol/ft^3 or 55.0 kmol/m^3. Thus, the mole fractions at equilibrium are

y_{SO_2}	x_{SO_2}
0.0191	0.000563
0.0303	0.000846
0.0546	0.001408
0.0850	0.001971

These data are fitted with average and maximum absolute deviations of 0.91% and 1.16%, respectively, by the quadratic equation

$$y_{SO_2} = 29.74x_{SO_2} + 6{,}733x_{SO_2}^2 \qquad (1)$$

Thus, differentiating, the slope of the equilibrium curve is given by

$$m = \frac{dy}{dx} = 29.74 + 13{,}466x_{SO_2} \qquad (2)$$

The given mass transfer coefficients can be converted to k_x and k_y by (3-216) and (3-217): $k_x = k_c c = 0.18(55.0) = 9.9 \frac{kmol}{h\text{-}m^2}$ and $k_y = 0.040(2)(101.3) = 8.1 \frac{kmol}{h\text{-}m^2}$.

(a) From (1) for $x_{A_b} = 0.001$, $y_A^* = 29.74(0.001) + 6{,}733(0.001)^2 = 0.0365$. From (1), for $y_{A_b} = 0.085$, we solve the quadratic equation to obtain $x_A^* = 0.001975$.

The average slope in this range is $m = \dfrac{0.085 - 0.0365}{0.001975 - 0.001} = 49.7$. From an examination of (3-231) and (3-232), the liquid-phase resistance is controlling because the term in k_x is much larger than the term in k_y. Therefore, from (3-232), using $m = m_x$,

$$\frac{1}{K_x} = \frac{1}{9.9} + \frac{1}{49.7(8.1)} = 0.1010 + 0.0025 = 0.1035 \text{ or } K_x = 9.66 \frac{kmol}{h\text{-}m^2}$$

From (3-212), $N_A = 9.66(0.001975 - 0.001) = 0.00942 \frac{kmol}{h\text{-}m^2}$.

(b) From part (a), the gas-phase resistance is almost negligible. Therefore, $y_{A_i} \approx y_{A_b}$ and $x_{A_i} \approx x_A^*$.

From (3-230), the slope m_y must, therefore, be taken at the point $y_{A_b} = 0.085$ and $x_A^* = 0.001975$ on the equilibrium line.

From (2), $m_y = 29.74 + 13{,}466(0.001975) = 56.3$. From (3-232),

$$K_x = \frac{1}{(1/9.9) + [1/(56.3)(8.1)]} = 9.69 \frac{kmol}{h\text{-}m^2},$$

giving $N_A = 0.00945$ kmol/h-m^2. This is only a slight change from part (a).

(c) From the results of parts (a) and (b), we have

$$y_{A_b} = 0.085, y_{A_i} = 0.085, x_{A_i} = 0.1975, x_{A_b} = 0.001$$

$$(y_B)_{LM} = 1.0 - 0.085 = 0.915 \text{ and } (x_B)_{LM} \approx 0.9986$$

From (3-218), $\quad k_x' = \dfrac{9.9}{0.9986} = 9.9 \dfrac{kmol}{h\text{-}m^2}$ and $k_y' = \dfrac{8.1}{0.915} = 8.85 \dfrac{kmol}{h\text{-}m^2}$

From (3-232), $\quad K_x = \dfrac{1}{(1/9.9) + [1/56.3(8.85)]} = 9.71 \dfrac{kmol}{h\text{-}m^2}$

From (3-212), $\quad N_A = 9.71(0.001975 - 0.001) = 0.00947 \dfrac{kmol}{h\text{-}m^2}$

which is only a very slight change from parts (a) and (b), where the bulk-flow effect was ignored. The effect is very small because here it is important only in the gas phase; but the liquid-phase resistance is controlling.

(d) The relative magnitude of the mass transfer resistances can be written as

$$\frac{1/m_y k_y'}{1/k_x'} = \frac{1/(56.3)(8.85)}{1/9.9} = 0.02$$

Thus, the gas-phase resistance is only 2% of the liquid-phase resistance. The interface vapor mole fraction can be obtained from (3-223), after accounting for the bulk-flow effect:

$$y_{A_i} = y_{A_b} - \frac{N_A}{k_y'} = 0.085 - \frac{0.00947}{8.85} = 0.084$$

Similarly,
$$x_{A_i} = \frac{N_A}{k_x'} + x_{A_b} = \frac{0.00947}{9.9} + 0.001 = 0.00196 \qquad \blacksquare$$

SUMMARY

1. Mass transfer is the net movement of a component in a mixture from one region to another region of different concentration, often between two phases across an interface. Mass transfer occurs by molecular diffusion, eddy diffusion, and bulk flow. Molecular diffusion occurs because of a number of driving forces, including concentration (the most important), pressure, temperature, and external force fields.

2. Fick's first law for steady-state conditions states that the mass transfer flux by ordinary molecular diffusion is equal to the product of the diffusion coefficient (diffusivity) and the negative of the concentration gradient.

3. Two limiting cases of mass transfer are equimolar counterdiffusion (EMD) and unimolecular diffusion (UMD). The former is also a good approximation for dilute conditions. The latter must include the bulk-flow effect.

4. When experimental data are not available, diffusivities in gas and liquid mixtures can be estimated. Diffusivities in solids, including porous solids, crystalline solids, metals, glass, ceramics, polymers, and cellular solids are best measured. For some solids—for example, wood—diffusivity is an anisotropic property.

5. Diffusivity values vary by orders of magnitude. Typical values are 0.10, 1×10^{-5}, and 1×10^{-9} cm^2/s for ordinary molecular diffusion of a solute in a gas, liquid, and solid, respectively.

6. Fick's second law for unsteady-state diffusion is readily applied to semiinfinite and finite stagnant media, including certain anisotropic materials.

7. Molecular diffusion under laminar-flow conditions can be determined from Fick's first and second laws, provided that velocity profiles are available. Common cases include falling liquid-film flow, boundary-layer flow on a flat plate, and fully developed flow in a straight, circular tube. Results are often expressed in terms of a mass transfer coefficient embedded in a dimensionless group called the Sherwood number. The mass transfer flux is given by the product of the mass transfer coefficient and a concentration-driving force.

8. Mass transfer in turbulent flow is often predicted by analogy to heat transfer. Of particular importance, is the Chilton–Colburn analogy, which utilizes empirical j-factor correlations and the dimensionless Stanton number for mass transfer. Semi-theoretical analogies, such as extensions of the Reynolds analogy, are also sometimes useful.

9. A number of models have been developed for mass transfer across a two-fluid interface and into a liquid. These include the film theory, penetration theory, surface-renewal theory, and the film-penetration theory. These theories predict mass transfer

coefficients that are proportional to the diffusivity raised to an exponent that varies from 0.5 to 1.0. Most experimental data provide exponents ranging from 0.5 to 0.75.

10. The two-film theory of Whitman (more properly referred to as a two-resistance theory) is widely used to predict the mass transfer flux from one fluid phase, across an interface, and into another fluid phase, assuming equilibrium at the interface. One resistance is often controlling. The theory defines an overall mass transfer coefficient that is determined from the separate coefficients for each of the two phases and the equilibrium relationship at the interface.

REFERENCES

1. Taylor, R., and R. Krishna, *Multicomponent Mass Transfer*, John Wiley and Sons, New York (1993).

2. Reid, R.C., J.M. Prausnitz, and B.E. Poling, *The Properties of Liquids and Gases*, 4th ed., McGraw-Hill, New York (1987).

3. Fuller, E.N., P.D. Schettler, and J.C. Giddings, *Ind. Eng. Chem.*, **58** (5), 18–27 (1966).

4. Takahashi, S., *J. Chem. Eng. Jpn.*, **7**, 417–420 (1974).

5. Slattery, J.C., M.S. thesis, University of Wisconsin, Madison (1955).

6. Wilke, C.R., and P. Chang, *AIChE J.*, **1**, 264–270 (1955).

7. Hayduk, W., and B.S. Minhas, *Can. J. Chem. Eng.*, **60**, 295–299 (1982).

8. Quayle, O.R., *Chem. Rev.*, **53**, 439–589 (1953).

9. Vignes, A., *Ind. Eng. Chem. Fundam.*, **5**, 189–199 (1966).

10. Sorber, H.A., *Handbook of Biochemistry, Selected Data for Molecular Biology*, 2nd ed., Chemical Rubber Co., Cleveland, OH (1970).

11. Geankoplis, C.J., *Transport Processes and Unit Operations*, 3rd ed., Prentice-Hall, Englewood Cliffs, NJ (1993).

12. Friedman, L., and E.O. Kraemer, *J. Am. Chem. Soc.*, **52**, 1298–1304, 1305–1310, 1311–1314, (1930).

13. Boucher, D.F., J.C. Brier, and J.O. Osburn, *Trans. AIChE*, **38**, 967–993 (1942).

14. Barrer, R.M., *Diffusion in and through Solids*, Oxford University Press, London (1951).

15. Swets, D.E., R.W. Lee, and R.C. Frank, *J. Chem. Phys.*, **34**, 17–22 (1961).

16. Lee, R. W., *J. Chem, Phys.*, **38**, 44–455 (1963).

17. Williams, E.L., *J. Am. Ceram. Soc.*, **48**, 190–194 (1965).

18. Sucov, E.W., *J. Am. Ceram. Soc.*, **46**, 14–20 (1963).

19. Kingery, W.D., H.K. Bowen, and D.R. Uhlmann, *Introduction to Ceramics*, 2nd ed., John Wiley and Sons, New York (1976).

20. Ferry, J.D., *Viscoelastic Properties of Polymers*, John Wiley and Sons, New York (1980).

21. Rhee, C.K., and J.D. Ferry, *J. Appl. Polym. Sci.*, **21**, 467–476 (1977).

22. Brandrup, J., and E.H. Immergut, Eds., *Polymer Handbook*, 3rd ed., John Wiley and Sons, New York (1989).

23. Gibson, L.J., and M.F. Ashby, *Cellular Solids, Structure and Properties*, Pergamon Press, Elmsford, NY (1988).

24. Stamm, A.J., *Wood and Celluose Science*, Ronald Press, New York (1964).

25. Sherwood, T.K., *Ind. Eng. Chem.*, **21**, 12–16 (1929).

26. Carslaw, H.S., and J.C. Jaeger, *Heat Conduction in Solids*, 2nd ed., Oxford University Press, London (1959).

27. Crank, J., *The Mathematics of Diffusion*, Oxford University Press, London (1956).

28. Bird, R.B., W.E. Stewart, and E.N. Lightfoot, *Transport Phenomena*, John Wiley and Sons, New York (1960).

29. Churchill, R.V., *Operational Mathematics*, 2nd ed., McGraw-Hill, New York (1958).

30. Abramowitz, M., and I.A. Stegun, Eds., *Handbook of Mathematical Functions*, National Bureau of Standards, Applied Mathematics Series 55, Washington, DC (1964).

31. Newman, A.B., *Trans. AIChE*, **27**, 310–333 (1931).

32. Grimley, S.S., *Trans. Inst. Chem. Eng.* (London), **23**, 228–235 (1948).

33. Johnstone, H.F., and R.L. Pigford, *Trans. AIChE*, **38**, 25–51 (1942).

34. Olbrich, W.E., and J.D. Wild, *Chem Eng. Sci.*, **24**, 25–32 (1969).

35. Churchill, S.W., *The Interpretation and Use of Rate Data: The Rate Concept*, McGraw-Hill, New York (1974).

36. Churchill, S.W., and R. Usagi, *AIChE J.*, **18**, 1121–1128 (1972).

37. Emmert, R.E., and R.L. Pigford, *Chem. Eng. Prog.*, **50**, 87–93 (1954).

38. Prandtl, L., *Proc. 3rd Int. Math. Congress*, Heidelberg (1904); reprinted in *NACA Tech. Memo 452* (1928).

39. Blasius, H., *Z. Math Phys.*, **56**, 1–37 (1908); reprinted in *NACA Tech. Memo 1256.*

40. Schlichting, H., *Boundary Layer Theory*, 4th ed., McGraw-Hill, New York (1960).

41. Pohlhausen, E., *Z. Angew. Math Mech.*, **1**, 252 (1921).

42. Pohlhausen, E., *Z. Angew. Math Mech.*, **1**, 115–121 (1921).

43. Langhaar, H.L., *Trans. ASME*, **64**, A-55 (1942).

44. Graetz, L., *Ann. d. Physik*, **25**, 337–357 (1885).

45. Sellars, J.R., M. Tribus, and J.S. Klein, *Trans. ASME*, **78**, 441–448 (1956).

46. Leveque, J., *Ann. Mines*, [12], **13**, 201, 305, 381 (1928).

47. Knudsen, J.G., and D.L. Katz, *Fluid Dynamics and Heat Transfer*, McGraw-Hill, New York (1958).

48. Hausen, H., *Verfahrenstechnik Beih. z. Ver. deut. Ing.*, **4**, 91 (1943).

49. Linton, Jr., W.H., and T.K. Sherwood, *Chem. Eng. Prog.*, **46**, 258–264 (1950).

50. Reynolds, O., *Trans. Roy. Soc.* (London), **174A**, 935–982 (1883).

51. Boussinesq, J., *Mem. Pre. par. div. Sav.*, XXIII, Paris (1877).

52. Prandtl, L., *Z. Angew, Math Mech.*, **5**, 136 (1925); reprinted in *NACA Tech. Memo 1231* (1949).

53. Reynolds, O., *Proc. Manchester Lit. Phil. Soc.*, **14**, 7 (1874).

54. Colburn, A.P., *Trans. AIChE*, **29**, 174–210 (1933).

55. Chilton, T.H., and A.P. Colburn, *Ind. Eng. Chem.*, **26**, 1183–1187 (1934).

56. Prandtl, L., *Physik. Z.*, **11**, 1072 (1910).

57. Friend, W.L., and A.B. Metzner, *AIChE J.,* **4,** 393–402 (1958).
58. Nernst, W., *Z. Phys. Chem.,* **47,** 52 (1904).
59. Higbie, R., *Trans. AIChE,* **31,** 365–389 (1935).
60. Danckwerts, P.V., *Ind. Eng. Chem.,* **43,** 1460–1467 (1951).

61. Levenspiel, O., *Chemical Reaction Engineering,* 2nd ed., John Wiley and Sons, New York (1972).
62. Toor, H.L., and J.M. Marchello, *AIChE J.,* **4,** 97–101 (1958).
63. Whitman, W.G., *Chem. Met. Eng.,* **29,** 146–148 (1923).

EXERCISES

Section 3.1

3.1 A beaker filled with an equimolar liquid mixture of ethyl alcohol and ethyl acetate evaporates at 0°C into still air at 101 kPa (1atm) total pressure. Assuming Raoult's law applies, what will be the composition of the liquid remaining when half the original ethyl alcohol has evaporated, assuming that each component evaporates independently of the other? Also assume that the liquid is always well mixed. The following data are available:

	Vapor Pressure, kPa at 0°C	Diffusivity in Air m²/s
Ethyl acetate (AC)	3.23	6.45×10^{-6}
Ethyl alcohol (AL)	1.62	9.29×10^{-6}

3.2 An open tank, 10 ft in diameter and containing benzene at 25°C, is exposed to air in such a manner that the surface of the liquid is covered with a stagnant air film estimated to be 0.2 in.-thick. If the total pressure is 1 atm and the air temperature is 25°C, what loss of material in pounds per day occurs from this tank? The specific gravity of benzene at 60°F is 0.877. The concentration of benzene at the outside of the film is so low that it may be neglected. For benzene, the vapor pressure at 25°C is 100 torr, and the diffusivity in air is 0.08 cm²/s.

3.3 An insulated glass tube and condenser are mounted on a reboiler containing benzene and toluene. The condenser returns liquid reflux so that it runs down the wall of the tube. At one point in the tube the temperature is 170°F, the vapor contains 30 mol% toluene, and the liquid reflux contains 40 mol% toluene. The effective thickness of the stagnant vapor film is estimated to be 0.1 in. The molar latent heats of benzene and toluene are equal. Calculate the rate at which toluene and benzene are being interchanged at this point in the tube in lbmol/h-ft².

Diffusivity of toluene in benzene = 0.2 ft²/h.
Pressure = 1 atm total pressure (in the tube).
Vapor pressure of toluene at 170°F = 400 torr.

3.4 Air at 25°C with a dew-point temperature of 0°C flows past the open end of a vertical tube filled with liquid water maintained at 25°C. The tube has an inside diameter of 0.83 in., and the liquid level was originally 0.5 in. below the top of the tube. The diffusivity of water in air at 25°C is 0.256 cm²/s.
(a) How long will it take for the liquid level in the tube to drop 3 in.?

(b) Make a plot of the liquid level in the tube as a function of time for this period.

3.5 Two bulbs are connected by a tube, 0.002 m in diameter and 0.20 m in length. Initially bulb 1 contains argon, and bulb 2 contains xenon. The pressure and temperature are maintained at 1 atm and 105°C, at which the diffusivity is 0.180 cm²/s. At time t = 0, diffusion is allowed to occur between the two bulbs. At a later time, the argon mole fraction in the gas at end 1 of the tube is 0.75, and 0.20 at the other end. Determine at the later time:
(a) The rates and directions of mass transfer of argon and xenon
(b) The transport velocity of each species
(c) The molar average velocity of the mixture

Section 3.2

3.6 The diffusivity of toluene in air was determined experimentally by allowing liquid toluene to vaporize isothermally into air from a partially filled vertical tube 3 mm in diameter. At a temperature of 39.4°C, it took 96×10^4 s for the level of the toluene to drop from 1.9 cm below the top of the open tube to a level of 7.9 cm below the top. The density of toluene is 0.852 g/cm³, and the vapor pressure is 57.3 torr at 39.4°C. The barometer reading was 1 atm. Calculate the diffusivity and compare it with the value predicted from (3-36). Neglect the counterdiffusion of air.

3.7 An open tube, 1 mm in diameter and 6 in. long, has pure hydrogen blowing across one end and pure nitrogen blowing across the other. The temperature is 75°C.
(a) For equimolar counterdiffusion, what will be the rate of transfer of hydrogen into the nitrogen stream (mol/s)? Estimate the diffusivity from (3-36).
(b) Assuming the flow is uniform at all points in any cross section of the tube, calculate the net flow of gas (mol/s) if the number of moles of hydrogen passing into the nitrogen is maintained at 10 times the number of moles of nitrogen passing into the hydrogen. What is the direction of the molar flow? Of the mass flow?
(c) For parts (a) and (b), plot the mole fraction of hydrogen against distance from the end of the tube past which nitrogen is blown.

3.8 Some HCl gas diffuses across a film of air 0.1 in. thick at 20°C. The partial pressure of HCl on one side of the film is 0.08 atm and it is zero on the other. Estimate the rate of diffusion, as mol HCl/s-cm², if the total pressure is
(a) 10 atm, (b) 1 atm, (c) 0.1 atm.
The diffusivity of HCl in air at 20°C and 1 atm is 0.145 cm²/s.

3.9 Estimate the diffusion coefficient for the gaseous binary system nitrogen (A)/toluene (B) at 25°C and 3 atm using the method of Fuller et al.

3.10 For the mixture of Example 3.3, estimate the diffusion coefficient if the pressure is increased to 100 atm, using the method of Takahashi.

3.11 Estimate the diffusivity of carbon tetrachloride at 25°C in a dilute solution of: (a) Methanol, (b) Ethanol, (c) Benzene, and (d) n-Hexane by the method of Wilke–Chang and Hayduk–Minhas. Compare the estimated values with the following experimental observations:

Solvent	Experimental D_{AB}, cm²/s
Methanol	1.69×10^{-5} cm²/s at 15°C
Ethanol	1.50×10^{-5} cm²/s at 25°C
Benzene	1.92×10^{-5} cm²/s at 25°C
n-Hexane	3.70×10^{-5} cm²/s at 25°C

3.12 Estimate the liquid diffusivity of benzene (A) in formic acid (B) at 25°C and infinite dilution. Compare the estimated value to that of Example 3.6 for formic acid at infinite dilution in benzene.

3.13 Estimate the liquid diffusivity of acetic acid at 25°C in a dilute solution of: (a) Benzene, (b) Acetone, (c) Ethyl acetate, and (d) Water by an appropriate method. Compare the estimated values with the following experimental values:

Solvent	Experimental D_{AB}, cm²/s
Benzene	2.09×10^{-5} cm²/s at 25°C
Acetone	2.92×10^{-5} cm²/s at 25°C
Ethyl acetate	2.18×10^{-5} cm²/s at 25°C
Water	1.19×10^{-5} cm²/s at 20°C

3.14 Water in an open dish exposed to dry air at 25°C is found to vaporize at a constant rate of 0.04 g/h-cm². Assuming the water surface to be at the wet-bulb temperature of 11.0°C, calculate the effective gas-film thickness (i.e., the thickness of a stagnant air film that would offer the same resistance to vapor diffusion as is actually encountered at the water surface).

3.15 Isopropyl alcohol is undergoing mass transfer at 35°C and 2 atm under dilute conditions through water, across a phase boundary, and then through nitrogen. Based on the date given below, estimate for isopropyl alcohol:
(a) The diffusivity in water using the Wilke-Chang equation
(b) The diffusivity in nitrogen using the Fuller et al. equation
(c) The product, $D_{AB}\rho_M$, in water
(d) The product, $D_{AB}\rho_M$, in air
where ρ_M is the molar density of the mixture.
   Using the above results, compare:
(e) The diffusivities in parts (a) and (b)
(f) The diffusivity-molar density products in Parts (c) and (d)

Lastly:
(g) What conclusions can you come to about molecular diffusion in the liquid phase versus the gaseous phase.
   Data:

Component	T_c, °R	P_c, psia	Z_c	v_L, cm³/mol
Nitrogen	227.3	492.9	0.289	—
Isopropyl alcohol	915	691	0.249	76.5

3.16 Experimental liquid-phase activity coefficient data are given in Exercise 2.23 for the ethanol/benzene system at 45°C. Estimate and plot diffusion coefficients for both ethanol and benzene over the entire composition range.

3.17 Estimate the diffusion coefficient of NaOH in a 1 M aqueous solution at 25°C.

3.18 Estimate the diffusion coefficient of NaCl in a 2 M aqueous solution at 18°C. Compare your estimate with the experimental value of 1.28×10^{-5} cm²/s.

3.19 Estimate the diffusivity of N_2 in H_2 in the pores of a catalyst at 300°C and 20 atm if the porosity is 0.45 and the tortuosity is 2.5. Assume ordinary molecular diffusion in the pores.

3.20 Gaseous hydrogen at 150 psia and 80°F is stored in a small spherical steel pressure vessel having an inside diameter of 4 in. and a wall thickness of 0.125 in. At these conditions, the solubility of hydrogen in steel is 0.094 lbmol/ft³ and the diffusivity of hydrogen in steel is 3.0×10^{-9} cm²/s. If the inner surface of the vessel remains saturated at the existing hydrogen pressure and the hydrogen partial pressure at the outer surface is assumed to be zero, estimate:
(a) The initial rate of mass transfer of hydrogen through the metal wall
(b) The initial rate of pressure decrease inside the vessel
(c) The time in hours for the pressure to decrease to 50 psia, assuming the temperature stays constant at 80°F

3.21 A polyisoprene membrane of 0.8-μm thickness is to be used to separate a mixture of methane and H_2. Using the data in Table 14.9 and the following compositions, estimate the mass transfer flux of each of the two species.

	Partial Pressures, MPa	
	Membrane Side 1	Membrane Side 2
Methane	2.5	0.05
Hydrogen	2.0	0.20

Section 3.3

3.22 A 3-ft depth of stagnant water at 25°C lies on top of a 0.10-in. thickness of NaCl. At time < 0, the water is pure. At time = 0, the salt begins to dissolve and diffuse into the water. If the concentration of salt in the water at the solid–liquid interface is maintained at saturation (36 g NaCl/

100 g H_2O) and the diffusivity of NaCl in water is 1.2×10^{-5} cm^2/s, independent of concentration, estimate, by assuming the water to act as a semiinfinite medium, the time and the concentration profile of salt in the water when
(a) 10% of the salt has dissolved
(b) 50% of the salt has dissolved
(c) 90% of the salt has dissolved

3.23 A slab of dry wood of 4-in. thickness and sealed edges is exposed to air of 40% relative humidity. Assuming that the two unsealed faces of the wood immediately jump to an equilibrium moisture content of 10 lb H_2O per 100 lb of dry wood, determine the time for the moisture to penetrate to the center of the slab (2 in. from either face). Assume a diffusivity of water in the wood as 8.3×10^{-6} cm^2/s.

3.24 A wet clay brick measuring $2 \times 4 \times 6$ in. has an initial uniform moisture content of 12 wt%. At time $= 0$, the brick is exposed on all sides to air such that the surface moisture content is maintained at 2 wt%. After 5 h, the average moisture content is 8 wt%. Estimate:
(a) The diffusivity of water in the clay in cm^2/s
(b) The additional time for the average moisture content to reach 4 wt%. All moisture contents are on a dry basis.

3.25 A spherical ball of clay, 2 in. in diameter, has an initial moisture content of 10 wt%. The diffusivity of water in the clay is 5×10^{-6} cm^2/s. At time $t = 0$, the surface of the clay is brought into contact with air such that the moisture content at the surface is maintained at 3 wt%. Estimate the time for the average moisture content in the sphere to drop to 5 wt%. All moisture contents are on a dry basis.

Section 3.4

3.26 Estimate the rate of absorption of pure oxygen at 10 atm and 25°C into water flowing as a film down a vertical wall 1 m high and 6 cm in width at a Reynolds number of 50 without surface ripples. Assume the diffusivity of oxygen in water is 2.5×10^{-5} cm^2/s and that the mole fraction of oxygen in water at saturation for the above temperature and pressure is 2.3×10^{-4}.

3.27 For the conditions of Example 3.13, determine at what height from the top, the average concentration of CO_2 would correspond to 50% of saturation.

3.28 Air at 1 atm flows at 2 m/s across the surface of a 2-in.-long surface that is covered with a thin film of water. If the air and water are maintained at 25°C, and the diffusivity of water in air at these conditions is 0.25 cm^2/s, estimate the mass flux for the evaporation of water at the middle of the surface assuming laminar boundary-layer flow. Is this assumption reasonable?

3.29 Air at 1 atm and 100°C flows across a thin flat plate of napthalene that is 1 m long, causing the plate to sublime. The Reynolds number at the trailing edge of the plate is at the upper limit for a laminar boundary layer. Estimate:
(a) The average rate of sublimation in kmol/s-m²

(b) The local rate of sublimation at a distance of 0.5 m from the leading edge of the plate.
Physical properties are given in Example 3.14.

3.30 Air at 1 atm and 100°C flows through a straight, 5-cm-diameter, circular tube, cast from napthalene, at a Reynolds number of 1,500. Air entering the tube has an established laminar-flow velocity profile. Properties are given in Example 3.14. If pressure drop through the tube is negligible, calculate the length of tube needed for the average mole fraction of napthalene in the exiting air to be 0.005.

3.31 A spherical water drop is suspended from a fine thread in still, dry air. Show:
(a) That the Sherwood number for mass transfer from the surface of the drop into the surroundings has a value of 2, if the characteristic length is the diameter of the drop.
  If the initial drop diameter is 1 mm, the air temperature is 38°C, the drop temperature is 14.4°C, and the pressure is 1 atm, calculate:
(b) The initial mass of the drop in grams.
(c) The initial rate of evaporation in grams per second.
(d) The time in seconds for the drop diameter to be reduced to 0.2 mm.
(e) The initial rate of heat transfer to the drop. If the Nusselt number is also 2, is the rate of heat transfer sufficient to supply the heat of vaporization and sensible heat of the evaporated water? If not, what will happen?

Section 3.5

3.32 Water at 25°C flows at 5 ft/s through a straight, cylindrical tube cast from benzoic acid, of 2 in. inside diameter. If the tube is 10 ft long, and fully developed turbulent flow is assumed, estimate the average concentration of benzoic acid in the water leaving the tube. Physical properties are given in Example 3.15.

3.33 Air at 1 atm flows at a Reynolds number of 50,000 normal to a long, circular, 1-in. diameter cylinder made of napthalene. Using the physical properties of Example 3.14 for a temperature of 100°C, calculate the average sublimation flux in kmol/s-m².

3.34 For the conditions of Exercise 3.33, calculate the initial average rate of sublimation in kmol/s-m² for a spherical particle of 1-in. initial diameter. Compare this result to that for a bed packed with napthalene spheres with a void fraction of 0.5.

Section 3.6

3.35 Carbon dioxide is stripped from water by air in a wetted-wall tube. At a certain location, where the pressure is 10 atm and the temperature is 25°C, the mass transfer flux of CO_2 is 1.62 lbmol/h-ft². The partial pressures of CO_2 are 8.2 atm at the interface and 0.1 atm in the bulk gas. The diffusivity of CO_2 in air at these conditions is 1.6×10^{-2} cm^2/s. Assuming turbulent flow of the gas, calculate by the film theory, the mass transfer coefficient k_c for the gas phase and the film thickness.

3.36 Water is used to remove CO_2 from air by absorption in a column packed with Pall rings. At a certain region of the column where the partial pressure of CO_2 at the interface is 150 psia and the concentration in the bulk liquid is negligible, the absorption rate is 0.017 lbmol/h-ft². The diffusivity of CO_2 in water is 2.0×10^{-5} cm²/s. Henry's law for CO_2 is $p = Hx$, where $H = 9,000$ psia. Calculate:
(a) The liquid-phase mass transfer coefficient and the film thickness
(b) Contact time for the penetration theory
(c) Average eddy residence time and the probability distribution for the surface renewal theory

3.37 Determine the diffusivity of H_2S in water, using the penetration theory, from the following data for the absorption of H_2S into a laminar jet of water at 20°C.

Jet diameter = 1 cm, Jet length = 7 cm, and Solubility of H_2S in water = 100 mol/m³

The average rate of absorption varies with the flow rate of the jet as follows:

Jet Flow Rate, cm³/s	Rate of Absorption, mol/s × 10⁶
0.143	1.5
0.568	3.0
1.278	4.25
2.372	6.15
3.571	7.20
5.142	8.75

Section 3.7

3.38 In a test on the vaporization of H_2O into air in a wetted-wall column, the following data were obtained:

Tube diameter, 1.46 cm, Wetted-tube length, 82.7 cm
Air rate to tube at 24°C and 1 atm, 720 cm³/s
Temperature of inlet water, 25.15°C, Temperature of outlet water, 25.35°C
Partial pressure of water in inlet air, 6.27 torr, and in outlet air, 20.1 torr

The value for the diffusivity of water vapor in air is 0.22 cm²/s at 0°C and 1 atm. The mass velocity of air is taken relative to the pipe wall. Calculate:
(a) Rate of mass transfer of water into the air
(b) K_G for the wetted-wall column

3.39 The following data were obtained by Chamber and Sherwood [*Ind. Eng. Chem., 29,* 1415 (1937)] on the absorption of ammonia from an ammonia/air system by a strong acid in a wetted-wall column 0.575 in. in diameter and 32.5 in. long:

Inlet acid (2 N H_2SO_4) temperature, °F	76
Outlet acid temperature, °F	81
Inlet air temperature, °F	77
Outlet air temperature, °F	84
Total pressure, atm	1.00
Partial pressure NH_3 in inlet gas, atm	0.0807
Partial pressure NH_3 in outlet gas, atm	0.0205
Air rate, lbmol/h	0.260

The operation was countercurrent, with the gas entering at the bottom of the vertical tower and the acid passing down in a thin film on the inner wall. The change in acid strength was inappreciable, and the vapor pressure of ammonia over the liquid may be assumed to have been negligible because of the use of a strong acid for absorption. Calculate the mass transfer coefficient, k_g from the data.

3.40 A new type of cooling-tower packing is being tested in a laboratory column. At two points in the column, 0.7 ft apart, the following data have been taken. Calculate the overall mass transfer coefficient K_y that can be used to design a large, packed-bed cooling tower.

	Bottom	Top
Water temperature, °F	120	126
Water vapor pressure, psia	1.69	1.995
Mole fraction H_2O in air	0.001609	0.0882
Total pressure, psia	14.1	14.3
Air rate, lbmol/h	0.401	0.401
Column area, ft²	0.5	0.5
Water rate, lbmol/h (approximate)	20	20

Chapter 4

Single Equilibrium Stages and Flash Calculations

The simplest separation process is one in which two phases in contact are brought to physical equilibrium, followed by phase separation. If the separation factor between two species in the two phases is very large, a single contacting stage may be sufficient to achieve a desired separation between them; if not, multiple stages are required. For example, if a vapor phase is in equilibrium with a liquid phase, the separation factor is the relative volatility, α, of a volatile component called the light key, LK, with respect to a less-volatile component called the heavy key, HK, where $\alpha_{LK,HK} = K_{LK}/K_{HK}$. If the separation factor is 10,000, an almost perfect separation is achieved in a single stage. If the separation factor is only 1.10, an almost perfect separation requires hundreds of stages. In this chapter, only a single equilibrium stage is considered, but a wide spectrum of separation operations is described. In all cases, the calculations are made by combining material balances with phase equilibria relations. When a phase change such as vaporization occurs, or when heat of mixing effects are large, an energy balance must be added. In the next chapter, arrangements of multiple stages, called cascades, are described.

4.1 THE GIBBS PHASE RULE AND DEGREES OF FREEDOM

The description of a single-stage system at physical equilibrium involves *intensive variables*, which are independent of the size of the system, and *extensive variables,* which depend on system size. Intensive variables are temperature, pressure, and phase compositions (mole fractions, mass fractions, concentrations, etc.). Extensive variables include mass or moles and energy for a batch system, and mass or molar flow rates and energy transfer rates for a flow system.

Regardless of whether only intensive variables or both intensive and extensive variables are considered, only a few of the variables are independent; when these are specified, all other variables become fixed. The number of independent variables is the *variance,* or the number of *degrees of freedom, $\mathscr{F}$,* for the system.

The phase rule of J. Willard Gibbs, which applies only to the intensive variables at equilibrium, states that

$$\mathscr{F} = C - \mathscr{P} + 2 \tag{4-1}$$

where C is the number of components and $\mathscr{P}$ is the number of phases at equilibrium. Equation (4-1) is derived by counting, at physical equilibrium, the number of intensive variables and the number of independent equations that relate these variables. The number of intensive variables, $\mathcal{V}$, is

$$\mathcal{V} = C\mathcal{P} + 2 \qquad\qquad \textbf{(4-2)}$$

where the 2 refers to the equilibrium temperature and pressure, while the term $C\mathcal{P}$ is the total number of composition variables (e.g., mole fractions) for components distributed among $\mathcal{P}$ equilibrium phases. The number of independent equations, $\mathcal{E}$, relating the intensive variables is

$$\mathcal{E} = \mathcal{P} + C(\mathcal{P} - 1) \qquad\qquad \textbf{(4-3)}$$

where the first term, $\mathcal{P}$, refers to the requirement that mole or mass fractions sum to one for each phase and the second term, $C(\mathcal{P} - 1)$, refers to the number of independent K-value equations of the general form

$$K_i = \text{mole fraction of } i \text{ in phase (1)/mole fraction of } i \text{ in phase (2)}$$

where (1) and (2) refer to equilibrium phases. For two phases, there are C independent expressions of this type; for three phases, $2C$; for four phases, $3C$; and so on. For example, for three phases $(V, L^{(1)}, L^{(2)})$, we can write $3C$ different K-value equations:

$$
\begin{aligned}
K_i^{(1)} &= y_i/x_i^{(1)} & i &= 1 \text{ to } C \\
K_i^{(2)} &= y_i/x_i^{(2)} & i &= 1 \text{ to } C \\
K_{D_i} &= x_i^{(1)}/x_i^{(2)} & i &= 1 \text{ to } C
\end{aligned}
$$

However, only $2C$ of these equations are independent, because

$$K_{D_i} = K_i^{(2)}/K_i^{(1)}$$

Thus, the term for the number of independent K-value equations is $C(\mathcal{P} - 1)$, not $C\mathcal{P}$.

Degrees-of-Freedom Analysis

The degrees of freedom is the number of intensive variables, $\mathcal{V}$, less the number of equations, $\mathcal{E}$. Thus, from (4-2) and (4-3),

$$\mathcal{F} = \mathcal{V} - \mathcal{E} = (C\mathcal{P} + 2) - [\mathcal{P} + C(\mathcal{P} - 1)] = C - \mathcal{P} + 2$$

which completes the derivation of (4-1). When the number, $\mathcal{F}$, of intensive variables is specified, the remaining $\mathcal{P} + C(\mathcal{P} - 1)$ intensive variables are determined from the $\mathcal{P} + C(\mathcal{P} - 1)$ equations.

As an example, consider the vapor–liquid equilibrium ($\mathcal{P} = 2$) shown in Figure 4.1a, where the equilibrium intensive variables are labels on the sketch located above the list of independent equations relating these variables. Suppose there are $C = 3$ components. From (4-1), $\mathcal{F} = 3 - 2 + 2 = 3$. The equilibrium intensive variables are T, P, x_1, x_2, x_3, y_1, y_2, and y_3. If values are specified for T, P, and one of the mole fractions, the remaining five mole fractions are fixed and can be computed from the five independent equations listed in Figure 4.1a. Irrational specifications lead to infeasible results. For example, if the components are H_2O, N_2, and O_2, and $T = 100°F$ and $P = 15$ psia are specified, a specification of $x_{N_2} = 0.90$ is not feasible because nitrogen is not nearly this soluble in water.

In using the Gibbs phase rule, it should be noted that the K-values are not variables, but are thermodynamic functions that depend on the intensive variables discussed in Chapter 2.

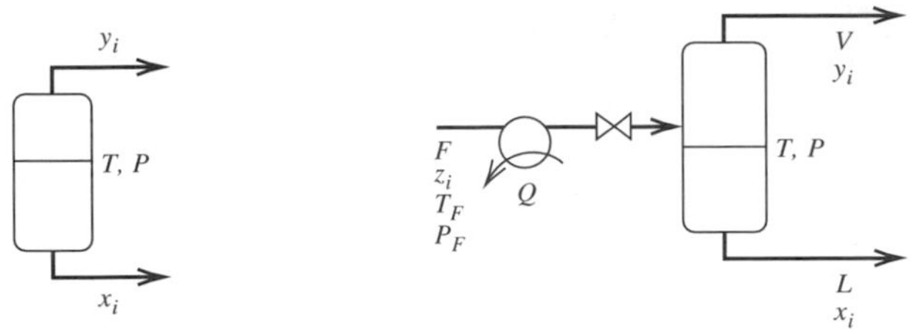

Independent equations:

$$\sum_{i=1}^{C} y_i = 1$$

$$\sum_{i=1}^{C} x_i = 1$$

$$K_i = \frac{y_i}{x_i}, \quad i = 1 \text{ to } C$$

(a)

Independent equations:
Same as for (a) plus

$$F_{z_i} = V_{y_i} + L_{x_i}, \quad i = 1 \text{ to } C$$

$$F h_F + Q = V h_V + L h_L$$

(b)

Figure 4.1 Different treatments of degrees of freedom for vapor–liquid phase equilibria: (a) Gibbs phase rule (considers equilibrium intensive variables only); (b) general analysis (considers all intensive and extensive variables).

The Gibbs phase rule is limited because it does not deal with feed streams sent to the equilibrium stage nor with extensive variables used when designing or analyzing separation operations. However, the phase rule can be extended for process applications, by adding the feed stream and extensive variables, and additional independent equations relating feed variables, extensive variables, and the intensive variables already considered by the rule.

Consider the single-stage, vapor–liquid ($\mathscr{P} = 2$) equilibrium separation process shown in Figure 4.1b. By comparison with Figure 4.1a, the additional variables are z_i, T_F, P_F, F, Q, V, and L, or $C + 6$ variables, all of which are indicated in the diagram. In general, for $\mathscr{P}$ phases, the additional variables number $C + \mathscr{P} + 4$. The additional independent equations, listed below the diagram, are the C component material balances and the energy balance, or $C + 1$ equations. Note that, like K-values, stream enthalpies are not counted as variables because they are thermodynamic functions that depend on intensive variables.

For the general degrees-of-freedom analysis for phase equilibrium, with C components, $\mathscr{P}$ phases, and a single feed phase, (4-2) and (4-3) are extended by adding the number of additional variables and equations, respectively:

$$\mathscr{V} = (C\mathscr{P} + 2) + (C + \mathscr{P} + 4) = \mathscr{P} + C\mathscr{P} + C + 6$$

$$\mathscr{E} = [\mathscr{P} + C(\mathscr{P} - 1)] + (C + 1) = \mathscr{P} + C\mathscr{P} + 1 \qquad \textbf{(4-4)}$$

$$\mathscr{F} = \mathscr{V} - \mathscr{E} = C + 5$$

For example, if the $C + 5$ degrees of freedom are used to specify all z_i and the five variables F, T_F, P_F, T, and P, the remaining variables are computed from the equations shown in Figure 4.1. To apply the Gibbs phase rule, (4-1), the number of phases must be known. When applying (4-4), the determination of the number of equilibrium phases, $\mathscr{P}$, is implicit in the computational procedure as illustrated in later sections of this chapter.

In the following sections, the Gibbs phase rule, (4-1), and the equation for the number of degrees of freedom of a flow system, (4-4), are applied to (1) tabular equilibrium data, (2) graphical equilibrium data, or (3) thermodynamic equations for K-values and enthalpies for vapor–liquid, liquid–liquid, solid–liquid, gas–liquid, gas–solid, vapor–liquid–solid, and vapor–liquid–liquid systems at equilibrium.

4.2   BINARY VAPOR–LIQUID SYSTEMS

Experimental vapor–liquid equilibrium data for systems containing two components, A and B, are widely available. Sources include *Perry's Handbook* [1] and Gmehling and Onken [2]. Because $y_B = 1 - y_A$ and $x_B = 1 - x_A$, the data are presented in terms of just four intensive variables: T, P, y_A, and x_A. Most commonly T, y_A, and x_A are tabulated for a fixed P for ranges of y_A and x_A from 0 to 1, where A is the more volatile component $(y_A > x_A)$. However, if an azeotrope (see Section 4.3) forms, B becomes the more volatile component on one side of the azeotropic point. By the Gibbs phase rule, (4-1), $\mathscr{F} = 2 - 2 + 2 = 2$. Thus, with pressure fixed, phase compositions are completely defined if temperature is also fixed, and the separation factor, that is, the relative volatility in the case of vapor–liquid equilibria,

$$\alpha_{A,B} = \frac{K_A}{K_B} = \frac{(y_A/x_A)}{(y_B/x_B)} = \frac{(y_A/x_A)}{(1 - y_A)/(1 - x_A)} \tag{4-5}$$

is also fixed.

Vapor–liquid equilibria data of the form $T - y_A - x_A$ for 1 atm pressure of three binary systems of industrial importance are given in Table 4.1. Included are values of relative volatility computed from (4-5). As discussed in Chapter 2, $\alpha_{A,B}$ depends on T, P, and the compositions of the equilibrium vapor and liquid. At 1 atm, where $\alpha_{A,B}$ is approximated well by $\gamma_A P_A^s / \gamma_B P_B^s$, $\alpha_{A,B}$ depends only on T and x_A, since vapor-phase nonidealities are small. Because of the dependency on x_A, $\alpha_{A,B}$ is not a constant, but varies from point to point. For the three binary systems in Table 4.1, the vapor and liquid phases become richer in the less volatile component, B, as temperature increases. For $x_A = 1$, the temperature is the normal boiling point of A; for $x_A = 0$, the temperature is the normal boiling point of B. For the three systems, all other data points are at temperatures between the two boiling points. Except for the pure components ($x_A = 1$ or 0), $y_A > x_A$ and $\alpha_{A,B} > 1$.

For the water–glycerol system, the difference in normal boiling points is 190°C. Therefore, relative volatility values are very high, making it possible to achieve a reasonably good separation in a single equilibrium stage. Industrially, the separation is often conducted in an evaporator, which produces a nearly pure water vapor and a glycerol-rich liquid. For example, from Table 4.1, at 207°C, a vapor of 98 mol% water is in equilibrium with a liquid phase containing more than 90 mol% glycerol.

For the methanol–water system, the difference in normal boiling points is 35.5°C. As a result, the relative volatility is an order of magnitude lower than for the water–glycerol system. A sharp separation cannot be made with a single stage. About 30 trays are required in a distillation operation to obtain a 99 mol% methanol distillate and a 98 mol% water bottoms, an acceptable industrial separation.

For the aromatic paraxylene-metaxylene isomer system, the normal boiling-point difference is only 0.8°C. Thus, the relative volatility is very close to 1.0, making the separation by distillation impractical because about 1,000 trays are required to produce nearly pure products. Instead, crystallization and adsorption, which have much higher separation factors, are used commercially to make the separation.

Experimental vapor–liquid equilibrium data for the methanol–water system are given in Table 4.2 in the form of $P-y_A-x_A$ for fixed temperatures of 50, 150, and 250°C. The three sets of data cover a pressure range of 1.789 to 1,234 psia, with the higher pressures corresponding to the higher temperatures. At 50°C, relative volatilities are moderately high at an average value of 4.94 over the composition range. At 150°C, the average relative volatility is only 3.22; and at 250°C, it decreases to 1.75. Thus, as the temperature and pressure increase, the relative volatility decreases significantly. In Table 4.2, for the data set at 250°C, it is seen that as the compositions become richer in methanol, a point is reached in the neighborhood of 1,219 psia, at a methanol mole fraction of 0.772, where

Table 4.1 Vapor–Liquid Equilibrium Data
for Three Common Binary Systems at
1 atm Pressure

a. Water (A)–Glycerol (B) System
$P = 101.3$ kPa
Data of Chen and Thompson, *J. Chem. Eng.
Data,* **15,** 471 (1970)

Temperature, °C	y_A	x_A	$\alpha_{A,B}$
100.0	1.0000	1.0000	
101.7	0.9998	0.9534	250
104.6	0.9996	0.8846	333
109.8	0.9991	0.7731	332
122.0	0.9982	0.5610	440
128.8	0.9980	0.4742	544
140.4	0.9974	0.3622	683
148.2	0.9964	0.3077	627
158.6	0.9940	0.2460	507
175.2	0.9898	0.1756	456
207.0	0.9804	0.0945	481
244.5	0.9341	0.0491	275
282.5	0.8308	0.0250	191
290.0	0.0000	0.0000	

b. Methanol (A)–Water (B) System
$P = 101.3$ kPa
Data of J.G. Dunlop, M.S. thesis, Brooklyn
Polytechnic Institute (1948)

Temperature, °C	y_A	x_A	$\alpha_{A,B}$
64.5	1.000	1.000	
65.0	0.979	0.950	2.45
66.0	0.958	0.900	2.53
67.5	0.915	0.800	2.69
69.3	0.870	0.700	2.87
71.2	0.825	0.600	3.14
73.1	0.779	0.500	3.52
75.3	0.729	0.400	4.04
78.0	0.665	0.300	4.63
81.7	0.579	0.200	5.50
84.4	0.517	0.150	6.07
87.7	0.418	0.100	6.46
89.3	0.365	0.080	6.61
91.2	0.304	0.060	6.84
93.5	0.230	0.040	7.17
96.4	0.134	0.020	7.58
100.0	0.000	0.000	

c. *Para*-xylene (A)–*Meta*-xylene (B) System
$P = 101.3$ kPa
Data of Kato, Sato, and Hirata, *J. Chem.
Eng. Jpn.,* **4,** 305 (1970)

Temperature, °C	y_A	x_A	$\alpha_{A,B}$
138.335	1.0000	1.0000	
138.414	0.9019	0.9000	1.0021
138.491	0.8033	0.8000	1.0041
138.568	0.7043	0.7000	1.0061
138.644	0.6049	0.6000	1.0082
138.720	0.5051	0.5000	1.0102
138.795	0.4049	0.4000	1.0123
138.869	0.3042	0.3000	1.0140
138.943	0.2032	0.2000	1.0160
139.016	0.1018	0.1000	1.0180
139.088	0.0000	0.0000	

Table 4.2 Vapor–Liquid Equilibrium Data
for the Methanol–Water
System at Temperatures of 50,
150, and 250°C

a. Methanol (A)–Water (B) System
   $T = 50°C$
   Data of McGlashan and Williamson, *J.
   Chem. Eng. Data,* **21,** 196 (1976)

Pressure, psia	y_A	x_A	$\alpha_{A,B}$
1.789	0.0000	0.0000	
2.373	0.2661	0.0453	7.64
2.838	0.4057	0.0863	7.23
3.369	0.5227	0.1387	6.80
3.764	0.5898	0.1854	6.32
4.641	0.7087	0.3137	5.32
5.163	0.7684	0.4177	4.63
5.771	0.8212	0.5411	3.90
6.122	0.8520	0.6166	3.58
6.811	0.9090	0.7598	3.16
7.280	0.9455	0.8525	3.00
7.800	0.9817	0.9514	2.74
8.072	1.0000	0.0000	

b. Methanol (A)–Water (B) System
   $T = 150°C$
   Data of Griswold and Wong, *Chem.
   Eng. Prog. Symp. Ser.,* **48** (3), 18 (1952)

Pressure, psia	y_A	x_A	$\alpha_{A,B}$
73.3	0.060	0.009	7.03
79.0	0.135	0.022	6.94
85.7	0.213	0.044	5.88
93.9	0.286	0.079	4.67
114.9	0.459	0.186	3.71
139.7	0.610	0.374	2.62
148.6	0.662	0.459	2.31
160.4	0.731	0.578	1.98
177.4	0.832	0.748	1.67
193.5	0.929	0.893	1.57
194.5	0.943	0.913	1.58
196.5	0.960	0.936	1.64
197.7	0.972	0.953	1.71
199.2	0.982	0.969	1.75

c. Methanol (A)–Water (B) System
   $T = 250°C$
   Data of Griswold and Wong, *Chem.
   Eng. Prog. Symp. Ser.,* **48** (3), 18 (1952)

Pressure, psia	y_A	x_A	$\alpha_{A,B}$
681	0.163	0.066	2.76
764	0.280	0.132	2.56
818	0.344	0.180	2.39
889	0.423	0.254	2.15
949	0.487	0.331	1.92
994	0.542	0.404	1.75
1049	0.596	0.483	1.58
1099	0.643	0.553	1.46
1159	0.698	0.631	1.35
1204	0.756	0.732	1.13
1219	0.772	0.772	1.00
1234	0.797	0.797	1.00

the relative volatility is 1.0 and no separation by distillation is possible because the compositions of the vapor and liquid are identical and the two phases become one phase. This is the critical point of a mixture of this composition. It is intermediate between the critical points of pure methanol and pure water:

$y_A = x_A$	T_c, °C	P_c, psia
0.000	374.1	3,208
0.772	250	1,219
1.000	240	1,154

A set of critical conditions exists for each binary-mixture composition. In industry, distillation columns operate at pressures well below the critical pressure of the mixture to be separated to avoid relative volatilities that approach a value of 1.

The data of Tables 4.1 and 4.2 for the methanol–water system are plotted in three different ways in Figure 4.2: (a) T versus y_A or x_A at $P = 1$ atm; (b) y_A versus x_A at $P = 1$ atm; and (c) P versus x_A at $T = 150$°C. These three plots all satisfy the requirement of the Gibbs phase rule that when two intensive variables are fixed, all other variables are fixed by the governing equilibrium equations and mole-fraction-summation constraints. Of the three diagrams in Figure 4.2, only (a) contains the complete data; (b) does not contain temperatures; and (c) does not contain vapor-phase mole fractions. Although mass fractions could be used in place of mole fractions, the latter are preferred because theoretical phase equilibrium relations are based on molar properties.

Plots like Figure 4.2a are useful for determining phase states, phase-transition temperatures, equilibrium phase compositions, and equilibrium phase amounts for a given feed of known composition. Consider the T–y–x plot in Figure 4.3 for the normal hexane (H)–normal octane (O) system at 101.3 kPa. Because normal hexane is the more volatile species, the mole fractions are for that component. The upper curve, labeled "saturated vapor," gives the dependency on the dew-point temperature of the vapor mole fraction y_H; the lower curve, labeled "saturated liquid," gives the dependency of the bubble-point temperature on the liquid-phase mole fraction, x_H. The two curves converge at $x_H = 0$, the normal boiling point of pure normal octane (258.2°F), and at $x_H = 1$, the normal boiling point of normal hexane (155.7°F). In order for two phases to exist, a point representing the overall composition of the two-phase binary mixture at a given temperature must be located in the two-phase region between the two curves. If the point lies above the saturated vapor curve, only a superheated vapor is present; if the point lies below the saturated liquid curve, only a subcooled liquid exists.

Suppose we have a mixture of 30 mol% H at 150°F. From Figure 4.3, at point A we have a subcooled liquid with $x_H = 0.3$ ($x_O = 0.7$). When this mixture is heated at a constant pressure of 1 atm, the liquid state is maintained until a temperature of 210°F is reached, which corresponds to point B on the saturated liquid curve. Point B is the *bubble point* because the first bubble of vapor appears. This bubble is a saturated vapor in equilibrium with the liquid at the same temperature. Thus, its composition is determined by following a *tie line,* BC from $x_H = 0.3$ to $y_H = 0.7$ ($y_O = 0.3$). The tie line is horizontal because the temperatures of the two equilibrium phases are the same. As the temperature of the two-phase mixture is increased to point E, on horizontal tie line DEF at 225°F, the mole fraction of H in the liquid phase decreases to $x_H = 0.17$ (because it is more volatile than O and preferentially vaporizes) and correspondingly the mole fraction of H in the vapor phase increases to $y_H = 0.55$. Throughout the two-phase region, the vapor is at its dew point, while the liquid is at its bubble point. The overall composition of the two phases remains at a mole fraction of 0.30 for hexane. At point E, the relative molar amounts of the two equilibrium phases is determined by the inverse lever-arm rule based on the lengths of the line segments DE and EF. Thus, referring to Figures 4.1b and 4.3, $V/L = DE/EF$

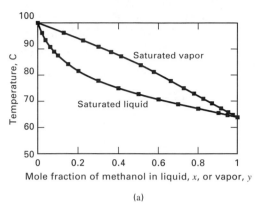

(a)

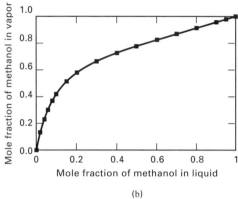

(b)

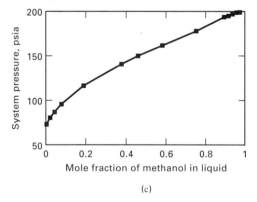

(c)

Figure 4.2 Vapor–liquid equilibrium conditions for the methanol–water system: (a) T–x diagram for 1 atm pressure; (b) y–x diagram for 1 atm pressure; (c) P–x diagram for 150°C.

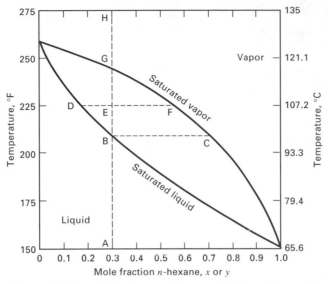

Figure 4.3 Use of the *T–y–x* phase equilibrium diagram for the normal hexane–normal octane system at 1 atm.

or $V/F = $ DE/DEF. When the temperature is increased to 245°F, point G, the dew point for $y_H = 0.3$ is reached, where only one droplet of equilibrium liquid remains with a composition from the tie line FG at point $\overline{F}$ of $x_H = 0.06$. A further increase in temperature—say, to point H at 275°F—gives a superheated vapor with $y_H = 0.30$. The steps are reversible starting from point H and moving down to point A.

Constant-pressure *x–y* plots like Figure 4.2b are also useful because the equilibrium vapor and liquid compositions are represented by points on the equilibrium curve. However, no phase–temperature information is included. Such plots usually include a 45° reference line, $y = x$. Consider the *y–x* plot in Figure 4.4 for H–O at 101.3 kPa. This plot is convenient for determining equilibrium phase compositions for various values of mole percent vaporization of a feed mixture of a given composition by geometric constructions.

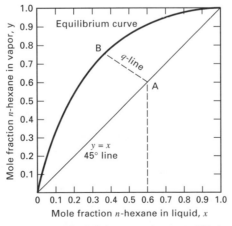

Figure 4.4 Use of the *y–x* phase equilibrium diagram for the normal hexane–normal octane system at 1 atm.

Suppose we have a feed mixture, F, shown in Figure 4.1b, of overall composition $z_H = 0.6$. To determine the equilibrium phase compositions if, say, 60 mol% of the feed is vaporized, we develop the dashed-line construction in Figure 4.4. Point A on the 45° line represents z_H. Point B on the equilibrium curve is reached by extending a line, called the *q-line,* upward and to the left toward the equilibrium curve at a slope equal to $[(V/F) - 1]/(V/F)$. Thus, for 60 mol% vaporization, the slope = $(0.6 - 1)/0.6 = -\frac{2}{3}$. Point B at the intersection of line AB with the equilibrium curve gives the equilibrium composition as $y_H = 0.76$ and $x_H = 0.37$. This computation requires a trial-and-error placement of a horizontal line if we use Figure 4.3. The derivation of the slope of the q-line in Figure 4.4 follows by combining the mole balance equation of Figure 4.1a,

$$Fz_H = Vy_H + Lx_H$$

with the total mole balance,

$$F = V + L$$

to eliminate L, giving the equation for the q-line:

$$y_H = \left[\frac{(V/F) - 1}{(V/F)}\right] x_H + \left[\frac{1}{(V/F)}\right] z_H$$

Thus, the slope of the q-line passing through the equilibrium point (y_H, x_H), is $[(V/F) - 1]/(V/F)$.

Figure 4.2c is the least used of the three plots in Figure 4.2. However, such a plot does illustrate, for a fixed temperature, the extent to which the binary mixture deviates from an ideal solution. If Raoult's law applies, the total pressure above the liquid is

$$
\begin{aligned}
P &= P_A^s x_A + P_B^s x_B \\
&= P_A^s x_A + P_B^s (1 - x_A) \\
&= P_B^s + x_A (P_A^s - P_B^s)
\end{aligned}
\tag{4-6}
$$

Thus, a plot of P versus x_A is a straight line with intersections at the vapor pressure of B for $x_A = 0$ and the vapor pressure of A for $x_B = 0$ ($x_A = 1$). The greater the departure from a straight line, the greater is the deviation from the assumptions of an ideal gas and/ or an ideal liquid solution. If the pressures are sufficiently low that the equilibrium vapor phase is ideal and the curve is convex, deviations from Raoult's law are positive, and species liquid-phase activity coefficients are greater than 1; if the curve is concave, deviations are negative and activity coefficients are less than 1. In either case, the total pressure is given by

$$P = \gamma_A P_A^s x_A + \gamma_B P_B^s x_B \tag{4-7}$$

If the vapor does not obey the ideal gas law, (4-7) does not apply. In Figure 4.2c, system pressures are sufficiently high that some deviation from the ideal gas law occurs. However, the convexity is due mainly to activity coefficients that are greater than 1.

For relatively close (narrow)-boiling binary mixtures that exhibit ideal or nearly ideal behavior, the relative volatility, $\alpha_{A,B}$, varies little with pressure. If $\alpha_{A,B}$ is assumed constant over the entire composition range, the y–x phase equilibrium curve can be determined and plotted from a rearrangement of (4-5):

$$y_A = \frac{\alpha_{A,B} x_A}{1 + x_A (\alpha_{A,B} - 1)} \tag{4-8}$$

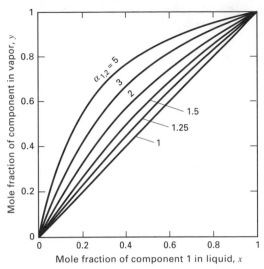

Figure 4.5 Vapor–liquid phase equilibrium curves for constant values of relative volatility.

For an ideal solution, $\alpha_{A,B}$ can be approximated with Raoult's law to give

$$\alpha_{A,B} = \frac{K_A}{K_B} = \frac{P_A^s/P}{P_B^s/P} = \frac{P_A^s}{P_B^s} \tag{4-9}$$

Thus, from a knowledge of just the vapor pressures of the two components at a temperature, say, midway between the two boiling points at the given pressure, a y–x phase equilibrium curve can be approximated using only one value of $\alpha_{A,B}$. Families of curves, as shown in Figure 4.5, can be used for preliminary calculations in the absence of detailed experimental data. The use of (4-8) and (4-9) is not recommended for wide-boiling or nonideal mixtures.

4.3 AZEOTROPIC SYSTEMS

Departures from Raoult's law frequently manifest themselves in the formation of *azeotropes,* particularly for mixtures of close-boiling species of different chemical types whose liquid solutions are nonideal. Azeotropes are formed by liquid mixtures exhibiting maximum or minimum boiling points. These represent, respectively, negative or positive deviations from Raoult's law. Vapor and liquid compositions are identical at the azeotropic composition; thus, all K-values are 1 and no separation of species can take place.

If only one liquid phase exists, the mixture forms a *homogeneous* azeotrope; if more than one liquid phase is present, the azeotrope is *heterogeneous*. In accordance with the Gibbs phase rule, at constant pressure in a two-component system, the vapor can coexist with no more than two liquid phases, while in a ternary mixture up to three liquid phases can coexist with the vapor.

Figures 4.6, 4.7, and 4.8 show three types of azeotropes that are commonly encountered with binary mixtures. The most common type by far is the minimum-boiling homogeneous azeotrope, illustrated in Figure 4.6 for the isopropyl ether–isopropyl alcohol system. In Figure 4.6a, for a temperature of 70°C, the maximum total pressure is greater than the vapor pressure of either component because activity coefficients are greater than 1. The y–x diagram in Figure 4.6b shows that for a pressure of 1 atm the azeotropic mixture occurs at 78 mol% ether. Figure 4.6c is a T–x diagram for a pressure of 101 kPa, where the azeotrope is seen to boil at 66°C. In Figure 4.6a, for 70°C, the azeotrope, at 123 kPa

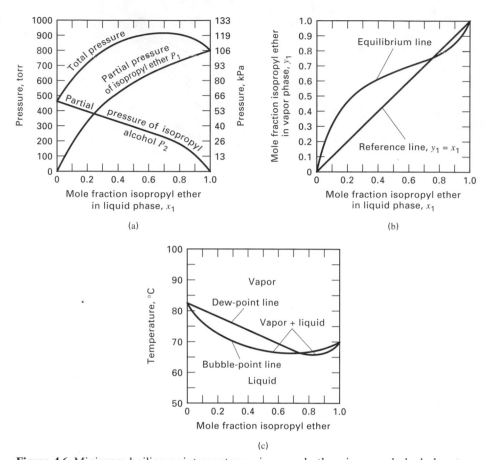

Figure 4.6 Minimum-boiling-point azeotrope, isopropyl ether–isopropyl alcohol system: (a) partial and total pressures at 70°C; (b) vapor–liquid equilibria at 101 kPa; (c) phase diagram at 101 kPa. [Adapted from O.A. Hougen, K.M. Watson, and R.A. Ragatz, *Chemical Process Principles. Part II,* 2nd ed., John Wiley and Sons, New York (1959).]

(923 torr), is 72 mol% ether. Thus, the azeotropic composition shifts with pressure. In distillation, the minimum-boiling azeotropic mixture is the overhead product.

For the maximum-boiling homogeneous azeotropic acetone–chloroform system in Figure 4.7a, the minimum total pressure is below the vapor pressures of the pure components because activity coefficients are less than 1. The azeotrope concentrates in the bottoms in a distillation operation.

Heterogeneous azeotropes are always minimum-boiling mixtures because activity coefficients must be significantly greater than 1 to cause splitting into two liquid phases. The region a–b in Figure 4.8a for the water–normal butanol system is a two-phase region where total and partial pressures remain constant as the relative amounts of the two phases change, but the phase compositions do not. The y–x diagram in Figure 4.8b shows a horizontal line over the immiscible region, and the phase diagram of Figure 4.8c shows a minimum constant temperature.

Azeotropes limit the separation achievable by ordinary distillation. It is possible to shift the equilibrium by changing the pressure sufficiently to "break" the azeotrope, or move it away from the region where the required separation must be made. For example, ethyl alcohol and water form a homogeneous minimum-boiling azeotrope of 95.6 wt% alcohol

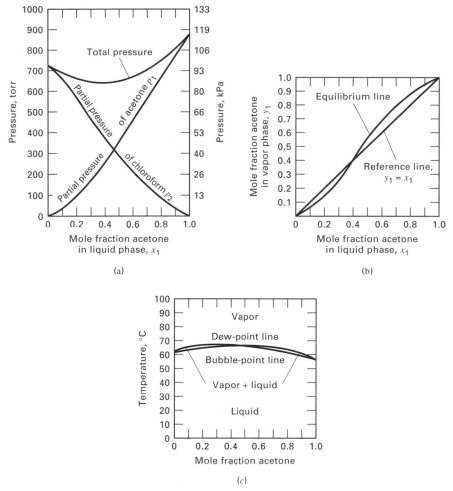

Figure 4.7 Maximum-boiling-point azeotrope, acetone–chloroform system: (a) partial and total pressures at 60°C; (b) vapor–liquid equilibria at 101 kPa; (c) phase diagram at 101 kPa pressure. [Adapted from O.A. Hougen, K.M. Watson, and R.A. Ragatz, *Chemical Process Principles. Part II,* 2nd ed., John Wiley and Sons, New York (1959).]

at 78.15°C and 101.3 kPa. However, at vacuums of less than 9.3 kPa, no azeotrope is formed. Ternary azeotropes also occur, and these offer the same barrier to complete separation as do binary azeotropes.

Azeotrope formation in general, and heterogeneous azeotropes in particular, can be employed to achieve difficult separations. As discussed in Chapter 1, an entrainer is added for the purpose of combining with one or more of the components in the feed to form a minimum-boiling azeotrope, which is then recovered as the distillate.

Figure 4.9 shows the Keyes process [3] for making pure ethyl alcohol by *heterogeneous azeotropic distillation.* Water and ethyl alcohol form a binary minimum-boiling azeotrope containing 95.6 wt% alcohol and boiling at 78.15°C at 101.3 kPa. Thus, it is impossible to obtain pure alcohol (boiling point = 78.40°C) by ordinary distillation at 1 atm. The addition of benzene to an alcohol–water mixture results in the formation of a minimum-boiling heterogeneous ternary azeotrope containing by weight, 18.5% alcohol, 74.1% benzene, and 7.4% water, boiling at 64.85°C. Upon condensation, the ternary azeotrope separates into

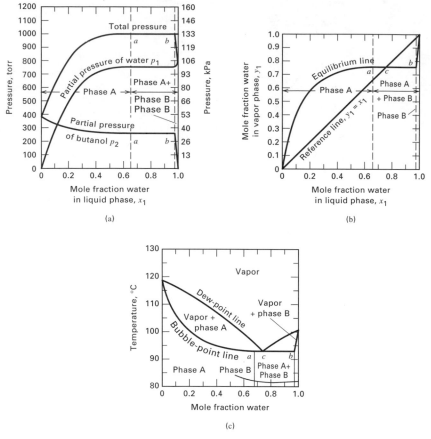

Figure 4.8 Minimum-boiling-point (two liquid phases) water/*n*-butanol system: (a) partial and total pressures at 100°C; (b) vapor–liquid equilibria at 101 kPa; (c) phase diagram at 101 kPa pressure. [Adapted from O.A. Hougen, K.M. Watson, and R.A. Ragatz, *Chemical Process Principles. Part II,* 2nd ed., John Wiley and Sons, New York (1959).]

two liquid layers: a top layer containing 14.5% alcohol, 84.5% benzene, and 1% water, and a bottoms layer of 53% alcohol, 11% benzene, and 36% water, all by weight. The benzene-rich layer is returned as reflux. The other layer is sent to a second distillation column for recovery and recycling of alcohol and benzene. Absolute alcohol, which has a boiling point above that of the ternary azeotrope, is removed at the bottom of the column.

In *extractive distillation,* as discussed in Chapter 1, a solvent is added, usually near the top of the column, to selectively alter the activity coefficients in order to increase the relative volatility between the two species to be separated. The solvent is generally a relatively polar, high-boiling constituent, such as phenol, aniline, or furfural, which concentrates at the bottom of the column.

4.4 MULTICOMPONENT FLASH, BUBBLE-POINT, AND DEW-POINT CALCULATIONS

A *flash* is a single-equilibrium-stage distillation in which a feed is partially vaporized to give a vapor richer in the more volatile components than the remaining liquid. In Figure 4.10a, a liquid feed is heated under pressure and flashed adiabatically across a valve to a

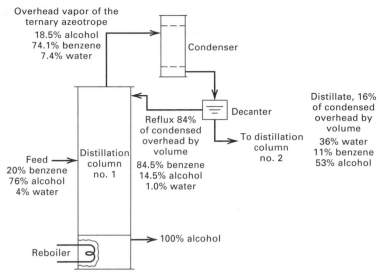

Figure 4.9 The Keyes process for absolute alcohol.

lower pressure, resulting in the creation of a vapor phase that is separated from the remaining liquid in a flash drum. If the valve is omitted, a low-pressure liquid can be partially vaporized in the heater and then separated into two phases in the flash drum. Alternatively, a vapor feed can be cooled and partially condensed, with phase separation in a flash drum, as in Figure 4.10b, to give a liquid that is richer in the less volatile components. In both cases, if the equipment is properly designed, the vapor and liquid leaving the drum are in equilibrium [4].

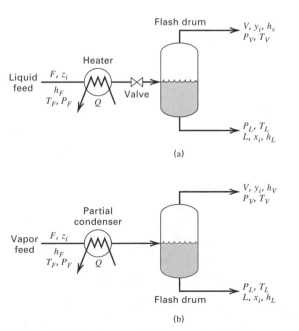

Figure 4.10 Continuous single-stage equilibrium separation: (a) flash vaporization (adiabatic flash with valve, isothermal flash without valve when T_V is specified); (b) partial condensation (analogous to isothermal flash when T_V is specified).

Unless the relative volatility is very large, the degree of separation achievable between two components in a single equilibrium stage is poor. Therefore, flashing (partial vaporization) or partial condensation are usually auxiliary operations used to prepare streams for further processing. Typically, the vapor phase is sent to a vapor separation system, while the liquid phase is sent to a liquid separation system. Computational methods for a single-stage flash calculation are of fundamental importance. Such calculations are used not only for the operations in Figure 4.10, but also to determine, anywhere in a process, the phase condition of a stream or batch of known composition, temperature, and pressure.

For the single-stage equilibrium operation with one feed stream and two product streams, shown in Figure 4.10, the $2C + 5$ equations listed in Table 4.3 apply. They relate the $3C + 10$ variables (F, V, L, z_i, y_i, x_i, T_F, T_V, T_L, P_F, P_V, P_L, Q) and leave $C + 5$ degrees of freedom. Assuming that $C + 3$ feed variables F, T_F, P_F, and C values of z_i are known, two additional variables can be specified. The most common sets of specifications are

T_V, P_V	Isothermal flash
$V/F = 0$, P_L	Bubble-point temperature
$V/F = 1$, P_V	Dew-point temperature
T_L, $V/F = 0$	Bubble-point pressure
T_V, $V/F = 0$	Dew-point pressure
$Q = 0$, P_V	Adiabatic flash
Q, P_V	Nonadiabatic flash
V/F, P_V	Percent vaporization flash

Calculation procedures, described in the following for all these cases, are well known and widely used.

Isothermal Flash

If the equilibrium temperature T_V (or T_L) and the equilibrium pressure P_V (or P_L) of a multicomponent mixture are specified, values of the remaining $2C + 5$ variables are determined from the same number of equations in Table 4.3. The computational procedure, referred to as the *isothermal flash calculation,* is not straightforward because Eq. (4) in Table 4.3 is a nonlinear equation in the unknowns V, L, y_i, and x_i. Many solution strategies have been developed, but the generally preferred procedure, as given in Table 4.4, is that

Table 4.3 Equations for Single-Stage Flash Vaporization and Partial Condensation Operations

Equation		Number of Equations
(1) $P_V = P_L$	(mechanical equilibrium)	1
(2) $T_V = T_L$	(thermal equilibrium)	1
(3) $y_i = K_i x_i$	(phase equilibrium)	C
(4) $Fz_i = Vy_i + Lx_i$	(component material balance)	C
(5) $F = V + L$	(total material balance)	1
(6) $h_F F + Q = h_V V + h_L L$	(energy balance)	1
(7) $\sum_i y_i - \sum_i x_i = 0$	(summations)	1
		$\mathscr{E} = 2C + 5$

$$K_i = K_i\{T_V, P_V, \mathbf{y}, \mathbf{x}\} \qquad h_F = h_F\{T_F, P_F, \mathbf{z}\}$$
$$h_V = h_V\{T_V, P_V, \mathbf{y}\} \qquad h_L = h_L\{T_L, P_L, \mathbf{x}\}$$

Table 4.4 Rachford–Rice Procedure for Isothermal
Flash Calculations When K-Values Are
Independent of Composition

Specified variables: $F, T_F, P_F, z_1, z_2, \ldots, z_C, T_V, P_V$
Steps
(1) $T_L = T_V$
(2) $P_L = P_V$
(3) Solve

$$f\{\Psi\} = \sum_{i=1}^{C} \frac{z_i(1 - K_i)}{1 + \Psi(K_i - 1)} = 0$$

for $\Psi = V/F$, where $K_i = K_i\{T_V, P_V\}$.
(4) $V = F\Psi$

(5) $x_i = \dfrac{z_i}{1 + \Psi(K_i - 1)}$

(6) $y_i = \dfrac{z_i K_i}{1 + \Psi(K_i - 1)} = x_i K_i$

(7) $L = F - V$
(8) $Q = h_V V + h_L L - h_F F$

---

of Rachford and Rice [5] when K-values are independent (or nearly independent) of equilibrium-phase compositions.

Equations containing only a single unknown are solved first. Thus, Eqs. (1) and (2) in Table 4.3 are solved respectively for P_L and T_L. The unknown Q appears only in Eq. (6), so Q is computed only after all other equations have been solved. This leaves Eqs. (3), (4), (5), and (7) in Table 4.3 to be solved for V, L, and all values of y and x. These equations can be partitioned so as to solve for the unknowns in a sequential manner by substituting Eq. (5) into Eq. (4) to eliminate L and combining the result with Eq. (3) to obtain Eqs. (5) and (6) in Table 4.4. Here (5) is in x_i, but not y_i, and (6) is in x_i but not y_i. Summing these two equations and combining them with $\Sigma y_i - \Sigma x_i = 0$ to eliminate y_i and x_i gives Eq. (3) in Table 4.4; a nonlinear equation in V (or $\Psi = V/F$) only. Upon solving this equation numerically in an iterative manner for Ψ and then V, from Eq. (4) of Table 4.4, one can obtain the remaining unknowns directly from Eqs. (5) through (8) in Table 4.4. When T_F and/or P_F are not specified, Eq. (6) of Table 4.3 is not solved for Q. By this isothermal flash procedure, the equilibrium phase condition of a mixture at a known temperature ($T_V = T_L$) and pressure ($P_V = P_L$) is determined.

Equation (3) of Table 4.4 can be solved iteratively by guessing values of Ψ between 0 and 1 until the function $f\{\Psi\} = 0$. A typical form of the function, as will be computed in Example 4.1, is shown in Figure 4.11. The most widely employed numerical method for solving Eq. (3) of Table 4.4 is Newtons method [6]. A predicted value of the Ψ root for iteration $k + 1$ is computed from the recursive relation

$$\Psi^{(k+1)} = \Psi^{(k)} - \frac{f\{\Psi^{(k)}\}}{f'\{\Psi^{(k)}\}} \tag{4-10}$$

where the superscript is the iteration index, and the derivative of $f\{\Psi\}$, from Eq. (3) in Table 4.4, with respect to Ψ is

$$f'\{\Psi^{(k)}\} = \sum_{i=1}^{C} \frac{z_i(1 - K_i)^2}{[1 + \Psi^{(k)}(K_i - 1)]^2} \tag{4-11}$$

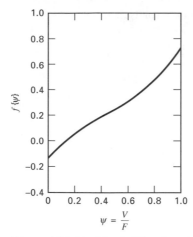

Figure 4.11 Rachford–Rice function for Example 4.1.

The iteration can be initiated by assuming $\Psi^{(1)} = 0.5$. Sufficient accuracy will be achieved by terminating the iterations when $|\Psi^{(k+1)} - \Psi^{(k)}|/\Psi^{(k)} < 0.0001$.

One should check the existence of a valid root ($0 \le \Psi \le 1$), before employing the procedure of Table 4.4, by checking to see if the equilibrium condition corresponds to subcooled liquid or superheated vapor rather than partial vaporization or partial condensation. A first estimate of whether a multicomponent feed gives a two-phase equilibrium mixture when flashed at a given temperature and pressure can be made by inspecting the K-values. If all K-values are greater than 1, the exit phase is superheated vapor above the dew point. If all K-values are less than 1, the single exit phase is a subcooled liquid below the bubble point. If one or more K-values are greater than 1 and one or more K-values are less than 1, the check is made as follows. First, $f\{\Psi\}$ is computed from Eq. (3) for $\Psi = 0$. If the resulting $f\{0\} > 0$, the mixture is below its bubble point (subcooled liquid). Alternatively if $f\{1\} < 0$, the mixture is above the dew point (superheated vapor).

EXAMPLE 4.1

A 100-kmol/h feed consisting of 10, 20, 30, and 40 mol% of propane (3), n-butane (4), n-pentane (5), and n-hexane (6), respectively, enters a distillation column at 100 psia (689.5 kPa) and 200°F (366.5°K). Assuming equilibrium, what mole fraction of the feed enters as liquid, and what are the liquid and vapor compositions?

SOLUTION

At flash conditions, from Figure 2.8, $K_3 = 4.2$, $K_4 = 1.75$, $K_5 = 0.74$, $K_6 = 0.34$, independent of compositions. Because some K-values are greater than 1 and some are less than 1, it is first necessary to compute values of $f\{0\}$ and $f\{1\}$ for Eq. (3) in Table 4.4 to determine if the mixture is between the bubble point and the dew point.

$$f\{0\} = \frac{0.1(1 - 4.2)}{1} + \frac{0.2(1 - 1.75)}{1} + \frac{0.3(1 - 0.74)}{1} + \frac{0.4(1 - 0.34)}{1} = -0.128$$

Since $f\{0\}$ is not greater than zero, the mixture is above the bubble point.

$$f\{1\} = \frac{0.1(1 - 4.2)}{1 + (4.2 - 1)} + \frac{0.2(1 - 1.75)}{1 + (1.75 - 1)} + \frac{0.3(1 - 0.74)}{1 + (0.74 - 1)} + \frac{0.4(1 - 0.34)}{1 + (0.34 - 1)} = 0.720$$

Since $f\{1\}$ is not less than zero, the mixture is below the dew point. Therefore, the mixture is part vapor and substitution of z_i and K_i values in Eq (3) of Table 4.4 gives

$$0 = \frac{0.1(1 - 4.2)}{1 + \Psi(4.2 - 1)} + \frac{0.2(1 - 1.75)}{1 + \Psi(1.75 - 1)} + \frac{0.3(1 - 0.74)}{1 + \Psi(0.74 - 1)} + \frac{0.4(1 - 0.34)}{1 + \Psi(0.34 - 1)}$$

Solution of this equation by Newton's method using an initial guess for Ψ of 0.50 gives the following iteration history:

k	$\Psi^{(k)}$	$f\{\Psi^{(k)}\}$	$f'\{\Psi^{(k)}\}$	$\Psi^{(k+1)}$	$\left\| \dfrac{\Psi^{(k+1)} - \Psi^{(k)}}{\Psi^{(k)}} \right\|$
1	0.5000	0.2515	0.6259	0.0982	0.8037
2	0.0982	−0.0209	0.9111	0.1211	0.2335
3	0.1211	−0.0007	0.8539	0.1219	0.0065
4	0.1219	0.0000	0.8521	0.1219	0.0000

For this example, convergence is very rapid, giving $\Psi = V/F = 0.1219$. From Eq. (4) of Table 4.4, the equilibrium vapor flow rate is $0.1219(100) = 12.19$ kmol/h, and the equilibrium liquid flow rate from Eq. (7) is $(100 - 12.19) = 87.81$ kmol/h. The liquid and vapor compositions computed from Eqs. (5) and (6) are

	x	y
Propane	0.0719	0.3021
n-Butane	0.1833	0.3207
n-Pentane	0.3098	0.2293
n-Hexane	0.4350	0.1479
	1.0000	1.0000

A plot of $f\{\Psi\}$ as a function of Ψ is shown in Figure 4.11. ∎

Bubble and Dew Points

Often, it is desirable to bring a mixture to the bubble point or the dew point. At the bubble point, $\Psi = 0$ and $f\{0\} = 0$. Therefore, from Eq. (3), Table 4.4,

$$f\{0\} = \sum_i z_i(1 - K_i) = \sum z_i - \sum z_i K_i = 0$$

However, $\Sigma z_i = 1$. Therefore, the bubble-point equation is

$$\sum_i z_i K_i = 1 \qquad \text{(4-12)}$$

At the dew point, $\Psi = 1$ and $f\{1\} = 0$. Therefore, from Eq. (3), Table 4.4,

$$f\{1\} = \sum_i \frac{z_i(1 - K_i)}{K_i} = \sum \frac{z_i}{K_i} - \sum z_i = 0$$

Therefore, the dew-point equation is

$$\sum_i \frac{z_i}{K_i} = 1 \qquad \text{(4-13)}$$

For a given feed composition, z_i, (4-12) or (4-13) can be used to find T for a specified P or to find P for a specified T.

Because of the K-values, the bubble- and dew-point equations are generally highly nonlinear in temperature, but only moderately nonlinear in pressure, except in the region of the *convergence pressure,* where K-values of very light or very heavy species change radically with pressure, as in Figure 2.10. Therefore, iterative procedures are required to solve for bubble- and dew-point conditions. One exception is where Raoult's law K-values are applicable. Substitution of $K_i = P_i^s/P$ into (4-12) leads to an equation for the direct calculation of bubble-point pressure:

$$P_{bubble} = \sum_{i=1}^{C} z_i P_i^s \qquad (4\text{-}14)$$

where P_i^s is the temperature-dependent vapor pressure of species i. Similarly, from (4-13), the dew-point pressure is

$$P_{dew} = \left(\sum_{i=1}^{C} \frac{z_i}{P_i^s} \right)^{-1} \qquad (4\text{-}15)$$

Another useful exception occurs for mixtures at the bubble point when K-values can be expressed by the modified Raoult's law, $K_i = \gamma_i P_i^s/P$. Substituting this equation into (4-12),

$$P_{bubble} = \sum_{i=1}^{C} \gamma_i z_i P_i^s \qquad (4\text{-}16)$$

Liquid-phase activity coefficients can be computed for a known temperature and composition, since $x_i = z_i$ at the bubble point.

Bubble- and dew-point calculations are used to determine saturation conditions for liquid and vapor streams, respectively. It is important to note that when vapor–liquid equilibrium is established, the vapor is at its dew point and the liquid is at its bubble point.

EXAMPLE 4.2

In Figure 1.9, the nC_4-rich bottoms product from column C3 has the composition given in Table 1.5. If the pressure at the bottom of the distillation column is 100 psia (689 kPa), estimate the temperature of the mixture.

SOLUTION

The bottoms product will be a liquid at its bubble point with the following composition:

Component	kmol/h	$z_i = x_i$
i-Butane	8.60	0.0319
n-Butane	215.80	0.7992
i-Pentane	28.10	0.1041
n-Pentane	17.50	0.0648
	270.00	1.0000

The bubble-point temperature can be estimated by finding the temperature that will satisfy (4-12), using K-values from Figure 2.8. Because the bottoms product is rich in nC_4, assume that the K-value of nC_4 is 1. From Figure 2.8, for 100 psia, $T = 150°F$. For this temperature, using Figure 2.8 to obtain the K-values of the other three hydrocarbons and substituting these values and the z-values into (4-12),

$$\sum z_i K_i = 0.0319(1.3) + 0.7992(1.0) + 0.1041(0.47) + 0.0648(0.38)$$

$$= 0.042 + 0.799 + 0.049 + 0.025 = 0.915$$

Because the sum is not 1.0, another temperature must be assumed and the summation repeated. To increase the sum, the K-values must be greater and, thus, the temperature must be higher. Because the sum is dominated by nC_4, assume that its K-value must be $1.000(1.00/0.915) = 1.09$. This corresponds to a temperature of 160°F, which results in a summation of 1.01. By linear interpolation, $T = 159°F$. ∎

EXAMPLE 4.3

Cyclopentane is to be separated from cyclohexane by liquid–liquid extraction with methanol at 25°C. In extraction it is important that the liquid mixtures be maintained at pressures greater than the bubble-point pressure. Calculate the bubble-point pressure using the following equilibrium liquid-phase compositions, activity coefficients, and vapor pressures:

	Methanol	**Cyclohexane**	**Cyclopentane**
Vapor pressure, psia	2.45	1.89	6.14
Methanol-rich layer:			
x	0.7615	0.1499	0.0886
γ	1.118	4.773	3.467
Cyclohexane-rich layer:			
x	0.1737	0.5402	0.2861
γ	4.901	1.324	1.074

SOLUTION

Because the bubble-point pressure is likely to be below ambient pressure, the modified Raoult's law in the form of (4-16) applies for either liquid phase. If the methanol-rich layer data are used:

$$P_{bubble} = 1.118(0.7615)(2.45) + 4.773(0.1499)(1.89) + 3.467(0.0886)(6.14)$$
$$= 5.32 \text{ psia } (36.7 \text{ kPa})$$

A similar calculation based on the cyclohexane-rich layer gives an identical result because the data are consistent with phase equilibrium theory such that $\gamma_{iL}^{(1)}x_i^{(1)} = \gamma_{iL}^{(2)}x_i^{(2)}$. A pressure higher than 5.32 psia will prevent formation of vapor at this location in the extraction process. Thus, operation at atmospheric pressure is a good choice. ∎

EXAMPLE 4.4

Propylene (P) is to be separated from 1-butene (B) by distillation into a vapor distillate containing 90 mol% propylene. Calculate the column operating pressure assuming the exit temperature from the partial condenser is 100°F (37.8°C), the minimum attainable temperature with cooling water. Determine the composition of the liquid reflux. In Figure 4.12, K-values estimated from Eq. (5), Table 2.3, using the Redlich–Kwong equation of state for the vapor fugacity, are plotted and compared to experimental data [7] and Raoult's law K-values.

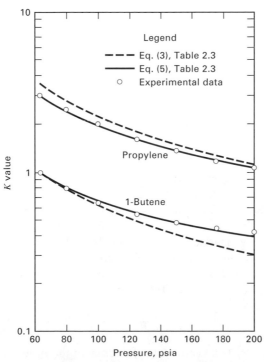

Figure 4.12 K-values for propylene/1-butene system at 100°F.

SOLUTION

The operating pressure corresponds to a dew-point condition for the vapor-distillate composition. The composition of the reflux corresponds to the liquid in equilibrium with the vapor distillate at its dew point. The method of false position [8] can be used to perform the iterative calculations by rewriting (4-13) in the form

$$f\{P\} = \sum_{i=1}^{C} \frac{z_i}{K_i} - 1 \qquad (1)$$

The recursion relationship for the method of false position is based on the assumption that $f\{P\}$ is linear in P such that

$$P^{(k+2)} = P^{(k+1)} - f\{P^{(k+1)}\}\left[\frac{P^{(k+1)} - P^{(k)}}{f\{P^{(k+1)}\} - f\{P^{(k)}\}}\right] \qquad (2)$$

This assumption is reasonable because, at low pressures, K-values in (2) are almost inversely proportional to pressure. Two values of P are required to initialize this formula. Choose 100 psia and 190 psia. At 100 psia, reading the K-values from the solid lines in Figure 4.12,

$$f\{P\} = \frac{0.90}{2.0} + \frac{0.10}{0.68} - 1.0 = -0.40$$

Subsequent iterations give

k	$P^{(k)}$, psia	K_P	K_B	$f\{P^{(k)}\}$
1	100	2.0	0.68	−0.40
2	190	1.15	0.42	+0.02
3	186	1.18	0.425	−0.0020

Iterations are terminated when $|P^{(k+2)} - P^{(k+1)}|/P^{(k+1)} < 0.005$.

An operating pressure of 186 psia (1,282 kPa) at the partial condenser outlet is indicated. The composition of the liquid reflux is obtained from $x_i = z_i/K_i$ with the result

Component	Equilibrium Mole Fraction	
	Vapor Distillate	Liquid Reflux
Propylene	0.90	0.76
1-Butene	0.10	0.24
	1.00	1.00

∎

Adiabatic Flash

When the pressure of a liquid stream of known composition, flow rate, and temperature (or enthalpy) is reduced adiabatically across a valve as in Figure 4.10a, an *adiabatic flash* calculation is made to determine the resulting temperature, compositions, and flow rates of the equilibrium liquid and vapor streams for a specified pressure downstream of the valve. For an adiabatic flash, the isothermal flash calculation procedure can be applied in the following iterative manner. A guess is made of the flash temperature, T_V. Then Ψ, V, x, y, and L are determined, as for an isothermal flash, from steps 3 through 7 in Table 4.4. The guessed value of T_V (equal to T_L) is next checked by an energy balance obtained by combining Eqs. (7) and (8) of Table 4.4 with $Q = 0$ to give

$$f\{T_V\} = \frac{\Psi h_V + (1 - \Psi)h_L - h_F}{1,000} = 0 \qquad \textbf{(4-17)}$$

where the division by 1,000 is to make the terms of the order of 1. If the computed value of $f\{T_V\}$ is not zero, the entire procedure is repeated for two or more guesses of T_V. A plot of $f\{T_V\}$ versus T_V is interpolated to determine the correct value of T_V. The procedure is tedious because it involves inner-loop iteration on Ψ and outer-loop iteration on T_V.

Outer-loop iteration on T_V is very successful when Eq. (3) of Table 4.4 is not sensitive to the guess of T_V. This is the case for wide-boiling mixtures. For close-boiling mixtures (e.g., isomers), the algorithm may fail because of extreme sensitivity to the value of T_V. In this case, it is preferable to do the outer-loop iteration on Ψ and solve Eq. (3) of Table 4.4 for T_V in the inner loop, using a guessed value for Ψ to initiate the process:

$$f\{T_V\} = \sum_{i=1}^{C} \frac{z_i(1 - K_i)}{1 + \Psi(K_i - 1)} = 0 \qquad \textbf{(4-18)}$$

Then, Eqs. (5) and (6) of Table 4.4 are solved for x and y, respectively. Equation (4-17) is then solved directly for Ψ, since

$$f\{\Psi\} = \frac{\Psi h_V + (1 - \Psi)h_L - h_F}{1,000} = 0 \qquad \textbf{(4-19)}$$

from which

$$\Psi = \frac{h_F - h_L}{h_V - h_L} \qquad \textbf{(4-20)}$$

If Ψ from (4-20) is not equal to the value of Ψ guessed to solve (4-18), the new value of Ψ is used to repeat the outer loop starting with (4-18).

Multicomponent isothermal flash, bubble-point, dew-point, and adiabatic flash calculations can be very tedious because of their iterative nature. They are unsuitable for manual calculations for nonideal vapor and liquid mixtures because of the complexity of the expressions for the thermodynamic properties, K, h_V, and h_L. However, robust algorithms for making such calculations are incorporated into widely used steady-state simulation computer programs such as ASPEN PLUS, ChemCAD DESIGN II, HYSIM, and PRO/II.

EXAMPLE 4.5

The equilibrium liquid from the flash drum at 120°F and 485 psia in Example 2.6 is fed to a distillation tower to remove the remaining hydrogen and methane. A tower for this purpose is often referred to as a *stabilizer*. Pressure at the feed plate of the stabilizer is 165 psia (1,138 kPa). Calculate the percent vaporization of the feed if the pressure is decreased adiabatically from 485 to 165 psia by valve and pipeline pressure drop.

SOLUTION

This problem is most conveniently solved by using a steady-state simulation program. If the ChemCAD program is used with K-values and enthalpies estimated from the P-R equation of state, the following results are obtained:

	kmol/h		
Component	Feed 120°F 485 psia	Vapor 112°F 165 psia	Liquid 112°F 165 psia
Hydrogen	1.0	0.7	0.3
Methane	27.9	15.2	12.7
Benzene	345.1	0.4	344.7
Toluene	113.4	0.04	113.36
Total	487.4	16.34	471.06
Enthalpy, kJ/h	−1,089,000	362,000	−1,451,000

This case involves a wide-boiling feed, so the procedure involving (4-17) is the best choice. The above results show that only a small amount of vapor ($\Psi = 0.0035$), predominantly H_2 and CH_4, is produced by the adiabatic flash. The computed flash temperature of 112°F is 8°F below the feed temperature. The enthalpy of the feed is equal to the sum of the vapor and liquid product enthalpies for this adiabatic operation. ∎

4.5 TERNARY LIQUID–LIQUID SYSTEMS

Ternary mixtures that undergo phase splitting to form two separate liquid phases can differ as to the extent of solubility of the three components in each of the two liquid phases. The simplest case is shown in Figure 4.13a, where only the solute, component B, has any appreciable solubility in either the carrier, A, or the solvent, C, both of which have negligible (although never zero), solubility in each other. In this case, the equations can be derived for a single equilibrium stage, using the variables F, S, $L^{(1)}$, and $L^{(2)}$ to refer, respectively, to the flow rates (or amounts) of the feed, solvent, exiting extract, and exiting raffinate. By definition the extract is the exiting liquid phase that contains the solvent and the extracted solute; the raffinate is the exiting liquid phase that contains the carrier, A, of the feed and the portion of the solute, B, that is not extracted. Although the extract is shown in Figure 4.13a as leaving from the top of the stage, this will only be so if the extract is the lighter (lower-density) exiting phase. Assuming that the entering solvent contains no solute, B, it is convenient to write material balance and phase equilibrium equations for the solute, B. These two equations may be written in terms of molar or mass flow rates. To obtain the simplest result, it is preferable to express compositions of the solute as mass or mole ratios instead of mass or mole fractions.

Let: F_A = feed rate of carrier A
S = flow rate of solvent C
X_B = ratio of mass (or moles) of solute B, to mass (or moles) of the other component in the feed (F), raffinate (R), or extract (E),

Then, the solute material balance is

$$X_B^{(F)} F_A = X_B^{(E)} S + X_B^{(R)} F_A \qquad (4\text{-}21)$$

and the distribution of solute at equilibrium is given by

$$X_B^{(E)} = K'_{D_B} X_B^{(R)} \qquad (4\text{-}22)$$

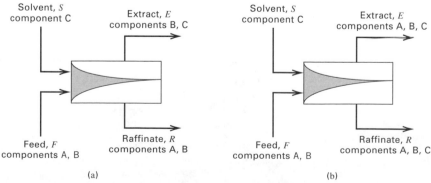

Figure 4.13 Phase splitting of ternary mixtures: (a) components A and C mutually insoluble; (b) components A and C partially soluble.

where K'_{D_B} is the distribution coefficient defined in terms of mass or mole ratios. Substituting (4-22) into (4-21) to eliminate $X_B^{(E)}$ gives

$$X_B^{(R)} = \frac{X_B^{(F)} F_A}{F_A + K'_{D_B} S} \tag{4-23}$$

It is convenient to define an extraction factor, E_B, for the solute B:

$$E_B = K'_{D_B} S / F_A \tag{4-24}$$

The larger the value of E, the greater the extent to which the solute is extracted. Large values of E result from large values of the distribution coefficient, K'_{D_B}, or large ratios of solvent to carrier. Substituting (4-24) in (4-23) gives the fraction of B that is not extracted as

$$X_B^{(R)} / X_B^{(F)} = \frac{1}{1 + E_B} \tag{4-25}$$

where it is clear that the larger the extraction factor, the smaller the fraction of B not extracted.

Values of mass (mole) ratios, X are related to mass (mole) fractions, x, by

$$X_i = x_i / (1 - x_i) \tag{4-26}$$

Values of the distribution coefficient, K'_D, in terms of ratios, are related to K_D in terms of fractions as given in (2-20) by

$$K'_{D_i} = \frac{x_i^{(1)}/(1 - x_i^{(1)})}{x_i^{(2)}/(1 - x_i^{(2)})} = K_{D_i}\left(\frac{1 - x_i^{(2)}}{1 - x_i^{(1)}}\right) \tag{4-27}$$

When values of x_i are small, K'_D approaches K_D. As discussed in Chapter 2, the distribution coefficient, K_{D_i}, which can be determined from activity coefficients using the expression $K_{D_B} = \gamma_B^{(2)}/\gamma_B^{(1)}$ when mole fractions are used, is a strong function of equilibrium phase compositions and temperature. However, when the raffinate and extract are both dilute in the solute, activity coefficients of the solute can be approximated by the values at infinite dilution so that K_{D_B} can be taken as a constant at a given temperature. An extensive listing of such K_{D_B} values in mass fraction units for various ternary systems is given in *Perry's Handbook* [9]. If values for F_B, $X_B^{(F)}$, S, and K_{D_B} are given, (4-25) can be solved for $X_B^{(R)}$.

EXAMPLE 4.6

A feed of 13,500 kg/h consists of 8 wt% acetic acid (B) in water (A). The removal of the acetic acid is to be accomplished by liquid–liquid extraction at 25°C with methyl isobutyl ketone solvent (C), because distillation of the feed would require vaporization of large amounts of water. If the raffinate is to contain only 1 wt% acetic acid, estimate the kilograms per hour of solvent required if a single equilibrium stage is used.

SOLUTION

Assume that the carrier (water) and the solvent are immiscible. From *Perry's Handbook*, $K_D = 0.657$ in mass-fraction units for this system. For the relatively low concentrations of acetic acid in this problem, assume that $K'_D = K_D$.

$$F_A = (0.92)(13,500) = 12,420 \text{ kg/h}$$

$$X_B^{(F)} = (13,500 - 12,420)/12,420 = 0.087$$

The raffinate is to contain 1 wt% B. Therefore,

$$X_B^{(R)} = 0.01/(1 - 0.01) = 0.0101$$

From (4-25), solving for E_B,

$$E_B = \frac{X_B^{(F)}}{X_B^{(R)}} - 1 = (0.087/0.0101) - 1 = 7.61$$

From (4-24), the definition of the extraction factor,

$$S = \frac{E_B F_A}{K_D'} = 7.61(12,420/0.657) = 144,000 \, \text{kg/h}$$

This is a very large solvent flow rate compared to the feed rate—more than a factor of 10! Multiple stages should be used to reduce the solvent rate or a solvent with a larger distribution coefficient should be sought.   ∎

In the ternary liquid–liquid system, shown in Figure 4.13b, components A and C are partially soluble in each other and component B again distributes between the extract and raffinate phases. Both of these exiting phases contain all components present in the feed and solvent. This case is by far the most commonly encountered, and a number of different phase diagrams and computational techniques have been devised to determine the equilibrium compositions. Examples of phase diagrams are shown in Figure 4.14 for the water (A)–ethylene glycol (B)–furfural (C) system at 25°C and a pressure of 101 kPa, which is above the bubble-point pressure, so no vapor phase exists. The pairs water–ethylene glycol and furfural–ethylene glycol are each completely miscible. The only partially miscible pair is furfural–water. In practice, furfural is used as a solvent to remove the solute, ethylene glycol, from water; the furfural-rich phase is the extract, and the water-rich phase is the raffinate.

Figure 4.14a, an equilateral triangular diagram, is the most common display of ternary liquid–liquid equilibrium data in the chemical literature. Any point located within or on an edge of the triangle represents a mixture composition. Such a diagram has the property that the sum of the lengths of the perpendicular lines drawn from any interior point to the sides equals the altitude of the triangle. Thus, if each of the three altitudes is scaled from 0 to 100, the percent of, say, furfural, at any point such as M, is simply the length of the line perpendicular to the base opposite the pure furfural apex, which represents 100% furfural. Figure 4.14a is constructed for compositions based on mass fractions (mole fractions and volume fractions are also sometimes used). Thus, the point M in Figure 4.14a represents a mixture of feed and solvent (before phase separation) containing 40 wt% water, 30 wt% ethylene glycol, and 30 wt% furfural.

The miscibility limits for the furfural–water binary system are at D and G. The miscibility boundary (saturation curve) DEPRG can be obtained experimentally by a *cloud-point titration;* water, for example, is added to a (clear) 50 wt% solution of furfural and glycol, and it is noted that the onset of cloudiness due to the formation of a second phase occurs when the mixture is 10% water, 45% furfural, and 45% glycol by weight. Other miscibility data are given in Table 4.5, from which the miscibility curve in Figure 4.14a was drawn.

Tie lines are used to connect points on the miscibility boundary, DEPRG, that represent equilibrium phase compositions. To obtain data to construct tie lines, such as ER, it is necessary to make a mixture such as M (30% glycol, 40% water, and 30% furfural), equilibrate it, and then chemically analyze the resulting extract and raffinate phases E and R (in this case, 41.8% glycol, 10% water, and 48.2% furfural; and 11.5% glycol, 81.5% water, and 7% furfural, respectively). At point P, the *plait point,* the two liquid phases have identical compositions. Therefore, the tie lines converge to a point and the two phases become one phase. Tie-line data for this system are given in Table 4.6, in terms of glycol composition.

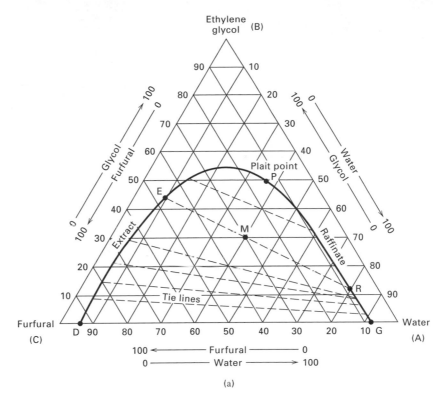

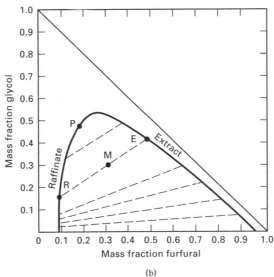

Figure 4.14 Liquid–liquid equilibrium, ethylene glycol–furfural–water, 25°C, 101 kPa: (a) equilateral triangular diagram; (b) right triangular diagram; (*continues*)

When there is mutual solubility between two phases, the thermodynamic variables necessary to define the equilibrium system are temperature, pressure, and the concentrations of the components in each phase. According to the phase rule, (4-1), for a three-component, two-liquid-phase system, there are three degrees of freedom. At constant temperature and pressure, specification of the concentration of one component in either

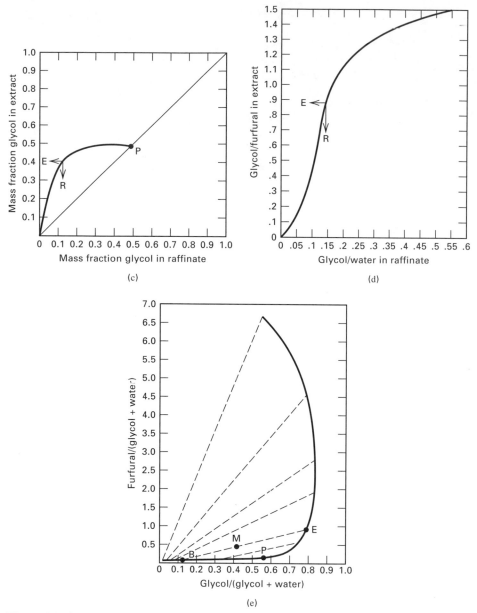

Figure 4.14 (*Continued*) (c) equilibrium solute diagram in mass fractions; (d) equilibrium solute diagram in mass ratios; (e) Janecke diagram.

of the phases suffices to completely define the state of the system. Thus, as shown in Figure 4.14a, one value for glycol weight percent on the miscibility boundary curve fixes the composition of the corresponding phase and, by means of the tie line, the composition of the other equilibrium phase.

Figure 4.14b is a representation of the same system on a right-triangular diagram. Here the concentrations in weight percent of any two of the three components (normally the solute and solvent) are given, the concentration of the third being obtained by difference from 100 wt%. Diagrams like this are easier to read than equilateral triangular diagrams.

Figures 4.14c and 4.14d are representations of the same ternary system in terms of weight fraction and weight ratios of the solute, respectively. Figure 4.14c is simply a plot of the

Table 4.5 Equilibrium Miscibility Data in Weight Percent for the Furfural–Ethylene Glycol–Water System at 25°C and 101 kPa

Furfural	Ethylene Glycol	Water
94.8	0.0	5.2
84.5	11.4	4.1
63.1	29.7	7.2
49.4	41.6	9.0
40.6	47.5	11.9
33.8	50.1	16.1
23.2	52.9	23.9
20.1	50.5	29.4
10.2	32.2	57.6
9.2	28.1	62.7
7.9	0.0	92.1

equilibrium (tie-line) data of Table 4.6 in terms of solute mass fraction. In Figure 4.14d, mass ratios of solute (ethylene glycol) to furfural and water for the extract and raffinate phases, respectively, are used. Such curves can be used to interpolate tie lines, since only a limited number of tie lines are shown on triangular graphs. Because of this, such diagrams are often referred to as *distribution diagrams*. When mole (rather than mass) fractions are used in a diagram like Figure 4.14c, a nearly straight line is often evident near the origin, where the slope is the distribution coefficient, K_D, for the solute at infinite dilution.

In 1906, Janecke [10] suggested the equilibrium data display shown as Figure 4.14e. Here, the mass of solvent per unit mass of solvent-free material, furfural/(water + glycol), is plotted as the ordinate versus the mass ratio, on a solvent-free basis, of glycol/(water + glycol) as abscissa. The ordinate and abscissa apply to both phases. Equilibrium conditions are related by tie lines. Mole ratios can be used also to construct Janecke diagrams.

Any of the five diagrams in Figure 4.14 can be used for solving problems involving material balances subject to liquid–liquid equilibrium constraints, as is demonstrated in the following example.

Table 4.6 Mutual Equilibrium (Tie Line) Data for the Furfural–Ethylene Glycol–Water System at 25°C and 101 kPa

Glycol in Water Layer, wt%	Glycol in Furfural Layer, wt%
49.1	49.1
32.1	48.8
11.5	41.8
7.7	28.9
6.1	21.9
4.8	14.3
2.3	7.3

EXAMPLE 4.7

Calculate the composition of the equilibrium extract and raffinate phases produced when a 45% by weight glycol (B)–55% water (A) solution is contacted with its own weight of pure furfural solvent (C) at 25°C and 101 kPa. Use each of the five diagrams in Figure 4.14, if possible. What is the composition of the water–glycol mixture obtained by removing all of the furfural from the extract?

SOLUTION

Assume a basis of 100 g of 45% glycol–water feed. Thus, in Figure 4.13b, the feed (F) is 55 g of A and 45 g of B. The solvent (S) is 100 g of C. Let $L^{(1)} = E$, the extract, and $L^{(2)} = R$, the raffinate.

(a) By an equilateral triangular diagram, Figure 4.15:

Step 1. Locate the feed and solvent compositions at points F and S, respectively.
Step 2. Define M, the mixing point, as $M = F + S = E + R$
Step 3. Apply the inverse lever rule to the equilateral triangular phase equilibrium diagram. Let $w_i^{(1)}$ be the mass fraction of species i in the extract, $w_i^{(2)}$ be the mass fraction of species i in the raffinate, and $w_i^{(M)}$ be the mass fraction of species i in the combined feed and solvent phases.

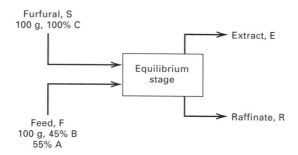

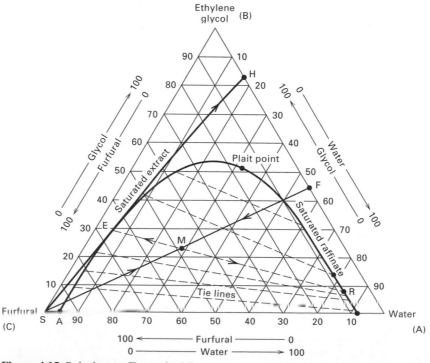

Figure 4.15 Solution to Example 4.7a.

From a balance on the solvent, C: $(F + S)w_C^{(M)} = Fw_C^{(F)} + Sw_C^{(S)}$.

Therefore,
$$\frac{F}{S} = \frac{w_C^{(S)} - w_C^{(M)}}{w_C^{(M)} - w_C^{(F)}}$$
(1)

Thus, points S, M, and F lie on a straight line, and, by the inverse lever rule, $\frac{F}{S} = \frac{SM}{MF} = 1$.

The composition at point M is 27.5% A, 22.5% B, and 50% C.

Step 4. Since M lies in the two-phase region, the mixture must separate along a tie line into the extract phase at point E (27.9% B, 6.5% A, and 65.6% C) and the raffinate at point R (8% B, 84% A, and 8% C).

Step 5. The inverse lever rule applies to points E, M, and R, so $E = M$ (RM/ER). Because $M = 100 + 100 = 200$ g, and from measurements of the line segments, $E = 200(49/67) = 146$ g and $R = M - E = 200 - 146 = 54$ g.

Step 6. The solvent-free extract composition is obtained by extending a straight line passing through points S and E to point H (83% B, 17% A, and 0% C), since it is easily shown that this line is the locus of all possible mixtures that can be obtained by adding or subtracting pure solvent to or from E.

(b) By a right-triangular diagram, Figure 4.16:

Step 1. Locate the points F and S for the two feed streams.

Step 2. Define the mixing point $M = F + S$.

Step 3. The inverse lever rule also applies to right-triangular diagrams, so MF/MS = 1, and the point M is located at the middle of line FS as shown.

Step 4. Points R and E are on the ends of the tie line passing through point M.

Step 5. Then point H, the furfural-stripped extract, is found by extending the line SE to the zero furfural axis.

The numerical results of part (b) are identical to those of part (a).

(c) By an equilibrium solute diagram, Figure 4.14c. A material balance on glycol, B,

$$Fw_B^{(F)} + Sw_B^{(S)} = 45 = Ew_B^{(E)} + Rw_B^{(R)}$$
(2)

must be solved simultaneously with a phase equilibrium relationship. It is not possible to do this graphically using Figure 4.14c in any straightforward manner unless the solvent (C) and carrier (A) are mutually insoluble. The outlet stream composition can be found, however, by the following iterative procedure.

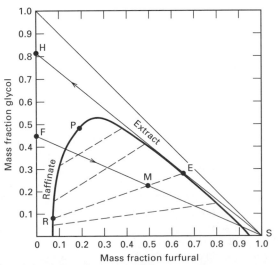

Figure 4.16 Solution to Example 4.7b.

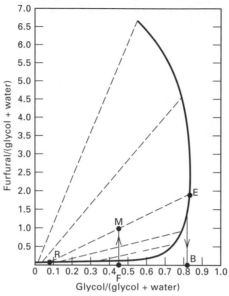

Figure 4.17 Solution to Example 4.7e.

Step 1. Guess a value for $w_B^{(R)}$ and read the equilibrium value, $w_B^{(E)}$ from Figure 4.14c.
Step 2. Substitute these two values into the equation obtained by combining (2) with the overall balance, $E + R = 200$ to eliminate R. Solve for E and then R.
Step 3. Check to see if the furfural (or water) balance is satisfied using the equilibrium data from Figures 4.14a, 4.14b, or 4.14e. If not, repeat steps 1 to 3 with a new guess for $w_B^{(R)}$. This procedure leads to the same results obtained in parts (a) and (b).

(d) By an equilibrium solute diagram in mass fractions, Figure 4.14d: This plot suffers from the same limitations as Figure 4.13c in that a solution must be achieved by an iterative procedure.

(e) By a Janecke diagram, Figure 4.17:

Step 1. The feed mixture is located at point F. With the addition of 100 g of pure furfural solvent, $M = F + S$ is located as shown, since the ratio of glycol to (glycol + water) remains the same.
Step 2. The mixture at point M separates into the two phases at points E and R, by tie-line construction, with the coordinates (1.91, 0.81) at E and (0.09, 0.09) at R.
Step 3. Let Z^E and Z^R equal the total mass of components A and B in the extract and raffinate, respectively. Then, the following balances apply:

Furfural: $1.91Z^E + 0.09Z^R = 100$
Glycol: $0.81Z^E + 0.09Z^R = 45$

Solving these two simultaneous equations, we obtain $Z^E = 50$ g, $Z^R = 50$ g.

Thus, the furfural in the extract = (1.91)(50 g) = 95.5 g, the furfural in the raffinate = (0.09)(50 g) = 4.5 g, the glycol in the extract = (0.81)(50 g) = 40.5 g, the glycol in the raffinate = (0.09)(50) = 4.5 g, the water in the raffinate = 50 − 4.5 = 45.5 g, and the water in the extract = 50 − 40.5 = 9.5 g. The total extract is (95.5 + 40.5 + 9.5) = 145.5 g, which is almost identical to the results obtained in part (a). The raffinate composition and amount can be obtained just as readily.

It should be noted on the Janecke diagram that ME/MR does not equal R/E; it equals the ratio of R/E on a solvent-free basis.

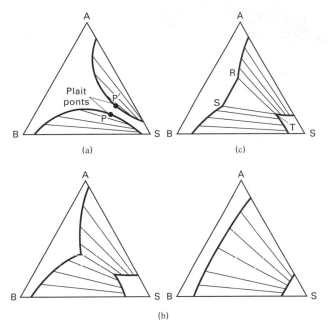

Figure 4.18 Equilibria for 3/2 systems: (a) miscibility boundaries are separate; (b) miscibility boundaries and tie-line equilibria merge; (c) tie lines do not merge and the three-phase region RST is formed.

Step 4. Point B, the furfural-free extract composition, is obtained by extrapolating a vertical line through E to the axis. The furfural-free extract mixture is 81% glycol and 19% water by weight. ∎

In Figure 4.14, two pairs of components are mutually soluble, while one pair is only partially soluble. Ternary systems where two pairs and even all three pairs are only partially soluble are also common. Figure 4.18 shows examples, taken from Francis [11] and Findlay [12], of four different cases where two pairs of components are only partially soluble.

In Figure 4.18a, two separate two-phase regions are formed, while in Figure 4.18c, in addition to the two-phase regions, a three-phase region, RST, is formed. In Figure 4.18b the two separate two-phase regions merge. For a ternary mixture, as temperature is reduced, phase behavior may progress from Figure 4.18a to 4.18b to 4.18c. In Figures 4.18a, 4.18b, and 4.18c, all tie lines slop in the same direction. In some systems of importance, *solutropy,* a reversal of tie-line slopes, occurs.

4.6 MULTICOMPONENT LIQUID–LIQUID SYSTEMS

Quarternary and higher multicomponent mixtures are often encountered in liquid–liquid extraction processes, particularly when two solvents are used for liquid–liquid extraction. As discussed in Chapter 2, multicomponent liquid–liquid equilibria are very complex and there is no compact graphical way of representing experimental phase equilibria. Accordingly, the computation of the equilibrium phase compositions is best made by a computer-assisted algorithm using activity coefficient equations from Chapter 2 that account for the effect of composition. One such algorithm is a modification of the Rachford–Rice algorithm for vapor–liquid equilibrium, given in Tables 4.3 and 4.4. To apply these tables to multicomponent liquid–liquid equilibria, the following symbol transformations are made, where all flow rates and compositions are in moles:

Vapor–Liquid Equilibria **(Tables 4.3, 4.4)**	**Liquid–Liquid Equilibria**
Feed, F	Feed, F, + solvent, S
Equilibrium vapor, V	Extract, E ($L^{(1)}$)
Equilibrium liquid, L	Raffinate, R ($L^{(2)}$)
Feed mole fractions, z_i	Mole fractions of combined F and S
Vapor mole fractions, y_i	Extract mole fractions, $x_i^{(1)}$
Liquid mole fractions, x_i	Raffinate mole fractions, $x_i^{(2)}$
K-value, K_i	Distribution coefficient, K_{D_i}
$\Psi = V/F$	$\Psi = E/F$

Most liquid–liquid equilibria are achieved under adiabatic conditions, thus necessitating consideration of an energy balance. However, if both feed and solvent enter the stage at identical temperatures, the only energy effect is the heat of mixing, which is often sufficiently small that only a very small temperature change occurs. Accordingly, the calculations are often made isothermally.

The modified Rachford–Rice algorithm is shown in the flow chart of Figure 4.19. This algorithm is applicable for either an isothermal vapor–liquid (V–L) or liquid–liquid (L–L)

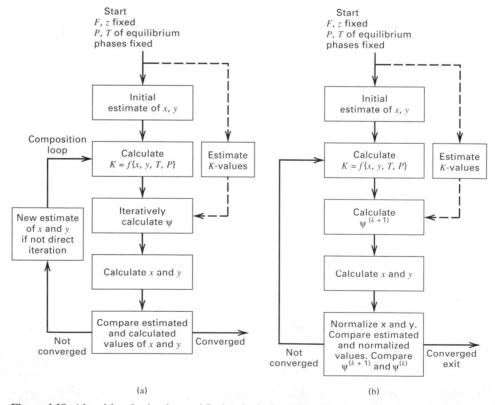

Figure 4.19 Algorithm for isothermal flash calculation when K-values are composition dependent: (a) separate nested iterations on Ψ and (x, y); (b) simultaneous iteration on Ψ and (x, y).

equilibrium-stage calculation when K-values depend strongly on phase compositions. For the L–L case, the algorithm assumes that the feed and solvent flow rates and compositions are fixed. The equilibrium pressure and temperature are also specified. An initial estimate is made of the phase compositions, $x_i^{(1)}$ and $x_i^{(2)}$, and corresponding estimates of the distribution coefficients are made from liquid-phase activity coefficients, using (2-30) with, for example, the NRTL or UNIQUAC equations discussed in Chapter 2. Equation 3 of Table 4.4 is then solved iteratively for $\Psi = E/(F + S)$, from which values of $x_i^{(2)}$ and $x_i^{(1)}$ are computed from Eqs. (5) and (6), respectively, of Table 4.4. The resulting values of $x_i^{(1)}$ and $x_i^{(2)}$ will not usually sum, respectively, to 1 for each liquid phase and are therefore normalized. The normalized values are obtained from equations of the form, $x_i' = x_i/\Sigma\, x_j$, where x_i' are the normalized values that force $\Sigma\, x_j'$ to equal 1. The normalized values replace the values computed from Eqs. (5) and (6). The iterative procedure is repeated until the compositions $x_i^{(1)}$ and $x_i^{(2)}$, to say three or four significant digits, no longer change from one iteration to the next. Multicomponent liquid–liquid equilibrium calculations are best carried out with a steady-state simulation computer program.

EXAMPLE 4.8

An azeotropic mixture of isopropanol, acetone, and water is being dehydrated with ethyl acetate in a distillation system of two columns. Benzene was previously used as the dehydrating agent, but recent legislation has made the use of benzene undesirable because it is carcinogenic. Ethyl acetate is far less toxic. The overhead vapor from the first column, with the composition given below, at a pressure of 20 psia and a temperature of 80°C is condensed and cooled to 35°C, without significant pressure drop, causing the formation of two liquid phases in equilibrium. Estimate the amounts of the two phases in kilograms per hour and the phase compositions in weight percent.

Component	kg/h
Isopropanol	4,250
Acetone	850
Water	2,300
Ethyl acetate	43,700

Note that the specification of this problem satisfies the degrees of freedom from (4-4), which for $C = 4$ is 9.

SOLUTION

This example was solved with the ChemCAD program using the UNIFAC group contribution method to estimate liquid-phase activity coefficients. The results are as follows:

	Weight Fraction	
Component	Organic-Rich Phase	Water-Rich Phase
Isopropanol	0.0843	0.0615
Acetone	0.0169	0.0115
Water	0.0019	0.8888
Ethyl acetate	0.8969	0.0382
	1.0000	1.0000
Flow rate, kg/h	48,617	2,483

It is of interest to compare the distribution coefficients computed from the above results based on the UNIFAC method to experimental values given in *Perry's Handbook* [1]:

	Distribution Coefficient (wt% Basis)	
Component	**UNIFAC**	***Perry's Handbook***
Isopropanol	1.37	1.205 (20°C)
Acetone	1.47	1.50 (30°C)
Water	0.0021	—
Ethyl acetate	23.5	—

Results for isopropanol and acetone are in reasonably good agreement at these relatively dilute conditions, considering that no temperature corrections were made. ∎

4.7 SOLID–LIQUID SYSTEMS

Solid–liquid separation operations include leaching, crystallization, and adsorption. In a leaching operation (solid–liquid extraction), a multicomponent solid mixture is separated by contacting the solid with a solvent that selectively dissolves some, but not all, components in the solid. Although this operation is quite similar to liquid–liquid extraction, two aspects of leaching make it a much more difficult separation operation in practice. Diffusion in solids is very slow compared to diffusion in liquids, thus making it difficult to achieve equilibrium. Also, it is virtually impossible to completely separate a solid phase from a liquid phase. In comparison, the separation of two liquid phases is fairly easy to accomplish.

A second solid–liquid system involves the crystallization of one or more, but not all, components from a liquid mixture. This operation is analogous to distillation. However, although equilibrium can be achieved, a sharp phase separation is again virtually impossible.

A third application of solid–liquid systems, adsorption, involves the use of a porous solid agent, which does not undergo phase change or composition change. The solid selectively adsorbs, on its exterior and interior surface, certain components of the liquid mixture. The adsorbed species are then desorbed and the solid adsorbing agent is regenerated. Variations of adsorption include ion exchange and chromatography. A solid–liquid system is also utilized in membrane separation operations, where the solid is a membrane that selectively absorbs and transports certain species, thus effecting a separation.

Solid–liquid separation processes, such as leaching and crystallization, almost always involve phase-separation operations such as gravity sedimentation, filtration, and centrifugation. These operations are not covered in this textbook, but are discussed in Section 19 of *Perry's Handbook* [1].

Leaching

A leaching stage for a ternary system is shown in Figure 4.20. The solid mixture to be separated consists of particles containing (inert) carrier A and solute B. The solvent, C, selectively dissolves B. The overflow from the stage is a solids-free liquid of solvent C, and dissolved B. The underflow is a wet solid or slurry of liquid and solid carrier A. In an ideal, equilibrium leaching stage, all of the solute is dissolved by the solvent; none of the carrier is dissolved. In addition, the composition of the liquid phase in the underflow is identical to the composition of the liquid overflow, and the overflow is free of solid. The mass ratio of solid to liquid in the underflow depends on the properties of the two phases and the type of equipment used, and is best determined from experience or tests with prototype equipment. Ideal leaching calculations result in diagrams like those shown

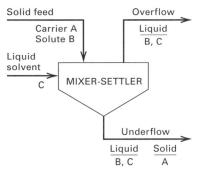

Figure 4.20 Leaching stage.

in Figure 4.21a, where the following nomenclature applies, with all quantities in either mass or moles:

X_s = solute/(solvent + solute) in the overflow
Y_s = solute/(solvent + solute) in the underflow
Y_I = carrier/(solvent + solute) in overflow or underflow
y = fraction solvent in overflow or underflow
x = fraction solute in overflow or underflow

Figure 4.21a depicts ideal leaching conditions. Thus, $Y_s = X_s$ because the equilibrium liquid solutions in both underflow and overflow are assumed to have the same composition. Also $Y_I = 0$ in the overflow because the carrier is not soluble in the solvent, and Y_I in the underflow equals a fixed value. In the y–x diagram, the underflow line AB is parallel to the overflow line CD for constant Y_I in the underflow, and the extrapolated tie lines (e.g., FE) pass through the origin for an insoluble carrier solid.

For nonideal leaching, Figure 4.21b represents typical data. The tie lines slant to the right, indicating that the solute is more highly concentrated in the underflow because of incomplete leaching. Also, curve CD does not coincide with the $Y_I = 0$ axis, indicating a partially soluble carrier or incomplete settling. In the right-triangular diagram, the tie line FE does not extrapolate to $y = 0$ and curves CFD and AEB are not parallel. If the solubility of solute in the solvent is limited, the underflow curve AEB dips down and intersects the abscissa before $X_s = 1$, and the Y_s–X_s curve is vertical at that point.

Construction of material-balance lines on solid–liquid diagrams depends on the data, assumptions, and coordinates used. In the following example, data are given as y–x mass fractions on a right-triangular diagram, assuming no solubility of the carrier in the solvent and with a constant ratio of solids to liquid in the underflow. Thus, diagrams of the type shown in Figure 4.21a apply. The method of solution, however, is identical if the diagrams resemble Figure 4.21b.

EXAMPLE 4.9

Oil is extracted from soybeans by leaching with benzene. If 1 kg of pure benzene is mixed with 1 kg of crushed soybean meal that contains 50% by weight of soluble oil and 50% inert (insoluble) solids, calculate the amounts and compositions of underflow and overflow leaving an ideal extractor. What percentage of the oil is recovered in the benzene overflow? Tests with a prototype extractor have verified the assumption of equilibrium and have determined that the underflow will contain 65 wt% solids.

SOLUTION

Step 1. Construct the y–x diagram shown in Figure 4.22 by drawing the overflow line so as to pass through the points ($y = 1$, $x = 0$) and ($y = 0$, $x = 1$), and the underflow line so as to be parallel to the overflow line and passing through the point ($y = 0.35$, $x = 0$). Tie lines pass through the origin. Locate the feed L_o at $y_{L_o} = 0$, $x_{L_o} = 0.5$; and solvent S_o at $y_{S_o} = 1$, $x_{S_o} = 0$.

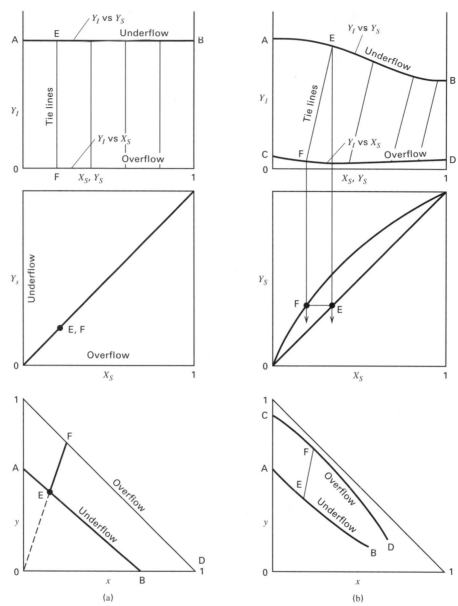

Figure 4.21 Underflow–overflow conditions for leaching: (a) ideal leaching conditions; (b) nonideal leaching conditions.

Step 2. Define the mixing point $M = L_o + S_o$, where quantities are in kilograms.
Step 3. Apply the inverse lever rule to an oil balance.

$$Mx_M = (L_o + S_o)x_M = L_o x_{L_o} + S_o x_{S_o}$$

Thus,
$$\frac{L_o}{S_o} = \frac{x_{S_o} - x_M}{x_M - x_{L_o}} \qquad (1)$$

By a solvent balance,
$$My_M = (L_o + S_o)y_M = L_o y_{L_o} + S_o y_{S_o} \qquad (2)$$

Thus,
$$\frac{L_o}{S_o} = \frac{y_{S_o} - y_M}{y_M - y_{L_o}}$$

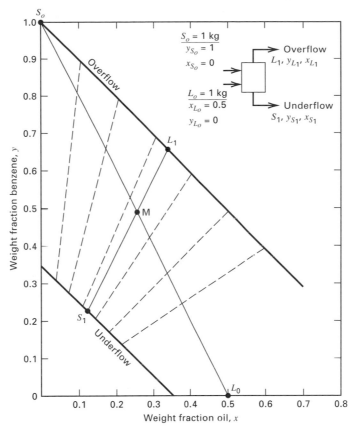

Figure 4.22 Experimental data for leaching of soybean oil with benzene. [Modified from W.L. Badger and J.T. Banchero, *Introduction to Chemical Engineering*, McGraw-Hill, New York, p. 347, (1955).]

The point M must lie at the middle of a straight line connecting S_o and L_o. Then, by (1) or (2), L_o/S_o = length of line S_oM/length of line ML_o.

Step 4. The mixture M is in the two-phase region between the overflow and underflow lines and must split into the two equilibrium streams: L_1 at $y = 0.667$ and $x = 0.333$, and S_1 at $y = 0.222$, and $x = 0.111$, which are found by drawing a straight tie line connecting point M to the origin.

Step 5. Because by material balance, $L_1 + S_1 = M$, the ratio L_1/M = length of line MS_1/ length of line $L_1S_1 = 0.625$, so $L_1 = 1.25$ kg and $S_1 = 2.00 - 1.25 = 0.75$ kg. The underflow consists of 0.50 kg of solid and 0.25 kg of solution adhering to the solid.

Step 6. Confirm the results by a solvent balance:

$$L_1 y_{L_1} + S_1 y_{S_1} = L_o y_{L_o} + S_o y_{S_o}$$

$$1.25(0.667) + 0.75(0.222) = 1.00 \text{ kg benzene}$$

Step 7. Percent extraction of oil = $(L_1 x_{L_1}/L_o x_{L_o})(100) = [(1.25)(0.333)]/[(1.0)(0.5)](100) = 83.25\%$. ∎

Crystallization

Crystallization may take place from aqueous or nonaqueous solutions. The simplest case is for a binary mixture of two organic chemicals such as naphthalene and benzene, whose

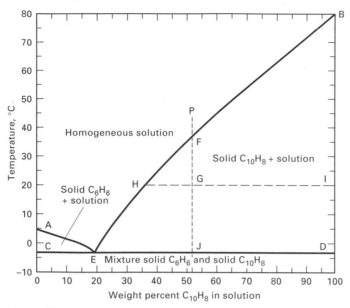

Figure 4.23 Solubility of naphthalene in benzene. [Adapted from O.A. Hougen, K.M. Watson, and R.A. Ragatz, *Chemical Process Principles. Part I,* 2nd ed., John Wiley and Sons, New York (1954).]

solubility or solid–liquid phase equilibrium diagram for a pressure of 1 atm is shown in Figure 4.23. Points A and B are the melting (freezing) points of pure benzene (5.5°C) and pure naphthalene (80.2°C), respectively. When benzene is dissolved in liquid naphthalene or naphthalene is dissolved in liquid benzene, the freezing point of the solvent is depressed. Point E is the *eutectic point,* corresponding to a eutectic temperature (−3°C) and eutectic composition (80 wt% benzene). The word "eutectic" is derived from a Greek word that means "easily fused," and in Figure 4.23 it represents the binary mixture of naphthalene and benzene with the lowest freezing (melting) point.

Temperature–composition points located above the curve AEB correspond to a homogeneous liquid phase. Curve AE is the solubility curve for benzene in naphthalene. For example, at 0°C the solubility is very high, 87 wt% benzene or 6.7 kg benzene/kg naphthalene. Curve EB is the solubility curve for naphthalene. At 25°C the solubility is 41 wt% naphthalene or 0.7 kg naphthalene/kg benzene. At 50°C, the solubility of naphthalene is much higher, 1.9 kg naphthalene/kg benzene. For this mixture, as with most mixtures, solubility increases with increasing temperature.

If a liquid solution of composition and temperature represented by point P is cooled along the vertical dashed line, it will remain a liquid until the line intersects the solubility curve at point F. If the temperature is lowered further, crystals of naphthalene form and the remaining liquid, called the *mother liquor,* becomes richer in benzene. For example, when point G is reached, pure naphthalene crystals and a mother liquor, given by point H on solubility curve EB, coexist at equilibrium, with the composition of the solution being 37 wt% naphthalene. This is in agreement with the Gibbs phase rule (4-1), because with $C = 2$ and $\mathscr{P} = 2$, $\mathscr{F} = 2$ and for fixed T and P, the phase compositions are fixed. The fraction of the solution crystallized can be determined by applying the inverse lever-arm rule. Thus, in Figure 4.23, the fraction is kilograms naphthalene crystals/kilograms original solution = length of line GH/length of line HI = $(52 - 37)/(100 - 37) = 0.238$.

As the temperature is lowered further until line CED, corresponding to the eutectic temperature, is reached at point J, the two-phase system consists of naphthalene crystals and a mother liquor of the eutectic composition given by point E. Any further removal of heat causes the eutectic solution to solidify.

EXAMPLE 4.10

A total of 8,000 kg/h of a liquid solution of 80 wt% naphthalene and 20 wt% benzene at 70°C is cooled to 30°C to form naphthalene crystals. Assuming that equilibrium is achieved, determine the amount of crystals formed and the composition of the equilibrium mother liquor.

SOLUTION

From Figure 4.23, at 30°C, the solubility of naphthalene is 45 wt% naphthalene. By the inverse lever-arm rule, for an original 80 wt% solution,

$$\frac{\text{kg naphthalene crystals}}{\text{kg original mixture}} = \frac{(80 - 45)}{(100 - 45)} = 0.636$$

The flow rate of crystals = 0.636 (8,000) = 5,090 kg/h.

The composition of the remaining 2,910 kg/h of mother liquor is 55 wt% benzene and 45 wt% naphthalene. ∎

Crystallization of a salt from an aqueous solution is frequently complicated by the formation of hydrates of the salt with water in certain definite molar proportions. These hydrates can be stable solid compounds within certain ranges of temperature as given in the solid–liquid phase equilibrium diagram. A rather extreme, but common, case is that of $MgSO_4$, which can form the stable hydrates $MgSO_4 \cdot 12H_2O$, $MgSO_4 \cdot 7H_2O$, $MgSO_4 \cdot 6H_2O$, and $MgSO_4 \cdot H_2O$. The high hydrate is stable at low temperatures, while the low hydrate is the stable form at higher temperatures.

A simpler example is that of Na_2SO_4 in mixtures with water. As seen in the phase diagram of Figure 4.24, only one stable hydrate is formed, $Na_2SO_4 \cdot 10H_2O$, commonly known as Glauber's salt. Not shown in Figure 4.24 is the metastable hydrate $Na_2SO_4 \cdot 7H_2O$. Since the molecular weights are 142.05 for Na_2SO_4 and 18.016 for H_2O, the weight percent Na_2SO_4 in the decahydrate is 44.1, which corresponds to the vertical line BFG.

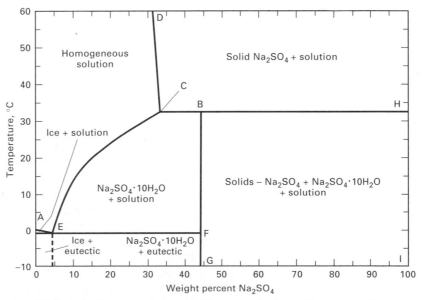

Figure 4.24 Solubility of sodium sulfate in water. [Adapted from O.A. Hougen, K.M. Watson, and R.A. Ragatz, *Chemical Process Principles. Part I,* 2nd ed., John Wiley and Sons, New York (1954).]

The freezing point of water, 0°C, is at A in Figure 4.24, but the melting point of Na_2SO_4, 884°C, is not shown because the temperature scale is terminated at 60°C. The decahydrate melts at 32.4°C, point B, to form solid Na_2SO_4 and a mother liquor, point C, of 32.5 wt% Na_2SO_4. As Na_2SO_4 is dissolved in water, the freezing point is depressed slightly along curve AE until the eutectic, point E, is reached.

Curves EC and CD represent the solubilities of the decahydrate crystals and anhydrous sodium sulfate, respectively, in water. Note that the solubility of Na_2SO_4 decreases slightly with increasing temperature.

For each region, the coexisting phases are indicated. For example, in the region below GFBHI, a solid solution of the anhydrous and decahydrate forms exists. The amounts of the coexisting phases can be determined by the inverse lever-arm rule.

EXAMPLE 4.11

A 30 wt% aqueous Na_2SO_4 solution of 5,000 lb/h enters a cooling-type crystallizer at 50°C. At what temperature will crystallization begin? Will the crystals be the decahydrate or anhydrous form? To what temperature will the mixture have to be cooled to crystallize 50% of the Na_2SO_4?

SOLUTION

From Figure 4.24, the original solution of 30 wt% Na_2SO_4 at 50°C corresponds to a point in the homogeneous liquid solution region. If a vertical line is dropped from that point, it intersects the solubility curve EC at 31°C. Below this temperature, the crystals formed are the decahydrate.

The feed contains $(0.30)(5,000) = 1,500$ lb/h of Na_2SO_4 and $(5,000 - 1,500) = 3,500$ lb/h of H_2O. Thus, $(0.5)(1,500) = 750$ lb/h are to be crystallized. The decahydrate crystals include water of hydration in an amount given by ratioing molecular weights or

$$750 \left[\frac{(10)(18.016)}{(142.05)} \right] = 950 \text{ lb/h}$$

Thus, the total amount of decahydrate is $750 + 950 = 1,700$ lb/h. The water remaining in the mother liquor is $3,500 - 950 = 2,550$ lb/h. The composition of the mother liquor is $\frac{750}{(2,550 + 750)} (100\%) = 22.7$ wt% Na_4SO_4. From Figure 4.24, the temperature corresponding to 22.7 wt% Na_2SO_4 on the solubility curve EC is 26°C.

The amount of crystals can be verified by applying the inverse lever-arm rule, which gives

$$5,000 \left(\frac{30 - 22.7}{44.1 - 22.7} \right) = 1,700 \text{ lb/h.} \qquad ∎$$

Liquid Adsorption

When a liquid mixture is brought into contact with a microporous solid, adsorption of certain components in the mixture takes place on the internal surface of the solid. The maximum extent of adsorption occurs when equilibrium is reached. The solid, which is essentially insoluble in the liquid, is the *adsorbent*. The component(s) being adsorbed are called *solutes* when in the liquid and constitute the *adsorbate* upon adsorption on the solid. In general, the higher the concentration of the solute, the higher is the equilibrium adsorbate concentration on the adsorbent. The component(s) of the liquid mixture other than the solute(s), that is, the solvent (carrier) are assumed not to adsorb.

No theory for predicting adsorption equilibrium curves, based on molecular properties of the solute and solid, is universally embraced. Instead, laboratory experiments must be performed at a fixed temperature for each liquid mixture and adsorbent to provide data for plotting curves, called *adsorption isotherms*. Figure 4.25, taken from the data of Fritz and Schuluender [13], is an isotherm for the adsorption of phenol from an aqueous solution onto activated carbon at 20°C. Activated, powdered, or granular carbon is a microcrystalline, nongraphitic form of carbon that has a microporous structure to give it a very high

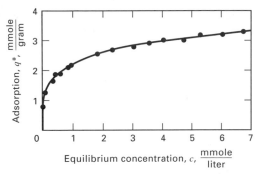

Figure 4.25 Adsorption isotherm for phenol from an aqueous solution in the presence of activated carbon at 20°C.

internal surface area per unit mass of carbon, and therefore a high capacity for adsorption. Activated carbon preferentially adsorbs organic compounds rather than water when contacted with an aqueous phase containing dissolved organics. As shown in Figure 4.25, as the concentration of phenol in the aqueous phase is increased, the extent of adsorption increases very rapidly at first, followed by a much slower increase. When the concentration of phenol is 1.0 mmol/L (0.001 mol/L of aqueous solution or 0.000001 mol/g of aqueous solution), the concentration of phenol on the activated carbon is somewhat more than 2.16 mmol/g (0.00216 mol/g of carbon or 0.203 g phenol/g of carbon). Thus, the affinity of this adsorbent for phenol is extremely high. The extent of adsorption depends markedly on the nature of the process used to produce the activated carbon. Adsorption isotherms like Figure 4.25 can be used to determine the amount of adsorbent required to selectively remove a given amount of solute from a liquid.

Consider the ideal single-stage adsorption process of Figure 4.26, where A is the carrier liquid, B is the solute, and C is the solid adsorbent. Let:

c_B = concentration of solute in the carrier liquid, mol/unit volume
q_B = concentration of adsorbate, mol/unit mass of adsorbent
Q = volume of liquid (assumed to remain constant during adsorption)
S = mass of adsorbent (solute-free basis)

A material balance on the solute, assuming that the entering adsorbent is free of solute and that adsorption equilibrium is achieved, as designated by the asterisk superscript on q, gives

$$c_B^{(F)}Q = c_B Q + q_B^* S \qquad (4\text{-}28)$$

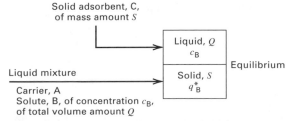

Figure 4.26 Equilibrium stage for liquid adsorption.

This equation can be rearranged to the form of a straight line that can be plotted on the graph of an adsorption isotherm of the type in Figure 4.25, to obtain a graphical solution at equilibrium for c_B and q_B^*. Thus, solving (4-28) for q_B^*,

$$q_B^* = -\frac{Q}{S}c_B + c_B^{(F)}\frac{Q}{S} \tag{4-29}$$

The intercept on the c_B axis is $c_B^{(F)}Q/S$, and slope is $-(Q/S)$. The intersection of (4-29) with the adsorption isotherm is the equilibrium condition, c_B and q_B^*.

Alternatively, an algebraic solution can be obtained. Adsorption isotherms for equilibrium liquid adsorption of a species i can frequently be fitted with the empirical Freundlich equation, discussed in Chapter 15.

$$q_i^* = Ac_i^{(1/n)} \tag{4-30}$$

where A and n depend on the solute, carrier, and particular adsorbent. The constant, n, is greater than 1, and A is a function of temperature. Freundlich developed his equation from experimental data on the adsorption on charcoal of organic solutes from aqueous solutions. Substitution of (4-30) into (4-29) gives

$$Ac_B^{(1/n)} = -\frac{Q}{S}c_B + c_B^{(F)}\frac{Q}{S} \tag{4-31}$$

which is a nonlinear equation in c_B that can be solved numerically by an iterative method, as illustrated in the following example.

EXAMPLE 4.12

One liter of an aqueous solution containing 0.010 mol of phenol is brought to equilibrium at 20°C with 5 g of activated carbon having the adsorption isotherm shown in Figure 4.25. Determine the percent adsorption of the phenol and the equilibrium concentrations of phenol on carbon by:

(a) A graphical method
(b) A numerical algebraic method

For the latter case, the curve of Figure 4.25 is fitted quite well with the Freundlich equation (4-30), giving

$$q_B^* = 2.16c_B^{(1/4.35)} \tag{1}$$

SOLUTION

From the data given, $c^{(F)} = 10$ mmol/L, $Q = 1$ L, and $S = 5$ g.

(a) *Graphical method.* From (4-29), $q_B^* = -(\frac{1}{5})c_B + 10(\frac{1}{5}) = -0.2c_B + 2$

This equation, with a slope of -0.2 and an intercept of 2, when plotted on Figure 4.25, yields an intersection with the equilibrium curve at $q_B^* = 1.9$ mmol/g and $c_B = 0.57$ mmol/liter. Thus, the percent adsorption of phenol is

$$\frac{c_B^{(F)} - c_B}{c_B^{(F)}} = \frac{10 - 0.57}{10} = 0.94 \quad \text{or} \quad 94\%$$

(b) *Numerical algebraic method.* Applying Eq. (1) from the problem statement and (4-31),

$$2.16c_B^{0.23} = -0.2c_B + 2 \tag{2}$$

or

$$f\{c_B\} = 2.16c_B^{0.23} + 0.2c_B - 2 = 0 \tag{3}$$

This nonlinear equation for c_B can be solved by any of a number of iterative numerical techniques. For example, Newton's method [14] can be applied to Eq. (3) by using the iteration rule:

$$c_B^{(k+1)} = c_B^{(k)} - f^{(k)}\{c_B\}/f'^{(k)}\{c_B\} \tag{4}$$

where k is the iteration index. For this example, $f\{c_B\}$ is given by Eq. (3) and $f'\{c_B\}$ is obtained by differentiating Eq. (3) with respect to c_B to give

$$f'^{(k)}\{c_B\} = 0.497 c_B^{-0.77} + 0.2$$

A convenient initial guess for c_B can be made by assuming almost 100% adsorption of phenol to give $q_B^* = 2$ mmol/g. Then, from (4-30),

$$c_B^{(0)} = (q_B^*/A)^n = (2/2.16)^{4.35} = 0.72 \text{ mmol/L}$$

where the (0) superscript designates the starting guess. The Newton iteration rule of Eq. (4) can now be applied, giving the following results:

k	$c_B^{(k)}$	$f^{(k)}\{c_B\}$	$f'^{(k)}\{c_B\}$	$c_B^{(k+1)}$
0	0.72	0.1468	0.8400	0.545
1	0.545	−0.0122	0.9928	0.558
2	0.558	−0.00009	0.9793	0.558

These results indicate convergence to $f\{c_B\} = 0$ for a value of $c_B = 0.558$ after only three iterations. From Eq. (1),

$$q_B^* = 2.16(0.558)^{(1/4.35)} = 1.89 \text{ mmol/g}$$

The result of the numerical method is within the accuracy of the graphical method. ∎

4.8 GAS–LIQUID SYSTEMS

Vapor–liquid systems were covered in Sections 4.2, 4.3, and 4.4. There, the vapor was a mixture of species, most or all of which were condensable. Although the terms *vapor* and *gas* are often used interchangeably, the term *gas* is used to designate a mixture for which the temperature is above the critical temperatures of most or all of the species in the mixture. Thus, the components of a gas mixture are not easily condensed to a liquid. In this section, the physical equilibrium of gas–liquid mixtures is considered.

Even though components of a gas mixture are at a temperature above critical, they can dissolve in an appropriate liquid solvent to an extent that depends on the temperature and their partial pressure in the gas mixture. With good mixing, equilibrium between the two phases can be achieved in a short time unless the liquid is very viscous.

Unlike equilibrium vapor–liquid mixtures, where, as discussed in Chapter 2, a number of theoretical relationships are in use for estimating K-values from molecular properties, no widely accepted theory exists for gas–liquid mixtures. Instead, experimental data, plots of experimental data, or empirical correlations are used.

Experimental solubility data for 13 common gases dissolved in water are plotted over a range of temperature from 0 to as high as 100°C in Figure 4.27. The ordinate is the equilibrium mole fraction of the gas (solute) in the liquid when the partial pressure of the gas is 1 atm. The curves of Figure 4.27 can be used to estimate the solubility in water at other pressures and for mixtures of gases by applying Henry's law, provided that mole-fraction solubilities are low and no chemical reactions occur among the gas species or with water. Henry's law, discussed briefly in Chapter 2, is written for use with Figure 4.27 as

$$x_i = \left(\frac{1}{H_i}\right) y_i P \tag{4-32}$$

where H_i = Henry's law constant, atm

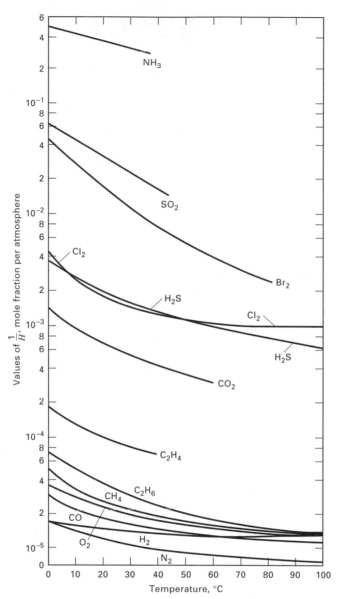

Figure 4.27 Henry's law constant for solubility of gases in water. [Adapted from O.A. Hougen, K.M. Watson, and R.A. Ragatz, *Chemical Process Principles. Part I,* 2nd ed., John Wiley and Sons, New York (1954).]

For gases with a high solubility, such as ammonia, Henry's law may not be applicable, even at low partial pressures. In that case, experimental data for the actual conditions of pressure and temperature are necessary. In either case, calculations of equilibrium conditions are made in the manner illustrated in previous sections of this chapter by combining material balances with equilibrium relationships or data. The following two examples illustrate single-stage, gas–liquid equilibria calculation methods.

EXAMPLE 4.13

An ammonia plant, located at the base of a 300-ft (91.44-m)-high mountain, employs a unique absorption system for disposing of by-product CO_2. The CO_2 is absorbed in water at a CO_2 partial pressure of 10 psi (68.8 kPa) above that required to lift water to the top of the mountain.

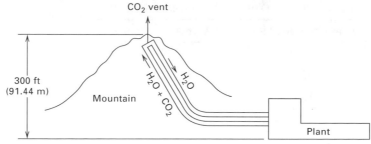

Figure 4.28 Flowsheet for Example 4.13.

The CO_2 is then vented to the atmosphere at the top of the mountain, the water being recirculated as shown in Figure 4.28. At 25°C, calculate the amount of water required to dispose of 1,000 ft³ (28.31 m³)(STP) of CO_2.

SOLUTION

Basis: 1,000 ft³ (28.31 m³) of CO_2 at 0°C and 1 atm (STP). From Figure 4.27, the reciprocal of the Henry's law constant for CO_2 at 25°C is 6×10^{-4} mole fraction/atm. The CO_2 pressure in the absorber (at the foot of the mountain) is

$$p_{CO_2} = \frac{10}{14.7} + \frac{300 \text{ ft H}_2\text{O}}{34 \text{ ft H}_2\text{O/atm}} = 9.50 \text{ atm} = 960 \text{ kPa}$$

At this partial pressure, the equilibrium concentration of CO_2 in the water is

$$x_{CO_2} = 9.50(6 \times 10^{-4}) = 5.7 \times 10^{-3} \text{ mole fraction CO}_2 \text{ in water}$$

The corresponding ratio of dissolved CO_2 to water is

$$\frac{5.7 \times 10^{-3}}{1 - 5.7 \times 10^{-3}} = 5.73 \times 10^{-3} \text{ mol CO}_2/\text{mol H}_2\text{O}$$

The total number of moles of CO_2 to be absorbed is

$$\frac{1,000 \text{ ft}^3}{359 \text{ ft}^3/\text{lbmol (at STP)}} = \frac{1,000}{359} = 2.79 \text{ lbmol}$$

or $(2.79)(44)(0.454) = 55.73$ kg

Assuming all the absorbed CO_2 is vented at the mountain top, the number of moles of water required is $2.79/(5.73 \times 10^{-3}) = 458$ lbmol = 8,730 lb = 3,963 kg

If one corrects for the fact that the pressure on top of the mountain is 101 kPa, so that not all of the CO_2 is vented, 4,446 kg (9,810 lb) of water are required. ∎

EXAMPLE 4.14

The partial pressure of ammonia (A) in air–ammonia mixtures in equilibrium with their aqueous solutions at 20°C is given in Table 4.7. Using these data, and neglecting the vapor pressure of

Table 4.7 Partial Pressure of Ammonia over Ammonia–Water Solutions at 20°C

NH_3 Partial Pressure, kPa	g NH_3/g H_2O
4.23	0.05
9.28	0.10
15.2	0.15
22.1	0.20
30.3	0.25

Table 4.8 Y–X Data for Ammonia–Water, 20°C

Y, mol NH_3/mol Air	X, mol NH_3/mol H_2O
0.044	0.053
0.101	0.106
0.176	0.159
0.279	0.212
0.426	0.265

water and the solubility of air in water, construct an equilibrium diagram at 101 kPa using mole ratios Y_A = mol NH_3/mol air, and X_A = mol NH_3/mol H_2O as coordinates. Henceforth, the subscript A is dropped. If 10 mol of gas, of composition $Y = 0.3$, are contacted with 10 mol of a solution of composition $X = 0.1$, what are the compositions of the resulting phases at equilibrium? The process is assumed to be isothermal and at atmospheric pressure.

SOLUTION

The equilibrium data given in Table 4.7 are recalculated in terms of mole ratios in Table 4.8 and plotted in Figure 4.29.

Mol NH_3 in entering gas = $10[Y/(1 + Y)] = 10(0.3/1.3) = 2.3$
Mol NH_3 in entering liquid = $10[X/(1 + X)] = 10(0.1/1.1) = 0.91$

A molar material balance for ammonia about the equilibrium stage is

$$GY_0 + LX_0 = GY_1 + LX_1 \tag{1}$$

where G = moles of air and L = moles of H_2O. Then $G = 10 - 2.3 = 7.7$ mol and $L = 10 - 0.91 = 9.09$ mol. Solving for Y_1 from Eq. (1),

$$Y_1 = -\frac{L}{G}X_1 + \left(\frac{L}{G}X_0 + Y_0\right) \tag{2}$$

This material balance relationship is an equation of a straight line of slope-$L/G = -9.09/7.7 = 1.19$, with an intercept of $(L/G)(X_0) + Y_0 = 0.42$.

The intersection of this material balance line with the equilibrium curve, as shown in Figure 4.29, gives the ammonia composition of the gas and liquid phases leaving the stage as $Y_1 =$

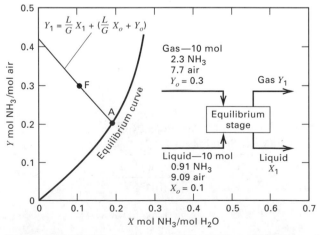

Figure 4.29 Equilibrium for air–NH_3–H_2O at 20°C, 1 atm, in Example 4.14.

0.195 and $X_1 = 0.19$. This result can be checked by an NH_3 balance, since the amount of NH_3 leaving is $(0.195)(7.70) + (0.19)(9.09) = 3.21$, which equals the total moles of NH_3 entering.

It is of importance to recognize that Eq. (2), the material balance line, called an *operating line* and discussed in great detail in Chapters 5 to 8, is the locus of all passing stream pairs; thus, X_0, Y_0 (point F) also lies on this operating line. ■

4.9 GAS–SOLID SYSTEMS

Systems consisting of gas and solid phases that tend to equilibrium are involved in sublimation, desublimation, and adsorption separation operations.

Sublimation and Desublimation

In sublimation, a solid vaporizes into a gas phase without passing through a liquid state. In desublimation, one or more components (solutes) in the gas phase are condensed to a solid phase, without passing through a liquid state. At low pressure, both sublimation and desublimation are governed by the solid vapor pressure of the solute. Sublimation of the solid takes place when the partial pressure of the solute in the gas phase is less than the vapor pressure of the solid at the system temperature. When the partial pressure of the solute in the gas phase exceeds the vapor pressure of the solid, desublimation occurs. At equilibrium, the vapor pressure of the species as a solid is equal to the partial pressure of the species as a solute in the gas phase. This is illustrated in the following example.

EXAMPLE 4.15

Ortho-xylene is partially oxidized in the vapor phase with air to produce phthalic anhydride, PA, in a catalytic reactor (fixed bed or fluidized bed) operating at about 370°C and 780 torr. However, a very large excess of air must be used to keep the xylene content of the reactor feed below 1 mol% to avoid an explosive mixture. In a typical plant, 8,000 lbmol/h of reactor effluent gas, containing 67 lbmol/h of PA and other amounts of N_2, O_2, CO, CO_2, and water vapor are cooled to separate the PA by desublimation to a solid at a total pressure of 770 torr. If the gas is cooled to 206°F, where the vapor pressure of solid PA is 1 torr, calculate the number of pounds of PA condensed per hour as a solid, and the percent recovery of PA from the gas if equilibrium is achieved. Assume that the xylene is converted completely to PA.

SOLUTION

At these conditions, only the PA condenses. At equilibrium, the partial pressure of PA is equal to the vapor pressure of solid PA or 1 torr. Thus, the amount of PA in the cooled gas is given by Dalton's law of partial pressures:

$$(n_{PA})_G = \frac{p_{PA}}{P} n_G \qquad (1)$$

where

$$n_G = (8{,}000 - 67) + (n_{PA})_G \qquad (2)$$

and n = lbmol/h. Combining Eqs. (1) and (2),

$$(n_{PA})_G = \frac{p_{PA}}{P}[(8{,}000 - 67) + (n_{PA})_G] = \frac{1}{770}[(8{,}000 - 67) + (n_{PA})_G] \qquad (3)$$

Solving this linear equation gives

$$(n_{PA})_G = 10.3 \text{ lbmol/h of PA}$$

The amount of PA desublimed is $67 - 10.3 = 56.7$ lbmol/h. The percent recovery of PA is $56.7/67 = 0.846$ or 84.6%. The amount of PA remaining in the gas is a very large quantity. In a typical plant, the gas is cooled to a much lower temperature, perhaps 140°F, where the vapor pressure of PA is less than 0.1 torr, bringing the recovery of PA to almost 99%. ■

Gas Adsorption

As with liquid mixtures, one or more components of a gas mixture can be adsorbed on the surface of a solid adsorbent. Data for a single solute can be represented by an adsorption isotherm of the type shown in Figure 4.25, or in similar diagrams, where the partial pressure of the solute in the gas is used in place of the concentration. However, when two components of a gas mixture are adsorbed and the purpose of adsorption is to separate these two components, other methods of representing the experimental data may be preferred. One such representation is shown in Figure 4.30, from the data of Lewis et al. [15], for the

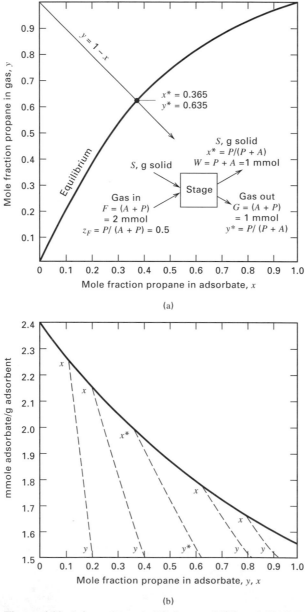

(a)

(b)

Figure 4.30 Adsorption equilibrium at 25°C and 101 kPa of propane and propylene on silica gel. [Adapted from W.K. Lewis, E.R. Gilliland, B. Chertow, and W. H. Hoffman, *J. Am. Chem. Soc.,* **72,** 1153 (1950).]

adsorption of a propane (P)–propylene (A) gas mixture with silica gel at 25°C and 101 kPa. At 25°C, a pressure of at least 1,000 kPa is required to initiate condensation (dew point) of a mixture of propylene and propane. However, in the presence of silica gel, significant amounts of the gas are adsorbed at 101 kPa.

Figure 4.30a is similar to a binary vapor–liquid equilibrium plot of the type discussed in Section 4.2. For adsorption equilibria, the liquid-phase mole fraction is replaced by the mole fraction in the adsorbate. For the propylene–propane mixture, propylene is adsorbed more strongly. For example, for an equimolar mixture in the gas phase, the adsorbate contains only 27 mol% propane. Figure 4.30b combines the data for the equilibrium mole fractions in the gas and adsorbate with the amount of adsorbate per unit of adsorbent. The mole fractions are obtained by reading the abscissa at the two ends of a tie line. For example, for equilibrium with $y_P = y^* = 0.50$, Figure 4.30b gives $x_P = x^* = 0.27$ and 2.08 mmol of adsorbate/g adsorbent. Therefore, $y_A = 0.50$, and $x_A = 0.73$. The separation factor analogous to the relative volatility for distillation is

$$(0.50/0.27)/(0.50/0.73) = 2.7$$

This value is much higher than the α-value for distillation, which, from Figure 2.8, at 25°C and 1,100 kPa is only 1.13. Accordingly, the separation of propylene and propane by adsorption has received some attention. Equilibrium calculations using data such as that shown in Figure 4.30 are made in the usual manner by combining such data with material balance equations, as illustrated in the following example.

EXAMPLE 4.16

Propylene (A) and propane (P), are to be separated by preferential adsorption on silica gel (S) at 25°C and 101 kPa.

Two millimoles of a gas containing 50 mol% P and 50 mol% A is equilibrated with silica gel at 25°C and 101 kPa. Manometric measurements show that 1 mmol of gas is adsorbed. If the data of Figure 4.30 apply, what is the mole fraction of propane in the equilibrium gas and adsorbate, and how many grams of silica gel are used?

SOLUTION

A pictorial representation of the process is included in Figure 4.30a, where W = millimoles of adsorbate, G = millimoles of gas leaving, and z_F = mole fraction of propane in the feed.

The propane mole balance is $\qquad Fz_F = Wx^* + Gy^*$ **(1)**

With $F = 2$, $z_F = 0.5$, $W = 1$, and $G = F - W = 1$, Eq. (1) becomes $1 = x^* + y^*$.

The operating (material balance) line $y^* = 1 - x^*$ is the locus of all solutions of the material balance equation, and is shown in Figure 4.30a. It intersects the equilibrium curve at $x^* = 0.365$, $y^* = 0.635$. From Figure 4.30b, at the point x^*, there must be 2.0 mmol adsorbate/g adsorbent; therefore there are $1.0/2 = 0.50$ g of silica gel in the system. ∎

4.10 MULTIPHASE SYSTEMS

In previous sections of this chapter, only two phases were considered to be in equilibrium. In some applications of multiphase systems, three or more phases coexist. Figure 4.31 is a schematic diagram of a photograph of a laboratory curiosity taken from Hildebrand [16], which shows seven phases in equilibrium at near-ambient temperature. The phase on top is air, followed by six liquid phases in order of increasing density: hexane-rich, aniline-rich, water-rich, phosphorous, gallium, and mercury. Each phase contains all components in the seven-phase mixture, but the mole fractions in many cases are extremely small. For example, the aniline-rich phase contains on the order of 10 mol% n-hexane, 20 mol% water, but much less than 1 mol% each of dissolved air, phosphous, gallium, and mercury. Note that even though the hexane-rich phase is not in direct contact with the water-rich phase, an equilibrium amount of water (approximately 0.06 mol%) is present in the hexane-

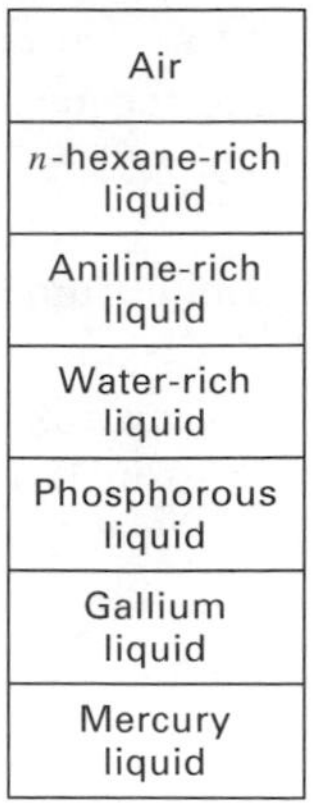

Air
n-hexane-rich liquid
Aniline-rich liquid
Water-rich liquid
Phosphorous liquid
Gallium liquid
Mercury liquid

Figure 4.31 Seven phases in equilibrium.

rich phase because each phase is in equilibrium with each of the other phases as attested by the equality of component fugacities:

$$f_i^{(1)} = f_i^{(2)} = f_i^{(3)} = f_i^{(4)} = f_i^{(5)} = f_i^{(6)} = f_i^{(7)}$$

More practical multiphase systems include the vapor–liquid–solid systems present in evaporative crystallization and pervaporation and the vapor–liquid–liquid systems that occur when distilling certain mixtures of water and hydrocarbons or other organic chemicals having a limited solubility in water. Actually, all of the two-phase systems considered in the previous sections of this chapter involve a third phase, the containing vessel. However, the material of the container is selected on the basis of its inertness to and lack of solubility in the phases it contains, and therefore the material of the container does not normally enter into phase equilibria calculations.

Although calculations of multiphase equilibrium are based on the same principles as for two-phase systems (material balances, energy balances, and phase equilibria criteria such as equality of fugacity), the computations can be quite complex unless simplifying assumptions are made, in which case approximate results are obtained. Rigorous calculations are best made with a computer algorithm. In this section both types of calculations are illustrated.

Approximate Method for a Vapor–Liquid–Solid System

The simplest case of multiphase equilibrium is that encountered in an evaporative crystallizer involving crystallization of an inorganic compound, B, from its aqueous solution at its bubble point in the presence of its vapor. Assume that only two components are present, B and water. In that case, it is common to assume that B has no vapor pressure and water is not present in the solid phase. Thus, the vapor is pure water (steam), the liquid is a mixture of water and B, and the solid phase is pure B. Then, the solubility of B in the liquid phase is not influenced by the presence of the vapor and the system pressure at a given temperature can be approximated by applying Raoult's law to the water in the liquid phase:

$$P = P_{H_2O}^s x_{H_2O} \tag{4-33}$$

where x_{H_2O} can be obtained from the solubility of B.

EXAMPLE 4.17

A 5,000-lb batch of 20 wt% aqueous $MgSO_4$ solution is fed to a vacuum evaporative crystallizer operating at 160°F. At this temperature, the stable solid phase is the monohydrate, with a $MgSO_4$ solubility of 36 wt%. If 75% of the water is evaporated, calculate:

(a) Pounds of water evaporated
(b) Pounds of monohydrate crystals, $MgSO_4 \cdot H_2O$
(c) Crystallizer pressure

SOLUTION

(a) The feed solution is 0.20(5,000) = 1,000 lb $MgSO_4$, and 5,000 − 1,000 = 4,000 lb H_2O. The amount of water evaporated is 0.75(4,000) = 3,000 lb H_2O.

(b) Let W = amount of $MgSO_4$ remaining in solution. Then
$MgSO_4$ in the crystals = 1,000 − W.
MW of H_2O = 18 and MW of $MgSO_4$ = 120.4.
Water of crystallization for the monohydrate = (1,000 − W)(18/120.4) = 0.15(1,000 − W).
Water remaining in solution = 4,000 − 3,000 − 0.15(1,000 − W) = 850 + 0.15W.
Total amount of solution remaining = 850 + 0.15W + W = 850 + 1.15W.

From the solubility of $MgSO_4$, $\qquad 0.36 = \dfrac{W}{850 + 1.15W}$

Solving: W = 522 pounds of dissolved $MgSO_4$.
$MgSO_4$ crystallized = 1,000 − 522 = 478 lb.
Water of crystallization = 0.15(1,000 − W) = 0.15(1,000 − 522) = 72 lb.
Total monohydrate crystals = 478 + 72 = 550 lb.

(c) Crystallizer pressure is given by (4-33). At 160°F, the vapor pressure of H_2O is 4.74 psia. Then water remaining in solution = (850 + 0.15W)/18 = 51.6 lbmol.
$MgSO_4$ remaining in solution = 522/120.4 = 4.3 lbmol.

Hence $\qquad\qquad x_{H_2O} = 51.6/(51.6 + 4.3) = 0.923$

By Raoult's law, $p_{H_2O} = P = 4.74(0.923) = 4.38$ psia ∎

Approximate Method for a Vapor–Liquid–Liquid System

Another case suitable for an approximate method is that of a mixture containing water and hydrocarbons (HCs), at conditions such that a vapor phase and two liquid phases, HC-rich (1) and water-rich (2) coexist. Often the solubilities of water in the liquid HC phase and HCs in the water phase are less than 0.1 mol% and may be neglected. In that case, if the liquid HC phase obeys Raoult's law, the total pressure of the system is given by the sum of the pressures exhibited by the separate phases:

$$P = P^s_{H_2O} + \sum_{HCs} P^s_i x^{(1)}_i \tag{4-34}$$

For more general cases, at low pressures where the vapor phase is ideal but the liquid HC phase may be nonideal,

$$P = P^s_{H_2O} + P \sum_{HCs} K_i x^{(1)}_i \tag{4-35}$$

which can be rearranged to

$$P = \frac{P^s_{H_2O}}{1 - \sum_{HCs} K_i x^{(1)}_i} \tag{4-36}$$

Equations (4-34) and (4-36) can be used directly to estimate the pressure for a given temperature and liquid-phase composition or iteratively to estimate the temperature for a given pressure. An important aspect of the calculation is the determination of the particular phases present from all six possible cases, namely, V, $V-L^{(1)}$, $V-L^{(1)}-L^{(?)}$, $V-L^{(2)}$, $L^{(1)}-L^{(2)}$ and L. It is not always obvious how many and which phases may be present. Indeed, if a $V-L^{(1)}-L^{(2)}$ solution to a problem exists, almost always $V-L^{(1)}$ and $V-L^{(2)}$ solutions also exist. In that case, the three-phase solution is the correct one. It is important, therefore, to seek the three-phase solution first.

EXAMPLE 4.18

A mixture of 1,000 kmol of 75 mol% water and 25 mol% n-octane is cooled under equilibrium conditions at a constant pressure 133.3 kPa (1,000 torr) from an initial temperature of 136°C to a final temperature of 25°C. Determine:

(a) The initial phase condition
(b) The temperature, phase amounts, and compositions when each phase change occurs

Assume that water and n-octane are immiscible liquids. The vapor pressure of octane is included in Figure 2.4.

SOLUTION

(a) Initial phase conditions are $T = 136°C = 276.8°F$ and $P = 133.3 \text{ kPa} = 19.34 \text{ psia}$. Vapor pressures at 276.8°F are $P^s_{H_2O} = 46.7 \text{ psia}$ and $P^s_{nC_8} = 19.5 \text{ psia}$. Because the initial pressure is less than the vapor pressure of each component, the initial phase condition is all vapor, with partial pressures

$$p_{H_2O} = y_{H_2O}P = 0.75(19.34) = 14.5 \text{ psia}$$

$$p_{nC_8} = y_{nC_8}P = 0.25(19.34) = 4.8 \text{ psia}$$

(b) As the temperature is decreased, the first phase change occurs when a temperature is reached where either $P^s_{H_2O} = p_{H_2O} = 14.5 \text{ psia}$ or $P^s_{nC_8} = p_{nC_8} = 4.8 \text{ psia}$. The corresponding temperatures where these vapor pressure occur are 211°F for H_2O and 194°F for nC_8. The highest temperature applies. Therefore, water condenses first when the temperature reaches 211°F. This is the dew-point temperature of the initial mixture at the system pressure. As the temperature is further reduced, the number of moles of water in the vapor decreases, causing the partial pressure of water to decrease below 14.5 psia and the partial pressure of nC_8 to increase above 4.8 psia. Thus, nC_8 begins to condense, forming a second liquid phase, at a temperature higher than 194°F but lower than 211°F. This temperature, referred to as the *secondary dew point*, must be determined iteratively. The calculation is simplified if the bubble point of the mixture is computed first.

From (4-34), $$P = 19.34 \text{ psi} = P^s_{H_2O} + P^s_{nC_8} \qquad \textbf{(1)}$$

Thus, a temperature is sought, as follows, to cause Eq. (1) to be satisfied:

T, °F	$P^s_{H_2O}$, psia	$P^s_{nC_8}$, psia	P, psia
194	10.17	4.8	14.97
202	12.01	5.6	17.61
206	13.03	6.1	19.13
207	13.30	6.2	19.50

By linear interpolation, $T = 206.7°F$ for $P = 19.34 \text{ psia}$. Below this temperature, the vapor phase disappears and only two immiscible liquid phases are present.

To determine the temperature at which one of the liquid phases disappears, which is the same condition as when the second liquid phase begins to appear (secondary dew point), it is noted for this case, with only pure water and a pure HC present, that vaporization, starting from the bubble point, is at a constant temperature until one of the two liquid phases is completely vaporized. Thus, the secondary dew-point temperature is the same as the bubble-point temperature or 206.7°F. At the secondary dew point, the partial pressures are $p_{H_2O} = 13.20$ psia and $p_{nC_8} = 6.14$ psia, with all of the nC_8 in the vapor phase. Therefore, the phase amounts and compositions are

Component	Vapor		H$_2$O-Rich Liquid
	kmol	*y*	kmol
H$_2$O	53.9	0.683	21.1
nC_8	25.0	0.317	0.0
	78.9	1.000	21.1

If desired, additional flash calculations can be made for conditions between the dew point and secondary dew point. The resulting flash curve is Figure 4.32a. If more than one HC species is present, the liquid HC phase does not evaporate at a constant composition and the secondary dew-point temperature is higher than the bubble-point temperature. In that case, the flash is described by Figure 4.32b. ■

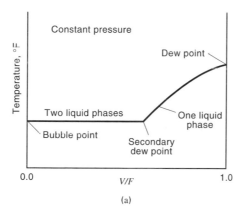

(a)

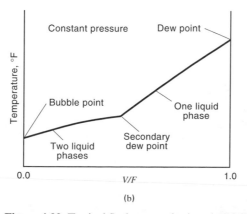

(b)

Figure 4.32 Typical flash curves for immiscible liquid mixtures of water and hydrocarbons at constant pressure: (a) only one hydrocarbon species present; (b) more than one hydrocarbon species present.

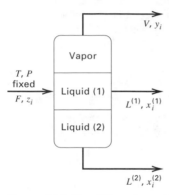

Figure 4.33 Conditions for a three-phase isothermal flash.

Rigorous Method for a Vapor–Liquid–Liquid System

The rigorous method for treating a vapor–liquid–liquid system at a given temperature and pressure is called a *three-phase isothermal flash.* As first presented by Henley and Rosen [17], it is analogous to the isothermal two-phase flash algorithm developed in Section 4.4. The system is shown schematically in Figure 4.33. The usual material balances and phase-equilibrium relations apply for each component:

$$Fz_i = Vy_i + L^{(1)}x_i^{(1)} + L^{(2)}x_i^{(2)} \tag{4-37}$$

$$K_i^{(1)} = y_i/x_i^{(1)} \tag{4-38}$$

$$K_i^{(2)} = y_i/x_i^{(2)} \tag{4-39}$$

Alternatively, the following relation can be substituted for (4-38) or (4-39):

$$K_{D_i} = x_i^{(1)}/x_i^{(2)} \tag{4-40}$$

These equations can be solved by a modification of the Rachford–Rice procedure if we let $\Psi = V/F$ and $\xi = L^{(1)}/(L^{(1)} + L^{(2)})$, where $0 \leq \Psi \leq 1$ and $0 \leq \xi \leq 1$. By combining (4-37), (4-38), and (4-39) with

$$\sum x_i^{(1)} - \sum y_i = 0 \tag{4-41}$$

and

$$\sum x_i^{(1)} - \sum x_i^{(2)} = 0 \tag{4-42}$$

to eliminate y_i, $x_i^{(1)}$, and $x_i^{(2)}$, two simultaneous equations in Ψ and ξ are obtained:

$$\sum_i \frac{z_i(1 - K_i^{(1)})}{\xi(1 - \Psi) + (1 - \Psi)(1 - \xi)K_i^{(1)}/K_i^{(2)} + \Psi K_i^{(1)}} = 0 \tag{4-43}$$

and

$$\sum_i \frac{z_i(1 - K_i^{(1)}/K_i^{(2)})}{\xi(1 - \Psi) + (1 - \Psi)(1 - \xi)K_i^{(1)}/K_i^{(2)} + \Psi K_i^{(1)}} = 0 \tag{4-44}$$

Values of Ψ and ξ are computed by solving the nonlinear equations (4-43) and (4-44) simultaneously by an appropriate numerical method such as that of Newton. Then the

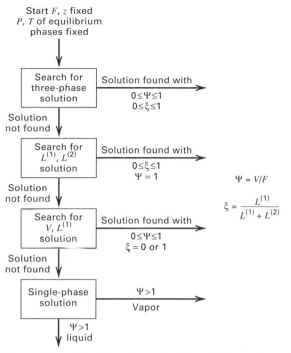

Figure 4.34 Algorithm for an isothermal three-phase flash.

amounts and compositions of the three phases are determined from

$$V = \Psi F \tag{4-45}$$

$$L^{(1)} = \xi(F{-}V) \tag{4-46}$$

$$L^{(2)} = F{-}V{-}L^{(1)} \tag{4-47}$$

$$y_i = \frac{z_i}{\xi(1 - \Psi)/K_i^{(1)} + (1 - \Psi)(1 - \xi)/K_i^{(2)} + \Psi} \tag{4-48}$$

$$x_i^{(1)} = \frac{z_i}{\xi(1 - \Psi) + (1 - \Psi)(1 - \xi)(K_i^{(1)}/K_i^{(2)}) + \Psi K_i^{(1)}} \tag{4-49}$$

$$x_i^{(2)} = \frac{z_i}{\xi(1 - \Psi)(K_i^{(2)}/K_i^{(1)}) + (1 - \Psi)(1 - \xi) + \Psi K_i^{(2)}} \tag{4-50}$$

Calculations for an isothermal three-phase flash are difficult and tedious because of the strong dependency of K-values on liquid-phase compositions when two immiscible liquid phases are present. In addition, it is usually not obvious that three phases will be present, and calculations may be necessary for other combinations of phases. A typical algorithm for determining the phase conditions is shown in Figure 4.34. Because of the complexity of the isothermal three-phase flash algorithm, calculations are best made with a steady-state-process simulation computer program. Such programs can also perform adiabatic or nonadiabatic three-phase flashes by iterating on temperature until the enthalpy balance,

$$h_F F + Q = h_V V + h_{L^{(1)}} L^{(1)} + h_{L^{(2)}} L^{(2)} = 0 \tag{4-51}$$

is satisfied.

EXAMPLE 4.19

In a process for producing styrene from toluene and methanol, the gaseous reactor effluent is as follows:

Component	kmol/h
Hydrogen	350
Methanol	107
Water	491
Toluene	107
Ethylbenzene	141
Styrene	350

If this stream is brought to equilibrium at 38°C and 300 kPa, compute the amounts and compositions of the phases present.

SOLUTION

Because water, hydrocarbons, and a light gas are present in the mixture, the possibility exists that a vapor and two liquid phases may be present, with the methanol being distributed among all three phases. The isothermal three-phase flash module of the ChemCAD simulation program was used with Henry's law for hydrogen and the UNIFAC method for estimating liquid-phase activity coefficients for the other components, to obtain the following results:

	kmol/h		
Component	V	$L^{(1)}$	$L^{(2)}$
Hydrogen	349.96	0.02	0.02
Methanol	9.54	14.28	83.18
Water	7.25	8.12	475.63
Toluene	1.50	105.44	0.06
Ethylbenzene	0.76	140.20	0.04
Styrene	1.22	348.64	0.14
Totals	370.23	616.70	559.07

As would be expected, little of the hydrogen is dissolved in either of the two liquid phases. Little of the other components is left uncondensed. The water-rich liquid phase contains little of the hydrocarbons, but much of the methanol. The organic-rich phase contains most of the hydrocarbons and small amounts of water and methanol.

Additional calculations at 300 kPa indicate that the organic phase condenses first with dew point = 143°C and secondary dew point = 106°C. ∎

SUMMARY

1. The phase rule of Gibbs, which applies to intensive variables at equilibrium, determines the number of independent variables that can be specified. This rule can be extended to the more general determination of the degrees of freedom (number of allowable specifications) for a flow system, including consideration of extensive variables. The intensive and extensive variables are related by material and energy balance equations together with phase equilibrium data in the form of equations, tables, and/or graphs.

2. Vapor–liquid equilibrium conditions for binary systems are conveniently represented and determined with $T-y-x$, $y-x$, and $P-x$ diagrams. The relative volatility for a binary system tends to 1.0 as the critical point is approached.

3. Minimum or maximum-boiling azeotropes, which are formed by close-boiling nonideal liquid mixtures, are conveniently represented by the same types of diagrams used for nonazeotropic (zeotropic) binary mixtures. Highly nonideal liquid mixtures can form heterogeneous azeotropes involving two liquid phases.

4. For multicomponent mixtures, vapor–liquid equilibrium phase compositions and amounts can be determined by isothermal flash, adiabatic flash, and bubble- and dew-point calculations. When the mixtures are nonideal, the computations are best done with process simulation computer programs.

5. Liquid–liquid equilibrium conditions for ternary mixtures are best determined graphically from triangular and other equilibrium diagrams, unless only one of the three components (called the solute) is soluble in the two liquid phases and the system is dilute in the solute. In that case, the conditions can be readily determined algebraically using phase distribution ratios for the solute.

6. Liquid–liquid equilibrium conditions for multicomponent mixtures of four or more components are best determined with process-simulation computer programs, particularly when the system is not dilute with respect to the solute(s).

7. Solid–liquid equilibrium commonly occurs in leaching, crystallization, and adsorption. Leaching calculations commonly assume that the solute is completely dissolved in the solvent and that the remaining solid leaving in the underflow is accompanied by a known fraction of liquid. Crystallization calculations are best made with a solid–liquid phase equilibrium diagram. For crystallization of inorganic salts from an aqueous solution, formation of hydrates must be considered. Equilibrium adsorption can be represented algebraically or graphically by adsorption isotherms.

8. Solubility of gases that are only sparingly soluble in a liquid are well represented by a Henry's law constant that depends on temperature.

9. Solid vapor pressure can be used to determine equilibrium sublimation and desublimation conditions for gas–solid systems. Adsorption isotherms and y–x diagrams are useful in determining adsorption equilibrium conditions for gas mixtures in the presence of a solid adsorbent.

10. Calculations of equilibrium when more than two phases are present are best made with simulation computer programs. However, approximate manual procedures are readily applied to vapor–liquid–solid systems when no component is found in all three phases and for vapor–liquid–liquid systems when only one component distributes in all three phases.

REFERENCES

1. Perry, R.H., D.W. Green, and J.O. Maloney, Eds., *Perry's Chemical Engineers' Handbook,* 6th ed., McGraw-Hill, New York, Section 13 (1984).

2. Gmehling, J., and U. Onken, *Vapor-Liquid Equilibrium Data Collection,* DECHEMA Chemistry Data Series, **1-8** (1977–1984).

3. Keyes, D.B., *Ind. Eng. Chem.,* **21,** 998–1001 (1929).

4. Hughes, R.R., H.D. Evans, and C.V. Sternling, *Chem. Eng. Progr.,* **49,** 78–87 (1953).

5. Rachford, H.H., Jr., and J.D. Rice, *J. Pet. Tech.,* **4** (10), Section 1, p. 19, and Section 2, p. 3 (Oct. 1952).

6. Press, W.H., S.A. Teukolsky, W.T. Vetterling, and B.P. Flannery, *Numerical Recipes in FORTRAN,* 2nd ed., Cambridge University Press, Cambridge, chap. 9 (1992).

7. Goff, G.H., P.S. Farrington, and B.H. Sage, *Ind. Eng. Chem.,* **42,** 735–743 (1950).

8. Constantinides, A., *Applied Numerical Methods with Personal Computers,* McGraw-Hill, New York, chap. 2 (1987).

9. Robbins, L.A., in R.H. Perry, D.H. Green, and J.O. Maloney, Eds., *Perry's Chemical Engineers' Handbook,* 6th ed., McGraw-Hill, New York, pp. 15-9 to 15-14 (1984).

10. Janecke, E., *Z. Anorg. Allg. Chem.,* **51,** 132–157 (1906).

11. Francis, A.W., *Liquid-Liquid Equilibriums,* Interscience, New York (1963).

12. Findlay, A., *Phase Rule,* Dover, New York (1951).

13. Fritz, W., and E.-U. Schuluender, *Chem. Eng. Sci.,* **29,** 1279–1282 (1974).

14. Felder, R.M., and R.W. Rousseau, *Elementary Principles of Chemical Processes,* 2nd ed., John Wiley and Sons, New York, pp. 596–599 (1986).

15. Lewis, W.K., E.R. Gilliland, B. Cherton, and W.H. Hoffman, *J. Am. Chem. Soc.,* **72,** 1153–1157 (1950).

16. Hildebrand, J.H., *Principles of Chemistry,* 4th ed., Macmillan, New York (1940).

17. Henley, E.J., and E.M. Rosen, *Material and Energy Balance Computations,* John Wiley and Sons, New York, pp. 351–353 (1969).

EXERCISES

Section 4.1

4.1 Consider the equilibrium stage shown in Figure 4.35. Conduct a degrees-of-freedom analysis by performing the following steps.
(a) List and count the variables.
(b) Write and count the equations relating the variables.
(c) Calculate the degrees of freedom.
(d) List a reasonable set of design variables.

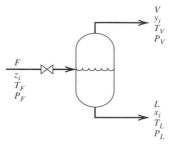

Figure 4.36 Conditions for Exercise 4.3.

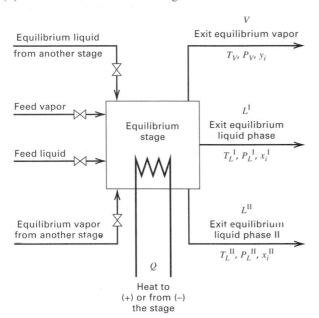

Figure 4.35 Conditions for Exercise 4.1.

4.2 Can the following problems be solved uniquely?
(a) The feed streams to an adiabatic equilibrium stage consist of liquid and vapor streams of known composition, flow rate, temperature, and pressure. Given the stage (outlet) temperature and pressure, calculate the composition and amounts of equilibrium vapor and liquid leaving the stage.
(b) The same as part (a), except that the stage is not adiabatic.
(c) A multicomponent vapor of known temperature, pressure, and composition is to be partially condensed in a condenser. The outlet pressure of the condenser and the inlet cooling water temperature are fixed. Calculate the cooling water required.

4.3 Consider an adiabatic equilibrium flash. The variables are all as indicated in Figure 4.36.
(a) Determine the number of variables.
(b) Write all the independent equations that relate the variables.
(c) Determine the number of equations.
(d) Determine the number of degrees of freedom.
(e) What variables would you prefer to specify in order to solve a typical adiabatic flash problem?

4.4. Determine the number of degrees of freedom for a nonadiabatic equilibrium flash for one liquid feed, one vapor stream product, and two immiscible liquid stream products as shown in Figure 4.33.

4.5 Consider the seven-phase equilibrium system shown in Figure 4.31. Assume that air consists of N_2, O_2, and argon. How many degrees of freedom are computed by the Gibbs phase rule? What variables might be specified to fix the system?

Section 4.2

4.6 A liquid mixture containing 25 mol% benzene and 75 mol% ethyl alcohol, which components are miscible in all proportions, is heated at a constant pressure of 1 atm (101.3 kPa, 760 torr) from a temperature of 60°C to 90°C. Using the following $T-x-y$ experimental data, perform calculations to determine the answers to parts (a) through (f).

EXPERIMENTAL $T-x-y$ DATA FOR BENZENE–ETHYL ALCOHOL AT 1 ATM

Temperature, °C:

78.4	77.5	75	72.5	70	68.5	67.7	68.5	72.5	75	77.5	80.1

Mole percent benzene in vapor:

0	7.5	28	42	54	60	68	73	82	88	95	100

Mole percent benzene in liquid:

0	1.5	5	12	22	31	68	81	91	95	98	100

(a) At what temperature does vaporization begin?
(b) What is the composition of the first bubble of equilibrium vapor formed?
(c) What is the composition of the residual liquid when 25 mol% has evaporated? Assume that all vapor formed is retained within the apparatus and that it is completely mixed and in equilibrium with the residual liquid.
(d) Repeat part (c) for 90 mol% vaporized.
(e) Repeat part (d) if, after 25 mol% vaporized as in part (c), the vapor formed is removed and an additional 35 mol% is vaporized by the same technique used in part (c).
(f) Plot the temperature versus the percent vaporized for parts (c) and (e).
(g) Use the following vapor pressure data in conjunction with Raoult's and Dalton's laws to construct a $T-x-y$ dia-

gram, and compare it for the answers obtained in parts (a) and (f) with those obtained using the experimental T–x–y data. What do you conclude about the applicability of Raoult's law to this binary system?

VAPOR PRESSURE DATA

Vapor pressure, torr:

20	40	60	100	200	400	760

Ethanol, °C:

8	19.0	26.0	34.9	48.4	63.5	78.4

Benzene, °C:

−2.6	7.6	15.4	26.1	42.2	60.6	80.1

4.7 Stearic acid is to be steam distilled at 200°C in a direct-fired still, heat-jacketed to prevent condensation. Steam is introduced into the molten acid in small bubbles, and the acid in the vapor leaving the still has a partial pressure equal to 70% of the vapor pressure of pure stearic acid at 200°C. Plot the kilograms of acid distilled per kilogram of steam added as a function of total pressure from 101.3 kPa down to 3.3 kPa at 200°C. The vapor pressure of stearic acid at 200°C is 0.40 kPa.

4.8 The relative volatility, α, of benzene to toluene at 1 atm is 2.5. Construct an x–y diagram for this system at 1 atm. Repeat the construction using vapor pressure data for benzene from Exercise 4.6 and for toluene from the following table in conjunction with Raoult's and Dalton's laws. Also construct a T–x–y diagram.
(a) A liquid containing 70 mol% benzene and 30 mol% toluene is heated in a container at 1 atm until 25 mol% of the original liquid is evaporated. Determine the temperature. The phases are then separated mechanically, and the vapors condensed. Determine the composition of the condensed vapor and the liquid residue.
(b) Calculate and plot the K-values as a function of temperature at 1 atm.

VAPOR PRESSURE OF TOLUENE

Vapor pressure, torr:

20	40	60	100	200	400	760	1,520

Temperature, °C:

18.4	31.8	40.3	51.9	69.5	89.5	110.6	136

4.9 The vapor pressure of toluene is given in Exercise 4.8, and that of n-heptane is given in the accompanying table.

VAPOR PRESSURE OF n-HEPTANE

Vapor pressure, torr:

20	40	60	100	200	400	760	1,520

Temperature, °C:

9.5	22.3	30.6	41.8	58.7	78.0	98.4	124

(a) Plot an x–y equilibrium diagram for this system at 1 atm by using Raoult's and Dalton's laws.

(b) Plot the T–x bubble-point curve at 1 atm.
(c) Plot α and K-values versus temperature.
(d) Repeat part (a) using the arithmetic average value of α, calculated from the two extreme values.
(e) Compare your x–y and T–x–y diagrams with the following experimental data of Steinhauser and White [*Ind. Eng. Chem.*, **41**, 2912 (1949)].

VAPOR–LIQUID EQUILIBRIUM DATA FOR n-HEPTANE/TOLUENE AT 1 ATM

$x_{n\text{-heptane}}$	$y_{n\text{-heptane}}$	T, °C
0.025	0.048	110.75
0.129	0.205	106.80
0.250	0.349	104.50
0.354	0.454	102.95
0.497	0.577	101.35
0.692	0.742	99.73
0.843	0.864	98.90
0.940	0.948	98.50
0.994	0.993	98.35

4.10 Saturated liquid feed, of $F = 40$ mol/h, containing 50 mol% A in B is supplied continuously to the apparatus shown in Figure 4.37. The condensate from the condenser is split so that half of it is returned to the still pot.
(a) If heat is supplied at such a rate that $W = 30$ mol/h and $\alpha = 2$, as subsequently defined, what will be the composition of the overhead and the bottom product?
(b) If the operation is changed so that no condensate is returned to the still pot and $W = 3D$ as before, what will be the composition of the products?

$$\alpha = \frac{P_A^s}{P_B^s} = \frac{y_A x_B}{y_B x_A}$$

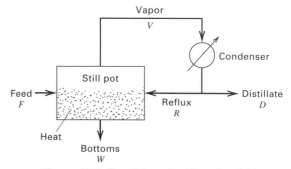

Figure 4.37 Conditions for Exercise 4.10.

4.11 It is required to design a fractionation tower to operate at 101.3 kPa to obtain a distillate consisting of 95 mol% acetone (A) and 5 mol% water, and a residue containing 1 mol% A. The feed liquid is at 125°C and 687 kPa and contains 57 mol% A. The feed is introduced to the column through an expansion valve so that it enters the column partially vaporized at 60°C. From the data below, determine the molar ratio of liquid to vapor in the partially vaporized feed. Enthalpy and equilibrium data are as follows:

Molar latent heat of A = 29,750 kJ/kmol (constant)
Molar latent heat of H_2O = 42,430 kJ/kmol (constant)
Molar specific heat of A = 134 kJ/kmol-K (constant)
Molar specific heat of H_2O = 75.3 kJ/kmol-K (constant)
Enthalpy of high-pressure, hot feed before adiabatic expansion = 0
Enthalpies of feed phases after expansion: h_V = 27,200 kJ/kmol, h_L = −5,270 kJ/kmol

VAPOR-LIQUID EQUILIBRIUM DATA FOR ACETONE-H_2O AT 101.3 kPa

	T, °C						
	56.7	57.1	60.0	61.0	63.0	71.7	100
Mol% A in liquid:	100	92.0	50.0	33.0	17.6	6.8	0
Mol% A in vapor:	100	94.4	85.0	83.7	80.5	69.2	0

4.12 Using vapor pressure data from Exercises 4.6 and 4.8 and the enthalpy data provided below:
(a) Construct an *h–x–y* diagram for the benzene–toluene system at 1 atm (101.3 kPa) based on the use of Raoult's and Dalton's laws.

	Saturated Enthalpy, kJ/kg			
	Benzene		Toluene	
T, °C	h_L	h_V	h_L	h_V
60	79	487	77	471
80	116	511	114	495
100	153	537	151	521

(b) Calculate the energy required for 50 mol% vaporization of a 30 mol% liquid solution of benzene in toluene, initially at saturation temperature. If the vapor is then condensed, what is the heat load on the condenser in kJ/kg of solution if the condensate is saturated and if it is subcooled by 10°C?

Section 4.3

4.13 Vapor–liquid equilibrium data at 101.3 kPa are given for the chloroform–methanol system on p. 13-11 of *Perry's Chemical Engineers' Handbook*, 6th ed. From these data, prepare plots like Figures 4.6b and 4.6c. From the plots, determine the azeotropic composition and temperature at 101.3 kPa. Is the azeotrope of the minimum- or maximum-boiling type?

4.14 Vapor–liquid equilibrium data at 101.3 kPa are given for the water–formic acid system on p. 13-14 of *Perry's Chemical Engineers' Handbook*, 6th ed. From these data, prepare plots like Figures 4.7b and 4.7c. From the plots, determine the azeotropic composition and temperature at

101.3 kPa. Is the azeotrope of the minimum- or maximum-boiling type?

4.15 Vapor–liquid equilibrium data for mixtures of water and isopropanol at 1 atm (101.3 kPa, 760 torr) are given below.
(a) Prepare *T–x–y* and *x–y* diagrams.
(b) When a solution containing 40 mol% isopropanol is slowly vaporized, what will be the composition of the initial vapor formed?
(c) If this same 40% mixture is heated under equilibrium conditions until 75 mol% has been vaporized, what will be the compositions of the vapor and liquid produced?

VAPOR–LIQUID EQUILIBRIUM FOR ISOPROPANOL AND WATER AT 1 ATM

	Mol% Isopropanol	
T, °C	Liquid	Vapor
93.00	1.18	21.95
89.75	3.22	32.41
84.02	8.41	46.20
83.85	9.10	47.06
82.12	19.78	52.42
81.64	28.68	53.44
81.25	34.96	55.16
80.62	45.25	59.26
80.32	60.30	64.22
80.16	67.94	68.21
80.21	68.10	68.26
80.28	76.93	74.21
80.66	85.67	82.70
81.51	94.42	91.60

Notes:

Composition of the azeotrope: $x = y = 68.54\%$.
Boiling point of azeotrope: 80.22°C.
Boiling point of pure isopropanol: 82.5°C.

(d) Calculate *K*-values and values of α at 80°C and 89°C.
(e) Compare your answers in parts (a), (b), and (c) to those obtained from *T–x–y* and *x–y* diagrams based on the following vapor pressure data and Raoult's and Dalton's laws. What do you conclude about the applicability of Raoult's law to this system?

VAPOR PRESSURES OF ISOPROPANOL AND WATER

Vapor pressure, torr	200	400	760
Isopropanol, °C	53.0	67.8	82.5
Water, °C	66.5	83	100

Section 4.4

4.16 Using the *y–x* and *T–y–x* diagrams in Figures 4.3 and 4.4, determine the temperature, amounts, and compositions of the equilibrium vapor and liquid phases at 101 kPa for the following conditions with a 100-kmol mixture of nC_6 (H) and nC_8 (C).

(a) $z_H = 0.5$, $\Psi = V/F = 0.2$
(b) $z_H = 0.4$, $y_H = 0.6$
(c) $z_H = 0.6$, $x_C = 0.7$
(d) $z_H = 0.5$, $\Psi = 0$
(e) $z_H = 0.5$, $\Psi = 1.0$
(f) $z_H = 0.5$, $T = 200°F$

4.17 For a binary mixture of components 1 and 2, show that the equilibrium phase compositions and amounts can be computed directly from the following reduced forms of Eqs. (5), (6), and (3) of Table 4.4.

$$x_1 = (1 - K_2)/(K_1 - K_2)$$

$$x_2 = 1 - x_1$$

$$y_1 = (K_1 K_2 - K_1)/(K_2 - K_1)$$

$$y_2 = 1 - y_1$$

$$\Psi = \frac{V}{F} = \frac{z_1[(K_1 - K_2)/(1 - K_2)] - 1}{K_1 - 1}$$

4.18 Consider the Rachford–Rice form of the flash equation,

$$\sum_{i=1}^{C} \frac{z_i(1 - K_i)}{1 + (V/F)(K_i - 1)} = 0$$

Under what conditions can this equation be satisfied?

4.19 A liquid containing 60 mol% toluene and 40 mol% benzene is continuously distilled in a single-equilibrium-stage unit at atmospheric pressure. What percent of benzene in the feed leaves in the vapor if 90% of the toluene entering in the feed leaves in the liquid? Assume a relative volatility of 2.3 and obtain the solution graphically.

4.20 Solve Exercise 4.19 by assuming an ideal solution and using vapor pressure data from Figure 2.4. Also determine the temperature.

4.21 A seven-component mixture is flashed at a specified temperature and pressure.

(a) Using the K-values and feed composition given below, make a plot of the Rachford–Rice flash function

$$f\{\Psi\} = \sum_{i=1}^{C} \frac{z_i(1 - K_i)}{1 + \Psi(K_i - 1)}$$

at intervals of Ψ of 0.1, and from the plot estimate the correct root of Ψ.

(b) An alternative form of the flash function is

$$f\{\Psi\} = \sum_{i=1}^{C} \frac{z_i K_i}{1 + \Psi(K_i - 1)} - 1$$

Make a plot of this equation also at intervals of Ψ of 0.1 and explain why the Rachford–Rice function is preferred.

Component	z_i	K_i
1	0.0079	16.2
2	0.1321	5.2
3	0.0849	2.6
4	0.2690	1.98
5	0.0589	0.91
6	0.1321	0.72
7	0.3151	0.28

4.22 One hundred kilomoles of a feed composed of 25 mol% n-butane, 40 mol% n-pentane, and 35 mol% n-hexane are flashed at steady-state conditions. If 80% of the hexane is to be recovered in the liquid at 240°F, what pressure is required, and what are the liquid and vapor compositions? Obtain K-values from Figure 2.8.

4.23 An equimolar mixture of ethane, propane, n-butane, and n-pentane is subjected to a flash vaporization at 150°F and 205 psia. What are the expected amounts and compositions of the liquid and vapor products? Is it possible to recover 70% of the ethane in the vapor by a single-stage flash at other conditions without losing more than 5% of nC_4 to the vapor? Obtain K-values from Figure 2.8.

4.24 The system shown in Figure 4.38 is used to cool the reactor effluent and separate the light gases from the heavier

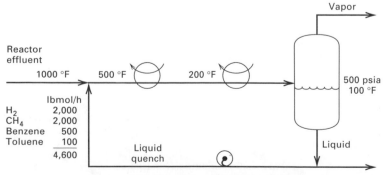

Figure 4.38 Conditions for Exercise 4.24.

hydrocarbons. K-values for the components at 500 psia and 100°F are

Component	K_i
H_2	80
CH_4	10
Benzene	0.010
Toluene	0.004

(a) Calculate the composition and flow rate of the vapor leaving the flash drum.
(b) Does the flow rate of liquid quench influence the result? Prove your answer analytically.

4.25 The mixture shown in Figure 4.39 is partially condensed and separated into two phases. Calculate the amounts and compositions of the equilibrium phases, V and L.

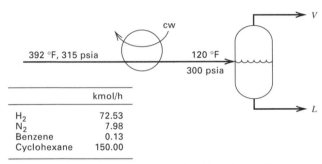

	kmol/h
H_2	72.53
N_2	7.98
Benzene	0.13
Cyclohexane	150.00

Figure 4.39 Conditions for Exercise 4.25.

4.26 The following stream is at 200 psia and 200°F. Determine whether it is a subcooled liquid or a superheated vapor, or whether it is partially vaporized, without making a flash calculation.

Component	lbmol/h	K-value
C_3	125	2.056
nC_4	200	0.925
nC_5	175	0.520

4.27 The overhead system for a distillation column is shown in Figure 4.40. The composition of the total distillates is indicated, with 10 mol% of it being taken as vapor. Determine the pressure in the reflux drum, if the temperature is 100°F. Use the following K-values by assuming that K is inversely proportional to pressure.

Component	K at 100°F, 200 psia
C_2	2.7
C_3	0.95
C_4	0.34

4.28 Determine the phase condition of a stream having the following composition at 7.2°C and 2,620 kPa.

Component	kmol/h
N_2	1.0
C_1	124.0
C_2	87.6
C_3	161.6
nC_4	176.2
nC_5	58.5
nC_6	33.7

Perform the calculations with a simulation computer program using at least three different options for K-values. Does the choice of K-value method influence the results?

4.29 A liquid mixture consisting of 100 kmol of 60 mol% benzene, 25 mol% toluene, and 15 mol% o-xylene is flashed at 1 atm and 100°C.
(a) Compute the amounts of liquid and vapor products and their composition.
(b) Repeat the calculation at 100°C and 2 atm.
(c) Repeat the calculation at 105°C and 0.1 atm.
(d) Repeat the calculation at 150°C and 1 atm.

Assume ideal solutions and use the vapor pressure curves of Figure 2.4 for benzene and toluene. For o-xylene, draw a vapor pressure line that goes through the points (100.2°C, 200 torr) and (144°C, 760 torr).

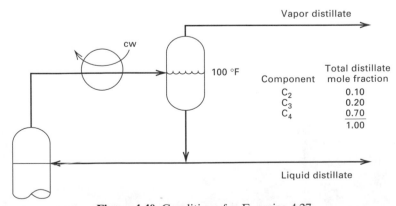

Component	Total distillate mole fraction
C_2	0.10
C_3	0.20
C_4	0.70
	1.00

Figure 4.40 Conditions for Exercise 4.27.

4.30 Prove that the vapor leaving an equilibrium flash is at its dew point and that the liquid leaving an equilibrium flash is at its bubble point.

4.31 The following mixture is introduced into a distillation column as saturated liquid at 1.72 MPa. Calculate the bubble-point temperature using the K-values of Figure 2.8.

Compound	kmol/h
Ethane	1.5
Propane	10.0
n-Butane	18.5
n-Pentane	17.5
n-Hexane	3.5

4.32 An equimolar solution of benzene and toluene is totally evaporated at a constant temperature of 90°C. What are the pressures at the beginning and end of the vaporization process? Assume an ideal solution and use the vapor pressure curves of Figure 2.4.

4.33 The following equations are given by Sebastiani and Lacquaniti [*Chem. Eng. Sci.*, **22**, 1155 (1967)] for the liquid-phase activity coefficients of the water (W)–acetic acid (A) system.

$$\log \gamma_W = x_A^2[A + B(4x_W - 1) + C(x_W - x_A)(6x_W - 1)]$$

$$\log \gamma_A = x_W^2[A + B(4x_W - 3) + C(x_W - x_A)(6x_W - 5)]$$

$$A = 0.1182 + \frac{64.24}{T(K)}$$

$$B = 0.1735 - \frac{43.27}{T(K)}$$

$$C = 0.1081$$

Find the dew point and bubble point of a mixture of composition $x_W = 0.5$, $x_A = 0.5$ at 1 atm. Flash the mixture at a temperature halfway between the dew point and the bubble point.

4.34 Find the bubble-point and dew-point temperatures of a mixture of 0.4 mole fraction toluene (1) and 0.6 mole fraction n-butanol (2) at 101.3 kPa. The K-values can be calculated from (2-72), the modified Raoult's law, using vapor pressure data, and γ_1 and γ_2 from the van Laar equation of Table 2.9 with $A_{12} = 0.855$ and $A_{21} = 1.306$. If the same mixture is flashed at a temperature midway between the bubble point and dew point, and 101.3 kPa, what fraction is vaporized, and what are the compositions of the two phases?

4.35 (a) For a liquid solution having a molar composition of ethyl acetate (A) of 80% and ethyl alcohol (E) of 20%, calculate the bubble-point temperature at 101.3 kPa and the composition of the corresponding vapor using (2-72) with vapor pressure data and the van Laar equation of Table 2.9 with $A_{AE} = 0.855$, $A_{EA} = 0.753$.
(b) Find the dew point of the mixture.

(c) Does the mixture form an azeotrope? If so, predict the temperature and composition.

4.36 A binary solution at 107°C contains 50 mol% water (W) and 50 mol% formic acid (F). Using (2-72) with vapor pressure data and the van Laar equation of Table 2.9 with $A_{WF} = -0.2935$ and $A_{FW} = -0.2757$, compute:
(a) The bubble-point pressure
(b) The dew-point pressure

Also determine whether the mixture forms a maximum- or minimum-boiling azeotrope. If so, predict the azeotropic pressure at 107°C and the azeotropic composition.

4.37 For a mixture consisting of 45 mol% n-hexane, 25 mol% n-heptane, and 30 mol% n-octane at 1 atm, use a simulation computer program to:
(a) Find the bubble- and dew-point temperatures
(b) Find the flash temperature, and the compositions and relative amounts of the liquid and vapor products if the mixture is subjected to a flash distillation at 1 atm so that 50 mol% of the feed is vaporized
(c) Find how much of the octane is taken off as vapor if 90% of the hexane is taken off as vapor

Repeat parts (a) and (b) at 5 atm and 0.5 atm.

4.38 In Figure 4.41, 150 kmol/h of a saturated liquid, L_1, at 758 kPa, of molar composition, propane 10%, n-butane 40%, and n-pentane 50%, enters the reboiler from stage 1. What are the compositions and amounts of V_B and B? What is Q_R, the reboiler duty? Use a simulation computer program to find the answers.

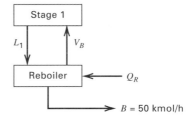

Figure 4.41 Conditions for Exercise 4.38.

4.39 (a) Find the bubble-point temperature of the following mixture at 50 psia, using K-values from Figure 2.8 or Figure 2.9.

Component	z_i
Methane	0.005
Ethane	0.595
n-Butane	0.400

(b) Find the temperature that results in 25% vaporization at this pressure. Determine the corresponding liquid and vapor compositions.

4.40 As shown in Figure 4.42, a hydrocarbon mixture is heated and expanded before entering a distillation column.

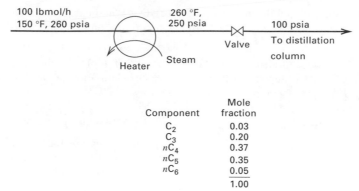

Component	Mole fraction
C_2	0.03
C_3	0.20
nC_4	0.37
nC_5	0.35
nC_6	0.05
	1.00

Figure 4.42 Conditions for Exercise 4.40.

Calculate, using a simulation computer program, the mole percent vapor phase and vapor and liquid phase mole fractions at each of the three locations indicated by a pressure specification.

4.41 Streams entering stage F of a distillation column are shown in Figure 4.43. What is the temperature of stage F and the compositions and amounts of streams V_F and L_F if the pressure is 785 kPa for all streams? Use a simulation computer program to obtain the answers.

4.42 Flash adiabatically, across a valve, a stream composed of the six hydrocarbons given below. The feed upstream of the valve is at 250°F and 500 psia. The pressure downstream of the valve is 300 psia.

Component	z_i
C_2H_4	0.02
C_2H_6	0.03
C_3H_6	0.05
C_3H_8	0.10
iC_4	0.20
nC_4	0.60

Compute using a simulation computer program:
(a) The phase condition upstream of the valve
(b) The temperature downstream of the valve
(c) The molar fraction vaporized downstream of the valve
(d) The mole fraction compositions of the vapor and liquid phases downstream of the valve

4.43 Propose a detailed algorithm like Figure 4.19a and Table 4.4 for a flash where the percent vaporized and the flash pressure are to be specified.

4.44 Determine algorithms for carrying out the following flash calculations, assuming that expressions for K-values and enthalpies are available.

Given	Find
h_F, P	Ψ, T
h_F, T	Ψ, P
h_F, Ψ	T, P
Ψ, T	h_F, P
Ψ, P	h_F, T
T, P	h_F, Ψ

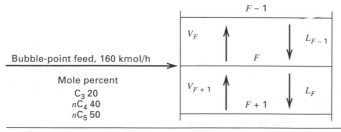

Stream	Total flow rate kmol/h	Composition, mol%		
		C_3	nC_4	nC_5
L_{F-1}	100	15	45	40
V_{F+1}	196	30	50	20

Figure 4.43 Conditions for Exercise 4.41.

Section 4.5

4.45 A feed of 13,500 kg/h consists of 8 wt% acetic acid (B) in water (A). The removal of the acetic acid is to be accomplished by liquid–liquid extraction at 25°C. The raffinate is to contain only 1 wt% acetic acid. The following four solvents, with accompanying distribution coefficients in mass fraction units, are being considered. Water and each solvent (C) can be considered immiscible. For each solvent, estimate the kilograms required per hour if a single equilibrium stage is used.

Solvent	K_D
Methyl acetate	1.273
Isopropyl ether	0.429
Heptadecanol	0.312
Chloroform	0.178

4.46 Forty-five kilograms of a solution containing 30 wt% ethylene glycol in water is to be extracted with furfural. Using Figures 4.14a and 4.14e, calculate:
(a) The minimum quantity of solvent
(b) The maximum quantity of solvent
(c) The weights of solvent-free extract and raffinate for 45 kg solvent, and the percent glycol extracted
(d) The maximum possible purity of glycol in the finished extract and the maximum purity of water in the raffinate for one equilibrium stage.

4.47 Prove that, in a triangular diagram, where each vertex represents a pure component, the composition of the system at any point inside the triangle is proportional to the length of the respective perpendicular drawn from the point to the side of the triangle opposite the vertex in question. It is not necessary to assume a special case (i.e., a right or equilateral triangle).

4.48 A mixture of chloroform ($CHCl_3$) and acetic acid at 18°C and 1 atm (101.3 kPa) is to be extracted with water to recover the acid.
(a) Forty-five kilograms of a mixture containing 35 wt% $CHCl_3$ and 65 wt% acid is treated with 22.75 kg of water at 18°C in a simple one-stage batch extraction. What are the compositions and weights of the raffinate and extract layers produced?
(b) If the raffinate layer from the above treatment is extracted again with one-half its weight of water, what will be the compositions and weights of the new layers?
(c) If all the water is removed from this final raffinate layer, what will its composition be?

Solve this exercise using the following equilibrium data to construct one or more of the types of diagrams in Figure 4.14.

LIQUID–LIQUID EQUILIBRIUM DATA FOR $CHCl_3$-H_2O-CH_3COOH AT 18°C AND 1 ATM

Heavy Phase (wt%)			Light Phase (wt%)		
$CHCl_3$	H_2O	CH_3COOH	$CHCl_3$	H_2O	CH_3COOH
99.01	0.99	0.00	0.84	99.16	0.00
91.85	1.38	6.77	1.21	73.69	25.10
80.00	2.28	17.72	7.30	48.58	44.12
70.13	4.12	25.75	15.11	34.71	50.18
67.15	5.20	27.65	18.33	31.11	50.56
59.99	7.93	32.08	25.20	25.39	49.41
55.81	9.58	34.61	28.85	23.28	47.87

4.49 Isopropyl ether (E) is used to separate acetic acid (A) from water (W). The liquid–liquid equilibrium data at 25°C and 1 atm (101.3 kPa) are presented below.
(a) One hundred kilograms of a 30 wt% A–W solution is contacted with 120 kg of ether in an equilibrium stage. What are the compositions and weights of the resulting extract and raffinate? What would be the concentration of acid in the (ether-rich) extract if all the ether were removed?
(b) A mixture containing 52 kg A and 48 kg W is contacted with 40 kg of E. What are the extract and raffinate compositions and quantities?

LIQUID-LIQUID EQUILIBRIUM DATA FOR ACETIC ACID (A), WATER (W), AND ISOPROPANOL ETHER (E) AT 25°C AND 1 ATM

Water-Rich Layer			Ether-Rich Layer		
Wt% A	Wt% W	Wt% E	Wt% A	Wt% W	Wt% E
1.41	97.1	1.49	0.37	0.73	98.9
2.89	95.5	1.61	0.79	0.81	98.4
6.42	91.7	1.88	1.93	0.97	97.1
13.30	84.4	2.3	4.82	1.88	93.3
25.50	71.1	3.4	11.4	3.9	84.7
36.70	58.9	4.4	21.6	6.9	71.5
45.30	45.1	9.6	31.1	10.8	58.1
46.40	37.1	16.5	36.2	15.1	48.7

Section 4.6

4.50 Diethylene glycol (DEG) is used as a solvent in the UDEX liquid–liquid extraction process [H.W. Grote, *Chem Eng. Progr.*, **54** (8), 43 (1958)] to separate paraffins from aromatics. If 280 lbmol/h of 42.86 mol% *n*-hexane, 28.57 mol% *n*-heptane, 17.86 mol% benzene, and 10.71 mol% toluene is contacted with 500 lbmol/h of 90 mol% aqueous DEG at 325°F and 300 psia, calculate, using a simulation computer program and the UNIFAC L/L method for estimating liquid-phase activity coefficients, the flow rates and molar compositions of the resulting two liquid phases. Is DEG more selective for the paraffins or the aromatics?

4.51 A feed of 110 lbmol/h includes 5, 3, and 2 lbmol/h, respectively, of formic acid, acetic acid, and propionic acid in water. If the acids are extracted in a single equilibrium stage with 100 lbmol/h of ethyl acetate (EA), calculate with a simulation computer program using the UNIFAC method, flow rates and molar compositions of the resulting two liquid phases. What is the order of selectivity of EA for the three organic acids?

Section 4.7

4.52 Repeat Example 4.9 for each of the following changes.
(a) Two kilograms of pure benzene are mixed with 1 kg of meal containing 50 wt% oil.
(b) One kilogram of pure benzene is mixed with 1 kg of meal containing 25 wt% oil.

4.53 Water is to be used in a single equilibrium stage to dissolve 1,350 kg/h of Na_2CO_3 from 3,750 kg/h of a solid, where the balance is an insoluble oxide. If 4,000 kg/h of water is used and the underflow from the stage is 40 wt% solvent on a solute-free basis, compute the flow rates and compositions of the overflow and the underflow.

4.54 Repeat Exercise 4.53 if the residence time is only sufficient to leach 80% of the carbonate.

4.55 A total of 6,000 lb/h of a liquid solution of 40 wt% benzene in naphthalene at 50°C is cooled to 15°C. Assuming that equilibrium is achieved, use Figure 4.23 to determine the amount of crystals formed, and the flow rate and composition of the mother liquor. Are the crystals benzene or naphthalene?

4.56 Repeat Example 4.10, except determine the temperature necessary to crystallize 80% of the naphthalene.

4.57 A total of 10,000 kg/h of a 10 wt% liquid solution of naphthalene in benzene is cooled from 30°C to 0°C. Assuming that equilibrium is achieved, determine the amount of crystals formed and the composition and flow rate of the mother liquid. Are the crystals benzene or naphthalene? Use Figure 4.23.

4.58 Repeat Example 4.11, except let the original solution be 20 wt% Na_2SO_4.

4.59 At 20°C, 1,000 kg of a mixture of 50 wt% $Na_2SO_4 \cdot 10H_2O$ and 50 wt% Na_2SO_4 crystals exists. How many kilograms of water must be added to just completely dissolve the crystals if the temperature is kept at 20°C and equilibrium is maintained? Use Figure 4.24.

4.60 Repeat Example 4.12, except determine the grams of activated carbon to achieve:
(a) 75% adsorption of phenol
(b) 90% adsorption of phenol
(c) 98% adsorption of phenol

4.61 A colored substance (B) is to be removed from a mineral oil by adsorption with clay particles at 25°C. The original oil has a color index of 200 units/100 kg oil, while the decolorized oil must have an index of only 20 units/100 kg oil. The following experimental adsorption equilibrium data have been measured in a laboratory:

c_B, color units/ 100 kg oil	200	100	60	40	10
q_B, color units/ 100 kg clay	10	7.0	5.4	4.4	2.2

(a) Fit the data to the Freundlich equation.
(b) Compute the kilograms of clay needed to treat 500 kg of oil if one equilibrium contact is used.

Section 4.8

4.62 Vapor–liquid equilibrium data in mole fractions for the system acetone–air–water at 1 atm (101.3 kPa) are as follows:

y, acetone in air:	0.004	0.008	0.014	0.017	0.019	0.020
x, acetone in water:	0.002	0.004	0.006	0.008	0.010	0.012

(a) Plot the data as (1) a graph of moles acetone per mole air versus moles acetone per mole water, (2) partial pressure of acetone versus g acetone per g water, and (3) y versus x.
(b) If 20 moles of gas containing 0.015 mole fraction acetone is brought into contact with 15 moles of water in an equilibrium stage, what would be the composition of the discharge streams? Solve graphically. For both parts, neglect partitioning of water and air.

4.63 It has been proposed that oxygen be separated from nitrogen by absorbing and desorbing air in water. Pressures from 101.3 to 10,130 kPa and temperatures between 0 and 100°C are to be used.
(a) Devise a workable scheme for doing the separation assuming the air is 79 mol% N_2 and 21 mol% O_2.
(b) Henry's law constants for O_2 and N_2 are given in Figure 4.27. How many batch absorption steps would be necessary to make 90 mol% pure oxygen? What yield of oxygen (based on total amount of oxygen feed) would be obtained?

4.64 A vapor mixture having equal volumes of NH_3 and N_2 is to be contacted at 20°C and 1 atm (760 torr) with water to absorb a portion of the NH_3. If 14 m³ of this mixture is brought into contact with 10 m³ of water and if equilibrium is attained, calculate the percent of the ammonia originally in the gas that will be absorbed. Both temperature and total pressure will be maintained constant during the absorption. The partial pressure of NH_3 over water at 20°C is as follows:

Partial Pressure of NH₃ in Air, torr	Grams of Dissolved NH₃/100 g of H₂O
470	40
298	30
227	25
166	20
114	15
69.6	10
50.0	7.5
31.7	5.0
24.9	4.0
18.2	3.0
15.0	2.5
12.0	2.0

Section 4.9

4.65 Repeat Example 4.15 for temperatures corresponding to the following vapor pressures for solid PA:
(a) 0.7 torr
(b) 0.4 torr
(c) 0.1 torr

Plot the percent recovery of PA versus the solid vapor pressure for the range from 0.1 torr to 1.0 torr.

4.66 Nitrogen at 760 torr and 300°C contains 10 mol% anthraquinone (A). If this gas is cooled to 200°C, calculate the percent desublimation of A. Vapor pressure data for solid A are as follows:

T, °C:	190.0	234.2	264.3	285.0
Vapor pressure, torr:	1	10	40	100

These data can be fitted to the Antoine equation (2-39) using the first three constants.

4.67 At 25°C and 101 kPa, 2 mol of a gas containing 35 mol% propylene in propane is equilibrated with 0.1 kg of silica gel adsorbent. Using the equilibrium data of Figure 4.30, calculate the moles and composition of the gas adsorbed and the equilibrium composition of the gas not adsorbed.

4.68 A gas containing 50 mol% propylene in propane is to be separated with silica gel having the equilibrium properties shown in Figure 4.30. The final products are to be 90 mol% propylene and 75 mol% propane. If 1,000 lb of silica gel/lbmol of feed gas or less is used, can the desired separation be made in one equilibrium stage? If not, what separation can be achieved?

Section 4.10

4.69 Repeat Example 4.17 for 90% evaporation of the water.

4.70 A 5,000-kg/h aqueous solution of 20 wt% Na₂SO₄ is fed to an evaporative crystallizer operating at 60°C. Equilibrium data are given in Figure 4.24. If 80% of the Na₂SO₄ is to be crystallized, calculate:
(a) The kilograms of water that must be evaporated per hour
(b) The crystallizer pressure in torr

4.71 Calculate the dew-point pressure, secondary dew-point pressure, and bubble-point pressure of the following mixtures at 50°C, assuming that the liquid aromatics and water are mutually insoluble:
(a) 50 mol% benzene and 50 mol% water
(b) 50 mol% toluene and 50 mol% water
(c) 40 mol% benzene, 40 mol% toluene, and 20 mol% water

4.72 Repeat Exercise 4.71, except compute temperatures for a pressure of 2 atm.

4.73 A liquid containing 30 mol% toluene, 40 mol% ethylbenzene, and 30 mol% water is subjected to a continuous flash distillation at a total pressure of 0.5 atm. Assuming that mixtures of ethylbenzene and toluene obey Raoult's law and that the hydrocarbons are completely immiscible in water and vice versa, calculate the temperature and composition of the vapor phase at the bubble-point temperature.

4.74 As shown in Figure 4.8, water (W) and *n*-butanol (B) can form a three phase system at 101 kPa. For a mixture of overall composition of 60 mol% W and 40 mol% B, use a simulation computer program and the UNIFAC method to estimate:
(a) Dew-point temperature and composition of the first drop of liquid
(b) Bubble-point temperature and composition of the first bubble of vapor
(c) Compositions and relative amounts of all three phases for 50 mol% vaporization

4.75 Repeat Example 4.19 for a temperature of 25°C. Are the changes significant?

Chapter 5

Cascades

A separation cascade is a collection of contacting stages arranged to (1) accomplish a separation that cannot be achieved in a single stage, and/or (2) reduce the required amount of the mass- or energy-separating agent. A typical cascade is shown in Figure 5.1, where, in each stage, an attempt is made to bring two or more process streams of different phase state into intimate contact to promote rapid mass and heat transfer, so as to approach physical equilibrium. The resulting phases are then separated and each phase is sent to another stage in the cascade, or withdrawn as a product. Although equilibrium conditions may not be achieved in each stage, it is common to design and analyze cascades using equilibrium-stage models.

5.1 CASCADE CONFIGURATIONS

Cascades can be configured in many ways as shown by the examples in Figure 5.2, where stages are represented by either boxes, as in Figure 5.1, or as horizontal lines in Figure 5.2d,e. Depending on the mechanical design of the stages, cascades may be arranged vertically or horizontally. The feed to be separated is designated by F; the mass-separating agent, if used, is designated by S; and products are designated by P_i.

In the *countercurrent cascade,* shown in Figures 5.1 and 5.2a, the two phases flow countercurrently to each other between stages. As will be shown in examples, this configuration is very efficient and is widely used in practice for absorption, stripping, liquid–liquid extraction, leaching, and washing. The *crosscurrent cascade,* shown in Figure 5.2b, is, in most cases, not as efficient as the countercurrent cascade, but it is much easier to apply in a batchwise manner. It differs from the countercurrent cascade in that the solvent is divided into portions fed individually to each stage.

A complex diamond variation of the crosscurrent cascade is shown in Figure 5.2c. Unlike the two former cascades, which are linear or one-dimensional, the diamond configuration is two-dimensional. One application is to batch crystallization. Feed F is separated in stage 1 into crystals, which pass to stage 2, and mother liquor, which passes to stage 4. In each of the other stages, partial crystallization or recrystallization occurs by processing crystals, mother liquor, or combinations of the two. Final products are purified crystals and impurity-bearing mother liquors.

The first three cascades in Figure 5.2 consist of single sections with streams entering and leaving only from the ends. Such cascades are used to recover components from a feed stream and are not generally useful for making a sharp separation between two selected feed components, called the *key components.* To do this, it is best to provide a cascade consisting of two sections. The countercurrent cascade of Figure 5.2d is often used. It consists of one section above the feed and one below. If two solvents are used, where

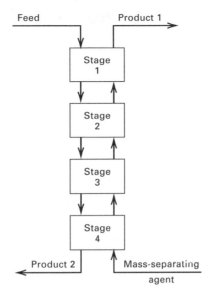

Figure 5.1 Cascade of contacting stages.

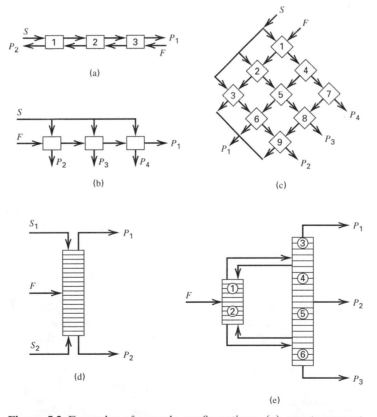

Figure 5.2 Examples of cascade configurations: (a) countercurrent cascade; (b) crosscurrent cascade; (c) two-dimensional diamond cascade; (d) two-section countercurrent cascade; (e) interlinked system of countercurrent cascades.

S_1 selectively dissolves certain components of the feed, while S_2 is more selective for the other components, the process, referred to as *fractional liquid–liquid extraction,* achieves a sharp separation. If S_1 is a liquid absorbent and S_2 is a vapor stripping agent, added to the cascade, as shown, or produced internally by condensation heat transfer at the top to give liquid reflux, and boiling heat transfer at the bottom to give vapor boilup, the process is simple distillation, for which a sharp split between two key components can be achieved if a reasonably high relative volatility exists between the two key components and if reflux, boilup, and the number of stages are sufficient.

Figure 5.2e shows an interlinked system of two distillation columns containing six counter-current cascade sections. Reflux and boilup for the first column are provided by the second column. This system is capable of taking a ternary (three-component) feed, F, and producing three relatively pure products, P_1, P_2, and P_3.

In this chapter, algebraic equations are developed for modeling idealized cascades to illustrate, quantitatively, their capabilities and advantages. First, a simple countercurrent, single-section cascade for solid–liquid leaching and/or washing process is considered. Then, cocurrent, crosscurrent, and countercurrent single-section cascades, based on simplified component distribution coefficients, are compared for a liquid–liquid extraction process. Finally, a two-section, countercurrent cascade is developed for a vapor–liquid distillation operation. In all three cases, a set of linear algebraic equations is reduced to a single relation for estimating the extent of separation as a function of the number of stages in the cascade, the separation factor, and the flow ratio of the mass- or energy-separating agent to the feed. More rigorous models for design and analysis purposes are developed in subsequent chapters. As will be seen, single relations cannot be obtained from rigorous models because of the nonlinear nature of rigorous models, making computer calculations a necessity.

5.2 SOLID–LIQUID CASCADES

Consider the N-stage, countercurrent leaching and/or washing process shown in Figure 5.3. This cascade is an extension of the single-stage systems discussed in Section 4.7. The solid feed, entering stage 1, consists of two components A and B, of mass flow rates F_A and F_B. Pure liquid solvent, C, which can dissolve B completely, but not A, enters stage N at a mass flow rate S. It is convenient to express liquid–phase concentrations of B, the solute, in terms of mass ratios of solute to solvent. The liquid *overflow* from each stage, j, contains Y_j mass of soluble material per mass of solute-free solvent, and no insoluble material. The *underflow* from each stage is a slurry consisting of a mass flow F_A of insoluble solids, a constant ratio of mass of solvent/mass of insoluble solids equal to R, and X_j mass of soluble material/mass of solute-free solvent. For a given solid feed, a relationship between the exiting underflow concentration of the soluble component, X_N, the solvent feed rate, and the number of stages is derived as follows.

If no solid equilibrium exists, and the solid carrier is insoluble in the solvent, then at each stage, the overflow solute concentration, Y_j, equals the underflow solute concentration in the liquid, X_j, which refers to liquid held in the interstices of the solid associated with the underflow. Then all soluble material is dissolved or leached in stage 1 and all other

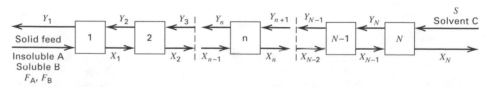

Figure 5.3 Countercurrent leaching or washing system.

stages are washing stages for reducing the amount of soluble material lost in the underflow leaving the last stage, N, and thereby increasing the amount of soluble material leaving in the overflow from stage 1. By solvent material balances, it is readily shown that for constant R, the flow rate of solvent leaving in the overflow from stages 2 to N is S. The flow rate of solvent leaving in the underflow from stages 1 to N is RF_A. Therefore, the flow rate of solvent leaving in the overflow from stage 1 is $S - RF_A$.

A material balance for the soluble material around any interior stage n from $n = 2$ to $N - 1$ is given by

$$Y_{n+1}S + X_{n-1}RF_A = Y_nS + X_nRF_A \tag{5-1}$$

For terminal stages 1 and N, the material balances on the soluble material are, respectively,

$$Y_2S + F_B = Y_1(S - RF_A) + X_1RF_A \tag{5-2}$$

$$X_{N-1}RF_A = Y_NS + X_NRF_A \tag{5-3}$$

Assuming equilibrium, the concentration of soluble material in the overflow from each stage is equal to the concentration of soluble material in the liquid part of the underflow from the same stage. Thus,

$$X_n = Y_n \tag{5-4}$$

In addition, it is convenient to define a washing factor, W, as

$$W = \frac{S}{RF_A} \tag{5-5}$$

If (5-1), (5-2), and (5-3) are each combined with (5-4) to eliminate Y, and the resulting equations are rearranged to allow the substitution of (5-5), the following equations result:

$$X_1 - X_2 = \left(\frac{F_B}{S}\right) \tag{5-6}$$

$$\left(\frac{1}{W}\right)X_{n-1} - \left(\frac{1+W}{W}\right)X_n + X_{n+1} = 0, \qquad n = 2 \text{ to } N - 1 \tag{5-7}$$

$$\left(\frac{1}{W}\right)X_{N-1} - \left(\frac{1+W}{W}\right)X_N = 0 \tag{5-8}$$

Equations (5-6) to (5-8) constitute a set of N linear algebraic equations in N unknowns, $X_n(n = 1 \text{ to } N)$. The equations are of a tridiagonal, sparse-matrix form, which—for example, with $N = 5$—is given by

$$\begin{bmatrix} 1 & -1 & 0 & 0 & 0 \\ \left(\frac{1}{W}\right) & -\left(\frac{1+W}{W}\right) & 1 & 0 & 0 \\ 0 & \left(\frac{1}{W}\right) & -\left(\frac{1+W}{W}\right) & 1 & 0 \\ 0 & 0 & \left(\frac{1}{W}\right) & -\left(\frac{1+W}{W}\right) & 1 \\ 0 & 0 & 0 & \left(\frac{1}{W}\right) & -\left(\frac{1+W}{W}\right) \end{bmatrix} \cdot \begin{bmatrix} X_1 \\ X_2 \\ X_3 \\ X_4 \\ X_5 \end{bmatrix} = \begin{bmatrix} \left(\frac{F_B}{S}\right) \\ 0 \\ 0 \\ 0 \\ 0 \end{bmatrix} \tag{5-9}$$

Equations of type (5-9) can be solved by Gaussian elimination by starting from the top and eliminating unknowns X_1, X_2, etc., in order to obtain

$$X_N = \left(\frac{F_B}{S}\right)\left(\frac{1}{W^{N-1}}\right) \tag{5-10}$$

By back-substitution, interstage values of X are given by

$$X_n = \left(\frac{F_B}{S}\right)\left(\frac{\sum\limits_{k=0}^{N-n} W^k}{W^{N-1}}\right) \tag{5-11}$$

For example, with $N = 5$,

$$X_1 = Y_1 = \left(\frac{F_B}{S}\right)\left(\frac{1 + W + W^2 + W^3 + W^4}{W^4}\right)$$

The purpose of the cascade, for any given S, is to maximize Y_1, the amount of soluble solids dissolved in the solvent leaving in the overflow from stage 1, and to minimize X_N, the amount of soluble solid dissolved in the solvent leaving the underflow with the insoluble material from stage N. Equation (5-10) indicates that this can be achieved for a given soluble solids feed rate, F_B, by specifying a large solvent feed rate, S, a large number of stages, N, and/or by employing a large washing factor, which can be achieved by minimizing the amount of liquid underflow compared to overflow. It should be noted that the minimum amount of solvent required corresponds to zero overflow from stage 1, or

$$S_{min} = RF_A \tag{5-12}$$

For this minimum value, $W = 1$ from (5-5) and all soluble solids leave in the underflow from the last stage, N, regardless of the number of stages. Therefore, it is best to specify a value of S significantly greater than S_{min}. Equations (5-10) and (5-5) show that the value of X_N is reduced exponentially by increasing the number of stages, N. Thus, the countercurrent cascade can be very effective. For two or more stages, X_N is also reduced exponentially by increasing the solvent rate, S. For three or more stages, the value of X_N is reduced exponentially by decreasing the underflow ratio R.

EXAMPLE 5.1

Pure water is to be used to dissolve 1,350 kg/h of Na_2CO_3 from 3,750 kg/h of a solid, where the balance is an insoluble oxide. If 4,000 kg/h of water is used as the solvent for the carbonate and the total underflow from each stage is 40 wt% solvent on a solute-free basis, compute and plot the percent recovery of the carbonate in the overflow product for one stage and for two to five countercurrent stages, as in Figure 5.3.

SOLUTION

Soluble solids feed rate $= F_B = 1,350$ kg/h
Insoluble solids feed rate $= F_A = 3,750 - 1,350 = 2,400$ kg/h
Solvent feed rate $= S = 4,000$ kg/h
Underflow ratio $R = 40/60 = 2/3$
Washing factor $W = S/RF_A = 4,000/[(2/3)(2,400)] = 2.50$
Overall fractional recovery of soluble solids $= Y_1(S - RF_A)/F_B$

By overall material balance on soluble solids for N stages,

$$F_B = Y_1(S - RF_A) + X_N RF_A$$

Solving for Y_1 and using (5-5) to introduce the washing factor,

$$Y_1 = \frac{(F_B/S) - (1/W)X_N}{(1 - 1/W)}$$

From the given data,

$$Y_1 = \frac{(1,350/4,000) - (1/2.50)X_N}{(1 - 1/2.50)} \quad \text{or} \quad Y_1 = 0.5625 - 0.6667 X_N \tag{1}$$

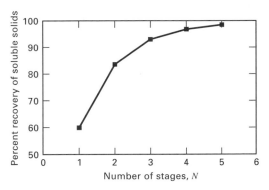

Figure 5.4 Effect of number of stages on percent recovery in Example 5.1.

where, from (5-10),

$$X_N = \left(\frac{1{,}350}{4{,}000}\right)\frac{1}{2.50^{N-1}} = \frac{0.3375}{2.50^{N-1}} \tag{2}$$

The percent recovery of soluble material is

$$Y_1(S - RF_A)/F_B = Y_1[4{,}000 - (2/3)(2{,}400)]/1{,}350 \times 100\% = 177.8Y_1 \tag{3}$$

Results for one to five stages, as computed from (1) to (3), are

No. of Stages in Cascade, N	X_N	Y_1	Percent Recovery of Soluble Solids
1	0.3375	0.3375	60.0
2	0.1350	0.4725	84.0
3	0.0540	0.5265	93.6
4	0.0216	0.5481	97.4
5	0.00864	0.5567	99.0

A plot of the percent recovery of soluble solids as a function of the number of stages is shown in Figure 5.4. Although only a 60% recovery is obtained with one stage, 99% recovery is achieved for five stages. To achieve 99% recovery with one stage, a water rate of 160,000 kg/h is required, which is 40 times that required for five stages. Thus, the use of multiple stages in a countercurrent cascade to increase recovery of soluble material can be much more effective than increased use of a mass-separating agent with a single stage. ■

5.3 SINGLE-SECTION LIQUID–LIQUID EXTRACTION CASCADES

Three possible two-stage liquid–liquid extraction cascades are the cocurrent, crosscurrent, and countercurrent arrangements in Figure 5.5. The countercurrent arrangement is generally preferred because, as will be shown in this section, that arrangement results in a higher degree of extraction for a given amount of solvent and number of equilibrium stages.

In Section 4.5, (4-25), for the fraction of solute, B, that is not extracted, was derived for a single liquid–liquid equilibrium extraction stage, assuming the use of pure solvent, and a constant value for the distribution coefficient, K'_{D_B}, for the solute, B, dissolved in components A and C, which are mutually insoluble. That equation is now extended to multiple stages for each type of cascade shown in Figure 5.5.

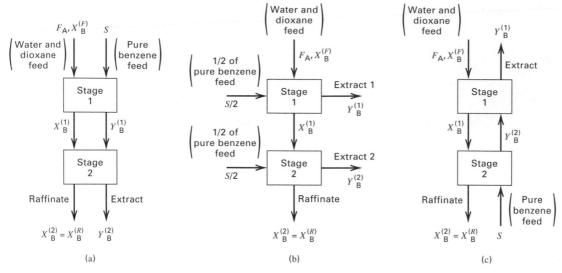

Figure 5.5 Two-stage arrangements: (a) cocurrent cascade; (b) crosscurrent cascade; (c) countercurrent cascade.

Cocurrent Cascade

If additional stages are added in the cocurrent arrangement in Figure 5.5a, the equation for the first stage is that of a single stage. That is, from (4-25) in mass ratio units,

$$X_B^{(1)}/X_B^{(F)} = \frac{1}{1 + E} \tag{5-13}$$

where E is the extraction factor, given by

$$E = K'_{D_B} S/F_A \tag{5-14}$$

Since $Y_B^{(1)}$ is in equilibrium with $X_B^{(1)}$ according to

$$K'_{D_B} = Y_B^{(1)}/X_B^{(1)} \tag{5-15}$$

the combination of (5-15) with (5-13) gives

$$Y_B^{(1)}/X_B^{(F)} = K'_{D_B}/(1 + E) \tag{5-16}$$

For the second stage, a material balance for B gives

$$X_B^{(1)}F_A + Y_B^{(1)}S = X_B^{(2)}F_A + Y_B^{(2)}S \tag{5-17}$$

$$\text{with } K'_{D_B} = Y_B^{(2)}/X_B^{(2)} \tag{5-18}$$

Combining (5-17) with (5-18), (5-13), and (5-16) to eliminate $X_B^{(1)}$, $Y_B^{(1)}$, and $Y_B^{(2)}$ gives

$$\frac{X_B^{(2)}}{X_B^{(F)}} = \frac{1}{1 + E} = \frac{X_B^{(N)}}{X_B^{(F)}} \tag{5-19}$$

Comparison of (5-19) with (5-13) shows that $X_B^{(2)} = X_B^{(1)}$. Thus, no additional extraction takes place in the second stage. This is as expected because the two streams leaving the first stage are at equilibrium and when they are recontacted in stage 2, no additional net mass transfer of B occurs. Accordingly, a cocurrent cascade of equilibrium stages has no merit other than to provide increased residence time.

Crosscurrent Cascade

For the crosscurrent cascade shown in Figure 5.5b, the feed progresses through each stage, starting with stage 1 and finishing with stage N. The solvent flow rate, S, is divided with portions sent to each stage. If the portions are equal, the following mass ratios are obtained by application of (5-13), where S is replaced by S/N, so that E is replaced by E/N:

$$X_B^{(1)}/X_B^{(F)} = 1/(1 + E/N)$$
$$X_B^{(2)}/X_B^{(1)} = 1/(1 + E/N)$$
$$\vdots$$
$$X_B^{(N)}/X_B^{(N-1)} = 1/(1 + E/N)$$

(5-20)

Combining the equations in (5-20) to eliminate all intermediate interstage variables, $X_B^{(n)}$, the final raffinate mass ratio is given by

$$X_B^{(N)}/X_B^{(F)} = X_B^{(R)}/X_B^{(F)} = \frac{1}{(1 + E/N)^N}$$

(5-21)

Interstage values of $X_B^{(n)}$ are obtained similarly from

$$X_B^{(n)}/X_B^{(F)} = \frac{1}{(1 + E/N)^n}$$

(5-22)

Thus, unlike the cocurrent cascade, the value of X_B decreases in each successive stage. For an infinite number of equilibrium stages, (5-21) becomes

$$X_B^{(\infty)}/X_B^{(F)} = 1/\exp(E)$$

(5-23)

Thus, even for an infinite number of stages, $X_B^{(R)} = X_B^{(\infty)}$ cannot be reduced to zero.

Countercurrent Cascade

In the countercurrent arrangement for two stages in Figure 5.5c, the feed liquid passes through the cascade countercurrently to the solvent. For a two-stage system, the material balance and equilibrium equations for solute, B, for each stage are as follows.

Stage 1:

$$X_B^{(F)}F_A + Y_B^{(2)}S = X_B^{(1)}F_A + Y_B^{(1)}S$$

(5-24)

$$K'_{D_B} = \frac{Y_B^{(1)}}{X_B^{(1)}}$$

(5-25)

Stage 2:

$$X_B^{(1)}F_A = X_B^{(2)}F_A + Y_B^{(2)}S$$

(5-26)

$$K'_{D_B} = \frac{Y_B^{(2)}}{X_B^{(2)}}$$

(5-27)

Combining (5-24) to (5-27) with (5-14) to eliminate $Y_B^{(1)}$, $Y_B^{(2)}$, and $X_B^{(1)}$ gives

$$X_B^{(2)}/X_B^{(F)} = X_B^{(R)}/X_B^{(F)} = \frac{1}{1 + E + E^2}$$

(5-28)

If the number of countercurrent stages is extended to N stages, the result is

$$X_B^{(N)}/X_B^{(F)} = X_B^{(R)}/X_B^{(F)} = 1 \bigg/ \sum_{n=0}^{N} E^n$$

(5-29)

Interstage values of $X_B^{(n)}$ are obtained in a similar fashion, giving

$$X_B^{(n)}/X_B^{(F)} = \sum_{k=0}^{N-n} E^k \Big/ \sum_{k=0}^{N} E^k \qquad (5\text{-}30)$$

As with the crosscurrent arrangement, the value of X_B decreases in each successive stage. The amount of decrease for the countercurrent arrangement is greater than for the crosscurrent arrangement, and the difference increases exponentially with increasing extraction factor, E. Therefore, the countercurrent cascade is the most efficient of the three linear cascades.

For an infinite number of equilibrium stages, the limit of (5-28) gives two results:

$$X_B^{(\infty)}/X_B^{(F)} = 0, \qquad 1 \le E \le \infty \qquad (5\text{-}31)$$

$$X_B^{(\infty)}/X_B^{(F)} = (1 - E), \qquad E \le 1 \qquad (5\text{-}32)$$

Thus, complete extraction can be achieved in a countercurrent cascade, but only for an extraction factor, E, greater than 1.

Ethylene glycol can be catalytically dehydrated completely to p-dioxane (a cyclic diether) by the reaction $2HOCH_2CH_2HO \rightarrow H_2CCH_2OCH_2CH_2O + 2H_2O$. Water and p-dioxane have normal boiling points of 100°C and 101.1°C, respectively, and cannot be separated by distillation. However, liquid–liquid extraction at 25°C (298.15 K), using benzene as a solvent, is reasonably effective. Assume that 4,536 kg/h (10,000 lb/h) of a 25 wt% solution of p-dioxane in water is to be separated continuously by using 6,804 kg/h (15,000 lb/h) of pure benzene. Assuming that benzene and water are mutually insoluble, determine the effect of the number and arrangement of stages on the percent extraction of p-dioxane. The flowsheet is shown in Figure 5.6.

SOLUTION

Three different arrangements of stages will be examined: (a) cocurrent cascade, (b) crosscurrent cascade, and (c) countercurrent cascade. Because water and benzene are almost mutually insoluble, (5-13), (5-21), and (5-29) can be used, respectively, to estimate $X_B^{(R)}/X_B^{(F)}$, the fraction of p-dioxane not extracted, as a function of the number of stages. From the equilibrium data of Berdt and Lynch [1], the distribution coefficient for p-dioxane, $K_{D_B}' = Y_B/X_B$, where Y refers to the benzene phase and X refers to the water phase, varies from 1.0 to 1.4 over the concentration range of interest. For this example, assume a constant value of 1.2. From the given data, $S = 6,804$ kg/h of benzene, $F_A = 4,536(0.75) = 3,402$ kg/h of water, and $X_B^{(F)} = 0.25/0.75 = 1/3$. From (5-14),

$$E = 1.2(6,804)/3,402 = 2.4$$

Single equilibrium stage

All three arrangements give identical results for a single stage. From (5-13),

$$X_B^{(1)}/X_B^{(F)} = 1/(1 + 2.4) = 0.294$$

The corresponding fractional extraction is

$$1 - X_B^{(1)}/X_B^{(F)} = 1 - 0.294 = 0.706 \quad \text{or} \quad 70.6\%$$

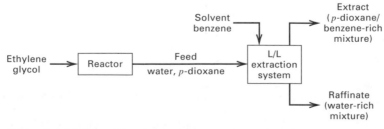

Figure 5.6 Flowsheet for Example 5.2.

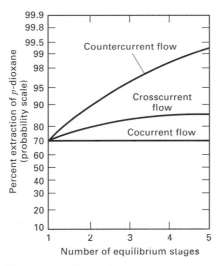

Figure 5.7 Effect of multiple-stage cascade arrangement on extraction efficiency.

More than one equilibrium stage

(a) Cocurrent cascade. For any number of stages, the percent extraction is the same as for one stage, 70.6%.

(b) Crosscurrent cascade. For any number of stages, (5-21) applies. For example, for two stages, assuming equal flow rates of solvent to each stage

$$X_B^{(2)}/X_B^{(F)} = 1/(1 + E/2)^2 = 1/(1 + 2.4/2)^2 = 0.207$$

and the percent extraction is 79.3%. Results for other numbers of stages are obtained in the same manner.

(c) Countercurrent cascade. For any number of stages, (5-29) applies. For example, for two stages,

$$X_B^{(2)}/X_B^{(F)} = 1/(1 + E + E^2) = 1/(1 + 2.4 + 2.4^2) = 0.109$$

and the percent extraction is 89.1%. Results for other numbers of stages are obtained in the same manner.

A plot of percent extraction as a function of the number of equilibrium stages for up to five stages is shown in Figure 5.7 for each of the three arrangements. The probability-scale ordinate is convenient because for the countercurrent arrangement, with $E > 1$, 100% extraction is approached as the number of stages approaches infinity. For the crosscurrent arrangement, a maximum percent extraction of 90.9% is computed from (5-23). For five stages, Figure 5.7 shows that the countercurrent cascade has already achieved 99% extraction. ∎

5.4 MULTICOMPONENT VAPOR–LIQUID CASCADES

Countercurrent cascades are used extensively for vapor–liquid separation operations, including absorption, stripping, and distillation. For absorption and stripping, a single-section cascade is used to recover designated components from the feed. For distillation, a two-section cascade is effective in achieving a separation between two selected components referred to as the key components. For both cases, approximate calculation procedures relate compositions of multicomponent vapor and liquid streams entering and exiting the cascade to the number of equilibrium stages required. These approximate procedures are called *group methods* because they provide only an overall treatment of the group of stages in the cascade, without considering detailed changes in temperature, phase compositions, and flows from stage to stage.

Single-Section Cascades by Group Methods

Kremser [2] originated the group method by deriving an equation for the fractional absorption of a species from a gas into a liquid absorbent for a multistage countercurrent absorber. The treatment presented here is similar to that of Edmister [3] for general application to vapor–liquid separation operations. An alternative treatment is given by Smith and Brinkley [4].

Consider first the countercurrent cascade of N adiabatic equilibrium stages used, as shown in Figure 5.8a, to absorb species present in the entering vapor. Assume that these species are absent in the entering liquid. Stages are numbered from top to bottom. It is convenient to express stream compositions in terms of component molar flow rates, v_i and l_i, in the vapor and liquid phases, respectively. However, in the following derivation, the subscript i is dropped. A material balance around the top of the absorber, including stages 1 through $N - 1$, for any absorbed species gives

$$v_N = v_1 + l_{N-1} \tag{5-33}$$

where
$$v = yV \tag{5-34}$$

$$l = xL \tag{5-35}$$

and $l_0 = 0$. From equilibrium considerations for stage N, the definition of the vapor–liquid equilibrium ratio or K-value can be employed to give

$$y_N = K_N x_N \tag{5-36}$$

Combining (5-34), (5-35), and (5-36), v_N becomes

$$v_N = \frac{l_N}{L_N/(K_N V_N)} \tag{5-37}$$

An absorption factor A, analogous to the extraction factor, E, for a given stage and component is defined by

$$A = \frac{L}{KV} \tag{5-38}$$

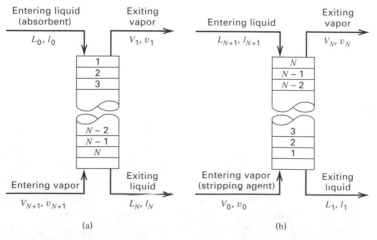

(a)   (b)

Figure 5.8 Countercurrent cascades of N adiabatic stages: (a) absorber; (b) stripper.

Combining (5-37) and (5-38),

$$v_N = \frac{l_N}{A_N} \tag{5-39}$$

Substituting (5-39) into (5-33),

$$l_N = (l_{N-1} + v_1)A_N \tag{5-40}$$

The internal flow rate, l_{N-1}, is eliminated by successive substitution using material balances around successively smaller sections of the top of the cascade. For stages 1 through $N - 2$,

$$l_{N-1} = (l_{N-2} + v_1)A_{N-1} \tag{5-41}$$

Substituting (5-41) into (5-40),

$$l_N = l_{N-2}A_{N-1}A_N + v_1(A_N + A_{N-1}A_N) \tag{5-42}$$

Continuing this process to the top stage, where $l_1 = v_1 A_1$, ultimately converts (5-42) into

$$l_N = v_1(A_1 A_2 A_3 \ldots A_N + A_2 A_3 \ldots A_N + A_3 \ldots A_N + \ldots + A_N) \tag{5-43}$$

A more useful form is obtained by combining (5-43) with the overall component balance

$$l_N = v_{N+1} - v_1 \tag{5-44}$$

to give an equation for the exiting vapor in terms of the entering vapor and a recovery fraction:

$$v_1 = v_{N+1}\phi_A \tag{5-45}$$

where, by definition, the recovery fraction is

$$\phi_A = \frac{1}{A_1 A_2 A_3 \ldots A_N + A_2 A_3 \ldots A_N + A_3 \ldots A_N + \ldots + A_N + 1} \tag{5-46}$$

$$= \text{fraction of species in entering vapor that is not absorbed}$$

In the group method, an average effective absorption factor, A_e, replaces the separate absorption factors for each stage. Equation (5-46) now becomes

$$\phi_A = \frac{1}{A_e^N + A_e^{N-1} + A_e^{N-2} + \cdots + A_e + 1} \tag{5-47}$$

When multiplied and divided by $(A_e - 1)$, (5-47) reduces to

$$\phi_A = \frac{A_e - 1}{A_e^{N+1} - 1} \tag{5-48}$$

Note that each component has a different A_e and, therefore, a different value of ϕ_A. Figure 5.9 from Edmister [3] is a plot of (5-48) with a probability scale for ϕ_A, a logarithmic scale for A_e, and N as a parameter. This plot, in linear coordinates, was first developed by Kremser [2].

Consider next the countercurrent stripper shown in Figure 5.8b. Assume that the components stripped from the liquid are absent in the entering vapor, and ignore condensation or absorption of the stripping agent. In this case, stages are numbered from bottom to top to facilitate the derivation. The pertinent stripping equations follow in a manner analogous to the absorber equations. The results are

$$l_1 = l_{N+1}\phi_S \tag{5-49}$$

$$\text{where } \phi_S = \frac{S_e - 1}{S_e^{N+1} - 1} = \text{fraction of species in entering liquid that is not stripped} \tag{5-50}$$

$$S = \frac{KV}{L} = \frac{1}{A} = \text{stripping factor} \tag{5-51}$$

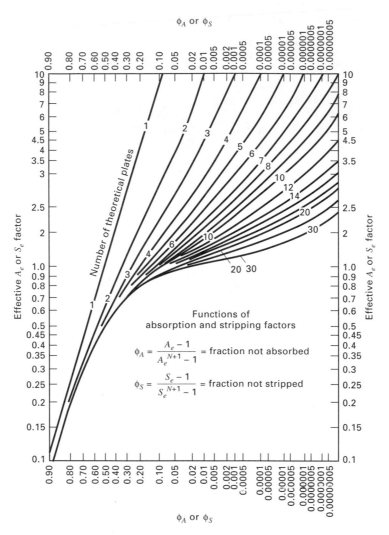

Figure 5.9 Plot of Kremser equation for a single-section countercurrent cascade. [From W.C. Edmister, *AIChE J.,* **3,** 165–171 (1957).]

Figure 5.9 also applies to (5-50). As shown in Figure 5.10, absorbers are frequently coupled with strippers or distillation columns to permit regeneration and recycle of absorbent. Since stripping action is not perfect, recycled absorbent entering the absorber contains species present in the vapor entering the absorber. Vapor passing up through the absorber can strip these as well as the absorbed species introduced in the makeup absorbent. A general absorber equation is obtained by combining (5-45) for absorption of species from the entering vapor with a modified form of (5-49) for stripping of the same species from the entering liquid. For stages numbered from top to bottom, as in Figure 5.8a, (5-49) becomes

$$l_N = l_0 \phi_S \tag{5-52}$$

or, since

$$l_0 = v_1 + l_N$$
$$v_1 = l_0(1 - \phi_S) \tag{5-53}$$

The total balance in the absorber for a component appearing in both entering vapor and entering liquid is obtained by adding (5-45) and (5-53) to give

$$v_1 = v_{N+1}\phi_A + l_0(1 - \phi_S) \tag{5-54}$$

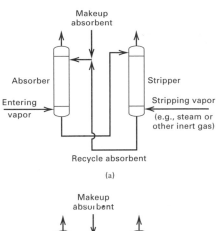

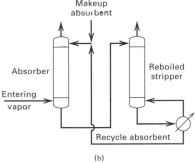

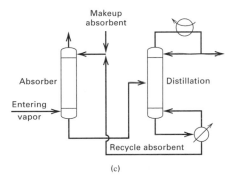

Figure 5.10 Various coupling schemes for absorbent recovery: (a) use of stream or inert gas stripper; (b) use of reboiled stripper; (c) use of distillation.

which is generally applied to each component in the vapor entering the absorber. Equation (5-52) is used for species that appear only in the entering liquid. The analogous equation for a stripper in Figure 5.8b is

$$l_1 = l_{N+1}\phi_S + v_0(1 - \phi_A) \tag{5-55}$$

EXAMPLE 5.3

In Figure 5.11, the heavier components in a slightly superheated hydrocarbon gas are to be removed by absorption at 400 psia (2,760 kPa) with a high-molecular-weight oil. Estimate exit vapor and exit liquid flow rates and compositions by the approximate group method of Kremser. Assume that effective absorption and stripping factors for each component can be estimated from the entering values of L, V, and the component K-values, as listed below based on an average entering temperature of $(90 + 105)/2 = 97.5°F$.

SOLUTION

From (5-38) and (5-51), $A_i = L/K_iV = 165/[K_i(800)] = 0.206/K_i$

$$S_i = 1/A_i = 4.85K_i$$

$$N = 6 \text{ stages}$$

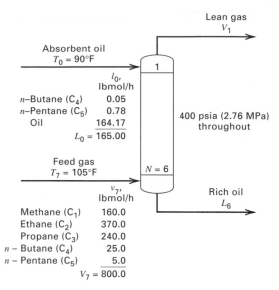

Figure 5.11 Specifications for absorber of Example 5.3.

Values of ϕ_A and ϕ_S are obtained from (5-48) and (5-50) or Figure 5.9. Values of $(v_i)_1$, the component flow rates in the exit vapor, are computed from (5-54). Values of $(l_i)_6$, the component flow rates in the exit liquid, are computed from an overall component material balance using Figure 5.8a:

$$(l_i)_6 = (l_i)_0 + (v_i)_7 - (v_i)_1 \qquad \textbf{(1)}$$

The computations, which are best made in tabular fashion with a spreadsheet computer program, give the following results:

Component	K @ 97.5°F, 400 psia	A	S	ϕ_A	ϕ_S	v_1	l_6
C_1	6.65	0.0310	—	0.969	—	155.0	5.0
C_2	1.64	0.126	—	0.874	—	323.5	46.5
C_3	0.584	0.353	—	0.647	—	155.4	84.6
nC_4	0.195	1.06	0.946	0.119	0.168	3.02	22.03
nC_5	0.0713	2.89	0.346	0.00112	0.654	0.28	5.5
Oil	0.0001	—	0.0005	—	0.9995	0.075	164.095
						637.275	327.725

The above results indicate that approximately 20% of the gas is absorbed. Less than 0.1% of the absorbent oil is stripped. ∎

Two-Section Cascades

A single-stage flash distillation produces a vapor that is somewhat richer in the lower-boiling constituents than the feed. Further enrichment can be achieved by a series of flash distillations in which the vapor from each stage is condensed, then reflashed. In principle, any desired product purity can be obtained by a multistage flash technique, provided a suitable volatility difference exists and a suitable number of stages is employed. In practice, however, the recovery of product is small, heating and cooling requirements are high, and relatively large quantities of various liquid products are produced.

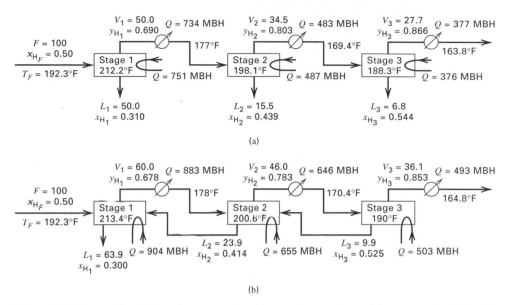

Figure 5.12 Successive flashes for recovering hexane from octane: (a) no recycle; (b) with recycle. Flow rates in lbmol/h. MBH = 1,000 Btu/h.

As an example, consider Figure 5.12a, where n-hexane (H) is separated from n-octane by a series of three flashes at 1 atm (pressure drop and pump needs are ignored). The feed to the first flash stage is an equimolar bubble-point liquid at a flow rate of 100 lbmol/h. A bubble-point temperature calculation yields 192.3°F. If the vapor rate leaving stage 1 is set equal to the amount of n-hexane in the feed to stage 1, the calculated equilibrium exit phases are as shown. The vapor V_1 is enriched to a hexane mole fraction of 0.690. The heating requirement is 751,000 Btu/h. Equilibrium vapor from stage 1 is condensed to bubble-point liquid with a cooling duty of 734,000 Btu/h. Repeated flash calculations for stages 2 and 3 gives the results shown. For each stage, the leaving molar vapor rate is set equal to the moles of hexane in the feed to the stage. The purity of n-hexane is increased from 50 mol% in the feed to 86.6 mol% in the final condensed vapor product, but the recovery of hexane is only 27.7(0.866)/50 or 48%. Total heating requirement is 1,614,000 Btu/h and liquid products total 72.3 lbmol/h.

In comparing feed and liquid products from two contiguous stages, we note that liquid from the later stage and the feed to the earlier stage are both leaner in hexane, the more volatile species, than the feed to the later state. Thus, if intermediate streams are recycled, intermediate recovery of hexane is improved. This processing scheme is depicted in Figure 5.12b, where again the molar fraction vaporized in each stage equals the mole fraction of hexane in the combined feeds to the stage. The mole fraction of hexane in the final condensed vapor product is 0.853, just slightly less than that achieved by successive flashes without recycle. However, the use of recycle increases recovery of hexane from 48% to 61.6%. As shown in Figure 5.12b, increased recovery of hexane is accompanied by approximately 28% increased heating and cooling requirements. If the same degree of heating and cooling is used for the no-recycle scheme in Figure 5.12a as in Figure 5.12b, the final hexane mole fraction is reduced from 0.866 to 0.815, but hexane recovery is increased to 36.1(0.815)/50 or 58.8%.

Both of the successive flash arrangements in Figure 5.12 involve a considerable number of heat exchangers and pumps. Except for stage 1, the heaters in Figure 5.12a can be eliminated if the two intermediate total condensers are converted to partial condensers with duties of 734 − 487 = 247 MBH (MBH = 1,000 Btu/h) and 483 − 376 = 107 MBH.

Total heating duty is now only 751,000 Btu/h, and total cooling duty is 731,000 Btu/h. Similarly, if heaters for stages 2 and 3 in Figure 5.12b are removed by converting the two total condensers to partial condensers, total heating duty is 904,000 Btu/h (20% greater than the no-recycle case), and cooling duty is 864,000 Btu/h (18% greater than the no-recycle case).

A considerable simplification of the successive flash technique with recycle is shown in Figure 5.13a. The total heating duty is provided by a feed boiler ahead of stage 1. The total cooling duty is utilized at the opposite end to condense totally the vapor leaving stage 3. Condensate in excess of distillate is returned as reflux to the top stage, from which it passes successively from stage to stage countercurrently to vapor flow. Vertically arranged adiabatic stages eliminate the need for interstage pumps, and all stages are contained within a single piece of inexpensive equipment. The set of stages is called a *rectifying section*. As discussed in Chapter 2, such an arrangement is thermodynamically inefficient because heat is added at the highest temperature level and removed at the lowest temperature level.

The number of degrees of freedom for the arrangement in Figure 5.13a is determined by the method of Chapter 4 to be $(C + 2N + 10)$. If all independent feed conditions,

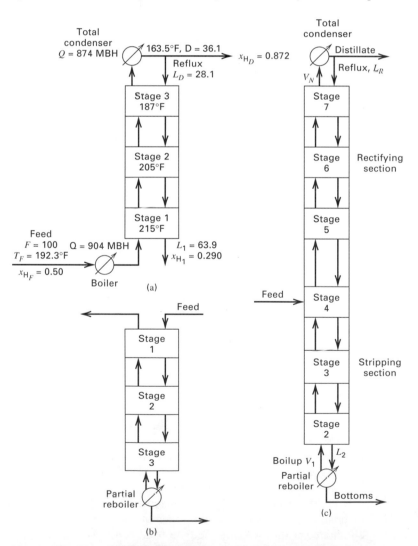

Figure 5.13 Successive adiabatic flash arrangements: (a) rectifying section; (b) stripping section; (c) multistage distillation.

number of stages (3), and all stage pressures (1 atm), bubble-point liquid leaving the condenser, and adiabatic stages are specified, two degrees of freedom remain. These are specified to be a heating duty for the boiler and a distillate rate equal to that of Figure 5.12b. Calculations result in a mole fraction of 0.872 for hexane in the distillate. This is somewhat greater than that shown in Figure 5.12b.

The same principles by which we have concluded that the adiabatic multistage countercurrent flow arrangement is advantageous for concentrating a light component in an overhead product can be applied to the concentration of a heavy component in a bottoms product, as in Figure 5.13b. Such a set of stages is called a *stripping section.*

Figure 5.13c, a combination of Figures 5.13a and 5.13b with a liquid feed, is a complete column for rectifying and stripping a feed to effect a sharper separation between a selected more volatile component, called the *light key,* and a less volatile component, called the *heavy key component,* than is possible with either a stripping or an enriching section alone. Adiabatic flash stages are placed above and below the feed. Recycled liquid reflux, L_R, is produced in the condenser and vapor boilup, V_1, in the reboiler. The reflux ratios are L_R/V_N and L_2/V_1 at the top and bottom of the apparatus, respectively. All interstage flows are countercurrent. Two-section cascades are widely used in industry for multistage distillation.

The rectifying stages above the point of feed introduction purify the light product by contacting upward flowing vapor with successively richer liquid reflux. Stripping stages below the feed increase light-product recovery because vapor relatively low in volatile constituents strips light components out of the liquid. For the heavy product, the functions are reversed: The stripping section increases purity; the enriching section increases recovery.

Edmister [3] applied the Kremser group method for absorbers and strippers to distillation where two cascades are coupled to a condenser, a reboiler, and a feed stage. In Figure 5.14, five separation zones are shown: (1) partial condenser, C; (2) absorption or rectifying cascade (enriching section), E; (3) feed-flash stage, F; (4) stripping cascade (exhausting section), X; and (5) partial reboiler, B.

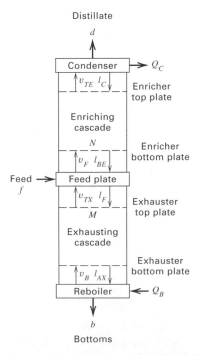

Figure 5.14 Countercurrent distillation cascade.

In Figure 5.14, N stages for the enricher are numbered from the top down and the overhead product is distillate; whereas for the exhauster, M stages are numbered from the bottom up. Component feeds to the enricher section are vapor, v_F, from the feed stage and liquid, l_C, from the condenser. Component feeds to the exhauster are liquid, l_F, from the feed stage and vapor, v_B, from the reboiler. Component flows leaving the enricher cascade are vapor, v_{TE}, from the top stage, 1, and liquid, l_{BE}, from the bottom stage, N. Component flows leaving the exhauster cascade are vapor, v_{TX}, from the top stage, M, and liquid, l_{BX}, from the bottom stage, 1. The recovery equations for the enricher are obtained from (5-54) by making the following substitutions, which are obtained from material balance and equilibrium considerations. For each component in the feed,

$$v_{TE} = l_C + d \tag{5-56}$$

$$v_F = l_{BE} + d \tag{5-57}$$

and

$$l_C = dA_C \tag{5-58}$$

where

$$A_C = \frac{L_C}{DK_C} \quad \text{(for a partial condenser)} \tag{5-59}$$

$$A_C = \frac{L_C}{D} \quad \text{(for a total condenser)} \tag{5-60}$$

The resulting enricher recovery equations for each species are

$$\frac{l_{BE}}{d} = \frac{A_C\phi_{SE} + 1}{\phi_{AE}} - 1 \tag{5-61}$$

or

$$\frac{v_F}{d} = \frac{A_C\phi_{SE} + 1}{\phi_{AE}} \tag{5-62}$$

where the additional subscript E on ϕ refers to the enricher.

The recovery equations for the exhauster are obtained in a similar manner, as

$$\frac{v_{TX}}{b} = \frac{S_B\phi_{AX} + 1}{\phi_{SX}} - 1 \tag{5-63}$$

$$\frac{l_F}{b} = \frac{S_B\phi_{AX} + 1}{\phi_{SX}} \tag{5-64}$$

where

$$S_B = K_B V_B / B$$

for a partial reboiler, and additional subscript X on ϕ denotes an exhauster.

For either an enricher or exhauster, ϕ_A and ϕ_E are given, from above, by (5-48) and (5-50), respectively, or from Figure 5.9.

To couple the enriching and exhausting cascades, a feed stage is employed for which the absorption factor is related to the streams leaving the feed stage by

$$A_F = \frac{L_F}{K_F V_F} = \frac{L_F}{(v_F L_F / V_F l_F)V_F} = \frac{l_F}{v_F} \tag{5-65}$$

For the distillation column of Figure 5.14, (5-62), (5-64), and (5-65) are combined to eliminate l_F and v_F. The result is

$$\frac{b}{d} = A_F \frac{[(A_C\phi_{SE} + 1)/\phi_{AE}]}{[(S_B\phi_{AX} + 1)/\phi_{SX}]} \tag{5-66}$$

To apply (5-66) for the calculation of component split ratios, b/d, it is necessary to establish values of absorption factors A_F and A_C, and the stripping factor S_B. Average values for factors A_E, A_X, S_E, and S_X for each component are also required for the two

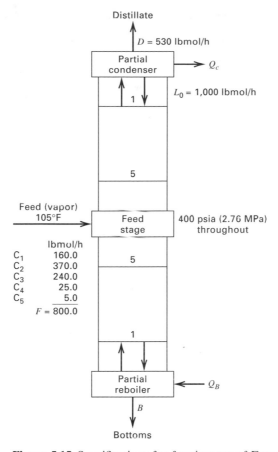

Figure 5.15 Specifications for fractionator of Example 5.4.

cascades to determine the corresponding ϕ values. To establish these values, it is necessary to estimate temperatures and molar vapor and liquid, V and L, flow rates. An approximate method for making these estimates is given in the following example.

EXAMPLE 5.4

The hydrocarbon gas of Example 5.3 is distilled at 400 psia (2.76 MPa), to separate ethane from propane, for the conditions shown in Figure 5.15. Estimate the distillate and bottoms compositions using (5-66). This example is best solved by using a spreadsheet computer program.

SOLUTION

Assume a feed stage temperature equal to the feed temperature, 105°F. To estimate the condenser and reboiler temperatures, assume a perfect split for the specified distillate rate of 530 lbmol/h, with all methane and ethane going to the distillate and all propane and heavier going to the bottoms. Thus the preliminary material balance is

| | | lbmol/h | |
Component	Feed, f	Assumed Distillate, d	Assumed Bottoms, b
C_1	160	160	0
C_2	370	370	0
C_3	240	0	240
nC_4	25	0	25
nC_5	5	0	5
	800	530	270

For these assumed products, applying procedures in Section 4.4, a distillate temperature of 12°F is obtained from a dew-point calculation and a bottoms temperature of 165°F from a bubble-point calculation. Average temperatures of $(12 + 105)/2 = 59°F$ and $(105 + 165)/2 = 135°F$ are estimated for the enriching and exhausting cascades, respectively. Assuming that total molar flow rates are constant in each cascade, the following vapor and liquid flow-rate estimates are obtained by working down from the top of the column, where the liquid reflux is specified:

Stage or Section	Average Temperature, °F	Average Flow Rates, lbmol/h	
		Vapor	Liquid
Condenser	12	530	1,000
Enricher	59	1,530	1,000
Feed	105	1,530	1,000
Exhauster	135	730	1,000
Reboiler	165	730	270

From the column pressure and the estimated temperature values, K-values are read from Figure 2.8. These values are then used to estimate absorption and stripping factors for the five sections, with the following results:

	Component				
	C_1	C_2	C_3	nC_4	nC_5
A_C	0.356	2.45	10.48	43.9	172
A_E	0.131	0.511	1.72	5.78	18.7
$S_E = 1/A_E$	7.65	1.96	0.581	0.173	0.0536
A_F	0.111	0.363	1.108	3.19	9.08
A_X	0.205	0.652	1.73	4.72	12.5
$S_X = 1/A_X$	4.89	2.26	0.577	0.212	0.080
S_B	19.5	7.03	2.78	1.08	0.433

From the values of A_E, A_X, S_E, and S_X, and the numbers of theoretical stages specified in Figure 5.15, the following values of ϕ are computed from (5-48) and (5-50) or read from Figure 5.9:

	Component				
	C_1	C_2	C_3	nC_4	nC_5
ϕ_{AE}	0.869	0.498	0.0289	0.0001	0.000
ϕ_{AX}	0.796	0.377	0.028	0.00034	0.000
ϕ_{SE}	0.0000	0.0173	0.435	0.827	0.946
ϕ_{SX}	0.0003	0.0444	0.439	0.788	0.920

From the values in the above two tables, values of (b/d) are computed for each component from (5-66). Since an overall balance for each component is given by $f = d + b$, values of d and b can then be computed from

$$d = \frac{f}{1 + (b/d)} \tag{1}$$

$$b = f - d \tag{2}$$

The following results are obtained:

	(b/d)	lbmol/h	
		d	b
C_1	0.000002	160	0
C_2	0.00924	366.6	3.4
C_3	86.8	2.7	237.3
nC_4	937,000	0	25
nC_5	Very large	0	5
Totals		529.3	270.7

The total distillate rate is somewhat less than the 530.0 lbmol/h specified. The values of d_i and b_i can be corrected to force the total to 530.0 by the method of Lyster et al. [5], which involves finding the positive root of θ in the relation

$$D = \sum_i \frac{f_i}{1 + \theta(b_i/d_i)}$$

followed by recalculation of d_i from

$$d_i = \frac{f_i}{1 + \theta(b_i/d_i)}$$

and b_i from $f_i - d_i$. The resulting value of θ is 0.8973, which gives $d_{C_2} = 367$, $b_{C_2} = 3$, $d_{C_3} = 3$, and $b_{C_3} = 237$, with no changes for the other components.

It is interesting to note that the separation achieved by distillation is considerably improved over the separation achieved by absorption in Example 5.3. Although the overhead total exit vapor flow rates are approximately the same (530 lbmol/h) in this example and in Example 5.3, a reasonably sharp split between ethane and propane occurs for distillation because of the two-section cascade, while the absorber, with only a one-section cascade, allows appreciable quantities of both ethane and propane to exit in the overhead exit vapor and the bottoms exit liquid. Even if the absorbent rate in Example 5.3 is doubled so that the recovery of propane in the bottoms exit liquid approaches 100%, more than 50% of the ethane also appears in the bottoms. ∎

5.5 DEGREES OF FREEDOM AND SPECIFICATIONS FOR COUNTERCURRENT CASCADES

The solution to a multicomponent, multiphase, multistage separation problem is found in the simultaneous or iterative solution of the material balance, energy balance, and phase equilibria equations. This implies that a sufficient number of design variables is specified so that the number of remaining unknown (output) variables exactly equals the number of independent equations. When this is done, a separation process is said to be specified. In this section, the degrees-of-freedom analysis discussed in Section 4.1 for a single equilibrium stage is extended to one- and multiple-section countercurrent cascades.

An intuitively simple, but operationally complex, method of finding N_D, the number of independent design variables, *degrees of freedom,* or *variance* in the process, is to enumerate all pertinent variables, N_V, and to subtract from these the total number of independent equations or relationships, N_E, relating the variables:

$$N_D = N_V - N_E \tag{5-67}$$

This approach to separation process design was developed by Kwauk [6], and a modification of his methodology forms the basis for this discussion.

Typically, the variables in a separation process are intensive variables such as composition, temperature, or pressure; extensive variables such as flow rate or heat transferred; and equipment parameters such as the number of equilibrium stages. Physical properties such as enthalpy or K-values are not counted because they are functions of the intensive variables. The variables are relatively easy to enumerate, but to achieve an unambiguous count of N_E it is necessary to carefully seek out all independent relationships due to material and energy conservations, phase equilibria restrictions, process specifications, and equipment configurations.

Separation equipment consists of physically identifiable elements (equilibrium stages, condensers, reboilers, etc.) as well as stream dividers and stream mixers. It is helpful to examine each element separately, before synthesizing the complete system.

Stream Variables

For each single-phase stream containing C components, a complete specification of intensive variables consists of C mole fractions (or other concentration variables) plus temperature and pressure, or $C + 2$ variables. However, only $C - 1$ of the feed mole fractions are independent, because the other mole fraction must satisfy the mole-fraction constraint:

$$\sum_{i=1}^{c} \text{mole fractions} = 1.0$$

Thus, only $C + 1$ intensive stream variables can be specified. This is in agreement with the phase rule, which states that, for a single-phase system, the intensive variables are specified by $C - \mathscr{P} + 2 = C + 1$ variables. To this number can be added the total flow rate of the stream, an extensive variable. Although the missing mole fraction is often treated implicitly, it is preferable for completeness to include the missing mole fraction in the list of stream variables and then to include in the list of equations the above mole-fraction constraint. Thus, associated with each stream are $C + 3$ variables. For example, for a liquid-phase stream, the variables are liquid mole fractions $x_1, x_2, \ldots, x_C$; total molar flow rate L; temperature T, and pressure P.

Adiabatic or Nonadiabatic Equilibrium Stage

For a single adiabatic or nonadiabatic equilibrium stage with two entering streams and two exit streams, as shown in Figure 5.16, the variables are those associated with the four streams plus the heat transfer rate to or from the stage. Thus:

$$N_V = 4(C + 3) + 1 = 4C + 13$$

The exiting streams V_{OUT} and L_{OUT} are in equilibrium, so there are equilibrium restrictions as well as component material balances, a total material balance, an energy balance,

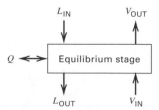

Figure 5.16 Equilibrium stage with heat addition.

and mole fraction constraints. Thus, the equations relating these variables and N_E are

Equations	Number of Equations
Pressure equality	1
$P_{V_{OUT}} = P_{L_{OUT}}$	
Temperature equality,	1
$T_{V_{OUT}} = T_{L_{OUT}}$	
Phase equilibrium relationships,	C
$(y_i)_{V_{OUT}} = K_i(x_i)_{L_{OUT}}$	
Component material balances,	$C - 1$
$L_{IN}(x_i)_{L_{IN}} + V_{IN}(y_i)_{V_{IN}} = L_{OUT}(x_i)_{L_{OUT}} + V_{OUT}(y_i)_{V_{OUT}}$	
Total material balance,	1
$L_{IN} + V_{IN} = L_{OUT} + V_{OUT}$	
Energy balance,	1
$Q + h_{L_{IN}}L_{IN} + h_{V_{IN}}V_{IN} = h_{L_{OUT}}L_{OUT} + h_{V_{OUT}}V_{OUT}$	
Mole fraction constraints in entering and exiting streams	4
e.g., $\sum\limits_{i=1}^{C} (x_i)_{L_{IN}} = 10$	
	$N_E = 2C + 7$

Alternatively, C, instead of $C-1$, component material balances can be written. The total material balance is then a dependent equation obtained by summing the component material balances and applying the mole-fraction constraints to eliminate the mole fractions. From (5-67),

$$N_D = (4C + 13) - (2C + 7) = 2C + 6$$

Several different sets of design variables can be specified. A typical set includes complete specification of the two entering streams as well as the stage pressure and heat transfer rate.

Variable Specification	Number of Variables
Component mole fractions, $(x_i)_{L_{IN}}$	$C - 1$
Total flow rate, L_{IN}	1
Component mole fractions, $(y_i)_{V_{IN}}$	$C - 1$
Total flow rate, V_{IN}	1
Temperature and pressure of L_{IN}	2
Temperature and pressure of V_{IN}	2
Stage pressure ($P_{V_{OUT}}$ or $P_{L_{OUT}}$)	1
Heat transfer rate, Q	1
	$N_D = 2C + 6$

Specification of these $(2C + 6)$ variables permits calculation of the unknown variables L_{OUT}, V_{OUT}, $(x_C)_{L_{IN}}$, $(y_C)_{V_{IN}}$, all $(x_i)_{L_{OUT}}$, T_{OUT}, and all $(y_i)_{V_{OUT}}$, where C denotes the missing mole fractions in the two entering streams.

Single-Section Countercurrent Cascade

Consider the N-stage, single-section countercurrent cascade unit shown in Figure 5.17. This cascade consists of N adiabatic or nonadiabatic equilibrium stage elements of the type shown in Figure 5.16. An algorithm is easily developed for enumerating variables, equations, and degrees of freedom for combinations of such elements to form a unit. The number of design variables for the unit is obtained by summing the variables associated with each element and then subtracting from the total variables the $C + 3$ variables for

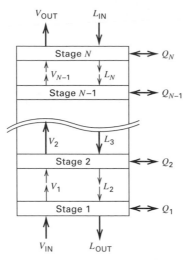

Figure 5.17 An N-stage cascade.

each of the N_R redundant interconnecting streams that arise when the output of one element becomes the input to another. Also, if an unspecified number of repetitions of any element occurs within the unit, an additional variable is added, one for each group of repetitions, giving a total of N_A additional variables. In a similar manner, the number of independent equations for the unit is obtained by summing the values of N_E for the units and then subtracting the N_R redundant mole-fraction constraints. The number of degrees of freedom is obtained as before, from (5-67). Thus:

$$(N_V)_{\text{unit}} = \sum_{\substack{\text{all} \\ \text{elements, } e}} (N_V)_e - N_R(C + 3) + N_A \tag{5-68}$$

$$(N_E)_{\text{unit}} = \sum_{\substack{\text{all} \\ \text{elements, } e}} (N_E)_e - N_R \tag{5-69}$$

Combining (5-67), (5-68), and (5-69), we have

$$(N_D)_{\text{unit}} = \sum_{\substack{\text{all} \\ \text{elements, } e}} (N_D)_e - N_R(C + 2) + N_A \tag{5-70}$$

or $$(N_D)_{\text{unit}} = (N_V)_{\text{unit}} - (N_E)_{\text{unit}} \tag{5-71}$$

For the N-stage cascade unit of Figure 5.17, with reference to the above degrees-of-freedom analysis for the single adiabatic or nonadiabatic equilibrium stage element, the total number of variables from (5-68) is

$$(N_V)_{\text{unit}} = N(4C + 13) - [2(N - 1)](C + 3) + 1 = 7N + 2NC + 2C + 7$$

since $2(N - 1)$ interconnecting streams exist. The additional variable is the total number of stages (i.e., $N_A = 1$).

The number of independent relationships from (5-69) is

$$(N_E)_{\text{unit}} = N(2C + 7) - 2(N - 1) = 5N + 2NC + 2$$

since $2(N - 1)$ redundant mole-fraction constraints exist.

The number of degrees of freedom from (5-71) is

$$(N_D)_{\text{unit}} = N_V - N_E = 2N + 2C + 5$$

One possible set of design variables is

Variable Specification	Number of Variables
Heat transfer rate for each stage (or adiabaticity)	N
Stage pressures	N
Stream V_{IN} variables	$C + 2$
Stream L_{IN} variables	$C + 2$
Number of stages	1
	$2N + 2C + 5$

Output variables for this specification include missing mole fractions for V_{IN} and L_{IN}, stage temperatures, and the variables associated with the V_{OUT} stream, L_{OUT} stream, and interstage streams. This N-stage cascade unit can represent simple absorbers, strippers, or liquid–liquid extractors.

Two-Section Countercurrent Cascades

Two-section countercurrent cascades can consist not only of adiabatic or nonadiabatic equilibrium stage elements, but also of other elements of the type shown in Table 5.1, including total and partial reboilers; total and partial condensers; equilibrium stages with a feed, F, or a side stream S; and stream mixers and dividers. These different elements can be combined into any of a number of complex cascades, by applying to (5-68) to (5-71) the values of N_V, N_E, and N_D given in Table 5.1 for the different elements.

The design or simulation of multistage separation operations involves solving the variable relationships for output variables after selecting values of design variables to satisfy the degrees of freedom. Two cases are commonly encountered. In case I, the design case, recovery specifications are made for one or two key components and the number of required equilibrium stages is determined. In case II, the simulation case, the number of equilibrium stages is specified and component separations are computed. For rigorous calculations involving multicomponent feeds, the second case is more widely applied because less computational complexity is involved with the number of stages fixed. Table 5.2 is a summary of possible variable specifications for each of these two cases for a number of separator types discussed in Chapter 1 and shown in Table 1.1. For all separators in Table 5.2, it is assumed that all inlet streams are completely specified (i.e., $C - 1$ mole fractions, total flow rate, temperature, and pressure) and all element and unit pressures and heat transfer rates (except for condensers and reboilers) are specified. Thus, only variable specifications for satisfying the remaining degrees of freedom are listed.

EXAMPLE 5.5

Consider a multistage distillation column with one feed, one side stream, a total condenser, a partial reboiler, and provisions for heat transfer to or from any stage. Determine the number of degrees of freedom and a reasonable set of specifications.

SOLUTION

This separator is assembled as shown in Figure 5.18, from the circled elements and units, which are all found in Table 5.1. The total variables are determined by summing the variables $(N_V)_e$ for each element from Table 5.1 and then subtracting the redundant variables due to interconnecting flows. As before, redundant mole-fraction constraints are subtracted from the summation of independent relationships for each element $(N_E)_e$. This problem was first treated by Gilliland and Reed [7] and more recently by Kwauk [6]. Differences in N_D obtained by various authors

Table 5.1 Degrees of Freedom for Separation Operation Elements and Units

	Schematic	Element or Unit Name	N_V, Total Number of Variables	N_E, Independent Relationships	N_D, Degrees of Freedom
(a)	$L \to \oslash \to V$, Q	Total boiler (reboiler)	$(2C + 7)$	$(C + 3)$	$(C + 4)$
(b)	$V \to \oslash \to L$, Q	Total condenser	$(2C + 7)$	$(C + 3)$	$(C + 4)$
(c)	$L_{in} \to \oslash \to V_{out}, L_{out}$, Q	Partial (equilibrium) boiler (reboiler)	$(3C + 10)$	$(2C + 6)$	$(C + 4)$
(d)	$V_{in} \to \oslash \to V_{out}, L_{out}$, Q	Partial (equilibrium) condenser	$(3C + 10)$	$(2C + 6)$	$(C + 4)$
(e)	$V_{out}\ L_{in}$ / $V_{in}\ L_{out}$	Adiabatic equilibrium stage	$(4C + 12)$	$(2C + 7)$	$(2C + 5)$
(f)	$V_{out}\ L_{in}$ / Q / $V_{in}\ L_{out}$	Equilibrium stage with heat transfer	$(4C + 13)$	$(2C + 7)$	$(2C + 6)$
(g)	$V_{out}\ L_{in}$ / F / Q / $V_{in}\ L_{out}$	Equilibrium feed stage with heat transfer and feed	$(5C + 16)$	$(2C + 8)$	$(3C + 8)$
(h)	$V_{out}\ L_{in}$ / $^a s$ / Q / $V_{in}\ L_{out}$	Equilibrium stage with heat transfer and side-stream	$(5C + 16)$	$(3C + 9)$	$(2C + 7)$
(i)	$V_{out}\ L_{in}$ / Stage N — Q_N, Q_{N-1} / Stage 1 — Q_2, Q_1 / $V_{in}\ L_{out}$	N connected equilibrium stages with heat transfer	$(7N + 2NC + 2C + 7)$	$(5N + 2NC + 2)$	$(2N + 2C + 5)$
(j)	L_1, L_2 / Q / L_3	Stream mixer	$(3C + 10)$	$(C + 4)$	$(2C + 6)$
(k)	Q, $^b L_1$ / L_2, L_3	Stream divider	$(3C + 10)$	$(2C + 5)$	$(C + 5)$

[a] Sidestream can be vapor or liquid. [b] Alternatively, all streams can be vapor.

Table 5.2 Typical Variable Specifications for Design Cases

Unit Operation		N_D	Case I, Component Recoveries Specified	Case II, Number of Equilibrium Stages Specified
			Variable Specification[a]	
(a) Absorption (two inlet streams)		$2N + 2C + 5$	1. Recovery of one key component	1. Number of stages
(b) Distillation (one inlet stream, total condenser, partial reboiler)		$2N + C + 9$	1. Condensate at saturation temperature 2. Recovery of light key component 3. Recovery of heavy key component 4. Reflux ratio (> minimum) 5. Optimum feed stage[b]	1. Condensate at saturation temperature 2. Number of stages above feed stage 3. Number of stages below feed stage 4. Reflux ratio 5. Distillate flow rate
(c) Distillation (one inlet stream, partial condenser, partial reboiler, vapor distillate only)		$(2N + C + 6)$	1. Recovery of light key component 2. Recovery of heavy key component 3. Reflux ratio (> minimum) 4. Optimum feed stage[b]	1. Number of stages above feed stage 2. Number of stages below feed stage 3. Reflux ratio 4. Distillate flow rate
(d) Liquid–liquid extraction with two solvents (three inlet streams)		$2N + 3C + 8$	1. Recovery of key component 1 2. Recovery of key component 2	1. Number of stages above feed 2. Number of stages below feed
(e) Reboiled absorption (two inlet streams)		$2N + 2C + 6$	1. Recovery of light key component 2. Recovery of heavy key component 3. Optimum feed stage[b]	1. Number of stages above feed 2. Number of stages below feed 3. Bottoms flow rate

(continued)

Table 5.2 (*Continued*)

Unit Operation		N_D	Variable Specification[a]	
			Case I, Component Recoveries Specified	Case II, Number of Equilibrium Stages Specified
(*f*) Reboiled stripping (one inlet stream)		$2N + C + 3$	1. Recovery of one key component 2. Reboiler heat duty[d]	1. Number of stages 2. Bottoms flow rate
(*g*) Distillation (one inlet stream, partial condenser, partial reboiler, both liquid and vapor distillates)		$2N + C + 9$	1. Ratio of vapor distillate to liquid distillate 2. Recovery of light key component 3. Recovery of heavy key component 4. Reflux ratio ($>$ minimum) 5. Optimum feed stage[b]	1. Ratio of vapor distillate to liquid distillate 2. Number of stages above feed stage 3. Number of stages below feed stage 4. Reflux ratio 5. Liquid distillate flow rate
(*h*) Extractive distillation (two inlet streams, total condenser, partial reboiler, single-phase condensate)		$2N + 2C + 12$	1. Condensate at saturation temperature 2. Recovery of light key component 3. Recovery of heavy key component 4. Reflux ratio ($>$ minimum) 5. Optimum feed stage[b] 6. Optimum MSA stage[b]	1. Condensate at saturation temperature 2. Number of stages above MSA stage 3. Number of stages between MSA and feed stages 4. Number of stages below feed stage 5. Reflux ratio 6. Distillate flow rate
(*i*) Liquid–liquid extraction (two inlet streams)		$2N + 2C + 5$	1. Recovery of one key component	1. Number of stages

Table 5.2 (*Continued*)

Unit Operation		N_D	Variable Specification[a]	
			Case I, Component Recoveries Specified	Case II, Number of Equilibrium Stages Specified
(*j*) Stripping (two inlet streams)		$2N + 2C + 5$	1. Recovery of one key component	1. Number of stages

[a] Does not include the following variables, which are also assumed specified: all inlet stream variables ($C + 2$ for each stream); all element and unit pressures; all element and unit heat transfer rates except for condensers and reboilers.

[b] Optimum stage for introduction of inlet stream corresponds to minimization of total stages.

[c] For case I variable specifications, MSA flow rates must be greater than minimum values for specified recoveries.

[d] For case I variable specifications, reboiler heat duty must be greater than minimum value for specified recovery.

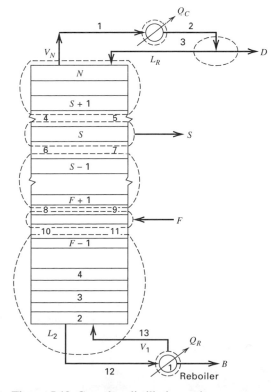

Figure 5.18 Complex distillation unit.

are due, in part, to their method of numbering stages. Here, the partial reboiler is the first equilibrium stage. From Table 5.1, element variables and relationships are as follows:

Element or Unit	$(N_V)_e$	$(N_E)_e$
Total condenser	$(2C + 7)$	$(C + 3)$
Reflux divider	$(3C + 10)$	$(2C + 5)$
$(N - S)$ stages	$[7(N - S) + 2(N - S)C + 2C + 7]$	$[5(N - S) + 2(N - S)C + 2]$
Side-stream stage	$(5C + 16)$	$(3C + 9)$
$(S - 1) - F$ stages	$[7(S - 1 - F) + 2(S - 1 - F)C$ $+ 2C + 7]$	$[5(S - 1 - F) + 2(S - 1 - F)C$ $+ 2]$
Feed stage	$(5C + 16)$	$(2C + 8)$
$(F - 1) - 1$ stages	$[7(F - 2) + 2(F - 2)C + 2C + 7]$	$[5(F - 2) + 2(F - 2)C + 2]$
Partial reboiler	$(3C + 10)$	$(2C + 6)$
	$\Sigma(N_V)_e = 7N + 2NC + 18C + 59$	$\Sigma(N_E)_e = 5N + 2NC + 4C + 22$

Subtracting $(C + 3)$ redundant variables for 13 interconnecting streams, according to (5-68), with $N_A = 0$ (no unspecified repetitions), gives

$$(N_V)_{\text{unit}} = \sum (N_E)_e - 13(C + 3) = 7N + 2NC + 5C + 20$$

Subtracting the corresponding 13 redundant mole-fraction constraints, according to (5-69),

$$(N_E)_{\text{unit}} = \sum (N_E)_e - 13 = 5N + 2NC + 4C + 9$$

Therefore, from (5-71),

$$N_D = (7N + 2NC + 5C + 20) - (5N + 2NC + 4C + 9) = 2N + C + 11$$

A set of feasible design variable specifications is

Variable Specification	Number of Variables
1. Pressure at each stage (including partial reboiler)	N
2. Pressure at reflux divider outlet	1
3. Pressure at total condenser outlet	1
4. Heat transfer rate for each stage (excluding partial reboiler)	$(N - 1)$
5. Heat transfer rate for divider	1
6. Feed mole fractions and total feed rate	C
7. Feed temperature	1
8. Feed pressure	1
9. Condensate temperature (e.g., saturated liquid)	1
10. Total number of stages, N	1
11. Feed stage location	1
12. Sidestream stage location	1
13. Sidestream total flow rate, S	1
14. Total distillate flow rate, D or D/F	1
15. Reflux flow rate, L_R, or reflux ratio, L_R/D	1
	$N_D = (2N + C + 11)$

In most separation operations, variables related to feed conditions, stage heat transfer rates, and stage pressure are known or set. Remaining specifications have proxies, provided that the variables are mathematically independent of each other and of those already known. Thus, in the above list, the first 9 entries are almost always known or specified. Variables 10 to 15, however, have surrogates. Some of these are

16. Condenser heat duty, Q_C
17. Reboiler heat duty, Q_R
18. Recovery or mole fraction of one component in bottoms
19. Recovery or mole fraction of one component in distillate

Heat duties Q_C and Q_R are not good design variables because they are difficult to specify. Condenser duty Q_C, for example, must be specified so that the condensate temperature lies between that corresponding to a saturated liquid and the freezing point of the condensate. Otherwise, a physically unrealizable (or no) solution to the problem is obtained. Similarly, it is much easier to calculate Q_R knowing the total flow rate and enthalpy of the bottom streams than vice versa. In general, Q_R and Q_C are so closely related that it is not advisable to specify both of them.

Other proxies are possible, such as a stage temperature, a flow rate leaving a stage, or any independent variable that characterizes the process. The problem of independence of variables requires careful consideration. Distillate product rate, Q_C, and L_R/D, for example, are not independent. It should also be noted that, for the design case, recoveries of no more than two species (items 18 and 19) are specified. These species are referred to as key components. Attempts to specify recoveries of three or four species will usually result in an unsuccessful solution of the equations.

The degrees of freedom for the complex distillation unit of Figure 5.18 can be determined quickly by modifying a similar unit operation in Table 5.2. The closest unit is (b), which differs from the unit in Figure 5.18 by only a sidestream. From Table 5.1, we see that an equilibrium stage with heat transfer but without a sidestream [element (f)] has $N_D = (2C + 6)$, while an equilibrium stage with heat transfer and with a sidestream [element (h)] has $N_D = (2C + 7)$ or one additional degree of freedom. In addition, when this sidestream stage is placed in a cascade, an additional degree of freedom is added for the location of the sidestream stage. Thus, two degrees of freedom are added to $N_D = 2N + C + 9$ for unit operation (b) in Table 5.2. The result is $N_D = 2N + C + 11$, which is identical to that determined in the above example.

In a similar manner, the above example can be readily modified to include a second feed stage. By comparing values of N_D for elements (f) and (g) in Table 5.1, it is seen that a feed adds $C + 2$ degrees of freedom. In addition, one more degree of freedom must be added for the location of this feed stage in a cascade. Thus, a total of $C + 3$ degrees of freedom are added, giving $N_D = 2N + 2C + 14$. ∎

SUMMARY

1. A cascade is a collection of contacting stages arranged to: (a) accomplish a separation that cannot be achieved in a single stage, and/or (b) reduce the amount of mass or energy separating agent.

2. Cascades are single- or multiple-sectioned and may be configured in cocurrent, crosscurrent, or countercurrent arrangements. Cascades are readily computed when governing equations are linear in component split ratios.

3. Stage requirements for a countercurrent solid–liquid leaching and/or washing cascade, involving constant underflow and mass transfer of one component, are given by (5-10).

4. Stage requirements for a single-section liquid–liquid extraction cascade assuming a constant distribution coefficient and immiscible solvent and carrier are given by (5-19), (5-22), and (5-29) for cocurrent, crosscurrent, and countercurrent flow arrangements, respectively. The countercurrent cascade is the most efficient.

5. Single-section stage requirements for a countercurrent cascade for absorption and stripping can be estimated with the Kremser equations, (5-48), (5-50), (5-54), and (5-55). A single-section countercurrent cascade is limited in its ability to achieve a separation between two components.

6. The Kremser equations can be combined for a two-section cascade to give (5-66), which is suitable for making approximate calculations of component splits for distillation. A two-section countercurrent cascade can achieve a sharp split between two key components. The rectifying section purifies the light components and increases recovery of heavy components. The stripping section provides the opposite function.

7. All of the cascade equations involve parameters referred to as washing W, extraction E, absorption A, and stripping S, factors that involve distribution coefficients, such as K, K_D, and R, and phase flow ratios, such as S/F and L/V.

8. The number of degrees of freedom (number of specifications) for a mathematical model of a cascade is the difference between the number of unique variables and the number of independent equations that relate the variables. For a single-section countercurrent cascade, the recovery of one component can be specified. For a two-section countercurrent cascade, the recoveries of two components can be specified.

REFERENCES

1. Berdt, R.J., and C.C. Lynch, *J. Am. Chem. Soc.,* **66,** 282–284 (1944).
2. Kremser, A., *Natl. Petroleum News,* **22** (21), 43–49 (May 21, 1930).
3. Edmister, W.C., *AIChE J.,* **3,** 165–171 (1957).
4. Smith, B.D., and W.K. Brinkley, *AIChE J.,* **6,** 446–450 (1960).

5. Lyster, W.N., S.L. Sullivan, Jr., D.S. Billingsley, and C.D. Holland, *Petroleum Refiner,* **38** (6), 221–230 (1959).
6. Kwauk, M., *AIChE J.,* **2,** 240–248 (1956).
7. Gilliland, E.R., and C.E. Reed, *Ind. Eng. Chem.,* **34,** 551–557 (1942).

EXERCISES

Section 5.1

5.1 Devise an interlinked cascade of the type shown in Figure 5.2e, but consisting of three columns for the separation of a four-component feed into four products.

5.2 A liquid–liquid extraction process is conducted batchwise as shown in Figure 5.19. The process begins in vessel 1 (original), where 100 mg each of solutes A and B are dissolved in 100 ml of water. After adding 100 ml of an organic solvent that is more selective for A than B, the distribution of A and B becomes that shown for equilibration 1 with vessel 1. The organic-rich phase is transferred to vessel 2 (transfer), leaving the water-rich phase in vessel 1 (transfer). Assume that water and the organic solvent are immiscible. Next, 100 ml of water is added to vessel 2, resulting in the phase distribution shown for vessel 2 (equilibration 2). Also, 100 ml of organic solvent is added to vessel 1 to give the phase distribution shown for vessel 1 (equilibration 2). The batch process is continued by adding vessel 3 and then 4 to obtain the results shown.

(a) Carefully study the process in Figure 5.19 and then draw a corresponding cascade diagram, labeled in a manner similar to Figure 5.2(b).

(b) Is the process of the cocurrent, countercurrent, or crosscurrent type?

(c) Compare the separation achieved with that for a single batch equilibrium step.

(d) How could the process be modified to make it a countercurrent cascade [see O. Post and L.C. Craig, *Anal. Chem.,* **35,** 641 (1963)].

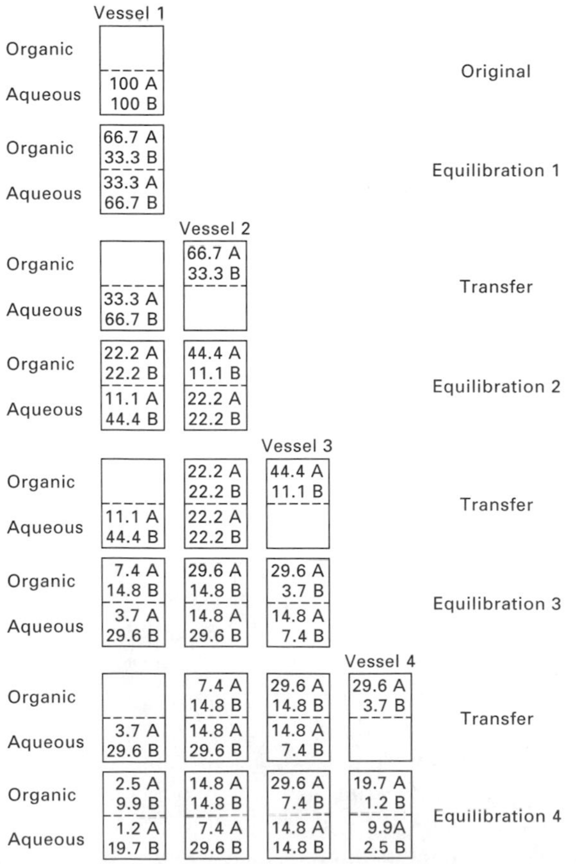

5.3 Nitrogen is to be removed from a gas mixture with methane by gas permeation (see Table 1.2) using a glassy polymer membrane that is selective for nitrogen. However, the desired degree of separation cannot be achieved in one stage. Draw sketches of two different two-stage membrane cascades that might be considered to perform the desired separation.

Section 5.2

5.4 In Example 4.9, 83.25% of the oil in soybeans is leached by benzene using a single equilibrium stage. Calculate the percent extraction of oil if:

(a) Two countercurrent equilibrium stages are used to process 5,000 kg/h of soybean meal with 5,000 kg/h of benzene.

(b) Three countercurrent equilibrium stages are used to process the same flows as in part (a).

Figure 5.19 Liquid–liquid extraction process for Exercise 5.2.

(c) Also, determine the number of countercurrent equilibrium stages required to extract 98% of the oil if a solvent rate of twice the minimum value is used.

5.5 For Example 5.1, involving the separation of sodium carbonate from an insoluble oxide, compute the minimum solvent feed rate in pounds per hour. What is the ratio of actual solvent rate to the minimum solvent rate? Determine and plot the percent recovery of soluble solids with a cascade of five countercurrent equilibrium stages for solvent flow rates from 1.5 to 7.5 times the minimum value.

5.6 Aluminum sulfate, commonly called alum, is produced as a concentrated aqueous solution from bauxite ore by reaction with aqueous sulfuric acid, followed by a three-stage countercurrent washing operation to separate soluble aluminum sulfate from the insoluble content of the bauxite ore, followed by evaporation. In a typical process, 40,000 kg/day of solid bauxite ore containing 50 wt% Al_2O_3 and 50% inert is crushed and fed together with the stoichiometric amount of 50 wt% aqueous sulfuric acid to a reactor, where the Al_2O_3 is reacted completely to alum by the reaction

$$Al_2O_3 + 3H_2SO_4 \rightarrow Al_2(SO_4)_3 + 3H_2O$$

The slurry effluent from the reactor (digester), consisting of solid inert material from the ore and an aqueous solution of aluminum sulfate is then fed to a three-stage countercurrent washing unit to separate the aqueous aluminum sulfate from the inert material. If the solvent is 240,000 kg/day of water and the underflow from each washing stage is 50 wt% water on a solute-free basis, compute the flow rates in kilograms per day of aluminum sulfate, water, and inert solid in each of the two product streams leaving the cascade. What is the percent recovery of the aluminum sulfate? Would the addition of one more stage be worthwhile?

5.7 (a) When rinsing clothes with a given amount of water, would one find it more efficient to divide the water and rinse several times; or should one use all the water in one rinse? Explain.
(b) Devise a clothes-washing machine that gives the most efficient rinse cycle for a fixed amount of water.

Section 5.3

5.8 An aqueous acetic acid solution containing 6.0 moles of acid per liter is to be extracted in the laboratory with chloroform at 25°C to recover the acid (B) from chloroform-insoluble impurities present in the water. The water (A) and chloroform (C) are essentially immiscible. If 10 liters of solution are to be extracted at 25°C, calculate the percent extraction of acid obtained with 10 liters of chloroform under the following conditions:
(a) Using the entire quantity of solvent in a single batch extraction
(b) Using three batch extractions with one-third of the total solvent used in each batch
(c) Using three batch extractions with 5 liters of solvent in the first, 3 liters in the second, and 2 liters in the third batch
Assume that the volumetric amounts of the feed and solvent

do not change during extraction. Also, assume the distribution coefficient for the acid, $K''_{D_B} = (c_B)_C/(c_B)_A = 2.8$, where $(c_B)_C$ = concentration of acid in chloroform and $(c_B)_A$ = concentration of acid in water, both in moles per liter.

5.9 A 20 wt% solution of uranyl nitrate (UN) in water is to be treated with tributyl phosphate (TBP) to remove 90% of the uranyl nitrate. All operations are to be batchwise equilibrium contacts. Assuming that water and TBP are mutually insoluble, how much TBP is required for 100 g of solution if at equilibrium (g UN/g TBP) = 5.5(g UN/g H_2O) and:
(a) All the TBP is used at once in one stage?
(b) Half is used in each of two consecutive stages?
(c) Two countercurrent stages are used?
(d) An infinite number of crosscurrent stages is used?
(e) An infinite number of countercurrent stages is used?

5.10 The uranyl nitrate (UN) in 2 kg of a 20 wt% aqueous solution is to be extracted with 500 g of tributyl phosphate. Using the equilibrium data in Exercise 5.9, calculate and compare the percentage recoveries for the following alternative procedures:
(a) A single-stage batch extraction
(b) Three batch extractions with one-third of the total solvent used in each batch (the solvent is withdrawn after contacting the entire UN phase)
(c) A two-stage cocurrent extraction
(d) A three-stage countercurrent extraction
(e) An infinite-stage countercurrent extraction
(f) An infinite-stage crosscurrent extraction

5.11 One thousand kilograms of a 30 wt% dioxane in water solution is to be treated with benzene at 25°C to remove 95% of the dioxane. The benzene is dioxane free, and the equilibrium data of Example 5.2 can be used. Calculate the solvent requirements for:
(a) A single batch extraction
(b) Two crosscurrent stages using equal amounts of benzene
(c) Two countercurrent stages
(d) An infinite number of crosscurrent stages
(e) An infinite number of countercurrent stages

5.12 Chloroform is to be used to extract benzoic acid from wastewater effluent. The benzoic acid is present at a concentration of 0.05 mol/liter in the effluent, which is discharged at a rate of 1,000 liter/h. The distribution coefficient for benzoic acid at process conditions is given by

$$c^I = K_D^{II} c^{II}$$

where $K_D^{II} = 4.2$, c^I = molar concentration of solute in solvent, and c^{II} = molar concentration of solute in water. Chloroform and water may be assumed immiscible. If 500 liters/h of chloroform is to be used, compare the fraction benzoic acid removed in
(a) A single equilibrium contact
(b) Three crosscurrent contacts with equal portions of chloroform
(c) Three countercurrent contacts

5.13 Repeat Example 5.2 with a solvent for which $E = 0.90$. Display your results in a plot like Figure 5.7. Does countercurrent flow still have a marked advantage over crosscurrent flow? Is it desirable to choose the solvent and solvent rate so that $E > 1$? Explain.

Section 5.4

5.14 Repeat Example 5.3 for $N = 1, 3, 10$, and 30 stages. Plot the percent absorption of each of the five hydrocarbons and the total feed gas, as well as the percent stripping of the oil versus the number of stages, N. What can you conclude about the effect of the number of stages on each component?

5.15 Solve Example 5.3 for an absorbent flow rate of 330 lbmol/h and three theoretical stages. Compare your results to the results of Example 5.3 and discuss the effect of trading stages for absorbent flow.

5.16 Estimate the minimum absorbent flow rate required for the separation calculated in Example 5.3 assuming that the key component is propane, whose flow rate in the exit vapor is to be 155.4 lbmol/h.

5.17 Solve Example 5.3 with the addition of a heat exchanger at each stage so as to maintain isothermal operation of the absorber at
(a) 125°F
(b) 150°F
What is the effect of temperature on absorption in the range of 100 to 150°F?

5.18 One million pound-moles per day of a gas of the following composition is to be absorbed by n-heptane at $-30°F$ and 550 psia in an absorber having 10 theoretical stages so as to absorb 50% of the ethane. Calculate the required flow rate of absorbent and the distribution, in lbmol/h, of all the components between the exiting gas and liquid streams.

Component	Mole Percent in Feed Gas	K-value @ $-30°F$ and 550 psia
C_1	94.9	2.85
C_2	4.2	0.36
C_3	0.7	0.066
nC_4	0.1	0.017
nC_5	0.1	0.004

5.19 A stripper operating at 50 psia with three equilibrium stages is used to strip 1,000 kmol/h of liquid at 300°F having the following molar composition: 0.03% C_1, 0.22% C_2, 1.82% C_3, 4.47% nC_4, 8.59% nC_5, 84.87% nC_{10}. The stripping agent is 1,000 kmol/h of superheated steam at 300°F and 50 psia. Use the Kremser equation to estimate the compositions and flow rates of the stripped liquid and exiting rich gas.

Assume a K-value for C_{10} of 0.20 and assume that no steam is absorbed. However, calculate the dew-point temperature of the exiting rich gas at 50 psia. If that temperature is above 300°F, what would you suggest be done?

5.20 In Figure 5.12, is anything gained by totally condensing the vapor leaving each stage? Alter the processes in Figure 5.12a and 5.12b so as to eliminate the addition of heat to stages 2 and 3 and still achieve the same separations.

5.21 Repeat Example 5.4 for external reflux flow rates L_0 of
(a) 1,500 lbmol/h
(b) 2,000 lbmol/h
(c) 2,500 lbmol/h
Plot d_{C_3}/b_{C_3} as a function of L_0 from 1,000 to 2,500 lbmol/h. In making the calculations, assume that stage temperatures do not change from the results of Example 5.4. Discuss the effect of reflux ratio on the separation.

5.22 Repeat Example 5.4 for the following numbers of equilibrium stages (see Figure 5.15):
(a) $M = 10, N = 10$
(b) $M = 15, N = 15$
Plot d_{C_3}/b_{C_3} as a function of $M + N$ from 10 to 30 stages. In making the calculations, assume that state temperatures and total flow rates do not change from the results of Example 5.4. Discuss the effect of the number of stages on the separation.

5.23 Use the Edmister group method to determine the compositions of the distillate and bottoms for the distillation operation shown in Figure 5.20. At column conditions, the feed is approximately 23 mol% vapor.

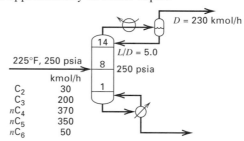

Figure 5.20 Conditions for Exercise 5.23.

5.24 A bubble-point liquid feed is to be distilled as shown in Figure 5.21. Use the Edmister group method to estimate the mole-fraction compositions of the distillate and bottoms. Assume initial overhead and bottoms temperatures are 150 and 250°F, respectively.

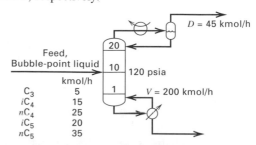

Figure 5.21 Conditions for Exercise 5.24.

Section 5.5

5.25 Verify the values given in Table 5.1 for N_V, N_E, and N_D for a partial reboiler and a total condenser.

5.26 Verify the values given in Table 5.1 for N_V, N_E, and N_D for a stream mixer and a stream divider.

5.27 A mixture of maleic anhydride and benzoic acid containing 10 mol% acid is a product of the manufacture of phthalic anhydride. The mixture is to be distilled continuously in a column with a total condenser and a partial reboiler at a pressure of 13.2 kPa (100 torr) with a reflux ratio of 1.2 times the minimum value to give a product of 99.5 mol% maleic anhydride and a bottoms of 0.5 mol% anhydride. Is this problem completely specified?

5.28 Verify N_D for the following unit operations in Table 5.2: (*b*), (*c*), and (*g*). How would N_D change if two feeds were used instead of one?

5.29 Verify N_D for unit operations (*e*) and (*f*) in Table 5.2. How would N_D change if a vapor side stream was pulled off some stage located between the feed stage and the bottom stage?

5.30 Verify N_D for unit operation (*h*) in Table 5.2. How would N_D change if a liquid side stream was added to a stage that was located between the feed stage and stage 2?

5.31 The following are not listed as design variables for the distillation unit operations in Table 5.2:
(a) Condenser heat duty
(b) Stage temperature
(c) Intermediate-stage vapor rate
(d) Reboiler heat load
Under what conditions might these become design variables? If so, which variables listed in Table 5.2 would you eliminate?

5.32 Show for distillation that, if a total condenser is replaced by a partial condenser, the number of degrees of freedom is reduced by 3, provided that the distillate is removed solely as a vapor.

5.33 Unit operation (*b*) in Table 5.2 is to be heated by injecting live steam directly into the bottom plate of the column instead of by using a reboiler, for a separation involving ethanol and water. Assuming a fixed feed, an adiabatic operation, atmospheric pressure throughout, and a top alcohol concentration specification:
(a) What is the total number of design variables for the general configuration?
(b) How many design variables are needed to complete the design? Which variables do you recommend?

5.34 (a) For the distillation column shown in Figure 5.22, determine the number of independent design variables.

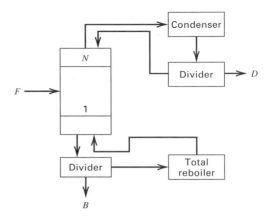

Figure 5.22 Conditions for Exercise 5.34.

(b) It is suggested that a feed consisting of 30% A, 20% B, and 50% C, all in moles, at 37.8°C and 689 kPa, be processed in the unit of Figure 5.22, consisting of a 15-plate, 3-m-diameter column that is designed to operate at vapor velocities of 0.3 m/s and an L/V of 1.2. The pressure drop per plate is 373 Pa at these conditions, and the condenser is cooled by plant water at 15.6°C.

The product specifications in terms of the concentration of A in the distillate and C in the bottoms have been set by the process department, and the plant manager has asked you to specify a feed rate for the column. Write a memorandum to the plant manager pointing out why you can't do this, and suggest some alternatives.

5.35 Calculate the number of degrees of freedom for the mixed-feed, triple effect evaporator system shown in Figure 5.23. Assume that the steam and all drain streams are at saturated conditions and the feed is an aqueous solution of a dissolved organic solid. Also, assume that all overhead streams are pure water vapor, with no entrainment.

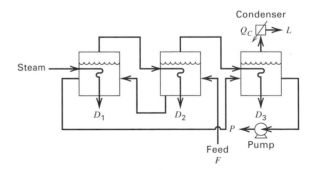

Figure 5.23 Conditions for Exercise 5.35.

If this evaporator system is used to concentrate a feed containing 2 wt% dissolved organic to a product with 25 wt% dissolved organic, using 689-kPa saturated steam, calculate the number of unspecified design variables and suggest likely candidates. Assume perfect insulation against heat loss for each effect.

5.36 A reboiled stripper as shown in Figure 5.24 is to be designed for the task shown. Determine
(a) The number of variables
(b) The number of equations relating the variables
(c) The number of degrees of freedom and indicate
(d) Which additional variables, if any, need to be specified

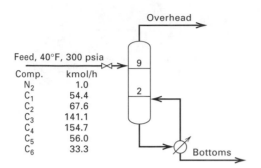

Figure 5.24 Conditions for Exercise 5.36.

5.37 The thermally coupled distillation system shown in Figure 5.25 is to be used to separate a mixture of three components into three products. Determine for the system
(a) The number of variables
(b) The number of equations relating the variables
(c) The number of degrees of freedom and propose
(d) A reasonable set of design variables

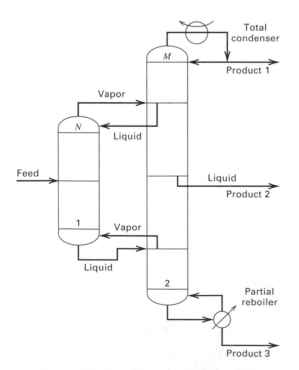

Figure 5.25 Conditions for Exercise 5.37.

5.38 When the feed to a distillation column contains a small amount of impurities that are much more volatile than the desired distillate, it is possible to separate the volatile impurities from the distillate by removing the distillate as a liquid sidestream from a stage located several stages below the top stage. As shown in Figure 5.26, this additional top section of stages is referred to as a pasteurizing section.
(a) Determine the number of degrees of freedom for the unit
(b) Determine a reasonable set of design variables

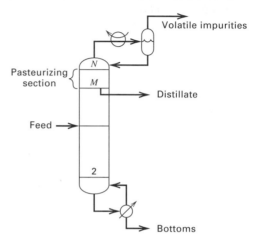

Figure 5.26 Conditions for Exercise 5.38.

5.39 A system for separating a mixture into three products is shown in Figure 5.27. For it, determine
(a) The number of variables
(b) The number of equations relating the variables
(c) The number of degrees of freedom and propose
(d) A reasonable set of design variables

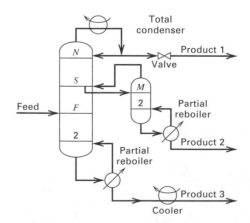

Figure 5.27 Conditions for Exercise 5.39.

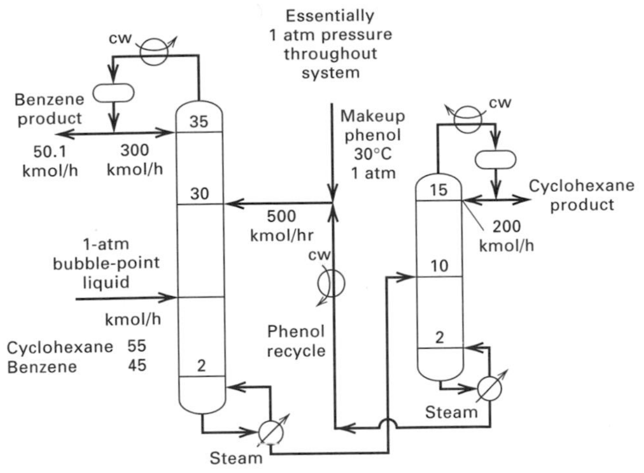

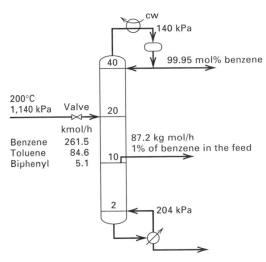

Figure 5.28 Conditions for Exercise 5.40.

5.40 A system for separating a binary mixture by extractive distillation, followed by ordinary distillation for recovery and recycle of the solvent, is shown in Figure 5.28 Are the design variables shown sufficient to specify the problem completely? If not, what additional design variable(s) would you select?

5.41 A single distillation column for separating a three-component mixture into three products is shown in Figure 5.29. Are the design variables shown sufficient to specify the problem completely? If not, what additional design variable(s) would you select?

Figure 5.29 Conditions for Exercise 5.41.

Chapter 6

Absorption and Stripping of Dilute Mixtures

In *absorption* (also called *gas absorption, gas scrubbing,* and *gas washing*), a gas mixture is contacted with a liquid (the *absorbent* or *solvent*) to selectively dissolve one or more components by mass transfer from the gas to the liquid. The components transferred to the liquid are referred to as *solutes* or *absorbate.* Absorption is used to separate gas mixtures; remove impurities, contaminants, pollutants, or catalyst poisons from a gas; or recover valuable chemicals. Thus, the species of interest in the gas mixture may be all components, only the component(s) not transferred, or only the component(s) transferred.

The opposite of absorption is *stripping* (also called *desorption*), wherein a liquid mixture is contacted with a gas to selectively remove components by mass transfer from the liquid to the gas phase. As discussed in Chapter 5, absorbers are frequently coupled with strippers to permit regeneration (or recovery) and recycling of the absorbent. Because stripping is not perfect, absorbent recycled to the absorber contains species present in the vapor entering the absorber. When water is used as the absorbent, it is more common to separate the absorbent from the solute by distillation rather than stripping.

A typical absorption operation is shown in Figure 6.1. The feed, which contains air (21% O_2, 78% N_2, and 1% Ar), water vapor, and acetone vapor, is the gas leaving a dryer where solid cellulose acetate fibers, wet with water and acetone, are dried. The purpose of the 30-tray (equivalent to 10 equilibrium stages) absorber is to remove the acetone by contacting the gas with a suitable absorbent, water. By using countercurrent flow of gas and liquid in a multiple-stage device, the material balance, shown in Figure 6.1, indicates that 99.5% of the acetone is absorbed. The gas leaving the absorber contains only 143 ppm (parts per million) by weight of acetone vapor and can be recycled to the dryer or exhausted to the atmosphere. Although the major component transferred between phases is acetone, the material balance indicates that small amounts of oxygen and nitrogen are also absorbed by the water solvent. Because water is present in both the feed gas and the absorbent, it can be both absorbed and stripped. As seen in Figure 6.1, the net effect is that water is stripped because more water appears in the exit gas than in the feed gas. The exit gas is almost saturated with water vapor and the exit liquid is almost saturated with air. The temperature of the absorbent decreases by 3°C to supply the energy of vaporization needed to strip the water, which in this example is greater than the energy of condensation liberated from the absorption of acetone.

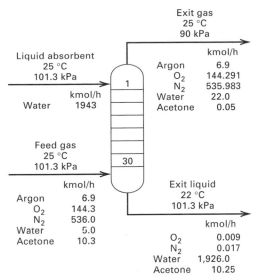

Figure 6.1 Typical absorption process.

As was shown in Figure 5.9, the fraction of a component absorbed in a countercurrent cascade depends on the number of equilibrium stages and the absorption factor, $A = L/KV$, for that component. For the conditions of Figure 6.1, estimated K-values and absorption factors, which range over many orders of magnitude, are

Component	$A = L/KV$	K-value
Water	11.7	0.031
Acetone	1.38	2.0
Oxygen	0.00006	45,000
Nitrogen	0.00003	90,000
Argon	0.00008	35,000

For acetone, the K-value is based on Eq. (4) of Table 2.3, the modified Raoult's law, $K = \gamma P^s/P$, with $\gamma = 6.7$ for a dilute solution of acetone in water at 25°C and 101.3 kPa. For oxygen and nitrogen, K-values are based on the use of Eq. (6) of Table 2.3, Henry's law, $K = H/P$, using constants from Figure 4.27 at 25°C. For water, the K-value is obtained from Eq. (3) of Table 2.3, Raoult's law, $K = P^s/P$, which applies because the mole fraction of water in the liquid phase is close to 1. For argon, the Henry's law constant at 25°C was obtained from the International Critical Tables [1].

Figure 5.9 shows that if the value of A is greater than 1, any degree of absorption can be achieved: the larger the value of A, the fewer the number of stages required to absorb a desired fraction of the solute. However, very large values of A can correspond to absorbent flow rates that are larger than necessary. From an economic standpoint, the value of A, for the main (key) species to be absorbed, should be in the range of 1.25 to 2.0, with 1.4 being a frequently recommended value. Thus, the above value of 1.38 for acetone is favorable.

For a given feed gas flow rate and choice of absorbent, factors that influence the value of A are absorbent flow rate, temperature, and pressure. Because $A = L/(KV)$, the larger the absorbent flow rate is, the larger will be the value of A. The required absorbent flow rate can be reduced by reducing the K-value of the solute. Because the K-value for many solutes varies exponentially with temperature and is inversely proportional to pressure, this reduction can be achieved by reducing the temperature and/or increasing the pressure. Increasing the pressure also serves to reduce the diameter of the equipment for a given gas throughput. However, temperature adjustment by feed-gas refrigeration and/or absorbent, and/or adjustment of the feed gas pressure by gas compression can be expensive. For these reasons, the absorber in Figure 6.1 operates at near-ambient conditions.

For a stripper, the stripping factor, $S = 1/A = KV/L$, is crucial. To reduce the required flow rate of stripping agent, operation of the stripper at a high temperature and/or a low pressure is desirable, with an optimum stripping factor in the vicinity of 1.4.

Absorption and stripping are technically mature separation operations. Design procedures are well developed and commercial processes are common. Table 6.1 lists representative commercial absorption applications. In most cases, the solutes are contained in gaseous effluents from chemical reactors. Passage of strict environmental standards with respect to pollution by emission of noxious gases has greatly increased the use of gas absorbers in the past decade.

When water and hydrocarbon oils are used as absorbents, no significant chemical reactions occur between the absorbent and the solute, and the process is commonly referred to as *physical absorption*. When aqueous sodium hydroxide (a strong base) is used as the absorbent to dissolve an acid gas, absorption is accompanied by a rapid and

Table 6.1 Representative Commercial Applications of Absorption

Solute	Absorbent	Type of Absorption
Acetone	Water	Physical
Acrylonitrile	Water	Physical
Ammonia	Water	Physical
Ethanol	Water	Physical
Formaldehyde	Water	Physical
Hydrochloric acid	Water	Physical
Hydrofluoric acid	Water	Physical
Sulfur dioxide	Water	Physical
Sulfur trioxide	Water	Physical
Benzene and toluene	Hydrocarbon oil	Physical
Butadiene	Hydrocarbon oil	Physical
Butanes and propane	Hydrocarbon oil	Physical
Naphthalene	Hydrocarbon oil	Physical
Carbon dioxide	Aq. NaOH	Irreversible chemical
Hydrochloric acid	Aq. NaOH	Irreversible chemical
Hydrocyanic acid	Aq. NaOH	Irreversible chemical
Hydrofluoric acid	Aq. NaOH	Irreversible chemical
Hydrogen sulfide	Aq. NaOH	Irreversible chemical
Chlorine	Water	Reversible chemical
Carbon monoxide	Aq. cuprous ammonium salts	Reversible chemical
CO_2 and H_2S	Aq. monoethanolamine (MEA) or diethanolamine (DEA)	Reversible chemical
CO_2 and H_2S	Diethyleneglycol (DEG) or triethyleneglycol (TEG)	Reversible chemical
Nitrogen oxides	Water	Reversible chemical

irreversible neutralization reaction in the liquid phase and the process is referred to as *chemical absorption* or *reactive absorption.* More complex examples of chemical absorption are processes for absorbing CO_2 and H_2S with aqueous solutions of monoethanolamine (MEA) and diethanolamine (DEA), where a reversible chemical reaction takes place in the liquid phase. Chemical reactions can increase the rate of absorption, increase the absorption capacity of the solvent, increase selectivity to preferential dissolve only certain components of the gas, and convert a hazardous chemical to a safe compound.

In this chapter, equipment for conducting absorption and stripping operations is discussed and fundamental *equilibrium-based* and *rate-based* calculation procedures, both graphical and algebraic, are presented for physical absorption and stripping of mainly dilute mixtures. The methods also apply to reactive absorption with irreversible and complete chemical reactions of the solute in the liquid phase. Calculations for concentrated mixtures and reactive absorption with reversible chemical reactions are best handled by computer-aided calculations, which are discussed in Chapters 10 and 11. An introduction to concentrated mixtures in packed columns is given in the last section of this chapter.

6.1 EQUIPMENT

Absorption and stripping are conducted in trayed towers (plate columns), packed columns, spray towers, bubble columns, and centrifugal contactors, as shown schematically in Figure 6.2. A trayed tower is a vertical, cylindrical pressure vessel in which vapor and liquid, which flow countercurrently, are contacted on a series of metal trays or plates, an example of which is shown in Figure 6.3. Liquid flows across each tray, over an outlet weir, and into a downcomer, which takes the liquid by gravity to the tray below. Gas flows upward through openings in each tray, bubbling through the liquid on the tray. When the openings are holes, any of the five two-phase-flow regimes shown in Figure 6.4, and considered in detail by Lockett [2], may occur. The most common and favored regime is the *froth regime,* in which the liquid phase is continuous and the gas passes through in the form of jets or a series of bubbles. The *spray regime,* in which the gas phase is continuous, occurs for low weir heights (low liquid depths) at high gas rates. For low gas rates, the *bubble regime* can occur, in which the liquid is fairly quiescent and bubbles rise in swarms. At high liquid rates, small gas bubbles may be undesirably emulsified. If bubble coalescence is hindered, an undesirable foam forms. Ideally, the liquid carries no vapor bubbles (*occlusion*) to the tray below, the vapor carries no liquid droplets (*entrainment*) to the tray above, and there is no *weeping* of liquid through the openings of the tray. With good contacting, equilibrium between the exiting vapor and liquid phases is approached on each tray.

As shown in Figure 6.5, openings in the tray for the passage of vapor are most commonly perforations, metal valves, and/or metal bubble caps. The simplest is perforations, usually $\frac{1}{8}$ to $\frac{1}{2}$ in. in diameter, used in a so-called *sieve tray* (also called a *perforated tray*). A *valve tray* has much larger openings, commonly from 1.5 to 2 in. in diameter. Each hole is fitted with a valve that consists of a cap, which overlaps the hole, with legs or a cage to limit the vertical rise while maintaining the horizontal location of the valve. With no vapor flow, each valve sits on the tray, over a hole. As the vapor rate is increased, the valve rises, providing a larger peripheral opening for vapor to flow into the liquid to create a froth. A *bubble-cap tray* has bubble caps that consist of a fixed cap, 3 to 6 in. in diameter, mounted over and above a riser of 2 to 3 in. in diameter. The cap has rectangular or triangular slots cut around its side. The vapor flows up through the tray opening into the riser, turns around, and passes out through the slots of the cap, into the liquid to form a

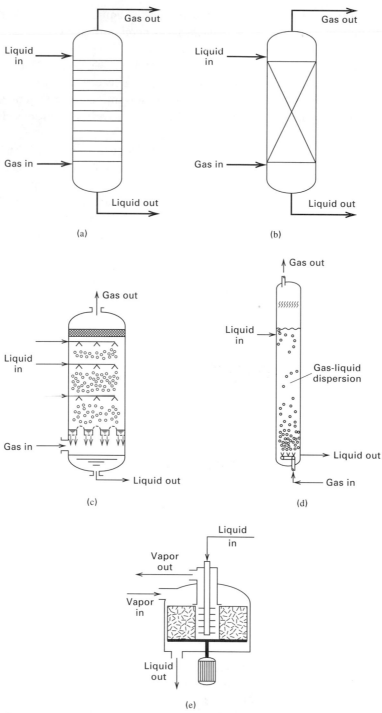

Figure 6.2 Industrial equipment for absorption and stripping: (a) trayed tower; (b) packed column; (c) spray tower; (d) bubble column; (e) centrifugal contactor.

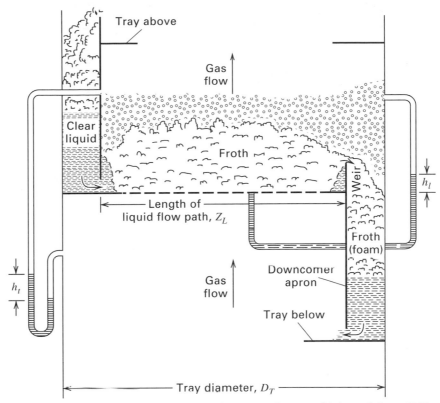

Figure 6.3 Details of a contacting tray in a trayed tower. [Adapted from B.F. Smith, *Design of Equilibrium Stage Processes,* McGraw-Hill, New York (1963).]

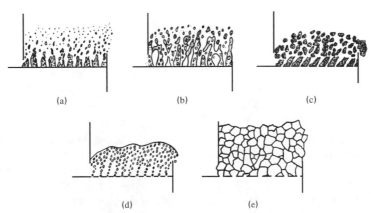

Figure 6.4 Possible vapor–liquid flow regimes for a contacting tray: (a) spray; (b) froth; (c) emulsion; (d) bubble; (e) cellular foam. [Reproduced by permission from M.J. Lockett, *Distillation Tray Fundamentals,* Cambridge University Press, London (1986).]

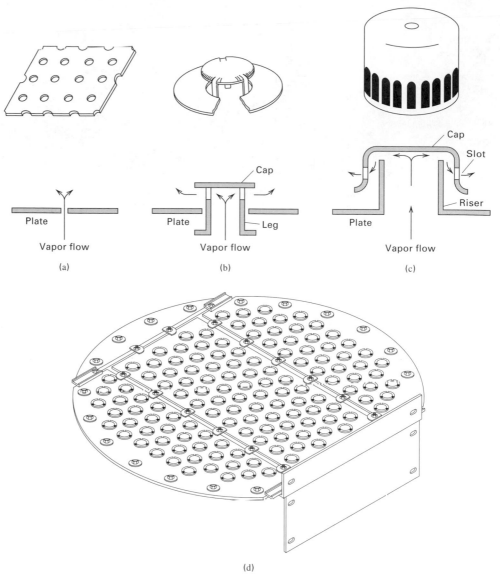

Figure 6.5 Three types of tray openings for passage of vapor up into liquid: (a) perforation; (b) valve cap; (c) bubble cap. (d) Tray with valve caps.

froth. An 11-ft-diameter column might have trays with 50,000 $\frac{3}{16}$-in.-diameter perforations, or 1,000 2-in.-diameter valve caps, or 500 4-in.-diameter bubble caps.

As listed in Table 6.2, tray types are compared on the basis of cost, pressure drop, mass transfer efficiency, vapor capacity, and flexibility in terms of turndown ratio (ratio of maximum to minimum vapor capacity). At the limiting vapor capacity, *flooding* of the column occurs because of excessive entrainment causing the liquid flow rate to exceed the capacity of the downcomer and, thus, go back up the column. At low vapor rates, weeping of liquid through the tray openings or vapor pulsation becomes excessive. Because of their low relative cost, sieve trays are preferred unless flexibility is required, in which case valve trays are best. Bubble-cap trays, which exist in many pre-1950 installations, are rarely specified for new installations, but may be preferred when the amount of liquid holdup on a tray must be controlled to provide adequate residence time for a chemical reaction or when weeping must be prevented.

Table 6.2 Comparison of Types of Trays

	Sieve Trays	Valve Trays	Bubble-Cap Trays
Relative cost	1.0	1.2	2.0
Pressure drop	Lowest	Intermediate	Highest
Efficiency	Lowest	Highest	Highest
Vapor capacity	Highest	Highest	Lowest
Typical Turndown ratio	2	4	5

A *packed column,* shown in detail in Figure 6.6, is a vertical cylindrical pressure vessel containing one or more sections of a packing material over whose surface the liquid flows downward by gravity, as a film or as droplets between packing elements. Vapor flows upward through the wetted packing, contacting the liquid. The sections of packing are contained between a lower gas-injection support plate, which holds the packing, and an upper grid or mesh hold-down plate, which prevents packing movement. A *liquid distributor,* placed above the hold-down plate, ensures uniform distribution of liquid as it enters the packed section. If the depth of packing is more than about 20 ft, liquid channeling may occur, causing the liquid to flow down the column mainly near the wall, and gas to flow mainly up the center of the column, thus greatly reducing the extent of vapor–liquid contact. In that case, a liquid *redistributor* should be installed.

Commercial packing materials include *random* (dumped) packings, some of which are shown in Figure 6.7a, and *structured* (also called arranged, ordered, or stacked packings), some of which are shown in Figure 6.7b. Among the random packings, which are poured into the column, are the old ceramic Raschig rings and Berl saddles, which are seldom specified for current installations. They have been largely replaced by metal and plastic Pall rings, metal Bialecki rings, and ceramic Intalox saddles, which provide more surface area for mass transfer, a higher flow capacity, and a lower pressure drop. Most recently, through-flow packings of a lattice-work design have been developed. These packings, which include metal Intalox IMTP; metal, plastic, and ceramic Cascade Mini-Rings; metal Levapak; metal Fleximax; metal, plastic, and ceramic Hiflow rings; metal tri-packs; and plastic Nor Pac rings, exhibit even lower pressure drop per unit height of packing and even higher mass transfer rates per unit volume of packing. Accordingly, they are called "high-efficiency" random packings. Most random packings are available in nominal diameters, ranging from 1 in. to 3.5 in. As packing size increases, mass-transfer efficiency and pressure drop decrease. Therefore, an optimal packing size exists that represents a compromise between these two factors, since low pressure drop and high mass transfer rates are both desirable. However, to minimize channeling of liquid, the nominal diameter of the packing should be less than one-eighth of the column diameter. Metal packings are usually preferred because of their superior strength and good wettability. Ceramic packings, which have superior wettability but inferior strength, are used only to resist corrosion at elevated temperatures, where plastics would fail. Plastic packings, usually of polypropylene, are inexpensive and have sufficient strength, but may experience poor wettability, particularly at low liquid rates.

Representative structured packings include the older corrugated sheets of metal gauze, such as Sulzer BX, Montz A, Gempak 4BG, and Intalox High-Performance Wire Gauze Packing. Newer and less expensive structured packings, which are fabricated from sheet metal and plastics and may or may not be perforated, embossed, or surface roughened, include metal and plastic Mellapak 250Y, metal Flexipac, metal and plastic Gempak 4A, metal Montz B1, and metal Intalox High-Performance Structured Packing. Structured packings come with different size openings between adjacent corrugated layers and are

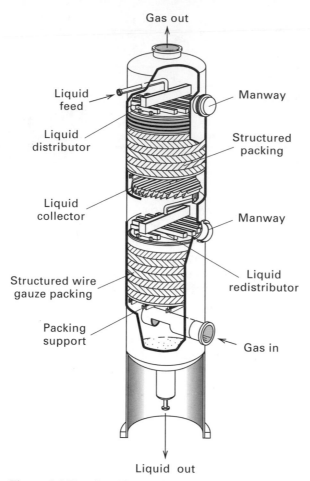

Figure 6.6 Details of internals used in a packed column.

stacked in the column. Although structured packings are considerably more expensive per unit volume than random packings, structured packings exhibit far less pressure drop per theoretical stage and have higher efficiency and capacity.

As shown in Table 6.3, packings are usually compared on the basis of the same factors used to compare tray types. However, the differences between random and structured packings are much greater than the differences among the three types of trays listed in Table 6.2.

If only one or two theoretical stages are required, only a very low pressure drop is allowed, and the solute is very soluble in the liquid phase, the use of a *spray tower* may be advantageous. As shown in Figure 6.2, a spray tower consists of a vertical, cylindrical vessel filled with gas into which liquid is sprayed. A *bubble column,* also shown in Figure 6.2, consists of a vertical cylindrical vessel partially filled with liquid into which the vapor is bubbled. Vapor pressure drop is high, and only one or two theoretical stages can be achieved. Such a device has a low vapor throughput and should not be considered unless the solute has a very low solubility in the liquid and/or a slow chemical reaction takes place in the liquid phase, thus requiring an appreciable residence time. A novel device is

Ceramic Raschig rings   Ceramic Berl saddle   Ceramic Intalox saddle

Plastic super
Intalox saddle   Metal Intalox IMTP   Metal Pall ring

Plastic Flexiring   Metal Bialecki ring   Metal Top-pak

Plastic Tellerette   Plastic Hackett   Plastic Igel

(a)

Figure 6.7 Typical materials used in a packed column: (a) random packing materials; (*continued*)

the *centrifugal contactor,* one example of which, as shown in Figure 6.2, consists of a stationary ringed housing, intermeshed with a ringed rotating section. The liquid phase is fed near the center of the packing, from which it is caused to flow outward by centrifugal force. The vapor phase flows inward by a pressure driving force. Very high mass transfer rates can be achieved with only moderately high rotation rates. It is possible to obtain the equivalent of several equilibrium stages in a very compact unit. This type of contact is

Table 6.3 Comparison of Types of Packing

	Random		Structured
	Raschig Rings and Saddles	"Through Flow"	
Relative cost	Low	Moderate	High
Pressure drop	Moderate	Low	Very low
Efficiency	Moderate	High	Very high
Vapor capacity	Fairly high	High	High
Typical turndown ratio	2	2	2

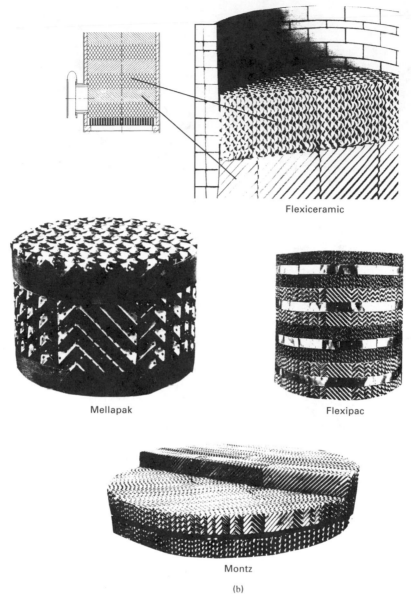

Flexiceramic

Mellapak

Flexipac

Montz

(b)

Figure 6.7 (*Continued*) (b) structured packing materials.

favored when headroom for a trayed tower or packed column is not available or when a short residence time is desired.

In most applications, the choice of contacting device is between a trayed tower and a packed column. The latter, using dumped packings, is almost always favored when a column diameter of less than 2 ft and a packed height of not more than 20 ft are sufficient. In addition, packed columns should be considered for corrosive services where ceramic or plastic materials are preferred over metals, in services where foaming may be severe if trays are used, and when pressure drop must be low, as in vacuum or near-ambient-pressure operations. Otherwise, trayed towers, which can be designed and scaled up more reliably, are preferred. Although structured packings are quite expensive, they may be the best choice for a new installation when pressure drop must be very low or for replacing existing trays (retrofitting) when a higher capacity or degree of separation is required in an existing column. Trayed towers are preferred when liquid velocities are low, while columns with random packings are best for high liquid velocities. In general, a continuous turbulent liquid flow is desirable if mass transfer is limiting in the liquid phase, while a continuous turbulent gas flow is desirable if mass transfer is limiting in the gas phase.

6.2 GENERAL DESIGN CONSIDERATIONS

Design or analysis of an absorber (or stripper) requires consideration of a number of factors, including:

1. Entering gas (liquid) flow rate, composition, temperature, and pressure
2. Desired degree of recovery of one or more solutes
3. Choice of absorbent (stripping agent)
4. Operating pressure and temperature, and allowable gas pressure drop
5. Minimum absorbent (stripping agent) flow rate and actual absorbent (stripping agent) flow rate as a multiple of the minimum rate needed to make the separation
6. Number of equilibrium stages
7. Heat effects and need for cooling (heating)
8. Type of absorber (stripper) equipment
9. Height of absorber (stripper)
10. Diameter of absorber (stripper)

The ideal absorbent should (a) have a high solubility for the solute(s) to minimize the need for absorbent, (b) have a low volatility to reduce the loss of absorbent and facilitate separation of absorbent from solute(s), (c) be stable to maximize absorbent life and reduce absorbent makeup requirement, (d) be noncorrosive to permit use of common materials of construction, (e) have a low viscosity to provide low pressure drop and high mass and heat transfer rates, (f) be nonfoaming when contacted with the gas so as to make it unnecessary to increase absorber dimensions, (g) be nontoxic and nonflammable to facilitate its safe use, and (h) be available, if possible, within the process, to make it unnecessary to provide an absorbent from external sources, or be inexpensive. As already indicated at the beginning of this chapter, the most widely used absorbents are water, hydrocarbon oils, and aqueous solutions of acids and bases. The most common stripping agents are water vapor, air, inert gases, and hydrocarbon gases.

In general, operating pressure should be high and temperature low for an absorber, to minimize stage requirements and/or absorbent flow rate and to lower the equipment volume required to accommodate the gas flow. Unfortunately, both compression and refrigeration of a gas are expensive. Therefore, most absorbers are operated at feed-gas pressure, which may be greater than ambient pressure, and ambient temperature, which can be achieved by cooling the feed gas and absorbent with cooling water, unless one or both streams already exist at a subambient temperature. Operating pressure should be low and temperature high for a stripper to minimize stage requirements or stripping agent flow rate. However, because maintenance of a vacuum is expensive, strippers are commonly operated at a pressure just above ambient. A high temperature can be used, but it should not be so high as to cause undesirable chemical reactions. Of course, operating temperature and pressure must be compatible with the necessary phase conditions of the streams being contacted. For example, an absorber should not be operated at a pressure and/or temperature that would condense the feed gas, and a stripper should not be operated at a pressure and/or temperature that would vaporize the feed liquid. The possibility of such conditions occurring can be checked by bubble-point and dew-point calculations, discussed in Chapter 4.

For given feed gas (liquid) flow rate, extent of solute absorption (stripping), operating pressure and temperature, and absorbent (stripping agent) composition, a minimum absorbent (stripping agent) flow rate exists that corresponds to an infinite number of countercurrent equilibrium contacts between the gas and liquid phases. In every design problem involving flow rates of the absorbent (stripping agent) and number of stages, a trade-off exists between the number of equilibrium stages and the absorbent (stripping agent) flow rates at rates greater than the minimum value. Graphical and analytical methods for computing the minimum flow rate and this trade-off are developed in the following sections for a mixture that is dilute in the solute(s). For this essentially isothermal case, the energy balance can be ignored. As discussed in Chapters 10 and 11, computer-aided methods are best used for concentrated mixtures, where multicomponent phase equilibrium and mass transfer effects can become complicated and it is necessary to consider the energy balance.

6.3 GRAPHICAL EQUILIBRIUM-STAGE METHOD FOR TRAYED TOWERS

Consider the countercurrent-flow, trayed tower for absorption (or stripping) operating under isobaric, isothermal, continuous, steady-state flow conditions shown in Figure 6.8. Phase equilibrium is assumed to be achieved at each tray between the vapor and liquid streams leaving the tray. That is, each tray is treated as an equilibrium stage. Assume that the only component transferred from one phase to the other is the solute. For application to an absorber, let:

L' = molar flow rate of solute-free absorbent
G' = molar flow rate of solute-free gas (carrier gas)
X = mole ratio of solute to solute-free absorbent in the liquid
Y = mole ratio of solute to solute-free gas in the vapor

Note that with these definitions, values of L' and G' remain constant through the tower, assuming no vaporization of absorbent into carrier gas or absorption of carrier gas by liquid. For the solute at any equilibrium stage, n,

$$K_n = \frac{y_n}{x_n} = \frac{Y_n/(1 + Y_n)}{X_n/(1 + X_n)} \tag{6-1}$$

The fixed temperature and pressure are estimated for a series of values of x and equilibrium values of y in the presence of the solute-free absorbent and solute-free gas by methods discussed in Chapter 2. From these values, an equilibrium curve of Y as a function of X is developed using (6-1), as shown in Figure 6.8. In general, this curve will not be a straight line, but it will pass through the origin. If the solute undergoes, in the liquid phase, a complete irreversible conversion by chemical reaction, to a nonvolatile solute, the equilibrium curve will be a straight line of zero slope passing through the origin.

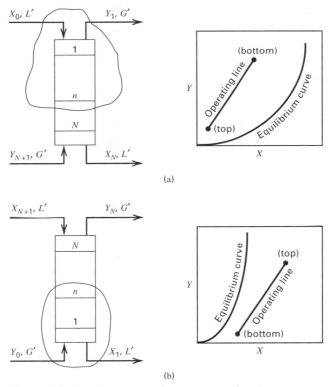

Figure 6.8 Continuous, steady-state operation in a countercurrent cascade with equilibrium stages: (a) absorber; (b) stripper.

At either end of the towers shown in Figure 6.8, entering and leaving streams and solute mole ratios are paired. For the absorber, the pairs are $(X_0, L'$ and $Y_1, G')$ at the top and $(X_N, L'$ and $Y_{N+1}, G')$ at the bottom; for the stripper, $(X_{N+1}, L'$ and $Y_N, G')$ at the top and $(X_1, L'$ and $Y_0, G')$ at the bottom. These terminal pairs can be related to intermediate pairs of passing streams by the following solute material balances for the envelopes shown in Figure 6.8. The balances are written around one end of the tower and an arbitrary intermediate equilibrium stage, n.

For the absorber,

$$X_0 L' + Y_{n+1} G' = X_n L' + Y_1 G' \tag{6-2}$$

or, solving for Y_{n+1},

$$Y_{n+1} = X_n (L'/G') + Y_1 - X_0 (L'/G') \tag{6-3}$$

For the stripper,

$$X_{n+1} L' + Y_0 G' = X_1 L' + Y_n G' \tag{6-4}$$

or, solving for Y_n,

$$Y_n = X_{n+1} (L'/G') + Y_0 - X_1 (L'/G') \tag{6-5}$$

Equations (6-3) and (6-5), which are called *operating-line equations,* are plotted in Figure 6.8. The terminal points of these lines represent the conditions at the top and bottom of the towers. For the absorber, the operating line is above the equilibrium line because, for a given solute concentration in the liquid, the solute concentration in the gas is always greater than the equilibrium value, thus providing the driving force for mass transfer of solute from the gas to the liquid. For the stripper, the operating line lies below the equilibrium line for the opposite reason. For the coordinate systems in Figure 6.8, the operating lines are straight with a slope of L'/G'.

For an absorber, the terminal point of the operating line at the top of the tower is fixed by the amount of solute, if any, in the entering absorbent, and the specified degree of absorption of the solute, which fixes the value of Y_1 in the leaving gas. The terminal point of the operating line at the bottom of the tower depends on the slope of the operating line and, thus, the flow rate, L', of solute-free absorbent.

Minimum Absorbent Flow Rate

Operating lines for four different absorbent flow rates are shown in Figure 6.9, where each operating line passes through the terminal point, (Y_1, X_0), at the top of the column, and corresponds to a different liquid absorbent rate and corresponding slope, L'/G'. To achieve the desired value of Y_1 for given Y_{N+1}, X_0, and G', the solute-free absorbent flow rate L', must lie in the range of ∞ (operating line 1) to $L'_{\min}$ (operating line 4). The value of the

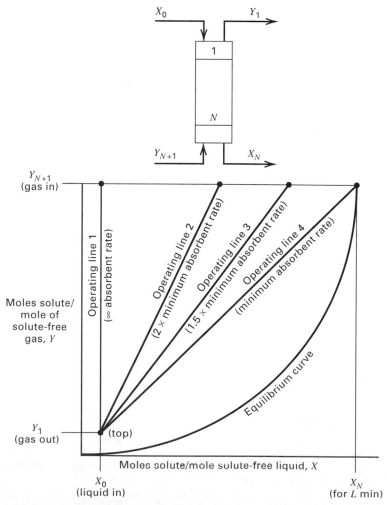

Figure 6.9 Operating lines for an absorber.

solute concentration in the outlet liquid, X_N, depends on L' by a material balance on the solute for the entire absorber. From (6-2), for $n = N$,

$$X_0 L' + Y_{N+1}G' = X_N L' + Y_1 G' \tag{6-6}$$

or

$$L' = \frac{G'(Y_{N+1} - Y_1)}{(X_N - X_0)} \tag{6-7}$$

Note that the operating line can terminate at the equilibrium line, as for operating line 4, but cannot cross it because that would be a violation of the second law of thermodynamics.

The value of L'_{min} corresponds to a value of X_N (leaving the bottom of the tower) in equilibrium with Y_{N+1}, the solute concentration in the feed gas. It takes an infinite number of stages for this equilibrium to be achieved. An expression for L'_{min} of an absorber can be derived from (6-7) as follows.

For stage N, (6-1) becomes, for the minimum absorbent rate,

$$K_N = \frac{Y_{N+1}/(1 + Y_{N+1})}{X_N/(1 + X_N)} \tag{6-8}$$

Solving (6-8) for X_N and substituting the result into (6-7) gives

$$L'_{min} = \frac{G'(Y_{N+1} - Y_1)}{\{Y_{N+1}/[Y_{N+1}(K_N - 1) + K_N]\} - X_0} \tag{6-9}$$

For dilute-solute conditions, where $Y \approx y$ and $X \approx x$, (6-9) becomes

$$L'_{min} = G'\left(\frac{y_{N+1} - y_1}{\frac{y_{N+1}}{K_N} - x_0}\right) \tag{6-10}$$

Furthermore, if the entering liquid contains no solute, that is, $X_0 \approx 0$, Eq. (6-10) becomes

$$L'_{min} = G'K_N \text{ (fraction of solute absorbed)} \tag{6-11}$$

This equation is reasonable because it would be expected that L'_{min} would increase with increasing G', K-value, and fraction of solute absorbed.

The selection of the actual operating absorbent flow rate is based on some multiple of L'_{min}, typically from 1.1 to 2. A value of 1.5 corresponds closely to the value of 1.4 for the optimal absorption factor mentioned earlier. In Figure 6-9, operating lines 2 and 3 correspond to 2.0 and 1.5 times L'_{min}, respectively. As the operating line moves from 1 to 4, the number of required equilibrium stages, N, increases from zero to infinity. Thus, a trade-off exists between L' and N, and an optimal value of L' exists.

A similar derivation of G'_{min}, for the stripper of Figure 6.8, results in an expression analogous to (6-11):

$$G'_{min} = \frac{L'}{K_N} \text{ (fraction of solute stripped)} \tag{6-12}$$

Number of Equilibrium Stages

As shown in Figure 6.10a, the operating line relates the solute concentration in the vapor passing upward between two stages to the solute concentration in the liquid passing downward between the same two stages. Figure 6.10b illustrates that the equilibrium curve relates the solute concentration in the vapor leaving an equilibrium stage to the solute concentration in the liquid leaving the same stage. This makes it possible, in the case of an absorber, to start from the top of the tower (at the bottom of the $Y-X$ diagram) and

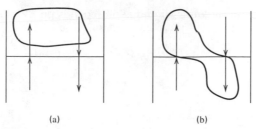

Figure 6.10 Vapor–liquid stream relationships: (a) operating line (passing streams); (b) equilibrium curve (leaving streams).

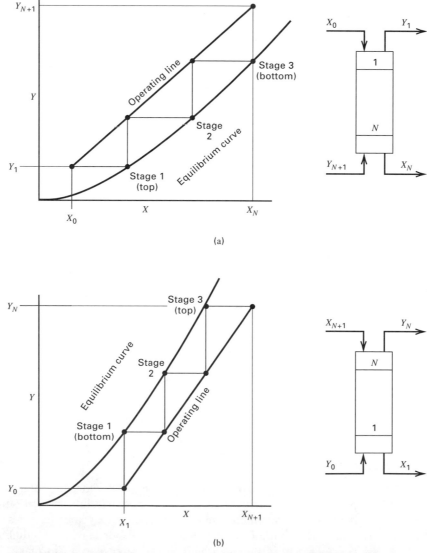

Figure 6.11 Graphical determination of the number of equilibrium stages for (a) absorber and (b) stripper.

move to the bottom of the tower (at the top of the Y–X diagram) by constructing a staircase confined between the operating line and the equilibrium curve, as shown in Figure 6.11a. The number of equilibrium stages required for a particular absorbent flow rate corresponding to the slope of the operating line, which in Figure 6.11a is for $(L'/G') = 1.5(L'_{min}/G')$, is stepped off by moving up the staircase, starting from the point (Y_1, X_0), on the operating line and moving horizontally to the right to the point (Y_1, X_1) on the equilibrium curve. From there, a vertical move is made to the point (Y_2, X_1) on the operating line. Proceeding in this manner, the staircase is climbed until the terminal point (Y_{N+1}, X_N) on the operating line is reached. As the slope (L'/G') is increased, fewer equilibrium stages are required. As (L'/G') is decreased, more stages are required until (L'_{min}/G') is reached, at which the operating line and equilibrium curve intersect at a so-called *pinch point*, for which an infinite number of stages is required. Operating line 4 in Figure 6.9 has a pinch point at Y_{N+1}, X_N. If (L'/G') is reduced below (L'_{min}/G'), the specified extent of absorption of the solute cannot be achieved.

The number of equilibrium stages required for stripping a solute is determined in a manner similar to that for absorption. An illustration is shown in Figure 6.11b, which refers to Figure 6.8b. For given specifications of Y_0, X_{N+1}, and the extent of stripping of the solute, which corresponds to a value of X_1, G'_{min} is determined from the slope of the operating line that passes through the points (Y_0, X_1) and (Y_N, X_{N+1}) on the equilibrium curve. The operating line in Figure 6.11b is for $G' = 1.5G'_{min}$ or a slope of $(L'/G') = (L'/G'_{min})/1.5$.

In Figure 6.11, the number of equilibrium stages for the absorber and stripper is exactly three each. These integer results are coincidental. Ordinarily, the result is some fraction above an integer number of stages, as is the case in the following example. In practice, the result is usually rounded to the next highest integer.

EXAMPLE 6.1

When molasses is fermented to produce a liquor containing ethyl alcohol, a CO_2-rich vapor containing a small amount of ethyl alcohol is evolved. The alcohol can be recovered by absorption with water in a sieve-tray tower. For the following conditions, determine the number of equilibrium stages required for countercurrent flow of liquid and gas, assuming isothermal, isobaric conditions in the tower and neglecting mass transfer of all components except ethyl alcohol.

Entering gas:
  180 kmol/h; 98% CO_2, 2% ethyl alcohol; 30°C, 110 kPa
Entering liquid absorbent:
  100% water; 30°C, 110 kPa
Required recovery (absorption) of ethyl alcohol: 97%

SOLUTION

From Section 5.5 for a single-section countercurrent cascade, the number of degrees of freedom is $2N + 2C + 5$. All stages operate adiabatically at a pressure of approximately 1 atm, taking $2N$ degrees of freedom. The entering gas is completely specified, taking $C + 2$ degrees of freedom. The entering liquid flow rate is not specified; thus, only $C + 1$ degrees of freedom are taken by the entering liquid. The recovery of ethyl alcohol takes one additional degree of freedom. Thus, the total number of degrees of freedom taken by the problem specification is $2N + 2C + 4$. This leaves one additional specification to be made, which in this example can be the entering liquid flow rate at, say, 1.5 times the minimum value.

The above application of the degrees of freedom analysis from Chapter 5 has assumed the use of an energy balance for each stage. The energy balances are assumed to result in the assumed isothermal operation at 30°C.

Assume that the exiting absorbent will be dilute in ethyl alcohol, whose K-value is determined from a modified Raoult's law, $K = \gamma P^s/P$. The vapor pressure of ethyl alcohol at 30°C is 10.5 kPa. At infinite dilution in water at 30°C, the liquid-phase activity coefficient of ethyl alcohol is taken as 6. Therefore, $K = (6)(10.5)/110 = 0.57$. The minimum solute-free absorbent rate is given by (6-11), where the solute-free gas rate, G', is $(0.98)(180) = 176.4$ kmol/h. Thus,

$$L'_{min} = (176.4)(0.57)(0.97) = 97.5 \text{ kmol/h}$$

The actual solute-free absorbent rate, at 50% above the minimum rate, is

$$L' = 1.5(97.5) = 146.2 \text{ kmol/h}$$

The amount of ethyl alcohol transferred from the gas to the liquid is 97% of the amount of alcohol in the entering gas or

$$(0.97)(0.02)(180) = 3.49 \text{ kmol/h}$$

The amount of ethyl alcohol remaining in the exiting gas is

$$(1.00 - 0.97)(0.02)(180) = 0.11 \text{ kmol/h}$$

We now compute the alcohol mole ratios at both ends of the operating line as follows, referring to Figure 6.8a:

$$\text{top} \left\{ X_0 = 0, \qquad\qquad\qquad Y_1 = \frac{0.11}{176.4} = 0.0006 \right.$$

$$\text{bottom} \left\{ Y_{N+1} = \frac{0.11 + 3.49}{176.4} = 0.0204, \qquad X_N = \frac{3.49}{146.2} = 0.0239 \right.$$

The equation for the operating line from (6-3) with $X_0 = 0$ is

$$Y_{N+1} = \left(\frac{146.2}{176.4}\right) X_N + 0.0006 = 0.829 X_N + 0.0006 \qquad (1)$$

It is clear that we are dealing with a dilute system. The equilibrium curve for ethyl alcohol can be determined from (6-1) using the value of $K = 0.57$ computed above. From (6-1),

$$0.57 = \frac{Y/(1 + Y)}{X/(1 + X)}$$

Solving for Y, we obtain $\qquad\qquad \dfrac{Y = 0.57X}{1 + 0.43\,X}.$ $\qquad\qquad$ (2)

To cover the entire column, the necessary range of X for a plot of Y vs X is 0 to almost 0.025. From the Y–X equation, (2),

Y	X
0.00000	0.000
0.00284	0.005
0.00569	0.010
0.00850	0.015
0.01130	0.020
0.01410	0.025

For this dilute system in ethyl alcohol, the maximum error in Y is 1.0% if Y is taken simply as $Y = KX = 0.57X$.

The equilibrium curve, which is almost straight in this example, and a straight operating line drawn through the terminal points (Y_1, X_0) and (Y_{N+1}, X_N) is given in Figure 6.12. The determination of points for the operating line and the equilibrium curve, as well as the plot of the points, is conveniently done with a spreadsheet program on a computer using Eqs. (1) and (2). The theoretical stages are stepped off as shown starting from the top stage (Y_1, X_0) located near the lower left corner of Figure 6.12. The required number of theoretical stages for 97% absorption of ethyl alcohol is just slightly more than six. Accordingly, it is best to provide seven theoretical stages. ∎

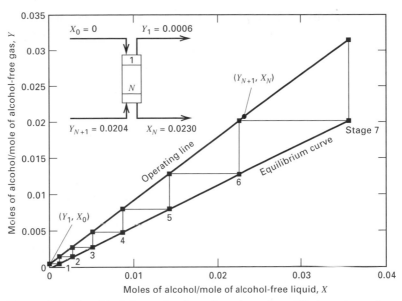

Figure 6.12 Graphical determination of number of equilibrium stages for an absorber.

6.4 ALGEBRAIC METHOD FOR DETERMINING THE NUMBER OF EQUILIBRIUM STAGES

Graphical methods for determining equilibrium stages have great educational value because a fairly complex multistage problem can be readily followed and understood. Furthermore, one can quickly gain visual insight into the phenomena involved. However, the application of a graphical method can become very tedious when (1) the problem specification fixes the number of stages rather than the percent recovery of solute, (2) when more than one solute is being absorbed or stripped, (3) when the best operating conditions of temperature and pressure are to be determined so that the location of the equilibrium curve is unknown, and/or (4) if very low or very high concentrations force the graphical construction to the corners of the diagram so that multiple y–x diagrams of varying sizes and dimensions are needed. Then, the application of an algebraic method may be preferred.

The Kremser method for single-section cascades, as developed in Section 5.4, is ideal for absorption and stripping of dilute mixtures. For example, (5-48) and (5-50) can be written in terms of the fraction of solute absorbed or stripped as

$$\text{Fraction of a solute, } i, \text{ absorbed} = \frac{A_i^{N+1} - A_i}{A_i^{N+1} - 1} \tag{6-13}$$

and

$$\text{Fraction of a solute, } i, \text{ stripped} = \frac{S_i^{N+1} - S_i}{S_i^{N+1} - 1} \tag{6-14}$$

where the solute absorption and stripping factors are, respectively,

$$A_i = L/(K_i V) \tag{6-15}$$

$$S_i = K_i V/L \tag{6-16}$$

Values of L and V in moles per unit time may be taken as entering values. Values of K_i depend mainly on temperature, pressure, and liquid-phase composition. Methods for estimating K-values are discussed in detail in Chapter 2. At near-ambient pressure, for dilute mixtures, some common expressions are

$$K_i = P_i^s/P \qquad \text{(Raoult's law)} \qquad \qquad \text{(6-17)}$$

$$K_i = \gamma_{iL}^\infty P_i^s/P \qquad \text{(modified Raoult's law)} \qquad \text{(6-18)}$$

$$K_i = H_i/P \qquad \text{(Henry's law)} \qquad \qquad \text{(6-19)}$$

$$K_i = P_i^s/x_i^* P \qquad \text{(solubility)} \qquad \qquad \text{(6-20)}$$

The first expression applies for ideal solutions involving solutes at subcritical temperatures. The second expression is useful for moderately nonideal solutions when activity coefficients are known at infinite dilution. For solutes at supercritical temperatures, the use of Henry's law may be preferable. For sparingly soluble solutes at subcritical temperatures, the fourth expression is preferred when solubility data in mole fractions, x_i^*, are available. This expression is derived by considering a three-phase system consisting of an ideal vapor containing solute, carrier vapor, and solvent; a pure or near-pure solute as liquid (1); and the solvent liquid (2) with dissolved solute. In that case, for solute, i, at equilibrium between the two liquid phases,

$$x_i^{(1)}\gamma_{iL}^{(1)} = x_i^{(2)}\gamma_{iL}^{(2)}$$

But,
$$x_i^{(1)} \approx 1, \qquad \gamma_{iL}^{(1)} \approx 1, \qquad x_i^{(2)} = x_i^*$$

Therefore,
$$\gamma_{iL}^{(2)} \approx 1/x_i^*$$

and from (6-18),
$$K_i^{(2)} = \gamma_{iL}^{(2)} P_i^s/P = P_i^s/(x_i^* P)$$

The advantage of (6-13) and (6-14) is that they can be solved directly for the percent absorption or stripping of a solute when the number of theoretical stages, N, and the absorption or stripping factor are known.

Example 6.2

As discussed by Okoniewski [3], volatile organic compounds (VOCs) can be stripped from wastewater by air. Such compounds are to be stripped at 70°F and 15 psia from 500 gpm of wastewater with 3,400 scfm of air (standard conditions of 60°F and 1 atm) in an existing tower containing 20 plates. A chemical analysis of the wastewater shows three organic chemicals in the amounts shown in the following table. Included are necessary thermodynamic properties from the 1966 *Technical Data Book–Petroleum Refining* of the American Petroleum Institute.

Organic Compound	Concentration in the Wastewater, mg/L	Solubility in Water at 70°F, mole fraction	Vapor Pressure at 70°F, psia
Benzene	150	0.00040	1.53
Toluene	50	0.00012	0.449
Ethylbenzene	20	0.000035	0.149

It is desirable that 99.9% of the total VOCs be stripped, but the plate efficiency of the tower is uncertain, with an estimated range of 5% to 20%, corresponding to one to four theoretical stages for the 20-plate tower. Calculate and plot the percent stripping of each of the three organic compounds for one, two, three, and four theoretical stages. Under what conditions can we expect to achieve the desired degree of stripping? What should be done with the exiting air?

SOLUTION   Because the wastewater is dilute in the VOCs, the Kremser equation may be applied independently to each of the three organic chemicals. We will ignore the absorption of air by the water and the stripping of water by the air. The stripping factor for each compound is given by $S_i = K_i V/L$, where V and L will be taken at entering conditions. The K-value may be computed from a modified Raoult's law, $K_i = \gamma_{iL} P_i^s/P$, where for a compound that is only slightly soluble, $\gamma_{iL} = 1/x_i^*$, where x_i^* is the solubility in mole fraction. Thus, from (6-20), $K_i = P_i^s/x_i^* P$

$$V = 3{,}400\,(60)/(379\,\text{scfm/lbmol}) \quad \text{or} \quad 538\,\text{lbmol/h}$$

$$L = 500(60)(8.33\,\text{lb/gal})/(18.02\,\text{lb/lbmol}) \quad \text{or} \quad 13{,}870\,\text{lbmol/h}$$

The corresponding K-values and stripping factors are

Component	K at 70°F, 15 psia	S
Benzene	255	9.89
Toluene	249	9.66
Ethylbenzene	284	11.02

From (6-14), $$\text{Fraction stripped} = \frac{S^{N+1} - S}{S^{N+1} - 1}$$

The calculations when carried out with a spreadsheet computer program give the following results:

Component	Percent Stripped			
	1 Stage	2 Stages	3 Stages	4 Stages
Benzene	90.82	99.08	99.91	99.99
Toluene	90.62	99.04	99.90	99.99
Ethylbenzene	91.68	99.25	99.93	99.99

The results are quite sensitive to the number of theoretical stages as shown in Figure 6.13. To achieve 99.9% removal of the total VOCs, three theoretical stages are needed, corresponding to the necessity for a 15% stage efficiency in the existing 20-tray tower.

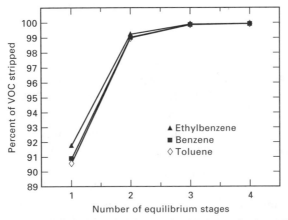

Figure 6.13 Results of Example 6.2 for stripping of VOCs from water with air.

It is best to process the exiting air to remove or destroy the VOCs, particularly the benzene, which is a carcinogen [4]. The amount of benzene stripped is

$$(500 \text{ gpm})(60 \text{ min/h})(3.785 \text{ liters/gal})(150 \text{ mg/liters}) = 17{,}030{,}000 \text{ mg/h} \quad \text{or} \quad 37.5 \text{ lb/h}$$

If benzene is valued at \$0.30/lb, the annual value is approximately \$100,000. It is doubtful that this would justify a recovery technique, such as carbon adsorption. It is perhaps preferable to destroy the VOCs by incineration. For example, the air can be sent to a utility boiler, a waste-heat boiler, or a catalytic incinerator. It is also to be noted that the amount of air was arbitrarily given as 3,400 scfm. To complete the design procedure, various air rates should be investigated. It will also be necessary to verify that, at the chosen air flow rates, no flooding or weeping will occur in the column.   ∎

6.5   STAGE EFFICIENCY

Graphical and algebraic methods for determining stage requirements for absorption and stripping assume equilibrium with respect to both heat and mass transfer at each stage. Thus, the number of *equilibrium stages* (*theoretical stages, ideal stages,* or *ideal plates*) is determined or specified when using those methods. Except when temperature changes significantly from stage to stage, the assumption that vapor and liquid phases leaving a stage are at the same temperature is often reasonable. The assumption of equilibrium with respect to mass transfer, however, is not often reasonable and, for streams leaving a stage, vapor-phase mole fractions are not related to liquid-phase mole fractions simply by thermodynamic K-values. To determine the actual number of plates, the number of equilibrium stages must be adjusted with a *stage efficiency* (*plate efficiency* or *tray efficiency*).

Stage efficiency concepts are applicable to devices in which the phases are contacted and then separated, that is, when discrete stages can be identified. This is not the case for packed columns or continuous-contact devices. For these, the efficiency is imbedded into an equipment- and system-dependent parameter, an example of which is the HETP (height of packing equivalent to a theoretical plate).

The simplest approach for staged columns, in preliminary design studies and in the evaluation of the performance of an existing column, is to apply an overall stage (or column) efficiency, defined by Lewis [5] as

$$E_o = N_t/N_a \qquad \qquad \textbf{(6-21)}$$

where E_o is the fractional overall stage efficiency, usually less than 1.0; N_t is the calculated number of equilibrium (theoretical) stages; and N_a is the actual number of contacting trays or plates (usually greater than N_t) required. Based on the results of extensive research conducted over a period of more than 60 years, the overall stage efficiency has been found to be a complex function of the

1. Geometry and design of the contacting trays
2. Flow rates and flow paths of vapor and liquid streams
3. Compositions and properties of vapor and liquid streams

For well-designed trays and for flow rates near the capacity limit, E_o depends mainly on the physical properties of the vapor and liquid streams.

Values of E_o can be predicted by any of the following four methods:

1. Comparison with performance data from industrial columns for the same or similar systems
2. Use of empirical efficiency models derived from data on industrial columns
3. Use of semitheoretical models based on mass and heat transfer rates
4. Scale-up from data obtained with laboratory or pilot-plant columns

These methods, which are discussed in some detail in the following four subsections, are applied to other vapor–liquid separation operations, such as distillation, as well as to absorption and stripping. Suggested correlations of mass transfer coefficients for trayed towers are deferred to Section 6.6, following the discussion of tray capacity.

Performance Data

Performance data obtained from industrial absorption and stripping columns equipped with trays generally include gas and liquid feed and product flow rates and compositions, average column pressure and temperature or pressures and temperatures at the bottom and top of the column, number of actual trays, N_a, column diameter, and type of tray with, perhaps, some details of the tray design. From these data, particularly if the system is dilute with respect to the solute(s), the graphical or algebraic methods, described in Sections 6.3 and 6.4, respectively, can be used to estimate the number of equilibrium stages, N_t, required. Then (6-21) can be applied to determine the overall stage efficiency, E_o. Values of E_o for absorbers and strippers are typically low, often less than 50%.

Table 6.4 presents performance data, from a study by Drickamer and Bradford [6], for five industrial hydrocarbon absorption and stripping operations using columns with bubble-cap trays. For the three absorbers, the stage efficiencies are based on the absorption of n-butane as the key component. For the two strippers, both of which use steam as the stripping agent, the key component is not given, but is probably n-heptane. Although the data cover a wide range of average pressure and temperature, the overall stage efficiencies, which cover a wide range of 10.4% to 57%, appear to depend primarily on the molar average liquid viscosity, a key factor for the rate of mass transfer in the liquid phase.

The gas feed to a hydrocarbon absorber contains a range of light hydrocarbons, each of which is absorbed to a different extent based on its K-value, as illustrated in Example 5.3. The data of Jackson and Sherwood [7] for a 9-ft-diameter hydrocarbon absorber equipped with 19 bubble-cap trays on 30-in. tray spacing and operating at 92 psia and 60°F, as analyzed by O'Connell [8] and summarized in Table 6.5, show that each component being absorbed has a different overall stage efficiency, which appears to increase with decreasing K-value (increasing solubility in the liquid absorbent). For the same molar average liquid viscosity (1.90 cps), the overall stage efficiency is seen to vary from as low as 10.3% for ethylene, the most volatile species considered, to 33.8% for butylene (presumably n-butene), the least volatile species considered.

Table 6.4 Performance Data for Absorbers and Strippers in Hydrocarbon Service

Service	Type Tray	Column Diameter, ft	No. of Trays	Tray Spacing, in.	Average Pressure, psia	Average Temp., °F	Molar Average Liquid Viscosity, CP	Overall Stage Efficiency, %
Absorption of butane	Bubble cap	4	24	18	260	120	0.48	36
Absorption of butane	Bubble cap	5	16	30	254	132	0.31	50
Absorption of butane	Bubble cap	4	16	24	94	117	1.41	10.4
Steam stripping of kerosene	Bubble cap	5	4	30	68	448	0.205	57
Steam stripping of gas oil	Bubble cap	5	6	30	60	507	0.250	49

Source: H.G. Drickamer and J.R. Bradford [6].

Table 6.5 Effect of Species on Overall
Stage Efficiency in a 9-ft-Diameter
Industrial Absorber Using
Bubble-Cap Trays

Component	Overall Stage Efficiency, %
Ethylene	10.3
Ethane	14.9
Propylene	25.5
Propane	26.8
Butylene	33.8

Source: H.E. O'Connell [8].

An even more dramatic effect of the species solubility in the absorbent on the overall stage efficiency is seen in Table 6.6, from a study by Walter and Sherwood [9] using small laboratory bubble-cap tray columns ranging in size from 2 to 18 in. in diameter. Stage efficiencies vary over a very wide range from 0.65% to 69%. Comparing the data for the water absorption of ammonia (a very soluble gas) and carbon dioxide (a slightly soluble gas), it is clear that the solubility of the gas (i.e., the K-value) has a large effect on stage efficiency. Thus, low stage efficiency can occur when the liquid viscosity is high and/or the gas solubility is low (high K-value); high stage efficiency can occur when the liquid viscosity is low and the gas solubility is high (low K-value).

Empirical Correlations

Using 20 sets of performance data from industrial hydrocarbon absorbers and strippers, including the data in Table 6.4, Drickamer and Bradford [6] correlated the overall stage

Table 6.6 Performance Data for Absorption in Laboratory Bubble-Cap Tray Columns

Service	Column Diameter, in.	No. of Trays	Tray Spacing, in.	Average Pressure, psia	Average Temp., °F	Overall Stage Efficiency, %
Absorption of ammonia in water	18	1	—	14.7	57	69
Absorption of isobutylene in heavy naphtha	2	1	—	66	78.8	36.4
Absorption of propylene in gas oil	2	1	—	66	118.4	13.1
Absorption of propylene in gas lube oil	2	1	—	66	105.8	4.7
Absorption of carbon dioxide in water	18	1	—	14.7	50.4	2.0
Desorption of carbon dioxide from 43.7 wt% aqueous glycerol	5	4	11	14.7	77	0.65

Source: J.F. Walter and T.K. Sherwood [9].

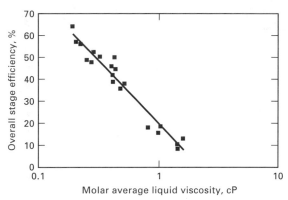

Figure 6.14 Drickamer and Bradford correlation for plate efficiency of hydrocarbon absorbers and strippers.

efficiency of the key component absorbed or stripped with just the molar average viscosity of the rich oil (liquid leaving an absorber or liquid entering a stripper) at the average tower temperature over a viscosity range of 0.19 to 1.58 cP. The empirical equation,

$$E_o = 19.2 - 57.8 \log \mu_L \tag{6-22}$$

where E_o is in percent and μ is in centipoise, fits the data with average and maximum percent deviations of 10.3% and 41%, respectively. A plot of the Drickamer and Bradford correlation, compared to performance data, is given in Figure 6.14. Equation (6-22) should not be used for absorption into nonhydrocarbon liquids and is restricted to the range of the data used to develop the correlation.

Mass transfer theory indicates that when the volatility of species being absorbed or stripped covers a wide range, the relative importance of liquid-phase and gas-phase mass transfer resistances can shift. Thus, O'Connell [8] found that the Drickamer-Bradford correlation, (6-22), was inadequate for absorbers and strippers when applied to species covering a wide range of volatility or K-values. This additional effect is indicated clearly in the performance data of Tables 6.5 and 6.6, where liquid viscosity alone cannot correlate the data. O'Connell obtained a more general correlation by using a parameter that included not only the liquid viscosity but also the liquid density and the Henry's law constant of the species being absorbed or stripped. Edmister [10] and Lockhart and Leggett [11] suggested slight modifications to the O'Connell correlation to permit its use with K-values. An O'Connell-type plot of overall stage efficiency for absorption or stripping in bubble-cap tray columns is given in Figure 6.15. The correlating parameter, suggested by Edmister, is $K_i M_L \mu_L / \rho_L$, where:

K_i = K-value of species being absorbed or stripped
M_L = molecular weight of the liquid
μ_L = viscosity of the liquid, cP
ρ_L = density of the liquid, lb/ft^3

Thus, the correlating parameter has the units of cP-ft^3/lb. A reasonable fit to the 33 data points is given by the empirical equation

$$\log E_o = 1.597 - 0.199 \log \left(\frac{K M_L \mu_L}{\rho_L} \right) - 0.0896 \left[\log \left(\frac{K M_L \mu_L}{\rho_L} \right) \right]^2 \tag{6-23}$$

The average and maximum deviations of (6-23) for the 33 data points of Figure 6.15 are 16.3% and 157%, respectively. More than 50% of the data points, including points for the highest and lowest observed efficiencies, are predicted to within 10%.

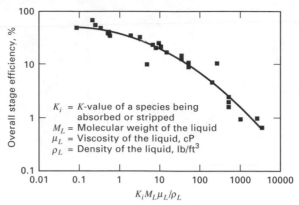

Figure 6.15 O'Connell correlation for plate efficiency of absorbers and strippers.

The 33 data points in Figure 6.15 cover a very wide range of conditions:

Column diameter:	2 in. to 9 ft
Average pressure:	14.7 to 485 psia
Average temperature:	60 to 138°F
Liquid viscosity:	0.22 to 21.5 cP
Overall stage efficiency:	0.65 to 69%

Absorbents include both hydrocarbons and water. For the absorption or stripping of more than one species, because of the effect of species K-value, different stage efficiencies are predicted, as observed from performance data of the type shown in Table 6.5. The inclusion of the K-value also permits the correlation to be used for aqueous systems where the solute may exhibit a very wide range of solubility (e.g., ammonia versus carbon dioxide) as included in Table 6.6. In using Figure 6.15 or Eq. (6-23), the K-value and absorbent properties are best evaluated at the end of the tower where the liquid phase is richest in solute(s). Prudent designs use the lowest predicted efficiency.

Most of the data used to develop the correlation of Figure 6.15 are for columns having a liquid flow path across the active tray area of from 2 to 3 ft. Theory and experimental data show that higher efficiencies are achieved for longer flow paths. For short liquid flow paths, the liquid flowing across the tray is usually completely mixed. For longer flow paths, the equivalent of two or more completely mixed, successive liquid zones may be present. The result is a greater average driving force for mass transfer and, thus, a higher efficiency— perhaps greater than 100%. For example, a column with a 10-ft liquid flow path may have an efficiency as much as 25% greater than that predicted by (6-23). However, at high liquid rates, long liquid path lengths are undesirable because they lead to excessive hydraulic gradients. When the effective height of a liquid on a tray is appreciably higher on the inflow side than at the overflow weir, vapor may prefer to enter the tray in the latter region, leading to nonuniform bubbling action. Multipass trays, as shown in Figure 6.16a, are used to prevent excessive liquid gradients. Estimation of the desired number of flow paths can be made with Figure 6.16b.

Based on estimates of the number of actual trays and tray spacing, the height of a column between the top tray and the bottom tray is computed. By adding an additional 4 ft above the top tray for removal of entrained liquid and 10 ft below the bottom tray for bottoms surge capacity, the total column height is estimated. If the height is greater than 212 ft (equivalent to 100 trays on 24-in. spacing, two or more columns arranged in

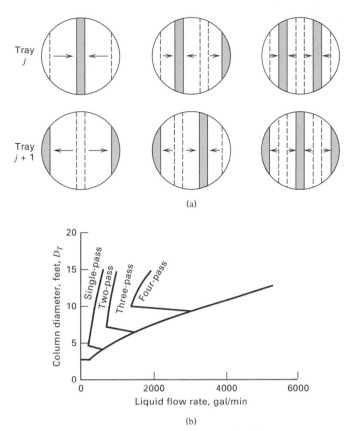

Figure 6.16 Estimation of number of required liquid flow passes. (a) Multipass trays: (1) two-pass; (2) three-pass; (3) four-pass. (b) Flow pass correlation. (Derived from *Koch Flexitray Design Manual, Bulletin 960,* Koch Engineering Co., Inc., Wichita, KA, 1960.)

series may be preferable to a single column. Perhaps the tallest column in the world, located at the Shell Chemical Company complex in Deer Park, Texas, stands 338 ft tall [*Chem. Eng.,* **84** (26), 84 (1977)].

EXAMPLE 6.3

Performance data, given below, for a bubble-cap tray absorber located in a Texas petroleum refinery, were reported by Drickamer and Bradford [6]. Based on these data, back-calculate the overall stage efficiency for *n*-butane and compare the result with both the Drickamer–Bradford and O'Connell correlations. Lean oil and rich gas enter the tower; rich oil and lean gas leave the tower.

PERFORMANCE DATA

Number of plates	16	Lean oil rate, lbmol/h	368
Plate spacing, in.	24	Rich oil rate, lbmol/h	525.4
Tower diameter, ft	4	Rich gas rate, lbmol/h	946
Tower pressure, psig	79	Lean gas rate, lbmol/h	786.9
Lean oil temperature, °F	102	Lean oil molecular weight	250
Rich oil temperature, °F	126	Lean oil viscosity at 116°F, cP	1.4
Rich gas temperature, °F	108	Lean oil gravity, °API	21
Lean gas temperature, °F	108		

STREAM COMPOSITIONS, MOL%

Component	Rich Gas	Lean Gas	Rich Oil	Lean Oil
C_1	47.30	55.90	1.33	
C_2	8.80	9.80	1.16	
$C_3^=$	5.20	5.14	1.66	
C_3	22.60	21.65	8.19	
$C_4^=$	3.80	2.34	3.33	
nC_4	7.40	4.45	6.66	
nC_5	3.00	0.72	4.01	
nC_6	1.90		3.42	
Oil absorbent			70.24	100
Totals	100.0	100.0	100.0	100

SOLUTION

Before computing the overall stage efficiency for n-butane, it is worthwhile to check the consistency of the plant data by examining the overall material balance and the material balance for each component. From the stream compositions, it is apparent that the compositions have been normalized to total 100%.

The overall material balance is

$$\text{Total flow into tower} = 368 + 946 = 1{,}314 \text{ lbmol/h}$$

$$\text{Total flow from tower} = 525.4 + 786.9 = 1{,}312.3 \text{ lbmol/h}$$

These two totals agree to within 0.13%. This is excellent agreement.

The component material balance for the oil absorbent is

$$\text{Total oil in} = 368 \text{ lbmol/h}$$

$$\text{Total oil out} = (0.7024)(525.4) = 369 \text{ lbmol/h}$$

These two totals agree to within 0.3%. Again, this is excellent agreement.

Component material balances for other hydrocarbons from spreadsheet calculations are as follows.

	lbmol/h			
Component	Lean Gas	Rich Oil	Total Out	Total In
C_1	439.9	7.0	446.9	447.5
C_2	77.1	6.1	83.2	83.2
$C_3^=$	40.4	8.7	49.1	49.2
C_3	170.4	43.0	213.4	213.8
$C_4^=$	18.4	17.5	35.9	35.9
nC_4	35.0	35.0	70.0	70.0
nC_5	5.7	21.1	26.8	28.4
nC_6	0.0	18.0	18.0	18.0
	786.9	156.4	943.3	946.0

Again, we see excellent agreement. The largest difference is 6% for pentanes. Plant data are not always so consistent.

For the back-calculation of stage efficiency from the performance data, The Kremser equation is applied to compute the number of equilibrium stages required for the measured absorption of n-butane.

$$\text{Fraction of } nC_4 \text{ absorbed} = \frac{35}{70} = 0.50$$

From (6-13), $0.50 = \dfrac{A^{N+1} - A}{A^{N+1} - 1}$ \qquad where A = absorption factor $= \dfrac{L}{KV}$

Let L = average liquid rate $= \dfrac{368 + 525.4}{2} = 446.7$ lbmol/h

and let V = average vapor rate $= \dfrac{946 + 786.9}{2} = 866.5$ lbmol/h

$$A = \frac{446.7}{(0.7)(866.5)} = 0.736$$

Therefore, $0.50 = \dfrac{0.736^{N+1} - 0.736}{0.736^{N+1} - 1}$

Solving, $N = N_t = 1.45$

From the performance data, $N_a - 16$

From (6-21), $E_o = \dfrac{1.45}{16} - 0.091$ \quad or \quad 9.1%

For the stage efficiency from the Drickamer-Bradford correlation, assume the ambient pressure is 14.7 psia. Then

Tower pressure $= 79 + 14.7 = 93.7$ psia

Assume average tower temperature = the average of inlet and outlet temperatures = $(102 + 126 + 108 + 108)/4 = 111°F$. Also assume that the viscosity of the lean oil at 116°F equals the viscosity of the rich oil at 111°F. Therefore, $\mu = 1.4$ cP.

Equation (6-22) is applicable to n-butane, because that component is absorbed to the extent of about 50% and thus can be considered one of the key components. Other possible key components are butenes and n-pentane.

From (6-22), $E_o = 19.2 - 57.8 \log(1.4) = 10.8%$

To find the stage efficiency from the O'Connell correlation assume that the lean oil properties can be used for the rich oil. Then,

$\mu_L = 1.4$ cP, \qquad $M_L = 250$

S.G. $= 141.5/(°API + 131.5) = 141.5/(21 + 131.5) = 0.928$

$\rho_L = 0.928\,(62.4) = 57.9$ lb/ft^3

From Figure 2.8, at 93.7 psia and 111°F, $K_{nC_4} = 0.7$. Thus,

$$\frac{KM_L\mu_L}{\rho_L} = \frac{0.7(250)(1.4)}{57.9} = 4.23$$

From (6-23), $\log E_o = 1.597 - 0.199 \log(4.23) - 0.0896[\log(4.23)]^2 = 1.437$

$E_o = 10^{1.437} = 27.4%$

For this hydrocarbon absorber, the Drickamer and Bradford correlation (10.8%) gives better agreement than the O'Connell correlation (27.4%) with the plant performance data (9.1%).

Semitheoretical Models

A third method for predicting the overall stage efficiency involves the application of a semitheoretical tray model based on mass and heat transfer rates. With this model, the fractional approach to equilibrium, called the *plate* or *tray efficiency,* is estimated for each component in the mixture for each tray in the column. These efficiency values are then utilized to determine conditions for each tray, or averaged for the column to obtain the overall plate efficiency.

Tray efficiency models, in order of increasing complexity, have been proposed by Holland [12], Murphree [13], Hausen [14], and Standart [15]. All four models are based on the assumption that vapor and liquid streams entering each tray are of uniform compositions. The *Murphree vapor efficiency,* which is the oldest and most widely used, is derived with the additional assumptions of (1) complete mixing of the liquid flowing across the tray such that the liquid is of a uniform concentration, equal to the composition of the liquid leaving the tray and entering the next tray below, and (2) plug flow of the vapor passing up through the liquid, as indicated in Figure 6.17. Considering species i, let:

n = rate of mass transfer for absorption from the gas to the liquid

K_G = overall gas mass transfer coefficient based on a partial pressure driving force

a = vapor–liquid interfacial area per volume of combined gas and liquid holdup (froth or dispersion) on the tray,

A_b = active bubbling area of the tray (total cross-sectional area minus liquid downcomer areas)

Z_f = height of combined gas and liquid holdup on the tray

P = total absolute pressure

y_i = mole fraction of i in the vapor rising up through the liquid

y_i^* = vapor mole fraction of i in equilibrium with the completely mixed liquid on the tray

Then the differential rate of mass transfer for a differential height of holdup on tray n, numbered down from the top, is

$$dn_i = K_G a(y_i - y_i^*) P A_b \, dZ \tag{6-24}$$

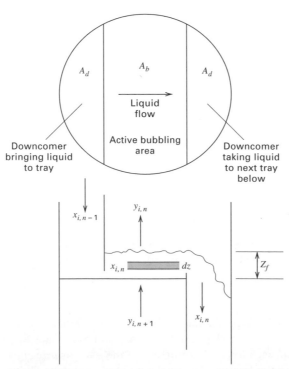

Figure 6.17 Schematic top and side views of tray for Murphree vapor-tray efficiency derivation.

where K_G takes into account both gas- and liquid-phase resistances to mass transfer. By material balance,

$$dn_i = -G \, dy_i \tag{6-25}$$

where G = molar gas flow rate up through the liquid on the tray.

Combining (6-24) and (6-25) to eliminate dn_i, separating variables, and converting to integral form,

$$A_b \int_0^{Z_f} \frac{K_G a P}{G} \, dZ = \int_{y_{i,n+1}}^{y_{i,n}} \frac{dy_i}{y_{i,n}^* - y_i} = N_{OG} \tag{6-26}$$

where a second subscript involving the tray number, n, has been added to the mole fraction of the vapor phase. The vapor enters tray n at $y_{i,n+1}$ and exits at $y_{i,n}$. By definition,

$$N_{OG} = \text{number of overall gas-phase mass transfer units}$$

Values of K_G, a, P, and G may vary somewhat as the gas flows up through the liquid, but if they as well as y_i^* are taken to be constant, (6-26) can be integrated to give

$$N_{OG} = \frac{K_G a P Z_f}{(G/A_b)} = \ln \left(\frac{y_{i,n+1} - y_{i,n}^*}{y_{i,n} - y_{i,n}^*} \right) \tag{6-27}$$

A rearrangement of (6-27) in terms of the fractional approach of y_i to equilibrium defines the Murphree vapor efficiency as

$$E_{MV} = \frac{y_{i,n+1} - y_{i,n}}{y_{i,n+1} - y_{i,n}^*} = 1 - e^{-N_{OG}} \tag{6-28}$$

or

$$N_{OG} = -\ln(1 - E_{MV}) \tag{6-29}$$

Suppose that measurements give

$$y_i \text{ entering tray } n = y_{i,n+1} = 0.64$$

$$y_i \text{ leaving tray } n = y_{i,n} = 0.61$$

and, from thermodynamics or phase equilibrium data, y_i^* in equilibrium with x_i on and leaving tray $n = 0.60$

Then, from (6-28), $E_{MV} = (0.64 - 0.61)/(0.64 - 0.60) = 0.75$

or a 75% approach to equilibrium. From (6-29), $N_{OG} = -\ln(1 - 0.75) = 1.386$

When $N_{OG} = 1$, $E_{MV} = 1 - e^{-1} = 0.632$.

The derivation of the Murphree vapor efficiency does not consider the exiting stream temperatures. However, it is implied that the completely mixed liquid phase is at its bubble-point temperature so that the equilibrium vapor phase mole fraction, $y_{i,n}^*$, can be computed.

For multicomponent mixtures, values of E_{MV} are component-dependent and can vary from tray to tray, but at each tray it can be shown that the number of independent values of E_{MV} is one less than the number of components. The dependent value of E_{MV} is determined by forcing $\Sigma \, y_i = 1$. It is thus possible that a negative value of E_{MV} can result for a component in a multicomponent mixture. Such negative efficiencies are possible because of mass transfer coupling among concentration gradients in a multicomponent mixture, which is discussed in Chapter 12. However, for a binary mixture, values of E_{MV} are always positive and identical for the two components.

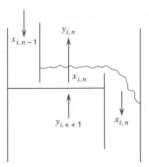

Figure 6.18 Schematic of tray for Murphree vapor-point efficiency.

Only if liquid travel distance across a tray is small will the liquid on a tray approach complete mixing. To handle the more general case of incomplete mixing, a *Murphree vapor-point efficiency* is defined by assuming that liquid composition varies with distance of travel across a tray, but is uniform in the vertical direction. Thus, for species i on tray n, at any horizontal distance from the downcomer that directs liquid onto tray n, as shown in Figure 6.18,

$$E_{OV} = \frac{y_{i,n+1} - y_{i,n}}{y_{i,n+1} - y_i^*} \qquad (6\text{-}30)$$

Because x_i varies across a tray, y_i^* and y_i also vary. However, the exiting vapor is then assumed to mix completely to give a uniform $y_{i,n}$ before entering the tray above. Because E_{OV} is a more fundamental quantity than E_{MV}, E_{OV} serves as the basis for semitheoretical estimates of tray efficiency and overall column efficiency.

Lewis [16] integrated E_{OV} over a tray for several cases. For complete mixing of liquid on a tray to give a uniform composition, $x_{i,n}$, it is obvious that

$$E_{OV} = E_{MV} \qquad (6\text{-}31)$$

For plug flow of liquid across a tray with no longitudinal diffusion (no mixing of liquid in the horizontal direction), Lewis derived

$$E_{MV} = \frac{1}{\lambda}(e^{\lambda E_{OV}} - 1) \qquad (6\text{-}32)$$

$$\text{with } \lambda = mG/L \qquad (6\text{-}33)$$

where G and L are gas and liquid molar flow rates, respectively, and $m = dy/dx = $ slope of the equilibrium line for a species, using the expression $y = mx + b$. If b is taken as zero, then m is the K-value, and for the key component, k, being absorbed,

$$\lambda = K_k G/L = 1/A_k$$

If A_k, the key-component absorption factor, is given the typical value of 1.4, $\lambda = 0.71$. Suppose the measured or predicted point efficiency is $E_{OV} = 0.25$. From (6-32),

$$E_{MV} = 1.4(e^{0.71(0.25)} - 1) = 0.27$$

which is only 9% higher than E_{OV}. However, if $E_{OV} = 0.9$, E_{MV} is 1.25, which is significantly higher and equivalent to more than a theoretical stage. This surprising result is due to the concentration gradient in the liquid across the length of travel on the tray, which allows the vapor to contact a liquid having an average concentration of species k that can be appreciably lower than that in the liquid leaving the tray.

Equations (6-31) and (6-32) represent extremes between complete mixing and no mixing of the liquid phase, respectively. A more realistic, but considerably more complex model that accounts for partial liquid mixing on the tray, as developed by Gerster et al. [17], is

$$\frac{E_{MV}}{E_{OV}} = \frac{1 - e^{-(\eta + N_{Pe})}}{(\eta + N_{Pe})\{1 + [(\eta + N_{Pe})/\eta]\}} + \frac{e^{\eta} - 1}{\eta\{1 + [\eta/(\eta + N_{Pe})]\}} \tag{6-34}$$

where

$$\eta = \frac{N_{Pe}}{2}\left[\left(1 + \frac{4\lambda E_{OV}}{N_{Pe}}\right)^{1/2} - 1\right] \tag{6-35}$$

The dimensionless Peclet number, N_{Pe}, which serves as a partial mixing parameter, is defined by

$$N_{Pe} = Z_L^2/D_F\theta_L = Z_L u/D_E \tag{6-36}$$

where Z_L is the length of liquid flow path across the tray as shown in Figure 6.3, D_E is the eddy diffusion coefficient in the direction of liquid flow, θ_L is the average liquid residence time on the tray, and $u = Z_L/\theta_L$ is the mean liquid velocity across the tray. Equation (6-34) is plotted in Figure 6.19 for wide ranges of N_{Pe} and λE_{OV}. When $N_{Pe} = 0$, Eq. (6-31) holds; when $N_{Pe} = \infty$, Eq. (6-32) holds.

From (6-36), the Peclet number can be viewed as the ratio of the mean liquid bulk velocity to the eddy diffusion velocity. When N_{Pe} is small, eddy diffusion is important and the liquid approaches a well-mixed condition. When N_{Pe} is large, bulk flow predominates and the liquid approaches plug flow. Experimental measurements of D_E in bubble-cap and sieve-plate columns [18–21] cover a range of 0.02 to 0.20 ft^2/s. Values of u/D_E typically range from 3 to 15 ft^{-1}. Based on the second form of (6-36), N_{Pe} increases directly with increasing Z_L and, therefore, column diameter. A typical value of N_{Pe} for a 2-ft-diameter column is 10; for a 6-ft-diameter column, N_{Pe} might be 30. For N_{Pe} values of this magnitude, Figure 6.19 shows that values of E_{MV} can be significantly larger than E_{OG} for large values of λ.

Lewis [16] showed that when the equilibrium and operating lines are straight, but not necessarily parallel, the overall stage efficiency, defined by (6-21), is related to the Murphree vapor efficiency by

$$E_o = \frac{\log[1 + E_{MV}(\lambda - 1)]}{\log \lambda} \tag{6-37}$$

When the two lines are not only straight but parallel, such that $\lambda = 1$, Eq. (6-38) becomes $E_o = E_{MV}$. Also, when $E_{MV} = 1$ then $E_o = 1$ regardless of the value of λ.

Scale-up from Laboratory Data

When vapor–liquid equilibrium data for a system are unavailable or not well known, and particularly if the system forms a highly nonideal liquid solution with possible formation of azeotropes, tray requirements are best estimated, and the feasibility of achieving the desired degree of separation verified, by conducting laboratory tests. A particularly useful apparatus is a small glass or metal sieve-plate column with center-to-side downcomers developed by Oldershaw [22] and shown schematically in Figure 6.20. Oldershaw columns are typically 1 to 2 in. in diameter and can be assembled with almost any number of sieve plates, usually containing 0.035- to 0.043-in. holes with a hole area of approximately 10%. A detailed study by Fair, Null, and Bolles [23] showed that overall plate efficiencies of Oldershaw columns operated over a pressure range of 3 to 165 psia are in conservative agreement with distillation data obtained from sieve-tray pilot-plant and industrial-size

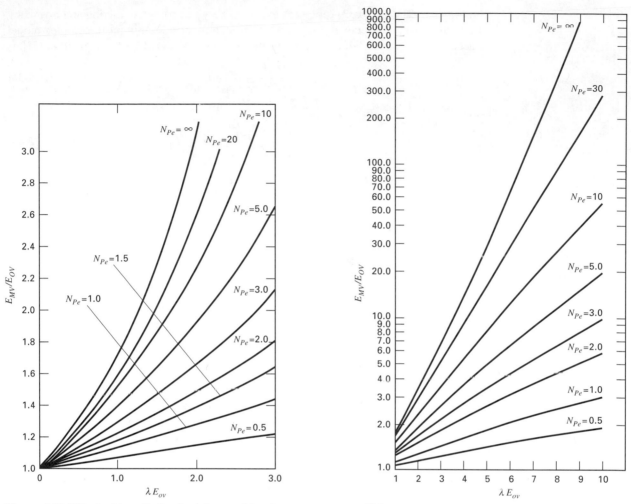

Figure 6.19 Effect of longitudinal mixing on Murphree vapor tray efficiency.

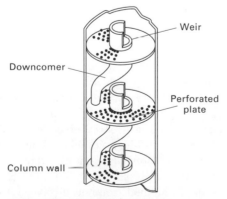

Figure 6.20 Oldershaw column.

columns ranging in size from 18 in. to 4 ft in diameter when operated in the range of 40% to 90% of flooding. It may be assumed that similar agreement might be realized for absorption and stripping.

It is believed that the small-diameter Oldershaw column achieves essentially complete mixing of liquid on each tray, thus permitting the measurement of a point efficiency. Somewhat larger efficiencies may be observed in much-larger-diameter columns due to incomplete liquid mixing, which results in a higher Murphree tray efficiency and, therefore, higher overall plate efficiency.

Fair et al. [23] recommend the following conservative scale-up procedure for the Oldershaw column:

1. Determine the flooding point.
2. Establish operation at about 60% of flooding (but 40 to 90% seems acceptable).
3. Run the system to find a combination of plates and flow rates that gives the desired degree of separation.
4. Assume that the commercial column will require the same number of plates for the same ratio of liquid to vapor molar flow rates.

If reliable vapor–liquid equilibrium data are available, they can be used with the Oldershaw data to determine the overall column efficiency, E_o. Then (6-37) and (6-34) can be used to estimate the average point efficiency. For the commercial-size column, the Murphree vapor efficiency can be determined from the Oldershaw column point efficiency using (6-34), which takes into account incomplete liquid mixing. In general, the tray efficiency of the commercial column, depending on the length of the liquid flow path, will be higher than for the Oldershaw column at the same percentage of flooding.

EXAMPLE 6.4	Assume that the column diameter for the absorption operation of Example 6.1 is 3 ft. If the overall stage efficiency, E_o, is 30% for the absorption of ethyl alcohol, estimate the average Murphree vapor efficiency, E_{MV}, and the possible range of the Murphree vapor-point efficiency, E_{OV}.
SOLUTION	For Example 6.1, the system is dilute in ethyl alcohol, the main component undergoing mass transfer. Therefore, the equilibrium and operating lines are essentially straight, and (6-37) can be applied. From the data of Example 6.1, $\lambda = KG/L = 0.57(180)/151.5 = 0.68$

Solving (6-37) for E_{MV}, using $E_o = 0.30$,

$$E_{MV} = (\lambda^{E_o} - 1)/(\lambda - 1) = (0.68^{0.30} - 1)/(0.68 - 1) = 0.34$$

For a 3-ft diameter column, the degree of liquid mixing probably lies intermediate between complete mixing and plug flow. From (6-31) for the former case, $E_{OV} = E_{MV} = 0.34$. From a rearrangement of (6-32) for the latter case, $E_{OV} = \ln(1 + \lambda E_{MV})/\lambda = \ln[1 + 0.68(0.34)]/0.68 = 0.31$. Therefore, E_{OV} lies in the range of 31% to 34%, probably closer to 34% for complete mixing. However, the differences between E_o, E_{MV}, and E_{OV} for this example are almost negligible.   ∎

6.6 TRAY CAPACITY, PRESSURE DROP, AND MASS TRANSFER

In the trayed tower shown in Figure 6.21, vapor flows vertically upward, contacting liquid in crossflow on each tray. When trays are designed properly, a stable operation is achieved wherein (1) vapor flows only through the perforations or open regions of the tray between the downcomers, (2) liquid flows from tray to tray only by means of the downcomers, (3) liquid neither weeps through the tray perforations nor is carried by the vapor as entrainment to the tray above, and (4) vapor is neither carried (occluded) down by the liquid in the downcomer to the tray below nor allowed to bubble up through the liquid in the downcomer.

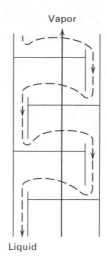

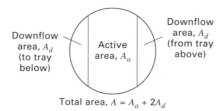

Figure 6.21 Vapor and liquid flow through a trayed tower.

Tray design includes the determination of tray diameter and the division of the tray cross-sectional area, A, as shown in Figure 6.21, into active vapor bubbling area, A_a, and liquid downcomer area, A_d. With the tray diameter fixed, vapor pressure drop and mass transfer coefficients can be estimated.

Tray Diameter

For a given liquid flow rate, as shown in Figure 6.22 for a sieve-tray column, a maximum vapor flow rate exists beyond which incipient column flooding occurs because of backup

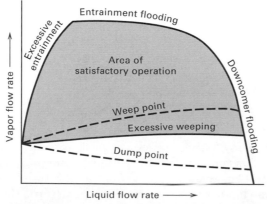

Figure 6.22 Limits of stable operation in a trayed tower. [Reproduced by permission from H.Z. Kister, *Distillation Design*, McGraw-Hill, New York (1992).]

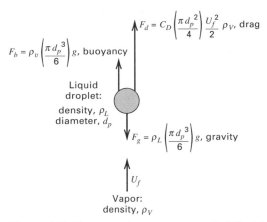

Figure 6.23 Forces acting on a suspended liquid droplet.

of liquid in the downcomer. This condition, if sustained, leads to carryout of liquid with the overhead vapor leaving the column. *Downcomer flooding* takes place when liquid backup is caused by downcomers of inadequate cross-sectional area, A_d, to carry the liquid flow, but rarely occurs if downcomer cross-sectional area is at least 10% of total column cross-sectional area and if tray spacing is at least 24 in. The usual design limit is *entrainment flooding,* which is caused by excessive carry-up of liquid, at the rate e, by vapor entrainment to the tray above. At incipient flooding, $(e + L) \gg L$ and downcomer cross-sectional area is inadequate for the excessive liquid load $(e + L)$. Tray diameter is determined as follows to avoid flooding.

Entrainment of liquid is due to carry up of suspended droplets by rising vapor or to throw up of liquid particles by vapor jets formed at tray perforations, valves, or bubble-cap slots. Souders and Brown [24] successfully correlated entrainment flooding data for 10 commercial trayed columns by assuming that carry-up of suspended droplets controls entrainment. At low vapor velocity, a droplet settles out; at high vapor velocity, it is entrained. At flooding or incipient entrainment velocity, U_f, the droplet is suspended such that the vector sum of the gravitational, buoyant, and drag forces acting on the droplet, as shown in Figure 6.23, are zero. Thus,

$$\Sigma F = 0 = F_g - F_b - F_d \tag{6-38}$$

In terms of droplet diameter, d_p,

$$\rho_L \left(\frac{\pi d_p^3}{6}\right) g - \rho_V \left(\frac{\pi d_p^3}{6}\right) g - C_D \left(\frac{\pi d_p^2}{4}\right) \frac{U_f^2}{2} \rho_V = 0 \tag{6-39}$$

where C_D is the drag coefficient. Solving for flooding velocity,

$$U_f = C \left(\frac{\rho_L - \rho_V}{\rho_V}\right)^{1/2} \tag{6-40}$$

where C = capacity parameter of Souders and Brown. According to the above theory,

$$C = \left(\frac{4 d_p g}{3 C_D}\right)^{1/2} \tag{6-41}$$

Parameter C can be calculated from (6-41) if the droplet diameter d_p is known. In practice, however, C is treated as an empirical parameter determined using experimental data obtained from operating equipment. Souders and Brown considered all the important

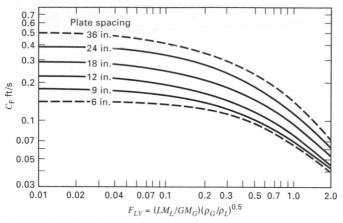

Figure 6.24 Entrainment flooding capacity in a trayed tower.

variables that could influence the value of C and obtained a correlation for commercial-size columns with bubble-cap trays. Data covered column pressures from 10 mmHg to 465 psia, plate spacings from 12 to 30 in., and liquid surface tensions from 9 to 60 dyne/cm. In accordance with (6-41), the value of C increases with increasing surface tension, which increases d_p. Also, C increases with increasing tray spacing, since this allowed more time for agglomeration to a larger d_p.

Using additional commercial operating data, Fair [25] produced the more general correlation of Figure 6.24, which is applicable to columns with bubble cap and sieve trays. Whereas Souders and Brown base the vapor velocity on the entire column cross-sectional area, Fair utilizes a net vapor flow area equal to the total inside column cross-sectional area minus the area blocked off by the downcomer, that is, $A - A_d$ in Figure 6.21. The value of C_F in Figure 6.24 depends on tray spacing and the ratio $F_{LV} = (LM_L/VM_V)(\rho_V/\rho_L)^{0.5}$ (where flow rates are in molar units), which is a kinetic energy ratio first used by Sherwood, Shipley, and Holloway [26] to correlate packed-column flooding data. The value of C in (6-41) is obtained from Figure 6.24 by correcting C_F for surface tension, foaming tendency, and the ratio of vapor hole area A_h to tray active area A_a, according to the empirical relationship

$$C = F_{ST}F_F F_{HA}C_F \tag{6-42}$$

where F_{ST} = surface tension factor = $(\sigma/20)^{0.2}$

$\quad F_F$ = foaming factor

$\quad F_{HA}$ = 1.0 for $A_h/A_a \geq 0.10$ and $5(A_h/A_a) + 0.5$ for $0.06 \geq A_h/A_a \leq 0.1$

$\quad \sigma$ = liquid surface tension, dyne/cm

For nonfoaming systems, $F_F = 1.0$; for many absorbers, F_F may be 0.75 or even less. The quantity A_h is the area open to the vapor as it penetrates into the liquid on a tray. It is the total cap slot area for bubble-cap trays and the perforated area for sieve trays.

Figure 6.24 appears to be conservative for valve trays. This is shown in Figure 6.25, where entrainment flooding data of Fractionation Research, Inc. (FRI) [27,28], for a 4-ft-diameter column equipped with Glitsch type A-1 and V-1 valve trays on 24-in. spacing are compared to the correlation in Figure 6.24. For valve trays, the slot area A_h is taken as the full valve opening through which vapor enters the frothy liquid on the tray at a 90° angle with the axis of the column.

Typically, column diameter D_T is based on fraction, f, of flooding velocity U_f, calculated from (6-40), using C from (6-42), based on C_F from Figure 6.24. By the continuity equation

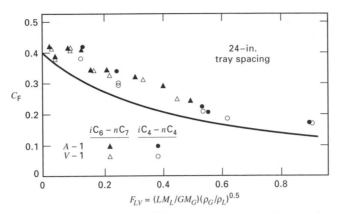

Figure 6.25 Comparison of flooding correlation with data for valve trays.

(flow rate = velocity × flow area × density), the molar vapor flow rate is related to the flooding velocity by

$$V = (fU_f)(A - A_d)\frac{\rho_V}{M_V} \qquad \text{(6-43)}$$

where A = total column cross-sectional area = $\pi D_T^2/4$. Thus,

$$D_T = \left[\frac{4VM_V}{fU_f\pi(1 - A_d/A)\rho_V}\right]^{0.5} \qquad \text{(6-44)}$$

Oliver [29] suggests that A_d/A be estimated from F_{LV} in Figure 6.24 by

$$\frac{A_d}{A} = \begin{cases} 0.1, & F_{LV} \le 0.1 \\ 0.1 + \dfrac{(F_{LV} - 0.1)}{9}, & 0.1 \le F_{LV} \le 1.0 \\ 0.2, & F_{LV} \ge 1.0 \end{cases}$$

Column diameter is calculated at both the top and bottom of the column, with the larger of the two diameters used for the entire column. Because of the need for internal access to columns with trays, a packed column is generally used if the calculated diameter from (6-44) is less than 2 ft.

Tray spacing must be specified to compute column diameter using Figure 6.24. As spacing is increased, column height is increased but column diameter is reduced. A spacing of 24 in., which provides ease of maintenance, is optimum for a wide range of conditions; however, a smaller spacing may be desirable for small-diameter columns with a large number of stages; and larger spacing is frequently used for large-diameter columns with a small number of stages.

As shown in Figure 6.22, a minimum vapor rate exists below which liquid weeps (dumps) through tray perforations or risers instead of flowing completely across the active area and into the downcomer. Below this minimum, the degree of contacting of liquid with vapor is reduced, causing tray efficiency to decline. The ratio of the vapor rate at flooding to the minimum vapor rate is the *turndown ratio,* which is approximately 8 for bubble-cap trays, 5 for valve trays, but only about 2 for sieve trays.

When vapor and liquid flow rates change appreciably from tray to tray, column diameter, tray spacing, or hole area can be varied to reduce column cost and ensure stable operation at high efficiency. Variation of tray spacing is particularly applicable to columns with sieve trays because of their low turndown ratio.

EXAMPLE 6.5

Estimate the required tray diameter for the absorber of Example 6.1, assuming a tray spacing of 24 in., a foaming factor of $F_F = 0.90$, a fraction flooding of $f = 0.80$, and a surface tension of $\sigma = 70$ dynes/cm.

SOLUTION

Because tower conditions are almost the same at the top and bottom, the calculation of column diameter is made only at the bottom, where the gas rate is highest. From Example 6.1,

$$T = 30°C \qquad P = 110 \, \text{kPa}$$
$$V = 180 \, \text{kmol/h}, \qquad L = 151.5 + 3.5 = 155.0 \, \text{kmol/h}$$
$$M_V = 0.98(44) + 0.02(46) = 44.0, \qquad M_L = \frac{151.5(18) + 3.5(46)}{155} = 18.6$$
$$\rho_V = \frac{PM}{RT} = \frac{(110)(44)}{(8.314)(303)} = 1.92 \, \text{kg/m}^3, \qquad \rho_L = (0.986)(1,000) = 986 \, \text{kg/m}^3$$
$$F_{LV} = \frac{(155)(18.6)}{(180)(44.0)}\left(\frac{1.92}{986}\right)^{0.5} = 0.016$$

For tray spacing = 24 in., from Figure 6.24, $C_F = 0.39$ ft/s,

$$F_{ST} = \left(\frac{\sigma}{20}\right)^{0.2} = \left(\frac{70}{20}\right)^{0.2} = 1.285, \qquad F_F = 0.90$$

Because $F_{LV} < 0.1$, $A_d/A = 0.1$ and $F_{HA} = 1.0$. Then, from (6-42),

$$C = 1.285 \, (0.90)(1.0)(0.39) = 0.45 \, \text{ft/s}$$

From (6-40), $U_f = 0.45\left(\frac{986 - 1.92}{1.92}\right)^{0.5} = 10.2 \, \text{ft/s}$

From (6-44),

$$D_T = \left[\frac{4(180/3,600)(44.0)}{(0.80)(10.2/3.28)(3.14)(1 - 0.1)(1.92)}\right]^{0.5} = 0.81 \, \text{m} = 2.65 \, \text{ft}$$

∎

Tray Vapor Pressure Drop

Typical tray pressure drop for flow of vapor in a tower is from 0.05 to 0.15 psi/tray. Referring to Figure 6.3, pressure drop (head loss) for a sieve tray is due to friction for vapor flow through the tray perforations, holdup of the liquid on the tray, and a loss due to surface tension:

$$h_t = h_d + h_l + h_\sigma \tag{6-45}$$

where h_t = total pressure drop/tray, in. of liquid
h_d = dry tray pressure drop, in. of liquid
h_l = equivalent head of clear liquid on tray, in. of liquid
h_σ = pressure drop due to surface tension, in. of liquid

The dry sieve tray pressure drop is given by a modified orifice equation,

$$h_d = 0.186\left(\frac{u_o^2}{C_o^2}\right)\left(\frac{\rho_V}{\rho_L}\right) \tag{6-46}$$

where u_o = hole velocity (ft/s) and C_o depends on the percent hole area and the ratio of tray thickness to hole diameter. For a typical 0.078-in.-thick tray with $\frac{3}{16}$-in.-diameter holes and a percent hole area (based on the cross-sectional area of the tower) of 10%, C_o may be taken as 0.73. Otherwise, C_o lies between about 0.65 and 0.85.

The equivalent height of clear liquid holdup on a tray depends on weir height, liquid and vapor densities and flow rates, and downcomer weir length, as given by the following empirical expression developed from experimental data by Bennett, Agrawal, and Cook [30]:

$$h_l = \phi_e \left[h_w + C \left(\frac{q_L}{L_w \phi_e} \right)^{2/3} \right] \tag{6-47}$$

where h_w = weir height, in.

ϕ_e = effective relative froth density (height of clear liquid/froth height)
$= \exp(-4.257 K_s^{0.91})$ \hfill (6-48)

K_s = capacity parameter, ft/s $= U_a \left(\dfrac{\rho_V}{\rho_L - \rho_V} \right)^{1/2}$ \hfill (6-49)

U_a = superficial vapor velocity based on active bubbling area,
$A_a = (A - 2A_d)$, of the tray, ft/s, \qquad L_w = weir length, in.
q_L = liquid flow rate across tray, gal/min
$C = 0.362 + 0.317 \exp(-3.5 h_w)$ \hfill (6-50)

The second term in (6-47) is related to the Francis weir equation for a straight segmental weir, taking into account the froth nature of the liquid flow over the weir. For $A_d/A = 0.1$, $L_w = 73\%$ of the tower diameter.

As the gas emerges from the tray perforations, the bubbles must overcome surface tension. The pressure drop due to surface tension is given by the difference between the pressure inside the bubble and that of the liquid, according to the theoretical relation

$$h_\sigma = \frac{6\sigma}{g \rho_L D_{B(\max)}} \tag{6-51}$$

where, except for tray perforations much smaller than $\frac{3}{16}$-in. in diameter, $D_{B(\max)}$, the maximum bubble size, may be taken as the perforation diameter, D_H.

Methods for estimating vapor pressure drop for bubble-cap trays and valve trays are given by Smith [31] and Klein [32], respectively, and are discussed by Kister [33] and Lockett [34].

EXAMPLE 6.6

Estimate the tray vapor pressure drop for the absorber of Example 6.1, assuming use of sieve trays with a tray diameter of 1 m, a weir height of 2 in., and a hole diameter of $\frac{3}{16}$ in.

SOLUTION

From Example 6.5,

$$\rho_V = 1.92 \text{ kg/m}^3 \qquad \rho_L = 986 \text{ kg/m}^3$$

At the bottom of the tower, vapor velocity based on the total cross-sectional area of the tower is

$$\frac{(180/3,600)(44)}{(1.92)[3.14(1)^2/4]} = 1.46 \text{ m/s}$$

For a 10% hole area, based on the total cross-sectional area of the tower,

$$u_o = \frac{1.46}{0.10} = 14.6 \text{ m/s} \quad \text{or} \quad 47.9 \text{ ft/s}$$

Using the above densities, (6-46) gives $h_d = 0.186 \left(\dfrac{47.9^2}{0.73^2} \right) \left(\dfrac{1.92}{986} \right) = 1.56$ in. of liquid

Take weir length as 73% of tower diameter, with $A_d/A = 0.10$. Then

$$L_w = 0.73(1) = 0.73 \text{ m} \quad \text{or} \quad 28.7 \text{ in.}$$

$$\text{Liquid flow rate in gpm} = \frac{(155/60)(18.6)}{986(0.003785)} = 12.9 \text{ gpm}$$

with
$$A_d/A = 0.1 \qquad A_a/A = (A - 2A_d)/A = 0.8$$

$$\text{Therefore, } U_a = 1.46/0.8 = 1.83 \text{ m/s} = 5.99 \text{ ft/s}$$

From (6-49),
$$K_s = 5.99[1.92/(986 - 1.92)]^{0.5} = 0.265 \text{ ft/s}$$

From (6-48),
$$\phi_e = \exp[-4.257(0.265)^{0.91}] = 0.28$$

From (6-50),
$$C = 0.362 + 0.317 \exp[-3.5(2)] = 0.362$$

From (6-47),
$$h_l = 0.28[2 + 0.362(12.9/28.7/0.28)^{2/3}] = 0.28(2 + 0.50) = 0.70 \text{ in.}$$

From (6-51), in metric units, using $D_{B(\max)} = D_H = \frac{3}{16}$ in. $= 0.00476$ m

$$\sigma = 70 \text{ dynes/cm} = 0.07 \text{ N/m} = 0.07 \text{ kg/s}^2, g = 9.8 \text{ m/s}^2, \text{ and } \rho_L = 986 \text{ kg/m}^3$$

$$h_\sigma = \frac{6(0.07)}{9.8(986)(0.00476)} = 0.00913 \text{ m} = 0.36 \text{ in.}$$

From (6-45), the total tray head loss is $h_t = 1.56 + 0.70 + 0.36 = 2.62$ in.

For $\rho_L = 986 \text{ kg/m}^3 = 0.0356 \text{ lb/in}^3$,

$$\text{tray vapor pressure drop} = h_t \rho_L = 2.62(0.0356) = 0.093 \text{ psi/tray} \qquad ■$$

Mass Transfer Coefficients and Transfer Units

Following the determination of tower diameter and major details of the tray layout, an estimate of the Murphree vapor point efficiency, defined by (6-30), can be made using empirical correlations for mass transfer coefficients, based on experimental data. For a vertical path for vapor flow up through the froth from a point on the bubbling area of the tray, (6-29) applies to the Murphree vapor-point efficiency:

$$N_{OG} = -\ln(1 - E_{OV}) \tag{6-52}$$

$$\text{where } N_{OG} = \frac{K_{OG} a P Z_f}{(G/A_b)} \tag{6-53}$$

The overall volumetric mass transfer coefficient, $K_G a$, is related to the individual volumetric mass transfer coefficients by the sum of the mass transfer resistances, which from equations in Section 3.7 can be shown to be

$$\frac{1}{K_G a} = \frac{1}{k_G a} + \frac{(K P M_L/\rho_L)}{k_L a} \tag{6-54}$$

where the two terms on the right-hand side are the gas- and liquid-phase resistances, respectively, and the symbols k_g for the gas and k_c for the liquid used in Chapter 3 have been replaced by k_G and k_L, respectively. In terms of individual transfer units, defined by

$$N_G = \frac{k_G a P Z_f}{(G/A_b)} \tag{6-55}$$

$$\text{and } N_L = \frac{k_L a \rho_L Z_f}{M_L(L/A_b)} \tag{6-56}$$

we obtain from (6-53) and (6-54)

$$\frac{1}{N_{OG}} = \frac{1}{N_G} + \frac{(KV/L)}{N_L} \tag{6-57}$$

Important empirical mass transfer correlations have been published by the AIChE [35] for bubble-cap trays, Chan and Fair [36,37] for sieve trays, and Scheffe and Weiland [38] for one type of valve tray (Glitsch V-1). These correlations have been developed in terms of N_L, N_G, k_L, k_G, a, and N_{Sh} for either the gas or liquid phase. In this section, we present only correlations for sieve trays, as given for binary systems by Chan and Fair [36], who used a correlation for the liquid phase based on the work of Foss and Gerster [39] as reported by the AIChE [40], and who developed a correlation for the vapor phase from a fairly extensive experimental data bank of 143 points for towers 2.5 to 4.0 ft in diameter, operating at pressures from 100 mmHg to 400 psia.

Experimental data for sieve trays have validated the assumed direct dependence of mass transfer on the interfacial area between the gas and liquid phases, and on the residence times in the froth of the gas and liquid phases. Accordingly, Chan and Fair give the following modifications of (6-55) and (6-56):

$$N_G = k_G \bar{a} \bar{t}_G \tag{6-58}$$

$$N_L = k_L \bar{a} \bar{t}_L \tag{6-59}$$

where $\bar{a}$ is the interfacial area per unit volume of equivalent clear liquid, $\bar{t}_G$ is the average residence time of the gas in the froth, and $\bar{t}_L$ is the average residence time of the liquid in the froth.

Average residence times are estimated from the following dimensionally consistent, theoretical continuity equations, using (6-47) for the equivalent head of clear liquid on the tray and (6-48) for the effective relative density of the froth:

$$\bar{t}_L = \frac{h_l A_a}{q_L} \tag{6-60}$$

and

$$\bar{t}_G = \frac{(1 - \phi_e) h_l}{(\phi_e U_a)} \tag{6-61}$$

where $(1 - \phi_e) h_l / \phi_e$ is the equivalent height of vapor holdup in the froth, and the residence times are usually computed in seconds.

Empirical expressions for $k_G \bar{a}$ and $k_L \bar{a}$ in units of s^{-1} are

$$k_G \bar{a} = \frac{1{,}030 \, D_V^{0.5} \, (f - 0.842 f^2)}{(h_l)^{0.5}} \tag{6-62}$$

and

$$k_L \bar{a} = 78.8 \, D_L^{0.5} \, (F + 0.425) \tag{6-63}$$

where the variables and their units are

D_V, D_L = diffusion coefficients, cm^2/s
h_l = clear liquid height, cm
$f = U_a/U_f$, fractional approach to flooding
$F = F$-factor $= U_a \rho_G^{0.5}$, (kg/m)$^{0.5}$/s

From (6-62), it is seen that an important factor influencing the value of $k_G \bar{a}$ is the fractional approach to flooding, $f = U_a/U_f$. This effect is shown in Figure 6.26, where (6-62) is compared to experimental data. At gas rates corresponding to a fractional approach to flooding of greater than 0.60, the mass transfer factor decreases with increasing value of f. This may be due to entrainment, which is discussed in the next sub-section. On an entrainment-free basis, the curve in Figure 6.26 might be expected to at least remain at its peak value for conditions above $f = 0.60$.

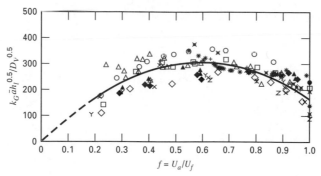

Figure 6.26 Comparison of experimental data to the correlation of Chan and Fair for gas-phase mass transfer. [From H. Chan and J.R. Fair, *Ind. Eng. Chem. Process Des. Dev.,* **23,** 817 (1984) with permission.]

From (6-63), it is seen that the *F*-factor is an important consideration for liquid-phase mass transfer. Experimental data that support this are shown in Figure 6.27, where $k_L\bar{a}$ depends strongly on *F* but is almost independent of liquid flow rate and weir height. The Murphree vapor-point efficiency model of (6-52), (6-57), (6-60), (6-61), (6-62), and (6-63) correlates the 143 points of the Chan and Fair [36] data bank with an average absolute deviation of 6.27%. Lockett [34] pointed out that (6-63) implies that $k_L\bar{a}$ depends on tray spacing, which seems unreasonable. However, the data bank did include data for tray spacings from 6 to 24 in.

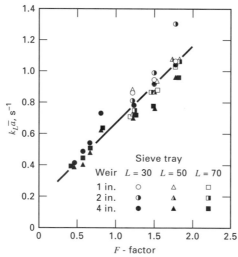

Figure 6.27 Effect of the *F*-factor on the liquid-phase volumetric mass transfer coefficient for desorption of oxygen from water with air at 1 atm. and 25°C, where L = gal/(min)/(ft of average flow width).

EXAMPLE 6.7

Estimate the Murphree vapor-point efficiency for the absorber of Example 6.1, using results from Examples 6.5 and 6.6, for the tray of Example 6.6. In addition, determine the controlling resistance to mass transfer.

SOLUTION

Pertinent data for the two phases are as follows.

	Gas	Liquid
Molar flow rate, kmol/h	180.0	155.0
Molecular Weight	44.0	18.6
Density, kg/m^3	1.92	986
Ethanol diffusivity, cm^2/s	7.86×10^{-2}	1.81×10^{-5}

Pertinent tray dimensions from Example 6.6 are $D_T = 1$ m, and $A = 0.785$ m^2; $A_a = 0.80\,A = 0.628$ m$^2 = 6,280$ cm^2; $L_w = 28.7$ in. $= 0.73$ m.

From Example 6.6, $\phi_e = 0.28$; $h_l = 0.70$ in. $= 1.78$ cm; $U_a = 5.99$ ft/s $= 183$ cm/s

From Example 6.5, $U_f = 10.2$ ft/s; $f = U_a/U_f = 5.99/10.2 = 0.59$

$$F = 1.83(1.92)^{0.5} = 2.54 \text{ (kg/m)}^{0.5}/\text{s}$$

$$q_L = \frac{(155.0)(18.6)}{986}\left(\frac{10^6}{3,600}\right) = 812 \text{ cm}^3/\text{s}$$

From (6-60), $\bar{t}_L = (1.78)(6,280)/812 = 13.8$ s

From (6-61), $\bar{t}_G = (1 - 0.28)(1.78)/[(0.28)(183)] = 0.025$ s

From (6-63), $k_L\bar{a} = 78.8(1.81 \times 10^{-5})^{0.5}(2.54 + 0.425) = 0.99$ s^{-1}

From (6-62), $k_G\bar{a} = 1,030(7.86 \times 10^{-2})^{0.5}[0.59 - 0.842(0.59)^2]/(1.78)^{0.5} = 64.3$ s^{-1}

From (6-59), $N_L = (0.99)(13.8) = 13.7$

From (6-58), $N_G = (64.3)(0.025) = 1.61$

From Example 6.1, $K = 0.57$. Therefore, $KV/L = (0.57)(180)/155 = 0.662$

From (6-57), $N_{OG} = \dfrac{1}{(1/1.62) + (0.662/13.7)} = \dfrac{1}{0.621 + 0.048} = 1.49$

and the mass transfer of ethanol is seen to be controlled by the vapor phase resistance. From (6-52), solving for E_{OV},

$$E_{OV} = 1 - \exp(-N_{OG}) = 1 - \exp(-1.49) = 0.77 = 77\% \qquad ■$$

Weeping, Entrainment, and Downcomer Backup

For a tray to operate at high efficiency, (1) weeping of liquid through the tray perforations must be small compared to flow over the outlet weir and into the downcomer, (2) entrainment of liquid by the gas must not be excessive, and (3) froth height in the downcomer must not approach tray spacing. The tray must operate in the stable region shown schematically in Figure 6.22. Weeping is associated with the lower limit of gas velocity, while entrainment flooding is associated with the upper limit.

Weeping occurs at low vapor velocities and/or high liquid rates when the clear liquid height on the tray exceeds the sum of the dry (no liquid flow) tray pressure drop, due to vapor flow, and the surface tension effect. Thus, to prevent weeping, it is necessary that

$$h_d + h_\sigma > h_l \qquad \textbf{(6-64)}$$

everywhere on the active area of the tray. If weeping occurs uniformly over the tray active area or mainly near the downcomer, a ratio of weep rate to downcomer liquid rate as high as 0.1 may not cause an unacceptable decrease in tray efficiency. Methods for estimating weep rates are discussed by Kister [33].

The prediction of fractional liquid entrainment by the vapor, defined as $\psi = e/(L + e)$, can be made by the correlation of Fair [41], given in Figure 6.28. As shown, entrainment

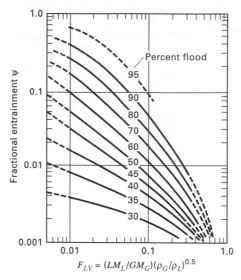

Figure 6.28 Correlation of Fair for fractional entrainment for sieve trays. [Reproduced by permission from B.D. Smith, *Design of Equilibrium Stage Processes,* McGraw-Hill, New York (1963).]

becomes excessive at high values of fraction of flooding, $f = U_a/U_f$, particularly for small values of the kinetic energy ratio, F_{LV}. The effect of entrainment on the Murphree vapor efficiency can be estimated by the following relation derived by Colburn [42], where E_{MV} is the usual "dry" efficiency and $E_{MV,\text{wet}}$ is the "wet" efficiency:

$$\frac{E_{MV,\text{wet}}}{E_{MV}} = \frac{1}{1 + eE_{MV}/L} = \frac{1}{1 + E_{MV}[\psi/(1 - \psi)]} \tag{6-65}$$

Equation (6-65) assumes that $\lambda = KV/L = 1$ and that the liquid is well mixed on the tray such that the composition of the entrained liquid is that of the liquid flowing to the tray below. For a given value of the entrainment ratio, ψ, the larger the value of E_{MV}, the greater is the effect of entrainment. For $E_{MV} = 1.0$ and $\psi = 0.10$, the "wet" efficiency is 0.90. An equation similar to (6-65) for the effect of weeping is not available, because this effect depends greatly on the degree of liquid mixing on the tray and on the distribution of weeping over the active area of the tray. If weeping occurs only in the vicinity of the downcomer, no decrease in the value of E_{MV} is observed.

The height of clear liquid in the downcomer, h_{dc}, is always greater than the height of clear liquid on the tray because, by reference to Figure 6.3, the pressure difference across the froth in the downcomer is equal to the total pressure drop across the tray from which liquid enters the downcomer, plus the height of clear liquid on the tray below to which the liquid flows, and plus the head loss for liquid flow under the downcomer apron. Thus, the clear liquid head in the downcomer is

$$h_{dc} = h_t + h_l + \mathrm{h}_{da} \tag{6-66}$$

where h_t is given by (6-45) and h_l by (6-47), and the hydraulic gradient is assumed to be negligible. The head loss for liquid flow under the downcomer, h_{da}, in inches of liquid can be estimated from an empirical orifice-type equation:

$$h_{da} = 0.03 \left(\frac{q_L}{100 \, A_{da}}\right)^2 \tag{6-67}$$

where q_L is the liquid flow in gpm and A_{da} is the area in ft^2 for liquid flow under the downcomer apron. If the height of the opening under the apron, (typically 0.5 in. less than h_w) is h_a, then $A_{da} = L_w h_a$. The height of the froth in the downcomer is

$$h_{df} = h_{dc}/\phi_{df} \qquad \qquad \textbf{(6-68)}$$

where the froth density, ϕ_{df}, can be taken conservatively as 0.5.

EXAMPLE 6.8

Using data from Examples 6.5, 6.6, and 6.7, estimate the entrainment rate, the froth height in the downcomer, and whether weeping occurs.

SOLUTION

Weeping criterion: From Example 6.6,

$$h_d = 1.56 \text{ in.} \qquad h_\sigma = 0.36 \text{ in.} \qquad h_l = 0.70 \text{ in.}$$

$$\text{From (6-64)}, 1.56 + 0.36 > 0.70$$

Therefore, if the liquid level is uniform across the active area, no weeping occurs.
Entrainment: From Example 6.5,

$$F_{LV} = 0.016 \qquad \text{From Example 6.7}, f = 0.59$$

From Fig. 6.28, $\psi = 0.06$. Therefore, for $L = 155$ from Example 6.7, the entrainment rate is $0.06(155) = 9.3$ kmol/h. Assuming that (6-65) is reasonably accurate for $\lambda = 0.662$ from Example 6.7, and that $E_{MV} = 0.78$, the effect of ψ on E_{MV}, is given by

$$\frac{E_{MV,wet}}{E_{MV}} = \frac{1}{1 + 0.78(0.06/0.94)} = 0.95 \qquad \text{or} \qquad E_{MV} = 0.95(0.78) = 0.74$$

Downcomer backup:

From Example 6.6, $h_t = 2.62$ in. $\qquad$ From Example 6.7, $L_w = 28.7$ in.

From Example 6.6, $h_w = 2.0$ in. $\qquad$ Assume that $h_a = 2.0 - 0.5 = 1.5$ in. Then

$$A_{da} = L_w h_a = 28.7(1.5) = 43.1 \text{ in.}^2 = 0.299 \text{ ft}^2 \qquad \text{From Example 6.6, } q_L = 12.9 \text{ gpm}$$

$$\text{From (6-67)}, h_{da} = 0.03 \left[\frac{12.9}{(100)(0.299)}\right]^2 = 0.006 \text{ in.}$$

$$\text{From (6-66)}, h_{dc} = 2.62 + 0.70 + 0.006 = 3.33 \text{ in. of clear liquid backup}$$

$$\text{From (6-68)}, h_{df} = \frac{3.33}{0.5} = 6.66 \text{ in. of froth in the downcomer}$$

Based on these results, neither weeping nor downcomer backup appear to be problems. An estimated 5% loss in tray efficiency occurs due to entrainment. ■

6.7 RATE-BASED METHOD FOR PACKED COLUMNS

Absorption and stripping are frequently conducted in packed columns, particularly when (1) the required column diameter is less than 2 ft; (2) the pressure drop must be low, as for a vacuum service; (3) corrosion considerations favor the use of ceramic or polymeric materials; and/or (4) low liquid holdup is desirable. Structured packing is often favored over random packing for revamps to overcome capacity limitations of trayed towers.

Packed columns are continuous differential contacting devices that do not have the physically distinguishable stages found in trayed towers. Thus, packed columns are best analyzed by mass transfer considerations rather than by the equilibrium-stage concept described in earlier sections of this chapter for trayed towers. Nevertheless, in practice, packed-tower performance is often analyzed on the basis of equivalent equilibrium stages using a packed height equivalent to a theoretical (equilibrium) plate (stage), called the

HETP or HETS and defined by the equation

$$\text{HETP} = \frac{\text{packed height}}{\text{number of equivalent equilibrium stages}} = \frac{l_T}{N_t} \qquad \textbf{(6-69)}$$

The HETP concept, unfortunately, has no theoretical basis. Accordingly, although HETP values can be related to mass transfer coefficients, such values are best obtained by back-calculation from (6-69) using experimental data from laboratory or commercial-size columns. To illustrate the application of the HETP concept, consider Example 6.1, which involves the recovery of ethyl alcohol from a CO_2-rich vapor by absorption with water. The required number of equilibrium stages is found to be just slightly more than 6, say, 6.1. Suppose that experience shows that if 1.5-in. metal Pall rings are used in a packed tower, an average HETP of 2.25 ft can be achieved. From (6-69), the required packed height, l_T, is $l_T = (\text{HETP})N_t = 2.25(6.1) = 13.7$ ft. With metal Intalox IMTP #40 random packing, the HETP might be 2.0 ft, giving $l_T = 12.3$ ft. With Mellapak 250Y corrugated sheet metal structured packing, the HETP might be only 1.2 ft, giving $l_T = 7.3$ ft.

For packed columns, it is preferable to determine packed height from a more theoretically based method involving mass-transfer coefficients for the liquid and vapor phases. As with cascades of equilibrium stages, countercurrent flow of vapor and liquid is generally preferred over cocurrent flow. Consider the countercurrent-flow packed columns shown in Figure 6.29, which is analogous to Figure 6.8 for trayed towers. For packed absorbers and strippers, operating-line equation, that are analogous to those of Section 6.3 can be derived in terms of mole fractions and total molar flow rates. Thus, for the absorber in Figure 6.29a, a material balance around the upper envelope, for the solute, gives

$$x_{in}L_{in} + yG_l = xL_l + y_{out}G_{out} \qquad \textbf{(6-70)}$$

or solving for y, assuming dilute solutions such that $G_l = G_{in} = G_{out} = G$ and $L_l = L_{in} = L_{out} = L$

$$y = x\left(\frac{L}{G}\right) + y_{out} - x_{in}\left(\frac{L}{G}\right) \qquad \textbf{(6-71)}$$

Similarly for the stripper in Figure 6.29b,

$$y = x\left(\frac{L}{G}\right) + y_{in} - x_{out}\left(\frac{L}{G}\right) \qquad \textbf{(6-72)}$$

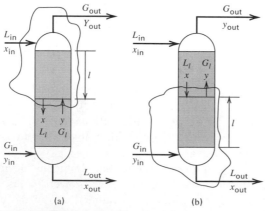

(a) (b)

Figure 6.29 Packed columns with countercurrent flow: (a) absorber; (b) stripper.

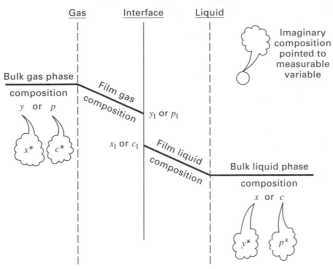

Figure 6.30 Interface properties in terms of bulk properties.

In Equations (6-70) to (6-72), mole fractions y and x represent, respectively, bulk compositions of the gas and liquid streams in contact with each other at any elevation of the packed part of the column. For the case of absorption, with mass transfer of the solute from the gas stream to the liquid stream, the two-film theory, developed in Section 3.7, can be applied as illustrated in Figure 6.30. A concentration gradient exists in each film. At the interface between the two phases, physical equilibrium is assumed to exist. Thus, as with trayed towers, an operating line and an equilibrium line are of great importance in a packed column. For a given problem specification, the location of the two lines is independent of whether the tower is trayed or packed. Thus, the method for determining the minimum absorbent liquid or stripping vapor flow rates in a packed column is identical to the method for trayed towers, as presented in Section 6.3 and illustrated in Figure 6.9.

The rate of mass transfer for absorption or stripping in a packed column can be expressed in terms of mass transfer coefficients for each phase. Coefficients, k, based on a unit area for mass transfer could be used, but the area for mass transfer in a packed bed is difficult to determine. Accordingly, as with mass transfer in the froth of a trayed tower, it is more common to use volumetric mass transfer coefficients, ka, where the quantity a represents the area for mass transfer per unit volume of packed bed. Thus, ka is based on a unit volume of packed bed. At steady state in an absorber, in the absence of chemical reactions, and since species moles are conserved, the rate of solute mass transfer across the gas-phase film must equal the rate across the liquid-phase film. If the system is dilute with respect to the solute, unimolecular diffusion (UMD) may be approximated by the simpler equations for equimolar counterdiffusion (EMD) discussed in Chapter 3. The rate of mass transfer per unit volume of packed bed, r, may be written in terms of mole-fraction driving forces in each of the two phases or in terms of a partial-pressure driving force in the gas phase and a concentration driving force in the liquid phase, as indicated in Figure 6.30. Using the former with the subscript I to denote the interface:

$$r = k_y a(y - y_I) = k_x a(x_I - x) \qquad \text{(6-73)}$$

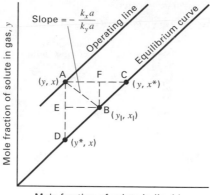

Figure 6.31 Interface composition in terms of the ratio of mass transfer coefficients.

The composition at the interface depends on the ratio, $k_x a / k_y a$, of the volumetric mass transfer coefficients, because (6-73) can be rearranged to

$$\frac{y - y_I}{x - x_I} = -\frac{k_x a}{k_y a} \tag{6-74}$$

Thus, a straight line of slope $-k_x a / k_y a$ drawn from the operating line at point (y, x) intersects the equilibrium curve at (y_I, x_I). This result is shown graphically in Figure 6.31.

The slope $-k_x a / k_y a$ determines the relative resistances of the two phases to mass transfer. In Figure 6.31 the distance AE is the gas-phase driving force $(y - y_I)$, while AF is the liquid-phase driving force $(x_I - x)$. If the mass transfer resistance in the gas phase is very low, y_I is approximately equal to y. Then, the resistance resides entirely in the liquid phase. This situation occurs in the absorption of a solute that is only slightly soluble in the liquid phase (i.e., a solute with a high K-value) and is referred to as a liquid-film resistance-controlling process. Alternatively, if the resistance in the liquid phase is very low, x_I is approximately equal to x. This situation occurs in the absorption of a solute that is very soluble in the liquid phase (i.e., a solute with a low K-value) and is referred to as a gas-film resistance-controlling process. It is important to know if one of the two resistances is controlling. If so, the rate of mass transfer can be increased by promoting turbulence in and/or increasing the dispersion of the controlling phase.

To avoid the need to determine the composition at the interface between the two phases, overall volumetric mass transfer coefficients can be defined in terms of overall driving forces for either the gas phase or the liquid phase. Thus,

$$r = K_y a (y - y^*) = K_x a (x^* - x) \tag{6-75}$$

where, as shown in Figure 6.31, y^* is the ficticious vapor mole fraction that is in equilibrium with the mole fraction, x, in the bulk liquid; and x^* is the ficticious liquid mole fraction that is in equilibrium with the mole fraction, y, in the bulk vapor. By combining (6-73) to (6-75), the overall coefficients can be expressed in terms of the separate coefficients for the two phases. Thus,

$$\frac{1}{K_y a} = \frac{1}{k_y a} + \frac{1}{k_x a} \left(\frac{y_I - y^*}{x_I - x} \right) \tag{6-76}$$

and

$$\frac{1}{K_x a} = \frac{1}{k_x a} + \frac{1}{k_y a}\left(\frac{x^* - x_\mathrm{I}}{y - y_\mathrm{I}}\right)$$ **(6-77)**

However, from Figure 6.31, for dilute solutions when the equilibrium curve is approximately a straight line through the origin,

$$\frac{y_\mathrm{I} - y^*}{x_\mathrm{I} - x} = \frac{\mathrm{ED}}{\mathrm{BE}} = \mathrm{K}$$ **(6-78)**

and

$$\frac{x^* - x_\mathrm{I}}{y - y_\mathrm{I}} = \frac{\mathrm{CF}}{\mathrm{FB}} = \frac{1}{K}$$ **(6-79)**

where K is the K-value for the solute. Combining (6-76) with (6-78) and (6-77) with (6-79),

$$\frac{1}{K_y a} = \frac{1}{k_y a} + \frac{K}{k_x a}$$ **(6-80)**

and

$$\frac{1}{K_x a} = \frac{1}{k_x a} + \frac{1}{K k_y a}$$ **(6-81)**

Determination of the packed height of a column most commonly involves the overall gas-phase coefficient, $K_y a$ because the liquid usually has a strong affinity for the solute so that resistance to mass transfer is mostly in the gas. This is analogous to a trayed tower, where the tray efficiency from mass transfer considerations is commonly based on $K_{OG} a$ or N_{OG}. Consider the countercurrent flow absorption column in Figure 6.32. For a dilute system, a differential material balance for the solute over a differential height of packing dl, gives:

$$-G\, dy = K_y a(y - y^*)S\, dl$$ **(6-82)**

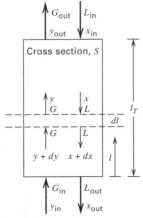

Figure 6.32 Differential contact in a countercurrent-flow packed absorption column.

where S is the cross-sectional area of the tower. In integral form, with nearly constant terms placed outside the integral, (6-82) becomes

$$\frac{K_y aS}{G} \int_0^{l_T} dl = \frac{K_y aSl_T}{G} = \int_{y_{out}}^{y_{in}} \frac{dy}{y - y^*} \tag{6-83}$$

Solving for the packed height gives

$$l_T = \frac{G}{K_y aS} \int_{y_{out}}^{y_{in}} \frac{dy}{y - y^*} \tag{6-84}$$

Chilton and Colburn [43] suggest that the right-hand side of (6-84) be written as the product of two terms:

$$l_T = H_{OG} N_{OG} \tag{6-85}$$

where

$$H_{OG} = \frac{G}{K_y aS} \tag{6-86}$$

and

$$N_{OG} = \int_{y_{out}}^{y_{in}} \frac{dy}{y - y^*} \tag{6-87}$$

The term H_{OG} is called the *overall height of a transfer unit* (HTU) based on the gas phase. Experimental data show that the HTU varies less with G than $K_y a$. The smaller the HTU, the more efficient is the contacting. The term N_{OG} is called the *overall number of transfer units* (NTU) based on the gas phase. It represents the overall change in solute mole fraction divided by the average mole fraction driving force. The larger the NTU, the greater is the extent of contacting required.

Equation (6-87) was first integrated by Colburn [44]. By using the linear equilibrium condition $y^* = Kx$ to eliminate y^* and using the linear solute material balance operating line, (6-71), to eliminate x, the result is

$$\int_{y_{out}}^{y_{in}} \frac{dy}{y - y^*} = \int_{y_{out}}^{y_{in}} \frac{dy}{(1 - KG/L)y + y_{out}(KG/L) - Kx_{in}} \tag{6-88}$$

Letting $L/KG = A$, the absorption factor, and integrating (6-88), gives

$$N_{OG} = \frac{\ln\{[(A - 1)/A][(y_{in} - Kx_{in})/(y_{out} - Kx_{in})] + (1/A)\}}{(A - 1)/A} \tag{6-89}$$

By applying (6-89) and (6-86), the required packed height, l_T, can be determined from (6-85).

The NTU (e.g., N_{OG}) and the HTU (e.g., H_{OG}) should not be confused with the number of equilibrium (theoretical) stages, N_t, and the HETP, respectively. However, when the operating and equilibrium lines are not only straight but also parallel, NTU $= N_t$ and HTU = HETP. Otherwise, the NTU is greater than or less than N_t as shown in Figure 6.33 for the case of absorption. When the operating and equilibrium lines are straight but not parallel, then

$$\text{HETP} = H_{OG} \frac{\ln(1/A)}{(1 - A)/A} \tag{6-90}$$

and

$$N_{OG} = N_t \frac{\ln(1/A)}{(1 - A)/A} \tag{6-91}$$

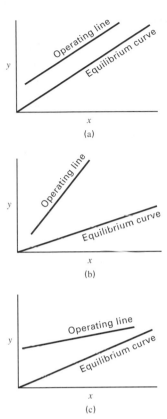

Figure 6.33 Relationship between the NTU and the number of theoretical stages N_t: (a) NTU = N_t; (b) NTU > N_t; (c) NTU < N_t.

Although the most common applications of the HTU and NTU are based on (6-85) to (6-87) and (6-89), a number of alternative groupings have been used, depending on the selected driving force for mass transfer and whether the overall basis is the gas phase, as above, or the liquid phase, where H_{OL} and N_{OL} apply. These groupings are summarized in Table 6.7. Included are driving forces based on partial pressures, p; mole ratios, X, Y; and concentrations, c; as well as mole fractions, x, y. Also included in Table 6.7 for later reference in the last section of this chapter are groupings for unimolecular diffusion (UMD) when solute concentration is not dilute.

EXAMPLE 6.9

Repeat Example 6.1 for absorption in a tower packed with 1.5-in. metal Pall rings. If $H_{OG} = 2.0$ ft, compute the required packed height.

SOLUTION

From Example 6.1, $G = 180$ kmol/h, $L = 151.5$ kmol/h, $y_{in} = 0.020$, $x_{in} = 0.0$, and $K = 0.57$. For 97% recovery of ethyl alcohol, by material balance,

$$y_{out} = \frac{(0.03)(0.02)(180)}{180 - (0.97)(0.02)(180)} = 0.000612$$

$$A = \frac{L}{KG} = \frac{51.5}{(0.57)(180)} = 1.477$$

$$\frac{y_{in}}{y_{out}} = \frac{0.020}{0.000612} = 32.68$$

Table 6.7 Alternative Mass Transfer Coefficient Groupings

		Height of a Transfer Unit, HTU			Number of Transfer Units, NTU	
Driving Force	Symbol	EM Diffusion or Dilute UM Diffusion	UM Diffusion	Symbol	EM Diffusion[a] or Dilute UM Diffusion	UM Diffusion
1. $(y - y^*)$	H_{OG}	$\dfrac{G}{K_y a S}$	$\dfrac{G}{K'_y a(1 - y)_{LM} S}$	N_{OG}	$\displaystyle\int \dfrac{dy}{(y - y^*)}$	$\displaystyle\int \dfrac{(1 - y)_{LM}\, dy}{(1 - y)(y - y^*)}$
2. $(p - p^*)$	H_{OG}	$\dfrac{G}{K_G a P S}$	$\dfrac{G}{K'_G a(1 - y)_{LM} P S}$	N_{OG}	$\displaystyle\int \dfrac{dp}{(p - p^*)}$	$\displaystyle\int \dfrac{(P - p)_{LM}\, dp}{(P - p)(p - p^*)}$
3. $(Y - Y^*)$	H_{OG}	$\dfrac{G'}{K_Y a S}$	$\dfrac{G'}{K_Y a S}$	N_{OG}	$\displaystyle\int \dfrac{dY}{(Y - Y^*)}$	$\displaystyle\int \dfrac{dY}{(Y - Y^*)}$
4. $(y - y_I)$	H_G	$\dfrac{G}{k_y a S}$	$\dfrac{G}{k'_y a(1 - y)_{LM} S}$	N_G	$\displaystyle\int \dfrac{dy}{(y - y_I)}$	$\displaystyle\int \dfrac{(1 - y)_{LM}\, dy}{(1 - y)(y - y_I)}$
5. $(p - p_I)$	H_G	$\dfrac{G}{k_g a P S}$	$\dfrac{G}{k'_g a(P - p)_{LM} S}$	N_G	$\displaystyle\int \dfrac{dp}{(p - p_I)}$	$\displaystyle\int \dfrac{(P - p)_{LM}\, dp}{(P - p)(p - p_I)}$
6. $(x^* - x)$	H_{OL}	$\dfrac{L}{K_x a S}$	$\dfrac{L}{K'_x a(1 - x)_{LM} S}$	N_{OL}	$\displaystyle\int \dfrac{dx}{(x^* - x)}$	$\displaystyle\int \dfrac{(1 - x)_{LM}\, dx}{(1 - x)(x^* - x)}$
7. $(c^* - c)$	H_{OL}	$\dfrac{L}{K_L a\,(\rho/M)\,S}$	$\dfrac{L}{K'_L a(\rho/M - c)_{LM} S}$	N_{OL}	$\displaystyle\int \dfrac{dc}{(c^* - c)}$	$\displaystyle\int \dfrac{(\rho/M - c)_{LM}\, dx}{(\rho/M - c)(c^* - c)}$
8. $(X^* - X)$	H_{OL}	$\dfrac{L'}{K_X a S}$	$\dfrac{L'}{K_X a S}$	N_{OL}	$\displaystyle\int \dfrac{dX}{(X^* - X)}$	$\displaystyle\int \dfrac{dX}{(X^* - X)}$
9. $(x_I - x)$	H_L	$\dfrac{L}{k_x a S}$	$\dfrac{L}{k'_x a(1 - x)_{LM} S}$	N_L	$\displaystyle\int \dfrac{dx}{(x_I - x)}$	$\displaystyle\int \dfrac{(1 - x)_{LM}\, dx}{(1 - x)(x_I - x)}$
10. $(c_I - c)$	H_L	$\dfrac{L}{k_L a\,(\rho/M)\,S}$	$\dfrac{L}{k'_L a(\rho/M - c)_{LM} S}$	N_L	$\displaystyle\int \dfrac{dc}{(c_I - c)}$	$\displaystyle\int \dfrac{(\rho/M - c)_{LM}\, dC}{(\rho/M - c)(c_I - c)}$

[a] The substitution $K_y = K'_y y_{B_{LM}}$ or its equivalent can be made.

From (6-89),

$$N_{OG} = \frac{\ln\{[(1.477 - 1)/1.477](32.68) + (1/1.477)\}}{(1.477 - 1)/1.477} = 7.5 \text{ transfer units}$$

The packed height, from (6-85), is

$$l_T = 2.0(7.5) = 15 \text{ ft}$$

Note that N_t for this example was determined in Example 6.1 to be about 6.1. The value of 7.5 for N_{OG} is greater than N_t because the slope of the operating line, L/G, is greater than the slope of the equilibrium line, K, so Figure 6.33b applies. ∎

EXAMPLE 6.10

Experimental data have been obtained for air containing 1.6% by volume SO_2 being scrubbed with pure water in a packed column of 1.5 m^2 in cross-sectional area and 3.5 m in packed height. Entering gas and liquid flow rates are 0.062 and 2.2 kmol/s, respectively. If the outlet mole fraction of SO_2 in the gas is 0.004 and column temperature is near ambient with $K_{SO_2} = 40$, calculate from the data:

(a) The N_{OG} for absorption of SO_2
(b) The H_{OG} in meters
(c) The volumetric overall mass transfer coefficient, $K_y a$ for SO_2 in kmol/m^3-s-(Δy).

SOLUTION

(a) Assume a straight operating line because the system is dilute in SO_2.

$$A = \frac{L}{KG} = \frac{2.2}{(40)(0.0672)} = 0.89 \qquad y_{in} = 0.016 \qquad y_{out} = 0.004$$

From (6-89),

$$N_{OG} = \frac{\ln\{[(0.89 - 1)/0.89](0.016/0.004) + (1/0.89)\}}{(0.89 - 1)/0.89} = 3.75$$

(b) $l_T = 3.5$ m. From (6-85), $H_{OG} = l_T/N_{OG} = 3.5/3.75 = 0.93$ m

(c) $G = 0.062$ kmol/s, $S = 1.5$ m^2.

From (6-86), $K_y a = G/H_{OG}S = 0.062/[(0.93)(1.5)] = 0.044$ kmol/m^3-s-(Δy) ∎

EXAMPLE 6.11

A gaseous reactor effluent consisting of 2 mol% ethylene oxide in an inert gas is scrubbed with water at 30°C and 20 atm. The total gas feed rate is 2,500 lbmol/h, and the water rate entering the scrubber is 3,500 lbmol/h. The column, with a diameter of 4 ft, is packed in two 12-ft-high sections with 1.5-in. metal Pall rings. A liquid redistributor is located between the two packed sections. Under the operating conditions for the scrubber, the K-value for ethylene oxide is 0.85 and estimated values of $k_y a$ and $k_x a$ are 200 lbmol/h-ft^3-Δy and 165 lbmol/h-ft^3-Δx, respectively.
Calculate: (a) $K_y a$ and (b) H_{OG}.

SOLUTION

(a) From (6-80), $K_y a = \dfrac{1}{(1/k_y a) + (K/k_x a)} = \dfrac{1}{(1/200) + (0.85/165)} = 98.5$ lbmol/h-ft^3-Δy

(b) $S = 3.14(4)^2/4 = 12.6$ ft^2

From (6-86), $H_{OG} = G/K_y aS = 2,500/[(98.5)(12.6)] = 2.02$ ft.

Note that in this example, both gas-phase and liquid-phase resistances are important.
The value of H_{OG} can also be computed from values of H_G and H_L using equations in Table 6.6:

$$H_G = G/k_y aS = 2,500/[(200)(12.6)] - 1.0 \text{ ft}$$
$$H_L = L/k_x aS = 3,500/[(165)(12.6)] = 1.68 \text{ ft}$$

Substituting these two expressions and (6-86) into (6-80) gives the following relationship for H_{OG} in terms of H_G and H_L:

$$H_{OG} = H_G + H_L/A$$
$$A = L/KG = 3,500/[(0.85)(2,500)] = 1.65 \tag{6-92}$$
$$H_{OG} = 1.0 + 1.68/1.65 = 2.02 \text{ ft}$$

∎

6.8 PACKED COLUMN EFFICIENCY, CAPACITY, AND PRESSURE DROP

Values of volumetric mass transfer coefficients and corresponding HTUs depend on gas and/or liquid flow rates per unit inside cross-sectional area of the packed column. Therefore, column diameter must be estimated before determining required height of packing. The estimation of a suitable column diameter for a given system, packing, and operating conditions requires consideration of liquid holdup, flooding, and pressure drop.

Liquid Holdup

Typical experimental curves, taken from Billet [45] and shown also by Stichlmair, Bravo, and Fair [46], for specific pressure drop in meters of water head per meter of packed height, and specific liquid holdup in cubic meters per cubic meter of packed bed as a

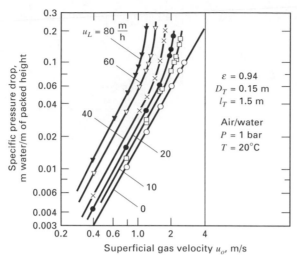

Figure 6.34 Specific pressure drop for dry and irrigated 25-mm metal Bialecki rings. [From R. Billet, *Packed Column Analysis and Design,* Ruhr-University Bochum (1989) with permission.]

function of superficial gas velocity for different values of superficial water velocity are shown in Figures 6.34 and 6.35, respectively, for a 0.15-m-diameter column packed with random 1-in. metal Bialecki rings to a height of 1.5 m and operated at 25C and 1 bar. In Figure 6.34, the lowest curve corresponds to zero liquid flow, that is, the dry pressure drop. Over an almost 10-fold range of air velocity, the pressure drop for air flowing up through the packing is proportional to air velocity to the 1.86 power. As liquid flows down through the packing at an increasing rate, gas-phase pressure drop for a given gas velocity increases. However, below a certain limiting gas velocity, the curve for each liquid velocity is a straight line parallel to the dry pressure drop curve. In this region, the liquid holdup for each liquid velocity is constant, as shown in Figure 6.35. Thus, for a liquid velocity of 40 m/h, specific liquid holdup is 0.08 m^3/m^3 of packed bed until a superficial gas velocity of 0.8 m/h is reached. Instead of a void fraction, ϵ, of 0.94 for the gas to flow through, the

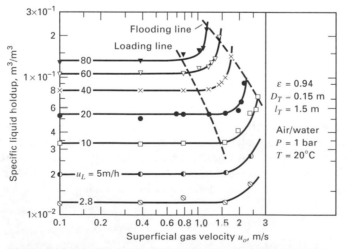

Figure 6.35 Specific liquid holdup for irrigated 25-mm metal Bialecki rings. [From R. Billet, *Packed Column Analysis and Design,* Ruhr-University Bochum (1989) with permission.]

effective void fraction is reduced by the liquid holdup to 0.94 − 0.08 = 0.86, causing an increased pressure drop. For a given liquid velocity, the upper limit to the gas velocity for a constant liquid holdup is termed the *loading point*. Below this point, the gas phase is the continuous phase. Above this point, liquid begins to accumulate or load the bed, replacing gas holdup and causing a sharp increase in pressure drop. Finally, a gas velocity is reached at which the liquid surface is continuous across the top of the packing and the column is flooded. At the *flooding point,* the pressure drop increases infinitely with increasing gas velocity. Approximate loci of both loading and flooding points are included in Figure 6.35.

The region between the loading point and the flooding point is the *loading region;* significant liquid entrainment is observed, liquid holdup increases sharply, mass transfer efficiency decreases, and column operation is unstable. Typically, according to Billet [45], the superficial gas velocity at the loading point is approximately 70% of that at the flooding point. Although a packed column can operate in the loading region, most packed columns are designed to operate below the loading point, in the *preloading region.*

The specific liquid holdup in the preloading region has been found, from extensive experiments by Billet and Schultes [47] for a wide variety of random and structured packings and, for a number of gas–liquid systems, to depend on packing characteristics, and the viscosity, density, and superficial velocity of the liquid according to the dimensionless expression

$$h_L = \left(12 \frac{N_{Fr_L}}{N_{Re_L}} \right)^{1/3} \left(\frac{a_h}{a} \right)^{2/3}$$ (6-93)

where N_{Re_L} = liquid Reynolds number = $\dfrac{\text{inertial force}}{\text{viscous force}} = \dfrac{u_L \rho_L}{a \mu_L} = \dfrac{u_L}{a v_L}$ (6-94)

where v_L is the kinematic viscosity.

$$N_{Fr_L} = \text{liquid Froude number} = \frac{\text{inertial force}}{\text{gravitational force}} = \frac{u_L^2 a}{g}$$ (6-95)

and the ratio of specific hydraulic area of packing, a_h, to specific surface area of packing, a, is given by

$$a_h/a = C_h N_{Re_L}^{0.15} N_{Fr_L}^{0.1} \qquad \text{for } N_{Re_L} < 5$$ (6-96)

$$a_h/a = 0.85\, C_h N_{Re_L}^{0.25} N_{Fr_L}^{0.1} \qquad \text{for } N_{Re_L} \geq 5$$ (6-97)

Values of a and C_h are characteristic of the particular type and size of packing, as listed, together with packing void fraction, ϵ, and other packing constants in Table 6.8. Because the specific liquid holdup is constant in the preloading region, as seen in Fig. 6.35, Eq. (6-93) does not involve gas-phase properties or gas velocity.

At low liquid velocities, liquid holdup can become so small that the packing is no longer completely wetted. When this occurs, packing efficiency decreases dramatically, particularly for aqueous systems of high surface tension. To ensure complete wetting of packing, proven liquid distributors and redistributors should be used and superficial liquid velocities should exceed the following values:

Type of Packing Material	$u_{L,min}$, m/s
Ceramic	0.00015
Oxidized or etched metal	0.0003
Bright metal	0.0009
Plastic	0.0012

Table 6.8 Characteristics of Packings

Packing	Material	Size	F_P, ft²/ft³	a, m²/m³	ε, m³/m³	C_h	C_p	C_L	C_V
					Characteristics from Billet:				
Random Packings									
Berl saddle	Ceramic	25 mm	110	260.0	0.680	0.620		1.246	0.387
Berl saddle	Ceramic	13 mm	240	545.0	0.650	0.833		1.364	0.232
Bialecki ring	Metal	50 mm		121.0	0.966	0.798	0.719	1.721	0.302
Bialecki ring	Metal	35 mm		155.0	0.967	0.787	1.011	1.412	0.390
Bialecki ring	Metal	25 mm		210.0	0.956	0.692	0.891	1.461	0.331
DINPAC ring	Plastic	70 mm		110.7	0.938	0.991			
DINPAC ring	Plastic	45 mm		182.9	0.922	1.173			
Envi Pac ring	Plastic	80 mm, no. 3		60.0	0.955	0.641	0.358	1.603	0.257
Envi Pac ring	Plastic	60 mm, no. 2		98.4	0.961	0.794	0.338	1.522	0.296
Envi Pac ring	Plastic	32 mm, no. 1		138.9	0.936	1.039	0.549	1.517	0.459
Cascade miniring	Metal	30 PMK		180.2	0.975	0.930			
Cascade miniring	Metal	30 P		168.9	0.958	0.851			
Cascade miniring	Metal	1.5″ CMR; T		188.0	0.972	0.870			
Cascade miniring	Metal	1.5″ CMR	29	174.9	0.974	0.935			
Cascade miniring	Metal	1.0″ CMR	40	232.5	0.971	1.040			
Cascade miniring	Metal	0.5″ CMR		356.0	0.955	1.338			
Hackette	Plastic	45 mm		133.4	0.931	0.643	0.399		
Hiflow ring	Ceramic	75 mm	15	54.1	0.868		0.435		
Hiflow ring	Ceramic	50 mm	29	89.7	0.809		0.538	1.377	0.379
Hiflow ring	Ceramic	35 mm	37	108.3	0.833		0.621		
Hiflow ring	Ceramic	20 mm, 6 stg.		265.8	0.776	0.958			
Hiflow ring	Ceramic	20 mm, 4 stg.		261.2	0.779	1.167	0.628	1.744	0.465
Hiflow ring	Metal	50 mm	16	92.3	0.977	0.876	0.421	1.168	0.408
Hiflow ring	Metal	25 mm	42	202.9	0.962	0.799	0.689	1.641	0.402
Hiflow ring	Plastic	90 mm	9	69.7	0.968		0.276		
Hiflow ring	Plastic	50 mm, hydr.		118.4	0.925		0.311	1.553	0.369
Hiflow ring	Plastic	50 mm	20	117.1	0.924	1.038	0.327	1.487	0.345
Hiflow ring	Plastic	25 mm		194.5	0.918		0.741	1.577	0.390
Hiflow ring, super	Plastic	50 mm		82.0	0.942	0.414		1.219	0.342
Hiflow saddle	Plastic	50 mm		86.4	0.938		0.454		
Intalox saddle	Ceramic	50 mm	40	114.6	0.761		0.747		
Intalox saddle	Plastic	50 mm	28	122.1	0.908		0.758		
NORPAC ring	Plastic	50 mm	14	86.8	0.947	0.651	0.350	1.080	0.322
NORPAC ring	Plastic	35 mm	21	141.8	0.944	0.587	0.371	0.756	0.425
NORPAC ring	Plastic	25 mm, type B		202.0	0.953		0.397	0.883	0.366
NORPAC ring	Plastic	25 mm, 10 stg.		197.9	0.920		0.383	0.976	0.410
NORPAC ring	Plastic	25 mm	31	180.0	0.927	0.601			
NORPAC ring	Plastic	22 mm		249.0	0.913		0.397		
NORPAC ring	Plastic	15 mm		311.4	0.918	0.343	0.365		
Pall ring	Ceramic	50 mm	43	116.5	0.783	1.335	0.662	1.227	0.415
Pall ring	Metal	50 mm	27	112.6	0.951	0.784	0.763	1.192	0.410
Pall ring	Metal	35 mm	40	139.4	0.965	0.644	0.967	1.012	
Pall ring	Metal	25 mm	56	223.5	0.954	0.719	0.957	1.440	0.336
Pall ring	Metal	15 mm	70	368.4	0.933	0.590	0.990		
Pall ring	Plastic	50 mm	26	111.1	0.919	0.593	0.698	1.239	0.368
Pall ring	Plastic	35 mm	40	151.1	0.906	0.718	0.927	0.856	0.380
Pall ring	Plastic	25 mm	55	225.0	0.887	0.528	0.865	0.905	0.446
Raflux ring	Plastic	15 mm		307.9	0.894	0.491	0.595		
Ralu ring	Plastic	50 mm, hydr.		94.3	0.939	0.439		1.481	0.341
Ralu ring	Plastic	50 mm		95.2	0.938	0.468		1.520	0.341

Table 6.8 (*Continued*)

Packing	Material	Size	F_P, ft²/ft³	a, m²/m³	ϵ, m³/m³	C_h	C_p	C_L	C_V
						Characteristics from Billet:			
Raschig ring	Carbon	25 mm		202.2	0.720	0.623		1.379	0.471
Raschig ring	Ceramic	25 mm	179	190.0	0.680	0.577	1.329	1.361	0.412
Raschig ring	Ceramic	15 mm	380	312.0	0.690	0.648		1.276	0.401
Raschig ring	Ceramic	10 mm	1,000	440.0	0.650	0.791		1.303	0.272
Raschig ring	Ceramic	6 mm	1,600	771.9	0.620	1.094		1.130	
Raschig ring	Metal	15 mm	170	378.4	0.917	0.455			
Tellerette	Plastic	25 mm	40	190.0	0.930	0.588	0.538	0.899	
Top-Pak ring	Aluminum	50 mm		105.5	0.956	0.881	0.604	1.326	0.389
VSP ring	Metal	50 mm, no. 2		104.6	0.980	1.135	0.773	1.222	0.420
VSP ring	Metal	25 mm, no. 1		199.6	0.975	1.369	0.782	1.376	0.405
Structured Packings									
Euroform	Plastic	PN-110		110.0	0.936	0.511	0.250	0.973	0.167
Gempak	Metal	A2 T-304		202.0	0.977	0.678			
Impulse	Ceramic	100		91.4	0.838	1.900	0.417	1.317	0.327
Impulse	Metal	250		250.0	0.975	0.431		0.983	0.270
Koch-Sulzer	Metal	CY	70						
Koch-Sulzer	Metal	BX	21						
Mellapak	Plastic	250 Y	22	250.0	0.960	0.554			
Montz	Metal	B1-100		100.0	0.987	0.626			
Montz	Metal	B1-200		200.0	0.979	0.547	0.355	0.971	0.390
Montz	Metal	B1-300	33	300.0	0.930	0.482	0.295	1.165	0.422
Montz	Plastic	C1-200		200.0	0.954		0.453	1.006	0.412
Montz	Plastic	C2-200		200.0	0.900		0.481	0.739	
Ralu Pak	Metal	YC-250		250.0	0.945		0.191	1.334	0.385

EXAMPLE 6.12

An absorption column is to be designed using oil absorbent with a kinematic viscosity of three times that of water at 20°C. The superficial liquid velocity will be 0.01 m/s, which is safely above the minimum value for good wetting. The superficial gas velocity will be such that operation will be in the preloading region. Two packing materials are being considered: (1) randomly packed 50-mm metal Hiflow rings and (2) metal Montz B1-200 structured packing. Estimate the specific liquid holdup for each of these two packings.

SOLUTION

From Table 6.8,

Packing	a, m²/m³	ϵ	C_h
50-mm metal Hiflow rings	92.3	0.977	0.876
Montz metal B1-200	200.0	0.979	0.547

At 20°C for water, kinematic viscosity, $\nu = \mu/\rho = 1 \times 10^{-6}$ m²/s. Therefore, for the oil, $\mu/\rho = 3 \times 10^{-6}$ m²/s. From (6-94) and (6-95),

$$N_{Re_L} = \frac{0.01}{3 \times 10^{-6}a} \qquad N_{Fr_L} = \frac{(0.01)^2 a}{9.8}$$

Therefore,

Packing	N_{Re_L}	N_{Fr_L}
Hiflow	36.1	0.000942
Montz	16.67	0.00204

From (6-97), since $N_{Re_L} > 5$, for the Hiflow packing, $a_h/a = (0.85)(0.876)(36.1)^{0.25}(0.000942)^{0.1} = 0.909$. For the Montz packing, $a_h/a = 0.85(0.547)(16.67)^{0.25}(0.00204)^{0.10} = 0.506$.

$$\text{From (6-93), for the Hiflow packing, } h_L = \left[\frac{12(0.000942)}{36.1}\right]^{1/3}(0.909)^{2/3} = 0.0637 \text{ m}^3/\text{m}^3$$

$$\text{For the Montz packing, } h_L = \left[\frac{12(0.0204)}{16.67}\right]^{1/3}(0.506)^{2/3} = 0.0722 \text{ m}^3/\text{m}^3$$

Note that for the Hiflow packing, the void fraction available for gas flow is reduced by the liquid flow from $\epsilon = 0.977$ (Table 6.8) to $0.0977 - 0.064 = 0.913$ m^3/m^3. For the Montz packing, the reduction is from 0.979 to 0.907 m^3/m^3. ∎

Capacity and Pressure Drop

Most packed columns consist of cylindrical vertical vessels. The column diameter is determined so as to safely avoid flooding and operate in the preloading region with a pressure drop of no greater than 1.5 in. of water head per foot of packed height (equivalent to 0.054 psi/ft of packing). In addition, for random packings, a nominal packing diameter not greater than one-eighth of the diameter of the column is selected; otherwise, poor distribution of liquid and vapor flow over the cross-sectional area of the column can occur, with liquid tending to migrate to the wall of the column.

Flooding data for packed columns with countercurrent flow of liquid and gas were first correlated successfully by Sherwood et al. [26], who used the same liquid-to-gas kinetic energy ratio, $F_{LG} = (LM_L/GM_G)(\rho_G/\rho_L)^{0.5}$, already discussed for the correlation of flooding and entrainment in trayed towers, as shown in Figures 6.24 and 6.28, respectively. The superficial gas velocity at flooding was embedded in the dimensionless term $u_o^2 a/g\epsilon^3$, which was arrived at by considering the square of the gas velocity, $u_G^2 = u_o^2/\epsilon^2$, the hydraulic radius, $r_H = \epsilon/a$, which is the volume available for flow divided by the wetted surface area of the packing, and the gravitational acceleration, g, to give the dimensionless expression, $u_G^2/gr_H = u_o^2 a/g\epsilon^3 = u_o^2 F_P/g$. The ratio, a/ϵ^3, is a function of the packing only, and is known as the packing factor, F_P. Values of a, ϵ, and F_P are included in Table 6.8. In some cases, F_P is a modified packing factor, treated as an empirical constant, backed out from experimental data so as to fit a generalized correlation. Additional factors were added by Sherwood et al. to account for liquid density and viscosity, and gas density.

In 1954, Leva [48] used experimental data on ring and saddle packings to extend the Sherwood et al. [26] flooding correlation to include lines of constant pressure drop, with the resulting chart becoming known as the generalized pressure drop correlation (GPDC).

A modern version of the GPDC chart is that of Leva [49], as shown in Figure 6.36a. The abscissa is the same F_{LG} parameter, but the ordinate is given by

$$Y = \frac{u_o^2 F_P}{g}\left(\frac{\rho_g}{\rho_{H_2O_{(L)}}}\right)f\{\rho_L\}f\{u_L\} \tag{6-98}$$

where the density of H$_2$O is taken as 62.4 lb/ft^3 with ρ_G in the same units. The functions $f\{\rho_L\}$ and $f\{\mu_L\}$ are corrections for liquid properties as given by Figures 6.36b and 6.36c, respectively.

For given fluid flow rates and properties, and a given packing material, the GPDC chart is used to compute u_o, the superficial gas velocity at flooding. Then a fraction of flooding, f, is selected (usually from 0.5 to 0.7), followed by calculation of the tower diameter from an equation similar to (6-44):

$$D_T = \left(\frac{4GM_G}{fu_o\pi\rho_G}\right)^{0.5} \tag{6-99}$$

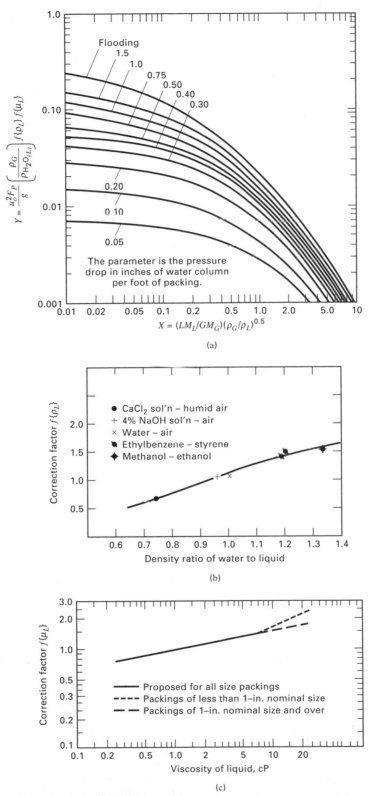

Figure 6.36 (a) Generalized pressure drop correlation of Leva for packed columns. (b) Correction factor for liquid density. (c) Correction factor for lqiuid viscosity. [From M. Leva, *Chem. Eng. Prog.,* **88** (1), 65–72 (1992) with permission.]

EXAMPLE 6.13

Air containing 5 mol% NH_3 at a total flow rate of 40 lbmol/h, enters a packed column operating at 20°C and 1 atm, where 90% of the ammonia is scrubbed by a countercurrent flow of 3,000 lb/h of water. Use the GPDC chart of Figure 6.36 to estimate the superficial gas flooding velocity, the column inside diameter for operation at 70% of flooding, and the pressure drop per foot of packing for two packing materials:

(a) One-inch ceramic Raschig rings ($F_P = 179$ ft^2/ft^3)
(b) One-inch metal IMTP packing ($F_P = 41$ ft^2/t^3)

SOLUTION

Because the superficial gas velocity is highest at the bottom of the column, calculations are made for conditions there.

Inlet gas:

$$M_G = 0.95(29) + 0.05(17) = 28.4 \qquad G = 40 \text{ lbmol/h}$$

$$\rho_G = PM_G/RT = (1)(28.4)/[(0.730)(293)(1.8)] = 0.0738 \text{ lb/ft}^3$$

Exiting liquid:

Ammonia absorbed = 0.90(0.05)(40)(17) = 30.6 lb/h or 1.8 lbmol/h
Water rate (neglecting any stripping by the gas) = 3,000 lb/h or 166.7 lbmol/h
Mole fraction of ammonia = 1.8/(166.7 + 1.8) = 0.0107

$$M_L = 0.0107(17) + (0.9893)(18) = 17.9 \qquad L = 1.8 + 166.7 = 168.5 \text{ lbmol/h}$$

Take: $\rho_L = 62.4$ lb/ft^3 and $\mu_L = 1.0$ cP

Now,

$$X = F_{LG}(\text{abscissa in Figure 6.36a}) = \frac{(168.5)(17.99)}{(40)(28.4)} \left(\frac{0.0738}{62.4}\right)^{0.5} = 0.092$$

From Figure 6.36a, $Y = 0.125$

From Figure 6.36b, $f\{\rho_L\} = 1.14$

From Figure 6.36c, $f\{\mu_L\} = 1.0$

From (6-98), $u_o^2 = 0.125 \left(\dfrac{g}{F_P}\right) \dfrac{62.4}{(0.0738)(1.14)(1.0)} = 92.7g/F_P$

Using $g = 32.2$ ft/s^2,

Packing Material	F_P, ft^2/ft^3	u_o, ft/s
Raschig rings	179	4.1
IMTP packing	41	8.5

For $f = 0.70$, using (6-99),

Packing Material	$u_G = u_o f$, ft/s	D_T, in.
Raschig rings	2.87	16.5
IMTP packing	5.95	11.5

From Figure 6.36a, for $F_{LG} = 0.092$ and $Y = 0.70^2(0.125) = 0.0613$, the pressure drop is 0.88 in. of water head per foot of packed height for both packings.

Based on these results, the IMTP packing has a much greater capacity than the Raschig rings, since the required column cross-sectional area is reduced by about 50%. ∎

Experimental flooding-point data for a variety of packing materials are in reasonable agreement with the upper curve of the GPDC chart of Figure 6.36. Unfortunately, such good agreement is not always the case for pressure drop, particularly for operation at superficial vapor velocities above 50% of flooding, where pressure drop is greater than 0.5 in. of water head per foot of packed height. Reasons for the difficulty of achieving a simple generalization of pressure drop measurements are discussed in detail by Kister [33]. As an example of the possible magnitude of the disparity, the predicted pressure drop of 1.5 in. of water per foot in Example 6.13 for operation with IMTP packing at 70% of flooding is in poor agreement with the value of 0.63 in. of water head per foot determined from data supplied by the packing manufacturer.

If Figure 6.36a is crossplotted as pressure drop versus Y for constant values of F_{LG}, it is found that a pressure drop of from 2.5 to 3 in. of water head per foot is predicted at the flooding condition for all packings. However, studies by Kister and Gill [33,50] for both random and structured packings show that the pressure drop at flooding is strongly dependent on the packing factor, F_P, by the empirical expression

$$\Delta P_{\text{flood}} = 0.115 F_P^{0.7} \qquad \text{(6-100)}$$

where ΔP_{flood} has units of inches of water head per foot of packed height and F_P has units of ft^2/ft^3. As seen in Table 6.8, the range of F_P is from about 10 to 100. Thus, (6-99) predicts pressure drops at flooding from as low as 0.6 to as high as 3 in. of water head per foot of packed height. Kister and Gill also give an interpolation procedure for estimating pressure drop, which utilizes experimental data in conjunction with a GPDC-type plot.

Theoretically based models for predicting pressure drop in packed beds with countercurrent gas/liquid flows have been presented by Stichlmair et al. [46], who use a particle model, and Billet and Schultes [51], who use a channel model. Both models extend well accepted equations for dry-bed pressure drop to account for the effect of liquid holdup. When a gas flows through a packed column under conditions of no liquid flow, a correlation for the pressure drop can be obtained in a manner similar to that for flow through an empty, straight pipe, by plotting a modified friction factor against a modified Reynolds number as shown in Figure 6.37 from the widely used study by Ergun [52]. In this plot,

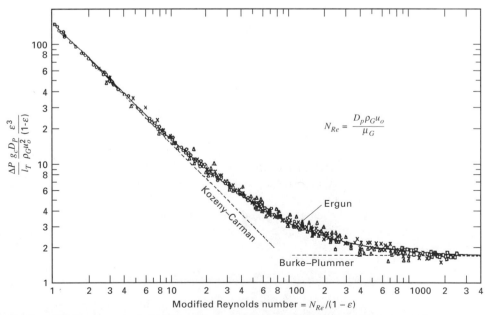

Figure 6.37 Ergun correlation for dry-bed pressure drop. [From S. Ergun, *Chem. Eng. Prog.* **48** (2), 89–94 (1952) with permission.]

in which D_P is an effective packing material diameter, it can be seen that at low superficial gas velocities (modified $N_{Re} < 10$), typical of laminar flow, the pressure drop per unit height is proportional to the gas velocity. At high gas velocities, typical of turbulent flow, the pressure drop per height approaches a dependency of the square of the gas velocity. Most packed columns used for separations operate in the turbulent region (modified $N_{Re} > 1,000$). Thus, dry pressure drop data shown in Figure 6.34 for Bialecki rings show an exponential dependency on gas velocity of about 1.86. Also, as shown in Figure 6.34, when liquid flows countercurrent to the gas in the preloading region, this same dependency continues, but at a higher pressure drop because the volume for gas flow decreases due to liquid holdup.

Based on extensive experimental studies using 54 different packing materials, including structured packings, Billet and Schultes [51] developed a correlation for dry-gas pressure drop, ΔP_o, similar in form to that of Figure 6.37. Their dimensionally consistent correlating equation is

$$\frac{\Delta P_o}{l_T} = \Psi_o \frac{a}{\epsilon^3} \frac{u_o^2 \rho_G}{2g_c} \frac{1}{K_W} \tag{6-101}$$

where
$$l_T = \text{height of packing}$$
$$K_W = \text{a wall factor}$$

K_W can be important for columns with an inadequate ratio of effective packing diameter to inside column diameter, and is given by

$$\frac{1}{K_W} = 1 + \frac{2}{3}\left(\frac{1}{1-\epsilon}\right)\frac{D_P}{D_T} \tag{6-102}$$

where the effective packing diameter, D_P, is determined from

$$D_P = 6\left(\frac{1-\epsilon}{a}\right) \tag{6-103}$$

The dry-packing resistance coefficient (a modified friction factor), Ψ_o, is given by the empirical expression

$$\Psi_o = C_p\left(\frac{64}{N_{Re_G}} + \frac{1.8}{N_{Re_G}^{0.08}}\right) \tag{6-104}$$

where
$$N_{Re_G} = \frac{u_o D_P \rho_G}{(1-\epsilon)\mu_G} K_W \tag{6-105}$$

and C_p is a packing constant, determined from experimental data, and tabulated for a number of packings in Table 6.8. In (6-104), the laminar-flow region is characterized by the term $64/N_{Re_G}$, while the next term characterizes the more common turbulent-flow regime.

When the packed tower is irrigated, the liquid holdup causes the pressure drop to increase. The experimental data are reasonably well correlated by

$$\frac{\Delta P}{\Delta P_o} = \left(\frac{\epsilon - h_L}{\epsilon}\right)^{1.5} \exp\left(\frac{N_{Re_L}}{200}\right) \tag{6-106}$$

where N_{Re_L} is defined by (6-94). For operation in the preloading region, the liquid holdup can be estimated from (6-93) to (6-97).

Mass Transfer Efficiency

The mass transfer efficiency of a packed column is incorporated in the HETP or the more theoretically based HTUs and volumetric mass transfer coefficients. Although the HETP concept lacks a sound theoretical basis, its simplicity, coupled with the relative ease with which equilibrium-stage calculations can be made with computer-aided simulation programs, has made it a widely used method for estimating packing height. In the preloading region and where good distribution of vapor and liquid is initiated and maintained, values of the HETP depend mainly on packing type and size, liquid viscosity, and surface tension. For rough estimates the following relations, taken from Kister [33], can be used.

1. Pall rings and similar high-efficiency random packings with low-viscosity liquids:

$$\text{HETP, ft} = 1.5 D_P, \text{in.} \tag{6-107}$$

2. Structured packings at low-to-moderate pressure with low-viscosity liquids:

$$\text{HETP, ft} = 100/a, \text{ft}^2/\text{ft}^3 + 4/12 \tag{6-108}$$

3. Absorption with viscous liquid:

$$\text{HETP} = 5 \text{ to } 6 \text{ ft}$$

4. Vacuum service:

$$\text{HETP, ft} = 1.5 D_P, \text{in.} + 0.5 \tag{6-109}$$

5. High-pressure service (> 200 psia):

$$\text{HETP for structured packings may be greater than predicted by (6-108)}$$

6. Small-diameter columns, $D_T < 2$ ft:

$$\text{HETP, ft} = D_T, \text{ft, but not less than 1 ft}$$

In general, lower values of HETP are achieved with smaller-size random packings, particularly in small-diameter columns, and with structured packings, particularly those with large values of a, the packing surface area per packed volume. The experimental data of Figure 6.38 for no. 2 (2-in.-diameter) Nutter rings from Kunesh [53] show that in the preloading region, the HETP is relatively independent of the vapor-flow F-factor:

$$F = u_o (\rho_G)^{0.5} \tag{6-110}$$

provided that the ratio L/G is maintained constant as the superficial gas velocity, u_o, is increased. Beyond the loading point, and as the flooding point is approached, the HETP can increase dramatically like the pressure drop and liquid holdup.

Experimental mass transfer data for packed columns are usually correlated in terms of volumetric mass transfer coefficients and/or HTUs, rather than in terms of HETPs. The data are obtained from experiments in which either the liquid-phase or the gas-phase mass transfer resistance is negligible, so that the other resistance can be studied and correlated independently. For applications where both resistances may be important, the two resistances are added together according to the two-film theory of Whitman [54], as discussed in Chapter 3, to obtain the overall resistance. This theory assumes the absence of any mass transfer resistance at the interface between the gas and liquid phases. Thus, the two phases are in equilibrium at the interface.

The two-film theory defines an overall coefficient in terms of the individual volumetric mass transfer coefficients discussed in Section 6.7. Most commonly, reference is made to

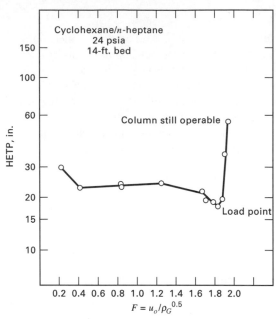

Figure 6.38 Effect of F-factor on HETP.

the overall gas-phase resistance, (6-80),

$$\frac{1}{K_y a} = \frac{1}{k_y a} + \frac{K}{k_x a}$$

for mass transfer rates expressed in terms of mole-fraction driving forces by (6-73),

$$r = k_y a(y - y_I) = k_x a(x_I - x) = K_y a(y - y^*)$$

where K is the vapor–liquid equilibrium ratio.

Alternatively, as summarized in Table 6.7, mass transfer rates can be expressed in terms of liquid-phase concentrations and gas-phase partial pressure

$$r = k_g a(p - p_I) = k_L a(c_I - c) = K_G a(p - p^*) \tag{6-111}$$

If we define a Henry's law constant at the equilibrium interface between the two phases by

$$p_I = H' c_I \tag{6-112}$$

and let

$$p^* = H' c \tag{6-113}$$

then

$$\frac{1}{K_G a} = \frac{1}{k_g a} + \frac{H'}{k_L a} \tag{6-114}$$

Alternatively, expressions can be derived for $K_x a$ and $K_L a$.

It should be noted that the units of various mass transfer coefficients differ:

	SI Units	**American Engineering Units**
r	mol/m^3-s	lbmol/ft^3-h
$k_y a, k_x a, K_x a, K_y a$	mol/m^3-s	lbmol/ft^3-h
$k_g a, K_G a$	mol/m^3-s-kPa	lbmol/ft^3-h-atm
$k_L a, k_G a$	s^{-1}	h^{-1}
k_L, k_G	m/s	ft/h

Instead of using mass transfer coefficients directly for column design, the transfer unit concept of Chilton and Colburn [43,44] is often employed because HTUs: (1) have only one dimension (length), (2) generally vary with column conditions less than mass transfer coefficients, and (3) are related to an easily understood geometrical quantity, namely, height per theoretical stage. Definitions of individual and overall HTUs are included in Table 6.7 for the dilute case. By substituting these definitions into (6-80),

$$H_{OG} = H_G + (KG/L)H_L \qquad \textbf{(6-115)}$$

Alternatively, an expression can be derived for H_{OL}. In the absorption or stripping of very insoluble gases, the solute K-value or Henry's law constant, H' is very large, making the last terms in (6-80), (6-114) and (6-115) large such that the resistance of the gas phase is negligible and the rate of mass transfer is controlled by the liquid phase. Such data can then be used to study the effect of the variables on the volumetric liquid-phase mass transfer coefficient and HTU. Typical data are shown in Figure 6.39 for three different size Berl-saddle packings for the stripping of oxygen from water by air, in a 20-in.-I.D. column operated at near-ambient temperature and pressure in the preloading region, as reported in an early study by Sherwood and Holloway [55]. The effect of liquid velocity on k_La is seen to be quite pronounced, with k_La increasing at about the 0.75 power of the liquid mass velocity. Gas velocity was observed to have no effect on k_La in the preloading region. Also included in Figure 6.39 are the data plotted in terms of H_L, where

$$H_L = \frac{M_L L}{\rho_L k_L a S} \qquad \textbf{(6-116)}$$

As seen, H_L does not depend as strongly as k_La on liquid velocity.

Another system for which the rate of mass transfer is controlled by the liquid phase is CO_2–air–H_2O, where CO_2 can be either absorbed or stripped. Measurements on this system for a variety of modern metal, ceramic, and plastic packings are reported by Billet [45]. Data on the effect of liquid velocity on k_La in the preloading region for two different size ceramic Hiflow ring packings are shown in Figure 6.40. The effect of gas velocity on

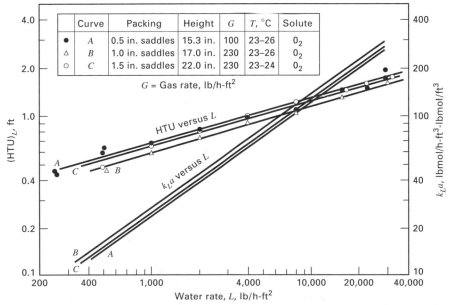

Curve	Packing	Height	G	T, °C	Solute
A	0.5 in. saddles	15.3 in.	100	23–26	O_2
B	1.0 in. saddles	17.0 in.	230	23–26	O_2
C	1.5 in. saddles	22.0 in.	230	23–24	O_2

G = Gas rate, lb/h-ft^2

Figure 6.39 Effect of liquid rate on liquid-phase mass transfer of O_2. [From T.K. Sherwood and F.A.L. Hollowan *Trans. AIChE.,* **36,** 39–70 (1940) with permission.]

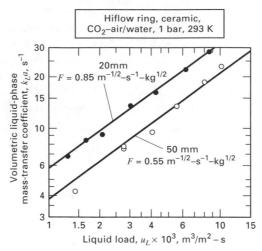

Figure 6.40 Effect of liquid rate on liquid-phase mass transfer of CO_2. [From R. Billet, *Packed Column Analysis and Design*, Ruhr-University Bochum (1989) with permission.]

$k_L a$ in terms of the F-factor at a constant liquid rate is shown in Figure 6.41 for the same system, but with 50-mm plastic Pall rings and Hiflow rings. Up to an F-factor value of about 1.8 $m^{-1/2}$-s^{-1}-$kg^{1/2}$, which is in the preloading region, no effect of gas velocity is observed. Above the loading limit, $k_L a$ increases with increasing gas velocity because of increased liquid holdup, which increases interfacial surface area for mass transfer. Although it is not illustrated in Figures 6.39 to 6.41, another major factor that influences the rate of mass transfer in the liquid phase is the solute molecular diffusivity in the solvent. For a given packing, experimental data on different systems in the preloading region can usually be correlated satisfactorily by the following empirical expression, which includes only the liquid velocity and liquid diffusivity:

$$k_L a = C_1 D_L^{1/2} u_L^n \tag{6-117}$$

where n has been observed by different investigators to vary from about 0.6 to 0.95, with 0.75 being a typical value. The exponent on the diffusivity is consistent with the penetration theory discussed in Chapter 3.

A convenient system for studying gas-phase mass transfer is NH_3–air–H_2O. The high solubility of NH_3 in H_2O, corresponds to a relatively low K-value. Accordingly, the last

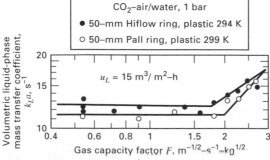

Figure 6.41 Effect of gas rate on liquid-phase mass transfer of CO_2. [From R. Billet, *Packed Column Analysis and Design*, Ruhr-University Bochun (1989) with permission.]

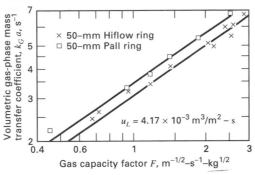

Figure 6.42 Effect of gas rate on gas-phase mass transfer of NH_3. [From R. Billet, *Packed Column Analysis and Design*, Ruhr-University Bochum (1989) with permission.]

terms in (6-80), (6-114), and (6-115) may be negligible so that the gas-phase resistance controls the rate of mass transfer. The small effect of the liquid-phase resistance can be backed out using a correlation such as (6-117). The typical effect of superficial vapor velocity, expressed in terms of the F-factor of (6-110), on the volumetric gas-phase mass transfer coefficient in the preloading region is shown in Figure 6.42 for two different plastic packings at the same liquid velocity. The coefficients are proportional to about the 0.75 power of F. Figure 6.43 shows that the liquid velocity also affects the gas-phase mass transfer coefficient, probably because as the liquid rate is increased, the holdup increases and more interfacial surface is created.

The volumetric gas-phase mass transfer coefficients, $k_G a$, plotted in Figures 6.42 and 6.43, are based on gas-phase molar concentrations. Thus, they have the same units as $k_L a$. For a given packing, experimental data on $k_g a$ or $k_G a$ for different systems in the preloading region can usually be correlated satisfactorily with empirical correlations of the form

$$k_g a = C_2 D_G^{0.67} F^{m'} u_L^{n'} \tag{6-118}$$

where D_G is the gas diffusivity and m' and n' have been observed by different investigators to vary from 0.65 to 0.85 and from 0.25 to 0.5, respectively, a typical value for m' being 0.8.

The development of separate generalized correlations for gas- and liquid-phase mass transfer coefficients and/or HTUs, which began with the study of Sherwood and Holloway [55] on the liquid phase, has led to a significant number of empirical and semitheoretical

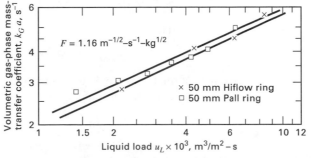

Figure 6.43 Effect of liquid rate on gas-phase mass transfer of NH_3. [From R. Billet, *Packed Column Analysis and Design*, Ruhr-University Bochum (1989) with permission.]

Table 6.9 Generalized Correlations for Mass Transfer in Packed Columns

Investigator	Year	Ref. No.	Type of Correlations	Packings
Shulman et al.	1955	64	k_g, k_L, a	Raschig rings, Berl saddles
Cornell et al.	1960	56, 57	H_G, H_L	Raschig rings, Berl saddles
Onda et al.	1968	65	k_g, k_L, a	Raschig rings, Berl saddles
Bolles and Fair	1979, 1982	58, 59	H_G, H_L	Raschig rings, Berl saddles, Pall rings
Bravo and Fair	1982	60	a	Raschig rings, Berl saddles, Pall rings
Bravo et al.	1985	61	k_G, k_L	Sulzer
Fair and Bravo	1987	62	k_G, k_L, a	Sulzer, Gempak, Mellapak, Montz, Ralu Pak
Fair and Bravo	1991	63	k_G, k_L, a	Flexipac, Gempak, Intalox 2T Montz, Mellapak, Sulzer
Billet and Schultes	1991	67	$k_G a, k_L a$	14 random packings and 4 structured packings

equations, most of which are based on the application of the two-film theory by Fair and co-workers [56–63] and others [64,65]. In some cases, values of k_G and k_L are correlated separately from a; in others, the combinations $k_G a$ and $k_L a$ are used. Important features of some of these correlations are summarized in Table 6.9. The development of such correlations from experimental data is difficult because, as shown by Billet [66] in a comprehensive study with metal Pall rings, values of the mass transfer coefficients are significantly affected by the technique used to pack the column and the number of liquid feed distribution points per unit of column cross section, when this number is less than 10 points/ft^2. When 25 points/ft^2 are used and $D_T/D_P > 10$, column diameter has little, if any, effect on mass transfer coefficients for packed heights up to 20 ft.

In an extensive investigation, Billet and Schultes [67] measured and correlated volumetric mass transfer coefficients and HTUs for 31 different binary and ternary chemical systems with 67 different types and sizes of packings in columns of diameter ranging from 2.4 in. to 4.6 ft. The systems include some for which mass transfer resistance resides mainly in the liquid phase and others for which resistance in the gas phase is predominant. They assume uniform distribution of gas and liquid over the cross-sectional area of the column and apply the two-film theory of mass transfer discussed in Chapter 3. For the liquid-phase resistance, they assume that the liquid flows in a thin film through the irregular channels of the packing, with continual remixing of the liquid at points of contact with the packing such that Higbie's penetration theory of diffusion [68], as developed in Chapter 3, can be applied. Thus, for the diffusing component, in terms of concentration units, the volumetric mass transfer coefficient is defined by

$$r = (k_L a_h)(c_{L_1} - c_L) \tag{6-119}$$

From the penetration theory of Higbie,

$$k_L = 2(D_L/\pi t_L)^{0.5} \tag{6-120}$$

where t_L = time of exposure of the liquid film before remixing. Billet and Schultes assume that this time is governed by a length of travel equal to the hydraulic diameter of the packing:

$$t_L = h_L d_H/u_L \tag{6-121}$$

where d_H, the hydraulic diameter, is equal to $4r_H$ or $4\epsilon/a$. Thus, in terms of the height of a liquid transfer unit, (6-120) and (6-121) give

$$H_L = \frac{u_L}{k_L a_h} = \frac{\sqrt{\pi}}{2}\left(\frac{4h_L\epsilon}{D_L a u_L}\right)^{1/2}\frac{u_L}{a_h} \tag{6-122}$$

Equation (6-122) was modified to include an empirical constant, C_L, which is back-calculated for each packing to fit the experimental data. The final predictive equation given by Billet and Schultes is

$$H_L = \frac{1}{C_L}\left(\frac{1}{12}\right)^{1/6}\left(\frac{4h_L\epsilon}{D_Lau_L}\right)^{1/2}\frac{u_L}{a_h} \tag{6-123}$$

where values of C_L are included in Table 6.8.

A similar development was made by Billet and Schultes for the gas-phase resistance, except that the time of exposure of the gas between periods of mixing was determined empirically, to give

$$H_G = \frac{1}{C_V}(\epsilon - h_L)^{1/2}\left(\frac{4\epsilon}{a^4}\right)^{1/2}(N_{Re_G})^{-3/4}(N_{Sc_G})^{-1/3}\left(\frac{u_oa}{D_Ga_h}\right) \tag{6-124}$$

where C_V is included in Table 6.8 and

$$N_{Re_G} = \frac{u_o\rho_G}{a\mu_G} \tag{6-125}$$

$$N_{Sc_G} = \frac{\mu_G}{\rho_GD_G} \tag{6-126}$$

Following the estimation of H_L and H_G from (6-123) and (6-124), respectively, the overall HTU value can be determined from (6-115), followed by the determination of packed height from

$$l_T = H_{OG}N_{OG} \tag{6-127}$$

where the determination of N_{OG} is discussed in Section 6.7.

EXAMPLE 6.14

For the absorption of ethyl alcohol from CO_2 with water, as considered in Example 6.1, a 2.5-ft-I.D. tower, packed with 1.5-in. metal Pall-like Rings, is to be used. It is estimated that the tower will operate in the preloading region with a pressure drop of approximately 1.5 in. of water head per foot of packed height. From Example 6.9, the required number of overall transfer units based on the gas phase is 7.5. Estimate H_G, H_L, H_{OG}, HETP, and the required packed height in feet using the following estimates of flow conditions and physical properties at the bottom of the packing:

	Vapor	Liquid
Flow rate, lb/h	17,480	6,140
Molecular weight	44.05	18.7
Density, lb/ft^3	0.121	61.5
Viscosity, cP	0.0145	0.63
Surface tension, dynes/cm	—	101
Diffusivity of ethanol, cm^2/s	0.0775	1.82×10^{-5}
Kinematic viscosity, m^2/s	0.75×10^{-5}	0.64×10^{-6}

SOLUTION

Cross-sectional area of tower = $(3.14)(2.5)^2/4 = 4.91$ ft^2.
Volumetric liquid flow rate = $6,140/61.5 = 99.8$ ft^3/h.
u_L = superficial liquid velocity = $99.8/[(4.91)(3,600)] = 0.0056$ ft/s or 0.0017 m/s.
From Section 6.8, $u_L > u_{L,min}$.
u_o = superficial gas velocity = $17,480/[(0.121)(4.91)(3,600)] = 8.17$ ft/s = 2.49 m/s.
Let the packing characteristics for the 1.5-inch metal Pall-like rings be as follows (somewhat different from values for Pall rings in Table 6.8):
$a = 149.6$ m^2/m^3 $\epsilon = 0.952$
C_h = approximately 0.7 $C_L = 1.227$ $C_V = 0.341$

Estimation of specific liquid holdup, h_L:

$$\text{From (6-94), } N_{\text{Re}_L} = \frac{0.0017}{(0.64 \times 10^{-6})(149.6)} = 17.8.$$

$$\text{From (6-95), } N_{\text{Fr}_L} = \frac{(0.01)^2(149.6)}{9.8} = 0.00153$$

$$\text{From (6-97), } \frac{a_h}{a} = 0.85\,(0.7)(17.8)^{0.25}(0.00153)^{0.10} = 0.64$$

$$a_h = 0.64(149.6) = 95.7 \text{ m}^2/\text{m}^3$$

$$\text{From (6-93), } h_L = \left[\frac{12(0.00153)}{17.8}\right]^{1/3}(0.64)^{2/3} = 0.075 \text{ m}^3/\text{m}^3$$

Estimation of H_L: From (6-123), using consistent SI units,

$$H_L = \frac{1}{1.227}\left(\frac{1}{12}\right)^{1/6}\left[\frac{(4)(0.075)(0.952)}{(1.82 \times 10^{-9})(149.6)(0.0017)}\right]^{1/2}\left(\frac{0.0017}{95.7}\right) = 0.24 \text{ m} = 0.79 \text{ ft}$$

Estimation of H_G:

$$\text{From (6-125), } N_{\text{Re}_G} = 2.49/[(149.6)(0.75 \times 10^{-5})] = 2,220$$

$$\text{From (6-126), } N_{\text{Sc}_G} = 0.75 \times 10^{-5}/0.0775 \times 10^{-4} = 0.968$$

From (6-124), using consistent SI units,

$$H_G = \frac{1}{0.341}(0.952 - 0.075)^{1/2}\left[\frac{(4)(0.952)}{(149.6)^4}\right]^{1/2}(2,220)^{-3/4}(0.968)^{-1/3}\left[\frac{(2.49)(149.6)}{(0.0775 \times 10^{-4})(95.7)}\right]$$

$$= 0.38 \text{ m} \quad \text{or} \quad 1.25 \text{ ft}$$

Estimation of H_{OG}: From Example 6.1, the K-value for ethyl alcohol = 0.57, $G = 17,480/44.05 = 397$ lbmol/h, $L = 6,140/18.7 = 328$ lbmol/h, and $1/A = KG/L = (0.57)(397)/328 = 0.69$.

$$\text{From (6-115), } H_{OG} = 1.25 + 0.69\,(0.79) = 1.55 \text{ ft}$$

Therefore, mass transfer resistances in both phases are important, but the resistance in the gas phase is larger.

Estimation of Packed Height:

$$\text{From (6-127), } l_T = 1.55(7.5) = 11.6 \text{ ft}$$

Estimation of HETP: From (6-90), for straight operating and equilibrium lines,

and
$$A = 1/069 = 1.45,$$

$$\text{HETP} = 1.55\,\frac{\ln(0.69)}{(1 - 1.45)/1.45} = 1.85 \text{ ft} \qquad \blacksquare$$

6.9 CONCENTRATED SOLUTIONS IN PACKED COLUMNS

When the solute concentration in the gas and/or liquid is concentrated so that the operating line and/or equilibrium line are noticeably curved, then the procedure given in Section 6.7 for determining N_{OG} and l_T cannot be used because (6-87) cannot be analytically integrated to give (6-89). Instead, alternative methods can be employed or the computer-aided methods discussed in Chapters 10 and 11 can be applied.

For concentrated solutions, the two columns in Table 6.7 labeled UM (unimolecular) diffusion apply. To obtain these columns from the two columns labeled EM (equimolar) diffusion, we let:

$$L' = L(1 - x) \quad \text{and} \quad G' = G(1 - y)$$

where L' and G' are the constant flow rates of the inert (solvent) liquid and (carrier) gas, respectively on a solute-free basis. Then

$$d(Gy) = G'd\left(\frac{y}{1-y}\right) = G'\frac{dy}{(1-y)^2} = G\frac{dy}{(1-y)} \tag{6-128}$$

$$d(Lx) = L'd\left(\frac{x}{1-x}\right) = L'\frac{dx}{(1-x)^2} = L\frac{dx}{(1-x)} \tag{6-129}$$

Equation (6-84) now becomes

$$l_T = \int_{y_2}^{y_1}\left(\frac{G}{K_y'aS}\right)\frac{dy}{(1-y)(y-y^*)} = \frac{G}{K_y'aS}\int_{y_2}^{y_1}\frac{dy}{(1-y)(y-y^*)} \tag{6-130}$$

where 1 refers to inlet and 2 refers to outlet conditions. Based on the liquid phase,

$$l_T = \int_{x_1}^{x_2}\left(\frac{L}{K_x'aS}\right)\frac{dx}{(1-x)(x^*-x)} = \frac{L}{K_x'aS}\int_{x_1}^{x_2}\frac{dx}{(1-x)(x^*-x)} \tag{6-131}$$

where the mass transfer coefficients are primed to signify UM diffusion.

If the numerators and denominators of (6-130) and (6-131) are multiplied by $(1-y)_{LM}$ and $(1-x)_{LM}$, respectively, where $(1-y)_{LM}$ is the log mean of $(1-y)$ and $(1-y^*)$, and $(1-x)_{LM}$ is the log mean of $(1-x)$ and $(1-x^*)$, we obtain the expressions in rows 1 and 6 of columns 4 and 7 in Table 6.7:

$$l_T = \int_{y_2}^{y_1}\left[\frac{G}{K_y'a(1-y)_{LM}S}\right]\frac{(1-y)_{LM}\,dy}{(1-y)(y-y^*)} = \frac{G}{K_y'a(1-y)_{LM}S}\int_{y_2}^{y_1}\frac{(1-y)_{LM}\,dy}{(1-y)(y-y^*)} \tag{6-132}$$

$$l_T = \int_{x_1}^{x_2}\left[\frac{L}{K_x'a(1-x)_{LM}S}\right]\frac{(1-x)_{LM}\,dx}{(1-x)(x^*-x)} = \frac{L}{K_x'a(1-x)_{LM}S}\int_{x_1}^{x_2}\frac{(1-x)_{LM}\,dx}{(1-x)(x^*-x)} \tag{6-133}$$

In these equations $K_y'(1-y)_{LM}$ is equal to the concentration-independent K_y, and $K_x'(1-x)_{LM}$ is equal to the concentration-independent K_x. If there is appreciable absorption, G decreases from the bottom to the top of the absorber. However, the values of Ka are also a function of flow rate, such that the ratio G/Ka is approximately constant and HTU groupings, $[L/K_x'a(1-x)_{LM}S]$ and $[G/K_y'a(1-y)_{LM}S]$, can often be taken out of the integral sign without incurring errors larger than those inherent in experimental measurements of Ka. Usually, average values of G, L, and $(1-y)_{LM}$ are used.

Another approach is to leave all of the terms in (6-132) or (6-133) under the integral sign and evaluate l_T by a stepwise or graphical integration. In either case, to obtain the terms $(y-y^*)$ or (x^*-x), the equilibrium and operating lines must be established. The equilibrium curve is determined from appropriate thermodynamic data or correlations. To establish the operating line, which will not be straight if the solutions are concentrated, the appropriate material balance equations must be developed. With reference to Figure 6.29, an overall balance around the upper part of the absorber gives

$$G + L_{in} = G_{out} + L \tag{6-134}$$

Similarly a balance around the upper part of the absorber for the component being absorbed, assuming a pure liquid absorbent, gives:

$$Gy = G_{out}y_{out} + Lx \tag{6-135}$$

An absorbent balance around the upper part of the absorber is:

$$L_{in} = L(1-x) \tag{6-136}$$

Combining (6-134) to (6-136) to eliminate G and L gives

$$y = \frac{G_{out}y_{out} + [L_{in}x/(1-x)]}{G_{out} + [L_{in}x/(1-x)]} \tag{6-137}$$

Equation (6-137) allows the y–x operating line to be calculated from a knowledge of terminal conditions only.

A simpler approach to the problem of concentrated gas or liquid mixtures is to linearize the operating line by expressing all concentrations in mole ratios, and the gas and liquid flows on a solute-free basis, that is, $G' = (1 - y)G$, $L' = (1 - x)L$. Then, in place of (6-132) and (6-133), we have

$$l_T = \int_{Y_2}^{Y_1} \left(\frac{G'}{K_Y a S}\right) \frac{dY}{(Y - Y^*)} = \frac{G'}{K_Y a S} \int_{Y_2}^{Y_1} \frac{dY}{(Y - Y^*)} \tag{6-138}$$

$$l_T = \int_{X_1}^{X_2} \left(\frac{L'}{K_X a S}\right) \frac{dX}{(X^* - X)} = \frac{L'}{K_X a S} \int_{X_1}^{X_2} \frac{dX}{(X^* - X)} \tag{6-139}$$

This set of equations is listed in rows 3 and 8 of Table 6.7.

EXAMPLE 6.15

To remove 95% of the ammonia from an air stream containing 40% ammonia by volume, 488 lbmol/h of an absorbent per 100 lbmol/h of entering gas are to be used, which is greater than the minimum requirement.

Equilibrium data are given in Figure 6.44. Pressure is 1 atm and temperature is assumed constant at 298 K. Calculate the number of transfer units by:

(a) Equation (6-132) using a curved operating line determined from (6-137)
(b) Equation (6-138) using mole ratios.

SOLUTION

(a) Take as a basis $L_{in} = 488$. Then $G_{out} = 100 - (40)(0.95) = 62$, and $y_{out} = (0.05)(40)/62 = 0.0323$. From (6-137), it is possible to construct the curved operating line of Figure 6.44. For example, if $x = 0.04$,

$$y = \frac{(62)(0.0323) + [(488)(0.04)/(1 - 0.04)]}{62 + [(488)(0.04)/(1 - 0.04)]} = 0.27$$

It is now possible to calculate the following values of y, y^*, $(1 - y)_{LM} = [(1 - y) - (1 - y^*)]/\ln[(1 - y)/(1 - y^*)]$, and $(1 - y)_{LM}/[(1 - y)(y - y^*)]$ for use in (6-132).

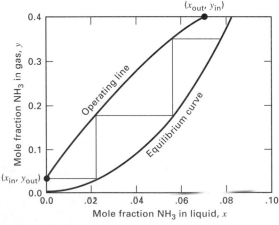

Figure 6.44 Determination of the number of theoretical stages for Example 6.15.

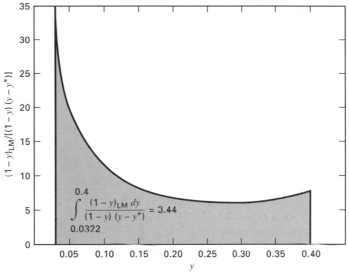

Figure 6.45 Determination of the number of transfer units for Example 6.15.

y	y^*	$(y - y^*)$	$(1 - y)$	$(1 - y)_{LM}$	$\dfrac{(1 - y)_{LM}}{(1 - y)(y - y^*)}$
0.03	0.002	0.028	0.97	0.99	36.47
0.05	0.005	0.045	0.95	0.97	22.68
0.10	0.01	0.09	0.90	0.94	11.60
0.15	0.025	0.125	0.85	0.91	8.56
0.20	0.04	0.16	0.80	0.89	6.95
0.25	0.08	0.17	0.75	0.85	6.66
0.30	0.12	0.18	0.70	0.82	6.51
0.35	0.17	0.18	0.65	0.73	6.24
0.40	0.26	0.14	0.60	0.67	7.97

Note that since $(1 - y) \approx (1 - y)_{LM}$, these two terms frequently cancel out of the NTU equations, particularly when y is small.

Figure 6.45 is a plot of y versus $(1 - y)_{LM}/[(1 - y)(y - y^*)]$ to determine N_{OG}. The integral on the right-hand side of (6-132), between $y = 0.4$ and $y = 0.0322$, is $3.44 = N_{OG}$. This is approximately 1 more than the number of equilibrium stages of 2.6, as seen in the steps of Figure 6.44.

(b) It is a simple matter to obtain the following values for $Y = y/(1 - y)$, $Y^* = y^*/(1 - y^*)$, $(Y - Y^*)$, and $(Y - Y^*)^{-1}$.

y	Y	y^*	Y^*	$(Y - Y^*)^{-1}$
0.03	0.031	0.002	0.002	34.48
0.05	0.053	0.005	0.005	20.83
0.1	0.111	0.01	0.010	9.9
0.15	0.176	0.025	0.026	6.66
0.20	0.250	0.04	0.042	4.8
0.25	0.333	0.08	0.087	4.06
0.30	0.43	0.12	0.136	3.40
0.35	0.54	0.17	0.205	2.98
0.40	0.67	0.26	0.310	2.78

Graphical integration of the right-hand-side integral of (6-138) is carried out by determining the area under the curve of $(Y - Y^*)^{-1}$ versus Y between $Y = 0.67$ and $Y = 0.033$. The result is $N_{OG} = 3.46$. Alternatively, the numerical integration can be performed on a computer with a spreadsheet.

It must be pointed out that for concentrated solutions, the assumption of constant temperature may not be valid and can result in a large error. If an overall energy balance indicates a temperature change that alters the equilibrium curve significantly, it is best to use a computer-aided method that includes the energy balance. Such methods are presented in Chapter 10.

∎

SUMMARY

1. A liquid can be used to selectively absorb one or more components from a gas mixture. A gas can be used to selectively desorb or strip one or more components from a liquid mixture.

2. The fraction of a component that can be absorbed or stripped in a countercurrent cascade depends on the number of equilibrium stages and the absorption factor, $A = L/KV$, or the stripping factor, $S = KV/L$, respectively.

3. Absorption and stripping are most commonly conducted in trayed towers equipped with sieve or valve trays, or in towers packed with random or structured packings.

4. Absorbers are most effectively operated at high pressure and low temperature. The reverse is true for stripping. However, high costs of gas compression, refrigeration, and vacuum often preclude operation at the most thermodynamically favorable conditions.

5. For a given gas flow rate and composition, a desired degree of absorption of one or more components, a choice of absorbent, and an operating temperature and pressure, there is a minimum absorbent flow rate, given by (6-9) to (6-11), that corresponds to the use of an infinite number of equilibrium stages. For the use of a finite and reasonable number of stages, an absorbent rate of 1.5 times the minimum is typical. A similar criterion, (6-12), holds for a stripper.

6. The number of equilibrium stages required for a selected absorbent or stripping agent flow rate for the absorption or stripping of a dilute solution can be determined from the equilibrium line, (6-1), and an operating line, (6-3) or (6-5), using graphical, algebraic, or numerical methods. Graphical methods, such as Figure 6.11, offer considerable visual insight into stage-by-stage changes in compositions of the gas and liquid streams.

7. Rough estimates of overall stage efficiency, defined by (6-21), can be made with the correlations of Drickamer and Bradford, (6-22), O'Connell, (6-23), and Figure 6.15. More accurate and reliable procedures involve the use of a small Oldershaw column or semitheoretical equations, e.g., of Chan and Fair, based on mass transfer considerations, to determine a Murphree vapor-point efficiency, (6-30), from which a Murphree vapor tray efficiency can be estimated from (6-31) to (6-34), which can then be related to the overall efficiency using (6-37).

8. Tray diameter can be determined from (6-44) based on entrainment flooding considerations using Figure 6.24. Tray vapor pressure drop, the weeping constraint, entrainment, and downcomer backup can be estimated from (6-45), (6-64), (6-65), and (6-66), respectively.

9. Packed column height can be estimated using the HETP, (6-69), or HTU/NTU, (6-85), concepts, with the latter having a more fundamental theoretical basis in the two-film theory of mass transfer. For straight equilibrium and operating lines, HETP is related to the HTU by (6-90), and the number of equilibrium stages is related to the NTU by (6-91).

10. Below a so-called loading point, in a preloading region, the liquid holdup in a packed column is independent of the vapor velocity. The loading point is typically about 70% of the flooding point and most packed columns are designed to operate in the preloading region at from 50% to 70% of flooding. From the GPDC chart of Figure 6.36, the flooding point can be estimated, from which the column diameter can be determined with (6-98).

11. One significant advantage of a packed column is its relatively low pressure drop per unit of packed height, as compared to a trayed tower. Packed column pressure drop can be roughly estimated from Figure 6.36 or more accurately from (6-100).

12. Numerous rules of thumb are available for estimating the HETP of packed columns. However, the preferred approach is to estimate H_{OG} from separate semitheoretical mass transfer correlations for the liquid and gas phases, such as those of (6-123) and (6-124) based on the extensive experimental work of Billet and Schultes.

13. Determination of theoretical stages for concentrated solutions involves numerical integration because of curved equilibrium and/or operating lines.

REFERENCES

1. Washburn, E.W., Ed.-in-Chief, *International Critical Tables*, McGraw-Hill, New York, Vol. III, p. 255 (1928).

2. Lockett, M., *Distillation Tray Fundamentals,* Cambridge University Press, Cambridge, UK, p. 13 (1986).

3. Okoniewski, B.A., *Chem. Eng. Prog.,* **88** (2), 89–93 (1992).

4. Sax, N.I., *Dangerous Properties of Industrial Materials,* 4th ed., Van Nostrand Reinhold, New York, pp. 440–441 (1975).

5. Lewis, W.K., *Ind. Eng. Chem.,* **14,** 492–497 (1922).

6. Drickamer, H.G., and J.R. Bradford, *Trans. AIChE,* **39,** 319–360 (1943).

7. Jackson, R.M., and T.K. Sherwood, *Trans. AIChE,* **37,** 959–978 (1941).

8. O'Connell, H.E., *Trans. AIChE,* **42,** 741–755 (1946).

9. Walter, J.F., and T.K. Sherwood, *Ind. Eng. Chem.,* **33,** 493–501 (1941).

10. Edmister, W.C., *The Petroleum Engineer,* C45–C54 (Jan. 1949).

11. Lockhart, F.J., and C.W. Leggett, in K.A. Kobe and J.J. McKetta, Jr., Ed., *Advances in Petroleum Chemistry and Refining,* Vol. 1, Interscience, New York, Vol. 1, pp. 323–326 (1958).

12. Holland, C.D., *Multicomponent Distillation,* Prentice-Hall, Englewood Cliffs, NJ, 1963.

13. Murphree, E.V., *Ind. Eng. Chem.,* **17,** 747 (1925).

14. Hausen, H., *Chem. Ing. Tech.,* **25,** 595 (1953).

15. Standart, G., *Chem Eng. Sci.,* **20,** 611 (1965).

16. Lewis, W.K., *Ind. Eng. Chem.,* **28,** 399 (1936).

17. Gerster, J.A., A.B. Hill, N.H. Hochgraf, and D.G. Robinson, "Tray Efficiencies in Distillation Columns," Final Report from the University of Delaware, American Institute of Chemical Engineers, New York (1958).

18. *Bubble-Tray Design Manual,* AIChE, New York (1958).

19. Gilbert, T.J., *Chem. Eng. Sci.,* **10,** 243 (1959).

20. Barker, P.E., and M.F. Self, *Chem. Eng. Sci.,* **17,** 541 (1962).

21. Bennett, D.L., and H.J. Grimm, *AIChE J.,* **37,** 589 (1991).

22. Oldershaw, C.F., *Ind. Eng. Chem. Anal. Ed.,* **13,** 265 (1941).

23. Fair, J.R., H.R. Null, and W.L. Bolles, *Ind. Eng. Chem. Process Des. Dev.,* **22,** 53–58 (1983).

24. Souders, M., and G.G. Brown, *Ind. Eng. Chem.,* **26,** 98–103 (1934).

25. Fair, J.R., *Petro/Chem. Eng.,* **33,** 211–218 (Sept. 1961).

26. Sherwood, T.K., G.H. Shipley, and F.A.L. Holloway, *Ind. Eng. Chem.,* **30,** 765–769 (1938).

27. *Glitsch Ballast Tray, Bulletin No. 159,* Fritz W. Glitsch and Sons, Dallas, TX (from FRI report of Sept. 3, 1958).

28. *Glitsch V-1 Ballast Tray, Bulletin No. 160,* Fritz W. Glitsch and Sons, Dallas, TX (from FRI report of Sept. 25, 1959).

29. Oliver, E.D., *Diffusional Separation Processes. Theory, Design, and Evaluation,* John Wiley and Sons, New York, pp. 320–321 (1966).

30. Bennett, D.L., R. Agrawal, and P.J. Cook, *AIChE J.,* **29,** 434–442 (1983).

31. Smith, B.D., *Design of Equilibrium Stage Processes,* McGraw-Hill, New York (1963).

32. Klein, G.F., *Chem. Eng.,* **89** (9), 81–85 (1982).

33. Kister, H.Z., *Distillation Design,* McGraw-Hill, New York (1992).

34. Lockett, M.J., *Distillation Tray Fundamentals,* Cambridge University Press, Cambridge, UK, p. 146 (1986).

35. American Institute of Chemical Engineers (AIChE), *Bubble-Tray Design Manual,* AIChE, New York, (1958).

36. Chan, H., and J.R. Fair, *Ind. Eng. Chem. Process Des. Dev.,* **23,** 814–819 (1984).

37. Chan, H., and J.R. Fair, *Ind. Eng. Chem. Process Des. Dev.,* **23,** 820–827 (1984).

38. Scheffe, R.D., and R.H. Weiland, *Ind. Eng. Chem. Res.,* **26,** 228–236 (1987).

39. Foss, A.S., and J.A. Gerster, *Chem. Eng. Prog.,* **52,** 28-J to 34-J (Jan. 1956).

40. Gerster, J.A., A.B. Hill, N.N. Hochgraf, and D.G. Robinson, "Tray Efficiencies in Distillation Columns," Final Report from University of Delaware, American Institute of Chemical Engineers (AIChE), New York (1958).

41. Fair, J.R., *Petro./Chem. Eng.,* **33** (10), 45 (1961).

42. Colburn, A.P., *Ind. Eng. Chem.,* **28,** 526 (1936).

43. Chilton, T.H., and A.P. Colburn, *Ind. Eng. Chem.,* **27,** 255–260, 904 (1935).

44. Colburn, A.P., *Trans. AIChE,* **35,** 211–236, 587–591 (1939).

45. Billet, R., *Packed Column Analysis and Design,* Ruhr-University Bochum (1989).

46. Stichlmair, J., J.L. Bravo, and J.R. Fair, *Gas Separation and Purification,* **3,** 19–28 (1989).

47. Billet, R., and M. Schultes, *Packed Towers in Processing and Environmental Technology,* translated by J. W. Fullarton, VCH Publishers, New York (1995).

48. Leva, M., *Chem. Eng. Prog. Symp. Ser.,* **50** (10), 51 (1954).

49. Leva, M., *Chem. Eng. Prog.,* **88,** (1) 65–72 (1992).

50. Kister, H.Z., and D.R. Gill, *Chem. Eng. Prog.,* **87** (2), 32–42 (1991).

51. Billet, R., and M. Schultes, *Chem. Eng. Technol.,* **14,** 89–95 (1991).

52. Ergun, S., *Chem. Eng. Prog.,* **48** (2), 89–94 (1952).

53. Kunesh, J.G., *Can. J. Chem. Eng.,* **65,** 907–913 (1987).

54. Whitman, W.G., *Chem. and Met. Eng.,* **29,** 146–148 (1923).

55. Sherwood, T.K., and F.A.L. Holloway, *Trans. AIChE.,* **36,** 39–70 (1940).

56. Cornell, D., W.G. Knapp, and J.R. Fair, *Chem. Eng. Prog.,* **56** (7) 68–74 (1960).

57. Cornell, D., W.G. Knapp, and J.R. Fair, *Chem. Eng., Prog.,* **56** (8), 48–53 (1960).

58. Bolles, W.L., and J.R. Fair, *Inst. Chem. Eng. Symp. Ser.,* **56,** 3/35 (1979).

59. Bolles, W.L., and J.R. Fair, *Chem. Eng.,* **89** (14), 109–116 (1982).

60. Bravo, J.L., and J.R. Fair, *Ind. Eng. Chem. Process Des. Devel.* **21,** 162–170 (1982).

61. Bravo, J.L., J.A. Rocha, and J.R. Fair, *Hydrocarbon Processing,* **64** (1), 56–60 (1985).

62. Fair, J.R., and J.L. Bravo, *I. Chem. E. Symp. Ser.,* **104,** A183–A201 (1987).

63. Fair, J.R., and J.L. Bravo, *Chem. Eng. Prog.,* **86** (1), 19–29 (1990).

64. Shulman, H.L., C.F. Ullrich, A.Z. Proulx, and J.O. Zimmerman, *AIChE J.,* **1,** 253–258 (1955).

65. Onda, K., H. Takeuchi, and Y.J. Okumoto, *J. Chem. Eng. Jpn.,* **1,** 56–62 (1968).

66. Billet, R., *Chem. Eng. Prog.,* **63** (9), 53–65 (1967).

67. Billet, R., and M. Schultes, *Beitrage zur Verfahrens-Und Umwelttechnik,* Ruhr-Universitat Bochum, pp. 88–106 (1991).

68. Higbie, R., *Trans. AIChE,* **31,** 365–389 (1935).

EXERCISES

Section 6.1

6.1 In any absorption operation, the absorbent is stripped to some extent depending on the *K*-value of the absorbent. In any stripping operation, the stripping agent is absorbed to some extent depending on its *K*-value. In Figure 6.1, it is seen that both absorption and stripping occur. Which occurs to the greatest extent in terms of kilomoles per hour? Should the operation be called an absorber or a stripper? Why?

6.2 Prior to 1950, only two types of commercial random packings were in common use: Raschig rings and Berl saddles. Starting in the 1950s, a wide variety of commercial random packings began to appear. What advantages do these newer packings have? By what advances in packing design and fabrication techniques were these advantages achieved? Why were structured packings introduced?

6.3 Bubble-cap trays were widely used in the design of trayed towers prior to the 1960s. Today sieve and valve trays are favored. However, bubble-cap trays are still occasionally specified, especially for operations that require very high turndown ratios. What characteristics of bubble-cap trays make it possible for them to operate satisfactorily at low vapor and liquid rates?

Section 6.2

6.4 In Example 6.3, a lean oil of 250 MW is used as the absorbent. Consideration is being given to the selection of a new absorbent. Available streams are:

	Rate, gpm	Density, lb/gal	MW
C$_5$s	115	5.24	72
Light oil	36	6.0	130
Medium oil	215	6.2	180

Which stream would you choose? Why? Which streams, if any, are unacceptable?

6.5 Volatile organic chemicals (VOCs) can be removed from water effluents by stripping in packed towers. Possible stripping agents are steam and air. Alternatively, the VOCs can be removed by carbon adsorption. The U.S. Environmental Protection Agency (EPA) has identified air stripping as the best available technology from an economic standpoint. What are the advantages and disadvantages of air compared to steam?

6.6 Prove by equations why, in general, absorbers should be operated at high pressure and low temperature, while strippers should be operated at low pressure and high temperature. Also prove, by equations, why a trade-off exists between number of stages and flow rate of the separating agent.

Section 6.3

6.7 The exit gas from an alcohol fermenter consists of an air–CO_2 mixture containing 10 mol% CO_2 that is to be absorbed in a 5.0 N solution of triethanolamine, containing 0.04 mol of carbon dioxide per mole of amine solution. If the column operates isothermally at 25°C, if the exit liquid contains 78.4% of the CO_2 in the feed gas to the absorber, and if the absorption is carried out in a six-theoretical-plate column, calculate:

(a) Moles of amine solution required per mole of feed gas
(b) Exit gas composition

Equilibrium Data

Y	0.003	0.008	0.015	0.023	0.032	0.043
X	0.01	0.02	0.03	0.04	0.05	0.06

Y	0.055	0.068	0.083	0.099	0.12	
X	0.07	0.08	0.09	0.10	0.11	

Y = moles CO_2/mole air; X = moles CO_2/mole amine solution

6.8 Ninety-five percent of the acetone vapor in an 85 vol% air stream is to be absorbed by countercurrent contact with pure water in a valve-tray column with an expected overall tray efficiency of 50%. The column will operate essentially at 20C and 101 kPa pressure. Equilibrium data for acetone–water at these conditions are:

Mole percent acetone in water	3.30	7.20	11.7	17.1
Acetone partial pressure in air, torr	30.00	62.80	85.4	103.0

Calculate:
(a) The minimum value of L'/G', the ratio of moles of water per mole of air.
(b) The number of equilibrium stages required using a value of L'/G' of 1.25 times the minimum.
(c) The concentration of acetone in the exit water. From Table 5.2 for N connected equilibrium stages, there are $2N + 2C + 5$ degrees of freedom. Specified in this problem are

Stage pressures (101 kPa)	N
Stage temperatures (20°C)	N
Feed stream composition	$C-1$
Water stream composition	$C-1$
Feed stream T, P	2
Water stream, T, P	2
Acetone recovery	1
L/G	1
	$\overline{2N + 2C + 4}$

The remaining specification is the feed flow rate, which can be taken on a basis of 100 kmol/h.

6.9 A solvent recovery plant consists of a plate column absorber and a plate column stripper. Ninety percent of the benzene (B) in the gas stream is recovered in the absorption column. Concentration of benzene in the inlet gas is 0.06 mol B/mol B-free gas. The oil entering the top of the absorber contains 0.01 mol B/mol pure oil. In the leaving liquid, X = 0.19 mol B/mol pure oil. Operating temperature is 77°F (25°C).

Open, superheated steam is used to strip benzene out of the benzene-rich oil at 110°C. Concentration of benzene in the oil = 0.19 and 0.01 (mole ratios) at inlet and outlet, respectively. Oil (pure)-to-steam (benzene-free) flow rate ratio = 2.0. Vapors are condensed, separated, and removed.

MW oil = 200   MW benzene = 78   MW gas = 32

Equilibrium Data at Column Pressures

X in Oil	Y in Gas, 25°C	Y in Steam, 110°C
0	0	0
0.04	0.011	0.1
0.08	0.0215	0.21
0.12	0.032	0.33
0.16	0.042	0.47
0.20	0.0515	0.62
0.24	0.060	0.795
0.28	0.068	1.05

Calculate:
(a) The molar flow rate ratio of B-free oil to B-free gas in the absorber; (b) The number of theoretical plates in the absorber; and (c) The minimum steam flow rate required to remove the benzene from 1 mol of oil under given terminal conditions, assuming an infinite-plates column.

6.10 A straw oil used to absorb benzene from coke-oven gas is to be steam-stripped in a sieve-plate column at atmospheric pressure to recover the dissolved benzene. Equilibrium conditions at the operating temperature are approximated by Henry's law such that, when the oil phase contains 10 mol% C_6H_6, the C_6H_6 partial pressure above the oil is 5.07 kPa. The oil may be considered nonvolatile. The oil enters containing 8 mol% benzene, 75% of which is to be recovered. The steam leaving contains 3 mol% C_6H_6. (a) How many theoretical stages are required? (b) How many moles of steam are required per 100 mol of oil–benzene mixture? (c) If 85% of the benzene is to be recovered with the same oil and steam rates, how many theoretical stages are required?

Section 6.4

6.11 Groundwater at a flow rate of 1,500 gpm, containing three volatile organic chemicals (VOCs), is to be stripped in a trayed tower with air to produce drinking water that will meet EPA standards. Relevant data are given below. Determine the maximum air flow rate in scfm (60F, 1 atm) and the number of equilibrium stages required if an air flowrate of twice the minimum is used and the tower operates at 25°C and 1 atm. Also determine the composition in parts per million for each VOC in the resulting drinking water.

		Concentration, ppm	
Component	**K-value**	**Groundwater**	**Max. for Drinking water**
1,2- Dichloroethane (DCA)	60	85	0.005
Trichloroethylene (TCE)	650	120	0.005
1,1,1-Trichloroethane (TCA)	275	145	0.200

Note: ppm = parts per million by weight.

6.12 Sulfur dioxide and butadienes (B3 and B2) are to be stripped with nitrogen from the liquid stream as shown in Figure 6.46 so that butadiene sulfone (BS) product will contain less than 0.05 mol% SO_2 and less than 0.5 mol% butadienes. Estimate the flow rate of nitrogen, N_2, and the number of equilibrium stages required. At 70°C, K-values for SO_2, B2, B3, and BS are, respectively, 6.95, 3.01, 4.53, and 0.016.

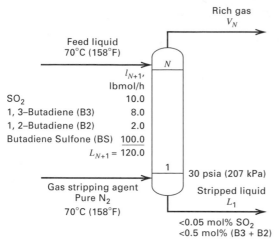

Figure 6.46 Data for Exercise 6.12.

	lbmol/h
SO_2	10.0
1, 3–Butadiene (B3)	8.0
1, 2–Butadiene (B2)	2.0
Butadiene Sulfone (BS)	100.0
L_{N+1} =	120.0

6.13 Determine by the Kremser method the separation that can be achieved for the absorption operation indicated in Figure 6.47 for the following combinations of conditions: (a) Six equilibrium stages and 75 psia operating pressure, (b) Three equilibrium stages and 150 psia operating pressure, (c) Six equilibrium stages and 150 psia operating pressure. At 90°F and 75 psia, the K-value of nC_{10} = 0.0011.

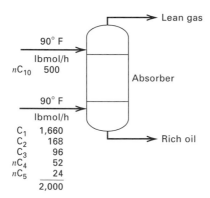

Figure 6.47 Data for Exercise 6.13.

6.14 One thousand kilomoles per hour of rich gas at 70°F with 25% C_1, 15% C_2, 25% C_3, 20% nC_4, and 15% nC_5 by moles is to be absorbed by 500 kmol/h of nC_{10} at 90°F in an absorber operating at 4 atm. Calculate by the Kremser method the percent absorption of each component for four, ten, and thirty theoretical stages. What do you conclude from the results? (Note: The K-value of nC_{10} at 80°F and 4 atm is 0.0014.)

Section 6.5

6.15 Using the performance data of Example 6.3, back-calculate the overall stage efficiency for propane and compare the result with estimates from the Drickamer–Bradford and O'Connell correlations.

6.16 Several hydrogenation processes are being considered that will require hydrogen of 95% purity. A refinery stream of 800,000 scfm (at 32°F, 1 atm), currently being used for fuel and containing 72.5% H_2, 25% CH_4, and 2.5% C_2H_6 is available. To convert this gas to the required purity, oil absorption, activated charcoal adsorption, and membrane separation are being considered. For oil absorption, an available n-octane stream can be used as the absorbent. Because the 95% H_2 must be delivered to a hydrogenation process at not less than 375 psia, it is proposed to operate the absorber at 400 psia and 100°F. If at least 80% of the hydrogen fed to the absorber is to leave in the exit gas, determine:
(a) The minimum absorbent rate in gallons per minute.
(b) The actual absorbent rate if 1.5 times the minimum amount is used.
(c) The number of theoretical stages.
(d) The stage efficiency for each of the three species in the feed gas, using the O'Connell correlation.
(e) The number of trays actually required.
(f) The composition of the exit gas, taking into account the stripping of octane.
(g) If the octane lost to the exit gas is not recovered, estimate the annual cost of this lost oil if the process operates 7,900 h/year and the octane is valued at $1.00/gal.

6.17 The absorption operation of Examples 6.1 and 6.4 is being scaled up by a factor of 15, such that a column with an 11.5-ft diameter will be needed. In addition, because of the low efficiency of 30% for the original operation, a new tray design has been developed and tested in an Oldershaw-type column. The resulting Murphree vapor-point efficiency, E_{OV}, for the new tray design for the system of interest is estimated to be 55%. Estimate E_{MV} and E_o. (To estimate the length of the liquid flow path, Z_L, use Figure 6.16. Also, assume that $u/D_E = 6$ ft^{-1}.)

Section 6.6

6.18 Conditions at the bottom tray of a reboiled stripper arc as shown in Figure 6.48. If valve trays are used with a 24-in. tray spacing, estimate the required column diameter for operation at 80% of flooding.

6.19 Determine the flooding velocity and column diameter for the following conditions at the top tray of a hydrocarbon absorber equipped with valve trays:

Pressure	400 psia
Temperature	128°F
Vapor rate	530 lbmol/h
Vapor MW	26.6
Vapor density	1.924 lb/ft^3
Liquid rate	889 lbmol/h

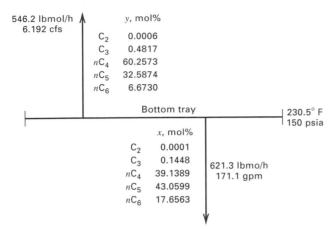

546.2 lbmol/h
6.192 cfs

	y, mol%
C_2	0.0006
C_3	0.4817
nC_4	60.2573
nC_5	32.5874
nC_6	6.6730

Bottom tray

230.5° F
150 psia

	x, mol%
C_2	0.0001
C_3	0.1448
nC_4	39.1389
nC_5	43.0599
nC_6	17.6563

621.3 lbmo/h
171.1 gpm

Figure 6.48 Data for Exercise 6.18.

Liquid MW	109
Liquid density	41.1 lb/ft^3
Liquid surface tension	18.4 dynes/cm
Foaming factor	0.75
Tray spacing	24 in.
Fraction flooding	0.85
Valve trays	

6.20 For Exercise 6.16, if a flow rate of 40,000 gpm of octane is used to carry out the absorption in a sieve-tray column using 24-in. tray spacing, a weir height of 2.5 in., and holes of $\frac{1}{4}$-in. diameter, determine for a foaming factor of 0.80 and a fraction flooding of 0.70:
(a) The column diameter based on conditions near the bottom of the column
(b) The vapor pressure drop per tray
(c) Whether weeping will occur
(d) The entrainment rate
(e) The fractional decrease in E_{MV} due to entrainment.
(f) The froth height in the downcomer.

6.21 Repeat the calculations of Examples 6.5, 6.6, and 6.7 for a column diameter corresponding to 40% of flooding.

6.22 For the acetone absorber of Figure 6.1, assuming the use of sieve trays with a 10% hole area and $\frac{3}{16}$-in. holes with an 18-in. tray spacing, estimate:
(a) The column diameter for a foaming factor of 0.85 and a fraction of flooding of 0.75
(b) The vapor pressure drop per tray
(c) The number of transfer units, N_G and N_L, from (6-58) and (6-59), respectively
(d) N_{OG} from (6-57)
(e) The controlling resistance to mass transfer
(f) E_{OV} from (6-52)
From your results, determine if 30 actual trays are adequate.

6.23 Design a VOC stripper for the flow conditions and separation of Example 6.2 except that the wastewater and air flow rates are twice as much. To develop the design, determine:

(a) The number of equilibrium stages required
(b) The column diameter for sieve trays
(c) The vapor pressure drop per tray
(d) Murphree vapor-point efficiencies using the Chan and Fair method
(e) The number of trays actually required.

Section 6.7

6.24 Air containing 1.6 vol% sulfur dioxide is scrubbed with pure water in a packed column of 1.5-m^2 cross-sectional area and 3.5-m height packed with no. 2 plastic Super Intalox saddles, at a pressure of 1 atm. Total gas flow rate is 0.062 kmol/s, the liquid flow rate is 2.2 kmol/s, and the outlet gas SO$_2$ concentration is $y = 0.004$. At the column temperature, the equilibrium relationship is given by $y^* = 40x$.
(a) What is L/L_{min}?
(b) Calculate N_{OG} and compare your answer to that for the number of theoretical stages required.
(c) Determine H_{OG} and the HETP from the operating data.
(d) Calculate $K_G a$ from the data.

6.25 An SO$_2$–air mixture is being scrubbed with water in a countercurrent-flow packed tower operating at 20°C and 1 atm. Solute-free water enters the top of the tower at a constant rate of 1,000 lb/h and is well distributed over the packing. The liquor leaving contains 0.6 lb SO$_2$/100 lb of solute-free water. The partial pressure of SO$_2$ in the spent gas leaving the top of the tower is 23 torr. The mole ratio of water to air is 25. The necessary equilibrium data are given below.
(a) What percent of the SO$_2$ in the entering gases is absorbed in the tower?
(b) In operating the tower it was found that the rate coefficients k_g and k_L remained substantially constant throughout the tower at the following values:

$k_L = 1.3$ ft/h
$k_g = 0.195$ lbmol/h-ft^2-atm

At a point in the tower where the liquid concentration is 0.001 lbmol SO$_2$ per lbmol of water, what is the liquid concentration at the gas–liquid interface in lbmol/ft^3? Assume that the solution has the same density as H$_2$O.

Solubility of SO$_2$ in H$_2$O at 20°C

$\dfrac{\text{lb SO}_2}{\text{100 lb H}_2\text{O}}$	Partial Pressure of SO$_2$ in Air, torr
0.02	0.5
0.05	1.2
0.10	3.2
0.15	5.8
0.20	8.5
0.30	14.1
0.50	26.0
0.70	39.0
1.0	59

6.26 A wastewater stream of 600 gpm, containing 10 ppm (by weight) of benzene, is to be stripped with air in a packed column operating at 25°C and 2 atm to produce water containing 0.005 ppm of benzene. The packing is 2-in. Flexirings made of polypropylene. The vapor pressure of benzene at 25°C is 95.2 torr. The solubility of benzene in water at 25°C is 0.180 g/100 g. An expert in VOC stripping with air has suggested use of 1,000 scfm of air (60°F, 1 atm), at which condition one should achieve for the mass transfer of benzene:

$$k_L a = 0.067 \text{ s}^{-1} \quad \text{and} \quad k_g a = 0.80 \text{ s}^{-1}$$

Determine:
(a) The minimum air stripping rate in scfm. Is it less than the rate suggested by the expert? If not, use 1.4 times your minimum value.
(b) The stripping factor based on the air rate suggested by the expert.
(c) The number of transfer units, N_{OG} required.
(d) The overall mass transfer coefficient, $K_G a$, in units of mol/m³-s-kPa and s⁻¹. Which phase controls mass transfer?
(e) The volume of packing in cubic meters.

Section 6.8

6.27 Germanium tetrachloride ($GeCl_4$) and silicon tetrachloride ($SiCl_4$) are used in the production of optical fibers. Both chlorides are oxidized at high temperature and converted to glasslike particles. However, the $GeCl_4$ oxidation is quite incomplete and it is necessary to scrub the unreacted $GeCl_4$ from its air carrier in a packed column operating at 25°C and 1 atm with a dilute caustic solution. At these conditions, the dissolved $GeCl_4$ has no vapor pressure and mass transfer is controlled by the gas phase. Thus, the equilibrium curve is a straight line of zero slope. Why? The entering gas is 23,850 kg/day of air containing 288 kg/day of $GeCl_4$. The air also contains 540 kg/day of Cl_2, which, when dissolved, also will have no vapor pressure. The two liquid-phase reactions are

$$GeCl_4 + 5OH^- \rightarrow HGeO_3^- + 4Cl^- + 2H_2O$$
$$Cl_2 + 2OH^- \rightarrow ClO^- + Cl^- + H_2O$$

It is desired to absorb 99% of both $GeCl_4$ and Cl_2 in an existing 2-ft-diameter column that is packed to a height of 10 ft with $\frac{1}{2}$-in. ceramic Raschig rings. The liquid rate should be set so that the column operates at 75% of flooding. For the packing: $\epsilon = 0.63$, $F_P = 580 \text{ ft}^{-1}$, and $D_P = 0.01774$ m.

Gas-phase mass transfer coefficients for $GeCl_4$ and Cl_2 can be estimated from the following empirical equations developed from experimental studies, where μ, ρ, and D_i are gas-phase properties:

$$K_y a = k_y a$$
$$\frac{k_y}{(G/S)} = 1.195 \left[\frac{D_p G'}{\mu(1 - \epsilon_o)} \right]^{-0.36} (N_{Sc})^{-2/3}$$
$$\epsilon_o = \epsilon - h_L$$

$$h_L = 0.03591(L')^{0.331}$$
$$a - \frac{14.69(808\, G'/\rho^{1/2})^n}{(L')^{0.111}}$$
$$n = 0.01114L' + 0.148$$

where S = column cross sectional area, m²
k_y = kmol/m²-s
G = molar gas rate, kmol/s
D_p = equivalent packing diameter, m
μ = gas viscosity, kg/m-s
ρ = gas density, kg/m³
N_{Sc} = Schmidt number = $\mu/\rho D_i$
D_i = molecular diffusivity of component i in the gas, m²/s
a = interfacial area for mass transfer, m²/m³ of packing
L' = liquid mass velocity, kg/m²-s
G' = gas mass velocity, kg/m²-s

For the two diffusing species, take

$$D_{GeCl_4} = 0.000006 \text{ m}^2/\text{s}$$
$$D_{Cl_2} = 0.000013 \text{ m}^2/\text{s}$$

Determine:
(a) The dilute caustic flow rate in kilograms per second.
(b) The required packed height in feet based on the controlling species ($GeCl_4$ or Cl_2). Is the 10 ft of packing adequate?
(c) The percent absorption of $GeCl_4$ and Cl_2 based on the available 10 ft of packing. If the 10 ft of packing is not sufficient, select an alternative packing that is adequate.

6.28 For the VOC stripping task of Exercise 6.26, the expert has suggested that we use a tower diameter of 0.80 m for which we can expect a pressure drop of 500 N/m²-m of packed height (0.612 in. H_2O/ft). Verify the information from the expert by estimating:
(a) The fraction of flooding using the GPDC chart of Figure 6.36 with $F_P = 24 \text{ ft}^2/\text{ft}^3$
(b) The pressure drop at flooding
(c) The pressure drop at the operating conditions of Exercise 6.26 using the GPDC chart
(d) The pressure drop at operating conditions using the correlation of Billet and Schultes by assuming that 2-in. plastic Flexiring packing has the same characteristics as 2-in. plastic Pall rings.

6.29 For the VOC stripping task of Exercise 6.26, the expert suggested certain mass transfer coefficients. Check this information by estimating the coefficients from the correlations of Billet and Schultes by assuming that 2-in. plastic Flexiring packing has the same characteristics as 2-in. plastic Pall rings.

6.30 A 2 mol% NH_3-in-air mixture at 68°F and 1 atm is to be scrubbed with water in a tower packed with 1.5-in. ceramic Berl saddles. The inlet water mass velocity will be 2400 lb/h-ft², and the inlet gas mass velocity 240 lb/h-ft². Assume that the tower temperature remains constant at 68°F, at which the gas solubility relationship follows Henry's law, $p = Hx$, where p is the partial pressure of ammonia over the solution, x is the mole fraction of ammonia in the liquid, and H is the Henry's law constant, equal to 2.7 atm/mole fraction.

(a) Calculate the required packed height for absorption of 90% of the NH_3.

(b) Calculate the minimum water mass velocity in lb/h-ft^2 for absorbing 98% of the NH_3.

(c) The use of 1.5-in. ceramic Hiflow rings rather than the Berl saddles has been suggested. What changes would this cause in K_Ga, pressure drop, maximum liquid rate, K_La, column height, column diameter, H_{OG}, and N_{OG}?

6.31 You are to design a packed column to absorb CO_2 from air into fresh dilute caustic solution. The entering air contains 3 mol% CO_2, and a 97% recovery of CO_2 is desired. The gas flow rate is 5,000 ft^3/min at 60°F, 1 atm. It may be assumed that in the range of operation, the equilibrium curve is $Y^* = 1.75X$, where Y and X are mole ratios of CO_2 to carrier gas and liquid, respectively. A column diameter of 30 in. with 2-in. Intalox saddle packing can be assumed for the initial design estimates. Assume the caustic solution has the properties of water. Calculate:

(a) The minimum caustic solution-to-air molar flow-rate ratio.

(b) The maximum possible concentration of CO_2 in the caustic solution.

(c) The number of theoretical stages at $L/G = 1.4$ times minimum.

(d) The caustic solution rate.

(e) The pressure drop per foot of column height. What does this result suggest?

(f) The overall number of gas transfer units, N_{OG}.

(g) The height of packing, using a K_Ga of 2.5 lbmol/h-ft^3-atm.

Section 6.9

6.32 At a point in an ammonia absorber using water as the absorbent and operating at 101.3 kPa and 20°C, the bulk gas phase contains 10 vol% NH_3. At the interface, the partial pressure of NH_3 is 2.26 kPa. The concentration of the ammonia in the body of the liquid is 1 wt%. The rate of ammonia absorption at this point is 0.05 kmol/h-m^2.

(a) Given this information and the equilibrium curve in Figure 6.49, calculate, X, Y, Y_I, X_I, X^*, Y^*, K_Y, K_X, k_Y, and k_X.

(b) What percent of the mass transfer resistance is in each phase?

(c) Verify for these data that $1/K_Y = 1/k_Y + H'/k_X$.

6.33 One thousand cubic feet per hour of a 10 mol% NH_3 in air mixture is required to produce nitrogen oxides. This mixture is to be obtained by desorbing an aqueous 20 wt% NH_3 solution with air at 20°C. The spent solution should not contain more than 1 wt% NH_3.

Calculate the volume of packing required for the desorption column. Vapor–liquid equilibrium data for Exercise 6.32 can be used and $K_Ga = 4$ lbmol/h-ft^3-atm partial pressure.

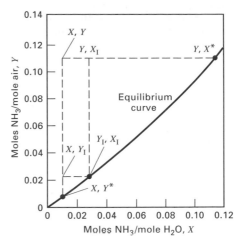

Figure 6.49 Data for Exercise 6.32.

6.34 Ammonia, present at a partial pressure of 12 torr in an air stream saturated with water vapor at 68°F and 1 atm, must be removed to the extent of 99.6% by water absorption at the same temperature and pressure. Two thousand pounds of dry air per hour are to be handled.

(a) Calculate the minimum amount of water necessary using the equilibrium data for Exercise 6.32 in Fig. 6.49.

(b) Assuming an operation at 2 times the minimum water flow and at one-half the flooding gas velocity, compute the dimensions of a column packed with 38-mm ceramic Berl Saddles.

(c) Repeat part (b) for 50-mm Pall rings.

(d) Which of the two packings would you recommend?

6.35 Exit gas from a chlorinator consists of a mixture of 20 mol% chlorine in air. This concentration is to be reduced to 1% chlorine by water absorption in a packed column to operate isothermally at 20°C and atmospheric pressure. Using the following equilibrium x–y data, calculate for 100 kmol/h of feed gas:

(a) The minimum water rate in kilograms per hour.

(b) N_{OG} for twice the minimum water rate

Data for x–y at 20°C (in chlorine mole fractions):

x	0.0001	0.00015	0.0002	0.00025	0.0003
y	0.006	0.012	0.024	0.04	0.06

6.36 Calculate the diameter and height for the column of Example 6.15 if the tower is packed with 1.5-in. metal Pall rings. Assume that the absorbing solution has the properties of water and use conditions at the bottom of the tower, where flow rates are highest.

6.37 You are asked to design a packed column to recover acetone from air continuously, by absorption with water at 60°F. The air contains 3 mol% acetone, and a 97% recovery is desired. The gas flow rate is 50 ft^3/min at 60°F, 1 atm. The maximum allowable gas superficial velocity in the column is

2.4 ft/s. It may be assumed that in the range of operation, $Y^* = 1.75X$, where Y and X are mole ratios for acetone.

Calculate:

(a) The minimum water-to-air molar flow rate ratio

(b) The maximum acetone concentration possible in the aqueous solution

(c) The number of theoretical stages for a flow rate ratio of 1.4 times the minimum

(d) The corresponding number of overall gas transfer units

(e) The height of packing, assuming $K_Y a = 12.0$ lbmol/h-ft³-molar ratio difference.

(f) The height of packing as a function of the molar flow rate ratio, assuming that G and HTU remain constant

6.38 Determine the diameter and packed height of a countercurrently operated packed tower required to recover 99% of the ammonia from a gas mixture that contains 6 mol% NH_3 in air. The tower, packed with 1-in. metal Pall rings, must handle 2,000 ft³/min of gas as measured at 68°F and 1 atm. The entering water absorbent rate will be twice the theoretical minimum, and the gas velocity will be such that it is 50% of the flooding velocity. Assume isothermal operation at 68°F and 1 atm. Equilibrium data are given in Figure 6.49.

6.39 A tower, packed with Montz B1-200 metal structured packing, is to be designed to absorb SO_2 from air by scrubbing with water. The entering gas, at an SO_2-free flow rate of 6.90 lbmol/h-ft² of bed cross section, contains 80 mol% air and 20 mol% SO_2. Water enters at a flow rate of 364 lbmol/h-ft² of bed cross section. The exiting gas is to contain only 0.5 mol% SO_2. Assume that neither air nor water will be transferred between phases and that the tower operates at 2 atm and 30°C. Equilibrium data in mole fractions for SO_2 solubility in water at 30°C and 2 atm (*Perry's Chemical Engineers' Handbook*, 4th ed., Table 14.31, p. 14-6) have been fitted by a least-squares method to the equation

$$y = 12.697x + 3148.0x^2 - 4.724 \times 10^5 x^3 + 3.001 \times 10^7 x^4 - 6.524 \times 10^8 x^5$$

(a) Derive the following molar material balance operating line for SO_2 mole fractions:

$$x = 0.0189 \left(\frac{y}{1 - y} \right) - 0.00010$$

(b) Write a computer program or use a spreadsheet program to calculate the number of required transfer units based on the overall gas-phase resistance.

Chapter 7

Distillation of Binary Mixtures

In *distillation* (*fractionation*), a feed mixture of two or more components is separated into two or more products, including, and often limited to, an overhead distillate and a bottoms, whose compositions differ from that of the feed. Most often, the feed is a liquid or a vapor–liquid mixture. The bottoms product is almost always a liquid, but the distillate may be a liquid or a vapor or both. The separation requires that (1) a second phase be formed so that both liquid and vapor phases are present and can contact each other on each stage within a separation column, (2) the components have different volatilities so that they will partition between the two phases to different extents, and (3) the two phases can be separated by gravity or other mechanical means. Distillation differs from absorption and stripping in that the second fluid phase is created by thermal means (vaporization and condensation) rather than by the introduction of a second phase that usually contains an additional component or components not present in the feed mixture.

According to Forbes [1], the art of distillation dates back to at least the first century A.D. By the eleventh century, distillation was being used in Italy to produce alcoholic beverages. At that time, distillation was probably a batch process based on the use of just a single stage, the boiler. The feed to be separated, a liquid, was placed in a vessel to which heat was applied, causing part of the liquid to evaporate. The vapor passed out of the heating vessel and was cooled in another chamber by transfer of heat through the wall of the chamber to water, producing condensate that dripped into a product receiver. The word *distillation* is derived from the Latin word *destillare*, which means dripping or trickling down. By at least the sixteenth century, it was known that the extent of separation could be improved by providing multiple vapor–liquid contacts (stages) in a so-called Rectificatorium. The term *rectification* is derived from the Latin words *recte facere*, meaning to improve. Modern distillation derives its ability to produce almost pure products from the use of multistage contacting.

Throughout the twentieth century, multistage distillation has been by far the most widely used method for separating liquid mixtures of chemical components. Unfortunately, distillation is a very energy-intensive technique, especially when the relative volatility, α, of the components being separated is low (<1.50). Mix et al. [2] report that the energy consumption for distillation in the United States for 1976 totaled 2×10^{15} Btu (2 quads), which was nearly 3% of the entire national energy consumption. Approximately two-thirds of the distillation energy was consumed by petroleum refining, where distillation is widely used to separate crude oil into petroleum fractions, light hydrocarbons (C_2's to C_5's), and aromatic chemicals. The separation of other organic chemicals, often in the presence of water, is widely practiced in the chemical industry.

The fundamentals of distillation are best understood by the study of binary distillation, the separation of a two-component mixture, which is the subject of this chapter. The more general and much more difficult case of a multicomponent mixture is covered in Chapters 10 and 11. A representative binary distillation operation is shown in Figure 7.1 for the separation of 620 lbmol/h (0.0781 kmol/s) of a binary mixture of 46 mol% benzene (the more volatile component) and 54 mol% toluene. The purpose of the 25-sieve-tray (equivalent to 20 theoretical stages plus a partial reboiler that acts as an additional theoretical stage) distillation column is to separate the feed into a liquid distillate of 99 mol% benzene and a liquid bottoms product of 98 mol% toluene. The column operates at a pressure in the reflux drum of 18 psia (124 kPa), just slightly above ambient pressure. For a negligible pressure drop across the condenser and a vapor pressure drop of 0.1 psi/tray (0.69 kPa/tray), the pressure in the reboiler is $18 + 0.1(25) = 20.5$ psia (141 kPa). In this range of pressure, benzene and toluene form nearly ideal mixtures with a relative volatility of from 2.26 at the bottom tray to 2.52 at the top tray, as determined from Raoult's law by (2-44). The reflux ratio (reflux rate to distillate rate) is 2.215. If an infinite number of stages were used, the required reflux ratio would be a

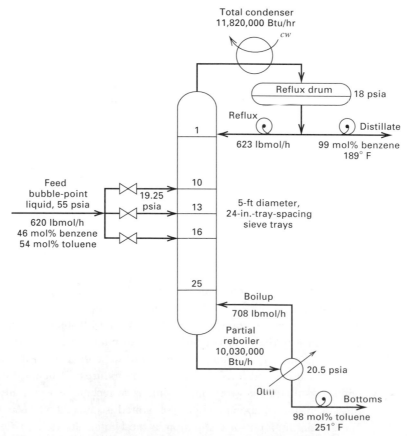

Figure 7.1 Distillation of a binary mixture of benzene and toluene.

minimum value of 1.708. Thus, the ratio of reflux rate to minimum reflux rate for this example is 1.297. Most distillation columns are designed to operate with optimal-reflux-to-minimum-reflux ratios of 1.1 to 1.5. If an infinite ratio of reflux to minimum reflux were used, only 10.76 theoretical stages would be required. Thus, the ratio of theoretical stages to minimum theoretical stages for this example is $21/10.7 = 1.96$. For most distillation columns, this ratio is approximately 2. The stage efficiency is 20/25 or 80%. This is close to the average efficiency observed for distillation.

The feed to the separation operation of Figure 7.1 is a saturated liquid at 55 psia (379 kPa). A bubble-point calculation gives a temperature of 294°F (419 K). When this feed is flashed adiabatically across the feed valve to the feed tray pressure of 19.25 psia (133 kPa), the feed temperature drops to 220°F (378 K), causing 23.4 mol% of the feed to be vaporized. A total condenser is used to obtain saturated liquid reflux and liquid distillate at a bubble-point temperature of 189°F (360 K) at 18 psia (124 kPa). The duty of the condenser is 11,820,000 Btu/h (3.46 MW). At the bottom of the column, a partial reboiler is used to produce vapor boilup and a liquid bottoms product. Assuming that the boilup and bottoms are in physical equilibrium, the partial reboiler functions as an additional theoretical stage, giving a total of 21 theoretical stages. Because the bottoms product is a saturated liquid, its temperature of 251°F (395 K) corresponds to the bubble point of the bottoms at 20.5 psia (141 kPa). The duty of the reboiler is 10,030,000 Btu/h (2.94 MW), which is within 15% of the condenser duty.

The inside diameter of the distillation column in Figure 7.1 is a constant 5 ft (1.53 m). At the top tray this diameter corresponds to 84% of flooding, while at the bottom tray the percent flooding is 81%. As shown, the column can be fed at any one of three trays. For the design conditions, the optimal feed entry is between trays 12 and 13. However, should the feed composition or product specifications change, one of the other two feed trays could become optimum.

Distillation columns similar to that of Figure 7.1 have been built for diameters up to at least 30 ft (9.14 m). With a 24-in. (0.61-m) tray spacing, the maximum number of trays included in a single column is usually no greater than 150. In general, for the sharp separation of a binary mixture with a relative volatility less than 1.05, distillation can require many hundreds of trays, so a more efficient separation technique should be sought. Even when distillation is the most economical separation technique, its second-law efficiency, using the calculational procedure developed in Chapter 2, can be less than 10%.

Technically, distillation is the most mature separation operation. Design and operation procedures are well established; for example, see Kister [3,4]. Only when vapor–liquid equilibrium or other data are uncertain is a laboratory and/or pilot-plant study necessary prior to the design of a commercial unit. Table 7.1, taken partially from the study of Mix et al. [2], lists just some of the more common commercial binary distillation operations in decreasing order of difficulty of separation. Included are representative nominal values of relative volatility, number of trays, column operating pressure, and reflux-to-minimum-reflux ratio. Although the data in Table 7.1 refer to trayed towers, distillation can also be carried out in packed columns. More and more frequently, additional distillation capacity is being achieved with existing trayed towers by replacing all or some of the trays with sections of random or structured packing.

In this chapter, equipment for conducting distillation operations is discussed and fundamental equilibrium-based and rate-based calculational procedures are developed for binary mixtures. Trayed and packed distillation columns are identical in most respects

Table 7.1 Representative Commercial Binary Distillation Operations [2]

Binary Mixture	Average Relative Volitility	Number of Trays	Typical Operating Pressure, psia	Reflux-to-Minimum-Reflux Ratio
1,3-Butadiene/vinyl acetylene	1.16	130	75	1.70
Vinyl acetate/ethyl acetate	1.16	90	15	1.15
o-Xylene/m-xylene	1.17	130	15	1.12
Isopentane/n-pentane	1.30	120	30	1.20
Isobutane/n-butane	1.35	100	100	1.15
Ethylbenzene/styrene	1.38	34	1	1.71
Propylene/propane	1.40	138	280	1.06
Methanol/ethanol	1.44	75	15	1.20
Water/acetic acid	1.83	40	15	1.35
Ethylene/ethane	1.87	73	230	1.07
Acetic acid/acetic anhydride	2.02	50	15	1.13
Toluene/ethylbenzene	2.15	28	15	1.20
Propyne/1,3-butadiene	2.18	40	120	1.13
Ethanol azeotrope/water	2.21	60	15	1.35
Isopropanol/water	2.23	12	15	1.28
Benzene/toluene	3.09	34	15	1.15
Methanol/water	3.27	60	45	1.31
Cumene/phenol	3.76	38	1	1.21
Benzene/ethylbenzene	6.79	20	15	1.14
HCN/water	11.20	15	50	1.36
Ethylene oxide/water	12.68	50	50	1.19
Formaldehyde/methanol	16.70	23	50	1.17
Water/ethylene glycol	81.20	16	4	1.20

to the absorption and stripping columns discussed in the previous chapter. Therefore, where appropriate, reference is made to Chapter 6 and only important differences are discussed in this chapter.

7.1 EQUIPMENT AND DESIGN CONSIDERATIONS

Industrial distillation operations are most commonly conducted in trayed towers, but packed columns are finding increasing use. Occasionally, distillation columns contain both trays and packing. Types of trays and packings are identical to those used for absorption and stripping, as described in Section 6.1, shown in Figures 6.2 to 6.7, and compared in Tables 6.2 and 6.3.

Factors that influence the design or analysis of a binary distillation operation include:

1. Feed flow rate, composition, temperature, pressure, and phase condition
2. Desired degree of separation between two components
3. Operating pressure (which must be below the critical pressure of the mixture)
4. Vapor pressure drop, particularly for vacuum operation
5. Minimum reflux ratio and actual reflux ratio
6. Minimum number of equilibrium stages and actual number of equilibrium stages (stage efficiency)
7. Type of condenser (total, partial, or mixed)
8. Degrees of subcooling, if any, of the liquid reflux
9. Type of reboiler (partial or total)

10. Type of contacting (trays or packing or both)
11. Height of the column
12. Feed entry stage
13. Diameter of the column
14. Column internals.

The phase condition (also called thermal condition) of the feed is determined at the feed-tray pressure by an adiabatic flash calculation across the feed valve. As the molar fraction of vapor in the feed increases, the required reflux ratio (L/D) increases, but the corresponding boilup ratio (V/B) decreases. The column operating pressure in the reflux drum should correspond to a distillate temperature somewhat higher (e.g., 10 to 50°F or 6 to 28°C) than the supply temperature of the cooling water used as the coolant in the overhead condenser. However, if this pressure approaches the critical pressure of the more volatile component, then a lower operating pressure must be used and a refrigerant is required as coolant. For example, in Table 7.1, the separation of ethylene/ethane is conducted at 230 psia (1,585 kPa), giving a column top temperature of −40°F (233 K), which requires a refrigerant. Water at 80°F (300 K) cannot be used in the condenser because the critical temperature of ethylene is 48.6°F (282 K). If the estimated pressure is less than atmospheric pressure, the operating pressure at the top of the column is often set just above atmospheric pressure to avoid vacuum operation, unless the temperature at the bottom of the column is found to exceed a bottoms temperature limited by decomposition, polymerization, excessive corrosion, or other chemical reaction. In that case, vacuum operation is necessary. In Table 7.1, vacuum operation is required for the separation of ethylbenzene from styrene to maintain a bottoms temperature sufficiently low to prevent polymerization of styrene.

For given (1) feed, (2) desired degree of separation, and (3) operating pressure, a minimum reflux ratio exists that corresponds to an infinite number of theoretical stages; and a minimum number of theoretical stages exists that corresponds to an infinite reflux ratio. A design trade-off is usually made between the number of stages and the reflux ratio. A graphical method for determining the data needed to make this trade-off and to determine the optimal feed-stage location is developed in the next section.

7.2   MCCABE–THIELE GRAPHICAL EQUILIBRIUM-STAGE METHOD FOR TRAYED TOWERS

Consider the general countercurrent-flow, multistage, binary distillation operation shown in Figure 7.2. The operation consists of a column containing the equivalent of N theoretical stages; a total condenser in which the overhead vapor leaving the top stage is totally condensed to give a liquid distillate product and liquid reflux that is returned to the top stage; a partial reboiler in which liquid from the bottom stage is partially vaporized to give a liquid bottoms product and vapor boilup that is returned to the bottom stage, and an intermediate feed stage. By means of multiple, countercurrent contacting stages arranged in a two-section cascade with reflux and boilup, as discussed in Section 5.4, it is possible to achieve a sharp separation between the two components in the feed unless an azeotrope is formed, in which case one of the two products will approach the azeotropic composition.

The feed, which contains a more volatile (light) component (the *light key*, LK), and a less volatile (heavy) component (the *heavy key*, HK), enters the column at a feed stage, *f*. At the feed-stage pressure, the feed may be liquid, vapor, or a mixture of liquid and vapor, with its overall mole-fraction composition with respect to the light component denoted by z_F. The mole fraction of the light key in the distillate is x_D, while the mole fraction of the light key in the bottoms product is x_B. Corresponding compositions with respect to the heavy key are $1 - z_F$, $1 - x_D$, and $1 - x_B$.

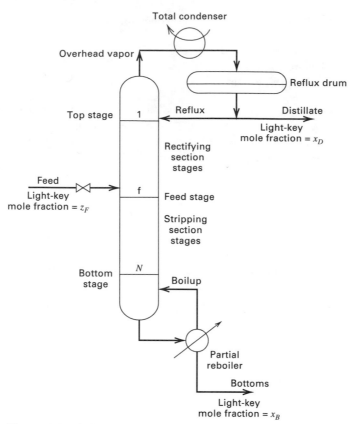

Figure 7.2 Distillation operation using a total condenser and partial reboiler.

The goal of distillation is to produce from the feed a distillate rich in the light key (i.e., x_D approaching 1.0) and a bottoms product rich in the heavy key (i.e., x_B approaching 0.0). The ease or difficulty with which the separation can be achieved depends on the relative volatility, α, of the two components (LK = 1 and HK = 2), where

$$\alpha_{1,2} = K_1/K_2 \qquad \text{(7-1)}$$

Methods for estimating K-values are discussed in Chapter 2.

If the two components form ideal solutions and follow the ideal gas law in the vapor phase, Raoult's law applies to give

$$K_1 = P_1^s/P \quad \text{and} \quad K_2 = P_2^s/P$$

and from (7-1), the relative volatility is given simply by the ratio of vapor pressures, $\alpha_{1,2} = P_1^s/P_2^s$ and thus is a function only of temperature. As discussed in Section 4.2, as the temperature (and therefore the pressure) increases, $\alpha_{1,2}$ decreases. At the convergence pressure of the mixture, $\alpha_{1,2} = 1.0$ and a separation cannot be achieved at this or any higher pressure.

The relative volatility can be expressed in terms of equilibrium vapor and liquid compositions from the definition of the K-value as $K_i = y_i/x_i$. For a binary mixture,

$$\alpha_{1,2} = \frac{y_1/x_1}{y_2/x_2} = \frac{y_1(1 - x_1)}{x_1(1 - y_1)} \qquad \text{(7-2)}$$

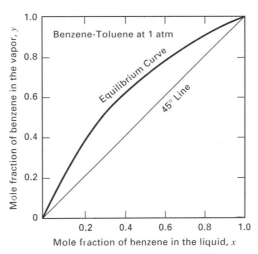

Figure 7.3 Equilibrium curve for benzene–toluene at 1 atm.

Solving (7-2) for y_1,

$$y_1 = \frac{\alpha_{1,2}x_1}{1 + x_1(\alpha_{1,2} - 1)} \tag{7-3}$$

For ideal binary mixtures of components with close boiling points, the temperature change over the column is small and $\alpha_{1,2}$ is almost constant. In any case, for a given pressure P and liquid-phase composition x_1, the Gibbs phase rule, discussed in Chapter 4, fixes the temperature and equilibrium vapor composition. An equilibrium curve for the benzene–toluene system is shown in Figure 7.3, where y and x correspond to the light key, benzene, and the pressure is 1 atm, at which pure benzene and pure toluene boil at 176 and 231°F, respectively. Thus, these two components are not close-boiling. Using (7-3) with this curve, α varies from about 2.6 at the bottom of the curve to about 2.35 at the top of the curve. Representative equilibrium curves for some average values of α are shown in Figure 4.5. The higher the average value of α, the easier it is to achieve the desired separation. Average values of α for the distillation operations in Table 7.1 range from 1.16 to 81.2.

In 1925, McCabe and Thiele [5] published an approximate graphical method for combining the equilibrium curve of Figure 7.3 with operating-line curves to estimate, for a given binary feed mixture and column operating pressure, the number of equilibrium stages and the amount of reflux required for a desired degree of separation of the feed. Although computer-aided methods, discussed later in Chapter 10, are more accurate and easier to apply, the graphical construction of the McCabe–Thiele method greatly facilitates the visualization of many of the important aspects of multistage distillation, and therefore the effort required to learn the method is well justified.

Typical problem specifications for and results from the McCabe–Thiele method are summarized in Table 7.2. This table applies to a simple binary distillation operation, like that in Figure 7.2, for a single feed and two products. The distillate can be a liquid from a total condenser, as shown in Figure 7.2, or a vapor from a partial condenser. The feed phase condition must be known at the column pressure, which is assumed to be uniform throughout the column for the McCabe–Thiele method. The type of condenser and reboiler must be specified, as well as the ratio of reflux to minimum reflux. From the specification of x_D and x_B for the light key, the distillate and bottoms flow rates, D and B, are fixed by material balance, since

$$Fz_F = x_D D + x_B B$$

Table 7.2 Specifications for and Results from the McCabe–Thiele Method
for Binary Distillation

Specifications	
F	Total feed rate
z_F	Mole fraction composition of the feed
P	Column operating pressure (assumed uniform throughout the column)
	Phase condition of the feed at column pressure
	Vapor–liquid equilibrium curve for the binary mixture at column pressure
	Type of overhead condenser (total or partial)
	Type of reboiler (usually partial)
x_D	Mole fraction composition of the distillate
x_B	Mole fraction composition of the bottoms
R/R_{min}	Ratio of reflux to minimum reflux
Results	
D	Distillate flow rate
B	Bottoms flow rate
N_{min}	Minimum number of equilibrium stages
R_{min}	Minimum reflux ratio, L_{min}/D
R	Reflux ratio, L/D
V_B	Boilup ratio, $\overline{V}/B$
N	Number of equilibrium stages
	Optimal feed-stage location
	Stage vapor and liquid compositions

But, $B = F - D$ and therefore

$$Fz_F = x_D D + x_B(F - D)$$

$$\text{or } D = F\left(\frac{z_F - x_B}{x_D - x_B}\right)$$

This result requires $x_B < z_F < x_D$.

The McCabe–Thiele method determines not only N, the number of equilibrium stages, but also N_{min}, R_{min}, and the optimal stage for feed entry. Following the application of the McCabe–Thiele method, energy balances are applied to estimate condenser and reboiler heat duties.

Besides the equilibrium curve, the McCabe–Thiele method involves a 45° reference line, separate operating lines for the upper *rectifying* (enriching) section of the column and the lower *stripping* (exhausting) section of the column, and a fifth line (the *q-line* or feed line) for the phase or thermal condition of the feed. A typical set of these lines is shown in Figure 7.4. Equations for these lines are derived in the following subsection.

Rectifying Section

As shown in Figure 7.2, the rectifying section of equilibrium stages extends from the top stage, 1, to just above the feed stage, f. Consider a top portion of the rectifying stages, including the total condenser. A material balance for the light key over the envelope shown in Figure 7.5a for the total condenser and stages 1 to n is as follows, where y and

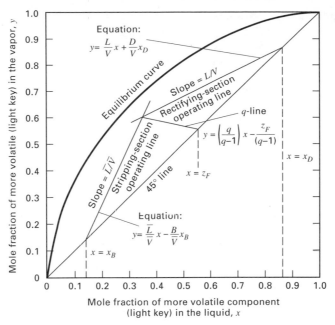

Figure 7.4 Construction lines for McCabe–Thiele method.

x refer to vapor and liquid mole fractions, respectively, for the light key:

$$V_{n+1}y_{n+1} = L_n x_n + D x_D \tag{7-4}$$

Solving for y_{n+1}:

$$y_{n+1} = \frac{L_n}{V_{n+1}} x_n + \frac{D}{V_{n+1}} x_D \tag{7-5}$$

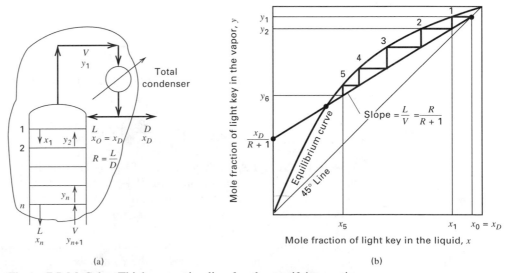

Figure 7.5 McCabe–Thiele operating line for the rectifying section.

Equation (7-5) relates the compositions, y_{n+1} and x_n, of two *passing* streams, V_{n+1} and L_n, respectively. For (7-5) to plot as a straight line of the form $y = mx + b$, which is the locus of compositions of all passing streams in the rectifying section, total molar flow rates L and V must not vary from stage to stage. This is the case if:

1. The two components have equal and constant molar enthalpies of vaporization (latent heats).
2. Component sensible enthalpy changes $(C_P \Delta T)$ and heat of mixing are negligible compared to latent heat changes.
3. The column is well insulated so that heat loss is negligible.
4. The pressure is uniform throughout the column (no pressure drop).

These assumptions are referred to as the *McCabe–Thiele assumptions* leading to the condition of *constant molar overflow* in the rectifying section, which refers to a molar liquid flow rate that remains constant as the liquid overflows each weir from one stage to the next. Since a total material balance for the rectifying-section envelope in Figure 7.5a gives $V_{n+1} = L_n + D$, if L is constant, then V is also constant for a particular value of D. Thus, (7-5) can be rewritten as

$$y = \frac{L}{V}x + \frac{D}{V}x_D \tag{7-6}$$

as shown in Figure 7.4. Thus, the slope of the operating line is L/V, which is constant. Because $V > L$, $L/V < 1$ in the rectifying section, as seen in Figure 7.5a.

For constant molar overflow, it is not necessary to consider energy balances in either the rectifying or stripping sections; only material balances and a vapor–liquid equilibrium curve are required. However, energy balances are needed to determine condenser and reboiler duties.

The liquid entering the top stage is the external reflux rate, L_0, and its ratio to the distillate rate, L_0/D, is the reflux ratio, R. Because of the assumption of constant molar overflow, R is a constant in the rectifying section, equal to L/D. Since $V = L + D$, the slope of the operating line is readily related to the reflux ratio:

$$\frac{L}{V} = \frac{L}{L+D} = \frac{L/D}{L/D + D/D} = \frac{R}{R+1} \tag{7-7}$$

Similarly,

$$\frac{D}{V} = \frac{D}{L+D} = \frac{1}{R+1} \tag{7-8}$$

Combining (7-6), (7-7), and (7-8) produces the most useful form of the operating line for the rectifying section:

$$y = \left(\frac{R}{R+1}\right)x + \left(\frac{1}{R+1}\right)x_D \tag{7-9}$$

If values of R and x_D are specified, (7-9) plots as a straight line with an intersection at $y = x_D$ on the 45° line, a slope of $L/V = R/(R+1)$, and an intersection at $y = x_D/(R+1)$ for $x = 0$, as shown in Figure 7.5b, which also contains a 45° line and an equilibrium curve. The equilibrium stages are stepped off in the manner described in Section 6.3 for absorption. Starting from the point $(y_1 = x_D, x_0 = x_D)$ on the operating line and the 45° line, a horizontal line is drawn to the left until it intersects the equilibrium curve at (y_1, x_1), that is, the compositions of the *equilibrium phases* leaving the top equilibrium stage. A vertical line is now dropped until it intersects the operating line at the point (y_2, x_1), the compositions of the two *phases passing* each other between stages 1 and 2.

The horizontal and vertical line constructions are continued down the rectifying section in the manner shown in Figure 7.5b to give the staircase construction shown, which is arbitrarily terminated at stage 5.

Stripping Section

As shown in Figure 7.2, the stripping section of equilibrium stages extends from the feed to the bottom stage. In Figure 7.6a, consider a bottom portion of the stripping stages, including the partial reboiler and extending up from stage N to stage $m + 1$, located somewhere below the feed. A material balance for the light key over the envelope shown in Figure 7.6a results in

$$\overline{L}x_m = \overline{V}y_{m+1} + Bx_B \tag{7-10}$$

Solving for y_{m+1}:

$$y_{m+1} = \frac{\overline{L}}{\overline{V}}x_m - \frac{B}{\overline{V}}x_B$$

$$\text{or } y = \frac{\overline{L}}{\overline{V}}x - \frac{B}{\overline{V}}x_B \tag{7-11}$$

where $\overline{L}$ and $\overline{V}$ are the total molar flows, which by the constant-molar-overflow assumption remain constant from stage to stage. The slope of this operating line for the compositions of passing streams in the stripping section is seen to be $\overline{L}/\overline{V}$. Because $\overline{L} > \overline{V}$, $\overline{L}/\overline{V} > 1$, as seen in Figure 7.6b. This is the reverse of conditions in the rectifying section.

The vapor leaving the partial reboiler is assumed to be in equilibrium with the liquid bottoms product. Thus, the partial reboiler acts as an additional equilibrium stage. The vapor rate leaving it is called the *boilup*, $\overline{V}_{N+1}$, and its ratio to the bottoms product rate, $V_B = \overline{V}_{N+1}/B$, is the *boilup ratio*. Because of the constant-molar-overflow assumption, V_B is constant in the stripping section. Since $\overline{L} - \overline{V} + B$,

$$\frac{\overline{L}}{\overline{V}} = \frac{\overline{V} + B}{\overline{V}} = \frac{V_B + 1}{V_B} \tag{7-12}$$

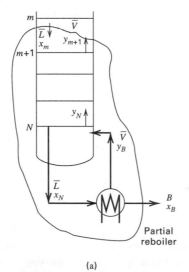

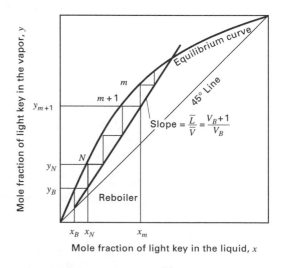

(a)   (b)

Figure 7.6 McCabe–Thiele operating line for the stripping section.

Similarly,

$$\frac{B}{\overline{V}} = \frac{1}{V_B} \tag{7-13}$$

Combining (7-11), (7-12), and (7-13), the operating-line equation for the stripping section becomes

$$y = \left(\frac{V_B + 1}{V_B}\right) x - \left(\frac{1}{V_B}\right) x_B \tag{7-14}$$

If values of V_B and x_B are known, (7-14) can be plotted, together with the equilibrium curve and a 45° line, as a straight line with an intersection at $y = x_B$ on the 45° line and a slope of $\overline{L}/\overline{V} = (V_B + 1)/V_B$, as shown in Figure 7.6b. The equilibrium stages are stepped off, in a manner similar to that described for the rectifying section, starting from the point $(y = x_B, x = x_B)$ on the operating and 45° lines and moving upward on a vertical line until the equilibrium curve is intersected at $(y = y_B, x = x_B)$, which represents the equilibrium mole fractions in the vapor and liquid leaving the partial reboiler. From that point, the staircase is constructed by drawing horizontal and then vertical lines, moving back and forth between the operating line and equilibrium curve, as observed in Figure 7.6b, where the staircase is arbitrarily terminated at stage m.

Feed-Stage Considerations

Thus far, the McCabe–Thiele construction has not considered the feed to the column. In determining the operating lines for the rectifying and stripping sections, it is very important to note that although x_D and x_B can be selected independently, R and V_B are related by the feed phase condition.

Consider the five possible feed conditions shown in Figure 7.7, which assumes that the feed has been flashed adiabatically to the feed-stage pressure. If the feed is a bubble-point liquid, it adds to the reflux, L, coming from the stage above to give $\overline{L} = L + F$. If the feed is a dew-point vapor, it adds to the boilup vapor, $\overline{V}$, coming from the stage below to give $V = \overline{V} + F$. For a partially vaporized feed, as shown in Figure 7.7c, $F = L_F + V_F$ and $\overline{L} = L + L_F$ and $V = \overline{V} + V_F$. If the feed is a subcooled liquid, it will cause a portion of the boilup, $\overline{V}$, to condense giving $\overline{L} > L + F$ and $V < \overline{V}$. If the feed is a superheated vapor, it will cause a portion of the reflux, L, to vaporize giving $\overline{L} < L$ and $V > \overline{V} + F$.

For cases (b), (c), and (d) of Figure 7.7, covering a range of feed conditions from a saturated liquid to a saturated vapor, the boilup $\overline{V}$ is related to the reflux L by the material balance:

$$\overline{V} = L + D - V_F \tag{7-15}$$

and the boilup ratio, $V_B = \overline{V}/B$, is

$$V_B = \frac{L + D - V_F}{B} \tag{7-16}$$

Alternatively, the reflux can be determined from the boilup by

$$L = \overline{V} + B - L_F \tag{7-17}$$

Although distillation operations can be specified by either the reflux ratio R or the boilup ratio V_B, by tradition R or R/R_{min} is used because the distillate product is most often the more important product.

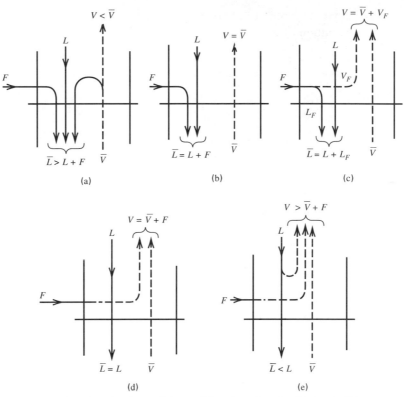

Figure 7.7 Possible feed conditions: (a) subcooled liquid feed; (b) bubble-point liquid feed; (c) partially vaporized feed; (d) dew-point vapor feed; (e) superheated vapor feed. [Adapted from W.L. McCabe, J.C. Smith, and P. Harriott, *Unit Operations of Chemical Engineering,* 5th ed., McGraw-Hill, New York (1993).]

For the other two cases, (a) and (e) of Figure 7.7, V_B and R cannot be related by simple material balances alone. It is necessary to consider an energy balance to convert sensible enthalpy into latent enthalpy of phase change. This is most conveniently done by defining a parameter, q, as the ratio of the increase in molar reflux rate across the feed stage to the molar feed rate,

$$q = \frac{\overline{L} - L}{F} \qquad \text{(7-18)}$$

or by material balance around the feed stage,

$$q = 1 + \frac{\overline{V} - V}{F} \qquad \text{(7-19)}$$

Values of q for the five feed conditions are

Feed condition	q
Subcooled liquid	>1
Bubble-point liquid	1
Partially vaporized	$L_F/F = 1 -$ molar fraction vaporized
Dew-point vapor	0
Superheated vapor	<0

To determine values of q for subcooled liquid and superheated vapor, a more general definition of q is applied:

q = enthalpy change to bring the feed to a dew-point vapor divided by enthalpy of vaporization of the feed (dew-point vapor minus bubble-point liquid), that is,

$$q = \frac{(h_F)_{\text{sat'd vapor temperature}} - (h_F)_{\text{feed temperature}}}{(h_F)_{\text{sat'd vapor temperature}} - (h_F)_{\text{sat'd liquid temperature}}} \tag{7-20}$$

For a subcooled liquid feed, (7-20) becomes

$$q = \frac{\Delta H^{\text{vap}} + C_{P_L}(T_b - T_F)}{\Delta H^{\text{vap}}} \tag{7-21}$$

For a superheated vapor, (7-20) becomes

$$q = \frac{C_{P_V}(T_d - T_F)}{\Delta H^{\text{vap}}} \tag{7-22}$$

where C_{P_L} and C_{P_V} are the liquid and vapor molar heat capacities, respectively, ΔH^{vap} is the molar enthalpy change from the bubble point to the dew point, and T_F, T_d, and T_b are the feed, dew-point, and bubble-point temperatures, respectively, of the feed at the column operating pressure.

Instead of using (7-14) to locate the stripping operating line on the McCabe–Thiele diagram, it is more common to use an alternative method that involves a q-line, which is included in Figure 7.4. The q-line, one point of which is the intersection of the rectifying and stripping operating lines, is derived in the following manner. Subtracting (7-11) from (7-6) gives

$$y(V - \overline{V}) = (L - \overline{L})x + Dx_D + Bx_B \tag{7-23}$$

But

$$Dx_D + Bx_B = Fz_F \tag{7-24}$$

and a material balance around the feed stage gives

$$F + \overline{V} + L = V + \overline{L} \tag{7-25}$$

Combining (7-23) to (7-25) with (7-18) gives

$$y = \left(\frac{q}{q-1}\right)x - \left(\frac{z_F}{q-1}\right) \tag{7-26}$$

which is the equation for the q-line. This line is located on the McCabe–Thiele diagram by noting that when $x = z_F$, Eq. (7-26) reduces to the point $y = z_F = x$, which lies on the 45° line. From (7-26), the slope of the line is $q/(q-1)$. This construction is shown in Figure 7.4 for a partially vaporized feed, for which $0 < q < 1$ and $-\infty < [q/(q-1)] < 0$. Following the placement of the rectifying section operating line and the q-line, the stripping-section operating line is located by drawing a straight line from the point ($y = x_B, x = x_B$) on the 45° line to and through the point of intersection of the q-line and the rectifying-section operating line as shown in Figure 7.4. The point of intersection must lie somewhere between the equilibrium curve and the 45° line.

As q changes from a value greater than 1 (subcooled liquid) to a value less than 0 (superheated vapor), the slope of the q-line, $q/(q-1)$, changes from a positive value to a negative value and back to a positive value, as shown in Figure 7.8. For a saturated liquid feed, the q-line is vertical; for a saturated vapor, the q-line is horizontal.

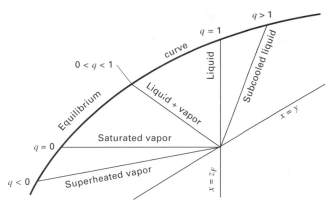

Figure 7.8 Effect of thermal condition of feed on slope of q-line.

Determination of Number of Equilibrium Stages and Feed-Stage Location

Following the construction of the five lines shown in Figure 7.4, the number of equilibrium stages required for the entire column, as well as the location of the feed stage, are determined by stepping off stages by any of several ways. The stages can be stepped off first from the top down and then from the bottom up, as described above, until a point of merger is found for the feed stage. Alternatively, the stages can be stepped off from the bottom all the way to the top, or vice versa. Hardly ever will an integer number of stages result, but rather a fractional stage will appear near the middle, at the top, or at the bottom. Usually the staircase is stepped off from the top and continued all the way to the bottom, starting from the point $(y = x_D, x = x_D)$ on the 45° line, as shown in Figure 7.9 for the case of a partially vaporized feed. In that figure, point P is the intersection of the q-line with the two operating lines. The transfer point for stepping off stages between the rectifying-section operating line and the equilibrium curve to stepping off stages between the stripping-section operating line and the equilibrium curve occurs at the feed stage. In Figure 7.9a, the feed stage is stage 3 from the top and a fortuitous total of exactly five stages is required, where the last stage is the partial reboiler. In Figure 7.9b the feed stage is stage 5 and a total of about 6.4 stages is required. In Figure 7.9c, the feed stage is stage 2 and a total of about 5.9 stages is required. In Figure 7.9b, the stepping off of stages in the rectifying section can be continued indefinitely, finally approaching, but never reaching, point K. In Figure 7.9c, if the stepping off of stages had started from the partial reboiler at the point $(y = x_B, x = x_B)$ and proceeded upward, the staircase in the stripping section could have been continued indefinitely, finally approaching, but never reaching, point R. In Figure 7.9, it is seen that the smallest number of total stages occurs when the transfer is made at the first opportunity after a horizontal line of the staircase passes over point P, as in Figure 7.9a. This feed-stage location is optimal.

Limiting Conditions

For a given specification (Table 7.2), a reflux ratio can be selected anywhere from the minimum, $R_{\min}$, to an infinite value (total reflux) where all of the overhead vapor is condensed and returned to the top stage (thus, no distillate is withdrawn). As shown in Figure 7.10, the minimum corresponds to the need for an infinite number of stages, while the infinite value corresponds to the minimum number of equilibrium stages. The McCabe–Thiele graphical method can quickly determine the two limits, $N_{\min}$ and $R_{\min}$. Then, for a practical operation, $N_{\min} < N < \infty$ and $R_{\min} < R < \infty$.

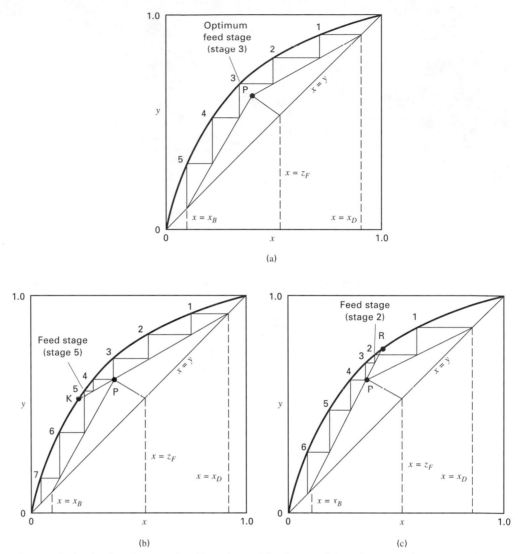

Figure 7.9 Optimal and nonoptimal locations of feed stage: (a) optimum feed-stage location; (b) feed-stage location below optimum stage; (c) feed-stage location above optimum stage.

Minimum Number of Equilibrium Stages

As the reflux ratio is increased, the slope of the rectifying section operating line, given by (7-7), increases from $L/V < 1$ to a limiting value of $L/V = 1$. Correspondingly, the boilup ratio increases and the slope of the stripping section operating line, given by (7-12), decreases from $\overline{L}/\overline{V} > 1$ to a limiting value of $\overline{L}/\overline{V} = 1$. Thus, at this limiting condition, both the rectifying and stripping operating lines coincide with the 45° line and neither the feed composition, z_F, nor the q-line influences the staircase construction. This is total reflux because when $L = V$, $D = B = 0$, and the total condensed overhead is returned to the column as reflux. Furthermore, all liquid leaving the bottom stage is vaporized and returned as boilup to the column. If both distillate and bottoms flow rates are zero, the feed to the column is also zero, which is consistent with the lack of influence of the feed condition. It is possible to operate a column at total reflux, and such an operation is

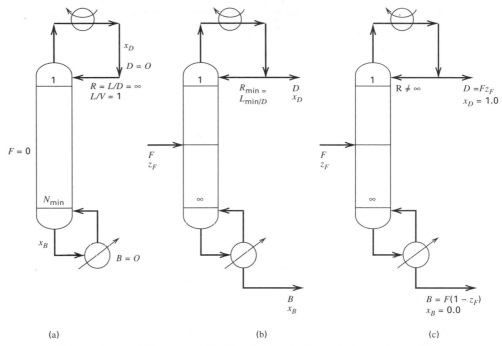

(a)   (b)   (c)

Figure 7.10 Limiting conditions for distillation: (a) total reflux, minimum stages; (b) minimum reflux, infinite stages; (c) perfect separation for nonazeotropic system.

convenient for measuring tray efficiency because a steady-state operating condition is readily achieved.

A simple example of the McCabe–Thiele construction for this limiting condition is shown in Figure 7.11 for two equilibrium stages. Because the operating lines are located as far away as possible from the equilibrium curve, a minimum number of stages is required.

Minimum Reflux Ratio

As the reflux ratio decreases from the limiting case of infinity (i.e., total reflux), the intersection of the two operating lines and the q-line moves from the 45° line toward the equilibrium curve. The number of equilibrium stages required increases because the

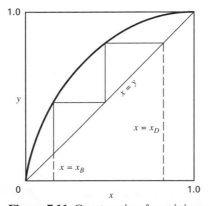

Figure 7.11 Construction for minimum stages at total reflux.

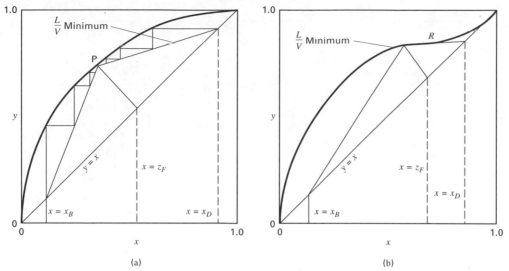

Figure 7.12 Construction for minimum reflux at infinite stages: (a) typical ideal or near-ideal system, pinch point at the feed stage; (b) typical nonideal system, pinch point above the feed stage.

operating lines move closer and closer to the equilibrium curve, thus requiring more and more stairs to move from the top of the column to the bottom. Finally a limiting condition is reached when the point of intersection is on the equilibrium curve, as shown in Figure 7.12. For binary mixtures that are not highly nonideal, the typical case is shown in Figure 7.12a, where the intersection, P, is at the feed stage. To reach that stage from either the rectifying section or the stripping section, an infinite number of stages is required. The point P is called a *pinch point* because the two operating lines each pinch the equilibrium curve.

For a highly nonideal binary system, the pinch point may occur at a stage above or below the feed stage. The former case is illustrated in Figure 7.12b, where the operating line for the rectifying section intersects the equilibrium curve before the feed stage is reached. The slope of this operating line cannot be reduced further because it would then cross over the equilibrium curve and thereby violate the second law of thermodynamics because of a reversal in the direction of mass transfer. This would require spontaneous mass transfer from a region of low concentration to a region of high concentration. This is similar to a second-law violation by a temperature crossover in a heat exchanger. Now, the pinch point occurs entirely in the rectifying section, where an infinite number of stages exists; the stripping section contains a finite number of stages.

From the slope of the limiting operating line for the rectifying section, the minimum reflux ratio can be determined. From (7-7), the minimum feasible slope is

$$(L/V)_{min} = R_{min}/(R_{min} + 1)$$

$$\text{or } R_{min} = (L/V)_{min}/[1 - (L/V)_{min}] \qquad \textbf{(7-27)}$$

Alternatively, the limiting condition of infinite stages corresponds to a minimum boilup ratio for $(\overline{L}/\overline{V})_{max}$. From (7-14),

$$(V_B)_{min} = 1/[(\overline{L}/\overline{V})_{max} - 1] \qquad \textbf{(7-28)}$$

Perfect Separation

A third limiting condition of interest involves the degree of separation. As a perfect split ($x_D = 1$, $x_B = 0$) is approached, for a reflux ratio at or greater than the minimum value, the number of stages required near the top and near the bottom of the column increases rapidly and without limit until pinches are encountered at $x_D = 1$ and $x_B = 0$. Thus, a perfect separation of a binary mixture that does not form an azeotrope requires an infinite number of stages in both sections of the column. However, this is not the case for the reflux ratio. In Figure 7.12, as x_D is moved from, say, 0.90 toward 1.0, the slope of the operating line at first increases, but in the range of x_D from 0.99 to 1.0 the slope changes only slightly. Furthermore, the value of the slope, and therefore the value of R, is finite for a perfect separation. For example, if the feed is a saturated liquid, application of (7-4) and (7-7) gives the following equation for the minimum reflux of a perfect binary separation:

$$R_{\min} = \frac{1}{z_F(\alpha - 1)} \tag{7-29}$$

where the relative volatility, α, is evaluated at the feed condition.

EXAMPLE 7.1

A trayed tower is to be designed to continuously distill 450 lbmol/h (204 kmol/h) of a binary mixture of 60 mol% benzene and 40 mol% toluene. A liquid distillate and a liquid bottoms product of 95 mol% and 5 mol% benzene, respectively, are to be produced. The feed is preheated so that it enters the column with a molar percent vaporization equal to the distillate-to-feed ratio. Use the McCabe–Thiele method to compute the following, assuming a uniform pressure of 1 atm (101.3 kPa) throughout the column: (a) Minimum number of theoretical stages, $N_{\min}$; (b) Minimum reflux ratio, $R_{\min}$; and (c) Number of equilibrium stages N, for a reflux-to-minimum reflux ratio, $R/R_{\min}$, of 1.3 and the optimal location of the feed stage.

SOLUTION

Calculate D and B. An overall material balance on benzene gives

$$0.60(450) = 0.95D + 0.05B \tag{1}$$

A total balance gives $450 = D + B$ (2)

Combining (1) and (2) to eliminate B, followed by solving the resulting equation for D and (2) for B gives $D = 275$ lbmol/h, $B = 175$ lbmol/h, and $D/F = 0.611$

Calculate the slope of the q-line:
$V_F/F = D/F$ for this example $= 0.611$ and q for a partially vaporized feed is

$$\frac{L_F}{F} = \frac{(F - V_F)}{F} - 1 - \frac{V_F}{F} = 0.389$$

From (7-26), the slope of the q-line is $\dfrac{q}{q-1} = \dfrac{0.389}{0.389 - 1} = -0.637$

(a) In Figure 7.13, where y and x refer to benzene, the more volatile component, with $x_D = 0.95$ and $x_B = 0.05$, the number of minimum equilibrium stages is stepped off between the equilibrium curve and the 45°, starting from the top, giving $N_{\min} = 6.7$.

(b) In Figure 7.14, a q-line is drawn that has a slope of -0.637 and passes through the feed composition ($z_F = 0.60$) on the 45° line. For the minimum reflux condition, an operating line for the rectifying section passes through the point $x = x_D = 0.95$ on the 45° line and through the point of intersection of the q-line and the equilibrium curve ($y = 0.684$, $x = 0.465$). The slope of this operating line is 0.55, which from (7-9) equals $R/(R + 1)$. Therefore, $R_{\min} = 1.22$.

(c) The operating reflux ratio is $1.3R_{\min} = 1.3(1.22) = 1.59$
From (7-9), the slope of the operating line for the rectifying section is

$$\frac{R}{R + 1} = \frac{1.59}{1.59 + 1} = 0.614$$

The construction for the resulting two operating lines, together with the q-line, is shown in Figure 7.15, where the operating line for the stripping section is drawn to pass through the point $x = x_B = 0.05$ on the 45° line and the point of intersection of the q-line and the operating line for the stripping section. The number of equilibrium stages is stepped off between first, the rectifying-section operating line and the equilibrium curve and then the stripping-section

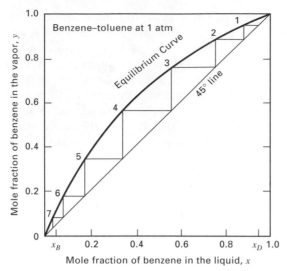

Figure 7.13 Determination of minimum stages for Example 7.1.

operating line and the equilibrium curve, starting from point A (at $x = x_D = 0.95$) and finishing at point B (to the left of $x = x_B = 0.05$). For the optimal feed-stage location, the transfer from the rectifying-section operating line to the stripping-section operating line takes place at point P. The result is $N = 13.2$ equilibrium stages, with stage 7 from the top being the feed stage. Thus, for this example, $N/N_{min} = 13.2/6.7 = 1.97$. The bottom stage is the partial reboiler, leaving 12.2 equilibrium stages contained in the column. If the plate efficiency were 0.8, 16 trays would be needed. ∎

Column Operating Pressure and Condenser Type

For preliminary design, column operating pressure and condenser type are established by the procedure shown in Figure 7.16, which is formulated to achieve, if possible, a reflux

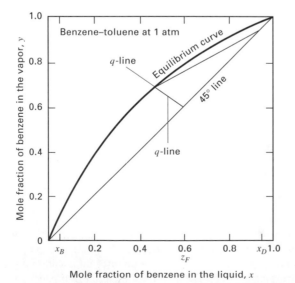

Figure 7.14 Determination of minimum reflux for Example 7.1.

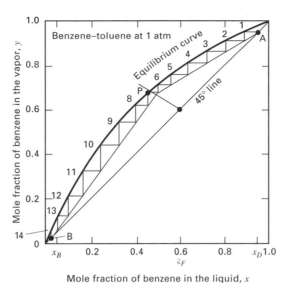

Figure 7.15 Determination of number of equilibrium stages and feed-stage location for Example 7.1.

drum pressure, P_D, between 0 and 415 psia (2.86 MPa) at a minimum temperature of 120°F (49°C) (corresponding to the use of water as the coolant in the condenser). The pressure and temperature limits are representative only and depend on economic factors. A condenser pressure drop of 0 to 2 psi (0 to 14 kPa) and an overall column pressure drop of 5 psi (35 kPa) may be assumed. However, when column tray requirements are known, more refined computations should result in at approximately 0.1 psi/tray (0.7 kPa/tray) pressure drop for atmospheric and superatmospheric pressure operation and 0.05 psi/tray (0.35 kPa/tray) pressure drop for vacuum column operation. Column bottom

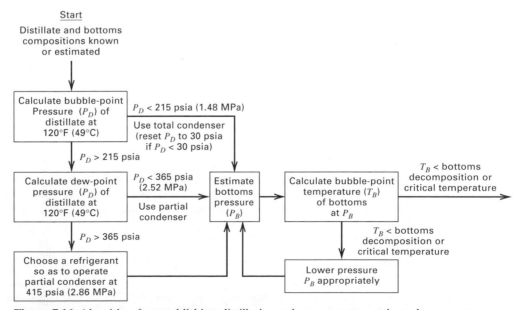

Figure 7.16 Algorithm for establishing distillation column pressure and condenser type.

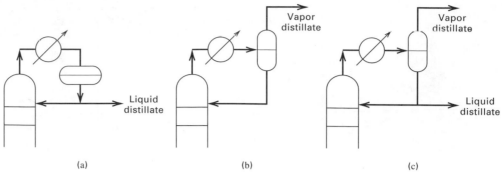

Figure 7.17 Condenser types: (a) total condenser; (b) partial condenser; (c) mixed condenser.

temperature must not result in bottoms decomposition or correspond to a near-critical condition.

A total condenser is recommended for reflux drum pressures to 215 psia (1.48 MPa). A partial condenser is appropriate from 215 psia to 365 psia (2.52 MPa). However, a partial condenser can be used below 215 psia when a vapor distillate is desired. A mixed condenser can provide both vapor and liquid distillates. The three types of condenser configurations are shown in Figure 7.17. A refrigerant is used as condenser coolant if pressure tends to exceed 365 psia.

When a partial condenser is specified, the McCabe–Thiele staircase construction for the case of a total condenser must be modified, as will be illustrated in Example 7.2, to account for the fact that the first equilibrium stage, counted down from the top, is the partial condenser. This is based on the assumption that the liquid reflux leaving the reflux drum is in equilibrium with the vapor distillate.

Subcooled Reflux

Although most distillation columns are designed so that the reflux is a saturated (bubble-point) liquid, such is not always the case for operating columns. If the condenser type is partial or mixed, the reflux is a saturated liquid unless heat losses cause its temperature to decrease. For a total condenser, however, the operating reflux is often a subcooled liquid at column pressure, particularly if the condenser is not tightly designed and the distillate bubble-point temperature is significantly higher than the inlet cooling water temperature. If the condenser outlet pressure is lower than the top-tray pressure of the column, the reflux is subcooled for any of the three types of condensers.

When subcooled reflux enters the top tray, its temperature rises and causes vapor entering the tray to condense. The latent enthalpy of condensation of the vapor provides the sensible enthalpy to heat the subcooled reflux to the bubble point. In that event, the internal reflux ratio within the rectifying section of the column is higher than the external reflux ratio from the reflux drum. The McCabe–Thiele construction should be based on the internal reflux ratio, which can be estimated by the following equation derived from an approximate energy balance around the top tray:

$$R_{internal} = R \left(1 + \frac{C_{P_L} \Delta T_{subcooling}}{\Delta H^{vap}} \right) \tag{7-30}$$

where C_{P_L} and ΔH^{vap} are per mole and $\Delta T_{subcooling}$ is the degrees of subcooling. The internal reflux ratio replaces R, the external reflux ratio, in (7-9). If a correction is not made for subcooled reflux, the calculated number of equilibrium stages is somewhat more than required.

EXAMPLE 7.2

One thousand kilomoles per hour of a feed containing 30 mol% n-hexane and 70% n-octane is to be distilled in a column consisting of a partial reboiler, one equilibrium (theoretical) plate, and a partial condenser, all operating at 1 atm (101.3 kPa). Thus, hexane is the light key and octane is the heavy key. The feed, a bubble-point liquid, is fed to the reboiler, from which a liquid bottoms product is continuously withdrawn. Bubble-point reflux is returned from the partial condenser to the plate. The vapor distillate, in equilibrium with the reflux, contains 80 mol% hexane, and the reflux ratio, L/D, is 2. Assume that the partial reboiler, plate, and partial condenser each function as equilibrium stages.

(a) Using the McCabe–Thiele method, calculate the bottoms composition and kilomoles per hour of distillate produced.
(b) If the relative volatility α is assumed constant at a value of 5 over the composition range (the relative volatility actually varies from approximately 4.3 at the reboiler to 6.0 at the condenser), calculate the bottoms composition analytically.

SOLUTION

First determine whether the problem is completely specified. From Table 5.2c, we have $N_D = C + 2N + 6$ degrees of freedom, where N includes the partial reboiler and the stages in the column, but not the partial condenser. With $N = 2$ and $C = 2$, $N_D = 12$. Specified in this problem are

Feed-stream variables	4
Plate and reboiler pressures	2
Condenser pressure	1
Q (=0) for plate	1
Number of stages	1
Feed-stage location	1
Reflux ratio, L/D	1
Distillate composition	1
Total	12

Thus, the problem is fully specified and can be solved.

(a) *Graphical solution.* A diagram of the separator is given in Figure 7.18 as is the McCabe–Thiele graphical solution, which is constructed in the following manner.

1. The point $y_D = 0.8$ at the partial condenser is located on the $x = y$ line.
2. Conditions in the condenser are fixed because x_R (reflux composition) is in equilibrium with y_D. Hence, the point (x_R, y_D) is located on the equilibrium curve.
3. Noting that $(L/V) = 1 - 1/[+(L/D)] = 2/3$, the operating line with slope $L/V = 2/3$ is drawn through the point $y_D = 0.8$ on the 45° line until it intersects the equilibrium curve. Because the feed is introduced into the partial reboiler, there is no stripping section.
4. Three theoretical stages (partial condenser, plate 1, and partial reboiler) are stepped off and the bottoms composition $x_B = 0.135$ is read.

The amount of distillate is determined from overall material balances. For hexane, $z_F F = y_D D + x_B B$. Therefore, $(0.3)(1,000) = (0.8)D + (0.135)B$. For the total flow, $B = 1,000 - D$. Solving these two equations simultaneously, $D = 248$ kmol/h.

(b) *Analytical solution.* For constant α, equilibrium liquid compositions for the light key, in terms of α and y are given by a rearrangement of (7-3): $x = \dfrac{y}{y + \alpha(1 - y)}$ (1)

where α is assumed constant at a value of 5.

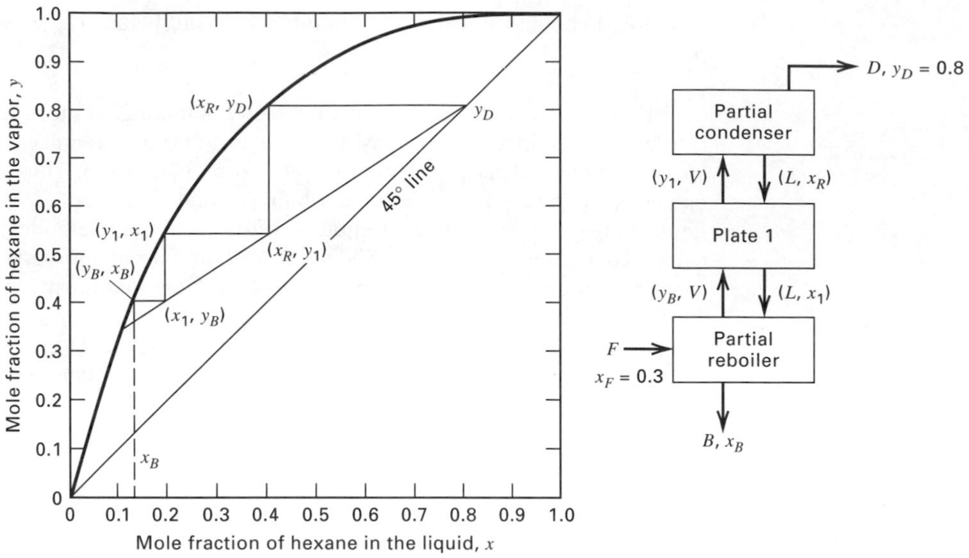

Figure 7.18 Solution to Example 7.2.

The steps in the solution are as follows:

1. The liquid leaving the partial condenser at x_R is calculated from (1), for $y = y_D = 0.8$:

$$x_R = \frac{0.8}{0.8 + 5(1 - 0.8)} = 0.44$$

2. Then y_1 is determined by a material balance about the partial condenser:

$$Vy_1 = Dy_D + Lx_R \quad \text{with} \quad D/V = 1/3 \quad \text{and} \quad L/V = 2/3$$

$$y_1 = (1/3)(0.8) + (2/3)(0.44) = 0.56$$

3. From (1), for plate 1, $x_1 = \dfrac{0.56}{0.56 + 5(1 - 0.56)} = 0.203$

4. By material balance around plate 1 and the partial condenser,

$$Vy_B = Dy_D + Lx_1 \quad \text{and} \quad y_B = (1/3)(0.8) + (2/3)(0.203) = 0.402$$

5. From (1), for the partial reboiler, $x_B = \dfrac{0.402}{0.402 + 5(1 - 0.402)} = 0.119$.

EXAMPLE 7.2

By approximating the equilibrium curve with $\alpha = 5$, an answer of 0.119 rather than 0.135 for x_B is obtained. Note that for a larger number of theoretical plates, part (b) can be readily computed with a spreadsheet program. ∎

EXAMPLE 7.3

Consider Example 7.2. (a) Solve it graphically, assuming that the feed is introduced on plate 1, rather than into the reboiler. (b) Determine the minimum number of stages required to carry out the separation. (c) Determine the minimum reflux ratio.

SOLUTION

(a) The flowsheet and solution given in Figure 7.19 are obtained as follows.

1. The point x_R, y_D is located on the equilibrium line.
2. The operating line for the enriching section is drawn through the point $y = x = 0.8$, with a slope of $L/V = 2/3$.
3. The intersection of the q-line, $x_F = 0.3$ (which, for a saturated liquid, is a vertical line), with the enriching-section operating line is located at point P. The stripping-section operating line must also pass through this point, but its slope and the point x_B are not known initially.

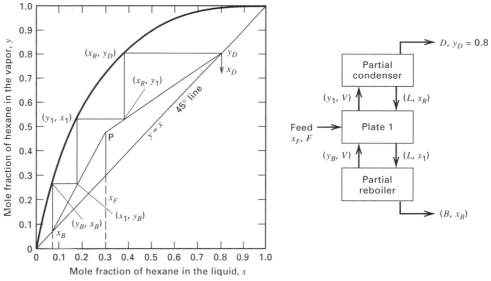

Figure 7.19 Solution to Example 7.3.

4. The slope of the stripping-section operating line is found by trial and error because there are three equilibrium contacts in the column, with the middle stage involved in the switch from one operating line to the other. If the middle stage is the optimum feed-stage location, the result is $x_B = 0.07$, as shown in Figure 7.19. The amount of distillate is obtained from the combined total and hexane overall material balances to give $(0.3)(1,000) = (0.8D) + 0.07(1,000 - D)$. Solving, $D = 315$ kmol/h

Comparing this result to that obtained in Example 7.2, we find that the bottoms purity and distillate yield are improved by introduction of the feed to plate 1, rather than to the reboiler. This improvement could have been anticipated if the q-line had been constructed in Figure 7.18.

(b) The construction corresponding to total reflux ($L/V = 1$, no products, no feed, minimum equilibrium stages) is shown in Figure 7.20. Slightly more than two stages are required for an x_B of 0.07, compared to the three stages previously required.

(c) To determine the minimum reflux ratio, the vertical q-line in Figure 7.19 is extended from point P until the equilibrium curve is intersected, which is determined to be the point

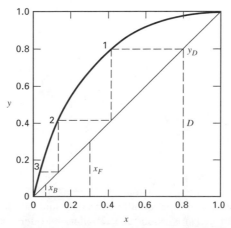

Figure 7.20 Solution for total reflux in Example 7.3.

(0.68, 0.3). The slope, $(L/V)_{min}$ of the operating line for the rectifying section, which connects this point to the point (0.8, 0.8) on the 45° line is 0.24. Thus $(L/D)_{min} = (L/V_{min})/[1 - (L/V_{min})] = 0.32$. This is considerably less than the $L/D = 2$ specified. ∎

Reboiler Type

Different types of reboilers are used to provide boilup vapor to the stripping section of a distillation column. For small laboratory and pilot-plant-size columns, the reboiler consists of a reservoir of liquid located just below the bottom plate to which heat is supplied from (1) a jacket or mantle that is heated by an electrical current or by condensing steam, or (2) tubes that pass through the liquid reservoir carrying condensing steam. Both of these types of reboilers have limited heat transfer surface and are not suitable for industrial applications.

For plant-size distillation columns, the reboiler is usually an external heat exchanger, as shown in Figure 7.21, of either the kettle or vertical thermosyphon type. Both can provide the amount of heat transfer surface required for large installations. In the former case, liquid leaving the sump (reservoir) at the bottom of the column enters the kettle,

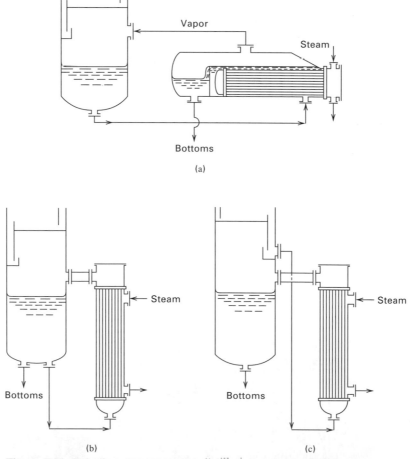

Figure 7.21 Reboilers for plant-size distillation columns: (a) kettle-type reboiler; (b) vertical thermosyphon-type reboiler, reboiler liquid withdrawn from bottom sump; (c) vertical thermosyphon-type reboiler, reboiler liquid withdrawn from bottom-tray downcomer.

where it is partially vaporized by the transfer of heat from tubes carrying condensing steam or some other heating medium. The bottoms product liquid leaving the reboiler is assumed to be in equilibrium with the vapor returning to the bottom tray of the column. Thus the kettle reboiler is a partial reboiler equivalent to one equilibrium stage. The kettle reboiler is sometimes located in the bottom of the column to avoid piping.

The vertical thermosyphon reboiler may be of the type shown in Figure 7.21b or 7.21c. In the former, both the bottoms product and the reboiler feed are withdrawn from the column bottom sump. Circulation through the tubes of the reboiler occurs because of the difference in static heads of the supply liquid and the column of partially vaporized fluid flowing through the reboiler tubes. The partial vaporization provides enrichment of the exiting vapor in the more volatile component. However, the exiting liquid is then mixed with liquid leaving the bottom tray, which contains a higher percentage of the more volatile component. The result is that this type of reboiler arrangement provides only a fraction of an equilibrium stage and it is best to take no credit for it.

A more complex and less common vertical thermosyphon reboiler is that of Figure 7.21c, where the reboiler liquid is withdrawn from the downcomer of the bottom tray. Partially vaporized liquid is returned to the column, where the bottoms product from the bottom sump is withdrawn. This type of reboiler does function as an equilibrium stage.

Kettle reboilers are common, but thermosyphon reboilers are favored when (1) the bottoms product contains thermally sensitive compounds, (2) bottoms pressure is high, (3) only a small ΔT is available for heat transfer, and (4) heavy fouling occurs. Horizontal thermosyphon reboilers are sometimes used in place of the vertical types when only small static heads are needed for circulation, surface area requirement is very large, and/or when frequent cleaning of the tubes is anticipated. A pump may be added for either thermosyphon type to improve circulation. Liquid residence time in the column bottom sump should be at least 1 minute and perhaps as much as 5 minutes or more.

Condenser and Reboiler Duties

Following the determination of the feed condition, reflux ratio, and number of theoretical stages by the McCabe–Thiele method, estimates of the heat duties of the condenser and reboiler are made. An energy balance for the entire column gives

$$Fh_F + Q_R = Dh_D + Bh_B + Q_C + Q_{\text{loss}} \tag{7-31}$$

Except for small and/or uninsulated distillation equipment, Q_{loss} is negligible and can be ignored. We can approximate the energy balance of (7-31) by applying the assumptions of the McCabe–Thiele method. An energy balance for a total condenser is

$$Q_C = D(R + 1)\,\Delta H^{\text{vap}} \tag{7-32}$$

where ΔH^{vap} = average molar heat of vaporization of the two components being separated. For a partial condenser,

$$Q_C = DR\,\Delta H^{\text{vap}} \tag{7-33}$$

For a partial reboiler,

$$Q_R = BV_B\,\Delta H^{\text{vap}} \tag{7-34}$$

When the feed is at the bubble point and a total condenser is used, (7-16) can be arranged to:

$$BV_B = L + D = D(R + 1) \tag{7-35}$$

Comparing this to (7-34) and (7-32), note that $Q_R = Q_C$. When the feed is partially vaporized and a total condenser is used, the heat required by the reboiler is less than the condenser duty and is given by

$$Q_R = Q_C \left[1 - \frac{V_F}{D(R + 1)} \right] \tag{7-36}$$

If saturated steam is the heating medium for the reboiler, the steam rate required is given by an energy balance:

$$m_s = \frac{M_s Q_R}{\Delta H_s^{\text{vap}}} \tag{7-37}$$

where,

m_s = mass flow rate of steam

Q_R = reboiler duty (rate of heat transfer)

M_s = molecular weight of steam

ΔH_s^{vap} = molar enthalpy of vaporization of steam

The cooling water rate for the condenser is

$$m_{\text{cw}} = \frac{Q_C}{C_{P_{\text{H}_2\text{O}}}(T_{\text{out}} - T_{\text{in}})} \tag{7-38}$$

where, m_{cw} = mass flow rate of cooling water

Q_C = condenser duty (rate of heat transfer)

$C_{P_{\text{H}_2\text{O}}}$ = specific heat of water

$T_{\text{out}}, T_{\text{in}}$ = temperature of cooling water out of and into the condenser, respectively

Because the annual cost of reboiler steam can be an order of magnitude higher than the annual cost of cooling water, the feed to a distillation column is frequently preheated and partially vaporized to reduce Q_R, in comparison to Q_C, as indicated by (7-36).

Feed Preheat

The feed to a distillation column is usually a process feed, an effluent from a reactor, or a liquid product from another separator. The feed pressure must be greater than the pressure in the column at the feed-tray location. If so, any excess feed pressure is dropped across a valve, which may cause the feed to partially vaporize before entering the column; if not, additional pressure is added with a pump.

The temperature of the feed as it enters the column does not necessarily equal the temperature in the column at the feed-tray location. However, such equality will increase second-law efficiency. It is usually best to avoid a subcooled liquid or superheated vapor feed and supply a partially vaporized feed. This is achieved by preheating the feed in a heat exchanger with the bottoms product or some other process stream that possesses a suitably high temperature, to ensure a reasonable ΔT driving force for heat transfer, and a sufficient available enthalpy.

Optimal Reflux Ratio

An industrial distillation column must be operated between the two limiting conditions of minimum reflux and total reflux. As shown in Table 7.3, for a typical case adapted from

Table 7.3 Effect of Reflux Ratio on Annualized Cost of a Distillation Operation

R/R_{min}	Actual N	Diam., ft	Reboiler Duty, Btu/h	Condenser Duty, Btu/h	Annualized Cost, $/yr			Total Annualized Cost, $/yr
					Equipment	Cooling Water	Steam	
1.00	Infinite	6.7	9,510,160	9,416,000	Infinite	17,340	132,900	Infinite
1.05	29	6.8	9,776,800	9,680,000	44,640	17,820	136,500	198,960
1.14	21	7.0	10,221,200	10,120,000	38,100	18,600	142,500	199,200
1.23	18	7.1	10,665,600	10,560,000	36,480	19,410	148,800	204,690
1.32	16	7.3	11,110,000	11,000,000	35,640	20,220	155,100	210,960
1.49	14	7.7	11,998,800	11,880,000	35,940	21,870	167,100	224,910
1.75	13	8.0	13,332,000	13,200,000	36,870	24,300	185,400	246,570

(Adapted from an example by Peters and Timmerhaus [6].)

Peters and Timmerhaus [6], as the reflux ratio is increased from the minimum value, the number of plates decreases, the column diameter increases, and the reboiler steam and condenser cooling water requirements increase. When the annualized fixed investment costs for the column, condenser, reflux drum, reflux pump, and reboiler are added to the annual cost of steam and cooling water, an optimal reflux ratio is established, as shown, for the conditions of Table 7.3, in Figure 7.22. For this example the optimal R/R_{min} is 1.1. The data in Table 7.3 show that although the condenser and reboiler duties are almost identical for a given reflux ratio, the annual cost of steam for the reboiler is almost eight times that of the cost of condenser cooling water. The total annual cost is dominated by the cost of steam except at the minimum-reflux condition. At the optimal reflux ratio, the cost of steam is 70% of the total annualized cost. Because the cost of steam is dominant,

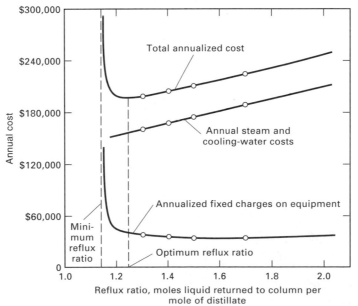

Figure 7.22 Optimum reflux ratio for a representative distillation operation. [Adapted from M.S. Peters and K.D. Timmerhaus, *Plant Design and Economics for Chemical Engineers,* 4th ed., McGraw-Hill, New York (1991).]

the optimal reflux ratio is sensitive to the steam cost. For example, at the extreme of zero cost for steam, the optimal R/R_{min} for this example is shifted from 1.1 to 1.32. This example assumes that the heat removed by cooling water in the condenser has no value.

The range of optimal ratio of reflux to minimum reflux often is from 1.05 to 1.50, with the lower value applying to a difficult separation (e.g., $\alpha = 1.2$) and the higher value applying to an easy separation (e.g., $\alpha = 5$). However, as seen in Figure 7.22, the optimal reflux ratio is not sharply defined. Accordingly, to achieve greater operating flexibility, columns are often designed for reflux ratios greater than the optimum.

Large Number of Stages

The McCabe–Thiele graphical construction is difficult to apply when conditions of relative volatility and/or product purities are such that a large number of stages must be stepped off. In that event, one of the following techniques can be used to determine the stage requirements.

1. Separate plots of expanded scales and/or larger dimensions are used for stepping off stages at the ends of the y–x diagram. For example, the additional plots might cover just the regions (1) 0.95 to 1.0 and (2) 0 to 0.05.
2. As described by Horvath and Schubert [7] and shown in Figure 7.23, a plot based on logarithmic coordinates is used for the low (bottoms) end of the y–x diagram, while for the high (distillate) end, the log–log graph is turned upside down and rotated 90°. Unfortunately, as seen in Figure 7.23, the operating lines become curved, but they can be plotted from a few points computed from (7-9) and (7-14). The 45° line remains straight and the normally curved equilibrium curve becomes nearly straight at the two ends.
3. The stages at the two ends are computed algebraically in the manner of part (b) of Example 7.2. This is readily done with a spreadsheet computer program.
4. If the equilibrium data are given in analytical form, commercially available McCabe–Thiele computer programs can be used
5. The stages are determined by combining the McCabe–Thiele graphical construction, for a suitable region in the middle, with the Kremser equations of Section 5.4 for the low and/or high ends, where absorption and stripping factors are almost constant. This technique, which is often preferred, is illustrated in the following example.

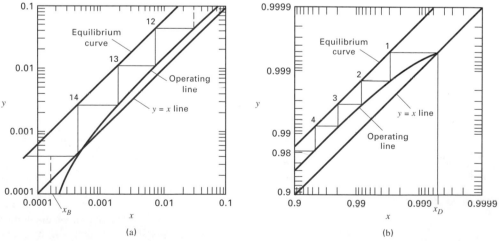

Figure 7.23 Use of log–log coordinates for McCabe–Thiele construction: (a) bottoms end of column; (b) distillate end of column.

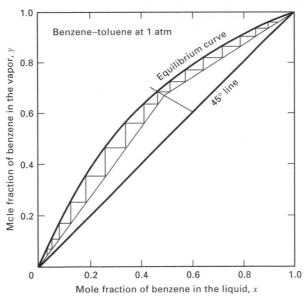

Figure 7.24 McCabe–Thiele construction for Example 7.4 from $x = 0.028$ to $x = 0.956$.

EXAMPLE 7.4

Repeat part (c) of Example 7.1 for distillate and bottoms purities of 99.9 and 0.1 mol%, respectively, using a reflux ratio of 1.88, which is about 30% higher than the minimum reflux of 1.44 for these new purities.

SOLUTION

Figure 7.24 shows the McCabe–Thiele construction for the region of x from 0.028 to 0.956, where the stages have been stepped off in two directions starting from the feed stage. In this middle region, nine stages are stepped off above the feed stage and eight below the feed stage, for a total of 18 stages, including the feed stage. The Kremser equations can now be applied to determine the remaining stages needed to achieve the desired high purities for the distillate and bottoms.

Additional stages for the rectifying Section. With respect to Figure 5.8a, counting stages from the top down, from Figure 7.24:

$$(y_{N+1})_{benzene} = 0.982 \quad \text{and} \quad (y_{N+1})_{toluene} = 0.018$$

Also, $(x_0)_{benzene} = (y_1)_{benzene} = 0.999$ and $(x_0)_{toluene} = (y_1)_{toluene} = 0.001$

Combining the Kremser equations (5.55), (5-34), (5-35), (5-48), and (5-50) and performing a number of algebraic manipulations:

$$N_R = \frac{\log\left[\frac{1}{A} + \left(1 - \frac{1}{A}\right)\left(\frac{y_{N+1} - x_0 K}{y_1 - x_0 K}\right)\right]}{\log A} \tag{7-39}$$

where N_R = additional equilibrium stages for the rectifying section. For that section, which is like an absorption section, it is best to apply (7-39) to toluene, the heavy key. Because $\alpha = 2.52$ at the top of the column, where $K_{benzene}$ is close to one, take $K_{toluene} = 1/2.52 = 0.408$. Since $R = 1.88$, $L/V = R/(R + 1) = 0.653$

Therefore, the absorption factor for toluene is $A_{\text{toluene}} = L/VK_{\text{toluene}} = 0.653/0.408 = 1.60$ which is assumed to remain constant in the uppermost part of the rectifying section. Therefore, from (7-39) for toluene,

$$N_R = \frac{\log\left[\dfrac{1}{1.60} + \left(1 - \dfrac{1}{1.60}\right)\left(\dfrac{0.018 - 0.001(0.408)}{0.001 - 0.001(0.408)}\right)\right]}{\log 1.60} = 5.2$$

Additional stages for the stripping section. With respect to Figure 5.8b, counting stages from the bottom up, we have from Figure 7.24: $(y_N)_{\text{benzene}} = 0.065$ and $(x_{N+1})_{\text{benzene}} = 0.048$. Also, $(x_B)_{\text{benzene}} = 0.001$. Combining the Kremser equations for a stripping section gives

$$N_S = \frac{\log\left[\overline{A} + (1 - \overline{A})\left(\dfrac{x_{N+1} - x_1/K}{x_1 - x_1/K}\right)\right]}{\log(1/\overline{A})} \tag{7-40}$$

where N_S = additional equilibrium stages for the stripping section

$\overline{A}$ = absorption factor in the stripping section = $\overline{L}/K\overline{V}$

Because benzene is being stripped in the stripping section, it is best to apply (7-40) to the benzene. At the bottom of the column, where K_{toluene} is approximately 1.0, $\alpha = 2.26$, and therefore $K_{\text{benzene}} = 2.26$. By material balance, with flows in lbmol/h, $D = 270.1$. For $R = 1.88$, $L = 507.8$, and $V = 270.1 + 507.8 = 777.9$. From Example 7.1, $V_F = 0.611(450) = 275$ and $L_F = 450 - 275 = 175$. Therefore, $\overline{L} = L + L_F = 507.8 + 175 = 682.8$ lbmol/h and $\overline{V} = V - V_F = 777.9 - 275 = 502.9$ lbmol/h.

$$\overline{L}/\overline{V} = 682.8/502.9 = 1.358; \overline{A}_{\text{benzene}} = \overline{L}/K\overline{V} = 1.358/2.26 = 0.601; x_1 = x_{B_{\text{benzene}}} = 0.001$$

Substitution into (7-40) gives

$$N_S = \frac{\log\left[0.601 + (1 - 0.601)\left(\dfrac{0.048 - 0.001/2.26}{0.001 - 0.001/2.26}\right)\right]}{\log(1/0.601)} = 7.0$$

This value includes the partial reboiler. Accordingly, the total number of equilibrium stages starting from the bottom is: partial reboiler + 7.0 + 8 + feed stage + 9 + 5.2 = 31.2. ∎

Use of Murphree Efficiency

The McCabe–Thiele method assumes that the two phases leaving each stage are in thermodynamic equilibrium. In industrial countercurrent, multistage equipment, it is not always practical to provide the combination of residence time and intimacy of contact required to approach equilibrium closely. Hence, concentration changes for a given stage are usually less than predicted by equilibrium.

As discussed in Section 6.5, a stage efficiency frequently used to describe individual tray performance for individual components is the Murphree plate efficiency. This efficiency can be defined on the basis of either phase and, for a given component, is equal to the change in actual composition in the phase, divided by the change predicted by equilibrium. This definition applied to the vapor phase can be expressed in a manner similar to (6-28):

$$E_{MV} = \frac{y_n - y_{n+1}}{y_n^* - y_{n+1}} \tag{7-41}$$

Where E_{MV} is the Murphree vapor efficiency for stage n, where $n + 1$ is the stage below and y_n^* is the composition in the hypothetical vapor phase in equilibrium with the liquid composition leaving stage n. Values of E_{MV} can be less than or somewhat more than 100%.

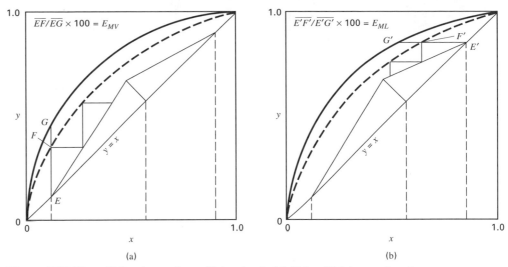

Figure 7.25 Use of Murphree plate efficiencies in McCabe–Thiele construction.

The component subscript in (7-41) is dropped because values of E_{MV} are equal for the two components of a binary mixture.

In stepping off stages, the Murphree vapor efficiency, if known, can be used to dictate the percentage of the distance taken from the operating line to the equilibrium line; only E_{MV} of the total vertical path is traveled. This is shown in Figure 7.25a for the case of Murphree efficiencies based on the vapor phase. Figure 7.25b shows the case when the Murphree tray efficiency is based on the liquid. In effect, the dashed curve for actual exit-phase composition replaces the thermodynamic equilibrium curve for a particular set of operating lines.

Multiple Feeds, Side Streams, and Open Steam

The McCabe–Thiele method for a single feed and two products is readily extended to the case of multiple feeds and/or side streams by adding one additional operating line for each additional feed or side stream. A multiple-feed arrangement is shown in Figure 7.26. In the absence of side stream L_S, this arrangement has no effect on the material balance associated with the rectifying section of the column above the upper feed point, F_1. The section of column between the upper feed point and the lower feed point F_2 (in the absence of feed F) is represented by an operating line of slope L'/V', this line intersecting the rectifying-section operating line. A similar argument holds for the stripping section of the column. Hence it is possible to apply the McCabe–Thiele graphical construction shown in Figure 7.27a, where feed F_1 is a dew-point vapor, while feed F_2 is a bubble-point liquid. Feed F_3 and side stream L_S of Figure 7.26 are not present. Thus, between the two feed points for this example, the molar vapor flow rate is $V' = V - F_1$ and $\overline{L} = L' + F_2 = L + F_2$. For given x_B, z_{F_2}, z_{F_1}, x_D, and L/D, the three operating lines in Figure 7.27a are readily constructed.

A side stream may be withdrawn from the rectifying section, the stripping section, or between multiple feed points, as a saturated vapor or saturated liquid. Within material-balance constraints, L_S and x_S can both be specified. In Figure 7.27b, a saturated liquid side stream of composition x_S and molar flow rate L_S is withdrawn from the rectifying section. In the section of stages between the side stream-withdrawal stage and the feed stage, $L' = L - L_S$, while $V' = V$. The McCabe–Thiele constructions determine the

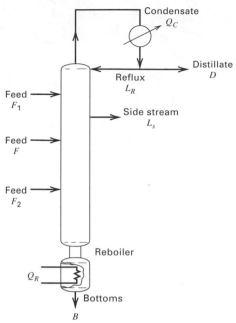

Figure 7.26 Complex distillation column with multiple feeds and side stream.

location of the side stream stage. However, if it is not located directly above x_S, the reflux ratio must be varied until it does.

For certain types of distillation, an inert hot gas is introduced directly into the base of the column. Open steam, for example, can be used if one of the components in the mixture is water, or if water can form a second liquid phase, thereby reducing the boiling point, as in the steam distillation of fats, where heat is supplied by live superheated steam and no reboiler is used. In this application, Q_R of Figure 7.26 is replaced by a stream of composition $y = 0$, which, with $x = x_B$, becomes a point on the operating line, since the passing streams at this point actually exist at the end of the column. The use of open steam rather than a reboiler for the operating condition $F_1 = F_2 = L_S = 0$ is represented graphically in Figure 7.27c.

EXAMPLE 7.5

A complex distillation column, equipped with a partial reboiler and total condenser, and operating at steady state with a saturated liquid feed, has a liquid side stream draw-off in the enriching (rectifying) section. Making the usual simplifying assumptions of the McCabe–Thiele method: (a) Derive an equation for the two operating lines in the enriching section. (b) Find the point of intersection of these operating lines. (c) Find the intersection of the operating line between F and L_S with the diagonal. (d) Show the construction on a y–x diagram.

SOLUTION

(a) By material balance over section 1 in Figure 7.28, $V_{n-1}y_{n-1} = L_n x_n + D x_D$. About section 2, $V_{s-2}y_{s-2} = L'_{s-1}x_{s-1} + L_s x_s + D x_D$. Invoking the usual simplifying assumptions, the two operating lines are

$$y = \frac{L'}{V}x + \frac{L_s x_s + D x_D}{V} \quad \text{and} \quad y = \frac{L}{V}x + \frac{D}{V}x_D$$

(b) Equating the two operating lines, the intersection occurs at $(L - L')x = L_s x_s$ and since $L - L' = L_S$, the point of intersection becomes $x = x_S$.

(c) The intersection of the lines $y = \frac{L'}{V}x + \frac{L_s x_s + D x_D}{V}$ and $y = x$ occurs at $x = \frac{L_s x_s + D x_D}{L_s + D}$

(d) The y–x diagram is shown in Figure 7.29.   ∎

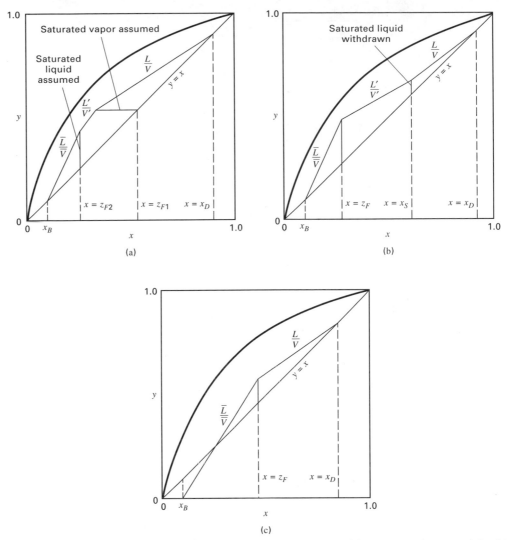

Figure 7.27 McCabe–Thiele construction for complex columns: (a) two feeds (saturated liquid and saturated vapor); (b) one feed, one side stream (saturated liquid); (c) use of open steam.

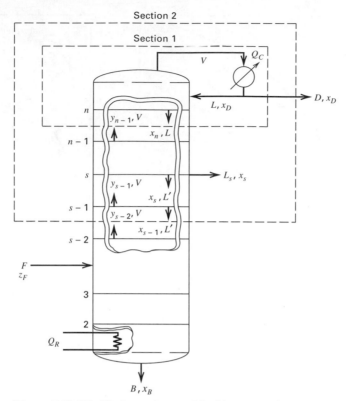

Figure 7.28 Distillation column with side stream for Example 7.5.

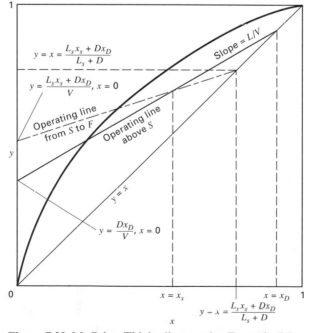

Figure 7.29 McCabe–Thiele diagram for Example 7.5.

7.3 ESTIMATION OF STAGE EFFICIENCY

Methods for estimating the stage efficiency for binary distillation are analogous to those for absorption and stripping, presented in Section 6.5. The efficiency is a complex function of tray design, fluid properties, and flow patterns. However, in hydrocarbon absorption and stripping, the liquid phase is often rich in heavy components so that liquid viscosity is high and mass transfer rates are relatively low. This leads to low stage efficiencies, usually less than 50%. In contrast, for binary distillation, particularly of close-boiling mixtures, liquid viscosity is low, with the result that stage efficiencies, for well-designed trays and optimal operating conditions, are often higher than 70% and can be even higher than 100% for large-diameter columns where a crossflow effect is present.

Performance Data

As discussed in *AIChE Equipment Testing Procedure* [8], performance data for an industrial distillation column are best obtained at conditions of total reflux (no feed or products) so as to avoid possible column-feed fluctuations, simplify location of the operating line, and avoid discrepancies between feed and feed-tray compositions. However, as shown by Williams, Stigger, and Nichols [9], efficiency measured at total reflux can differ markedly from that at design reflux ratio. Ideally, the column is operated in the range of 50% to 85% of flooding. If liquid samples are taken from the top and bottom of the column, the overall plate efficiency, E_o, can be determined from (6-21), where the number of theoretical stages required is determined by applying the McCabe–Thiele method at total reflux, as in Figure 7.11. If liquid samples are taken from the downcomers of intermediate trays, Murphree vapor efficiencies, E_{MV}, can be determined using (6-28). If liquid samples are withdrawn from different points on one tray, (6-30) can be applied to obtain point efficiencies, E_{OV}. Reliable values for these efficiencies require the availability of accurate vapor–liquid equilibrium data. For that reason, efficiency data for binary mixtures that form ideal solutions are preferred.

Table 7.4, from Gerster et al. [10], lists plant data, obtained from Eastman Kodak Company in Rochester, New York, for the distillation at total reflux of a methylene chloride (MC)–ethylene chloride (EC) mixture in a 5.5-ft-diameter column containing 60 bubble-cap trays on 18-in. tray spacing and operating at 85% of flooding at total reflux.

EXAMPLE 7.6

Using the performance data of Table 7.4, estimate: (a) the overall tray efficiency for the section of trays from 35 to 29 and (b) the Murphree vapor efficiency for tray 32. Assume the following values for relative volatility:

x_{MC}	$\alpha_{MC,EC}$	y_{MC} from (7-3)
0.00	3.55	0.00
0.10	3.61	0.286
0.20	3.70	0.481
0.30	3.76	0.617
0.40	3.83	0.719
0.50	3.91	0.796
0.60	4.00	0.857
0.70	4.03	0.904
0.80	4.09	0.942
0.90	4.17	0.974
1.00	4.25	1.00

Table 7.4 Performance Data for the Distillation of a Mixture of Methylene Chloride and Ethylene Chloride

Company	Eastman Kodak
Location	Rochester, New York
Column diameter	5.5 ft (65.5 in. I.D.)
No. of trays	60
Tray spacing	18 in.
Type tray	10 rows of 3-in.-diameter bubble caps on 4-7/8-in. triangular centers. 115 caps/tray
Bubbling area	20 ft^2
Length of liquid travel	49 in.
Outlet weir height	2.25 in.
Downcomer clearance	1.5 in.
Liquid rate	24.5 gal/min-ft = 1,115.9 lb/min
Vapor *F*-factor	1.31 ft/s (lb/ft^3)$^{0.5}$
Percent of flooding	85
Pressure, top tray	33.8 psia
Pressure, bottom tray	42.0 psia
Liquid composition, mole % methylene chloride:	
From tray 33	89.8
From tray 32	72.6
From tray 29	4.64

Source: J.A. Gerster, A.B. Hill, N.H. Hochgrof, and D.B. Robinson, *Tray Efficiencies in Distillation Columns, Final Report from the University of Delaware,* AIChE, New York (1958).

SOLUTION

(a) The above x–α–y data are plotted in Figure 7.30. Four theoretical stages are stepped off from $x_{33} = 0.898$ to $x_{29} = 0.0464$ for total reflux. Since the actual number of stages is also 4, the overall stage efficiency from (6-21) is 100%.

(b) At total reflux conditions, passing vapor and liquid streams have the same composition. That is, the operating line is the 45° line. Using this together with the above performance data and the equilibrium curve in Figure 7.30, we obtain for methylene chloride, with trays counted from the bottom up:

$$y_{32} \, x_{33} = 0.898 \text{ and } y_{31} = x_{32} = 0.726$$

$$\text{From (6-28)}, (E_{MV})_{32} = \frac{y_{32} - y_{31}}{y_{32}^* - y_{31}}$$

$$\text{From Figure 7.30, for } x_{32} = 0.726, y_{32}^* = 0.917, (E_{MV})_{32} = \frac{0.898 - 0.726}{0.917 - 0.726} = 0.90 \quad \text{or} \quad 90\%$$ ■

Empirical Correlations

Based on 41 sets of performance data for bubble-cap-tray and sieve-tray columns, distilling mainly hydrocarbon mixtures and a few water and miscible organic mixtures, Drickamer and Bradford [11] correlated the overall stage efficiency for the separation of the two key components in terms of the molar average liquid viscosity of the tower feed at the average tower temperature. The data covered average temperatures from 157 to 420°F, pressures from 14.7 to 366 psia, liquid viscosities from 0.066 to 0.355 cP, and overall tray efficiencies from 41% to 88%. The empirical equation

$$E_o - 13.3 - 66.8 \log \mu \tag{7-42}$$

where E_o is in percent and μ is in centipoise, fits the data with average and maximum percent deviations of 5.0% and 13.0%, respectively. A plot of the Drickamer and Bradford

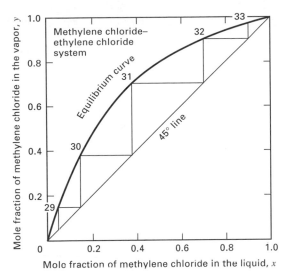

Figure 7.30 McCabe–Thiele diagram for Example 7.6.

correlation, compared to performance data for distillation, is given in Figure 7.31. Equation (7-42) is restricted to the range of the data and is intended mainly for hydrocarbon mixtures.

Mass transfer theory, discussed in Section 6.5, indicates that, when the relative volatility covers a wide range, the relative importance of liquid-phase and gas-phase mass transfer resistances can shift. Thus, as might be expected, O'Connell [12] found that the Drickamer–Bradford correlation correlates data inadequately for fractionators operating on key components with large relative volatilities. Separate correlations in terms of a viscosity–volatility product were developed for fractionators and for absorbers and strippers by O'Connell. However, as shown in Figure 7.32, Lockhart and Leggett [13] were able to obtain a single correlation by using the product of liquid viscosity and an appropriate volatility as the correlating variable. For fractionators, the relative volatility of the key components was

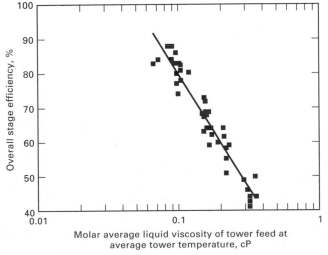

Figure 7.31 Drickamer and Bradford correlation for plate efficiency of distillation columns.

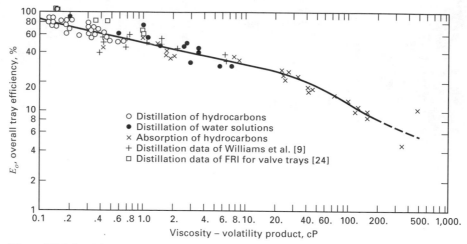

Figure 7.32 Lockhart and Leggett version of the O'Connell correlation for overall tray efficiency of fractionators, absorbers, and strippers. [Adapted from F.J. Lockhart and C.W. Leggett, in *Advances in Petroleum Chemistry and Refining*, Vol. 1, Eds., K.A. Kobe and John J. McKetta, Jr., Interscience, New York, pp. 323–326 (1958).

used; for hydrocarbon absorbers, the volatility was taken as 10 times the K-value of a selected key component, which must be one that is reasonably distributed between top and bottom products. The data used by O'Connell cover a range of relative volatility from 1.16 to 20.5. A comprehensive study of the effect on E_o of the ratio of liquid-to-vapor molar flow rates, L/V, for eight different binary systems in a 10-in.-diameter column with bubble-cap trays was reported by Williams, et al. [9]. The systems include water, hydrocarbons, and other organic compounds. For fractionation with L/V nearly equal to 1.0 (i.e., total reflux), their distillation data, which are included in Figure 7.32, are in reasonable agreement with the O'Connell correlation. For the distillation of hydrocarbons in a column having a diameter of 0.45 m, Zuiderweg, Verburg, and Gilissen [14] found differences in E_o among bubble-cap, sieve, and valve trays to be insignificant at 85% of flooding. Accordingly, Figure 7.32 is assumed to be applicable to all three tray types, but may be somewhat conservative for well-designed trays. For example, data of Fractionation Research Incorporated (FRI) for valve trays operating with the cyclohexane/n-hexane and isobutane/n-butane systems are also included in Figure 7.32 and show efficiencies 10% to 20% higher than the correlation.

For the distillation data plotted in Figure 7.32, which cover a viscosity–relative volatility range for distillation of from 0.1 to 10 cP, the O'Connell correlation fits the empirical equation

$$E_o = 50.3(\alpha\mu)^{-0.226} \tag{7-43}$$

Where E_o is in percent and μ is in centipoise. The relative volatility is determined for the two key components at average column conditions.

Most of the data for developing the correlation of Figure 7.32 are for columns having a liquid flow path across the active tray area of from 2 to 3 ft. Gautreaux and O'Connell [15], using theory and experimental data, showed that higher efficiencies are achieved for longer flow paths. For short liquid flow paths, the liquid flowing across the tray is usually mixed completely. For longer flow paths, the equivalent of two or more completely mixed, successive liquid zones may be present. The result is a greater average driving force for mass transfer, and, thus, a higher efficiency—sometimes even greater than 100%. Provided

Table 7.5 Correction to Overall Tray Efficiency for Length of Liquid Flow Path ($0.1 \leq \mu\alpha \leq 1.0$)

Length of Liquid Flow Path, ft	Factor to Be Added to E_o from Figure 7.32, %
3	0
4	10
5	15
6	20
8	23
10	25
15	27

Source: F.J. Lockhart and C.W. Leggett, in K.A. Kobe and J.J. McKetta, Jr., Eds., *Advances in Petroleum Chemistry and Refining,* Vol. 1, Interscience, New York, pp. 323–326 (1958).

that the viscosity–volatility product lies between 0.1 and 1.0, Lockhart and Leggett [13] recommend addition of the increments in Table 7.5 to the value of E_o from Figure 7.32 when the liquid flow path is greater than 3 ft. However, at large liquid rates, long liquid path lengths are undesirable because they lead to excessive liquid gradients, causing maldistribution of vapor flow. The use of multipass trays, shown in Figure 6.16, to prevent excessive liquid gradients is discussed in Section 6.5.

EXAMPLE 7.7

For the benzene–toluene distillation of Figure 7.1, use the Drickamer–Bradford and O'Connell correlations to estimate the overall stage efficiency and number of actual plates required. Calculate the height of the tower assuming 24-in. tray spacing, with 4 ft above the top tray for removal of entrained liquid and 10 ft below the bottom tray for bottoms surge capacity. The separation requires 20 equilibrium stages plus a partial reboiler that acts as an equilibrium stage.

SOLUTION

For estimating overall stage efficiency, the liquid viscosity is determined at the feed-stage condition of 220°F, assuming a liquid composition of 50 mol% benzene.

μ of benzene 0.10 cP; μ of toluene $= 0.12$ cP; Average $\mu = 0.11$ cP.

From Fig. 7.3, take the average relative volatility as

$$\text{Average } \alpha = \frac{\alpha_{\text{top}} + \alpha_{\text{bottom}}}{2} = \frac{2.52 + 2.26}{2} = 2.39$$

From the Drickamer-Bradford correlation (7-42), $E_o = 13.3 - 66.8 \log(0.11) = 77\%$

This is close to the value given in the description of this problem. Therefore, 26 actual trays are required and column height $= 4 + 2(26 - 1) + 10 = 64$ ft.

From the O'Connell correlation (7-43), $E_o = 50.3[(2.39)(0.11)]^{-0.226} = 68\%$.

For a 5-ft-diameter column, the length of the liquid flow path is about 3 ft for a single-pass tray and even less for a two-pass tray. From Table 7.5, the efficiency correction is zero. Therefore, the actual number of trays required is $20/0.68 = 29.4$ or call it 30 trays. Column height $= 4 + 2(30 - 1) + 10 = 72$ ft. ∎

Semitheoretical Models

In Section 6.5, semitheoretical tray models based on the Murphree vapor efficiency and the Murphree vapor-point efficiency are applied to absorption and stripping. These same relationships are valid for distillation. However, because the equilibrium line is curved for distillation, λ must be taken as mV/L (not $KV/L = 1/A$), where m = local slope of the equilibrium curve = dy/dx. In Section 6.6, the method of Chan and Fair [16] is used for estimating the Murphree vapor-point efficiency from mass transfer considerations. The Murphree vapor efficiency can then be estimated. The Chan and Fair correlation is specifically applicable to binary distillation because it was developed from experimental data that includes six different binary systems.

Scale-up from Laboratory Data

When binary mixtures form ideal or nearly ideal solutions, it is rarely necessary to obtain laboratory distillation data. Where nonideal solutions are formed and/or the possibility of azeotrope formation exists, use of a small laboratory Oldershaw column, of the type discussed in Section 6.5, should be used to verify the desired degree of separation and to obtain an estimate of the Murphree vapor point efficiency. The ability to predict the efficiency of an industrial-size sieve-tray column from measurements with 1-in. glass and 2-in. metal diameter Oldershaw columns is shown in Figure 7.33, from the work of Fair, Null, and Bolles [17]. The measurements were made for the cyclohexane/n-heptane system at vacuum conditions (Figure 7.33a) and at near-atmospheric conditions (Figure 7.33b) and for the isobutane/n-butane system at 11.2 atm (Figure 7.33c). The Oldershaw data are correlated by the solid lines. Data for the 4-ft-diameter column with sieve trays of 8.3% and 13.7% open area were obtained by Sakata and Yanagi [18] and Yanagi and Sakata [19], respectively, of FRI. The Oldershaw column is assumed to measure point efficiency. The FRI column measured overall efficiency, but the relations of Section 6.5 were used to convert the FRI data to the point efficiencies shown in Figure 7.33. The data

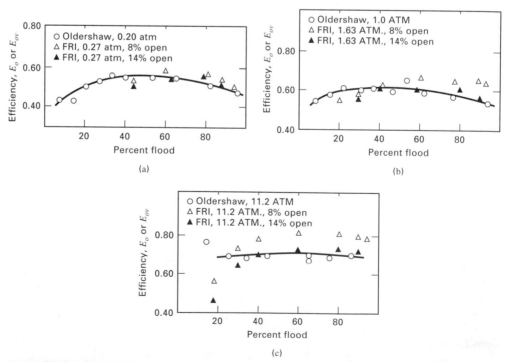

Figure 7.33 Comparison of Oldershaw column efficiency with point-efficiency in 4-ft-diameter FRI column: (a) cyclohexane/n-heptane system; (b) cyclohexane/n-heptane systems; (c) isobutane/n-butane system.

cover a percent of flooding range from about 10% to 95%. Data from the Oldershaw column are in reasonable agreement with the FRI data for 14% open area, except at the lower part of the flooding range. In Figures 7.33b and 7.33c, the FRI data for 8% open area show efficiencies as much as 10 percentage points higher.

7.4 CAPACITY OF TRAYED TOWERS AND REFLUX DRUMS

In Section 6.6, methods for estimating tray capacity and pressure drop for absorbers and strippers are presented. These same methods apply to distillation columns. Calculations of column diameter are usually made for conditions at the top and bottom trays of the tower. If the diameters differ by 1 ft or less, the larger diameter is used for the entire column. If the diameters differ by more than 1 ft, it is often more economical to swage the column, using the different diameters for the sections above and below the feed.

Reflux Drums

Almost all commercial towers are provided with a cylindrical reflux drum, as shown in Figure 7.1. This drum is usually located near ground level, necessitating a pump to lift the reflux to the top of the column. If a partial condenser is used, the drum is often installed vertically to facilitate the separation of vapor from liquid—in effect, acting as a flash drum. Vertical reflux and flash drums are sized by calculating a minimum drum diameter, D_T, to prevent liquid carryover by entrainment, using (6-44) in conjunction with the curve for 24-in. tray spacing in Figure 6.24 and a value of $F_{HA} = 1.0$ in (6-42). To absorb process upsets and fluctuations, and otherwise facilitate control, vessel volume, V_V, is determined on the basis of liquid residence time, t, which should be at least 5 min, with the vessel half full of liquid [20]:

$$V_V = \frac{2LM_Lt}{\rho_L} \tag{7-44}$$

where L is the molar liquid flow rate leaving the vessel. Assuming a vertical, cylindrical vessel and neglecting the volume associated with the heads, the height H of the vessel is

$$H = \frac{4V_V}{\pi D_T^2} \tag{7-45}$$

However, if $H > 4D_T$, it is generally preferable to increase D_T and decrease H to give $H = 4D$. Then

$$D_T = \frac{H}{4} = \left(\frac{V_V}{\pi}\right)^{1/3} \tag{7-46}$$

A height above the liquid level of at least 4 ft is necessary for feed entry and disengagement of liquid droplets from the vapor. Within this space, it is common to install a wire mesh pad, which serves as a mist eliminator.

When vapor is totally condensed, a cylindrical, horizontal reflux drum is commonly employed to receive the condensate. Equations (7-44) and (7-46) permit estimates of the drum diameter, D_T, and length, H, by assuming a near-optimal value for H/D_T of 4, and the same liquid residence time suggested for a vertical drum. A horizontal drum is also used following a partial condenser when the liquid flow rate is appreciably greater than the vapor flow rate.

EXAMPLE 7.8

Equilibrium vapor and liquid streams leaving a flash drum are as follows:

Component	Vapor	Liquid
Pound-moles per hour:		
HC1	49.2	0.8
Benzene	118.5	81.4
Monochlorobenzene	71.5	178.5
Total	239.2	260.7
Pounds per hour	19,110	26,480
T, °F	270	270
P, psia	35	35
Density, lb/ft^3	0.371	57.08

Determine the dimensions of the flash drum.

SOLUTION

$$\text{Using Figure 6.24, } F_{LV} = \frac{26,480}{19,110} \left(\frac{0.371}{57.08}\right)^{0.5} = 0.112$$

C_F at a 24-in. tray spacing is 0.34. Assume, in (6-24), that $C = C_F$. From (6-40),

$$U_f = 0.34 \left(\frac{57.08 - 0.371}{0.371}\right)^{0.5} = 4.2 \text{ ft/s} = 15,120 \text{ ft/h}$$

From (6-44) with $A_d/A = 0$, $D_T = \left[\dfrac{(4)(19,110)}{(0.85)(15,120)(3.14)(1)(0.371)}\right]^{0.5} = 2.26 \text{ ft}$

From (7-44), with $t = 5 \text{ min} = 0.0833 \text{ h}$, $V_V = \dfrac{(2)(26,480)(0.0833)}{(57.08)} = 77.3 \text{ ft}^3$

From (7-43), $H = \dfrac{(4)(77.3)}{(3.14)(2.26)^2} = 19.3 \text{ ft}$

However, $H/D_T = 19.3/2.26 = 8.54 > 4$. Therefore, redimension V_V for $H/D_T = 4$.

From (7-46), $D_T = \left(\dfrac{77.3}{3.14}\right)^{1/3} = 2.91 \text{ ft and } H = 4D_T = (4)(2.91) = 11.64 \text{ ft}$

Height above the liquid level is $11.64/2 = 5.82$ ft, which is adequate.

Alternatively, with a height of twice the minimum disengagement height, $H = 8$ ft and $D_T = 3.5$ ft. ■

7.5   RATE-BASED METHOD FOR PACKED COLUMNS

With the availability of economical and efficient packings, packed towers are finding increasing use in new distillation processes and for retrofitting existing trayed towers. Methods in Section 6.8 for estimating packed column efficiency, capacity, and pressure drop for absorbers are applicable to distillation. Methods for determining packed height are similar to those presented in Section 6.7 and are extended here for use in conjunction with the McCabe–Thiele diagram. Both the HETP and the HTU methods are discussed and illustrated. Unlike the case of absorption or stripping of dilute solutions, where values of HETP and HTU may be constant throughout the packed height, values of HETP and HTU can vary over the packed height of a distillation column, especially across the feed entry, where appreciable changes in vapor and liquid traffic occur. Also, because the

Table 7.6 Modified Efficiency and Mass Transfer Equations for Binary Distillation

$$\lambda = mV/L$$

$$m = dy/dx = \text{local slope of equilibrium curve} \tag{7-47}$$

Efficiency
   Equations (6.31) to (6.37) hold if λ is defined by (7.47)
Mass transfer

$$\frac{1}{N_{OG}} = \frac{1}{N_G} + \frac{\lambda}{N_L} \tag{7-48}$$

$$\frac{1}{K_{OG}} = \frac{1}{k_G a} + \frac{mPM_L/\rho_L}{k_L a} \tag{7-49}$$

$$\frac{1}{K_y a} = \frac{1}{k_y a} + \frac{m}{k_x a} \tag{7-50}$$

$$\frac{1}{K_x a} = \frac{1}{k_x a} + \frac{1}{mk_y a} \tag{7-51}$$

$$H_{OG} = H_G + \lambda H_L \tag{7-52}$$

$$\text{HETP} = H_{OG} \ln \lambda/(\lambda - 1) \tag{7-53}$$

equilibrium line for distillation is curved rather than straight, the mass transfer equations of Section 6.8 must be modified by replacing $\lambda = \dfrac{KG}{L} = \dfrac{1}{A}$ with

$$\lambda = \frac{mV}{L} = \frac{\text{slope of equilibrium curve}}{\text{slope of operating line}}$$

where $m = dy/dx$ varies with location in the tower. The modified efficiency and mass transfer relationships are summarized in Table 7.6.

HETP Method

In the HETP method, the equilibrium stages are first stepped off on a McCabe–Thiele diagram. The case of equimolar counterdiffusion (EMD) applies to distillation. At each stage, the temperature, pressure, phase flow ratio, and phase compositions are noted. A suitable packing material is selected and the column diameter is estimated for operation at, say, 70% of flooding by one of the methods of Section 6.8. Mass transfer coefficients for the individual phases are estimated for the conditions at each stage from correlations also discussed in Section 6.8. From these coefficients, values of H_{OG} and HETP are estimated for each stage. The latter values are then summed to obtain the separate packed heights of the rectifying and stripping sections. If experimental values of HETP are available, they are used directly. In computing values of H_{OG} from H_G and H_L, or K_y from k_y and k_x, (6-92) and (6-80) must be modified because for binary distillation where the mole fraction of the light key may range from almost 0 at the bottom of the column to almost 1 at the top of the column, the ratio $(y_I - y^*)/(x_I - x)$ in (6-76) is no longer a constant equal to the K-value, but is dy/dx equal to the slope, m, of the equilibrium curve. The modified equations are included in Table 7.6.

EXAMPLE 7.9

For the benzene–toluene distillation of Example 7.1, determine packed heights of the rectifying and stripping sections based on a column diameter and packing material with the following values for the individual HTUs. Included are the L/V values for each section from Example 7.1.

	H_G, ft	H_L, ft	L/V
Rectifying section	1.16	0.48	0.62
Stripping section	0.90	0.53	1.40

SOLUTION

Slopes dy/dx of the equilibrium curve are obtained from Figure 7.15 and values of λ from (7-47). H_{OG} for each stage is determined from (7-52) in Table 7.6. HETP for each stage is determined from (7-53) in Table 7.6. The results are given in Table 7.7, where only 0.2 of stage 13 is needed and stage 14 is the partial reboiler.

Based on the results in Table 7.7, 10 ft of packing should be used in each of the two sections. ∎

HTU Method

In the HTU method, equilibrium stages are not stepped off on a McCabe–Thiele diagram. Instead, the diagram provides data to perform an integration over the packed height of each section using either mass transfer coefficients or transfer units.

Consider the schematic diagram of a packed distillation column and its accompanying McCabe–Thiele diagram in Figure 7.34. Assume that V, L, $\overline{V}$, and $\overline{L}$ are constant in their respective sections. For equimolar countercurrent diffusion (EMD), the rate of mass transfer of the light-key component from the liquid phase to the vapor phase is

$$n = k_x a(x - x_I) = k_y a(y_I - y) \tag{7-54}$$

Rearranging:

$$-\frac{k_x a}{k_y a} = \frac{y_I - y}{x_I - x} \tag{7-55}$$

Thus, as shown in Figure 7.34b, for any point (x, y) on the operating line, the corresponding point (x_I, y_I) on the equilibrium curve is obtained by drawing a line of slope $(-k_x a/k_y a)$ from the point (x, y) to the point where it intersects the equilibrium curve.

Table 7.7 Results for Example 7.9

Stage	m	$\lambda = \dfrac{mV}{L}$	H_{OG}, ft	HETP, ft
1	0.47	0.76	1.52	1.74
2	0.53	0.85	1.56	1.70
3	0.61	0.98	1.62	1.64
4	0.67	1.08	1.68	1.62
5	0.72	1.16	1.71	1.59
6	0.80	1.29	1.77	1.56
Total for rectifying section:				9.85
7	0.90	0.64	1.32	1.64
8	0.98	0.70	1.28	1.52
9	1.15	0.82	1.34	1.47
10	1.40	1.00	1.43	1.43
11	1.70	1.21	1.53	1.40
12	1.90	1.36	1.62	1.38
13	2.20	1.57	1.73	1.37(0.2) = 0.27
Total for stripping section:				9.11
Total packed height:				18.96

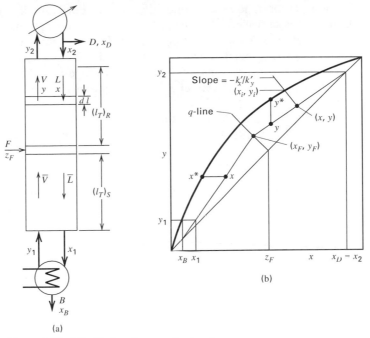

Figure 7.34 Distillation in a packed column.

By material balance over an incremental section of packed height, assuming constant molar overflow,

$$V \, dy = k_y a(y_I - y)S \, dl \tag{7-56}$$

$$L \, dx = k_x a(x - x_I)S \, dl \tag{7-57}$$

where S is the cross sectional area of the packed section. Integrating over the rectifying section,

$$(l_T)_R = \int_0^{(l_T)_R} dl = \int_{y_F}^{y_2} \frac{V \, dy}{k_y a S(y_I - y)} = \int_{x_F}^{x_D} \frac{L \, dx}{k_x a S(x - x_I)} \tag{7-58}$$

or

$$(l_T)_R = \int_{y_F}^{y_2} \frac{H_G \, dy}{(y_I - y)} = \int_{x_F}^{x_D} \frac{H_L \, dx}{(x - x_I)} \tag{7-59}$$

Integrating over the stripping section,

$$(l_T)_S = \int_0^{(l_T)_S} dl = \int_{y_1}^{y_F} \frac{V \, dy}{k_y a S(y_I - y)} = \frac{L \, dx}{k_x a S(x - x_I)} \tag{7-60}$$

$$\text{or } (l_T)_S = \int_{y_1}^{y_F} \frac{H_G \, dy}{(y_I - y)} = \int_{x_1}^{x_F} \frac{H_L \, dx}{(x - x_I)} \tag{7-61}$$

In general, values of k_y and k_x vary over the packed height causing the slope ($-k_x a/k_y a$) to vary. If $k_x a > k_y a$, the main resistance to mass transfer resides in the vapor and it is most accurate to evaluate the integrals in y. For $k_y a > k_x a$, the integrals in x are used. Usually, it is sufficient to evaluate k_y and k_x at just three points in each section, from which their variation with x can be determined. Then by computing and plotting their ratios from (7-55), a locus of points P can be found, from which values of $(y_I - y)$ for any value of y, or $(x - x_I)$ for any value of x can be read for use in integrals (7-58) to (7-61). These integrals can be evaluated either graphically or numerically to determine the packed heights.

EXAMPLE 7.10

Suppose that 250 kmol/h of a mixture of 40 mol% isopropyl ether in isopropanol is distilled in a packed column operating at 1 atm to obtain a distillate of 75 mol% isopropyl ether and a bottoms of 95 mol% isopropanol. At the feed entry, the mixture is a saturated liquid. A reflux ratio of 1.5 times minimum is used and the column is equipped with a total condenser and a partial reboiler. For the packing and column diameter, mass transfer coefficients given below have been estimated from empirical correlations of the type discussed in Section 6.8. Compute the required packed heights of the rectifying and stripping sections.

SOLUTION

The distillate and bottoms rates are computed by an overall material balance on isopropyl ether:

$$0.40(250) = 0.75D + 0.05(250 - D)$$

$$\text{Solving, } D = 125 \text{ kmol/h and } B = 250 - 125 = 125 \text{ kmol/h}$$

The equilibrium curve for this mixture at 1 atm is shown in Figure 7.35, where it is noted that isopropyl ether is the light key and an azeotrope is formed at 78 mol% isopropyl ether. The distillate composition of 75 mol% is safely below the azeotropic composition. Also shown in Figure 7.35 are the q-line and the rectification section operating line for the condition of minimum reflux. The slope of the latter line is measured to be $(L/V)_{min} = 0.39$. From (7-27),

$$R_{min} = 0.39/(1 - 0.39) = 0.64 \text{ and } R = 1.5 R_{min} = 0.96$$

$$L = RD = 0.96(125) = 120 \text{ kmol/h and } V = L + D = 120 + 125 = 245 \text{ kmol/h}$$

$$\overline{L} = L + L_F = 120 + 250 = 370 \text{ kmol/h and } \overline{V} = V - V_F = 245 - 0 = 245 \text{ kmol/h}$$

$$\text{Slope of rectification section operating line} = L/V = 120/245 = 0.49$$

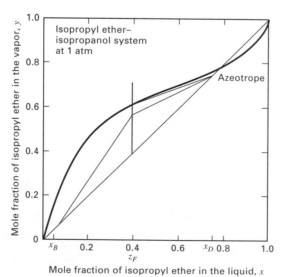

Figure 7.35 Operating lines and minimum reflux line for Example 7.10.

This line and the stripping-section operating line are plotted in Figure 7.35. The partial reboiler is stepped off in Figure 7.35 to give the following end points for determining the packed heights of the two sections, where the symbols refer to Figure 7.34a:

	Stripping Section	**Rectifying Section**
Top	($x_F = 0.40$, $y_F = 0.577$)	($x_2 = 0.75$, $y_2 = 0.75$)
Bottom	($x_1 = 0.135$, $y_1 = 0.18$)	($x_F = 0.40$, $y_F = 0.577$)

Mass transfer coefficients at three values of x in each section are as follows:

x	$k_y a$ kmol/m^3-h-(mole fraction)	$k_x a$ kmol/m^3-h-(mole fraction)
Stripping section:		
0.15	305	1,680
0.25	300	1,760
0.35	335	1,960
Rectifying section:		
0.45	185	610
0.60	180	670
0.75	165	765

The slopes of the ratios of these coefficients, given by (7-55), are plotted in Figure 7.36. From these, dashed locus lines AB and BC are drawn. Since $k_x a \gg k_y a$, the integrals in terms of y are evaluated using the locus lines to determine $(y_I - y)$ for a number of points sufficient to obtain satisfactory accuracy. Corresponding values of $k_y a$ can be interpolated from the above values. Since the diameter of the column is not given, the packed volumes are determined from the following rearrangements of (7-58) and (7-60), where $V = Sl_T$:

$$V_R = \int_{y_F}^{y_2} \frac{V\,dy}{k_y a(y_I - y)} \tag{7-62}$$

$$V_S = \int_{y_1}^{y_F} \frac{V\,dy}{k_y a(y_I - y)} \tag{7-63}$$

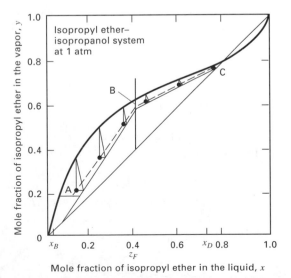

Figure 7.36 Mass-transfer driving forces for Example 7.10.

y	$(y_I - y)$	$k_y a$	$\dfrac{V (\text{or } \overline{V})}{k_y a (y_I - y)}, \text{m}^3$
Stripping section:			
0.18	0.145	307	5.5
0.25	0.150	303	5.4
0.35	0.143	300	5.7
0.45	0.103	320	7.4
0.577	0.030	350	23.3
Rectifying section:			
0.577	0.030	187	43.7
0.60	0.033	185	40.1
0.65	0.027	182	49.9
0.70	0.017	175	82.3
0.75	0.010	165	148.5

By numerical integration, $V_S = 3.6 \text{ m}^3$ and $V_R = 12.3 \text{ m}^{3.}$ ■

7.6 PONCHON-SAVARIT GRAPHICAL EQUILIBRIUM-STAGE METHOD FOR TRAYED TOWERS

The McCabe–Thiele method assumes that molar vapor and liquid flow rates are constant in each section of the column. This assumption (constant molar overflow) eliminates the need to make an energy balance around each stage. For nonideal binary mixtures, such an assumption may not be valid and the McCabe–Thiele method may not be accurate. A graphical method that includes energy balances as well as material balances and phase equilibrium relations is the Ponchon–Savarit method [21,22], which utilizes an enthalpy–composition diagram of the type shown in Figure 7.37 for the *n*-hexane/*n*-octane system at 1 atm. This diagram includes curves for the enthalpies of saturated vapor and liquid mixtures. Terminal points of tie lines connecting these two curves represent the equilibrium vapor and liquid compositions for the given temperature. Curves above the saturated vapor curve represent enthalpies of the superheated vapor, while curves below the saturated liquid curve represent the subcooled liquid.

The application of the enthalpy–concentration diagram to equilibrium-stage calculations may be illustrated by considering a single equilibrium stage, $n - 1$, where vapor from stage $n - 2$ is mixed adiabatically with liquid from stage n to give an overall mixture, denoted by mole-fraction z, and then brought to equilibrium. The process is represented schematically in two steps at the top of Figure 7.38. The energy-balance equations for stage $n - 1$ are

$$V_{n-2}H_{n-2} + L_n h_n = (V_{n-2} + L_n)h_z \tag{7-64}$$

$$(V_{n-2} + L_n)h_z = V_{n-1}H_{n-1} + L_{n-1}h_{n-1} \tag{7-65}$$

where H and h are vapor and liquid molar enthalpies, respectively. The governing material-balance equations for the light component are:

$$y_{n-2}V_{n-2} + x_n L_n = z(V_{n-2} + L_n) \tag{7-66}$$

$$z(V_{n-2} + L_n) = y_{n-1}V_{n-1} + x_{n-1}L_{n-1} \tag{7-67}$$

Simultaneous solution of (7-64) and (7-66) gives

$$\frac{H_{n-2} - h_z}{y_{n-2} - z} = \frac{h_z - h_n}{z - x_n} \tag{7-68}$$

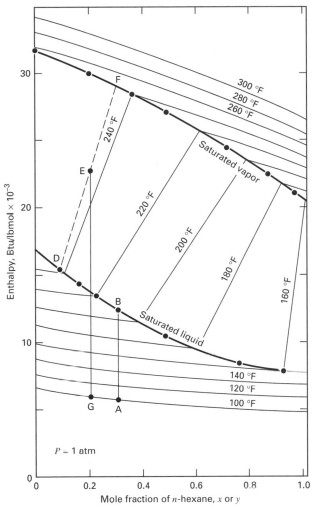

Figure 7.37 Enthalpy–concentration diagram for *n*-hexane/
n-octane.

which is the three-point form of a straight line plotted in Figure 7.38. Similarly, the simultaneous solution of (7-65) and (7-67) gives:

$$\frac{H_{n-1} - h_z}{y_{n-1} - z} = \frac{h_z - h_{n-1}}{z - x_{n-1}} \tag{7-69}$$

Which is also the equation for a straight line. However, in this case y_{n-1} and x_{n-1} are in equilibrium and, therefore, the points (H_{n-1}, y_{n-1}) and (h_{n-1}, x_{n-1}) must lie on the opposite ends of the tie line that passes through the mixing point (h_z, z), as shown in Figure 7.38.

The Ponchon–Savarit method for binary distillation is an extension of the construction in Figure 7.38 to countercurrent cascades above and below the feed stage, with consideration of the condenser and reboiler. A detailed description of the method is not given here because the method has been largely superseded by the rigorous computer-aided calculation procedures, discussed in Chapter 10, which can be applied to multicomponent as well as binary mixtures. A detailed presentation of the Ponchon–Savarit method is given by Henley and Seader [23].

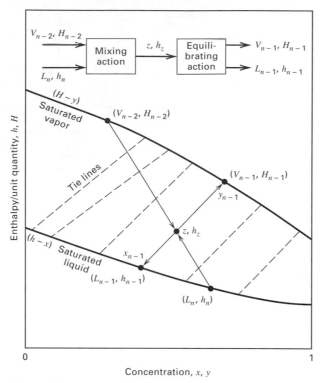

Figure 7.38 Two-phase mixing on an enthalpy–concentration diagram.

SUMMARY

1. A binary liquid and/or vapor binary mixture can be separated into two nearly pure products (distillate and bottoms) by distillation, provided that the value of the relative volatility of the two components is high enough, usually greater than 1.05.

2. Distillation is the most mature and widely used separation operation, with design procedures and operation practices well established.

3. The purities of the products from distillation depend on the number of equilibrium stages in the rectifying section above the feed entry and in the stripping section below the feed entry, and on the reflux ratio. Both the number of stages and the reflux ratio must be greater than the minimum values corresponding to total reflux and infinite stages, respectively. The optimal reflux-to-minimum-reflux ratio is usually in the range of 1.10 to 1.50.

4. Distillation is most commonly conducted in trayed towers equipped with sieve or valve trays, or in columns packed with random or structured packings. Many older towers are equipped with bubble-cap trays.

5. Most distillation towers are equipped with a condenser, cooled with cooling water, to provide reflux, and a reboiler, heated with steam, to provide boilup.

6. When the assumption of constant molar overflow is valid in each of the two sections of the distillation tower, the McCabe–Thiele graphical method is convenient for determining stage and reflux requirements. This method facilitates the visualization of many aspects of distillation and provides a procedure for locating the optimal feed-stage location.

7. Miscellaneous considerations involved in the design of a distillation tower include selection of operating pressure, type of condenser, degree of reflux subcooling, type of reboiler, and extent of feed preheat.

8. The McCabe–Thiele method can be extended to handle Murphree stage efficiency, multiple feeds, side streams, open steam, and use of interreboilers and intercondensers.

9. Rough estimates of overall stage efficiency, defined by (6-21), can be made with the Drickamer and Bradford, (7-42), or O'Connell, (7-43), correlations. More accurate and reliable procedures use data from a small Oldershaw column or the same semitheoretical equations for mass transfer in Chapter 6 that are used for absorption and stripping.

10. Tray diameter, pressure drop, weeping, entrainment, and downcomer backup can all be estimated by the procedures in Chapter 6.

11. Reflux and flash drums are sized by a procedure based on avoidance of entrainment and provision for adequate liquid residence time.

12. Packed column diameter and pressure drop are determined by the same procedures presented in Chapter 6 for absorption and stripping.

13. The height of a packed column may be determined by the HETP method, or preferably from the HTU method. Application of the latter method is similar to that of Chapter 6 for absorbers and strippers, but differs in the manner in which the curved equilibrium curve must be handled, as given by (7-47).

14. The Ponchon–Savarit graphical method removes the assumption of constant molar overflow in the McCabe–Thiele method by employing energy balances with an enthalpy-concentration diagram. However, the Ponchon–Savarit method has largely been supplanted by rigorous computer-aided methods.

REFERENCES

1. Forbes, R.J., *Short History of the Art of Distillation,* E.J. Brill, Leiden (1948).

2. Mix, T W., J.S. Dweck, M. Weinberg, and R.C. Armstrong, *Chem. Eng. Prog.,* **74** (4), 49–55 (1978).

3. Kister, H.Z., *Distillation Design,* McGraw-Hill, New York (1992).

4. Kister, H.Z., *Distillation Operation,* McGraw-Hill, New York (1990).

5. McCabe, W.L., and E.W. Thiele, *Ind. Eng. Chem.,* **17,** 605–611 (1925).

6. Peters, M.S., and K.D. Timmerhaus, *Plant Design and Economics for Chemical Engineers,* 4th ed., McGraw-Hill, New York (1991).

7. Horvath, P.J., and R.F. Schubert, *Chem. Eng.,* **65** (3), 129–132 (1958).

8. *AIChE Equipment Testing Procedure, Tray Distillation Columns,* 2nd ed., AIChE, New York (1987).

9. Williams, G.C., E.K. Stigger, and J.H. Nichols, *Chem. Eng. Progr.,* **46** (1), 7–16 (1950).

10. Gerster, J.A., A.B. Hill, N.H. Hochgrof, and D.B. Robinson, *Tray Efficiencies in Distillation Columns, Final Report from the University of Delaware,* AIChE, New York (1958).

11. Drickamer, H.G., and J.R. Bradford, *Trans. AIChE,* **39,** 319–360 (1943).

12. O'Connell, H.E., *Trans. AIChE,* **42,** 741–755 (1946).

13. Lockhart, F.J., and C.W. Leggett, in K.A. Kobe and John J. McKetta, Jr., Eds., *Advances in Petroleum Chemistry and Refining,* Vol. 1, Interscience, New York, pp. 323–326 (1958).

14. Zuiderweg, F.J., H. Verburg, and F.A.H. Gilissen, *Proc. International Symposium on Distillation,* Institution of Chem. Eng., London, 202–207 (1960).

15. Gautreaux, M.F., and H.E. O'Connell, *Chem. Eng. Prog.,* **51** (5) 232 237 (1955).

16. Chan, H., and J.R. Fair, *Ind. Eng. Chem. Process Des. Dev.,* **23,** 814–819 (1984).

17. Fair, J.R., H.R. Null, and W.L. Bolles, *Ind. Eng. Chem. Process Des. Dev.,* **22,** 53–58 (1983).

18. Sakata, M., and T. Yanagi, *I. Chem. E. Symp. Ser.,* **56,** 3.2/21 (1979).

19. Yanagi, T., and M. Sakata, *Ind. Eng. Chem. Process Des. Devel.,* **21,** 712 (1982).

20. Younger, A.H., *Chem. Eng.,* **62** (5), 201–202 (1955).

21. Ponchon, M., *Tech. Moderne,* **13,** 20, 55 (1921).

22. Savarit, R., *Arts et Metiers,* pp. 65, 142, 178, 241, 266, 307 (1922).

23. Henley, E.J., and J.D. Seader, *Equilibrium-Stage Separation Operations in Chemical Engineering,* John Wiley and Sons, New York (1981).

24. Glitsch Ballast Tray, Bulletin 159, Fritz W. Glitsch and Sons, Dallas (from FRI report of September 3, 1958).

EXERCISES

Unless otherwise stated, the usual simplifying assumptions of saturated liquid reflux, optimal feed-stage location, no heat losses, steady state, and constant molar liquid and vapor flows apply to each of the following problems.

Section 7.1

7.1 List as many differences between absorption and distillation as you can. List as many differences between stripping and distillation as you can.

7.2 Prior to the 1980s, packed columns were rarely used for distillation unless column diameter was less than 2.5 ft. Explain why, in recent years, some existing trayed towers are being retrofitted with packing and some new large-diameter columns are being designed for packing rather than trays.

7.3 A mixture of methane and ethane is to be separated by distillation. Explain why water cannot be used as the coolant in the condenser. What would you choose as the coolant?

7.4 A mixture of ethylene and ethane is to be separated by distillation. Determine the maximum operating pressure of the column. What operating pressure would you suggest? Why?

7.5 Under what circumstances would it be advisable to conduct laboratory or pilot-plant tests of a proposed distillation separation?

7.6 Explain why an economic tradeoff exists between the number of trays and the reflux ratio.

Section 7.2

7.7 Following the development by Sorel in 1894 of a mathematical model for continuous, steady-state, equilibrium-stage distillation, a number of methods were proposed for solving the equations graphically or algebraically during an 18-year period from 1920 to 1938, prior to the availability of digital computers. Today, the only method from that era that remains in widespread use is the McCabe–Thiele method. What are the attributes of this method that are responsible for its continuing popularity?

7.8 (a) For the cascade shown in Figure 7.39a, calculate the compositions of streams V_4 and L_1. Assume atmospheric pressure, saturated liquid and vapor feeds, and the vapor–liquid equilibrium data given below. Compositions are in mole percent.
(b) Given the feed compositions in cascade (a), how many equilibrium stages are required to produce a V_4 containing 85 mol% alcohol?
(c) For the cascade configuration shown in Figure 7.39b, with $D = 50$ mol, what are the compositions of D and L_1?
(d) For the configuration of cascade (b), how many equilibrium stages are required to produce a D of 50 mol% alcohol?

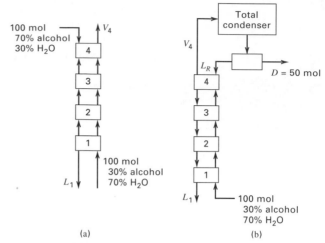

Figure 7.39 Data for Exercise 7.8.

EQUILIBRIUM DATA, MOLE FRACTION ALCOHOL

x	0.1	0.3	0.5	0.7	0.9
y	0.2	0.5	0.68	0.82	0.94

7.9 Liquid air is fed to the top of a perforated-tray reboiled stripper operated at substantially atmospheric pressure. Sixty percent of the oxygen in the feed is to be drawn off in the bottoms vapor product from the still. This product is to contain 0.2 mol% nitrogen. Based on the assumptions and data given below, calculate:
(a) The mole percent of nitrogen in the vapor leaving the top plate.
(b) The moles of vapor generated in the still per 100 mol of feed.
(c) The number of theoretical plates required.
  Notes: To simplify the problem, assume constant molar overflow equal to the moles of feed. Liquid air contains 20.9 mol% of oxygen and 79.1 mol% of nitrogen. The equilibrium data [*Chem. Met. Eng.*, **35**, 622 (1928)] at atmospheric pressure are

Temperature, K	Mole Percent N_2 in Liquid	Mole Percent N_2 in Vapor
77.35	100.00	100.00
77.98	90.00	97.17
78.73	79.00	93.62
79.44	70.00	90.31
80.33	60.00	85.91
81.35	50.00	80.46
82.54	40.00	73.50
83.94	30.00	64.05
85.62	20.00	50.81
87.67	10.00	31.00
90.17	0.00	0.00

7.10 A mixture of A (more volatile) and B is being separated in a plate distillation column. In two separate tests run with a saturated liquid feed of 40 mol% A, the following compositions, in mol% A, were obtained for samples of liquid and vapor streams from three consecutive stages between the feed and total condenser at the top:

	Mol% A			
	Test 1		Test 2	
Stage	Vapor	Liquid	Vapor	Liquid
$M + 2$	79.5	68.0	75.0	68.0
$M + 1$	74.0	60.0	68.0	60.5
M	67.9	51.0	60.5	53.0

Determine the reflux ratio and overhead composition in each case, assuming that the column has more than three stages.

7.11 A saturated liquid mixture containing 70 mol% benzene and 30 mol% toluene is to be distilled at atmospheric pressure to produce a distillate of 80 mol% benzene. Five procedures, described below, are under consideration. For each of the procedures, calculate and tabulate: (a) Moles of distillate per 100 moles of feed, (b) Moles of total vapor generated per mole of distillate, (c) Mole percent benzene in the residue, and

(d) For each part, construct a y–x diagram. On this, indicate the compositions of the overhead product, the reflux, and the composition of the residue.

(e) If the objective is to maximize total benzene recovery, which, if any, of these procedures is preferred?

Note: Assume that the relative volatility equals 2.5.

The procedures are as follows:

1. Continuous distillation followed by partial condensation. The feed is sent to the direct-heated still pot, from which the residue is continuously withdrawn. The vapors enter the top of a helically coiled partial condenser that discharges into a trap. The liquid is returned (refluxed) to the still, while the residual vapor is condensed as a product containing 80 mol% benzene. The molar ratio of reflux to product is 0.5.

2. Continuous distillation in a column containing one equilibrium plate. The feed is sent to the direct-heated still, from which residue is withdrawn continuously. The vapors from the plate enter the top of a helically coiled partial condenser that discharges into a trap. The liquid from the trap is returned to the plate, while the uncondensed vapor is condensed to form a distillate containing 80 mol% benzene. The molar ratio of reflux to product is 0.5.

3. Continuous distillation in a column containing the equivalent of two equilibrium plates. The feed is sent to the direct-heated still, from which residue is withdrawn continuously. The vapors from the top plate enter the top of a helically coiled partial condenser that dis-

charges into a trap. The liquid from the trap is returned to the top plate (refluxed) while the uncondensed vapor is condensed to form a distillate containing 80 mol% benzene. The molar ratio of reflux to product is 0.5.

4. The operation is the same as that described for procedure 3 with the exception that the liquid from the trap is returned to the bottom plate.

5. Continuous distillation in a column containing the equivalent of one equilibrium plate. The feed at its boiling point is introduced on the plate. The residue is withdrawn continuously from the direct-heated still pot. The vapors from the plate enter the top of a helically coiled partial condenser that discharges into a trap. The liquid from the trap is returned to the plate while the uncondensed vapor is condensed to form a distillate containing 80 mol% benzene. The molar ratio of reflux to product is 0.5.

7.12 A saturated liquid mixture of benzene and toluene containing 50 mol% benzene is distilled in an apparatus consisting of a still pot, one theoretical plate, and a total condenser. The still pot is equivalent to one equilibrium stage, and the pressure is 101 kPa.

The still is supposed to produce a distillate containing 75 mol% benzene. For each of the following procedures, calculate, if possible, the number of moles of distillate per 100 moles of feed. Assume a relative volatility of 2.5.

(a) No reflux with feed to the still pot.

(b) Feed to the still pot, reflux ratio $L/D = 3$.

(c) Feed to the plate with a reflux ratio of 3.

(d) Feed to the plate with a reflux ratio of 3. However, in this case, a partial condenser is employed.

(e) Part (b) using minimum reflux.

(f) Part (b) using total reflux.

7.13 A fractionation column operating at 101 kPa is to separate 30 kg/h of a solution of benzene and toluene containing 0.6 mass fraction toluene into an overhead product containing 0.97 mass fraction benzene and a bottoms product containing 0.98 mass fraction toluene. A reflux ratio of 3.5 is to be used. The feed is liquid at its boiling point, feed is to the optimal tray, and the reflux is at saturation temperature.

(a) Determine the quantity of top and bottom products.

(b) Determine the number of stages required.

EQUILIBRIUM DATA IN MOLE FRACTION BENZENE, 101 kPa

y	0.21	0.37	0.51	0.64	0.72	0.79	0.86	0.91	0.96	0.98
x	0.1	0.2	0.3	0.4	0.5	0.6	0.7	0.8	0.9	0.95

7.14 A mixture of 54.5 mol% benzene in chlorobenzene at its bubble point is fed continuously to the bottom plate of a column containing two theoretical plates. The column is equipped with a partial reboiler and a total condenser. Sufficient heat is supplied to the reboiler to give $V/F = 0.855$, and the reflux ratio L/V in the top of the column is kept constant at 0.50. Under these conditions, what quality of product and bottoms (x_D, x_B) can be expected?

EQUILIBRIUM DATA AT COLUMN PRESSURE, MOLE FRACTION BENZENE

x	0.100	0.200	0.300	0.400	0.500	0.600	0.700	0.800
y	0.314	0.508	0.640	0.734	0.806	0.862	0.905	0.943

7.15 A continuous distillation operation with a reflux ratio (L/D) of 3.5 yields a distillate containing 97 wt% B (benzene) and bottoms containing 98 wt% T (toluene). Due to weld failures, the 10 plates in the bottom section of the column are ruined, but the 14 upper plates are intact. It is suggested that the column still be used, with the feed (F) as saturated vapor at the dew point, with F = 13,600 kg/h containing 40 wt% B and 60 wt% T. Assuming that the plate efficiency remains unchanged at 50%: (a) Can this column still yield a distillate containing 97 wt% B, (b) How much distillate can we get, and (c) What will the composition of the residue be in mole percent?
For vapor–liquid equilibrium data, see Exercise 7.13.

7.16 A distillation column having eight theoretical stages (seven in the column + partial reboiler + total condenser) is being used to separate 100 kmol/h of a saturated liquid feed containing 50 mol% A into a product stream containing 90 mol% A. The liquid-to-vapor molar ratio at the top plate is 0.75. The saturated liquid feed is introduced on plate 5 from the top. Determine: (a) The composition of the bottoms, (b) The L/V ratio in the stripping section, and (c) The moles of bottoms per hour.
Unbeknown to the operators, the bolts holding plates 5, 6, and 7 rust through, and the plates fall into the still pot. If no adjustments are made, what is the new bottoms composition?

It is suggested that, instead of returning reflux to the top plate, an equivalent amount of liquid product from another column be used as reflux. If this product contains 80 mol% A, what now is the composition of: (a) The distillate, and (b) The bottoms.

EQUILIBRIUM DATA, MOLE FRACTION OF A

y	0.19	0.37	0.5	0.62	0.71	0.78	0.84	0.9	0.96
x	0.1	0.2	0.3	0.4	0.5	0.6	0.7	0.8	0.9

7.17 A distillation unit consists of a partial reboiler, a column with seven equilibrium plates, and a total condenser. The feed consists of a 50 mol% mixture of benzene in toluene. It is desired to produce a distillate containing 96 mol% benzene, when operating at 101 kPa.
(a) With saturated liquid feed fed to the fifth plate from the top, calculate: (1) Minimum reflux ratio $(L_R/D)_{min}$, (2) The bottoms composition, using a reflux ratio (L_R/D) of twice the minimum, and (3) Moles of product per 100 moles of feed.
(b) Repeat part (a) for a saturated vapor feed fed to the fifth plate from the top.

(c) With saturated vapor feed fed to the reboiler and a reflux ratio (L/V) of 0.9, calculate: (1) Bottoms composition, (2) Moles of product per 100 mole of feed.
Equilibrium data are given in Exercise 7.13.

7.18 A valve-tray fractionating column containing eight theoretical plates, a partial reboiler equivalent to one theoretical plate, and a total condenser is in operation separating a benzene–toluene mixture containing 36 mol% benzene at 101 kPa. Under normal operating conditions, the reboiler generates 100 kmol of vapor per hour. A request has been made for very pure toluene, and it is proposed to operate this column as a stripper, introducing the feed on the top plate as a saturated liquid, employing the same boilup at the still, and returning no reflux to the column. Equilibrium data are given in Exercise 7.13.
(a) What is the minimum feed rate under the proposed conditions, and what is the corresponding composition of the liquid in the reboiler at the minimum feed?
(b) At a feed rate 25% above the minimum, what is the rate of production of toluene, and what are the compositions in mole percent of the product and distillate?

7.19 A solution of methanol and water at 101 kPa containing 50 mol% methanol is continuously rectified in a seven-theo-retical-plate, perforated-tray column, equipped with a total condenser and a partial reboiler heated by steam.

During normal operation, 100 kmol/h of feed is introduced on the third plate from the bottom. The overhead product contains 90 mol% methanol, and the bottoms product contains 5 mol% methanol. One mole of liquid reflux is returned to the column for each mole of overhead product.

Recently it has been impossible to maintain the product purity in spite of an increase in the reflux ratio. The following test data were obtained:

Stream	kmol/h	mol% alcohol
Feed	100	51
Waste	62	12
Product	53	80
Reflux	94	—

What is the most probable cause of this poor performance? What further tests would you make to establish definitely the reason for the trouble? Could some 90% product be obtained by further increasing the reflux ratio, while keeping the vapor rate constant?

Vapor–liquid equilibrium data at 1 atm [*Chem. Eng. Prog.* **48**, 192 (1952)] in mole fraction methanol are

x	0.0321	0.0523	0.075	0.154	0.225	0.349	0.813	0.918
y	0.1900	0.2940	0.352	0.516	0.593	0.703	0.918	0.963

7.20 A fractionating column equipped with a partial reboiler heated with steam, as shown in Figure 7.40, and with a total condenser, is operated continuously to separate a mixture

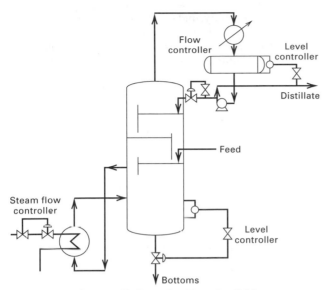

Figure 7.40 Data for Exercise 7.20.

of 50 mol% A and 50 mol% B into an overhead product containing 90 mol% A and a bottoms product containing 20 mol% A. The column has three theoretical plates, and the reboiler is equivalent to one theoretical plate. When the system is operated at an $L/V = 0.75$ with the feed as a saturated liquid to the bottom plate of the column, the desired products can be obtained. The system is instrumented as shown. The steam to the reboiler is controlled by a flow controller so that it remains constant. The reflux to the column is also on a flow controller so that the quantity of reflux is constant. The feed to the column is normally 100 kmol/h, but it was inadvertently cut back to 25 kmol/h. What would be the composition of the reflux, and what would be the composition of the vapor leaving the reboiler under these new conditions? Assume that the vapor leaving the reboiler is not superheated. Relative volatility for the system is 3.0.

7.21 A saturated vapor mixture of maleic anhydride and benzoic acid containing 10 mol% acid is a by-product of the manufacture of phthalic anhydride. This mixture is distilled continuously at 13.3 kPa to give a product of 99.5 mol% maleic anhydride and a bottoms of 0.5 mol% anhydride. Using the data below, calculate the number of theoretical plates needed using an L/D of 1.6 times the minimum.

VAPOR PRESSURE, TORR:

	10	50	100	200	400
Temperature, °C:					
Maleic anhydride	78.7	116.8	135.8	155.9	179.5
Benzoic acid	131.6	167.8	185.0	205.8	227

7.22 A bubble-point binary mixture containing 5 mol% A in B is to be distilled to give a distillate containing 35 mol% A and a bottoms product containing 0.2 mol% A. If the relative volatility is constant at a value of 6, calculate the

following algebraically, assuming that the column will be equipped with a partial reboiler and a partial condenser.
(a) The minimum number of equilibrium stages
(b) The minimum boilup ratio V/B leaving the reboiler
(c) The actual number of equilibrium stages for a boilup ratio equal to 1.2 times the minimum value

7.23 Methanol (M) is to be separated from water (W) by distillation as shown in Figure 7.41. The feed is subcooled such that $q = 1.12$. Determine the feed stage location and the number of theoretical stages required. Vapor–liquid equilibrium data are given in Exercise 7.19.

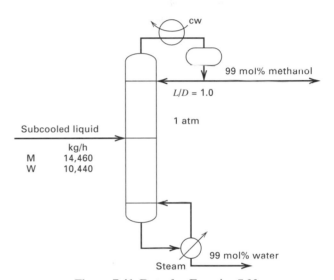

Figure 7.41 Data for Exercise 7.23.

7.24 A saturated liquid mixture of 69.4 mol% benzene (B) in toluene (T) is to be continuously distilled at atmospheric pressure to produce a distillate containing 90 mol% benzene, with a yield of 25 moles of distillate per 100 moles of feed. The feed is sent to a steam-heated still (reboiler), where residue is to be withdrawn continuously. The vapors from the still pass directly to a partial condenser. From a liquid separator following the condenser, reflux is returned to the still. Vapors from the separator, which are in equilibrium with the liquid reflux, are sent to a total condenser and are continuously withdrawn as distillate. At equilibrium the mole ratio of B to T in the vapor may be taken as 2.5 times the mole ratio of B to T in the liquid. Calculate analytically and graphically the total moles of vapor generated in the still per 100 mol of feed.

7.25 A plant has a batch of 100 kmol of a liquid mixture containing 20 mol% benzene and 80 mol% chlorobenzene. It is desired to rectify this mixture at 1 atm to obtain bottoms containing only 0.1 mol% benzene. The relative volatility may be assumed constant at 4.13. There are available a suitable still to vaporize the feed, a column containing the equivalent of four theoretical plates, a total condenser, and a reflux drum to collect the condensed overhead. The run is to be made at total reflux. While the steady state is being approached, a finite amount of distillate is held in a reflux trap. When the steady state is reached, the bottoms contain 0.1 mol% benzene. With this apparatus, what yield of bottoms can be obtained? The liquid holdup in the column is negligible compared to that in the still and in the reflux drum.

7.26 A mixture of acetone and isopropanol containing 50 mol% acetone is to be distilled continuously to produce an overhead product containing 80 mol% acetone and a bottoms containing 25 mol% acetone. If a saturated liquid feed is employed, if the column is operated with a reflux ratio of 0.5, and if the Murphree vapor efficiency is 50%, how many trays will be required? Assume a total condenser, partial reboiler, saturated liquid reflux, and optimal feed stage. The vapor–liquid equilibrium data for this system are

EQUILIBRIUM DATA, MOLE PERCENT ACETONE

Liquid	0	2.6	5.4	11.7	20.7	29.7	34.1	44.0	52.0
Vapor	0	8.9	17.4	31.5	45.6	55.7	60.1	68.7	74.3

Liquid	63.9	74.6	80.3	86.5	90.2	92.5	95.7	100.0
Vapor	81.5	87.0	89.4	92.3	94.2	95.5	97.4	100.0

7.27 A mixture of 40 mol% carbon disulfide (CS_2) in carbon tetrachloride (CCl_4) is continuously distilled. The feed is 50% vaporized ($q = 0.5$). The top product from a total condenser is 95 mol% CS_2, and the bottoms product from a partial reboiler is a liquid of 5 mol% CS_2.

The column operates with a reflux ratio, L/D, of 4 to 1. The Murphree vapor efficiency is 80%.
(a) Calculate graphically the minimum reflux, the minimum boilup ratio from the reboiler, V/B, and the minimum number of stages (including reboiler).
(b) How many trays are required for the actual column at 80% efficiency by the McCabe–Thiele method.

The vapor–liquid equilibrium data at column pressure for this mixture in terms of CS_2 mole fraction are

x	0.05	0.1	0.2	0.3	0.4	0.5	0.6	0.7	0.8	0.9
y	0.135	0.245	0.42	0.545	0.64	0.725	0.79	0.85	0.905	0.955

7.28 A distillation unit consists of a partial reboiler, a bubble-cap column, and a total condenser. The overall plate efficiency is 65%. The feed is a liquid mixture, at its bubble point, consisting of 50 mol% benzene in toluene. This liquid is fed to the optimal plate. The column is to produce a distillate containing 95 mol% benzene and a bottoms of 95 mol% toluene. Calculate for an operating pressure of 1 atm:
(a) Minimum reflux ratio $(L/D)_{min}$, (b) Minimum number of actual plates to carry out the desired separation, (c) Using a reflux ratio (L/D) of 50% more than the minimum, the number of actual plates needed, (d) The kilograms per hour of product and residue, if the feed is 907.3 kg/h, (e) The saturated steam at 273.7 kPa required in kilograms per hour for heat to the reboiler using enthalpy data below and any

assumptions necessary, and (f) A rigorous enthalpy balance on the reboiler, using the enthalpy data tabulated below and assuming ideal solutions. Enthalpies in Btu/lbmol at reboiler temperature:

	h_L	h_V
benzene	4,900	18,130
toluene	8,080	21,830

Vapor–liquid equilibrium data are given in Exercise 7.13.

7.29 A continuous distillation unit, consisting of a perforated-tray column together with a partial reboiler and a total condenser, is to be designed to operate at atmospheric pressure to separate ethanol and water. The feed, which is introduced into the column as liquid at its bubble point, contains 20 mol% alcohol. The distillate is to contain 85 mol% alcohol, and the alcohol recovery is to be 97%.
(a) What is the molar concentration of the bottoms?
(b) What is the minimum value of:
  (1) The reflux ratio L/V?
  (2) The reflux ratio L/D?
  (3) The boilup ratio V/B from the reboiler?
(c) What is the minimum number of theoretical stages and the corresponding number of actual plates if the overall plate efficiency is 55%?
(d) If the reflux ratio L/V used is 0.80, how many actual plates will be required?

Vapor–liquid equilibrium for ethanol–water at 1 atm in terms of mole fractions of ethanol are [*Ind. Eng. Chem.*, **24**, 881 (1932)]:

x	y	$T, °C$	x	y	$T, °C$
0.0190	0.1700	95.50	0.3273	0.5826	81.50
0.0721	0.3891	89.00	0.3965	0.6122	80.70
0.0966	0.4375	86.70	0.5079	0.6564	79.80
0.1238	0.4704	85.30	0.5198	0.6599	79.70
0.1661	0.5089	84.10	0.5732	0.6841	79.30
0.2337	0.5445	82.70	0.6763	0.7385	78.74
0.2608	0.5580	82.30	0.7472	0.7815	78.41
			0.8943	0.8943	78.15

7.30 A solvent A is to be recovered by distillation from its water solution. It is necessary to produce an overhead product containing 95 mol% A and to recover 95% of the A in the feed. The feed is available at the plant site in two streams, one containing 40 mol% A and the other 60 mol% A. Each stream will provide 50 kmol/h of component A, and each will be fed into the column as saturated liquid. Since the less volatile component is water, it has been proposed to supply the necessary heat in the form of open steam. For the preliminary design, it has been suggested that the operating reflux ratio, L/D, be 1.33 times the minimum value. A total condenser will be employed. For this system, it is estimated that the overall plate efficiency will be 70%. How many plates will be required, and what will be the bottoms composition? The relative volatility may be assumed to be constant at 3.0. Determine analytically the points necessary to locate the operating lines. Each feed should enter the column at its optimal location.

7.31 A saturated liquid feed stream containing 40 mol% n-hexane (H) and 60 mol% n-octane is fed to a plate column. A reflux ratio L/D equal to 0.5 is maintained at the top of the column. An overhead product of 0.95 mole fraction H is required, and the column bottoms is to be 0.05 mole fraction H. A cooling coil submerged in the liquid of the second plate from the top removes sufficient heat to condense 50 mol% of the vapor rising from the third plate down from the top.
(a) Derive the equations needed to locate the operating line.
(b) Locate the operating lines and determine the required number of theoretical plates if the optimal feed plate location is used.

7.32 One hundred kilogram-moles per hour of a saturated liquid mixture of 12 mol% ethyl alcohol in water is distilled continuously by direct steam at 1 atm introduced directly to the bottom plate. The distillate required is 85 mol% alcohol, representing 90% recovery of the alcohol in the feed. The reflux is saturated liquid with $L/D = 3$. Feed is on the optimal stage. Vapor–liquid equilibrium data are given in Exercise 7.29. Calculate:
(a) Steam requirement, kmol/h
(b) Number of theoretical stages
(c) The feed stage (optimal)
(d) Minimum reflux ratio, $(L/D)_{\min}$

7.33 A water–isopropanol mixture at its bubble point containing 10 mol% isopropanol is to be continuously rectified at atmospheric pressure to produce a distillate containing 67.5 mol% isopropanol. Ninety-eight percent of the isopropanol in the feed must be recovered. If a reflux ratio L/D of 1.5 times the minimum is used, how many theoretical stages will be required: (a) If a partial reboiler is used? (b) If no reboiler is used and saturated steam at 101 kPa is introduced below the bottom plate? (c) How many stages are required at total reflux?

Vapor–liquid equilibrium data in mole fraction of isopropanol at 101 kPa are

T, °C	93.00	84.02	82.12	81.25	80.62	80.16	80.28	81.51
y	0.2195	0.4620	0.5242	0.5686	0.5926	0.6821	0.7421	0.9160
x	0.0118	0.0841	0.1978	0.3496	0.4525	0.6794	0.7693	0.9442

Notes: Composition of the azeotrope is $x = y = 0.6854$. Boiling point of azeotrope = 80.22°C

7.34 An aqueous solution containing 10 mol% isopropanol is fed at its bubble point to the top of a continuous stripping column, operated at atmospheric pressure, to produce a va-

por containing 40 mol% isopropanol. Two procedures are under consideration, both involving the same heat expenditure with V/F (moles of vapor generated/mole of feed) = 0.246 in each case. Scheme 1 uses a partial reboiler at the bottom of a plate-type stripping column, generating vapor by the use of steam condensing inside a closed coil. In Scheme 2, the reboiler is omitted and live steam is injected directly below the bottom plate. Determine the number of stages required in each case.

Equilibrium data for the system isopropanol–water are given in Exercise 7.33. The usual simplifying assumptions may be made.

7.35 Determine the optimal stage location for each feed and the number of theoretical stages required for the distillation separation shown in Figure 7.42 using the following equilibrium data in mole fractions.

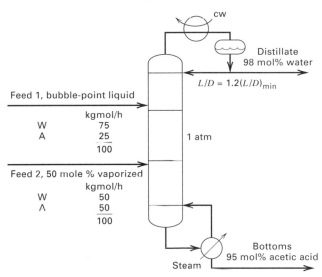

Figure 7.42 Data for Exercise 7.35.

WATER (W)/ACETIC ACID (A), 1 atm

x_W	0.0055	0.053	0.125	0.206	0.297	0.510	0.649	0.803	0.9594
y_W	0.0112	0.133	0.240	0.338	0.437	0.630	0.751	0.866	0.9725

7.36 Determine the number of theoretical stages required and the optimal stage locations for the feed and liquid side stream for the distillation process shown in Figure 7.43 assuming that methanol (M) and ethanol (E) form an ideal solution.

7.37 A mixture of n-heptane (H) and toluene (T) is separated by extractive distillation with phenol (P). Distillation is then used to recover the phenol for recycle as shown in Figure 7.44a, where the small amount of n-heptane in the feed is ignored. For the conditions shown in Figure 7.44a, determine the number of theoretical stages required. Note that heat will have to be supplied to the reboiler at a high temperature because of the high boiling point of phenol.

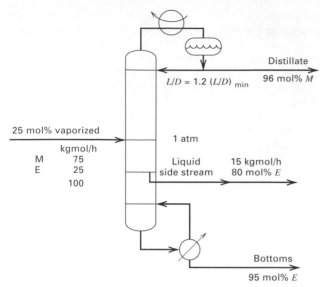

Figure 7.43 Data for Exercise 7.36.

x_T	y_T	T, °C	x_T	y_T	T, °C
0.0435	0.3410	172.70	0.6512	0.9260	120.00
0.0872	0.5120	159.40	0.7400	0.9463	119.70
0.1186	0.6210	153.80	0.7730	0.9536	119.40
0.1248	0.6250	149.40	0.8012	0.9545	115.60
0.2190	0.7850	142.20	0.8840	0.9750	112.70
0.2750	0.8070	133.80	0.9108	0.9796	112.20
0.4080	0.8725	128.30	0.9394	0.9861	113.30
0.4800	0.8901	126.70	0.9770	0.9948	111.10
0.5898	0.9159	122.20	0.9910	0.9980	111.10
0.6348	0.9280	120.20	0.9939	0.9986	110.50
			0.9973	0.9993	110.50

7.38 A distillation column for the separation of *n*-butane from n-pentane was recently put into operation in a petroleum refinery. Apparently, an error was made in the design because the column fails to make the desired separation as shown in the following table [*Chem. Eng. Prog.,* **61** (8), 79 (1965)]:

	Design Specification	Actual Operation
Mol% nC_5 in distillate	0.26	13.49
Mol% nC_4 in bottoms	0.16	4.28

In order to correct the situation, it is proposed to add an intercondenser in the rectifying section to generate more reflux and an interreboiler in the stripping section to produce additional boilup. Show by use of a McCabe–Thiele diagram how such a proposed change can improve the operation.

Therefore, consider the alternative scheme in Figure 7.44b, where an interreboiler, located midway between the bottom plate and the feed stage, is used to provide 50% of the boilup used in Figure 7.44a. The remainder of the boilup is provided by the reboiler. Determine the number of theoretical stages required for the case with the interreboiler and the temperature of the interreboiler stage. Unsmoothed vapor–liquid equilibrium data at 1 atm are [*Trans. AIChE,* **41,** 555 (1945)]:

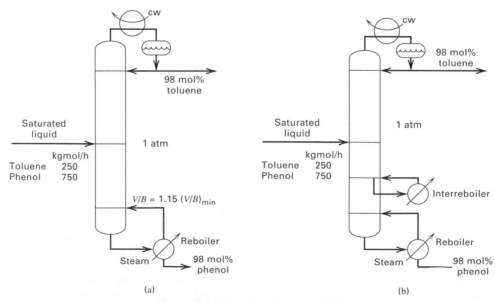

Figure 7.44 Data for Exercise 7.37.

7.39 In the production of chlorobenzenes by chlorination of benzene, two close-boiling isomers, *para*-dichlorobenzene (P) and *ortho*-dichlorobenzene (O), are separated by distillation. The feed to the column consists of 62 mol% of the para isomer and 38 mol% of the ortho isomer. Assume that the pressures at the bottom and top of the column are 20 psia (137.9 kPa) and 15 psia (103.4 kPa), respectively. The distillate is a liquid containing 98 mol% para isomer. The bottoms product is to contain 96 mol% ortho isomer. At column pressure, the feed is slightly vaporized with $q = 0.9$. Calculate the number of theoretical stages required for a reflux ratio equal to 1.15 times the minimum reflux ratio. Base your calculations on a constant relative volatility obtained as the arithmetic average between the column top and column bottom using appropriate vapor pressure data and the assumption of Raoult's and Dalton's laws. The McCabe–Thiele construction should be supplemented at the two ends by use of the Kremser equations as illustrated in Example 7.4.

7.40 Relatively pure oxygen and nitrogen can be obtained by the distillation of air using the Linde double column, which, as shown in Figure 7.45, consists of a lower column operating at elevated pressure surmounted by an atmospheric-pressure column. The boiler of the upper column is at the same time the reflux condenser for both columns.

Gaseous air plus enough liquid to take care of heat leak into the column (more liquid, of course, if liquid-oxygen product is withdrawn) enters the exchanger at the base of the lower column and condenses, giving up heat to the boiling liquid and thus supplying the vapor flow for this column. The liquid air enters an intermediate point in this column, as shown in Figure 7.45. The vapors rising in this column are partially condensed to form the reflux, and the uncondensed vapor passes to an outer row of tubes and is totally condensed, the liquid nitrogen collecting in an annulus, as shown. By operating this column at 4 to 5 atm, the liquid oxygen boiling at 1 atm is cold enough to condense pure nitrogen. The liquid that collects in the bottom of the lower column contains about 45 mol% O_2 and forms the feed for the upper column. Such a double column can produce very pure oxygen with high oxygen recovery, and relatively pure nitrogen. On a single McCabe–Thiele diagram—using equilibrium lines, operating lines, q-lines, 45°line, stepped-off stages, and other illustrative aids—show qualitatively how the stage requirements of the double column can be computed.

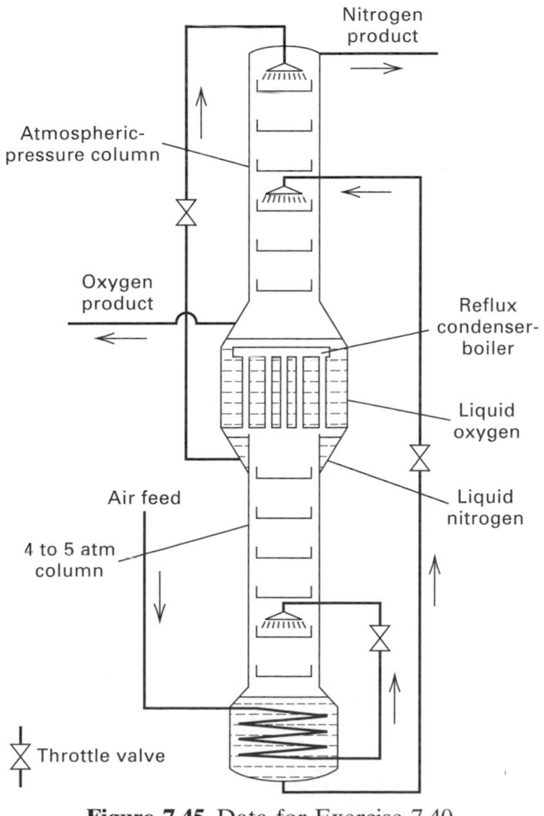

Section 7.3

7.41 The following performance data have been obtained for a distillation tower separating a 50/50 by weight percent mixture of methanol and water:

Feed rate = 45,438 lb/h, Feed condition = bubble-point liquid at feed-tray pressure, Wt% methanol in distillate = 95.04, and Wt% methanol in bottoms = 1.00
Reflux ratio = 0.947; Reflux condition = saturated liquid
Boilup ratio = 1.138; Pressure in reflux drum = 14.7 psia
Type condenser = total; Type reboiler = partial
Condenser pressure drop = 0.0 psi; Tower pressure drop = 0.8 psi
Trays above feed tray = 5; Trays below feed tray = 6
Total trays = 12; Tray diameter = 6 ft
Type tray = single-pass sieve tray; Flow path length = 50.5 in.
Weir length = 42.5 in.; Hole area = 10%; Hole size = 3/16 in.
Weir height = 2 in.; Tray spacing = 24 in.
Viscosity of feed = 0.34 cP
Surface tension of distillate = 20 dyne/cm; Surface tension of bottoms = 58 dyne/cm
Temperature of top tray = 154°F; Temperature of bottom tray = 207°F

Vapor–liquid equilibrium data at column pressure in mole fraction of methanol are

y	0.0412	0.156	0.379	0.578
x	0.00565	0.0246	0.0854	0.205

y	0.675	0.729	0.792	0.915
x	0.315	0.398	0.518	0.793

Figure 7.45 Data for Exercise 7.40.

Figure 7.46 Data for Exercise 7.43.

Based on the above data:
(a) Determine the overall tray efficiency from the data, assuming that the reboiler is the equivalent of a theoretical stage.
(b) Estimate the overall tray efficiency from the Drickamer–Bradford correlation.
(c) Estimate the overall tray efficiency from the O'Connell correlation, accounting for length of flow path.
(d) Estimate the Murphree vapor tray efficiency by the method of Chan and Fair.

7.42 For the conditions of Exercise 7.41, a laboratory Oldershaw column measures an average Murphree vapor point efficiency of 65%. Estimate E_{MV} and E_o.

Section 7.4

7.43 Conditions for the top tray of a distillation column are as shown in Figure 7.46. Determine the column diameter corresponding to 85% of flooding if a valve tray is used. Make whatever assumptions necessary.

7.44 A separation of propylene from propane is achieved by distillation as shown in Figure 7.47, where two columns in series are used because a single column would be too tall. The tray numbers refer to equilibrium stages. Determine the column diameters, tray efficiency using the O'Connell correlation, number of actual trays, and column heights if perforated trays are used.

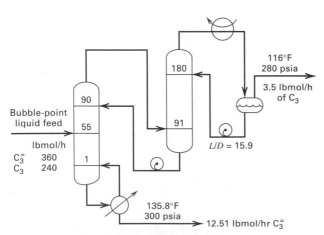

Figure 7.47 Data for Exercise 7.44.

7.45 Determining the height and diameter of a vertical flash drum for the conditions shown in Figure 7.48.

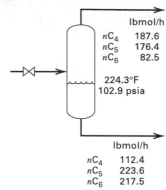

Figure 7.48 Data for Exercise 7.45.

7.46 Determine the length and diameter of a horizontal reflux drum for the conditions shown in Figure 7.49.

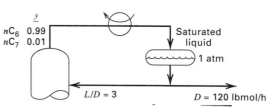

Figure 7.49 Data for Exercise 7.46.

7.47 Results of design calculations for a methanol-water distillation operation are given in Figure 7.50.

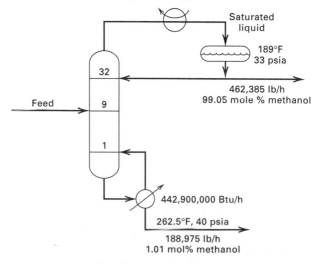

Figure 7.50 Data for Exercise 7.47.

(a) Calculate the column diameter at the top tray and at the bottom tray for sieve trays. Should the column be swaged?
(b) Calculate the length and diameter of the horizontal reflux drum.

7.48 For the conditions given in Exercise 7.41, estimate for the top tray and the bottom tray: (a) Percent of flooding, (b) Tray pressure drop in psi, (c) Whether weeping will occur, (d) Entrainment rate, and (e) Froth height in the downcomer.

7.49 If the feed rate to the tower of Exercise 7.41 is increased by 30% with all other conditions except for tower pressure drop remaining the same, estimate for the top and bottom trays: (a) Percent of flooding, (b) Tray pressure drop in psi, (c) Entrainment rate, (d) Froth height in the downcomer. Will the new operation be acceptable? If not, should you consider a retrofit with packing? If so, should both sections of the column be packed or could just one section be packed to achieve an acceptable operation?

Section 7.5

7.50 A mixture of benzene and dichloroethane is used to test the efficiency of a packed column that contains 10 ft of packing and operates adiabatically at atmospheric pressure. The liquid is charged to the reboiler, and the column is operated at total reflux until equilibrium is established. At equilibrium, liquid samples from the distillate and reboiler, as analyzed by refractive index, give the following compositions for benzene: $x_D = 0.653$, $x_B = 0.298$.

Calculate the value of HETP in inches for this packing. What are the limitations on using this calculated value for design?

Data for x–y at 1 atm (in benzene mole fractions) are

x	0.1	0.2	0.3	0.4	0.5	0.6	0.7	0.8	0.9
y	0.11	0.22	0.325	0.426	0.526	0.625	0.720	0.815	0.91

7.51 Consider a distillation column for separating ethanol from water at 1 atm. The following specifications are set:

Feed: 10 mol% ethanol (bubble-point liquid)
Bottoms: 1 mol% ethanol
Distillate: 80 mol% ethanol (saturated liquid)
Reflux ratio: 1.5 times the minimum

Constant molar overflow may be assumed and vapor–liquid equilibrium data are given in Exercise 7.29.
(a) How many theoretical plates are required above and below the feed if a plate column is used?
(b) How many transfer units are required above and below the feed if a packed column is used?
(c) Assuming that the plate efficiency is approximately 80% and the plate spacing is 18 in., how high is the plate column?
(d) Using an H_{OG} value of 1.2 ft., how high is the packed column?

(e) Assuming that you had HTU data available only on the benzene–toluene system, how would you go about applying the data to obtain the HTU for the ethanol–water system?

7.52 Plant capacity for the methanol–water distillation of Exercise 7.41 is to be doubled. Rather than installing a second trayed tower identical to the one in operation, a packed column is to be considered for the new installation. This column will have a feed location identical to the present trayed tower and will be expected to achieve the same product purities with the same top pressure and reflux ratio. Two packings are being considered:

1. 50-mm plastic NOR PAC rings (a random packing)
2. Montz metal B1-300 (a structured packing)

For each of these two packings, design a packed column to operate at 70% of flooding by calculating for each section: (a) Liquid holdup, (b) Column diameter, (c) H_{OG}, (d) Packed height, (e) Pressure drop.
What are the advantages, if any, of each of the packed-column designs over a second trayed tower? Which packing, if either, is preferable?

7.53 For the specifications of Example 7.1, design a packed column using 50-mm metal Hiflow rings and operating at 70% of flooding by calculating for each section: (a) Liquid holdup, (b) Column diameter, (c) H_{OG}, (d) Packed height, and (e) Pressure drop.
What are the advantages and disadvantages of a packed column as compared to a trayed tower for this service?

Section 7.6

7.54 An enthalpy–concentration diagram is given in Figure 7.37 for a mixture of n-hexane (H), and n-octane (O) at 101 kPa. Using this diagram, determine the following:
(a) The mole fraction composition of the vapor when a liquid containing 30 mol% H is heated from point A to the bubble-point temperature at point B.
(b) The energy required to vaporize 60 mol% of a mixture initially at 100°F and containing 20 mol% H (point G).
(c) The compositions of the equilibrium vapor and liquid resulting from part (b).

7.55 Using the enthalpy–concentration diagram of Figure 7.37, determine the following for a mixture of n-hexane (H) and n-octane (O) at 1 atm:
(a) The temperature and compositions of equilibrium liquid and vapor resulting from adiabatic mixing of 950 lb/h of a mixture of 30 mol% H in O at 180°F with 1,125 lb/h of a mixture of 80 mol% H in O at 240°F.
(b) The energy required to partially condense, by cooling, a mixture of 60 mol% H in O from an initial temperature of 260°F to 200°F. What are the compositions and amounts of the resulting vapor and liquid phases per pound-mole of original mixture?

Table 7.8 Methanol–Water Vapor–Liquid Equilibrium and Enthalpy Data for 1 atm (MeOH = Methyl Alcohol)

Mol% MeOH y or x	Enthalpy above 0°C, Btu/lbmol Solution				Vapor Liquid Equilibrium Data		
	Saturated Vapor		Saturated Liquid		Mol% MeOH in		Boiling Point, °C
	T, °C	h_V	T, °C	h_L	Liquid	Vapor	
0	100	20,720	100	3,240	0	0	100
5	98.9	20,520	92.8	3,070	2.0	13.4	96.4
10	97.7	20,340	87.7	2,950	4.0	23.0	93.5
15	96.2	20,160	84.4	2,850	6.0	30.4	91.2
20	94.8	20,000	81.7	2,760	8.0	36.5	89.3
30	91.6	19,640	78.0	2,620	10.0	41.8	87.7
40	88.2	19,310	75.3	2,540	15.0	51.7	84.4
50	84.9	18,970	73.1	2,470	20.0	57.9	81.7
60	80.9	18,650	71.2	2,410	30.0	66.5	78.0
70	76.6	18,310	69.3	2,370	40.0	72.9	75.3
80	72.2	17,980	67.6	2,330	50.0	77.9	73.1
90	68.1	17,680	66.0	2,290	60.0	82.5	71.2
100	64.5	17,390	64.5	2,250	70.0	87.0	69.3
					80.0	91.5	67.6
					90.0	95.8	66.0
					95.0	97.9	65.0
					100.0	100.0	64.5

Source: J.G. Dunlop, "Vapor–Liquid Equilibrium Data," M.S. thesis, Brooklyn Polytechnic Institute, Brooklyn, NY (1948).

(c) If the equilibrium vapor from part (b) is further cooled to 180°F, determine the compositions and relative amounts of the resulting vapor and liquid.

7.56 One hundred pound-moles per hour of a mixture of 60 mol% methanol in water at 30°C and 1 atm is to be separated by distillation at the same pressure into a liquid distillate containing 98 mol% methanol and a bottoms liquid product containing 96 mol% water. Enthalpy and equilibrium data for the mixture at 1 atm are given in Table 7.8. The enthalpy of the feed mixture is 765 Btu/lbmol.

(a) Using the given data, plot an enthalpy–concentration diagram.
(b) Devise a procedure to determine, from the diagram of part (a), the minimum number of equilibrium stages for the condition of total reflux and the required separation.
(c) From the procedure developed in part (b), determine N_{min}. Why is the value independent of the feed condition?
(d) What are the temperatures of the distillate and the bottoms?

Chapter 8

Liquid–Liquid Extraction with Ternary Systems

In *liquid–liquid extraction,* a liquid feed of two or more components to be separated is contacted with a second liquid phase, called the *solvent,* which is immiscible or only partly miscible with one or more components of the liquid feed and completely or partially miscible with one or more of the other components of the liquid feed. Thus, the solvent, which is a single chemical species or a mixture, partially dissolves certain components of the liquid feed, effecting at least a partial separation of the feed. Liquid–liquid extraction is sometimes called *extraction, solvent extraction,* or *liquid extraction.* These, as well as the term *solid–liquid extraction,* are also applied to the recovery of substances from a solid by contact with a liquid solvent, such as the recovery of oil from seeds by an organic solvent. Solid-liquid extraction is covered in Chapter 16.

According to Derry and Williams [1], liquid extraction has been practiced since at least the time of the Romans, who separated gold and silver from molten copper by extraction using molten lead as a solvent. This was followed by the discovery that sulfur could selectively dissolve silver from an alloy with gold. However, it was not until the early 1930s that the first large-scale liquid–liquid extraction process began operation. In that industrial process, named after its inventor L. Edeleanu, aromatic and sulfur compounds were selectively removed from liquid kerosene by liquid–liquid extraction with liquid sulfur dioxide at 10 to 20°F. Removal of aromatic compounds resulted in a cleaner-burning kerosene. Liquid–liquid extraction has grown in importance in recent years because of the growing demand for temperature-sensitive products, higher-purity requirements, more efficient equipment, and availability of solvents with higher selectivity.

The simplest liquid–liquid extraction involves only a ternary system. The feed consists of two miscible components, the *carrier, C,* and the *solute, A.* solvent, *S,* is a pure compound. Components *C* and *S* are at most only partially soluble in each other. Solute *A* is soluble in *C* and completely or partially soluble in *S.* During the extraction process, mass transfer of *A* from the feed to the solvent occurs, with less transfer of *C* to the solvent, or *S* to the feed. However, complete or nearly complete transfer of *A* to the solvent is seldom achieved in just one stage, as discussed in Chapter 4. In practice, a number of stages are used in one- or two-section countercurrent cascades, as discussed in Chapter 5.

Acetic acid is produced by methanol carbonylation or oxidation of acetaldehyde, or as a by-product of cellulose acetate manufacture. In all three cases, a mixture of acetic acid (normal b.p. = 118.1°C) and water (normal b.p. = 100°C) must be separated to give glacial acetic acid (99.8 wt% min). When the mixture contains less than 50% acetic acid, separation by distillation is expensive because of the need to vaporize large amounts of the more volatile water, with its very high heat of vaporization. Accordingly, an alternative, less expensive liquid–liquid extraction process is often used. A typical implementation is shown in Figure 8.1. In this process, it is important to note that two additional distillation separation steps are required to recover the solvent for recycle to the extractor. These additional separation steps are common to almost all extraction processes.

In the process of Figure 8.1, a feed of 30,260 lb/h, of 22 wt% acetic acid in water, is sent to a single-section extraction column, operating at near-ambient conditions, where the feed is countercurrently contacted with 71,100 lb/h of ethyl acetate solvent (normal

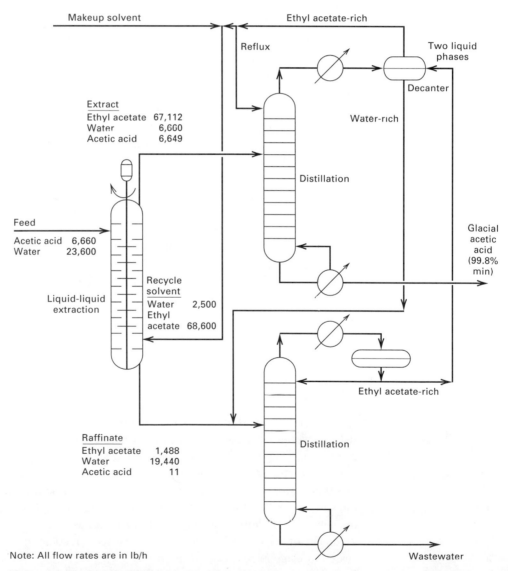

Note: All flow rates are in lb/h

Figure 8.1 Typical liquid–liquid extraction process.

b.p. = 77.1°C), saturated with water. The extract (solvent-rich product), being the low-density liquid phase, exits from the top of the extractor with 99.8% of the acetic acid originally contained in the feed. The raffinate (carrier-rich product), being the high-density liquid phase, exits from the bottom of the extractor and contains only 0.05 wt% acetic acid. The extract is sent to a distillation column, where glacial acetic acid is the bottoms product. The overhead vapor, which is rich in ethyl acetate but which also contains appreciable water vapor, splits into two liquid phases upon condensation. The two phases are separated by gravity in the decanter. The lighter ethyl acetate-rich phase is divided into two streams. One is used for reflux for the distillation operation and the other is used for solvent recycle to the extractor.

The water-rich phase from the decanter is sent, together with the raffinate from the extractor, to a second distillation column, where wastewater is removed from the bottom and the ethyl acetate-rich overhead distillate is recycled to the decanter. Makeup ethyl acetate solvent is provided for solvent losses to the glacial acetic acid and wastewater products. The fact that there are solvent losses implies environmental degradation. This, plus the fact that extraction requires solvent recovery equipment in addition to extraction equipment, makes it more costly and less desirable than distillation.

At an average extraction temperature of 100°F, six equilibrium stages are required to transfer 99.8% of the acetic acid from the feed to the extract using a solvent-to-feed ratio of 2.35 on a weight basis, where the recycled solvent is saturated with water. For six theoretical stages, a mechanically assisted extractor is preferred and a rotating-disk contactor (RDC), in a column configuration, is shown in Figure 8.1. The organic-rich phase is dispersed into droplets by rotating disks, while the water-rich phase is a continuous phase throughout the column. Dispersion and subsequent coalescence and settling takes place easily because at extractor operating conditions, liquid-phase viscosities are less than 1 cP, the phase-density difference is more than 0.08 g/cm^3, and the interfacial tension between the two phases is appreciable, at more than 30 dyne/cm.

The column has an inside diameter of 5.5 ft and a total height from the tangent of the top head to the tangent of the bottom head of 28 ft. The column is divided into 40 compartments, each 7.5 in. high and each containing a 40-in. diameter rotor disk located between a pair of stator (donut) rings of 46-in. inside diameter. Above the top stator ring and below the bottom stator ring are settling zones. Because the light liquid phase is dispersed, the liquid–liquid interface is maintained near the top of the column. The rotors are mounted on a centrally located, single shaft driven at a nominal 60 rpm by a 5-hp motor, equipped with a speed changer, the optimal disk speed being determined during plant operation. The HETP for the extractor is 50 in., equivalent to 6.67 compartments per theoretical stage. The HETP would be only 33 in. if axial (longitudinal) mixing did not occur. Because of the corrosive nature of aqueous acetic acid solutions, the extractor is constructed of stainless steel. Since 1948, hundreds of extraction columns similar to that of Figure 8.1, with diameters ranging up to at least 25 ft, have been built. As discussed in Section 8.1, a number of other extraction devices are suitable for the process in Figure 8.1.

Liquid–liquid extraction is a reasonably mature separation operation, although not as mature or as widely applied as distillation, absorption, and stripping. Since the 1930s, more than 1,000 laboratory, pilot-plant, and industrial extractors have been installed. Procedures for determining the number of theoretical stages to achieve a desired solute recovery are well established. However, in the thermodynamics of liquid–liquid extraction, no simple limiting theory, such as that of ideal solutions for vapor–liquid equilibrium, exists. In many cases, experimental equilibrium data are preferred over predictions based on activity-coefficient correlations. However, such data can often be correlated well by semitheoretical activity-coefficient equations such as the NRTL or UNIQUAC equations discussed in Chapter 2. Also, considerable laboratory effort may be required just to find an acceptable and efficient solvent. Furthermore, as will be discussed in the next section, a wide variety of industrial extraction equipment is available, making it necessary to consider many alternatives before making a final selection. Unfortunately, no generalized capacity and efficiency correlations are available for all equipment types. Often, equipment vendors must be relied upon to determine equipment size, or pilot-plant tests must be performed, followed by application of scale-up procedures recommended by the vendor or taken from sources such as this text.

Since the introduction of industrial liquid–liquid extraction processes, a large number of applications have been proposed and developed. The petroleum industry represents the largest-volume application for liquid–liquid extraction. By the late 1960s, more than 100,000 m^3/day of liquid feedstocks were being processed with physically selective solvents [2]. Extraction processes are well suited to the petroleum industry because of the need to separate heat-sensitive liquid feeds according to chemical type (e.g., aliphatic, aromatic, naphthenic) rather than by molecular weight or vapor pressure. Table 8.1

Table 8.1 Industrial Liquid–Liquid Extraction Processes

Solute	Carrier	Solvent
Acetic acid	Water	Ethyl acetate
Acetic acid	Water	Isopropyl acetate
Aconitic acid	Molasses	Methyl ethyl ketone
Ammonia	Butenes	Water
Aromatics	Paraffins	Diethylene glycol
Aromatics	Paraffins	Furfural
Aromatics	Kerosene	Sulfur dioxide
Aromatics	Paraffins	Sulfur dioxide
Asphaltenes	Hydrocarbon oil	Furfural
Benzoic acid	Water	Benzene
Butadiene	1-Butene	*aq.* Cuprammonium acetate
Ethylene cyanohydrin	Methyl ethyl ketone	Brine liquor
Fatty acids	Oil	Propane
Formaldehyde	Water	Isopropyl ether
Formic acid	Water	Tetrahydrofuran
Glycerol	Water	High alcohols
Hydrogen peroxide	Anthrahydroquinone	Water
Methyl ethyl ketone	Water	Trichloroethane
Methyl borate	Methanol	Hydrocarbons
Naphthenes	Distillate oil	Nitrobenzene
Naphthenes/aromatics	Distillate oil	Phenol
Phenol	Water	Benzene
Phenol	Water	Chlorobenzene
Penicillin	Broth	Butyl acetate
Sodium chloride	*aq.* Sodium hydroxide	Ammonia
Vanilla	Oxidized liquors	Toluene
Vitamin A	Fish-liver oil	Propane
Vitamin E	Vegetable oil	Propane
Water	Methyl ethyl ketone	*aq.* Calcium chloride

shows some representative industrial extraction processes. Other major applications exist in the biochemical industry, where emphasis is on the separation of antibiotics and protein recovery from natural substrates; in the recovery of metals, such as copper from ammoniacal leach liquors, and in separations involving rare metals and radioactive isotopes from spent fuel elements; and in the inorganic chemical industry, where high-boiling constituents such as phosphoric acid, boric acid, and sodium hydroxide need to be recovered from aqueous solutions.

In general, extraction is preferred to distillation for the following applications:

1. In the case of dissolved or complexed inorganic substances in organic or aqueous solutions.
2. For the removal of a component present in small concentrations, such as a color former in tallow or hormones in animal oil.
3. When a high-boiling component is present in relatively small quantities in a waste stream, as in the recovery of acetic acid from cellulose acetate. Extraction becomes competitive with distillation because of the expense of evaporating large quantities of water with its very high heat of vaporization.
4. In the recovery of heat-sensitive materials, where extraction may be less expensive than vacuum distillation.
5. In the case of the separation of a mixture according to chemical type rather than relative volatility.
6. In the case of the separation of close-melting or close-boiling liquids, where solubility differences can be exploited.
7. In the case of mixtures that form azeotropes.

The key to an effective extraction process is the discovery of a suitable solvent. In addition to being nontoxic, inexpensive, and easily recoverable, a good solvent should be relatively immiscible with feed components(s) other than the solute and have a different density from the feed to facilitate phase separation. Also, it must have a very high affinity for the solute, from which it should be easily separated by distillation, crystallization, or other means. Ideally, the distribution coefficient for the solute between the two liquid phases should be greater than 1; otherwise a large solvent-to-feed ratio is required. When the degree of solute extraction is not particularly high and/or when a large extraction factor can be achieved, an extractor will not require many stages. This is fortunate because mass transfer resistance in liquid–liquid systems is often high and stage efficiency is low in commercial contacting devices, unless mechanical agitation is provided.

In this chapter, equipment for conducting liquid–liquid extraction operations is discussed and fundamental equilibrium-based and rate-based calculation procedures are presented mainly for extraction in ternary systems. The use of graphical methods is emphasized. Except for systems dilute in solute(s), calculations for higher-order multicomponent systems are best conducted with computer-aided methods discussed in Chapter 10.

8.1 EQUIPMENT

Given the wide diversity of applications, one might expect a correspondingly large variety of liquid–liquid extraction devices. Indeed, such is the case. Equipment similar to that used for absorption, stripping, and distillation is sometimes used, but such devices are inefficient unless liquid viscosities are low and the difference in phase density is high. For that reason, centrifugal and mechanically agitated devices are often preferred. Regardless of the type of equipment, the necessary number of theoretical stages is computed. Then the size of the device for a continuous countercurrent process is obtained from experimental HETP or mass transfer performance data characteristic of the particular piece of equipment. In extraction, some authors use the acronym HETS, height equivalent to a theoretical stage, rather than HETP. Also, the *dispersed phase* is sometimes referred to as the *discontinuous phase,* the other phase being the *continuous phase.*

Mixer-Settlers

In mixer-settlers, the two liquid phases are first mixed and then separated by settling. Any number of mixer-settler units may be connected together to form a multistage, countercurrent cascade. During mixing, one of the liquids is dispersed in the form of small droplets into the other liquid phase. The dispersed phase may be either the heavier or the lighter of the two phases. The mixing step is commonly conducted in an agitated vessel, with sufficient agitation and residence time so that a reasonable approach to equilibrium (e.g., 80% to 90% of a theoretical stage) is attained. The vessel may be compartmented as shown in Figure 8.2, and is usually agitated by means of impellers of the type shown in Figure 8.3. If dispersion is easily achieved and equilibrium is rapidly approached, as with liquids of low interfacial tension and viscosity, the mixing step can be carried out by impingement in a jet mixer; by turbulence in a nozzle mixer, orifice mixer, or other in-line mixing device; by shearing action if both phases are fed simultaneously into a centrifugal pump; or by injectors, wherein the flow of one liquid is induced by another.

The settling step is by gravity in a second vessel called a settler or decanter. In the configuration shown in Figure 8.4, a horizontal vessel, with an impingement baffle to prevent the jet of the entering two-phase dispersion from disturbing the gravity-settling process, is used. Vertical and inclined vessels are also used. A major problem in settlers is emulsification, which may occur if the agitation is so intense that the dispersed droplet size falls below 1 to 1.5 micrometers. When this happens, coalescers, separator membranes,

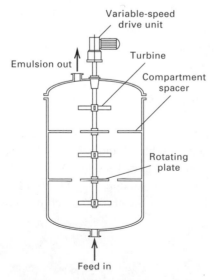

Figure 8.2 Compartmented mixing vessel with variable-speed turbine agitators [Adapted from R.E. Treybal, *Mass Transfer,* 3rd ed., McGraw-Hill, New York (1980).]

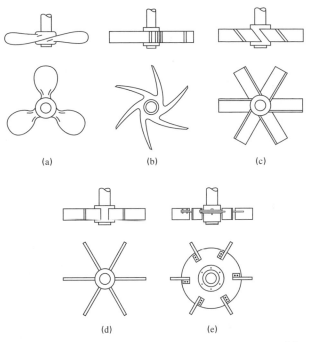

Figure 8.3 Some common types of mixing impellers: (a) marine-type propeller; (b) centrifugal turbine; (c) pitched-blade turbine; (d) flat-blade paddle; (e) flat-blade turbine. [From R.E. Treybal, *Mass Transfer,* 3rd ed., McGraw-Hill, New York (1980), with permission.]

meshes, electrostatic forces, ultrasound, chemical treatment, or other ploys are required to speed settling. The rate of settling can also be increased by substituting centrifugal for gravitational force. This may be necessary if the phase-density difference is small.

A large number of commercial single- and multi-stage mixer-settler units are available, many of which are described by Bailes, Hanson, and Hughes [3] and by Lo, Baird, and Hanson [4]. Particularly worthy of mention is the Lurgi extraction tower [4], which was originally developed for extracting aromatics from hydrocarbon mixtures. In this device, the phases are mixed by centrifugal mixers stacked vertically outside the column and driven from a single shaft. Settling takes place in the column, with phases flowing interstage-wise, guided by a complex baffle design located within the settling zones.

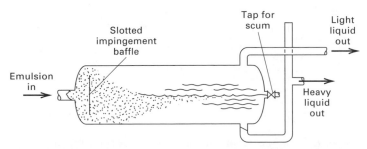

Figure 8.4 Horizontal gravity-settling vessel. [Adapted from R.E. Treybal, *Liquid Extraction,* 2nd ed., McGraw-Hill, New York (1963), with permission.]

Spray Columns

The simplest and one of the oldest extraction devices is the spray column. Either the heavy phase or the light phase can be dispersed, as shown in Figure 8.5. The droplets of the dispersed phase are generated only at the inlet, usually by spray nozzles. Because of lack of column internals, throughputs are large, depending upon phase-density difference and phase viscosities. As in gas absorption, *axial dispersion* (*backmixing*) in the continuous phase limits these devices to applications where only one or two stages are required. Axial dispersion is so serious for columns with large diameter-to-length ratio that the continuous phase may be completely mixed. Therefore, spray columns are rarely used, despite their very low cost.

Packed Columns

Axial mixing in a spray column can be substantially reduced, but not eliminated, by packing the column. The packing also improves mass transfer by breaking up large drops to increase interfacial area and promotes mixing in drops by distorting droplet shape. With the exception of Raschig rings [5], the same packings used in distillation and absorption are employed for liquid–liquid extraction. The choice of packing material, however, is somewhat more critical. A material preferentially wetted by the continuous phase is preferred. Figure 8.6 shows performance data, in terms of HTU, for Intalox Saddles in an extraction service as a function of continuous, U_C, and discontinuous, U_D, phase superficial velocities. Because of backmixing, the HETP is generally larger than for staged devices. For that reason, packed columns are used only where few stages are needed.

Plate Columns

Sieve plates in a column also reduce axial mixing and achieve a more stagewise type of contact. The dispersed phase may be the light or the heavy phase. In the former case, the dispersed phase, analogous to vapor bubbles in distillation, flows vertically up the column, with redispersion at each tray. The heavy phase is the continuous phase, flowing at each

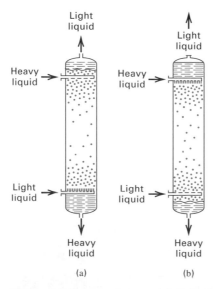

Figure 8.5 Spray columns: (a) light liquid dispersed, heavy liquid continuous; (b) heavy liquid dispersed, light liquid continuous.

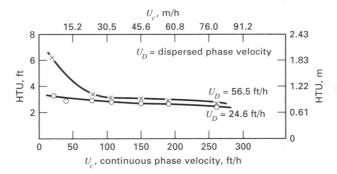

Figure 8.6 Efficiency of 1-in. Intalox saddles in a column 60 in. high with MEK–water–kerosene. [From R.R. Neumatis, J.S. Eckert, E.H. Foote, and L.R. Rollinson, *Chem. Eng. Progr., 67*(1), 60 (1971), with permission.]

stage through a *downcomer* and then across the tray the way a liquid does in a trayed distillation tower. If the heavy phase is dispersed, *upcomers* are used for the light phase. Columns have been built and successfully operated for diameters larger than 4.5 m. Holes from 0.64 to 0.32 cm in diameter and 1.25 to 1.91 cm apart are commonly used. Tray spacings are much closer than in distillation—10 to 15 cm in most applications involving low-interfacial-tension liquids. Plates are usually built without outlet weirs on the downspouts. A variation of the simple sieve column is the Koch Kascade Tower, where perforated plates are set in vertical arrays of complex designs.

If operated in the proper hydrodynamic flow regime, extraction rates in sieve-plate columns are high because the dispersed-phase droplets coalesce and re-form on each stage. This helps destroy concentration gradients, which develop if a droplet passes through the entire column without disturbance. Sieve-plate columns in extraction service are subject to the same limitations as distillation columns: flooding, entrainment, and, to a lesser extent, weeping. Additional problems, such as scum formation at interfaces due to small amounts of impurities, are frequently encountered in all types of extraction devices.

Columns with Mechanically Assisted Agitation

If the surface tension is high, and/or the density difference between the two liquid phases is low, and/or liquid viscosities are high, gravitational forces are inadequate for proper phase dispersal and the creation of turbulence. In that case, some type of mechanical agitation is necessary to increase interfacial area per unit volume and/or decrease mass transfer resistance. For packed and plate columns, agitation is provided by an oscillating pulse to the liquid, either by mechanical or pneumatic means. Pulsed perforated-plate columns found considerable application in the nuclear industry in the 1950s, but their popularity declined because of mechanical problems and the difficulty of propagating a pulse through a large volume [6]. The most important mechanically agitated columns are those that employ rotating agitators, driven by a shaft that extends axially through the column. The agitators create shear mixing zones, which alternate with settling zones in the column. Differences among the various agitated columns lie primarily in the mixers and settling chambers used. Eleven of the more popular arrangements are shown in Figure 8.7. Agitation can also be induced in a column by moving the plates back and forth in a reciprocating motion (Figure 8.7j) or in a novel horizontal contactor (Figure 8.7k). Such devices are also included in Figure 8.7. These devices answer the plea of Fenske, Carlson, and Quiggle [7] in 1947 for equipment that can efficiently provide large numbers of equilibrium stages in a compact device without large numbers of pumps and motors,

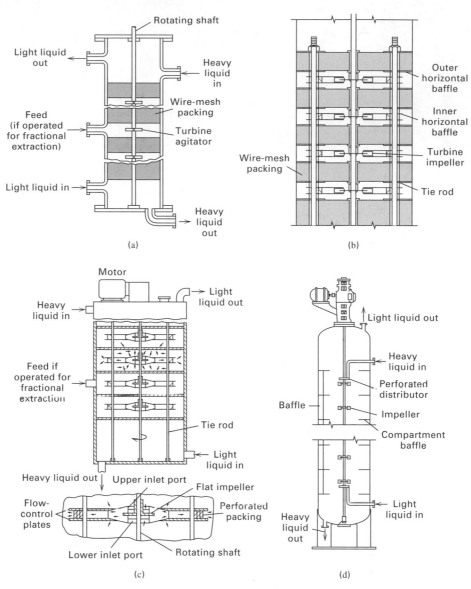

Figure 8.7 Commercial extractors with mechanically assisted agitation: (a) Scheibel column—first design; (b) Scheibel column—second design; (c) Scheibel column—third design; (d) Oldshue–Rushton (Mixco) column; (e) rotating-disk-contactor (RDC); (f) asymmetric rotating-disk contactor (ARD); (g) section of ARD contactor; (h) Kuhni column; (i) flow pattern in Kuhni column; (j) Karr reciprocating-plate column (RPC); (k) Graesser raining-buckct (RTL) extractor.

and extensive piping. They stated, "Despite . . . advantages of liquid–liquid separational processes, the problems of accumulating twenty or more theoretical stages in a small compact and relatively simple countercurrent operation have not yet been fully solved." Indeed, in 1946 it was considered impractical to design for more than seven theoretical stages, which represented the number of mixer-settler units in the only large-scale commercial liquid extraction process in use at that time.

Perhaps the first mechanically agitated column of importance was the Scheibel column [8] (Figure 8.7a), in which countercurrent liquid phases are contacted at fixed intervals by

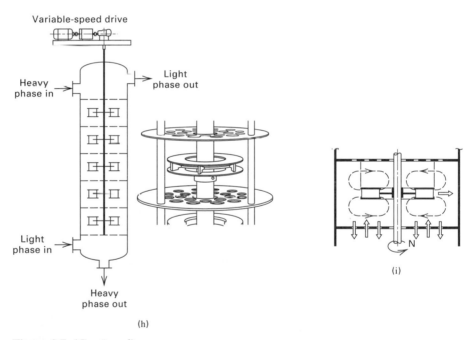

Figure 8.7 *(Continued)*

unbaffled, flat-bladed, turbine-type agitators (Figure 8.3) mounted on a vertical shaft. In the unbaffled separation or calming zones, located between the mixing zones, knitted wire-mesh packing is installed to prevent backmixing between mixing zones and to induce coalescence and settling of drops. The mesh material must be wetted by the dispersed phase. For more economical designs for larger-diameter installations (>1 m), Scheibel [9] (Figure 8.7b) added outer and inner horizontal annular baffles to divert the vertical flow of the phases in the mixing zone and to ensure complete mixing. For systems with high interfacial surface tension and viscosities, the wire mesh is removed. The first two Scheibel

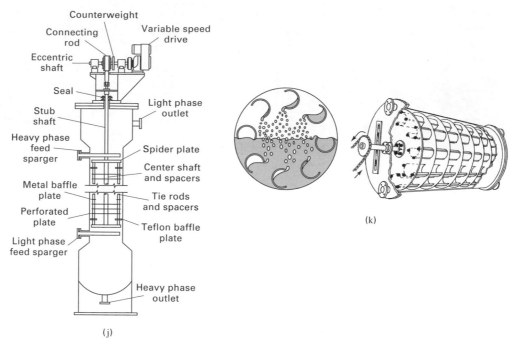

(j)

(k)

Figure 8.7 *(Continued)*

designs did not permit removal of the agitator shaft for inspection and maintenance. Instead the entire internal assembly (called the cartridge) had to be removed. To permit removal of just the agitator assembly shaft, especially for very-large-diameter columns (e.g., >1.5 m), and allow an access way through the column for any necessary inspection, cleaning, and repair, Scheibel [10] offered a third design, shown in Figure 8.7c. Here the agitator assembly shaft can be removed because it has a smaller diameter than the opening in the inner baffle.

The Oldshue–Rushton extractor [11] (Figure 8.7d) consists of a column with a series of compartments separated by annular outer stator-ring baffles, each with four vertical baffles attached to the wall. The centrally mounted vertical shaft drives a flat-bladed turbine impeller in each compartment.

A third type of column with rotating agitators that appeared about the same time as the Scheibel and Oldshue–Rushton columns is the rotating-disk contactor (RDC) [12,13] (Figure 8.7e), an example of which is described at the beginning of this chapter and shown in Figure 8.1. On a worldwide basis, it is probably the most extensively used liquid–liquid extraction device, with hundreds of units in use by 1983 [4]. Horizontal disks, mounted on a centrally located rotating shaft, are the agitation elements. Mounted at the column wall are annular stator rings with an opening larger than the agitator disk diameter. Thus, the agitator assembly shaft is easily removed from the column. Because the rotational speed of the rotor controls the drop size, the rotor speed can be continuously varied over a wide range.

A modification of the RDC concept is the asymmetric rotating-disk contactor (ARD) [14], which has been in industrial use since 1965. As shown in Figure 8.7f, the contactor consists of a column, a baffled stator, and an offset multistage agitator fitted with disks. The asymmetric arrangement, shown in more detail in Figure 8.7g, provides contact and transport zones that are separated by a vertical baffle, to which is attached a series of horizontal baffles. Compared to the RDC, this design retains the efficient shearing action, but reduces backmixing because of the separate mixing and settling compartments.

Another extractor based on the Scheibel concept is the Kuhni extraction column [15]. As shown in Figure 8.7h, the column is compartmented by a series of stator disks made of perforated plates. On a centrally positioned shaft is mounted a series of double-entry, radial-flow, shrouded turbine mixers, which promote, in each compartment, the circulation action shown in Figure 8.7i. For columns of diameter greater than 3 m, three turbine-mixer shafts on parallel axes are normally provided to preserve scale-up. Three hundred of these extractors were in use, mainly in Europe, by 1983 [4].

Rather than provide agitation by rotating impellers on a vertical shaft, or by pulsing the liquid phases, Karr [16,17] devised a reciprocating perforated-plate extractor, also called the Karr column, in which the plates move up and down approximately two times per second with a stroke length of 0.75 inch. As shown in Figure 8.7j, annular baffle plates are provided periodically in the plate stack to minimize axial mixing. The perforated plates use large holes (typically 9/16-in. diameter) and a high hole area (typically 58%). The central shaft, which supports both sets of plates, is reciprocated by a drive mechanism located at the top of the column. A modification of the Karr column is the vibrating-plate extractor (VPE) of Prochazka et al. [18], which uses perforated plates of smaller hole size and smaller percent hole area than the Karr column. The small holes provide passage for the dispersed phase, while one or more large holes on each plate provide passage for the continuous phase. Some VPE columns operate like the Karr column with uniform motion of all plates; others are provided with two shafts to obtain countermotion of alternate plates.

Another novel device for providing agitation is the Graesser raining-bucket contactor (RTL), which was developed in the late 1950s [4], primarily for extraction processes involving liquids of small density difference, low interfacial tension, and a tendency to form emulsions As shown in Figure 8.7k, a series of disks is mounted inside a shell on a central, horizontal rotating shaft, with a series of horizontal C-shaped buckets fitted between and around the periphery of the disks. An annular gap between the disks and the inside perphery of the shell allows countercurrent longitudinal flow of the phases. Dispersing action is very gentle, with each phase cascading through the other in opposite directions toward the two-phase interface, which is maintained close to the equatorial position.

A number of industrial centrifugal extractors have been available since 1944, when the Podbielniak (POD) extractor, with its short residence time, was successfully applied to penicillin extraction [19]. In the POD, several concentric sieve trays are arranged around

Table 8.2 Maximum Size and Loading for Commercial Liquid–Liquid Extraction Columns

Column Type	Approximate Maximum Liquid Throughout, m^3/m^2-h	Maximum Column Diameter, m
Lurgi tower	30	8.0
Pulsed packed	40	3.0
Pulsed sieve tray	60	3.0
Scheibel	40	3.0
RDC	40	8.0
ARD	25	5.0
Kuhni	50	3.0
Karr	100	1.5
Graesser	<10	7.0

Above data apply to systems of:
1. High interfacial surface tension (30 to 40 dyne/cm).
2. Viscosity of approximately 1 cP.
3. Volumetric phase ratio of 1:1.
4. Phase-density difference of approximately 0.6 g/cm^3.

Table 8.3 Advantages and Disadvantages of Different Extraction Equipment

Class of Equipment	Advantages	Disadvantages
Mixer-settlers	Good contacting Handles wide flow ratio Low headroom High efficiency Many stages available Reliable scale-up	Large holdup High power costs High investment Large floor space Interstage pumping may be required
Continuous counterflow contactors (no mechanical drive)	Low initial cost Low operating cost Simplest construction	Limited throughput with small density difference Cannot handle high flow ratio High headroom Sometimes low efficiency Difficult scale-up
Continuous counterflow contactors (mechanical agitation)	Good dispersion Reasonable cost Many stages possible Relatively easy scale-up	Limited throughput with small density difference Cannot handle emulsifying systems Cannot handle high flow ratio
Centrifugal extractors	Handles low density difference between phases Low holdup volume Short holdup time Low space requirements Small inventory of solvent	High initial costs High operating cost High maintenance cost Limited number of stages in single unit

a horizontal axis through which the two liquid phases flow countercurrently. Liquid inlet pressures of 4 to 7 atm are required to overcome pressure drop and centrifugal force. As many as five theoretical stages can be achieved in one unit.

Many of the commercial extractors described above have seen numerous industrial applications. Maximum loadings and sizes for column-type equipment, as given by Reissinger and Schroeter [5,20] and Lo et al. [4], are listed in Table 8.2. As seen, the Lurgi tower, RDC, and Graesser extractors have been built in very large sizes. Throughputs per unit cross-sectional area are highest for the Karr extractor and lowest for the Graesser extractor.

The selection of an appropriate extractor is based on a large number of factors. Table 8.3 lists the advantages and disadvantages of the various types of extractors. Figure 8.8 shows a selection scheme for commercial extractors. For example, if only a small number of stages is required, a mixer-settler unit might be selected. If more than five theoretical stages, a high throughput, and a large load range (m³/m²-h) are needed, and floor space is limited, an RDC or ARD contactor should be considered.

8.2 GENERAL DESIGN CONSIDERATIONS

The design and analysis of a liquid–liquid extractor involves more factors than for vapor–liquid operations because of complications introduced by the two liquid phases. One of the three different cascade arrangements in Figure 8.9, or a more complex arrangement,

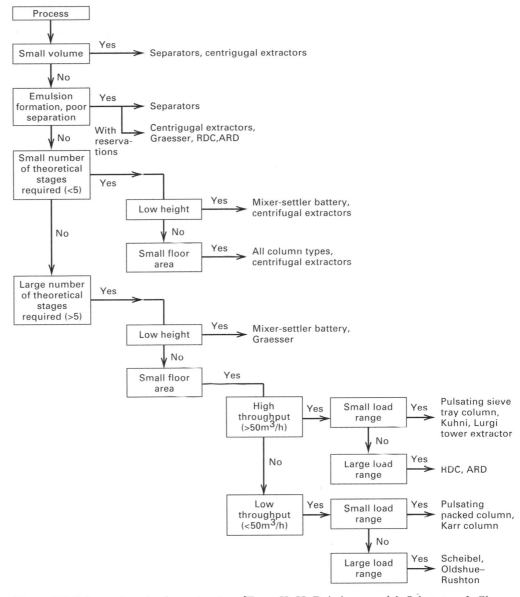

Figure 8.8 Scheme for selecting extractors. [From K.-H. Reissinger and J. Schroeter, *I. Chem. E. Symp. Ser. No. 54*, 33–48 (1978).]

must be selected. The single-section cascade of Figure 8.9a is similar to that used for absorption and stripping. It is designed to transfer to the solvent a certain percentage of the solute in the feed. The two-section cascade of Figure 8.9b is similar to distillation. Solvent enters at one end and reflux, derived from the extract, enters at the other end. The feed enters between the two sections. With two sections, depending on solubility considerations, it is sometimes possible to achieve a reasonably sharp separation between components of the feed; if not, a dual solvent arrangement with two sections, as in Figure 8.9c, with or without reflux at the two ends, may be advantageous. For the latter configuration, which involves a minimum of four components (two in the feed and two solvents), computer-aided calculations are preferred, as discussed in Chapter 10. Although the con-

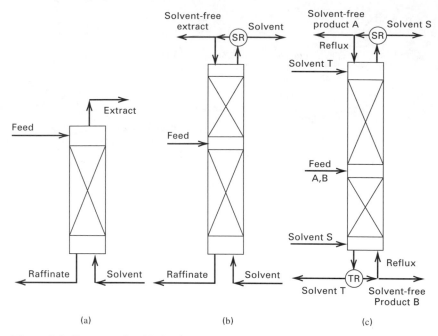

Figure 8.9 Common liquid–liquid extraction cascade configurations: (a) single-section cascade; (b) two-section cascade; (c) dual solvent with two-section cascade.

figurations in Figure 8.9 are shown with packed sections, any of the extractors discussed in Section 8.1 may be considered. The factors influencing extraction include:

1. Entering feed flow rate, composition, temperature, and pressure
2. Type of stage configuration (one-section or two-section)
3. Desired degree of recovery of one or more solutes for one-section cascades
4. Degree of separation of the feed for two-section cascades
5. Choice of liquid solvent(s)
6. Operating temperature
7. Operating pressure (greater than the bubble point of the system)
8. Minimum solvent flow rate and actual solvent flow rate as a multiple of the minimum rate for one-section cascades or reflux rate and minimum reflux ratio for two-section cascades
9. Number of equilibrium stages
10. Emulsification and scum-formation tendency
11. Interfacial tension
12. Phase-density difference
13. Type of extractor
14. Extractor size and horsepower requirement

The ideal solvent has:

1. A high selectivity for the solute relative to the carrier, so as to minimize the need to recover carrier from the solvent
2. A high capacity for dissolving the solute, so as to minimize the solvent-to-feed ratio
3. A minimal solubility in the carrier
4. A volatility sufficiently different from the solute that recovery of the solvent can be achieved by distillation, but the vapor pressure should not be so high that a high

extractor pressure is needed or so low that a high temperature is needed if the solvent is recovered by distillation

5. Stability to maximize the solvent life and minimize the solvent make-up requirement
6. Inertness to permit use of common materials of construction
7. A low viscosity to promote phase separation, minimize pressure drop, and provide a high solute mass transfer rate
8. Nontoxic and nonflammable characteristics to facilitate its safe use
9. Availability at a relatively low cost
10. A moderate interfacial tension to balance the ease of dispersion and the promotion of phase separation
11. A large difference in density relative to the carrier to achieve a high capacity in the extractor
12. Compatibility with the solute and carrier to avoid contamination
13. A lack of tendency to form a stable rag or scum layer at the phase interface
14. Desirable wetting characteristics with respect to extractor internals

Solvent selection is frequently a compromise among all the properties listed above. However, initial consideration is usually given first to selectivity and environmental concerns, and second to capacity. From Chapter 2, the distribution coefficient for solute A between solvent S and carrier C is given by:

$$(K_A)_D = (x_A)^{II}/(x_A)^{I} = (\gamma_A)^{I}/(\gamma_A)^{II} \tag{8-1}$$

where II is the extract phase rich in S and I is the raffinate phase rich in C. Similarly, for the carrier and the solvent, respectively,

$$(K_C)_D = (x_C)^{II}/(x_C)^{I} = (\gamma_C)^{I}/(\gamma_C)^{II} \tag{8-2}$$

$$(K_S)_D = (x_S)^{II}/(x_S)^{I} = (\gamma_S)^{I}/(\gamma_S)^{II} \tag{8-3}$$

The relative selectivity of the solute with respect to the carrier is obtained by taking the ratio of (8-1) to (8-2), giving

$$\beta_{AC} = (K_A)_D/(K_C)_D = \frac{(x_A)^{II}/(x_A)^{I}}{(x_C)^{II}/(x_C)^{I}} = \frac{(\gamma_A)^{I}/(\gamma_A)^{II}}{(\gamma_C)^{I}/(\gamma_C)^{II}} \tag{8-4}$$

For high selectivity, the value of β_{AC} should be high, that is, there should be a high concentration of A and a low concentration of C in the solvent. A first estimate of β_{AC} is made from available values or predictions of the activity coefficients $(\gamma_A)^{I}$, $(\gamma_A)^{II}$, and $(\gamma_C)^{II}$, at infinite dilution where $(\gamma_C)^{I} = 1$, or by using liquid–liquid equilibrium data for the lowest tie line on a triangular diagram of the type discussed in Chapter 4. If A and C form a nearly ideal solution, the value of $(\gamma_A)^{I}$ in (8-4) can also be taken as 1.

For high solvent capacity, the value of $(K_A)_D$ should be high. From (8-2) it is seen that this is difficult to achieve if A and C form nearly ideal solutions, such that $(\gamma_A)^{I} = 1.0$, unless A and S have a great affinity for each other, which would result in a negative deviation from Raoult's law to give $(\gamma_A)^{II} < 1$. Unfortunately, such systems are rare.

For ease in solvent recovery, $(K_S)_D$ should be as large as possible and $(K_C)_D$ as small as possible to minimize the presence of solvent in the raffinate and carrier in the extract. This will generally be the case if activity coefficients $(\gamma_S)^{I}$ and $(\gamma_C)^{II}$ at infinite dilution are large.

If a water-rich feed is to be separated, it is common to select an organic solvent; for an organic-rich feed, an aqueous solvent is often selected. In either case, it is desirable to select a solvent that lowers the activity coefficient of the solute. Consideration of molecule group interactions can help narrow the search for such a solvent before activity coefficients are estimated or liquid–liquid equilibrium data are sought. A table of interactions for solvent-screening purposes, as given by Cusack et al. [21], based on a modification of the

Table 8.4 Group Interactions for Solvent Selection

Group	Solute	Solvent 1	2	3	4	5	6	7	8	9
1	Acid, aromatic OH (phenol)	0	−	−	−	−	0	+	+	+
2	Paraffinic OH(alcohol), water, imide or amide with active H	−	0	+	+	+	+	+	+	+
3	Ketone, aromatic nitrate, tertiary amine, pyridine, sulfone, trialkyl phosphate, or phosphine oxide	−	+	0	+	+	−	0	+	+
4	Ester, aldehyde, carbonate, phosphate, nitrite or nitrate, amide without active H; intramolecular bonding, e.g., *o*-nitrophenol	−	+	+	0	+	−	+	+	+
5	Ether, oxide, sulfide, sulfoxide, primary and secondary amine or imine	−	+	+	+	0	−	0	+	+
6	Multihaloparaffin with active H	0	+	−	−	−	0	0	+	0
7	Aromatic, halogenated aromatic, olefin	+	+	0	+	0	0	0	0	0
8	Paraffin	+	+	+	+	+	+	0	0	0
9	Monohaloparaffin or olefin	+	+	+	+	+	0	0	+	0

(+) Plus sign means that compounds in the column group tend to raise activity coefficients of compounds in the row group.
(−) Minus sign means a lowering of activity coefficients.
(0) Zero means no effect.
Choose a solvent that lowers the activity coefficient.
Source: Cusack, R.W., P. Fremeaux, and D. Glate, *Chem Eng*, **98**(2), 66–76 (1991).

work of Robbins [22], is shown as Table 8.4. In this table, a minus (−) sign for a given solute–solvent pair means that the solvent will lower the value of the activity coefficient of the solute relative to its value in the feed solution. For example, suppose it is desired to extract acetone from water. Acetone, the solute, is a ketone. Thus, in Table 8.4, group 3 applies for the solute and desirable solvents are of the type given in groups 1 and 6. In particular, trichloroethane, a group 6 compound, is known to be a highly selective solvent with high capacity for acetone over a wide range of feed compositions. However, if the compound is environmentally objectionable, it must be rejected. A more sophisticated solvent-selection method, based on the UNIFAC group-contribution method for estimating activity coefficients and utilizing a computer-aided constrained optimization approach, has been developed by Naser and Fournier [23].

In Chapter 4, ternary diagrams were introduced for representing liquid–liquid equilibrium data for three-component systems at constant temperature. Such diagrams are available for a large number of systems, as discussed by Humphrey et al. [6]. For liquid–liquid extraction with ternary systems, the most common diagram is type I shown in Figure 8.10a; much less common is type II shown in Figure 8.10b. Examples of type II systems are (1) *n*-heptane/aniline/methyl cyclohexane, (2) styrene/ethylbenzene/diethylene glycol, and (3) chlorobenzene/water/methylethyl ketone. For type I, the solute and solvent are miscible in all proportions, while in type II they are not. For type I systems, the greater the two-phase region on line $\overline{CS}$, the greater will be the immiscibility of carrier and solvent. The closer the top of the two-phase region is to apex A, the greater will be the range of feed composition, along line $\overline{AC}$, that can be separated with solvent S. In Figure 8.11, it is possible to separate feed solutions only in the composition range from C to F because, regardless of the amount of solvent added, two liquid phases are not formed in the feed composition range of $\overline{FA}$ (i.e., $\overline{FS}$ does not pass through the two-phase region). For type II systems, a high degree of insolubility of S in C and C in S will produce a desirable high

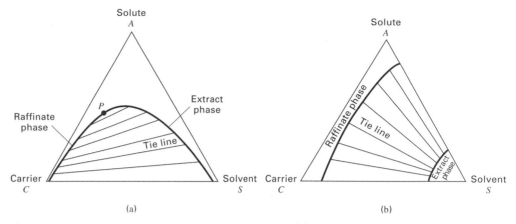

(a)   (b)

Figure 8.10 Most common classes of ternary systems: (a) type I, one immiscible pair; (b) type II, two immiscible pairs.

relative selectivity, but at the expense of solvent capacity. Thus, solvents that result in type I systems are more desirable.

Whether a ternary system is of type I or type II often depends on the temperature. For example, data of Darwent and Winkler [24] for the ternary system n-hexane (H)/ methylcyclopentane (M)/aniline (A) for temperatures of 25, 34.5, and 45°C are shown in Figure 8.12. At the lowest temperature, 25°C, we have a type II system because both H and M are only partially miscible in the aniline solvent. As the temperature increases, the solubility of M in aniline increases more rapidly than the solubility of H in aniline until at 34.5°C, the critical solution temperature for M in aniline is reached. At this temperature, the system is at the border of type II and type I. At 45°C, the system is clearly of type I, with aniline more selective for M than H. Type I systems have a plait point; type II systems do not.

Except in the near-critical region, pressure has little if any effect on liquid-phase activity coefficients and, therefore, on liquid–liquid equilibrium. It is only necessary to select an operating pressure of at least ambient, and greater than the bubble-point pressure of the two-liquid-phase mixture at any location in the extractor. Most extractors operate at near-ambient temperature. If feed and solvent enter the extractor at the same temperature, the operation will be nearly isothermal because the only thermal effect is the heat of mixing, which is usually small.

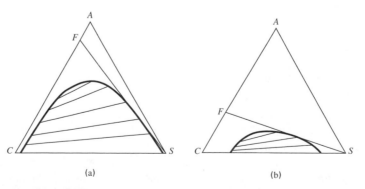

(a)   (b)

Figure 8.11 Effect of solubility on range of feed composition that can be extracted.

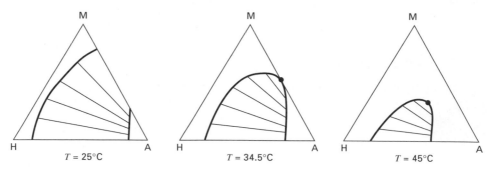

Figure 8.12 Effect of temperature on solubility for the system n-hexane (H)/methylcyclopentane (M)/aniline (A).

Laboratory or pilot-plant work, using actual or expected plant feed and solvent, is almost always necessary to ascertain dispersion and coalescence properties of the liquid–liquid system. Although rapid coalescence of drops is desirable, this reduces interfacial area, leading to reduced mass transfer rates. Thus, compromises are necessary. Coalescence is enhanced when the solvent phase is continuous and mass transfer of solute is from the droplets. This phenomenon, called the *Marangoni effect,* is due to a lowering of interfacial tension by a significant presence of the solute in the interfacial film. When the solvent is the dispersed phase, the interfacial film is depleted of solute, causing an increase in interfacial tension and inhibition of coalescence.

For a given (1) feed liquid, (2) degree of solute extraction, (3) operating pressure and temperature, and (4) choice of solvent for a single-section cascade, a minimum solvent-to-feed flow-rate ratio exists that corresponds to an infinite number of countercurrent equilibrium contacts. A trade-off exists between the number of equilibrium stages and the solvent-to-feed flow-rate ratio. For a two-section cascade, as for distillation, the trade-off involves the reflux ratio and the number of stages. Algebraic methods, similar to those for absorption and stripping described in Chapter 6, for computing the minimum ratios and the trade-off are rapid, but are useful only for very dilute solutions, where values of the solute activity coefficients are essentially those at infinite dilution. When the carrier and the solvent are mutually insoluble, the algebraic method of Sections 5.3 and 5.4 can be used. For more general applications, use of the graphical methods described in this chapter is preferred for ternary systems. Computer-aided methods discussed in Chapter 10 are necessary for higher-order multicomponent systems.

8.3 HUNTER AND NASH GRAPHICAL EQUILIBRIUM-STAGE METHOD

Stagewise extraction calculations for ternary systems of type I and type II (Figure 8.10) are most conveniently carried out with equilibrium diagrams [25]. In this section, procedures are developed and illustrated, using mainly triangular diagrams. The use of other diagrams is covered in the next section.

Consider a countercurrent-flow, N-stage contactor for liquid–liquid extraction of a ternary system operating under isothermal, continuous, steady-state flow conditions at a pressure sufficient to prevent vaporization, as shown in Figure 8.13. Stages are numbered from the feed end. Thus, the final extract is E_1 and the final raffinate is R_N. Equilibrium is assumed to be achieved at each stage, so that for any stage, n, the extract, E_n, and the raffinate, R_n, are in equilibrium with all three components. Mass transfer of all components occurs at each stage. The feed, F, contains the carrier, C, and the solute, A, and can also

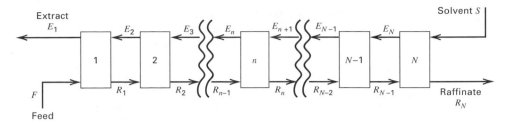

Figure 8.13 Countercurrent-flow, N-stage liquid–liquid extraction cascade.

contain solvent, S, up to the solubility limit. The entering solvent, S, can contain C and A, but preferably contains little, if any. Because most liquid–liquid equilibrium data are given in mass rather than mole concentrations, let:

F = mass flow rate of feed to the contactor
S = mass flow rate of solvent to the contactor
E_n = mass flow rate of extract leaving stage n
R_n = mass flow rate of raffinate leaving stage n
$(y_i)_n$ = mass fraction of species i in extract leaving stage n
$(x_i)_n$ = mass fraction of species i in raffinate leaving stage n

Although Figure 8.13 might seem to imply that the extract is the light phase, either phase can be the light phase. Phase equilibrium may be represented, as discussed in Chapter 4, on an equilateral-triangle diagram, as proposed by Hunter and Nash [26], or on a right-triangle diagram as proposed by Kinney [27]. Assume, for illustration purposes, that the ternary system is ethylene glycol, water, and furfural at a particular temperature, T, such that the liquid–liquid equilibrium data are represented on the equilateral-triangle diagram of Figure 8.14, where the dashed lines are the tie lines that connect the equilibrium phases of the equilibrium curve (also called the *binodal curve* because the plait point separates

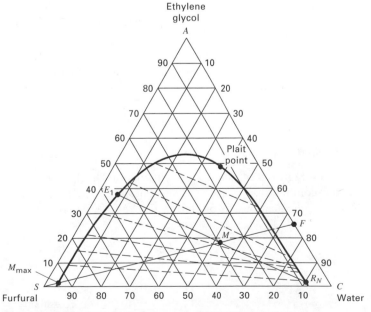

Figure 8.14 Construction 1: Location of product points.

the curve into an extract and a raffinate). Because ethylene glycol is completely soluble in both water and furfural, let glycol be the solute, A. The equilibrium tie lines slope upward from the water side of the diagram toward the furfural side. Therefore, at equilibrium the glycol has a concentration higher in furfural than in water. Thus, furfural is an effective solvent for extracting glycol from a mixture with water. On the other hand, because the tie lines slope downward from the furfural side toward the water side, water is not an effective solvent for extracting glycol from furfural.

Some ternary systems, such as isopropanol–water–benzene, exhibit a phenomenon called *solutropy,* wherein moving from the plait point, the tie lines first slope in one direction. However, the slope diminishes until an intermediate tie line becomes horizontal. Below that tie line, the remaining tie lines slope in the other direction. Sometimes, the solutropy phenomenon disappears if mole fraction coordinates, rather than mass fraction coordinates, are used.

Number of Equilibrium Stages

From the degrees-of-freedom discussion in Chapters 4 and 5, the following sets of specifications, for the cascade of Figure 8.13 with a ternary system, can be made, where all sets include the specification of F, $(x_i)_F$, $(y_i)_S$, and T:

Set 1. S and $(x_i)_{R_N}$ Set 4. N and $(x_i)_{R_N}$
Set 2. S and $(y_i)_{E_1}$ Set 5. N and $(y_i)_{E_1}$
Set 3. $(x_i)_{R_N}$ and $(y_i)_{E_1}$ Set 6. S and N

where values of $(x_i)_{R_N}$ and $(y_i)_{E_1}$ and all exiting phases must lie on the equilibrium curve.

Calculations for sets 1 to 3, which involve the determination of N, are made directly using the triangular diagram. Sets 4 to 6, which involve a specified N, require an iterative procedure. We first consider the calculational procedure for set 1, with the procedures for sets 2 and 3 being just minor modifications.   The procedure, sometimes referred to as the *Hunter–Nash method* [26], involves three kinds of construction on the triangular diagram and is somewhat more difficult than the McCabe–Thiele staircase-type construction for distillation. Although the procedure is illustrated only for the type I system, the principles are readily extended to a type II system. The constructions for the ethylene glycol (A)–water (C)–furfural (S) system are shown in Figure 8.14, where A is the solute, C is the carrier, and S is the solvent. On the binodal curve, all extract compositions lie to the left of the plait point, while all raffinate compositions lie to the right. The problem posed is to determine the number of stages given the flow rates and compositions of the feed and solvent, and the desired raffinate composition.

Construction 1 (Product Points)

First, on Figure 8.14, we locate mixing point M, which represents the overall composition of the combination of feed, F, and entering solvent, S. Assume the following feed and solvent specifications, as plotted in Figure 8.14, where furfural is the solvent:

	Feed	Solvent
	$F = 250$ kg	$S = 100$ kg
	$(x_A)_F = 0.24$	$(x_A)_S = 0.00$
	$(x_C)_F = 0.76$	$(x_C)_S = 0.00$
	$(x_S)_F = 0.00$	$(x_S)_S = 1.00$

The overall composition M for combined F and S is obtained from the following material balances:

$$M = F + S = 250 + 100 = 350 \text{ kg}$$
$$(x_A)_M M = (x_A)_F F + (x_A)_S S = 0.24(250) + 0(100) = 60 \text{ kg}$$
$$(x_A)_M = 60/350 = 0.171$$

$$(x_C)_M M = (x_C)_F F + (x_C)_S S = 0.76(250) + 0(100) = 190 \text{ kg}$$
$$(x_C)_M = 190/350 = 0.543$$

$$(x_S)_M M = (x_S)_F F + (x_S)_S S = 0(250) + 1(100) = 100 \text{ kg}$$
$$(x_S)_M = 100/350 = 0.286$$

From any two of these $(x_i)_M$ values, point M is located, as shown, in Figure 8.14. Based on the properties of the triangular diagram, presented in Chapter 4, point M must be located somewhere on the straight line connecting F and S. Therefore, M can be located knowing just one value of $(x_i)_M$, say $(x_S)_M$. Also, the ratio S/F is given by the inverse lever-arm rule as

$$S/F = \overline{MF}/\overline{MS} = 100/250 = 0.400$$
$$\text{or } S/M = \overline{MF}/\overline{SF} = 100/350 = 0.286$$

Thus, point M can be located by two points or by measurement, employing either of these ratios.

With point M located, the composition of extract, E_1, exiting from a countercurrent, multistage extractor, can be determined from overall material balances:

$$M = R_N + E_1 = 350 \text{ kg}$$
$$(x_A)_M M = 60 = (x_A)_{R_N} R_N + (x_A)_{E_1} E_1$$
$$(x_C)_M M = 190 = (x_C)_{R_N} R_N + (x_C)_{E_1} E_1$$
$$(x_S)_M M = 100 = (x_S)_{R_N} R_N + (x_S)_{E_1} E_1$$

Because the raffinate, R_N, is assumed to be at equilibrium, its composition must lie on the equilibrium curve of Figure 8.14. Therefore, if we specify the value $(x_A)_{R_N} = 0.025$, we can locate the point R_N, and the values of $(x_C)_{R_N}$ and $(x_S)_{R_N}$ can be read from Figure 8.14. A straight line drawn from R_N through M will locate E_1 at the intersection of the equilibrium curve, from which, in Figure 8.14, the composition of E_1 can be read. Values of the flow rates R_N and E_1 can then be determined from the overall material balances above or from Figure 8.14 by the inverse lever-arm rule:

$$E_1/M = \overline{MR_N}/\overline{E_1R_N}$$
$$R_N/M = \overline{ME_1}/\overline{E_1R_N}$$

with $M = 350$ kg for this illustration. By either method, we find:

Raffinate Product	Extract Product
$R_N = 198$ kg	$E_1 = 152$ kg
$(x_A)_{R_N} = 0.025$	$(x_A)_{E_1} = 0.365$
$(x_C)_{R_N} = 0.90$	$(x_C)_{E_1} = 0.075$
$(x_S)_{R_N} = 0.075$	$(x_S)_{E_1} = 0.560$

Construction 2 (Operating Point and Lines)

An operating line is the locus of passing streams. Referring to Figure 8.13, material balances around groups of stages from the feed end are

$$F - E_1 = \cdots = R_{n-1} - E_n = \cdots = R_N - S = P \qquad \text{(8-5)}$$

Because the passing streams are differenced, point P defines a *difference point* rather than a *mixing point*. From the same geometric considerations as apply to a mixing point, a difference point also lies on a straight line drawn through the points involved. However, while the mixing point always lies inside the triangular diagram and between the two end points, the difference point usually lies outside the triangular diagram along an extrapolation of the line through two points such as F and E_1, R_N and S, and so on.

To locate the difference point, two straight lines are drawn, respectively, through the point pairs (E_1, F) and (S, R_N), which are established by Construction 1 and shown in Figure 8.15. These lines are extrapolated until they intersect at point P. Figure 8.15 shows these lines and the difference point, P. From (8-5), straight lines drawn through points on the triangular diagram for any other pair of passing streams, such as (E_n, R_{n-1}), must also pass through point P. Thus, we refer to the difference point as an *operating point,* and the lines drawn through pairs of points for passing streams and extrapolated to Point P as *operating lines.*

The difference point has properties similar to those of the mixing point. If $F - E_1 = P$ is rewritten as $F = E_1 + P$, we see that F can be interpreted as the mixing point for P and E_1. Therefore, by the inverse lever-arm rule, the length of line $\overline{E_1 P}$ relative to the length of the line $\overline{FP}$ is given by

$$\frac{\overline{E_1 P}}{\overline{FP}} = \frac{E_1 + P}{E_1} = \frac{F}{E_1} \qquad \text{(8-6)}$$

Thus, point P can be located, if desired, by measurement with a ruler using either pair of feed-product passing streams.

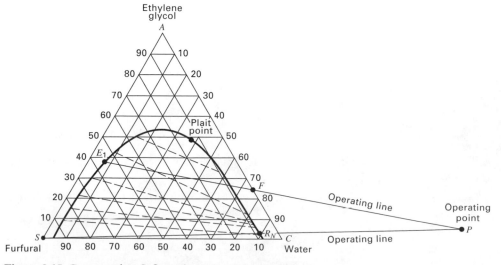

Figure 8.15 Construction 2: Location of operating point.

The operating point, P, lies on the feed or raffinate side of the triangular diagram in the illustration of Figure 8.15. Depending on the relative amounts of feed and solvent and the slope of the tie lines, point P may be located on the solvent or feed side of the diagram, and inside or outside the diagram.

Construction 3 (Equilibrium Lines)

The third type of construction involves *tie lines* that connect opposite sides of the equilibrium curve, which is divided into the two sides by the plait point, which for type I diagrams is the point where the two equilibrium phases become one phase. A material balance around any stage n for any of the three components is

$$(x_i)_{n-1}R_{n-1} + (y_i)_{n+1}E_{n+1} = (x_i)_n R_n + (y_i)_n E_n \tag{8-7}$$

Because R_n and E_n are in equilibrium, their composition points are on the triangular diagram at the two ends of a tie line. Typically a diagram will not contain all the tie lines needed. Tie lines may be added by centering them between existing experimental or predicted tie lines, or by using either of two interpolation procedures illustrated in Figure 8.16. In Figure 8.16a, conjugate line $\overline{PJ}$ is determined from four tie lines and the plait point, P. From tie line $\overline{DE}$, lines $\overline{DG}$ and $\overline{EF}$ are drawn parallel to triangle sides $\overline{CB}$ and $\overline{AC}$, respectively. The intersection at point H gives a second point on the conjugate curve. Subsequent intersections, using the other tie lines, establish additional points from which the conjugate curve is drawn. Then, using the curve, additional tie lines are drawn by reversing the procedure. If it is desired to keep the conjugate curve inside the two liquid-phase region of the triangular diagram, the procedure illustrated in Figure 8.16b is used, where lines are drawn parallel to triangle sides $\overline{AB}$ and $\overline{AC}$.

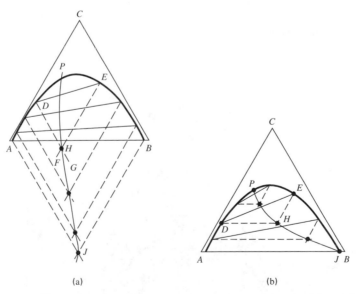

(a)   (b)

Figure 8.16 Use of conjugate curves to interpolate tie lines: (a) method of *International Critical Tables,* Vol. III, McGraw-Hill, New York, p. 393 (1928); (b) method of T.K. Sherwood, *Absorption and Extraction,* McGraw-Hill, New York, p. 242 (1937). [From R.E. Treybal, *Liquid Extraction,* 2nd ed., McGraw-Hill, New York (1963), with permission.]

Stepping off Stages

Equilibrium stages are stepped off on the triangular diagram by alternate use of equilibrium and operating lines, as shown in Figure 8.17, where Constructions 1 and 2 have already been employed to locate the five points F, E, S, R_1, and P. We start at the feed end from point E_1. Referring to Figure 8.13, we see that R_1 is in equilibrium with E_1. Therefore, by Construction 3, R_1 in Figure 8.17 must be at the opposite end of a tie line (shown as a dashed line) connecting to E_1. From Figure 8.13, R_1 passes E_2. Therefore, by Construction 2, E_2 must lie at the intersection of the operating line, drawn through points R_1 and P, and the extract side of the equilibrium curve. From E_2, we locate R_2 by Construction 3; from R_2, we locate E_3 by Construction 2. Continuing in this fashion by alternating between equilibrium tie lines and operating lines, we finally reach or pass the known point R_N. If the latter, a fraction of the last stage is taken. In Figure 8.17 approximately 2.8 equilibrium stages are required.

Procedures for problem specification sets 2 and 3 are very similar to that for set 1. Sets 4 and 5 can be handled by iteration on assumed values for S and following the above procedure for set 1. Set 6 can also use the procedure of set 1 by iterating on E_1.

From (8-6), we see that if the ratio F/E_1 approaches a value of 1, the operating point, P, will be located at a large distance from the triangular diagram. In that case, using an arbitrary rectangular-coordinate system superimposed over the triangular diagram, the coordinates of P can be calculated from (8-6) using the equations for the two straight lines established in Construction 2 . Operating lines for intermediate stages can then be located on the triangular diagram so as to pass through P. Details of this procedure are given by Treybal [25].

Minimum and Maximum Solvent-to-Feed Flow-Rate Ratios

The graphical procedure just described for determining the number of equilibrium stages to achieve a desired solute extraction for a given solvent-to-feed flow-rate ratio presupposes that this ratio is greater than the minimum ratio corresponding to the need for an infinite number of stages. In practice, one usually determines the minimum ratio before solving

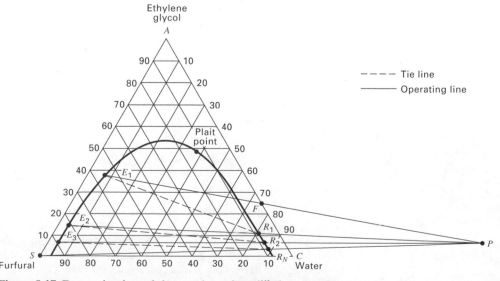

Figure 8.17 Determination of the number of equilibrium stages.

specification sets 1 or 2. In essence, we must solve set 4 with $N = \infty$, where, as in distillation, absorption, and stripping, the infinity of stages occurs at a pinch point of the equilibrium curve and the operating line(s). In ternary systems, the pinch point occurs when a tie line is coincident with an operating line. The calculation is somewhat involved because the location of the pinch point is not always at the feed end of the cascade. Consider the ethylene glycol (A)–water (C)–furfural (S) system shown in Figure 8.18. The points F, S, and R_N are specified, but E_1 is not because the solvent rate has not yet been specified. The operating line $\overline{OL}$ is drawn through the points S and R_N and extended to the left and right of the diagram. This line is the locus of all possible material balances determined by adding S to R_N. Each tie line is then assumed to be a pinch point by extending each tie line until it intersects the line $\overline{OL}$. In this manner, a sequence of intersections, P_1, P_2, P_3, and so on, is found. If these points lie on the raffinate side of the diagram, as in Figure 8.18, the pinch point corresponds to the point $P_{\min}$ located at the greatest distance from R_N. If the triangular diagram does not have a sufficient number of tie lines to determine that point accurately, additional tie lines are introduced by a method described previously and illustrated in Figure 8.16. If we assume in Figure 8.18 that no other tie line gives a point P_i farther away from R_N than P_1, then $P_1 = P_{\min}$.

With $P_{\min}$ known, an operating line can be drawn through point F and extended to E_1 at an intersection with the extract side of the equilibrium curve. From the compositions of the four points, S, R_N, F, and E_1, the mixing point M can be found and the following material balances can then be used to solve for $S_{\min}/F$:

$$F + S_{\min} = R_N + E_1 = M \tag{8-8}$$

$$(x_A)_F F + (x_A)_S S_{\min} = (x_A)_M M \tag{8-9}$$

from which

$$\frac{S_{\min}}{F} = \frac{(x_A)_F - (x_A)_M}{(x_A)_M - (x_A)_S} \tag{8-10}$$

A solvent flow rate greater than $S_{\min}$ must be selected for the extraction to be conducted in a finite number of stages. In Figure 8.18, such a solvent rate results in an operating point P to the right of $P_{\min}$, that is, at a location farther away from R_N. A reasonable

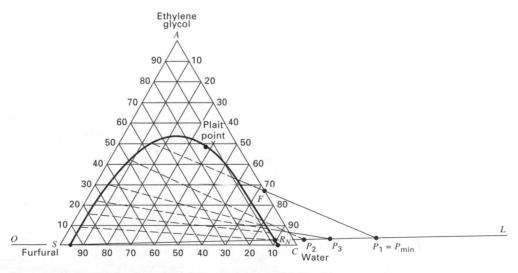

Figure 8.18 Determination of minimum solvent-to-feed ratio.

value for S might be 1.5 S_{min}. From Figure 8.18, we find $(x_A)_M = 0.185$, from which, by (8-10), $S_{min}/F = 0.30$. In our example, we used $S/F = 0.40$, giving $S/S_{min} = 1.33$.

In Figure 8.18 the tie lines slope downward toward the raffinate side of the diagram. If the tie lines slope downward toward the extract side of the diagram, the above procedure for finding S_{min}/F must be modified. The sequence of points P_1, P_2, P_3, and so on, is now found on the other side of the diagram. However, the pinch point now corresponds to that point, P_{min}, that is closest to point S and an operating point, P, must be chosen between points P and S. In the case of a system that exhibits solutropy, intersections P_1, P_2, and so on, will be found on both sides of the diagram. Those on the extract side will determine the minimum solvent-to-feed ratio.

In Figure 8.14, the mixing point M must lie in the two-phase region. As this point is moved along the line $\overline{SF}$ toward S, the ratio S/F increases according to the inverse lever-arm rule. In the limit, a maximum S/F ratio is reached when $M = M_{max}$ arrives at the equilibrium curve on the extract side. At this point, all of the feed is dissolved in the solvent, no raffinate is obtained, and only one stage is required. To avoid this impractical condition, as well as the other extreme of infinite stages, we must select a solvent ratio, S/F, such that $(S/F)_{min} < (S/F) < (S/F)_{max}$. In Figure 8.14, the mixing point M_{max} is located as shown, from which $(S/F)_{max}$ is determined to be about 16.

EXAMPLE 8.1

Acetone is to be extracted from a feed mixture of 30 wt% acetone (A) and 70 wt% ethyl acetate (E) at 30°C by using pure water (S) as the solvent. The final raffinate is to contain 5 wt% acetone on a water-free basis. Determine the minimum and maximum solvent-to-feed ratios and the number of equilibrium stages required for two intermediate S/F ratios. The equilibrium data, which are shown in Figure 8.19 and are taken from Venkaratnam and Rao [28], correspond to a type I system, but with tie lines sloping downward toward the extract side of the diagram. Thus, although water is a convenient solvent, it does not have a high capacity, relative to ethyl

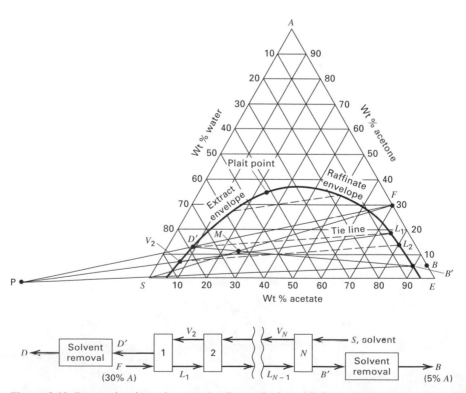

Figure 8.19 Determination of stages for Example 8.1 with $S/F = 1.75$.

acetate, for dissolving acetone. Figure 8.19 also includes a flow diagram of the extractor with the nomenclature to be used for this example. Also, determine for the feed the maximum weight percent acetone that can enter the extractor. This example, as well as Example 8.2 later, are taken largely from an analysis by Sawistowski and Smith [29].

SOLUTION

Point B represents the solvent-free final raffinate. By drawing a straight line from B to S, the intersection with the equilibrium curve on the raffinate side, B', is the actual raffinate composition leaving stage N.

Minimum S/F. Because the tie lines slope downward toward the extract side of the diagram, we seek the extrapolated tie line that intersects the extrapolated line $\overline{SB}$ closest to S. This tie line, leading to P_{min} is shown in Figure 8.20. Because this tie line is at the feed end of the extractor, the location of the extract composition, D'_{min}, is determined as shown in Figure 8.20. The mixing point, M_{min}, for $(S/F)_{min}$ is the intersection of lines $\overline{B'D'_{min}}$ and $\overline{SF}$. By the inverse lever-arm rule, $(S/F)_{min} = \overline{FM_{min}}/\overline{SM_{min}} = 0.75$.

Maximum S/F. If M in Figure 8.20 is moved along line $\overline{FS}$ toward S, the intersection for $(S/F)_{max}$ occurs at the point shown on the extract side of the binodal curve. By the inverse lever-arm rule, using line $\overline{FS}$, $(S/F)_{max} = \overline{FM_{max}}/\overline{SM_{max}} = 8.0$.

Equilibrium stages for other S/F ratios. First consider $S/F = 1.75$. In Figure 8.19, the composition of the saturated extract D' is obtained from a material balance about the extractor,

$$S + F = D' + B' = M$$

For $S/F = 1.75$, point M can be located such that $\overline{FM}/\overline{MS} = 1.75$. A straight line must also pass through D', B', and M. Therefore, D' can be located by extending $\overline{B'M}$ to the extract envelope.

The flow difference point P is located to the left of the triangular diagrams. Therefore, $P = S - B' = D' - F$. It is located at the intersection of extensions of lines $\overline{FD'}$ and $\overline{B'S}$.

Stepping off stages poses no problem. Starting at D', we follow a tie line to L_1. Then V_2 is located by noting the intersection of the operating line $\overline{L_1P}$ with the phase envelope. Additional stages are stepped off in the same manner by alternating between the tie lines and operating lines. For the sake of clarity, only the first stage is shown; four are required.

For $S/F = 4(S/F)_{min} = 3.0$, M is determined and the stages are stepped off in a similar manner

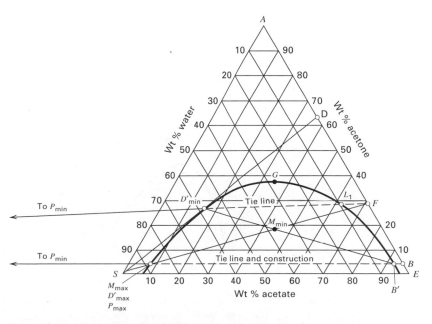

Figure 8.20 Minimum and maximum S/F for Example 8.1.

to give two equilibrium stages. In summary, for the countercurrent cascade, we have

S/F (solvent/feed ratio)	0.75	1.75	3	8
N (equilibrium stages)	∞	4	2	1
x_D (wt% acetone)	64	62	50	30

If the wt% acetone in the feed mixture is increased from the base value of 30%, a feed composition will be reached that cannot be extracted because two liquid phases in equilibrium will not form (no phase splitting). This feed composition is determined by extending a line through S, tangent to the equilibrium curve, until it intersects $\overline{AE}$. This is shown as point D in Figure 8.20. The feed composition is 64 wt% acetone. Feed mixtures with a higher acetone content cannot be extracted with water. ∎

Use of Right-Triangle Diagrams

As discussed in Chapter 4, diagrams other than the equilateral-triangle diagram are used for calculations involving ternary liquid–liquid systems. Ternary, countercurrent extraction calculations can also be made on a right-triangle diagram as shown by Kinney [27]; no new principles are involved. The disadvantage is that compositions of only two of the components are plotted; the third being determined, when needed, by difference from 100%. The advantage of right-triangle diagrams is that ordinary rectangular-coordinates graph paper can be used and either one of the coordinates can be expanded, if necessary, to increase the accuracy of the constructions.

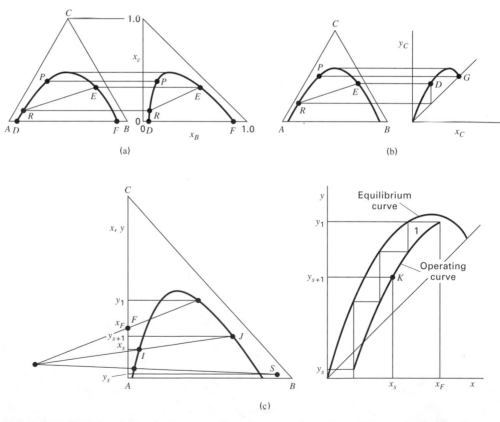

Figure 8.21 Development of other coordinate systems from the equilateral-triangle diagram: (a) to right-triangle diagram; (b) to auxiliary distribution curve. (c) Location of operating point on auxiliary distribution curve. [From R.E. Treybal, *Liquid Extraction,* 2nd ed., McGraw-Hill, New York (1963), with permission.]

A right-triangle diagram can be developed from an equilateral-triangle diagram as shown in Figure 8.21a, where the coordinates in both diagrams are in mass fractions or in mole fractions. Point P on the equilibrium curve and tie line $\overline{RE}$ in the equilateral triangle become point P and tie line $\overline{RE}$ in the right-triangle diagram, which uses rectangular coordinates, x_C and x_B, where C is the solute and B is the solvent.

Consider the right triangle diagram in Figure 8.22 for the ethylene glycol (G), furfural (F), and water (H) system at 25°C and 101.3 kPa. The compositions of F (the solvent) and G (the solute) are plotted in weight (mass) fractions, x_i. For example, point A represents a liquid mixture of overall composition ($x_G = 0.43$, $x_F = 0.28$). By difference, x_H, which is not shown on the diagram, is $1 - 0.43 - 0.28 = 0.29$. Although lines of constant x_H are included on the right triangle of Figure 8.22, such lines are usually omitted because they clutter the diagram. As with the equilateral-triangle diagram, Figure 8.22 for a right triangle includes the binodal curve, with extract and raffinate sides, tie lines connecting compositions of equilibrium phases, and the plait point, P, at $x_G = 0.49$.

Because point M falls within the phase envelope, the mixture separates into two liquid phases, whose compositions are given by points A' and A'' at the ends of the tie line that passes through point M. In this case, the extract at A'' is richer in glycol and furfural than the raffinate at A'.

Point M might be the result of mixing a feed, point B, consisting of 26,100 kg/h of 60 wt% glycol in water ($x_G = 0.6$, $x_F = 0$) with 10,000 kg/h of pure furfural, point C. At equilibrium, the mixture splits into the phases represented by A' and A''. The location of point M and the amounts of extract and raffinate are given by the same mixing rule and inverse lever-arm rule used for equilateral-triangle diagrams. Thus, we find that M is located at ($x_F = 0.278$, $x_G = 0.434$). The mixture separates spontaneously into 11,600 kg/h of raffinate ($x_F = 0.08$, $x_G = 0.32$) and 24,500 kg/h of extract ($x_F = 0.375$, $x_G = 0.48$).

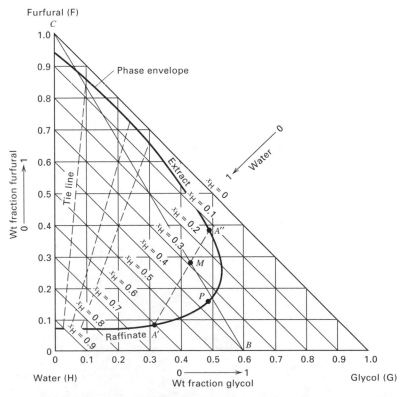

Figure 8.22 Right-triangle diagram for furfural–ethylene glycol–water system.

Figure 8.23 represents the portion of an n-stage, countercurrent-flow cascade, where x and y are weight fractions of glycol in the raffinate and extract, respectively, and L and V are total amounts of raffinate and extract, respectively. Glycol is to be extracted from water by furfural. The feed to stage n is $L_{n+1} = 180$ kg of 35 wt% glycol in a saturated mixture with water and furfural ($x_{n+1} = 0.35$), and the solvent to stage 1 is $V_W = 100$ kg of pure furfural ($y_W = 0.0$). Thus, the solvent-to-feed ratio is $100/180 = 0.556$. These two points are shown on the right-triangle diagram of Figure 8.24. The mixing point for L_{n+1} and V_W is shown as point M_1, as determined by the inverse lever-arm rule. Suppose that the final raffinate, L_W, leaving stage 1 is to contain 0.05 weight-fraction glycol ($x_W = 0.05$). By an overall balance,

$$M_1 = V_W + L_{n+1} = V_n + L_W \qquad \text{(8-11)}$$

Applying the mixing rule, since V_W, L_{n+1}, and M_1 lie on a straight line, V_n, L_W, and M_1 must also lie on a straight line. Furthermore, because V_n leaves stage n at equilibrium and L_W leaves stage 1 at equilibrium, these two streams must lie on the extract and raffinate sides, respectively, of the equilibrium curve. The resulting points are shown in Figure 8.24, where it is seen that the weight fraction of glycol in the final extract is $y_n = 0.34$.

Figures 8.23b and 8.23c, and 8.24, include two additional cases of solvent-to-feed ratio, each with the same compositions for the solvent and the feed and the same value for x_W:

Case	Feed L_{n+1}, kg	Solvent, V_W, kg	Solvent-to-Feed Ratio	Extract Designation	Mixing Point
1	180	100	0.556	V_n	M_1
2	55	100	1.818	V_{n2}	M_2
3	600	100	0.167	V_{n3}	M_3

For case 2, a difference point, P_2, may be defined, as with the equilateral-triangle diagram, in terms of passing streams, as

$$P_2 = V_{n2} - L_{n+1} = V_W - L_W \qquad \text{(8-12)}$$

This point, shown in Figure 8.24, is located at the top of the diagram, where the lines $\overline{L_W V_W}$ and $\overline{L_{n+1} V_{n2}}$ intersect. For case 3, the difference point, P_3, falls at the bottom of the diagram, where the lines $\overline{L_W V_W}$ and $\overline{L_{n+1} V_{n3}}$ intersect.

Equilibrium stages for Figure 8.24 are stepped off in a manner similar to that for an equilateral-triangle diagram by alternating use of equilibrium tie lines and operating lines passing through the difference point. For example, considering case 2, with the high solvent-to-feed ratio of 1.818, and stepping off stages from stage n, a tie line from the point y_{n2} gives a value of $x_n = 0.04$. But this is less than the specified value of $x_w = 0.05$. Therefore, less than one equilibrium stage is required.

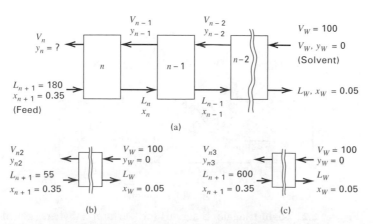

Figure 8.23 Multistage countercurrent contactors.

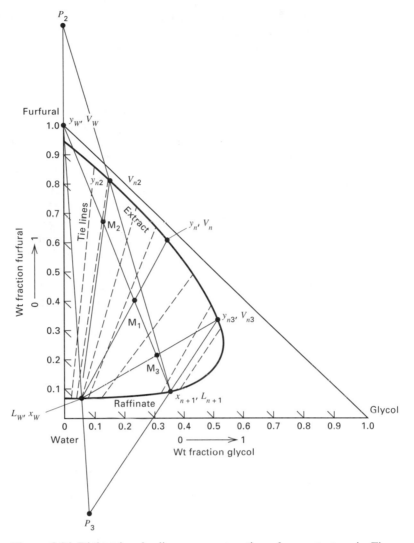

Figure 8.24 Right-triangle diagram constructions for contactors in Figure 8.23.

If we step off stages for case 3, starting from y_{n3}, the tie line and the operating line coincide. That is, we have a pinch point. Thus, the solvent-to-feed ratio of 0.167 for this case is the minimum value corresponding to an infinite number of equilibrium stages.

For case 1, where the solvent-to-feed ratio is in between that of cases 2 and 3, the required number of equilibrium stages lies between 1 and ∞. Construction of the difference point and the steps for this case are not shown in Figure 8.14. The difference point is found to be located at a very large distance from the triangle because lines $\overline{L_W V_W}$ and $\overline{L_{n+1} V_n}$ are almost parallel. When the stages are stepped off, using operating lines parallel to $\overline{L_W V_W}$, it is found that between one and two stages are required.

Use of an Auxiliary Distribution Curve

As the number of equilibrium stages to be stepped off on either one of the two types of triangular diagrams becomes more than a few, the diagram becomes cluttered. In that event, the use of a triangular diagram in conjunction with a y–x distribution diagram becomes attractive. This is a method devised by Varteressian and Fenske [30]. The y–x diagram, discussed in Section 4.5 and illustrated in Figure 8.21b, is simply a plot of the tie-line data in terms of mass fractions (or mole fractions) of the solute in the extract (y_C) and equilibrium raffinate (x_C). The curve begins at the origin and terminates at the plait point, P, where the curve intersects the 45° line. A tie line, such as $\overline{RE}$, in the triangle diagram becomes a point, such as D, in the distribution curve.

If an operating line is added to the distribution curve in Figure 8.21b, a staircase construction of the type used in the McCabe–Thiele method of Chapter 7 can rapidly determine the number of equilibrium stages. However, unlike distillation, where the operating line is straight because of the assumption of constant molal overflow, the operating line for liquid–liquid extraction in a ternary system will always be curved except in the low-solute-concentration region. Fortunately, the curved operating line is quite readily drawn using the technique of Varteressian and Fenske [30]. In Figure 8.19 for the equilateral-triangle diagram, or in Figure 8.24 for the right-triangle diagram, the intersections of the equilibrium curve with a line drawn through a difference (operating) point represent the compositions of the passing streams. Thus, for each such operating line on the triangular diagram, one point of the operating line for the y–x plot is determined. The operating lines passing through the difference point can be drawn at random; they need not coincide with passing streams of actual equilibrium-stage operating lines. Usually five or six such fictitious operating-line intersections, covering the expected range of compositions in the extraction cascade, is sufficient to establish the curved operating line in the y–x plot. For example, in Figure 8.21c, the arbitrary operating line that intersects the equilibrium curve at I and J in the right-triangle diagram becomes a point K on the operating line of the y–x distribution curve. The y–x plot of Figure 8.25 for the glycol–furfural–water system includes an operating line established in this manner, based on the data of Figure 8.24, but with a solvent-to-feed ratio of 0.208, that is, $V_W = 100$, $L_{n+1} = 480$ (25% greater than the minimum ratio of 0.167). The stages are stepped off in the McCabe–Thiele manner starting from the feed end. The result is seen to be almost exactly three equilibrium stages.

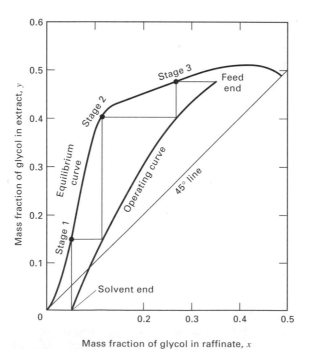

Figure 8.25 Stepping-off stages on an auxiliary distribution curve for the glycol–furfural–water system.

Extract and Raffinate Reflux

The simple single-section, countercurrent, equilibrium-stage extraction cascade shown in Figure 8.13 can be refluxed, as in Figure 8.26a, to resemble distillation. In Figure 8.26a, L is used for raffinate flows, V is used for extract flows, and stages are numbered from the solvent end of the process. Extract reflux, L_R, is provided by sending the extract, V_N, to a solvent-recovery step, which removes most of the solvent, to give a solute-rich solution, $L_R + D$, which is divided into extract reflux, L_R, which is returned to stage N, and solute product, D. At the other end of the cascade, a portion of the raffinate, L_1, is withdrawn in a stream divider and added as raffinate reflux, V_B, to fresh solvent, S. The remaining raffinate is sent to a solvent-removal step to produce a carrier-rich raffinate product. When using extract reflux, minimum and total reflux conditions, corresponding to infinite and minimum number of stages, bracket the optimum extract reflux ratio. Raffinate reflux is not processed through the solvent-removal unit because fresh solvent is added at this end of the cascade. It is necessary, however, to remove solvent from extract reflux at the enriching end of the cascade.

The analogy between a two-section liquid–liquid extractor with feed entering a middle stage, and distillation, is considered in some detail by Randall and Longtin [32]. Different

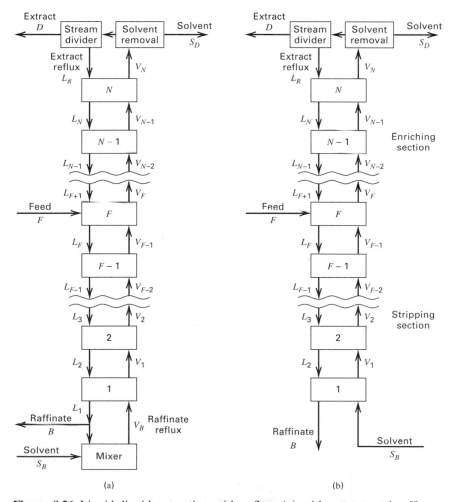

Figure 8.26 Liquid–liquid extraction with reflux: (a) with extract and raffinate reflux; (b) with extract reflux only.

Table 8.5 Analogy between Distillation and Extraction

Distillation	Extraction
Addition of heat	Addition of solvent
Reboiler	Solvent mixer
Removal of heat	Removal of solvent
Condenser	Solvent separator
Vapor at the boiling point	Solvent-rich solution saturated with solvent
Superheated vapor	Solvent-rich solution containing more solvent than that required to saturate it
Liquid below the boiling point	Solvent-lean solution, containing less solvent than that required to saturate it
Liquid at the boiling point	Solvent-lean solution saturated with solvent
Mixture of liquid and vapor	Two-phase liquid mixture
Relative volatility	Relative selectivity
Change of pressure	Change of temperature
D = distillate	D = extract product (solute on a solvent-free basis)
B = bottoms	B = raffinate (solvent-free basis)
L = saturated liquid	L = saturated raffinate (solvent-free)
V = saturated vapor	V = saturated extract (solvent-free)
A = more volatile component	A = solute to be recovered
C = less volatile component	C = carrier from which A is extracted
F = feed	F = feed
x = mole fraction A in liquid	X = mole or weight ratio of A (solvent-free), $A/(A + C)$
y = mole fraction A in vapor	Y = $S/(A + C)$

aspects of the analogy are listed in Table 8.5. The most important analogy is that the solvent in extraction serves the same purpose as heat in distillation.

The use of raffinate reflux has been judged to be of little, if any, benefit by Skelland [31], who shows that the amount of raffinate reflux does not affect the number of stages required. Accordingly, we will consider a two-section countercurrent cascade that includes only extract reflux, as shown in Figure 8.26b.

Analysis of a refluxed extractor, such as that of Figure 8.26b, involves relatively straightforward extensions of the procedures already developed. As will be shown, however, results for a type I system depend critically on the feed composition and the nature of the equilibrium-phase diagram, and it is very difficult to draw any general conclusions with respect to the effect (or even feasibility) of reflux.

For the two-section cascade with extract reflux shown in Figure 8.26b, a degrees-of-freedom analysis can be performed as described in Chapter 5. The result, using as elements two countercurrent cascades, a feed stage, a splitter, and a divider, is $N_D = 2N + 3C + 13$. All but four of the specifications will usually be

Variable Specification	Number of Variables
Pressure at each stage	N
Temperature for each stage	N
Feed-stream flow rate, composition, temperature, and pressure	$C + 2$
Solvent composition, temperature, and pressure	$C + 1$
Split of each component in the splitter	C
Temperature and pressure of the two streams leaving the splitter	4
Pressure and temperature of the divider	2
	$\overline{2N + 3C + 9}$

The four additional specifications can be taken from one of the following sets:

Set 1	**Set 2**	**Set 3**
Solvent rate	Reflux ratio	Solvent rate
Solute concentration in extract (solvent free)	Solute concentration in extract (solvent-free)	Reflux ratio
Solute concentration in raffinate (solvent-free)	Solute concentration in raffinate (solvent-free)	Number of stages
Optimal feed-stage location	Optimal feed-stage location	Feed-stage location

Sets 1 and 2 are of particular interest in the design of a new extractor because two of the specifications deal with the split of the feed into two products of designated purities, on a solvent-free basis. Set 2 is analogous to the design of a binary distillation column using the McCabe–Thiele graphical method, where the purities of the distillate and bottoms, the reflux ratio, and the optimal feed-stage location are specified. For a single-section cascade, it is not feasible to specify the split of the feed with respect to two key components. Instead, as in absorption and stripping, the recovery of just one component in the feed is specified.

It will be recalled for binary distillation that the purity of one of the products may be limited by the formation of an azeotrope. A similar limitation can occur for a type I system when using a two-section cascade with extract reflux, because of the plait point, which separates the two-liquid phase region from the homogeneous, single-phase region. This limitation can be determined from a triangular diagram, but it is most readily observed on a Janecke diagram, of the type described in Chapter 4. In Figure 8.27, liquid–liquid equilibrium data are given for the ethylene glycol–water–furfural system, where glycol is the solute and furfural is the solvent. In the triangular representation of Figure 8.27a, the maximum solvent-free solute concentration that can be achieved in the extract by a countercurrent cascade with extract reflux is determined by the intersection of line $\overline{SE'}$, drawn tangent to the binodal curve, with the solvent-free composition line $\overline{AC}$, giving, in this case, 83 wt% glycol. The same value is read from the binodal curve of the Janecke diagram of Figure 8.27b as the value of the abscissa for the point E' farthest to the right.

Without extract reflux, the maximum solvent-free solute concentration that can be achieved corresponds to an extract that is in equilibrium with the feed, when saturated with the extract. If this maximum value is close to the maximum value determined as in Figure 8.27, then the use of extract reflux will be of little value. This is often the case for type I systems, as illustrated in the following example.

EXAMPLE 8.2

In Example 8.1, a feed mixture of 30 wt% acetone and 70 wt% ethyl acetate was extracted in a single-section countercurrent cascade with pure water to obtain a raffinate of 5 wt% acetone on a water (solvent)-free basis. The maximum solvent-free solute concentration in the extract was found to be 64 wt%, as shown in Figure 8.20 at point D, corresponding to the condition of minimum $S/F = 0.75$ at infinite stages. For an actual $S/F = 1.75$ with four equilibrium stages, the extract contains 62 wt% acetone on a solvent-free basis. Thus, use of extract reflux for the purpose of producing a more pure (solvent-free) extract is not very attractive, given the particular phase equilibrium diagram and feedstock composition. However, to demonstrate the technique, the calculation for extract reflux is carried out nevertheless. Also, the minimum number of equilibrium stages at total reflux and the minimum reflux ratio are determined.

SOLUTION

For the case of the single-section countercurrent cascade, the extract pinch point is at 63 wt% water, 24 wt% acetone, and 13 wt% acetate, as shown in Figure 8.20 at point D'_{min}. If stages are added above the feed point, as in the two-section refluxed cascade of Figure 8.26b, it is possible, theoretically, to reduce the water content of the extract to about 32 wt%, as shown by point G in Figure 8.20. However, the solvent (water)-free extract would not be as rich in acetone (53 wt%), which is determined from the line drawn through points S and G and extended to where it intersects the solvent-free line $\overline{AE}$.

To make this example more interesting, assume that a saturated extract containing 50 wt% water is required elsewhere in the process. Thus, the extraction cascade is that shown in Figure

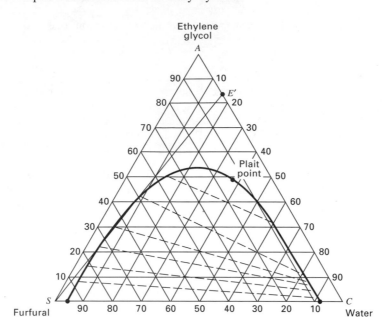

(a)

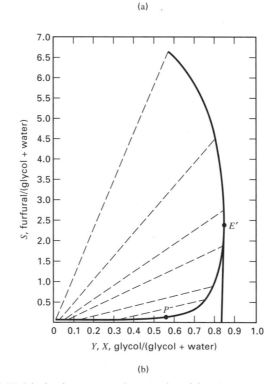

(b)

Figure 8.27 Limitation on product purity: (a) using an equilateral-triangle diagram; (b) using a Janecke diagram.

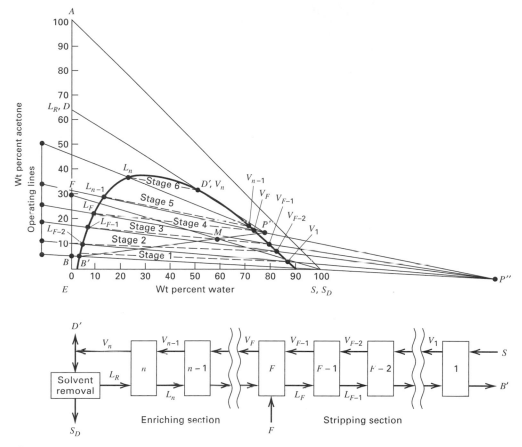

Figure 8.28 Equilibrium stages for Example 8.2.

8.28, rather than that of Figure 8.26b. The difference lies in the location of the solvent-removal step. This saturated extract product is shown as point D' in Figure 8.28. Assume the ratio S/F to be 1.43, which is almost twice the minimum ratio found in Example 8.1. The desired raffinate composition is again 5 wt% acetone on a water-free basis (point B in Figure 8.28), which maps to point B' on the raffinate side of the binodal curve on a line connecting points B and S.

As with single-section cascades, the mixing point, M, in Figure 8.28, for the two streams entering the cascade, S and F, is determined by applying the inverse lever-arm rule, using the S/F ratio, or by computing the overall composition of M, which in this case is 59 wt% water, 29 wt% acetate, and 12 wt% acetone.

The cascade in Figure 8.28 consists of an enriching section to the left of the feed point and a stripping section to the right, where extract is enriched in solute and raffinate is stripped of solute, respectively. A difference or operating point is needed for each section. We will let these be P' and P'', respectively, for the enriching and stripping sections. In the enriching section, referring to the cascade in Figure 8.28, $P' = V_n - L_R = V_{n-1} - L_n$. But, by material balance, $V_n - L_R = D' + S_D$. Therefore, $P'' = D' + S_D$, that is, the total flow leaving the extract end of the cascade. Also, by overall material balance, $M = F + S = B' + D' + S_D = B' + P'$. Thus, P' must lie on a line drawn through points B' and M. To locate the position of P' on that line, we also note that V_n has the same composition as D', and L_R is simply D' with the solvent removed (point D). From above, however, $P' = V_n - L_R$ or $V_n = P' + L_R$. Thus, point P' must also lie on a line drawn through points $V_n(D')$ and $L_R(D)$. Thus, point P' is located at the intersection of the extended lines $\overline{B'M}$ and $\overline{DD'}$ as shown in Figure 8.28.

The difference point, P'', for the stripping section is located in a similar manner, if we note that $P'' = B' - S = F - D' - S_D = F - P'$. Thus, P'' must be the intersection of extended lines $\overline{FP'}$ and $\overline{B'S}$ as shown to the right of the triangular diagram of Figure 8.28.

The stages can now be stepped off in the usual manner by starting from V_n and using the difference point P' until an operating line crosses the feed line $\overline{FP'}$. From there, the stages are stepped off using the difference point P'' until the raffinate composition is reached or exceeded. In Figure 8.28, it is seen that six equilibrium stages are required, with two in the enriching section and four in the stripping section. The feed enters the third stage from the left.

The reflux ratio, defined for this example as $(V_n - D')/D' = (L_R + S_D)/D'$, can be determined as follows. From above, $P' = D' + S_D$. Therefore, by the mixing rule, $S_D/D' = \overline{D'P'}/\overline{S_D P'}$. By material balance, $V_n - D' = L_R + S_D$. Therefore,

$$\frac{V_n - D'}{S_D} = \frac{L_R + S_D}{S_D} = \frac{\overline{L_R S_D}}{\overline{L_R D'}} \quad \text{and} \quad \frac{V_n - D'}{D'}\left(\frac{S_D}{D'}\right)\left(\frac{V_n - D'}{S_D}\right) = \left(\frac{\overline{D'P'}}{\overline{S_D P'}}\right)\left(\frac{\overline{L_R S_D}}{\overline{L_R D'}}\right)$$

By measurement from Figure 8.28, $(V_n - D')/D' = (1.2)(2.0) = 2.4$. The reflux ratio is valid only for the selected solvent-to-feed ratio of 1.43.

Next, we consider the case of total reflux, corresponding to the minimum number of stages. With reference to the equilateral-triangle diagram of Figure 8.29, compositions of exiting streams are as previously specified or computed. With respect to acetone, we have 30 wt% in F, 4.9 wt%, in B', 33 wt%, in D', and 62 wt% in L_R. As in the case of the single-section cascade of Figure 8.20, as the solvent-to-feed ratio is increased, the mixing point $M = F + S$ moves toward the pure-solvent apex. At the maximum solvent addition, M lies at the intersection of the line through F and S with the extract side of the binodal curve. Difference points P' and P'' also move toward S because $P' = D' + S_D$ approaches S_D at total reflux and $P'' = F - P'$ approaches P', recalling that at total reflux $F = D' = 0$. As shown in Figure 8.29, the minimum number of equilibrium stages is three, as stepped off from the S apex.

Lastly, we consider the case of minimum reflux ratio at infinite stages, which also corresponds to the minimum solvent ratio. As the solvent ratio is reduced, point M moves toward the feed point, F, and point P'' moves away from the binodal curve. Also, point P' moves toward V_n. Ultimately, a value of the S/F ratio is reached where an operating line in either the enriching

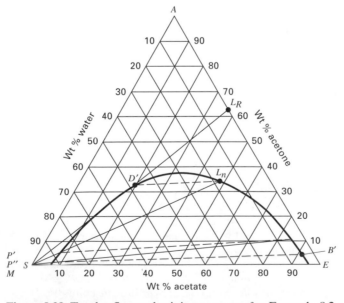

Figure 8.29 Total reflux and minimum stages for Example 8.2.

section or the stripping section coincides with a tie line, giving a pinch point and, therefore, an infinite number of stages. Often, this occurs for the extended tie line that passes through the feed point. Such is the case here, giving a minimum reflux ratio of about 0.6 and a corresponding minimum S/F ratio of 0.75. ■

8.4 MALONEY AND SCHUBERT GRAPHICAL EQUILIBRIUM-STAGE METHOD

For type II ternary systems, use of a two-section cascade with extract reflux is particularly desirable. Without a plait point, the two-phase region extends all the way across the solute composition . Thus, while maximum solvent-free solute concentration in the extract is limited for a type I system as was shown in Figure 8.27, no such limit exists with a type II system. Accordingly, it is possible with extract reflux to achieve as sharp a separation as desired between the solute (A) and carrier (C).

When reflux is used, many stages may be required and the use of triangular diagrams is often not convenient. Instead, use can be made of a McCabe–Thiele-type diagram. Alternatively, a Janecke diagram, often in conjunction with a distribution diagram, has proved to be useful. In Janecke diagrams, which use convenient rectangular coordinates, solvent concentration on a solvent-free basis is plotted as the ordinate against solute concentration on a solvent-free basis as the abscissa, that is, $\%S/(\%A + \%C)$ against $\%A/(\%A + \%C)$, either with mass or mole percents. The Janecke diagram is analogous to an enthalpy–concentration diagram and is consistent with the distillation-extraction analogy of Table 8.5. The application of such a diagram to liquid–liquid extraction of a type II system with the use of reflux is considered in detail by Maloney and Schubert [33], who use an auxiliary distribution diagram of the McCabe–Thiele type, but on a solvent-free basis, to facilitate visualization of the stages. This method is also referred to as the *Ponchon–Savarit method for extraction*. Unlike the analogous method for distillation mentioned briefly at the end of Chapter 7 and which requires both enthalpy and vapor–liquid equilibrium data, the method for extraction requires only ternary liquid–liquid solubility data, which are far more common than combined vapor–liquid enthalpy and equilibrium data. Accordingly, despite the development of rigorous computer-aided methods, the Ponchon-Savarit method for extraction has remained useful, while the analogous method for distillation has rapidly declined in popularity. Although the Janecke diagram can also be applied to type I systems, it becomes difficult to use when the carrier and the solvent are highly immiscible, because the resulting values of the ordinate can become very large.

With the Janecke diagram, construction of tie lines, mixing points, operating points, and operating lines are all made in a manner similar to that for a triangular diagram [33]. Consider the case of extraction for a type II system with extract reflux, shown in Figure 8.26b. A representative Janecke diagram is shown in Figure 8.30, where all flow rates are on a solvent-free mass basis and the following solvent-free concentrations for extract and raffinate phases are based on Janecke coordinates:

$$Y = \frac{\text{mass solvent}}{\text{mass of solvent-free liquid phase}}$$

$$X = \frac{\text{mass solute}}{\text{mass of solvent-free liquid phase}}$$

Values of the ordinate, Y, especially for the saturated extract phase, can vary over a wide range depending on the solubility of the solute and carrier in the solvent. Values of the abscissa, X, vary from 0 to 1 (pure carrier to pure solute). Equilibrium tie lines relate concentrations in the saturated extract to the saturated raffinate. The Y location of the feed, F, is somewhere between zero and the saturated raffinate curve. The extract, V_N,

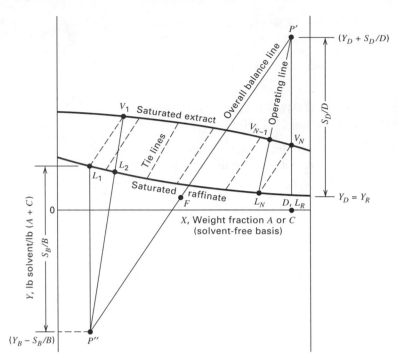

Figure 8.30 Construction of equilibrium stages on a Janecke diagram.

leaving stage N and prior to solvent removal, is rich in solute and lies on the saturated extract curve. Upon solvent removal, the extract, D, and extract reflux, L_R, with identical compositions, lie on the $Y = 0$ horizontal line. The solvent, S_D, is assumed to be pure. The raffinate, L_1, leaving stage 1 is rich in the carrier and lies on the saturated raffinate line. Solvent S_B is assumed to be pure.

Points P' and P'' are difference or operating points for the enriching and stripping sections, respectively, that are used to draw operating lines. The locations of P' and P'' are derived as follows: Solvent-free and solvent material balances around the solvent-recovery step give, in terms of passing streams,

$$V_N - L_R = D \tag{8-13}$$

$$Y_{V_N}V_N - Y_D L_R = S_D + Y_D D \tag{8-14}$$

For a solvent difference balance around a section of top stages down to stage n, located above stage F in Figure 8.26b, we obtain

$$Y_{V_n}V_n - Y_{L_{n+1}}L_{n+1} = S_D + Y_D D \tag{8-15}$$

Thus, any solvent flow difference between passing streams in the enriching section above the feed stage is given by $S_D + Y_D D$. If (8-13) and (8-14) are combined to eliminate V_N, we obtain

$$\frac{L_R}{D} = \frac{(Y_D + S_D/D) - Y_{V_N}}{Y_{V_N} - Y_D} \tag{8-16}$$

In Figure 8.30, however, $(Y_{V_N} - Y_D)$ is the vertical distance between the points V_N and (D, L_R). Therefore, the difference point, P', must be $(S_D + Y_D D)$ divided by D to give

$(Y_D + S_D/D)$. That is,

$$P' = Y_D + S_D/D \qquad \text{(8-17)}$$

$$\text{and} \quad L_R/D = \overline{P'V_N}/\overline{V_N L_R} \qquad \text{(8-18)}$$

Similarly, it can be shown that,

$$P'' = Y_B - S_B/B \qquad \text{(8-19)}$$

$$D/B = \overline{FP''}/\overline{P'F} \qquad \text{(8-20)}$$

Stages are stepped off in a manner analogous to that for the triangular diagram, starting from either the extract, D, or the raffinate, B, alternating between operating lines and tie lines. For example, in Figure 8.30, we can start from the top of the cascade at the extract D and step off stages in the enriching section. The solute compositions of D, L_R, and V_n on a solvent-free basis are identical. Thus, the operating line for passing streams L_R and V_N is a vertical line passing through the difference point P'. From point V_N on the saturated extract curve, a tie line is followed to the equilibrium raffinate phase, L_N. An operating line connecting points P' and L_N intersects the extract curve at the passing stream V_{N-1}. Subsequent stages in the enriching section are stepped off in a similar manner until the feed stage is reached. The optimal location of this stage is determined in a manner analogous to the intersection of the rectification and stripping operating lines in the McCabe–Thiele method. On the Janecke diagram, this intersection is the line $\overline{P'P''}$, which passes through the feed point. Thus, the transition from the enriching section (where the difference point P' is used) to the stripping section (where P'' is used) is made when an equilibrium tie line for a stage crosses the line $\overline{P'P''}$. Following the location of the feed stage, the remaining stripping-section stages are stepped off until the desired product raffinate solvent-free concentration is reached or crossed over.

The Janecke diagram can also be used to determine the two limiting conditions of total reflux (minimum stages) and minimum reflux (infinite stages). For total reflux, the difference points P' and P'' lie at $+\infty$ and $-\infty$, respectively, because $F = B = D = 0$. Thus, all operating lines become vertical lines and the minimum number of stages are stepped off in the manner illustrated in Figure 8.31a.

For the condition of minimum reflux, a pinch condition is sought either at the feed stage or some other stage location. In Figure 8.31b, where the pinch is assumed to be at the feed stage, an operating line is drawn coincident with a tie line and the feed point, F, to determine points P' and P''. To determine if the pinch does occur at the feed stage, tie lines to the right of the feed-stage tie line are extended to an intersection with the vertical line through D. If a higher intersection occurs, then that P' may be the correct P'_{min} difference point for minimum reflux. In a similar manner, ties lines to the left of the feed-stage tie line are extended to an intersection with the vertical line through B. If a lower intersection occurs, then that P'' may be the factor that determines the minimum reflux. The former case is shown in Figure 8.31c, where P' is higher than P'_1. Thus, $P' = P'_{min}$. In any case, once the controlling P' or P'' is determined, a line through F determines the other difference point and the minimum reflux is computed from (8-18) using P'_{min}.

EXAMPLE 8.3

As shown in Figure 8.32, a countercurrent-extraction cascade equipped with a perfect solvent separator to provide extract reflux is used to separate methylcyclopentane (A) and n-hexane (C) into a final extract and raffinate containing, on a solvent-free basis, 95 wt% and 5 wt% A, respectively, using aniline (S) as the solvent. The feed rate is 1,000 kg/h with 55 wt% A, and the mass ratio of solvent to feed is 4.0. The feed contains no aniline and the fresh solvent is pure. Recycle solvent is also assumed to be pure.

(a) Determine the reflux ratio and number of stages. Equilibrium data at extractor temperature and pressure are shown for mass units in the Janecke diagram of Figure 8.33. Feed is to enter at the optimal stage.

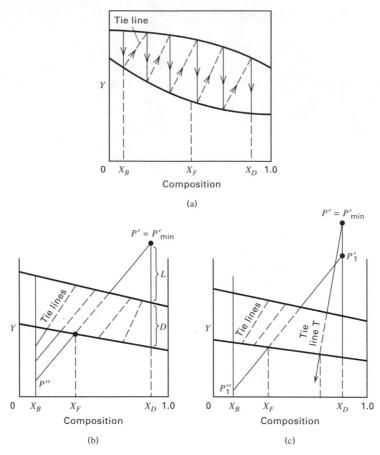

Figure 8.31 Limiting conditions on a Janecke diagram: (a) minimum number of stages at total reflux; (b) minimum reflux determined by a tie line through the feed point; (c) minimum reflux determined by a tie line to the right of the feed point.

(b) Determine the minimum number of stages for the specified solvent-free extract and raffinate compositions.

(c) Determine the minimum reflux ratio for the specified feed and product compositions.

SOLUTION

(a) First, determine all product rates by material-balance calculations. An overall balance on solute plus carrier gives $D + B = 1,000$ kg/h

A solute balance gives $0.95D + 0.05B = (0.55)(1,000) = 550$ kg/h

Solving these two equations simultaneously gives $D = 556$ kg/h and $B = 444$ kg/h. Since $S_B = 4,000$ kg/h, $S_B/B = 9.0$. In Figure 8.30, point P'' is located at a distance of S_B/B below the raffinate composition, X_B, at point B. Since Y at point B, from Figure 8.33, is approximately 0.3, point P'' is located at $0.3 - 9.0 = -8.7$.

A line drawn through P'' and F, extended to the intersection with the vertical line through D, gives $P' = 6.7$. By measurement from Figure 8.33, using (8-16), $L_R/D = 3.0$.

In Figure 8.33, stages are stepped off starting from point D. At the third stage $(N - 2)$, the tie line crosses line $\overline{P''FP'}$. Thus, this is the optimal feed stage. Three more stages are required to reach B, giving a total of six equilibrium stages.

(b) If the construction for minimum stages, shown in Figure 8.31a, is used in Figure 8.33, just less than five stages are determined.

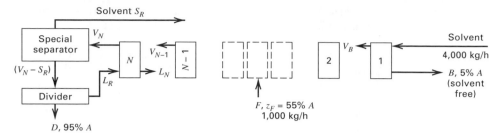

Figure 8.32 Countercurrent extraction cascade with extract reflux for Example 8.3.

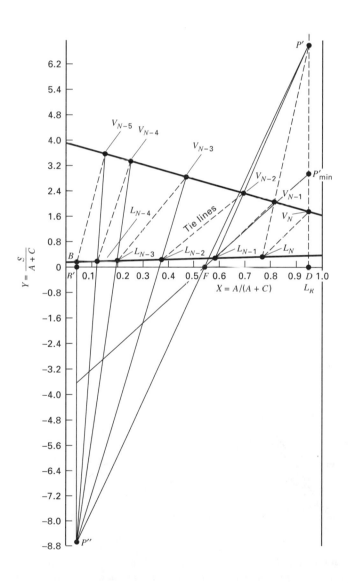

Figure 8.33 Maloney–Schubert constructions on Janecke diagram for Example 8.3.

(c) If the construction for minimum reflux, shown in Figure 8.31b for a pinch at the feed stage, is used in Figure 8.33, a value of $P' = 2.90$. No other tie line in either section gives a larger value. Therefore, $P_{min} = 2.9$. By measurement, using (8-18), $(L_R/D)_{min} = 0.73$. Using the construction indicated in Figure 8.30, the corresponding $(S_B/B)_{min}$ is found to be 4.2. Thus, $(S_B)_{min} = 4.2(444) = 1,865$ kg/h or $(S_B/F)_{min} = 1,865/1,000 = 1.865$.

For this example, a relatively high reflux ratio and corresponding solvent-to-feed ratio is employed to keep the required number of equilibrium stages small. When the number of equilibrium stages is large, the Janecke diagram becomes cluttered with operating lines and tie lines. In that case, an auxiliary distribution plot of solute mass fraction in the extract layer versus solute mass fraction in the raffinate layer, both on a solvent-free basis, as in the Janecke

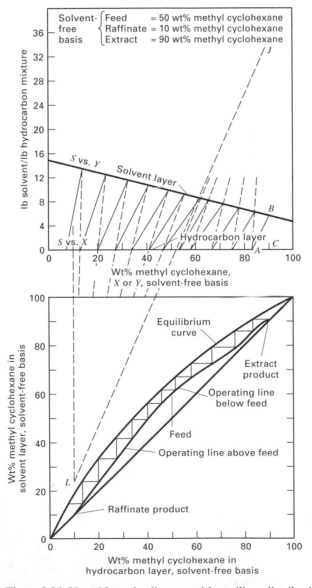

Figure 8.34 Use of Janecke diagram with auxiliary distribution diagram. [From *Chemical Engineers' Handbook*, 5th ed., R.H. Perry and C.H. Chilton, Eds., McGraw-Hill, New York (1973).]

diagram, can be drawn, with points on the enriching and stripping operating lines determined, as discussed above, from arbitrary operating lines on the Janecke diagram. Stages are then stepped off on the distribution diagram in the McCabe–Thiele manner. An example of the Janecke diagram with such an auxiliary distribution diagram is shown in Figure 8.34, taken from Maloney and Schubert [33]. ∎

8.5 THEORY AND SCALE-UP OF EXTRACTOR PERFORMANCE

Following the estimation, by methods described in Sections 8.3 and 8.4, of the number of equilibrium stages, suitable extraction equipment can be selected using the scheme of Figure 8.8. Often, the choice is between a cascade of mixer-settler units or a multicompartment column-type extractor with mechanical agitation, the main considerations being the number of stages required, and the floor space and head room available. Methods for estimating size and power requirements of these two general types of extractors are presented next. Column devices with no mechanical agitation are also considered.

Mixer-Settler Units

Sizing of mixer-settler units is done most accurately by scale-up from batch or continuous runs in laboratory or pilot-plant equipment. However, preliminary sizing calculations can be made using available theory and empirical correlations. Experimental data of Flynn and Treybal [34] show that when liquid-phase viscosities are less than 5 cP and the specific-gravity difference between the two liquid phases is greater than about 0.10, the average residence time required of the two liquid phases in the mixing vessel to achieve at least 90% stage efficiency may be as low as 30 s and is usually not more than 5 min, when an agitator-power input per mixer volume of 1,000 ft-lbf/min-ft^3 (4 hp/1,000 gal) is used. Based on experiments reported by Ryan, Daley, and Lowrie [35], the capacity of a settler vessel can be expressed in terms of C gal/min of combined extract and raffinate per square foot of phase-disengaging area. For a horizontal, cylindrical vessel of length L and diameter D_T, the economic ratio of L to D_T is approximately 4. Thus, if the phase interface is located at the middle of the vessel, the disengaging area is $D_T L$ or $4D_T^2$. A typical value of C given by Happel and Jordan [36] is about 5. Frequently, the settling vessel will be larger than the mixing vessel, as is the case in the following example.

EXAMPLE 8.4

Benzoic acid is to be continuously extracted from a dilute solution in water with a solvent of toluene in a series of discrete mixer-settler vessels operated in countercurrent flow. The flow rates of the feed and solvent are 500 and 750 gal/min, respectively. Assuming a residence time, t_{res}, of 2 min in each mixer and a settling vessel capacity of 5 gal/min-ft^2, estimate:

(a) Diameter and height of a mixing vessel, assuming $H/D_T = 1$
(b) Agitator horsepower for a mixing vessel
(c) Diameter and length of a settling vessel, assuming $L/D_T = 4$
(d) Residence time in a settling vessel in minutes

SOLUTION

(a) Q = total flow rate = 500 + 750 = 1,250 gal/min
   V = volume = Qt_{res} = 1,250(2) = 2,500 gal or 2,500/7.48 = 334 ft^3

$$V = \pi D_T^2 H/4, \ H = D_T, \text{ and } V = \pi D_T^3/4$$

$$D_T = (4V/\pi)^{1/3} = [(4)(334)/3.14]^{1/3} = 7.52 \text{ ft and } H = 7.52 \text{ ft}$$

(b) Horsepower = 4(2,500/1,000) = 10 hp
(c) $D_T L$ = 1,250/5 = 250 ft^2; D_T^2 = 250/4 = 62.5 ft^2

$$D_T = 7.9 \text{ ft}; \ L = 4D_T = 4(7.9) = 31.6 \text{ ft}$$

(d) Volume of settler $= \pi D_T^2 L/4 = 3.14(7.9)^2(31.6)/4 = 1,548$ ft^3 or $1,548(7.48) = 11,580$ gal

$$t_{res} = V/Q = 11,580/1,250 = 9.3 \text{ min}$$

■

A typical single-compartment mixing tank for liquid–liquid extraction is shown in Figure 8.35. The vessel is closed with the two liquid phases entering at the bottom and the effluent, in the form of a two-phase emulsion, leaving at the top. Although flat tank heads are shown in Figure 8.35, rounded heads of the type in Figure 8.2 are preferred to eliminate stagnant fluid regions. Air or other gases must be evacuated from the vessel so no gas–liquid interface exists.

Mixing is accomplished by an appropriate, centrally located impeller selected from the many types available, some of which are shown in Figure 8.3. For example, a flat-blade turbine might be chosen as in Figure 8.35. A single turbine is adequate unless the vessel height is greater than the vessel diameter, in which case a compartmented vessel with two or more impellers might be employed. When the vessel is open, vertical side baffles are mandatory to prevent vortex formation at the gas–liquid interface. For closed vessels that run full of liquid, vortexing will not occur. Nevertheless, it is common to install baffles, even in closed tanks, to minimize swirling and improve circulation patterns. Although no standards exist for vessel and turbine geometry, the following, with reference to Figure 8.35, give good dispersion performance in liquid–liquid agitation:

Number of turbine blades = 6; Number of vertical baffles = 4

$H/D_T = 1$; $D_i/D_T = 1/3$; $W/D_T = 1/12$ and $H_i/H = 1/2$

To achieve a high stage efficiency for extraction in a mixing vessel—say, between 90 and 100%—it is necessary to provide fairly vigorous agitation. For a given type of impeller and vessel–impeller geometry, the agitator power, P, can be estimated from an empirical correlation in terms of a power number, N_{Po}, which depends on an impeller Reynolds number, N_{Re}, where

$$N_{Po} = \frac{Pg_c}{N^3 D_i^5 \rho_M} \tag{8-21}$$

$$N_{Re} = \frac{D_i^2 N \rho_M}{\mu_M} \tag{8-22}$$

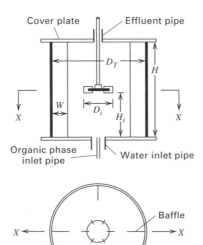

Figure 8.35 Agitated vessel with flat-blade turbine and baffles.

The impeller Reynolds number is the ratio of the inertial force to the viscous force:

$$\text{Inertial force} \propto (ND_i)^2 \rho_M D_i^2$$

$$\text{Viscous force} \propto \frac{\mu_M(ND_i)D_i^2}{D_i}$$

where N = rate of impeller rotation. Thus, the characteristic length in the impeller Reynolds number is the impeller diameter and the characteristic velocity is ND_i = impeller peripheral velocity.

The agitator power is proportional to the product of the volumetric liquid flow produced by the impeller and the applied kinetic energy per unit volume of fluid. The result is

$$P \propto (ND_i^3)[\rho_M(ND_i)^2/2g_c]$$

which can be rewritten as (8-21), where the constant of proportionality is $2N_{Po}$. Both the impeller Reynolds number and the power number (also called the Newton number) are dimensionless groups. Thus, any consistent set of units can be used. The power number for an agitated vessel serves the same purpose as the friction factor for the flow of a fluid through a pipe. This is illustrated, over a wide range of impeller Reynolds number, for a typical mixing impeller in Figure 8.36a, taken from the work of Rushton and Oldshue [37].

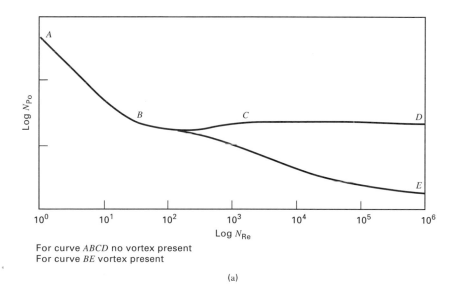

For curve *ABCD* no vortex present
For curve *BE* vortex present

(a)

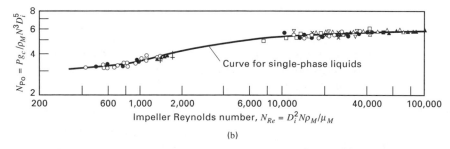

(b)

Figure 8.36 Power consumption of agitated vessels. (a) Typical power characteristics. [From J.H. Rushton and J.Y. Oldshue, *Chem. Eng. Prog.*, **49**, 161–168 (1953).] (b) Power correlation for six-bladed, flat-blade turbines with no vortex. [From D.S. Laity and R.E. Treybal, *AIChE J.*, **3**, 176–180 (1957).]

The upper curve, ABCD, pertains to a vessel with baffles, while the lower curve, ABE, pertains to the same tank with no baffles. In the low-Reynolds-number region, AB, viscous forces dominate and the impeller power is proportional to $\mu_M N^2 D_i^3$. Somewhere beyond a Reynolds number of about 200, a vortex appears if no baffles are present and the power-number relation is given by curve BE. In this region, the Froude number, $N_{Fr} = N^2 D_i/g$, which is the ratio of inertial to gravitational forces, also becomes a factor. With baffles present, and the Reynolds number greater than about 1,000, a region, CD, is reached where fully developed turbulent flow exists. Now, inertial forces dominate and the power is proportional to $\rho_M N^3 D_i^5$. It is clear that the addition of baffles greatly increases power requirements in the turbulent flow region.

A correlation of experimental data for baffled vessels with six-bladed, flat-blade turbines is shown in Figure 8.36b, from a study by Laity and Treybal [38]. The range of impeller Reynolds number covers only the turbulent flow region, where efficient liquid–liquid mixing is achieved. The solid line represents batch mixing of single–phase liquids. The data points represent liquid–liquid mixing, where agreement is achieved with the single-phase curve by computing two-phase mixture properties from

$$\rho_M = \rho_C \phi_C + \rho_D \phi_D \tag{8-23}$$

$$\mu_M = \frac{\mu_C}{\phi_C}\left(1 + \frac{1.5 \mu_D \phi_D}{\mu_C + \mu_D}\right) \tag{8-24}$$

where ϕ is the volume fraction of holdup in the tank, with subscripts C for the continuous phase and D the dispersed phase, such that $\phi_D + \phi_C = 1$. When measurements were made for continuous flow from inlets at the bottom of the vessel to an outlet for the emulsion from the top of the vessel and with the impeller located at a position above the liquid–liquid interface when at rest, the data were correlated with the curve of Figure 8.36b.

With fully developed turbulent flow, the volume fraction of dispersed phase in the vessel closely approximates that in the feed to the vessel; otherwise the volume fraction may be different from that in the total feed to the vessel. That is, the residence times of the two phases in the vessel may not be the same. At best, spheres of uniform size can pack tightly to give a void fraction of 0.26. Therefore, $\phi_C > 0.26$ and $\phi_D < 0.74$. For continuous flow, the vessel is first filled with the phase to be continuous. Following initiation of agitation, the two-feed liquids are then introduced into the vessel in their desired volume ratio.

Based on the work of Skelland and Ramsay [39] and Skelland and Lee [40], a minimum impeller rate of rotation is required for complete and uniform dispersion of one liquid into another. For a flat-blade turbine in a baffled vessel of the type discussed above, this minimum rotation rate can be estimated from

$$\frac{N_{min}^2 \rho_M D_i}{g\,\Delta\rho} = 1.03\left(\frac{D_T}{D_i}\right)^{2.76} \phi_D^{0.106}\left(\frac{\mu_M^2 \sigma}{D_i^5 \rho_M g^2(\Delta\rho)^2}\right)^{0.084} \tag{8-25}$$

where $\Delta\rho$ is the absolute value of the difference in density and σ is the interfacial tension between the two liquid phases. The dimensionless group on the left-hand side of (8-25) is the two-phase Froude number. The dimensionless group at the far right of (8-25) is a ratio of forces:

$$\frac{(\text{viscous})^2(\text{interfacial tension})}{(\text{inertial})(\text{gravitational})^2}$$

EXAMPLE 8.5

Furfural is to be continuously extracted from a dilute solution in water by toluene at 25°C in an agitated vessel of the type shown in Figure 8.35. The feed enters at a flow rate of 20,400 lb/h, while the solvent enters at 11,200 lb/h. For a residence time in the vessel of 2 min, estimate for either phase as the dispersed phase:

(a) The dimensions of the mixing vessel and the diameter of the flat-blade turbine impeller
(b) The minimum rate of rotation of the impeller for complete and uniform dispersion
(c) The power requirement of the agitator at the minimum rotation rate

SOLUTION

Mass flow rate of feed = 20,400 lb/h; feed density = 62.3 lb/ft^3
Volumetric flow rate of feed = Q_F = 20,400/62.3 = 327 ft^3/h
Mass flow rate of solvent = 11,200 lb/h
Solvent density = 54.2 lb/ft^3; volumetric flow rate of solvent = Q_S = 11,200/54.2 = 207 ft^3/h

Because of the dilute concentration of solute in the feed and sufficient agitation to achieve complete and uniform dispersion, assume fractional volumetric holdups of raffinate and extract in the vessel are equal to the corresponding volume fractions in the combined feed and solvent entering the mixer:

$$\phi_R = 327/(327 + 207) = 0.612; \phi_E = 1 - 0.612 = 0.388$$

(a) Mixer volume = $(Q_F + Q_S)t_{res}$ = V = (327 + 207)(2/60) = 17.8 ft^3. Assume a cylindrical vessel with $D_T = H$ and neglect the volume of the bottom and top heads and the volume occupied by the agitator and the baffles. Then

$$V = (\pi D_T^2/4)H = \pi D_T^3/4$$

$$D_T = [(4/\pi)V]^{1/3} = [(4/3.14)17.8]^{1/3} = 2.83 \text{ ft}$$

$$H = D_T = 2.83 \text{ ft}$$

Make the vessel 3 ft in diameter by 3 ft high. Assume that

$$D_i/D_T = 1/3; D_i = D_T/3 = 3/3 = 1 \text{ ft.}$$

(b) *Case 1—Raffinate phase dispersed:*

$$\phi_D = \phi_R = 0.612; \phi_C = \phi_E = 0.388$$

$$\rho_D = \rho_R = 62.3 \text{ lb/ft}^3; \rho_C = \rho_E = 54.2 \text{ lb/ft}^3$$

$$\mu_D = \mu_R = 0.89 \text{ cP} = 2.16 \text{ lb/h-ft}; \mu_C = \mu_E = 0.59 \text{ cP} = 1.43 \text{ lb/h-ft}$$

$$\Delta\rho = 62.3 - 54.2 - 8.1 \text{ lb/ft}^3; \sigma = 25 \text{ dyne/cm} = 719,000 \text{ lb/h}^2$$

From (8-23), ρ_M = (54.2)(0.388) + (62.3)(0.612) = 59.2 lb/ft^3

From (8-24), $\mu_M = \dfrac{1.43}{0.388}\left[1 + \dfrac{1.5(2.16)(0.612)}{1.43 + 2.16}\right] = 5.72$ lb/h-ft

From (8-25), using American engineering units, with $g = 4.17 \times 10^8$ ft/h^2,

$$\frac{\mu_M^2\sigma}{D_i^5\rho_M g^2(\Delta\rho)^2} = \frac{(5.72)^2(719,000)}{(1)^5(59.2)(4.17 \times 10^8)^2(8.1)^2} = 3.47 \times 10^{-14}$$

$$N_{min}^2 = 1.03\left(\frac{g\,\Delta\rho}{\rho_M D_i}\right)\left(\frac{D_T}{D_i}\right)^{2.76}\phi_D^{0.106}(3.47 \times 10^{-14})^{0.084}$$

$$= 1.03\left[\frac{(4.17 \times 10^8)(8.1)}{(59.2)(1)}\right]\left(\frac{3}{1}\right)^{2.76}(0.612)^{0.106}(0.0740) = 8.56 \times 10^7 (\text{rph})^2$$

$$N_{min} = 9,250 \text{ rph} = 155 \text{ rpm}$$

Case 2—Extract phase dispersed: Calculations similar to case 1 result in N_{min} = 8,820 rph = 147 rpm

(c) *Case 1—Raffinate phase dispersed:*

From (8-22), $\qquad\qquad N_{Re} = \dfrac{(1)^2(9,250)(59.2)}{(5.72)} = 9.57 \times 10^4$

From Figure 8.36b, it is seen that a fully turbulent flow exists, with the power number given by its asymptotic value of $N_{Po} = 5.7$.

From (8-21), $P = N_{Po}N^3D_i^5\rho_M/g_c$

$$= (5.7)(9,250)^3(1)^5(59.2)/(4.17 \times 10^8) = 640,000 \text{ ft-lbf/h} = 0.323 \text{ hp}$$
$$P/V = 640,000/17.8 = 36,000 \text{ ft-lbf/h-ft}^3 = 600 \text{ ft-lbf/min-ft}^3$$

Case 2—Extract phase dispersed:

Calculations similar to case 1 result in $P = 423,000 \text{ ft-lbf/h} = 0.214 \text{ hp}.$ ∎

Mass Transfer Efficiency

When dispersion is complete and uniform, the contents of the vessel are perfectly mixed with respect to both phases. In that case, the concentration of the solute in each of the two phases in the mixing vessel is uniform and equal to the concentrations in the two-phase emulsion leaving the mixing vessel. This is the so-called ideal CFSTR or CSTR (continuous-flow stirred-tank reactor) model, sometimes called the completely back-mixed or perfectly mixed model, first discussed by MacMullin and Weber [41]. The Murphree dispersed-phase efficiency for liquid–liquid extraction, based on the raffinate as the dispersed phase, can be expressed as the fractional approach to equilibrium. In terms of bulk molar concentrations of the solute,

$$E_{MD} = \frac{c_{D,in} - c_{D,out}}{c_{D,in} - c_D^*} \tag{8-26}$$

where c_D^* is the solute concentration in equilibrium with the bulk solute concentration in the exiting continuous phase, $c_{C,out}$. The rate of mass transfer of the solute from the dispersed phase to the continuous phase can be expressed as

$$n = K_{OD}a(c_{D,out} - c_D^*)V \tag{8-27}$$

where the concentration driving force for mass transfer is uniform throughout the well-mixed vessel and is equal to the driving force based on the exit concentrations, a is the interfacial area for mass transfer per unit volume of liquid phases, V is the total volume of liquid phases in the vessel, and K_{OD} is the overall mass transfer coefficient based on the dispersed phase, which is given in terms of the separate resistances of the dispersed and continuous phases by

$$\frac{1}{K_{OD}} = \frac{1}{k_D} + \frac{1}{mk_C} \tag{8-28}$$

where equilibrium is assumed at the interface between the two phases and m = the slope of the equilibrium curve for the solute plotted as c_C versus c_D:

$$m = dc_C/dc_D \tag{8-29}$$

For dilute solutions, changes in volumetric flow rates of the raffinate and extract are small, and thus the rate of mass transfer based on the change in solute concentration in the dispersed phase is given by material balance:

$$n = Q_D(c_{D,in} - c_{D,out}) \tag{8-30}$$

where Q_D is the volumetric flow rate of the dispersed phase.

To obtain an expression for E_{MD} in terms of $K_{OD}a$, (8-26), (8-27), and (8-30) are combined in the following manner. From (8-26),

$$\frac{E_{MD}}{1 - E_{MD}} = \frac{c_{D,in} - c_{D,out}}{c_{D,out} - c_D^*} \tag{8-31}$$

Equating (8-27) and (8-30), and noting that the right-hand side of (8-31) is the number of dispersed-phase transfer units for a perfectly mixed vessel with $c_D = c_{D,\text{out}}$,

$$N_{OD} = \int_{c_{D,\text{out}}}^{c_{D,\text{in}}} \frac{dc_D}{c_D - c_D^*} = \frac{c_{D,\text{in}} - c_{D,\text{out}}}{c_{D,\text{out}} - c_D^*} = \frac{K_{OD}aV}{Q_D} \tag{8-32}$$

Combining (8-31) and (8-32) and solving for E_{MD},

$$E_{MD} = \frac{K_{OD}aV/Q_D}{1 + K_{OD}aV/Q_D} = \frac{N_{OD}}{1 + N_{OD}} \tag{8-33}$$

When $N_{OD} = (K_{OD}aV/Q_D) \gg 1$, $E_{MD} = 1$.

Drop Size and Interfacial Area

From (8-33) and (8-28), it is seen that an estimate of E_{MD} requires generalized correlations of experimental data for the interfacial area for mass transfer, a, and the dispersed- and continuous-phase mass transfer coefficients, k_D and k_C, respectively. The population of dispersed-phase droplets in an agitated vessel will cover a range of sizes and shapes. For each droplet, it is useful to define d_e, the equivalent diameter of a spherical drop, using the method of Lewis, Jones, and Pratt [42],

$$d_e = (d_1^2 d_2)^{1/3} \tag{8-34}$$

where d_1 and d_2 are the major and minor axes, respectively, of an ellipsoidal drop image. For a spherical drop, d_e is simply the diameter of the drop. For the population of drops, it is useful to define an average or mean drop diameter. A number of different definitions are available depending on whether weight-mean, mean-volume, surface-mean, mean-surface, length-mean, or mean-length diameter is appropriate [43]. For mass transfer calculations, the surface-mean diameter, d_{vs} (also called the *Sauter mean diameter*) is most appropriate because it is the mean drop diameter that gives the same interfacial surface area as the entire population of drops for the same mass of drops. It is determined from experimental drop-size distribution data for N drops by the definition:

$$\frac{\pi d_{vs}^2}{(\pi/6)d_{vs}^3} = \frac{\pi \sum_N d_e^2}{(\pi/6) \sum_N d_e^3}$$

which, when solved for d_{vs}, gives

$$d_{vs} = \frac{\sum_N d_e^3}{\sum_N d_e^2} \tag{8-35}$$

With this definition, the interfacial surface area per unit volume of a two-phase mixture is

$$a = \frac{\pi N d_{vs}^2 \phi_D}{\pi N d_{vs}^3/6} = \frac{6\phi_D}{d_{vs}} \tag{8-36}$$

Equation (8-36) is used to estimate the interfacial area, a, from a measurement of d_{vs} or vice versa. Early experimental investigations, such as those of Vermeulen, Williams, and Langlois [44], found that d_{vs} is dependent on a Weber number:

$$N_{\text{We}} = \frac{(\text{inertial force})}{(\text{interfacial tension force})} = \frac{D_i^3 N^2 \rho_C}{\sigma} \tag{8-37}$$

High Weber numbers give small droplets and high interfacial areas. Gnanasundram, Dega-leesan, and Laddha [45] correlated d_{vs} over a wide range of N_{We}. Below a critical value of $N_{We} = 10,000$, d_{vs} is dependent on dispersed-phase holdup, ϕ_D, because of coalescence effects. For $N_{We} > 10,000$, inertial forces dominate so that coalescence effects are much less prominent and d_{vs} is almost independent of holdup up to $\phi_D = 0.5$. The recommended correlations are

$$\frac{d_{vs}}{D_i} = 0.052(N_{We})^{-0.6}e^{4\phi_D}, \quad N_{We} < 10,000 \tag{8-38}$$

$$\frac{d_{vs}}{D_i} = 0.39(N_{We})^{-0.6}, \quad N_{We} > 10,000 \tag{8-39}$$

Typical values of N_{We} for industrial extractors are less than 10,000, so (8-38) applies. Values of d_{vs}/D_i are frequently in the range of 0.0005 to 0.01.

Experimental studies, for example, those of Chen and Middleman [46] and Sprow [47], show that the dispersion produced in an agitated vessel is a dynamic phenomenon. Droplet breakup by turbulent pressure fluctuations dominates in the vicinity of the impeller blades, while for reasonable dispersed-phase holdup, coalescence of drops by collisions dominates away from the impeller. Thus, a distribution of drop sizes is found in the vessel, with smaller drops in the vicinity of the impeller blades and larger drops elsewhere. Typically, when both drop breakup and coalescence occur, the drop-size distribution is such that $d_{min} \approx d_{vs}/3$ and $d_{max} \approx 3d_{vs}$. Thus, the drop size varies over about a 10-fold range, and the distribution approximates a normal Gaussian distribution.

EXAMPLE 8.6

For the conditions and results of Example 8.5, with the extract phase as the dispersed phase, estimate the Sauter mean drop diameter, the range of drop sizes, and the interfacial area.

SOLUTION

$$D_i = 1 \text{ ft}; \quad N = 147 \text{ rpm} = 8,820 \text{ rph}$$
$$\rho_C = 62.3 \text{ lb/ft}^3; \quad \sigma = 718,800 \text{ lb/h}^2$$

From (8-37);

$$N_{We} = (1)^3(8,820)^2(62.3)/718,800 = 6,742; \quad \phi_D = 0.388$$

From (8-38),

$$d_{vs} = (1)(0.052)(6,742)^{-0.6}\exp[4(0.388)] = 0.00124 \text{ ft} \quad \text{or} \quad (0.00124)(12)(25.4) = 0.38 \text{ mm}$$
$$d_{min} = d_{vs}/3 = 0.126 \text{ mm}; \, d_{max} = 3d_{vs} = 1.134 \text{ mm}$$

From (8-36),

$$a = 6(0.388)/0.00124 = 1,880 \text{ ft}^2/\text{ft}^3$$

■

Mass Transfer Coefficients

Experimental studies, conducted since the early 1940s, show that mass transfer in mechanically agitated liquid–liquid systems is very complex. This is true for mass transfer in (1) the dispersed-phase droplets, (2) the continuous phase, and (3) at the interface. The reasons for this complexity are many. The magnitude of k_D depends on drop diameter, solute diffusivity, and fluid motion within the drop. When drop diameter is small (less than 1 mm according to Davies [48]), interfacial tension is high (say > 15 dyne/cm), and trace amounts of surface-active agents are present, droplets are rigid (internally stagnant), and they behave like solids. As droplets become larger, interfacial tension decreases, surface-active agents become relatively ineffective, and internal toroidal fluid circulation patterns, caused by viscous drag of the continuous phase, appear within the drops. For larger-diameter drops, the shape of the drop may oscillate between spheroid and ellipsoid or other shapes.

Mass-transfer coefficients in the continuous phase depend on the relative motion between the droplets and the continuous phase, and whether the drops are forming or breaking, or are coalescing. Interfacial movements or turbulence, called *Marangoni effects,* occur due to interfacial tension gradients. Such effects can induce substantial increases in mass transfer rates.

A relatively conservative estimate of the overall mass-transfer coefficient, K_{OD}, in (8-28), can be made from estimates of k_D and k_C, by assuming rigid drops, the absence of Marangoni effects, and a stable drop size (i.e., no drop forming, breaking, or coalescing). For k_D, the asymptotic steady-state solution for mass transfer in a rigid sphere with negligible resistance of the surroundings is given by Treybal [25] as

$$(N_{\text{Sh}})_D = \frac{k_D d_{vs}}{D_D} = \frac{2}{3}\pi^2 = 6.6 \tag{8-40}$$

where N_{Sh} is the Sherwood number. Exercise 3.31 in Chapter 3 for diffusion from the surface of a sphere into an infinite, quiescent fluid gives the following result for the continuous-phase Sherwood number:

$$(N_{\text{Sh}})_C = \frac{k_C d_{vs}}{D_C} = 2 \tag{8-41}$$

However, if other spheres of equal diameter are located near the sphere of interest, $(N_{\text{Sh}})_C$ may decrease to a value as low as 1.386, according to Cornish [49]. In an agitated vessel, the continuous-phase Sherwood number will usually be much greater than 1.386. A reasonable estimate can be made with the semi-theoretical correlation of Skelland and Moeti [50]. They fitted 180 data points for three different solutes, three different dispersed organic solvents, and water as the continuous phase. Mass transfer was from the dispersed phase to the continuous phase, but only for $\phi_D = 0.01$. Skelland and Moeti assumed an equation of the form

$$(N_{\text{Sh}})_C \propto (N_{\text{Re}})_C^y (N_{\text{Sc}})_C^x \tag{8-42}$$

where

$$(N_{\text{Sh}})_C = k_C d_{vs}/D_C \tag{8-43}$$

$$(N_{\text{Sc}})_C = \mu_C/\rho_C D_C \tag{8-44}$$

For the Reynolds number, they assumed that the characteristic velocity is the square root of the mean-square local fluctuating velocity in the vicinity of the droplet, based on the theory of local isotropic turbulence of Batchelor [51]:

$$\overline{u}^2 \propto \left(\frac{Pg_c}{V}\right)^{2/3}\left(\frac{d_{vs}}{\rho_C}\right)^{2/3} \tag{8-45}$$

Thus,

$$(N_{\text{Re}})_C = \frac{(\overline{u}^2)^{1/2} d_{vs}\rho_C}{\mu_C} \tag{8-46}$$

Combining (8-45) and (8-46), with omission of the proportionality constant:

$$(N_{\text{Re}})_C = \frac{d_{vs}^{4/3}\rho_C^{2/3}(Pg_c/V)^{1/3}}{\mu_C} \tag{8-47}$$

As discussed previously, in the turbulent-flow region,

$$Pg_c \propto \rho N^3 D_i^5 \quad \text{or} \quad Pg_c/V \propto \rho_C N^3 D_i^5/D_T^3$$

Thus,

$$(N_{Re})_C = \frac{d_{vs}^{4/3} \rho_C^{5/3} N D_i^{5/3}}{\mu_C D_T} \tag{8-48}$$

Skellend and Moeti correlated their mass transfer coefficient data with

$$k_C \propto D_C^{2/3} \mu_C^{-1/3} N^{3/2} d_{vs}^0$$

The exponents in this proportionality are used to determine the exponents y and x in (8-42) as $\frac{2}{3}$ and $\frac{1}{3}$, respectively. In addition, based on the work of previous investigators, a droplet Eotvos number,

$$N_{Eo} = \rho_D d_{vs}^2 g / \sigma \tag{8-49}$$

where N_{Eo} = (gravitational force)/(surface tension force)

and the dispersed-phase holdup, ϕ_D, are incorporated into the following final correlation, which predicts 180 experimental data points to an average absolute deviation of 19.71%:

$$(N_{Sh})_C = \frac{k_C d_{vs}}{D_C} = 1.237 \times 10^{-5} \left(\frac{\mu_C}{\rho_C D_C} \right)^{1/3} \left(\frac{D_i^2 N \rho_C}{\mu_C} \right)^{2/3} \phi_D^{-1/2} \left(\frac{D_i N^2}{g} \right)^{5/12} \times$$
$$\left(\frac{D_i}{d_{vs}} \right)^2 \left(\frac{d_{vs}}{D_T} \right)^{1/2} \left(\frac{\rho_D d_{vs}^2 g}{\sigma} \right)^{5/4} \tag{8-50}$$

EXAMPLE 8.7

For the system, conditions, and results of Examples 8.5 and 8.6, with the extract as the dispersed phase, estimate:

(a) The dispersed-phase mass transfer coefficient, k_D
(b) The continuous-phase mass transfer coefficient, k_C
(c) The Murphree dispersed-phase efficiency, E_{MD}
(d) The fractional extraction of furfural

The molecular diffusivities of furfural in toluene (dispersed) and water are, respectively,

$$D_D = 8.32 \times 10^{-5} \text{ ft}^2/\text{h and } D_C = 4.47 \times 10^{-5} \text{ ft}^2/\text{h}$$

The distribution coefficient for dilute conditions is $m = dc_C/dc_D = 0.0985$.

SOLUTION

(a) From (8-40), $k_D = 6.6(D_D)/d_{vs} = 6.6(8.32 \times 10^{-5})/0.00124 = 0.44$ ft/h
(b) To apply (8-50) to the estimation of k_C, first compute each of the dimensionless groups in that equation:

$$N_{Sc} = \mu_C/\rho_C D_C = 2.165/[(62.3)(4.47 \times 10^{-5})] = 777$$
$$N_{Re} = D_i^2 N \rho_C/\mu_C = (1)^2(8,820)(62.3)/2.165 = 254,000$$
$$N_{Fr} = D_i N^2/g = (1)(8,820)^2/(4.17 \times 10^8) = 0.187$$
$$D_i/d_{vs} = 1/0.00124 = 806; \, d_{vs}/D_T = 0.00124/3 = 0.000413$$
$$N_{Eo} = \rho_D d_{vs}^2 g/\sigma = (54.2)(0.00124)^2(4.17 \times 10^8)/718,800 = 0.0483$$

From (8-50),

$$N_{Sh} = 1.237 \times 10^{-5}(777)^{1/3}(254,000)^{2/3}(0.388)^{-1/2}(0.187)^{5/12}(806)^2(0.000413)^{1/2}(0.0483)^{5/4} = 109$$

Note that this Sherwood number is much greater than the value of 2 for a single droplet in an infinite, quiescent fluid.

$$k_C = N_{Sh} D_C/d_{vs} = (109)(4.47 \times 10^{-5})/0.00124 = 3.93 \text{ ft/h}$$

(c) From (8-28), $K_{OD} = \dfrac{1}{(1/0.44) + 1/[(0.0985)(3.93)]} = 0.206$ ft/h. From the results of Example 8.6, $K_{OD}a = (0.206)(1,880) = 387$ h^{-1}. From (8-33), $N_{OD} = K_{OD}aV/Q_D = 387(21.2)/(207) = 39.6$ where $V = \pi D_T^2 H/4 = (3.14)(3)^2(3)/4 = 21.2$ ft^3. From (8-33), $E_{MD} = 39.6/(1 + 39.6) = 0.975$ or 97.5%.

(d) The fractional extraction of furfural for dilute solutions is given by

$$f_{\text{Extracted}} = \frac{c_{C,\text{in}} - c_{C,\text{out}}}{c_{C,\text{in}}} = 1 - \frac{c_{C,\text{out}}}{c_{C,\text{in}}}$$

By material balance,

$$Q_C(c_{C,\text{in}} - c_{C,\text{out}}) = Q_D c_{D,\text{out}} \tag{1}$$

From (8-26),

$$E_{MD} = c_{D,\text{out}}/c_D^* = m c_{D,\text{out}}/c_{C,\text{out}} \tag{2}$$

Combining (1) and (2) to eliminate $c_{D,\text{out}}$ gives

$$\frac{c_{C,\text{out}}}{c_{C,\text{in}}} = \frac{1}{1 + Q_D E_{MD}/(Q_C m)} \tag{3}$$

$$\text{or } f_{\text{Extracted}} = \frac{Q_D E_{MD}/(Q_C m)}{1 + Q_D E_{MD}/(Q_C m)}$$

$$\frac{Q_D}{Q_C}\frac{E_{MD}}{m} = \frac{(207)(0.975)}{(327)(0.0985)} = 6.27$$

$$\text{Thus, } f_{\text{Extracted}} = \frac{6.27}{1 + 6.27} = 0.862 \quad \text{or} \quad 86.2\%$$

∎

Multicompartment Columns

Sizing extraction columns, which may or may not include mechanical agitation, involves the determination of column diameter and column height. The diameter must be sufficiently large to permit the two phases to flow countercurrently through the column without flooding. The column height must be sufficient to achieve the number of equilibrium stages corresponding to the desired degree of extraction.

For small-diameter columns, rough estimates of the diameter and height can be made using the results of a study by Stichlmair [52] with the toluene–acetone–water system for $Q_D/Q_C = 1.5$. Typical ranges of 1/HETS and the sum of the superficial phase velocities for a number of extractor types are given in Table 8.6.

Because of the large number of important variables, an accurate estimation of column diameter for liquid–liquid contacting devices is far more complex and more uncertain than for vapor–liquid contactors. These variables include individual phase flow rates, density difference between the two phases, interfacial tension, direction of mass transfer, viscosity and density of the continuous phase, rotating or reciprocating speed, and geometry of internals. Column diameter is best determined by scale-up from tests run in standard laboratory or pilot plant test units with a diameter of 1 in. or larger. The sum of the measured superficial velocities of the two liquid phases in the test unit can then be assumed to hold for larger commercial units. This sum is often expressed in total gallons per hour per square foot of empty column cross section.

In the absence of laboratory data, preliminary estimates of diameter for some columns can be made by a simplification of the theory of Logsdail, Thornton, and Pratt [53], which is compared to other procedures by Landau and Houlihan [54] in the case of the rotating-

Table 8.6 Performance of Several Types of Column Extractors

Extractor Type	1/HETS, m^{-1}	$U_D + U_C$, m/h
Packed column	1.5–2.5	12–30
Pulsed packed column	3.5–6	17–23
Sieve-plate column	0.8–1.2	27–60
Pulsed-plate column	0.8–1.2	25–35
Scheibel column	5–9	10–14
RDC	2.5–3.5	15–30
Kuhni column	5–8	8–12
Karr column	3.5–7	30–40
RTL contactor	6–12	1–2

Source: J. Stichlmair, *Chemie-Ingenieur-Technik,* **52,** 253 (1980).

disk contactor. Because the relative motion between a dispersed droplet phase and a continuous phase is involved, this theory is based on a concept that is similar to that developed in Chapter 6 for liquid droplets dispersed in a vapor phase.

Consider the case of liquid droplets of the lower-density phase rising through the denser, downward-flowing continuous liquid phase, as shown in Figure 8.37. If the average superficial velocities of the discontinuous (droplet) phase and the continuous phase are U_D in the upward direction and U_C in the downward direction (i.e., both of these velocities are positive), respectively, the corresponding average actual velocities relative to the column wall are

$$\bar{u}_D = \frac{U_D}{\phi_D} \tag{8-51}$$

$$\text{and } \bar{u}_D = \frac{U_C}{1 - \phi_D} \tag{8-52}$$

The average droplet rise velocity relative to the continuous phase is the sum of (8-51) and (8-52):

$$\bar{u}_r = \frac{U_D}{\phi_D} + \frac{U_C}{1 - \phi_D} \tag{8-53}$$

This relative velocity (also called *slip velocity*) can be expressed in terms of a modified form of (6-40) where the continuous-phase density in the buoyancy term is replaced by the density of the two-phase mixture, ρ_M. Thus, after noting for the case here that the drag force, F_d, and gravitational force, F_g, act downward while buoyancy, F_b, acts upward, we obtain

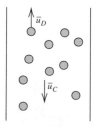

Figure 8.37 Countercurrent flows of dispersed and continuous liquid phases in a column.

$$\bar{u}_r = C\left(\frac{\rho_M - \rho_D}{\rho_C}\right)^{1/2} f\{1 - \phi_D\} \tag{8-54}$$

where C is the same parameter as in (6-41) and $f\{1 - \phi_D\}$ is a factor that allows for the hindered rising effect of neighboring droplets. The density ρ_M is a volumetric mean given by

$$\rho_M = \phi_D\rho_D + (1 - \phi_D)\rho_C \tag{8-55}$$

$$\rho_M - \rho_D = (1 - \phi_D)(\rho_C - \rho_D) \tag{8-56}$$

Substitution of (8-56) into (8-54) yields

$$\bar{u}_r = C\left(\frac{\rho_C - \rho_D}{\rho_C}\right)^{1/2}(1 - \phi_D)^{1/2}f\{1 - \phi_D\} \tag{8-57}$$

From experimental data, Gayler, Roberts, and Pratt [55] found that, for a given liquid–liquid system, the right-hand side of (8-57) can be expressed empirically as

$$\bar{u}_r = u_0(1 - \phi_D) \tag{8-58}$$

where u_0 is a characteristic rise velocity for a single droplet, which depends on all the variables discussed above, except those on the right-hand side of (8-53). Thus, for a given liquid–liquid system, column design, and operating conditions, the combination of (8-53) and (8-58) gives

$$\frac{U_D}{\phi_D} + \frac{U_C}{1 - \phi_D} = u_0(1 - \phi_D) \tag{8-59}$$

where u_0 is a constant. Equation (8-59) is cubic in ϕ_D, with a typical solution shown in Figure 8.38 for $U_C/u_0 = 0.1$. Thornton [56] argues that, with U_C fixed, an increase in U_D results in an increased value of the holdup ϕ_D, until the flooding point is reached, at which $(\partial U_D/\partial \phi_D)_{U_C} = 0$. Thus, in Figure 8.38, only that portion of the curve for $\phi_D = 0$ to $(\phi_D)_f$, the holdup at the flooding point, is realized in practice. Alternatively, with U_D fixed, $(\partial U_C/\partial \phi_D)_{U_D} = 0$ at the flooding point. If these two derivatives are applied to (8-59), we

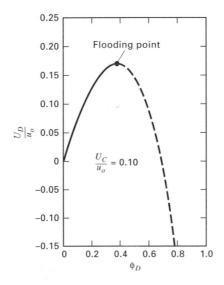

Figure 8.38 Typical holdup curve for liquid–liquid extraction column.

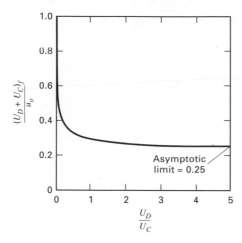

Figure 8.39 Effect of phase ratio on total capacity of liquid–liquid extraction column.

obtain, respectively,

$$U_C = u_0[1 - 2(\phi_D)_f][1 - (\phi_D)_f]^2 \tag{8-60}$$

$$U_D = 2u_0[1 - (\phi_D)_f](\phi_D)_f^2 \tag{8-61}$$

where the subscript f denotes flooding. Combining (8-60) and (8-61) to eliminate u_0 gives the following expression for $(\phi_D)_f$:

$$(\phi_D)_f = \frac{[1 + 8(U_C/U_D)]^{0.5} - 3}{4[(U_C/U_D) - 1]} \tag{8-62}$$

This equation predicts values of $(\phi_D)_f$ ranging from zero at $U_D/U_C = 0$ to 0.5 at $U_C/U_D = 0$. At $U_D/U_C = 1$, $(\phi_D)_f = \frac{1}{3}$. The simultaneous solution of (8-59) and (8-62) results in Figure 8.39 for the variation of total capacity as a function of phase flow ratio. The largest total capacities are achieved, as might be expected, at the smallest ratios of dispersed-phase flow rate to continuous-phase flow rate.

For fixed values of column geometry and rotor speed, experimental data of Logsdail et al. [53] for a laboratory-scale RDC indicate that the dimensionless group $(u_0\mu_C\rho_C/\sigma \Delta\rho)$ is approximately constant. Data of Reman and Olney [57] and Strand, Olney, and Ackerman [58] for well-designed and efficiently operated commercial RDC columns ranging from 8 to 42 in. in diameter indicate that this dimensionless group has a value of roughly 0.01 for systems involving water as either the continuous or dispersed phase. This value is suitable for preliminary calculations of RDC and Karr column diameters, when the sum of the actual superficial phase velocities is taken as 50% of the estimated sum at flooding conditions.

EXAMPLE 8.8

Estimate the diameter of an RDC to extract acetone from a dilute toluene–acetone solution into water at 20°C. The flow rates for the dispersed organic and continuous aqueous phases are 27,000 and 25,000 lb/h, respectively.

SOLUTION

The necessary physical properties are

$$\mu_C = 1.0 \text{ cP } (0.000021 \text{ lbf-s/ft}^2) \text{ and } \rho_C = 1.0 \text{ g/cm}^3$$

$$\Delta\rho = 0.14 \text{ g/cm}^3 \text{ and } \sigma = 32 \text{ dyne/cm } (0.00219 \text{ lbf/ft})$$

$$\frac{U_D}{U_C} = \left(\frac{27,000}{25,000}\right)\left(\frac{\rho_C}{\rho_D}\right) = \left(\frac{27,000}{25,000}\right)\left(\frac{1.0}{0.86}\right) = 1.26$$

From Figure 8.39, $(U_D + U_C)_f/u_0 = 0.29$. Assume that $\dfrac{u_0 \mu_C \rho_C}{\sigma\,\Delta\rho} = 0.01$.

Therefore,
$$u_0 = \frac{(0.01)(0.00219)(0.14)}{(0.000021)(1.0)} = 0.146 \text{ ft/s}$$

$$(U_D + U_C)_f = 0.29(0.146) = 0.0423 \text{ ft/s}$$

$$(U_D + U_C)_{50\% \text{ of flooding}} = \left(\frac{0.0423}{2}\right)(3{,}600) = 76.1 \text{ ft/h}$$

$$\text{Total ft}^3/\text{h} = \frac{27{,}000}{(0.86)(62.4)} + \frac{25{,}000}{(1.0)(62.4)} = 904 \text{ ft}^3/\text{h}$$

$$\text{Column cross-sectional area} = A_c = \frac{904}{76.1} = 11.88 \text{ ft}^2$$

$$\text{Column diameter} = D_T = \left(\frac{4A_c}{\pi}\right)^{0.5} = \left[\frac{(4)(11.88)}{3.14}\right]^{0.5} = 3.9 \text{ ft}$$

Note that from Table 8.6, a typical $(U_D + U_C)$ for an RDC is 15 to 30 m/h or 49 to 98.4 ft/h.
∎

Despite their compartmentalization, mechanically assisted liquid-liquid extraction columns, such as the RDC and Karr columns, operate more nearly like differential contacting devices than like staged contactors. Therefore, it is more common to consider stage efficiency for such columns in terms of HETS (height equivalent to a theoretical stage) or as some function of mass transfer parameters, such as HTU (height of a transfer unit). Although it is not on as sound a theoretical basis as the HTU, the HETS is preferred here because it can be applied directly to determine column height from the number of equilibrium stages.

Because of the great complexity of liquid–liquid systems and the large number of variables that influence contacting efficiency, general correlations for HETS have been difficult to develop. However, for well-designed and efficiently operated columns, the available experimental data indicate that the dominant physical properties influencing HETS are the interfacial tension, the phase viscosities, and the density difference between the phases. In addition, it has been observed by Reman [59] for RDC units and by Karr and Lo [60] for Karr columns that HETS increases with increasing column diameter because of axial mixing effects discussed in the next section.

It is preferred to obtain values of HETS by conducting small-scale laboratory experiments with systems of interest. These values are scaled to commercial-size columns by assuming that HETS varies with column diameter D_T, raised to an exponent, which may vary from 0.2 to 0.4 depending on the system.

In the absence of experimental data, the crude correlation of Figure 8.40 can be used for preliminary design if phase viscosities are no greater than 1 cP. The data points correspond to minimum reported HETS values for RDC and Karr units with the exponent on column diameter set arbitrarily to $\frac{1}{3}$. The points represent values of HETS that vary from as low as 6 in. for a 3-in.-diameter laboratory-size column operating with a low-interfacial-tension/low-viscosity system such as methyl–isobutyl ketone/acetic acid/water, to as high as 25 in. for a 36-in.-diameter commercial column operating with a high-interfacial-tension/low-viscosity system such as xylenes–acetic acid–water. For systems having one phase of high viscosity, values of HETS can be 24 in. or more, even for a small, laboratory-size column.

Estimate HETS for the conditions of Example 8.8.

EXAMPLE 8.9

SOLUTION

Because toluene has a viscosity of approximately 0.6 cP, this is a low-viscosity system. From Example 8.8 the interfacial tension is 32 dyne/cm. From Figure 8.40, HETS/$D_T^{1/3} = 6.9$. For

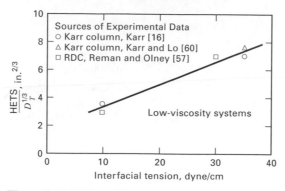

Figure 8.40 Effect of interfacial tension on HETS for RDC and Karr columns.

D_T = 3.9 ft, HETS = $6.9[(3.9)(12)]^{1/3}$ = 24.8 in. Note that from Table 8.6, HETS for an RDC varies from 0.29 to 0.40 m or 11.4 to 15.7 in. for a small column. ∎

More accurate estimates of flooding and HETS are discussed in detail by Lo et al. [4] and by Thornton [61]. Packed column design is considered by Strigle [62].

Axial Dispersion

In this and previous chapters covering liquid–liquid and vapor–liquid countercurrent-flow contactors, plug flow of each phase has been assumed. Each element of a phase is assumed to have the same residence time in the contactor, while each phase has a different residence time. Because axial concentration gradients in the direction of bulk flow are established in each phase, diffusion of a species is superimposed on the bulk flow of the species in that phase. Axial diffusion degrades the efficiency of multistage separation equipment, and in the limit, a multistage separator behaves like a single well-mixed stage. In Figure 8.41, solute concentration profiles for the extract and raffinate phases of a liquid–liquid extraction column are shown for plug flow (dashed lines) and for flow with significant axial

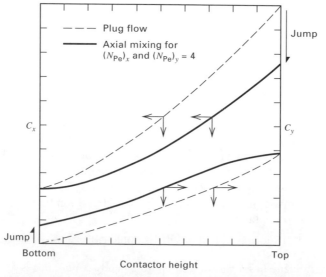

Figure 8.41 Solute concentration profiles for continuous countercurrent extraction with and without axial mixing.

diffusion in each of the two phases (solid lines). The continuous phase is the feed/raffinate (x subscript), which enters the contactor at the top ($z = 0$) The dispersed phase is the solvent/extract (y subscript), which enters the contactor at the bottom ($z = H$). Solute transfer is from the continuous phase to the dispersed phase. Two effects of axial diffusion are seen: (1) The concentration curves in the presence of axial diffusion are closer together than for plug flow and (2) these close proximities are due partially to concentrations at the two ends, which are different from those in the original feed and solvent. These differences are called *jumps* and are due to axial diffusion outside the region in the contactor where the two liquid phases are in contact. The jump at the top is caused by axial diffusion, superimposed on the bulk flow, in the feed liquid before it enters the contactor. This causes the concentration of solute in the feed just as it enters the contactor to be less than its concentration in the original feed liquid. Similarly, diffusion of solute into the incoming solvent causes the concentration of solute in the solvent just entering the bottom of the contactor to be greater than the concentration in the original solvent, which in Figure 8.41 is zero. The overall effect of axial diffusion is a reduction in the average driving force for mass transfer of the solute between the two phases, necessitating a taller column to accomplish the desired separation.

The effects shown in Figure 8.41 are actually due to a number of factors besides diffusion, which are lumped together into one overall effect, commonly referred to as *axial dispersion, axial mixing, longitudinal dispersion,* or *back-mixing.* These factors include:

1. Molecular and turbulent diffusion of the continuous phase along concentration gradients
2. Circulatory motion of the continuous phase due to the droplets of the dispersed phase
3. Transport and shedding of the continuous phase in the wakes attached to the rear of droplets of the dispersed phase
4. Circulation of continuous and dispersed phases in mechanically agitated columns
5. Channeling and nonuniform velocity profiles leading to distributions of residence times in the two phases

In general, the effect of axial dispersion is most pronounced when (1) a high recovery of solute is necessary, (2) the contactor is short in height, (3) large circulation patterns occur, (4) a wide range of droplet sizes is present, and/or (5) the feed-to-solvent flow ratio is very small or very large. Although axial dispersion effects are generally negligible in extractors where phase separation occurs between stages, such as in mixer-settler cascades and sieve-plate columns with downcomers, axial dispersion can be significant in spray columns, packed columns, and RDCs. Although axial dispersion can occur in packed absorbers, packed strippers, and packed distillation columns, it is significant only when operating at very high liquid-to-gas ratios. However, axial dispersion can be significant in spray and bubble columns used for absorption.

Two types of models have been developed for predicting the extent and effect of axial mixing: (1) diffusion models for differential-type contactors, due to Sleicher [63] and Miyauchi and Vermeulen [64]; and (2) backflow models for staged extractors without complete phase separation between stages, due to Sleicher [65] and Miyauchi and Vermeulen [66]. Both types are discussed by Vermeulen et al. [67]. Diffusion models, which have received the most attention and have been most applied more frequently, are convenient for studying the complex nature of axial dispersion.

Consider a differential height, dz, of a differential contactor with countercurrent two-phase flow, as shown in Figure 8.42. Feed enters the top of the column at $z = 0$, while solvent enters the bottom of the column at $z = H$. Assume that: (1) axial dispersion in each phase is characterized by a constant turbulent diffusion coefficient, E; (2) phase superficial velocities are each uniform over the cross section and constant in the axial direction; (3) the volumetric overall mass transfer coefficients for the solute are constant; (4) only the solute undergoes mass transfer between the two phases; and (5) the phase

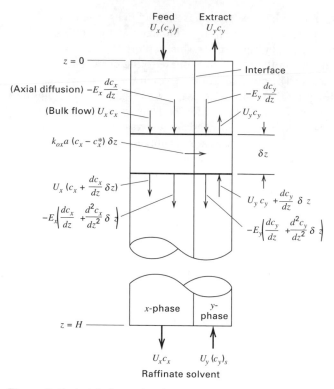

Figure 8.42 Axial dispersion in an extraction column. [From J.D. Thornton, *Science and Practice of Liquid–Liquid Extraction,* Vol. 1, Clarendon Oxford, (1992), with permission.]

equilibrium ratio for the solute is constant. Then the solute mass balance equations for the feed/raffinate (x) and solvent/extract (y) phases, respectively, are

$$E_x \frac{d^2c_x}{dz^2} - U_x \frac{dc_x}{dz} - K_{Ox}a(c_x - c_x^*) = 0 \tag{8-63}$$

$$E_y \frac{d^2c_y}{dz^2} + U_y \frac{dc_y}{dz} + K_{Ox}a(c_x - c_x^*) = 0 \tag{8-64}$$

where c_x^* is the concentration of the solute in the raffinate that is equilibrium with the solute concentration in the bulk extract. For these two differential equations, the boundary conditions, which were first proposed by Danckwerts [68] and were further elucidated by Wehner and Wilhelm [69], are
At $z = 0$,

$$U_x c_{x_f} - U_x c_{x_0} = -E_x \frac{dc_x}{dz} \tag{8-65}$$

and

$$dc_y/dz = 0 \tag{8-66}$$

At $z = H$,

$$U_y c_{y_H} - U_y c_{y_s} = E_y \frac{dc_y}{dz} \tag{8-67}$$

and

$$dc_x/dz = 0 \tag{8-68}$$

where: c_{x_f} = concentration of solute in the original feed

c_{x_0} = concentration of solute in the feed at $z = 0$

c_{y_H} = concentration of solute in the solvent at $z = H$

c_{y_s} = concentration of solute in the original solvent

The two terms on the left-hand sides of (8-65) and (8-67) are the jumps shown in Figure 8.41.

It is customary to convert (8-63) and (8-64) to alternative forms in terms of pertinent dimensionless groups. This is readily done by defining

$$Z = z/H \tag{8-69}$$

$$N_{\mathrm{Pe}_y} = U_y H / E_y = \text{axial turbulent column Peclet number for the extract phase} \tag{8-70}$$

$$N_{\mathrm{Pe}_x} = U_x H / E_x = \text{axial turbulent column Peclet number for the raffinate phase} \tag{8-71}$$

$$N_{Ox} = K_{Ox} a H / U_x = K_{Ox} a V / Q_x \tag{8-72}$$

Equations (8-63) and (8-64) then become

$$\frac{d^2 c_x}{dZ^2} - N_{\mathrm{Pe}_x} \frac{dc_x}{dZ} - N_{Ox} N_{\mathrm{Pe}_x} (c_x - c_x^*) = 0 \tag{8-73}$$

$$\frac{d^2 c_y}{dZ^2} - N_{\mathrm{Pe}_y} \frac{dc_y}{dZ} - \left(\frac{U_x}{U_y}\right) N_{Ox} N_{\mathrm{Pe}_y} (c_x - c_x^*) = 0 \tag{8-74}$$

The boundary conditions are transformed in a similar way. For a straight equilibrium curve, $c_x^* = mc_y$. Thus, we have a coupled set of ordinary differential equations, whose solutions for c_y and c_x are functions of c_{y_f}, c_{y_s}, m, N_{Pe_x}, N_{Pe_y}, N_{Ox}, U_x/U_y, and Z.

Further algebraic manipulations involving the substitution of dimensionless solute concentrations can reduce the number of variables from 10 to 7. In either case, the solution of the axial dispersion equations as obtained by Sleicher [65] and Miyauchi and Vermeulen [66] is very difficult to display in tabular or graphical form. However, the possible importance of axial dispersion is most commonly judged by the magnitudes of the Peclet numbers. A Peclet number of 0 corresponds to complete back-mixing, such that at most only one equilibrium stage is achieved; the entire column functions like a single mixer stage. A Peclet number of ∞ corresponds to an absence of axial dispersion. Experimental data on several different types of liquid-liquid extraction columns indicate that N_{Pe} for the dispersed phase is frequently greater than 50, while N_{Pe} for the continuous phase may be in the range of 5 to 30. Thus, as a first approximation, axial dispersion in the dispersed phase can be largely ignored. This effect was observed experimentally by Geankoplis and Hixson [70] in a spray extraction column and by Gier and Hougen [71] in spray and packed extraction columns. They reported end concentration changes of significant magnitude at the continuous-phase entrance, but not at the dispersed-phase entrance.

A number of approximate solutions to the axial dispersion equations, (8-63) to (8-68), have been published, including one by Sleicher [63]. Alternatively, if the original solvent is free of solute, a rapid and somewhat conservative estimate of the effect of axial dispersion can be made by the method of Watson and Cochran [72] from an empirical relation for the column efficiency:

$$\frac{H_{\text{plug flow}}}{H_{\text{actual}}} = \frac{(\mathrm{HTU}_{Ox})(\mathrm{NTU}_{Ox})}{H}$$

$$= 1 - \frac{1}{1 + N_{\mathrm{Pe}_x}(\mathrm{HTU}_{Ox}/H) - E + (1/\mathrm{NTU}_{Ox})}$$

$$- \frac{E}{N_{\mathrm{Pe}_y}(\mathrm{HTU}_{Ox}/H) - 1 + E + (1/\mathrm{NTU}_{Ox})} \tag{8-75}$$

where $H_{\text{actual}} = H =$ height of column taking into account axial dispersion

$\text{HTU}_{Ox} =$ height of an overall transfer unit based on the raffinate phase for plug flow

$\text{NTU}_{Ox} =$ number of overall transfer units based on the raffinate phase for plug flow

$E =$ extraction factor $= mU_x/U_y$

$m = dc_x/dc_y$

The product of HTU_{Ox} and NTU_{Ox} is the column height for plug flow, which is $<H$. Thus, the ratio on the left-hand side of (8-75) is a column efficiency. The NTU_{Ox} is approximated by:

$$\text{NTU}_{Ox} = \ln\left(\frac{X}{XE + 1 - E}\right)\Big/(E - 1) \qquad \textbf{(8-76)}$$

$$\text{where } X = (c_x)_{\text{out}}/(c_x)_{\text{in}} \qquad \textbf{(8-77)}$$

The HTU_{Ox} is defined by:

$$\text{HTU}_{Ox} = U_x/K_{Ox}a \qquad \textbf{(8-78)}$$

For given values of HTU_{Ox}, NTU_{Ox}, E, N_{Pe_x}, and N_{Pe_y}, (8-75) is solved for H. Caution must be exercised in using (8-75) because of its empirical nature. The equation is limited to $NTU_{Ox} \geq 2$, $E > 0.25$, $N_{\text{Pe}_x}(\text{HTU}_{Ox}/H) > 1.5$, and the calculated value of the column efficiency, $H_{\text{plug flow}}/H_{\text{actual}}$, must be ≥ 0.20. Within these restrictions, an extensive comparison by Watson and Cochran with the exact solution of (8-63) to (8-68) gives conservative efficiency values that deviate by no more than 0.07 (7%), with the highest accuracy for estimated efficiencies greater than 0.5 (50%).

EXAMPLE 8.10

Experiments conducted for a dilute system under laboratory conditions approximating plug flow give $\text{HTU}_{Ox} = 3$ ft. If a commercial column is to be designed for $\text{NTU}_{Ox} = 4$ and N_{Pe_x} and N_{Pe_y} are estimated to be 19 and 50, respectively, determine the necessary column height if $E = 0.5$.

SOLUTION

For plug flow, column height is $(\text{HTU}_{Ox})(\text{NTU}_{Ox})$ Substitution of the data into (8-75) gives

$$\frac{12}{H} = 1 - \frac{1}{(57/H) + 0.75} - \frac{0.5}{(180/H) - 0.25}$$

This is a nonlinear algebraic equation in H. Solving by an iterative method,

$$H = 17 \text{ ft}$$
$$\text{Efficiency} = (\text{HTU}_{Ox})(\text{NTU}_{Ox})/H = 12/17 = 0.706 \qquad (70.6\%)$$

■

SUMMARY

1. A solvent can be used to selectively extract one or more components from a liquid mixture.

2. Although liquid–liquid extraction is a reasonably mature separation operation, considerable experimental effort is often needed to find a suitable solvent and to determine residence time requirements or values of HETS, NTU, or mass transfer coefficients.

3. Compared to vapor–liquid separation operations, extraction has a higher overall mass transfer resistance. Stage efficiencies in columns are frequently low.

4. A wide variety of commercial extractors are available, as shown in Figures 8.2 to 8.7, ranging from simple columns with no mechanical agitation to centrifugal devices that may spin at several thousand revolutions per minute. A selection scheme, given in Table 8.3, is useful for choosing the most suitable extractors for a given separation.

5. Solvent selection is facilitated by consideration of a number of chemical factors given in Table 8.4 and physical factors discussed in Section 8.2.

6. For liquid–liquid extraction with ternary mixtures, phase equilibrium is conveniently represented on equilateral- or right-triangle diagrams for both type I (solute and solvent completely miscible) and the less common type II (solute and solvent not completely miscible) systems.

7. For determining equilibrium-stage requirements of single-section countercurrent cascades for ternary systems, the graphical methods of Hunter and Nash (equilateral-triangle diagram), Kinney (right-triangle diagram), or Varteressian and Fenske (distribution diagram of McCabe–Thiele type) can be applied, as described in Section 8.3. These methods can also determine minimum and maximum solvent requirements.

8. A two-section countercurrent cascade with extract reflux can be employed with a type II ternary system to enable a sharp separation of a binary feed mixture. The calculation of stage requirements of such a two-section cascade is conveniently carried out by the graphical method of Maloney and Schubert using a Janecke equilibrium diagram, as discussed in Section 8.4. The addition of raffinate reflux to such a cascade is of little value. The Maloney–Schubert method can also be applied to single-section cascades.

9. When only a few equilibrium stages are required, a cascade of mixer-settler units may be attractive because each mixer can be designed to closely approach an equilibrium stage. With many ternary and higher-order systems the residence-time requirement may be only a few minutes for a 90% approach to equilibrium using an agitator input of approximately 4 hp/1,000 gal. Adequate phase-disengaging area for the settlers may be estimated from the rule of 5 gal of combined extract and raffinate per minute per square foot of disengaging area.

10. For mixers utilizing a six-flat-bladed turbine in a closed vessel with side vertical baffles, as shown in Figure 8.35, useful extractor design correlations are available for estimating, for a given extraction, the mixing vessel dimensions, minimum impeller rotation rate for complete and uniform dispersion, impeller horsepower, mean droplet size, range of droplet sizes, interfacial area per unit volume, dispersed-phase and continuous-phase mass transfer coefficients, and Murphree efficiency.

11. For column-type extractors, with and without mechanical agitation, correlations for determining column diameter, to avoid flooding, and column height are suitable only for very preliminary sizing calculations. For final extractor selection and design, recommendations of equipment vendors based on experimental data from pilot-size equipment are highly desirable.

12. Sizing of column-type extractors must consider axial dispersion, which can significantly reduce mass transfer-driving forces and thus increase the required column height. Axial dispersion effects are often most significant in the continuous phase.

REFERENCES

1. Derry, T.K., and T.I. Williams, *A Short History of Technology,* Oxford University Press, New York (1961).

2. Bailes, P.J., and A. Winward, Trans. Inst. Chem. Eng., **50,** 240–258 (1972).

3. Bailes, P.J., C. Hanson, and M.A. Hughes, *Chem. Eng.,* **83** (2), 86–100 (1976).

4. Lo, T.C., M.H.I. Baird, and C. Hanson, Eds., *Handbook of Solvent Extraction,* Wiley-Interscience, New York (1983).

5. Reissinger, K.-H., and J. Schroeter, "Alternatives to Distillation," *I. Chem. E. Symp. Ser. No. 54,* 33–48 (1978).

6. Humphrey, J.L., J.A. Rocha, and J.R. Fair, *Chem. Eng.,* **91** (19), 76–95 (1984).

7. Fenske, M.R., C.S. Carlson, and D. Quiggle, *Ind. Eng. Chem.,* **39,** 1932 (1947).

8. Scheibel, E.G., *Chem. Eng. Prog.,* **44,** 681 (1948).

9. Scheibel, E.G., *AIChE J.,* **2,** 74 (1956).

10. Scheibel, E.G., U. S. Patent 3,389,970 (June 25, 1968).

11. Oldshue, J., and J. Rushton, *Chem. Eng. Prog.,* **48** (6), 297 (1952).

12. Reman, G.H., *Proceedings of the 3rd World Petroleum Congress,* The Hague, Netherlands, Sec. III, 121 (1951).

13. Reman, G.H., *Chem. Eng. Prog.,* **62** (9), 56 (1966).

14. Misek, T., and J. Marek, *Br. Chem. Eng.,* **15,** 202 (1970).

15. Fischer, A., *Verfahrenstechnik,* **5,** 360 (1971).

16. Karr, A.E., *AIChE J.,* **5,** 446 (1959).

17. Karr, A.E., and T.C. Lo, *Chem. Eng. Prog.,* **72** (11), 68 (1976).

18. Prochazka, J., J. Landau, F. Souhrada, and A. Heyberger, *Br. Chem. Eng.,* **16,** 42 (1971).

19. Barson, N., and G.H. Beyer, *Chem. Eng. Prog.,* **49** (5), 243–252 (1953).

20. Reissinger, K.-H., and J. Schroeter, "Liquid–Liquid Extraction, Equipment Choice," in J.J. McKetta and W.A. Cunningham, Eds., *Encyclopedia of Chemical Processing and Design,* Vol. 21, Marcel Dekker, New York (1984).

21. Cusack, R.W., P. Fremeaux, and D. Glatz, *Chem. Eng.,* **98** (2), 66–76 (1991).

22. Robbins, L.A., *Chem. Eng. Prog.,* **76** (10), 58–61 (1980).

23. Naser, S.F., and R.L. Fournier, *Comput. Chem. Eng.,* **15,** 397–414 (1991).

24. Darwent, B., and C.A. Winkler, *J. Phys. Chem.,* **47,** 442–454 (1943).

25. Treybal, R.E., *Liquid Extraction,* 2nd ed., McGraw-Hill, New York (1963).

26. Hunter, T.G., and A.W. Nash, *J. Soc. Chem. Ind.,* **53,** 95T–102T (1934).

27. Kinney, G.F., *Ind. Eng. Chem.,* **34,** 1102–1104 (1942).

28. Venkataranam, A., and R.J. Rao, *Chem. Eng. Sci.,* **7,** 102–110 (1957).

29. Sawistowski, H., and W. Smith, *Mass Transfer Process Calculations,* Interscience, New York (1963).

30. Varteressian, K.A., and M.R. Fenske, *Ind. Eng. Chem.,* **28,** 1353–1360 (1936).

31. Skelland, A.H.P., *Ind. Eng. Chem.,* **53,** 799–800 (1961).

32. Randall, M., and B. Longtin, *Ind. Eng. Chem.,* **30,** 1063, 1188, 1311 (1938); **31,** 908, 1295 (1939); **32,** 125 (1940).

33. Maloney, J.O., and A.E. Schubert, *Trans. AIChE,* **36,** 741 (1940).

34. Flynn, A.W. and R.E. Treybal, *AIChE J.,* **1,** 324–328 (1955).

35. Ryon, A.D., F.L. Daley, and R.S. Lowrie, *Chem. Eng. Prog.,* **55** (10), 70–75 (1959).

36. Happel, J., and D.G. Jordan, *Chemical Process Economics,* 2nd ed., Marcel Dekker, New York (1975).

37. Rushton, J.H., and J.Y. Oldshue, *Chem Eng. Prog.,* **49,** 161–168 (1953).

38. Laity, D.S., and R.E. Treybal, *AIChE J.,* **3,** 176–180 (1957).

39. Skelland, A.H.P., and G.G. Ramsey, *Ind. Eng. Chem. Res.,* **26,** 77–81 (1987).

40. Skelland, A.H.P., and J.M. Lee, *Ind. Eng. Chem. Process Des. Dev.,* **17,** 473–478 (1978).

41. MacMullin, R.B., and M. Weber, *Trans. AIChE,* **31,** 409–458 (1935).

42. Lewis, J.B., I. Jones, and H.R.C. Pratt, *Trans. Inst. Chem. Eng.,* **29,** 126 (1951).

43. Coulson, J.M., and J.F. Richardson, *Chemical Engineering,* Vol. 2, 4th ed., Pergamon, Oxford (1991).

44. Vermuelen, T., G.M. Williams, and G.E. Langlois, *Chem. Eng. Prog.,* **51,** 85F (1955).

45. Gnanasundaram, S., T.E. Degaleesan, and G.S. Laddha, *Can. J. Chem. Eng.,* **57,** 141–144 (1979).

46. Chen, H.T., and S. Middleman, *AIChE J.,* **13,** 989–995 (1967).

47. Sprow, F.B., *AIChE J.,* **13,** 995–998 (1967).

48. Davies, J.T., *Turbulence Phenomena,* Academic Press, New York, p. 311 (1978).

49. Cornish, A.R.H., *Trans. Inst. Chem. Eng.,* **43,** T332–T333 (1965).

50. Skelland, A.H.P., and L.T. Moeti, *Ind. Eng. Chem. Res.,* **29,** 2258–2267 (1990).

51. Batchelor, G.K., *Proc. Cambridge Phil. Soc.,* **47,** 359–374 (1951).

52. Stichlmair, J., *Chemie-Ingenieur-Technik,* **52,** 253 (1980).

53. Logsdail, D.H., J.D. Thornton, and H.R.C. Pratt, *Trans. Inst. Chem. Eng.,* **35,** 301–315 (1957).

54. Landau, J., and R. Houlihan, *Can. J. Chem. Eng.,* **52,** 338–344 (1974).

55. Gayler, R., N.W. Roberts, and H.R.C. Pratt, *Trans. Inst. Chem. Eng.,* **31,** 57–68 (1953).

56. Thornton, J.D., *Chem. Eng. Sci.,* **5,** 201–208 (1956).

57. Reman, G.H., and R.B. Olney, *Chem. Eng. Prog.,* **52** (3), 141–146 (1955).

58. Strand, C.P., R.B. Olney, and G.H. Ackerman, *AIChE J.,* **8,** 252–261 (1962).

59. Reman, G.H., *Chem. Eng. Prog.,* **62** (9), 56–61 (1966).

60. Karr, A.E., and T.C. Lo, "Performance of a 36-inch Diameter Reciprocating-Plate Extraction Column," paper presented at the 82nd National Meeting of AIChE, Atlantic City, NJ (Aug. 29–Sept. 1, 1976).

61. Thornton, J.D., *Science and Practice of Liquid–Liquid Extraction,* Vol. 1, Clarendon Press, Oxford (1992).

62. Strigle, R.F., Jr., *Random Packings and Packed Towers,* Gulf Publishing Company, Houston, TX (1987).

63. Sleicher, C.A., Jr., *AIChE J.,* **5,** 145–149 (1959).

64. Miyauchi, T., and T. Vermeulen, *Ind. Eng. Chem. Fund.,* **2,** 113–126 (1963).

65. Sleicher, C.A., Jr., *AIChE J.,* **6,** 529–531 (1960).

66. Miyauchi, T., and T. Vermeulen, *Ind. Eng. Chem. Fund.,* **2,** 304–310 (1963).

67. Vermeulen, T., J.S. Moon, A. Hennico, and T. Miyauchi, *Chem. Eng Prog.,* **62** (9), 95–101 (1966).

68. Danckwerts, P.V., *Chem. Eng. Sci.,* **2,** 1–13 (1953).

69. Wehner, J.F., and R.H. Wilhelm, *Chem. Eng. Sci.,* **6,** 89–93 (1956).

70. Geankoplis, C.J., and A.N. Hixson, *Ind. Eng. Chem.,* **42,** 1141–1151 (1950).

71. Gier, T.E., and J.O. Hougen, *Ind. Eng. Chem.,* **45,** 1362–1370 (1953).

72. Watson, J.S., and H.D. Cochran, Jr., *Ind. Eng. Chem. Process Des. Dev.,* **10,** 83–85 (1971).

EXERCISES

Section 8.1

8.1 Explain why it is preferable to separate a dilute mixture of benzoic acid in water by liquid–liquid extraction rather than distillation.

8.2 Why is liquid–liquid extraction preferred over distillation for the separation of a mixture of formic acid and water?

8.3 Based on the information in Table 8.3 and the selection scheme in Figure 8.8, is the choice of an RDC appropriate for the extraction of acetic acid from water by ethyl acetate in the process described in the introduction to this chapter and shown in Figure 8.1? What other types of extractors might be considered?

8.4 What is the major advantage of the ARD over the RDC? What is the disadvantage of the ARD compared to the RDC?

8.5 Under what conditions is a cascade of mixer-settler units probably the best choice of extraction equipment?

8.6 A petroleum reformate stream of 4,000 bbl/day is to be contacted with diethylene glycol to extract the aromatics from the paraffins. The ratio of solvent volume to reformate volume is 5. It is estimated that eight theoretical stages will be needed. Using Tables 8.2 and 8.3, and Fig. 8.8, which types of extractors would be most suitable?

Section 8.2

8.7 Using Table 8.4, select possible liquid–liquid extraction solvents for separating the following mixtures: (a) water–ethyl alcohol, (b) water–aniline, and (c) water–acetic acid. For each case, indicate clearly which of the two components should be the solute.

8.8 Using Table 8.4, select possible liquid–liquid extraction solvents for removing the solute from the carrier in the following cases:

	Solute	Carrier
(a)	Acetone	Ethylene glyol
(b)	Toluene	*n*-Heptane
(c)	Ethyl alcohol	Glycerine

8.9 For the extraction of acetic acid (A) from a dilute solution in water (C) into ethyl acetate (S) at 25°C, estimate or obtain data for $(K_A)_D$, $(K_C)_D$, $(K_S)_D$, and β_{AC}. Does this system exhibit: (a) High selectivity, (b) High solvent capacity and (c) Ease in recovering the solvent? Can you select a solvent that would exhibit better factors than ethyl acetate?

8.10 Interfacial tension can be an important factor in liquid–liquid extraction. Very low values of interfacial tension result in stable emulsions that are difficult to separate, while very high values require large energy inputs to form the dispersed phase. It is best to measure the interfacial tension for the two-phase mixture of interest. However, in the absence of experimental data, propose a method for estimating the interfacial tension of a ternary system using only the compositions of the equilibrium phases and the values of surface tension in air for each of the three components.

Section 8.3

8.11 One thousand kilograms per hour of a 45 wt% acetone-in-water solution is to be extracted at 25°C in a continuous countercurrent system with pure 1,1,2-trichloroethane to obtain a raffinate containing 10 wt% acetone. Using the following equilibrium data, determine with an equilateral triangle diagram:
(a) the minimum flow rate of solvent,
(b) the number of stages required for a solvent rate equal to 1.5 times the minimum, and
(c) the flow rate and composition of each stream leaving each stage.

	Acetone, Weight Fraction	Water, Weight Fraction	Trichloroethane, Weight Fraction
Extract	0.60	0.13	0.27
	0.50	0.04	0.46
	0.40	0.03	0.57
	0.30	0.02	0.68
	0.20	0.015	0.785
	0.10	0.01	0.89
Raffinate	0.55	0.35	0.10
	0.50	0.43	0.07
	0.40	0.57	0.03
	0.30	0.68	0.02
	0.20	0.79	0.01
	0.10	0.895	0.005

The tie-line data are:

Raffinate, Weight Fraction Acetone	Extract, Weight Fraction Acetone
0.44	0.56
0.29	0.40
0.12	0.18

8.12 Solve Exercise 8.11 with a right-triangle diagram.

8.13 A distillate containing 45 wt% isopropyl alcohol, 50 wt% diisopropyl ether, and 5 wt% water is obtained from the heads column of an isopropyl alcohol finishing unit. The company desires to recover the ether from this stream by liquid–liquid extraction in a column, with water, as the solvent, entering the top and the feed entering the bottom so as to produce an ether containing no more than 2.5 wt% alcohol and to obtain the extracted alcohol at a concentration of at least 20 wt%. The unit will operate at 25°C and 1 atm. Using the method of Varteressian and Fenske with a McCabe–Thiele diagram, find how many theoretical stages are required.

Is it possible to obtain an extracted alcohol composition of 25 wt%? Equilibrium data are given below.

PHASE EQUILIBRIUM DATA AT 25°C, 1 ATM

Ether Phase			Water Phase		
Wt% Alcohol	Wt% Ether	Wt% Water	Wt% Alcohol	Wt% Ether	Wt% Water
2.4	96.7	0.9	8.1	1.8	90.1
3.2	95.7	1.1	8.6	1.8	89.6
5.0	93.6	1.4	10.2	1.5	88.3
9.3	88.6	2.1	11.7	1.6	86.7
24.9	69.4	5.7	17.5	1.9	80.6
38.0	50.2	11.8	21.7	2.3	76.0
45.2	33.6	21.2	26.8	3.4	69.8

ADDITIONAL POINTS ON PHASE BOUNDARY

Wt% Alcohol	Wt% Ether	Wt% Water
45.37	29.70	24.93
44.55	22.45	33.00
39.57	13.42	47.01
36.23	9.66	54.11
24.74	2.74	72.52
21.33	2.06	76.61
0	0.6	99.4
0	99.5	0.5

8.14 Benzene and trimethylamine (TMA) are to be separated in a three-stage liquid–liquid extraction column using water as the solvent. If the solvent-free extract and raffinate products are to contain, respectively, 70 and 3 wt% TMA, find the original feed composition and the water-to-feed ratio with a right-triangle diagram. There is no reflux and the solvent is pure water. Equilibrium data are as follows:

TRIMETHYLAMINE–WATER–BENZENE COMPOSITIONS ON PHASE BOUNDARY

Extract, wt%			Raffinate, wt%		
TMA	H₂O	Benzene	TMA	H₂O	Benzene
5.0	94.6	0.04	5.0	0.0	95.0
10.0	89.4	0.06	10.0	0.0	90.0
15.0	84.0	1.0	15.0	1.0	84.0
20.0	78.0	2.0	20.0	2.0	78.0
25.0	72.0	3.0	25.0	4.0	71.0
30.0	66.4	3.6	30.0	7.0	63.0
35.0	58.0	7.0	35.0	15.0	50.0
40.0	47.0	13.0	40.0	34.0	26.0

The tie-line data are:

Extract, wt% TMA	Raffinate, wt% TMA
39.5	31.0
21.5	14.5
13.0	9.0
8.3	6.8
4.0	3.5

8.15 The system docosane–diphenylhexane (DPH)–furfural is representative of more complex systems encountered in the solvent refining of lubricating oil. Five hundred kilograms per hour of a 40 wt% mixture of DPH in docosane are to be continuously extracted in a countercurrent system with 500 kg/h of a solvent containing 98 wt% furfural and 2 wt% DPH to produce a raffinate that contains only 5 wt% DPH. Calculate with a right-triangle diagram the number of theoretical stages required and the number of kilograms per hour of DPH in the extract at 45°C and at 80°C. Equilibrium data are as follows.

EQUILIBRIUM DATA: BINODAL CURVES IN DOCOSANE–DIPHENYLHEXANE–FURFURAL SYSTEM [*IND. ENG. CHEM.*, 35, 711 (1943)]

Wt% at 45°C			Wt% at 80°C		
Docosane	DPH	Furfural	Docosane	DPH	Furfural
96.0	0.0	4.0	90.3	0.0	9.7
84.0	11.0	5.0	50.5	29.5	20.0
67.0	26.0	7.0	34.2	35.8	30.0
52.5	37.5	10.0	23.8	36.2	40.0
32.6	47.4	20.0	16.2	33.8	50.0
21.3	48.7	30.0	10.7	29.3	60.0
13.2	46.8	40.0	6.9	23.1	70.0
7.7	42.3	50.0	4.6	15.4	80.0
4.4	35.6	60.0	3.0	7.0	90.0
2.6	27.4	70.0	2.2	0.0	97.8
1.5	18.5	80.0			
1.0	9.0	90.0			
0.7	0.0	99.3			

The tie lines in the docosane-diphenylhexane-furfural system are:

Docosane Phase Composition, wt%			Furfural Phase Composition, wt%		
Docosane	DPH	Furfural	Docosane	DPH	Furfural
Temperature, 45°C:					
85.2	10.0	4.8	1.1	9.8	89.1
69.0	24.5	6.5	2.2	24.2	73.6
43.9	42.6	13.3	6.8	40.9	52.3
Temperature, 80°C:					
86.7	3.0	10.3	2.6	3.3	94.1
73.1	13.9	13.0	4.6	15.8	79.6
50.5	29.5	20.0	9.2	27.4	63.4

8.16 For each of the ternary systems shown in Figure 8.43, indicate whether: (a) simple countercurrent extraction, or (b) countercurrent extraction with extract reflux, or (c) countercurrent extraction with raffinate reflux, or (d) countercurrent extraction with both extract and raffinate reflux would be expected to yield the most economical process.

8.17 Two solutions, feed F at the rate of 7,500 kg/h containing 50 wt% acetone and 50 wt% water, and feed F' at the rate of 7,500 kg/h containing 25 wt% acetone and 75 wt% water, are to be extracted in a countercurrent system with 5,000 kg/h of 1,1,2-trichloroethane at 25°C to give a raffinate containing 10 wt% acetone. Calculate the number of equilibrium stages required and the stage to which each feed should be introduced, using a right-triangle diagram. Equilibrium data are given in Exercise 8.11.

8.18 The three-stage extractor shown in Figure 8.44 is used to extract the amine from a fluid consisting of 40 wt% benzene (B) and 60 wt% trimethylamine (T). The solvent (water)

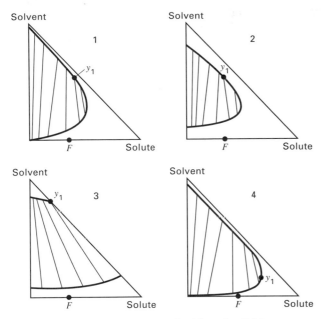

Figure 8.43 Data for Exercise 8.16.

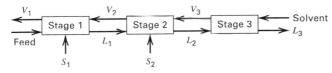

Figure 8.44 Data for Exercise 8.18.

flow to stage 3 is 5,185 kg/h and the feed flow rate is 10,000 kg/h. On a solvent-free basis V_1 is to contain 76 wt% T and L_3 is to contain 3 wt% T. Determine the required solvent flow rates S_1 and S_2 using an equilateral-triangle diagram. Equilibrium data are given in Exercise 8.14.

8.19 The extraction process shown Figure 8.45 is conducted in a multiple-feed countercurrent unit without extract or raffinate reflux. Feed F' is composed of solvent and solute, and is an extract-phase feed. Feed F'' is composed of unextracted raffinate and solute and is a raffinate phase feed. Derive the equations required to establish the three reference points needed to step off the theoretical stages in the extraction column. Show the graphical determination of these points on a right-triangle graph.

8.20 A mixture containing 50 wt% methylcyclohexane (MCH) in *n*-heptane is fed to a countercurrent stage-type extractor at 25°C. Aniline is used as solvent. Reflux is used on both ends of the column.

An extract containing 95 wt% MCH and a raffinate containing 5 wt% MCH (both on solvent-free basis) are required. The minimum extract reflux ratio is 3.49. Using a right-triangle diagram with the equilibrium data of Exercise 8.22 below, calculate: (a) the raffinate reflux ratio, (b) the amount of aniline that must be removed at the separator "on top" of the column, and (c) the amount of solvent that must be added to the solvent mixer at the bottom of the column.

8.21 In its natural state, zirconium, which is an important material of construction for nuclear reactors, is associated with hafnium, which has an abnormally high neutron-absorp-

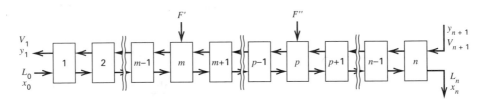

Figure 8.45 Data for Exercise 8.19.

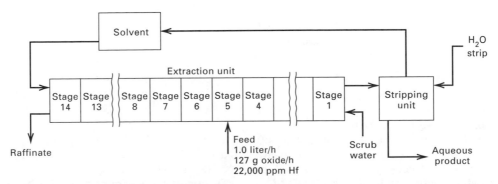

Figure 8.46 Data for Exercise 8.21.

tion cross section and must be removed before the zirconium can be used. Refer to Figure 8.46 for a proposed liquid–liquid extraction process wherein tributyl phosphate (TBP) is used as a solvent for the separation of hafnium from zirconium.

One liter per hour of 5.10 N HNO_3 containing 127 g of dissolved Hf and Zr oxides per liter is fed to stage 5 of the 14-stage extraction unit. The feed contains 22,000 g Hf per million g of Zr. Fresh TBP enters stage 14, while scrub water is fed to stage 1. Raffinate is removed at stage 14, while the organic extract phase which is removed at stage 1 goes to a stripping unit. The stripping operation consists of a single contact between fresh water and the organic phase. The following table gives experimental data. (a) Use these data to fashion a complete material balance for the process. (b) Check the data for consistency in as many ways as you can. (c) What is the advantage of running the extractor as shown? Would you recommend that all the stages be used?

STAGEWISE ANALYSES OF MIXER-SETTLER RUN

	Organic Phase			Aqueous Phase		
Stage	g oxide/ liter	N HNO_3	(Hf/Zr) ×(100)	g oxide/ liter	N HNO_3	(Hf/Zn) ×(100)
1	22.2	1.95	<0.010	17.5	5.21	<0.010
2	29.3	2.02	<0.010	27.5	5.30	<0.010
3	31.4	2.03	<0.010	33.5	5.46	<0.010
4	31.8	2.03	0.043	34.9	5.46	0.24
5	32.2	2.03	0.11	52.8	5.15	3.6
6	21.1	1.99	0.60	30.8	5.15	6.8
7	13.7	1.93	0.27	19.9	5.05	9.8
8	7.66	1.89	1.9	11.6	4.97	20
9	4.14	1.86	4.8	8.06	4.97	36
10	1.98	1.83	10	5.32	4.75	67
11	1.03	1.77	23	3.71	4.52	110
12	0.66	1.68	32	3.14	4.12	140
13	0.46	1.50	42	2.99	3.49	130
14	0.29	1.18	28	3.54	2.56	72
Stripper		0.65		76.4	3.96	<0.01

[Data From R.P. Cox, H.C. Peterson, and C.H. Beyer, *Ind. Eng. Chem.,* **50** (2), 141 (1958). Exercise adapted from E.J. Henley and H. Bieber, *Chemical Engineering Calculations,* McGraw-Hill, New York, p. 298 (1959).]

Section 8.4

8.22 A feed mixture containing 50 wt% n-heptane and 50 wt% methylcyclohexane (MCH) is to be separated by liquid–liquid extraction into one product containing 92.5 wt% methylcyclohexane and another containing 7.5 wt% methylcyclohexane, both on a solvent-free basis. Aniline will be used as the solvent. Using the equilibrium data given below and the graphical method of Maloney and Schubert: (a)

What is the minimum number of theoretical stages necessary to effect this separation? (b) What is the minimum extract reflux ratio? (c) If the reflux ratio is 7.0, how many theoretical contacts are required?

LIQUID–LIQUID EQUILIBRIUM DATA FOR THE SYSTEM n-HEPTANE/METHYLCYCLOHEXANE/ANILINE AT 25°C AND AT 1 ATM (101 kPa)

Hydrocarbon Layer		Solvent Layer	
Weight Percent MCH, Solvent-Free Basis	Pounds Aniline/ Pound Solvent-Free Mixture	Weight Percent MCH, Solvent-Free Basis	Pounds Aniline/ Pound Solvent-Free Mixture
0.0	0.0799	0.0	15.12
9.9	0.0836	11.8	13.72
20.2	0.087	33.8	11.5
23.9	0.0894	37.0	11.34
36.9	0.094	50.6	9.98
44.5	0.0952	60.0	9.0
50.5	0.0989	67.3	8.09
66.0	0.1062	76.7	6.83
74.6	0.1111	84.3	6.45
79.7	0.1135	88.8	6.0
82.1	0.116	90.4	5.9
93.9	0.1272	96.2	5.17
100.0	0.135	100.0	4.92

8.23 Two liquids, A and B, which have nearly identical boiling points, are to be separated by liquid–liquid extraction with solvent C. The following data represent the equilibrium between the two liquid phases at 95°C.

EQUILIBRIUM DATA, WT%

Extract Layer			Raffinate Layer		
A, %	B, %	C, %	A, %	B, %	C, %
0	7.0	93.0	0	92.0	8.0
1.0	6.1	92.9	9.0	81.7	9.3
1.8	5.5	92.7	14.9	75.0	10.1
3.7	4.4	91.9	25.3	63.0	11.7
6.2	3.3	90.5	35.0	51.5	13.5
9.2	2.4	88.4	42.0	41.0	17.0
13.0	1.8	85.2	48.1	29.3	22.6
18.3	1.8	79.9	52.0	20.0	28.0
24.5	3.0	72.5	47.1	12.9	40.0
31.2	5.6	63.2		Plait point	

[Adapted from McCabe and Smith, *Unit Operations of Chemical Engineering,* 4th ed., McGraw-Hill, New York, p. 557 (1985).]

Determine the minimum amount of reflux that must be returned from the extract product to produce an extract containing 83% A and 17% B (on a solvent-free basis) and a raffinate product containing 10% A and 90% B (solvent-free basis). The feed contains 35% A and 65% B on a solvent-free basis and is a saturated raffinate. The raffinate is the heavy liquid. Determine the number of ideal stages on both sides of the feed required to produce the same end products from the same feed when the reflux ratio of the extract, expressed as pounds of extract reflux per pound of extract product (including solvent), is twice the minimum. Calculate the masses of the various streams per 1,000 lb of feed, all on a solvent-free basis. Solve the problem using equilateral-triangle coordinates, right-triangle coordinates, and solvent-free coordinates. Which method is best for this exercise?

8.24 Solve Exercise 8.20 by the graphical method of Maloney and Schubert.

Section 8.5

8.25 Acetic acid is continuously extracted from a 3 wt% dilute solution in water with a solvent of isopropyl ether in a mixer-settler unit. The flow rates of the feed and solvent are 12,400 and 24,000 lb/h, respectively. Assuming a residence time of 1.5 min in the mixer and a settling vessel capacity of 4 gal/min-ft^2, estimate: (a) Diameter and height of the mixing vessel, assuming $H/D_T = 1$, (b) Agitator horsepower for the mixing vessel, (c) Diameter and length of the settling vessel, assuming $L/D_T = 4$, and (d) Residence time in minutes in the settling vessel.

8.26 A cascade of six mixer-settler units is available, each unit consisting of a 10-ft-diameter by 10-ft-high mixing vessel equipped with a 20-hp agitator, and a 10-ft diameter by 40-ft-long settling vessel. If this cascade is used for the acetic acid extraction described in the introduction to this chapter, estimate the pounds per hour of feed that could be processed.

8.27 Acetic acid is to be extracted from a dilute aqueous solution with isopropyl ether at 25°C in a countercurrent cascade of mixer-settler units. In one of the units, the following conditions apply:

	Raffinate	Extract
Flow rate, lb/h	21,000	52,000
Density, lb/ft^3	63.5	45.3
Viscosity, cP	3.0	1.0

Interfacial tension = 13.5 dyne/cm. If the raffinate is the dispersed phase and the mixer residence time is 2.5 minutes, estimate for the mixer: (a) The dimensions of a closed, baffled vessel, (b) The diameter of a flat-bladed impeller, (c) The minimum rate of rotation in revolutions per minute of the impeller for complete and uniform dispersion and (d) The power requirement of the agitator at the minimum rate of rotation.

8.28 For the conditions of Exercise 8.27, estimate: (a) Sauter mean drop size, (b) Range of drop sizes, and (c) Interfacial area of the two-phase liquid–liquid emulsion.

8.29 For the conditions of Exercises 8.27 and 8.28, and the additional data given below, estimate: (a) The dispersed-phase mass transfer coefficient, (b) The continuous-phase mass-transfer coefficient, (c) The Murphree dispersed-phase efficiency, and (d) The fraction of acetic acid extracted.
Additional data:
Diffusivity of acetic acid: in the raffinate, 1.3×10^{-9} m^2/s and in the extract, 2.0×10^{-9} m^2/s.
Distribution coefficient for acetic acid: $c_D/c_C - 2.7$

8.30 For the conditions and results of Example 8.4, involving the extraction of benzoic acid, from a dilute solution in water with toluene, determine the following when using a six-flat-blade turbine impeller in a closed vessel with baffles and with the extract phase dispersed, based on the physical properties given: (a) The minimum rate of rotation of the impeller for complete and uniform dispersion, (b) The power requirement of the agitator at the minimum rotation rate, (c) The Sauter mean droplet diameter, (d) The interfacial area, (e) The overall mass transfer coefficient, K_{OD}, (f) The number of overall transfer units, N_{OD}, (g) The Murphree efficiency, E_{MD}, and (h) The fractional extraction of benzoic acid.
Liquid properties are:

	Raffinate Phase	**Extract Phase**
Density, g/cm^3	0.995	0.860
Viscosity, cP	0.95	0.59
Diffusivity of benzoic acid, cm^2/s	2.2×10^{-5}	1.5×10^{-5}

Interfacial tension = 22 dyne/cm
Distribution coefficient for benzoic acid = $c_D/c_C = 21$

8.31 Estimate the diameter of an RDC column to extract acetic acid from water with isopropyl ether for the conditions and data of Exercises 8.25 and 8.27.

8.32 Estimate the diameter of a Karr column to extract benzoic acid from water with toluene for the conditions of Exercise 8.30.

8.33 Estimate the value of HETS for an RDC column operating under the conditions of Exercise 8.31.

8.34 Estimate the value of HETS for a Karr column operating under the conditions of Exercise 8.32.

Chapter 9

Approximate Methods for Multicomponent, Multistage Separations

Although rigorous computer methods, discussed in Chapter 10, are available for solving multicomponent separation problems, approximate methods continue to be used in practice for various purposes, including preliminary design, parametric studies to establish optimum design conditions, and process synthesis studies to determine optimal separation sequences.

In Section 5.4, the approximate methods of Kremser [1] for absorbers, and Edmister [2] for distillation are discussed. This chapter presents an additional approximate method that is widely used for making preliminary designs and optimization of simple distillation. The method is commonly referred to as the *Fenske–Underwood–Gilliland* or FUG method. In addition, application of the Kremser method is extended to and illustrated for strippers and liquid–liquid extraction. Although these methods can be applied fairly readily by manual calculation if physical properties are independent of composition, computer calculations are preferred, and FUG models are included in most computer-aided process design programs.

9.1 FENSKE–UNDERWOOD–GILLILAND METHOD

An algorithm for the empirical Fenske–Underwood–Gilliland method, named after the authors of the three important steps in the procedure, is shown in Figure 9.1 for a simple distillation column of the type shown in Figure 9.3. The column can be equipped with a partial or total condenser. From Table 5.2, the number of degrees of freedom with a total condenser is $2N + C + 9$. In this case, the following variables are generally specified with the partial reboiler counted as a theoretical stage:

Number of Specifications

Feed flow rate	1
Feed mole fractions	$C - 1$
Feed temperature[1]	1
Feed pressure[1]	1
Adiabatic stages (excluding reboiler)	$N - 1$
Stage pressures (including reboiler)	N
Split of light key component	1
Split of heavy key component	1
Feed stage location	1
Reflux ratio (as multiple of minimum reflux)	1
Reflux temperature	1
Adiabatic reflux divider	1
Pressure of total condenser	1
Pressure at reflux divider	1
	$2N + C + 9$

Similar specifications can be written for columns with a partial condenser.

Selection of Two Key Components

For multicomponent feeds, specification of two key components and their distribution between distillate and bottoms is accomplished in a variety of ways. Preliminary estimation of the distribution of nonkey components can be sufficiently difficult to require the iterative procedure indicated in Figure 9.1. However, generally only two and seldom more than three iterations are necessary.

Consider the multicomponent hydrocarbon feed in Figure 9.2. This mixture is typical of the feed to the recovery section of an alkylation plant [3]. Components are listed in order of decreasing volatility. A sequence of distillation columns including a deisobutanizer and a debutanizer is to be used to separate this mixture into the three products indicated. In case 1 of Table 9.1, the deisobutanizer is selected as the first column in the sequence. Since the allowable quantities of n-butane in the isobutane recycle, and isobutane in the n-butane product, are specified, isobutane is the light key and n-butane is the heavy key. These two keys are adjacent in order of volatility. Because a fairly sharp separation between these two keys is indicated and the nonkey components are not close in volatility to the butanes, as a preliminary estimate we can assume the separation of the nonkey components to be perfect.

Alternatively, in case 2, if the debutanizer is placed first in the sequence, specifications in Figure 9.2 require that n-butane be selected as the light key. However, selection of the heavy key is uncertain because no recovery or purity is specified for any component less volatile than n-butane. Possible heavy key components for the debutanizer are iC_5, nC_5, or C_6. The simplest procedure is to select iC_5 so that the two keys are again adjacent.

For example, suppose we specify that 13 lbmol/h of iC_5 in the feed is allowed to appear in the distillate. Because the split of iC_5 is not sharp and nC_5 is close in volatility to iC_5, it is probable that the quantity of nC_5 in the distillate will not be negligible. A preliminary estimate of the distributions of the nonkey components for case 2 is given in Table 9.1. Although iC_4 may also distribute, a preliminary estimate of zero is made for the bottoms quantity.

---

[1] Feed temperature and pressure may correspond to known stream conditions leaving the previous piece of equipment.

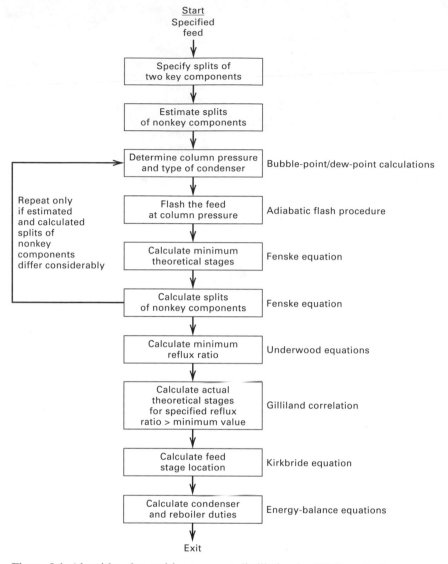

Figure 9.1 Algorithm for multicomponent distillation by FUG method.

Finally, in case 3, we select C_6 as the heavy key for the debutanizer at a specified rate of 0.01 lbmol/h in the distillate, as shown in Table 9.1. Now iC_5 and nC_5 will distribute between the distillate and bottoms in amounts to be determined; as a preliminary estimate, we assume the same distribution as in case 2.

In practice, the deisobutanizer is usually placed first in the sequence. In Table 9.1, the bottoms for case 1 then becomes the feed to the debutanizer, for which, if nC_4 and iC_5 are selected as the key components, component separation specifications for the debutanizer are as indicated in Figure 9.3 with preliminary estimates of the separation of nonkey components shown in parentheses. This separation has been treated by Bachelor [4]. Because nC_4 and C_8 comprise 82.2 mol% of the feed and differ widely in volatility, the temperature difference between distillate and bottoms is likely to be large. Furthermore, the light key split is rather sharp, but the heavy key split is not. As will be shown later, this case provides a relatively severe test of the empirical design procedure discussed in this section.

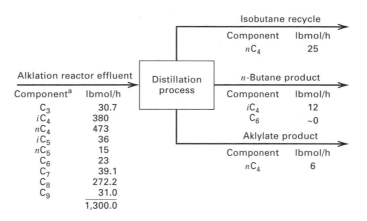

^{a} C_6, C_7, C_8, C_9 are taken as normal paraffins.

Figure 9.2 Separation specifications for alkylation reactor effluent.

Table 9.1 Specifications of Key Component Splits and Preliminary Estimation of Nonkey Component Splits for Alkylation Reactor Effluent

Component	Feed, lbmol/h	Case 1, Deisobutanizer Column First, lbmol/h		Case 2, Debutanizer Column First (iC_5 is HK), lbmol/h		Case 3, Debutanizer Column First (C_6 is HK), lbmol/h	
		Distillate	Bottoms	Distillate	Bottoms	Distillate	Bottoms
C_3	30.7	(30.7)	(0)	(30.7)	(0)	(30.7)	(0)
iC_4	380	368^a	12^b	(380.0)	(0)	(380.0)	(0)
nC_4	473	25^b	448^a	467^a	6^b	467^a	6^b
iC_5	36	(0)	(36)	13^b	23^a	(13)	(23)
nC_5	15	(0)	(15)	(1)	(14)	(1)	(14)
C_6	23	(0)	(23)	(0)	(23)	0.01^b	22.99^a
C_7	39.1	(0)	(39.1)	(0)	(39.1)	(0)	(39.1)
C_8	272.2	(0)	(272.2)	(0)	(272.2)	(0)	(272.2)
C_9	31.0	(0)	(31.0)	(0)	(31.0)	(0)	(31.0)
	1,300.0	423.7	876.3	891.7	408.3	891.71	408.29

^{a} By material balance.

^{b} Specification.

(Preliminary estimate.)

Column Operating Pressure

For preliminary design, column operating pressure and type of condenser can be established by the procedure discussed in Section 7.2 and shown in Figure 7.16, as illustrated in the following example. With column operating pressure established, the column feed can be flashed adiabatically at an estimated feed-tray pressure to determine feed-phase condition.

EXAMPLE 9.1

Determine column operating pressures and type of condenser for the debutanizer of Figure 9.3.

SOLUTION

Using the estimated distillate composition in Figure 9.3, we compute the distillate bubble-point pressure at 120°F (48.9°C) iteratively from (4-12) in a manner similar to Example 4.4. This

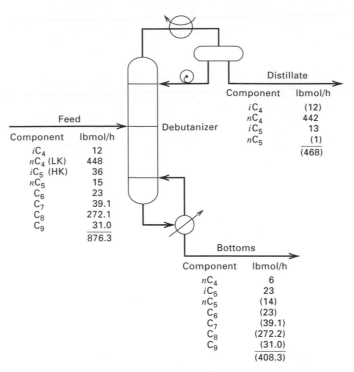

Figure 9.3 Specifications for debutanizer.

procedure gives 79 psia as the reflux drum pressure. Thus, a total condenser is indicated. Allowing a 2-psi condenser pressure drop, column top pressure is (79 + 2) = 81 psia; and, allowing a 5-psi pressure drop through the column, the bottoms pressure is (81 + 5) = 86 psia. Assume a feed tray pressure midway between the column top and bottom pressures or 83.5 psia.

Bachelor [4] sets column pressure at 80 psia throughout. He obtains a distillate temperature of 123°F. A bubble-point calculation for the bottoms composition at 80 psia gives 340°F. This temperature is sufficiently low to prevent decomposition.

Feed to the debutanizer is presumably bottoms from a deisobutanizer operating at a pressure of perhaps 100 psia or more. Results of an adiabatic flash of this feed, by the procedure of Section 4.4, to 80 psia are given by Bachelor [4] as follows.

	Pound-Moles per hour	
Component	**Vapor Feed**	**Liquid Feed**
iC_4	3.3	8.7
nC_4	101.5	346.5
iC_5	4.6	31.4
nC_5	1.6	13.4
nC_6	1.3	21.7
nC_7	1.2	37.9
nC_8	3.2	269.0
nC_9	0.2	30.8
	116.9	759.4

The temperature of the flashed feed is 180°F (82.2°C). From above, the feed mole fraction vaporized is (116.9/876.3) = 0.1334. ∎

Fenske Equation for Minimum Equilibrium Stages

For a specified separation between two key components of a multicomponent mixture, an exact expression is easily developed for the required minimum number of equilibrium stages, which corresponds to total reflux. This condition can be achieved in practice by charging the column with feedstock and operating it with no further input of feed and no withdrawal of distillate or bottoms, as illustrated in Figure 9.4. To facilitate derivation of the Fenske equation, stages are numbered from the bottom up. All vapor leaving stage N is condensed and returned to stage N as reflux. All liquid leaving stage 1 is vaporized and returned to stage 1 as boilup. For steady-state operation within the column, heat input to the reboiler and heat output from the condenser are made equal (assuming no heat losses). Then, by a material balance, vapor and liquid streams passing between any pair of stages have equal flow rates and compositions, for example, $V_{N-1} = L_N$ and $y_{i,N-1} = x_{i,N}$. However, vapor and liquid flow rates will change from stage to stage unless the assumption of constant molal overflow is valid.

Derivation of an exact equation for the minimum number of equilibrium stages involves only the definition of the K-value and the mole-fraction equality between stages. For component i at stage 1 in Figure 9.4,

$$y_{i,1} = K_{i,1}x_{i,1} \tag{9-1}$$

But for passing streams

$$y_{i,1} = x_{i,2} \tag{9-2}$$

Combining these two equations,

$$x_{i,2} = K_{i,1}x_{i,1} \tag{9-3}$$

Similarly, for stage 2,

$$y_{i,2} = K_{i,2}x_{i,2} \tag{9-4}$$

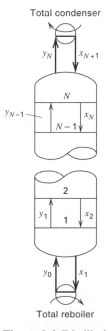

Figure 9.4 Distillation column operation at total reflux.

Combining (9-3) and (9-4), we have

$$y_{i,2} = K_{i,2}K_{i,1}x_{i,1} \tag{9-5}$$

Equation (9-5) is readily extended in this fashion to give

$$y_{i,N} = K_{i,N}K_{i,N-1} \cdots K_{i,2}K_{i,1}x_{i,1} \tag{9-6}$$

Similarly, for component j,

$$y_{j,N} = K_{j,N}K_{j,N-1} \cdots K_{j,2}K_{j,1}x_{j,1} \tag{9-7}$$

Combining (9-6) and (9-7), we find that

$$\frac{y_{i,N}}{y_{j,N}} = \alpha_N \alpha_{N-1} \cdots \alpha_2 \alpha_1 \left(\frac{x_{i,1}}{x_{j,1}}\right) \tag{9-8}$$

or

$$\left(\frac{x_{i,N+1}}{x_{i,1}}\right)\left(\frac{x_{j,1}}{x_{j,N+1}}\right) = \prod_{k=1}^{N_{min}} \alpha_k \tag{9-9}$$

where $\alpha_k = K_{i,k}/K_{j,k}$, the relative volatility between components i and j. Equation (9-9) relates the relative enrichments of any two components i and j over a cascade of N theoretical stages to the stage relative volatilities between the two components. Although (9-9) is exact, it is rarely used in practice because the conditions of each stage must be known to compute the set of relative volatilities. However, if the relative volatility is assumed constant, (9-9) simplifies to

$$\left(\frac{x_{i,N+1}}{x_{i,1}}\right)\left(\frac{x_{j,1}}{x_{j,N+1}}\right) = \alpha^N \tag{9-10}$$

or

$$N_{min} = \frac{\log\{[(x_{i,N+1})/x_{i,1}][x_{j,1}/(x_{j,N+1})]\}}{\log \alpha_{i,j}} \tag{9-11}$$

Equation (9-11) is extremely useful. It is referred to as the *Fenske equation* [5]. When i = the light key (LK) and j = the heavy key (HK), the minimum number of equilibrium stages is influenced by the nonkey components only by their effect (if any) on the value of the relative volatility between the key components.

Equation (9-11) permits a rapid estimation of minimum equilibrium stages. A more convenient form of (9-11) is obtained by replacing the product of the mole-fraction ratios by the equivalent product of mole-distribution ratios in terms of component distillate and bottoms flow rates d and b, respectively,[2] and by replacing the relative volatility by a geometric mean of the top-stage and bottom-stage values. Thus,

$$N_{min} = \frac{\log[(d_i/d_j)(b_j/b_i)]}{\log \alpha_m} \tag{9-12}$$

where the mean relative volatility is approximated by

$$\alpha_m = [(\alpha_{i,j})_N(\alpha_{i,j})_1]^{1/2} \tag{9-13}$$

Thus, the minimum number of equilibrium stages depends on the degree of separation of the two key components and their relative volatility, but is independent of feed-phase condition. Equation (9-12) in combination with (9-13) is exact for two minimum stages.

---

[2] This substitution is valid even though no distillate or bottoms products are withdrawn at total reflux.

For one stage, it is equivalent to the equilibrium flash equation. In practice, distillation columns are designed for separations corresponding to as many as 150 minimum equilibrium stages.

The Fenske equation is quite reliable except when the relative volatility varies appreciably over the column, and/or when the mixture forms nonideal liquid solutions. In those cases, if the Fenske equation is applied with (9-13), it should be done with great caution, and should be followed by rigorous calculations of the type in Chapter 10.

EXAMPLE 9.2

For the debutanizer shown in Figure 9.3 and considered in Example 9.1, estimate the minimum equilibrium stages by the Fenske equation. Assume uniform operating pressure of 80 psia (552 kPA) throughout and utilize the ideal K-values given by Bachelor [4] as plotted in Figure 9.5.

SOLUTION

The two key components are n-butane and isopentane. Distillate and bottoms conditions based on the estimated product distributions for nonkey components in Figure 9.3 are

Component	$x_{N+1} = x_D$	$x_1 = x_B$
iC_4	0.0256	~ 0
nC_4 (LK)	0.9445	0.0147
iC_5 (HK)	0.0278	0.0563
nC_5	0.0021	0.0343
nC_6	~ 0	0.0563
nC_7	~ 0	0.0958
nC_8	~ 0	0.6667
nC_9	~ 0	0.0759
	1.0000	1.0000

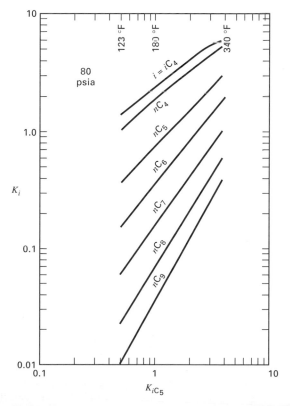

Figure 9.5 Ideal K-values for hydrocarbons at 80 psia.

From Figure 9.5, at 123°F, the assumed top-stage temperature is

$$(\alpha_{nC_4,iC_5})_N = 1.03/0.495 = 2.08$$

At 340°F, the assumed bottom-stage temperature is

$$(\alpha_{nC_4,iC_5})_1 = 5.20/3.60 = 1.44$$

From (9-13),

$$\alpha_m = [(2.08)(1.44)]^{1/2} = 1.73$$

Noting that $(d_i/d_j) = (x_{D_i}/x_{D_j})$ and $(b_i/b_j) = (x_{B_i}/x_{B_j})$, (9-12) becomes

$$N_{\min} = \frac{\log[(0.9445/0.0278)(0.0563/0.0147)]}{\log 1.73} = 8.88 \text{ stages} \qquad \blacksquare$$

Distribution of Nonkey Components at Total Reflux

The Fenske equation is not restricted to the two key components. Once $N_{\min}$ is known, (9-12) can be used to calculate mole fractions x_{N+1} and x_1 for all nonkey components. These values provide a first approximation to the actual product distribution when more than the minimum number of stages is employed.

Let i = a nonkey component and j = the heavy key or reference component denoted by r. Then (9-12) becomes

$$\left(\frac{d_i}{b_i}\right) = \left(\frac{d_r}{b_r}\right)(\alpha_{i,r})_m^{N_{\min}} \qquad \textbf{(9-14)}$$

Substituting $f_i = d_i + b_i$ in (9-14) gives

$$b_i = \frac{f_i}{1 + (d_r/b_r)(\alpha_{i,r})_m^{N_{\min}}} \qquad \textbf{(9-15)}$$

$$\text{or } d_i = \frac{f_i(d_r/b_r)(\alpha_{i,r})_m^{N_{\min}}}{1 + (d_r/b_r)(\alpha_{i,r})_m^{N_{\min}}} \qquad \textbf{(9-16)}$$

Equations (9-15) and (9-16) give the distribution of nonkey components at total reflux as predicted by the Fenske equation.

For accurate calculations, (9-15) and (9-16) should be used to compute the smaller of the two quantities b_i and d_i. The other quantity is best obtained by overall material balance.

EXAMPLE 9.3

Estimate the product distributions for nonkey components by the Fenske equation for the conditions of Example 9.2.

SOLUTION

All nonkey relative volatilities are calculated relative to isopentane using the K-values of Figure 9.5.

| | α_{i,iC_5} | | |
Component	123°F	340°F	Geometric Mean
iC_4	2.81	1.60	2.12
nC_5	0.737	0.819	0.777
nC_6	0.303	0.500	0.389
nC_7	0.123	0.278	0.185
nC_8	0.0454	0.167	0.0870
nC_9	0.0198	0.108	0.0463

Based on $N_{min} = 8.88$ stages from Example 9.2 and the above geometric-mean relative volatilities, values of $(\alpha_{i,r})_{m}^{N_{min}}$ are computed relative to isopentane as tabulated below.

From (9-15), using the feed rate specifications in Figure 9.3 for f_i,

$$b_{iC_4} = \frac{12}{1 + (13/23)790} = 0.0268 \text{ lbmol/h}$$

$$d_{iC_4} = f_{iC_4} - b_{iC_4} = 12 - 0.0268 = 11.9732 \text{ lbmol/h}$$

Results of similar calculations for the other nonkey components are included in the following table.

Component	$(\alpha_{i,iC_5})_{m}^{N_{min}}$	d_i	b_i
iC_4	790	11.9732	0.0268
nC_4	130	442.0	6.0
iC_5	1.00	13.0	23.0
nC_5	0.106	0.851	14.149
nC_6	0.000228	0.00297	22.99703
nC_7	3.11×10^{-7}	6.87×10^{-6}	39.1
nC_8	3.83×10^{-10}	5.98×10^{-8}	272.2
nC_9	1.41×10^{-12}	2.48×10^{-11}	31.0
		467.8272	408.4728

∎

Underwood Equations for Minimum Reflux

Minimum reflux is based on the specifications for the degree of separation between two key components. The minimum reflux is finite and feed product withdrawals are permitted. However, a column cannot operate under this condition because of the accompanying requirement of infinite stages. Nevertheless, minimum reflux is a useful limiting condition.

For binary distillation at minimum reflux, as shown in Figure 7.12a, most of the stages are crowded into a constant-composition zone that bridges the feed stage. In this zone, all vapor and liquid streams have compositions essentially identical to those of the flashed feed. This zone constitutes a single *pinch point* or *point of infinitude* as shown in Figure 9.6a. If nonideal phase conditions are such as to create a point of tangency between the equilibrium curve and the operating line in the rectifying section, as shown in Figure 7.12b, the pinch point will occur in the rectifying section as in Figure 9.6b. Alternatively, the single pinch point can occur in the stripping section.

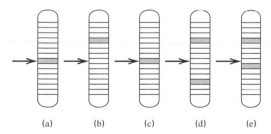

(a)   (b)   (c)   (d)   (e)

Figure 9.6 Location of pinch-point zones at minimum reflux: (a) binary system; (b) binary system, nonideal conditions giving point of tangency; (c) multicomponent system, all components distributed (class 1); (d) multicomponent system, not all LLK and HHK distributing (class 2); (e) multicomponent system, all LLK, if any, distributing, but not all HHK distributing (class 2). (LLK = lighter than light key; HHK = heavier than heavy key.)

Shiras, Hanson, and Gibson [6] classified multicomponent systems as having one (class 1) or two (class 2) pinch points. For class 1 separations, all components in the feed distribute to both the distillate and bottoms products. Then the single pinch point bridges the feed stage as shown in Figure 9.6c. Class 1 separations can occur when narrow-boiling-range mixtures are distilled or when the degree of separation between the key components is not sharp.

For class 2 separations, one or more of the components appear in only one of the products. If neither the distillate nor the bottoms product contains all feed components, two pinch points occur away from the feed stage as shown in Figure 9.6d. Stages between the feed stage and the rectifying-section pinch point remove heavy components that do not appear in the distillate. Light components that do not appear in the bottoms are removed by the stages between the feed stage and the stripping-section pinch point. However, if all feed components appear in the bottoms, the stripping-section pinch point moves to the feed stage as shown in Figure 9.6e.

Consider the general case of a rectifying-section pinch point at or away from the feed stage as shown in Figure 9.7. A component material balance over all stages gives

$$y_{i,\infty}V_\infty = x_{i,\infty}L_\infty + x_{i,D}D \tag{9-17}$$

A total balance over all stages is

$$V_\infty = L_\infty + D \tag{9-18}$$

Since phase compositions do not change in the pinch zone, the phase equilibrium relation is

$$y_{l,\infty} = K_{i,\infty}x_{l,\infty} \tag{9-19}$$

Combining (9-17) to (9-19) for components i and j to eliminate $y_{i,\infty}$, $y_{j,\infty}$ and V_∞; solving for the internal reflux ratio at the pinch point; and substituting $(\alpha_{i,j})_\infty = K_{i,\infty}/K_{j,\infty}$, we have

$$\frac{L_\infty}{D} = \frac{[(x_{i,D}/x_{i,\infty}) - (\alpha_{i,j})_\infty(x_{j,D}/x_{j,\infty})]}{(\alpha_{i,j})_\infty - 1} \tag{9-20}$$

For class 1 separations, flashed feed- and pinch-zone compositions arc identical.[3] Therefore, $x_{i,\infty} = x_{i,F}$ and (9-20) for the light key (LK) and the heavy key (HK) becomes

$$\frac{(L_\infty)_{min}}{F} = \frac{(L_F/F)[(Dx_{LK,D})/(L_Fx_{LK,F}) - (\alpha_{LK,HK})_F(Dx_{HK,D}/L_Fx_{HK,F})]}{(\alpha_{LK,HK})_F - 1} \tag{9-21}$$

---

[3] Assuming the feed is neither subcooled nor superheated.

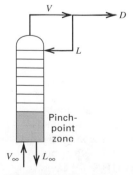

Figure 9.7 Rectifying-section pinch-point zone.

This equation is attributed to Underwood [7] and can be applied to subcooled liquid or superheated vapor feeds by using fictitious values of L_F and $x_{i,F}$ computed by making a flash calculation outside the two-phase region. As with the Fenske equation, (9-21) applies to components other than the key components. Therefore, for a specified split of two key components, the distribution of nonkey components is obtained by combining (9-21) with the analogous equation for component i in place of the light key to give

$$\frac{Dx_{i,D}}{L_F x_{i,F}} = \left[\frac{(\alpha_{i,HK})_F - 1}{(\alpha_{LK,HK})_F - 1}\right]\left(\frac{Dx_{LK,D}}{L_F x_{LK,F}}\right) + \left[\frac{(\alpha_{LK,HK})_F - (\alpha_{i,HK})_F}{(\alpha_{LK,HK})_F - 1}\right]\left(\frac{Dx_{HK,D}}{L_F x_{HK,F}}\right) \quad \textbf{(9-22)}$$

For a class 1 separation,

$$0 < \left(\frac{Dx_{i,D}}{Fx_{i,F}}\right) < 1$$

for all nonkey components. If that is so, the external reflux ratio is obtained from the internal reflux by an enthalpy balance around the rectifying section in the form

$$\frac{(L_{min})_{external}}{D} = (R_{min})_{external} = \frac{(L_\infty)_{min}(h_{V_\infty} - h_{L_\infty}) + D(h_{V_\infty} - h_V)}{D(h_V - h_L)} \quad \textbf{(9-23)}$$

where subscripts V and L refer to vapor leaving the top stage and external liquid reflux, respectively. For conditions of constant molar overflow,

$$(R_{min})_{external} = (L_\infty)_{min}/D$$

Even when (9-21) is invalid, it is useful because, as shown by Gilliland [8], the minimum reflux ratio computed by assuming a class 1 separation is equal to or greater than the true minimum. This is because the presence of distributing nonkey components in the pinch-point zones increases the difficulty of the separation, thus increasing the reflux requirement.

EXAMPLE 9.4

Calculate the minimum internal reflux for the conditions of Example 9.2 assuming a class 1 separation. Check the validity of this assumption.

SOLUTION

From Figure 9.5, the relative volatility between nC_4(LK) and iC_5(HK) at the feed temperature of 180°F is 1.93. Feed liquid and distillate quantities are given in Figure 9.3 and Example 9.1. From (9-21),

$$(L_\infty)_{min} = \frac{759.4[(442/346.5) - 1.93(13/31.4)]}{1.93 - 1} = 389 \text{ lbmol/h}$$

Distribution of nonkey components in the feed is determined by (9-22). The most likely nonkey component to distribute is nC_5 because its volatility is close to that of iC_5(HK), which does not undergo a sharp separation. For nC_5, using data for K-values from Figure 9.5, we have

$$\frac{Dx_{nC_5,D}}{L_F x_{nC_5,F}} = \left[\frac{0.765 - 1}{1.93 - 1}\right]\left(\frac{442}{346.5}\right) + \left[\frac{1.93 - 0.765}{1.93 - 1}\right]\left(\frac{13}{31.4}\right)$$

$$= 0.1963$$

Therefore, $Dx_{nC_5,D} = 0.1963(13.4) = 2.63$ lbmol/h of nC_5 in the distillate. This is less than the quantity of nC_5 in the total feed. Therefore, nC_5 distributes between the distillate and the bottoms. However, similar calculations for the other nonkey components give negative distillate flow rates for the other heavy components and, in the case of iC_4, a distillate flow rate greater than the feed rate. Thus, the computed reflux rate is not valid. However, as expected, it is greater than the true internal value of 298 lbmol/h reported by Bachelor [4]. ∎

For class 2 separations, (9-17) to (9-20) still apply. However, (9-20) cannot be used directly to compute the internal minimum reflux ratio because values of $x_{i,\infty}$ are not simply related to feed composition for class 2 separations. Underwood [9] devised an ingenious algebraic procedure to overcome this difficulty. For the rectifying section, he defined a quantity Φ by

$$\sum \frac{(\alpha_{i,r})_\infty x_{i,D}}{(\alpha_{i,r})_\infty - \Phi} = 1 + (R_\infty)_{\min} \qquad \textbf{(9-24)}$$

Similarly, for the stripping section, Underwood defined Φ' by

$$\sum \frac{(\alpha'_{i,r})_\infty x_{i,B}}{(\alpha'_{i,r})_\infty - \Phi'} = 1 - (R'_\infty)_{\min} \qquad \textbf{(9-25)}$$

where $R'_\infty = L'_\infty/B$ and the prime refers to conditions in the stripping-section pinch-point zone. In his derivation, Underwood assumed that relative volatilities are constant in the region between the two pinch-point zones and that $(R_\infty)_{\min}$ and $(R'_\infty)_{\min}$ are related by the assumption of constant molar overflow in the region between the feed entry and the rectifying-section pinch point and in the region between the feed entry and the stripping-section pinch point. Hence,

$$(L'_\infty)_{\min} - (L_\infty)_{\min} = qF \qquad \textbf{(9-26)}$$

With these two critical assumptions, Underwood showed that at least one common root θ (where $\theta = \Phi = \Phi'$) exists between (9-24) and (9-25).

Equation (9-24) is analogous to the following equation derived from (9-19), and the relation $\alpha_{i,r} = K_i/K_r$,

$$\sum \frac{(\alpha_{i,r})_\infty x_{i,D}}{(\alpha_{i,r})_\infty - L_\infty/[V_\infty(K_r)_\infty]} = 1 + (R_\infty)_{\min} \qquad \textbf{(9-27)}$$

where $L_\infty/[V_\infty(K_r)_\infty]$ is called the *absorption factor* for a reference component in the rectifying-section pinch-point zone. Although Φ is analogous to the absorption factor, a different root of Φ is used to solve for $(R_\infty)_{\min}$, as discussed by Shiras et al. [6].

The common root θ may be determined by multiplying (9-24) and (9-25) by D and B, respectively, adding the two equations, substituting (9-25) to eliminate $(R'_\infty)_{\min}$ and $(R_\infty)_{\min}$, and utilizing the overall component balance $z_{i,F}F = x_{i,D}D + x_{i,B}B$ to obtain

$$\sum \frac{(\alpha_{i,r})_\infty z_{i,F}}{(\alpha_{i,r})_\infty - \theta} = 1 - q \qquad \textbf{(9-28)}$$

where q is the thermal condition of the feed from (7-20) and r is conveniently taken as the heavy key, HK. When only the two key components distribute, (9-28) is solved iteratively for a root of θ that satisfies $\alpha_{LK,HK} > \theta > 1$. The following modification of (9-24) is then solved for the internal reflux ratio $(R_\infty)_{\min}$:

$$\sum \frac{(\alpha_{i,r})_\infty x_{i,D}}{(\alpha_{i,r})_\infty - \theta} = 1 + (R_\infty)_{\min} \qquad \textbf{(9-29)}$$

If any nonkey components are suspected of distributing, estimated values of $x_{i,D}$ cannot be used directly in (9-29). This is particularly true when nonkey components are intermediate in volatility between the two key components. In this case, (9-28) is solved for m roots of θ, where m is one less than the number of distributing components. Furthermore, each root of θ lies between an adjacent pair of relative volatilities of distributing components. For instance, in Example 9.4, it was found the nC_5 distributes at minimum reflux, but nC_6 and heavier do not and iC_4 does not. Therefore, two roots of θ are necessary, where

$$\alpha_{nC_4,iC_5} > \theta_1 > 1.0 > \theta_2 > \alpha_{nC_5,iC_5}$$

With these two roots, (9-29) is written twice and solved simultaneously to yield $(R_\infty)_{min}$ and the unknown value of $x_{nC_5,D}$. The solution must, of course, satisfy the condition $\Sigma x_{i,D} = 1.0$.

With the internal reflux ratio $(R_\infty)_{min}$ known, the external reflux ratio is computed by enthalpy balance with (9-23). This requires a knowledge of the rectifying-section pinch-point compositions. Underwood [9] shows that

$$x_{i,\infty} = \frac{\theta x_{i,D}}{(R_\infty)_{min}[(\alpha_{i,r})_\infty - \theta]} \tag{9-30}$$

with $y_{i,\infty}$ given by (9-17). The value of θ to be used in (9-30) is the root of (9-29) satisfying the inequality

$$(\alpha_{HNK,r})_\infty > \theta > 0$$

where HNK refers to the heaviest nonkey component in the distillate at minimum reflux. This root is equal to $L_\infty/[V_\infty(K_r)_\infty]$ in (9-27). With wide-boiling feeds, the external reflux can be significantly higher than the internal reflux. Bachelor [4] cites a case where the external reflux rate is 55% greater than the internal reflux.

For the stripping-section pinch-point composition, Underwood obtains

$$x'_{i,\infty} = \frac{\theta x_{i,B}}{[(R'_\infty)_{min} + 1][(\alpha_{i,r})_\infty - \theta]} \tag{9-31}$$

where, in this case, θ is the root of (9-29) satisfying the inequality

$$(\alpha_{HNK,r})_\infty > \theta > 0$$

where HNK refers to the heaviest nonkey in the bottoms product at minimum reflux.

Because of their relative simplicity, the Underwood minimum reflux equations for class 2 separations are widely used, but too often without examining the possibility of nonkey distribution. In addition, the assumption is frequently made that $(R_\infty)_{min}$ equals the external reflux ratio. When the assumptions of constant relative volatility and constant molar overflow in the regions between the two pinch-point zones are not valid, values of the minimum reflux ratio computed from the Underwood equations for class 2 separations can be appreciably in error because of the sensitivity of (9-28) to the value of q, as will be shown in Example 9.5. When the Underwood assumptions appear to be valid and a negative minimum reflux ratio is computed, this may be interpreted to mean that a rectifying section is not required to obtain the specified separation. The Underwood equations show that the minimum reflux depends mainly on the feed condition and relative volatility and, to a lesser extent, on the degree of separation between the two key components. A finite minimum reflux ratio exists even for a perfect separation.

An extension of the Underwood method for distillation columns with multiple feeds is given by Barnes, Hanson, and King [10]. Exact computer methods for determining minimum reflux are available [11]. For making rigorous distillation calculations at actual reflux conditions by the computer methods of Chapter 10, knowledge of the minimum reflux is not essential, but the minimum number of equilibrium stages is very useful.

EXAMPLE 9.5

Repeat Example 9.4 assuming a class 2 separation and utilizing the corresponding Underwood equations. Check the validity of the Underwood assumptions. Also calculate the external reflux ratio.

SOLUTION

From the results of Example 9.4, assume that the only distributing nonkey component is n-pentane. Assuming that the feed temperature of 180°F is reasonable for computing relative volatilities in the pinch zone, the following quantities are obtained from Figures 9.3 and 9.5:

Species i	$z_{i,F}$	$(\alpha_{i,HK})_\infty$
iC_4	0.0137	2.43
nC_4 (LK)	0.5113	1.93
iC_5 (HK)	0.0411	1.00
nC_5	0.0171	0.765
nC_6	0.0262	0.362
nC_7	0.0446	0.164
nC_8	0.3106	0.0720
nC_9	0.0354	0.0362
	1.0000	

The q for the feed is assumed to be the mole fraction of liquid in the flashed feed. From Example 9.1, $q = 1 - 0.1334 = 0.8666$. Applying (9-28), we have

$$\frac{2.43(0.0137)}{2.43 - \theta} + \frac{1.93(0.5113)}{1.93 - \theta} + \frac{1.00(0.0411)}{1.00 - \theta} + \frac{0.765(0.0171)}{0.765 - \theta}$$
$$+ \frac{0.362(0.0262)}{0.362 - \theta} + \frac{0.164(0.0446)}{0.164 - \theta} + \frac{0.072(0.3106)}{0.072 - \theta} + \frac{0.0362(0.0354)}{0.0362 - \theta}$$
$$= 1 - 0.8666$$

Solving this equation by a bounded Newton method for two roots of θ that satisfy

$$\alpha_{nC_4,iC_5} > \theta_1 > \alpha_{iC_5,iC_5} > \theta_2 > \alpha_{nC_5,iC_5}$$

or

$$1.93 > \theta_1 > 1.00 > \theta_2 > 0.765$$

$\theta_1 = 1.04504$ and $\theta_2 = 0.78014$. Because distillate rates for nC_4 and iC_5 are specified (442 and 13 lbmol/h, respectively), the following form of (9-29) is preferred:

$$\sum_i \frac{(\alpha_{i,r})_\infty(x_{i,D}D)}{(\alpha_{i,r})_\infty - \theta} = D + (L_\infty)_{\min} \qquad \textbf{(9-32)}$$

with the restriction that

$$\sum_i (x_{i,D}D) = D \qquad \textbf{(9-33)}$$

Assuming that $x_{i,D}D$ equals 0.0 for components heavier than nC_5 and 12.0 lbmol/h for iC_4, we find that these two relations give the following three linear equations:

$$D + (L_\infty)_{\min} = \frac{2.43(12)}{2.43 - 1.04504} + \frac{1.93(442)}{1.93 - 1.04504} + \frac{1.00(13)}{1.00 - 1.04504} + \frac{0.765(x_{nC_5,D}D)}{0.765 - 1.04504}$$

$$D + (L_\infty)_{\min} = \frac{2.43(12)}{2.43 - 0.78014} + \frac{1.93(442)}{1.93 - 0.78014} + \frac{1.00(13)}{1.00 - 0.78014} + \frac{0.765(x_{nC_5,D}D)}{0.765 - 0.78014}$$

$$D = 12 + 442 + 13 + (x_{nC_5,D}D)$$

Solving these three equations gives

$$x_{nC_5,D}D = 2.56 \text{ lbmol/h}$$

$$D = 469.56 \text{ lbmol/h}$$

$$(L_\infty)_{\min} = 219.8 \text{ lbmol/h}$$

The distillate rate for nC_5 is very close to the value of 2.63 computed in Example 9.4, if we assume a class 1 separation. The internal minimum reflux ratio at the rectifying pinch point is considerably less than the value of 389 computed in Example 9.4 and is also much less than the true internal value of 298 reported by Bachelor [4]. The main reason for the discrepancy

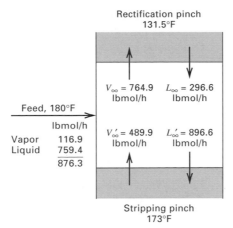

Figure 9.8 Pinch-point region conditions for Example 9.5 from computations by Bachelor. [From J.B. Bachelor, *Petroleum Refiner,* **36**(6), 161–170 (1957).]

between the value of 219.8 and the true value of 298 is the invalidity of the assumption of constant molar overflow. Bachelor computed the pinch-point region flow rates and temperatures shown in Figure 9.8. The average temperature of the region between the two pinch regions is 152°F (66.7°C), which is appreciably lower than the flashed feed temperature. The relatively hot feed causes additional vaporization across the feed zone. The effective value of q in the region between the pinch points is obtained from (7-18):

$$q_{\text{eff}} = \frac{L'_\infty - L_\infty}{F} = \frac{896.6 - 296.6}{876.3} = 0.685$$

This is considerably lower than the value of 0.8666 for q based on the flashed feed condition. On the other hand, the value of $\alpha_{\text{LK,HK}}$ at 152°F (66.7°C) is not much different from the value at 180°F (82.2°C). If this example is repeated using q equal to 0.685, the resulting value of $(L_\infty)_{\min}$ is 287.3 lbmol/h, which is only 3.6% lower than the true value of 298. Unfortunately, in practice, this corrected procedure cannot be applied because the true value of q cannot be readily determined.

To compute the external reflux ratio from (9-23), rectifying pinch-point compositions must be calculated from (9-30) and (9-17). The root of θ to be used in (9-30) is obtained from the version of (9-29) used above. Thus,

$$\frac{2.43(12)}{2.43 - \theta} + \frac{1.93(442)}{1.93 - \theta} + \frac{1.00(13)}{1.00 - \theta} + \frac{0.765(2.56)}{0.765 - \theta} = 469.56 + 219.8$$

where $0.765 > \theta > 0$. Solving, $\theta = 0.5803$. Liquid pinch-point compositions are obtained from the following form of (9-30):

$$x_{i,\infty} = \frac{\theta(x_{i,D}D)}{(L_\infty)_{\min}[(\alpha_{i,r})_\infty - \theta]}$$

with $(L_\infty)_{\min} = 219.8$ lbmol/h.

For iC_4,

$$x_{iC_4,\infty} = \frac{0.5803(12)}{219.8(2.43 - 0.5803)} = 0.0171$$

From a combination of (9-17) and (9-18),

$$y_{i,\infty} = \frac{x_{i,\infty}L_\infty + x_{i,D}D}{L_\infty + D}$$

For iC_4,

$$y_{iC_4,\infty} = \frac{0.0171(219.8) + 12}{219.8 + 469.56} = 0.0229$$

Similarly, the mole fractions of the other components appearing in the distillate are

Component	$x_{i,\infty}$	$y_{i,\infty}$
iC_4	0.0171	0.0229
nC_4	0.8645	0.9168
iC_5	0.0818	0.0449
nC_5	0.0366	0.0154
	1.0000	1.0000

The temperature of the rectifying-section pinch point is obtained from either a bubble-point temperature calculation on $x_{i,\infty}$ or a dew-point temperature calculation on $y_{i,\infty}$. The result is 126°F. Similarly, the liquid-distillate temperature (bubble point) and the temperature of the vapor leaving the top stage (dew point) are both computed to be approximately 123°F. Because rectifying-section pinch-point temperature and distillate temperatures are very close, it is expected that $(R_\infty)_{min}$ and $(R_{min})_{external}$ will be almost identical. Bachelor [4] obtained a value of 292 lbmol/h for the external reflux rate, compared to 298 lbmol/h for the internal reflux rate.   ∎

Gilliland Correlation for Actual Reflux Ratio and Theoretical Stages

To achieve a specified separation between two key components, the reflux ratio and the number of theoretical stages must be greater than their minimum values. The actual reflux ratio is generally established by economic considerations at some multiple of minimum reflux. The corresponding number of theoretical stages is then determined by suitable analytical or graphical methods or, as discussed in this section, by an empirical equation. However, there is no reason why the number of theoretical stages cannot be specified as a multiple of minimum stages and the corresponding actual reflux computed by the same empirical relationship. As shown in Figure 9.9, from studies by Fair and Bolles [12], the optimum value of R/R_{min} is approximately 1.05. However, near-optimal conditions extend over a relatively broad range of mainly larger values of R/R_{min}. In practice, superfractiona-

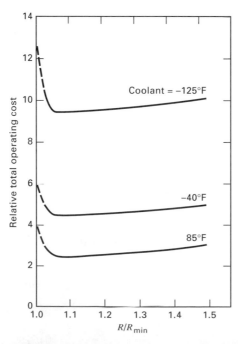

Figure 9.9 Effect of reflux ratio on cost. [From J.R. Fair and W.L. Bolles, *Chem. Eng.*, **75**(9), 156–178 (1968).]

tors requiring a large number of stages are frequently designed for a value of $R/R_{\min}$ of approximately 1.10, while separations requiring a small number of stages are designed for a value of $R/R_{\min}$ of approximately 1.50. For intermediate cases, a commonly used rule of thumb is that $R/R_{\min}$ equal to 1.30.

The number of equilibrium stages required for the separation of a binary mixture assuming constant relative volatility and constant molar overflow depends on $z_{i,F}$, $x_{i,D}$, $x_{i,B}$, q, R, and α. From (9-11), for a binary mixture, $N_{\min}$ depends on $x_{i,D}$, $x_{i,B}$, and α, while $R_{\min}$ depends on $z_{i,F}$, $x_{i,D}$, q, and α. Accordingly, a number of investigators have assumed empirical correlations of the form

$$N = N\{N_{\min}\{x_{i,D}, x_{i,B}, \alpha\}, R_{\min}\{z_{i,F}, x_{i,D}, q, \alpha\}, R\}$$

Furthermore, they have assumed that such a correlation might exist for nearly ideal multi-component systems where the additional feed composition variables and nonkey relative volatilities also influence the value of $R_{\min}$.

The most successful and simplest empirical correlation of this type is the one developed by Gilliland [13] and slightly modified in a later version by Robinson and Gilliland [14]. The correlation is shown in Figure 9.10, where the three sets of data points, which are based on accurate calculations, are the original points from Gilliland [13] and the multicomponent data points of Brown and Martin [15] and Van Winkle and Todd [16]. The 61 data points cover the following ranges of conditions:

1. Number of components: 2 to 11
2. q: 0.28 to 1.42
3. Pressure: vacuum to 600 psig
4. α: 1.11 to 4.05
5. $R_{\min}$: 0.53 to 9.09
6. $N_{\min}$: 3.4 to 60.3

The line drawn through the data represents the equation developed by Molokanov et al. [17].

$$Y = \frac{N - N_{\min}}{N + 1} = 1 - \exp\left[\left(\frac{1 + 54.4X}{11 + 117.2X}\right)\left(\frac{X - 1}{X^{0.5}}\right)\right] \tag{9-34}$$

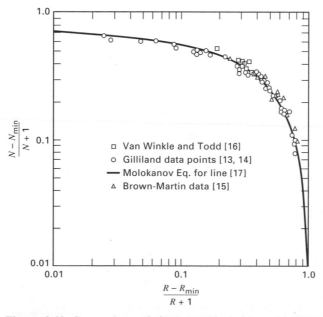

Figure 9.10 Comparison of rigorous calculations with Gilliland correlation.

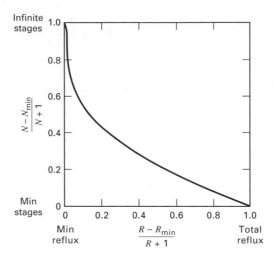

Figure 9.11 Gilliland correlation with linear coordinates.

where
$$X = \frac{R - R_{min}}{R + 1}$$

This equation satisfies the end points ($Y = 0$, $X = 1$) and ($Y = 1$, $X = 0$). At a value of R/R_{min} near the optimum of 1.3, Figure 9.10 predicts an optimum ratio for N/N_{min} of approximately 2. The value of N includes one stage for a partial reboiler and one stage for a partial condenser, if any.

The Gilliland correlation is very useful for preliminary exploration of design variables. Although it was never intended for final design, the Gilliland correlation was used to design many distillation columns for multicomponent separations without benefit of accurate stage-by-stage calculations. In Figure 9.11, a replot of the correlation in linear coordinates shows that a small initial increase in R above R_{min} causes a large decrease in N, but further changes in R have a much smaller effect on N. The knee in the curve of Figure 9.11 corresponds closely to the optimum value of R/R_{min} in Figure 9.9.

Robinson and Gilliland [14] state that a more accurate correlation should utilize a parameter involving the feed condition q. This effect is shown in Figure 9.12 using data

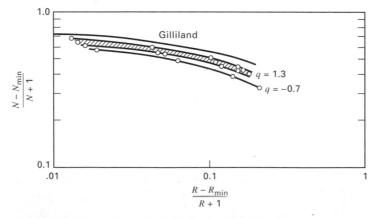

Figure 9.12 Effect of feed condition on Gilliland correlation. [From G. Guerreri, *Hydrocarbon Processing,* **48**(8), 137–142 (1969).]

points for the sharp separation of benzene–toluene mixtures from Guerreri [18]. The data, which cover feed conditions ranging from subcooled liquid to superheated vapor (q equals 1.3 to -0.7), show a trend toward decreasing theoretical stage requirements with increasing feed vaporization. The Gilliland correlation appears to be conservative for feeds having low values of q. Donnell and Cooper [19] state that this effect of q is important only when the α between the key components is high or when the feed is low in volatile components.

A serious problem with the Gilliland correlation can occur when stripping is much more important than rectification. For example, Oliver [20] considers a fictitious binary case with specifications of $z_F = 0.05$, $x_D = 0.40$, $x_B = 0.001$, $q = 1$, $\alpha = 5$, $R/R_{min} = 1.20$, and constant molar overflow. By exact calculations, $N = 15.7$. From the Fenske equation, $N_{min} = 4.04$. From the Underwood equation, $R_{min} = 1.21$. From (9-32) for the Gilliland correlation, $N = 10.3$. This is 34% lower than the exact value. This limitation, which is caused by ignoring boilup, is discussed further by Strangio and Treybal [21], who present a more accurate, but far more tedious, method for such cases.

EXAMPLE 9.6

Use the Gilliland correlation to estimate the theoretical stage requirements for the debutanizer of Examples 9.1, 9.2, and 9.5 for an external reflux of 379.6 lbmol/h (30% greater than the exact value of the minimum reflux rate from Bachelor).

SOLUTION

From the examples cited, values of R_{min} and $[(R - R_{min})/(R + 1)]$ are obtained using a distillate rate from Example 9.5 of 469.56 lbmol/h. Thus, $R = 379.6/469.56 = 0.808$. With $N_{min} = 8.88$,

$$R_{min} = 0.479, \text{ and } \quad \frac{R - R_{min}}{R + 1} = X = 0.182$$

From (9-34),

$$\frac{N - N_{min}}{N + 1} = 1 - \exp\left[\left(\frac{1 + 54.4(0.182)}{11 + 117.2(0.182)}\right)\left(\frac{0.182 - 1}{0.182^{0.5}}\right)\right] = 0.476$$

$$N = \frac{8.88 + 0.476}{1 - 0.476} = 17.85$$

$$N - 1 = 16.85$$

where $N - 1$ corresponds to the equilibrium stages in the tower allowing one theoretical stage for the reboiler, but no stage for the total condenser.

It should be kept in mind that, had the exact value of R_{min} not been known and a value of R equal to 1.3 times R_{min} from the Underwood method been used, the value of R would have been 292 lbmol/h. But this, by coincidence, is only the true minimum reflux. Therefore, the desired separation would not be achieved. ∎

Feed-Stage Location

Implicit in the application of the Gilliland correlation is the specification that the theoretical stages be distributed optimally between the rectifying and stripping sections. As suggested by Brown and Martin [15], the optimum feed stage can be located by assuming that the ratio of stages above the feed to stages below the feed is the same as the ratio determined by simply applying the Fenske equation to the separate sections at total reflux conditions to give

$$\frac{N_R}{N_S} \approx \frac{(N_R)_{min}}{(N_S)_{min}} = \frac{\log\left[(x_{LK,D}/z_{LK,F})(z_{HK,F}/x_{HK,D})\right]}{\log\left[(z_{LK,F}/x_{LK,B})(x_{HK,B}/z_{HK,F})\right]} \frac{\log\left[(\alpha_B\alpha_F)^{1/2}\right]}{\log\left[(\alpha_D\alpha_F)^{1/2}\right]} \tag{9-35}$$

Unfortunately, (9-35) is not reliable except for fairly symmetrical feeds and separations.

A reasonably good approximation of optimum feed-stage location can be made by employing the empirical equation of Kirkbride [22]:

$$\frac{N_R}{N_S} = \left[\left(\frac{z_{HK,F}}{z_{LK,F}}\right)\left(\frac{x_{LK,B}}{x_{HK,D}}\right)^2\left(\frac{B}{D}\right)\right]^{0.206} \tag{9-36}$$

An extreme test of both these equations is provided by the fictitious binary-mixture problem of Oliver [20] cited in the previous section. Exact calculations by Oliver and calculations using (9-35) and (9-36) give the following results:

Method	N_R/N_S
Exact	0.08276
Kirkbride (9-34)	0.1971
Fenske ratio (9-33)	0.6408

Although the result from the Kirkbride equation is not very satisfactory, the result from the Fenske ratio method is much worse.

EXAMPLE 9.7

Use the Kirkbride equation to determine the feed-stage location for the debutanizer of Example 9.1, assuming an equilibrium-stage requirement of 18.27.

SOLUTION

Assume that the product distribution computed in Example 9.3 for total reflux conditions is a good approximation to the distillate and bottoms compositions at actual reflux conditions. Then

$$x_{nC_4,B} = \frac{6.0}{408.5} = 0.0147 \qquad x_{iC_5,D} = \frac{13}{467.8} = 0.0278$$

$$D = 467.8 \text{ lbmol/h} \qquad B = 408.5 \text{ lbmol/h}$$

From Figure 9.3,

$$z_{nC_4,F} = 448/876.3 = 0.5112 \quad \text{and} \quad z_{iC_5,F} = 36/876.3 = 0.0411$$

From (9-36),

$$\frac{N_R}{N_S} = \left[\left(\frac{0.0411}{0.5112} \right) \left(\frac{0.0147}{0.0278} \right)^2 \left(\frac{408.5}{467.8} \right) \right]^{0.206} = 0.445$$

Therefore, $N_R = \dfrac{0.445}{1.445} (18.27) = 5.63$ stages   and $N_S = 18.27 - 5.63 = 12.64$ stages

Rounding the estimated stage requirements leads to 1 stage as a partial reboiler, 12 stages below the feed, and 6 stages above the feed.   ∎

Distribution of Nonkey Components at Actual Reflux

For multicomponent mixtures, all components distribute to some extent between distillate and bottoms at total reflux conditions. However, at minimum reflux conditions, none or only a few of the nonkey components distribute. Distribution ratios for these two limiting conditions are shown in Figure 9.13 for the debutanizer example. For total reflux conditions, results from the Fenske equation in Example 9.3 plot as a straight line for the log–log coordinates. For minimum reflux, results from the Underwood equation in Example 9.5 are shown as a dashed line.

It might be expected that a product distribution curve for actual reflux conditions would lie between the two limiting curves. However, as shown by Stupin and Lockhart [23], product distributions in distillation are complex. A typical result is shown in Figure 9.14. For a reflux ratio near minimum, the product distribution (curve 3) lies between the two limits (curves 1 and 4). However, for a high reflux ratio, the product distribution for a nonkey component (curve 2) may actually lie outside the limits, so that an inferior separation results.

For the behavior of the product distribution in Figure 9.14, Stupin and Lockhart provide an explanation that is consistent with the Gilliland correlation of Figure 9.10. As the reflux

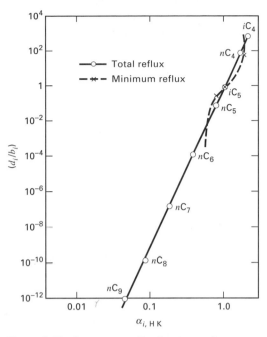

Figure 9.13 Component distribution ratios at extremes of distillation operating conditions.

ratio is decreased from total reflux while maintaining the specified splits of the two key components, equilibrium-stage requirements increase only slowly at first, but then rapidly as minimum reflux is approached. Initially, large decreases in reflux cannot be adequately compensated for by increasing stages. This causes inferior nonkey component distributions. However, as minimum reflux is approached, comparatively small decreases in reflux are

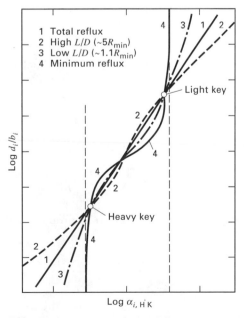

Figure 9.14 Component distribution ratios at various reflux ratios.

more than compensated for by large increases in equilibrium stages; and the separation of nonkey components becomes superior to that at total reflux. It appears reasonable to assume that, at a near-optimal reflux ratio of 1.3, nonkey component distribution is close to that estimated by the Fenske equation for total reflux conditions.

9.2 KREMSER GROUP METHOD

Many multicomponent separators are cascades of stages where the two contacting phases flow countercurrently. Approximate calculation procedures have been developed to relate compositions of streams entering and exiting cascades to the number of equilibrium stages required. These approximate procedures are called *group methods* because they provide only an overall treatment of the stages in the cascade without considering detailed changes in temperature, flow rates, and composition in the individual stages. In this section, single cascades used for absorption, stripping, and liquid–liquid extraction are considered.

Kremser [1] originated the group method. He derived overall species material balances for a multistage countercurrent absorber. Subsequent articles by Souders and Brown [24] Horton and Franklin [25] and Edmister [26] improved the method. The Kremser equations are derived and applied to absorption in Section 5.4. These equations are further extended to and illustrated for strippers and extractors here. Another treatment by Smith and Brinkley [27] emphasizes liquid–liquid separations.

Strippers

The vapor entering a stripper is often steam or another inert gas. When the stripping agent contains none of the species in the feed liquid, is not present in the entering liquid, and is not absorbed or condensed in the stripper, the only direction of mass transfer is from the liquid to the gas phase. Then, only values of the effective stripping factor, S_e, as defined by (5-51), are needed to apply the group method via (5-49) and (5-50). The equations for strippers are analogous to those for absorbers. However, the Kremser approximation can be improved somewhat by using the modification of Edmister [26]. With the stages of a stripper numbered as in Figure 5.8b, Edmister gives

$$S_e = [S_N(S_1 + 1) + 0.25]^{1/2} - 0.5 \qquad \textbf{(9-37)}$$

To estimate S_1 and S_N, total flow rates can be approximated by the following relations from Horton and Franklin [25]:

$$L_2 = L_1 \left(\frac{L_{N+1}}{L_1}\right)^{1/N} \qquad \textbf{(9-38)}$$

$$V_1 = V_0 + L_2 - L_1 \qquad \textbf{(9-39)}$$

$$L_N = L_{N+1} \left(\frac{L_1}{L_{N+1}}\right)^{1/N} \qquad \textbf{(9-40)}$$

Taking the temperature change of the liquid to be proportional to the liquid contraction, we have

$$\frac{T_{N+1} - T_N}{T_{N+1} - T_1} = \frac{L_{N+1} - L_N}{L_{N+1} - L_1} \qquad \textbf{(9-41)}$$

This equation is solved simultaneously with an overall enthalpy balance for T_1 and T_N, the terminal-stage temperatures. Often $(T_{N+1} - T_N)$ ranges from 0 to 20°F, depending on the fraction of entering liquid stripped.

For optimal stripping, temperatures should be high and pressures low. However, temperatures should not be so high as to cause decomposition, and vacuum should be used only if necessary. The minimum stripping-agent flow rate, for a specified value of ϕ_S for a key component K corresponding to an infinite number of stages, can be estimated from an equation obtained from (5-50) with $N = \infty$,

$$(V_0)_{\text{min}} = \frac{L_{N+1}}{K_K}(1 - \phi_{S_K}) \tag{9-42}$$

This equation assumes that $A_K < 1$ and the fraction of liquid feed stripped is small.

EXAMPLE 9.8

Sulfur dioxide and butadienes (B3 and B2) are to be stripped with nitrogen from the liquid stream given in Figure 9.15 so that butadiene sulfone (BS) product will contain less than 0.05 mol% SO_2 and less than 0.5 mol% butadienes. Estimate the flow rate of nitrogen, N_2, and the number of equilibrium stages required.

SOLUTION

Neglecting stripping of BS, the stripped liquid must have the following component flow rates, and corresponding values for ϕ_S:

Species	l_1, lbmol/h	$\phi_S = \dfrac{l_1}{l_{N+1}}$
SO_2	<0.0503	<0.00503
B3 + B2	<0.503	<0.0503
BS	100.0	—

Thermodynamic properties can be computed based on ideal solutions at low pressures. For butadiene sulfone, the vapor pressure is given by

$$P_{\text{BS}}^s = \exp\left(17.30 - \frac{11{,}142}{T + 459.67}\right)$$

where P_{BS}^s is in pounds force per square inch absolute and T is in degrees Fahrenheit. The liquid enthalpy of BS is

$$(h_L)_{\text{BS}} = 50T$$

where $(h_L)_{\text{BS}}$ is in British thermal units per pound-mole and T is in degrees Fahrenheit.

The entering flow rate of the stripping agent V_0 is not specified. The minimum rate at infinite stages can be computed from (9-42), provided that a key component is selected. Suppose we

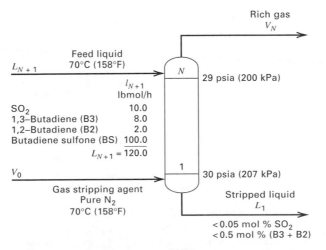

Figure 9.15 Specifications for stripper of Example 9.8.

choose B2, which is the heaviest component to be stripped to a specified extent. At 70°C, the vapor pressure of B2 is 90.4 psia. From Raoult's law at 30 psia total pressure,

$$K_{B2} = \frac{90.4}{30} = 3.01$$

From (9-42), using $(\phi_S)_{B2} = 0.0503$, we have

$$(V_0)_{min} = \frac{120}{3.01}(1 - 0.0503) = 37.9 \text{ lbmol/h}$$

For this value of $(V_0)_{min}$, (9-42) can now be used to determine ϕ_S for B3 and SO$_2$. The K-values for these two species are 4.53 and 6.95, respectively. From (9-42), at infinite stages with $V_0 = 37.9$ lbmol/h,

$$(\phi_S)_{B3} = 1 - \frac{4.53(37.9)}{120} = -0.43 \quad \text{and} \quad (\phi_S)_{SO_2} = 1 - \frac{6.95(37.9)}{120} = -1.19$$

These negative values indicate complete stripping of B3 and SO$_2$. Therefore, the total butadienes in the stripped liquid would be only $(0.0503)(2.0) = 0.1006$, compared to the specified value of 0.503. We can obtain a better estimate of $(V_0)_{min}$ by assuming that all of the butadiene content of the stripped liquid is due to B2. Then $(\phi_S)_{B2} = 0.503/2 = 0.2515$, and $(V_0)_{min}$ from (9-42) is 29.9 lbmol/h. Values of $(\phi_S)_{B3}$ and $(\phi_S)_{SO_2}$ are still negative.

The actual entering flow rate for the stripping vapor must be greater than the minimum value. To estimate the effect of V_0 on the theoretical stage requirements and ϕ_S values for the nonkey components, the Kremser approximation is used with K-values at 70°C and 30 psia, $L = L_{N+1} = 120$ lbmol/h, and $V = V_0$ equal to a series of multiples of 29.9 lbmol/h. The calculations are greatly facilitated if values of N are selected and values of V are determined from (5-51), where S is obtained from Figure 5.9. Because B3 will be found to some extent in the stripped liquid, $(\phi_S)_{B2}$ will be held below 0.2515. By making iterative calculations, one can choose $(\phi_S)_{B2}$ so that $(\phi_S)_{B2+B3}$ satisfies the specification of, say, 0.05. For 10 theoretical stages, assuming essentially complete stripping of B3 such that $(\phi_S)_{B2} \approx 0.25$, $S_{B2} = 0.76$ from Figure 5.9. From (5-51),

$$V = V_0 = \frac{(120)(0.76)}{3.01} = 30.3 \text{ lbmol/h}$$

For B3, from (5-51), $S_{B3} = \frac{(4.53)(30.3)}{120} = 1.143$

From Figure 5.9, $(\phi_S)_{B3} = 0.04$. Thus $(\phi_S)_{B2+B3} = \frac{0.25(2) + 0.04(8)}{10} = 0.082$

This is considerably above the specification of 0.05. Therefore, repeat the calculations with, say, $(\phi_S)_{B2} = 0.09$ and continue to repeat until the specified value of $(\phi_S)_{B2+B3}$ is obtained. In this manner, calculations for various numbers of theoretical stages are carried out with converged results as shown.

N	V_0, lbmol/h	$V_0/(V_0)_{min}$	ϕ_{SO_2}	ϕ_{B3}	ϕ_{B2}	ϕ_{B2+B3}	ϕ_{BS}
			\multicolumn{5}{c}{**Fraction Not Stripped**}				
∞	29.9	1.00	0.0	0.0	0.2515	0.0503	0.9960
10	33.9	1.134	0.0005	0.017	0.18	0.050	0.9955
5	45.4	1.518	0.0050	0.029	0.117	0.040	0.9940
3	94.3	3.154	0.0050	0.016	0.045	0.022	0.9874

These results show that the specification on SO$_2$ cannot be met for $N < 6$ stages. Therefore, for five and three stages, SO$_2$ must be the key component for determining the value of V_0.

From the above table, initial estimates of L_1 can be determined from (5-49) for given values of V_0 and N. For example, for $N = 5$,

	Pound-Moles per Hour			
Component	l_{N+1}	v_0	l_1	v_5
SO_2	10.0	0.0	0.05	9.95
B3	8.0	0.0	0.23	7.77
B2	2.0	0.0	0.23	1.77
BS	100.0	0.0	99.40	0.60
N_2	0.0	45.4	0.00	45.40
	120.0	45.4	99.91	64.59

To refine the estimates of l_1, use (9-38) through (9-40).

$$L_2 = 99.91 \left(\frac{120}{99.91} \right)^{1/5} = 103.64 \, \text{lbmol/h}; \quad V_1 = 45.4 + 103.64 - 99.91 = 49.13 \, \text{lbmol/h}$$

$$L_5 = 120 \left(\frac{99.91}{120} \right)^{1/5} = 115.68 \, \text{lbmol/h}$$

From (9-39), with T in degrees Fahrenheit,

$$\frac{158 - T_5}{158 - T_1} = \frac{120 - 115.68}{120 - 99.91} = 0.215; \quad \text{or } T_5 - 0.215 T_1 = 124$$

Solving this equation simultaneously with the overall enthalpy balance, we have $T_1 = 124.6°F$ (51.4°C) and $T_5 = 150.8°F$ (66.0°C).

Stripping factors can now be computed for the two terminal stages, where

$$\frac{V_1}{L_1} = \frac{49.13}{99.91} = 0.4917 \quad \text{and} \quad \frac{V_5}{L_5} = \frac{65.49}{115.68} = 0.5661$$

Also, it will be assumed that stripper top pressure is 29 psia (199.9 kPa).

	Bottom Stage, 124.6°F, 30 psia		**Top Stage, 150.8°F, 29 psia**	
Component	K_1	S_1	K_5	S_5
SO_2	4.27	2.10	6.50	3.68
B3	2.89	1.42	4.27	2.42
B2	1.84	0.905	2.82	1.60
BS	0.0057	0.0028	0.0133	0.0075

Effective stripping factors are computed from (9-35). Values of ϕ_S are read from Figure 5.9 or computed from (5-50), followed by calculations of $(l_i)_1$ from (5-49) and $(v_i)_5$ from an overall species material balance.

			Pound-Moles per Hour	
Component	S_e	ϕ_S	l_1	v_5
SO_2	2.91	0.0032	0.032	9.968
B3	1.97	0.017	0.136	7.864
B2	1.32	0.075	0.150	1.850
BS	0.0075	0.9925	99.24	0.75
N_2	—	—	0.00	45.40
			99.568	65.832

The calculated values of L_1 and V_5 are very close to the assumed values from the Kremser approximation. Therefore, the calculations need not be repeated. However, the stripped liquid contains 0.032 mol% SO_2 and 0.29 mol% butadienes, which are considerably less than limiting specifications. The entering stripping gas flow rate could therefore be reduced somewhat for five theoretical stages. ∎

Liquid–Liquid Extraction

A schematic representation of a countercurrent extraction cascade is shown in Figure 9.16, with stages numbered from the top down and solvent V_{N+1} entering at the bottom.[4] The group method of calculation can be applied, with the equations written by analogy to absorbers. In place of the K-value, the distribution coefficient is used:

$$K_{D_i} = \frac{y_i}{x_i} = \frac{v_i/V}{l_i/L}$$

(9-43)

Here, y_i is the mole fraction of i in the solvent or extract phase and x_i is the mole fraction in the feed or raffinate phase. Also, in place of the absorption factor, an *extraction factor*, E, is used, where

$$E_i = \frac{K_{D_i}V}{L}$$

(9-44)

The reciprocal of E is

$$U_i = \frac{1}{E_i} = \frac{L}{K_{D_i}V}$$

(9-45)

The working equations for each component are

$$v_1 = v_{N+1}\phi_U + l_0(1 - \phi_E)$$

(9-46)

$$l_N = l_0 + v_{N+1} - v_1$$

(9-47)

---

[4] In a vertical extractor, solvent would have to enter at the top if of greater density than the feed.

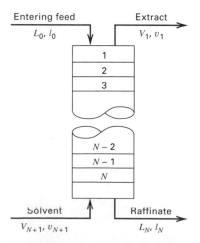

Entering feed — L_0, l_0

Extract — V_1, v_1

1

2

3

$N-2$

$N-1$

N

Solvent — V_{N+1}, v_{N+1}

Raffinate — L_N, l_N

Figure 9.16 Countercurrent liquid–liquid extraction cascade.

where

$$\phi_U = \frac{U_e - 1}{U_e^{N+1} - 1} \tag{9-48}$$

$$\phi_E = \frac{E_e - 1}{E_e^{N+1} - 1} \tag{9-49}$$

$$V_2 = V_1 \left(\frac{V_{N+1}}{V_1}\right)^{1/N} \tag{9-50}$$

$$L_1 = L_0 + V_2 - V_1 \tag{9-51}$$

$$V_N = V_{N+1} \left(\frac{V_1}{V_{N+1}}\right)^{1/N} \tag{9-52}$$

$$E_e = [E_1(E_N + 1) + 0.25]^{1/2} - 0.5 \tag{9-53}$$

$$U_e = [U_N(U_1 + 1) + 0.25]^{1/2} - 0.5 \tag{9-54}$$

If desired, (9-41) through (9-52) can be applied using mass units rather than mole units. No enthalpy balance equations are required because ordinarily temperature changes in an adiabatic extractor are not great unless the feed and solvent enter at appreciably different temperatures or the heat of mixing is large. Unfortunately, the group method is not always reliable for liquid–liquid extraction cascades because the distribution coefficient, as discussed in Chapter 2, is a ratio of activity coefficients, which can vary drastically with composition.

EXAMPLE 9.9

Countercurrent liquid–liquid extraction with methylene chloride is to be used at 25°C to recover dimethylformamide from an aqueous stream as shown in Figure 9.17. Estimate flow rates and compositions of extract and raffinate streams by the group method using mass units. Distribution coefficients for all components except DMF are essentially constant over the expected composition range and on a mass-fraction basis are

Component	K_{D_i}
MC	40.2
FA	0.005
DMA	2.2
W	0.003

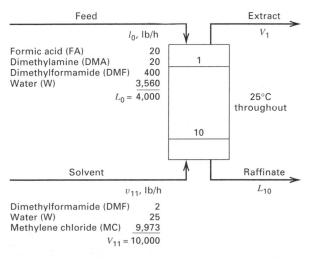

Figure 9.17 Specifications for extractor of Example 9.9.

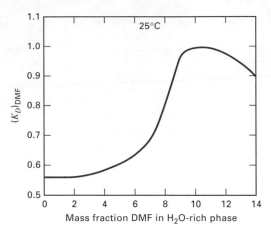

Figure 9.18 Distribution coefficient for dimethylformamide between water and methylene chloride.

The distribution coefficient for DMF depends on concentration in the water-rich phase as shown in Figure 9.18.

SOLUTION

Although the Kremser approximation could be applied for the first trial calculation, the following values will be assumed from guesses based on the magnitudes of the K_D-values.

Component	Pounds per Hour			
	Feed, l_0	Solvent, v_{11}	Raffinate, l_{10}	Extract, v_1
FA	20	0	20	0
DMA	20	0	0	20
DMF	400	2	2	400
W	3,560	25	3,560	25
MC	0	9,973	88	9,885
	4,000	10,000	3,670	10,330

From (9-50) through (9-52), we have

$$V_2 = 10,330 \left(\frac{10,000}{10,330}\right)^{1/10} = 10,297 \text{ lb/h}; \quad L_1 = 4,000 + 10,297 - 10,330 = 3,967 \text{ lb/h};$$

$$V_{10} = 10,000 \left(\frac{10,330}{10,000}\right)^{1/10} = 10,033 \text{ lb/h}$$

From (9-44), (9-45), (9-53), and (9-54), assuming a mass fraction of 0.09 for DMF in L_1 in order to obtain $(K_D)_{DMF}$ for stage 1, we have

Component	E_1	E_{10}	U_1	U_{10}	E_e	U_e
FA	0.013	0.014	—	—	0.013	—
DMA	5.73	6.01	—	—	5.86	—
DMF	2.50	1.53	0.400	0.653	2.06	0.579
W	0.0078	0.0082	128	122	0.0078	125
MC	—	—	0.0096	0.0091	—	0.0091

From (9-49), (9-48), (9-46), and (9-47), we have

Component	ϕ_E	ϕ_U	Pounds per Hour Raffinate, l_{10}	Extract, v_1
FA	0.9870	—	19.7	0.3
DMA	0.0	—	0.0	20.0
DMF	0.000374	0.422	1.3	400.7
W	0.9922	0.0	3,557.2	37.8
MC	—	0.9909	90.8	9,882.2
			3,669.0	10,331.0

The calculated total flow rates L_{10} and V_1 are almost exactly equal to the assumed rates. Therefore, an additional iteration is not necessary. The degree of extraction of DMF is very high. It would be worthwhile to calculate additional cases with less solvent and/or fewer equilibrium stages. ■

SUMMARY

1. The Fenske–Underwood–Gilliland (FUG) method for simple distillation of ideal and nearly ideal multicomponent mixtures is useful for making preliminary estimates of stage and reflux requirements.

2. Based on a specified split of two key components in the feed mixture, the theoretical Fenske equation is used to determine the minimum number of equilibrium stages at total reflux. The theoretical Underwood equations are used to determine the minimum reflux ratio for an infinite number of stages. The empirical Gilliland correlation relates the minimum stages and minimum reflux ratio to the actual reflux ratio and the actual number of equilibrium stages.

3. Estimates of the distribution of nonkey components and the feed-stage location can be made with the Fenske and Kirkbride equations, respectively.

4. The Underwood equations are more restrictive than the Fenske equation and must be used with care and caution.

5. The Kremser group method can be applied to simple strippers and liquid–liquid extractors to make approximate estimates of component recoveries for specified values of entering flow rates and number of equilibrium stages.

REFERENCES

1. Kremser, A., *Natl. Petroleum News*, **22**(21), 43–49 (1930).
2. Edmister, W.C., *AIChE J.*, **3**, 165–171 (1957).
3. Kobe, K.A., and J.J. McKetta, Jr., Eds, *Advances in Petroleum Chemistry and Refining*, Vol. 2, Interscience, New York, 315–355 (1959).
4. Bachelor, J.B., *Petroleum Refiner*, **36**(6), 161–170 (1957).
5. Fenske, M.R., *Ind. Eng. Chem.*, **24**, 482–485 (1932).
6. Shiras, R.N., D.N. Hanson, and C.H. Gibson, *Ind. Eng. Chem.*, **42**, 871–876 (1950).
7. Underwood, A.J.V., *Trans. Inst. Chem. Eng.*, **10**, 112–158 (1932).
8. Gilliland, E.R., *Ind. Eng. Chem.*, **32**, 1101–1106 (1940).
9. Underwood, A.J.V., *J. Inst. Petrol.*, **32**, 614–626 (1946).
10. Barnes, F.J., D.N. Hanson, and C.J. King, *Ind. Eng. Chem., Process Des. Dev.*, **11**, 136–140 (1972).
11. Tavana, M., and D.N. Hanson, *Ind. Eng. Chem., Process Des. Dev.*, **18**, 154–156 (1979).

12. Fair, J.R., and W.L. Bolles, *Chem. Eng.*, **75**(9), 156–178 (1968).
13. Gilliland, E.R., *Ind. Eng. Chem.*, **32**, 1220–1223 (1940).
14. Robinson, C.S., and E.R. Gilliland, *Elements of Fractional Distillation*, 4th ed., McGraw-Hill, New York, pp. 347–350 (1950).
15. Brown, G.G., and H.Z. Martin, *Trans. AIChE*, **35**, 679–708 (1939).
16. Van Winkle, M., and W.G. Todd, *Chem. Eng.*, **78**(21), 136–148 (1971).
17. Molokanov, Y.K., T.P. Korablina, N.I. Mazurina, and G.A. Nikiforov, *Int. Chem. Eng.*, **12**(2), 209–212 (1972).
18. Guerreri, G., *Hydrocarbon Processing*, **48**(8), 137–142 (1969).
19. Donnell, J.W., and C.M. Cooper, *Chem. Eng.*, **57**, 121–124 (1950).
20. Oliver, E.D., *Diffusional Separation Processes: Theory, Design, and Evaluation*, John Wiley and Sons, New York, pp. 104–105 (1966).

21. Strangio, V.A., and R.E. Treybal, *Ind. Eng. Chem., Process Des. Dev.*, **13**, 279–285 (1974).

22. Kirkbride, C.G., *Petroleum Refiner,* **23**(9), 87–102 (1944).

23. Stupin, W.J., and F.J. Lockhart, "The Distribution of Non-Key Components in Multicomponent Distillation," presented at the 61st Annual Meeting of the AIChE, Los Angeles, CA, December 1–5, 1968.

24. Souders, M., and G.G. Brown, *Ind. Eng. Chem.*, **24**, 519–522 (1932).

25. Horton, G., and W.B. Franklin, *Ind. Eng. Chem.*, **32**, 1384–1388 (1940).

26. Edmister, W.C., *Ind. Eng. Chem.*, **35**, 837–839 (1943).

27. Smith, B.D., and W.K. Brinkley, *AIChE J.*, **6**, 446–450 (1960).

EXERCISES

Section 9.1

9.1 A mixture of propionic and *n*-butyric acids, which can be assumed to form ideal solutions, is to be separated by distillation into a distillate containing 95 mol% propionic acid and a bottoms product containing 98 mol% *n*-butyric acid. Determine the type of condenser to be used and estimate the distillation column operating pressure.

9.2 A sequence of two distillation columns is to be used to produce the products indicated in Figure 9.19. Establish the type of condenser and an operating pressure for each column for: (a) The direct sequence (C_2/C_3 separation first) and (b) The indirect sequence (C_3/nC_4 separation first). Use *K*-values from Figures 2.8 and 2.9.

9.3 For each of the two distillation separations (*D*-1 and *D*-2) indicated in Figure 9.20, establish the type of condenser and an operating pressure.

9.4. A deethanizer is to be designed for the separation indicated in Figure 9.21. Estimate the number of equilibrium stages required, assuming it is equal to 2.5 times the minimum number of equilibrium stages at total reflux.

9.5 For the complex distillation operation shown in Figure 9.22, use the Fenske equation to determine the minimum number of stages required between: (a) The distillate and feed, (b) The feed and the side stream, and (c) The side stream and bottoms. The *K*-values can be obtained from Raoult's law.

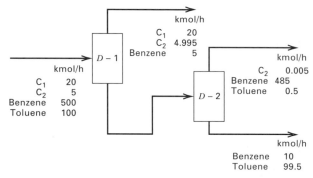

Figure 9.20 Data for Exercise 9.3.

9.6 A 25 mol% mixture of acetone (A) in water (W) is to be separated by distillation at an average pressure of 130 kPa into a distillate containing 95 mol% acetone and a bottoms containing 2 mol% acetone. The infinite-dilution activity coefficients are

$$\gamma_A^\infty = 8.12 \qquad \gamma_W^\infty = 4.13$$

Calculate by the Fenske equation the number of equilibrium stages required. Compare the result to that calculated from the McCabe–Thiele method. Is the Fenske equation reliable for this separation?

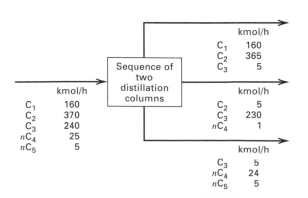

Figure 9.19 Data for Exercise 9.2.

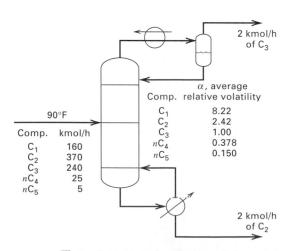

Figure 9.21 Data for Exercise 9.4.

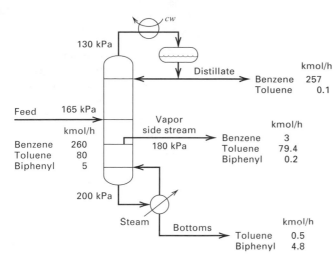

Figure 9.22 Data for Exercise 9.5.

9.7 For the distillation operation indicated in Figure 9.23, calculate the minimum number of equilibrium stages and the distribution of the nonkey components by the Fenske equation, using Figures 2.8 and 2.9 for K-values.

9.8 For the distillation operation shown in Figure 9.24, establish the type of condenser and an operating pressure, calculate the minimum number of equilibrium stages, and estimate the distribution of the nonkey components. Obtain K-values from Figures 2.8 and 2.9.

9.9 For 15 minimum equilibrium stages at 250 psia, calculate and plot the percent recovery of C_3 in the distillate as a function of distillate flow rate for the distillation of 1,000 lbmol/h of a feed containing 3% C_2, 20% C_3, 37% nC_4, 35% nC_5, and 5% nC_6 by moles. Obtain K-values from Figures 2.8 and 2.9.

9.10 Use the Underwood equations to estimate the minimum external reflux ratio for the separation by distillation

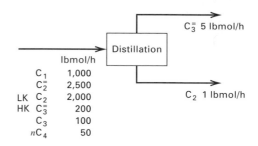

Figure 9.24 Data for Exercise 9.8.

of 30 mol% propane in propylene to obtain 99 mol% propylene and 98 mol% propane, if the feed condition at a column operating pressure of 300 psia is: (a) Bubble-point liquid, (b) Fifty mole percent vaporized, and (c) Dew-point vapor. Use K-values from Figures 2.8 and 2.9.

9.11 For the conditions of Exercise 9.7, compute the minimum external reflux rate and the distribution of the nonkey components at minimum reflux by the Underwood equation if the feed is a bubble-point liquid at column pressure.

9.12 Calculate and plot the minimum external reflux ratio and the minimum number of equilibrium stages against percent product purity for the separation by distillation of an equimolar bubble-point liquid feed of isobutane/n-butane at 100 psia. The distillate is to have the same iC_4 purity as the bottoms is to have nC_4 purity. Consider percent purities from 90% to 99.99%. Discuss the significance of the results.

9.13 Use the Fenske–Underwood–Gilliland shortcut method to determine the reflux ratio required to conduct the distillation operation indicated in Figure 9.25 if $N/N_{min} = 2.0$, the average relative volatility = 1.11, and the feed is at the bubble-point temperature at column feed-stage pressure. Assume that external reflux equals internal reflux at the upper pinch zone. Assume a total condenser and a partial reboiler.

9.14. A feed consisting of 62 mol% *para*-dichlorobenzene in *ortho*-dichlorobenzene is to be separated by distillation at near atmospheric pressure into a distillate containing 98 mol% para isomer and bottoms containing 96 mol% ortho isomer.

If a total condenser and partial reboiler are used, $q = 0.9$, average relative volatility = 1.154, and reflux/minimum

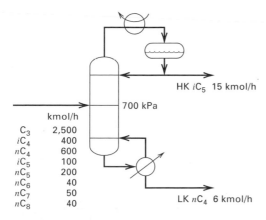

Figure 9.23 Data for Exercise 9.7.

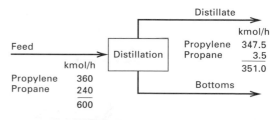

Figure 9.25 Data for Exercise 9.13.

reflux = 1.15, use the Fenske–Underwood–Gilliland procedure to estimate the number of theoretical stages required.

9.15 Explain why the Gilliland correlation can give erroneous results for cases where the ratio of rectifying to stripping stages is small.

9.16 The hydrocarbon feed to a distillation column is a bubble-point liquid at 300 psia with the mole fraction composition, $C_2 = 0.08$, $C_3 = 0.15$, $nC_4 = 0.20$, $nC_5 = 0.27$, $nC_6 = 0.20$, and $nC_7 = 0.10$.
(a) For a sharp separation between nC_4 and nC_5, determine the column pressure and type of condenser if condenser outlet temperature is 120°F.
(b) At total reflux, determine the separation for eight theoretical stages overall, specifying 0.01 mole fraction nC_4 in the bottoms product.
(c) Determine the minimum reflux ratio for the separation in part (b).
(d) Determine the number of theoretical stages at $L/D = 1.5$ times minimum using the Gilliland correlation.

9.17 The following feed mixture is to be separated by ordinary distillation at 120 psia so as to obtain 92.5 mol% of the nC_4 in the liquid distillate and 82.0 mol% of the iC_5 in the bottoms.

Component	lbmol/h
C_3	5
iC_4	15
nC_4	25
iC_5	20
nC_5	35
	100

(a) Estimate the minimum number of equilibrium stages required by applying the Fenske equation. Obtain K-values from Figures 2.8 and 2.9.
(b) Use the Fenske equation to determine the distribution of nonkey components between distillate and bottoms.
(c) Assuming that the feed is at its bubble point, use the Underwood method to estimate the minimum reflux ratio.
(d) Determine the number of theoretical stages required by the Gilliland correlation assuming $L/D = 1.2(L/D)_{min}$, a partial reboiler, and a total condenser.
(e) Estimate the feed-stage location.

9.18 Consider the separation by distillation of a chlorination effluent to recover C_2H_5Cl. The feed is a bubble-point liquid at the column pressure of 240 psia with the following composition and K-values for the column conditions:

Component	Mole Fraction	K
C_2H_4	0.05	5.1
HCl	0.05	3.8
C_2H_6	0.10	3.4
C_2H_5Cl	0.80	0.15

Specifications are: (x_D/x_B) for C_2H_5Cl = 0.01
(x_D/x_B) for C_2H_6 = 75

Calculate the product distribution, the minimum theoretical stages, the minimum reflux, and the theoretical stages at 1.5 times minimum L/D and locate the feed state. The column is to have a partial condenser and a partial reboiler.

9.19 One hundred kilogram-moles per hour of a three-component bubble-point mixture to be separated by distillation has the following composition:

Component	Mole Fraction	Relative Volatility
A	0.4	5
B	0.2	3
C	0.4	1

(a) For a distillate rate of 60 kmol/h, five theoretical stages, and total reflux, calculate the distillate and bottoms compositions by the Fenske equation.
(b) Using the separation in part (a) for components B and C, determine the minimum reflux and minimum boilup ratio by the Underwood equation.
(c) For an operating reflux ratio of 1.2 times the minimum, determine the number of theoretical stages and the feed-stage location.

9.20 For the conditions of Exercise 9.6, determine the ratio of rectifying to stripping equilibrium stages by: (a) Fenske equation, (b) Kirkbride equation, and (c) McCabe–Thiele diagram. Discuss your results.

Section 9.2

9.21 Derive (9-37) in detail starting with equations like (5-46) and (5-47).

9.22 Determine by the Kremser group method the separation that can be achieved for the absorption operation indicated in Figure 9.26 for the following combinations of conditions: (a) Six equilibrium stages and 75-psia operating pressure, (b) Three equilibrium stages and 150-psia op-

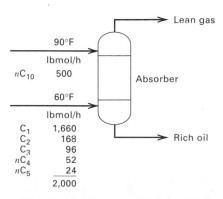

Figure 9.26 Data for Exercise 9.22.

erating pressure, and (c) Six equilibrium stages and 150-psia operating pressure

9.23 One thousand kilogram-moles per hour of rich gas at 70°F with 25% C_1, 15% C_2, 25% C_3, 20% nC_4, and 15% nC_5 by moles is to be absorbed by 500 kmol/h of nC_{10} at 90°F in an absorber operating at 4 atm. Calculate by the Kremser group method the percent absorption of each component for: (a) Four theoretical stages, (b) Ten theoretical stages, and (c) Thirty theoretical stages. Use Figures 2.8 and 2.9 for K-values.

9.24 For the flashing and stripping operation indicated in Figure 9.27, determine by the Kremser group method the kilogram-moles per hour of steam if the stripper is operated at 2 atm and has five theoretical stages.

9.25 A stripper operating at 50 psia with three equilibrium stages is used to strip 1,000 kmol/h of liquid at 250°F having the following molar composition: 0.03% C_1, 0.22% C_2, 1.82% C_3, 4.47% nC_4, 8.59% nC_5, 84.87% nC_{10}. The stripping agent is 100 kmol/h of superheated steam at 300°F and 50 psia. Use the group method to estimate the compositions and flow rates of the stripped liquid and rich gas.

9.26 One hundred kilogram-moles per hour of an equimolar mixture of benzene (B), toluene (T), n-hexane (C_6), and n-heptane (C_7) is to be extracted at 150°C by 300 kmol/h of diethylene glycol (DEG) in a countercurrent liquid–liquid extractor having five equilibrium stages. Estimate the flow rates and compositions of the extract and raffinate streams by the group method. In mole-fraction units, the distribution coefficients for the hydrocarbon can be assumed essentially constant at the following values:

Component	K_{D_i} = y(solvent phase)/x(raffinate phase)
B	0.33
T	0.29
C_6	0.050
C_7	0.043

For diethylene glycol, assume $K_D = 30$. [E.D. Oliver, *Diffusional Separation Processes*, John Wiley and Sons, New York, p. 432 (1966).]

9.27 A reboiled stripper in a natural-gas plant is to be used to remove mainly propane and lighter components from the feed shown in Figure 9.28. Determine by the group method the compositions of the vapor and liquid products.

9.28 A mixture of ethylbenzene and xylenes is to be distilled as shown in Figure 9.29. Assuming the applicability of Raoult's and Dalton's laws:
(a) Use the Fenske–Underwood–Gilliland method to estimate the number of stages required for a reflux-to-minimum reflux ratio of 1.10. Estimate the feed stage location by the Kirkbride equation.
(b) From the results of part (a) for reflux, stages, and distillate rate, use the Edmister group method of Section 5.4 to predict the compositions of the distillate and bottoms. Compare the results with the specifications.

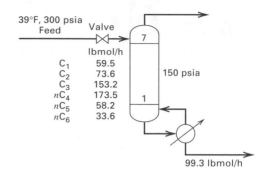

Figure 9.28 Data for Exercise 9.27.

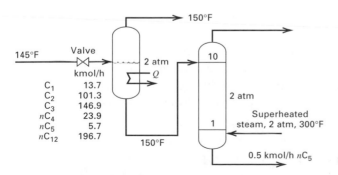

Figure 9.27 Data for Exercise 9.24.

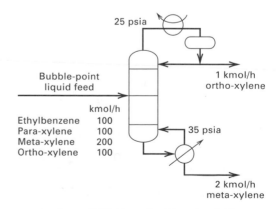

Figure 9.29 Data for Exercise 9.28.

Chapter 10

Equilibrium-Based Methods for Multicomponent Absorption, Stripping, Distillation, and Extraction

Previous chapters have considered graphical, empirical, and approximate group methods for the solution of multistage separation problems involving equilibrium stages. Except for simple cases, such as binary distillation, these methods are suitable only for preliminary design studies. Final design of multistage equipment for conducting multicomponent separations requires rigorous determination of temperatures, pressures, stream flow rates, stream compositions, and heat transfer rates at each stage. (However, rigorous calculational procedures may not be justified when multicomponent physical properties or stage efficiencies are not reasonably well known.) This determination is made by solving material balance, energy (enthalpy) balance, and equilibrium relations for each stage. Unfortunately, these relations are nonlinear algebraic equations that interact strongly. Consequently, solution procedures are relatively difficult and tedious. However, once the procedures are programmed for a high-speed digital computer, solutions are achieved fairly rapidly and almost routinely. Such programs are readily available and widely used. This chapter discusses the solution methods used by such programs, with applications to absorption, stripping, distillation, and liquid–liquid extraction. Applications to extractive, azeotropic, and reactive distillation are covered in Chapter 11.

10.1 THEORETICAL MODEL FOR AN EQUILIBRIUM STAGE

Consider a general, continuous, steady-stage vapor–liquid or liquid–liquid separator consisting of a number of stages arranged in a countercurrent cascade. Assume that: (1) phase equilibrium is achieved at each stage, (2) no chemical reactions occur, and (3) entrainment of liquid drops in vapor and occlusion of vapor bubbles in liquid are negligible. A general

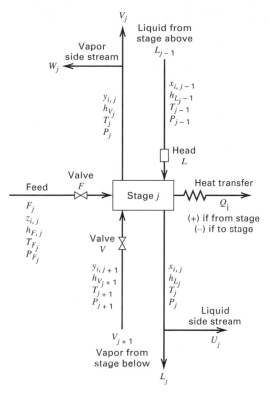

Figure 10.1 General equilibrium stage.

schematic representation of an equilibrium stage j is shown in Fig. 10.1 for a vapor–liquid separator, where the stages are numbered down from the top. The same representation applies to a liquid–liquid separator if the higher-density liquid phases are represented by liquid streams and the lower-density liquid phases are represented by vapor streams.

Entering stage j can be one single- or two-phase feed of molal flow rate F_j, with overall composition in mole fractions $z_{i,j}$ of component i, temperature T_{F_j}, pressure P_{F_j}, and corresponding overall molar enthalpy h_{F_j}. Feed pressure is assumed equal to or greater than stage pressure P_j. Any excess feed pressure $(P_F - P_j)$ is reduced to zero adiabatically across valve F.

Also entering stage j can be interstage liquid from stage $j - 1$ above, if any, of molar flow rate L_{j-1}, with composition in mole fractions $x_{i,j-1}$, enthalpy $h_{L_{j-1}}$, temperature T_{j-1}, and pressure P_{j-1}, which is equal to or less than the pressure of stage j. Pressure of liquid from stage $j - 1$ is increased adiabatically by hydrostatic head change across head L.

Similarly, from stage $j + 1$ below, interstage vapor of molar flow rate V_{j+1}, with composition in mole fractions $y_{i,j+1}$, enthalpy $h_{V_{j+1}}$, temperature T_{j+1}, and pressure P_{j+1} can enter stage j. Any excess pressure $(P_{j+1} - P_j)$ is reduced to zero adiabatically across valve V.

Leaving stage j is vapor of intensive properties $y_{i,j}$, h_{V_j}, T_j, and P_j. This stream can be divided into a vapor side stream of molar flow rate W_j and an interstage stream of molar flow rate V_j to be sent to stage $j - 1$ or, if $j = 1$, to leave the separator as a product. Also leaving stage j is liquid of intensive properties $x_{i,j}$, h_{L_j}, T_j, and P_j, which is in equilibrium with vapor $(V_j + W_j)$. This liquid can be divided also into a liquid side stream of molar flow rate U_j and an interstage or product stream of molar flow rate L_j to be sent to stage $j + 1$ or, if $j = N$, to leave the multistage separator as a product.

Heat can be transferred at a rate Q_j from (+) or to (−) stage j to simulate stage intercoolers, interheaters, condensers, or reboilers as shown in Fig. 1.8. The model in Figure

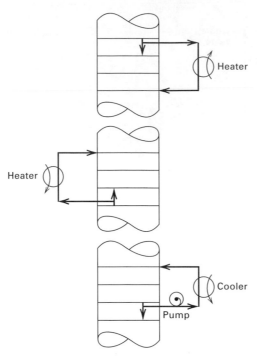

Figure 10.2 Pumparounds.

10.1 does not allow for pumparounds of the type shown in Fig. 10.2. Such pumparounds are often used in columns having side streams in order to conserve energy and balance column vapor loads.

Associated with each general theoretical stage are the following indexed equations expressed in terms of the variable set in Figure 10.1. However, variables other than those shown in Figure 10.1 can be used. For example, component flow rates can replace mole fractions, and side-stream flow rates can be expressed as fractions of interstage flow rates. The equations are similar to those of Section 5.5[1] and are often referred to as the MESH equations after Wang and Henke [1]

1. *M* equations–*M*aterial balance for each component (*C* equations for each stage).

$$M_{i,j} = L_{j-1}x_{i,j-1} + V_{j+1}y_{i,j+1} + F_jz_{i,j} - (L_j + U_j)x_{i,j} - (V_j + W_j)y_{i,j} = 0 \qquad \textbf{(10-1)}$$

2. *E* equations—phase *E*quilibrium relation for each component (*C* equations for each stage).

$$E_{i,j} = y_{i,j} - K_{i,j}x_{i,j} = 0 \qquad \textbf{(10-2)}$$

where $K_{i,j}$ is the phase equilibrium ratio.

---

[1] Unlike the treatment in Section 5.5, all *C* component material balances are included here, and the total material balance is omitted. Also, the separate but equal temperature and pressure of the equilibrium phases are replaced by the stage temperature and pressure.

3. *S* equations—mole fraction *S*ummations (one for each stage),

$$(S_y)_j = \sum_{i=1}^{C} y_{i,j} - 1.0 = 0 \qquad \textbf{(10-3)}$$

$$(S_x)_j = \sum_{i=1}^{C} x_{i,j} - 1.0 = 0 \qquad \textbf{(10-4)}$$

4. *H* equation—energy balance (one for each stage).

$$H_j = L_{j-1}h_{L_{j-1}} + V_{j+1}h_{V_{j+1}} + F_j h_{F_j} - (L_j + U_j)h_{L_j} - (V_j + W_j)h_{V_j} - Q_j = 0 \quad \textbf{(10-5)}$$

where kinetic and potential energy changes are ignored.

A total material balance equation can be used in place of (10-3) or (10-4). It is derived by combining these two equations and $\sum_i z_{i,j} = 1.0$ with (10-1) summed over the *C* components and over stages 1 through *j* to give

$$L_j = V_{j+1} + \sum_{m=1}^{j} (F_m - U_m - W_m) - V_1 \qquad \textbf{(10-6)}$$

In general, $K_{i,j} = K_{i,j}\{T_j, P_j, \mathbf{x}_j, \mathbf{y}_j\}$, $h_{V_j} = h_{V_j}\{T_j, P_j, \mathbf{y}_j\}$, and $h_{L_j} = h_{L_j}\{T_j, P_j, \mathbf{x}_j\}$. If these relations are not counted as equations and the three properties are not counted as variables, each equilibrium stage is defined only by the $2C + 3$ MESH equations. A countercurrent cascade of *N* such stages, as shown in Figure 10.3, is represented by $N(2C + 3)$ such equations in $[N(3C + 10) + 1]$ variables. If *N* and all F_j, $z_{i,j}$, T_{F_j}, P_{F_j}, P_j, U_j, W_j, and Q_j are specified, the model is represented by $N(2C + 3)$ simultaneous

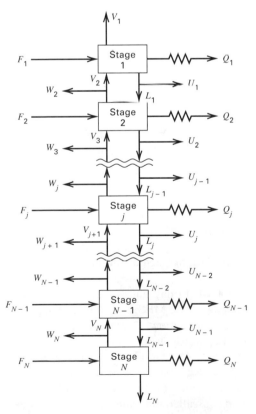

Figure 10.3 General countercurrent cascade of *N* stages.

algebraic equations in $N(2C + 3)$ unknown (output) variables comprising all $x_{i,j}$, $y_{i,j}$, L_j, V_j, and T_j, where the M, E, and H equations are nonlinear. If other variables are specified, as they often are, corresponding substitutions are made to the list of output variables. Regardless of the specifications, the result is a set of nonlinear equations that must be solved by iterative techniques.

10.2 GENERAL STRATEGY OF MATHEMATICAL SOLUTION

A wide variety of iterative solution procedures for solving nonlinear algebraic equations has appeared in the literature. In general, these procedures make use of equation partitioning in conjunction with equation tearing and/or linearization by Newton–Raphson techniques, which are described in detail by Myers and Seider [2]. The equation-tearing method was applied in Section 4.4 for computing an adiabatic flash.

Early attempts to solve (10-1) to (10-5) or equivalent forms of these equations resulted in the classical *stage-by-stage, equation-by-equation* calculational procedures of Lewis–Matheson [3] in 1932 and Thiele–Geddes [4] in 1933 based on equation tearing for solving simple fractionators with one feed and two products. Composition-independent K-values and component enthalpies were generally employed. The Thiele–Geddes method was formulated to handle the Case II variable specification in Table 5.2 wherein the number of equilibrium stages above and below the feed, the reflux ratio, and the distillate flow rate are specified, and stage temperatures and interstage vapor (or liquid) flow rates are the iteration (tear) variables. Although widely used for hand calculations in the years immediately following its appearance in the literature, the Thiele–Geddes method was found often to be numerically unstable when attempts were made to program it for a digital computer. However, Holland and co-workers [5] developed an improved Thiele–Geddes procedure called the *theta method*, which in various versions has been applied with considerable success.

The Lewis–Matheson method is also an equation-tearing procedure. It was formulated according to the Case I variable specification in Table 5.2 to determine stage requirements for specifications of the separation of two key components, a reflux ratio and a feed-stage location criterion. Both outer and inner iterations are required. The outer loop tear variables are the mole fractions or flow rates of nonkey components in the products. The inner loop tear variables are the interstage vapor (or liquid) flow rates. The Lewis–Matheson method was widely used for hand calculations, but it also proved often to be numerically unstable when implemented on a digital computer.

Rather than using an equation-by-equation solution procedure, Amundson and Pontinen [6] in a significant development in 1958, showed that (10-1), (10-2), and (10-6) of the MESH equations for a Case II specification could be combined and solved component-by-component from simultaneous linear equation sets for all N stages by an equation-tearing procedure using the same tear variables as the Thiele–Geddes method. Although too tedious for hand calculations, such equation sets are readily solved with a digital computer.

In a classic study in 1964, Friday and Smith [7] systematically analyzed a number of tearing techniques for solving the MESH equations. They carefully considered the choice of output variable for each equation. They showed that no one technique could solve all types of problems. For separators where the feed(s) contains only components of similar volatility (narrow-boiling case), a modified Amundson–Pontinen approach termed the bubble-point (BP) method was recommended. For a feed(s) containing components of widely different volatility (wide-boiling case) or solubility, the BP method was shown to be subject to failure and a so-called sum-rates (SR) method was suggested. For intermediate cases, the equation-tearing technique may fail to converge; in that case, Friday and Smith indicated that either a Newton–Raphson method or a combined tearing and Newton–Raphson technique was necessary. Boston and Sullivan [8] in 1974 presented an alternative robust approach to obtaining a solution to the MESH equations. They defined energy and volatility parameters, which are used as the primary successive approximation variables. A third parameter, which is a combination of the phase flow rates and temperature at each stage, is employed to iterate on the primary variables; thus the name *inside-out method.* Current practice is based mainly on the BP, SR, Newton–Raphson, and inside-out methods, all of which are treated in this chapter. The latter two methods permit considerable flexibility in the choice of specified variables and generally are capable of solving most problems.

10.3 EQUATION-TEARING PROCEDURES

In general the modern equation-tearing procedures are readily programmed, are rapid, and require a minimum of computer storage. Although they can be applied to a much wider variety of problems than the classical Thiele–Geddes tearing procedure, they are usually limited to the same choice of specified variables. Thus, neither product purities, species recoveries, interstage flow rates, nor stage temperatures can be specified.

Tridiagonal Matrix Algorithm

The key to the success of the BP and SR tearing procedures is the tridiagonal matrix that results from a modified form of the M equations (10-1) when they are torn from the other equations by selecting T_j and V_j as the tear variables, which leaves the modified M equations linear in the unknown liquid mole fractions. This set of equations for each component is solved by a highly efficient and reliable modified Gaussian elimination algorithm due to Thomas as applied by Wang and Henke [1]. The modified M equations are obtained by substituting (10-2) into (10-1) to eliminate y and by substituting (10-6) into (10-1) to eliminate L. Thus, equations for calculating y and L are partitioned from the other equations. The result for each component and each stage is as follows, where the i subscripts have been deleted from the B, C, and D terms.

$$A_j x_{i,j-1} + B_j x_{i,j} + C_j x_{i,j+1} = D_j \tag{10-7}$$

where

$$A_j = V_j + \sum_{m=1}^{j-1} (F_m - W_m - U_m) - V_1, \quad 2 \le j \le N \tag{10-8}$$

$$B_j = - \left[V_{j+1} + \sum_{m=1}^{j} (F_m - W_m - U_m) - V_1 + U_j + (V_j + W_j) K_{i,j} \right], \quad 1 \le j \le N \tag{10-9}$$

$$C_j = V_{j+1} K_{i,j+1}, \quad 1 \le j \le N - 1 \tag{10-10}$$

$$D_j = -F_j z_{i,j}, \quad 1 \le j \le N \tag{10-11}$$

with $x_{i,0} = 0$, $V_{N+1} = 0$, $W_1 = 0$, and $U_N = 0$, as indicated in Fig. 10-3. If the modified M equations are grouped by component, they can be partitioned by writing them as a series of C separate tridiagonal matrix equations where the output variable for each matrix equation is x_i over the entire countercurrent cascade of N stages.

$$\begin{bmatrix} B_1 & C_1 & 0 & 0 & 0 & .. & .. & & & .. & 0 \\ A_2 & B_2 & C_2 & 0 & 0 & .. & .. & & & .. & 0 \\ 0 & A_3 & B_3 & C_3 & 0 & .. & .. & & & .. & 0 \\ .. & .. & & & & & & & & & .. \\ .. & & & & & & & & & & .. \\ .. & & & & & & & & & & .. \\ .. & & & & & & & & & & .. \\ .. & & & & & & & & & & .. \\ 0 & .. & & & .. & .. & 0 & A_{N-2} & B_{N-2} & C_{N-2} & 0 \\ 0 & .. & & & .. & .. & 0 & 0 & A_{N-1} & B_{N-1} & C_{N-1} \\ 0 & .. & & & .. & .. & 0 & 0 & 0 & A_N & B_N \end{bmatrix} \begin{bmatrix} x_{i,1} \\ x_{i,2} \\ x_{i,3} \\ ... \\ ... \\ ... \\ ... \\ ... \\ x_{i,N-2} \\ x_{i,N-1} \\ x_{i,N} \end{bmatrix} = \begin{bmatrix} D_1 \\ D_2 \\ D_3 \\ .. \\ .. \\ .. \\ .. \\ .. \\ D_{N-2} \\ D_{N-1} \\ D_N \end{bmatrix}$$

$$(10\text{-}12)$$

Constants B_j and C_j for each component depend only on tear variables T and V provided that K-values are composition independent. If not, compositions from the previous iteration may be used to estimate the K-values.

The Thomas algorithm for solving the linearized equation set (10-12) is a Gaussian elimination procedure that involves forward elimination starting from stage 1 and working toward stage N to finally isolate $x_{i,N}$. Other values of $x_{i,j}$ are then obtained starting with $x_{i,N-1}$ by backward substitution. For five stages, the matrix equations at the beginning, middle, and end of the procedure are as shown in Figure 10-4.

$$\begin{bmatrix} B_1 & C_1 & 0 & 0 & 0 \\ A_2 & B_2 & C_2 & 0 & 0 \\ 0 & A_3 & B_3 & C_3 & 0 \\ 0 & 0 & A_4 & B_4 & C_4 \\ 0 & 0 & 0 & A_5 & B_5 \end{bmatrix} \begin{bmatrix} x_1 \\ x_2 \\ x_3 \\ x_4 \\ x_5 \end{bmatrix} = \begin{bmatrix} D_1 \\ D_2 \\ D_3 \\ D_4 \\ D_5 \end{bmatrix}$$

(a)

$$\begin{bmatrix} 1 & p_1 & 0 & 0 & 0 \\ 0 & 1 & p_2 & 0 & 0 \\ 0 & 0 & 1 & p_3 & 0 \\ 0 & 0 & 0 & 1 & p_4 \\ 0 & 0 & 0 & 0 & 1 \end{bmatrix} \begin{bmatrix} x_1 \\ x_2 \\ x_3 \\ x_4 \\ x_5 \end{bmatrix} = \begin{bmatrix} q_1 \\ q_2 \\ q_3 \\ q_4 \\ q_5 \end{bmatrix}$$

(b)

$$\begin{bmatrix} 1 & 0 & 0 & 0 & 0 \\ 0 & 1 & 0 & 0 & 0 \\ 0 & 0 & 1 & 0 & 0 \\ 0 & 0 & 0 & 1 & 0 \\ 0 & 0 & 0 & 0 & 1 \end{bmatrix} \begin{bmatrix} x_1 \\ x_2 \\ x_3 \\ x_4 \\ x_5 \end{bmatrix} = \begin{bmatrix} r_1 \\ r_2 \\ r_3 \\ r_4 \\ r_5 \end{bmatrix}$$

(c)

Figure 10.4 The coefficient matrix for the modified M-equations of a given component at various steps in the Thomas algorithm for five equilibrium stages (Note that the i subscript is deleted from x.) (a) Initial matrix. (b) Matrix after forward elimination. c) Matrix after backward substitution.

The equations used in the Thomas algorithm are as follows:

For stage 1, (10-7) is $B_1 x_{i,1} + C_1 x_{i,2} = D_1$, which can be solved for $x_{i,1}$ in terms of unknown $x_{i,2}$ to give

$$x_{i,1} = \frac{D_1 - C_1 x_{i,2}}{B_1}$$

Let
$$p_1 = \frac{C_1}{B_1} \quad \text{and} \quad q_1 = \frac{D_1}{B_1}$$

Then
$$x_{i,1} = q_1 - p_1 x_{i,2} \tag{10-13}$$

Thus, the coefficients in the matrix become $B_1 \leftarrow 1$, $C_1 \leftarrow p_1$, and $D_1 \leftarrow q_1$, where $\leftarrow$ means "is replaced by." Only values for p_1 and q_1 need be stored.

For stage 2, (10-7) can be combined with (10-13) and solved for $x_{i,2}$ to give

$$x_{i,2} = \frac{D_2 - A_2 q_1}{B_2 - A_2 p_1} - \left(\frac{C_2}{B_2 - A_2 p_1} \right) x_{i,3}$$

Let
$$q_2 = \frac{D_2 - A_2 q_1}{B_2 - A_2 p_1} \quad \text{and} \quad p_2 = \frac{C_2}{B_2 - A_2 p_1}$$

Then
$$x_{i,2} = q_2 - p_2 x_{i,3}$$

Thus, $A_2 \leftarrow 0$, $B_2 \leftarrow 1$, $C_2 \leftarrow p_2$, and $D_2 \leftarrow q_2$. Only values for p_2 and q_2 need be stored.

In general, we can define

$$p_j = \frac{C_j}{B_j - A_j p_{j-1}} \tag{10-14}$$

$$q_j = \frac{D_j - A_j q_{j-1}}{B_j - A_j p_{j-1}} \tag{10-15}$$

Then
$$x_{i,j} = q_j - p_j x_{i,j+1} \tag{10-16}$$

with $A_j \leftarrow 0$, $B_j \leftarrow 1$, $C_j \leftarrow p_j$, and $D_j \leftarrow q_j$. Only values of p_j and q_j need be stored. Thus, starting with stage 1, values of p_j and q_j are computed recursively in the order $p_1, q_1, p_2, q_2, \ldots, p_{N-1}, q_{N-1}, q_N$. For stage N, (10-16) isolates $x_{i,N}$ as

$$x_{i,N} = q_N \tag{10-17}$$

Successive values of x_i are computed recursively by backward substitution from (10-16) in the form

$$x_{i,j-1} = q_{j-1} - p_{j-1} x_{i,j} = r_{j-1} \tag{10-18}$$

Equation (10-18) corresponds to the identity coefficient matrix.

The Thomas algorithm, when applied in this fashion, generally avoids buildup of computer truncation errors because usually none of the steps involves subtraction of nearly equal quantities. Furthermore, computed values of $x_{i,j}$ are almost always positive. The algorithm is highly efficient, requires a minimum of computer storage as noted previously, and is superior to alternative matrix-inversion routines. A modified Thomas algorithm for difficult cases is given by Boston and Sullivan [9]. Such cases can occur for columns having large numbers of equilibrium stages and with components whose absorption factors [see (5-38)] are less than unity in one section of stages and greater than unity in another section.

Bubble-Point Method for Distillation

Frequently, distillation involves species that cover a relatively narrow range of vapor–liquid equilibrium ratios (K-values). A particularly effective solution procedure for this case was suggested by Friday and Smith [7] and developed in detail by Wang and Henke [1]. It is referred to as the bubble-point (BP) method because a new set of stage temperatures is computed during each iteration from bubble-point equations. In the method, all equations are partitioned and solved sequentially except for the modified M equations, which are solved separately for each component by the tridiagonal matrix technique.

The algorithm for the Wang–Henke BP method is shown in Figure 10.5. A FORTRAN computer program for the method is available [10]. Problem specifications consist of conditions and stage location of all feeds, pressure at each stage, total flow rates of all side streams (note that liquid distillate flow rate, if any, is designated as U_1), heat transfer rates to or from all stages except stage 1 (condenser) and stage N (reboiler), total number of stages, external bubble-point reflux flow rate, and vapor distillate flow rate. A sample problem specification is shown in Figure 10.6.

To initiate the calculations, values for the tear variables are assumed. For most problems, it is sufficient to establish an initial set of V_j values based on the assumption of constant molar interstage flows using the specified reflux, distillate, feed, and side-stream flow rates. A generally adequate initial set of T_j values can be provided by computing or assuming both the bubble-point temperature of an estimated bottoms product and the dew-point

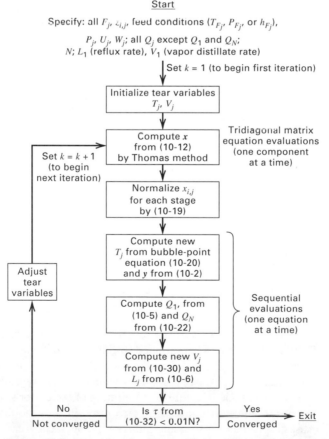

Figure 10.5 Algorithm for Wang–Henke BP method for distillation.

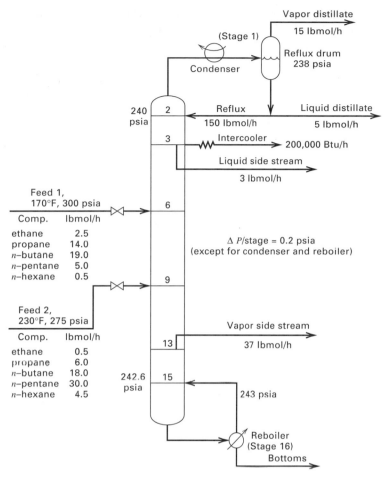

Figure 10.6 Sample specification for application of Wang–Henke BP method to distillation.

temperature of an assumed vapor distillate product; or computing or assuming bubble-point temperature if distillate is liquid or a temperature in-between the dew-point and bubble-point temperatures if distillate is mixed (both vapor and liquid); and then determining the other stage temperatures by assuming a linear variation of temperature with stage location.

To solve (10-12) for x_i by the Thomas method, $K_{i,j}$ values are required. When they are composition dependent, initial assumptions for all $x_{i,j}$ and $y_{i,j}$ values are also needed unless ideal K-values are employed for the first iteration. For each iteration, the computed set of $x_{i,j}$ values for each stage will, in general, not satisfy the summation constraint given by (10-4). Although not mentioned by Wang and Henke, it is advisable to normalize the set of computed $x_{i,j}$ values by the relation

$$(x_{i,j})_{\text{normalized}} = \frac{x_{i,j}}{\displaystyle\sum_{i=1}^{C} x_{i,j}} \tag{10-19}$$

These normalized values are used for all subsequent calculations involving $x_{i,j}$ during the iteration.

A new set of temperatures T_j is computed stage by stage by computing bubble-point temperatures from the normalized $x_{i,j}$ values. Friday and Smith [7] showed that bubble-point calculations for stage temperatures are particularly effective for mixtures having a narrow range of K-values because temperatures are not then sensitive to composition. For example, in the limiting case where all components have identical K-values, the temperature corresponds to the conditions of $K_{i,j} = 1$ and is not dependent on $x_{i,j}$ values. At the other extreme, however, bubble-point calculations to establish stage temperatures can be very sensitive to composition. For example, consider a binary mixture containing one component with a high K-value that changes little with temperature. The second component has a low K-value that changes very rapidly with temperature. Such a mixture is methane and n-butane at 400 psia. The effect on the bubble-point temperature of small quantities of methane dissolved in liquid n-butane is very large, as indicated by the following results.

Liquid Mole Fraction of Methane	Bubble-Point Temperature, °F
0.000	275
0.018	250
0.054	200
0.093	150

Thus, the BP method is best when components have a relatively narrow range of K-values.

The necessary bubble-point equation is obtained in the manner described in Chapter 4 by combining (10-2) and (10-3) to eliminate $y_{i,j}$, giving

$$\sum_{i=1}^{C} K_{i,j}x_{i,j} = 1.0 = 0 \tag{10-20}$$

which is nonlinear in T_j and must be solved iteratively. Wang and Henke prefer to use Muller's iterative method [11] because it is reliable and does not require the calculation of derivatives. Muller's method requires three initial assumptions of T_j. For each assumption, the value of S_j is computed from

$$S_j = \sum_{i=1}^{C} K_{i,j}x_{i,j} - 1.0 \tag{10-21}$$

The three sets of (T_j, S_j) are fitted to a quadratic equation for S_j in terms of T_j. The quadratic equation is then employed to predict T_j for $S_j = 0$, as required by (10-20). The validity of this value of T_j is checked by using it to compute S_j in (10-21). The quadratic fit and S_j check are repeated with the three best sets of (T_j, S_j) until some convergence tolerance is achieved, say $|T_j^{(n)} - T_j^{(n-1)}|/T_j^{(n)} \le 0.0001$, with T in absolute degrees, where n is the iteration number for the temperature loop in the bubble-point calculation, or one can use $S_j \le 0.0001\ C$, which is preferred.

Values of $y_{i,j}$ are determined along with the calculation of stage temperatures using the E equations, (10-2). With a consistent set of values for $x_{i,j}$, T_j, and $y_{i,j}$, molar enthalpies are computed for each liquid and vapor stream leaving a stage. Since F_1, V_1, U_1, W_1, and L_1 are specified, V_2 is readily obtained from (10-6), and the condenser duty, a (+) quantity, is obtained from (10-5). Reboiler duty, a (−) quantity, is determined by summing (10-5) for all stages to give

$$Q_N = \sum_{j=1}^{N} (F_jh_{F_j} - U_jh_{L_j} - W_jh_{v_j}) - \sum_{j=1}^{N-1} Q_j - V_1h_{V_1} - L_Nh_{L_N} \tag{10-22}$$

A new set of V_j tear variables is computed by applying the following modified energy balance, which is obtained by combining (10-5) and (10-6) twice to eliminate L_{j-1} and L_j. After rearrangement,

$$\alpha_j V_j + \beta_j V_{j+1} = \gamma_j \tag{10-23}$$

where

$$\alpha_j = h_{L_{j-1}} - h_{V_j} \tag{10-24}$$

$$\beta_j = h_{V_{j+1}} - h_{L_j} \tag{10-25}$$

$$\gamma_j = \left[\sum_{m=1}^{j-1} (F_m - W_m - U_m) - V_1 \right] (h_{L_j} - h_{L_{j-1}}) + F_j (h_{L_j} - h_{F_j}) + W_j (h_{V_j} - h_{L_j}) + Q_j \tag{10-26}$$

and enthalpies are evaluated at the stage temperatures last computed rather than at those used to initiate the iteration. Written in didiagonal matrix form (10-23) applied over stages 2 to $N - 1$ is:

$$
\begin{bmatrix}
\beta_2 & 0 & 0 & 0 & . & . & & & 0 \\
\alpha_3 & \beta_3 & 0 & 0 & . & . & & & 0 \\
0 & \alpha_4 & \beta_4 & 0 & . & . & & & 0 \\
. & . & & & & & & & . \\
. & . & & & & & & & . \\
0 & . & . & 0 & \alpha_{N-3} & \beta_{N-3} & 0 & 0 \\
0 & . & . & 0 & 0 & \alpha_{N-2} & \beta_{N-2} & 0 \\
0 & . & . & 0 & 0 & 0 & \alpha_{N-1} & \beta_{N-1}
\end{bmatrix}
\cdot
\begin{bmatrix}
V_3 \\
V_4 \\
V_5 \\
. \\
. \\
V_{N-2} \\
V_{N-1} \\
V_N
\end{bmatrix}
=
\begin{bmatrix}
\gamma_2 - \alpha_2 V_2 \\
\gamma_3 \\
\gamma_4 \\
. \\
. \\
\gamma_{N-3} \\
\gamma_{N-2} \\
\gamma_{N-1}
\end{bmatrix}
\tag{10-27}
$$

Matrix equation (10-27) is readily solved one equation at a time by starting at the top where V_2 is known and working down recursively. Thus,

$$V_3 = \frac{\gamma_2 - \alpha_2 V_2}{\beta_2} \tag{10-28}$$

$$V_4 = \frac{\gamma_3 - \alpha_3 V_3}{\beta_3} \tag{10-29}$$

or, in general

$$V_j = \frac{\gamma_{j-1} - \alpha_{j-1} V_{j-1}}{\beta_{j-1}} \tag{10-30}$$

and so on. Corresponding liquid flow rates are obtained from (10-6).

The solution procedure is considered to be converged when sets of $T_j^{(k)}$ and $V_j^{(k)}$ values are within some prescribed tolerance of corresponding sets of $T_j^{(k-1)}$ and $V_j^{(k-1)}$ values, where k is the iteration index. One possible convergence criterion is

$$\sum_{j=1}^{N} \left[\frac{T_j^{(k)} - T_j^{(k-1)}}{T_j^{(k)}} \right]^2 + \sum_{j=1}^{N} \left[\frac{V_j^{(k)} - V_j^{(k-1)}}{V_j^{(k)}} \right]^2 \le \epsilon \tag{10-31}$$

where T is the absolute temperature and ϵ is some prescribed tolerance. However, Wang and Henke suggest that the following simpler criterion, which is based on successive sets of T_j values only, is adequate.

$$\tau = \sum_{j=1}^{N} [T_j^{(k)} - T_j^{(k-1)}]^2 \le 0.01 N \tag{10-32}$$

Successive substitution is often employed for iterating the tear variables; that is, values of T_j and V_j generated from (10-20) and (10-30), respectively, during an iteration are used

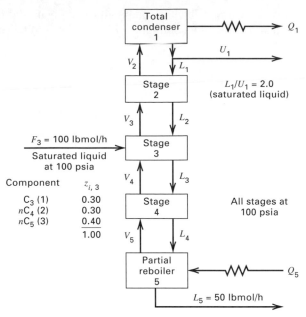

Figure 10.7 Specifications for distillation column of Example 10.1.

directly to initiate the next iteration. However, experience indicates that it is desirable frequently to adjust the values of the generated tear variables prior to beginning the next iteration. For example, upper and lower bounds should be placed on stage temperatures, and any negative values of interstage flow rates should be changed to near-zero positive values. Also, to prevent oscillation of the iterations, damping can be employed to limit changes in the values of V_j and absolute T_j from one iteration to the next to, say, 10%.

EXAMPLE 10.1

For the distillation column shown in Figure 10.7, do one iteration of the BP method up to and including the calculation of a new set of T_j values from (10-20). Use composition-independent K-values.

SOLUTION

By overall total material balance

$$\text{Liquid distillate} = U_1 = F_3 - L_5 = 100 - 50 = 50 \text{ lbmol/h}$$

Then

$$L_1 = (L_1/U_1)U_1 = (2)(50) = 100 \text{ lbmol/h}$$

By total material balance around the total condenser

$$V_2 = L_1 + U_1 = 100 + 50 = 150 \text{ lbmol/h}$$

Initial guesses of tear variables are

Stage j	V_j, lbmol/h	T_j, °F
1	(Fixed at 0 by specifications)	65
2	(Fixed at 150 by specifications)	90
3	150	115
4	150	140
5	150	165

At 100 psia, the estimated K-values at the assumed stage temperatures are

| | $K_{i,j}$ | | | | |
Stage	1	2	3	4	5
$C_3(1)$	1.23	1.63	2.17	2.70	3.33
$nC_4(2)$	0.33	0.50	0.71	0.95	1.25
$nC_5(3)$	0.103	0.166	0.255	0.36	0.49

The matrix equation (10-12) for the first component C_3 is developed as follows. From (10-8) with $V_1 = 0$, $W = 0$,

$$A_j = V_j + \sum_{m=1}^{j-1} (F_m - U_m)$$

Thus, $A_5 = V_5 + F_3 - U_1 = 150 + 100 - 50 = 200$ lbmol/h. Similarly, $A_4 = 200$, $A_3 = 100$, and $A_2 = 100$ in the same units.

From (10-9) with $V_1 = 0$, $W = 0$,

$$B_j = -\left[V_{j+1} + \sum_{m=1}^{j} (F_m - U_m) + U_j + V_j K_{i,j} \right]$$

Thus, $B_5 = -[F_3 - U_1 + V_5 K_{1,5}] = -[100 - 50 + (150)3.33] = -549.5$ lbmol/h. Similarly, $B_4 = -605$, $B_3 = -525.5$, $B_2 = -344.5$, and $B_1 = -150$ in the same units.

From (10-10), $C_j = V_{j+1} K_{1,j+1}$. Thus, $C_1 = V_2 K_{1,2} = 150(1.63) = 244.5$ lbmol/h. Similarly, $C_2 = 325.5$, $C_3 = 405$, and $C_4 = 499.5$ in the same units.

From (10-11), $D_j = -F_j z_{1,j}$. Thus, $D_3 = -100(0.30) = -30$ lbmol/h. Similarly, $D_1 = D_2 = D_4 = D_5 = 0$.

Substitution of the above values in (10-7) gives

$$
\begin{bmatrix}
-150 & 244.5 & 0 & 0 & 0 \\
100 & -344.5 & 325.5 & 0 & 0 \\
0 & 100 & -525.5 & 405 & 0 \\
0 & 0 & 200 & -605 & 499.5 \\
0 & 0 & 0 & 200 & -549.5
\end{bmatrix}
\begin{bmatrix}
x_{1,1} \\
x_{1,2} \\
x_{1,3} \\
x_{1,4} \\
x_{1,5}
\end{bmatrix}
=
\begin{bmatrix}
0 \\
0 \\
-30 \\
0 \\
0
\end{bmatrix}
$$

Using (10-14) and (10-15), we apply the forward step of the Thomas algorithm as follows.

$$p_1 = \frac{C_1}{B_1} = 244.5/(-150) = -1.630$$

$$q_1 = \frac{D_1}{B_1} = 0/(-150) = 0$$

$$p_2 = \frac{C_2}{B_2 - A_2 p_1} = \frac{325.5}{-344.5 - 100(-1.630)} = -1.793$$

By similar calculations, the matrix equation after the forward elimination procedure is

$$
\begin{bmatrix}
1 & -1.630 & 0 & 0 & 0 \\
0 & 1 & -1.793 & 0 & 0 \\
0 & 0 & 1 & -1.170 & 0 \\
0 & 0 & 0 & 1 & -1.346 \\
0 & 0 & 0 & 0 & 1
\end{bmatrix}
\begin{bmatrix}
x_{1,1} \\
x_{1,2} \\
x_{1,3} \\
x_{1,4} \\
x_{1,5}
\end{bmatrix}
=
\begin{bmatrix}
0 \\
0 \\
0.0867 \\
0.0467 \\
0.0333
\end{bmatrix}
$$

Applying the backward steps of (10-17) and (10-18) gives

$$x_{1,5} = q_5 = 0.0333$$

$$x_{1,4} = q_4 - p_4 x_{1,5} = 0.0467 - (-1.346)(0.0333) = 0.0915$$

Similarly, $\qquad x_{1,3} = 0.1938 \quad x_{1,2} = 0.3475 \quad x_{1,1} = 0.5664$

The matrix equations for nC_4 and nC_5 are solved in a similar manner to give

			$x_{i,j}$		
Stage	1	2	3	4	5
C_3	0.5664	0.3475	0.1938	0.0915	0.0333
nC_4	0.1910	0.3820	0.4483	0.4857	0.4090
nC_5	0.0191	0.1149	0.3253	0.4820	0.7806
$\sum_i x_{i,j}$	0.7765	0.8444	0.9674	1.0592	1.2229

After these compositions are normalized, bubble-point temperatures at 100 psia are computed iteratively from (10-20) and compared to the initially assumed values,

Stage	$T^{(2)}$, °F	$T^{(1)}$, °F
1	66	65
2	94	90
3	131	115
4	154	140
5	184	165

∎

The rate of convergence of the BP method is unpredictable, and, as shown in Example 10.2, it can depend drastically on the assumed initial set of T_j values. In addition, cases with high reflux ratios can be more difficult to converge than cases with low reflux ratios. Orbach and Crowe [12] describe a generalized extrapolation method for accelerating convergence based on periodic adjustment of the tear variables when their values form geometric progressions during at least four successive iterations.

EXAMPLE 10.2

Calculate stage temperatures, interstage vapor and liquid flow rates and compositions, reboiler duty, and condenser duty by the BP method for the distillation column specifications given in Example 5.4.

SOLUTION

The computer program of Johansen and Seader [10] based on the Wang–Henke procedure was used. In this program, no adjustments to the tear variables are made prior to the start of each iteration, and the convergence criterion is (10-32). The K-values and enthalpies are computed

	Assumed Temperatures, °F		Number of Iterations
Case	Distillate	Bottoms	for Convergence
1	11.5	164.9	29
2	0.0	200.0	5
3	20.0	180.0	12
4	50.0	150.0	19

from correlations for hydrocarbons. The only initial assumptions required are distillate and bottoms temperatures shown previously for four cases.

The significant effect of initially assumed distillate and bottoms temperatures on the number of iterations required to satisfy (10-32) is indicated by the results shown previously. The terminal temperatures of Case 1 were within a few degrees of the exact values and were much closer estimates that those of the other three cases. Nevertheless, Case 1 required the largest number of iterations. Figure 10.8 is a plot of τ from (10-32) as a function of the number of iterations for each of the four cases. Case 2 converged rapidly to the criterion of $\tau < 0.13$. Cases 1, 3, and 4 converged rapidly for the first three or four iterations, but then moved only slowly toward the criterion. This was particularly true of Case 1, for which application of a convergence acceleration method would be particularly desirable. In none of the four cases did oscillations of values of the tear variables occur; rather the values approached the converged results in a monotonic fashion.

The overall results of the converged calculations, as taken from Case 2, are shown in Fig. 10.9. Product component flow rates were not quite in material balance with the feed. Therefore, adjusted values that do satisfy overall material balance equations were determined by averaging the calculated values and are included in Figure 10.9. A smaller value of τ would have improved the overall material balance. Figures 10.10 to 10.13 are plots of converged values for stage temperatures, interstage flow rates, and mole fraction compositions from the results of Case 2. Results from the other three cases were almost identical to those of Case 2. Included in Figure 10.10 is the initially assumed linear temperature profile. Except for the bottom stages, it does not deviate significantly from the converged profile. A jog in the profile is seen at the feed stage. This is a common occurrence.

In Figure 10.11, it is seen that the assumption of constant interstage molar flow rates does not hold in the rectifying section. Both liquid and vapor flow rates decrease in moving down from the top stage toward the feed stage. Because the feed is vapor near the dew point, the liquid rate changes only slightly across the feed stage. Correspondingly, the vapor rate decreases

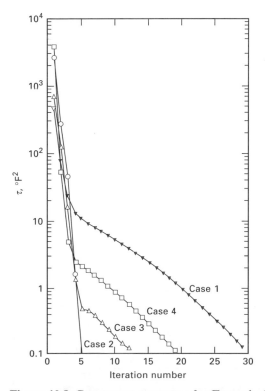

Figure 10.8 Convergence patterns for Example 10.2.

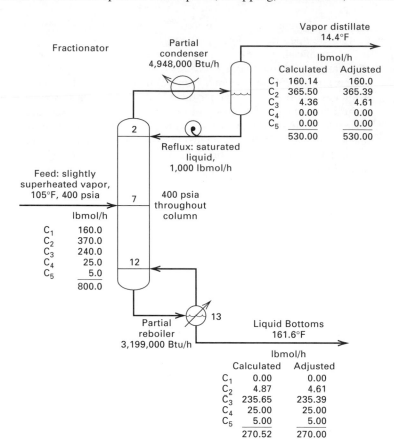

Figure 10.9 Specifications and overall results for Example 10.2.

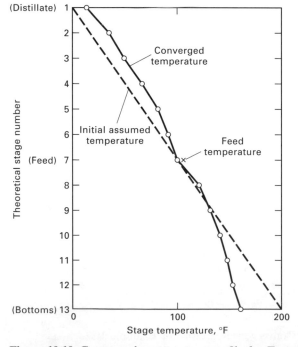

Figure 10.10 Converged temperature profile for Example 10.2.

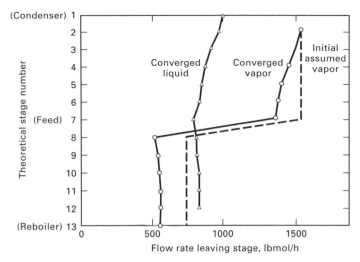

Figure 10.11 Converged interstage flow rate profiles for Example 10.2.

across the feed stage by an amount equal to the feed rate. For this problem, the interstage molar flow rates are almost constant in the stripping section. However, the assumed vapor flow rate in this section based on adjusting the rectifying section rate across the feed zone is approximately 33% higher than the average converged vapor rate. A much better initial estimate of the vapor rate in the stripping section can be made by first computing the reboiler duty from the condenser duty based on the specified reflux rate and then determining the corresponding vapor rate leaving the partial reboiler.

For this problem, the separation is between C_2 and C_3. Thus, these two components can be designated as the light key (LK) and heavy key (HK), respectively. Thus C_1 is a lighter-than-light key (LLK), and C_4 and C_5 are heavier than the heavy key (HHK). Each of these four designations exhibits a different type of composition profile curve as shown in Figures 10.12 and 10.13. Except at the feed zone and at each end of the column, both liquid and vapor mole fractions of the light key (C_2) decrease smoothly and continuously from the top of the column

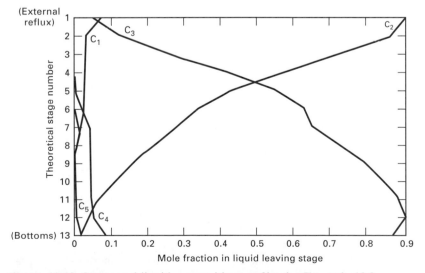

Figure 10.12 Converged liquid composition profiles for Example 10.2.

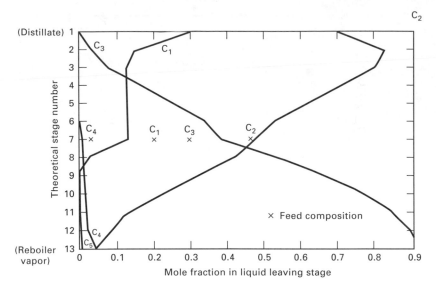

Figure 10.13 Converged vapor composition profiles for Example 10.2.

to the bottom. The inverse occurs for the heavy key (C_3). Mole fractions of methane (LLK) are almost constant over the rectifying section except near the top. Below the feed zone, methane rapidly disappears from both vapor and liquid streams. The inverse is true for the two HHK components. In Fig. 10.13, it is seen that the feed composition is somewhat different from the composition of either the vapor entering the feed stage from the stage below or the vapor leaving the feed stage.   ■

For problems where a specification is made of the distillate flow rate and the number of theoretical stages, it is difficult to specify the feed-stage location that will give the highest degree of separation. However, once the results of a rigorous calculation are available, a modified McCabe–Thiele plot based on the key components [13] can be constructed to determine whether the feed stage is optimally located or whether it should be moved. For this plot, mole fractions of the light-key component are computed on a nonkey-free basis. The resulting diagram for Example 10.2 is shown in Figure 10.14. It is seen that the trend toward a pinched-in region is more noticeable in the rectifying section just above stage 7 than in the stripping section just below stage 7. This suggests that a better separation between the key components might be made by shifting the feed entry to stage 6. The effect of feed-stage location on the percent loss of ethane to the bottoms product is shown in Fig. 10.15. As predicted from Figure 10.14, the optimum feed stage is stage 6.

Sum-Rates Method for Absorption and Stripping

The chemical components present in most absorbers and strippers cover a relatively wide range of volatility. Hence, the BP method of solving the MESH equations will fail because calculation of stage temperature by bubble-point determination (10-20) is too sensitive to liquid-phase composition and the stage energy balance (10-5) is much more sensitive to stage temperatures than to interstage flow rates. In this case, Friday and Smith [7] showed that an alternative procedure devised by Sujata [14] could be successfully applied. This procedure, termed the *sum-rates* (SR) *method,* was further developed in conjunction with the tridiagonal matrix formulation for the modified *M* equations by Burningham and Otto [15].

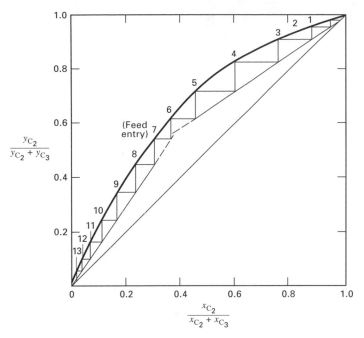

Figure 10.14 Modified McCabe–Thiele diagram for Example 10.2.

Figure 10.16 shows the algorithm for the Burningham–Otto SR method. A FORTRAN computer program for the method is available [16]. Problem specifications consist of conditions and stage locations for all feeds, pressure at each stage, total flow rates of any side streams, heat transfer rates to or from any stages, and total number of stages.

An initial set of tear variables T_j and V_j is assumed to initiate the calculations. For most problems it is sufficient to assume a set of V_j values based on the assumption of constant molar interstage flows, working up from the bottom of the absorber using specified vapor feeds and any vapor side-stream flows. Generally, an adequate initial set of T_j values can be derived from assumed top-stage and bottom-stage values and a linear variation with stages in-between.

Values of $x_{i,j}$ are obtained by solving (10-12) by the Thomas algorithm. However, the

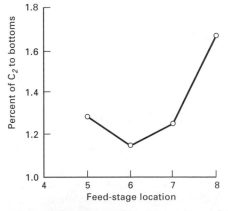

Figure 10.15 Effect of feed-stage location on separation for Example 10.2.

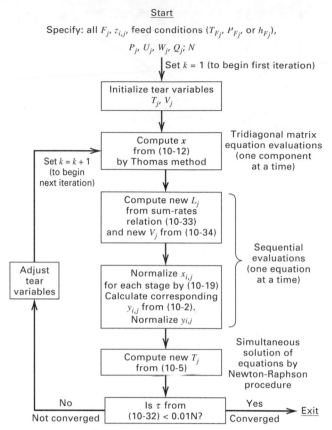

Figure 10.16 Algorithm for Burningham–Otto SR method for absorption/stripping.

values obtained are not normalized at this step but are utilized directly to produce new values of L_j by applying (10-4) in the form referred to as the *sum-rates equation*.

$$L_j^{(k+1)} = L_j^{(k)} \sum_{i=1}^{C} x_{i,j} \tag{10-33}$$

where values of $L_j^{(k)}$ are obtained from values of $V_j^{(k)}$ by (10-6). Corresponding values of $V_j^{(k+1)}$ are obtained from a total material balance, which is derived by summing (10-1) over the C components, combining the result with (10-3) and (10-4), and summing that result over stages j through N to give

$$V_j = L_{j-1} - L_N + \sum_{m=j}^{N} (F_m - W_m - U_m) \tag{10-34}$$

Normalized values of $x_{i,j}$ are next calculated from (10-19). Corresponding values of $y_{i,j}$ are computed from (10-2).

A new set of values for stage temperatures T_j is obtained by solving the simultaneous set of energy balance relations for the N stages given by (10-5). The temperatures are embedded in the specific enthalpies corresponding to the unspecified vapor and liquid flow rates. In general, these enthalpies are nonlinear in temperature. Therefore, an iterative solution procedure is required, such as the Newton–Raphson method [10].

In the Newton–Raphson method, the simultaneous nonlinear equations are written in the form

$$f_i\{x_1, x_2, \ldots, x_n\} = 0, \quad i = 1, 2, \ldots, n \tag{10-35}$$

Initial guesses, marked by asterisks, are provided for the n variables and each function is expanded about these guesses in a Taylor's series that is terminated after the first derivatives to give

$$0 = f_i\{x_1, x_2, \ldots, x_n\} \tag{10-36}$$

$$\approx f_i\{x_1^*, x_2^*, \ldots, x_n^*\} + \left.\frac{\partial f_i}{\partial x_1}\right|^* \Delta x_1 + \left.\frac{\partial f_i}{\partial x_2}\right|^* \Delta x_2 + \cdots + \left.\frac{\partial f_i}{\partial x_n}\right|^* \Delta x_n \tag{10-37}$$

where $\Delta x_j = x_j - x_j^*$.

Equations (10-36) are linear and can be solved directly for the corrections Δx_i. If the corrections are all found to be zero, the guesses are correct and equations (10-35) have been solved; if not, the corrections are added to the guesses to provide a new set of guesses that are applied to (10-36). The procedure is repeated until all the corrections, and thus the functions, become zero to within some tolerance. In recursion form (10-36) and (10-37) are

$$\sum_{j=1}^{n} \left[\left(\frac{\partial f_i}{\partial x_j}\right)^{(r)} \Delta x_j^{(r)} \right] = -f_i^{(r)}, \quad i = 1, 2, \ldots, n \tag{10-38}$$

$$x_j^{(r+1)} = x_j^{(r)} + \Delta x_j^{(r)}, \quad j = 1, 2, \ldots, n \tag{10-39}$$

EXAMPLE 10.3

Solve the simultaneous nonlinear equations

$$x_1 \ln x_2 + x_2 \exp(x_1) = \exp(1)$$
$$x_2 \ln x_1 + 2x_1 \exp(x_2) = 2 \exp(1)$$

for x_1 and x_2 to within ± 0.001, by the Newton–Raphson method.

SOLUTION

In the form of (10-35), the two equations are

$$f_1\{x_1, x_2\} = x_1 \ln x_2 + x_2 \exp(x_1) - \exp(1) = 0$$
$$f_2\{x_1, x_2\} = x_2 \ln x_1 + 2x_1 \exp(x_2) - 2 \exp(1) = 0$$

From (10-38), the linearized recursive form of these equations is

$$\left(\frac{\partial f_1}{\partial x_1}\right)^{(r)} \Delta x_1^{(r)} + \left(\frac{\partial f_1}{\partial x_2}\right)^{(r)} \Delta x_2^{(r)} = -f_1^{(r)}$$

$$\left(\frac{\partial f_2}{\partial x_1}\right)^{(r)} \Delta x_1^{(r)} + \left(\frac{\partial f_2}{\partial x_2}\right)^{(r)} \Delta x_2^{(r)} = -f_2^{(r)}$$

The solution of these two equations is readily obtained by the method of determinants to give

$$\Delta x_1^{(r)} = \frac{\left[f_2^{(r)} \left(\frac{\partial f_1}{\partial x_2}\right)^{(r)} - f_1^{(r)} \left(\frac{\partial f_2}{\partial x_2}\right)^{(r)} \right]}{D}$$

and

$$\Delta x_2^{(r)} = \frac{\left[f_1^{(r)} \left(\frac{\partial f_2}{\partial x_1} \right)^{(r)} - f_2^{(r)} \left(\frac{\partial f_1}{\partial x_1} \right)^{(r)} \right]}{D}$$

where

$$D = \left(\frac{\partial f_1}{\partial x_1} \right)^{(r)} \left(\frac{\partial f_2}{\partial x_2} \right)^{(r)} - \left(\frac{\partial f_1}{\partial x_2} \right)^{(r)} \left(\frac{\partial f_2}{\partial x_1} \right)^{(r)}$$

and the derivatives as obtained from the equations are

$$\left(\frac{\partial f_1}{\partial x_1} \right)^{(r)} = \ln(x_2^{(r)}) + x_2^{(r)} \exp(x_1^{(r)}) \qquad \left(\frac{\partial f_2}{\partial x_1} \right)^{(r)} = \frac{x_2^{(r)}}{x_1^{(r)}} + 2 \exp(x_2^{(r)})$$

$$\left(\frac{\partial f_1}{\partial x_2} \right)^{(r)} = \frac{x_1^{(r)}}{x_2^{(r)}} + \exp(x_1^{(r)}) \qquad \left(\frac{\partial f_2}{\partial x_2} \right)^{(r)} = \ln(x_1^{(r)}) + 2x_1^{(r)} \exp(x_2^{(r)})$$

As initial guesses, take $x_1^{(1)} = 2$, $x_2^{(1)} = 2$. Applying the Newton–Raphson procedure, one obtains the following results where at the sixth iteration values of $x_1 = 1.0000$ and $x_2 = 1.0000$ correspond closely to the required values of zero for f_1 and f_2.

r	$x_1^{(r)}$	$x_2^{(r)}$	$f_1^{(r)}$	$f_2^{(r)}$	$(\partial f_1/\partial x_1)^{(r)}$	$(\partial f_1/\partial x_2)^{(r)}$	$(\partial f_2/\partial x_1)^{(r)}$	$(\partial f_2/\partial x_2)^{(r)}$	$\Delta x_1^{(r)}$	$\Delta x_2^{(r)}$
1	2.0000	2.0000	13.4461	25.5060	15.4731	8.3891	15.7781	30.2494	−0.5743	−0.5436
2	1.4247	1.4564	3.8772	7.3133	6.4354	5.1395	9.6024	12.5880	−0.3544	−0.3106
3	1.0713	1.1457	0.7720	1.3802	3.4806	3.8541	7.3591	6.8067	−0.0138	−0.1878
4	1.0575	0.9579	−0.0059	0.1290	2.7149	3.9830	6.1183	5.5679	−0.0591	0.0417
5	0.9984	0.9996	−0.0057	−0.0122	2.7126	3.7127	6.4358	5.4244	0.00159	0.000368
6	1.0000	1.0000	5.51×10^{-6}	2.86×10^{-6}	2.7183	3.7183	6.4366	5.4366	12.1×10^{-6}	-3.0×10^{-6}
7	1.0000	1.0000	0.0	-2×10^{-9}	2.7183	3.7183	6.4366	5.4366	—	—

■

As applied to the solution of a new set of T_j values from the energy equation (10-5), the recursion equation for the Newton–Raphson method is

$$\left(\frac{\partial H_j}{\partial T_{j-1}} \right)^{(r)} \Delta T_{j-1}^{(r)} + \left(\frac{\partial H_j}{\partial T_j} \right)^{(r)} \Delta T_j^{(r)} + \left(\frac{\partial H_j}{\partial T_{j+1}} \right)^{(r)} \Delta T_{j+1}^{(r)} = -H_j^{(r)} \tag{10-40}$$

where

$$\Delta T_j^{(r)} = T_j^{(r+1)} - T_j^{(r)} \tag{10-41}$$

$$\frac{\partial H_j}{\partial T_{j-1}} = L_{j-1} \frac{\partial h_{L_{j-1}}}{\partial T_{j-1}} \tag{10-42}$$

$$\frac{\partial H_j}{\partial T_j} = -(L_j + U_j) \frac{\partial h_{L_j}}{\partial T_j} - (V_j + W_j) \frac{\partial h_{V_j}}{\partial T_j} \tag{10-43}$$

$$\frac{\partial H_j}{\partial T_{j+1}} = V_{j+1} \frac{\partial h_{V_{j+1}}}{\partial T_{j+1}} \tag{10-44}$$

The partial derivatives depend upon the enthalpy correlations that are utilized. For example, if composition-independent polynomial equations in temperature are used, then

$$h_{V_j} = \sum_{i=1}^{C} y_{i,j}(A_i + B_i T + C_i T^2) \tag{10-45}$$

$$h_{L_j} = \sum_{i=1}^{C} x_{i,j}(a_i + b_i T + c_i T^2) \tag{10-46}$$

and the partial derivatives are

$$\frac{\partial h_{V_j}}{\partial T_j} = \sum_{i=1}^{C} y_{i,j}(B_i + 2C_i T) \tag{10-47}$$

$$\frac{\partial h_{L_j}}{\partial T_j} = \sum_{i=1}^{C} x_{i,j}(b_i + 2c_i T) \tag{10-48}$$

The N relations given by (10-40) form a tridiagonal matrix equation that is linear in $\Delta T_j^{(r)}$. The form of the matrix equation is identical to (10-12) where, for example, $A_2 = (\partial H_2/\partial T_1)^{(r)}$, $B_2 = (\partial H_2/\partial T_2)^{(r)}$, $C_2 = (\partial H_2/\partial T_3)^{(r)}$, $x_{i,2} \leftarrow \Delta T_2^{(r)}$, and $D_2 = -H_2^{(r)}$. The matrix of partial derivatives is called the *Jacobian correction matrix*. The Thomas algorithm can be employed to solve for the set of corrections $\Delta T_j^{(r)}$. New guesses of T_j are then determined from

$$T_j^{(r+1)} = T_j^{(r)} + t \Delta T_j^{(r)} \tag{10-49}$$

where t is a scalar attenuation factor that is useful when initial guesses are not reasonably close to the true values. Generally, as in (10-39), t is taken as 1, but an optimal value can be determined at each iteration to minimize the sum of the squares of the functions,

$$\sum_{j=1}^{N} [H_j^{(r+1)}]^2$$

When all the corrections $\Delta T_j^{(r)}$ have approached zero, the resulting values of T_j are used with criteria such as (10-31) or (10-32) to determine whether convergence has been achieved. If not, before beginning a new k iteration, one can adjust values of V_j and T_j as indicated in Figure 10.16 and previously discussed for the BP method. Rapid convergence is generally observed for the sum-rates method.

EXAMPLE 10.4

Calculate stage temperatures and interstage vapor and liquid flow rates and compositions by the rigorous SR method for the absorber column specifications given in Figure 5.11.

SOLUTION

The digital computer program of Shinohara et al., [16] based on the Burningham–Otto solution procedure, was used. Initial assumptions for the top-stage and bottom-stage temperatures were 90°F (32.2°C) (entering liquid temperature) and 105°F (40.6°C) (entering gas temperature), respectively. The corresponding number of iterations to satisfy the convergence criterion of (10-32) was seven. Values of τ were as follows.

Iteration Number	τ, (°F)^2
1	9,948
2	2,556
3	46.0
4	8.65
5	0.856
6	0.124
7	0.0217

The overall results of the converged calculations are shown in Figure 10.17. Adjusted values of product component flow rates that satisfy overall material balance equations are included. Figures 10.18 to 10.20 are plots of converged values for stage temperatures, interstage total flow rates, and interstage component vapor flow rates, respectively. Figure 10.18 shows that the initial assumed linear temperature profile is grossly in error. Because of the substantial degree of absorption and accompanying high heat of absorption, stage temperatures are consider-

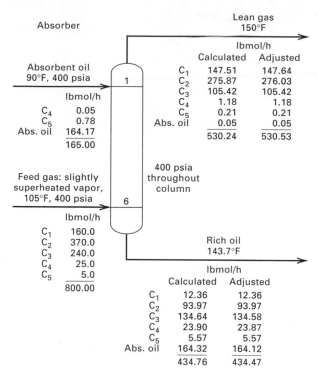

Figure 10.17 Specifications and overall results for Example 10.4.

ably greater than the two entering stream temperatures. The heat is absorbed by both the vapor and liquid streams. The peak stage temperature is essentially at the midpoint of the column. Figure 10.19 shows that the bulk of the overall absorption occurs at the two terminal stages. In Figure 10.20, it is seen that absorption of C_1 and C_2 occurs almost exclusively at the top and bottom stages. Absorption of C_3 occurs throughout the column, but mainly at the two terminal stages. Absorption of C_4 and C_5 also occurs throughout the column, but mainly at the bottom where vapor first contacts absorption oil. ■

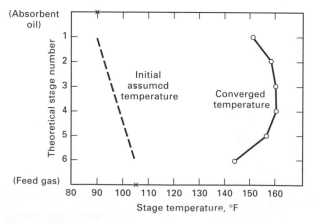

Figure 10.18 Converged temperature profile for Example 10.4.

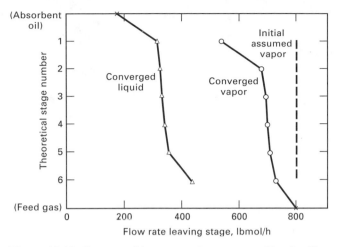

Figure 10.19 Converged interstage flow rate profiles for Example 10.4.

Isothermal Sum-Rates Method for Liquid-Liquid Extraction

Multistage liquid–liquid extraction equipment is operated frequently in an adiabatic manner. When entering streams are at the same temperature and heat of mixing is negligible, the operation is also isothermal. For this condition, or when stage temperatures are specified, as indicated by Friday and Smith [7] and shown in detail by Tsuboka and Katayama [17], a simplified isothermal version of the sum-rates method (ISR) can be applied. It is based on the same equilibrium-stage model presented in Section 10.1. However, with all stage temperatures specified, values of Q_j can be computed from stage energy balances, which can be partitioned from the other equations and solved in a separate step following the calculations discussed here. In the ISR method, particular attention is paid to the possibility that phase compositions may strongly influence K_{ij} values.

Figure 10.21 shows the algorithm for the Tsuboka–Katayama ISR method. Liquid-phase and vapor-phase symbols correspond to raffinate and extract, respectively. Problem

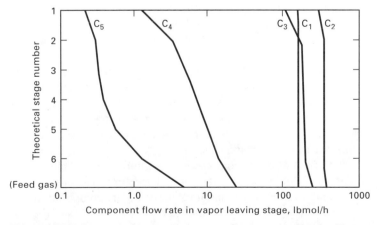

Figure 10.20 Converged component vapor flow rate profiles for Example 10.4.

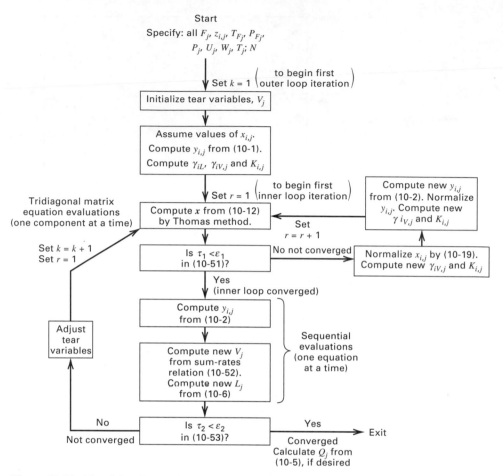

Figure 10.21 Algorithm for Tsuboka–Katayama ISR method for liquid–liquid extraction.

specifications consist of flow rates, compositions, and stage locations for all feeds; stage temperatures (frequently all equal); total flow rates of any side streams; and total number of stages. Stage pressures need not be specified but are understood to be greater than corresponding stage bubble-point pressures to prevent vaporization.

With stage temperatures specified, the only tear variables are V_j values. An initial set is obtained by assuming a perfect separation among the components of the feed and neglecting mass transfer of the solvent to the raffinate phase. This gives approximate values for the flow rates of the exiting raffinate and extract phases. Intermediate values of V_j are obtained by linear interpolation over the N stages. Modifications to this procedure are necessary for side streams or intermediate feeds. As shown in Fig. 10.21, the tear variables are reset in an outer iterative loop.

The effect of phase compositions is often considerable on K-values (distribution coefficients) for liquid–liquid extraction. Therefore, it is best also to provide initial estimates of $x_{i,j}$ and $y_{i,j}$ from which initial values of $K_{i,j}$ are computed. Initial values of $x_{i,j}$ are obtained by linear interpolation, with stage, of the compositions of the known entering and assumed exit streams. Corresponding values of $y_{i,j}$ are computed by material balance from (10-1). Values of $\gamma_{iL,j}$ and $\gamma_{iV,j}$ are determined from an appropriate correlation—for example, the van Laar, NRTL, UNIQUAC, or UNIFAC equations discussed in Chapter 2. Corresponding K-values are obtained from the following equation, which is equivalent to (2-30).

$$K_{i,j} = \frac{\gamma_{iL,j}}{\gamma_{iV,j}} \tag{10-50}$$

A new set of $x_{i,j}$ values is obtained by solving (10-12) by the Thomas algorithm. These values are compared to the assumed values by computing

$$\tau_1 = \sum_{j=1}^{N} \sum_{i=1}^{C} |x_{i,j}^{(r-1)} - x_{i,j}^{(r)}| \tag{10-51}$$

where r is an inner loop index. If $\tau_1 > \epsilon_1$, where, for example, the convergence criterion ϵ_1 might be taken as 0.01 NC, the inner loop is used to improve values of $K_{i,j}$ by using normalized values of $x_{i,j}$ and $y_{i,j}$ to compute new values of $\gamma_{iL,j}$ and $\gamma_{iV,j}$.

When the inner loop is converged, values of $x_{i,j}$ are used to calculate new values of $y_{i,j}$ from (10-2). A new set of tear variables V_j is then computed from the sum-rates relation

$$V_j^{(k+1)} = V_j^{(k)} \sum_{i=1}^{C} y_{i,j} \tag{10-52}$$

where k is an outer loop index. Corresponding values of $L_j^{(k+1)}$ are obtained from (10-6).

The outer loop is converged when

$$\tau_2 = \sum_{j=1}^{N} \left(\frac{V_j^{(k)} - V_j^{(k-1)}}{V_j^{(k)}} \right)^2 \leq \epsilon_2 \tag{10-53}$$

where, for example, the convergence criterion ϵ_2 may be taken as 0.01 N.

Before beginning a new k iteration, we can adjust values of V_j as previously discussed for the BP method. Convergence of the ISR method is generally rapid but is subject to the extent to which $K_{i,j}$ depends upon composition.

EXAMPLE 10.5

The separation of benzene (B) from n-heptane (H) by ordinary distillation is difficult. At atmospheric pressure, the boiling points differ by 18.3°C. However, because of liquid-phase nonideality, the relative volatility decreases to a value less than 1.15 at high benzene concentrations [18]. An alternative method of separation is liquid–liquid extraction with a mixture of dimethylformamide (DMF) and water [19]. The solvent is much more selective for benzene than for n-heptane at 20°C. For two different solvent compositions, calculate interstage flow rates and compositions by the rigorous ISR method for the countercurrent liquid–liquid extraction cascade, which contains five equilibrium stages and is shown schematically in Figure 10.22.

SOLUTION

Experimental phase equilibrium data for the quaternary system [19] were fitted to the NRTL equation by Cohen and Renon [20]. The resulting binary pair constants in (2-92) and (2-93) are

Binary Pair, ij	τ_{ij}	τ_{ji}	α_{ji}
DMF, H	2.036	1.910	0.25
Water, H	7.038	4.806	0.15
B, H	1.196	−0.355	0.30
Water, DMF	2.506	−2.128	0.253
B, DMF	−0.240	0.676	0.425
B, Water	3.639	5.750	0.203

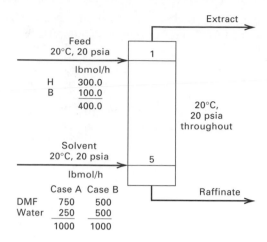

Figure 10.22 Specifications for Example 10.5.

For Case A, estimates of V_j (the extract phase), $x_{i,j}$, and $y_{i,j}$ are as follows, based on a perfect separation and linear interpolation by stage.

Stage		$y_{i,j}$				$x_{i,j}$			
j	V_j	H	B	DMF	Water	H	B	DMF	Water
1	1100	0.0	0.0909	0.6818	0.2273	0.7895	0.2105	0.0	0.0
2	1080	0.0	0.0741	0.6944	0.2315	0.8333	0.1667	0.0	0.0
3	1060	0.0	0.0566	0.7076	0.2359	0.8824	0.1176	0.0	0.0
4	1040	0.0	0.0385	0.7211	0.2404	0.9375	0.0625	0.0	0.0
5	1020	0.0	0.0196	0.7353	0.2451	1.0000	0.0	0.0	0.0

The converged solution is obtained by the ISR method with the following corresponding stage flow rates and compositions

Stage		$y_{i,j}$				$x_{i,j}$			
j	V_j	H	B	DMF	Water	H	B	DMF	Water
1	1113.1	0.0263	0.0866	0.6626	0.2245	0.7586	0.1628	0.0777	0.0009
2	1104.7	0.0238	0.0545	0.6952	0.2265	0.8326	0.1035	0.0633	0.0006
3	1065.6	0.0213	0.0309	0.7131	0.2347	0.8858	0.0606	0.0532	0.0004
4	1042.1	0.0198	0.0157	0.7246	0.2399	0.9211	0.0315	0.0471	0.0003
5	1028.2	0.0190	0.0062	0.7316	0.2432	0.9438	0.0125	0.0434	0.0003

Computed products for the two cases are:

	Extract, lbmol/h		Raffinate, lbmol/h	
	Case A	Case B	Case A	Case B
H	29.3	5.6	270.7	294.4
B	96.4	43.0	3.6	57.0
DMF	737.5	485.8	12.5	14.2
Water	249.9	499.7	0.1	5.0
	1113.1	1034.1	286.9	365.9

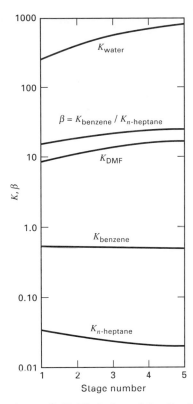

Figure 10.23 Variation of distribution coefficient and relative selectivity for Example 10.5, Case A.

On a percentage extraction basis, the results are:

	Case A	Case B
Percent of benzene feed extracted	96.4	43.0
Percent of n-heptane feed extracted	9.8	1.87
Percent of solvent transferred to raffinate	1.26	1.45

Thus, the solvent with 75% DMF extracts a much larger percentage of the benzene, but the solvent with 50% DMF is more selective between benzene and n-heptane.

For Case A, the variations with stage of K-values and the relative selectivity are shown in Figure 10.23, where the relative selectivity is $\beta_{B,H} = K_B/K_H$. The distribution coefficient for n-heptane varies by a factor of almost 1.75 from stage 5 to stage 1, while the coefficient for benzene is almost constant. The relative selectivity varies by a factor of almost 2. ∎

10.4 SIMULTANEOUS CORRECTION PROCEDURES

The BP and SR methods for vapor–liquid contacting converge only with difficulty or not at all for separations involving very nonideal liquid mixtures or for cases where the separator is like an absorber or stripper in one section and a fractionator in another section (e.g., a reboiled absorber). Furthermore, BP and SR methods are generally restricted to the very limited specifications stated previously. More general procedures capable of solving all types of multicomponent, multistage separation problems are based on the solution of all the MESH equations, or combinations thereof, by simultaneous correction (SC) techniques, often using the Newton–Raphson method.

In order to develop an SC procedure that uses the Newton–Raphson method, one must select and order the unknown variables and the corresponding functions (MESH equations) that contain them. As discussed by Goldstein and Stanfield [21] grouping of the functions by type is computationally most efficient for problems involving a large number of components, but few stages. Alternatively, it is most efficient to group the functions according to stage location for problems involving many stages, but relatively few components. The latter grouping is described by Naphtali [22] and was implemented by Naphtali and Sandholm [23].

The SC procedure of Naphtali and Sandholm is developed in detail because it utilizes many of the mathematical techniques presented in Section 10.3 on tearing methods. A computer program for their method is given by Fredenslund, Gmehling, and Rasmussen [24]. However, that program does not have the flexibility of specifications in Newton–Raphson implementations found in commercial simulators for computer-aided process design.

The equilibrium stage model of Figs. 10.1 and 10.3 is again employed. However, rather than solving the $N(2C + 3)$ MESH equations simultaneously, we combine (10-3) and (10-4) with the other MESH equations to eliminate $2N$ variables and thus reduce the problem to the simultaneous solution of $N(2C + 1)$ equations. This is done by first multiplying (10-3) and (10-4) by V_j and L_j, respectively, to give

$$V_j = \sum_{i=1}^{C} v_{i,j} \tag{10-54}$$

$$L_j = \sum_{i=1}^{C} l_{ij} \tag{10-55}$$

where we have used the mole fraction definitions

$$y_{i,j} = \frac{v_{i,j}}{V_j} \tag{10-56}$$

$$x_{i,j} = \frac{l_{i,j}}{L_j} \tag{10-57}$$

Equations (10-54) to (10-57) are now substituted into (10-1), (10-2), and (10-5) to eliminate V_j, L_j, $y_{i,j}$ and $x_{i,j}$ and introduce component flow rates $v_{i,j}$ and $l_{i,j}$. As a result, the following $N(2C + 1)$ equations are obtained, where $s_j = U_j/L_j$ and $S_j = W_j/V_j$ are dimensionless side-stream flow rates.

Material Balance

$$M_{i,j} = l_{i,j}(1 + s_j) + v_{i,j}(1 + S_j) - l_{i,j-1} - v_{i,j+1} - f_{i,j} = 0 \tag{10-58}$$

Phase Equilibria

$$E_{i,j} = K_{i,j} l_{i,j} \frac{\sum\limits_{\kappa=1}^{C} v_{\kappa,j}}{\sum\limits_{\kappa=1}^{C} l_{\kappa,j}} - v_{i,j} = 0 \tag{10-59}$$

Energy Balance

$$H_j = h_{L_j}(1 + s_j) \sum_{i=1}^{C} l_{i,j} + h_{V_j}(1 + S_j) \sum_{i=1}^{C} v_{i,j} - h_{L_{j-1}} \sum_{i=1}^{C} l_{i,j-1}$$
$$- h_{V_{j+1}} \sum_{i=1}^{C} v_{i,j+1} - h_{F_j} \sum_{i=1}^{C} f_{i,j} - Q_j = 0 \tag{10-60}$$

where $f_{i,j} = F_j z_{i,j}$.

If N and all $f_{i,j}$, T_{F_j}, P_{F_j}, P_j, s_j, S_j, and Q_j are specified, the M, E, and H functions are nonlinear in the $N(2C + 1)$ unknown (output) variables $v_{i,j}$, $l_{i,j}$, and T_j for $i = 1$ to C and $j = 1$ to N. Although other sets of specified and unknown variables are possible, we consider these sets first.

Equations (10-58), (10-59), and (10-60) are solved simultaneously by the Newton–Raphson iterative method in which successive sets of the output variables are produced until the values of the M, E, and H functions are driven to within some tolerance of zero. During the iterations, nonzero values of the functions are called *discrepancies* or *errors*. Let the functions and output variables be grouped by stage in order from top to bottom. As will be shown, this is done to produce a block tridiagonal structure for the Jacobian matrix of partial derivatives so that the Thomas algorithm can be applied. Let

$$\mathbf{X} = [\mathbf{X}_1, X_2, \ldots, \mathbf{X}_j, \ldots, \mathbf{X}_N]^T \tag{10-61}$$

and

$$\mathbf{F} = [\mathbf{F}_1, \mathbf{F}_2, \ldots, \mathbf{F}_j, \ldots, \mathbf{F}_N]^T \tag{10-62}$$

where $\mathbf{X}_j$ is the vector of output variables for stage j arranged in the order

$$\mathbf{X}_j = [v_{1,j}, v_{2,j}, \ldots, v_{i,j}, \ldots, v_{C,j}, T_j, l_{1,j}, l_{2,j}, \ldots, l_{i,j}, \ldots, l_{C,j}]^T \tag{10-63}$$

and $\mathbf{F}_j$ is the vector of functions for stage j arranged in the order

$$\mathbf{F}_j = [H_j, M_{1,j}, M_{2,j}, \ldots, M_{i,j}, \ldots, M_{C,j}, E_{1,j}, E_{2,j}, \ldots, E_{i,j}, \ldots, E_{C,j}]^T \tag{10-64}$$

The Newton–Raphson iteration is performed by solving for the corrections $\Delta\mathbf{X}$ to the output variables from (10-38), which in matrix form becomes

$$\Delta\mathbf{X}^{(k)} = -\left[\left(\frac{\partial\mathbf{F}}{\partial\mathbf{X}}\right)^{-1}\right]^{(k)} \mathbf{F}^{(k)} \tag{10-65}$$

These corrections are used to compute the next approximation to the set of output variables from

$$\mathbf{X}^{(k+1)} = \mathbf{X}^{(k)} + t\,\Delta\mathbf{X}^{(k)} \tag{10-66}$$

The quantity $(\partial\mathbf{F}/\partial\mathbf{X})$ is the following Jacobian or $(N \times N)$ matrix of blocks of partial derivatives of all the functions with respect to all the output variables.

$$\frac{d\mathbf{F}}{d\mathbf{X}} = \begin{bmatrix} \overline{\mathbf{B}}_1 & \overline{\mathbf{C}}_1 & 0 & 0 & .. & & 0 \\ \overline{\mathbf{A}}_2 & \overline{\mathbf{B}}_2 & \overline{\mathbf{C}}_2 & 0 & .. & & 0 \\ 0 & \overline{\mathbf{A}}_3 & \overline{\mathbf{B}}_3 & \overline{\mathbf{C}}_3 & .. & & 0 \\ .. & & & & & & .. \\ .. & & & & & & .. \\ 0 & .. & & & & & 0 \\ 0 & .. & & & 0 & \overline{\mathbf{A}}_{N-1} & \overline{\mathbf{B}}_{N-1} & \overline{\mathbf{C}}_{N-1} \\ 0 & .. & & & 0 & 0 & \overline{\mathbf{A}}_N & \overline{\mathbf{B}}_N \end{bmatrix} \tag{10-67}$$

This Jacobian is of a block tridiagonal form like (10-12) because functions for stage j are only dependent on output variables for stages $j - 1$, j, and $j + 1$. Each $\overline{\mathbf{A}}$, $\overline{\mathbf{B}}$, or $\overline{\mathbf{C}}$ block in (15-67) represents a $(2C + 1)$ by $(2C + 1)$ submatrix of partial derivatives, where the arrangements of output variables and functions are given by (10-63) and (10-64), respectively. Blocks $\overline{\mathbf{A}}_j$, $\overline{\mathbf{B}}_j$, and $\overline{\mathbf{C}}_j$ correspond to submatrices of partial derivatives of the functions on stage j with respect to the output variables on stages $j - 1$, j, and $j + 1$, respectively. Thus, using (10-58), (10-59), and (10-60), and denoting only the nonzero partial derivatives by $+$, or by row or diagonal strings of $+ \cdots +$, or by the following square or rectangular blocks enclosed by connected strings,

we find that the blocks have the following form, where $+$ is replaced by a numerical value (-1 or 1) in the event that the partial derivative has only that value.

$$
\overline{\mathbf{A}}_j = \frac{\overline{\partial \mathbf{F}_j}}{\partial \mathbf{X}_{j-1}} =
\begin{array}{c}
\\
\begin{array}{cc} & \mathbf{X}_{j-1} \\ \mathbf{F}_j & \end{array} \quad \text{\textit{Output variables}}
\end{array}
$$

$$
\overline{\mathbf{A}}_j = \frac{\overline{\partial \mathbf{F}_j}}{\partial \mathbf{X}_{j-1}} =
\begin{array}{c} \textit{Functions} \end{array}
\begin{array}{c}
H_j \\ M_{1,j} \\ \cdot \\ \cdot \\ \cdot \\ M_{C,j} \\ E_{1,j} \\ \cdot \\ \cdot \\ \cdot \\ E_{C,j}
\end{array}
\overset{\displaystyle v_{1,j-1} \cdots v_{C,j-1} \quad T_{j-1} \quad l_{1,j-1} \cdots l_{C,j-1}}{
\begin{bmatrix}
& & + & + \cdots \cdots + \\
& & -1 & \\
& & & \ddots \\
& & & & -1 \\
& & & \\
& & & \\
& & & \\
& & & \\
& & &
\end{bmatrix}}
\tag{10-68}
$$

$$
\overline{\mathbf{B}}_j = \frac{\overline{\partial \mathbf{F}_j}}{\partial \mathbf{X}_j} =
\begin{array}{c} \textit{Functions} \end{array}
\begin{array}{c}
H_j \\ M_{1,j} \\ \cdot \\ \cdot \\ M_{C,j} \\ E_{1,j} \\ \cdot \\ \cdot \\ E_{C,j}
\end{array}
\overset{\displaystyle v_{1,j} \cdots v_{C,j} \quad T_j \quad l_{1,j} \cdots l_{C,j}}{
\begin{bmatrix}
+ \cdots \cdots \cdots \cdots \cdots \cdots + \\
+ \qquad\qquad + \\
\ddots \qquad \ddots \\
+ \qquad + \\
+ \cdots \cdots \cdots \cdots \cdots + \\
\vdots \qquad\qquad \vdots \\
+ \cdots \cdots \cdots \cdots \cdots +
\end{bmatrix}}
\tag{10-69}
$$

$$
\overline{\mathbf{C}}_j = \frac{\partial \mathbf{F}_j}{\partial \mathbf{X}_{j+1}} =
\begin{array}{c} \textit{Functions} \end{array}
\begin{array}{c}
H_j \\ M_{1,j} \\ \cdot \\ M_{C,j} \\ E_{1,j} \\ \cdot \\ \cdot \\ \cdot \\ E_{C,j}
\end{array}
\overset{\displaystyle v_{1,j+1} \cdots v_{C,j+1} \quad T_{j+1} \quad l_{1,j+1} \cdots l_{C,j+1}}{
\begin{bmatrix}
+ \cdots \cdots + \\
-1 \\
\ddots \\
-1 \\
+ \cdots + \\
\vdots \\
\vdots \\
+ \cdots + \\
\end{bmatrix}}
\tag{10-70}
$$

Thus, (10-65) consists of a set of $N(2C + 1)$ simultaneous linear equations in the $N(2C + 1)$ corrections $\Delta\mathbf{X}$. For example, the $2C + 2$ equation in the set is obtained by expanding function H_2 (10-60) into a Taylor's series like (10-36) around the $N(2C + 1)$ output variables. The result is as follows after the usual truncation of terms involving derivatives of order greater than one:

$$0(\Delta v_{1,1} + \cdots + \Delta v_{C,1}) - \frac{\partial h_{L_1}}{\partial T_1}\sum_{i=1}^{C} l_{i,1}(\Delta T_1) - \left(\frac{\partial h_{L_1}}{\partial l_{1,1}}\sum_{i=1}^{C} l_{i,1} + h_{L_1}\right)\Delta l_{1,1}$$

$$- \cdots - \left(\frac{\partial h_{L_1}}{\partial l_{C,1}}\sum_{i=1}^{C} l_{i,1} + h_{L_1}\right)\Delta l_{C,1} + \left[\left(\frac{\partial h_{V_2}}{\partial v_{1,2}}\right)(1 + S_2)\sum_{i=1}^{C} v_{i,2} + h_{V_2}(1 + S_2)\right]\Delta v_{1,2}$$

$$+ \cdots + \left[\left(\frac{\partial h_{V_2}}{\partial v_{C,2}}\right)(1 + S_2)\sum_{i=1}^{C} v_{i,2} + h_{V_2}(1 + S_2)\right]\Delta v_{C,2}$$

$$+ \left[\left(\frac{\partial h_{L_2}}{\partial T_2}\right)(1 + s_2)\sum_{i=1}^{C} l_{i,2} + \left(\frac{\partial h_{V_2}}{\partial T_2}\right)(1 + S_2)\sum_{i=1}^{C} v_{i,2}\right]\Delta T_2$$

$$+ \left[\left(\frac{\partial h_{L_2}}{\partial l_{1,2}}\right)(1 + s_2)\sum_{i=1}^{C} l_{i,2} + h_{L_2}(1 + s_2)\right]\Delta l_{1,2}$$

$$+ \cdots + \left[\left(\frac{\partial h_{L_2}}{\partial l_{C,2}}\right)(1 + s_2)\sum_{i=1}^{C} l_{i,2} + h_{L_2}(1 + s_2)\right]\Delta l_{C,2}$$

$$- \left(\frac{\partial h_{V_3}}{\partial v_{1,3}}\sum_{i=1}^{C} v_{i,3} + h_{V_3}\right)\Delta v_{1,3} - \cdots - \left(\frac{\partial h_{V_3}}{\partial v_{C,3}}\sum_{i=1}^{C} v_{i,3} + h_{V_3}\right)\Delta v_{C,3}$$

$$- \frac{\partial h_{V_3}}{\partial T_3}\sum_{i=1}^{C} v_{i,3}\,\Delta T_3 + 0(\Delta l_{1,3} + \cdots + \Delta l_{C,N}) = -H_2$$

(10-71)

Although lengthy, equations such as (10-71) are handled readily in computer programs.

As a further example, the entry in the Jacobian matrix for row $(2C + 2)$ and column $(C + 3)$ is obtained from (10-71) as

$$\frac{\partial H_2}{\partial l_{2,1}} = -\frac{\partial h_{L_1}}{\partial l_{2,1}}\sum_{i=1}^{C} l_{i,1} + h_{L_1}$$

(10-72)

All partial derivatives are stated by Naphtali and Sandholm [23].

Partial derivatives of enthalpies and K-values depend upon the particular correlation utilized for these properties and are sometimes simplified by including only the dominant terms. For example, suppose that the Chao–Seader correlation is to be used for K-values. In general,

$$K_{i,j} = K_{i,j}\left\{P_j, T_j, \frac{l_{i,j}}{\sum_{\kappa=1}^{C} l_{\kappa,j}}, \frac{v_{i,j}}{\sum_{\kappa=1}^{C} v_{\kappa,j}}\right\}$$

In terms of the output variables, the partial derivatives $\partial K_{i,j}/\partial T_j$; $\partial K_{i,j}/\partial l_{i,j}$; and $\partial K_{i,j}/\partial v_{i,j}$ all exist and can be expressed analytically or evaluated numerically if desired. However, for some problems, the terms that include the first and second of these three groups of derivatives may be the dominant terms so that the third group may be taken as zero.

Derive an expression for $(\partial h_V/\partial T)$ from the Redlich–Kwong equation of state.

EXAMPLE 10.6

SOLUTION

From (2-53), $\qquad h_V = \sum_{i=1}^{C}(y_i h_{iV}^\circ) + RT\left[Z_V - 1 - \frac{3A}{2B}\left(1 + \frac{B}{Z_V}\right)\right]$

where h_{iV}°, Z_V, A, and B all depend on T, as determined from (2-36) and (2-46) to (2-50). Thus,

$$\frac{\partial h_V}{\partial T} = \sum_{i=1}^{C} \left[y_i \left(\frac{\partial h_{iV}^{\circ}}{\partial T} \right) \right] + R \left[Z_V - 1 - \frac{3A}{2B} \left(1 + \frac{B}{Z_V} \right) \right]$$

$$+ RT \left\{ \left(\frac{\partial Z_V}{\partial T} \right) - \frac{3}{2} \left(\frac{\partial (A/B)}{\partial T} \right) \left(1 + \frac{B}{Z_V} \right) \right.$$

$$\left. - \frac{3A}{2B} \left[\frac{1}{Z_V} \left(\frac{\partial B}{\partial T} \right) - \frac{B}{Z_V^2} \left(\frac{\partial Z_V}{\partial T} \right) \right] \right\}$$

From (2-36) and (2-35),

$$\left(\frac{\partial h_{iV}^{\circ}}{\partial T} \right) = \sum_{k=1}^{4} (a_k)_i T^{k-1} = (C_{P_V}^{\circ})_i$$

From (2-48) and Table 2.5,

$$B = \frac{bP}{RT} \text{ and } b = \frac{0.08664 RT_c}{P_c}$$

Thus,

$$\frac{\partial B}{\partial T} = -\frac{B}{T}$$

From (2-47) to (2-50) and Table 2.5,

$$\frac{A}{B} = \frac{a}{bRT} \text{ and } a = \frac{0.42748 R^2 T_c^{2.5}}{P_c T^{0.5}}$$

Thus,

$$\frac{\partial (A/B)}{\partial T} = -1.5 \frac{A}{BT}$$

From (2-46),

$$Z_V^3 - Z_V^2 + (A - B - B^2) Z_V - AB = 0$$

By implicit differentiation,

$$3Z_V^2 \frac{\partial Z_V}{\partial T} - 2Z_V \frac{\partial Z_V}{\partial T} + (A - B - B^2) \frac{\partial Z_V}{\partial T}$$

$$+ \frac{Z_V}{T} (-2.5A + B + 2B^2) + \frac{3.5AB}{T} = 0$$

which when combined with the preceding expressions for $(\partial B/\partial T)$ and $\partial(A/B)/\partial T$ gives

$$\frac{\partial Z_V}{\partial T} = \frac{\dfrac{Z_V}{T} (2.5A - B - 2B^2 - 3.5AB)}{3Z_V^2 - 2Z_V + (A - B - B^2)}$$ ∎

Because the Thomas algorithm can be applied to the block tridiagonal structure of (10-67), submatrices of partial derivatives are computed only as needed. The solution of (10-65) follows the scheme in Section 10.3, given by (10-13) to (10-18) and represented in Figure 10-4, where matrices and vectors $\overline{\mathbf{A}}_j$, $\overline{\mathbf{B}}_j$, $\overline{\mathbf{C}}_j$, $-\overline{\mathbf{F}}_j$, and $\Delta\mathbf{X}_j$ correspond to variables A_j, B_j, C_j, D_j, and x_j, respectively. However, the simple multiplication and division operations in Section 10.3 are changed to matrix multiplication and inversion, respectively. The steps are as follows:

Starting at stage 1, $\overline{\mathbf{C}}_1 \leftarrow (\overline{\mathbf{B}}_1)^{-1}\overline{\mathbf{C}}_1$, $\mathbf{F}_1 \leftarrow (\overline{\mathbf{B}}_1)^{-1}\mathbf{F}_1$, and $\overline{\mathbf{B}}_1 \leftarrow \mathbf{I}$ (the identity submatrix). Only $\overline{\mathbf{C}}_1$ and $\mathbf{F}_1$ are saved. For stages j from 2 to $(N-1)$, $\overline{\mathbf{C}}_j \leftarrow (\overline{\mathbf{B}}_j - \overline{\mathbf{A}}_j \overline{\mathbf{C}}_{j-1})^{-1} \overline{\mathbf{C}}_j$, $\mathbf{F}_j \leftarrow (\overline{\mathbf{B}}_j - \overline{\mathbf{A}}_j \overline{\mathbf{C}}_{j-1})^{-1}(\mathbf{F}_j - \overline{\mathbf{A}}_j \mathbf{F}_{j-1})$. Then $\overline{\mathbf{A}}_j \leftarrow 0$, and $\overline{\mathbf{B}}_j \leftarrow \mathbf{I}$. Save $\overline{\mathbf{C}}_j$ and $\mathbf{F}_j$ for each stage. For the last stage, $\mathbf{F}_N \leftarrow (\mathbf{B}_N - \mathbf{A}_N \mathbf{C}_{N-1})^{-1}(\mathbf{F}_N - \mathbf{A}_N \mathbf{F}_{N-1})$, $\overline{\mathbf{A}}_N \leftarrow 0$, $\overline{\mathbf{B}}_N \leftarrow \mathbf{I}$, and therefore $\Delta\mathbf{X}_N = -\mathbf{F}_N$. This completes the forward steps. Remaining values of $\Delta\mathbf{X}$ are obtained by successive backward substitution from $\Delta\mathbf{X}_j = -\mathbf{F}_j \leftarrow -(\mathbf{F}_j - \overline{\mathbf{C}}_j \mathbf{F}_{j+1})$. This procedure is illustrated by the following example.

For the last stage, $\mathbf{F}_N \leftarrow (\mathbf{B}_N - \mathbf{A}_N \overline{\mathbf{C}}_{N-1})^{-1}(\mathbf{F}_N - \mathbf{A}_N \mathbf{F}_{N-1})$, $\overline{\mathbf{A}}_N \leftarrow 0$, $\overline{\mathbf{B}}_N \leftarrow \mathbf{I}$, and therefore $\Delta \mathbf{X}_N = -\mathbf{F}_N$. This completes the forward steps. Remaining values of $\Delta \mathbf{X}$ are obtained by successive backward substitution from $\Delta \mathbf{X}_j = -\mathbf{F}_j \leftarrow -(\mathbf{F}_j - \overline{\mathbf{C}}_j \mathbf{F}_{j+1})$. This procedure is illustrated by the following example.

EXAMPLE 10.7

Solve the following matrix equation, which has a block tridiagonal structure, by the Thomas algorithm.

$$
\begin{bmatrix}
1 & 2 & 1 & 2 & 2 & 1 & 0 & 0 & 0 \\
2 & 1 & 1 & 2 & 1 & 0 & 0 & 0 & 0 \\
1 & 2 & 2 & 1 & 2 & 0 & 0 & 0 & 0 \\
0 & 1 & 3 & 1 & 2 & 1 & 1 & 2 & 1 \\
0 & 0 & 1 & 2 & 2 & 0 & 1 & 2 & 0 \\
0 & 0 & 2 & 2 & 1 & 1 & 1 & 1 & 0 \\
0 & 0 & 0 & 0 & 1 & 2 & 2 & 1 & 1 \\
0 & 0 & 0 & 0 & 0 & 2 & 1 & 1 & 1 \\
0 & 0 & 0 & 0 & 0 & 1 & 2 & 1 & 2
\end{bmatrix}
\cdot
\begin{bmatrix}
\Delta x_1 \\
\Delta x_2 \\
\Delta x_3 \\
\Delta x_4 \\
\Delta x_5 \\
\Delta x_6 \\
\Delta x_7 \\
\Delta x_8 \\
\Delta x_9
\end{bmatrix}
=
\begin{bmatrix}
9 \\
7 \\
8 \\
12 \\
8 \\
8 \\
7 \\
5 \\
6
\end{bmatrix}
$$

SOLUTION

The matrix equation is in the form

$$
\begin{bmatrix}
\overline{\mathbf{B}}_1 & \overline{\mathbf{C}}_1 & 0 \\
\overline{\mathbf{A}}_2 & \overline{\mathbf{B}}_2 & \overline{\mathbf{C}}_2 \\
0 & \overline{\mathbf{A}}_3 & \overline{\mathbf{B}}_3
\end{bmatrix}
\cdot
\begin{bmatrix}
\Delta \mathbf{X}_1 \\
\Delta \mathbf{X}_2 \\
\Delta \mathbf{X}_3
\end{bmatrix}
= -
\begin{bmatrix}
\mathbf{F}_1 \\
\mathbf{F}_2 \\
\mathbf{F}_3
\end{bmatrix}
$$

Following the procedure just given, starting at the first block row,

$$
\overline{\mathbf{B}}_1 = \begin{bmatrix} 1 & 2 & 1 \\ 2 & 1 & 1 \\ 1 & 2 & 2 \end{bmatrix}, \overline{\mathbf{C}}_1 = \begin{bmatrix} 2 & 2 & 1 \\ 2 & 1 & 0 \\ 1 & 2 & 0 \end{bmatrix}, \mathbf{F}_1 = \begin{bmatrix} -9 \\ -7 \\ -8 \end{bmatrix}
$$

By standard matrix inversion

$$
(\overline{\mathbf{B}}_1)^{-1} = \begin{bmatrix} 0 & 2/3 & -1/3 \\ 1 & -1/3 & -1/3 \\ -1 & 0 & 1 \end{bmatrix}
$$

By standard matrix multiplication

$$
(\overline{\mathbf{B}}_1)^{-1}(\overline{\mathbf{C}}_1) = \begin{bmatrix} 1 & 0 & 0 \\ 1 & 1 & 1 \\ -1 & 0 & -1 \end{bmatrix}
$$

which replaces $\overline{\mathbf{C}}_1$, and

$$
(\overline{\mathbf{B}}_1)^{-1}(\mathbf{F}_1) = \begin{bmatrix} -2 \\ -4 \\ 1 \end{bmatrix}
$$

which replaces $\mathbf{F}_1$. Also

$$
\mathbf{I} = \begin{bmatrix} 1 & 0 & 0 \\ 0 & 1 & 0 \\ 0 & 0 & 1 \end{bmatrix} \text{ replaces } \overline{\mathbf{B}}_1
$$

For the second block row

$$\overline{\mathbf{A}}_2 = \begin{bmatrix} 0 & 1 & 3 \\ 0 & 0 & 1 \\ 0 & 0 & 2 \end{bmatrix}, \overline{\mathbf{B}}_2 = \begin{bmatrix} 1 & 2 & 1 \\ 2 & 2 & 0 \\ 2 & 1 & 1 \end{bmatrix}, \overline{\mathbf{C}}_2 = \begin{bmatrix} 1 & 2 & 1 \\ 1 & 2 & 0 \\ 1 & 1 & 0 \end{bmatrix}, \mathbf{F}_2 = \begin{bmatrix} -12 \\ -8 \\ -8 \end{bmatrix}$$

By matrix multiplication and subtraction

$$(\overline{\mathbf{B}}_2 - \overline{\mathbf{A}}_2\overline{\mathbf{C}}_1) = \begin{bmatrix} 3 & 1 & 3 \\ 3 & 2 & 1 \\ 4 & 1 & 3 \end{bmatrix}$$

which upon inversion becomes

$$(\overline{\mathbf{B}}_2 - \overline{\mathbf{A}}_2\overline{\mathbf{C}}_1)^{-1} = \begin{bmatrix} -1 & 0 & 1 \\ 1 & 3/5 & -6/5 \\ 1 & -1/5 & -3/5 \end{bmatrix}$$

By multiplication

$$(\overline{\mathbf{B}}_2 - \overline{\mathbf{A}}_2\overline{\mathbf{C}}_1)^{-1}\overline{\mathbf{C}}_2 = \begin{bmatrix} 0 & -1 & -1 \\ 2/5 & 2 & 1 \\ 1/5 & 1 & 1 \end{bmatrix}$$

which replaces $\overline{\mathbf{C}}_2$. In a similar manner, the remaining steps for this and the third block row are carried out to give

$$\begin{bmatrix} \begin{bmatrix} 1 & 0 & 0 \\ 0 & 1 & 0 \\ 0 & 0 & 1 \end{bmatrix} & \begin{bmatrix} 1 & 0 & 0 \\ 1 & 1 & 1 \\ -1 & 0 & -1 \end{bmatrix} & \begin{bmatrix} 0 & 0 & 0 \\ 0 & 0 & 0 \\ 0 & 0 & 0 \end{bmatrix} \\ \begin{bmatrix} 0 & 0 & 0 \\ 0 & 0 & 0 \\ 0 & 0 & 0 \end{bmatrix} & \begin{bmatrix} 1 & 0 & 0 \\ 0 & 1 & 0 \\ 0 & 0 & 1 \end{bmatrix} & \begin{bmatrix} 0 & -1 & -1 \\ 2/5 & 2 & 1 \\ 1/5 & 1 & 1 \end{bmatrix} \\ \begin{bmatrix} 0 & 0 & 0 \\ 0 & 0 & 0 \\ 0 & 0 & 0 \end{bmatrix} & \begin{bmatrix} 0 & 0 & 0 \\ 0 & 0 & 0 \\ 0 & 0 & 0 \end{bmatrix} & \begin{bmatrix} 1 & 0 & 0 \\ 0 & 1 & 0 \\ 0 & 0 & 1 \end{bmatrix} \end{bmatrix} \cdot \begin{bmatrix} \Delta X_1 \\ \Delta X_2 \\ \Delta X_3 \\ \Delta X_4 \\ \Delta X_5 \\ \Delta X_6 \\ \Delta X_7 \\ \Delta X_8 \\ \Delta X_9 \end{bmatrix} = - \begin{bmatrix} -2 \\ -4 \\ +1 \\ +1 \\ -22/5 \\ -16/5 \\ -1 \\ -1 \\ -1 \end{bmatrix}$$

Thus, $\Delta X_7 = \Delta X_8 = \Delta X_9 = 1$.

The remaining backward steps begin with the second block row where

$$\overline{\mathbf{C}}_2 = \begin{bmatrix} 0 & -1 & -1 \\ 2/5 & 2 & 1 \\ 1/5 & 1 & 1 \end{bmatrix}, \overline{\mathbf{F}}_2 = \begin{bmatrix} 1 \\ -22/5 \\ -16/5 \end{bmatrix}$$

$$(\mathbf{F}_2 - \overline{\mathbf{C}}_2\mathbf{F}_3) = \begin{bmatrix} -1 \\ -1 \\ -1 \end{bmatrix}$$

Thus, $\Delta X_4 = \Delta X_5 = \Delta X_6 = 1$. Similarly, for the first block row, the result is

$$\Delta X_1 = \Delta X_2 = \Delta X_3 = 1 \qquad \blacksquare$$

Usually, it is desirable to specify certain top- and bottom-stage variables other than the condenser duty and/or reboiler duty. (In fact, the condenser and reboiler duties are usually so interdependent that specification of both values is not recommended.) Specifying the variables is readily accomplished by removing heat balance functions H_1 and/or H_N from the simultaneous equation set and replacing them with discrepancy functions depending

Table 10.1 Alternative Functions for H_1 and H_N

Specification	Replacement for H_1	Replacement for H_N
Reflux or reboil (boilup) ratio, (L/D) or (V/B)	$\Sigma\, l_{i,1} - (L/D)\,\Sigma\, v_{i,1} = 0$	$\Sigma\, v_{i,N} - (V/B)\,\Sigma\, l_{i,N} = 0$
Stage temperature, T_D or T_B	$T_1 - T_D = 0$	$T_N - T_B = 0$
Product flow rate, D or B	$\Sigma\, v_{i,1} - D = 0$	$\Sigma\, l_{i,N} - B = 0$
Component flow rate in product, d_i or b_i	$v_{i,1} - d_i = 0$	$l_{i,N} - b_i = 0$
Component mole fraction in product, y_{iD} or x_{iB}	$v_{i,1} - (\Sigma\, v_{i,1})\,y_{iD} = 0$	$l_{i,N} - (\Sigma\, l_{i,N})\,x_{iB} = 0$

upon the desired specification(s). Functions for alternative specifications for a column with a partial condenser are listed in Table 10.1.

If desired, (10-54) can be modified to permit real rather than theoretical stages to be computed. Values of the Murphree vapor-phase plate efficiency must then be specified. These values are related to phase compositions by the definition

$$\eta_j = \frac{y_{i,j} - y_{i,j+1}}{K_{i,j}x_{i,j} - y_{i,j+1}} \tag{10-73}$$

In terms of component flow rates, (10-73) becomes the following discrepancy function, which replaces (10-59).

$$E_{i,j} = \frac{\eta_j K_{i,j} l_{i,j} \sum\limits_{\kappa=1}^{C} v_{\kappa,j}}{\sum\limits_{\kappa=1}^{C} l_{\kappa,j}} - v_{i,j} + \frac{(1 - \eta_j)v_{i,j+1} \sum\limits_{\kappa=1}^{C} v_{\kappa,j}}{\sum\limits_{\kappa=1}^{C} v_{\kappa,j+1}} = 0 \tag{10-74}$$

If a total condenser with subcooling is desired, it is necessary to specify the degrees of subcooling, if any, and to replace (10-59) or (10-74) with functions that express identity of reflux and distillate compositions as discussed by Naphtali and Sandholm [23].

The algorithm for the Naphtali–Sandholm implementation of the Newton–Raphson or SC method is shown in Figure 10.24. Problem specifications are quite flexible. Pressure, compositions, flow rates, and stage locations are necessary specifications for all feeds. The thermal condition of each feed can be given in terms of enthalpy, temperature, or molar fraction vaporized. If a feed is found to consist of two phases, the phases can be sent to the same stage or the vapor can be directed to the stage above the designated feed stage. Stage pressures and stage efficiencies can be designated by specifying top- and bottom-stage values. Remaining values are obtained by linear interpolation. By default, intermediate stages are assumed to be adiabatic unless Q_j or T_j values are specified. Vapor and/or liquid side streams can be designated in terms of total flow rate or flow rate of a specified component, or by the ratio of the side-stream flow rate to the flow rate remaining and passing to the next stage. The top- and bottom-stage specifications are selected from Q_1 or Q_N, and/or more generally from the other specifications listed in Table 10.1.

In order to achieve convergence, the Newton–Raphson procedure requires that reasonable guesses be provided for the values of all output variables. Rather than provide all these guesses a priori, we can generate them if T, V, and L are guessed for the bottom

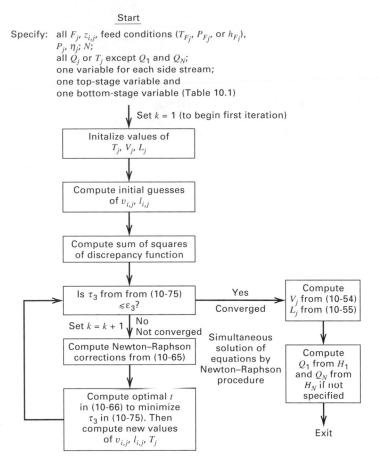

Figure 10.24 Algorithm for Naphtali–Sandholm method for all V/L separators.

and top stages and, perhaps, for one or more intermediate stages. Remaining guessed values of T_j, V_j, and L_j are readily obtained by linear interpolation of the given T_j values and computed (V_j/L_j) values. Initial values for $v_{i,j}$ and $l_{i,j}$ are then obtained by either of two techniques. If K-values are composition independent or can be approximated as such, one technique is to compute $x_{i,j}$ values and corresponding $y_{i,j}$ values from (10-12) and (10-2) as in the first iteration of the BP or SR method. A much cruder estimate is obtained by flashing the combined feeds at some average column pressure and a V/L ratio that approximates the ratio of overheads to bottoms products. The resulting mole fraction compositions of the equilibrium vapor and liquid phases are assumed to hold for each stage. The second technique works surprisingly well, but the first technique is preferred for difficult cases. For either technique, the initial component flow rates are computed by using the $x_{i,j}$ and $y_{i,j}$ values to solve (10-56) and (10-57) for $l_{i,j}$ and $v_{i,j}$, respectively.

Based on initial guesses for all output variables, the sum of the squares of the discrepancy functions is computed and compared to a convergence criterion

$$\tau_3 = \sum_{j=1}^{N} \left\{ (H_j)^2 + \sum_{i=1}^{C} \left[(M_{i,j})^2 + (E_{i,j})^2 \right] \right\} \le \epsilon_3 \tag{10-75}$$

In order that the values of all discrepancies be of the same order of magnitude, it is necessary to divide energy balance functions H_j by a scale factor approximating the latent heat of vaporization (e.g., 1,000 Btu/lbmol). If the convergence criterion is computed from

$$\epsilon_3 = N(2C + 1) \left(\sum_{j=1}^{N} F_j^2 \right) 10^{-10} \tag{10-76}$$

resulting converged values of the output variables will generally be accurate, on the average, to four or more significant figures. When employing (10-76), most problems are converged in 10 iterations or less.

Generally, the convergence criterion is far from satisfied during the first iteration when guessed values are assumed for the output variables. For each subsequent iteration, the Newton–Raphson corrections are computed from (10-65). These corrections can be added directly to the present values of the output variables to obtain a new set of values for the output variables. Alternatively, (10-66) can be employed where t is a nonnegative scalar step factor. At each iteration, a single value of t is applied to all output variables. By permitting t to vary from, say, slightly greater than zero up to 2, it can serve to dampen or accelerate convergence, as appropriate. For each iteration, an optimal value of t is sought to minimize the sum of the squares given by (10-75). Generally, optimal values of t proceed from an initial value for the second iteration at between 0 and 1 to a value nearly equal to or slightly greater than 1 when the convergence criterion is almost satisfied. An efficient optimization procedure for finding t at each iteration is the Fibonacci search [25]. If no optimal value of t can be found within the designated range, t can be set to 1, or some smaller value, and the sum of squares can be allowed to increase. Generally, after several iterations, the sum of squares will decrease for every iteration.

If the application of (10-66) results in a negative component flow rate, Naphtali and Sandholm recommend the following mapping equation, which reduces the value of the unknown variable to a near-zero, but nonnegative, quantity.

$$X^{(k+1)} = X^{(k)} \exp \left[\frac{t \, \Delta X^{(k)}}{X^{(k)}} \right] \tag{10-77}$$

In addition, it is advisable to limit temperature corrections at each iteration.

The Naphtali–Sandholm SC method is readily extended to staged separators involving two liquid phases (e.g., extraction) and three coexisting phases (e.g., three-phase distillation), as shown by Block and Hegner [26], and to interlinked separators as shown by Hofeling and Seader [27].

EXAMPLE 10.8

A reboiled absorber is to be designed to separate the hydrocarbon vapor feed of Examples 10.2 and 10.4. Absorbent oil of the same composition as that of Example 10.4 will enter the top stage. Complete specifications are given in Figure 10.25. The 770 lbmol/h (349 kmol/h) of bottoms product corresponds to the amount of C_3 and heavier in the two feeds. Thus, the column is to be designed as a deethanizer. Calculate stage temperatures, interstage vapor and liquid flow rates and compositions, and reboiler duty by the rigorous SC method. Assume all stage efficiencies are 100%. Compare the degree of separation of the feed to that achieved by ordinary distillation in Example 10.2.

SOLUTION

A digital computer program for the method of Naphtali and Sandholm was used. The K-values and enthalpies were assumed independent of composition and were computed by linear interpolation between tabular values given at 100°F increments from 0 to 400°F (-17.8 to 204.4°C).

From (10-76), the convergence criterion is

$$\epsilon_3 = 13[2(6) + 1](500 + 800)^2 10^{-10} = 2.856 \times 10^{-2}$$

Figure 10.26 shows the reduction in the sum of the squares of the 169 discrepancy functions from iteration to iteration. Seven iterations were required to satisfy the convergence criterion.

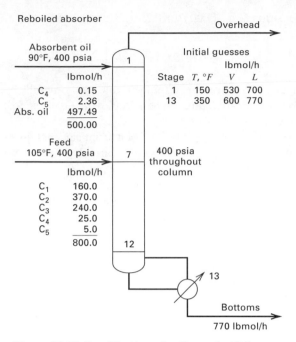

Figure 10.25 Specifications for Example 10.8.

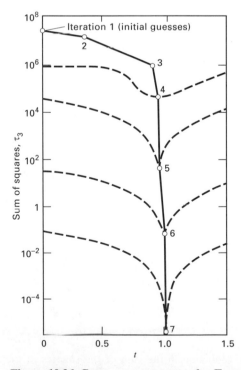

Figure 10.26 Convergence pattern for Example 10.8.

The initial iteration was based on values of the unknown variables computed from interpolation of the initial guesses shown in Fig. 10.25 together with a flash of the combined feeds at 400 psia (2.76 MPa) and a V/L ratio of 0.688 (530/770). Thus, for the first iteration, the following mole fraction compositions were computed and were assumed to apply to every stage.

Species	y	x
C_1	0.2603	0.0286
C_2	0.4858	0.1462
C_3	0.2358	0.1494
C_4	0.0153	0.0221
C_5	0.0025	0.0078
Abs. Oil	0.0003	0.6459
	1.0000	1.0000

The corresponding sum of squares of the discrepancy functions τ_3 of 2.865×10^7 was very large. Subsequent iterations employed the Newton–Raphson method. For iteration 2, the optimal value of t was found to be 0.34. However, this caused only a moderate reduction in the sum of squares. The optimal value of t increased to 0.904 for iteration 3, and the sum of squares was reduced by an order of magnitude. For the fourth and subsequent iterations, the effect of t on the sum of squares is included in Figure 10.26. Following iteration 4, the sum of squares was reduced by at least two orders of magnitude for each iteration. Also, the optimal value of t was rather sharply defined and corresponded closely to a value of 1. An improvement of τ_3 was obtained for every iteration.

In Figures 10.27 and 10.28, converged temperature and V/L profiles are compared to the initially guessed profiles. In Figure 10.27, the converged temperatures are far from linear with respect to stage number. Above the feed stage, the temperature profile increases from the top down in a gradual and declining manner. The relatively cold feed causes a small temperature drop from stage 6 to stage 7. Temperature also increases from stage 7 to stage 13. A particularly dramatic increase occurs in moving from the bottom stage in the column to the reboiler, where heat is added. In Figure 10.28, the V/L profile is also far from linear with respect to stage number. Dramatic changes in this ratio occur at the top, middle, and bottom of the column.

Component flow rate profiles for the two key components (ethane vapor and propane liquid) are shown in Figure 10.29. The initial guessed values are in very poor agreement with the converged values. The propane liquid profile is quite regular except at the bottom, where a large decrease occurs because of vaporization in the reboiler. The ethane vapor profile has

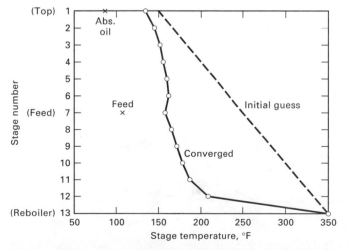

Figure 10.27 Converged temperature profile for Example 10.8.

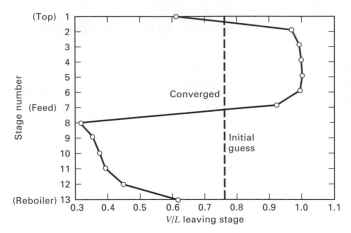

Figure 10.28 Converged vapor–liquid ratio profile for Example 10.8.

large changes at the top, where entering oil absorbs appreciable ethane, and at the feed stage, where substantial ethane vapor is introduced.

Converged values for the reboiler duty and overhead and bottoms compositions are given in Table 10.2. Also included are converged results for two additional solutions that used the Chao–Seader and Soave–Redlich–Kwong correlations for K-values and enthalpies in place of interpolation of composition-independent tabular properties. With the Soave–Redlich–Kwong correlation, a somewhat sharper separation between the two key components is predicted. In addition, the Soave–Redlich–Kwong correlation predicts a substantially higher bottoms temperature and a much larger reboiler duty. As discussed in Chapter 4, the effect of physical properties on equilibrium stage calculations can be significant.

It is interesting to compare the separation achieved with the reboiled absorber of this example to the separation achieved by ordinary distillation of the same feed in Example 10.2 as shown in Figure 10.9. The latter separation technique results in a much sharper separation and a much lower bottoms temperature and reboiler duty for the same number of stages. However, refrigeration is necessary for the overhead condenser, and the reflux flow rate is twice the absorbent oil flow rate. If the absorbent oil flow rate for the reboiled absorber is made equal

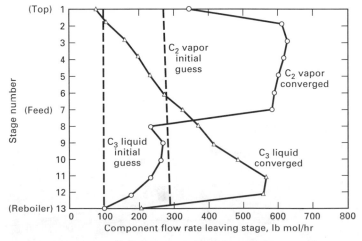

Figure 10.29 Converged flow rates for key components in Example 10.8.

Table 10.2 Product Compositions and Reboiler Duty for Example 10.8

	Composition-Independent Tabular Properties	Chao–Seader Correlation	Soave–Redlich–Kwong Correlation
Overhead component flow rates, lbmol/h			
C_1	159.99	159.98	159.99
C_2	337.96	333.52	341.57
C_3	31.79	36.08	28.12
C_4	0.04	0.06	0.04
C_5	0.17	0.21	0.18
Abs. oil	0.05	0.15	0.10
	530.00	530.00	530.00
Bottoms component flow rates, lbmol/h			
C_1	0.01	0.02	0.01
C_2	32.04	36.4	28.43
C_3	208.21	203.92	211.88
C_4	25.11	25.09	25.11
C_5	7.19	7.15	7.18
Abs. oil	497.44	497.34	497.39
	770.00	770.00	770.00
Reboiler duty, Btu/h	11,350,000	10,980,000	15,640,000
Bottoms temperature, °F	346.4	338.5	380.8

to the reflux flow rate, calculations give a separation almost as sharp as for ordinary distillation. However, the bottoms temperature and reboiler duty are increased to almost 600°F (315.6°C) and 60,000,000 Btu/h (63.3 GJ/h), respectively. ■

10.5 INSIDE-OUT METHOD

In the bubble-point (BP) and sum-rates (SR) methods described in Section 10.3 and the simultaneous correction (SC) method described in Section 10.4, a large percentage of the computational effort is expended in calculating K-values, vapor-phase enthalpies, and liquid-phase enthalpies, particularly when rigorous thermodynamic property models (e.g. Soave–Redlich–Kwong, Peng–Robinson, Wilson, NRTL, UNIQUAC) are utilized. As seen in Figure 10.30a and b, these property calculations are made at each iteration. Furthermore, at each iteration, derivatives are required of: (1) all three thermodynamic properties with respect to temperature and compositions of both phases, for the SC method; (2) K-values with respect to temperature for the BP method, unless Mullers method is used to compute bubble points; and (3) vapor and liquid enthalpies with respect to temperature for the SR method.

In 1974, Boston and Sullivan [28] presented an algorithm designed to significantly reduce the time spent in computing thermodynamic properties when designing steady-state, multicomponent separation operations. As shown in Figure 10.30c, two sets of thermodynamic property models are employed: (1) a simple approximate empirical set used frequently to converge inner loop calculations, and (2) the rigorous and complex set used less often in the outer loop. The MESH equations are always solved in the inner loop with the approximate set. The parameters in the empirical equations for the approximate set are updated in the outer loop by the rigorous equations, but only at infrequent intervals. A distinguishing feature of the Boston–Sullivan method is these inner and outer loops; hence the name *inside-out* for this class of methods. Another name, less frequently used, is *two-tier* methods.

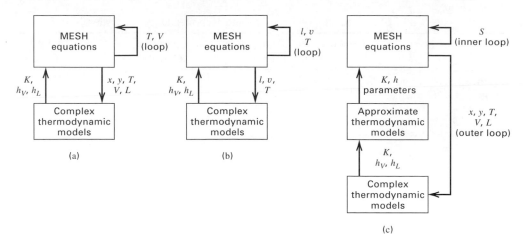

Figure 10.30 Incorporation of thermodynamic property correlations into interactive loops. (a) BP and SR methods. (b) SC method. (c) Inside-out method.

Another difference that distinguishes the inside-out method, as shown in Figure 10.30, is the choice of iteration variables. For the SC method, the iteration variables are $l_{i,j}$, v_{ij}, T_j. For the BP and SR methods, the choice is $x_{i,j}$, $y_{i,j}$, T_j, L_j, and V_j. For the inside-out method, the iteration variables for the outer loop are the parameters in the approximate equations for the thermodynamic properties. The iteration variables for the inner loop are related to the stripping factors, $S_{i,j} = K_{i,j} V_j / L_j$.

In the original presentation of the inside-out method in 1974, the development and application of the method was restricted to hydrocarbon distillation (moderately nonideal systems) for the Case II variable specification in Table 5.2, but with multiple feeds, sidestreams and intermediate heat exchangers. For these applications, the inside-side method was shown to be rapid and robust. Since 1974, the method has been extended and improved in a number of published articles [29,30,31,32,33,34] and proprietary implementations in simulation computer programs. These extensions permit the inside-out method to be applied to almost any type of steady-state multicomponent, multistage vapor–liquid separation operation. In the extensive implementation of the inside-out method by ASPEN technology in ASPEN PLUS, in computer programs called RADFRAC and MULTIFRAC, these applications include:

1. Absorption, stripping, reboiled absorption, reboiled stripping, extractive distillation, and azeotropic distillation
2. Three-phase (vapor-liquid-liquid) systems
3. Reactive systems
4. Highly nonideal systems requiring activity-coefficient models
5. Interlinked systems of separation units, including pumparounds, bypasses, and external heat exchangers
6. Narrow-boiling, wide-boiling, and dumbbell (mostly heavy and light components with little in between) feeds
7. Presence of free water
8. Wide variety of specifications other than Case II of Table 5.2 for the reflux ratio and product rates (e.g. product purities)
9. Use of Murphree-stage efficiencies

The inside-out method takes advantage of the following characteristics of the iterative calculations:

1. Component relative volatiles vary much less than component K-values.
2. Enthalpy of vaporization varies less than phase enthalpies.
3. Component stripping factors combine effects of temperature and liquid and vapor flows at each stage.

The inner loop of the inside-out method uses relative volatility, energy, and stripping factors to improve stability and reduce computing time. A widely used implementation of the inside-out method is that of Russell [31], which is described here together with further refinements suggested and tested by Jelinek [33].

MESH Equations

As with the BP, SR, and SC methods, the equilibrium stage model of Figures 10.1 and 10.3 is again employed. The form of the equations is similar to the SC method in that component flow rates are utilized. However, in addition, the following inner-loop variables are defined:

$$\alpha_{i,j} = K_{i,j}/K_{b,j} \tag{10-78}$$

$$S_{b,j} = K_{b,j}V_j/L_j \tag{10-79}$$

$$R_{Lj} = 1 + U_j/L_j \tag{10-80}$$

$$R_{Vj} - 1 + W_j/V_j \tag{10-81}$$

where K_b is the K-value for a base or hypothetical reference component, $S_{b,j}$ is the stripping factor for the base component, R_{Lj} is a liquid-phase withdrawal factor, and R_{Vj} is a vapor-phase withdrawal factor. For stages without sidestreams, R_{Lj} and R_{Vj} reduce to 1. With the defined variables of (10-78) to (10-81), (10-54) to (10-57) still apply, but the MESH equations, (10-58) to (10-60), become as follows, where (10-83) results from the use of (10-80) to (10-82) to eliminate the variables in V and the sidestream ratios s and S:

Phase equilibria:

$$v_{i,j} = \alpha_{i,j}S_{b,j}l_{i,j}, \qquad i = 1 \text{ to } C, \quad j = 1 \text{ to } N \tag{10-82}$$

Component material balance:

$$l_{i,j-1} - (R_{Lj} + \alpha_{i,j}S_{b,j}R_{Vj})l_{i,j} + (\alpha_{i,j+1}S_{b,j+1})l_{i,j+1} = -f_{i,j}, \quad i = 1 \text{ to } C, \quad j = 1 \text{ to } N \tag{10-83}$$

Energy balance:

$$H_j = h_{Lj}R_{Lj}L_j + h_{V_j}R_{Vj}V_j - h_{L_{j-1}}L_{j-1} - h_{V_{j+1}}V_{j+1} - h_{F_j}F_j - Q_j = 0, \quad j = 1 \text{ to } N \tag{10-84}$$

where $S_{i,j} = \alpha_{i,j}S_{b,j}$.

In addition, discrepancy functions of the type shown in Table 10.1 for the SC method can be added to the MESH equations to permit any reasonable set of product specifications.

Rigorous and Complex Thermodynamic Property Models

The complex thermodynamic models referred to in Figure 10.30 can include any of the types of models discussed in Chapter 2, including those based on $P-v-T$ equations of state (e.g. Soave–Redlich–Kwong and Peng–Robinson) and those based on free-energy models for predicting liquid-phase activity coefficients (e.g. Wilson, NRTL, and UNIQUAC). These models are used to generate parameters in the approximate thermodynamic property models. In general, the rigorous property models are of the form:

$$K_{i,j} = K_{i,j}\{P_j, T_j, \mathbf{x}_j, \mathbf{y}_j\} \tag{10-85}$$

$$h_{V_j} = h_{V_j}\{P_j, T_j, \mathbf{y}_j\} \tag{10-86}$$

$$h_{L_j} = h_{L_j}\{P_j, T_j, \mathbf{x}_j\} \tag{10-87}$$

Approximate Thermodynamic Property Models

K-Values

The approximate models used in the inside-out method are designed to facilitate the calculation of stage temperatures and stripping factors. The approximate K-value model of Russell [31] and Jelinek [33], which differs only slightly from the model of Boston and Sullivan [28] and originates from a proposal in the classic text book by Robinson and Gilliland [35], is (10-78) combined with

$$K_{b,j} = \exp(A_j - B_j/T_j) \tag{10-88}$$

Either a component in the feed or a hypothetical reference component can be selected as the base, b, component, with the latter preferred. For that case, the base component is determined from a vapor-composition weighting using the following relations:

$$K_{b,j} = \exp\left(\sum_i w_{i,j} \ln K_{i,j}\right) \tag{10-89}$$

where, $w_{i,j}$ are weighting functions given by,

$$w_{i,j} = \frac{y_{i,j}[\partial \ln K_{i,j}/\partial(1/T)]}{\sum_i y_{i,j}[\partial \ln K_{i,j}/\partial(1/T)]} \tag{10-90}$$

A unique K_b model and values of $\alpha_{i,j}$ in (10-78) are derived for each stage j from values of $K_{i,j}$ determined from the rigorous model. At the top stage, the base component will be close to one of the light components, while at the bottom stage, the base component will be close to a heavy component. The derivatives in (10-90) are obtained numerically or analytically from the rigorous model. To determine the values of A_j and B_j in (10-88), two temperatures must be selected for each stage. For example, the estimated or current temperatures of the two adjacent stages, $j - 1$ and $j + 1$, might be selected. Calling these two temperatures T_1 and T_2 and using (10-88) at each stage, b:

$$B = \frac{\ln(K_{b_{T_1}}/K_{b_{T_2}})}{\left(\dfrac{1}{T_2} - \dfrac{1}{T_1}\right]} \tag{10-91}$$

and

$$A = \ln K_{b_{T_1}} + B/T_1 \tag{10-92}$$

If highly nonideal liquid solutions are involved, it is advisable to separate the rigorous K-value into two parts, as in (2-27). Thus,

$$K_i = \gamma_{iL}(\phi_{iL}/\overline{\phi}_{iV}) \tag{10-93}$$

Then, $(\phi_{iL}/\overline{\phi}_{iV})$ is used to determine K_b and, as proposed by Boston [30], values of γ_{iL} at each stage are fitted at a reference temperature T^*, to the liquid-phase mole fraction by the linear function

$$\gamma_{iL}^* = a_i + b_i x_i \qquad (10\text{-}94)$$

to obtain the approximate estimates, γ_{iL}^*. Equation (10-83) is then modified by replacing $\alpha_{i,j}$ with $\alpha_{i,j}\gamma_{iL}^*$, where

$$\alpha_{i,j} = \frac{(\phi_{iL}/\overline{\phi}_{iV})_j}{K_{b,j}} \qquad (10\text{-}95)$$

rather than the $\alpha_{i,j}$ given by (10-78).

Enthalpies

Boston and Sullivan [28] and Russell [31] employ the same approximate enthalpy models. Jelinek [33] does not use approximate enthalpy models, because the additional complexity involved in the use of two enthalpy models may not always be justified to the extent that the use of both approximate and rigorous K-value models are justified.

The basis for the enthalpy calculations is the same as for the rigorous equations discussed in Chapter 2. Thus, for either phase, from Table 2.6,

$$h = h_V^{\circ} + (h - h_V^{\circ}) = h_V^{\circ} + \Delta H \qquad (10\text{-}96)$$

where h_V° is the ideal gas mixture enthalpy, as given by the polynomial equations, (2-35) and (2-36), based on the vapor-phase composition for h_V and the liquid-phase composition for h_L. The ΔH term is the enthalpy departure, $\Delta H_V = (h_V - h_V^{\circ})$ for the vapor phase, which accounts for the effect of pressure, and $\Delta H_L = (h_L - h_V^{\circ})$ for the liquid phase, which accounts for the enthalpy of vaporization and the effect of pressure on both liquid and vapor phases, as indicated in (2-57). Of particular importance is the enthalpy of vaporization, which dominates the ΔH_L term. The time-consuming parts of the enthalpy calculations are the two enthalpy departure terms, which are complex when an equation of state is used. Therefore, in the approximate enthalpy equations, the rigorous enthalpy departures are replaced by the simple linear functions,

$$\Delta H_{Vj} = c_j - d_j(T_j - T^*) \qquad (10\text{-}97)$$

and

$$\Delta H_{Lj} = e_j - f_j(T_j - T^*) \qquad (10\text{-}98)$$

where the departures are modeled in terms of enthalpy per unit mass instead of per unit mole, and T^* is a reference temperature. The parameters c, d, e, and f are evaluated from the rigorous models at each iteration of the outer loop.

Inside-Out Algorithm

The inside-out algorithm of Russell [31] involves an initialization procedure, inner-loop iterations, and outer-loop iterations.

Initialization Procedure

Before inner or outer loop calculations can begin, it is necessary to provide reasonably good estimates of all stage values of $x_{i,j}$, $y_{i,j}$, T_j, V_j, and L_j. Boston and Sullivan [28] suggest the following procedure:

1. Specify the number of theoretical stages, conditions of all feeds, feed-stage locations, and column pressure profile.
2. Specify stage locations for each product withdrawal (including side streams) and for each heat exchanger.
3. Provide an additional specification for each product and each intermediate heat exchanger.
4. If not specified, estimate each product withdrawal rate, and estimate each value of V_j. Estimate values of L_j from the total material balance equation, (10-6).
5. Estimate an initial temperature profile, T_j, by combining all feed streams (composite feed) and determining the bubble and dew point temperatures at the average column pressure. The dew-point temperature is taken as the top-stage temperature, T_1, whereas the bubble-point temperature is taken as the bottom-stage temperature, T_N. Intermediate stage temperatures are estimated by linear interpolation. Reference temperatures T^* for use with (10-94), (10-97), and (10-98) are set equal to T_j.
6. Flash the composite feed isothermally at the average column pressure and average column temperature. The resulting vapor and liquid compositions, y_i and x_i, are the estimated compositions for each stage.
7. Using the initial estimates from Steps 1 through 7, use the selected complex thermodynamic property correlation to determine values of the stagewise outside-loop K and h parameters A_j, B_j, $a_{i,j}$, $b_{i,j}$, c_j, d_j, e_j, f_j, $K_{b,j}$, and $\alpha_{i,j}$ of the approximate models.
8. Compute initial values of $S_{b,j}$, R_{Lj}, and R_{Vj} from (10-79), (10-80), and (10-81).

Inner-Loop Calculation Sequence

An iterative sequence of inner loop calculations begins with a set of values for the outside-loop parameters listed in Step 7, obtained initially from the initialization procedure and later from outer-loop calculations, using results from the inner loop, as shown in Figure 10.30c.

9. Compute component liquid flow rates, $l_{i,j}$, from the set of N equations (10-83) for each of the C components by the tridiagonal-matrix algorithm.
10. Compute component vapor flow rates, v_{ij}, from (10-82).
11. Compute a revised set of total flow rates, V_j and L_j, from the component flow rates by (10-54) and (10-55), respectively.
12. To calculate a revised set of stage temperatures, T_j, as follows, compute a set of x_i values for each stage from (10-57), then a revised set of $K_{b,j}$ values from a combination of the bubble-point equation, (4-12), $\sum_i K_i x_i = 1$, with (10-78), which gives

$$K_{b,j} = 1 \bigg/ \sum_{i=1}^{C} (\alpha_{i,j} x_{i,j}) \tag{10-99}$$

From this new set of $K_{b,j}$ values, compute a new set of stage temperatures from the following rearrangement of (10-88):

$$T_j = \frac{B_j}{A_j - \ln K_{b,j}} \tag{10-100}$$

At this point in the inner-loop iterative sequence, we have a revised set of values for $v_{i,j}$, $l_{i,j}$, and T_j, which satisfy the component material balance and phase equilibria equations for the estimated thermodynamic properties. However, these values do not satisfy the energy balance and specification equations unless the estimated base-component stripping factors and product withdrawal rates are correct.
13. Select inner-loop iteration variables as

$$\ln S_{b,j} = \ln(K_{b,j} V_j / L_j) \tag{10-101}$$

together with any other iteration variables. For a simple distillation column of the type shown in Figure 10.9, no other inner-loop iteration variables would be needed if the condenser and reboiler duties were specified. If the reflux ratio (L/D) and bottoms flow rate (B) are specified in place of the two duties, which is the more common situation, one adds, in place of the two (10-84) equations for H_1 and H_N, the following two specification equations from Table 10.1 in the form of discrepancy functions, D_1 and D_2:

$$D_1 = L_1 - (L/D)V_1 = 0 \tag{10-102}$$

$$D_2 = L_N - B = 0 \tag{10-103}$$

For each side stream, a side-stream withdrawal factor is added as an inner-loop iteration variable, e.g., $\ln(U_j/L_j)$ and $\ln(W_j/V_j)$, together with a specification equation on purity or some other variable.

14. Compute enthalpies of all streams from (10-96) to (10-98).

15. Compute normalized discrepancies of H_j, D_1, D_2, etc., from the energy balances (10-84) and (10-102), (10-103), etc., except compute Q_1 from H_1 and Q_N from H_N where appropriate. A typical normalization is discussed in Section 10.4 for the SC method.

16. Compute the Jacobian of partial derivatives of H_j, D_1, D_2, etc., with respect to the iteration variables of (10-101), etc. This is done by successive perturbation of each iteration variable and recalculation of the discrepancies through Steps 9 to 15, numerically or by differentiation.

17. Compute corrections to the inner-loop iteration variables by a Newton–Raphson iteration of the type discussed for the SR method in Section 10.3 and the SC method in Section 10.4.

18. Compute new values of the iteration variables from the sum of the previous values and the corrections with (10-66), using damping if necessary to reduce the sum of the squares of the normalized discrepancies.

19. Check whether the sum of the squares is sufficiently small. If so, proceed to the outer-loop calculation procedure given next. If not, repeat Steps 15 to 18 using the latest values of the iteration variables. For any subsequent cycles through Steps 15 to 18, Russell [31] uses Broyden [36] updates to avoid reestimation of the Jacobian partial derivatives, whereas Jelinek [33] recommends the standard Newton–Raphson method of recalculating of the partial derivatives for each inner-loop iteration.

20. Upon convergence of Steps 15 to 19, Steps 9 through 12 will have produced an improved set of primitive variables $x_{i,j}$, $v_{i,j}$, $l_{i,j}$, T_j, V_j, and L_j. From (10-56), corresponding values of $y_{i,j}$ can be computed. The values of these variables are not correct until the approximate thermodynamic properties are in agreement with the properties from the rigorous models. The primitive variables are input to the outer-loop calculations to bring the approximate and complex models into successively better agreement.

Outer-Loop Calculation Sequence

21. Using the values of the primitive variables from Step 20, compute relative volatilities and stream enthalpies from the complex thermodynamic models. If they are in close agreement with the previous values used to initiate a set of inner-loop iterations, both the outer-loop and inner-loop iterations are converged and the problem is solved. If not, proceed to Step 22.

22. Determine values of the stagewise outside-loop K and h parameters of the approximate models from the complex models as in initialization Step 7.

23. Compute values of $S_{b,j}$, R_{Lj}, and R_{Vj}, as in initialization Step 8.

24. Repeat the inner-loop calculation sequence of Steps 9 through 20.

Convergence of the inside-out method is not guaranteed. However, for most problems, the method is robust and rapid. Convergence can encounter difficulty because of poor initial estimates, resulting in negative or zero flow rates at certain locations in the column. To counteract this tendency, all component stripping factors are scaled with a scalar multiplier, S_b, sometimes called the base stripping factor, to give

$$S_{i,j} = S_b \alpha_{i,j} S_{b,j} \qquad (10\text{-}104)$$

The value of S_b is initially chosen to force the results of the initialization procedure to give a reasonable distribution of component flows throughout the column. Russell recommends that S_b be chosen only once, whereas Boston and Sullivan compute a new value of S_b for each new set of $S_{b,j}$ values.

For highly nonideal liquid mixtures, use of the inside-out method may become quite difficult. When that occurs, the SC method may be preferred. If the SC method also fails to converge, relaxation or continuation methods, described by Kister [37] are usually successful, but computing time may be an order of magnitude longer than that for similar problems converged successfully with the inside-out method.

EXAMPLE 10.9

For the conditions of the distillation column shown in Figure 10.7, obtain a converged solution by the inside-out method, using the SRK equation-of-state for thermodynamic properties.

SOLUTION

A computer solution was obtained with the equipment module TOWR (an inside-out method) of the ChemCAD process simulation program of Chemstations, Inc. The only initial assumptions are a condenser outlet temperature of 65°F and a bottoms-product temperature of 165°F. The bubble-point temperature of the feed is computed as 123.5°F. In the initialization procedure, the constants A and B in (10-88), with T in °R, are determined from the SRK equation, with the following results:

Stage	T, °F	A	B	K_b
1	65	6.870	3708	0.8219
2	95	6.962	4031	0.7374
3	118	7.080	4356	0.6341
4	142	7.039	4466	0.6785
5	165	6.998	4576	0.7205

Values of the enthalpy coefficients c, d, e, and f in (10-97) and (10-98) are not tabulated here but are also computed for each stage, based on the initial temperature distribution.

In the inner-loop calculation sequence, component flow rates are computed from (10-83) by the tridiagonal matrix method. The resulting bottoms-product flow rate deviates somewhat from the specified value of 50 lbmol/h. However, by modifying the component stripping factors with a base stripping factor, S_b, in (10-104) of 1.1863, the error in the bottoms flow rate is reduced to 0.73%.

The initial inside-loop error from the solution of the normalized energy-balance equations, (10-84), is found to be only 0.04624. This is reduced to 0.000401 after two iterations through the inner loop.

At this point in the inside-out method, the revised column profiles of temperature and phase compositions are used in the outer loop with the complex SRK thermodynamic models to compute updates of the approximate K and h constants. Only one inner-loop iteration is required to obtain satisfactory convergence of the energy equations. The K and h constants are again updated in the outer loop. After one inner-loop iteration, the approximate K and h values are found to be sufficiently close to the SRK values that overall convergence is achieved. Thus, a total of only three outer-loop iterations and four inner-loop iterations are required.

To illustrate the efficiency of the inside-out method to converge this example, the results from each of the three outer-loop iterations are summarized in the following tables:

Outer-Loop Iteration	Stage Temperatures, °F				
	T_1	T_2	T_3	T_4	T_5
Initial guess	65	—	—	—	165
1	82.36	118.14	146.79	172.66	193.20
2	83.58	119.50	147.98	172.57	192.53
3	83.67	119.54	147.95	172.43	192.43

Outer Loop Iteration	Total Liquid Flows, lbmol/h				
	L_1	L_2	L_3	L_4	L_5
Specification	100	—	—	—	—
1	100.00	89.68	187.22	189.39	50.00
2	100.03	89.83	188.84	190.59	49.99
3	100.0	89.87	188.96	190.56	50.00

Outer-Loop Iteration	Component Flows in Bottoms Product, lbmol/h			
	C_3	nC_4	nC_5	L_5
1	0.687	12.045	37.268	50.000
2	0.947	12.341	36.697	49.985
3	0.955	12.363	36.683	50.001

From these tables it is seen that the stage temperatures and total liquid flows are already close to the converged solution after only one outer-loop iteration. However, the composition of the bottoms product, specifically with respect to the lightest component, C_3, is not close to the converged solution until after two iterations. The inside-out method does not always converge so dramatically but is usually quite efficient as shown in the following table.

Problem	Total Number of Inner Loops	Number of Outer-Loop Iterations
Exercise 10.11	7	6
Exercise 10.25	6	3
Exercise 10.37	17	9
Exercise 10.41	16	5

Computing times for each of these four exercises was less than 5 seconds on a 386 DX/20MHz PC with a 80387 coprocessor. ■

SUMMARY

1. Rigorous methods are readily available for computer-solution of equilibrium-based models for multicomponent, multistage absorption, stripping, distillation, and liquid–liquid extraction.

2. The equilibrium-based model for a countercurrent-flow cascade provides for multiple feeds, vapor side streams, liquid side streams, and intermediate heat exchangers. Thus, the model can handle almost any type of column configuration.

3. The model equations include component material balances, total material balances, phase equilibria relations, and energy balances.

4. Some or all of the model equations can usually be grouped so as to obtain tridiagonal matrix equations, for which an efficient solution algorithm is available.

5. Widely used methods for iteratively solving all of the model equations are the bubble-point (BP) method, the sum-rates (SR) method, the simultaneous correction (SC) method, and the inside-out method.

6. The BP method is generally restricted to distillation problems involving narrow-boiling feed mixtures.

7. The SR method is generally restricted to absorption and stripping problems involving wide-boiling feed mixtures or in the ISR form to extraction problems.

8. The SC and inside-out methods are designed to solve any type of column configuration for any type of feed mixture. Because of its computational efficiency, the inside-out method is often the method of choice; however, it may fail to converge when highly nonideal liquid mixtures are involved, in which case the slower SC method should be tried. Both methods permit considerable flexibility in specifications.

9. When both the SC and inside-out methods fail, resort can be made to the much slower relaxation and continuation methods.

REFERENCES

1. Wang, J.C., and G.E. Henke, *Hydrocarbon Processing* **45**(8), 155–163 (1966).
2. Myers, A.I., and W.D. Seider, *Introduction to Chemical Engineering and Computer Calculations*, Prentice-Hall, Englewood Cliffs, NJ, 484–507 (1976).
3. Lewis, W.K., and G.L. Matheson, *Ind. Eng. Chem.* **24**, 496–498 (1932).
4. Thiele, E.W., and R.L. Geddes, *Ind. Eng. Chem.* **25**, 290 (1933).
5. Holland, C.D., *Multicomponent Distillation*. Prentice-Hall, Englewood Cliffs, NJ (1963).
6. Amundson, N.R., and A.J. Pontinen, *Ind. Eng. Chem.* **50**, 730–736 (1958).
7. Friday, J.R., and B.D. Smith, *AIChE J.* **10**, 698–707 (1964).
8. Boston, J.F., and S.L. Sullivan, Jr., *Can. J. Chem. Eng.* **52**, 52–63 (1974).
9. Boston, J.F., and S.L. Sullivan, Jr., *Can. J. Chem. Eng.* **50**, 663–669 (1972).
10. Johanson, P.J., and J.D. Seader, *Stagewise Computations—Computer Programs for Chemical Engineering Education* (ed. by J. Christensen), Aztec Publishing, Austin, TX pp. 349–389, A-16 (1972).
11. Lapidus, L., *Digital Computation for Chemical Engineers*, McGraw-Hill, New York pp. 308–309 (1962).
12. Orbach, O., and C.M. Crowe, *Can. J. Chem. Eng.* **49**, 509–513 (1971).
13. Scheibel, E.G., *Ind. Eng. Chem* **38**, 397–399 (1946).
14. Sujata, A.D., *Hydrocarbon Processing* **40**(12), 137–140 (1961).
15. Burningham, D.W., and F.D. Otto, *Hydrocarbon Processing* **46**(10), 163–170 (1967).
16. Shinohara, T., P.J. Johansen, and J.D. Seader, *Stagewise Computations—Computer Programs for Chemical Engineering Education*, J. Christensen, Ed., Aztec Publishing, Austin, TX pp. 390–428, A-17 (1972).
17. Tsuboka, T., and T. Katayama, *J. Chem. Eng. Japan* **9**, 40–45 (1976).
18. Hála, E., I. Wichterle, J. Polak, and T. Boublik, *Vapor–Liquid Equilibrium Data at Normal Pressures*, Pergamon, Oxford p. 308 (1968).
19. Steib, V.H., *J. Prakt. Chem.* **4**, Reihe, Bd. 28, 252–280 (1965).
20. Cohen, G., and H. Renon, *Can. J. Chem. Eng.* **48**, 291–296 (1970).
21. Goldstein, R.P., and R.B. Stanfield, *Ind. Eng. Chem., Process Des. Develop.* **9**, 78–84 (1970).
22. Naphtali, L.M., "The distillation column as a large system," paper presented at the AIChE 56th National Meeting, San Francisco, May 16–19, 1965.
23. Naphtali, L.M., and D.P. Sandholm, *AIChE J.* **17**, 148–153 (1971).
24. Fredenslund, A., J. Gmehling, and P. Rasmussen, *Vapor–Liquid Equilibria Using UNIFAC, A Group Contribution Method*. Elsevier, Amsterdam (1977).
25. Beveridge, G.S.G., and R.S. Schechter, *Optimization: Theory and Practice*, McGraw-Hill, New York pp. 180–189 (1970).
26. Block, U., and B. Hegner, *AIChE J.* **22**, 582–589 (1976).
27. Hofeling, B., and J.D. Seader, *AIChE J.* **24**, 1131–1134 (1978).
28. Boston, J.F., and S.L. Sullivan, Jr., *Can. J. Chem. Engr.* **52**, 52–63 (1974).
29. Boston, J.F., and H.I. Britt, *Comput. Chem. Engng.* **2**, 109–122 (1978).
30. Boston, J.F., *ACS Symp. Ser*, No. 124, 135–151 (1980).
31. Russell, R.A., *Chem. Eng.* **90**(20), 53–59 (1983).
32. Trevino-Lozano, R.A., T.P. Kisala, and J.F. Boston, *Comput. Chem. Engng.* **8**, 105–115 (1984).
33. Jelinek, J., *Comput. Chem. Engng.* **12**, 195–198 (1988).
34. Venkataraman, S., W.K. Chan, and J.F. Boston, *Chem. Eng. Prog.* **86**(8), 45–54 (1990).
35. Robinson, C.S., and E.R. Gilliland, *Elements of Fractional Distillation*, 4th edition, pp. 232–236. McGraw-Hill, New York (1950).
36. Broyden, C.G., *Math Comp.* **19**, 577–593 (1965).
37. Kister, H. Z., "Distillation Design", McGraw-Hill, Inc., NY (1992).

EXERCISES

The exercises for this chapter are most conveniently divided into two groups: (1) those that can be solved manually, and (2) those that are best solved with computer implementation of the methods discussed in this chapter. The first group is referenced to section numbers of this chapter. The second group of problems follows the first group and is referenced to the type of separator. Computer implementations for use with the second group are found in the following widely available programs and simulators:

ASPEN PLUS of Aspen Technology
ChemCAD of Chemstations
HYSIM of Hyprotech
PRO/II of Simulation Sciences

Section 10.1

10.1 Show mathematically that (10-6) is not independent of (10-1), (10-3), and (10-4).

10.2 Revise the MESH equations to account for entrainment, occlusion, and chemical reaction.

Section 10.2

10.3 Revise the MESH equations (10-1) to (10-6) to allow for pumparounds of the type shown in Figure 10.2 and discussed by Bannon and Marple [*Chem. Eng. Prog.* **74**(7), 41–45 (1978)] and Huber [*Hydrocarbon Processing* **56**(8), 121–125 (1977)]. Combine the equations to obtain modified *M* equations similar to (10.7). Can these equations still be partitioned in a series of *C* tridiagonal matrix equations?

10.4 Use the Thomas algorithm to solve the following matrix equation for x_1, x_2, and x_3.

$$\begin{bmatrix} -160 & 200 & 0 \\ 50 & -350 & 180 \\ 0 & 150 & -230 \end{bmatrix} \cdot \begin{bmatrix} x_1 \\ x_2 \\ x_3 \end{bmatrix} = \begin{bmatrix} 0 \\ -50 \\ 0 \end{bmatrix}$$

10.5 Use the Thomas algorithm to solve the following tridiagonal matrix equation for the **x** vector.

$$\begin{bmatrix} -6 & 3 & 0 & 0 & 0 \\ 3 & -4.5 & 3 & 0 & 0 \\ 0 & 1.5 & -7.5 & 3 & 0 \\ 0 & 0 & 4.5 & -7.5 & 3 \\ 0 & 0 & 0 & 4.5 & -4.5 \end{bmatrix} \cdot \begin{bmatrix} x_1 \\ x_2 \\ x_3 \\ x_4 \\ x_5 \end{bmatrix} = \begin{bmatrix} 0 \\ 0 \\ 100 \\ 0 \\ 0 \end{bmatrix}$$

Section 10.3

10.6 On page 162 of their article, Wang and Henke [1] claim that their method of solving the tridiagonal matrix for the liquid-phase mole fractions does not involve subtraction of nearly equal quantities. Prove or disprove their statement.

10.7 Derive an equation similar to (10-7), but with $v_{i,j} = y_{i,j}V_j$ as the variables instead of the liquid-phase mole fractions. Can the resulting equations still be partitioned into a series of *C* tridiagonal matrix equations?

10.8 In a computer program for the Wang–Henke bubble-point method, 10,100 storage locations are wastefully set aside for the four indexed coefficients of the tridiagonal matrix solution of the component material balances for a 100-stage distillation column.

$$A_j x_{i,j-1} + B_j x_{i,j} + C_j x_{i,j+1} - D_j = 0$$

Determine the minimum number of storage locations required if the calculations are conducted in the most efficient manner.

10.9 Solve by the Newton–Raphson method the simultaneous nonlinear equations

$$x_1^2 + x_2^2 = 17$$

$$(8x_1)^{1/3} + x_2^{1/2} = 4$$

for x_1 and x_2 to within $\pm$ 0.001. As initial guesses, assume
(a) $x_1 = 2$, $x_2 = 5$.
(b) $x_1 = 4$, $x_2 = 5$,
(c) $x_1 = 1$, $x_2 = 1$.
(d) $x_1 = 8$, $x_2 = 1$.

10.10 Solve by the Newton–Raphson method the simultaneous nonlinear equations

$$\sin(\pi x_1 x_2) - \frac{x_2}{2} - x_1 = 0$$

$$\exp(2x_1)\left[1 - \frac{1}{4\pi}\right] + \exp(1)\left[\frac{1}{4\pi} - 1 - 2x_1 + x_2\right] = 0$$

for x_1 and x_2 to within $\pm$ 0.001. As initial guesses, assume
(a) $x_1 = 0.4$, $x_2 = 0.9$.
(b) $x_1 = 0.6$, $x_2 = 0.9$.
(c) $x_1 = 1.0$, $x_2 = 1.0$.

10.11 One thousand kilogram-moles per hour of a saturated liquid mixture of 60 mol% methanol, 20 mol% ethanol, and 20 mol% n-propanol is fed to the middle stage of a distillation column having three equilibrium stages, a total condenser, a partial reboiler, and an operating pressure of 1 atm. The distillate rate is 600 kmol/h, and the external reflux rate is 2,000 kmol/h of saturated liquid. Assuming that ideal solutions are formed such that K-values can be obtained from vapor pressures and assuming constant molar overflow such that the vapor rate leaving the reboiler and each stage is 2,600 kmol/h, calculate one iteration of the BP method up to and including a new set of T_j values. To initiate the iteration, assume a linear temperature profile based on a distillate temperature equal to the normal boiling point of methanol and a bottoms temperature equal to the arithmetic average of the normal boiling points of the other two alcohols.

Section 10.4

10.12 Solve the following nine simultaneous linear equations, which have a block tridiagonal matrix structure, by the Thomas algorithm.

$$x_2 + 2x_3 + 2x_4 + x_6 = 7$$
$$x_1 + x_3 + x_4 + 3x_5 = 6$$
$$x_1 + x_2 + x_3 + x_5 + x_6 = 6$$
$$x_4 + 2x_5 + x_6 + 2x_7 + 2x_8 + x_9 = 11$$
$$x_4 + x_5 + 2x_6 + 3x_7 + x_9 = 8$$
$$x_5 + x_6 + x_7 + 2x_8 + x_9 = 8$$
$$x_1 + 2x_2 + x_3 + x_4 + x_5 + 2x_6 + 3x_7 + x_8 = 13$$
$$x_2 + 2x_3 + 2x_4 + x_5 + x_6 + x_7 + x_8 + 3x_9 = 14$$
$$x_3 + x_4 + 2x_5 + x_6 + 2x_7 + x_8 + x_9 = 10$$

10.13 Naphtali and Sandholm group the $N(2C + 1)$ equations by stage. Instead, group the equations by type (i.e., enthalpy balances, component balances, and equilibrium relations). Using a three-component, three-stage example, show whether the resulting matrix structure is still block tridiagonal.

10.14 Derivatives of properties are needed in the Naphtali–Sandholm SC method. For the Chao–Seader correlation, determine analytical derivatives for

$$\frac{\partial K_{i,j}}{\partial T_j}, \quad \frac{\partial K_{i,j}}{\partial v_{i,k}}, \quad \frac{\partial K_{i,j}}{\partial l_{i,k}}$$

10.15 A rigorous partial SC method for multicomponent, multistage vapor–liquid separations can be devised that is midway between the complexity of the BP/SR methods on the one hand and the SC methods on the other hand. The first major step in the procedure is to solve the modified M equations for the liquid-phase mole fractions by the usual tridiagonal matrix algorithm. Then, in the second major step, new sets of stage temperatures and total vapor flow rates leaving a stage are computed simultaneously by a Newton–Raphson method. These two major steps are repeated until a sum-of-squares criterion is satisfied. For this partial SC method:
(a) Write the two indexed equations you would use to simultaneously solve for a new set of T_j and V_j.
(b) Write the truncated Taylor series expansions for the two indexed equations in the T_j and V_j unknowns, and derive complete expressions for all partial derivatives, except that derivatives of physical properties with respect to temperature can be left as such. These derivatives are subject to the choice of physical property correlations.
(c) Order the resulting linear equations and the new variables ΔT_j and ΔV_j into a Jacobian matrix that will permit a rapid and efficient solution.

10.16 Revise equations (10-58) to (10-60) to allow two interlinked columns of the type shown in Figure 10.31 to be solved simultaneously by the SC method. Does the matrix equation

that results from the Newton–Raphson procedure still have a block tridiagonal structure?

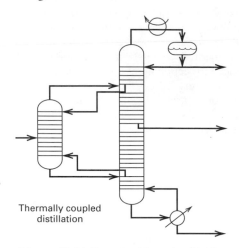

Figure 10.31 Data for Exercise 10.15.

10.17 In Equation (10-63), why is the variable order selected as v, T, l? What would be the consequence of changing the order to l, v, T? In Equation (10-64), why is the function order selected as H, M, E? What would be the consequence of changing the order to E, M, H?

Section 10.5

10.18 Suggest in detail a method for determining the scalar multiplier, S_b, in (10-104).

10.19 Suggest in detail an error function, similar to (10-75), that could be used to determine convergence of the inner-loop calculations for the inside-out method.

Distillation Problems

10.20 Calculate, product compositions, stage temperatures, interstage vapor and liquid flow rates and compositions, reboiler duty, and condenser duty for the following distillation column specifications.

Feed (bubble-point liquid at 250 psia and 213.9°F):

Component	Lbmol/h
Ethane	3.0
Propane	20.0
n-Butane	37.0
n-Pentane	35.0
n-Hexane	5.0

Column pressure = 250 psia
Partial condenser and partial reboiler
Distillate rate = 23.0 lbmol/h
Reflux rate = 150.0 lbmol/hr
Number of equilibrium stages (exclusive of condenser and reboiler) = 15
Feed is sent to middle stage

For this system at 250 psia, *K*-values and enthalpies may be computed by the Soave–Redlich–Kwong equations.

10.21 Determine the optimum feed stage location for Exercise 10.20.

10.22 Revise Exercise 10.20 so as to withdraw a vapor side stream at a rate of 37.0 lbmol/h from the fourth stage from the bottom.

10.23 Revise Exercise 10.20 so as to provide an intercondenser on the fourth stage from the top with a duty of 200,000 Btu/h and an interreboiler on the fourth stage from the bottom with a duty of 300,000 Btu/h.

10.24 Using the Peng–Robinson equations for thermodynamic properties, calculate the product compositions, stage temperatures, interstage vapor and liquid flow rates and compositions, reboiler duty, and condenser duty for the following multiple-feed distillation column, which has 30 equilibrium stages exclusive of a partial condenser and a partial reboiler and operates at 250 psia.

Feeds (both bubble-point liquids at 250 psia):

	Pound-moles per Hour	
	Feed 1	**Feed 2**
	to Stage 15	**to Stage 6**
Component	**from the Bottom**	**from the Bottom**
Ethane	1.5	0.5
Propane	24.0	10.0
n-Butane	16.5	22.0
n-Pentane	7.5	14.5
n-Hexane	0.5	3.0

Distillate rate = 36.0 lbmol/hr.
Reflux rate = 150.0 lbmol/hr.

Determine whether the feed locations are optimal.

10.25 Use the Chao–Seader or Grayson–Streed correlation for thermodynamic properties to calculate product compositions, stage temperatures, interstage flow rates and compositions, reboiler duty, and condenser duty for the distillation specifications in Figure 10.32.

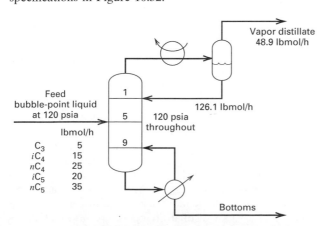

Figure 10.32 Data for Exercise 10.25.

Compare your results with those given in the *Chemical Engineers' Handbook*, Sixth Edition, pp. 13-42 to 13-45. Why do the two solutions differ?

10.26 Solve Exercise 10.11 using the UNIFAC method for *K*-values and obtain the converged solution.

10.27 Calculate with the Peng–Robinson equations for thermodynamic properties, the product compositions, stage temperatures, interstage flow rates and compositions, reboiler duty, and condenser duty for the distillation specifications in Figure 10.33, which represent an attempt to obtain four nearly pure products from a single distillation operation. Reflux is a saturated liquid. Why is such a high reflux ratio required?

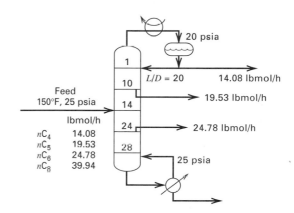

Figure 10.33 Data for Exercise 10.27.

10.28 Repeat Exercise 10.25, but substitute the following specifications for the specifications of vapor distillate rate and reflux rate:

Recovery of *n*C$_4$ in distillate = 98%
Recovery of *i*C$_5$ in bottoms = 98%

If the calculations fail to converge, the number of stages may be less than the minimum value. If so, increase the number of stages, revise the feed location, and repeat until convergence is achieved.

10.29 A saturated liquid feed at 125 psia contains 200 lbmol/h of 5 mol% *i*C$_4$, 20 mol% *n*C$_4$, 35 mol% *i*C$_5$, and 40 mol% *n*C$_5$. This feed is to be distilled at 125 psia with a column equipped with a total condenser and partial reboiler. The distillate is to contain 95% of the *n*C$_4$ in the feed, and the bottoms is to contain 95% of the *i*C$_5$ in the feed. Use the SRK equation for thermodynamic properties to determine a suitable design. Twice the minimum number of equilibrium stages, as estimated by the Fenske equation in Chapter 9, should provide a reasonable number of equilibrium stages.

10.30 A depropanizer distillation column is designed to operate at an average total pressure of 315 psia for separating a feed into distillate and bottoms with the flow rates shown next:

	lbmol/h		
	Feed	**Distillate**	**Bottoms**
Methane (C_1)	26	26	
Ethane (C_2)	9	9	
Propane (C_3)	25	24.6	0.4
n-Butane (C_4)	17	0.3	16.7
n-Pentane (C_5)	11		11
n-Hexane (C_6)	12		12
Totals	100	59.9	40.1

The thermal condition of the feed is such that it is 66 mol% vapor at tower pressure. Steam at 315 psia and cooling water at 65°F are available for the reboiler and condenser. The total pressure drop across the column may be taken to be 2 psi as a first approximation.
(a) Should a total condenser be used for this column?
(b) What are the feed temperature, K-values, and relative volatilities (with reference to C_3) at the feed temperature and pressure?
(c) If the reflux ratio is 1.3 times the minimum reflux, what is the actual reflux ratio? How many theoretical plates are needed in the rectifying and stripping sections?
(d) Compute the separation of species. How will the separation differ, if a reflux ratio of 1.5, 15 theoretical plates, and feed at the 9th plate are chosen?
(e) For part (c), compute the temperature and concentrations on each stage. What is the effect of feed plate location? How will the results differ if a reflux ratio of 1.5 and 15 theoretical plates are used?

10.31 Toluene is to be separated from biphenyl by ordinary distillation. The specifications for the separation are as follows:

	lbmol/h		
	Feed	**Distillate**	**Bottoms**
Benzene	3.4		
Toluene	84.6		2.1
Biphenyl	5.1	1.0	

Temperature = 264°F; Pressure = 37.1 psia for the feed
Reflux ratio = 1.3 times minimum reflux with total condenser
Top pressure = 36 psia; bottom pressure = 38.2 psia

(a) Determine the actual reflux ratio and the number of theoretical trays in the rectifying and stripping sections.
(b) For a D/F ratio of (3.4 + 82.5 + 1.0)/93.1, compute the separation of species. Compare the results to the preceding specifications.
(c) If the separation of species computed in part (b) is not sufficiently close to the specified split, adjust the reflux ratio to achieve the specified toluene flow in the bottoms.

10.32 The following stream at 100°F and 480 psia is to be separated by two ordinary distillation columns into the indicated products.

	lbmol/h			
Species	**Feed**	**Product 1**	**Product 2**	**Product 3**
H_2	1.5	1.5		
CH_4	19.3	19.2	0.1	
C_6H_6 (benzene)	262.8	1.3	258.1	3.4
C_7H_8 (toluene)	84.7		0.1	84.6
$C_{12}H_{10}$ (biphenyl)	5.1			5.1

Two different distillation sequences are to be examined. In the first sequence, CH_4 is removed in the first column. In the second sequence, toluene is removed in the first column. Compute the two sequences in the following manner: Estimate the actual reflux ratio and theoretical tray requirements for both sequences. Specify a reflux ratio equal to 1.3 times the minimum. Adjust isobaric column pressures so as to obtain distillate temperatures of about 130°F; however, no column pressure should be less than 20 psia. Specify total condensers, except that a partial condenser should be used when methane is taken overhead.

10.33 A process for the separation of a propylene–propane mixture to produce 99 mol% propylene and 95 mol% propane is shown in Figure 10.34. Because of the high product purities and the low relative volatility, 200 stages may be required. Assuming a tray efficiency of 100% and tray spacing of 24 inches, this will necessitate the two columns shown in series, because a single tower would be too tall. Assume a vapor distillate pressure of 280 psia, a pressure drop of 0.1 psi per tray, and a 2-psi drop through the condenser. The stage numbers and reflux ratio shown are only approximate. Determine the necessary reflux ratio for the stage numbers shown. Pay close attention to the determination of the proper feed-stage location so as to avoid pinch or near-pinch conditions wherein several adjacent trays may not be accomplishing anything.

10.34 So-called stabilizers are distillation columns that are often used in the petroleum industry to perform relatively easy separations between light components and considerably heavier components when one or two single-stage flashes are inadequate. An example of a stabilizer is shown in Figure 10.35 for the separation of H_2, methane, and ethane from benzene, toluene, and xylenes. Such columns can be difficult to calculate because a purity specification for the vapor distillate cannot be readily determined. Instead, it is more likely that the designer will be told to provide a column with 20 to 30 trays and a water-cooled partial condenser to provide 100°F reflux at a rate that will provide sufficient boilup at the bottom of the column to meet the purity specification there. It is desired to more accurately design the stabilizer column. The number of theoretical stages shown are just a first approximation and may be varied. Strive to achieve a bottoms product with no more than 0.05 mol% methane plus ethane and a vapor distillate temperature of about 100°F.

Exercises **583**

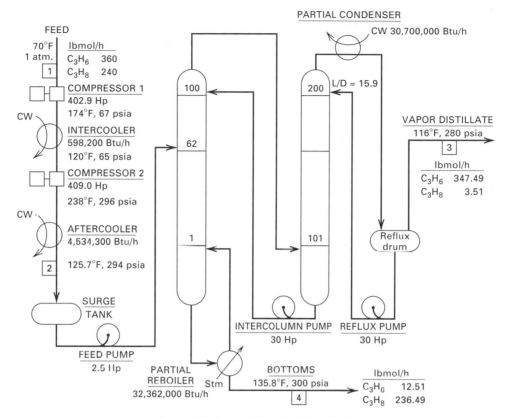

Figure 10.34 Data for Exercise 10.33.

These specifications may be achieved by varying the distillate rate and the reflux ratio. Reasonable initial estimates for these two quantities are 49.4 lbmol/h and 2. Assume a tray efficiency of 70%.

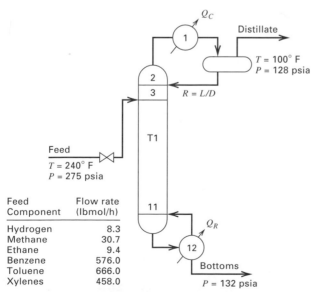

Feed Component	Flow rate (lbmol/h)
Hydrogen	8.3
Methane	30.7
Ethane	9.4
Benzene	576.0
Toluene	666.0
Xylenes	458.0

Figure 10.35 Data for Exercise 10.34.

10.35 A multiple recycle-loop problem formulated by Cavett[1] and shown in Figure 10.36 has been used extensively to test tearing, sequencing, and convergence procedures. The flowsheet is the equivalent of a four-theoretical-stage, near-isothermal distillation (rather than the conventional near-isobaric type), for which a patent by Gunther[2] exists. The flowsheet does not include necessary mixers, compressors, pumps, valves, and heat exchangers to make it a practical system. For the specifications shown on the drawing, determine the component flow rates for all streams in the process.

10.36 An absorber is to be designed for a pressure of 75 psia to handle 2,000 lbmol/h of gas at 60°F having the following composition.

Component	Mole Fraction
Methane	0.830
Ethane	0.084
Propane	0.048
n-Butane	0.026
n-Pentane	0.012

[1] R. H. Cavett, *Proc. Am. Petrol. Inst.* **43,** 57 (1963).
[2] A. Gunther, U.S. Patent 3,575,077 (April 13, 1971).

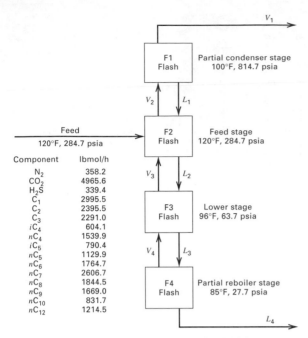

Figure 10.36 Data for Exercise 10.35.

The absorbent is an oil, which can be treated as a pure component having a molecular weight of 161. Calculate product rates and compositions, stage temperatures, and interstage vapor and liquid flow rate and compositions for the following conditions.

	Number of Equilibrium Stages	Entering Absorbent Flow Rate lbmol/h	Entering Absorbent Temperature, °F
(a)	6	500	90
(b)	12	500	90
(c)	6	1,000	90
(d)	6	500	60

10.37 Calculate product rates and compositions, stage temperatures, and interstage vapor and liquid flow rates and compositions for an absorber having four equilibrium stages with the specifications in Figure 10.37. Assume the oil is nC_{10}.

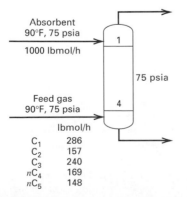

Figure 10.37 Data for Exercise 10.37.

10.38 In Example 10.4, temperatures of the gas and oil, as they pass through the absorber, increase substantially. This limits the extent of absorption. Repeat the calculations with a heat exchanger that removes 500,000 Btu/h from:
(a) Stage 2.
(b) Stage 3.
(c) Stage 4.
(d) Stage 5.

How effective is the intercooler? Which stage is the preferred location for the intercooler? Should the duty of the intercooler be increased or decreased assuming that the minimum stage temperature is 100°F using cooling water? Assume the absorber oil is nC_{12}.

10.39 Calculate product rates and compositions, stage temperatures, and interstage vapor and liquid flow rates and compositions for the absorber shown in Figure 10.38.

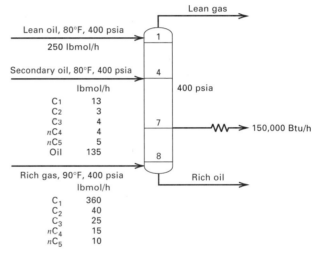

Figure 10.38 Data for Exercise 10.39.

10.40 Determine product compositions, stage temperatures, interstage flow rates and compositions, and reboiler duty for the reboiled absorber shown in Figure 10.39. Repeat the calculations without the interreboiler and compare both sets

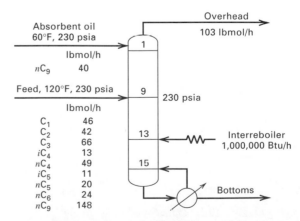

Figure 10.39 Data for Exercise 10.40.

of results. Is the interreboiler worthwhile? Should an intercooler in the top section of the column be considered?

10.41 Calcuate the product compositions, stage temperatures, interstage flow rates and compositions, and reboiler duty for the reboiled stripper shown in Figure 10.40.

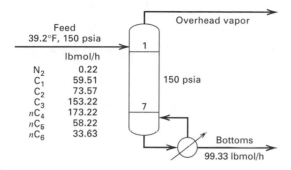

Figure 10.40 Data for Exercise 10.41.

Liquid–Liquid Extraction Problems

10.42 A mixture of cyclohexane and cyclopentane is to be separated by liquid–liquid extraction at 25°C with methanol. Phase equilibria for this system may be predicted by the NRTL or UNIQUAC equations. Calculate product rates and compositions and interstage flow rates and compositions for the conditions in Figure 10.41 with:
(a) $N = 1$ equilibrium stage.
(b) $N = 2$ equilibrium stages.
(c) $N = 5$ equilibrium stages.
(d) $N = 10$ equilibrium stages.

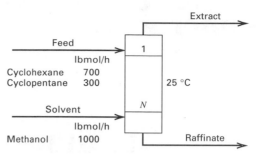

Figure 10.41 Data for Exercise 10.42.

10.43 The liquid–liquid extractor in Figure 8.1 operates at 100°F and a nominal pressure of 15 psia. For the feed and solvent flows shown, determine the number of equilibrium stages to extract 99.5% of the acetic acid, using the NRTL equation for activity coefficients. The NRTL constants may be taken as follows:

1 = ethyl acetate
2 = water
3 = acetic acid

I	J	B_{IJ}	B_{JI}	α_{IJ}
1	2	166.36	1190.1	0.2
1	3	643.30	−702.57	0.2
2	3	−302.63	−1.683	0.2

Compare the computed compositions of the raffinate and extract products to those of Figure 8.1.

Chapter 11

Enhanced Distillation and Supercritical Extraction

When two or more components differ in boiling point by less than approximately 50°C and form a nonideal liquid solution, the relative volatility may be below 1.10. Then, ordinary distillation may be uneconomic, and if an azeotrope forms even impossible. In that event, the following separation techniques, referred to as *enhanced distillation* by Stichlmair, Fair, and Bravo [1], should be explored:

1. *Extractive Distillation:* A method that uses a large amount of a relatively high-boiling solvent to alter the liquid-phase activity coefficients of the mixture, so that the relative volatility of the key components becomes more favorable. The solvent enters the column above the feed entry and a few trays below the top, and exits from the bottom of the column without causing an azeotrope to be formed. If the feed to the column is an azeotrope, the solvent breaks it. Also, the solvent may reverse volatilities.

2. *Salt Distillation:* A variation of extractive distillation in which the relative volatility of the key components is altered by dissolving a soluble, ionic salt in the top reflux. Because the salt is nonvolatile, it stays in the liquid phase as it passes down the column.

3. *Pressure-Swing Distillation:* A method for separating a pressure-sensitive azeotrope that utilizes two columns operated in sequence at two different pressures.

4. *Homogeneous Azeotropic Distillation:* A method of separating a mixture by adding an entrainer that forms a homogeneous minimum- or maximum-boiling azeotrope with one or more feed components. The entrainer is added near the top of the column, to the feed, or near the bottom of the column, depending upon whether the azeotrope is removed from the top or bottom.

5. *Heterogeneous Azeotropic Distillation:* A more useful azeotropic distillation in which a minimum-boiling heterogeneous azeotrope is formed by the entrainer. The azeotrope splits into two liquid phases in the overhead condensing system. One liquid phase is sent back to the column as reflux, while the other liquid phase is sent to another separation step or is a product.

6. *Reactive Distillation:* A method that adds a separating agent to react selectively and reversibly with one or more of the constituents of the feed. The reaction product is subsequently distilled from the nonreacting components. The reaction is then

reversed to recover the separating agent and the other reacting components. Reactive distillation also refers to the case where a chemical reaction and multistage distillation are conducted simultaneously in the same apparatus to produce other chemicals. This combined operation, sometimes referred to as *catalytic distillation* if a catalyst is used, is especially suited to chemical reactions limited by equilibrium constraints, since one or more of the products of the reaction are continuously separated from the reactants.

For ordinary distillation of multicomponent mixtures, the determination of feasible distillation sequences, the design of the columns in the sequence by rigorous methods described in Chapters 10 and 12, and the optimization of the column operating conditions are tedious, but are relatively straightforward. In contrast, determining and optimizing feasible enhanced distillation sequences is a considerably more difficult task. In particular, rigorous calculations of enhanced distillation frequently fail because of liquid-solution nonidealities and/or the difficulty of specifying feasible separations. To significantly reduce the chances of failure, especially for ternary systems, graphical techniques, described by Partin [2] and developed largely by Doherty and co-workers, and by Stichlmair and co-workers, as referenced later, provide valuable guidance for the development of feasible enhanced-distillation sequences. This chapter presents an introduction to the principles of these graphical methods and applies them to enhanced distillation.

Also included is a discussion of supercritical extraction, which differs considerably from conventional liquid–liquid extraction because of strong nonideal effects, and also requires considerable care in the development of an optimal system. The principles and techniques in this chapter are largely restricted to ternary systems; enhanced distillation and supercritical extraction are most commonly applied to such systems because the expense of these operations often requires that a multicomponent mixture first be reduced, by distillation or other means, to a binary or ternary system.

11.1 USE OF TRIANGULAR GRAPHS

When a binary mixture at a given pressure is separated by continuous distillation in equilibrium stages, all possible equilibrium compositions are uniquely located on a vapor–liquid (y–x) equilibrium curve. Figure 11.1 shows typical isobaric vapor–liquid equilibrium curves in terms of the mole fractions of the lowest-boiling component, A. In Figure 11.1a, possible compositions of the distillate and bottoms cover the entire range from pure B to pure A for a *zeotropic* (nonazeotropic) system. Temperatures, although not shown on the isobaric equilibrium curve, range from the boiling point of A to the boiling point of B. As the liquid and vapor compositions change from pure B to pure A, the temperature decreases.

In Figure 11.1b, a minimum-boiling azeotrope is formed at C and divides the plot into two regions. For distillation in Region 1, distillate and bottoms compositions can only vary from pure B to azeotrope C, and in Region 2, only from pure A to azeotrope C. For Region 1, as the composition changes from pure B to azeotrope C, the temperature decreases, as shown for example in Figure 4.8, where B is isopropyl alcohol, A is isopropyl ether, and the minimum-boiling azeotrope occurs at 78 mol% isopropyl ether and 66°C at 1 atm. In Region 2, the temperature also decreases as the composition changes from pure A to azeotrope C. Thus, a single distillation column operating at 1 atm cannot separate A from B. Depending upon whether the feed composition lies in Region 1 or 2, the column, at best, can only produce a distillate of azeotrope C and a bottoms of either pure B or

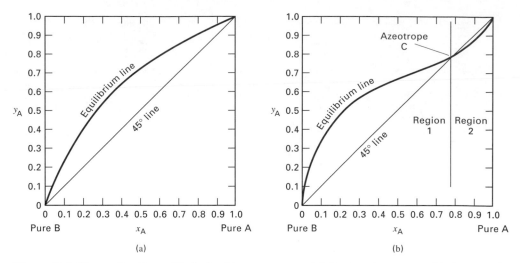

Figure 11.1 Vapor–liquid equilibria for binary systems. (a) Zeotropic system. (b) Azeotropic system.

pure A. However, all possible equilibrium compositions still lie on the equilibrium curve. These results are consistent with the Gibbs phase rule, as discussed in Chapter 4. From (4-1), for two components and two phases, that rule gives two degrees of freedom. Thus, if the pressure and temperature are fixed, the equilibrium vapor and liquid compositions arc fixed. However, as shown in Figure 11.2, for the case of an azeotrope-forming binary mixture, two feasible solutions exist within a certain temperature range. The particular solution observed depends on the overall composition of the two phases.

In the distillation of a ternary system, possible equilibrium compositions do not lie uniquely on a single isobaric equilibrium curve because the Gibbs phase rule gives an additional degree of freedom. The other compositions are determined only if the temperature, pressure, and composition of one component in one phase are fixed.

As discussed in Chapters 4 and 8, the composition of a ternary mixture can be represented on a triangular diagram, either equilateral or right, where the three apexes of the triangle represent the pure components. Although Stichlmair [3] shows that vapor–liquid phase

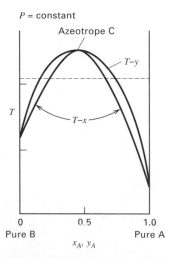

Figure 11.2 Multiple equilibrium solutions for an azeotropic system.

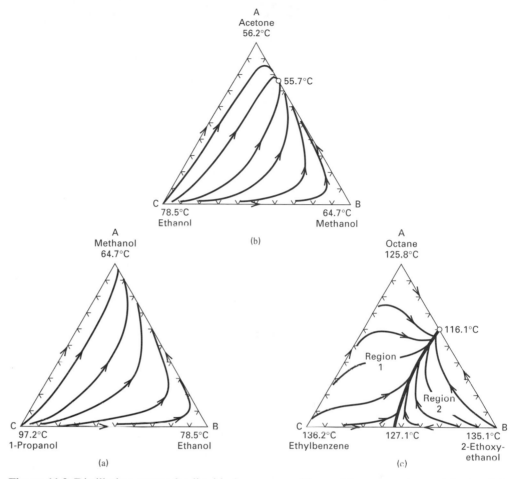

Figure 11.3 Distillation curves for liquid-phase compositions of ternary systems at 1 atm. (a) Mixture not forming an azeotrope. (b) Mixture forming one minimum-boiling azeotrope. (c) Mixture forming two minimum-boiling azeotropes.

equilibria at a fixed pressure can be plotted by letting the triangular grid represent the liquid phase and superimposing lines of constant equilibrium-vapor composition for two of the three components, this representation is confusing. It is more useful, when developing a feasible separation process, to plot only equilibrium liquid-phase compositions on the triangular diagram. Typical plots of this type at 1 atm, for three different ternary systems, are shown in Figure 11.3, where compositions are in mole fractions. Each curve in each diagram is the locus of possible equilibrium liquid-phase compositions that occur during distillation of a mixture, starting from any point on the curve. The boiling points of the three components and their binary and/or ternary azeotropes are included on the diagrams. The zeotropic alcohol system of Fig. 11.3a does not form any azeotropes. If a mixture of these three alcohols is distilled, there is only one distillation region, similar to the binary system of Fig. 11.1a. Accordingly, the distillate product can be nearly pure methanol (A) or the bottoms product can be nearly pure 1-propanol (C). However, nearly pure ethanol (B), the intermediate-boiling component, cannot be produced either as a distillate or bottoms. To separate this ternary mixture into the three components, a sequence of two ordinary distillation columns is used, as shown in Figure 11.4, where the feed, distillate, and bottoms product compositions must lie on a straight (total material-balance) line within the triangular diagram. Thus, in the so-called *direct sequence* of Figure 11.4a, the feed, F, is first separated into distillate A and a bottoms of B and C; then B is separated

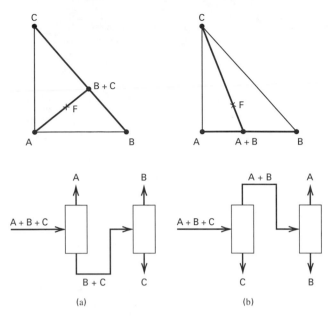

Figure 11.4 Distillation sequences for ternary zeotropic mixtures. (a) Direct sequence. (b) Indirect sequence.

from C in the second column. In the *indirect sequence* of Figure 11.4b, a distillate of A and B and a bottoms of C is produced in the first column, followed by the separation of A from B in the second column.

When a ternary mixture forms an azeotrope, the possible products from a single ordinary distillation column depend on the feed composition, as for a binary mixture. However, unlike the case of the binary mixture, where two distillation regions, shown in Figure 11.1b, are simple and well defined, the determination of possible distillation regions for azeotrope-forming ternary mixtures is complex. Consider first the example of Figure 11.3b, for a mixture of acetone (A), methanol (B), and ethanol (C), which are in the order of increasing boiling point. The only azeotrope formed at 1 atm is a minimum-boiling binary azeotrope, at 55.7°C, of the two lower-boiling components, acetone and methanol. The azeotrope contains 78.4 mol% acetone. For this type of system, as will be shown later, no distillation boundaries for the ternary mixture exist, even though an azeotrope is present. Thus, a feed composition located within the triangular diagram can be separated into two binary products, consistent with the straight (total material-balance) line. That is, ternary distillate or bottoms products can be avoided if the column split is properly selected. For example, the following five feed compositions can all produce, at a high reflux ratio and for a large number of stages, a distillate of the minimum-boiling azeotrope of acetone and methanol, and a bottoms product containing methanol and ethanol. That is, little or no ethanol will be in the distillate and little or no acetone in the bottoms.

Case	Feed:		Distillate:		Bottoms:	
	$x_{acetone}$	$x_{methanol}$	$x_{acetone}$	$x_{methanol}$	$x_{acetone}$	$x_{methanol}$
1	0.1667	0.1667	0.7842	0.2158	0.0000	0.1534
2	0.1250	0.3750	0.7837	0.2163	0.0000	0.4051
3	0.2500	0.2500	0.7837	0.2163	0.0000	0.2658
4	0.3750	0.1250	0.7837	0.2163	0.0000	0.0412
5	0.3333	0.3333	0.7837	0.2163	0.0000	0.4200

Alternatively, the column split can be selected to obtain a bottoms of nearly pure ethanol and a distillate of acetone and methanol. For either split, the straight, total material-balance line passing through the feed point can extend to the sides of the triangle.

Next, consider the more complex case of the ternary mixture of *n*-octane (A), 2-ethoxy-ethanol (B), and ethylbenzene (C), shown in Figure 11.3c. A and B form a minimum-boiling binary azeotrope at 116.1°C, and B and C do the same at 127.1°C. A triangular diagram for this type of system is separated by a *distillation boundary* (shown as a bold curved line) into regions 1 and 2. A material balance line connecting the feed to the distillate and bottoms cannot cross this distillation boundary, thus restricting the possible products of ordinary distillation of the ternary feed mixture. For example, a mixture with a feed composition inside Region 2 cannot produce a bottoms of ethylbenzene, the highest-boiling component in the mixture. It can be distilled to produce a distillate of the A–B azeotrope and a bottoms of a mixture of B and C, or a bottoms of B and a distillate of all three components. If the feed composition lies in Region 1 of Figure 11.3c, ordinary distillation can produce the A–B azeotrope and a bottoms of a mixture of A and C, or a bottoms of C and a distillate of a mixture of A and B. Thus, each region produces unique products.

To illustrate the restriction in product compositions caused by a distillation boundary, consider ordinary distillation with a feed mixture of 15 mol% A, 70 mol% B, and 15 mol% C. If the mixture is that in Figure 11.3a or b, a bottoms product of nearly pure C, the highest-boiling component, is obtained with a feed split corresponding to a distillate-to-bottoms ratio of 85/15. If, however, the mixture is that in Figure 11.3c, the same feed split ratio results in a bottoms of nearly pure B, the second-highest-boiling component.

Thus, products of a ternary mixture cannot be predicted merely from the boiling points of the components and azeotropes and a specified distillate-to-bottoms molar ratio, when distillation boundaries are present. These boundaries, as well as the mappings of distillation curves in the ternary plots of Figure 11.3, can be determined by either of the methods described in the next two sections.

Residue-Curve Maps

Consider the simple Rayleigh batch or differential distillation (no trays, packing, or reflux) shown schematically in Fig. 13.1. For any component of a ternary mixture, a material balance for its vaporization from the liquid in the still, assuming that the liquid is perfectly mixed and at its bubble-point temperature, is given by (13-1), which can be written as

$$\frac{dx_i}{dt} = (y_i - x_i)\frac{dW}{Wdt} \tag{11-1}$$

where x_i = mole fraction of component i in W moles of perfectly mixed liquid residue in the still

y_i = mole fraction of component i in the vapor (instantaneous distillate) in equilibrium with x_i

Because W changes (decreases) with time, t, it is possible to combine W and t into a single variable. Following the treatment of Doherty and Perkins [4], let this variable be ξ, such that

$$\frac{dx_i}{d\xi} = x_i - y_i \tag{11-2}$$

Combining (11-1) and (11-2) to eliminate $dx_i/(x_i - y_i)$:

$$\frac{d\xi}{dt} = -\frac{1}{W}\frac{dW}{dt} \tag{11-3}$$

Let the initial condition be $x = 0$ and $W = W_0$ at $t = 0$. Then the solution to (11-3) for ξ at time t is

$$\xi\{t\} = \ln[W_0/W\{t\}] \tag{11-4}$$

Because $W\{t\}$ decreases monotonically with time, $\xi\{t\}$ must increase monotonically with time and is considered a dimensionless, warped time. Thus, for the ternary mixture, the simple distillation process can be modeled by the following set of differential-algebraic equations (DAEs), assuming that a second liquid phase does not form:

$$\frac{dx_i}{d\xi} = x_i - y_i, \quad i = 1, 2 \tag{11-5}$$

$$\sum_{i=1}^{3} x_i = 1 \tag{11-6}$$

$$y_i = K_i x_i, \quad i = 1, 2, 3 \tag{11-7}$$

and the bubble-point-temperature equation:

$$\sum_{i=1}^{3} K_i x_i = 1 \tag{11-8}$$

where, in the general case, $K_i = K_i\{T, P, \mathbf{x}, \mathbf{y}\}$.

Thus, the system consists of seven equations in nine variables: P, T, x_1, x_2, x_3, y_1, y_2, y_3, and ξ. If the pressure is fixed, the next seven variables can be computed from (11-5) to (11-8) as a function of ξ, from a specified initial condition. The calculations can proceed in either the forward or backward direction of ξ. The results, when plotted on a triangular graph, are called a *residue curve* because the plot follows, with time, the liquid-residue composition in the still. A collection of residue curves, for a given ternary system at a fixed pressure, is a *residue-curve map*. A simple, but inefficient, procedure for calculating a residue curve is illustrated in the following example. Better, but more elaborate, procedures are given by Doherty and Perkins [4] and Bossen, Jørgensen, and Gani [5]. The last procedure is applicable when two separate liquid phases form, as is a procedure by Pham and Doherty [6].

EXAMPLE 11.1

Calculate and plot a portion of a residue curve for the ternary system, n-propanol (1), isopropanol (2), and benzene (3) at 1 atm, starting from a bubble-point liquid with a composition of 20 mol% each of 1 and 2, and 60 mol% of component 3. For K-values, use the modified Raoult's law (see Table 2.3) with regular-solution theory [see (2-64)] for estimating the liquid-phase activity coefficient as a function of composition and temperature. The normal boiling points of the three components in °C are 97.3, 82.3, and 80.1, respectively. Minimum-boiling azeotropes are formed at 77.1°C for components 1,3 and at 71.7°C for 2,3.

SOLUTION

A bubble-point calculation, using (11-7) and (11-8), gives starting values of $\mathbf{y}$ of 0.1437, 0.2154, and 0.6409, respectively, and a value of 79.07°C for the starting temperature, from the ChemSep program of Taylor and Kooijman [7].

For a specified increment in the dimensionless time, ξ, the differential equations (11-5) can be solved for x_1 and x_2 using Euler's method with a spreadsheet. Then x_3 is obtained from (11-6). The corresponding values of $\mathbf{y}$ and T are then obtained from (11-7) and (11-8). This procedure is repeated for the next increment in ξ. Thus, from (11-5):

$$x_1^{(1)} = x_1^{(0)} + (x_1^{(0)} - y_1^{(0)})\Delta\xi = 0.2000 + (0.2000 - 0.1437)0.1 = 0.2056$$

where superscripts (0) indicate starting values and a superscript of (1) indicates the value after the first increment in ξ. The value of 0.1 for $\Delta\xi$ gives reasonable accuracy, since the change in x_1 is seen to be only 2.7%. Similarly:

$$x_2^{(1)} = 0.2000 + (0.2000 - 0.2154)0.1 = 0.1985$$

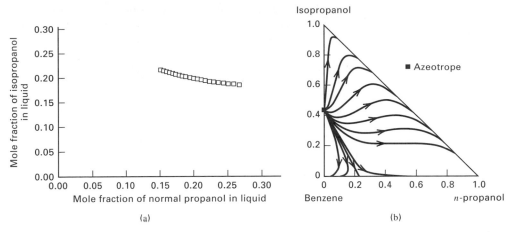

(a)   (b)

Figure 11.5 Residue curves for the normal propanol–isopropanol–benzene system at 1 atm for Example 11.1. (a) Calculated partial residue curve. (b) Residue-curve map.

From (11-6):

$$x_3^{(1)} = 1 - x_1^{(1)} - x_2^{(1)} = 1 - 0.2056 - 0.1985 = 0.5959$$

From a bubble-point calculation using (11-7) and (11-8), from ChemSep,

$$y^{(1)} = [0.1474, 0.2134, 0.6392]^T \text{ and } T^{(1)} = 79.14°C$$

The calculations are continued in the forward direction of ξ to $\xi = 1.0$. The calculations are also carried out in the backward direction back to $\xi = -1.0$. The results are in the table below, and the partial residue curve is plotted in Fig. 11.5a. For comparison, the complete residue-curve map for this system, from Doherty [8], is given on a right-triangle diagram in Fig. 11.5b.

ξ	x_1	x_2	y_1	y_2	T, °C
−1.0	0.1515	0.2173	0.1112	0.2367	78.67
−0.9	0.1557	0.2154	0.1141	0.2344	78.71
−0.8	0.1600	0.2135	0.1171	0.2322	78.75
−0.7	0.1644	0.2117	0.1201	0.2300	78.79
−0.6	0.1690	0.2099	0.1232	0.2278	78.83
−0.5	0.1737	0.2081	0.1264	0.2256	78.87
−0.4	0.1786	0.2064	0.1297	0.2235	78.91
−0.3	0.1837	0.2047	0.1331	0.2214	78.95
−0.2	0.1889	0.2031	0.1365	0.2194	79.00
−0.1	0.1944	0.2015	0.1401	0.2173	79.05
0.0	0.2000	0.2000	0.1437	0.2154	79.07
0.1	0.2056	0.1985	0.1474	0.2134	79.14
0.2	0.2115	0.1970	0.1512	0.2115	79.19
0.3	0.2175	0.1955	0.1550	0.2095	79.24
0.4	0.2237	0.1941	0.1589	0.2076	79.30
0.5	0.2302	0.1928	0.1629	0.2058	79.24
0.6	0.2369	0.1915	0.1671	0.2041	79.41
0.7	0.2439	0.1902	0.1714	0.2023	79.48
0.8	0.2512	0.1890	0.1758	0.2006	79.54
0.9	0.2587	0.1878	0.1804	0.1989	79.61
1.0	0.2665	0.1867	0.1850	0.1973	79.68

The residue-curve map in Figure 11.5b shows residue curves with arrows. The curves include the three border sides of the triangular diagram. The arrow on each curve points from a lower-boiling component or azeotrope to a higher-boiling component or azeotrope. In Figure 11.5b, all residue curves of the ternary mixture originate from the isopropanol–benzene azeotrope (lowest boiling point of 71.7°C). One of the curves terminates at the other azeotrope (*n*-propanol–benzene, which has a higher boiling point of 77.1°C). This is a special residue curve, called a *simple distillation boundary* because it divides the ternary region into two separate distillation regions. All residue curves lying above and to the right of the distillation boundary terminate at the *n*-propanol apex, which has the highest boiling point (97.3°C) for that region. All residue curves lying below and to the left of the distillation boundary are deflected to the benzene apex, whose boiling point of 80.1°C is the highest for this second region. ■

On the triangular diagram, all pure-component vertices and azeotropic points, whether binary azeotropes on the borders of the triangle, as in Figure 11.5b, or a ternary azeotrope within the triangle, are singular or fixed points of the residue curves because at these points, $dx/d\xi = 0$. In the vicinity of these points, the behavior of a residue curve depends upon the two eigenvalues of (11-5). At each pure-component vertex, the two eigenvalues are identical. At each azeotropic point, the two eigenvalues are different. Three cases, illustrated by each of three pattern groups in Figure 11.6, are possible:

Case 1: Both eigenvalues are negative. This is the point reached as ξ tends to infinity. It is the point at which all residue curves in a given region terminate. Thus, it is the component or azeotrope with the highest boiling point in the region. This point is a *stable node* because it is like the low point of a valley, in which a rolling ball finds a stable position.

Case 2: Both eigenvalues are positive. This is the point from which all residue curves in a given region originate. Thus, it is the component or azeotrope with the lowest boiling point in the region. This point is an *unstable node* because it is like the top of a peaked mountain from which a ball rolls toward a stable position.

Case 3: One eigenvalue is positive and one is negative. Residue curves within the triangle move toward and then away from such points, which are *saddles*. For a given distillation region, all pure components and azeotropes intermediate in boiling point between the stable node and the unstable node are saddles. In Figure 11.5b, the upper distillation region has one at the isopropanol vertex and the other at the normal propanol–benzene azeotrope.

From Example 11.1, it is clear that calculation of a residue-curve map requires a considerable effort. However, computer-aided simulation programs such as ASPEN PLUS [9] compute residue maps. Alternatively, as developed by Doherty and Perkins [10] and Doherty [8], the classification of singular points as stable nodes, unstable nodes, and saddles provides a rapid method for approximating a residue-curve map, including simple distillation boundaries, from just the pure-component boiling points and azeotrope boiling points and compositions. Boiling points of pure components are readily found in handbooks and component data banks of computer-aided simulation programs. Extensive listings of binary azeotropes are found in Horsley [11] and Gmehling [12]. The former lists more than 1,000 binary azeotropes. The latter includes experimental data for 15,300 binary azeotropes and 903 ternary azeotropes. The listings of ternary azeotropes are undoubtedly quite incomplete. However, in lieu of experimental data, a homotopy-continuation method for estimating all azeotropes of a multicomponent mixture from a thermodynamic model (e.g., Wilson, NRTL, UNIQUAC, UNIFAC) has been developed by Fidkowski, Malone, and Doherty [13].

Based on experimental evidence, for ternary mixtures, with very few exceptions, there are at most three binary azeotropes and one ternary azeotrope. Accordingly, the following set of restrictions apply to a ternary system:

$$N_1 + S_1 = 3 \tag{11-9}$$

$$N_2 + S_2 = B \le 3 \tag{11-10}$$

$$N_3 + S_3 = 1 \text{ or } 0 \tag{11-11}$$

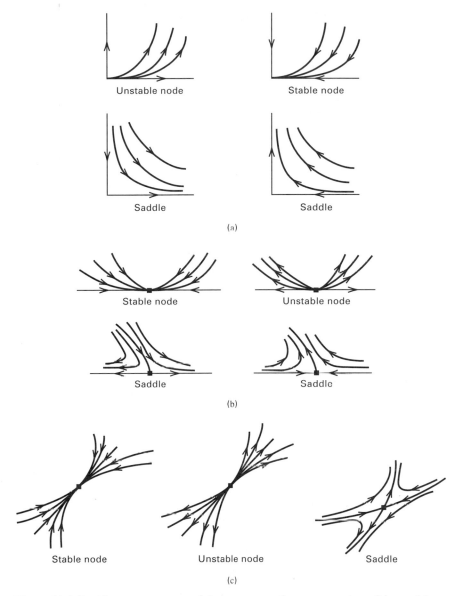

Figure 11.6 Residue curve patterns (a) near pure-component vertices; (b) near binary azeotropes; (c) near ternary azeotropes. [From M.F. Doherty and G.A. Caldarola, *IEC Fundam.,* **24,** 477 (1985) with permission.]

where, N is the number of stable and unstable nodes, S is the number of saddles, B is the number of binary azeotropes, and the subscript is the number of components at the node (stable or unstable) or saddle. Thus, S_2 is the number of binary azeotrope saddles. Doherty and Perkins [10] developed the following topological relationship among N and S:

$$2N_3 - 2S_3 + 2N_2 - B + N_1 = 2 \qquad \textbf{(11-12)}$$

For the system of Figure 11.5b, which has no ternary azeotrope, we see that $N_1 = 2$, $N_2 = 1$, $N_3 = 0$, $S_1 = 1$, $S_2 = 1$, $S_3 = 0$, and $B = 2$. Applying (11-12) gives $0 - 0 + 2 - 2 + 2 = 2$. Equation (11-9) gives $2 + 1 = 3$, (11-10) gives $1 + 1 = 2$, and (11-11) gives $0 + 0 = 0$. Thus, all four relations are satisfied.

The topological relationships are especially useful for rapidly sketching, on a ternary diagram, an approximate residue-curve map, including distillation boundaries, as described in detail by Foucher, Doherty, and Malone [14]. Their procedure involves the following nine steps (0–8), which are partly illustrated by a hypothetical example taken from their article and shown in Figure 11.7. The procedure is summarized in Figure 11.8. Approximate maps are usually developed from data at 1 atm. In the description of the steps, the term *species* refers to pure components and azeotropes.

Step 0: Label the ternary diagram with the pure-component normal-boiling-point temperatures. It is preferable to designate the top vertex of the triangle as the low boiler (L), the bottom-right vertex as the high boiler (H), and the bottom-left vertex as the intermediate boiler (I). Plot composition points for the binary and ternary azeotropes and add labels for their normal boiling points. This determines the value of B. See Figure 11.7, Step 0, where two minimum-boiling and one maximum-boiling binary azeotropes and one ternary azeotrope are designated by filled square markers. Thus, $B = 3$.

Step 1: Draw arrows on the edges of the triangle, in the direction of increasing temperature, for each pair of adjacent species. See Figure 11.7, Step 1, where there are six species on the edges of the triangle and six arrows have been added.

Step 2: Determine the type of singular point for each pure-component vertex, by using Fig. 11.6 with the arrows drawn in Step 1. This determines the values for N_1 and S_1. If a ternary azeotrope exists, go to Step 3; if not, go to Step 5. In Figure 11.7, Step 2, L is a saddle because one arrow points toward L and one points away from L; H is a stable node because both arrows point toward H, and I is a saddle. Therefore, $N_1 = 1$ and $S_1 = 2$.

Step 3 (for a ternary azeotrope): Determine the type of singular point for the ternary azeotrope, if one exists. The point is a node if (a) $N_1 + B < 4$, and/or (b) excluding the pure-component saddles, the ternary azeotrope has the highest, second-highest, lowest, or second-lowest boiling point of all species. Otherwise, the point is a saddle. This determines the values for N_3 and S_3. If the point is a node, go to Step 5; if a saddle, go to Step 4. In Figure 11.7, Step 3, $N_1 + B = 1 + 3 = 4$. However, excluding L and I because they are saddles, the ternary azeotrope has the second-lowest boiling point. Therefore, the point is a node, and $N_3 = 1$ and $S_3 = 0$. The type of node is still to be determined.

Step 4 (for a ternary saddle): Connect the ternary saddle, by straight lines, to all binary azeotropes and to all pure-component nodes (but not to pure-component saddles) and draw arrows on the lines to indicate the direction of increasing temperature. Determine the type of singular point for each binary azeotrope, by using Figure 11.6 with the arrows drawn in this step. This determines the values for N_2 and S_2. These values should be consistent with (11-10) and (11-12). This completes the development of the approximate residue-curve map, with no further steps needed. However, if $N_1 + B = 6$, then special checks must be made, as given in detail by Foucher, Doherty, and Malone [14]. This step does not apply to the example in Figure 11.7, because the ternary azeotrope is not a saddle.

Step 5 (for a ternary node or no ternary azeotrope): Determine the number of binary nodes, N_2, and binary saddles, S_2, from (11-10) and (11-12), where (11-12) can be solved for N_2 to give

$$N_2 = (2 - 2N_3 + 2S_3 + B - N_1)/2 \qquad \textbf{(11-13)}$$

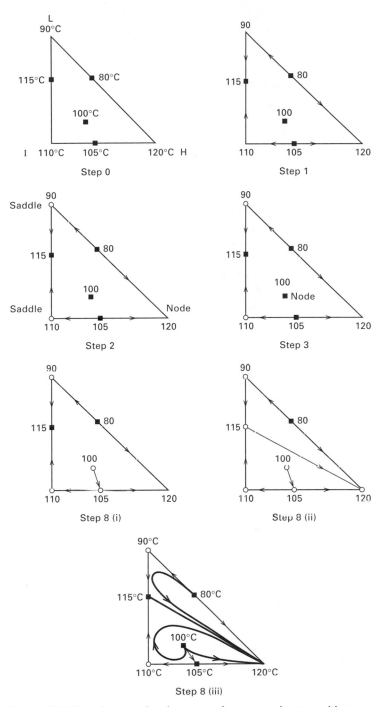

Figure 11.7 Step-by-step development of an approximate residue-curve map for a hypothetical system with two minimum-boiling binary azeotropes, one maximum-boiling binary azeotrope and one ternary azeotrope. [From E.R. Foucher, M.F. Doherty, and M.F. Malone, *IEC Res.* **30,** 764 (1991) with permission.]

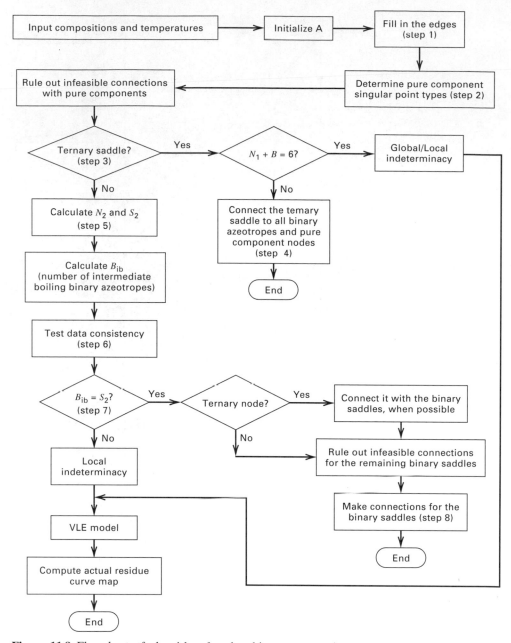

Figure 11.8 Flowchart of algorithm for sketching an approximate residue-curve map. [From E.R. Foucher, M.F. Doherty, and M.F. Malone, *IEC Res.* **30,** 763 (1991) with permission.]

For the example of Figure 11.7, $N_2 = (2 - 2 + 0 + 3 - 1)/2 = 1$. From (11-10), $S_2 = 3 - 1 = 2$.

Step 6: Count the binary azeotropes that are intermediate boilers (i.e. that are not the highest- or the lowest-boiling species), (and call that number B_{ib}). Make the following two data consistency checks: (a) The number of binary azeotropes, B, less B_{ib} must equal N_2, and (b) S_2 must be $\leq B_{ib}$. For the system in Figure 11.7, both checks are satisfied because $B_{ib} = 2$, $B - B_{ib} = 1$, $N_2 = 1$, and $S_2 = 2$. If these two consistency checks are not satisfied, one or more of the species' boiling points may be in error.

Step 7: If $S_2 \neq B_{ib}$, this procedure cannot determine a unique residue-curve-map structure, which therefore must be computed from (11-5) to (11-8). If $S_2 = B_{ib}$, there is a unique structure, which is completed in Step 8. For the example in Figure 11.7, $S_2 = B_{ib} = 2$; therefore, there is a unique map.

Step 8: In this final step for a ternary node or no ternary azeotrope, the distillation boundaries (connections), if any, are determined and entered on the triangular diagram as straight lines, and, if desired, one or more representative residue curves are sketched as curved lines within each distillation region. This step applies to cases of $S_3 = 0$, $N_3 = 0$ or 1, and $S_2 = B_{ib}$. In all cases, the number of distillation boundaries equals the number of binary saddles, S_2. Each binary saddle must be connected to a node (pure component, binary, or ternary). A ternary node must be connected to at least one binary saddle. Thus, a pure-component node cannot be connected to a ternary node, and an unstable node cannot be connected to a stable node. The connections are made by determining a connection for each binary saddle such that (a) a minimum-boiling binary saddle connects to an unstable node that boils at a lower temperature and (b) a maximum-boiling binary saddle connects to a stable node that boils at a higher temperature. It is best to first consider connections with the ternary node and then examine the possible connections for the remaining binary saddles. In the example of Figure 11.7, $S_2 = 2$, with these saddles denoted as L-I, a maximum-boiling azeotrope at 115°C, and as I-H, a minimum-boiling azeotrope at 105°C. Therefore, we make two connections to establish two distillation boundaries. The ternary node at 100°C cannot connect to L-I because 100°C is not greater than 115°C. The ternary node can, however, connect, as shown in Step 8 (i), to I-H because 100°C is lower than 105°C. This marks the ternary node as unstable. The connection for L-I can only be to H, as shown in Step 8 (ii) because it is a node (stable), and 120°C is greater than 115°C. This completes the connections. Finally, as shown in Step 8 (iii) of Figure 11.7, three typical, but approximate, residue curves are added to the diagram. These curves originate from unstable nodes and terminate at stable nodes.

Residue-curve maps are used to determine feasible distillation sequences for nonideal ternary systems. Matsuyama and Nishimura [15] show that the topological constraints just discussed limit the number of possible maps to about 113. Doherty and Caldarola [16] provide sketches of 87 maps that contain at least one minimum-boiling binary azeotrope. These maps cover most of the cases found in industrial applications, since minimum-boiling azeotropes are much more common than maximum-boiling azeotropes.

Distillation-Curve Maps

A residue curve represents the liquid-residue composition with time as the result of a simple, one-stage batch distillation. The curve is pointed in the direction of increasing time, from a lower-boiling state to a higher-boiling state. An alternative representation for distillation on a ternary diagram is a *distillation curve* for continuous, rather than batch, distillation. The curve is most readily determined for total reflux (infinite reflux ratio) at a constant pressure, usually 1 atm. The calculations are made down or up the column starting from any composition. Suppose we choose to make the calculations by moving up the column, starting from a stage designated as Stage 1, and numbering the stages upward. At any location between equilibrium stages j and $j + 1$, it will be recalled, from the McCabe–Thiele method for binary mixtures in Chapter 7 or the Fenske equation for multicomponent systems from Chapter 9, that passing vapor and liquid streams have the same composition. Thus:

$$x_{i,j+1} = y_{i,j} \qquad \textbf{(11-14)}$$

Also, liquid and vapor streams leaving the same stage are in equilibrium. Thus:

$$y_{i,j} = K_{i,j}x_{i,j} \tag{11-15}$$

To calculate a distillation curve for a fixed pressure, an initial liquid-phase composition, $x_{i,1}$, is assumed. This liquid is at its bubble-point temperature, which is determined from (11-8), which also gives the equilibrium-vapor composition, $y_{i,1}$ in agreement with (11-15). The composition, $x_{i,2}$, of the passing liquid stream is equal to $y_{i,1}$ by (11-14). The process is then repeated to obtain $x_{i,3}$, then $x_{i,4}$, and so forth. The sequence of liquid-phase compositions, which corresponds to the operating line for the total reflux condition, is plotted on the triangular diagram. The procedure is essentially that of Fenske for the determination of the minimum number of equilibrium stages for operation at total reflux to achieve a specified split of two key components, as discussed in Chapter 9. The distillation curve is analogous to the 45° line on a McCabe–Thiele diagram for a binary mixture. The calculation of a portion of a distillation curve is illustrated next.

EXAMPLE 11.2

Calculate and plot a portion of a distillation curve for the same starting conditions as Example 11.1.

SOLUTION

The starting values, $x^{(1)}$, are 0.2000, 0.2000, and 0.8000 for components 1,2, and 3, respectively. From Example 11.1, the bubble-point calculation gives a temperature of 79.07°C and the following values for $y^{(1)}$: 0.1437, 0.2154, and 0.6409. From (11-14), values of $x^{(2)}$ are 0.1437, 0.2154, and 0.6409. A bubble-point calculation for this composition gives $T^{(2)} = 78.62$°C and $y^{(2)} = 0.1063$, 0.2360, and 0.6577. Subsequent calculations are summarized in the following table:

Equilibrium Stage	x_1	x_2	y_1	y_2	T, °C
1	0.2000	0.2000	0.1437	0.2154	79.07
2	0.1437	0.2154	0.1063	0.2360	78.62
3	0.1063	0.2360	0.0794	0.2597	78.29
4	0.0794	0.2597	0.0592	0.2846	78.02
5	0.0592	0.2846	0.0437	0.3091	77.80

The resulting distillation curve is plotted in Figure 11.9, where points represent equilibrium stages and are connected by straight lines. ■

Distillation curves can be computed more rapidly than residue curves, and closely approximate them for reasons noted by Fidkowski, Doherty, and Malone [17]. If (11-5), which

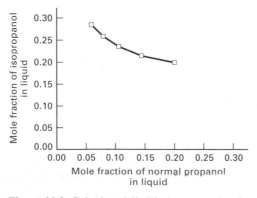

Figure 11.9 Calculated distillation curve for the normal propanol–isopropanol–benzene system at 1 atm for Example 11.2.

must be solved numerically as in Example 11.1, is written in a forward-finite-difference form, we obtain

$$(x_{i,j+1} - x_{i,j})/\Delta\xi = x_{i,j} - y_{i,j} \tag{11-16}$$

In Example 11.1, $\Delta\xi$ was set to $+0.1$ for calculations that give increasing values of T and to -0.1 to give decreasing values of T. If we choose the latter direction to be consistent with the direction used in Example 11.2, but set $\Delta\xi$ equal to -1.0, (11-16) becomes identical to (11-14). Thus, residue curves (which are true continuous curves) are equal to distillation curves (which are discrete points, through which a smooth curve is drawn), when the residue curves are approximated by a crude forward-finite-difference formulation.

A collection of distillation curves, including lines for distillation boundaries, is a *distillation-curve map,* an example of which from Fidkowski et al. [17] is shown in Figure 11.10 for the acetone–chloroform–methanol system at 1 atm. The Wilson equation was used to compute liquid-phase activity coefficients for the system. The dashed lines are the distillation curves; they approximate the residue curves, which are solid lines. This system has two minimum-boiling binary azeotropes, one maximum-boiling binary azeotrope, and a ternary saddle azeotrope. The map shows four distillation boundaries, designated by A, B, C, and D, consistent with Step 4 earlier. These computed boundaries are all curved lines rather than the approximate straight lines in the sketches of Figure 11.7.

Distillation-curve maps have been used extensively by Stichlmair and associates [1,3,18] for the development of feasible distillation sequences. In their maps, arrows on the distillation curves are directed toward the lower-boiling species, rather than the higher-boiling species as in residue-curve maps.

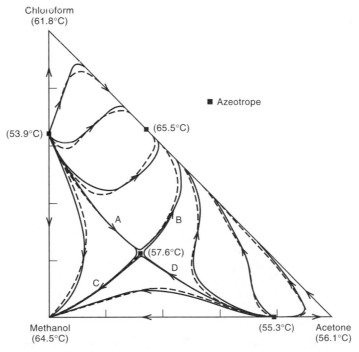

Figure 11.10 Comparison of residue curves to distillation curves. [From Z.T. Fidkowski, M.F. Malone, and M.F. Doherty, *AIChE J.,* **39,** 1303 (1993) with permission.]

Product-Composition Regions at Total Reflux

Residue-curve maps and distillation-curve maps are used to make preliminary estimates of regions of *feasible product compositions* for distillation of nonideal ternary mixtures. The product regions are determined by superimposing a column material-balance line on the curve-map diagram. Consider first the zeotropic ternary system in Figure 11.11a, which shows a typical isobaric residue-curve map with three residue curves. Assume that this map is identical to a corresponding distillation-curve map for total-reflux conditions and to a map for a finite, but very high reflux ratio. Suppose a ternary feed, denoted by F in Figure 11.11a, is to be continuously distilled in a column, operating isobarically at a high reflux ratio, to produce a distillate, D, and a bottoms, B. As shown in Chapters 4 and 8, if a straight line is drawn that connects distillate and bottoms compositions, the line must pass through the feed composition at some intermediate point to satisfy overall and component material-balance equations. This line is a *material-balance line,* three of which are included on Figure 11.11a. For a given material-balance line, the D and B composition points, designated by open circles, must lie on the same distillation curve. This causes the material-balance line to intersect the distillation curve at these two points and be a chord to the distillation curve.

The limiting distillate-composition point for this zeotropic system is pure low-boiling component, L. From the material-balance line passing through F, as shown in Figure 11.11b, the corresponding bottoms composition with the least amount of component L is point B. At the other extreme, the limiting bottoms-composition point is pure high-boiling component, H. A material-balance line from this point, through feed point F, ends at D. These two lines and the distillation curve define the feasible product-composition regions, shown shaded. Note that because, for a given feed, both the distillate and bottoms compositions must lie on the same distillation curve, the shaded feasible regions lie on the convex side of the distillation curve that passes through the feed point. Because of its appearance, the feasible product-composition region is referred to as a *bow-tie region.*

For an azeotropic system, where distillation boundaries are present, a feasible product-composition region can be found for each distillation region. Two examples are shown in Figure 11.12. The first, in Figure 11.12a, has two distillation regions caused by two minimum-boiling binary azeotropes. In the left-hand distillation region, distillate compositions are confined to the shaded region, $\dot{D}2$, whereas bottoms compositions are confined to $\dot{B}2$. For a feed composition in this distillation region, a distillate product of pure octane cannot

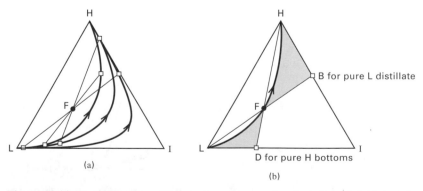

(a)

(b)

Figure 11.11 Product-composition regions for a zeotropic system. (a) Material-balance lines and distillation curves. (b) Product-composition regions. [From S. Widagdo and W.D. Seider, *AIChE J.,* **42,** 96–130 (1996) with permission.]

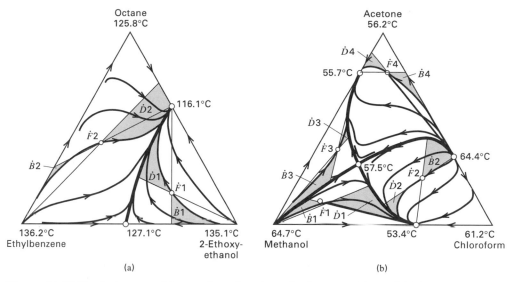

Figure 11.12 Product-composition regions for given feed compositions. (a) Ternary mixture with two minimum-boiling binary azeotropes at 1 atm. (b) Ternary mixture with three binary and one ternary azeotrope at 1 atm.

even be approached. For a feed composition in the right-hand distillation region, a bottoms product of ethylbenzene cannot be obtained. A more complex distillation-curve map, with four distillation regions, is shown in Figure 11.12b for a system with two minimum-boiling binary azeotropes, one maximum-boiling binary azeotrope, and one ternary azeotrope. One shaded bow-tie region is present for each distillation region. For this system, feasible product-composition regions are highly restricted.

In Figures 11.11b, 11.12a, and 11.12b, each bow-tie region is confined to its distillation region, as defined by the distillation boundaries. In all cases, the feed, distillate, and bottoms points on the material-balance line lie within a distillation region, with the feed point between the distillate and bottoms points. The material-balance lines do not cross the distillation-boundary lines. Is this always so? The answer is no! Under conditions where the distillation-boundary line is highly curved, it can be crossed by material-balance lines to obtain feasible product compositions. That is, a feed point can be on one side and the distillate and bottoms points on the other side of the distillation-boundary line. Consider the example in Figure 11.13, taken from Widagdo and Seider [19]. The distillation-boundary line, which is highly curved, extends from a minimum-boiling azeotrope K of H-I to the pure component L. This line divides the triangular diagram into two distillation regions, 1 and 2. Feed F_1 can be separated into products D_1 and B_1, which lie on distillation curve (a). In this case, the material-balance line and the distillation curve are both on the convex side of the distillation-boundary line. However, because the feed point F_1 lies close to the highly curved boundary line, F_1 can also be separated into D_2 and B_2 (or B_3), which lie on a distillation curve in region 2 on the concave side of the boundary. Thus, the material-balance line crosses the boundary from the convex to the concave side. Feed F_2 can be separated into D_4 and B_4, but not into D and B. In the latter case, the material-balance line cannot cross the boundary from the concave to the convex side, because the point F_2 does not lie between D and B on the material-balance line. The determination of the feasible product-composition regions for Figure 11.13 is left for an exercise at the end of this chapter. A detailed treatment of product composition regions is given by Wahnschafft et al. [20].

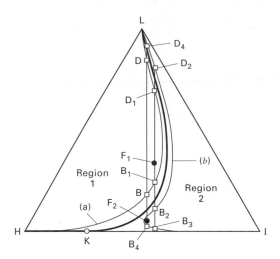

Figure 11.13 Feasible and infeasible crossings of distillation boundaries for an azeotropic system. [From S. Widagdo and W.D. Seider, *AIChE J.*, **42**, 96–130 (1996) with permission.]

11.2 EXTRACTIVE DISTILLATION

Extractive distillation is used to separate azeotropes and other mixtures that have key components with a relative volatility below about 1.1 over an appreciable range of concentration. If the feed is a minimum-boiling azeotrope, a solvent, with a lower volatility than the key components of the feed mixture, is added to a tray above the feed stage and a few trays below the top of the column so that (1) the solvent is present in the down-flowing liquid phase to the bottom of the column, and (2) little solvent is stripped and lost to the overhead vapor. If the feed is a maximum-boiling azeotrope, the solvent enters the column with the feed. The components in the feed must have different affinities for the solvent so that the solvent causes an increase in the relative volatility of the key components, to the extent that separation becomes feasible and economical. The solvent should not form an azeotrope with any components in the feed. Generally, a molar ratio of solvent-to-feed on the order of 1 is required to achieve this goal. The bottoms from the extractive distillation column is processed further to recover the solvent for recycle and complete the feed separation. The name, extractive distillation, was introduced by Dunn et al. [21] in connection with the commercial separation of toluene from a paraffin–hydrocarbon mixture using phenol as solvent.

Table 11.1 lists a number of industrial applications of extractive distillation. Consider the case of the acetone–methanol system. At 1 atm, acetone (nbp = 56.2°C) and methanol (nbp = 64.7°C) form a minimum-boiling azeotrope of 80 mol% acetone at a temperature of 55.7°C. The UNIFAC program was used to predict the vapor–liquid equilibria for this system at 1 atm. The azeotrope was estimated to occur at 55.2°C with 77.1 mol% acetone. At infinite dilution with respect to methanol, the relative volatility of acetone (A) with respect to methanol (M), $\alpha_{A,M}$, is predicted to be 0.74, with a liquid-phase activity coefficient for methanol of 1.88. At infinite dilution with respect to acetone, $\alpha_{A,M}$ is 2.48; by coincidence, the liquid-phase activity coefficient for acetone is 1.88 also. Water is a possible solvent for the system because at 1 atm, it does not form a binary or ternary azeotrope with acetone and/or methanol. The resulting residue-curve map with arrows directed from the azeotrope to pure water, computed by ASPEN PLUS using UNIFAC, is shown in Figure 11.14, where it is seen that no distillation boundaries exist. As discussed by Doherty and Caldarola [16], this is an ideal situation for the selection of an extractive distillation process. Their

Table 11.1 Some Industrial Applications of Extractive Distillation

Key Components in Feed Mixture	Solvent
Acetone–methanol	Aniline, ethylene glycol, water
Benzene–cyclohexane	Aniline
Butadienes–butanes	Acetone
Butadiene–butene-1	Furfural
Butanes–butenes	Acetone
Butenes–isoprene	Dimethylformamide
Cumene–phenol	Phosphates
Cyclohexane–heptanes	Aniline, phenol
Cyclohexanone–phenol	Adipic acid diester
Ethanol–water	Glycerine, ethylene glycol
Hydrochloric acid–water	Sulfuric acid
Isobutane–butene-1	Furfural
Isoprene–pentanes	Acetonitrile, furfural
Isoprene–pentenes	Acetone
Methanol–methylene bromide	Ethylene bromide
Nitric acid–water	Sulfuric acid
n-Butane–butene-2s	Furfural
Propane–propylene	Acrylonitrile
Pyridine–water	Bisphenol
Tetrahydrofuran–water	Dimethylformamide, propylene glycol
Toluene–heptanes	Aniline, phenol

schematic residue-curve map for this type system (designated 100) is included as an insert in Figure 11.14.

Ternary mixtures of acetone, methanol, and water at 1 atm give the following separation factors, estimated from the UNIFAC equation:

Mol% Water	Relative Volatility, $\alpha_{A,M}$			Liquid-Phase Activity Coefficient at Infinite Dilution	
	Methanol-rich	Acetone-rich	Equimolar	Acetone	Methanol
40	2.48	2.57	2.03	2.12	0.70
50	2.56	2.86	2.29	2.41	0.72

Thus, the presence of appreciable water increases the liquid-phase activity coefficient of acetone and decreases that of methanol, with the result that, over the entire concentration range of acetone and methanol, the relative volatility of acetone to methanol is at least 2.0. This makes it possible, with extractive distillation, to obtain a distillate of acetone and a bottoms of methanol and water. Furthermore, the relative volatilities of acetone to water and methanol to water average about 4.5 and 2.0, respectively. Thus, it is relatively easy to prevent an appreciable amount of water from reaching the distillate, and, in subsequent operations, to separate methanol from water by ordinary distillation.

EXAMPLE 11.3

Forty moles per second of a bubble-point mixture of 75 mol% acetone and 25 mol% methanol at 1 atm is separated by an extractive-distillation process, using water as the solvent, to produce an acetone product of not less than 95 mol% acetone, a methanol product of not less than 98 mol% methanol, and a water stream for recycle of at least 99.9 mol% purity. Prepare a preliminary process design, using the traditional sequence consisting of ordinary distillation followed by

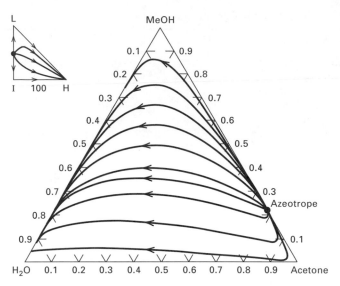

Figure 11.14 Residue-curve map for acetone–methanol–water system at 1 atm.

extractive distillation, and then ordinary distillation to recover the solvent, as shown for another system in Figure 11.15.

SOLUTION

In the first column, the feed mixture of acetone and methanol would be partially separated by ordinary distillation, where the distillate composition approaches that of the binary azeotrope. The bottoms would be nearly pure acetone or nearly pure methanol depending upon whether the feed contains more or less than 80 mol% acetone, respectively. However, in this example, the feed composition is already close to the azeotrope composition; therefore, the first column is omitted.

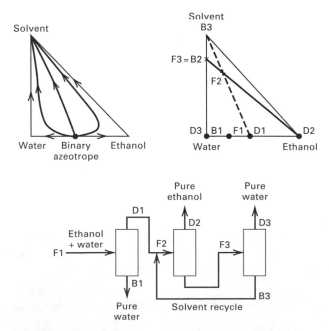

Figure 11.15 Distillation sequence for extractive distillation. [From M.F. Doherty and G.A. Caldarola, *IEC Fundam.,* **24,** 479 (1985) with permission.]

Table 11.2 Material and Energy Balances for Extractive Distillation Process of Example 11.3

	Material Balances					
Species	Flow rate, mol/s: Column 2 Feed	Column 2 Solvent	Column 2 Distillate	Column 2 Bottoms	Column 3 Distillate	Column 3 Bottoms
Acetone	30	0	29.86	0.14	0.14	0.0
Methanol	10	0	0.016	9.984	9.926	0.058
Water	0	60	1.35	58.65	0.06	58.59
Total	40	60	31.226	68.774	10.126	58.648

Energy Balances		
	Column 1	Column 2
Condenser duty, MW	4.71	1.07
Reboiler duty, MW	4.90	1.12

Accordingly, the acetone–methanol feed is sent to the second column, an extractive-distillation column equipped with a total condenser and a partial reboiler to produce a distillate of at least 95 mol% acetone. The acetone recovery is better than approximately 99% to achieve methanol purity in the third column. The bottoms from the extractive-distillation column is pumped to that column, where methanol and water are separated by ordinary distillation to achieve the specified purities.

The ChemSep and ChemCAD programs were used to make the calculations, with the UNIFAC method for activity coefficients. Equilibrium-stage, feed-stage, solvent entry-stage, solvent flow-rate, and reflux-ratio requirements were varied until a satisfactory design was achieved. The resulting material and energy balances are summarized in Table 11.2.

For the extractive-distillation column, a solvent flow rate of 60 mol/s of water is suitable. Using 28 theoretical trays, a 50°C solvent entry at Tray 6 from the top, a feed entry at Tray 12 from the top, and a reflux ratio of 4, a distillate composition of 95.6 mol% acetone is achieved. The impurity is mainly water. The acetone recovery is 99.5%. A 6-ft-diameter column with 60 sieve trays on 2-ft tray spacing is adequate for the operation. A liquid-phase composition profile is shown in Figure 11.16. The mole fraction of water (the solvent) in the liquid phase is appreciable, at least 0.35 for all of the stages below the solvent-entry stage. A distillation-curve

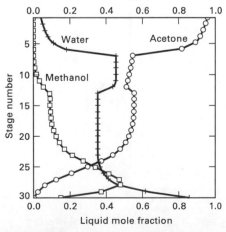

Figure 11.16 Liquid composition profile for extractive-distillation column of Example 11.3.

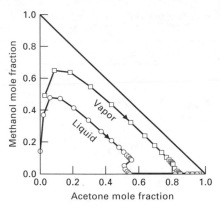

Figure 11.17 Distillation-curve map for Example 11.3. Data points are for theoretical stages.

map for the actual extractive distillation operation is given in Figure 11.17, where both the vapor and liquid curves are plotted. The arrows are directed from the bottom of the column to the top.

For the ordinary-distillation column, operation with 16 theoretical stages, a bubble-point feed-stage location of Stage 11 and a reflux ratio of 2 is adequate to achieve a methanol distillate of 98.1 mol% purity and a water bottoms, suitable for recycle, of 99.9 mol% purity. A McCabe–Thiele diagram in Figure 11.18 shows the locations of the theoretical stages. The feed stage is optimally located. Water makeup is less than 1.5 mol/s. A 2.5-ft-diameter column packed with 48 feet of 50-mm-diameter metal Pall rings is suitable for the separation. ∎

One unfortunate aspect in the design of the extractive-distillation column in Example 11.3 is caused by the relatively low boiling point of water. With a solvent entry point of Tray 6 from the top, 1.35 mol/s (2.25% of the water solvent) is stripped from the liquid into the distillate. The use of two other higher-boiling solvents listed in Table 11.1, aniline (nbp = 184°C) or ethylene glycol (nbp = 198°C), results in far less stripping of solvent. Other possible solvents for the separation of acetone from methanol by extractive distillation include methylethylketone (MEK) and ethanol. MEK behaves in a fashion opposite to that of water: MEK causes the volatility of methanol to be greater than that of acetone.

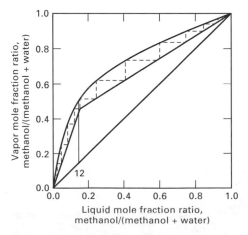

Figure 11.18 McCabe–Thiele diagram for methanol–water distillation in Example 11.3.

Thus, methanol becomes the distillate in the extractive-distillation column, leaving acetone to be separated from MEK in the following column.

In selecting a solvent for extractive distillation, a number of factors are considered, including availability, cost, corrosivity, vapor pressure, thermal stability, heat of vaporization, reactivity, toxicity, infinite-dilution activity coefficients in the solvent of the components to be separated, and ease of recovery for recycle. In addition, the solvent should not form azeotropes. Initial screening is based on the measurement or prediction of infinite-dilution activity coefficients. Berg [22] discusses, in detail, the selection of separation agents for both extractive and azeotropic distillation. He points out that all successful solvents for extractive distillation are highly hydrogen-bonded liquids, such as (1) water, amino alcohols, amides, and phenols that form three-dimensional networks of strong hydrogen bonds, and (2) alcohols, acids, phenols, and amines that are composed of molecules containing both active hydrogen atoms and donor atoms (oxygen, nitrogen, and fluorine). In general, it is very difficult or impossible to find a suitable solvent to economically separate components having the same functional groups.

Extractive distillation is also used to separate binary mixtures that form a maximum-boiling azeotrope, as shown in the following example.

EXAMPLE 11.4

Acetone (nbp = 56.16°C) and chloroform (nbp = 61.10°C) form a maximum-boiling homogeneous azeotrope at 1 atm and 64.43°C that contains 37.8 mol% acetone. Thus, they cannot be separated by ordinary distillation at 1 atm. Instead, it is proposed to separate them using extractive distillation in a two-column sequence, shown in Figure 11.19, with benzene (nbp = 80.24°C) as the solvent. Benzene does not form azeotropes with either of the feed components.

In the first column, the feed, blended with recycled solvent, is distilled to produce a distillate of 99 mol% acetone. The bottoms is sent to the second column, where 99 mol% chloroform leaves as distillate and the bottoms, which is rich in benzene, is recycled to the inlet of the first column with a small flow of makeup benzene. If the fresh feed is 21.8858 mol/s of 54.83 mol% acetone, with the balance chloroform, design a feasible two-column system using a ratio of 3.1667 moles of benzene per mole of acetone + chloroform in the combined feed to the first column. Both columns operate at a nominal pressure of 1 atm with total condensers, saturated liquid reflux, and partial reboilers. Use the UNIFAC method for estimating activity coefficients. The combined feed to the first column is brought to the bubble point before entering the feed stage.

SOLUTION

The residue-curve map for the ternary system acetone–chloroform–benzene at 1 atm is shown in Figure 11.20. The only azeotrope is that formed by acetone and chloroform. A curved distillation boundary extending from that azeotrope to the pure benzene apex divides the

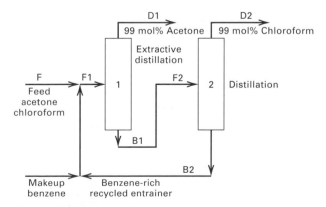

Figure 11.19 Process for the separation of acetone and chloroform in Example 11.4.

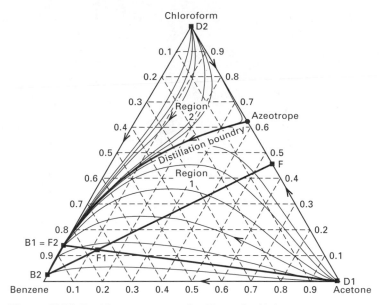

Figure 11.20 Residue-curve map for Example 11.4.

diagram into two distillation regions. The first column, which produces nearly pure acetone, operates in Region 1, whereas the second column operates in upper Region 2.

This ternary system was studied in detail by Fidkowski, Doherty, and Malone [17]. A design, based on their studies and using the ChemCAD process simulator, is summarized in Table 11.3. The first column contains 65 theoretical stages with the combined feed entering Stage 30 from the top. With a reflux ratio of 10, the acetone distillate purity is achieved with an acetone recovery of better than 99.95%. In Column 2, which contains 50 theoretical stages with the feed entering at Stage 30 from the top, a reflux ratio of 11.783 gives the required chloroform purity in the distillate, but with a recovery of only 82.23%. However, this is not serious because the remaining chloroform leaving in the bottoms is recycled with the benzene to Column 1. The benzene makeup rate is 0.1141 mol/s. Feed, distillate, and bottoms compositions are designated in Fig. 11.20. ∎

Table 11.3 Material and Energy Balances for Homogeneous Azeotropic Distillation of Example 11.4

Material Balances with Flows in mol/s						
Species	F	F_1	D_1	$B_1 = F_2$	D_2	B_2
Acetone	12.0000	12.0000	11.9948	0.0052	0.0052	0.0000
Chloroform	9.8858	12.0000	0.1046	11.8954	9.7812	2.1142
Benzene	0.0000	76.0000	0.0207	75.9793	0.0934	75.8859

Energy Balances		
Heat duty, kcal/h	Column 1	Column 2
Condenser	950,000	891,600
Reboiler	958,400	1,102,000

11.3 SALT DISTILLATION

The use of water as a solvent in the extractive distillation of acetone and methanol in Example 11.3 has the two disadvantages that a large amount of water is required to adequately alter the relative volatility and, even though the solvent is introduced into the column several trays below the top tray, enough water is stripped by vapor traffic into the distillate to reduce the acetone purity to 95.6 mol%. The water vapor pressure can be lowered, and thus the purity of acetone distillate increased, by use of an aqueous inorganic-salt solution as the solvent. For example, a 1927 patent application by Othmer [23] describes the use of a concentrated calcium chloride brine. Not only does calcium chloride, which is highly soluble in water, reduce the volatility of water, but it also has a strong affinity for methanol. Thus, the relative volatility of acetone with respect to methanol is further enhanced. The separation of the brine solution from methanol is easily accommodated in the subsequent distillation step, with the brine solution recycled to the extractive distillation column. The vapor pressure of the dissolved salt is so small that it never enters the vapor phase. An even earlier patent by Van Raymbeke [24] describes the extractive distillation of ethanol from water by using solutions of calcium chloride, zinc chloride, or potassium carbonate in glycerol.

Rather that using a solvent that contains a dissolved salt, the salt can be added as a solid or melt directly into the column by dissolving it in the liquid reflux before it enters the column. This technique was demonstrated experimentally by Cook and Furter [25] in a 4-inch-diameter, 12-tray rectifying column with bubble caps for the separation of ethanol from water using potassium acetate. At salt concentrations below saturation and between 5 and 10 mol%, an almost pure ethanol distillate was achieved. The salt, which must be soluble in the reflux, is recovered from the aqueous bottoms by evaporation and crystallization.

In aqueous alcohol solutions, both *salting out* and *salting in* have been observed by Johnson and Furter [26], as shown in the vapor–liquid equilibrium data in Figure 11.21: in (a), sodium nitrate salts out methanol, but in (b), mercuric chloride salts in methanol. Even low concentrations of potassium acetate can eliminate the ethanol–water azeotrope,

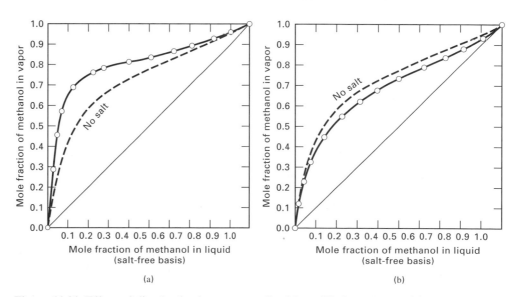

(a)

(b)

Figure 11.21 Effect of dissolved salts on vapor–liquid equilibria at 1 atm. (a) Salting-out of methanol by saturated aqueous sodium nitrate. (b) Salting-in of methanol by saturated aqueous mercuric chloride.

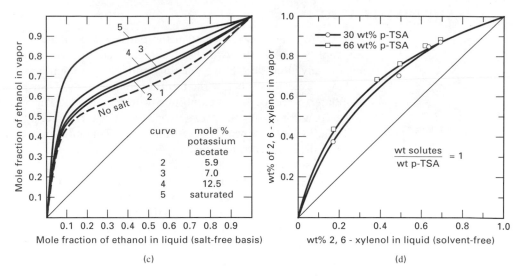

Figure 11.21 (c) Effect of salt concentration on ethanol–water equilibria. (d) Effect of *p*-toluenesulfonic acid (*p*-TSA) on phase equilibria of 2,6 xylenol-*p*-cresol. [From A.I. Johnson and W.F. Furter, *Can. J. Chem. Eng.,* **43,** 356–358 (1965), with permission.]

as shown in Figure 11.21c. Mixed acetate salts have been used in Germany for the separation of ethanol and water.

Surveys of the use of inorganic salts for extractive distillation, including effects on vapor–liquid equilibria, are given by Johnson and Furter [27], Furter and Cook [28], and Furter [29,30]. A survey of methods for predicting the effect of inorganic salts on vapor–liquid equilibria is given by Kumar [31].

Salt distillation can also be applied to the separation of organic compounds that have little capacity alone for dissolving inorganic salts by using a special class of organic salts called *hydrotropes.* Typical hydrotropic salts are alkali and alkaline-earth salts of the sulfonates of toluene, xylene, or cymene, and the alkali benzoates, thiocyanates, and salicylates. For example, Mahapatra, Gaikar, and Sharma [32] showed that the addition of aqueous solutions of 30 and 66 wt% *p*-toluenesulfonic acid to mixtures of 2,6-xylenol and *p*-cresol at 1 atm increased the relative volatility from approximately 1 to about 3, as shown in Figure 11.21d. Hydrotropes can also be used to enhance separations by liquid–liquid extraction, as shown by Agarwal and Gaikar [33].

11.4 PRESSURE-SWING DISTILLATION

When a binary azeotrope disappears at some pressure or changes composition by 5 mol% or more over a moderate range of pressure, consideration should be given to using, without a solvent, two distillation columns operating in series at different pressures. This process is referred to as *pressure-swing distillation* or *two-column distillation*. Knapp and Doherty [34] list 36 pressure-sensitive binary azeotropes, taken mainly from the compilation of Horsley [11]. The effect of pressure on the temperature and composition of two minimum-boiling azeotropes is shown in Figure 11.22. The mole fraction of ethanol in the ethanol–water azeotrope increases from 0.8943 at 760 torr to more than 0.9835 at 90 torr. Although not shown in Figure 11.22b, the azeotrope finally disappears at below about 70 torr. A much more dramatic change in azeotropic composition with pressure is seen in Figure 11.22b for the ethanol–benzene system, which forms a minimum-boiling azeotrope at 44.8 mol% ethanol and 1 atm. Applications of pressure-swing distillation, which was first noted

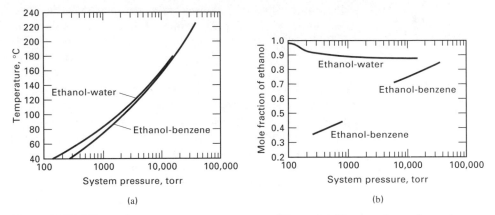

Figure 11.22 Effect of pressure on azeotrope conditions. (a) Temperature of azeotrope. (b) Composition of azeotrope. [From *Perry's Chemical Engineers' Handbook* 6th ed., R.H. Perry and D.W. Green, Eds. McGraw-Hill, (1984) with permission.]

by Lewis [35] in a 1928 patent, include the separations of the minimum-boiling azeotrope of tetrahydrofuran–water and the maximum-boiling azeotropes of hydrochloric acid–water and formic acid–water.

Consider the case, described by Van Winkle [36], of a minimum-boiling azeotrope for the mixture A–B, with T–y–x curves as shown in Fig. 11.23a. As the pressure is decreased from P_2 to P_1, the azeotropic composition moves toward a smaller percentage of A. An operable pressure-swing sequence is shown in Figure 11.23b. The total feed, F_1, to Column 1, operating at the lower pressure, P_1, is the sum of the fresh feed, F, whose composition is richer in A than the azeotrope, and the recycled distillate, D_2, whose composition is close to that of the azeotrope at pressure, P_2. The compositions of D_2 and, consequently, F_1 are both richer in A than the azeotrope composition at P_1. The bottoms, B_1, leaving Column 1 is almost pure A. The distillate, D_1, which is slightly richer in A than the azeotrope, but less rich in A than the azeotrope at P_2, is fed to Column 2, where the bottoms, B_2, is almost pure B. Robinson and Gilliland [37] provide an example of the separation of ethanol and water, where the fresh-feed composition is less rich in ethanol than the azeotrope. For that case, the products are still removed as bottoms, but nearly pure B is taken from the first column and A from the second.

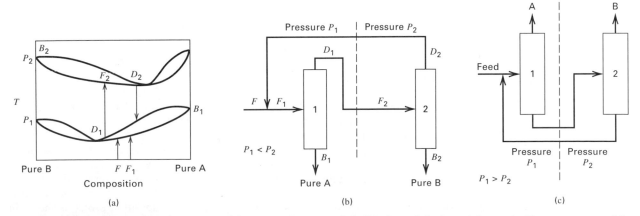

Figure 11.23 Pressure-swing distillation. (a) T–y–x curves at pressures P_1 and P_2 for minimum-boiling azeotrope. (b) Distillation sequence for minimum-boiling azeotrope. (c) Distillation sequence for maximum-boiling azeotrope.

Pressure-swing distillation can also be applied to the separation of the less-common maximum-boiling binary azeotropes. The sequence is shown in Figure 11.23c, where both products are withdrawn as distillates, rather than as bottoms. In this case, the composition of the azeotrope becomes richer in A as the pressure is decreased. The fresh feed, which is richer in A than the azeotrope at the higher pressure, is first distilled in Column 1 at the higher pressure, P_1, to produce a distillate of nearly pure A and a bottoms slightly richer in A than the azeotrope at the higher pressure. The bottoms is fed to Column 2, operating at the lower pressure, P_2, where the azeotrope composition is richer in A than the feed to that column. Accordingly, the distillate is nearly pure B, while the recycled bottoms from Column 2 is slightly less rich in A than the azeotrope at the lower pressure.

For all pressure-swing distillation sequences, the recycle ratio is an important factor in the design and depends upon the difference in azeotropic composition for the column pressures. The following example illustrates the calculation and importance of the recycle stream.

EXAMPLE 11.5

Ninety mol/s of a mixture of two-thirds by moles ethanol and one-third benzene at the bubble point at 101.3 kPa is to be separated into 99 mol% ethanol and 99 mol% benzene. Ordinary distillation is not feasible because the mixture forms a minimum-boiling azeotrope at 760 torr with a composition of 44.8 mol% ethanol and a temperature of 68°C. If the pressure is reduced to 200 torr, as shown in Figure 11.22b, the azeotrope composition shifts to 36 mol% ethanol at 35°C. This magnitude of shift makes this a candidate for pressure-swing distillation.

Apply the sequence shown in Figure 11.23b. Let the first column operate with a top tray pressure of 30 kPa (225 torr). Because the feed composition is greater than the azeotrope composition at the pressure of this column, the distillate composition approaches the minimum-boiling azeotrope at the top-tray pressure, and 99 mol% ethanol can be withdrawn as bottoms. The distillate is sent to the second column, which operates with a top tray pressure of 106 kPa. The feed to this column has an ethanol content greater than that of the azeotrope at the pressure of the second column. Accordingly, the distillate composition approaches the azeotrope at the top-tray pressure, and 99 mol% benzene can be withdrawn as bottoms. The distillate is recycled to the first column. Design a pressure-swing distillation system for this separation.

SOLUTION

For the first column, which operates under vacuum, the reflux-drum and reboiler pressures are set at 26 and 40 kPa, respectively. For the second column, which operates just slightly above ambient pressure, the reflux-drum and reboiler pressures are set at 101.3 and 120 kPa, respectively. The bottoms compositions are specified at the required purities. The distillate composition for the first column is set at 37 mol% ethanol, slightly greater than the azeotrope composition at 30 kPa. The distillate composition for the second column is set at 44 mol% ethanol, slightly less than the azeotrope composition at 106 kPa. With these composition specifications, material-balance calculations on ethanol and benzene give the following flow rates in moles per second.

Component	F	D_2	F_1	B_1	D_1	B_2
Ethanol	60.0	67.3	127.3	59.7	67.6	0.3
Benzene	30.0	85.6	115.6	0.6	115.0	29.4
Totals:	90.0	152.9	242.9	60.3	182.6	29.7

It is seen that the recycle molar flow rate, D_2, is about 10% greater than that of the fresh feed, F.

Equilibrium-stage calculations for the two columns were made with the ChemSep program, using total condensers and partial reboilers. For Column 1, a number of runs were made in an attempt to find optimal feed-tray locations for the fresh feed and the recycle, using a reasonable reflux rate that avoided any near-pinch conditions. The selected design uses seven theoretical trays (not counting the partial reboiler), with the recycle stream, at a temperature of 68°C, sent to Tray 3 from the top and the fresh feed to Tray 5 from the top. A reflux ratio of 0.5 is sufficient to achieve specifications. The resulting liquid-phase composition profile is shown in Figure 11.24a, where the desirable lack of composition pinch points is observed. The McCabe-Thiele diagram for Column 1 is given in Figure 11.24b, where the three operating lines are evident and optimal feed locations are indicated. Because of the azeotrope, the operating lines

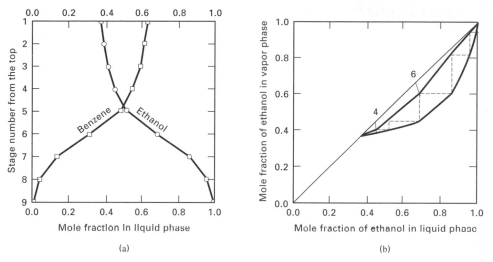

Figure 11.24 Computed results for Column 1 of pressure-swing distillation system in Example 11.5. (a) Liquid composition profiles. (b) McCabe–Thiele diagram.

and equilibrium curve all lie below the 45° line. The condenser duty is 9.88 MW, while the reboiler duty is 8.85 MW. The bottoms temperature is 56°C. This column was sized with the ChemCAD program for sieve trays on 24-inch tray spacing and a 1-inch weir height to minimize pressure drop. The resulting diameter is 3.2 meters (10.5 ft). A tray efficiency of about 47% is predicted, making the required number of trays equal to 15.

For the design of Column 2, a similar procedure was used to establish the optimal feed tray, total trays, and reflux ratio. The selected design turned out to be a refluxed stripper with only three theoretical stages (not counting the partial reboiler). A reflux rate of only 25.5 mol/s achieves the product specifications, with most of the liquid traffic in the stripper coming from the feed. The resulting liquid-phase composition profile is shown in Figure 11.25a, where, again, no composition pinches are evident. The McCabe–Thiele diagram for Column 2 is given in Figure 11.25b, where an optimal feed location is indicated. The condenser duty is 6.12 MW, while the reboiler duty is 7.07 MW. The bottoms temperature is 84°C. This column was sized for the same conditions as Column 1, resulting in a column diameter of 2.44 meters (8 ft). A tray efficiency of 50% results in 6 actual trays. ∎

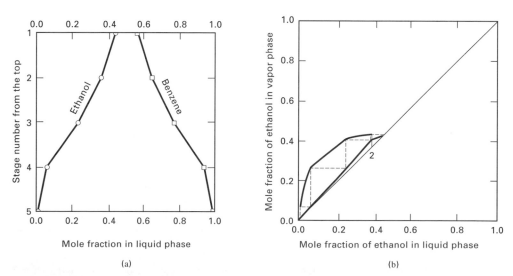

Figure 11.25 Computed results for Column 2 of pressure-swing distillation system in Example 11.5. (a) Liquid composition profiles. (b) McCabe–Thiele diagram.

11.5   HOMOGENEOUS AZEOTROPIC DISTILLATION

As discussed earlier, an azeotrope can be separated by extractive distillation, using a solvent that is higher boiling (less volatile) than the components in the feed, and that does not form an azeotrope with any of them. Alternatively, the separation can be made by *homogeneous azeotropic distillation,* using an *entrainer* that is not subject to such restrictions. Like extractive distillation, a sequence of two or three distillation columns is used. Alternatively, the sequence includes separation operations other than distillation, such as liquid–liquid extraction.

The conditions that a potential entrainer must satisfy for homogeneous azeotropic distillation to be feasible has been a subject of study by a number of investigators, including Doherty and Caldarola [16], Stichlmair, Fair, and Bravo [1], Foucher, Doherty, and Malone [14], Stichlmair and Herguijuela [18], Fidkowski, Malone, and Doherty [13], Wahnschafft and Westerberg [38], and Laroche, Bekiaris, Andersen, and Morari [39]. If it is assumed that a distillation boundary, if any, of a residue-curve map is straight or cannot be crossed, the conditions of Doherty and Caldarola apply. These conditions are based on the rule that for a potential entrainer, E, the two components, A and B, to be separated, or any product azeotrope, must lie in the same distillation region of the residue-curve map. Thus, a distillation boundary cannot be connected to the A-B azeotrope. Furthermore, A or B, but not both, must be a saddle. The maps suitable for a sequence that includes homogeneous azeotropic distillation together with ordinary distillation are classified into the five groups illustrated in Figure 11.26a, b, c, d, and e. The figure for each group includes the applicable residue-curve maps and the sequence of separation columns used to separate A from B and recycle the entrainer. For all groups, the residue-curve map is drawn in the manner of Doherty and Caldarola [16], with the lowest-boiling component, L, at the top vertex; the intermediate-boiling component, I, at the bottom-left vertex; and the highest-boiling component, H, at the bottom-right vertex. Component A is the lower-boiling component of the binary mixture and B the higher-boiling. For the first three groups, A and B form a minimum-boiling azeotrope; for the other two groups, they form a maximum-boiling azeotrope.

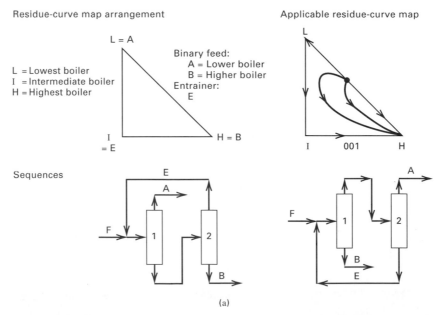

(a)

Figure 11.26 Residue-curve maps and distillation sequences for homogeneous azeotropic distillation. (a) Group 1: A and B form a minimum-boiling azeotrope, I = E, E forms no azeotropes.

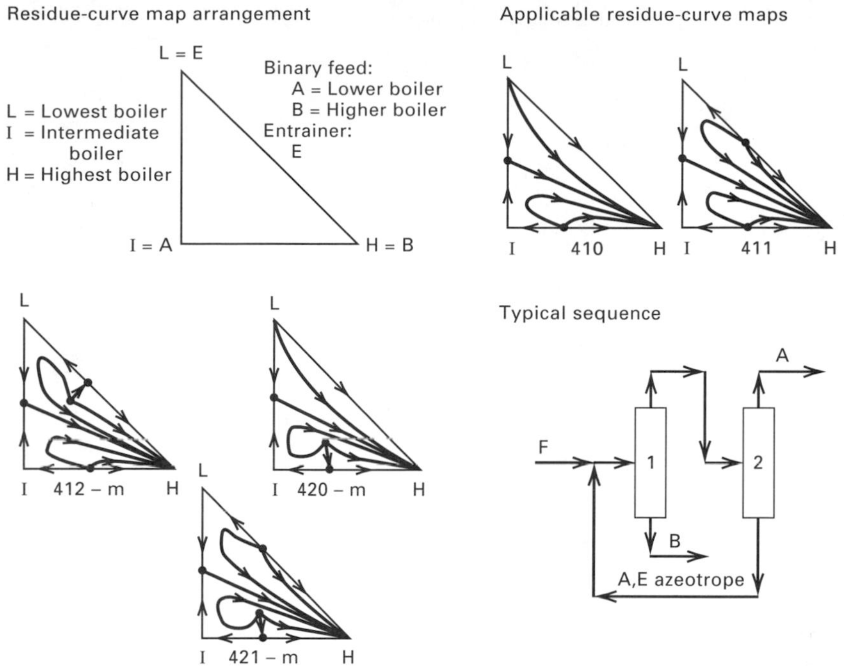

Residue-curve map arrangement

Applicable residue-curve maps

Typical sequence

(b)

Residue-curve map arrangement

Applicable residue-curve maps

Typical sequence

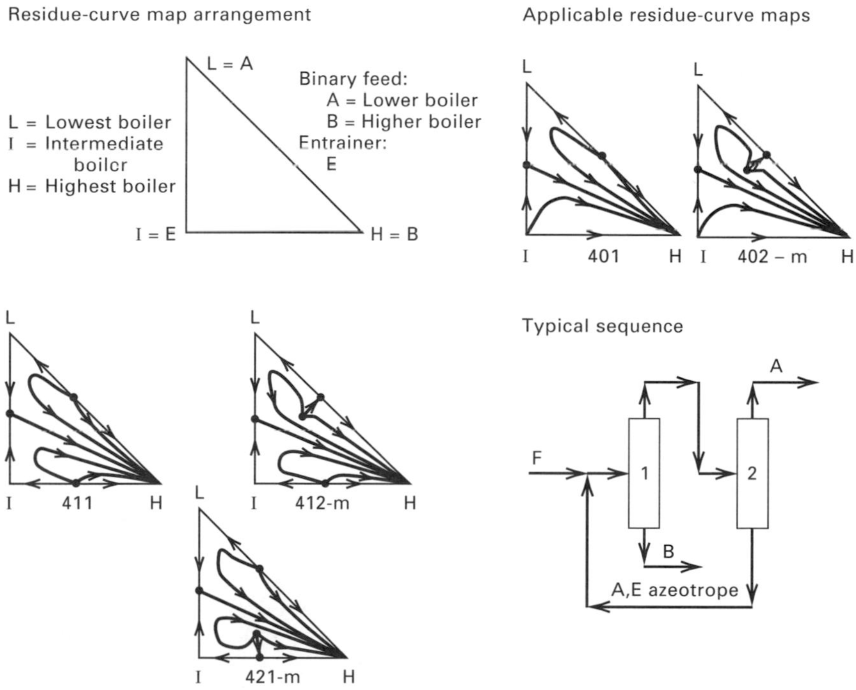

(c)

Figure 11.26 (b) Group 2: A and B form a minimum-boiling azeotrope, L = E, E forms a maximum-boiling azeotrope with A. (c) Group 3: A and B form a minimum boiling azeotrope, I = E, E forms a maximum-boiling azeotrope with A.

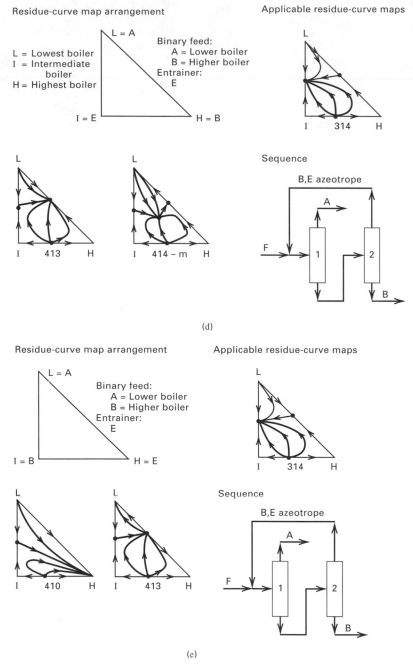

Figure 11.26 (d) Group 4: A and B form a maximum-boiling azeotrope, I = E, E forms a minimum-boiling azeotrope with B. (e) Group 5: A and B form a maximum-boiling azeotrope, H = E, E forms a minimum-boiling azeotrope with B.

In Group 1, the intermediate boiler, I, is E, which forms no azeotropes with A and/or B. As shown in Figure 11.26a, this case, like extractive distillation, involves no distillation boundary. Two sequences are shown, both of which assume that the fresh feed, F, of A and B, as fed to Column 1, is close to the azeotropic composition. Thus, this feed may be the distillate from a previous column used to produce the azeotrope from the original mixture of A and B. Either the *direct sequence,* in which Column 2 is fed by the bottoms from Column 1, or the *indirect sequence,* in which Column 2 is fed by the distillate from Column 1, may be used. In the first sequence, the entrainer is recovered as distillate from Column 2 and recycled to Column 1. In the second sequence, the entrainer is recovered as bottoms from Column 2 and recycled to Column 1. Although both sequences show the entrainer being combined with the fresh feed before being fed to Column 1, the fresh feed and the recycled entrainer can be fed to different trays to enhance the separation.

In Group 2, the low boiler, L, is E, which forms a maximum-boiling azeotrope with A. Entrainer E may also form a minimum-boiling azeotrope with B, and/or a minimum-boiling (unstable node) ternary azeotrope. Thus, in Figure 11.26b, any of the five residue-curve maps shown may apply. In all five cases, a distillation boundary exists, which is directed from the maximum-boiling azeotrope of A–E to pure B, the high boiler. A feasible indirect or direct sequence is restricted to the subtriangle bounded by the vertices of pure components A, B, and the binary azeotrope of A–E. An example of an indirect sequence, is included in Figure 11.26b. In this case, the azeotrope of A–E is recycled to Column 1 from the bottoms of Column 2. Alternatively, as shown in Figure 11.26c for Group 3, A and E may be switched to make A the low boiler and E the intermediate boiler, which again forms a maximum-boiling azeotrope with A. All sequences for Group 3 are confined to the same subtriangle as for Group 2.

Groups 4 and 5, shown in Figs. 11.26d and e, respectively, are similar to Groups 2 and 3. However, A and B now form a maximum-boiling azeotrope. In Group 4, the entrainer is the intermediate boiler, which forms a minimum-boiling azeotrope with B. The entrainer may also form a maximum-boiling azeotrope with A, and/or a maximum-boiling (stable node) ternary azeotrope. A feasible sequence is restricted to the subtriangle formed by vertices A, B, and the B–E azeotrope. Although just a direct sequence is shown, the indirect sequence can also be used. Alternatively, as shown in Figure 11.26e for Group 5, B and E may be switched to make E the high boiler. Otherwise, the other conditions and the sequences are the same as for Group 4.

The distillation boundaries for the hypothetical ternary systems in Figure 11.26 are shown as straight lines. When a distillation boundary is curved, it may be crossed, provided that both the distillate and bottoms products lie on the same side of the boundary.

It is often difficult to find an entrainer for a sequence involving homogeneous azeotropic distillation and ordinary distillation. However, azeotropic distillation can also be incorporated into a sequence involving separation operations other than distillation. In that case, some of the restrictions for the entrainer and the resulting residue-curve map may not apply. For example, the separation of the very close-boiling and minimum-azeotrope-forming system of benzene and cyclohexane using acetone as the entrainer violates the restrictions for a distillation-only sequence because the ternary system involves only two minimum-boiling binary azeotropes. However, the separation can be achieved by the sequence shown in Figure 11.27, which involves: (1) homogeneous azeotropic distillation with acetone entrainer to produce a bottoms product of nearly pure benzene and a distillate close in composition to the minimum-boiling binary azeotrope of acetone and cyclohexane; (2) liquid–liquid extraction of the distillate with water to give a raffinate of nearly pure cyclohexane and an extract of acetone and water; and (3) ordinary distillation of the extract to recover the acetone for recycle. As shown in the following example, the azeotropic distillation column is still subject to product-composition-region restrictions.

EXAMPLE 11.6

Benzene (nbp = 80.24°C) and cyclohexane (nbp = 80.64°C) form a minimum-boiling homogeneous azeotrope at 1 atm and 77.2°C that contains 54.2 mol% benzene. Thus, they cannot be separated by ordinary distillation at 1 atm. Instead, it is proposed to separate them by using acetone as the entrainer in the separation sequence shown in Figure 11.27. The fresh feed to the azeotropic column consists of 100 kmol/h of 25 mol% benzene and 75 mol% cyclohexane. Determine a feasible acetone addition rate to the feed so that nearly pure benzene can be

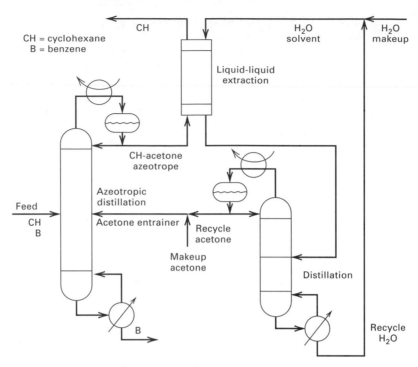

Figure 11.27 Separation sequence for separating cyclohexane and benzene using homogeneous azeotropic distillation with acetone entrainer. [From *Perry's Chemical Engineers' Handbook,* 6th ed., R.H. Perry and D.W. Green, Eds., McGraw-Hill, New York (1984) with permission.]

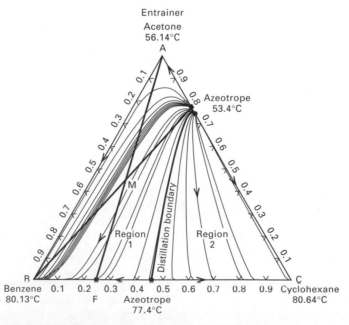

Figure 11.28 Residue-curve map for Example 11.6.

Table 11.4 Effect of Acetone Entrainer Flow Rate on
Benzene Purity for the Homogeneous Azeotropic Distillation
Process of Example 11.6

Case	Acetone Flow Rate, kmol/h	Benzene Purity in Bottoms, %
1	50	88.69
2	75	94.21
3	100	99.781
4	125	99.779

obtained as the bottoms product. Acetone (nbp = 56.14°C) forms a minimum-boiling azeotrope with cyclohexane (but not benzene) at 53.4°C and 1 atm at 74.6 mol% acetone. The residue-curve map at 1 atm is given in Figure 11.28.

SOLUTION

The residue-curve map shows a slightly curved distillation boundary connecting the two azeotropes and dividing the diagram into distillation regions, 1 and 2. The fresh feed composition is designated in Figure 11.28 by a filled-in box labeled F. If a straight line is drawn from F to the pure acetone apex, A, the mixture of the fresh feed and the acetone entrainer must lie somewhere in Region 1 on this line. Suppose that the 100 kmol/h of fresh feed is combined with an equal flow rate of entrainer. The mixing point, M, is located at the midpoint of the line connecting F and A. If a line is drawn from the benzene apex, B, through M and to the side of the triangle that connects the acetone apex to the cyclohexane apex, it does not cross the distillation boundary separating the two regions, but lies completely in Region 1. Thus, the separation into a nearly pure benzene bottoms and a distillate mixture containing mainly acetone and cyclohexane is possible. This is confirmed by calculations with the ASPEN PLUS process simulator for a column operating at 1 atm with 38 theoretical stages, a total condenser, a partial reboiler, a reflux ratio of 4, a bottoms product flow rate of 75 kmol/h (equivalent to the benzene flow rate in the feed to the column), and a bubble-point combined feed sent to Stage 19 from the top. The resulting product flow rates are listed in Table 11.4 as Case 3, where it is seen that a bottoms of 99.8 mol% benzene is achieved with a benzene recovery of the same value. A higher entrainer flow rate of 125 kmol/h, included in Table 11.4 as Case 4, is also successful in achieving high benzene bottoms-product purity and recovery. However, if only 75 kmol/h (Case 2) or 50 kmol/h (Case 1) of entrainer is used, a nearly pure benzene bottoms is not achieved because of the distillation boundary restriction. ■

11.6 HETEROGENEOUS AZEOTROPIC DISTILLATION

The requirement for a distillation sequence based on homogeneous azeotropic distillation, that A and B must lie in the same distillation region of the residue-curve map with entrainer E, is so restrictive that it is usually difficult, if not impossible, to find a feasible entrainer. The Group 1 map in Figure 11.26a requires that the entrainer not form an azeotrope but yet be the intermediate-boiling component, while the other two components form a minimum-boiling azeotrope. Such systems are rare, because most intermediate-boiling entrainers form an azeotrope with one or both of the other two azeotrope-forming components. The other four groups in Figure 11.26 all require that at least one maximum-boiling azeotrope be formed. However, such azeotropes are far less common than minimum-boiling azeotropes. The result is that industrial applications of distillation sequences based on homogeneous azeotropic distillation are not common.

An alternative technique that does find wide industrial application is heterogeneous azeotropic distillation, which is used to separate close-boiling binary mixtures and minimum-boiling binary azeotropes by employing an entrainer that forms a binary and/or ternary heterogeneous azeotrope. The overhead vapor from the column is close to the composition of the heterogeneous azeotrope. When condensed, two liquid phases form in a decanter downstream of the condenser. After separation in the decanter, most or all of the entrainer-rich liquid phase is returned to the column as reflux, while most or all of the other liquid phase is sent to the next column for further separation. Because these two phases usually lie in different distillation regions of the residue-curve map, the restriction that usually dooms distillation sequences based on homogeneous azeotropic distillation is overcome. Thus, in heterogeneous azeotropic distillation, the components to be separated need not lie in the same distillation region.

Heterogeneous azeotropic distillation has been practiced for almost a century, first by batch and then by continuous processing. Two of the most widely used applications are (1) the use of benzene or one of a number of other entrainers to separate the minimum-boiling azeotrope of ethanol and water, and (2) the use of ethyl acetate or one of a number of other entrainers to separate the close-boiling mixture of acetic acid and water. Other applications, cited by Widagdo and Seider [19], include dehydrations of isopropanol with isopropylether, *sec*-butyl-alcohol with *disec*-butyl-ether, chloroform with mesityl oxide, formic acid with toluene, and acetic acid with toluene. Also, dehydration of tanker-transported feedstocks such as styrene and benzene is a major application.

Consider the separation of the azeotrope of ethanol and water by heterogeneous azeotropic distillation. The two most widely used entrainers are benzene and diethyl ether. A number of other entrainers are feasible, including *n*-pentane, illustrated later in Example 11.7, and cyclohexane. In 1902, Young [40] discussed the use of benzene as an entrainer for the batch dehydration of ethanol, in perhaps the first application of heterogeneous azeotropic distillation. In 1928, Keyes obtained a patent [41] on a continuous process, discussed in a 1929 article [42]. A residue-curve map, computed by Bekiaris, Meski, and Morari [43] for the ethanol (E)–water (W)–benzene (B) system at 1 atm, using the UNI-QUAC equation (with parameters from ASPEN PLUS) for liquid-phase activity coefficients, is shown in Figure 11.29. The normal boiling points of E, W, and B are 78.4, 100, and 80.1°C, respectively. The UNIQUAC equation predicts that homogeneous minimum-boiling azeotropes AZ1 and AZ2 are formed by E and W at 78.2°C and 10.0 mol% W, and by E and B at 67.7°C and 44.6 mol% E, respectively. A heterogeneous minimum-boiling azeotrope AZ3 is predicted for W and B at 69.3°C, with a vapor composition of 29.8 mol% W. The overall composition of the two liquid phases is the same as that of the vapor, but each liquid phase is almost pure. The B-rich liquid phase is predicted to contain 0.55 mol% W, while the W-rich liquid phase contains only 0.061 mol% B. A ternary minimum-boiling heterogeneous azeotrope AZ4 is predicted at 64.1°C, with a vapor composition of 27.5 mol% E, 53.1 mol% B, and 19.4 mol% W. The overall composition of the two liquid phases of the ternary azeotrope is the same as that of the vapor, but the benzene-rich liquid phase contains 18.4 mol% E, 79.0 mol% B, and 2.6 mol% W, while the water-rich liquid phase contains 43.9 mol% E, 6.3 mol% B, and 49.8 mol% W.

In Figure 11.29, the map is divided into three distillation regions by three thick, solid-line distillation boundaries that each extend from the ternary azeotrope to a binary azeotrope. Each distillation region contains one pure component. Because the ternary azeotrope is the lowest-boiling azeotrope, it is an unstable node. Because all three binary azeotropes boil below the boiling points of the three pure components, the binary azeotropes are saddles and the pure components are stable nodes. Accordingly, all residue curves begin at the ternary azeotrope and terminate at a pure component apex. Liquid–liquid solubility is shown as a thick dashed curved line. However, this curve is not like the usual ternary solubility curve, because it is for isobaric, rather than isothermal, conditions. Superimposed on the distillation boundary that separates distillation regions 2 and 3 are thick dashes that represent the vapor composition in equilibrium with two liquid phases. The compositions of

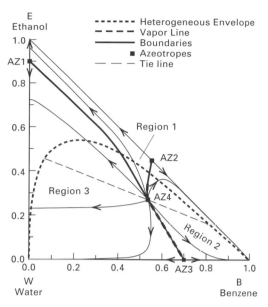

Figure 11.29 Residue-curve map for the ethanol–water–benzene system at 1 atm.

the two equilibrium liquid phases for a particular vapor composition are obtained from the two ends of the straight tie line that passes through the vapor composition point and terminates at the liquid solubility curve. The only tie line shown in Figure 11.29 is a thin dashed line that passes through the ternary azeotrope. Other tie lines, which would represent other temperatures, could be added; however, in most heterogeneous azeotropic distillation operations, an attempt is made to restrict the formation of two liquid phases to just the decanter downstream of the condenser where the composition approaches the ternary azeotrope.

Figure 11.29 clearly shows how a distillation boundary is crossed by the formation of two liquid phases in the decanter. This phase split is utilized in the following manner for a typical azeotropic-tower operation for the dehydration of ethanol by benzene. The tower is treated as a column with no condenser, a main feed that enters a few trays below the top of the column, and the reflux of benzene-rich liquid as a second feed. The composition of the combined two feeds lies in distillation region 1. Thus, from the directions of the residue curves, the products of the tower can be a bottoms of nearly pure ethanol and an overhead vapor approaching the composition of the ternary azeotrope. When that vapor is condensed, phase splitting occurs to give a water-rich phase that lies in distillation region 3 and an entrainer-rich phase in distillation region 2. Thus, if the water-rich phase is sent to a reboiled stripper, the residue curves indicate that a nearly pure-water bottoms can be produced, with the overhead vapor, rich in ethanol, recycled to the decanter. When the entrainer-rich phase in distillation region 2 is added to the main feed, which lies in distillation region 1, the overall composition lies in region 1.

It is preferable to restrict the formation of two liquid phases to the decanter. To avoid formation of two liquid phases on the top trays of the azeotropic tower, the composition of the vapor leaving the top tray must be such that the equilibrium liquid lies outside of the two-phase liquid region enclosed by the solubility curve and the base of the triangle in Figure 11.29. As shown in a comprehensive study by Prokopakis and Seider [44], vapor

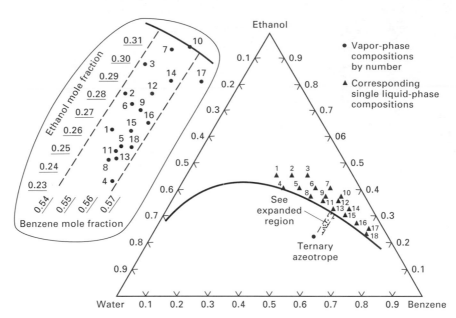

Figure 11.30 Overhead vapor compositions not in equilibrium with two liquid phases. [From J. Prokopakis and W.D. Seider, *AIChE J.*, **29**, 49–60 (1983) with permission.]

compositions that form two liquid phases when condensed, but are in equilibrium with only one liquid phase, are restricted to a very small window, as shown in Figure 11.30. Furthermore, that window can only be achieved by adding to the entrainer-rich liquid phase a portion of the water-rich liquid phase or a portion of the condensed vapor prior to separation in the decanter.

A variety of column sequences for heterogeneous azeotropic distillation have been proposed. Three of these that utilize only distillation, taken from a study by Ryan and Doherty [45], are shown in Figure 11.31. Most common is the three-column sequence, in which an aqueous feed dilute in ethanol is first preconcentrated in Column 1 to obtain a nearly pure water bottoms product and a distillate with composition approaching that of the binary azeotrope. The latter is fed to the azeotropic tower, Column 2, where nearly pure ethanol is recovered as the bottoms product and the tower is refluxed by most or all of the entrainer-rich liquid phase from the decanter. The water-rich phase, which contains ethanol and a small amount of entrainer, is sent to the entrainer-recovery column, which is a distillation column with both rectifying and stripping sections, or a stripper. The distillate from the recovery column is recycled to the azeotropic column. Alternatively, the distillate from Column 3 could be recycled to the decanter. As shown in all three sequences of Figure 11.31, portions of either liquid phase from the decanter can be returned to the azeotropic tower or to the next column in the sequence to control phase splitting on the top trays of the azeotropic tower.

A four-column distillation sequence is shown in Figure 11.31b. The first column is identical to the first column of the three-column sequence of Figure 11.31a. The second column is the azeotropic column, which is fed by the near-azeotrope distillate of ethanol and water from Column 1 and by a recycle distillate of about the same composition from Column 4. The purpose of Column 3 is to remove, as distillate, the entrainer from the water-rich liquid phase leaving the decanter and recycle it back to the decanter. Ideally, the composition of this distillate is identical to that of the vapor distillate from Column 2. The bottoms from Column 3 is separated in Column 4 into a bottoms of nearly pure water, and a distillate that approaches the ethanol–water azeotrope and is therefore recycled to

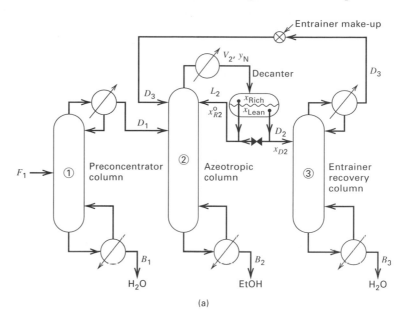

(a)

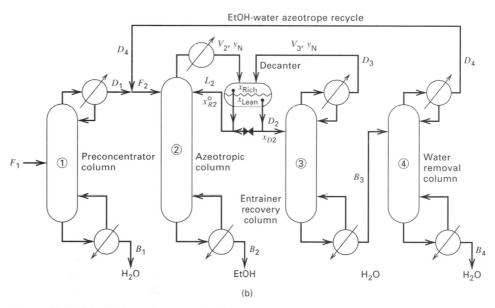

(b)

Figure 11.31 Distillation sequences for heterogeneous azeotropic distillation: (a) Three-column sequence; (b) four-column sequence; (c) two-column sequence. [From P.J. Ryan and M.F. Doherty, *AIChE J., **35,*** 1592–1601 (1989) with permission.]

the feed to Column 2. A study by Pham and Doherty [46] found no advantage for the four-column sequence over the three-column sequence.

A novel two-column distillation sequence, due to Lynn and described by Ryan and Doherty [45], is shown in Figure 11.31c. The feed is sent to Column 2, which is a combined preconcentrator and entrainer recovery column. The distillate from this column is the feed to the azeotropic column. The bottoms from Column 1 is nearly pure ethanol, while Column 2 produces a bottoms of nearly pure water. For feeds that are very dilute in ethanol, Ryan and Doherty found that the two-column sequence has a lower investment

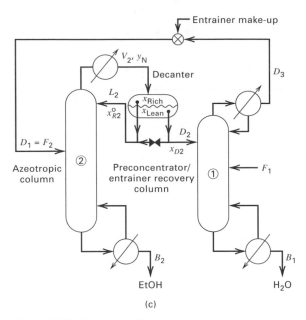

(c)

Figure 11.31 (*continued*)

cost, but a higher operating cost, than the three-column sequence. For feeds that are richer in ethanol, these two sequences are economically comparable.

The ethanol–benzene–water residue curve map of Figure 11.29 is only one of a number of different residue-curve maps that can lead to feasible distillation sequences that include heterogeneous azeotropic distillation. Pham and Doherty [46] note that a feasible entrainer is one that causes phase splitting over a portion of the three-component composition region, but does not cause the two components of the feed to be placed in different distillation regions. Figure 11.32 shows seven such maps, where the dash-dot lines are the liquid–liquid solubility curves.

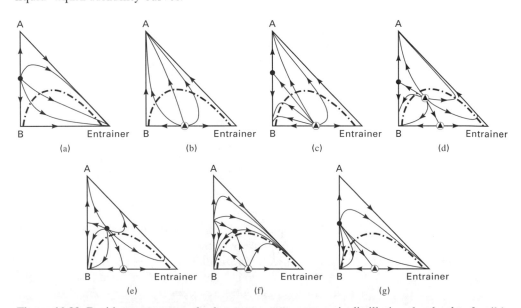

Figure 11.32 Residue-curve maps for heterogeneous azeotropic distillation that lead to feasible distillation sequences. [H.N. Pham and M.F. Doherty, *Chem. Eng. Sci.*, **45,** 1845–1854 (1990) with permission.]

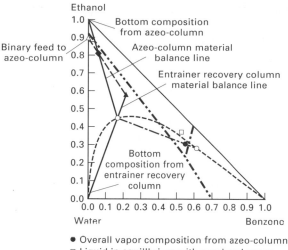

- ● Overall vapor composition from azeo-column
- □ Liquid in equilibrium with overhead vapor composition from azeo-column
- ▲ Distillate composition from entrainer recovery column
- ◆ Overall feed composition to azeo-column

Figure 11.33 Material balance lines for the three-column sequence of Figure 11.31a. [From P.J. Ryan and M.F. Doherty, *AIChE J.*, **35,** 1592–1601 (1989) with permission.]

The convergence of rigorous calculations for heterogeneous azeotropic distillation columns by the methods described in Chapter 10 can be extremely difficult, especially when the convergence of the entire sequence is attempted. For calculation purposes, it is preferable to uncouple the columns by using a residue-curve map to establish, by material-balance calculations, the flow rates and compositions of the feeds and products for each column. This procedure is illustrated for a three-column sequence in Figure 11.33, where the dash-dot lines separate the three distillation regions, the short-dash line is the liquid–liquid solubility curve, and the remaining lines are material-balance lines. Each column in the sequence can be computed separately. Even then, the calculations can be so sensitive, because of nonidealities in the liquid phase and possible phase splitting, that it may be necessary to use more robust methods. Among the most successful approaches for the most difficult cases are the boundary-value tray-by-tray method of Ryan and Doherty [45], the homotopy-continuation method of Kovach and Seider [47], and the collocation method of Swartz and Stewart [48].

Multiplicity

Solutions to mathematical models for operations of interest to chemical engineers are not always unique. The existence of multiple steady-state solutions for the continuous stirred-tank reactor (CSTR) has been known since at least 1922 and is described in detail in a number of textbooks on chemical reaction engineering. The existence of multiplicity in steady-state separation problems is a relatively new discovery. Gani and Jørgensen [49] define the following three types of multiplicity, all of which can, under certain conditions, occur in distillation simulations:

1. *Output multiplicity,* where all input variables are specified and more than one solution for the set of output variables is found. For example, for a distillation column, the feed condition, number of stages, feed-stage location, distillate flow rate, reflux ratio,

type condenser and reboiler, and column pressure profile might be specified and two or more sets of product compositions and column profiles found.

2. *Input multiplicity,* where one or more output variables are specified and multiple solutions are found for the unknown input variables.

3. *Internal-state multiplicity,* where multiple sets of internal conditions or profiles are found for the same values of the input and output variables.

Of particular interest here is output multiplicity for azeotropic distillation, which was first discovered by Shewchuk [50] in 1974. With different starting guesses, he found two steady-state solutions for the dehydration of ethanol by azeotropic distillation with benzene. In a more detailed study for the same system, Magnussen, Michelsen, and Fredenslund [51] found, with difficulty, for a rather narrow range of ethanol flow rate in the top feed to the column, three steady-state solutions, two of which are stable. The unstable solution can not be achieved in an operating plant. A similar multiplicity was found when pentane was used as the entrainer. One of the two stable solutions predicts a far purer ethanol bottoms product than the other stable solution. Thus, from a practical standpoint it is important to obtain all stable solutions when more than one exists. Subsequent studies, some contradictory, show that multiple solutions usually persist only over a narrow range of distillate or bottoms flow rate specifications, but may exist over a wide range of reflux rate provided that a sufficient number of stages are present. Composition profiles of five multiple solutions found by Kovach and Seider [47] for a 40-tray ethanol–water–benzene heterogeneous azeotropic distillation are shown in Figure 11.34. The variation in the profiles is extremely large. Again, when multiple solutions exist, it is important to locate them. Unfortunately, the use of current steady-state computer-aided process design and simulation programs to find multiple solutions is fraught with a number of difficulties because: (1) azeotropic columns are difficult to converge to even one solution, (2) multiple solutions may exist only in a very restricted range, (3) the multiple solutions can only be found in these programs by changing the initial guesses of the composition profiles, and (4) the choice of activity-coefficient correlation and interaction parameters can be crucial. Accordingly, the best results have been obtained when more advanced techniques such as continuation and bifurcation analysis are employed. These methods are described and applied by Kovach and Seider [47], Widagdo and Seider [19], Bekiaris, Meski, Radu, and Morari [52], and Bekiaris, Meski, and Morari [43]. The last two articles provide reasons why multiple solutions occur in homogeneous and heterogeneous azeotropic distillation.

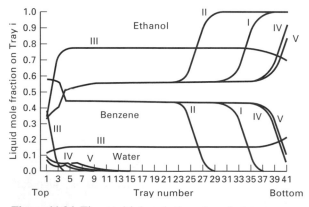

Figure 11.34 Five multiple solutions for a heterogeneous distillation operation. [From J.W. Kovach III and W.D. Seider, *Computer Chem. Engng.*, **11,** 593 (1987) with permission.]

EXAMPLE 11.7

Design and economic studies by Black and Ditsler [53] and Black, Golding, and Ditsler [54] show that *n*-pentane is a superior entrainer for the dehydration of ethanol. Like benzene, *n*-pentane forms a minimum-boiling heterogeneous ternary azeotrope with ethanol and water. Design a separation system for the dehydration of 16.8176 kmol/h of 80.937 mol% ethanol and 19.063 mol% water as a liquid at 344.3 K and 333 kPa, using *n*-pentane as an entrainer, to produce 99.5 mol% ethanol, and water with less than 100 ppm (by weight) of combined ethanol and *n*-pentane.

SOLUTION

A heterogeneous azeotropic distillation process for this ternary system has been studied extensively by Black [55], who proposed the two-column process flow diagram shown in Figure 11.35. The process consists of an 18-equilibrium-stage heterogeneous azeotropic distillation column (C-1) equipped with a total condenser and a partial reboiler, a decanter (D-1), a 4-equilibrium-stage reboiled stripper (C-2), and a condenser (E-1) to condense the overhead vapor from C-2. Each reboiler adds the equivalent of another equilibrium stage. Column C-1 operates at a bottoms pressure of 344.6 kPa with a column pressure drop of 13.1 kPa. Column C-2 operates at a top pressure of 308.9 kPa, with a column pressure drop of 3.0 kPa. These pressures permit the use of cooling water in the condensers. Purity specifications are placed on the bottoms products. The feed enters C-1 at Stage 3 from the top. The ethanol product is withdrawn from the bottom of C-1. A small *n*-pentane makeup stream, not shown in Figure 11.35, enters Stage 2 from the top. The overhead vapor from C-1 is condensed and sent to D-1, where a pentane-rich liquid phase and a water-rich liquid phase are formed. The pentane-rich phase is returned to C-1 as reflux, while the water-rich phase is sent to C-2, where the water is stripped of residual pentane and ethanol to produce a bottoms of the specified water purity. Twenty percent of the condensed vapor from C-2 is returned to D-1. To ensure that two liquid phases form in the decanter but not on the trays of C-1, the remaining 80% of the condensed vapor from C-2 is combined with the pentane-rich phase from D-1 for use as additional reflux to C-1. The specifications for the problem are included on Figure 11.35.

A very important step in the design of a heterogeneous azeotropic distillation column is the selection of a suitable method for predicting liquid-phase activity coefficients and the determination of the binary interaction parameters. The latter usually involves the regression

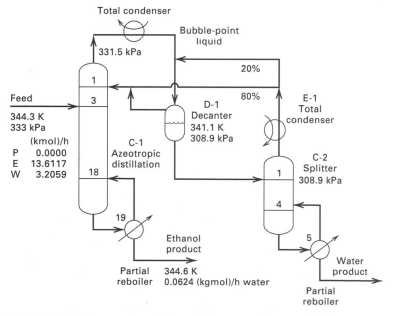

Figure 11.35 Process flow diagram for Example 11.7. [From *Perry's Chemical Engineers' Handbook,* 6th ed., R.H. Perry and D.W. Green, Eds., McGraw-Hill, New York (1984) with permission.]

Table 11.5 Converged Material Balance for Example 11.7

	Flow rate, kmol/h			
Stream	*n*-Pentane	Ethanol	Water	Total
C-1 feed	0.0000	13.6117	3.2059	16.8176
C-1 overhead	22.5565	2.1298	10.7269	35.4132
C-1 bottoms	0.0000	13.6117	0.0624	13.6741
C-1 reflux	22.5565	2.1298	7.5834	32.2697
D-1 *n*C5-rich	22.5500	1.0637	0.1129	23.7266
D-1 water-rich	0.0081	1.3326	12.4816	13.8223
C-2 overhead	0.0081	1.3326	9.3381	10.6788
C-2 bottoms	0.0000	0.0000	3.1435	3.1435

of both vapor–liquid (VLE) and liquid–liquid (LLE) experimental equilibrium data for all binary pairs. If available, ternary data can also be included in the regression. Unfortunately, for most activity-coefficient prediction methods, it is difficult to simultaneously fit VLE and LLE data. Accordingly, often different binary-interaction parameters are used for the azeotropic column where VLE is important and for the decanter where LLE is important. This has been found to be particularly desirable for the ethanol–water–benzene system. For this example, however, the use of a single set of binary-interaction parameters with the modification by Black [56] of the Van Laar equation was deemed adequate. The binary-interaction parameters are listed by Black, Golding, and Ditsler [54].

The calculations were made with the Process simulation program of Simulation Sciences, Inc., using their rigorous distillation routine to model the columns and a three-phase-flash routine to model the decanter. Because the entrainer was internal to the system, except for a very small makeup rate, it was necessary to provide a reasonable initial guess for the component flow rates in the combined feed to the decanter. The guessed values in kilomoles per hour were 25.0 for *n*-pentane, 3.0 for ethanol, and 7.5 for water. The converged material balance is given in Table 11.5, where it is seen that the product specifications are met and approximately 22.6 kmol/h of *n*-pentane circulates through the top trays of the azeotropic distillation column. The computed condenser and reboiler duties for Column C-1 are 1,116.5 and 1,135.0 MJ/h, respectively. The reboiler duty for Column C-2 is 486 MJ/h and the duty for Condenser E-1 is 438 MJ/h.

Because of the large effect of composition on liquid-phase activity coefficients, column profiles for azeotropic columns often show steep fronts. In Figure 11.36a to c, stage temperatures, total vapor and liquid flow rates, and liquid-phase compositions for Column C-1 vary only slightly from the reboiler (Stage 19) up to Stage 13. In this region, the liquid phase is greater than 99 mol% ethanol, whereas the *n*-pentane concentration slowly builds up from a negligible concentration in the bottoms to just less than 0.02 mol% at Stage 13. From Stage 13 to Stage 8, the *n*-pentane mole fraction in the liquid increases very rapidly to 53.8 mol%. In the same region, the temperature decreases sharply from 385.6 K to 348.4 K. Continuing up the column from Stage 8 to Stage 3, where the feed enters, the most significant change is the mole fraction of water in the liquid. Rather drastic changes in all variables take place about Stage 3. The large effects of *n*-pentane concentration on the relative volatility of water to ethanol and of water concentration on the relative volatility of *n*-pentane to ethanol are shown in Figure 11.36d, where the variation over the column is about 10-fold for each pair.

No phase splitting occurs in either column, but two liquid phases of drastically different composition are formed and separated in the decanter. The light phase, which is almost twice the quantity of the heavy phase, is 95 mol% *n*-pentane, whereas the heavy phase is 90 mol% water. These extremely different compositions are due to the small amount of ethanol in the overhead vapor from C-1. Because of the high concentration of water in the feed to the stripper, C-2, the concentrations of ethanol and *n*-pentane in the liquid phase are quickly reduced to parts-per-million levels. Temperatures, vapor flow rates, and liquid flow rates in the stripper, C-2, are almost constant at 408 K, 15.6 kmol/h, and 12.4 kmol/h, respectively. Because of the

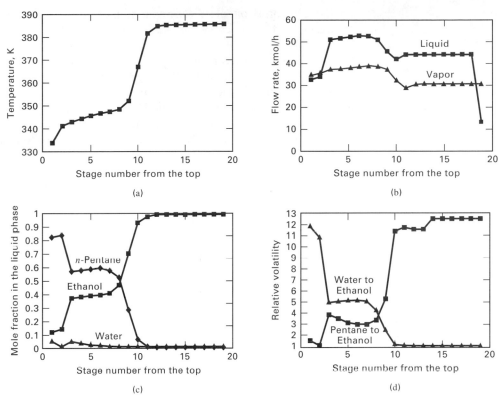

Figure 11.36 Results for Azeotropic Distillation Column of Example 11.7. (a) Temperature profile. (b) Vapor and liquid traffic profiles. (c) Liquid-phase composition profiles. (d) Relative volatility profiles.

large relative volatility of ethanol with respect to water (approximately 9) under the dilute ethanol conditions in C-2, the ethanol mole fraction decreases by almost an order of magnitude for each equilibrium stage. The extremely large relative volatility of n-pentane to water (more than 1,000) causes the n-pentane to be entirely stripped in just two stages. ∎

11.7 REACTIVE DISTILLATION

Reactive distillation involves simultaneous chemical reaction and distillation. The chemical reaction usually takes place in the liquid phase or at the surface of a solid catalyst in contact with the liquid phase. One general application of reactive distillation, described by Terrill, Sylvestre, and Doherty [57], is the separation of a close-boiling or azeotropic mixture of components A and B, where a reactive entrainer E is introduced into the distillation column. If A is the lower-boiling component, it is preferable that E be higher boiling than B and that it react selectively and reversibly with B to produce reaction product C, which also has a higher boiling point than component A and does not form an azeotrope with A, B, or E. Component A is removed from the distillation column as distillate, and components B and C, together with any excess E, are removed as bottoms. Components B and E are recovered from C in a separate distillation step, where the reaction is reversed to completely react C back to B and E; B is taken off as distillate, and E is taken off as bottoms and recycled to the first column. Terrill, Sylvestre, and Doherty [57] discuss the application of reactive entrainers to the separation of mixtures of p-xylene and m-xylene, whose normal boiling points differ by only 0.8°C, resulting in a relative volatility of only 1.029. Separation by ordinary distillation is impractical because, for example, to produce 99 mol% pure products from an equimolar feed, more than 500 theoretical stages are required. By reacting the *m*-xylene with a reactive entrainer such as *tert*-butylbenzene accompanied by a solid aluminum chloride catalyst, or chelated sodium *m*-xylene dissolved in cumene, the stage requirements are drastically reduced.

Closely related to the use of reactive entrainers in distillation is the use of reactive absorbents in absorption, which finds wide application in industry. For example, sour natural gas is sweetened by the removal of hydrogen sulfide and carbon dioxide acid gases by absorption into aqueous alkaline solutions of mono- and diethanolamines. Fast and reversible reactions occur to form soluble salt complexes such as carbonates, bicarbonates, sulfides, and mercaptans. The rich solution leaving the absorber is sent to a reboiled stripper where the reactions are reversed at higher temperatures to regenerate the amine solution as the bottoms and deliver the acid gases as overhead vapor.

A second application of reactive distillation involves taking into account undesirable chemical reactions that may occur during distillation. For example, Robinson and Gilliland [58] present an example involving the separation of cyclopentadiene from C_7 hydrocarbons. During distillation, cyclopentadiene dimerizes. The more volatile cyclopentadiene is taken overhead as distillate, but a small amount dimerizes in the lower section of the column and leaves in the bottoms with the C_7s. Alternatively, the cyclopentadiene can be dimerized to facilitate its separation by distillation from other constituents of a mixture. Then the dicyclopentadiene is removed as bottoms from the distillation column. However, during distillation, it is also necessary to account for possible depolymerization to produce cyclopentadiene, which would migrate to the distillate.

The most interesting application of reactive distillation, and the only one considered in detail in this section, involves combining chemical reaction(s) and separation by distillation in a single distillation apparatus. This concept appears to have been first pronounced by Backhaus, who, starting in 1921 [59], obtained a series of patents for esterification reactions in a distillation column. This concept of continuous and simultaneous chemical reaction and distillation in a single vessel was verified experimentally by Leyes and Othmer [60] for the esterification of acetic acid with an excess of *n*-butanol in the presence of sulfuric acid catalyst to produce butyl acetate and water. This type of reactive distillation should be considered as an alternative to the use of separate reactor and distillation vessels whenever the following hold:

1. The chemical reaction occurs in the liquid phase, in the presence or absence of a homogeneous catalyst, or at the interface of a liquid and a solid catalyst.
2. Feasible temperature and pressure for the reaction and distillation are the same. That is, reaction rates and distillation rates are of the same order of magnitude.
3. The reaction is equilibrium-limited such that if one or more of the products formed can be removed, the reaction can be driven to completion; thus, a large excess of a reactant is not necessary to achieve a high conversion. This is particularly advantageous when recovery of the excess reagent is difficult because of azeotrope formation. For reactions that are irreversible, it is more economical to take the reactions to completion in a reactor and then separate the products in a separate distillation column. In general, reactive distillation is not attractive for supercritical conditions, for gas-phase reactions, and for reactions that must take place at high temperatures and pressures, and/or that involve solid reactants or products.

Careful consideration must be given to the configuration of the distillation column when employing reactive distillation. Important factors are feed entry and product removal stages, the possible need for intercoolers and interheaters when the heat of reaction is appreciable, and the method for obtaining required residence time for the liquid phase.

In the following ideal cases, it is possible, as shown by Belck [61] and others for several two-, three-, and four-component systems, to obtain the desired products without the need for additional distillation.

Case 1: The reactions A ↔ R or A ↔ 2R, where R has a higher volatility than A. In this case, only a reboiled rectification section is needed. Pure A is sent to the column reboiler where all or most of the reaction takes place. As R is produced, it is vaporized, passing to the rectification column where it is purified. Overhead vapor from the column is condensed, with part of the condensate returned to the column as reflux. Chemical reaction may also take place in the column. If A and R form a maximum-boiling azeotrope, this configuration is still applicable if, under steady-state conditions, the mole fraction of R in the reboiler is greater than the azeotropic composition.

Case 2: The reaction A ↔ R or 2A ↔ R, where A has the lower boiling point or higher volatility. In this case, only a stripping section is needed. The feed of pure liquid A is sent to the top of the column, from which it flows down the column, reacting to produce R. The column is provided with a total condenser and a partial reboiler. No product is withdrawn from the top of the column. Product R is withdrawn from the reboiler. This configuration requires close examination because at a certain location in the column, chemical equilibrium may be achieved, and if the reaction is allowed to proceed below that point, the reverse reaction can occur.

Case 3: The reactions 2A ↔ R + S or A + B ↔ R + S, where A and B are intermediate in volatility to R and S, and R has the highest volatility. In this case, the feed enters an ordinary distillation column somewhere near the middle, with R withdrawn as distillate and S withdrawn as bottoms. If B is less volatile than A, then B may enter the column separately and at a higher level than A.

Commercial applications of reactive distillation include the following

1. The esterification of acetic acid with ethanol to produce ethyl acetate and water
2. The reaction of formaldehyde and methanol to produce methylal and water, using a solid acid catalyst, as described by Masamoto and Matsuzaki [62]
3. The esterification of acetic acid with methanol to produce methyl acetate and water, using sulfuric acid catalyst, as patented by Agreda and Partin [63], and described by Agreda, Partin, and Heise [64]
4. The reaction of isobutene with methanol to produce methyl-*tert*-butyl ether (MTBE), using a solid, strong-acid ion-exchange resin catalyst, as patented by Smith [65–67] and further developed by DeGarmo, Parulekar, and Pinjala [68]

The first widely studied example of reactive distillation is the esterification of acetic acid (A) with ethanol (B) to produce ethyl acetate (R) and water (S). The respective normal boiling points in °C are 118.1, 78.4, 77.1, and 100. Also, minimum-boiling binary homogeneous azeotropes are formed by B–S at 78.2°C with 10.57 mol% B, and by B–R at 71.8°C with 46 mol% B. A minimum-boiling binary heterogeneous azeotrope is formed by R–S at 70.4°C with 24 mol% S, and a ternary minimum-boiling azeotrope is formed by B–R–S at 70.3°C with 12.4 mol% B and 60.1 mol% R. Thus, this system is exceedingly complex and nonideal. A number of studies, both experimental and computational, have been published, many of which are cited by Chang and Seader [69], who developed a robust computational procedure for reactive distillation based on a homotopy continuation method. More recently, other computational procedures, used in computer-aided process design programs, have been reported by Venkataraman, Chan, and Boston [70] and Simandl and Svrcek [71]. Kang, Lee, and Lee [72] obtained binary interaction parameters for the UNIQUAC equation by fitting experimental data simultaneously for vapor–liquid equilibrium and liquid-phase chemical equilibrium.

In all of the computational procedures, a reaction-rate term must be added to the component material balance for a stage. For example, following the development of Chang and Seader [69], (10-58) for the simultaneous correction procedure is modified to include a reaction-rate source term for the liquid phase, assuming that at each stage, the liquid phase is completely mixed:

$$M_{i,j} = l_{i,j}(1 + s_j) + v_{i,j}(1 + S_j) - l_{i,j-1} - v_{i,j+1} - f_{i,j} - (V_{LH})_j \sum_{n=1}^{NRX} v_{i,n} r_{j,n}, i = 1, \ldots C$$

(11-17)

where $(V_{LH})_j$ = the volumetric liquid holdup at stage j

$v_{i,n}$ = stoichiometric coefficient for component i and reaction n using the customary convention of positive values for products and negative values for reactants

$r_{j,n}$ = reaction rate for reaction n on stage j, as the increase in moles of a reference reactant per unit time per unit volume of liquid phase

NRX = number of reversible and irreversible chemical reactions.

Typically, each reaction rate is expressed in a power-law form with liquid molar concentrations (where the n subscript is omitted in the following equation):

$$r_j = \sum_{p=1}^{2} k_p \prod_{q=1}^{NRC} c_{j,q}^m = \sum_{p=1}^{2} A_p \exp\left(-\frac{E_p}{RT_j}\right) \prod_{q=1}^{NRC} c_{j,q}^m$$

(11-18)

where $c_{j,q}$ = concentration of component q on stage j

k_p = reaction rate constant for the pth term, where $p = 1$ indicates the forward reaction and $p = 2$ indicates the reverse reaction; k_1 is positive and k_2 is negative

m = the exponent on the concentration

NRC = number of components in the power-law expression

A_p = preexponential (frequency) factor

E_p = activation energy

With (11-17) and (11-18), a reaction may be treated as irreversible ($k_2 = 0$), reversible (k_2 negative and not equal to zero), or at equilibrium. The last can be achieved by using very large values for the volumetric liquid holdup at each stage in the case of a single reversible reaction, or by multiplying each of the two frequency factors, A_1 and A_2, by the same large number, thus greatly increasing the forward and backward reactions, but maintaining the correct value for the chemical-reaction equilibrium constant. For equilibrium reactions, it is important that the power-law expression for the backward reaction be derived from the power-law expression for the forward reaction and the reaction stoichiometry so as to be consistent with the expression for the chemical-reaction equilibrium constant. The volumetric liquid holdup for a stage, when using a trayed tower, depends on the active bubbling area of the tray, the height of the froth on the tray as influenced by the weir height, and the liquid-volume fraction of the froth. These factors are all considered in the section on pressure-drop calculations in Chapter 6. In general, the liquid backup in the downcomer is not included in the estimate of volumetric liquid holdup. When large holdups are necessary, bubble-cap trays are preferred because they do not allow weeping. When the chemical reaction is in the reboiler, a large liquid holdup can be provided. The following example illustrates the application of the computational procedure to the esterification of acetic acid with ethanol to produce ethyl acetate and water. In this example, the single reversible chemical reaction is assumed to reach chemical equilibrium at each stage. Thus, no estimate of liquid holdup is needed. In a subsequent example, chemical equilibrium is not achieved and holdup estimates are made, which necessitates an estimate of tower diameter.

EXAMPLE 11.8

A reactive-distillation column containing the equivalent of 13 theoretical stages and equipped with a total condenser and partial reboiler is used to produce ethyl acetate (R) at 1 atm. A saturated liquid feed of 90 lbmol/h of acetic acid (A) enters Stage 2 from the top, while 100 lbmol/h of a saturated liquid of 90 mol% ethanol (B) and 10 mol% water (S) (close to the azeotropic composition) enters Stage 9 from the top. Thus, the acetic acid and ethanol are in stoichiometric ratio for esterification. The other specifications are a reflux ratio of 10 and a distillate rate of 90 lbmol/h in the hope that complete conversion to ethyl acetate (the low boiler) will occur. Kinetic data for the homogeneous reaction are given by Izarraraz, Bentzen, Anthony, and Holland [73], in terms of the rate law:

$$r = k_1 c_A c_B - k_2 c_R c_S$$

with $k_1 = 29{,}000 \exp(-14{,}300/RT)$ in L/(mol-min) with T in kelvins, and $k_2 = -7{,}380 \exp(-14{,}300/RT)$ in L/(mol-min) with T in kelvins. Because the activation energies for the forward and backward steps are the same, the chemical equilibrium constant is independent of temperature and equal to $k_1/k_2 = 3.93$. Assume that chemical equilibrium is achieved at each theoretical stage. Thus, very large values of liquid holdup are specified for each stage. Binary interaction parameters, for all six binary pairs, for predicting liquid-phase activity coefficients from the UNIQUAC equation are as follows, from Kang, Lee, and Lee [72]:

Components in Binary Pair, $i - j$	Binary Parameters	
	$u_{i,j}/R$, K	$u_{j,i}/R$, K
Acetic acid–ethanol	268.54	−225.62
Acetic acid–water	398.51	−255.84
Acetic acid–ethyl acetate	−112.33	219.41
Ethanol–water	−126.91	467.04
Ethanol–ethyl acetate	−173.91	500.68
Water–ethyl acetate	−36.18	638.60

Vapor-phase association of acetic acid is to be accounted for and the possible formation of two liquid phases is to be checked at each stage. Calculate the compositions of the distillate and bottoms products and determine the liquid-phase-composition and reaction-rate profiles.

SOLUTION

The calculations were made with the SCDS model (simultaneous-correction method) of the ChemCAD computer-aided process simulation program, where the total condenser is counted as the first stage. The only initial estimates provided were 163 and 198°F for the temperatures of the distillate and the bottoms, respectively. Convergence of the calculations required 17 iterations. A complete conversion to ethyl acetate was not achieved, as indicated by the following distillate and bottoms:

Component	Product Flow Rates, lbmol/h	
	Distillate	Bottoms
Ethyl acetate	49.52	6.39
Ethanol	31.02	3.07
Water	6.73	59.18
Acetic acid	2.73	31.36
Total	90.00	100.00

All four components appear in both products. The overall conversion to ethyl acetate is only 62.1%, with 88.6% of this going to the distillate. The distillate is 55 mol% acetate, while the bottoms is 59.2 mol% water. Only small changes in these compositions occur when the feed locations are varied. Two important factors in the failure to achieve a high conversion and nearly pure products are (1) the highly nonideal nature of the quaternary mixture, accompanied by the large number of azeotropes, and (2) the tendency of the reverse reaction to occur on certain stages. The former effect is shown in Fig. 11.37a, where the relative volatilities between ethyl acetate and water and between ethanol and water in the top section of the column are no greater than 1.25, making the separations difficult. The liquid-phase mole fraction distribution is shown in Figure 11.37b, where, in the section between the two feed points, compositions change slowly despite the esterification reaction. In Figure 11.37c, the reaction-rate profile is quite unusual. Above the upper feed stage (now Stage 3), the reverse reaction is dominant. From that feed point down to the second feed entry (now Stage 10), the forward reaction dominates, but mainly at the upper feed stage. The reverse reaction is dominant for Stages 11–13, whereas the forward reaction dominates at Stages 14 and 15 (the reboiler). The largest extents of forward reaction occur at Stages 3 and 15. Even when the number of stages is increased to 60, with the reaction confined to Stages 25 to 35, the distillate contains an appreciable fraction of ethanol and the bottoms contains a substantial fraction of acetic acid. For this example, the development of a reactive-distillation scheme for achieving a high conversion and nearly pure products represents a significant challenge.   ■

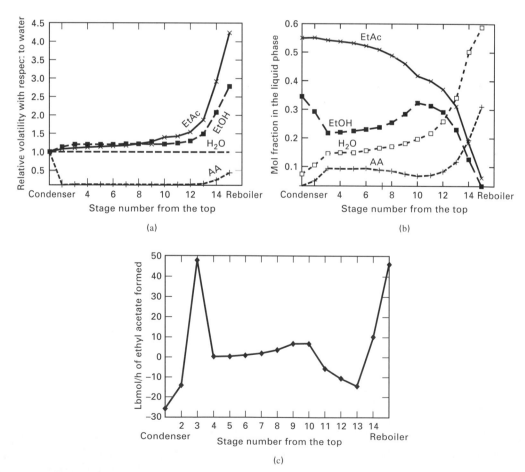

Figure 11.37 Profiles for reactive distillation in Example 11.8. (a) Relative volatility profile. (b) Liquid-phase mole fraction profiles. (c) Reaction-rate profile.

EXAMPLE 11.9

Using thermodynamic and kinetic data from Rehfinger and Hoffmann [74] for the formation of MTBE from methanol (MeOH) and isobutene (IB), in the presence of n-butene (NB), both Jacobs and Krishna [75] and Nijhuis, Kerkhof, and Mak [76] computed, for catalyzed reactive distillation, with the ASPEN PLUS simulator, multiple solutions having drastically different isobutene conversions when the feed stage for methanol was varied. An explanation for these multiple solutions is given by Hauan, Hertzberg, and Lien [77].

Compute a converged solution for the following conditions, taking into account the kinetics of the reaction, but assuming vapor–liquid equilibrium at each stage. The distillation column has a total condenser, a partial reboiler, and 15 equilibrium stages in the column, which operates at 11 bars. Stages are counted down from the top, with the total condenser numbered stage 1 when using a process simulator, even though it is not an equilibrium stage. The mixed butenes feed, consisting of 195.44 mol/s of IB and 353.56 mol/s of NB enters stage 11 as a vapor at 350 K and 11 bar. The methanol at a flow rate of 215.5 mol/s enters stage 10 as a liquid at 320 K and 11 bar. The reflux ratio is 7 and the bottoms flow rate is set at 197 mol/s. The catalyst is provided only for Stages 4 through 11 (8 stages total), with 204.1 kg of catalyst per stage. The catalyst is a strong-acid ion-exchange resin with 4.9 equivalents of acid groups per kilogram of catalyst. Thus, the equivalents per stage are 1,000 or 8,000 for the 8 stages. Compute the product compositions and column profiles using the RADFRAC model in ASPEN PLUS.

SOLUTION

The only chemical reaction considered is:

$$IB + MeOH \leftrightarrow MTBE$$

with NB inert.

For the forward reaction, the rate law is formulated in terms of mole-fraction concentrations, instead of activities (products of activity coefficient and mole fraction) as in Rehfinger and Hoffmann [74]:

$$r_{forward} = 3.67 \times 10^{12} \exp(-92,440/RT) x_{IB}/x_{MeOH} \qquad (1)$$

The corresponding backward rate law is

$$r_{backward} = 2.67 \times 10^{17} \exp(-134,454/RT) x_{MTBE}/x^2_{MeOH} \qquad (2)$$

where r is in moles per second per equivalent of acid groups, $R = 8.314$ J/mol-K, T is in kelvins, and x_i is liquid mole fraction.

The Redlich–Kwong equation of state is used to estimate vapor-phase fugacities and the UNIQUAC equation to estimate the liquid-phase activity coefficients. The UNIQUAC binary interaction parameters are as follows, where it is very important to include the inert NB in the system by assuming it has the same parameters as IB and that the two butenes form an ideal solution. The parameters are defined as follows, with all $a_{ij} = 0$.

$$T_{ij} = \exp\left(-\frac{u_{ij} - u_{jj}}{RT}\right) = \exp\left(a_{ij} + \frac{b_{ij}}{T}\right) \qquad (3)$$

Components in Binary Pair, *ij*	Binary Parameters	
	b_{ij}, **K**	b_{ji}, **K**
MeOH–IB	35.38	−706.34
MeOH–MTBE	88.04	−468.76
IB–MTBE	−52.2	24.63
MeOH–NB	35.38	−706.34
NB–MTBE	52.2	24.63

The only initial guesses provided to the computer program are temperatures of 350 and 420 K, respectively, for Stages 1 and 17; liquid-phase mole fractions of 0.05 for MeOH and 0.95 for MTBE leaving Stage 17; and vapor-phase mole fractions of 0.125 for MeOH and 0.875 for

MTBE leaving Stage 17. The ASPEN PLUS input data are listed in Table 11.6. The converged temperatures for Stages 1 and 17, respectively, are 347 and 420 K. Converged product flow rates are as follows:

	Flow Rate, mol/s	
Component	**Distillate**	**Bottoms**
MeOH	28.32	0.31
IB	7.27	1.31
NB	344.92	8.64
MTBE	0.12	186.74
Total	380.63	197.00

Table 11.6 ASPEN PLUS Input Data for Example 11.9

```
TITLE 'mtbe'

IN-UNITS MET VOLUME-FLOW='CUM/HR' ENTHALPY-FLO='MMKCAL/HR'   &
        HEAT-TRANS-C='KCAL/HR-SQM-K' PRESSURE=BAR TEMPERATURE=C   &
        VOLUME=CUM DELTA-T=C HEAD=METER MOLE-DENSITY='KMOL/CUM'   &
        MASS-DENSITY='KG/CUM' MOLE-ENTHALP='KCAL/MOL'   &
        MASS-ENTHALP='KCAL/KG' HEAT=MMKCAL MOLE-CONC='MOL/L'   &
        PDROP=BAR

DEF-STREAMS CONVEN ALL

DATABANKS PURECOMP  /  AQUEOUS  /  SOLIDS  /  INORGANIC  /  &
        NOASPENPCD

PROP-SOURCES PURECOMP  /  AQUEOUS  /  SOLIDS  /  INORGANIC

COMPONENTS
    MEOH CH4O MEOH /
    IB C4H8-5 IB /
    NB C4H8-1 NB /
    MTBE C5H12O-D2 MTBE

FLOWSHEET
    BLOCK B1 IN=1 2 OUT=3 4

PROPERTIES SYSOP11

PROP-REPLACE SYSOP11 UNIQ-RK
    PROP PHILMX PHILMX11
    PROP HLMX HLMX11
    PROP GLMX GLMX11
    PROP SLMX SLMX11
    PROP MUVMX MUVMX02
    PROP MULMX MULMX02
    PROP KVMX KVMX02
    PROP DV DV01

PROP-DATA UNIQ-1
    IN-UNITS MET VOLUME-FLOW='CUM/HR' ENTHALPY-FLO='MMKCAL/HR'   &
        HEAT-TRANS-C='KCAL/HR-SQM-K' PRESSURE=BAR TEMPERATURE=K   &
        VOLUME=CUM DELTA-T=C HEAD=METER MOLE-DENSITY='KMOL/CUM'   &
        MASS-DENSITY='KG/CUM' MOLE-ENTHALP='KCAL/MOL'   &
        MASS-ENTHALP='KCAL/KG' HEAT=MMKCAL MOLE-CONC='MOL/L'   &
        PDROP=BAR
    PROP-LIST UNIQ
    BPVAL MEOH IB 0.0 35.38 0.0 0.0 0.0 1000.000
```

```
      BPVAL IB MEOH 0.0 -706.34 0.0 0.0 0.0 1000.000
      BPVAL MEOH MTBE 0.0 88.04 0.0 0.0 0.0 1000.000
      BPVAL MTBE MEOH 0.0 -468.76 0.0 0.0 0.0 1000.000
      BPVAL IB MTBE 0.0 -52.2 0.0 0.0 0.0 1000.000
      BPVAL MTBE IB 0.0 24.63 0.0 0.0 0.0 1000.000
      BPVAL MEOH NB 0.0 35.38 0.0 0.0 0.0 1000.000
      BPVAL NB MEOH 0.0 -706.34 0.0 0.0 0.0 1000.000
      BPVAL NB MTBE 0.0 -52.2 0.0 0.0 0.0 1000.000
      BPVAL MTBE NB 0.0 24.63 0.0 0.0 0.0 1000.000

PROP-SET VLE PHIMX GAMMA PL SUBSTREAM=MIXED PHASE=V L

PROP-SET VLLE PHMIX GAMMA PL SUBSTREAM=MIXED PHASE=V L1 L2

STREAM 1
      SUBSTREAM MIXED TEMP=320 <K> PRES=11 NPHASE=1 PHASE=L
      MOLE-FLOW MEOH 215.5 <MOL/SEC> / IB 0. <MOL/SEC> / NB  &
          0. <MOL/SEC> / MTBE 0. <MOL/SEC>

STREAM 2
      SUBSTREAM MIXED TEMP=350 <K> PRES=11 NPHASE=1
      MOLE-FLOW MEOH 0. <MOL/SEC> / IB 195.44 <MOL/SEC> / NB  &
          353.56 <MOL/SEC> / MTBE 0. <MOL/SEC>

BLOCK B1 RADFRAC
      PARAM NSTAGE=17 MAXOL=50 MAXIL=50 ILMETH=NEWTON
      FEEDS 1 10 / 2 11 ON-STAGE
      PRODUCTS 3 1 L / 4 17 L
      P-SPEC 1 11
      COL-SPECS DP-COL=0 MOLE-RDV=0 MOLE-B=197 <MOL/SEC> MOLE-RR=7
      REAC-STAGES 4 11 r-1
      HOLD-UP 4 11 MASS-LHLDP=8000
      T-EST 1 350 <K> / 17 420 <K>
      X-EST 17 MEOH .05 / 17 MTBE .95
      Y-EST 17 MEOH .125 / 17 MTBE .875
      TRAY-REPORT TRAY-OPTION=ALL-TRAYS PROPERTIES=VLE VLLE

STREAM-REPOR PROPERTIES=VLE VLLE

REACTIONS R-1 REAC-DIST
      REAC-DATA 1 KINETIC CBASIS=MOLEFRAC
      REAC-DATA 2 KINETIC CBASIS=MOLEFRAC
      RATE-CON 1 PRE-EXP=3.67E12 ACT-ENERGY=92400 <KJ/KMOL>
      RATE-CON 2 PRE-EXP=2.67E17 ACT-ENERGY=134454 <KJ/KMOL>
      STOIC 1 MEOH -1 / IB -1 / NB 0 / MTBE 1
      STOIC 2 MTBE -1 / NB 0 / MEOH 1 / IB 1
      POWLAW-EXP 1 MEOH -1 / IB 1 / NB 0 / MTBE 0
      POWLAW-EXP 2 MTBE 1 / NB 0. / MEOH -2 / IB 0
   ;
   ;
   ;
```

The combined feeds to the reactive distillation contained a 10.3% mole excess of MeOH over IB. Therefore, IB was the limiting reactant and the preceding product distribution indicates that 95.6% of the IB, or 186.86 mol/s, reacted to form MTBE. The percent purity of the MTBE in the bottoms is 94.8%. Only 2.4% of the inert NB and 1.1% of the unreacted MeOH are found in the bottoms. The computed condenser and reboiler duties are, respectively, 53.2 and 40.4 MW.

Seven iterations were required to obtain a converged solution. The column profiles are in Figure 11.38. Figure 11.38a shows that most of the reaction occurs in a narrow temperature

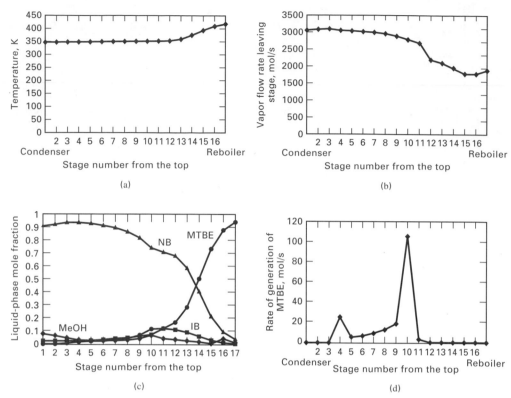

Figure 11.38 Profiles for reactive distillation in Example 11.9. (a) Temperature profile. (b) Vapor traffic profile. (c) Liquid-phase mole fraction profile. (d) Reaction-rate profile.

range of 348.6 to 353 K. The reaction temperature can be varied by adjusting the column pressure. Figure 11.38b shows that the vapor traffic in the column above the two feed entries changes by less than 11%, because of only small changes in temperature. As one moves down the column below the two feed entries, the temperature increases rapidly from 353 to 420 K, causing the vapor traffic to decrease by about 20%. In Figure 11.38c, the liquid composition profiles show that the liquid is dominated by NB from the top stage down to Stage 13, thus drastically reducing the driving force for the reaction. Below Stage 11, the liquid quickly becomes richer in MTBE as the mole fractions of the other components decrease because of increasing temperature. In the section of the column above the reaction zone, the mole fraction of MTBE quickly decreases as one moves to the top stage. These changes are due mainly to the large differences between the K-values for MTBE and those for the other three components. The relative volatility of MTBE with respect to any of the other components ranges from about 0.24 at the top stage to about 0.35 at the bottom. Nonideality in the liquid phase influences mainly MeOH, whose liquid-phase activity coefficient varies from a high of 10 at Stage 5 to a low of 2.6 at Stage 17. This causes the unreacted MeOH to leave mainly with the NB in the distillate rather than with MTBE in the bottoms. The profile for the rate of reaction is shown in Fig. 11.38d, where it is seen that the forward reaction dominates on every stage of the reaction section. However, 56% of the reaction occurs on Stage 10, which is the MeOH feed stage. The least amount of reaction occurs on Stage 11.

As mentioned earlier, the literature indicates that the percent conversion of IB to MTBE will vary depending upon the stage to which the MeOH is fed. Furthermore, in the range of MeOH feed stages from about 8 to 11, both a low-conversion and a high-conversion solution can be computed. This is shown in Figure 11.39, where the high-conversion solutions are mainly

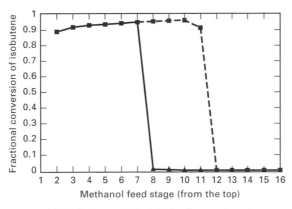

Figure 11.39 Effect of MeOH feed stage location on conversion of IB to MTBE.

in the 90+% range, while the low-conversion solutions are all less than 10%. However, if activities are used in the rate expressions, rather than mole fractions, the low-conversion solutions would be higher because of the large values for the activity coefficient for MeOH. The results in Figure 11.39 were computed starting with the MeOH feed entering Stage 2. The resulting profiles for this run were used as the initial guesses for the run with MeOH entering Stage 3. Subsequent runs were performed in a similar manner, increasing the MeOH feed stage by 1 each time and initializing with the results of the previous run. High-conversion solutions were obtained for each run until the MeOH feed stage was lowered to Stage 12, at which point the conversion decreased dramatically. Further lowering of the MeOH feed stage to Stage 16 also resulted in a low-conversion solution. However, when the direction of change to the MeOH feed stage was reversed starting from Stage 12, a low-conversion was obtained until the feed stage was decreased to Stage 9, at which point the conversion jumped back to the high-conversion result. ∎

11.8 SUPERCRITICAL-FLUID EXTRACTION

Solute extraction from a liquid or solid mixture is usually accomplished with a liquid solvent, as discussed in Chapters 8 and 16, respectively, at conditions of temperature and pressure that lie substantially below the critical temperature and pressure of the solvent. Following the extraction step, the solvent and dissolved solute are subjected to a subsequent separation step, such as distillation, to purify the solvent for recycle and recover the solute.

In 1879, Hannay and Hogarth [78] reported that solid potassium iodide could be dissolved in ethanol, as a dense gas, at supercritical conditions of $T > T_c = 516$ K and $P > P_c = 65$ atm. The iodide could then be precipitated from the ethanol by reducing the pressure. This process, which operates in the supercritical-fluid region shown in Figure 11.40, was later referred to as *supercritical-fluid extraction, supercritical-gas extraction, supercritical extraction* (SCE), *dense-gas extraction,* or *destraction* (a combination of distillation and extraction). By the 1940s, as chronicled by Williams [79], proposed practical applications of SCE began to appear in the patent and technical literature.

The solvent power of a compressed gas can undergo an enormous change in the vicinity of its critical point. Consider, for example, the solubility of *p*-iodochlorobenzene (pICB) in ethylene, as shown in Figure 11.41, at 298 K for pressures from 2 to 8 MPa. This temperature is 1.05 times the critical temperature of ethylene (283 K) and the pressure range straddles the critical pressure of ethylene (5.1 MPa). At 298 K, pICB is a solid

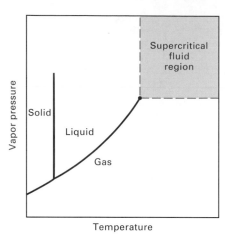

Figure 11.40 Supercritical-fluid region.

(melting point = 330 K) with a vapor pressure of the order of 0.1 torr. At 2 MPa, if pICB formed an ideal gas solution with ethylene, the mole fraction of pICB in the gas in equilibrium with pure, solid pICB would be extremely small at about 6.7×10^{-6} or a concentration of 0.00146 g/l. The experimental concentration from Figure 11.41 is 0.015 g/l, which is an order of magnitude higher because of nonideal gas effects. If the pressure is increased from 2 MPa to almost the critical pressure at 5 MPa (an increase by a factor of 2.5), the equilibrium concentration of pICB is increased about 10-fold to 0.15 g/l. At 8 MPa, the concentration begins to level out at 40 g/l, which is 2,700 times higher than predicted from the vapor pressure for an ideal gas solution. It is this dramatic increase in solubility of a solute at near-critical conditions of a solvent that makes SCE of interest.

Why such a dramatic increase in solvent power? The explanation lies in the change that occurs to the solvent density while the solubility of the solute increases. A pressure–enthalpy diagram for ethylene is shown in Figure 11.42, which includes the specific volume as a parameter, from which the density can be determined. The range of variables and

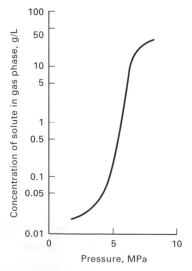

Figure 11.41 Effect of pressure on solubility of pICB in super-critical ethylene.

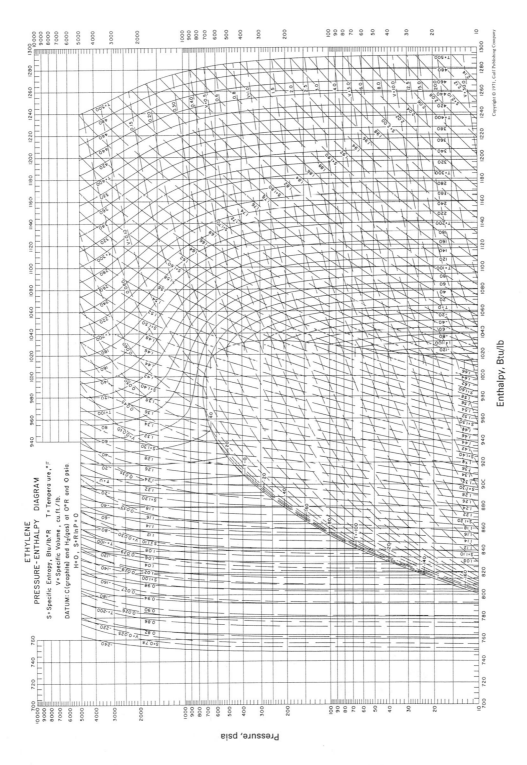

Figure 11.42 Pressure–enthalpy diagram for ethylene. [From K.E. Starling, "Fluid Thermodynamic Properties for Light Petroleum Systems," Gulf Publishing, Houston (1973) reprinted with permission.]

parameters straddles the critical point of ethylene. The density of ethylene compared to the solubility of pICB is as follows at 298 K:

Pressure, MPa	Ethylene Density, g/l	Solubility of pICB, g/l
2	25.8	0.015
5	95	0.15
8	267	40

Although there is far from a 1:1 correspondence in the increase of pICB solubility with ethylene over this range of pressure, there is a meaningful correlation. As the pressure is increased, closer packing of the solvent molecules allows them to surround and trap solute molecules. This phenomenon is most dramatic and useful at reduced temperatures from about 1.01 to 1.12.

Two other effects in the supercritical region are favorable for SCE. It will be recalled that the molecular diffusivity of a solute in an ambient-pressure gas is about four orders of magnitude higher than for a liquid. For a near-critical fluid, the diffusivity of solute molecules is usually one to two orders of magnitude higher than in a normal liquid solvent, thus resulting in a lower mass-transfer resistance in the solvent phase than might be expected. In addition, the viscosity of the supercritical fluid is about an order of magnitude less than that of a normal liquid solvent.

Many patents have been issued, proposals prepared, and experimental studies conducted on SCE as a possible alternative for distillation, enhanced distillation, and liquid–liquid extraction. However, in general, when these other techniques are feasible, SCE usually cannot compete economically because of high solvent compression costs. SCE is most favorable for the extraction of small amounts of large, relatively nonvolatile solutes in solid or liquid mixtures. Such applications are cited by Williams [79] and McHugh and Krukonis [80].

Solvent selection depends on the composition of the feed mixture. If only the chemical(s) to be extracted is (are) soluble in a potential solvent, then high solubility is a key factor. However, if a chemical besides the desired solute is soluble in the potential solvent, then the selectivity of the solvent becomes as important as solubility. A number of light gases and other low-molecular-weight chemicals, including the following, have received attention as solvents for SCE:

Solvent	Critical Temperature, K	Critical Pressure, MPa	Critical Density, kg/m³
Methane	192	4.60	162
Ethylene	283	5.03	218
Carbon dioxide	304	7.38	468
Ethane	305	4.88	203
Propylene	365	4.62	233
Propane	370	4.24	217
Ammonia	406	11.3	235
Water	647	22.0	322

Those solvents with a critical temperature below 373 K have been well studied. A particularly desirable solvent, particularly for the extraction of undesirable, valuable, or heat-sensitive chemicals from natural products such as foods, is carbon dioxide, which has a moderate critical pressure, a high critical density, and a critical temperature close to ambient temperature. Carbon dioxide is nonflammable, noncorrosive, nontoxic in low concentrations, readily available, inexpensive, and safe. Also, supercritical carbon dioxide has a relatively low viscosity and high molecular diffusivity. Separation of carbon dioxide from the solute is often possible by simply reducing the extract pressure. According to Williams [79], supercritical carbon dioxide has been used to extract caffeine from coffee, hops oil from beer, piperine from pepper, capsaicin from chilis, oil from nutmeg, and nicotine from tobacco.

Carbon dioxide is not a suitable solvent for all potential applications. McHugh and Krukonis [81] cite the energy crisis of the 1970s that led to substantial research on an energy-efficient separation of ethanol and water. The primary goal, which was to break the ethanol–water azeotrope, was not achieved by SCE with carbon dioxide because, although supercritical carbon dioxide has unlimited capacity to dissolve pure ethanol, water is also dissolved in significant amounts. A liquid–supercritical-fluid phase diagram for the ethanol–water–carbon dioxide ternary system at 308.2 K and 10.08 MPa, based on the experimental data of Takishima, Saiki, Arai, and Saito [82], is shown in Figure 11.43. These conditions correspond to $T_r = 1.014$ and $P_r = 1.366$ for carbon dioxide. For the binary mixture of water and carbon dioxide, two phases exist; a water-rich phase with about 2 mol% carbon dioxide and a carbon dioxide-rich phase with about 1 mol% water. Ethanol and carbon dioxide are completely soluble in each other. Ternary mixtures containing more than 40 mol% ethanol are completely miscible. If a near-azeotropic mixture of ethanol and water, say, 85 mol% ethanol and 15 mol% water, is extracted by carbon dioxide at the conditions of Figure 11.43, a mixing line drawn between this composition and a point for pure carbon dioxide does not appear to cross into the two-phase region. That is, regardless of the amount of solvent used, both water and ethanol are completely soluble in the carbon dioxide and no separation is possible at these temperatures and pressures. Alternatively, consider an ethanol–water broth from a fermentation reactor with 10 wt% (4.17 mol%) ethanol. If this mixture is extracted with supercritical carbon dioxide, complete dissolution will not occur and a modest degree of separation of ethanol from water can be achieved, as shown in the next example. The separation can be further enhanced by the use of a cosolvent, such as glycerol, that improves the selectivity, as shown by Inomata, Kondo, Ari, and Saito [83].

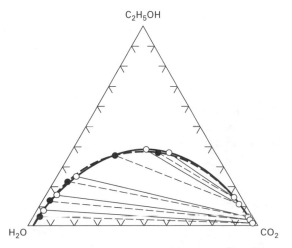

Figure 11.43 Liquid–fluid equilbria for CO_2–C_2H_5OH–H_2O at 308–313.2 K and 10.1–10.34 MPa.

When CO_2 is used as a solvent, it must be recovered and recycled. Three schemes discussed by McHugh and Krukonis [81] are shown in Figure 11.44. In the first scheme, shown for the separation of ethanol and water, the ethanol–water feed is pumped as a liquid to the pressure of the extraction column, where it is contacted with supercritical carbon dioxide. The raffinate leaving the extractor at the bottom is enriched with respect to water and can be sent to another part of the plant for further processing. The extract stream, which leaves from the top of the extractor and contains most of the carbon dioxide, some ethanol, and a smaller amount of water, is expanded across a valve to a lower pressure. In a flash drum downstream of the valve, ethanol–water condensate is collected and the CO_2-rich gas is recycled through a gas compressor back to the extractor. However, unless the pressure is greatly reduced across the valve, resulting in large compression costs, little of the ethanol is condensed.

A second CO_2 recovery scheme, due to de Filippi and Vivian [84], is shown in Figure 11.44b. The flash drum is replaced by a high-pressure distillation column, which operates at a pressure just below the pressure of the extraction column to produce a CO_2-rich distillate and an ethanol-rich bottoms. The distillate is compressed and recycled through the reboiler and back to the extractor. Both the raffinate and the distillate are flashed to recover dissolved CO_2. This scheme, although more complicated than the first, is more versatile.

A third CO_2 recovery scheme, due to Katz et al. [85] for the decaffeination of coffee, is shown in Figure 11.44c. In the extractor, green, wet coffee beans are mixed with supercritical CO_2 to extract caffeine. The extract is sent to a second extraction column, where the caffeine is extracted with water. The CO_2-rich raffinate from this column is recycled through a compressor (not shown) back to the first extraction column, from which the decaffeinated coffee leaves from the bottom and is sent to a roasting tower. The caffeine-rich water leaving the second column is sent to a reverse-osmosis unit, where the water is purified and recycled through a pump (not shown) to the water column. All three separation steps operate at high pressure. The concentrated caffeine–water mixture leaving the osmosis unit is sent to a crystallizer to produce caffeine crystals.

Multiple equilibrium stages in a countercurrent-flow contactor are generally needed to obtain the desired extent of extraction. A major problem in determining the number of stages required is the estimation of liquid–supercritical fluid phase-equilibrium constants. Most commonly, cubic-equation-of-state methods, such as the Soave–Redlich–Kwong (SRK) or Peng–Robinson (PR) equations, are used, but they have two shortcomings. First, their accuracy diminishes in the critical region of the solvent. Second, if the feed contains polar components that form a nonideal liquid mixture, an appropriate mixing rule, such as that of Wong and Sandler [86], that provides a correct bridge between equation-of-state methods and activity-coefficient methods must be employed.

As discussed in Section 2.5, the SRK and PR equations for pure components both contain two parameters, a and b, that are computed from critical constants. The SRK and PR equations are extended to liquid or vapor mixtures by a *mixing rule* for computing values of a_m and b_m for the mixture from values for the pure components. The simplest mixing rule, due to van der Waals, is:

$$a_m = \sum_{i=1}^{C} \sum_{j=1}^{C} x_i x_j a_{ij} \qquad \textbf{(11-19)}$$

$$b_m = \sum_{i=1}^{C} \sum_{j=1}^{C} x_i x_j b_{ij} \qquad \textbf{(11-20)}$$

where x is a mole fraction in the vapor or liquid mixture. Although these two mixing-rule equations are identical in form, the following *combining rules* for a_{ij} and b_{ij} are quite different, with the former being a geometric mean and the latter an arithmetic mean:

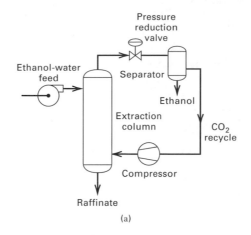

(a)

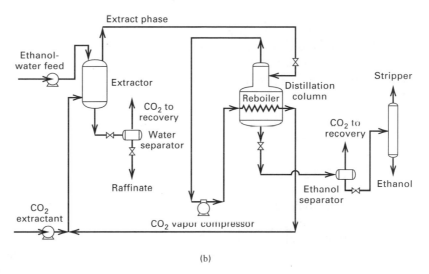

(b)

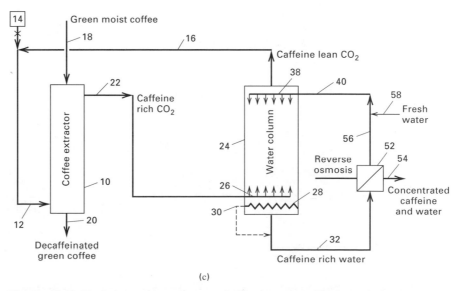

(c)

Figure 11.44 Techniques for recovery of CO_2 in supercritical extraction processes. (a) Pressure reduction. (b) High-pressure distillation. (c) High-pressure absorption with water.

$$a_{ij} = (a_i a_j)^{1/2} \tag{11-21}$$

$$b_{ij} = (b_i + b_j)/2 \tag{11-22}$$

As stated by Sandler, Orbey, and Lee [87], (11-19) to (11-22) are usually adequate for nonpolar mixtures of hydrocarbons and light gases when critical-temperature and/or size differences between the molecules are not very large.

Molecular size differences and/or modest degrees of polarity are handled by the following modified combining rules:

$$a_{ij} = (a_i a_j)^{1/2}(1 - k_{ij}) \tag{11-23}$$

$$b_{ij} = [(b_i + b_j)/2](1 - l_{ij}) \tag{11-24}$$

where k_{ij} and l_{ij} are binary interaction parameters back-calculated from experimental vapor–liquid equilibrium and/or density data. Often the latter parameter is set equal to zero. A tabulation of values of k_{ij}, suitable for use with the SRK and PR equations when the mixture contains hydrocarbons with CO_2, H_2S, N_2, and/or CO, is given by Knapp et al. [88]. In a study by Shibata and Sandler [89], using experimental phase-equilibria and phase-density data for the nonpolar binary system nitrogen–n-butane at 410.9 K over a pressure range of about 30 to 70 bar, reasonably good predictions, except in the critical region, were obtained using (11-19) and (11-20), with (11-23) and (11-24), and values of $k_{ij} = -0.164$ and $l_{ij} = -0.233$ in conjunction with the PR equation. Similar good agreement with experimental data was obtained for the systems nitrogen–cyclohexane, carbon dioxide–n-butane, and carbon dioxide–cyclohexane, and the ternary systems nitrogen–carbon dioxide–n-butane and nitrogen–carbon dioxide–cyclohexane.

For high pressures and mixtures containing one or more strongly polar components, the preceding rules are inadequate and it would seem desirable, in those cases, to combine the equation-of-state method with the activity-coefficient (free energy) method to handle strong nonidealities in the liquid phase. The following theoretically based mixing rule of Wong and Sandler [86] accomplishes such a bridge between a cubic equation of state and a free-energy or activity-coefficient equation. If, for example, the PR equation of state and the NRTL activity-coefficient equation are used, the Wong and Sandler mixing rule leads to the following expressions for computing a_m and b_m to be used in the PR equation:

$$a_m = RTQD/(1 - D) \tag{11-25}$$

$$b_m = Q/(1 - D) \tag{11-26}$$

where

$$Q = \sum_{i=1}^{C} \sum_{j=1}^{C} x_i x_j \left(b - \frac{a}{RT} \right)_{ij} \tag{11-27}$$

$$D = \sum_{i=1}^{C} x_i \frac{a_i}{b_i RT} + \frac{G^{ex}(x_i)}{\sigma RT} \tag{11-28}$$

$$\left(b - \frac{a}{RT} \right)_{ij} = \frac{1}{2}\left[\left(b_i - \frac{a_i}{RT} \right) + \left(b_j - \frac{a_j}{RT} \right) \right](1 - k_{ij}) \tag{11-29}$$

$$\sigma = \frac{1}{\sqrt{2}}[\ln(\sqrt{2} - 1)] \tag{11-30}$$

$$\frac{G^{ex}}{RT} = \sum_{i=1}^{C} x_i \left(\frac{\sum_{j=1}^{C} x_j \tau_{ji} g_{ji}}{\sum_{k=1}^{C} x_k g_{ki}} \right) \tag{11-31}$$

$$g_{ij} = \exp(-\alpha_{ij}\tau_{ij}) \tag{11-32}$$

with $a_{ij} = a_{ji}$.

From Equations (11-25) to (11-32), it is seen that for a binary system, using the NRTL equation, there are four adjustable binary interaction parameters (BIPs): k_{ij}, α_{ij}, τ_{ij}, and τ_{ji}. These four parameters, for a temperature and pressure range of interest, are best obtained by regression of experimental binary-pair data for VLE, LLE, or VLLE. The parameters can then be used to predict phase equilibria for ternary and higher multicomponent mixtures. However, Wong, Orbey, and Sandler [90] show that when values of the latter three parameters are already available, even at just near-ambient temperature and pressure conditions, from an experimental-data-compilation source such as that of Gmehling and Onken [91], those parameters can be assumed to be independent of temperature and used directly to make reasonably accurate predictions of phase equilibria, even at temperatures to at least 200°C and pressures to 200 bar. Furthermore, regression of experimental data to obtain a value of k_{ij} is not necessary either, because Wong, Orbey, and Sandler show that it can be determined from the other three parameters by choosing its value so that the excess Gibbs free energy computed from the equation of state matches that computed from the activity-coefficient model. Thus, the application of the Wong–Sandler mixing rule to supercritical extraction conditions is facilitated.

Another phase-equilibrium prediction method applicable to wide ranges of pressure, temperature, molecular size, and polarity is the group contribution equation of state (GC-EOS) of Skjold-Jørgensen [92]. This method, which combines features of the van der Waals equation of state, the Carnahan–Starling expression for hard spheres, the NRTL activity-coefficient equation, and the group-contribution principle, has been successfully applied to supercritical-extraction conditions. The GC-EOS method is particularly useful when all of the necessary binary data are not available to determine all binary interaction parameters.

When experimental K-values are available, or when the Wong–Sandler mixing rule or the GC-EOS can be applied, equilibrium-stage calculations for supercritical extraction can be made by conventional computer programs, as the following example illustrates.

EXAMPLE 11.10

One mol/s of 10 wt% ethanol in water is extracted with 3 mol/s of carbon dioxide at 305 K and 9.86 MPa in a countercurrent-flow extraction column with the equivalent of five equilibrium stages. Determine the flow rates and compositions of the exiting extract and raffinate.

SOLUTION

This problem, which was taken from Colussi, Fermeglia, Gallow, and Kikic [93], was solved with the Tower Plus model of the ChemCAD process simulator, under conditions of constant temperature and constant pressure, at which composition changes were small enough that K-values could be assumed constant. The following K-values, taken from Colussi et al., who used the GC-EOS method, and which are defined as the mole fraction in the extract divided by the mole fraction in the raffinate, are in reasonably good agreement with experimental data.

Component	K-Value
CO_2	34.5
Ethanol	0.115
Water	0.00575

The percent extraction of ethyl alcohol was computed to be 33.6%, with an extract of 69 wt% pure ethanol (solvent-free basis) and a raffinate containing 93 wt% water (solvent-free basis). The calculated stage-wise flow rates and component mole fractions are listed in Table 11.7, where stages are numbered from the feed end of the cascade. ∎

Table 11.7 Calculated Flow and Composition Profiles for Example 11.10

Leaving Streams	Extract Mole Fraction	Raffinate Mole Fraction
Stage 1		
Carbon dioxide	0.98999	0.02870
Ethanol	0.00466	0.04053
Water	0.00535	0.93077
Total flow, gmol/s	3.0013	1.0298
Stage 2		
Carbon dioxide	0.99002	0.02870
Ethanol	0.00463	0.04023
Water	0.00535	0.93107
Total flow, gmol/s	3.0311	1.0294
Stage 3		
Carbon dioxide	0.99012	0.02870
Ethanol	0.00452	0.03929
Water	0.00536	0.93201
Total flow, gmol/s	3.0308	1.0285
Stage 4		
Carbon dioxide	0.99043	0.02870
Ethanol	0.00419	0.03645
Water	0.00538	0.93485
Total flow, gmol/s	3.0298	1.0255
Stage 5		
Carbon dioxide	0.99138	0.02874
Ethanol	0.00319	0.02775
Water	0.00543	0.94351
Total flow, gmol/s	3.0268	0.9987

SUMMARY

1. Extractive distillation, salt distillation, pressure-swing distillation, homogeneous azeotropic distillation, heterogeneous azeotropic distillation, and reactive distillation are enhanced distillation techniques to be considered when separation by ordinary distillation is uneconomical or impossible. Reactive distillation can also be used to conduct, simultaneously and in the same apparatus, a chemical reaction and a separation by distillation.

2. For ternary systems, a composition plot on a triangular graph is very useful for finding feasible separations, especially when binary and ternary azeotropes form. With such a diagram, distillation paths, called residue curves or distillation curves, are readily tracked. The curves may be restricted to certain regions of the triangular diagram by distillation boundaries. Feasible product compositions at total reflux are readily determined.

3. Extractive distillation, using a low-volatility solvent that enters near the top of the column, is widely used to separate azeotropes and very close-boiling mixtures. Preferably, the solvent should not form an azeotrope with any component in the feed.

4. Certain salts, when added to a solvent, reduce the volatility of the solvent and increase the relative volatility between the two components to be separated. In this process, called salt distillation, the salt is dissolved in the solvent or added as a solid or melt to the reflux.

5. Pressure-swing distillation, utilizing two columns operating at different pressures, can be used to separate an azeotropic mixture when the azeotrope can be made to disappear at some pressure. If not, the technique may still be practical if the azeotropic composition changes by 5 mol% or more over a moderate range of pressure.

6. In homogeneous azeotropic distillation, an entrainer is added to a stage, usually above the feed stage. A minimum- or maximum-boiling azeotrope, formed by the entrainer with one or more feed components, is removed from the top or bottom of the column, respectively. Unfortunately, potential applications of this technique for difficult-to-separate mixtures are not common because of limitations due to distillation boundaries.

7. A more common and useful technique is heterogeneous azeotropic distillation, in which the entrainer forms, with one or more components of the feed, a minimum-boiling heterogeneous azeotrope. When condensed, the overhead vapor splits into organic-rich and water-rich phases. The azeotrope is broken by returning one liquid phase as reflux, with the other sent on as distillate for further processing.

8. A growing application of reactive or catalytic distillation is the combined operation of chemical reaction and distillation in one vessel. To be effective, it must be possible to carry out the reaction and phase separation at the same pressure and range of temperature, with reactants and products favoring different phases so that an equilibrium-limited reaction can go to completion.

9. Liquid–liquid or solid–liquid extraction can be carried out with a supercritical-fluid solvent at temperatures and pressures just above the critical because of favorable values for solvent density and viscosity, solute diffusivity, and solute solubility in the solvent. An attractive supercritical solvent is carbon dioxide, particularly for extraction of certain chemicals from natural products.

REFERENCES

1. Stichlmair, J., J.R. Fair, and J.L. Bravo, *Chem. Eng. Progress,* **85**(1), 63–69 (1989).

2. Partin, L.R., *Chem. Eng. Progress,* **89** (1), 43–48 (1993).

3. Stichlmair, J., "Distillation and Rectification," in *Ullmann's Encyclopedia of Industrial Chemistry,* 5th ed., VCH Verlagsgesellschaft Weinheim Vol. B3, pp. 4-1 to 4-94, (1988).

4. Doherty, M.F., and J.D. Perkins, *Chem. Eng. Sci.,* **33**, 281–301 (1978).

5. Bossen, B.S., S.B. Jørgensen, and R. Gani, *Ind. Eng. Chem. Res.,* **32**, 620–633 (1993).

6. Pham, H.N. and M.F. Doherty, *Chem. Eng. Sci.,* **45**, 1837–1843 (1990).

7. Taylor, R., and H.A. Kooijman, *CACHE News,* No. 41, 13–19 (1995).

8. Doherty, M.F., *Chem. Eng. Sci.,* **40**, 1885–1889 (1985).

9. ASPEN PLUS, "What's New in Release 9," Aspen Technology, Cambridge, MA (1994).

10. Doherty, M.F., and J.D. Perkins, *Chem. Eng. Sci.,* **34**, 1401–1414 (1979).

11. Horsley, L.H., "Azeotropic Data III," in *Advances in Chemistry Series,* American Chemical Society, Washington, D.C., Vol. 116 (1973).

12. Gmehling, J., *Azeotropic Data,* VCH Publishers, Deerfield Beach, FL (1994).

13. Fidkowski, Z.T., M.F. Malone, and M.F. Doherty, *Computers Chem. Engng.,* **17**, 1141–1155 (1993).

14. Foucher, E.R., M.F. Doherty, and M.F. Malone, *Ind. Eng. Chem. Res.,* **30**, 760–772 (1991) and **30**, 2364 (1991).

15. Matsuyama, H., and H.J. Nishimura, *J. Chem. Eng. Japan,* **10**, 181 (1977).

16. Doherty, M.F., and G.A. Caldarola, *Ind. Eng. Chem. Fundam.,* **24**, 474–485 (1985).

17. Fidkowski, Z.T., M.F. Doherty, and M.F. Malone, *AIChE J.,* **39**, 1303–1321 (1993).

18. Stichlmair, J.G., and J.-R. Herguijuela, *AIChE J.,* **38**, 1523–1535 (1992).

19. Widagdo, S., and W.D. Seider, *AIChE J.,* **42**, 96–130 (1996).

20. Wahnschafft, O.M., J.W. Koehler, E. Blass, and A.W. Westerberg, *Ind. Eng. Chem. Res.,* **31**, 2345–2362 (1992).

21. Dunn, C.L., R.W. Millar, G.J. Pierotti, R.N. Shiras, and M. Souders, Jr., *Trans. AIChE,* **41**, 631–644 (1945).

22. Berg, L., *Chem. Eng. Progress,* **65** (9), 52–57 (1969).

23. Othmer, D.F., *AIChE Symp. Series,* **235** (79), 90–117 (1983).

24. Van Raymbeke, U.S. Patent 1,474,216 (1922).

25. Cook, R.A., and W. F. Furter, *Can. J. Chem. Eng.,* **46**, 119–123 (1968).

26. Johnson, A.I., and W.F. Furter, *Can. J. Chem. Eng.,* **43**, 356–358 (1965).

27. Johnson, A.I., and W.F. Furter, *Can. J. Chem. Eng.,* **38**, 78–87 (1960).

28. Furter, W.F., and R.A. Cook, *Int. J. Heat Mass Transfer,* **10**, 23–36 (1967).

29. Furter, W.F., *Can. J. Chem. Eng.*, **55**, 229–239 (1977).

30. Furter, W.F., *Chem. Eng. Commun.*, **116**, 35 (1992).

31. Kumar, A., *Sep. Sci. and Tech.*, **28**, 1799–1818 (1993).

32. Mahapatra, A., V.G. Gaikar, and M.M. Sharma, *Sep. Sci. and Tech.*, **23**, 429–436 (1988).

33. Agarwal, M., and V.G. Gaikar, *Sep. Technol.*, **2**, 79–84 (1992).

34. Knapp, J.P., and M.F. Doherty, *Ind. Eng. Chem. Res.*, **31**, 346–357 (1992).

35. Lewis, W.K., U.S. Patent 1,676,700 (1928).

36. Van Winkle, M., *Distillation*, McGraw-Hill, New York (1967).

37. Robinson, C.S., and E.R. Gilliland, *Elements of Fractional Distillation*, 4th ed., McGraw-Hill, New York (1950).

38. Wahnschafft, O.M., and A.W. Westerberg, *Ind. Eng. Chem. Res.*, **32**, 1108 (1993).

39. Laroche, L., N. Bekiaris, H.W. Andersen, and M. Morari, *AIChE J.*, **38**, 1309 (1992).

40. Young, S., *J. Chem. Soc. Trans.*, **81**, 707–717 (1902).

41. Keyes, D.B., U.S. Patent 1,676,735 (1928).

42. Keyes, D.B., *Ind. Eng. Chem.*, **21**, 998–1001 (1929).

43. Bekiaris, N., G.A. Meski, and M. Morari, *Ind. Eng. Chem. Res.*, **35**, 207–217 (1996).

44. Prokopakis, G.J., and W.D. Seider, *AIChE J.*, **29**, 49–60 (1983).

45. Ryan, P.J., and M.F. Doherty, *AIChE J.*, **35**, 1592–1601 (1989).

46. Pham, H.N., and M.F. Doherty, *Chem. Eng. Sci.*, **45**, 1845–1854 (1990).

47. Kovach, III, J.W., and W.D. Seider, *Computers and Chem. Engng.*, **11**, 593 (1987).

48. Swartz, C.L.E., and W.E. Stewart, *AIChE J.*, **33**, 1977–1985 (1987).

49. Gani, R., and S.B. Jørgensen, *Computers Chem. Engng.*, **18**, Suppl., S55 (1994).

50. Shewchuk, C.F., "Computation of Multiple Distillation Towers," Ph.D. Thesis, University of Cambridge (1974).

51. Magnussen, T., M.L. Michelsen, and A. Fredenslund, *Inst. Chem. Eng. Symp. Series No. 56, Third International Symp. on Distillation*, Rugby, England (1979).

52. Bekiaris, N., G.A. Meski, C.M. Radu, and M. Morari, *Ind. Eng. Chem. Res.*, **32**, 2023–2038 (1993).

53. Black, C., and D.E. Ditsler, *Advances in Chemistry Series*, ACS, Washington, D.C., Vol. 115, pp. 1–15 (1972).

54. Black, C., R.A. Golding, and D.E. Ditsler, *Advances in Chemistry Series*, ACS, Washington, D.C., Vol. 115, pp. 64–92 (1972).

55. Black, C., *Chem. Eng. Progress*, **76** (9), 78–85 (1980).

56. Black, C., *Ind. Eng. Chem.*, **50**, 403–412 (1958).

57. Terrill, D.L., L.F. Sylvestre, and M.F. Doherty, *Ind. Eng. Chem. Proc. Des. Develop.*, **24**, 1062–1071 (1985).

58. Robinson, C.S., and E.R. Gilliland, *Elements of Fractional Distillation*, 4th ed., McGraw-Hill, New York (1950).

59. Backhaus, A.A., U.S. Patent 1,400,849 (1921).

60. Leyes, C.E., and D.F. Othmer, *Trans. AIChE*, **41**, 157–196 (1945).

61. Belck, L.H., *AIChE J.*, **1**, 467–470 (1955).

62. Masamoto, J., and K. Matsuzaki, *J. Chem. Eng. Japan*, **27**, 1–5 (1994).

63. Agreda, V.H., and L.R. Partin, U.S. Patent 4,435,595 (March 6, 1984).

64. Agreda, V.H., L.R. Partin, and W.H. Heise, *Chem. Eng. Prog.*, **86** (2), 40–46 (1990).

65. Smith, L.A., U.S. Patent 4,307,254 (Dec. 22, 1981).

66. Smith, L.A., U.S. Patent 4,443,559 (April 17, 1984).

67. Smith, L.A., U.S. Patent 4,978,807 (Dec. 18, 1990).

68. DeGarmo, J.L., V.N. Parulekar, and V. Pinjala, *Chem. Eng. Prog.*, **88** (3), 43–50 (1992).

69. Chang, Y.A., and J.D. Seader, *Computers Chem. Engng.*, **12**, 1243–1255 (1988).

70. Venkataraman, S., W.K. Chan, and J.F. Boston, *Chem. Eng. Progress*, **86** (8), 45–54 (1990).

71. Simandl, J., and W.Y. Svrcek, *Computers Chem. Engng.*, **15**, 337–348 (1991).

72. Kang, Y.W., and Y.Y. Lee, and W.K. Lee, *J. Chem. Eng. Japan*, **25**, 649–655 (1992).

73. Izarraraz, A., G.W. Bentzen, R. G. Anthony, and C.D. Holland, *Hydrocarbon Processing*, **59** (6), 195 (1980).

74. Rehfinger, A., and U. Hoffmann, *Chem. Eng. Sci.*, **45**, 1605–1617 (1990).

75. Jacobs, R., and R. Krishna, *Ind. Eng. Chem. Res.*, **32**, 1706–1709 (1993).

76. Nijhuis, S.A., F.P.J.M. Kerkhof, and N.S. Mak, *Ind. Eng. Chem. Res.*, **32**, 2767–2774 (1993).

77. Hauan, S., T. Hertzberg, and K.M. Lien, *Ind. Eng. Chem. Res.*, **34**, 987–991 (1995).

78. Hannay, J.B., and J. Hogarth, *Proc. Roy. Soc.* (*London*) *Sec. A*, **29**, 324 (1879).

79. Williams, D.F., *Chem. Eng. Sci.*, **36**, 1769–1788 (1981).

80. McHugh, M., and V. Krukonis, *Supercritical Fluid Extraction—Principles and Practice*, Butterworths, Boston (1986).

81. McHugh, M., and V. Krukonis, *Supercritical Fluid Extraction—Principles and Practice*, 2nd ed., Butterworth-Heinemann, Boston (1994).

82. Takishima, S., A. Saiki, K. Arai, and S. Saito, *J. Chem. Eng. Japan*, **19**, 48–56 (1986).

83. Inomata, H., A. Kondo, K. Arai, and S. Saito, *J. Chem. Eng. Japan*, **23**, 199–207 (1990).

84. de Fillipi, R.P., and J.E. Vivian, U.S. Patent 4,349,415 (1982).

85. Katz, S.N., J.E. Spence, M.J. O'Brian, R.H. Skiff, G.J. Vogel, and R. Prasad, U.S. Patent 4,911,941 (1990).

86. Wong, D.S.H., and S.I. Sandler, *AIChE J.*, **38**, 671–680 (1992).

87. Sandler, S.I., H. Orbey, and B-I. Lee, in *Models for Thermodynamic and Phase Equilibria Calculations*, S.I. Sandler, Ed., Marcel Dekker, New York, pp. 87–186 (1994).

88. Knapp, H., R. Doring, L. Oellrich, U. Plocker, and J.M. Prausnitz, *Vapor–Liquid Equilibria for Mixtures of Low Boiling Substances*, Chem. Data Ser., Vol. VI, DECHEMA, pp. 771–793 (1982).

89. Shibata, S.K., and S.I. Sandler, *Ind. Eng. Chem. Res.*, **28**, 1893–1898 (1989).

90. Wong, D.S.H., H. Orbey, and S.I. Sandler, *Ind. Eng. Chem. Res.*, **31**, 2033–2039 (1992).

91. Gmehling, J., and U. Onken, *Vapor–Liquid Equilibrium Data Compilation*, DECHEMA Data Series, DECHEMA, Frankfurt (1977).

92. Skjold-Jørgensen, S., *Ind. Eng. Chem. Res.*, **27**, 110–118 (1988).

93. Colussi, I.E., M. Fermeglia, V. Gallo, and I. Kikic, *Computers Chem. Engng.*, **16**, 211–224 (1992).

EXERCISES

Section 11.1

11.1 For the ternary system, normal hexane–methanol–methyl acetate at 1 atm find, in suitable references, all the binary and ternary azeotropes, sketch an approximate residue curve map on a right-triangular diagram, and indicate the distillation boundaries. Determine for each azeotrope and pure component whether it is a stable node, an unstable node, or a saddle.

11.2 For the same ternary system as in Exercise 11.1, use a process simulation program with the UNIFAC equation to calculate a portion of a residue curve at 1 atm starting from a bubble-point liquid with a composition of 20 mol% normal hexane, 60 mol% methanol, and 20 mol% methyl acetate.

11.3 For the same conditions as Exercise 11.2, use a process simulation program with the UNIFAC equation to calculate a portion of a distillation curve at 1 atm.

11.4 For the ternary system acetone, benzene, and *n*-heptane at 1 atm find, in suitable references, all the binary and ternary azeotropes, and sketch an approximate distillation-curve map on an equilateral-triangle diagram, and indicate the distillation boundaries. Determine for each azeotrope and pure component whether it is a stable node, an unstable node, or a saddle.

11.5 For the same ternary system as in Exercise 11.4, use a process simulation program with the UNIFAC equation to calculate a portion of a residue curve at 1 atm starting from a bubble-point liquid with a composition of 20 mol% acetone, 60 mol% benzene, and 20 mol% *n*-heptane.

11.6 For the same conditions as Exercise 11.5, use a process simulation program with the UNIFAC equation to calculate a portion of a distillation curve at 1 atm.

11.7 Develop the feasible product composition regions for the system of Figure 11.13, using Feed F_1.

11.8 Develop the feasible product composition regions for the system of Figure 11.10 if the feed composition is 50 mol% chloroform, 25 mol% methanol, and 25 mol% acetone.

Section 11.2

11.9 Repeat Example 11.3, but with ethanol as the solvent.

11.10 Repeat Example 11.3, but with MEK as the solvent.

11.11 Repeat Example 11.4, but with toluene as the solvent.

11.12 An equimolar mixture of n-heptane and toluene at 200°F, 20 psia, and a flow rate of 400 lbmol/h is to be separated by extractive distillation at 20 psia, using phenol at 220°F as the solvent, at a flow rate of 1200 lbmol/h. Design a suitable two-column system, obtaining reasonable product purities, with only a small loss of solvent

Section 11.4

11.13 Repeat Example 11.5, but with a feed of 100 mol/s of 55 mol% ethanol and 45 mol% benzene.

11.14 Determine the feasibility of separating 100 mol/s of a mixture of 20 mol% ethanol and 80 mol% benzene by pressure-swing distillation. If feasible, design such a system.

11.15 Design a pressure-swing distillation system to produce 99.8 mol% ethanol for 100 mol/s of an aqueous feed containing 30 mol% ethanol.

Section 11.5

11.16 In Example 11.6, a mixture of benzene and cyclohexane is separated in a separation sequence that begins with homogeneous azeotropic distillation using acetone as the entrainer. Can the same separation be achieved using methanol as the entrainer? If not, why not? [Ref.: Ratliff, R.A., and W.B. Strobel, *Petro. Refiner,* **33** (5), 151 (1954)].

11.17 Devise a separation sequence to separate 100 mol/s of an equimolar mixture of toluene and 2,5-dimethylhexane into nearly pure products. Include in the sequence a homogeneous azeotropic distillation column using methanol as the entrainer and determine a feasible design for that column. [Ref.: Benedict, M., and L.C. Rubin, *Trans. AIChE,* **41,** 353–392 (1945)].

11.18 A mixture of 55 wt% methyl acetate and 45 wt% methanol at a flow rate of 16,500 kg/h is to be separated into one product of 99.5 wt% methyl acetate and another product of 99 wt% methanol. It has been suggested that such a separation might be possible by using a sequence of one homogeneous azeotropic distillation column and one ordinary distillation column. Possible entrainers are *n*-hexane, cyclohexane, and toluene. Determine the feasibility of such a sequence. If feasible, prepare a process design. If not feasible, suggest an alternative process and prove its feasibility.

Section 11.6

11.19 Design a three-column distillation sequence to separate 150 mol/s of an azeotropic mixture of ethanol and water at 1 atm into nearly pure ethanol and nearly pure water using heterogeneous azeotropic distillation with benzene as the entrainer.

11.20 Design a three-column distillation sequence to separate 120 mol/s of an azeotropic mixture of isopropanol and water at 1 atm into nearly pure isopropanol and nearly pure water using heterogeneous azeotropic distillation with benzene as the entrainer. [Ref.: Pham, H.N., P.J. Ryan, and M.F. Doherty, *AIChE J.,* **35,** 1585–1591 (1989)].

11.21 Design a two-column distillation sequence to separate 1,000 kmol/h of 20 mol% aqueous acetic acid into nearly pure acetic acid and nearly pure water. The first column should use heterogeneous azeotropic distillation with *n*-propyl acetate as the entrainer.

Section 11.7

11.22 Repeat Example 11.9, with the entire range of methanol feed-stage locations. Compare your results for isobutene conversion with the values shown in Figure 11.39.

11.23 Repeat Exercise 11.22, but with activities, instead of mole fractions, in the reaction rate expressions. How do the results differ? Explain.

11.24 Repeat Exercise 11.22, but with the assumption of chemical equilibrium on stages where catalyst is employed. How do the results differ from Figure 11.39? Explain.

Section 11.8

11.25 Repeat Example 11.10, but with 10 equilibrium stages instead of 5. What is the effect of this change?

11.26 An important application of supercritical extraction is the removal of solutes from particles of porous natural materials. Such applications include the extraction of caffeine from coffee beans and the extraction of ginger oil from ginger root. When CO_2 is used as the solvent, the rate of extraction is found to be independent of the flow rate of CO_2 past the particles, but dependent upon the particle size. Develop a suitable mathematical model for the rate of extraction that is consistent with these observations. What parameter in the model would have to be determined by experiment?

11.27 Cygnarowicz and Seider [*Biotechnol. Prog.,* **6,** 82–91 (1990)] present a process design for the supercritical extraction of β-carotene from water using the GC-EOS method of Skjold-Jørgensen to estimate phase equilibria. Repeat the calculations for the conditions of their design using the Peng–Robinson EOS with the Wong–Sandler mixing rules. How do the two designs compare?

11.28 Cygnarowicz and Seider [*Ind. Eng. Chem. Res.,* **28,** 1497–1503 (1989)] present a process design for the supercritical extraction of acetone from water using the GC-EOS method of Skjold-Jørgensen to estimate phase equilibria. Repeat the calculations for the conditions of their design using the Peng–Robinson EOS with the Wong–Sandler mixing rules. How do the two designs compare?

Chapter 12

Rate-Based Models for Distillation

Chapter 10 contains rigorous equilibrium-based models for continuous-flow, steady-state, multicomponent, multistage distillation, absorption, stripping, and liquid–liquid extraction based on component material balances, energy balances, and thermodynamic correlations and criteria for phase equilibria. These models are extended in Chapter 11 to supercritical extraction and enhanced distillation, including extractive distillation, azeotropic distillation, and reactive distillation. The fundamental equations for the equilibrium-based models were first published by Sorel [1] in 1893. His equations consisted of total and component material balances around top and bottom sections of equilibrium stages (theoretical plates), including a total condenser and a reboiler, and corresponding energy balances that included provision for heat losses, which are an important factor for small columns. Sorel used graphs of phase-equilibrium data instead of equations. Because of the complexity of Sorel's model, it was not widely applied until 1921, when it was adapted to graphical solution techniques for binary systems, first by Ponchon and then by Savarit, who used an enthalpy–concentration diagram. In 1925, a much simpler, but less rigorous, graphical technique was developed by McCabe and Thiele, who eliminated the energy balances by assuming constant vapor and liquid molar flow rates from equilibrium stage to equilibrium stage except across feed or side-stream withdrawal stages. This is referred to as the constant-molar-overflow assumption. When applicable, the McCabe–Thiele graphical method, developed in detail in Chapter 7, is applied even today for binary distillation, because the method gives valuable insight into changes in phase compositions from stage to stage.

Because some of Sorel's equations are nonlinear, it is not possible to obtain algebraic solutions, unless simplifying assumptions are made. A notable achievement in this respect was made by Smoker [2] in 1938 for the distillation of a binary mixture by assuming not only constant molar overflow, but also constant relative volatility between the two components. Smoker's equation is still useful for superfractionators involving close-boiling binary mixtures, where that assumption is valid. Starting in 1932, two iterative, numerical methods were developed for obtaining a general solution to Sorel's model for the distillation of multicomponent mixtures. The Thiele–Geddes method [3] requires specification of the number of equilibrium stages, the feed stage, the reflux ratio, and the distillate flow rate, with the resulting distribution of the components between distillate and bottoms being calculated. The Lewis–Matheson method [4] computes the number of stages required and the location of the feed stage for a specified reflux ratio and split between two key components. These two methods were widely used for the simulation and design of single-feed multicomponent distillation columns prior to the 1960s.

Attempts in the late 1950s and early 1960s to adapt the Thiele–Geddes and Lewis–Matheson methods to computations with a digital computer had limited success. The real breakthrough in computerization of equilibrium-stage calculations occurred when Amundson and co-workers, starting in 1958, applied techniques of matrix algebra. This led to a number of successful computer-aided methods, based on sparse-matrix algebra, for Sorel's equilibrium-based model. The most important of these models are presented in Chapter 10. Today, computer-aided design and simulation programs abound for the rigorous, iterative numerical solution of Sorel's equilibrium-based model for a wide variety of column configurations and specifications. Although the iterative computations sometimes fail to converge, the methods are widely applied and have become more flexible and robust with each passing year.

The methods presented in Chapters 10 and 11 assume that equilibrium is achieved, at each stage, with respect to both heat and component mass transfer. Except when temperature changes significantly from stage to stage, the assumption of temperature equality for vapor and liquid phases leaving a stage is usually acceptable. However, in most industrial applications, the assumption of equilibrium with respect to exiting phase compositions is not reasonable. In general, exiting vapor-phase mole fractions are not related to exiting liquid-phase mole fractions by thermodynamic K-values. To overcome this limitation of equilibrium-based models, Lewis [5], in 1922, proposed the use of an overall stage efficiency for converting theoretical stages to actual stages. Unfortunately, experimental data show that this efficiency varies, depending on the application, over a range of about 5 to 120%, where the high values are achieved for distillation in large-diameter, single liquid-pass trays because of a cross-flow effect, whereas the lower values occur in absorption columns when a high-viscosity, high-molecular-weight absorbent is used.

A preferred procedure for accounting for nonequilibrium with respect to mass transfer, since its introduction by Murphree [6] in 1925, has been to incorporate the Murphree vapor-phase tray efficiency, $(E_{MV})_{i,j}$, directly into Sorel's model as a replacement for the equilibrium equation based on definition of the K-value. Thus, the equation

$$K_{i,j} = y_{i,j}/x_{i,j} \tag{12-1}$$

where i refers to the component and j the stage, is replaced by

$$(E_{MV})_{i,j} = (y_{i,j} - y_{i,j+1})/(y_{i,j}^* - y_{i,j+1}) \tag{12-2}$$

Thus, the efficiency is the ratio of the actual change in vapor-phase mole fraction to the change that would occur if equilibrium were achieved. The equilibrium value, $y_{i,j}^*$, is obtained from (12-1), with substitution into (12-2) giving

$$(E_{MV})_{i,j} = (y_{i,j} - y_{i,j+1})/(K_{i,j}x_{i,j} - y_{i,j+1}) \tag{12-3}$$

Equations (12-2) and (12-3) assume the following:

1. Uniform concentrations of vapor and liquid streams entering into and exiting a tray
2. Complete mixing throughout the liquid flowing across the tray
3. Plug flow of the vapor up through the liquid
4. Negligible resistance to mass transfer in the liquid phase

Application of the Murphree efficiency using empirical correlations has proved to be adequate for binary and close-boiling, ideal, and near-ideal multicomponent vapor–liquid mixtures. However, deficiencies of the Murphree efficiency for general multicomponent vapor–liquid mixtures have long been recognized. Murphree himself stated clearly the deficiencies of his development for multicomponent mixtures and for cases where the efficiency is low. He even stated that the theoretical plate should not be the basis of calculation for ternary mixtures.

When the equilibrium-based model is applied to multicomponent mixtures, a number of problems arise. Values of E_{MV} differ from component to component and vary from stage to stage. But, at each stage, the number of independent values of E_{MV} must be determined so as to force the sum of the mole fractions in the vapor phase to sum to 1. This introduces the possibility that negative values of E_{MV} can result. This is in contrast to binary mixtures for which the values of E_{MV} are always positive and are identical for the two components. When using the Murphree vapor-phase efficiency, the temperatures of the exiting vapor and liquid phases are assumed to be the same and equal to the bubble-point temperature of the exiting liquid phase. Because the vapor phase is not in equilibrium with the liquid phase, the vapor temperature does not correspond to the dew-point temperature. It is even possible, algebraically, for the vapor temperature to correspond to a value below its dew-point temperature, which is physically impossible.

Values of E_{MV} can be obtained from experimental data or correlations. These values, however, are more likely to be Murphree vapor point (rather than tray) efficiencies. Point efficiencies only apply to a particular location in the liquid on the tray. To convert these point efficiencies to tray efficiencies, vapor and liquid flow patterns must be assumed after the manner of Lewis [7], as discussed by Seader [8]. However, if the vapor and liquid phases are both completely mixed, the point efficiency equals the tray efficiency.

Walter and Sherwood [9] found that experimentally measured tray efficiencies covered an enormous range: 0.65 to 4.2% for absorption and stripping of carbon dioxide from water and glycerine solutions; 4.7 to 24% for absorption of olefins into oils; and 69 to 92% for absorption of ammonia, humidification of air, and rectification of alcohol.

In 1957, Toor [10] showed that diffusion in a ternary mixture is enormously more complex than in a binary mixture because of coupling among component concentration gradients, especially when components differ widely in size, shape, and polarity. Toor showed that, in addition to diffusion due to the conventional Fickian concentration driving force, the possible consequences of gradient coupling could result in: (1) diffusion against a driving force (reverse diffusion), (2) no diffusion even though a concentration driving force is present (diffusion barrier), and (3) diffusion with zero driving force (osmotic diffusion). Theoretical calculations by Toor and Burchard [11] predicted the possibility of negative values of E_{MV} in multicomponent systems, but values of E_{MV} for binary systems are restricted to the range from 0 to 100%.

In 1977, Krishna et al. [12] extended the theoretical work of Toor and Buchard and showed that when the vapor mole-fraction driving force of a component (call it A) is small compared to the other components in the mixture, the transport rate of A is controlled by the other components, with the result that E_{MV} for A is anywhere in the range from minus infinity to plus infinity. They confirmed this theoretical prediction by conducting experiments with the ethanol/*tert*-butanol/water system and obtained values of E_{MV} for *tert*-butanol ranging from −2,978% to +527%. In addition, values of E_{MV} for ethanol and water sometimes differed significantly.

Two other tray efficiencies are defined in the literature: the vaporization efficiency of Holland, which was first mentioned by McAdams, and the Hausen tray efficiency, which eliminates the assumption in E_{MV} that the exiting liquid is at its bubble point. The former cannot distinguish the Toor phenomena and can vary widely in a manner that is not ascribable to the particular component. The latter does appear to be superior to E_{MV}, but is considerably more complicated and difficult to use, and it has not found wide application.

Although the equilibrium-based model, modified to incorporate stage efficiency, is adequate for binary mixtures and for the major components in nearly ideal multicomponent mixtures, that model has serious deficiencies for more general cases and the development of a more realistic nonequilibrium, transport- or rate-based model has long been a desirable goal. In 1977, Waggoner and Loud [13] developed a rate-based, mass-transport model limited to nearly ideal, close-boiling, multicomponent systems. However, an energy-transport equation was not included (because thermal equilibrium would be closely approximated for a close-boiling mixture) and the coupling of component mass-transfer rates was ignored.

In 1979, Krishna and Standart [14] showed the possibility of applying rigorous multicomponent mass- and heat-transfer theory to calculations of simultaneous transport. The theory was further developed by Taylor and Krishna [15]. The availability of this theory led to the development in 1985 by Krishnamurthy and Taylor [16] of the first general rate-based, computer-aided model for application to trayed and packed columns for distillation and other continuous, countercurrent vapor–liquid separation operations. This model applies the two-film theory of mass transfer discussed in Chapter 3, with phase equilibria assumed at the interface of the two phases, and provides options for vapor and liquid flow configurations in trayed columns, including plug flow and perfectly mixed flow, on each tray. Although the model does not require tray efficiencies or values of HETP, correlations of mass-transfer and heat-transfer coefficients are needed for the particular type of trays or packing employed. The model was extended in 1994 by Taylor, Kooijman, and Hung [17] to include: (1) effects of entrainment of liquid droplets in the vapor and occlusion of vapor bubbles in the liquid; (2) estimation of the column pressure profile; (3) interlinking streams; and (4) axial dispersion in packed columns. In addition, unlike the 1985 model, which required the user to specify the column diameter and tray geometry or packing size, the 1994 version includes a design mode that estimates column diameter for a specified fraction of flooding or pressure drop. Rate-based models are implemented in several computer programs, including RATEFRAC [18] of Aspen Technology, and ChemSep Release 3.1 [19].

12.1 RATE-BASED MODEL

A schematic diagram of a nonequilibrium stage, consisting of a tray, a group of trays, or a segment of a packed section, is shown in Figure 12.1. Entering stage j, at pressure P_j, are liquid F_j^L and/or vapor F_j^V molar flow rates; component i molar flow rates, $f_{i,j}^L$ and $f_{i,j}^V$; and stream molar enthalpies, H_j^{LF} and H_j^{VF}. Also leaving from $(+)$ or entering to $(-)$ the liquid and/or vapor phases in the stage are heat transfer rates Q_j^V and Q_j^L, respectively. Also entering the stage from the stage above is liquid molar flow rate L_{j-1} at temperature T_{j-1}^L and pressure P_{j-1}, with molar enthalpy H_{j-1}^L and component mole fractions $x_{i,j-1}$; and entering the stage from the stage below is vapor molar flow rate V_{j+1} at temperature T_{j+1}^V and pressure P_{j+1}, with molar enthalpy H_{j+1}^V and component mole fractions $y_{i,j+1}$. Within the stage, mass transfer of components occurs across the phase boundary at molar

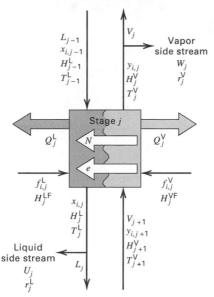

Figure 12.1 Nonequilibrium stage for rate-based method.

rates $N_{i,j}$ from the vapor phase to the liquid phase $(+)$ or vice versa $(-)$, and heat transfer occurs across the phase boundary at rates e_j from the vapor phase to the liquid phase $(+)$ or vice versa $(-)$. Leaving the stage is liquid at temperature T_j^L and pressure P_j, with molar enthalpy H_j^L; and vapor at temperature T_j^V and pressure P_j, with molar enthalpy H_j^V. A fraction, r_j^L, of the liquid exiting the stage may be withdrawn as a liquid side stream at molar flow rate U_j, leaving the molar flow rate L_j to enter the stage below or to exit the column. A fraction, r_j^V, of the vapor exiting the stage may be withdrawn as a vapor side stream at molar flow rate W_j, leaving the molar flow rate V_j to enter the stage above or to exit the column. If desired, entrainment, occlusion, interlink flows, a second immiscible liquid phase, and chemical reaction(s) can be added to the model.

Recall that the equilibrium-stage model of Chapter 10 utilizes the $2C + 3$ MESH equations for each stage:

C mass balances for components
C phase equilibria relations
2 summations of mole fractions
1 energy balance

In the rate-based model, the mass and energy balances around each equilibrium stage are each replaced by separate balances for each phase around a stage, which can be a tray, a collection of trays, or a segment of a packed section. In residual form, the equations are as follows, where the residuals are on the left-hand sides and become zero when the computations are converged. When not converged, the residuals are used to determine the proximity to convergence.

Liquid-phase component material balance:

$$M_{i,j}^L \equiv (1 + r_j^L)L_j x_{i,j} - L_{j-1}x_{i,j-1} - f_{i,j}^L - N_{i,j}^L = 0, \qquad i = 1, 2, \ldots, C \qquad \textbf{(12-4)}$$

Vapor-phase component material balance:

$$M_{i,j}^V \equiv (1 + r_j^V)V_j y_{i,j} - V_{j+1}y_{i,j+1} - f_{i,j}^V + N_{i,j}^V = 0, \qquad i = 1, 2, \ldots, C \qquad \textbf{(12-5)}$$

Liquid-phase energy balance:

$$E_j^L \equiv (1 + r_j^L)L_j H_j^L - L_{j-1}H_{j-1}^L - \left(\sum_{i=1}^{C} f_{i,j}^L\right)H_j^{LF} + Q_j^L - e_j^L = 0 \qquad \textbf{(12-6)}$$

Vapor-phase energy balance:

$$E_j^V \equiv (1 + r_j^V)V_jH_j^V - V_{j+1}H_{j+1}^V - \left(\sum_{i=1}^{C} f_{i,j}^V\right)H_j^{VF} + Q_j^V + e_j^V = 0 \tag{12-7}$$

where at the phase interface, I,

$$E_j^I = e_j^V - e_j^L = 0 \tag{12-8}$$

Equations (12-4) and (12-5) are coupled by the component mass-transfer rates:

$$R_{i,j}^L \equiv N_{i,j} - N_{i,j}^L = 0, \qquad i = 1, 2, \ldots, C - 1 \tag{12-9}$$

$$R_{i,j}^V \equiv N_{i,j} - N_{i,j}^V = 0, \qquad i = 1, 2, \ldots, C - 1 \tag{12-10}$$

The equations for the mole-fraction summation for each phase are applied at the vapor–liquid interface:

$$S_j^{LI} \equiv \sum_{i=1}^{C} x_{i,j}^I - 1 = 0 \tag{12-11}$$

$$S_j^{VI} \equiv \sum_{i=1}^{C} y_{i,j}^I - 1 = 0 \tag{12-12}$$

A hydraulic equation for stage pressure drop is given by

$$H_j \equiv P_{j+1} - P_j - (\Delta P_j) = 0, \qquad j = 1, 2, 3, \ldots, N - 1 \tag{12-13}$$

where the stage is assumed to be at mechanical equilibrium such that

$$P_j^L = P_j^V = P_j \tag{12-14}$$

and ΔP_j is the gas-phase pressure drop from stage $j + 1$ to stage j. Equation (12-13) is optional. It is only included when it is desired to compute one or more stage pressures from hydraulics, as discussed in Chapter 6. Phase equilibrium for each component is assumed to exist only at the phase interphase:

$$Q_{i,j}^I \equiv K_{i,j}x_{i,j}^I - y_{i,j}^I = 0, \qquad i = 1, 2, \ldots, C \tag{12-15}$$

Because only $C - 1$ equations are written for the component mass-transfer rates in (12-9) and (12-10), total phase material balances in terms of total mass-transfer rates, $N_{T,j}$, can be added to the system:

$$M_{T,j}^L \equiv (1 + r_j^L)L_j - L_{j-1} - \sum_{i=1}^{C} f_{i,j}^L - N_{T,j} = 0 \tag{12-16}$$

$$M_{T,j}^V \equiv (1 + r_j^V)V_j - V_{j+1} - \sum_{i=1}^{C} f_{i,j}^V + N_{T,j} = 0 \tag{12-17}$$

where

$$N_{T,j} = \sum_{i=1}^{C} N_{i,j} \tag{12-18}$$

Equations (12-4), (12-5), (12-9), (12-10), (12-16), (12-17), and (12-18) contain terms for component mass-transfer rates, estimated from diffusive and bulk-flow (convective) contributions. The former are based on interfacial area, average mole-fraction driving forces, and mass-transfer coefficients that account for component coupling effects through binary-pair coefficients. Empirical equations are used for the interfacial area and binary mass-transfer coefficients, based on correlations of experimental data for bubble-cap trays, sieve trays, valve trays, random packings, and structured packings. The average mole-fraction driving forces for diffusion depend upon the assumed vapor and liquid flow patterns. The simplest case is perfectly mixed flow for both the vapor and liquid phases,

Table 12.1 Summary of Independent Equations for Rate-Based Model

Equation	No. of Equations
$M_{i,j}^{L}$	C
$M_{i,j}^{V}$	C
$M_{T,j}^{L}$	1
$M_{T,j}^{V}$	1
E_{j}^{L}	1
E_{j}^{V}	1
E_{j}^{I}	1
$R_{i,j}^{L}$	$C-1$
$R_{i,j}^{V}$	$C-1$
S_{j}^{LI}	1
S_{j}^{VI}	1
H_{j}	(optional)
$Q_{i,j}^{I}$	C
	$5C+5$

which simulates small-diameter trayed columns. The case of countercurrent plug flow for vapor and liquid phases simulates a packed column with no axial dispersion.

Equations (12-6) to (12-8) contain terms for heat-transfer rates. These are estimated from convective and enthalpy flow contributions, where the former are based on interfacial area, average temperature driving forces, and convective heat-transfer coefficients, estimated from the Chilton–Colburn analogy for the vapor phase and the penetration theory for the liquid phase.

The K-values in (12-15) are estimated from the same equation-of-state or activity-coefficient models used with equilibrium-based models. Tray or packed-segment pressure drops are estimated from suitable correlations of the type discussed in Chapter 6.

The total number of independent equations, referred to as the MERSHQ equations, for each nonequilibrium stage, is $5C + 5$, as listed in Table 12.1. These equations apply for N stages, that is, $N_E = N(5C + 5)$ equations, in terms of $7NC + 14N + 1$ variables, listed in Table 12.2. The number of degrees of freedom is

$$N_D = N_V - N_E = (7NC + 14N + 1) - (5NC + 5N) = 2NC + 9N + 1$$

If variable types 1 to 7, 10, 14, and 16 to 18 in Table 12.2 are specified, a total of $2NC + 9N + 1$ variables are assigned values and the degrees of freedom are totally consumed. Then, the remaining $5C + 5$ independent variables in the $5C + 5$ equations are

$$x_{i,j}, y_{i,j}, x_{i,j}^{I}, y_{i,j}^{I}, N_{i,j}, T_{j}^{L}, T_{j}^{V}, T_{i}^{I}, L_{j}, \text{ and } V_{j}$$

which are the variables to be computed from the equations. Properties $K_{i,j}^{I}, H_{j}^{LF}, H_{j}^{VF}, H_{j}^{L}$, and H_{j}^{V} are computed from thermodynamic correlations in terms of the remaining independent variables. Transport rates, $N_{i,j}^{L}, N_{i,j}^{V}, e_{j}^{L}$, and e_{j}^{V} are computed from transport correlations and certain physical properties, in terms of the remaining independent variables. Stage pressures are computed from pressure drops, ΔP_{j}, stage geometry, fluid-mechanic equations, and certain physical properties, in terms of the remaining independent variables.

For a distillation column, it is preferable to specify $Q_{1}^{L} = 0$ and $Q_{N}^{V} = 0$. In that case, Q_{1}^{V} (heat-transfer rate from the vapor in the condenser) and Q_{N}^{L} (heat-transfer rate to the liquid in the reboiler) are not specified, but instead, as in the case of a column with a

Table 12.2 List of Variables for Rate-Based Model

Variable Type No.	Variable	No. of Variables
1	No. of stages, N	1
2	$f_{i,j}^{L}$	NC
3	$f_{i,j}^{V}$	NC
4	T_{j}^{LF}	N
5	T_{j}^{VF}	N
6	P_{j}^{LF}	N
7	P_{j}^{VF}	N
8	L_{j}	N
9	$x_{i,j}$	NC
10	r_{j}^{L}	N
11	T_{j}^{L}	N
12	V_{j}	N
13	$y_{i,j}$	NC
14	r_{j}^{V}	N
15	T_{j}^{V}	N
16	P_{j}	N
17	Q_{j}^{L}	N
18	Q_{j}^{V}	N
19	$x_{i,j}^{I}$	NC
20	$y_{i,j}^{I}$	NC
21	T_{j}^{I}	N
22	$N_{i,j}$	NC
		$N_V = 7NC + 14N + 1$

partial condenser, L_1 (reflux rate) and L_N (bottoms flow rate) are substitute specifications, which are sometimes referred to as standard specifications for ordinary distillation. For an adiabatic absorber or adiabatic stripper, however, all Q_j^L and Q_j^V are set equal to zero, with no substitution of specifications.

12.2 THERMODYNAMIC PROPERTIES AND TRANSPORT-RATE EXPRESSIONS

Rate-based models use the same K-value and enthalpy correlations as equilibrium-based models. However, the K-values apply only at the equilibrium interface between the vapor and liquid phases on trays or in packing. In general, the K-value correlation, whether based on an equation-of-state or an activity-coefficient model, is a function of phase-interface temperature and compositions, and tray pressure. Enthalpies are evaluated only at the conditions of the phases as they exit a tray. For the equilibrium-based model, the vapor is at the dew-point temperature and the liquid is at the bubble-point temperature, where both temperatures are equal and at the stage temperature. For the rate-based model, the liquid is subcooled and the vapor superheated.

The accuracy of enthalpies and, particularly, K-values is crucial to equilibrium-based models. For rate-based models, accurate predictions of heat-transfer rates and, particularly, mass-transfer rates are also required. These rates depend upon transport coefficients, interfacial area, and driving forces. It is important that mass-transfer rates account for component-coupling effects through binary-pair coefficients.

The general forms for component mass-transfer rates across the vapor and liquid films, respectively, on a tray or in a packed segment, are as follows, where both diffusive and convective (bulk-flow) contributions are included:

$$N_{i,j}^V = a_j^I J_{i,j}^V + y_{i,j} N_{T,j} \tag{12-19}$$

and

$$N_{i,j}^L = a_j^I J_{i,j}^L + x_{i,j} N_{T,j} \tag{12-20}$$

where a_j^I is the total interfacial area for the stage and $J_{i,j}^P$ is the molar diffusion flux relative to the molar average velocity, where P stands for the phase (V or L). For a binary mixture, as discussed in Chapter 3, these fluxes, in terms of mass-transfer coefficients, are given by

$$J_i^V = c_t^V k_i^V (y_i^V - y_i^I)_{avg} \tag{12-21}$$

and

$$J_i^L = c_t^L k_i^L (x_i^I - x_i^L)_{avg} \tag{12-22}$$

where c_t^P is the total molar concentration, k_i^P is the mass-transfer coefficient for a binary mixture based on a mole-fraction driving force, and the last terms in (12-21) and (12-22) are the mean mole-fraction driving forces over the stage. The positive direction of mass transfer is assumed to be from the vapor phase to the liquid phase. From the definition of the molar diffusive flux:

$$\sum_{i=1}^{C} J_i = 0 \tag{12-23}$$

Thus, for the binary system (1, 2), $J_1 = -J_2$.

As discussed in detail by Taylor and Krishna [15], the general multicomponent case for mass transfer is considerably more complex than the binary case because of component-coupling effects. For example, for the ternary system (1, 2, 3), from Taylor and Krishna [15], the fluxes for the first two components are

$$J_1^V = c_t^V \kappa_{11}^V (y_1^V - y_1^I)_{avg} + c_t^V \kappa_{12}^V (y_2 - y_2^I)_{avg} \tag{12-24}$$

$$J_2^V = c_t^V \kappa_{21}^V (y_1^V - y_1^I)_{avg} + c_t^V \kappa_{22}^V (y_2 - y_2^I)_{avg} \tag{12-25}$$

The flux for the third component is not independent of the other two, but is obtained from (12-23):

$$J_3^V = -J_1^V - J_2^V \tag{12-26}$$

In these equations, the binary-pair coefficients, κ^P, are complex functions related to inverse rate functions described below and are called Maxwell–Stefan mass-transfer coefficients in binary mixtures.

For the general multicomponent system (1, 2, . . . , C, the independent fluxes for the first $C - 1$ components are given in matrix equation form as

$$\boldsymbol{J}^V = c_t^V [\boldsymbol{\kappa}^V](\boldsymbol{y}^V - \boldsymbol{y}^I)_{avg} \tag{12-27}$$

$$\boldsymbol{J}^L = c_t^L [\boldsymbol{\kappa}^L](\boldsymbol{x}^I - \boldsymbol{x}^L)_{avg} \tag{12-28}$$

where $\boldsymbol{J}^P$, $(\boldsymbol{y}^V - \boldsymbol{y}^I)_{avg}$, and $(\boldsymbol{x}^I - \boldsymbol{x}^L)_{avg}$ are column vectors of length $C - 1$ and $[\boldsymbol{\kappa}^P]$ is a $(C - 1) \times (C - 1)$ square matrix. The method for determining the average mole-fraction driving forces depends, as discussed in the next section, upon the flow patterns of the vapor and liquid phases.

The most fundamental theory for multicomponent diffusion is that of Maxwell and Stefan, who, in the period from 1866 to 1871, applied the kinetic theory of ideal gases. Their theory is presented most conveniently in terms of rate coefficients, $\boldsymbol{B}$, which are defined in reciprocal diffusivity terms [15]. Likewise, it is convenient to determine $[\boldsymbol{\kappa}^P]$

from a reciprocal mass-transfer coefficient function, R, defined by Krishna and Standart [14]. For an ideal gas solution:

$$[\kappa^V] = [R^V]^{-1} \tag{12-29}$$

For a nonideal liquid solution:

$$[\kappa^L] = [R^L]^{-1}[\Gamma^L] \tag{12-30}$$

where the elements of R^P in terms of general mole fractions, z_i, are:

$$R_{ii}^P = \frac{z_i}{k_{iC}^P} + \sum_{\substack{k=1 \\ k \neq i}}^{C} \frac{z_k}{k_{ik}^P} \tag{12-31}$$

$$R_{ij}^P = -z_i \left(\frac{1}{k_{ij}^P} - \frac{1}{k_{iC}^P} \right) \tag{12-32}$$

where here, j refers to the jth component and not the jth stage and the values of k are binary-pair mass-transfer coefficients obtained from correlations of experimental data.

For the four-component vapor-phase system, the combination of (12-27) and (12-29) gives

$$\begin{bmatrix} J_1^V \\ J_2^V \\ J_3^V \end{bmatrix} = c_t^V \begin{bmatrix} R_{11}^V & R_{12}^V & R_{13}^V \\ R_{21}^V & R_{22}^V & R_{23}^V \\ R_{31}^V & R_{32}^V & R_{33}^V \end{bmatrix}^{-1} \begin{bmatrix} (y_1^V - y_1^I)_{\text{avg}} \\ (y_2^V - y_2^I)_{\text{avg}} \\ (y_3^V - y_3^I)_{\text{avg}} \end{bmatrix} \tag{12-33}$$

with

$$J_4^V = -(J_1^V + J_2^V + J_3^V) \tag{12-34}$$

and, for example, from (12-32) and (12-33), respectively:

$$R_{11}^V = \frac{y_1}{k_{14}^V} + \frac{y_2}{k_{12}^V} + \frac{y_3}{k_{13}^V} + \frac{y_4}{k_{14}^V} \tag{12-35}$$

$$R_{12}^V = -y_1 \left(\frac{1}{k_{12}^V} - \frac{1}{k_{14}^V} \right) \tag{12-36}$$

The term $[\Gamma^L]$ in (12-30) is a $(C-1) \times (C-1)$ matrix of thermodynamic factors that corrects for nonideality, which often is a necessary correction for the liquid phase. When an activity-coefficient model is used:

$$\Gamma_{ij}^L = \delta_{ij} + x_i \left(\frac{\partial \ln \gamma_i}{\partial x_j} \right)_{T,P,x_k,k \neq j=1,\ldots,C-1} \tag{12-37}$$

For a nonideal vapor, a $[\Gamma^V]$ term can be included in (12-29), but this is rarely necessary. For either phase, if an equation-of-state model is used, (12-37) can be rewritten by substituting $\overline{\phi}_i$, the mixture fugacity coefficient, for γ_i. The term δ_{ij} is the Kronecker delta, which is 1 if $i = j$ and 0 if not. The thermodynamic factor is required because it is generally accepted that the fundamental driving force for diffusion is the gradient of the chemical potential rather than the mole fraction or concentration gradient.

When mass-transfer fluxes are moderate to high, an additional correction term is needed in (12-29) and (12-30) to correct for distortion of the composition profiles. This correction, which can have a serious effect on the results, is discussed in detail by Taylor and Krishna [15]. The calculation of the low mass-transfer fluxes, according to (12-19) to (12-32), is illustrated by the following example.

EXAMPLE 12.1

This example is similar to Example 11.5.1 on page 283 of Taylor and Krishna [15]. The following results were obtained for tray n from a rate-based calculation of a ternary distillation at 14.7 psia, involving acetone (1), methanol (2), and water (3) in a 5.5-ft diameter column using sieve trays with a 2-inch-high weir. Vapor and liquid phases are assumed to be completely mixed.

Component	y_n	y_{n+1}	y_n^I	K_n^I	x_n
1	0.2971	0.1700	0.3521	2.759	0.1459
2	0.4631	0.4290	0.4677	1.225	0.3865
3	0.2398	0.4010	0.1802	0.3673	0.4676
	1.0000	1.0000	1.0000		1.0000

The computed products of the gas-phase binary mass-transfer coefficients and interfacial area, using the Chan–Fair correlations, are as follows in lbmol/(h-unit mole fraction):

$$k_{12} = k_{21} = 1,955; \quad k_{13} = k_{31} = 2,407; \quad k_{23} = k_{32} = 2,797$$

(a) Compute the molar diffusion rates.
(b) Compute the mass-transfer rates.
(c) Calculate the Murphree vapor tray efficiencies.

SOLUTION

Because rates instead of fluxes are given, the equations developed in this section are used with rates rather than fluxes.

(a) Compute the reciprocal rate functions, **R**, from (12-31) and (12-32), assuming linear mole fraction gradients such that z_i can be replaced by $(y_i + y_i^I)/2$.
Thus:

$$z_1 = (0.2971 + 0.3521)/2 = 0.3246$$
$$z_2 = (0.4631 + 0.4677)/2 = 0.4654$$
$$z_3 = (0.2398 + 0.1802)/2 = 0.2100$$

$$R_{11}^V = \frac{z_1}{k_{13}} + \frac{z_2}{k_{12}} + \frac{z_3}{k_{13}} = \frac{0.3246}{2,407} + \frac{0.4654}{1,955} + \frac{0.2100}{2,407} = 0.000460$$

$$R_{22}^V = \frac{z_2}{k_{23}} + \frac{z_1}{k_{21}} + \frac{z_3}{k_{23}} = \frac{0.4654}{2,797} + \frac{0.3246}{1,955} + \frac{0.2100}{2,797} = 0.000408$$

$$R_{12}^V = -z_1 \left(\frac{1}{k_{12}} - \frac{1}{k_{13}} \right) = -0.3246 \left(\frac{1}{1,955} - \frac{1}{2,407} \right) = -0.0000312$$

$$R_{21}^V = -z_2 \left(\frac{1}{k_{21}} - \frac{1}{k_{23}} \right) = -0.4654 \left(\frac{1}{1,955} - \frac{1}{2,797} \right) = -0.0000717$$

Thus, in matrix form: $[\mathbf{R}^V] = \begin{bmatrix} 0.000460 & -0.0000312 \\ -0.0000717 & 0.000408 \end{bmatrix}$

From (12-29), by matrix inversion: $[\mathbf{\kappa}^V] = [\mathbf{R}^V]^{-1} = \begin{bmatrix} 2,200 & 168.2 \\ 386.6 & 2,480 \end{bmatrix}$

Because the off-diagonal terms in the preceding 2 × 2 matrix are much smaller that the diagonal terms, the effect of coupling in this example is small.

From (12-27):

$$\begin{bmatrix} J_1^Y \\ J_2^Y \end{bmatrix} = \begin{bmatrix} \kappa_{11}^Y & \kappa_{12}^Y \\ \kappa_{21}^Y & \kappa_{22}^Y \end{bmatrix} \begin{bmatrix} (y_1 - y_1^I) \\ (y_2 - y_2^I) \end{bmatrix}$$

$$J_1^Y = \kappa_{11}^Y(y_1 - y_1^I) + \kappa_{12}^Y(y_2 - y_2^I)$$

$$= 2,200(0.2971 - 0.3521) + 168.2(0.4631 - 0.4677) = -121.8 \text{ lbmol/h}$$

$$J_2^Y = \kappa_{21}^Y(y_1 - y_1^I) + \kappa_{22}^Y(y_2 - y_2^I)$$

$$= 386.6(0.2971 - 0.3521) + 2,480(0.4631 - 0.4677) = -32.7 \text{ lbmol/h}$$

From (12-23): $J_3^Y = -J_1^Y - J_2^Y = 121.8 + 32.7 = 154.5 \text{ lbmol/h}$

(b) From (12-19), but with diffusion and mass-transfer rates instead of fluxes:

$$N_1^Y = J_1^Y + z_1 N_T^Y = -121.8 + 0.3246 N_T^Y \qquad (1)$$

Similarly:

$$N_2^Y = -32.7 + 0.4654 N_T^Y \qquad (2)$$

$$N_3^Y = 154.5 + 0.2100 N_T^Y \qquad (3)$$

To determine the component mass-transfer rates, it is necessary to know the total mass-transfer rate for the tray, N_T^Y. The problem of determining this quantity when the diffusion rates, J, are known is referred to as the *bootstrap problem* (p. 145 in Taylor and Krishna [15]). In chemical reaction with diffusion, N_T is determined by the stoichiometry. In distillation, N_T is determined by an energy balance, which gives the change in molar vapor rate across a tray. For the assumption of constant molar overflow, $N_T = 0$. In this example, that assumption is not valid, and the change is

$$N_T = V_{n+1} - V_n = -54 \text{ lbmol/h}$$

From (1), (2), and (3):

$$N_1^Y = -121.8 + 0.3246(-54) = -139.4 \text{ lbmol/h}$$

$$N_2^Y = -32.7 + 0.4654(-54) = -57.8 \text{ lbmol/h}$$

$$N_3^Y = 154.5 + 0.2100(-54) = 143.2 \text{ lbmol/h}$$

(c) Approximate values of the Murphree vapor tray efficiency are obtained from (12-3), with K-values at phase interface conditions:

$$E_{MV_i} = (y_{i,n} - y_{i,n+1})/(K_{i,n}^I x_{i,n} - y_{i,n+1}) \qquad (4)$$

From (4):

$$E_{MV_1} = \frac{(0.2971 - 0.1700)}{[(2.759)(0.1459) - 0.1700]} = 0.547$$

$$E_{MV_2} = \frac{(0.4631 - 0.4290)}{[(1.225)(0.3865) - 0.4290]} = 0.767$$

$$E_{MV_3} = \frac{(0.2398 - 0.4010)}{[(0.3673)(0.4676) - 0.4010]} = 0.703 \qquad \blacksquare$$

The general forms for rates of heat transfer across the vapor and liquid films of a stage, respectively, are

$$e_j^V = a_j^I h^V(T^V - T^I) + \sum_{i=1}^{C} N_{i,j}^V \overline{H}_{i,j}^V \qquad (12\text{-}38)$$

$$e_j^L = a_j^I h^L(T^I - T^L) + \sum_{i=1}^{C} N_{i,j}^L \overline{H}_{i,j}^L \qquad (12\text{-}39)$$

where $\overline{H}_{i,j}^{\mathrm{P}}$ are the partial molar enthalpies of component i for stage j and h^{P} are convective heat-transfer coefficients. The second terms on the right-hand sides of (12-38) and (12-39) account for the transfer of enthalpy by mass transfer. Temperatures T^{V} and T^{L} are the temperatures exiting the stage regardless of the assumed flow patterns for the vapor and liquid.

12.3 METHODS FOR ESTIMATING TRANSPORT COEFFICIENTS AND INTERFACIAL AREA

Equations (12-31) and (12-32) require binary-pair mass-transfer coefficients. In most rate-based model applications, the coefficients are estimated from empirical correlations of experimental data for different contacting devices. For trayed columns, correlations have been published by the AIChE [20], Harris [21], Hughmark [22], Zuiderweg [23], Chan and Fair [24], and Chen and Chuang [25]. A more fundamental and comprehensive model for sieve trays has been developed by Taylor and Krishna [15]. For sieve trays, the simplest correlations are those of Zuiderweg [24], but the correlations of Chan and Fair [25] appear to be more reliable and are more widely used.

Most of the mass-transfer correlations are presented in terms of the number of transfer units, N_V and N_L, where, by definition:

$$N_V \equiv k^{\mathrm{V}} a h_f / u_s \qquad \text{(12-40)}$$

$$N_L \equiv k^{\mathrm{L}} a h_f z / (Q_L / W) \qquad \text{(12-41)}$$

where

a = interfacial area/volume of froth on the tray
h_f = froth height
u_s = superficial vapor velocity based on the bubbling area of the tray
z = length of liquid flow path across the bubbling area of the tray
Q_L = volumetric liquid flow rate
W = weir length

The interfacial area for a tray, a^{I}, is related to a by

$$a^{\mathrm{I}} = a h_f A_b \qquad \text{(12-42)}$$

where A_b = bubbling area

Thus, k^{P} and a^{I} are obtained from correlations in terms of N_V and N_L.

For random (dumped) packings, empirical correlations for mass-transfer coefficients and interfacial area density (area/packed volume) have been published by Onda, Takeuchi, and Okumoto [26] and Bravo and Fair [27]. For structured packings, the empirical correlations of Bravo, Rocha, and Fair for gauze packings [28] and for a wide variety of structured packings [29] are available. A semitheoretical correlation by Billet and Schultes [30] requires five packing parameters and is applicable to both random and structured packings.

Heat-transfer coefficients for the vapor film are estimated from the Chilton–Colburn analogy between heat and mass transfer, described in Chapter 3. Thus:

$$h^{\mathrm{V}} = k^{\mathrm{V}} \rho^{\mathrm{V}} C_P^{\mathrm{V}} (N_{\mathrm{Le}})^{2/3} \qquad \text{(12-43)}$$

where:

$$N_{\mathrm{Le}} = \left(\frac{N_{\mathrm{Sc}}}{N_{\mathrm{Pr}}} \right) \qquad \text{(12-44)}$$

For the liquid-phase film, a penetration model is preferred, where

$$h^{\mathrm{L}} = k^{\mathrm{L}} \rho^{\mathrm{L}} C_P^{\mathrm{L}} (N_{\mathrm{Le}})^{1/2} \qquad \text{(12-45)}$$

12.4 VAPOR AND LIQUID FLOW PATTERNS

The simplest flow pattern for a stage corresponds to the assumption of a perfectly mixed vapor and a perfectly mixed liquid. Under these conditions, the mass-transfer driving forces in (12-27) and (12-28) are simplified to

$$(y^V - y^I)_{avg} = (y^V - y^I) \qquad \textbf{(12-46)}$$

$$(x^I - x^L)_{avg} = (x^I - x^L) \qquad \textbf{(12-47)}$$

where y^V and x^L are exiting stage mole fractions. These flow patterns are only valid for trayed towers with a short liquid flow path.

A plug-flow pattern for the vapor and/or liquid assumes that the phase moves through the froth without mixing. This pattern requires that the mass-transfer rates be integrated over the froth. An approximation of the integration is provided by Kooijman and Taylor [31], who assume constant mass-transfer coefficients and interface compositions. The resulting expressions for the average mole-fraction driving forces are the same as (12-46) and (12-47) except for a correction factor in terms of N^V or N^L, included on the right-hand side of each equation. Plug-flow patterns are generally more accurate for trayed towers than perfectly mixed flow patterns and are also applicable to packed towers.

The perfectly-mixed-flow and plug-flow patterns are the two patterns presented by Lewis [7] to convert Murphree vapor point efficiencies to Murphree vapor tray efficiencies. They represent the extreme situations. Fair, Null, and Bolles [32] recommend a more realistic partial mixing or dispersion model that utilizes a turbulent Peclet number, whose value can cover a wide range. This model provides a bridge between the two extremes.

12.5 METHOD OF CALCULATION

As indicated in Section 12.1, the number of equations to be solved for the rate-based model of Figure 12.1 is $N(5C + 5)$ when the pressure-drop equations are omitted, as summarized in Table 12.1. The equations contain the variables listed in Table 12.2. Other parameters in the equations are computed from these variables. When the number of equations is subtracted from the number of variables, the number of degrees of freedom is $2NC + 9N + 1$. If the total number of stages and all column feed conditions, including feed-stage locations ($2NC + 4N + 1$ variables) are specified, the number of remaining degrees of freedom, using the variable designations in Table 12.2, is $5N$. A computer program for the rate-based model would generally require the user to specify these $2NC + 4N + 1$ variables. The degree of flexibility provided to the user in the selection of the remaining $5N$ variables depends upon the particular rate-based computer algorithm, two of which are widely available: (1) Chem-Sep Release v3.1 from R. Taylor and H. A. Kooijman of Clarkson University, and (2) RATEFRAC in Release 9 of ASPEN PLUS from Aspen Technology, Cambridge, Massachusetts. Both algorithms provide a wide variety of correlations for thermodynamic and transport properties. Both programs also provide considerable flexibility in the selection of the remaining $5N$ specifications. The basic $5N$ specifications are

$$r_j^L \text{ or } U_j, r_j^V \text{ or } W_j, P_j, Q_j^L, \text{ and } Q_j^V$$

However, substitutions can be made as discussed next.

ChemSep Program

The ChemSep program applies the transport equations to trays or short heights (called segments) of packing. The condenser and reboiler stages are treated as equilibrium stages.

The specification options provided are as follows:

1. r_j^L and r_j^V: From each stage, either a liquid or a vapor side stream can be specified as (a) a side-stream flow rate or (b) a ratio of the side-stream flow rate to the flow rate of the remaining fluid passing to the next stage, that is,

$$r_j^L = U_j/L_j \quad \text{or} \quad r_j^V = W_j/V_j \text{ in Figure 12.1.}$$

2. P_j: Four options are available:

   (a) Condenser pressure (if any) and a constant, but different, pressure for all stages in the tower and for the reboiler, if any.
   (b) Condenser pressure (if any), top tower pressure, and bottom pressure (bottom tower stage or reboiler, if any). Pressures of stages intermediate between top and bottom are obtained by linear interpolation.
   (c) Condenser pressure (if any,), top tower pressure, and specified pressure drop per stage to obtain remaining stage pressures.
   (d) Condenser pressure (if any) and top tower pressure, with stage pressure drops estimated by ChemSep from hydraulic correlations.

3. Q_j^L and Q_j^V: The heat duty must be specified for all stage heaters and coolers except for the condenser and/or reboiler, if present. In addition, a heat loss for the tower can be specified that is divided equally over all stages. When a condenser (total without subcooling, total with subcooling, or partial) is present, one of the following specifications can replace the heat duty of the condenser: (a) molar reflux ratio, (b) condensate temperature, (c) distillate molar flow rate, (d) reflux molar flow rate, (e) component molar flow rate in distillate, (f) mole fraction of a component in distillate,(g) fractional recovery, from all feeds, of a component in the distillate,(h) molar fraction of all feeds to the distillate, and (i) molar ratio of two components in the distillate.

   For distillation, an often-used specification is the molar reflux ratio.

When a reboiler (partial, total with a vapor product, or total with a superheated vapor product) is present, the following list of specification options, similar to those just given for a condenser, can replace the heat duty of the reboiler: (a) molar boilup ratio, (b) reboiler temperature, (c) bottoms molar flow rate, (d) reboiled vapor (boilup) molar flow rate, (e) component molar flow rate in bottoms, (f) mole fraction of a component in bottoms, (g) fractional recovery, from all feeds, of a component in the bottoms, (h) molar fraction of all feeds to the bottoms, and (i) molar ratio of two components in the bottoms. For distillation, an often-used specification is the molar bottoms flow rate, which must be estimated if it is not specified.

The preceding number of optional specifications is considerable. In addition, ChemSep also provides so-called "flexible" specifications that can substitute for the condenser and/ or reboiler duties. These are advanced options supplied in the form of strings that contain values of certain allowable variables and/or combinations of these variables using the five common arithmetic operators ($+$, $-$, $*$, $/$, and exponentiation). The variables include stage variables (L, V, x, y, and T) and interface variables (x^I, y^I, and T^I) at any stage. Flow rates can be in mole or mass units.

Certain options and advanced options must be used with great care because values can be specified that cannot lead to a converged solution. For example, with a simple distillation column of a fixed number of stages, that number may be less than the minimum number to achieve specified distillate and bottoms purities. As always, it is generally wise to begin a simulation with a standard pair of top and bottom specifications, such as reflux ratio and a bottoms molar flow rate that corresponds to the desired distillate rate. These specifications are almost certain to converge unless interstage liquid or vapor flow rates tend to zero somewhere in the column. A study of the calculated results will provide valuable insight into possible limits in the use of other options.

The equations for the rate-based model, some linear and some nonlinear, are solved by Newton's method in a manner similar to that developed by Naphtali and Sandholm for the equilibrium-based model described in Chapter 10. Thus, the variables and equations are grouped by stage so that the Jacobian matrix is of the block-tridiagonal form. However, the equations to be solved number $5C + 6$ or $5C + 5$ per stage, depending upon whether stage pressures are computed or specified, compared to just $2C + 1$ for the equilibrium-based method.

Calculations of transport coefficients and pressure drops require column diameter and dimensions of column internals. These may be specified (simulation mode) or computed (design mode). In the latter case, default dimensions are selected for the internals, with column diameter computed from a specified value for percent of flooding for a trayed or packed column, or a specified pressure drop per unit height for a packed column.

Computing time per iteration for the design mode is only approximately twice that for the simulation mode, which usually requires less than twice the time for the equilibrium-based model. The number of iterations required for the design mode can be two to three times that for the equilibrium-based model. Overall, the total computing time for the design mode is usually less than an order of magnitude greater than that for the equilibrium-based model. With today's fast workstations and PCs, computing times for the design mode of the rate-based model are usually less than one minute.

Like the Naphtali–Sandholm method, the rate-based model utilizes mainly analytical partial derivations in the Jacobian matrix, and requires initial estimates of all variables. These estimates are generated automatically by the ChemSep program using a method of Powers et al. [33]. In this method, the usual assumptions of constant molar overflow and a linear temperature profile are employed. The initialization of the stage mole fractions is made by performing several iterations of the bubble-point method using ideal K-values for the first iteration and nonideal K-values thereafter. Initial interface mole fractions are set equal to estimated bulk values and initial mass-transfer rates are arbitrarily set to values of $\pm 10^{-3}$ kmol/h with the sign dependent upon the component K-value.

To prevent oscillations and promote convergence of the iterations, corrections to certain variables from iteration to iteration can be limited. Defaults are 10 K for temperature and 50% for flows. When a correction to a mole fraction would result in a value outside of the feasible range of 0 to 1, the default correction is one-half of the step that would take the value to a limit. For very difficult problems, homotopy continuation methods described by Powers et al. [33] can be applied to promote convergence.

Convergence of Newton's method is determined from values of the residuals of the functions, as in the Naphtali–Sandholm method, or from the corrections to the variables. ChemSep applies both criteria and terminates when either of the following are satisfied:

$$\left[\sum_{j=1}^{N} \sum_{k=1}^{N_j} f_{k,j}^2 \right]^{1/2} < \epsilon \qquad (12\text{-}48)$$

$$\sum_{j=1}^{N} \sum_{k=1}^{N_j} |\Delta X_{k,j}| / X_{k,j} < \epsilon \qquad (12\text{-}49)$$

where

$f_{k,j}$ = residuals in Table 12.1

N = number of stages

N_j = number of equations for the jth stage

$X_{k,j}$ = unknown variables from Table 12.2

ϵ = a small number with a default value of 10^{-4}

Table 12.3 Specifications for Example 12.2

Total condenser delivering saturated liquid
Partial reboiler
Pressure at condenser outlet = 14.7 psia
Pressure at condenser inlet = 15.0 psia
Reflux ratio = 1.5
Bottoms flow rate = 45 lbmol/h
Total number of trays = 20
Feed 1 to tray 10 from top:
   55 lbmol/h of n-heptane
   45 lbmol/h of toluene
   100 lbmol/h of methylethyl ketone (MEK)
   Saturated liquid at 20 psia
Feed 2 to tray 15 from top:
   100 lbmol/h of MEK
   Saturated liquid at 20 psia
UNIFAC for liquid-phase activity coefficients
Chan–Fair correlation for mass-transfer coefficients
Plug flow for vapor
Mixed flow for liquid
85% of flooding
Tray spacing = 0.5 m (19.7 inches)
Weir height = 2 inches

Unlike in the Naphtali–Sandholm method, the residuals are not scaled. Accordingly, the second criterion is usually satisfied first.

From the results of a converged solution, it is highly desirable to back-calculate Murphree vapor-tray efficiencies, component by component and tray by tray, from (12-3) for trayed columns, and HETP values for packed towers.

ChemSep can also perform rate-based calculations for liquid–liquid extraction.

EXAMPLE 12.2

A mixture of n-heptane and toluene cannot be separated at 1 atm by ordinary distillation. Accordingly, an enhanced-distillation scheme using methylethyl ketone as a solvent is used. As part of an initial design study, use the rate-based model of ChemSep with the specifications listed in Table 12.3 to calculate a sieve-tray column.

SOLUTION

The information in Table 12.3 was entered via the ChemSep menu and the program was executed. A converged solution was achieved in 8 iterations in 6 seconds on a PC with a Pentium 90 CPU, running the Windows 95 operating system. Initialization of all variables was done by the program.

The predicted separation is as follows:

Component	Distillate, lbmol/h	Bottoms, lbmol/h
n-Heptane	54.87	0.13
Toluene	0.45	44.55
Methylethyl ketone	199.68	0.32

Predicted column profiles for pressure, liquid-phase temperature, total vapor and liquid flow rates, component vapor and liquid mole fractions, component mass-transfer rates, and Murphree vapor tray efficiencies are shown in Figure 12.2, where stages are numbered from the top down

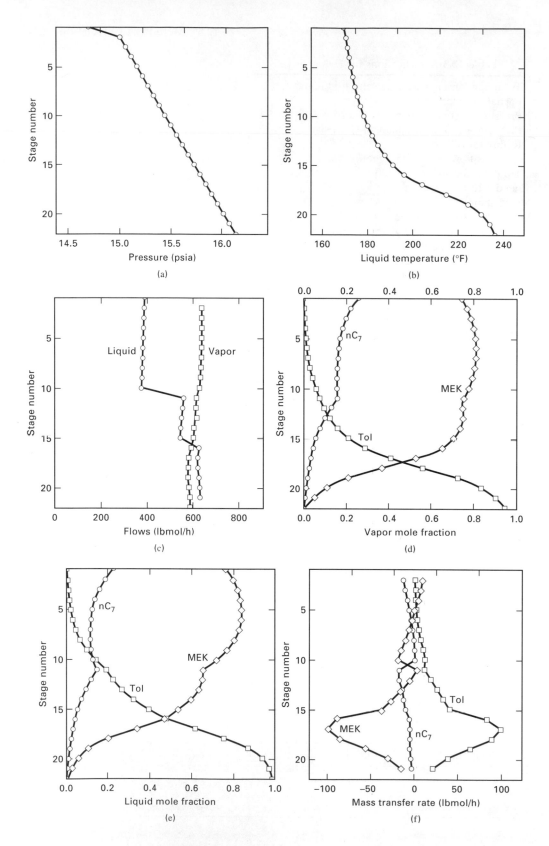

Figure 12.2 Column profiles for Example 12.2: (a) pressure profile; (b) liquid-phase temperature profile; (c) vapor and liquid flow rate profiles; (d) vapor mole-fraction profiles; (e) liquid mole-fraction profiles; (f) mass-transfer rate profiles; (g) Murphree vapor tray efficiencies.

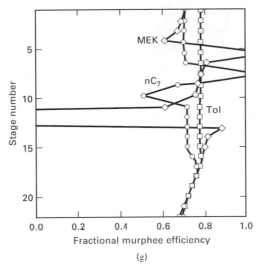

(g)

Figure 12.2 (*continued*)

and stages 2 to 21 are sieve trays. Back-calculated Murphree tray efficiencies are summarized as follows:

Component	Fractional Murphree Efficiencies	
	Range	Median
n-Heptane	0.52 to 1.10	0.73
Toluene	0.70 to 0.79	0.79
Methylethyl ketone	−3.23 to 1.14	0.76

The median values, based on experience, seem reasonable and give confidence in the rate-based method. The 20 trays are equivalent to approximately 15 equilibrium stages.

For purposes of sizing, the column was divided into three sections: 9 trays above the top feed, 5 trays from the top feed to the bottom feed, and 6 trays below the bottom feed. Computed column diameters are, respectively, 1.75 m (5.74 ft), 1.74 m (5.71 ft), and 1.83 m (6.00 ft). Thus, a 1.83 m (6.00 ft)-diameter column is a reasonable choice. Average predicted pressure drop per tray is 0.06 psi. Computed heat exchanger duties are as follows:

Condenser: 2.544 MW (8,680,000 Btu/h)
Reboiler: 2.482 MW (8,470,000 Btu/h)   ■

EXAMPLE 12.3

Repeat Example 12.2 for a tower packed with Flexipac 2 structured packing, operating at 75% of flooding. The packing heights are as follows:

Section	Packing Height, ft
Above top feed	13
Between top and bottom feeds	6.5
Below bottom feed	6.5

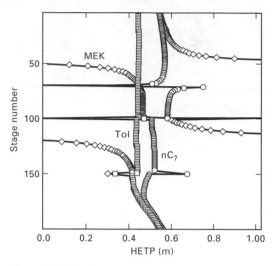

Figure 12.3 Column HETP profiles for Example 12.3.

SOLUTION

Each 6.5 feet of packing was simulated by 50 segments. Because of the large number of segments, mixed flow is assumed for both vapor and liquid. Newton's method could not converge the calculations. Therefore, the homotopy continuation option was selected. Then convergence was achieved in 73 s after a total of 26 iterations. The predicted separation, which is just slightly better than that in Example 12.2, is as follows:

Component	Distillate, lbmol/h	Bottoms, lbmol/h
n-Heptane	54.88	0.12
Toluene	0.40	44.60
Methylethyl ketone	199.72	0.28

The HETP profile is plotted in Figure 12.3. Median values for *n*-heptane, toluene, and methylethyl ketone, respectively, are approximately 0.55 m (21.7 inches), 0.45 m (17.7 inches), and 0.5 m (19.7 inches). The HETP values for the ketone are seen to vary widely.

Predicted column diameters for the three sections, starting from the top, are 1.65, 1.75, and 1.85 m, which are very close to the predicted sieve-tray diameters. ∎

RATEFRAC Program

The RATEFRAC program differs somewhat from the ChemSep program in that RATEFRAC is designed to model columns used for reactive distillation. The reactions can be equilibrium-based or kinetics-based, including reactions among electrolytes. For kinetically controlled reactions, built-in power-law expressions are selected, or the user supplies FORTRAN subroutines for the rate law(s). For equilibrium-based reactions, the user supplies a temperature-dependent equilibrium constant, or RATEFRAC computes reaction-equilibrium constants from free-energy values stored in the data bank. The user specifies the phase in which the reaction takes place. Flow rates of side streams and the column pressure profile must be provided. The heat duty must be specified for each intercooler or interheater. The standard specifications for the rating mode are the reflux ratio and the bottoms flow rate. However, these specifications can be manipulated in the design mode to achieve any of the following substitute specifications:

(a) Purity of a product or internal stream with respect to one component or a group of components.
(b) Recovery of a component or group of components in a product stream.
(c) Flow rate of a component or group of components in a product or internal stream.
(d) Temperature of a product or internal vapor or liquid stream.
(e) Heat duty of condenser or reboiler.
(f) Value of a product or internal stream physical property.
(g) Ratio or difference of any pair of product or internal stream physical properties, where the two streams can be the same or different.

Mass-transfer correlations are built into RATEFRAC for bubble-cap trays, valve trays, sieve trays, and random packings. Users may provide their own FORTRAN subroutines for transport coefficients and interfacial area. Newton's method is used to converge the calculations.

EXAMPLE 12.4

Use RATEFRAC to predict the column profiles for a 3.5-ft-diameter, 20-bubble-cap tray absorber operating at the conditions listed in Table 12.4.

Table 12.4 Specifications for Example 12.4

Column top pressure = 182 psia
Column bottom pressure = 185 psia
Weir height = 2 inches
Vapor completely mixed on each tray
Liquid completely mixed on each tray
AIChE correlations for binary mass-transfer coefficients and interfacial area
Chilton–Colburn analogy for heat transfer
Chao–Seader correlation for K-values

Vapor feed at 123°F and 184 psia:	
Component	lbmol/h
Hydrogen	218
Nitrogen	87
Methane	136
Ethane	139
Propane	118
Isobutane	6
n-Butane	2
Isopentane	43
n-Hexane	14
n-Heptane	4
	767

Liquid absorbent feed at 100°F and 182 psia:	
Component	lbmol/h
n-Dodecane	165
n-Tridecane	165

SOLUTION

No initial estimates of variables were provided. The program was run on a PC with a Pentium 90 CPU running under MS-DOS 6.22. Initial estimates of the values of the variables were provided by RATEFRAC. A total of five iterations were required following an initialization step. Total computing time, including translation, compiling, linking, and execution, was about 60 s. The following results were obtained for the product streams:

Component	Lean Vapor, lbmol/h	Rich Oil, lbmol/h
Hydrogen	216.6	1.4
Nitrogen	86.1	0.9
Methane	131.4	4.6
Ethane	120.0	19.0
Propane	73.4	44.6
Isobutane	1.1	4.9
n-Butane	0.0	2.0
Isopentane	0.0	43.0
n-Hexane	0.0	14.0
n-Heptane	0.0	4.0
n-Dodecane	0.0	165.0
n-Tridecane	0.0	165.0
	628.6	468.4

The back-calculated fractional Murphree vapor tray efficiencies are as follows:

Component	E_{MV} Range	E_{MV} Median Value
Hydrogen	−1.25 to 1.21	0.43
Nitrogen	−0.51 to 3.25	0.41
Methane	−0.13 to 0.87	0.41
Ethane	−0.03 to 1.02	0.30
Propane	−1.89 to 2.71	0.33
Isobutane	−6.51 to 1.16	0.42
n-Butane	−0.75 to 5.65	0.50
Isopentane	−3.17 to 1.15	0.57
n-Hexane	0.63 to 1.38	0.63
n-Heptane	0.64 to 0.88	0.64
n-Dodecane	−2.92 to 1.01	0.42
n-Tridecane	−5.04 to 0.97	0.42

It is seen that the efficiencies vary widely from component to component and from tray to tray. For absorber simulation and design, a rate-based model is clearly superior to an equilibrium-based model. ∎

As chemical engineers become more knowledgeable in the principles of mass and heat transfer, and improved correlations for mass-transfer and heat-transfer coefficients are developed for trays and packings, the use of rate-based models will accelerate. For best results, these models will also benefit from more realistic options for vapor and liquid flow patterns. More comparisons of rate-based models with industrial operating data are needed to gain confidence in the use of such models. Some recent comparisons are presented by Taylor, Kooijman, and Woodman [34]. Comparisons by Ovejero et al. [35], with distillation data obtained in a column packed with spheres and cylinders of known interfacial area, show very good agreement for three binary and two ternary systems.

SUMMARY

1. Rate-based models of multicomponent, multistage vapor–liquid separation operations became available in the late 1980s. These models are potentially superior to equilibrium-based models for all but near-ideal systems.

2. Rate-based models incorporate rigorous procedures for treating component-coupling effects in multicomponent mass transfer.

3. The number of equations for a rate-based model is much greater than for an equilibrium-based model because separate balances are needed for each of the two phases. In addition, rate-based models are influenced by the geometry of the column internals. Correlations are used to predict mass-transfer and heat-transfer rates. Tray or packing hydraulics are also incorporated into the rate-based model to enable prediction of column pressure profile. Equilibrium is assumed at the phase interface.

4. Computing time for a rate-based model is not generally more than an order of magnitude greater than that for an equilibrium-based model.

5. Both the ChemSep and RATEFRAC rate-based computer programs offer considerable flexibility in user specifications, so much so that inexperienced users can easily specify impossible conditions. Therefore, it is best to begin simulation studies with standard specifications.

REFERENCES

1. Sorel, Ernest, *La rectification de l' alcool,* Paris (1893).

2. Smoker, E.H., *Trans. AIChE,* **34,** 165 (1938).

3. Thiele, E.W., and R.L. Geddes, *Ind. Eng. Chem.,* **25,** 290 (1933).

4. Lewis, W.K., and G.L. Matheson, *Ind. Eng. Chem.,* **24,** 496–498 (1932).

5. Lewis, W.K., *Ind. Eng. Chem.,* **14,** 492 (1922).

6. Murphree, E.V., *Ind. Eng. Chem.,* **17,** 747–750, 960–964 (1925).

7. Lewis, W.K., *Ind. Eng. Chem.,* **28,** 399 (1936).

8. Seader, J.D., *Chem. Eng. Prog.,* **85** (10), 41–49 (1989).

9. Walter, J.F., and T.K. Sherwood, *Ind. Eng. Chem.,* **33,** 493–501 (1941).

10. Toor, H.L., *AIChE J.,* **3,** 198 (1957).

11. Toor, H.L., and J.K. Burchard, *AIChE J.,* **6,** 202 (1960).

12. Krishna, R., H.F. Martinez, R. Sreedhar, and G.L. Standart, *Trans. I. Chem. E.,* **55,** 178 (1977).

13. Waggoner, R.C., and G.D. Loud, *Comput. Chem. Engng.,* **1,** 49 (1977).

14. Krishna, R., and G.L. Standart, *Chem. Eng. Comm.,* **3,** 201 (1979).

15. Taylor, R., and R. Krishna, *Multicomponent Mass Transfer,* John Wiley and Sons, New York (1993).

16. Krishnamurthy, R., and R. Taylor, *AIChE J.,* **31,** 449, 456 (1985).

17. Taylor, R., H.A. Kooijman, and J.-S. Hung, *Comput. Chem. Engng.,* **18,** 205–217 (1994).

18. *ASPEN PLUS Reference Manual—Volume 1,* Aspen Technology, Cambridge, MA (1994).

19. Taylor, R., and H.A. Kooijman, *CACHE News,* No. 41, 13–19 (1995).

20. AIChE, *Bubble-Tray Design Manual,* New York (1958).

21. Harris, I.J., *British Chem. Engng.,* **10**(6), 377 (1965).

22. Hughmark, G.A., *Chem. Eng. Progress,* **61**(7), 97–100 (1965).

23. Zuiderweg, F.J., *Chem Eng. Sci.,* **37,** 1441 (1982).

24. Chan, H., and J.R. Fair, *Ind. Eng. Chem. Process Des. Dev.,* **23,** 814–827 (1984).

25. Chen, G.X., and K.T. Chuang, *Ind. Eng. Chem. Res.,* **32,** 701–708 (1993).

26. Onda, K., H. Takeuchi, and Y.J. Okumoto, *J. Chem. Eng. Japan,* **1,** 56–62 (1968).

27. Bravo, J.L., and J.R. Fair, *Ind. Eng. Chem. Process Des. Devel.,* **21,** 162–170 (1982).

28. Bravo, J.L., J.A. Rocha, and J.R. Fair, *Hydrocarbon Processing,* **64**(1), 56–60 (1985).

29. Bravo, J.L., J.A. Rocha, and J.R. Fair, *I. Chem. E. Symp. Ser.,* No. 128, A489–A507 (1992).

30. Billet, R., and M. Schultes, *I. Chem. E. Symp. Ser.,* No. 128, B129 (1992).

31. Kooijman, H.A., and R. Taylor, *Chem. Eng. J.,* **57**(2), 177–188 (1995).

32. Fair, J.R., H.R. Null, and W.L. Bolles, *Ind. Eng. Chem. Process Des. Dev.,* **22,** 53–58 (1983).

33. Powers, M.F., D.J. Vickery, A. Arehole, and R. Taylor, *Comput. Chem. Engng.,* **12,** 1229–1241 (1988).

34. Taylor, R., H.A. Kooijman, and M.R. Woodman, *I. Chem. E. Symp. Ser.,* No. 128, A415–A427 (1992).

35. Ovejero, G., R. Van Grieken, L. Rodriguez, and J.L. Valverde, *Sep. Sci. Tech.,* **29,** 1805–1821 (1994).

EXERCISES

Section 12.1

12.1 Modify the rate-based model of (12-4) to (12-18) to include entrainment and occlusion.

12.2 Modify the rate-based model of (12-4) to (12-18) to include a chemical reaction in the liquid phase under conditions of:

(a) Chemical equilibrium.

(b) Kinetic rate law.

12.3 Explain how the number of rate-based modeling equations can be reduced. Would this be worthwhile?

Section 12.2

12.4 The following results were obtained at tray n from a rate-based calculation at 14.7 psia, for a ternary mixture of acetone (1), methanol (2), and water (3) in a sieve-tray column assuming that both phases are perfectly mixed.

Component	y_n	y_{n+1}	y_n^I	K_n^I	x_n
1	0.4913	0.4106	0.5291	1.507	0.3683
2	0.4203	0.4389	0.4070	0.900	0.4487
3	0.0884	0.1505	0.0639	0.3247	0.1830

The products of the computed gas-phase binary mass-transfer coefficients and interfacial area from the Chan–Fair correlations are as follows in units of lbmol/(h-unit mole fractions).

$$k_{12} = k_{21} = 1,750$$
$$k_{13} = k_{31} = 2,154$$
$$k_{23} = k_{32} = 2,503$$

The computed vapor rates are $V_n = 1,200$ lbmol/h and $V_{n+1} = 1,164$ lbmol/h. Determine:
(a) The component molar diffusion rates.
(b) The mass-transfer rates.
(c) The Murphree vapor tray efficiencies.

12.5 Write all the expanded equations (12-31) and (12-32) for $\mathbf{R}^P$ for a five-component system.

12.6 Repeat the calculations of Example 12.1, but using 1 = methanol, 2 = water, and 3 = acetone. Are the results any different? If not, why not? Prove your conclusion mathematically.

Section 12.3

12.7 Compare and discuss the advantages and disadvantages of the available correlations for estimating binary-pair mass-transfer coefficients for trayed columns.

12.8 Compare and discuss the advantages and disadvantages of the available correlations for estimating binary-pair mass-transfer coefficients for columns with random (dumped) and structured packings.

Section 12.4

12.9 Discuss how the method of Fair, Null, and Bolles [32] might be used to model the flow patterns in a rate-based model. How would the mole-fraction driving forces be computed?

Section 12.5

12.10 A bubble-point mixture of 100 kmol/h of methanol, 50 kmol/h of isopropanol, and 100 kmol/h of water at 1 atm is sent to the 25th tray from the top of a 40-sieve-tray column equipped with a total condenser and partial reboiler, operating at a nominal pressure of 1 atm. If the reflux ratio is 5 and the bottoms flow rate is 150 kmol/h, determine the separation achieved if the UNIFAC method is used to estimate K-values and the Chan–Fair correlations are used for mass transfer. Assume that both phases are perfectly mixed on each tray and that operation is at about 80% of flooding.

12.11 A sieve-tray column, operating at a nominal pressure of 1 atm, is used to separate a mixture of acetone and methanol by extractive distillation using water. The column has 40 trays with a total condenser and partial reboiler. The feed of 50 kmol/h of acetone and 150 kmol/h of methanol at 60°C and 1 atm enters tray 35 from the top, while 50 kmol/h of water at 65°C and 1 atm enters tray 5 from the top. Determine the separation for a reflux ratio of 10 and a bottoms flow rate of 200 kmol/h. Use the UNIFAC method for K-values and the AIChE method for mass transfer. Assume a perfectly mixed liquid and a vapor in plug flow on each tray, with operation at 80% of flooding. Also determine the number of equilibrium stages (to the nearest stage) to achieve the same separation.

12.12 Repeat Exercise 12.10, if a column packed with 2-inch stainless-steel Pall rings is used with 25 feet of rings above the feed and 15 feet below. Be sure to use a sufficient number of segments for the calculations.

12.13 Repeat Exercise 12.10, if a column with structured packing is used with 25 feet above the feed and 15 feet below. Be sure to use a sufficient number of segments.

12.14 Solve Exercise 12.10 for combinations of the following values of percent flooding, weir height, and hole area, respectively:

40, 60, and 80%
1, 2, and 3 inches
6, 10, and 14%

12.15 The upper column of an air-separation system, of the type discussed and shown in Exercise 7.40, contains 48 sieve trays and operates at a nominal pressure of 131.7 kPa. A feed at 80 K and 131.7 kPa enters the top plate at 1,349 lbmol/h with a composition of 97.868 mol% nitrogen, 0.365 mol% argon, and 1.767 mol% oxygen. A second feed enters tray 12 from the top at 83 K and 131.7 kPa at 1,832 lbmol/h with a composition of 59.7 mol% nitrogen, 1.47 mol% argon, and 38.83 mol% oxygen. The column has no condenser, but has a split reboiler. Vapor distillate leaves the top plate at 2,487 lbmol/h, with remaining products leaving the reboiler as 50 mol% vapor and 50 mol% liquid. Assume that ideal solutions are formed. Determine the effect of percent flooding on the separation and the median Murphree vapor-tray efficiency for oxygen.

12.16 The following bubble-point organic-liquid mixture at 1.4 atm is distilled by extractive distillation with the following phenol-rich solvent at 1.4 atm and at the same temperature as the main feed:

Component	Feed, kmol/h	Solvent, kmol/h
Methanol	50	0
n-Hexane	20	0
n-Heptane	180	0
Toluene	150	10
Phenol	0	800

The column has 30 sieve trays, with a total condenser and a partial reboiler. The solvent enters the 5th tray and the feed enters tray 15, from the top. The pressure in the condenser is 1.1 atm; the pressure at the top tray is 1.2 atm, and the pressure at the bottom is 1.4 atm. The reflux ratio is 5 and the bottoms rate is 960 kgmol/h. Thermodynamic properties can be estimated with the UNIFAC method for the liquid phase and the SRK equation for the vapor phase. The Antoine equation is suitable for vapor pressure. Use the nonequilibrium model of the ChemSep program to estimate the separation. Assume that the vapor and liquid are both well mixed and that the trays operate at 75% of flooding. Specify the Chan–Fair correlation for calculating mass-transfer coefficients. In addition, determine from the tray-by-tray results the average Murphree vapor-tray efficiency for each component (after discarding values that appear to be much different than the majority of values). Try to improve the sharpness of the split by changing the feed and solvent entry tray locations. How can you increase the sharpness of the separation? List as many ideas as you have.

12.17 A bubble-cap tray absorber is designed to absorb 40% of the propane from a rich gas at 4 atm. The specifications for the entering rich gas and absorbent oil are as follows:

	Absorbent Oil	Rich Gas
Flow rate, kmol/s	11.0	11.0
Temperature, °C	32	62
Pressure, atm	4	4
Mole fraction:		
Methane	0	0.286
Ethane	0	0.157
Propane	0	0.240
n-Butane	0.02	0.169
n-Pentane	0.05	0.148
n-Dodecane	0.93	0

(a) Determine the number of equilibrium stages required and the splits of all components.
(b) Determine the actual number of trays required and the splits and Murphree vapor tray efficiencies of all components.
(c) Compare and discuss the equilibrium-based and rate-based results. What do you conclude?

12.18 A ternary mixture of methanol, ethanol, and water is distilled in a sieve-tray column to obtain a distillate with not more than 0.01 mol% water. The feed to the column is as follows:

Flow rate, kmol/h	142.46
Pressure, atm	1.3
Temperature, K	316
Mole fractions:	
Methanol	0.6536
Ethanol	0.0351
Water	0.3113

For a distillate rate of 93.10 kmol/h, a reflux ratio of 1.2, a condenser outlet pressure of 1.0 atm, and a top tray pressure of 1.1 atm, determine using the UNIFAC method for activity coefficients:
(a) The number of equilibrium stages required and the corresponding split, if the feed enters at the optimal stage.
(b) The number of actual trays required if the column operates at about 85% of flooding and the feed is introduced to the optimal tray. Compare the split to that in part (a). In addition, compute the component Murphree vapor-tray efficiencies. What do you conclude about the two methods of calculation?

12.19 Repeat Exercise 12.18 for a column packed with 2-inch stainless steel Pall rings.

12.20 It is required to absorb 96% of the benzene from a gas stream with absorption oil in a sieve-tray column at a nominal pressure of 1 atm. The feed conditions are as follows:

	Vapor	Liquid
Flow rate, kmol/s	0.01487	0.005
Pressure, atm	1.0	1.0
Temperature, K	300	300
Composition, mol fraction:		
Nitrogen	0.7505	0
Oxygen	0.1995	0
Benzene	0.0500	0.005
n-Tridecane (C_{13})	0	0.995

Tray geometry is as follows:

Tray spacing, m	0.5
Weir height, m	0.05
Hole diameter, m	0.003
Sheet thickness, m	0.002

Determine column diameter for 80% of flooding, the number of actual trays required, and the Murphree vapor-tray efficiency profile for benzene for the possible combinations of vapor and liquid flow patterns on a tray. Could the equilibrium-based method be used to obtain a reliable solution to this problem?

Chapter **13**

Batch Distillation

In batch separation operations, a feed mixture is charged to the equipment and one or more products are withdrawn. A familiar example is laboratory distillation, shown in Figure 13.1, where a liquid mixture is charged to a still pot, retort, or flask and heated to boiling. The vapor formed is continuously removed and condensed to produce a distillate.

The composition of both the initial charge and distillate change with time; there is no steady state. The still temperature increases and the relative amount of lower-boiling components in the charge decreases as distillation proceeds.

Batch operations can be used to advantage under the following circumstances:

1. The capacity of a facility is too small to permit continuous operation at a practical rate.
2. It is necessary, because of seasonal demands, to distill with one unit different feedstocks to produce different products.
3. It is desired to produce several new products with one distillation unit for evaluation by potential buyers.
4. Upstream process operations are batchwise and the composition of feedstocks for distillation vary with time or from batch to batch.
5. The feed contains solids or materials that form solids, tars, or resin that plug or foul a continuous distillation column.

13.1 DIFFERENTIAL DISTILLATION

The simplest case of batch distillation, as discussed by Lord Rayleigh [1], is *differential distillation*, which involves use of the apparatus shown in Figure 13.1. There is no reflux; at any instant, vapor leaving the still pot with composition y_D is assumed to be in equilibrium with perfectly mixed liquid in the still. For total condensation, $y_D = x_D$. Thus, there is only a single equilibrium stage, the still pot. This apparatus is useful for separating wide-boiling mixtures. The following nomenclature is used for variables that vary with time, t, assuming that all compositions refer to a particular species in the multicomponent mixture.

D = instantaneous distillate rate, mol/h
$y = y_D = x_D$ = instantaneous distillate composition, mole fraction
W = moles of liquid left in still
$x = x_W$ = composition of liquid left in still, mole fraction
0 = subscript referring to $t = 0$

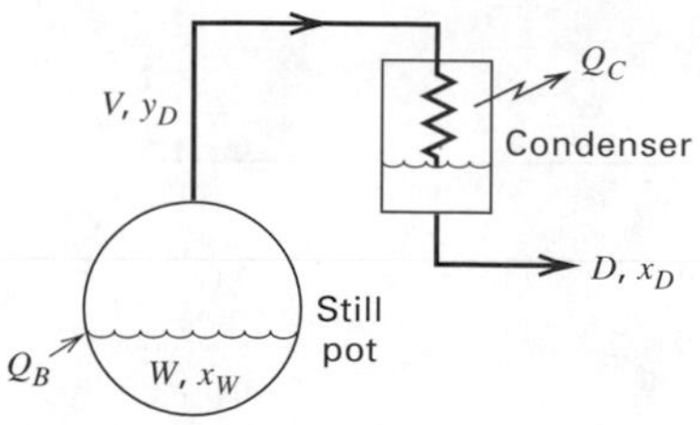

Figure 13.1 Differential distillation.

For any component in the mixture:

Instantaneous rate of output $= Dy_D$

$$
\left.\begin{array}{l}
\text{Instantaneous} \\
\text{rate of depletion} \\
\text{in the still}
\end{array}\right\} = -\frac{d}{dt}(Wx_W) = -W\frac{dx_W}{dt} - x_w\frac{dW}{dt}
$$

The distillate rate and, therefore, the rate of depletion of the liquid in the still depend on the rate of heat input to the still. By material balance at any instant:

$$
\frac{d}{dt}(Wx_W) = W\frac{dx_W}{dt} + x_W\frac{dW}{dt} = -Dy_D \tag{13-1}
$$

Multiplying by dt:

$$
Wdx_W + x_WdW = y_D(-Ddt) = y_DdW
$$

since by total balance $-Ddt = dW$. Separating variables and integrating from the initial charge condition:

$$
\int_{x_{W_0}}^{x_W}\frac{dx_W}{y_D - x_W} = \int_{W_0}^{W}\frac{dW}{W} = \ln\left(\frac{W}{W_0}\right) \tag{13-2}
$$

This is the well-known Rayleigh equation, which was first applied to the separation of wide-boiling mixtures such as $HCl–H_2O$, $H_2SO_4–H_2O$, and $NH_3–H_2O$. Without reflux, y_D and x_W are in equilibrium and (13-2) simplifies to

$$
\int_{x_0}^{x}\frac{dx}{y - x} = \ln\left(\frac{W}{W_0}\right) \tag{13-3}
$$

Equation (13-3) is easily integrated only when pressure is constant, temperature change in the still pot is relatively small (close-boiling mixture), and K-values are composition independent. Then $y = Kx$, where K is approximately constant, and (13-3) becomes

$$
\ln\left(\frac{W}{W_0}\right) = \frac{1}{K - 1}\ln\left(\frac{x}{x_0}\right) \tag{13-4}
$$

For a binary mixture, if the relative volatility α is assumed constant, substitution of (4-8) into (13-3), followed by integration and simplification, gives

$$
\ln\left(\frac{W_0}{W}\right) = \frac{1}{\alpha - 1}\left[\ln\left(\frac{x_0}{x}\right) + \alpha\ln\left(\frac{1 - x}{1 - x_0}\right)\right] \tag{13-5}
$$

If the equilibrium relationship $y = f\{x\}$ is in graphical or tabular form, integration of (13-3) can be performed graphically or numerically. The final liquid remaining in the still pot is often referred to as the *residue*.

EXAMPLE 13.1

A batch still is loaded with 100 kmol of a liquid containing a binary mixture of 50 mol% benzene in toluene. As a function of time, make plots of (a) still temperature, (b) instantaneous vapor composition, (c) still-pot composition, and (d) average total distillate composition. Assume a constant boilup rate of 10 kmol/h and a constant relative volatility of 2.41 at a pressure of 101.3 kPa (1 atm).

SOLUTION

Initially, $W_0 = 100$ kmol, $x_0 = 0.5$. Solving (13.5) for W at values of x from 0.5 in increments of 0.05, and determining corresponding values of time from $t = (W_0 - W)/10$, the following table is generated:

t, h	2.12	3.75	5.04	6.08	6.94	7.66	8.28	8.83	9.35
W, kmol	78.85	62.51	49.59	39.16	30.59	23.38	17.19	11.69	6.52
$x = x_W$	0.45	0.40	0.35	0.30	0.25	0.20	0.15	0.10	0.05

The instantaneous vapor composition, y, is obtained from (4-8), which is $y = 2.41x/(1 + 1.41x)$, the equilibrium relationship for constant α. The average value of y_D or x_D over the time interval 0 to t is related to x and W at time t by combining overall component and total material balances to give

$$(x_D)_{\text{avg}} = (y_D)_{\text{avg}} = \frac{W_0 x_0 - W x}{W_0 - W} \tag{13-6}$$

To obtain the temperature in the still, it is necessary to use experimental T–x–y data for benzene–toluene at 101.3 kPa as given in Table 13.1. The temperature and compositions as a function of time are shown in Figure 13.2. ∎

EXAMPLE 13.2

Repeat Example 13.1, except instead of using a constant value of 2.41 for the relative volatility, use the vapor-liquid equilibrium data for benzene-toluene at 101.3 kPa, given in Table 13.1, to solve the problem graphically or numerically with (13-3) rather than (13-5).

SOLUTION

Equation (13-3) can be solved by plotting $1/(y - x)$ versus x with a lower limit of $x_0 = 0.5$. Using the data of Table 13.1 for y as a function of x, points for the plot in terms of benzene are as follows:

x	0.5	0.4	0.3	0.2	0.1
$\dfrac{1}{y - x}$	4.695	4.717	4.831	5.814	9.259

Table 13.1 Vapor–Liquid Equilibrium Data for Benzene (B)–Toluene (T) at 101.3 kPa

x_B	y_B	T, °C
0.100	0.208	105.3
0.200	0.372	101.5
0.300	0.507	98.0
0.400	0.612	95.1
0.500	0.713	92.3
0.600	0.791	89.7
0.700	0.857	87.3
0.800	0.912	85.0
0.900	0.959	82.7
0.950	0.980	81.4

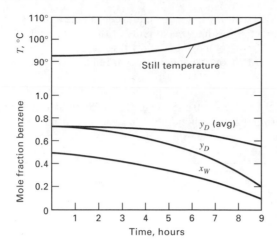

Figure 13.2 Distillation conditions for Example 13.1.

The area under the plotted curve from $x_0 = 0.5$ to a given value of x is equated to $\ln(W/W_0)$, and W is computed for $W_0 = 100$ kmol. In the region from $x = 0.5$ to 0.3, the value of $1/(y - x)$ changes only slightly. Therefore, a numerical integration by the trapezoidal rule is readily made:

$x = 0.4$:

$$\ln\left(\frac{W}{W_0}\right) = \int_{0.5}^{0.4} \frac{dx}{y - x} \approx \Delta x \left[\frac{1}{y - x}\right]_{\text{avg}} - (0.4 - 0.5)\left(\frac{4.695 + 4.717}{2}\right) = -0.4706$$

$$W/W_0 = 0.625, \ W = -0.625(100) = 62.5 \text{ kmol}$$

$x = 0.3$:

$$\ln\left(\frac{W}{W_0}\right) = \int_{0.5}^{0.3} \frac{dx}{y - x} \approx \Delta x \left[\frac{1}{y - x}\right]_{\text{ave}} = (0.3 - 0.5)\left[\frac{4.695 + 4.717 + 4.717 + 4.831}{4}\right] = -0.948$$

$$W/W_0 = 0.388, \ W = 0.388(100) = 38.8 \text{ kmol}$$

These two values are in good agreement with those in Example 13.1. A graphical integration from $x_0 = 0.4$ to $x = 0.1$ gives $W = 10.7$, which is approximately 10% less than the result in Example 13.1, which uses a constant value of the relative volatility. ∎

The Rayleigh equation (13-1) can be applied to any two components, i and j, of a multicomponent mixture. Thus, if we let

$$M_i = Wx_{W_i} \tag{13-6}$$

Then $$dM_i/dM_j = y_{D_i}/y_{D_j} \tag{13-7}$$

For constant $\alpha_{i,j} = y_{D_i}x_{W_j}/y_{D_j}x_{W_i}$, (13-7) becomes

$$dM_i/dM_j = \alpha_{i,j}(x_{W_i}/x_{W_j}) \tag{13-8}$$

Substitution of (13-6) for both i and j into (13-8) gives

$$dM_i/M_i = \alpha_{i,j}dM_j/M_j \tag{13-9}$$

Integration from the initial-charge condition gives

$$\ln(M_i/M_{i_0}) = \alpha_{i,j}\ln(M_j/M_{j_0}) \tag{13-10}$$

As shown in the following example, (13-10) is useful for determining the effect of relative volatility on the degree of separation that can be achieved by Rayleigh distillation.

EXAMPLE 13.3

The charge to a simple batch still consists of an equimolar binary mixture of A and B. For values of $\alpha_{A,B}$ of 2, 5, 10, 100, and 1,000, and 50% vaporization of A, determine the percent vaporization of B and the mole fraction of B in the total distillate.

SOLUTION

For $\alpha_{A,B} = 2$ and $M_A/M_{A_0} = 1 - 0.5 = 0.5$, (13-10) gives

$$M_B/M_{B_0} = (M_A/M_{A_0})^{1/\alpha_{A,B}} = (0.5)^{0.5} = 0.7071$$

Percent vaporization of B $= (1 - 0.7071)(100) = 29.29\%$.
For 200 moles of charge, the amounts of components in the distillate are $D_A = (0.5)(0.5)(200) = 50$ mol and $D_B = (0.2929)(0.5)(200) = 29.29$ mol

$$\text{Mole fraction of B in the total distillate} = \frac{29.29}{50 + 29.29} = 0.3694$$

Similar calculations for other values of $\alpha_{A,B}$ give the following results:

$\alpha_{A,B}$	% Vaporization of B	Mole Fraction of B in Total Distillate
2	29.29	0.3694
5	12.94	0.2057
10	6.70	0.1182
100	0.69	0.0136
1,000	0.07	0.0014

These results show that a sharp separation between A and B for 50% vaporization of A is only achieved if $\alpha_{A,B} \geq 100$. Furthermore, the purity achieved depends on the percent vaporization of A. For $\alpha_{A,B} = 100$, if 90% of A is vaporized, the mole fraction of B in the total distillate increases from 0.0136 to 0.0247. For this reason, as discussed in detail and illustrated by example in a later section of this chapter, it is common to conduct a binary batch distillation separation of light key (LK) and heavy key (HK) in the following manner:

1. Produce a distillate LK cut until the limit of impurity of HK in the total distillate is reached.
2. Continue the batch distillation to produce an intermediate cut of impure LK until the limit of impurity of LK in the liquid in the still is reached.
3. Empty the HK-rich cut from the still.
4. Recycle the intermediate cut to the next still charge.
   For desired purities of the LK cut and the HK cut, the fraction of intermediate cut increases as the LK-HK relative volatility decreases. ■

13.2 BINARY BATCH RECTIFICATION WITH CONSTANT REFLUX AND VARIABLE DISTILLATE COMPOSITION

To achieve a sharp separation and/or reduce the intermediate-cut fraction, a trayed or packed column, located above the still, and a means of sending reflux to the column, is provided as shown for the batch rectifier of Figure 13.3. For a column of a given diameter, the molar vapor boilup rate is usually fixed at a value safely below the column flooding point. If the reflux ratio R is fixed, distillate and still bottoms compositions vary with time. For a total condenser, negligible holdup of vapor and liquid in the column, phase equilibrium at each stage, and constant molar overflow, (13-2) still apply with $y_D = x_D$. The analysis of such a batch rectification for a binary system is facilitated by the McCabe–Thiele diagram using the method of Smoker and Rose [2].

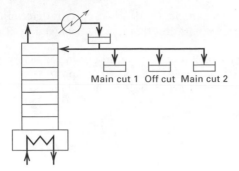

Figure 13.3 Batch rectification.

Initially, the composition of the light-key component in the liquid in the reboiler of the column in Figure 13.3 is the charge composition, x_0, which is given the value 0.45 in the McCabe–Thiele diagram of Figure 13.4. If there are two theoretical stages, the initial distillate composition x_0 at time 0 can be found by constructing an operating line of slope $L/V = R/(R + 1)$, such that exactly two stages are stepped off from x_0 to the $y = x$ line in Figure 13.4. At an arbitrary later time, say time 1, at still-pot composition $x_W < x_0$, the instantaneous distillate composition is x_D. A time-dependent series of points for x_D is thus established by trial and error, with L/V and the number of stages held constant.

Equation (13-2) cannot be integrated analytically because the relationship between y_D and x_W depends on the liquid-to-vapor ratio, the number of theoretical stages, and the phase equilibrium relationship. However, it can be integrated graphically with pairs of values for x_W and $y_D = x_D$ obtained from the McCabe–Thiele diagram.

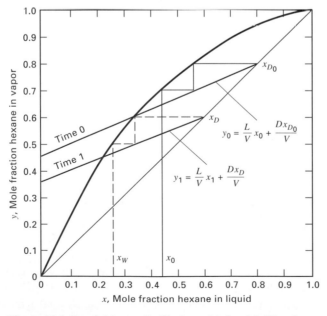

Figure 13.4 Batch binary distillation with fixed L/V and two theoretical stages.

The time t required for batch rectification at constant reflux ratio and negligible holdup in the column can be computed by a total material balance based on a constant boilup rate V to give the following equation due to Block [3]:

$$t = \frac{W_0 - W_t}{V\left(1 - \frac{L}{V}\right)} = \frac{R+1}{V}(W_0 - W_t) \tag{13-11}$$

With a constant-reflux policy, the distillate purity is above the specification at the beginning of distillation and below specification at the end of the run. By an overall material balance, the average mole fraction of the light-key component in the accumulated distillate at time t is given by

$$x_{D_{\text{ave}}} = \frac{W_0 x_0 - W_t x_{W_t}}{W_0 - W_t} \tag{13-12}$$

EXAMPLE 13.4

A three-theoretical-stage batch rectifier (first stage is the still pot) is charged with 100 kmol of a 20 mol% n-hexane in n-octane mixture. At a constant reflux ratio of 1 ($L/V = 0.5$), how many moles of charge must be distilled if an average product composition of 70 mol% C_6 is required? The phase equilibrium curve at column pressure is given in Figure 13.5. If the boilup rate is 10 kmol/h, calculate the distillation time.

SOLUTION

A series of operating lines and, hence, values of x_W are located by the trial-and-error procedure described earlier, as shown in Figure 13.5 for $x_0 = 0.20$ and $x_W = 0.09$. It is then possible to construct the following table:

$y_D = x_D$	0.85	0.60	0.5	0.35	0.3
x_W	0.2	0.09	0.07	0.05	0.035
$\dfrac{1}{y_D - x_W}$	1.54	1.96	2.33	3.33	3.77

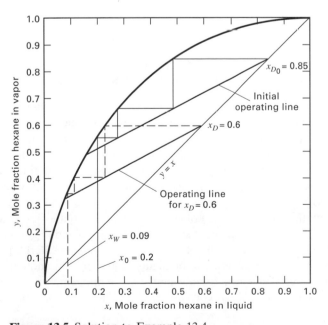

Figure 13.5 Solution to Example 13.4.

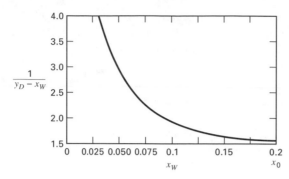

Figure 13.6 Graphical integration for Example 13.4.

The graphical integration is shown in Figure 13.6. Assuming a final value of $x_W = 0.1$, for instance, integration of (13-2) gives

$$\ln \frac{100}{W} = \int_{0.1}^{0.2} \frac{dx_W}{y_D - x_W} = 0.162$$

Hence,

$$W = 85 \text{ and } D = 15.$$

From (13-12):

$$(x_D)_{\text{ave}} = \frac{100(0.20) - 85(0.1)}{(100 - 85)} = 0.77$$

The $(x_D)_{\text{avg}}$ is higher than the desired value of 0.70; hence, another final x_W must be chosen. By trial, the correct answer is found to be $x_W = 0.06$, with $D = 22$, and $W = 78$, corresponding to a value of 0.25 for the integral.

From (13-11), the distillation time is $t = \dfrac{1+1}{10}(100 - 78) = 4.4\,\text{h}.$

When Rayleigh differential distillation is used, Figure 13.5 shows that a 70 mol% hexane distillate is not achievable because the initial distillate is only 56 mol% hexane. ∎

13.3 BINARY BATCH RECTIFICATION WITH CONSTANT DISTILLATE COMPOSITION AND VARIABLE REFLUX

The constant reflux policy described in the previous section is simple and easy to implement. For small batch rectification systems, it may be the least expensive policy. A more optimal operating policy is to maintain a constant molar vapor rate, but continuously vary the reflux ratio to achieve a constant distillate composition that meets the specified purity. This policy requires a more complex control system, including a composition monitor (or suitable substitute) on the distillate, which may be justified only for large batch rectification systems.

Calculations for the policy of constant distillate composition can also be made with the McCabe–Thiele diagram, as described by Bogart [4] and illustrated in Example 13.5. Other methods of operating batch columns are described by Ellerbe [5].

The Bogart method assumes negligible liquid holdup and constant molar overflow. An overall material balance for the light-key component, at any time, t, is given by a rearrangement of (13-12) at constant x_D, for W as a function of x_W.

$$W = W_0 \left[\frac{x_D - x_0}{x_D - x_W} \right] \tag{13-13}$$

Differentiating (13-13) with respect to t for varying W and x_W gives

$$\frac{dW}{dt} = W_0 \frac{(x_D - x_0)}{(x_D - x_W)^2} \frac{dx_W}{dt} \qquad \textbf{(13-14)}$$

For constant molar overflow, the rate of distillation is given by the rate of loss of charge, or

$$-\frac{dW}{dt} = (V - L) = \frac{dD}{dt} \qquad \textbf{(13-15)}$$

where D is now the amount of distillate, not the distillate rate. Substituting (13-15) into (13-14) and integrating:

$$t = \frac{W_0(x_D - x_0)}{V} \int_{x_{W_t}}^{x_0} \frac{dx_W}{(1 - L/V)(x_D - x_W)^2} \qquad \textbf{(13-16)}$$

For fixed values of W_0, x_0, x_D, V, and the number of equilibrium stages, the McCabe–Thiele diagram is used to determine values of L/V for a series of values of still composition between x_0 and the final value of x_W. These values are then used with (13-16) to determine, by graphical or numerical integration, the time for rectification or the time to reach any intermediate value of still composition. The required number of theoretical stages can be estimated by assuming total-reflux conditions for the final value of x_W. While rectification is proceeding, the instantaneous distillate rate will vary according to (13-15), which can be expressed in terms of L/V as

$$\frac{dD}{dt} = V(1 - L/V) \qquad \textbf{(13-17)}$$

EXAMPLE 13.5

A three-stage batch still (boiler and the equivalent of two equilibrium plates) is loaded with 100 kmol of a liquid containing a mixture of 50 mol% n-hexane in n-octane. A liquid distillate of 0.9 mole fraction hexane is to be maintained by continuously adjusting the reflux ratio, while maintaining a distillate rate of 20 kmol/h. What should the reflux ratio be after 1 h when the accumulated distillate is 20 kmol? Theoretically, when must accumulation of the distillate cut be stopped. Assume negligible holdup on the plates and constant molar overflow.

SOLUTION

When the accumulated distillate = 20 kmol, W = 80 kmol, and the still residue composition with respect to the light-key is given by a rearrangement of (13-13):

$$x_W = \frac{W x_D - W_0(x_D - x_0)}{W} = \frac{0.9(80) - 100(0.9 - 0.5)}{80} = 0.4$$

For $y_D = 0.9$, a series of operating lines of varying slope, $L/V = R/(R + 1)$, with three stages stepped off is used to determine the corresponding still residue composition. By trial and error, Line 1 in Figure 13.7 is found for $x_W = 0.4$, corresponding to an $L/V = 0.22$. The reflux ratio $= (L/V)/[1 - (L/V)] = 0.282$.

At the highest reflux rate possible, $L/V = 1$ (total reflux), and $x_W = 0.06$ according to the dashed-line construction shown in Figure 13.7. The corresponding time by material balance is given by $0.06(100 - 20t) = 50 - 20t(0.9)$. Solving, $t = 2.58$ h. ∎

13.4 BATCH STRIPPING AND COMPLEX BATCH DISTILLATION

A batch stripper consisting of a large accumulator, a trayed or packed stripping column, and a reboiler is shown in Figure 13.8. The initial charge is placed in the accumulator rather than the reboiler. The mixture in the accumulator is fed to the top of the column and the bottoms cut is removed from the reboiler. A batch stripper is useful for removing small quantities of volatile impurities. For binary mixtures, the McCabe–Thiele construc-

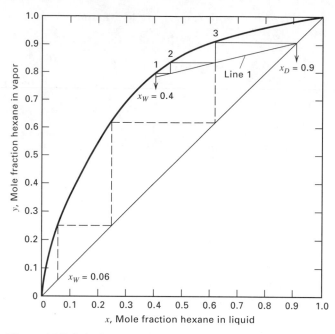

Figure 13.7 Solution to Example 13.5.

tion applies, and the graphical methods described in Sections 13.2 and 13.3 can be modified to follow with time the change in composition in the accumulator and the corresponding instantaneous and average composition of the bottoms cut.

A complex batch distillation unit, of the type described by Hasebe et al. [6], that permits considerable operating flexibility is shown in Figure 13.9. The charge in the feed tank is fed to a suitable column location. Holdups in the reboiler and condenser are kept to a minimum. Products or intermediate cuts are withdrawn from the condenser, the reboiler, or both. In addition, the liquid in the column at the feed location can be recycled to the feed tank if it is desirable to make the composition in the feed tank close to the composition of the liquid at the feed location.

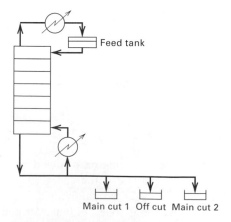

Figure 13.8 Batch stripping.

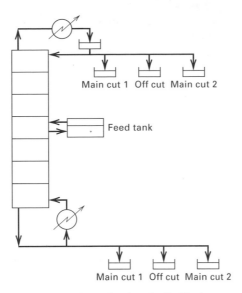

Figure 13.9 Complex batch distillation.

13.5   EFFECT OF LIQUID HOLDUP

Except at high pressure, vapor holdup in a rectifying column is negligible in batch distillation because of the small molar density of the vapor phase. However, the effect of liquid holdup on the trays and in the condensing and reflux system can be significant when the molar ratio of holdup to original charge is more than a few percent. This is especially true when a charge contains low concentrations of one or more of the components to be separated. In general, the effect of holdup in a trayed column is greater than in a packed column because of the lower amount of holdup in the latter. For either type of column, liquid holdup can be estimated by methods described in Chapter 6.

A batch rectifier is usually operated under total reflux conditions for an initial period of time prior to the withdrawal of distillate product. During this initial time period, liquid holdup in the column increases and approaches a value that is reasonably constant for the remainder of the distillation cycle. Because of the total-reflux concentration profile, the initial concentration of light components in the remaining charge to the still is less than in the original charge. At high liquid holdups, this causes the initial purity and degree of separation to be reduced from estimates based on methods that ignore liquid holdup. Liquid holdup can reduce the size of product cuts, increase the size of intermediate fractions that are recycled, increase the amount of residue, increase the batch cycle time, and increase the total energy input. Although approximate methods for predicting the effect of liquid holdup were developed in the past, the complexity of the holdup effect is such that it is now considered best to use the rigorous computer-based batch-distillation algorithms described later to study the effect on a case-by-case basis.

13.6   SHORTCUT METHOD FOR MULTICOMPONENT BATCH RECTIFICATION WITH CONSTANT REFLUX

The batch rectification method presented in Section 13.2 is limited to binary mixtures, under the assumptions of constant molar overflow, and negligible vapor and liquid holdup. Shortcut methods for handling multicomponent mixtures under the same assumptions have been developed by Diwekar and Madhaven [7] for the two cases of constant distillate composition and constant reflux, and by Sundaram and Evans [8] for constant reflux. Both methods avoid tedious stage-by-stage calculations of vapor and liquid compositions by employing the Fenske–Underwood–Gilliland (FUG) shortcut procedure for continuous distillation, described in Chapter 9, at successive time steps. In essence, they treat batch rectification as a sequence of continuous, steady-state rectifications. As in the FUG method, no estimations of compositions or temperatures are made for intermediate stages.

Sundaram and Evans [8] apply their shortcut method to a column of the type shown in Figure 13.3. An overall mole balance gives

$$-\frac{dW}{dt} = D \tag{13-18}$$

Therefore,
$$-\frac{dW}{dt} = \frac{V}{1+R} \tag{13-19}$$

For any component, i, an instantaneous mole balance around the column gives

$$\frac{d(x_{W_i}W)}{dt} = x_{D_i}\frac{dW}{dt} \tag{13-20}$$

Expanding the LHS of (13-20) and solving for dx_{W_i}:

$$dx_{W_i} = (x_{D_i} - x_{W_i})\frac{dW}{W} \tag{13-21}$$

In finite-difference form, using Euler's method, (13-19) and (13-21) become, respectively:

$$W^{(k+1)} = W^{(k)} - \left(\frac{V}{1+R}\right)\Delta t \tag{13-22}$$

$$x_{W_i}^{(k+1)} = x_{W_i}^{(k)} + (x_{D_i}^{(k)} - x_{W_i}^{(k)})\left[\frac{W^{(k+1)} - W^{(k)}}{W^{(k)}}\right] \tag{13-23}$$

where k is the time-increment index. For a given Δt time increment, $W^{(k+1)}$ is computed from (13-22) and then $x_{W_i}^{(k+1)}$ is computed for each component from (13-23). However, (13-23) requires values for $x_{D_i}^{(k)}$.

Calculations are initiated at $k = 0$. The initial charge to the still is $W^{(0)}$. Values of $x_{W_i}^{(0)}$ are equal to the mole fractions of the initial charge. Corresponding values of $x_{D_i}^{(0)}$ depend on the method used to start up the batch rectification still. If total-reflux operation is employed as the start-up method, as mentioned in Section 13.5, the Fenske equation of Chapter 9 can be applied to compute values of $x_{D_i}^{(0)}$ for given values of $x_{W_i}^{(0)}$ if column and condenser holdups are negligible. For a given number of equilibrium stages, N:

$$N = \frac{\log\left[\left(\frac{x_{D_i}}{x_{W_i}}\right)\left(\frac{x_{W_r}}{x_{D_r}}\right)\right]}{\log\alpha_{i,r}} \tag{13-24}$$

Solving,
$$x_{D_i} = x_{W_i}\left(\frac{x_{D_r}}{x_{W_r}}\right)\alpha_{i,r}^N \tag{13-25}$$

where r is an arbitrary reference component of the mixture, such as the least volatile component. Since

$$\sum_{i=1}^{C} x_{D_i} = 1.0 \tag{13-26}$$

substitution of (13-25) in (13-26) gives

$$x_{D_r} = \frac{x_{W_r}}{\sum\limits_{i=1}^{C} x_{W_i} \alpha_{i,r}^N} \tag{13-27}$$

The initial distillate composition, $x_{D_r}^{(0)}$, is computed from (13-27). The remaining values of $x_{D_i}^{(0)}$ are computed in turn from (13-25).

Using the initial set of values for $x_{D_i}^{(0)}$, values of $x_{W_i}^{(1)}$ are computed from (13-23) following the calculation of $W^{(1)}$ from (13-22). To compute each subsequent set of $x_{W_i}^{(k+1)}$ for $k > 0$, values of $x_{D_i}^{(k)}$ for $k > 0$ are needed. These are obtained by applying the FUG method. Equation (13-24) applies during batch rectification if N is replaced by $N_{\min} < N$ with $i =$ LK and $r =$ HK. But $N_{\min}$ is related to N by the Gilliland correlation. A convenient, but approximate, equation for that correlation, due to Eduljee [9], is

$$\frac{N - N_{\min}}{N + 1} = 0.75 \left[1 - \left(\frac{R - R_{\min}}{R + 1} \right)^{0.5668} \right] \tag{13-28}$$

An estimate of the minimum reflux ratio, $R_{\min}$, is provided by the Class I Underwood equation of Chapter 9, which assumes that all components in the charge distribute between the two products. Thus:

$$R_{\min} = \frac{\left(\dfrac{x_{D_{LK}}}{x_{W_{LK}}} \right) - \alpha_{LK,HK} \left(\dfrac{x_{D_{HK}}}{x_{W_{HK}}} \right)}{\alpha_{LK,HK} - 1} \tag{13-29}$$

If one or more components fail to distribute, then Class II Underwood equations should be used. Sundaran and Evans use only (13-29) with LK and HK equal to the lightest component, 1, and the heaviest component, C, in the mixture, respectively.

If (13-25), with $i = 1$, $r = C$, and $N = N_{\min}$, and (13-27) with $r = C$ are substituted into (13-29) with LK = 1 and HK = C, the result is

$$R_{\min} = \frac{\alpha_{1,C}^{N_{\min}} - \alpha_{1,C}}{(\alpha_{1,C} - 1) \sum\limits_{i=1}^{C} x_{W_i} \alpha_{i,C}^{N_{\min}}} \tag{13-30}$$

For specified values of N and R, (13-28) and (13-30) are solved for $R_{\min}$ and $N_{\min}$ simultaneously by an iterative method. The value of x_{D_C} is then computed from (13-27) with $N = N_{\min}$, followed by the calculation of the other values of x_{D_i} from (13-25). Values of $N_{\min}$ and $R_{\min}$ change with time.

The procedure of Sundaram and Evans involves an inner loop for the calculation of $x_{D_i}^{(k)}$, and an outer loop for $W^{(k+1)}$ and $x_{W_i}^{(k+1)}$. The inner loop requires iterations because of the nonlinear nature of (13-28) and (13-30). Calculations of the outer loop are direct because (13-22) and (13-23) are linear. Application of the method is illustrated in the following example, where relative volatilities are assumed constant.

EXAMPLE 13.6

A charge of 100 kmol of a ternary mixture of A, B, and C with composition, $x_{W_A}^{(0)} = 0.33$, $x_{W_B}^{(0)} = 0.33$, and $x_{W_C}^{(0)} = 0.34$ is distilled in a batch rectifier with $N = 3$ (including the reboiler), $R = 10$, and $V = 110$ kmol/h. Estimate the variation of the still, instantaneous distillate, and distillate accumulator compositions as a function of time for 2 h of operation following an initial start-up period during which a steady-state operation at total reflux is achieved. Use $\alpha_{AC} = 2.0$ and $\alpha_{BC} = 1.5$, and neglect column holdup.

SOLUTION

The method of Sandaram and Evans is applied with $D = V/(1 + R) = 110/(1 + 10) = 10$ kmol/h. Therefore, 10 h would be required to distill the entire charge.

Start-up Period:

From (13-27), with C as the reference r, $\quad x_{D_C}^{(0)} = \dfrac{0.34}{0.33(2)^3 + 0.33(1.5)^3 + 0.34(1)^3} = 0.0831$.

From (13-25), $\quad x_{D_A}^{(0)} = 0.33 \left(\dfrac{0.0831}{0.34} \right) 2^3 = 0.6449 \quad$ and $\quad x_{D_B}^{(0)} = 0.33 \left(\dfrac{0.0831}{0.34} \right) 1.5^3 = 0.2720$

Take time increments, Δt, of 0.5 h.

At t = 0.5 h for outer loop:

From (13-22), $\qquad\qquad W^{(1)} = 100 - \left(\dfrac{110}{1 + 10} \right) 0.5 = 95 \text{ kmol}$

From (13-23) with $k = 0$:

$$x_{W_A}^{(1)} = 0.33 + (0.6449 - 0.33) \left[\frac{95 - 100}{100} \right] = 0.3143$$

$$x_{W_B}^{(1)} = 0.33 + (0.2720 - 0.33) \left[\frac{95 - 100}{100} \right] = 0.3329$$

$$x_{W_C}^{(1)} = 0.34 + (0.0831 - 0.34) \left[\frac{95 - 100}{100} \right] = 0.3528$$

At t = 0.5 h for inner loop:

From (13-28) $\qquad\qquad \dfrac{3 - N_{\min}}{3 + 1} = 0.75 \left[1 - \left(\dfrac{10 - R_{\min}}{10 + 1} \right)^{0.5668} \right]$

Solving for $R_{\min}$, $\qquad\qquad R_{\min} = 10 - 1.5835 \, N_{\min}^{1.7643}$ $\qquad\qquad$ **(1)**

This equation holds for all values of time t.

From (13-30), $\qquad R_{\min} = \dfrac{2^{N_{\min}} - 2}{(2 - 1)[0.3143(2)^{N_{\min}} + 0.3329(1.5)^{N_{\min}} + 0.3528(1)^{N_{\min}}]}$ $\qquad$ **(2)**

Equations (1) and (2) are solved simultaneously for $R_{\min}$ and $N_{\min}$. This can be done by numerical or graphical methods including successive substitution, Newton's method, or with a spreadsheet by plotting each equation as $R_{\min}$ versus $N_{\min}$ and determining the intersection. The result is $R_{\min} = 1.2829$ and $N_{\min} = 2.6294$.

From (13-27), with $N = 2.6294$:

$$x_{D_C}^{(1)} = 0.3528/[0.3143(2)^{2.6294} + 0.3329(1.5)^{2.6294} + 0.3528] = 0.1081$$

Table 13.2 Results for Example 13.6

Time, h	W, kmol	x_W			$N_{\min}$	$R_{\min}$	x_D			x of Accumulated Distillate		
		A	B	C			A	B	C	A	B	C
0.0	100	0.3300	0.3300	0.3400	—	—	0.6449	0.2720	0.0831	—	—	—
0.5	95	0.3143	0.3329	0.3528	2.6294	1.2829	0.5957	0.2962	0.1081	0.6283	0.2749	0.0968
1.0	90	0.2995	0.3348	0.3657	2.6249	1.3092	0.5803	0.3048	0.1149	0.6045	0.2868	0.1087
1.5	85	0.2839	0.3365	0.3796	2.6199	1.3385	0.5633	0.3142	0.1225	0.5912	0.2932	0.1156
2.0	80	0.2675	0.3378	0.3947	2.6143	1.3709	0.5446	0.3242	0.1312	0.5800	0.2988	0.1212

$$x_{D_A} = 0.3143 \left(\frac{0.1081}{0.3528}\right) 2^{2.6924} = 0.5959$$

From (13-25):

$$x_{D_B} = 0.3329 \left(\frac{0.1081}{0.3528}\right) 1.5^{2.6294} = 0.2962$$

Subsequent, similar calculations give the results in Table 13.2. ∎

13.7 STAGE-BY-STAGE METHODS FOR MULTICOMPONENT BATCH RECTIFICATION

For final design studies or for the simulation of multicomponent batch rectification, complete stage-by-stage temperature, flow, and composition profiles as a function of time are required. Such calculations are tedious, but can be carried out conveniently with either of two types of computer-based methods. Both methods are based on the same differential-algebraic equations for the distillation model, but differ in the way the equations are solved.

Rigorous Model

Meadows [10] developed the first rigorous multicomponent batch distillation model, based on the assumptions of equilibrium stages, perfect mixing of liquid and vapor phases at each stage, negligible vapor holdup, constant molar liquid holdup, M, on a stage and in the condenser system, and adiabatic stages in the column. Distefano [11] extended the model and developed a computer-based method for solving the set of equations. A more efficient method for solving the equations is presented by Boston et al. [12]. For more rapid calculations, Galindez and Fredenslund [13] developed a quasi-steady state solution procedure that, at each step, utilizes the simultaneous-correction or inside-out methods for continuous distillation discussed in Chapter 10.

The Distefano model is based on the multicomponent batch rectification operation shown in Figure 13.10. The equipment consists of a partial reboiler, a column with N equilibrium stages or equivalent in packing, and a total condenser with a reflux drum. Also included, but not shown in Figure 13.10, are a number of accumulator or receiver drums equal to the desired number of overhead product and intermediate cuts. To initiate operation, the feed is charged to the reboiler, to which heat is supplied. Vapor leaving Stage 1 at the top of the column is totally condensed and passes to the reflux drum. At first, a total reflux condition is established for a steady-state fixed overhead vapor flow rate. Depending upon the amount of liquid holdup in the column and in the condenser system, the amount and composition of the liquid in the reboiler at total reflux differs to some extent from the original feed.

Starting at time $t = 0$, distillate is removed from the reflux drum and sent to a receiver (accumulator) at a constant molar rate, and a reflux ratio is established. The heat-transfer rate to the reboiler is adjusted so as to maintain the overhead vapor molar flow rate. Model equations are derived for the overhead condensing system, the column stages, and the reboiler, as illustrated for the overhead condensing system. For Section I, in Figure 13.10, component material balances, a total material balance, and an energy balance are given, respectively, by

$$V_1 y_{i,1} - L_0 x_{i,0} - D x_{i,D} = \frac{d(M_0 x_{i,0})}{dt} \tag{13-31}$$

$$V_1 - L_0 - D = \frac{dM_0}{dt} \tag{13-32}$$

$$V_1 h_{V_1} - (L_0 + D)h_{L_0} = Q_0 + \frac{d(M_0 h_{L_0})}{dt} \tag{13-33}$$

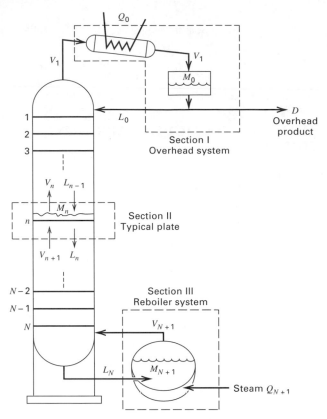

Figure 13.10 Multicomponent batch rectification operation. [From G.P. Distefano, *AIChE J.,* **140,** 190 (1968) with permission.]

where the derivative terms are accumulations due to holdup. Also, for phase equilibrium at Stage 1 of the column:

$$y_{i,1} = K_{i,1}x_{i,1} \qquad \text{(13-34)}$$

The working equations are obtained by combining (13-31) and (13-34) to obtain a revised component material balance in terms of liquid-phase compositions, and by combining (13-32) and (13-33) to obtain a revised energy balance that does not include dM_0/dt. Equations for Sections II and III in Figure 13.10 are derived in a similar manner. The resulting working model equations for $t = 0+$ are as follows, where i refers to the component and j refers to the stage, and M is molar liquid holdup.

1. Component mole balances for the overhead condensing system, column stages, and reboiler, respectively:

$$\frac{dx_{i,0}}{dt} = -\left[\frac{L_0 + D + \dfrac{dM_0}{dt}}{M_0}\right]x_{i,0} + \left[\frac{V_1 K_{i,1}}{M_0}\right]x_{i,1}, \quad i = 1 \text{ to } C \qquad \text{(13-35)}$$

$$\frac{dx_{i,j}}{dt} = \left[\frac{L_{j-1}}{M_j}\right]x_{i,j-1} - \left[\frac{L_j + K_{i,j}V_j + \dfrac{dM_j}{dt}}{M_j}\right]x_{i,j}$$

$$+ \left[\frac{K_{i,j+1}V_{j+1}}{M_j}\right]x_{i,j+1}, \quad i = 1 \text{ to } C, j = 1 \text{ to } N \qquad \text{(13-36)}$$

$$\frac{dx_{i,N+1}}{dt} = \left(\frac{L_N}{M_{N+1}}\right)x_{i,N} - \left[\frac{V_{N+1}K_{i,N+1} + \dfrac{dM_{N+1}}{dt}}{M_{N+1}}\right]x_{i,N+1}, \quad i = 1 \text{ to } C \qquad \textbf{(13-37)}$$

where $L_0 = RD$.

2. Total mole balances for overhead condensing system and column stages, respectively:

$$V_1 = D(R + 1) + \frac{dM_0}{dt} \qquad \textbf{(13-38)}$$

$$L_j = V_{j+1} + L_{j-1} - V_j - \frac{dM_j}{dt}, \quad j = 1 \text{ to } N \qquad \textbf{(13-39)}$$

3. Enthalpy balances around overhead condensing system, adiabatic column stages, and reboiler, respectively:

$$Q_0 = V_1(h_{V_1} - h_{L_0}) - M_0\frac{dh_{L_0}}{dt} \qquad \textbf{(13-40)}$$

$$V_{j+1} = \frac{1}{(h_{V_{j+1}} - h_{L_j})}\left[V_j(h_{V_j} - h_{L_j}) - L_{j-1}(h_{L_{j-1}} - h_{L_j}) + M_j\frac{dh_{L_j}}{dt}\right], \quad j = 1 \text{ to } N \qquad \textbf{(13-41)}$$

$$Q_{N+1} = V_{N+1}(h_{V_{N+1}} - h_{L_{N+1}}) - L_N(h_{L_N} - h_{L_{N+1}}) + M_{N+1}\left(\frac{dh_{L_{N+1}}}{dt}\right) \qquad \textbf{(13-42)}$$

4. Phase equilibrium on column stages and in the reboiler:

$$y_{i,j} = K_{i,j}x_{i,j}, \quad i = 1 \text{ to } C, \quad j = 1 \text{ to } N + 1 \qquad \textbf{(13-43)}$$

5. Mole fraction sums at column stages and in the reboiler:

$$\sum_{i=1}^{C} y_{i,j} = \sum_{i=1}^{C} K_{i,j}x_{i,j} = 1.0, \quad j = 0 \text{ to } N + 1 \qquad \textbf{(13-44)}$$

6. Molar holdups in the condenser system and on the column stages based on constant-volume holdups, G_j, respectively:

$$M_0 = G_0\rho_0 \qquad \textbf{(13-45)}$$

$$M_j = G_j\rho_j, \quad j = 1 \text{ to } N \qquad \textbf{(13-46)}$$

where ρ is liquid molar density.

7. Variation of molar holdup in the reboiler, where M_{N+1}^0 is the initial charge to the reboiler:

$$M_{N+1} = M_{N+1}^0 - \sum_{j=0}^{N} M_j - \int_0^t D\,dt \qquad \textbf{(13-47)}$$

Equations (13-35) through (13-47) constitute an initial-value problem for a system of ordinary differential and algebraic equations (DAEs). The total number of equations is $(2CN + 3C + 4N + 7)$. If variables N, D, $R = L_0/D$, M_{N+1}^0, and all G_j are specified, and if correlations are available for computing liquid densities, vapor and liquid enthalpies, and K-values, the number of unknown variables, distributed as follows, is equal to the number of equations.

$x_{i,j}$	$CN + 2C$
$y_{i,j}$	$CN + C$
L_j	N
V_j	$N + 1$
T_j	$N + 2$
M_j	$N + 2$
Q_0	1
Q_{N+1}	1
	$2CN + 3C + 4N + 7$

Initial values at $t = 0$ for all these variables are obtained from the steady-state total reflux calculation, which depends only on values of N, M_{N+1}^0, x_{N+1}^0, G_j, and V_1.

Equations (13-35) through (13-42) include first derivatives of $x_{i,j}$, M_j, and h_{L_j}. Except for M_{N+1}, derivatives of the latter two variables can be approximated with sufficient accuracy by incremental changes over the previous time step. If the reflux ratio is high, as it often is, the derivative of M_{N+1} can also be approximated in the same manner. This leaves only the $C(N + 2)$ ordinary differential equations (ODEs) for the component material balances to be integrated in terms of the $x_{i,j}$ dependent variables.

Rigorous Integration Method

The nonlinear equations (13-35) to (13-37) cannot be integrated analytically. Distefano [11] developed a numerical method of solution based on an investigation of 11 different numerical integration techniques that step in time. Of particular concern were the problems of *truncation* error and *stability.*

Local truncation errors result from using approximations for the functions on the right-hand side of the ODEs at each time step. These errors may be small, but they can propagate through subsequent time steps, resulting in global truncation errors sufficiently large to be unacceptable. As truncation errors become large, the number of significant digits in the computed dependent variables gradually decrease. Truncation errors can be reduced by decreasing the size of the time step.

Stability problems are much more serious. When instability occurs, the computed values of the dependent variables become totally inaccurate, with no significant digits at all. Reducing the time step does not eliminate instability until a time-step criterion, which depends on the numerical method, is satisfied. Even then, a further reduction in the time step is required to prevent oscillations of dependent variables.

Problems of stability and truncation error are conveniently illustrated by comparing results obtained by using the explicit and implicit Euler methods, both of which are first-order in accuracy, as discussed by Davis [15] and Riggs [16].

Consider the nonlinear, first-order ODE:

$$\frac{dy}{dt} = f\{t, y\} = ay^2te^y \tag{13-48}$$

for $y\{t\}$, where initially $y\{t_0\} = y_0$. The explicit (forward) Euler method approximates (13-48) with a sequence of discretizations of the form

$$\frac{y_{k+1} - y_k}{\Delta t} = ay_k^2 t_k e^{y_k} \tag{13-49}$$

where Δt is the time step and k the sequence index. The function $f\{t, y\}$ is evaluated at the beginning of the current time step. Solving for y_{k+1} gives the recursion equation:

$$y_{k+1} = y_k + ay_k^2 t_k e^{y_k} \Delta t \tag{13-50}$$

Regardless of the nature of $f\{t, y\}$ in (13-48), the recursion equation can be solved explicitly for y_{k+1} using results from the previous time step. However, as discussed later, this advantage is counterbalanced by a limitation on the magnitude of Δt to avoid instability and oscillations.

The implicit (backward) Euler method also utilizes a sequence of discretizations of (13-48), but the function, $f\{t, y\}$, is evaluated at the end of the current time step. Thus:

$$\frac{y_{k+1} - y_k}{\Delta t} = ay_{k+1}^2 t_{k+1} e^{y_{k+1}} \tag{13-51}$$

Because the function $f\{t, y\}$ is nonlinear in y, (13-51) cannot be solved explicitly for y_{k+1}. This disadvantage is counterbalanced by unconditional stability with respect to selection of Δt. However, too large a value can result in unacceptable truncation errors.

When the explicit Euler method is applied to (13-35) to (13-47) for batch rectification, as shown in the following example, the maximum value of Δt can be estimated from the maximum absolute eigenvalue, $|\lambda|_{max}$, of the Jacobian matrix of (13-35) to (13-37). To prevent instability, $\Delta t_{max} \leq 2/|\lambda|_{max}$. To prevent oscillations, $\Delta t_{max} \leq 1/|\lambda|_{max}$. Applications of the explicit and implicit Euler methods are compared in the following batch rectification example.

EXAMPLE 13.7

A charge of 100 kmol of an equimolar mixture of n-hexane (A) and n-heptane (B) is distilled at 15 psia in a batch rectifier consisting of a total condenser with a constant liquid holdup, M_0, of 0.10 kmol, a single equilibrium stage with a constant liquid holdup, M_1, of 0.01 kmol, and a reboiler. Initially the system is brought to the following total-reflux condition, with saturated liquid leaving the total condenser:

Stage	T, °F	x_A	K_A	K_B	M, kmol
Condenser	162.6	0.85935	—	—	0.1
Plate, 1	168.7	0.70930	1.212	0.4838	0.01
Reboiler, 2	178.6	0.49962	1.420	0.5810	99.89

Distillation begins ($t = 0$) with a reflux rate, L_0, of 10 kmol/h and a distillate rate, D, of 10 kmol/h. Calculate the mole fractions of n-hexane and n-heptane at $t = 0.05$ h (3 min), at each of the three rectifier locations, assuming constant molar overflow and constant K-values from above for this small period of elapsed time. Use both the explicit and implicit Euler methods to determine the influence of the choice of the time step, Δt.

SOLUTION

Based on the constant molar overflow assumption:

$$V_1 = V_2 = 20 \text{ kmol/h} \quad \text{and} \quad L_0 = L_1 = 10 \text{ kmol/h}$$

Using the K-values and liquid holdups given earlier, (13-35) to (13-37), with all $dM_j/dt = 0$, become as follows:

Condenser:
$$\frac{dx_{A,0}}{dt} = -200x_{A,0} + 242.4x_{A,1} \tag{1}$$

$$\frac{dx_{B,0}}{dt} = -200x_{B,0} + 96.76x_{B,1} \tag{2}$$

Plate:

$$\frac{dx_{A,1}}{dt} = 1{,}000 x_{A,0} - 3{,}424 x_{A,1} + 2{,}840 x_{A,2} \tag{3}$$

$$\frac{dx_{B,1}}{dt} = 1{,}000 x_{B,0} - 1{,}967 x_{B,1} + 1{,}162 x_{B,2} \tag{4}$$

Reboiler:

$$\frac{dx_{A,2}}{dt} = \left(\frac{10}{M_2}\right) x_{A,1} - \left(\frac{28.40}{M_2}\right) x_{A,2} \tag{5}$$

$$\frac{dx_{B,2}}{dt} = \left(\frac{10}{M_2}\right) x_{B,1} - \left(\frac{11.62}{M_2}\right) x_{B,2} \tag{6}$$

where

$$M_2(t = t) = M_2(t = 0) - (V_2 - V_1)t$$
$$\text{or } M_2 = 99.89 - 10t \tag{7}$$

Equations (1) through (6) can be grouped by component into the following two matrix equations:

Component A:
$$\begin{bmatrix} -200 & 242.2 & 0 \\ 1{,}000 & -3{,}424 & 2{,}840 \\ 0 & 10/M_2 & -28.40/M_2 \end{bmatrix} \cdot \begin{bmatrix} x_{A,0} \\ x_{A,1} \\ x_{A,2} \end{bmatrix} = \begin{bmatrix} dx_{A,0}/dt \\ dx_{A,1}/dt \\ dx_{A,2}/dt \end{bmatrix} \tag{8}$$

Component B:
$$\begin{bmatrix} -200 & 96.76 & 0 \\ 1{,}000 & -1{,}967 & 1{,}160 \\ 0 & 10/M_2 & -11.62/M_2 \end{bmatrix} \cdot \begin{bmatrix} x_{B,0} \\ x_{B,1} \\ x_{B,2} \end{bmatrix} = \begin{bmatrix} dx_{B,0}/dt \\ dx_{B,1}/dt \\ dx_{B,2}/dt \end{bmatrix} \tag{9}$$

Although (8) and (9) do not appear to be coupled, they are because at each time step the sums, $x_{A,j} + x_{B,j}$ do not equal 1. Accordingly, the mole fractions are normalized at each time step to force them to sum to 1. The initial eigenvalues of the Jacobian matrices (8) and (9) are computed from any of a number of computer programs, such as MathCad, Mathematica, MATLAB, or Maple, to be as follows, using $M_2 = 99.89$ kmol:

	Component A	Component B
λ_0	−126.54	−146.86
λ_1	−3,497.6	−2,020.2
λ_2	−0.15572	−0.03789

It is seen that $|\lambda|_{max} = 3{,}497.6$. Thus, for the explicit Euler method, instability and oscillations can be prevented by choosing:

$$\Delta t \le 1/3497.6 = 0.000286$$

If we select $\Delta t = 0.00025$ h (just slightly smaller than the criterion), it takes $0.05/0.00025 = 200$ time steps to reach $t = 0.05$ h (3 min). No such restriction applies to the implicit Euler method, but too large a Δt may result in an unacceptable truncation error.

Explicit Euler Method

According to Distefano [11], the maximum step size for integration using an explicit method is nearly always limited by stability considerations, and usually the truncation error is small. Assuming this to be true for this example, the following results were obtained using $\Delta t = 0.00025$ h with a spreadsheet program by converting (8) and (9) together with (7) for M_2, to the form of (13-50). Only the results for every 40 time steps are given.

Time, h	Normalized Mole Fractions in Liquid for n-Hexane			Normalized Mole Fractions in Liquid for n-Heptane		
	Distillate	Plate	Still	Distillate	Plate	Still
0.01	0.8183	0.6271	0.4993	0.1817	0.3729	0.5007
0.02	0.8073	0.6219	0.4991	0.1927	0.3781	0.5009
0.03	0.8044	0.6205	0.4988	0.1956	0.3795	0.5012
0.04	0.8036	0.6199	0.4985	0.1964	0.3801	0.5015
0.05	0.8032	0.6195	0.4982	0.1968	0.3805	0.5018

To show the instability effect, a time step of 0.001 h (four times the previous time step) gives the following unstable results during the first five time steps to an elapsed time of 0.005 h. Also included are values at 0.01 h for comparison to the preceding stable results.

Time, h	Normalized Mole Fractions in Liquid for n-Hexane			Normalized Mole Fractions in Liquid for n-Heptane		
	Distillate	Plate	Still	Distillate	Plate	Still
0.000	0.85935	0.7093	0.49962	0.14065	0.2907	0.50038
0.001	0.859361	0.559074	0.499599	0.140639	0.440926	0.500401
0.002	0.841368	0.75753	0.499563	0.158632	0.24247	0.500437
0.003	0.852426	0.00755	0.499552	0.147574	0.99245	0.500448
0.004	0.809963	0.884925	0.499488	0.190037	0.115075	0.500512
0.005	0.874086	1.154283	0.499546	0.125914	−0.15428	0.500454
0.01	1.006504	0.999254	0.493573	−0.0065	0.000746	0.506427

Much worse results are obtained if the time step is increased 10-fold to 0.01 h, as shown in the following table, where at $t = 0.01$ h, a very negative mole fraction has appeared.

Time, h	Normalized Mole Fractions in Liquid for n-Hexane			Normalized Mole Fractions in Liquid for n-Heptane		
	Distillate	Plate	Still	Distillate	Plate	Still
0.00	0.85935	0.7093	0.49962	0.14065	0.2907	0.50038
0.01	0.859456	−0.79651	0.49941	0.140544	1.796512	0.50059
0.02	2.335879	2.144666	0.497691	−1.33588	−1.14467	0.502309
0.03	1.284101	1.450481	0.534454	−0.2841	−0.45048	0.465546
0.04	1.145285	1.212662	8.95373	−0.14529	−0.21266	−7.95373
0.05	1.07721	1.11006	1.191919	−0.07721	−0.11006	−0.19192

Implicit Euler Method

If (8) and (9) are converted to implicit equations like (13-51), they can be rearranged into a linear tridiagonal set for each component. For example, the equation for component A on the plate becomes

$$(1{,}000\,\Delta t)x_{A,0}^{(k+1)} - (1 + 3{,}424\,\Delta t)x_{A,1}^{(k+1)} + (2{,}840\,\Delta t)x_{A,2}^{(k+1)} = -x_{A,1}^{(k)}$$

The two tridiagonal equation sets can be solved by the tridiagonal matrix algorithm or with a spreadsheet program using the iterative circular reference technique. For the implicit Euler method, the selection of the time step, Δt, is not restricted by stability considerations. However, too large a Δt can lead to unacceptable truncation errors. Normalized liquid mole fraction results at $t = 0.05$ h for just component A are as follows for a number of different choices of Δt, all of which are greater than the 0.00025 h used earlier to obtain stable and oscillation-free results with the explicit Euler method. Included for comparison is the explicit Euler result for $\Delta t = 0.00025$ h.

Time = 0.05 h:

	Normalized Mole Fractions in Liquid for n-Hexane		
Δt, h	Distillate	Plate	Still
	Explicit Euler		
0.00025	0.8032	0.6195	0.4982
	Implicit Euler		
0.0005	0.8042	0.6210	0.4982
0.001	0.8042	0.6210	0.4982
0.005	0.8045	0.6211	0.4982
0.01	0.8049	0.6213	0.4982
0.05	0.8116	0.6248	0.4982

The preceding data show acceptable results with the implicit Euler method using a time step about 200 times the Δt_{max} for the explicit Euler method. ∎

Another serious computational problem occurs when integrating the equations of batch distillation. Because the liquid holdups on the trays and in the condenser are small, the values of the corresponding liquid mole fractions, $x_{i,j}$, respond quickly to changes. The opposite holds for the reboiler (still) with its large liquid holdup. Hence, the required time step for accuracy is usually small, leading to a very slow response of the overall rectification system. Systems of ODEs having this characteristic constitute so-called stiff systems. For such a system, as discussed by Carnahan and Wilkes [17], an explicit method of solution must utilize a small time step for the entire period even though values of the dependent variables may all be changing slowly for a large portion of the time period. Accordingly, it is preferred to utilize special implicit integration techniques developed by Gear [14] and others, as contained in the public-domain software package called ODEPACK. Gear-type methods strive for accuracy, stability, and computational efficiency by using multistep, variable order, and variable step-size implicit techniques.

A commonly used measure of the degree of stiffness is the eigenvalue ratio $|\lambda|_{max}/|\lambda|_{min}$, where λ values are the eigenvalues of the Jacobian matrix of the set of ODEs. For the Jacobian matrix of (13-35) through (13-37), the Gerschgorin circle theorem, discussed by Varga [18], can be employed to estimate the eigenvalue ratio. The maximum absolute eigenvalue corresponds to the component with the largest K-value and the tray with the smallest liquid molar holdup. When the Gerschgorin theorem is applied to a row of the Jacobian matrix based on (13-36):

$$|\lambda|_{max} \leq \left[\left(\frac{L_{j-1}}{M_j} \right) + \left(\frac{L_j + K_{i,j} V_j}{M_j} \right) + \left(\frac{K_{i,j+1} V_{j+1}}{M_j} \right) \right] \approx 2 \left[\frac{L_j + K_{i,j} V_j}{M_j} \right] \quad \textbf{(13-52)}$$

where i refers to the most volatile component and j to the stage with the smallest liquid molar holdup. The minimum absolute eigenvalue almost always corresponds to a row of the Jacobian matrix for the reboiler. Thus, from (13-37):

$$|\lambda|_{min} \leq \left[\left(\frac{L_N}{M_{N+1}} \right) + \left(\frac{V_{N+1} K_{i,N+1}}{M_{N+1}} \right) \right] \approx \left[\frac{L_N + K_{i,N+1} V_{N+1}}{M_{N+1}} \right] \quad \textbf{(13-53)}$$

where i now refers to the least volatile component and $N + 1$ is the reboiler stage. The largest value of the reboiler holdup is M_{N+1}^0. The stiffness ratio, SR, is

$$\text{SR} = \frac{|\lambda|_{\max}}{|\lambda|_{\min}} \approx 2 \left(\frac{L + K_{\text{lightest}} V}{L + K_{\text{heaviest}} V} \right) \left(\frac{M_{N+1}^0}{M_{\text{tray}}} \right) \tag{13-54}$$

From (13-54), the stiffness ratio depends not only on the difference between tray and initial reboiler molar holdups, but also on the difference between K-values of the lightest and heaviest components in the charge to the still.

Davis [15] states SR = 20 is not stiff, SR = 1,000 is stiff, and SR = 1,000,000 is very stiff. For the conditions of Example 13.7, using (13-54):

$$\text{SR} \approx 2 \left[\frac{10 + (1.212)(20)}{10 + (0.581)(20)} \right] \left(\frac{100}{0.01} \right) = 31,700$$

which meets the criterion of a stiff problem. A modification of the computational procedure of Distefano [11], for solving (13-35) through (13-46), is as follows:

Initialization

1. Establish total-reflux conditions, based on vapor and liquid molar flow rates V_j^0 and L_j^0. V_{N+1}^0 is the desired boilup rate or L_0^0 is based on the desired distillate rate and reflux ratio such that $L_0^0 = D(R + 1)$.

2. At $t = 0$, reduce L_0^0 to begin distillate withdrawal, but maintain the boilup rate established or specified for the total-reflux condition. This involves replacing all L_j^0 with $L_j^0 - D$. Otherwise, the initial values of all variables are those established for total reflux.

Time Step

3. In (13-35) to (13-37), replace liquid holdup derivatives by total material balance equations:

$$\frac{dM_j}{dt} = V_{j+1} + L_{j-1} - V_j - L_j$$

Solve the resulting equations for the liquid mole fractions using an appropriate implicit integration technique and a suitable time step. Normalize the mole fractions if they do not sum to one at each stage.

4. Compute a new set of stage temperatures and corresponding vapor-phase mole fractions from (13-44) and (13-43), respectively.

5. Compute liquid densities and liquid holdups from (13-45) and (13-46), and liquid and vapor enthalpies. Then determine derivatives of enthalpies and liquid holdups with respect to time by forward-finite-difference approximations.

6. Compute a new set of liquid and vapor molar flow rates from (13-38), (13-39), and (13-41).

7. Compute the new reboiler molar holdup from (13-47).

8. Compute condenser and reboiler heat-transfer rates from (13-40) and (13-42).

Iteration to Completion of Operation

9. Repeat Steps 3 through 8 for additional time steps until a specified operation is complete. The specified operation might be for a desired amount of distillate, desired mole fraction of a particular component in the accumulated distillate, etc.

New Operation

10. Dump the accumulated distillate into a receiver, change operating conditions, and repeat Steps 2 through 9. Terminate calculations following the final operation.

The foregoing procedure is limited to narrow-boiling feeds and the simple configuration shown in Figure 13.10. A more flexible and efficient method, specifically designed to cope with stiffness, is that of Boston et al. [12], which uses a modified inside-out algorithm of the type discussed for continuous distillation in Chapter 10, which can handle feeds ranging from narrow-boiling to wide-boiling, even for nonideal liquid solutions. In addition, the Boston et al. method permits multiple feeds, side streams, tray heat transfer, vapor distillate, and considerable flexibility in operation specifications.

EXAMPLE 13.8

A charge of 100 kmol of 30 mol% acetone, 30 mol% methanol, and 40 mol% water at 60°C and 1 atm is to be distilled in a batch rectifier consisting of a reboiler, a column with five theoretical stages, a total condenser, a reflux drum, and three distillate accumulators. The molar liquid holdup of the condenser-reflux drum is 5 kmol, whereas the molar liquid holdup of each stage is 1 kmol. The pressure is assumed constant at 1 atm throughout the rectifier. The following four events are to occur, each with a reboiler duty of 1 million kcal h:

Event 1: Establishment of total reflux conditions.
Event 2: Rectification with a reflux ratio of 3 until the purity of the accumulated distillate in the first accumulator drops to 73 mol%
Event 3: Rectification with a reflux ratio of 3 and a second accumulator for 21 min
Event 4: Rectification with a reflux ratio of 3 and a third accumulator for 27 min

Determine accumulator and column conditions at the end of each event. Use the Wilson equation for computing K-values.

SOLUTION

The following results were obtained with the program DESIGN II/Batch of the ChemShare Corporation. The stiffness ratio, SR, may be computed from (13-54) based on total-reflux conditions at the end of Event 1. The conditions are as follows:

Event 1: Total Reflux Conditions:

Stage	T, °C	L, kmol/h	Acetone	Methanol	Water
			\multicolumn{3}{c}{Mole Fraction in Liquid}		
Condenser	55.6	138.9	0.770	0.223	0.007
1	55.6	138.6	0.761	0.227	0.012
2	55.7	138.0	0.747	0.235	0.018
3	55.9	137.0	0.722	0.247	0.031
4	56.2	134.8	0.673	0.269	0.058
5	57.3	128.7	0.560	0.306	0.134
Reboiler	62.2	—	0.252	0.307	0.441

The charge remaining in the still is $100 - 5 - 5(1) = 90$ kmol. The most volatile component is acetone, with a K-value at the bottom stage of 1.203. The least volatile component is water, with a corresponding K-value of 0.428. The stiffness ratio is

$$\text{SR} \approx 2\left[\frac{128.7 + (1.203)(134.8)}{128.7 + (0.428)(134.8)}\right]\left(\frac{90}{1}\right) \approx 281$$

Thus, this problem is not very stiff. A time step of 0.06 min is used.

Event 2:

The time required to complete Event 2 is computed to be 57.5 minutes. The accumulated distillate in Tank 1 is 32.0 kmol with a composition as follows: 73.0 mol% acetone, 26.0 mol%

methanol, and 1.0 mol% water. The liquid remaining in the reboiler is 58.0 kmol with the following composition: 2.8 mol% acetone, 30.0 mol% methanol, and 67.2 mol% water.

Event 3:

The time required to complete this event is specified as 21 min. The accumulated distillate in Tank 2 is 11.3 kmol of the following composition: 47.2 mol% acetone, 51.8 mol% methanol, and 1.0 mol% water. This intermediate cut is recycled for addition to the next charge.

Event 4:

At the end of the 27-min specification, the accumulated distillate in Tank 3 is 13.8 kmol of the following composition: 8.3 mol% acetone, 86.2 mol% methanol, and 5.5 mol% water. The composition of the remaining 32.9 kmol in the still is as follows: 0.0 mol% acetone, 0.4 mol% methanol, and 99.6 mol% water. ∎

Rapid Solution Method

As an alternative to integration of the stiff system of differential equations, the quasi-steady-state procedure of Galindez and Fredenslund [13] can be used. With this method, the transient conditions are simulated as a succession of a finite number of continuous steady states of short duration, typically 0.05 h (3 min). Holdup is taken into account, but the stiffness of the problem is of no consequence. Results compare favorably with those from the rigorous integration method.

Consider an intermediate theoretical stage, j, with molar holdup, M_j, in the batch rectifier in Figure 13.11a. A material balance for component i, in terms of component flow rates, rather than mole fractions, is

$$l_{i,j} + v_{i,j} - l_{i,j-1} - v_{i,j+1} + \frac{d(M_j x_{i,j})}{dt} = 0 \qquad \textbf{(13-55)}$$

Assume constant molar holdup. Also, assume that during a short time period, $dt = \Delta t = t_{k+1} - t_k$, the component flow rates given by the first four terms in (13-55) remain constant at values corresponding to time t_{k+1}. The component holdup term in (13-55) is

$$\frac{d(M_j x_{i,j})}{dt} = M_j \left[\frac{x_{i,j}\{t_{k+1}\} - x_{i,j}\{t_k\}}{\Delta t} \right] \qquad \textbf{(13-56)}$$

But, $x_{i,j} = l_{i,j}/L_j$. Therefore, (13-56) can be rewritten as

$$\frac{d(M_j x_{i,j})}{dt} = \frac{M_j l_{i,j}}{L_j \, \Delta t} - \frac{M_j l_{i,j}\{t_k\}}{L_j\{t_k\} \, \Delta t} \qquad \textbf{(13-57)}$$

If (13-57) is substituted into (13-55) and terms in the component flow rate $l_{i,j}$ are collected:

$$l_{i,j}\left(1 + \frac{M_j}{L_j \, \Delta t}\right) + v_{i,j} - l_{i,j-1} - v_{i,j+1} - \frac{M_j l_{i,j}\{t_k\}}{L_j\{t_k\} \, \Delta t} \qquad \textbf{(13-58)}$$

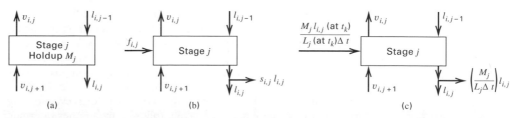

Figure 13.11 Simulation of holdup in a batch rectifier. (a) Stage in a batch rectifier with holdup. (b) Stage in a continuous fractionator. (c) Simulation of batch holdup in a continuous fractionator.

If (13-58) for unsteady-state (batch) distillation is compared to (10-58) for steady-state (continuous) distillation, we see that the term $M_j/(L_j \, \Delta t)$ in (13-58) corresponds to the liquid side-stream ratio in (10-58) or that $M_j/\Delta t$ corresponds to a liquid side-stream flow rate. We also see that the term $M_j l_{i,j}\{t_k\}/(L_j\{t_k\} \, \Delta t)$ in (13-58) corresponds to a component feed rate in (10-58). The analogy is shown in parts (b) and (c) of Figure 13.11. Thus, the change in component liquid holdup per unit time, $d(M_j x_{i,j})/dt$ in (13-56), is interpreted for a small, finite time difference as the difference between a component feed rate into the stage and a component flow rate in a liquid side stream leaving the stage. In a similar manner, the stage enthalpy holdup in the energy balance for the stage is interpreted as the difference over a small finite time interval between a heat input to the stage and an enthalpy output in a liquid side stream leaving the stage. The overall result is a system of steady-state equations, identical in form to the equations for the simultaneous-correction and inside-out methods of Chapter 10. Accordingly, as implemented in the Batch Column computer model of Chemstations, Inc., either of those two methods can be used to solve the system of component material-balance, phase-equilibrium, and energy-balance equations at each time step. The initial guesses used to initiate each time step are the values at the end of the previous time step. Because the variables generally change by only a small amount for each time step, convergence of the simultaneous correction or inside-out method is generally achieved in a small number of iterations.

EXAMPLE 13.9

A charge of 100 lbmol of a mixture of 25 mol% benzene (B), 50 mol% monochlorobenzene (MCB), and 25 mol% *ortho*-dichlorobenzene (DCB) is distilled in a batch rectifier consisting of a reboiler, 10 equilibrium stages, a reflux drum, and three distillate product accumulators. The condenser-reflux drum holdup is constant at 0.20 ft^3, and each stage in the column has a liquid holdup of 0.02 ft^3. Pressures are 17.5 psia in the reboiler and 14.7 psia in the reflux drum, with a linear pressure profile in the column from 15.6 psia at the top to 17 psia at the bottom. Following an initialization at total reflux, the batch is distilled in the following three operation steps, each with a vapor boilup rate of 200 lbmol/h and a reflux ratio of 3. Thus, the distillate rate is 50 lbmol/h.

Operation step 1: Terminate when the mole fraction of benzene in the distillate drops below 0.100.
Operation step 2: Terminate when the mole fraction of MCB in the distillate drops below 0.40.
Operation step 3: Terminate when the mole fraction of DCB in the reboiler rises above 0.98.

Assume ideal solutions and the ideal gas law.

SOLUTION

This problem is quite stiff, with a stiffness ratio of approximately 15,000. The quasi-steady-state procedure of Galindez and Fredenslund [13], as implemented in Batch Column, was used with a time increment of 0.005 h for each of the three operation steps. Although 0.05 h is normal for the Galindez and Fredenslund method, the high ratio of distillate rate to charge for this problem necessitated a smaller Δt. Computed results are given in Table 13.3, where it is seen that the accumulated distillate cuts from operation steps 1 and 3 are quite impure with respect to benzene and DCB, respectively. The cut from Step 2 is 95 mol% pure MCB. The residual left in the reboiler after Step 3 is quite pure in DCB. A plot of the instantaneous distillate composition as a function of total distillate accumulation for all steps is shown in Figure 13.12. Changes in mole fractions occur very rapidly at certain times during the batch rectification indicating that relatively pure cuts may be possible. This plot is useful in developing alternative schedules to obtain almost pure cuts. For example, suppose relatively rich distillate cuts of B, MCB, and DCB are desired. From Figure 13.12, an initial benzene-rich cut of, say, 18 lbmol might be taken, followed by an intermediate cut for recycle of, say, 18 lbmol. Then, an MCB-rich cut of 34 lbmol might be taken, followed by another intermediate cut of 8 lbmol, leaving a DCB-rich residual of 22 lbmol. For this series of operation steps, with the same vapor boilup rate of 200 lbmol/h and reflux ratio of 3, the computed results for each distillate accumulation (cut), using Batch Column with a time step of 0.005 h, are given in Table 13.4. As seen, all three

Table 13.3 Results at the End of Each Operation Step for Example 13.9

	Operation Step		
	1	2	3
Operation time, h	0.605	0.805	0.055
No. of time increments	121	161	11
Accumulated distillate:			
Total lbmol	33.65	41.96	2.73
Mole fractions:			
B	0.731	0.009	0.000
MCB	0.269	0.950	0.257
DCB	0.000	0.041	0.743
Reboiler holdup:			
Total lbmol	66.13	24.19	21.46
Mole fractions:			
B	0.006	0.000	0.000
MCB	0.616	0.044	0.018
DCB	0.378	0.956	0.982
Total heat duties, 10^6 Btu:			
Condenser	1.95	2.65	0.19
Reboiler	2.08	2.63	0.18

product cuts are better than 98 mol% pure. However, (18 + 8) = 26 lbmol of intermediate cuts, or about one-fourth of the original charge, would have to be recycled. Further improvements in purities of the cuts or reduction in the amounts of intermediate cuts for recycle can be made by increasing the reflux ratio and/or the number of theoretical stages. ∎

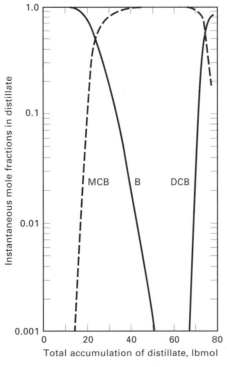

Figure 13.12 Instantaneous distillate–composition profile for Example 13.9. [*Perry's Chemical Engineers' Handbook,* 6th ed., R.H. Perry and D.W. Green, Eds., McGraw-Hill, New York (1984) with permission.]

Table 13.4 Results of Alternative Operating Schedule for Example 13.9

Distillate Cut	Amount, lbmol	Composition, Mole Fractions		
		B	MCB	DCB
Benzene-rich	18	0.993	0.007	0.000
Intermediate 1	18	0.374	0.626	0.000
MCB-rich	34	0.006	0.994	0.000
Intermediate 2	8	0.000	0.536	0.464
DCB-rich residual	22	0.000	0.018	0.982
Total	100			

SUMMARY

1. The simplest case of batch distillation corresponds to the condensation of a vapor rising from a boiling liquid, called differential or Rayleigh distillation. The vapor leaving the liquid surface is assumed to be in equilibrium with the liquid. The compositions of the liquid and vapor continually change as distillation proceeds. The instantaneous vapor and liquid compositions can be computed as a function of time for a given vaporization rate.

2. A batch rectifier system consists of a reboiler, a column with plates or packing that sits on top of the reboiler, a condenser, a reflux drum, and one or more distillate receivers.

3. For a binary system, a batch rectifier is usually operated at a constant reflux ratio or at a constant distillate composition. For either case, a McCabe–Thiele diagram can be used to follow the progress of the rectification, if the assumptions of constant molar overflow and negligible tray (or packing), condenser, and reflux drum liquid holdups are made.

4. A batch stripper is useful for removing small quantities of impurities from a charge. For complete flexibility, complex batch distillation involving both rectification and stripping can be employed.

5. Liquid holdup on the trays (or packing) and in the condenser and reflux drum can influence the course of batch rectification and the size and composition of the distillate cuts. The complexity of the holdup effect is such that it is best determined by rigorous calculations for specific cases.

6. For multicomponent batch rectification, with negligible liquid holdup except in the reboiler, the shortcut method of Sundaram and Evans, based on successive applications of the Fenske–Underwood–Gilliland (FUG) method at a sequence of time intervals, can be used to obtain approximate distillate and charge compositions and amounts as a function of time.

7. For accurate and detailed multicomponent batch rectification compositions, the model of Distefano as implemented by Boston et al. should be used. It accounts for liquid holdup and permits a sequence of operation steps to produce multiple distillate cuts. The model consists of algebraic and ordinary differential equations, which, when stiff, are best solved by Gear-type implicit integration methods. The Distefano model can also be solved by the method of Galindez and Fredenslund, which simulates the unsteady batch process by a succession of steady states of short duration, which are solved by either the simultaneous correction or the inside-out methods of Chapter 10.

REFERENCES

1. Rayleigh, J.W.S., *Phil. Mag. and J. Sci.,* Series 6, **4** (23), 521–537 (1902).

2. Smoker, E.H., and A. Rose, *Trans. AIChE,* **36,** 285–293 (1940).

3. Block, B., *Chem. Eng.,* **68** (3), 87–98 (1961).

4. Bogart, M.J.P., *Trans. AIChE,* **33,** 139–152 (1937).

5. Ellerbe, R.W., *Chem. Eng.,* **80,** 110–116 (1973).

6. Hasbe, S., B.B. Abdul Aziz, I. Hashimoto, and T. Watanabe, *Proc. IFAC Workshop, London, Sept. 7–8, 1992,* p. 177.

7. Diwekar, U.M., and K.P. Madhaven, *Ind. Eng. Chem. Res.*, **30**, 713–721 (1991).

8. Sundaram, S., and L.B. Evans, *Ind. Eng. Chem. Res.*, **32**, 511–518 (1993).

9. Eduljee, H.E., *Hydrocarbon Processing*, **56** (9), 120–122 (1975).

10. Meadows, E.L., *Chem. Eng. Progr. Symp. Ser. No. 46*, **59**, 48–55 (1963).

11. Distefano, G.P., *AIChE J.*, **14**, 190–199 (1968).

12. Boston, J.F., H.I. Britt, S. Jirapongphan, and V.B. Shah, in *Foundations of Computer-Aided Chemical Process Design*, Vol. II, AIChE, R.H.S. Mah and W.D. Seider, Eds., 203–237 (1981).

13. Galindez, H., and A. Fredenslund, *Comput. Chem. Eng.*, **12**, 281–288 (1988).

14. Gear, C.W., *Numerical Initial Value Problems in Ordinary Differential Equations*, Prentice-Hall, Englewood Cliffs, NJ (1971).

15. Davis, M.E., *Numerical Methods and Modeling for Chemical Engineers*, John Wiley and Sons, New York (1984).

16. Riggs, J.B., *An Introduction to Numerical Methods for Chemical Engineers*, Texas Tech. Univ. Press, Lubbock, TX (1988).

17. Carnahan, B., and J.O. Wilkes, "Numerical solution of differential equations—an overview," in *Foundations of Computer-Aided Chemical Process Design*, R.S.H. Mah and W.D. Seider, Eds., Engineering Foundation, New York, Vol. I, pp. 225–340 (1981).

18. Varga, R.S., *Matrix Iterative Analysis*, Prentice-Hall, Englewood Cliffs, NJ (1962).

EXERCISES

Section 13.1

13.1 (a) A bottle of pure *n*-heptane is accidentally poured into a drum of pure toluene in a commercial laboratory. One of the laboratory assistants, with almost no background in chemistry, suggests that, since heptane boils at a lower temperature than toluene, the following purification procedure can be used:

Pour the mixture (2 mol% *n*-heptane) into a simple still pot. Boil the mixture at 1 atm and condense the vapors until all heptane is boiled away. Obtain the pure toluene from the residue in the still pot.

You, being a chemical engineer, immediately realize that such a purification method will not work. Indicate this by a curve showing the composition of the material remaining in the pot after various quantities of the liquid have been distilled. What is the composition of the residue after 50 wt% of the original material has been distilled? What is the composition of the cumulative distillate?

(b) When one-half of the heptane has been distilled, what is the composition of the cumulative distillate and of the residue? What weight percent of the original material has been distilled?

Vapor–liquid equilibrium data at 1 atm [*Ind. Eng. Chem.*, **42**, 2912 (1949)] are as follows:

Mole Fraction *n*-Heptane

Liquid	Vapor	Liquid	Vapor
0.025	0.048	0.448	0.541
0.062	0.107	0.455	0.540
0.129	0.205	0.497	0.577
0.185	0.275	0.568	0.637
0.235	0.333	0.580	0.647
0.250	0.349	0.692	0.742
0.286	0.396	0.843	0.864
0.354	0.454	0.950	0.948
0.412	0.504	0.975	0.976

13.2 A mixture of 40 mol% isopropanol in water is to be distilled at 1 atm by a simple batch distillation until 70 mol% of the charge has been vaporized (equilibrium data are given in Exercise 7.33). What will be the compositions of the liquid residue remaining in the still pot and of the collected distillate?

13.3 A 30 mol% feed of benzene in toluene is to be distilled in a batch operation. A product having an average composition of 45 mol% benzene is to be produced. Calculate the amount of residue, assuming $\alpha = 2.5$ and $W_0 = 100$.

13.4 A charge of 250 lb of 70 mol% benzene and 30 mol% toluene is subjected to batch differential distillation at atmospheric pressure. Determine the compositions of the distillate and residue after one-third of the original mass is distilled off. Assume the mixture forms an ideal solution and apply Raoult's and Dalton's laws with vapor-pressure data.

13.5 A mixture containing 60 mol% benzene and 40 mol% toluene is subjected to batch differential distillation at 1 atm, under three different conditions:

1. Until the distillate contains 70 mol% benzene
2. Until 40 mol% of the feed is evaporated
3. Until 60 mol% of the original benzene leaves in the vapor phase

Using $\alpha = 2.43$, determine for each of the three cases:
(a) The number of moles in the distillate for 100 mol of feed.
(b) The compositions of distillate and residue.

13.6 A mixture consisting of 15 mol% phenol in water is to be batch distilled at 260 torr. What fraction of the original batch remains in the still when the total distillate contains 98 mol% water? What is the residue concentration?

Vapor–liquid equilibrium data at 260 torr [*Ind. Eng. Chem.*, **17**, 199 (1925)]:

x, wt% (H_2O):
   1.54   4.95   6.87   7.73   19.63   28.44   39.73   82.99   89.95   93.38   95.74

y, wt% (H_2O):
   41.10   79.72   82.79   84.45   89.91   91.05   91.15   91.86   92.77   94.19   95.64

13.7 A still is charged with 25 mol of a mixture of benzene and toluene containing 0.35 mole fraction benzene. Feed of the same composition is supplied at a rate of 7 mol/h, and the heat rate is adjusted so that the liquid level in the still remains constant. No liquid leaves the still pot, and $\alpha = 2.5$. How long will it be before the distillate composition falls to 0.45 mole fraction benzene?

13.8 A distillation system consisting of a reboiler and a total condenser (no column) is to be used to separate A and B from a trace of nonvolatile material. The reboiler initially contains 20 lbmol of feed of 30 mol% A. Feed is to be supplied to the reboiler at the rate of 10 lbmol/h, and the heat input is so adjusted that the total moles of liquid in the reboiler remains constant at 20. No residue is withdrawn from the still. Calculate the time required for the composition of the overhead product to fall to 40 mol% A. The relative volatility may be assumed constant at 2.50.

Section 13.2

13.9 Repeat Exercise 13.2 for the case of a batch distillation carried out in a two-stage column with a reflux ratio of $L/V = 0.9$.

13.10 Repeat Exercise 13.3 assuming the operation is carried out in a three-stage column with $L/V = 0.6$.

13.11 One kilomole of an equimolar mixture of benzene and toluene is fed to a batch still containing three equivalent stages (including the boiler). The liquid reflux is at its bubble point, and $L/D = 4$. What is the average composition and amount of product at a time when the instantaneous product composition is 55 mol% benzene? Neglect holdup, and assume $\alpha = 2.5$.

13.12 The fermentation of corn produces a mixture of 3.3 mol% ethyl alcohol in water. If 20 mol% of this mixture is distilled at 1 atm by a simple batch distillation, calculate and plot the instantaneous vapor composition as a function of mole percent of batch distilled. If reflux with three theoretical stages (including the reboiler) is used, what is the maximum purity of ethyl alcohol that can be produced by batch distillation?

Equilibrium data are given in Exercise 7.29.

Section 13.3

13.13 An acetone–ethanol mixture of 0.5 mole fraction acetone is to be separated by batch distillation at 101 kPa. Vapor–liquid equilibrium data at 101 kPa are as follows:

Mole Fraction Acetone

y	0.16	0.25	0.42	0.51	0.60	0.67	0.72	0.79	0.87	0.93
x	0.05	0.10	0.20	0.30	0.40	0.50	0.60	0.70	0.80	0.90

(a) Assuming an L/D of 1.5 times the minimum, how many stages should this column have if we want the composition of the distillate to be 0.90 mole fraction acetone at a time when the residue contains 0.1 mole fraction acetone?

(b) Assume the column has eight stages and the reflux rate is varied continuously so that the top product is maintained constant at 0.9 mole fraction acetone. Make a plot of the reflux ratio versus the still pot composition and the amount of liquid left in the still.

(c) Assume now that the same distillation is carried out at constant reflux ratio (and varying product composition). We wish to have a residue containing 0.1 and an (average) product containing 0.9 mole fraction acetone, respectively. Calculate the total vapor generated. Which method of operation is more energy intensive? Can you suggest operating policies other than constant reflux ratio and constant distillate compositions that might lead to equipment and/or operating cost savings?

13.14 A total of 2,000 gallons of 70 wt% ethanol in water, having a specific gravity of 0.871, is to be separated at 1 atm in a batch rectifier operating at constant distillate composition with a constant molar vapor boilup rate to obtain a distillate product of 85 mol% ethanol and a residual waste water containing 3 wt% ethanol. If the task is to be completed in 24 h, allowing 4 h for charging, start-up, shutdown, and cleaning, determine: (a) the number of theoretical plates required in the column, (b) the reflux ratio when the concentration of ethanol in the pot is 25 mol%, (c) the instantaneous distillate rate in lbmol/h when the concentration of ethanol in the pot is 15 mol%, (d) the lbmol of distillate product, and (e) the lbmol of residual wastewater.

Vapor–liquid equilibrium data are given in Exercise 7.29.

13.15 A charge of 1,000 kmol of a mixture of 20 mol% ethanol in water is to undergo batch rectification at 101.3 kPa at a vapor boilup rate of 100 kmol/h. If the column has the equivalent of six theoretical plates and the distillate composition is to be maintained at 80 mol% ethanol by varying the reflux ratio, determine: (a) the time in hours for the residue to reach an ethanol mole fraction of 0.05, (b) the kmol of distillate obtained when the condition of part (a) is achieved, (c) the minimum and maximum reflux ratios during the rectification period, and (d) the variation of the distillate rate in kmol/h during the rectification period. Assume constant molar overflow, neglect liquid holdup, and obtain vapor–liquid equilibrium data from Exercise 7.29.

13.16 A 500 lbmol mixture of 48.8 mol% A and 51.2 mol% B with a relative volatility $\alpha_{A,B}$ of 2.0 is to be separated in a batch rectifier consisting of a total condenser, a column with seven theoretical stages, and a partial reboiler. The reflux ratio is to be varied so as to maintain the distillation composition constant at 95 mol% A. The column can operate satisfactorily with a molar vapor boilup rate of 213.5 lbmol/h. The rectification is to be stopped when the mole fraction of A in the still drops to 0.192. Determine: (a) the time required for rectification, and (b) the total amount of distillate produced.

Section 13.4

13.17 Develop a procedure similar to that of Section 13.2 to calculate a binary batch stripping operation using the equipment arrangement of Figure 13.8.

13.18 A three-theoretical stage batch stripper (one stage is the reboiler) is charged to the feed tank (see Figure 13.8) with 100 kmol of 10 mol% *n*-hexane in *n*-octane mix. The boilup rate is 30 kmol/h. If a constant boilup ratio (V/L) of 0.5 is used, determine the instantaneous bottoms composition and the composition of the accumulated bottoms product at the end of 2 h of operation.

13.19 Develop a procedure similar to that of Section 13.2 to calculate a complex binary batch distillation operation using the equipment arrangement of Figure 13.9.

Section 13.5

13.20 For a batch rectifier with appreciable column holdup:
(a) Why is the composition of the charge to the still higher in the light component than the still composition at the start of rectification, assuming that total reflux conditions are established before rectification begins?
(b) Why will separation be more difficult than with zero holdup?

13.21 For a batch rectifier with appreciable column holdup, why do tray compositions change less rapidly compared to a rectifier with negligible column holdup, and why is the degree of separation improved?

13.22 Based on the statements in Exercises 13.20 and 13.21, why is it difficult to predict the effect of holdup?

Section 13.6

13.23 Use the shortcut method of Sandaram and Evans to solve Example 13.7, but with zero condenser and stage holdups.

13.24 A charge of 100 kmol of an equimolar mixture of A, B, and C, with $\alpha_{A,B} = 2$ and $\alpha_{A,C} = 4$, is distilled in a batch rectifier containing the equivalent of four theoretical stages, including the reboiler. If holdup can be neglected, use the shortcut method with $R = 5$ and $V = 100$ kmol/h to estimate the variation of the still and instantaneous distillate compositions as a function of time following a start-up period during which total reflux conditions are established.

13.25 A charge of 200 kmol of a mixture of 40 mol% A, 50 mol% B, and 10 mol% C with $\alpha_{A,C} = 2.0$ and $\alpha_{B,C} = 1.5$ is to be separated in a batch rectifier with a total of three theoretical stages and operating at a reflux ratio of 10, with a molar vapor boilup rate of 100 kmol/h. Holdup is negligible. Use the shortcut method to estimate instantaneous distillate and bottoms compositions as a function of time for the first hour of operation following start-up to achieve total reflux conditions.

Section 13.6

13.26 A charge of 100 lbmol of 35 mol% *n*-hexane, 35 mol% *n*-heptane, and 30 mol% *n*-octane is to be distilled at 1 atm in a batch rectifier, consisting of a partial reboiler, a column, and a total condenser, at a constant boilup rate of 50 lbmol/h and a constant reflux ratio of 5. Before rectification begins, total reflux conditions are to be established. Then, the following three operation steps are to be carried out to obtain an *n*-hexane-rich cut, an intermediate cut for recycle, an *n*-heptane-rich cut, and an *n*-octane-rich residue:

Step 1: Stop when the accumulated distillate purity drops below 95 mol% *n*-hexane.
Step 2: Empty the *n*-hexane-rich cut produced in Step 1 into a receiver and resume rectification until the instantaneous distillate composition reaches 80 mol% *n*-heptane.
Step 3: Empty the intermediate cut produced in Step 2 into a receiver and resume rectification until the accumulated distillate composition reaches 4 mol% *n*-octane.

For thermodynamic properties, assume ideal solutions and the ideal gas law.
Consider conducting the rectification in two different columns, each with the equivalent of 10 theoretical stages and a condenser-reflux drum liquid holdup of 1.0 lbmol. For each column, determine with a suitable batch distillation computer program the compositions and amounts in lbmol of each of the four products.

Column 1: A plate column with a total liquid holdup of 8 lbmol.
Column 2: A packed column with a total liquid holdup of 2 lbmol.

Discuss the effect of liquid holdup for the two columns. Are the results what you expected?

13.27 A charge of 100 lbmol of a hydrocarbon mixture containing 10 mol% propane, 30 mol% *n*-butane, 10 mol% *n*-pentane, and the balance *n*-hexane is to be separated in a batch rectifier equipped with a partial reboiler, a total condenser with a liquid holdup of 1.0 ft³, and a column with the equivalent of eight theoretical stages and a total holdup of 0.80 ft³. The pressure in the condenser is 50.0 psia and the column pressure drop is 2.0 psi. The rectification campaign or operating policy, given as follows, is designed to produce cuts of 98 mol% propane, 99.8 mol% *n*-butane, and a residual cut of 99 mol% *n*-hexane, and two intermediate cuts, one of which may be a relatively rich cut of *n*-pentane. All five operating steps are conducted at a molar vapor boilup rate of 40 lbmol/h. Use a suitable batch distillation computer program to determine the amounts and compositions of all cuts.

712 Chapter 13 Batch Distillation

Step	Reflux Ratio	Stop Criterion
1	5	98% propane in accumulator
2	20	95% n-butane in instantaneous distillate
3	25	99.8% n-butane in accumulator
4	15	80% n-pentane in instantaneous distillate
5	25	99% n-hexane in the pot

Make suggestions as to how you might alter the operation steps so as to obtain larger amounts of the product cuts and smaller amounts of the intermediate cuts.

13.28 A charge of 100 lbmol of benzene (B), monochlorobenzene (MCB), and o-dichlorobenzene (DCB) is being distilled in a batch rectifier that consists of a total condenser, a column with 10 theoretical stages, and a partial reboiler. Following the establishment of total-reflux, the first operation step begins for a boilup rate of 200 lbmol/h and a reflux ratio of about 3. At the end of 0.60 h, the following conditions exist for the top three stages in the column:

	Top Stage	Stage 2	Stage 3
Temperature, °F	267.7	271.2	272.5
V, lbmol/h	206.1	209.0	209.5
L, lbmol/h	157.5	158.0	158.1
M, lbmol	0.01092	0.01088	0.01087

Vapor Mole Fractions:

B	0.0994	0.0449	0.0331
MCB	0.9006	0.9551	0.9669
DCB	0.0000	0.0000	0.0000

Liquid Mole Fractions:

B	0.0276	0.0121	0.00884
MCB	0.9724	0.9879	0.99104
DCB	0.0000	0.0000	0.00012

In addition, reboiler and condenser holdups at 0.6 h are 66.4 and 0.1113 lbmol, respectively.

For benzene, use the preceding data with (13-36) and (13-39) to estimate the liquid-phase mole fraction of benzene leaving Stage 2 at 0.61 h by using the explicit Euler method with a Δt of 0.01 h. If the result is unreasonable, explain why with respect to stability and stiffness considerations.

Chapter 14

Membrane Separations

In a membrane separation process, a feed consisting of a mixture of two or more components is partially separated by means of a semipermeable barrier (the membrane) through which one or more species move faster than another or other species. The most general membrane process is shown in Figure 14.1 where the feed mixture is separated into a *retentate* (that part of the feed that does not pass through the membrane, i.e., is retained) and a *permeate* (that part of the feed that does pass through the membrane). Although the feed, retentate, and permeate are usually liquid or gas, they may also be solid. The barrier is most often a thin, nonporous polymeric film, but may also be porous polymer, ceramic, or metal materials, or even a liquid or gas. The barrier must not dissolve, disintegrate, or break. The optional sweep, shown in Fig. 14.1, is a liquid or gas, used to help remove the permeate. Many of the industrially important membrane separation operations are listed in Tables 1.2 and 14.1.

In membrane separations: (1) the two products are usually miscible, (2) the separating agent is a semipermeable barrier, and (3) a sharp separation is often difficult to achieve. Thus, membrane separations differ in two or three of these respects from the more common separation operations of absorption, stripping, distillation, and liquid–liquid extraction.

Although membranes as separating agents have been known for more than 100 years [1], large-scale applications have only appeared in the past 50 years. In the 1940s, porous fluorocarbons were used to separate $^{235}UF_6$ from $^{238}UF_6$ [2]. In the mid-1960s, reverse osmosis with cellulose acetate was first used to desalinize seawater to produce potable water (drinkable water with less than 500 ppm by weight of dissolved solids) [3]. In 1979, Monsanto Chemical Company introduced a hollow-fiber membrane of polysulfone to separate certain gas mixtures—for example, to enrich hydrogen- and carbon dioxide-containing streams [4]. Commercialization of alcohol dehydration by pervaporation began in the late 1980s, as did the large-scale application of emulsion liquid membranes for removal of metals and organics from wastewater.

The replacement of the more common separation operations with membrane separations has the potential to save large amounts of energy. This replacement requires the production of high mass-transfer flux, defect-free, long-life membranes on a large scale and the fabrication of the membrane into compact, economical modules of high surface area per unit volume.

A common application of membranes is to the separation of hydrogen from methane. Following World War II, during which large amounts of toluene were required to produce TNT (trinitrotoluene) explosives, petroleum refiners sought other markets for tolu-

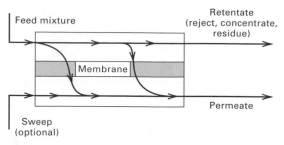

Figure 14.1 General membrane process.

ene. One potential market was the use of toluene as a feedstock for the manufacture of benzene, a precursor for nylon, and xylenes, precursors for a number of other chemicals, including polyesters. Toluene can be catalytically disproportionated to benzene and mixed xylenes in an adiabatic reactor with the feed entering at 950°F and a pressure greater than 500 psia. The main reaction is

$$2C_7H_8 \rightarrow C_6H_6 + C_8H_{10} \text{ isomers}$$

To suppress the formation of coke, which fouls the catalyst, the reactor feed must contain a substantial fraction of hydrogen at a partial pressure of at least 215 psia. Unfortunately, the hydrogen takes part in a side reaction for the hydrodealkylation of toluene to benzene and methane:

$$C_7H_8 + H_2 \rightarrow C_6H_6 + CH_4$$

Makeup hydrogen is usually not pure, but contains perhaps 15 mol% methane and 5 mol% ethane. Thus, typically, the reactor effluent contains H_2, CH_4, C_2H_6, C_6H_6, unreacted C_7H_8, and C_8H_{10} isomers. As shown in Figure 14.2a, for just the reaction section of the process, this effluent is cooled and partially condensed to 100°F at a pressure of 465 psia. At these conditions, a reasonably good separation is achieved between C_2H_6 and C_6H_6 in the flash drum. Thus, the vapor leaving the flash drum contains most of the H_2, CH_4, and C_2H_6, with most of the aromatic chemicals leaving in the liquid. Because of the large amount of hydrogen in the flash-drum vapor, it is important to recycle this stream to the reactor, rather than sending it to a flare or using it as a fuel. However, if all of the vapor were recycled, methane and ethane would build up in the recycle loop, since no other exit is provided. Before the development of acceptable membranes for the separation of H_2 from CH_4 by gas permeation, part of the vapor stream was customarily purged from the process, as shown in Figure 14.2a, to provide an exit for CH_4, and C_2H_6. With the introduction of a suitable membrane in 1979, it became possible to apply membrane separators, as shown in Figure 14.2b.

Table 14.2 is the steady-state material balance of the reaction section of Figure 14.2b for a plant designed to process 7,750 barrels (42 gal/bbl) per operating day of fresh toluene feed. The gas permeation membrane system separates the flash vapor (stream S11) into an H_2-enriched permeate (S14, the recycled hydrogen), and a methane-enriched retentate (S12, the purge). The flash vapor to the membrane system contains 89.74 mol% H_2 and 9.26 mol% CH_4. No sweep fluid is necessary. The permeate is enriched to 94.46 mol% in H_2. The retentate is enriched in CH_4 to 31.10 mol%. The recovery of H_2 in the permeate is 90%. Thus, only 10% of the H_2 in the vapor leaving the flash drum is lost to the purge. Before entering the membrane separator system, the vapor is heated to a

Table 14.1 Industrial Applications of Membrane
Separation Processes

1. Reverse osmosis:
    Desalinization of brackish water
    Treatment of wastewater to remove a wide variety of impurities
    Treatment of surface and ground water
    Concentration of foodstuffs
    Removal of alcohol from beer and wine
2. Dialysis:
    Separation of nickel sulfate from sulfuric acid
    Hemodialysis (removal of waste metabolites, excess body water, and restoration of electrolyte balance in blood)
3. Electrodialysis:
    Production of table salt from seawater
    Concentration of brines from reverse osmosis
    Treatment of wastewaters from electroplating
    Demineralization of cheese whey
    Production of ultrapure water for the semiconductor industry
4. Microfiltration:
    Sterilization of drugs
    Clarification and biological stabilization of beverages
    Purification of antibiotics
    Separation of mammalian cells from a liquid
5. Ultrafiltration:
    Preconcentration of milk before making cheese
    Clarification of fruit juice
    Recovery of vaccines and antibiotics from fermentation broth
    Color removal from Kraft black liquor in paper making
6. Prevaporation:
    Dehydration of ethanol–water azeotrope
    Removal of water from organic solvents
    Removal of organics from water
7. Gas permeation:
    Separation of CO_2 or H_2 from methane and other hydrocarbons
    Adjustment of the H_2/CO ratio in synthesis gas
    Separation of air into nitrogen- and oxygen-enriched streams
    Recovery of helium
    Reocvery of methane from biogas
8. Liquid membranes:
    Recovery of zinc from wastewater in the viscose fiber industry
    Recovery of nickel from electroplating solutions

temperature of at least 200°F (the dew-point temperature of the retentate) at a pressure of 450 psia (heater not shown). Because the hydrogen content of the feed is reduced in passing through the membrane separator, the retentate becomes more concentrated in the heavier components. Without the heater, undesirable condensation would occur in the separator. The retentate leaves the separator at about the same temperature and pressure as that of heated flash vapor entering the separator. The permeate leaves at

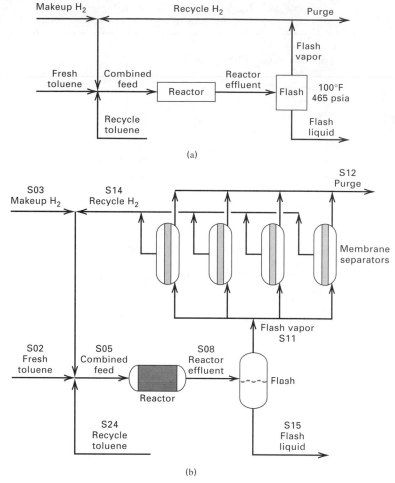

Figure 14.2 Reactor section of process to disproportionate toluene into benzene and xylene isomers. (a) Without a vapor separation step. (b) With a membrane separation step. Note: Heat exchangers, compressors, pump not shown.

Table 14.2 Material Balance for Membrane Separation Process in a Toluene Disproportionation Plant; Flow Rates in lbmol/h for Streams in Reactor Section of Figure 14.2b

Component	S02	S03	S24	S14	S05	S08	S15	S11	S12
Hydrogen		269.0		1,685.1	1,954.1	1,890.6	18.3	1,872.3	187.2
Methane		50.4		98.8	149.3	212.8	19.7	193.1	94.3
Ethane		16.8			16.8	16.8	5.4	11.4	11.4
Benzene			13.1		13.1	576.6	571.8	4.8	4.8
Toluene	1,069.4		1,333.0		2,402.4	1,338.9	1,334.7	4.2	4.2
p-Xylene			8.0		8.0	508.0	507.4	0.6	0.6
Total	1,069.4	336.3	1,354.1	1,783.9	4,543.7	4,543.7	2,457.4	2,086.3	302.4

the much lower pressure of 50 psia and a temperature somewhat lower than 200°F because of gas expansion.

The membrane is an aromatic polyamide polymer consisting of a 0.3-micron-thick, nonporous layer in contact with the feed, and a much thicker porous support backing to give the membrane strength and ability to withstand the pressure differential of 450 − 50 = 400 psi. This large pressure difference is needed to force the hydrogen through the nonporous membrane, which is in the form of a spiral-wound module made from flat membrane sheets. The average flux of hydrogen through the membrane is 40 scfh (standard ft^3/h at 60°F and 1 atm) per ft^2 of membrane surface area. From the material balance in Table 14.2, the total amount of H$_2$ transported through the membrane is

$$(1,685.1 \text{ lbmol/h})(379 \text{ scf/lbmol}) = 639,000 \text{ scfh}$$

Thus, the required membrane surface area is 639,000/40 = 16,000 ft^2. The membrane is packaged in pressure-vessel modules of 4,000 ft^2 each. Thus, four modules in parallel are used, as shown in Figure 14.2b. A disadvantage of the membrane separator in this application is the need to recompress the recycle hydrogen to the reactor inlet pressure. Unlike distillation, where the energy of separation is usually heat, the energy for gas permeation is the shaft work of gas compression.

Membrane separation is an emerging unit operation. Important progress is still being made in the development of efficient membrane materials and the packaging thereof for the processes listed in Table 14.1. Other novel methods for conducting separation with barriers for a wider variety of mixtures are being researched and developed. Applications covering wider ranges of temperature and types of membrane materials are being found. Already, membrane separation processes have found wide application in such diverse industries as the beverage, chemical, dairy, electronic, environmental, food, medical, paper, petrochemical, petroleum, pharmaceutical, and textile industries. Some of these applications are given in Table 1.2 and included in Table 14.1. Often, compared to other separation equipment, membrane separators are more compact, less capital intensive, and more easily operated, controlled, and maintained. However, membrane separators are usually modular in construction, with many parallel units required for large-scale applications, as contrasted with the more common separation techniques, where larger pieces of equipment are designed as plant size becomes larger.

The key to an efficient and economical membrane separation process is the membrane and the manner in which it is packaged and modularized. Desirable attributes of a membrane are (1) good permeability, (2) high selectivity, (3) chemical and mechanical compatibility with the processing environment, (4) stability, freedom from fouling, and reasonable useful life, (5) amenability to fabrication and packaging, and (6) ability to withstand large pressure differences across the membrane thickness. Research and development of membrane processes deals mainly with the discovery of suitable membrane materials and their fabrication.

This chapter discusses types of membrane materials, membrane modules, the theory of transport through membrane materials and modules, and the scale-up of membrane separators from experimental performance data. Emphasis is on dialysis, electrodialysis, reverse osmosis, gas permeation, and pervaporation, but many of the theoretical principles apply as well to emerging, but not yet commercialized, membrane processes such as membrane distillation, membrane gas absorption, membrane stripping, membrane solvent extraction, perstraction, and facilitated transport. The use of membranes in microfiltration and ultrafiltration is not discussed. The status of industrial membrane separation systems and directions in research to improve existing applications and make possible new applications are considered in detail by Baker et al. [5] in a study supported by the U.S. Department of Energy (DOE) and by a host of contributors in a recent handbook edited by Ho and Sirkar [6], which includes emerging membrane processes.

14.1 MEMBRANE MATERIALS

Almost all industrial membrane processes are made from natural or synthetic polymers (macromolecules). Natural polymers include wool, rubber, and cellulose. A wide variety of synthetic polymers has been developed and commercialized since 1930. Synthetic polymers are produced by polymerization of a monomer by condensation (step reactions) or addition (chain reactions), or by the copolymerization of two different monomers. The resulting polymer is categorized as having (1) a long linear chain, such as linear polyethylene; (2) a branched chain, such as polybutadiene; (3) a three-dimensional, highly cross-linked structure, such as phenol–formaldehyde; or (4) a moderately cross-linked structure, such as butyl rubber. The linear-chain polymers soften with an increase in temperature, are often soluble in organic solvents, and are referred to as *thermoplastic* polymers. At the other extreme, highly cross-linked polymers do not soften appreciably, are almost insoluble in most organic solvents, and are referred to as *thermosetting* polymers. Of more interest in the application of polymers to membranes is a classification based on the arrangement or conformation of the polymer molecules. At low temperatures, typically below 100C, idealized polymers can be classified as *glassy* or *crystalline*. The former refers to a polymer that is brittle and glassy in appearance and lacks any crystalline structure (i.e., *amorphous*), whereas the latter refers to a polymer that is brittle, hard, and stiff, with a crystalline structure. If the temperature of a glassy polymer is increased, a point, called the *glass-transition temperature, T_g*, may be reached where the polymer becomes *rubbery*. If the temperature of a crystalline polymer is increased, a point, called the *melting temperature, T_m*, is reached where the polymer becomes a melt. However, a thermosetting polymer never melts. Most polymers have both amorphous and crystalline regions, that is, a certain degree of crystallinity that varies from 5 to 90%, making it possible for some polymers to have both a T_g and a T_m. Membranes made of glassy polymers can operate below or above T_g; membranes of crystalline polymers must operate below T_m. Table 14.3 lists *repeat units* and values of T_g and/or T_m for several natural and synthetic polymers, from which membranes have been fabricated. Included are crystalline, glassy, and rubbery polymers. Cellulose triacetate is the reaction product of cellulose and acetic anhydride. Cellulose is the most readily available organic raw material in the world. The repeat unit of cellulose is identical to that shown for cellulose triacetate in Table 14.3, except that the acetyl, Ac (CH_3CO) groups are replaced by H. Typically, the number of repeat units (*degree of polymerization*) in cellulose is 1,000 to 1,500, whereas that in cellulose triacetate is around 300. Partially acetylated products are cellulose acetate and cellulose diacetate, with blends of two or three of the acetates being common. The triacetate is highly crystalline, of uniformly high quality, and hydrophobic.

Polyisoprene (natural rubber) is obtained from at least 200 different plants, with many of the rubber-producing countries being located in the Far East. Compared to the other polymers in Table 14.3, polisoprene has a very low glass-transition temperature. Natural rubber has a degree of polymerization of from about 3,000 to 40,000 and is hard and rigid when cold, but soft, easily deformed, and sticky when hot. Depending on the temperature, it slowly crystallizes. To increase the strength, elasticity and stability of rubber, it is vulcanized with sulfur, a process that introduces cross-links, but still allows unrestricted local motion of the polymer chain.

Table 14.3 Common Polymers Used in Membranes

Polymer	Type	Representative Repeat Unit	Glass Transition Temp., °C	Melting Temp., °C
Cellulose triace-tate	Crystalline	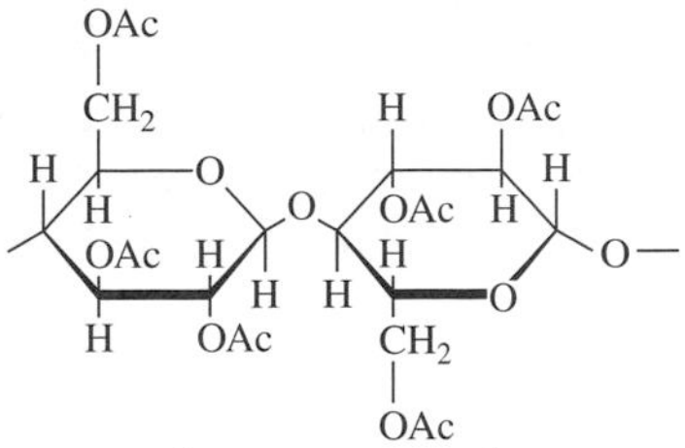		300
Polyisoprene (natural rubber)	Rubbery	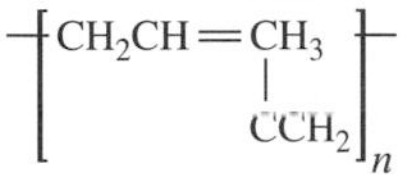 $\left[CH_2CH=CH_3 \atop \quad\quad CCH_2 \right]_n$	−70	
Aromatic poly-amide	Crystalline	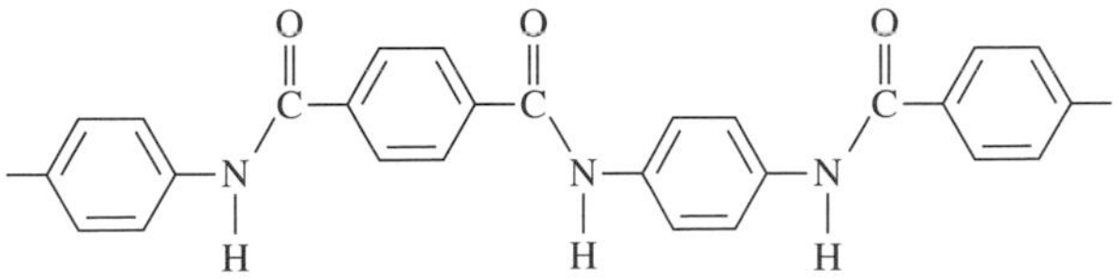		275
Polycarbonate	Glassy	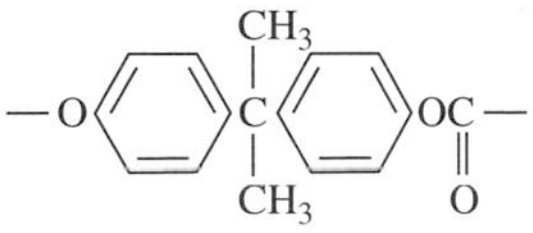	150	
Polyimide	Glassy	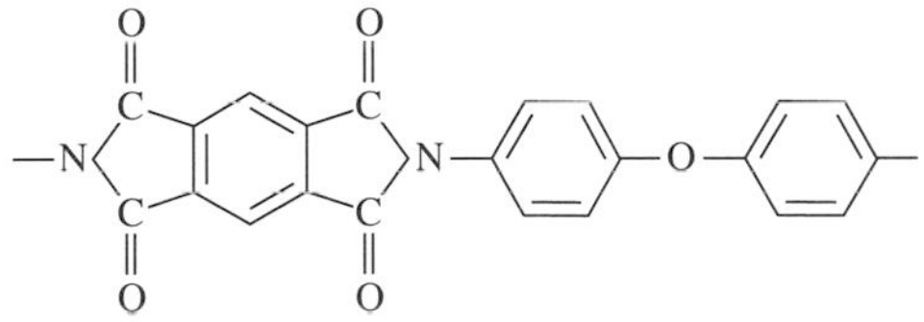	310–365	
Polystyrene	Glassy	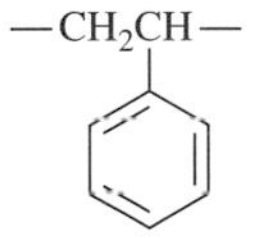 $-CH_2CH-$	74–110	
Polysulfone	Glassy	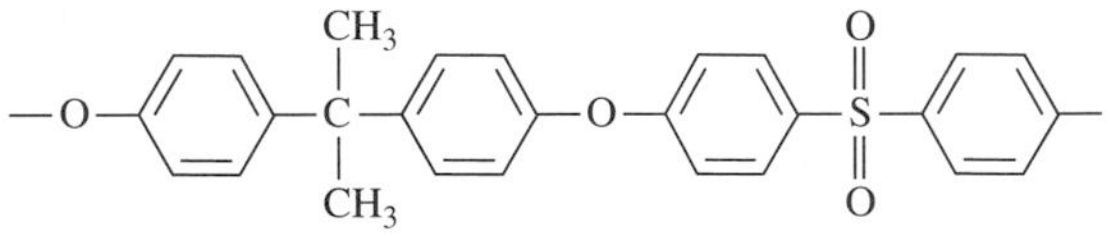	190	
Polytetrafluoro-ethylene (Teflon)	Crystalline	$-CF_2-CF_2-$		327

Aromatic polyamides (also called aramids) are high-melting crystalline polymers that have better long-term thermal stability and higher resistance to solvents than do aliphatic polyamides, such as nylon. Some aromatic polyamides are easily fabricated into fibers, films, and sheets. The polyamide structure shown in Table 14.3 is that of Kevlar, a trade name of DuPont.

Polycarbonates, which are characterized by the presence of the -OCOO- group in the chain, are mainly amorphous in structure. The polycarbonate shown in Table 14.3 is an aromatic form, but aliphatic forms also exist. Polycarbonates differ from most other amorphous polymers in that they possess ductility and toughness below T_g. Because polycarbonates are thermoplastic, they can be extruded into various shapes, including films and sheets.

Polyimides are characterized by the presence of aromatic rings and heterocyclic rings containing nitrogen and attached oxygen. The structure shown in Table 14.3 is only one of a number available. Polyimides are tough, amorphous polymers with high resistance to heat and excellent wear resistance. They can be fabricated into a wide variety of forms, including fibers, sheets, and films.

Polystyrene is a linear, amorphous, highly pure polymer of about 1,000 units of the structure shown in Table 14.3. Above a relatively low T_g, which depends on molecular weight, polystyrene becomes a viscous liquid that is easily fabricated by extrusion or injection molding. Like many other polymers, polystyrene can be annealed (heated and then cooled slowly) to convert it to a crystalline polymer with a melting point of 240°C. Styrene monomer can be copolymerized with a number of other organic monomers, including acrylonitrile and butadiene to form ABS copolymers.

Polysulfones are relatively new synthetic polymers, first introduced in 1966. The structure in Table 14.3 is just one of many, all of which contain the SO_2 group, which gives the polymers high strength. Polysulfones are easily spun into hollow fibers.

Polytetrafluoroethylene is a straight-chain, highly crystalline polymer with a very high degree of polymerization of the order of 100,000, which gives it considerable strength. It possesses exceptional thermal stability and can be formed into sheets, films, and tubing.

To be effective for separating a mixture of chemical components, a polymer membrane must possess high *permeance* and a high permeance ratio for the two species being separated by the membrane. The permeance for a given species diffusing through a membrane of given thickness is analogus to a mass transfer coefficient, i.e., the flow rate of that species per unit cross-sectional area of membrane per unit driving force (concentration, partial pressure, etc.). The molar transmembrane flux of species i is

$$N_i = \left(\frac{P_{M_i}}{l_M}\right)(\text{driving force}) = \overline{P}_{M_i}(\text{driving force}) \qquad \textbf{(14-1)}$$

where $\overline{P}_{M_i}$ is the permeance, which is defined as the ratio of P_{M_i}, the *permeability,* to l_M, the membrane thickness.

Polymer membranes can be dense or microporous. For dense amorphous membranes, no pores of microscopic dimensions are present, and diffusing species must dissolve into the polymer and then diffuse through the polymer between the segments of the macromolecular chains. Diffusion can be difficult, but highly selective for glassy polymers. If the polymer is partly crystalline, diffusion will occur almost exclusively through the amorphous regions, with the crystalline regions decreasing the diffusion area and increasing the diffusion path.

A microporous membrane contains interconnected pores that are small (on the order of 0.005–20 μm; 50–200,000 Å), but large in comparison to the size of small molecules. The pores are formed by a variety of proprietary techniques, some of which are described by Baker et al. [5]. Such techniques are especially valuable for producing symmetric, microporous, crystalline membranes. Permeability for microporous membranes is high, but selectivity is low for small molecules. However, when molecules both smaller and larger than the pore size are in the feed to the membrane, the molecules may be separated almost perfectly by size.

Thus, for the separation of small molecules, we seem to be presented with a dilemma. We can have high permeability or a high separation factor, but not both. The beginning of the resolution of this dilemma occurred in 1963 with the fabrication by Loeb and Sourirajan [7] of an asymmetric membrane of cellulose acetate by a novel casting procedure. As shown in Figure 14.3a, the resulting membrane consists of a thin dense skin about 0.1–1.0 μm in thick, called the *permselective* layer, formed over a much thicker microporous layer that provides support for the skin. The flux rate of a species is controlled by the permeance of the very thin permselective skin. From (14-1), the permeance of species i can be high because of the very small value of l_M even though the permeability, P_{M_i}, is

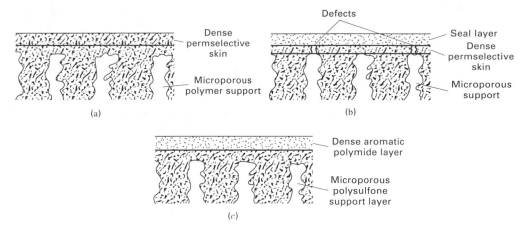

Figure 14.3 Polymer membranes: (a) asymmetric, (b) caulked asymmetric, and (c) typical thin-layer composite.

low because of the absence of pores. When large differences of P_{M_i} exist among molecules, both high permeance and high selectivity can be achieved with asymmetric membranes.

A very thin asymmetric membrane is subject to minute defects or pinholes in the permselective skin, which can render the membrane useless for the separation of a gas mixture. A practical solution to the defect problem for an asymmetric polysulfone membrane was patented by Henis and Tripodi [8] of the Monsanto Company in 1980. They pulled silicone rubber, from a coating on the surface of the skin, into the defects by applying a vacuum. The resulting membrane is sometimes referred to as a *caulked membrane,* as shown in Figure 14.3b.

Another patent by Wrasidlo [9] in 1977, introduced the thin-film composite membrane as an alternative to the asymmetric membrane. In the first application, as shown in Figure 14.3c, a very thin, dense film of polyamide polymer, 250 to 500 Å in thickness, was formed on a thicker microporous polysulfone support. Today asymmetric and thin-film composite membranes are fabricated from a variety of polymers by a variety of techniques.

The application of polymer membranes is generally limited to temperatures below about 200°C and to the separation of mixtures that are chemically inert. When operation at high temperatures and/or with chemically active mixtures is necessary, membranes made of inorganic materials can be used. These include mainly microporous ceramics, metals, and carbon; and dense metals, such as palladium, that allow the selective diffusion of very small molecules such as hydrogen and helium.

Some examples of inorganic membranes are (1) asymmetric microporous α-alumina tubes with 40–100 Å pores at the inside surface and 100,000 Å pores at the outside; (2) microporous glass tubes, the pores of which may or may not be filled with other oxides or the polymerization–pyrolysis product of trichloromethylsilane; (3) silica hollow fibers with extremely fine pores of 3–5 Å; (4) porous ceramic, glass, or polymer materials coated with a thin, dense film of palladium metal that is just a few microns thick; (5) sintered metal; (6) pyrolyzed carbon; and (7) zirconia on sintered carbon. Extremely fine pores (<10 Å) are necessary to separate gas mixtures. Larger pores (>50 Å) may be satisfactory for the separation of large molecules or solid particles from solutions containing small molecules.

EXAMPLE 14.1

A silica-glass membrane of 2 μm thickness and with very fine pores less than 10 Å in diameter has been developed for separating H_2 from CO at a temperature of 500°F. From laboratory data, the membrane permeabilities for hydrogen and carbon monoxide, respectively, are 200,000 and 700 barrer, where the barrer, a commonly used unit for gas permeation, is defined by:

$$1 \text{ barrer} = 10^{-10} \text{ cm}^3 \text{ (STP)-cm/(cm}^2\text{-s-cmHg)}$$

where cm^3 (STP)/cm^2-s) refers to the volumetric transmembrane flux of the diffusing species in terms of standard conditions of 0°C and 1 atm, cm refers to the membrane thickness, and cmHg refers to the transmembrane partial pressure driving force for the diffusing species.

The barrer unit is named for R. M. Barrer, who published an early article [10] on the nature of diffusion in a membrane, followed later by a widely referenced monograph on diffusion in and through solids [11].

If the transmembrane partial pressure driving forces for H$_2$ and CO, respectively, are 240 psi and 80 psi, calculate the transmembrane fluxes in kmol/(m^2-s). Compare the hydrogen flux to that for hydrogen in the commercial application discussed at the beginning of this chapter.

SOLUTION

At 0°C and 1 atm, 1 kmol of gas occupies 22.42 $\times$ 10^6 cm^3. Also, 2 μm thickness =2 $\times$ 10^{-4} cm and 1 cmHg ΔP = 0.1934 psi. Therefore, using (14-1):

$$N_{H_2} = \frac{(200{,}000)(10^{-10})(240/0.1934)(10^4)}{(22.42 \times 10^6)(2 \times 10^{-4})} = 0.0554 \frac{kmol}{m^2\text{-}s}$$

$$N_{CO} = \frac{(700)(10^{-10})(80/0.1934)(10^4)}{(22.42 \times 10^6)(2 \times 10^{-4})} = 0.000065 \frac{kmol}{m^2\text{-}s}$$

In the application discussed at the beginning of this chapter, the flux of H$_2$ for the polymer membrane is

$$\frac{(1685.1)(1/2.205)}{(16{,}000)(0.3048)^2(3600)} = 0.000143 \frac{kmol}{m^2\text{-}s}$$

Thus, the flux of H$_2$ through the ultramicroporous glass membrane is more than 100 times higher than the flux through the dense polymer membrane. Large differences in molar fluxes through different membranes are common. ∎

14.2 MEMBRANE MODULES

The asymmetric and thin-film composite polymer membrane materials described in the previous section are available in one or more of the three shapes shown in Figure 14.4a, b, and c. Flat sheets have typical dimensions of 1 m by 1 m by 200 μm thick, with a dense skin or thin, dense layer 500 to 5,000 Å in thickness. Tubular membranes are typically 0.5 to 5.0 cm in diameter and up to 6 m in length. The thin, dense layer ia on either the inside, as shown in Figure 14.4b, or the outside surface of the tube. The porous supporting part of the tube is fiberglass, perforated metal, or other suitable porous material. Very small-diameter hollow fibers, first reported by Mahon [12,13] in the 1960s, are typically 42 μm i.d. by 85 μm o.d. by 1.2 m long with a 0.1 to 1.0 μm-thick dense skin. Hollow fibers, shown in Figure 14.4c, provide a large membrane surface area per unit volume. A honeycomb monolithic element for inorganic oxide membranes is shown in Figure 14.4.d. Elements of both hexagonal and circular cross-section are available [14]. The circular flow channels are typically 0.3 to 0.6 cm in diameter, with a 20 to 40 mm-thick membrane layer. The hexagonal element in Figure 14.4d has 19 channels and is 0.85 m long. Both the bulk support and the thin membrane layer are porous, but the pores of the latter can be very small, down to 40 Å.

The membrane shapes of Figure 14.4 are incorporated into compact commercial modules and cartridges, some of which are shown in Figure 14.5. Flat sheets used in plate-and-frame modules are circular, square, or rectangular in cross-section. The sheets are separated by support plates that channel the permeate. In Figure 14.5a, a feed of brackish water flows across the surface of each membrane sheet in the stack. Pure water is the permeate product, whereas the retentate is a concentrated brine solution.

Flat sheets are also fabricated into spiral-wound modules shown in Figure 14.5b. A laminate, consisting of two membrane sheets separated by spacers for the flow of the feed and permeate, is wound around a central perforated collection tube to form a module that

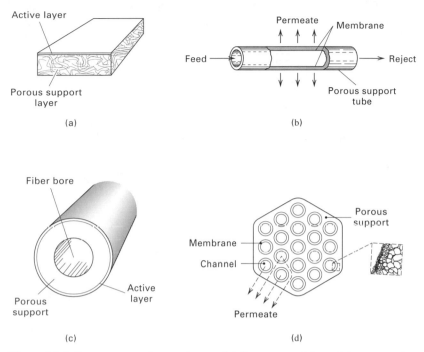

Figure 14.4 Common membrane shapes: (a) flat asymmetric or thin-composite sheet; (b) tubular; (c) hollow fiber; (d) monolithic.

is inserted into a pressure vessel. The feed flows axially in the channels created between the membranes by the porous spacers. Permeate passes through the membrane, traveling inward in a spiral path to the central collection tube. From there, the permeate flows in either axial direction through and out of the central tube. A typical spiral-wound module is 0.1 to 0.3 m in diameter and 3 m long. Six such modules are often placed in series. The four-leaf modification in Figure 14.5c minimizes the pressure drop of the permeate because the permeate travel is less for the same membrane area.

The hollow-fiber module shown in Figure 14.5d, for a gas permeation application, resembles a shell-and-tube heat exchanger. The pressurized feed enters the shell side at one end. While flowing over the fibers toward the other end, permeate passes through the fiber walls into the central fiber channels. Typically the fibers are sealed at one end and embedded into a tube sheet with epoxy resin at the other end. A commercial module might be 1 m long and 0.1 to 0.25 m in diameter and contain more than one million hollow fibers.

A tubular module is shown in Figure 14.5e. This module also resembles a shell-and-tube heat exchanger, but the feed flows through the tubes. Permeate passes through the wall of the tubes into the shell side of the module. Tubular modules contain up to 30 tubes.

The monolithic module in Figure 14.5f contains from 1 to 37 monolithic elements in a module housing. The feed flows through the circular channels and permeate passes through the membrane and porous support and into the open region between elements.

Table 14.4 is a comparison of the characteristics of four of the modules shown in Figure 14.5. The packing density is the membrane surface area per unit volume of module, for which the hollow-fiber membrane modules are clearly superior. Although the plate-and-frame module has a high cost and a moderate packing density, it finds use in all membrane applications except gas permeation. It is the only module widely used for pervaporation. The spiral-wound module is very popular for most applications because of its low cost and reasonable resistance to fouling. Tubular modules are only used for small applications or

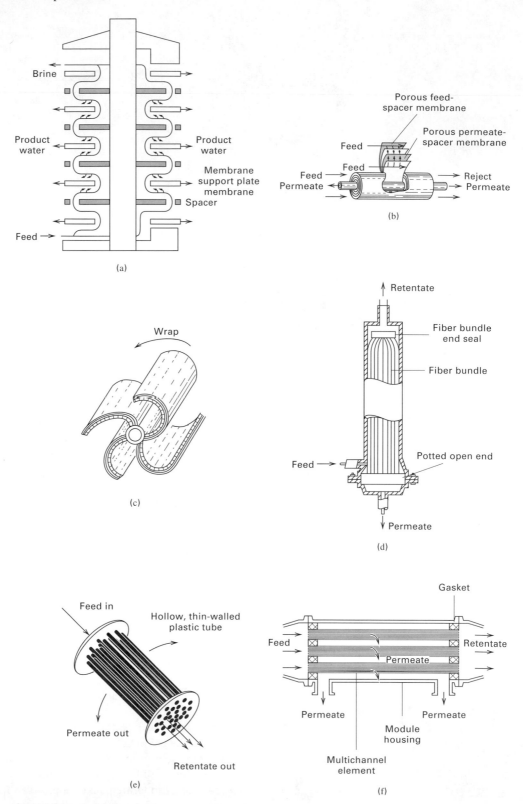

Figure 14.5 Common membrane modules: (a) plate-and frame, (b) spiral-wound, (c) four-leaf spiral-wound, (d) hollow-fiber, (e) tubular, (f) monolithic.

Table 14.4 Typical Characteristics of Membrane Modules

	Plate and Frame	Spiral-Wound	Tubular	Hollow-Fiber
Packing density, m^2/m^3	30 to 500	200 to 800	30 to 200	500 to 9,000
Resistance to fouling	Good	Moderate	Very good	Poor
Ease of cleaning	Good	Fair	Excellent	Poor
Relative cost	High	Low	High	Low
Main applications	D, RO, PV, UF, MF	D, RO, GP, UF, MF	RO, UF	D, RO, GP, UF

Note. D, dialysis; RO, reverse osmosis; GP, gas permeation; PV, pervaporation; UF, ultrafiltration; MF, microfiltration.

when a high resistance to fouling and/or ease of cleaning are essential. Hollow-fiber modules, with a very high packing density and low cost, are popular where fouling does not occur and cleaning is not necessary.

14.3 TRANSPORT IN MEMBRANES

For a given application, the calculation of the required membrane surface area is based on laboratory data for the selected membrane. Although permeation can occur by one or more of the mechanisms discussed in this section, these mechanisms are all consistent with (14-1) in either its permeance form or its permeability form, with the latter being applied more widely. However, because both the driving force and the permeability or permeance depend markedly on the mechanism of transport, it is important to understand the nature of transport in membranes, which is the subject of this section. Applications to dialysis, reverse osmosis, gas permeation, and pervaporation are presented in subsequent sections.

Membranes can be macroporous, microporous, or dense (nonporous). Only microporous or dense membranes are permselective. However, macroporous membranes are widely used to support thin microporous and dense membranes when significant pressure differences across the membrane are necessary to achieve a reasonable throughput. The theoretical basis for transport through microporous membranes is more highly developed than that for dense membranes, so porous-membrane transport is discussed first.

Porous Membranes

Mechanisms for the transport of liquid and gas molecules through a porous membrane are depicted in Figure 14.6a, b, and c. If the pore diameter is large compared to the molecular diameter, and a pressure difference exists across the membrane, bulk or convective flow through the pores occurs, as shown in Figure 14.6a. Such a flow is generally undesirable

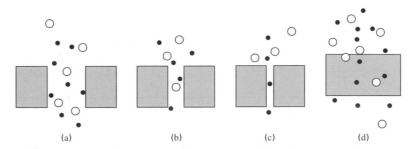

Figure 14.6 Mechanisms of transport in membranes. (Flow is downward.) (a) Bulk flow through pores; (b) diffusion through pores; (c) restricted diffusion through pores; (d) solution-diffusion through dense membranes.

because it is not permselective and, therefore, no separation between components of the feed occurs. If fugacity, activity, chemical potential, concentration, or partial pressure differences exist across the membrane for the various components, but the pressure is the same on both sides of the membrane, permselective diffusion of the components through the pores will take place, effecting a separation as shown in Figure 14.6b. If the pores are of the order of molecular size for at least some of the components in the feed mixture, the diffusion of those components will be restricted (hindered) as shown in Figure 14.6c, resulting in an enhanced separation. Molecules of size larger than the pores will be prevented altogether from diffusing through the pores. This special case is highly desirable and is referred to as *sieving*. Another special case exists for gas diffusion where the pore size and/or pressure (typically a vacuum) is such that the mean free path of the molecules is greater than the pore diameter, resulting in so-called *Knudsen diffusion,* which is dependent on molecular weight.

Bulk Flow

Consider the bulk flow of a fluid, due to a pressure difference, through an idealized straight, cylindrical pore. If the flow is in the laminar regime ($N_{Re} = Dv\rho/\mu < 2,100$), which is almost always the case for flow in small-diameter pores, the flow velocity, v, is given by the Hagen–Poiseuille law [15] as being directly proportional to the transmembrane pressure drop:

$$v = \frac{D^2}{32\mu L}(P_0 - P_L) \tag{14-2}$$

where D is the pore diameter, μ is the viscosity of the fluid, and L is the length of the pore. This law assumes that a parabolic velocity profile exists across the pore radius for the entire length of the pore, that the fluid is Newtonian, and if a gas, that the mean free path of the molecules is small compared to the pore diameter. If the membrane contains n such pores per unit cross-section of membrane surface area normal to flow, the porosity (void fraction) of the membrane is

$$\epsilon = n\pi D^2/4 \tag{14-3}$$

Then the superficial fluid bulk-flow flux (mass velocity), N, through the membrane is

$$N = \frac{\epsilon\rho D^2}{32\mu l_M}(P_0 - P_L) = \frac{n\pi\rho D^4}{128\mu l_M}(P_0 - P_L) \tag{14-4}$$

where l_M is the membrane thickness and ρ and μ are fluid properties.

In real porous membranes, pores may not be cylindrical and straight, making it necessary to modify (14-4). One procedure is that due to Carman and Kozeny, as extended by Ergun [16], where the pore diameter in (14-2) is replaced, as a rough approximation, by the hydraulic diameter:

$$d_H = 4\left(\frac{\text{Volume available for flow}}{\text{Total pore surface area}}\right)$$

$$= \frac{4\left(\dfrac{\text{Total pore volume}}{\text{Membrane volume}}\right)}{\left(\dfrac{\text{Total pore surface area}}{\text{Membrane volume}}\right)} = \frac{4\epsilon}{a} \tag{14-5}$$

where the membrane volume includes the volume of the pores. The specific surface area, a_v, which is the pore surface area per unit volume of membrane, is

$$a_v = a/(1 - \epsilon) \tag{14-6}$$

Pore length is longer than the membrane thickness and can be represented by $l_M\tau$, where τ is a tortuosity factor >1. Substituting (14-5), (14-6), and the tortuosity factor into (14-4) gives

$$N = \frac{\rho\epsilon^3(P_0 - P_L)}{2(1 - \epsilon)^2\tau a_v^2\mu l_M}$$ (14-7)

In terms of a bulk-flow permability, (14-7) becomes

$$N = \frac{P_M}{l_M}(P_0 - P_L)$$ (14-8)

where

$$P_M = \frac{\rho\epsilon^3}{2(1 - \epsilon)^2\tau a_v^2\mu}$$ (14-9)

Typically, τ is approximately 2.5, whereas a_v is inversely proportional to the average pore diameter, giving it values over a wide range.

Equation (14-7) may be compared to the following semitheoretical Ergun equation [16], which represents the best fit of experimental data for flow of a fluid through a packed bed:

$$\frac{P_0 - P_L}{l_M} = \frac{150\mu v_0(1 - \epsilon)^2}{D_P^2\epsilon^3} + \frac{1.75\rho v_0^2(1 - \epsilon)}{D_P\epsilon^3}$$ (14-10)

where D_P is the mean particle diameter and v_0 is the superficial fluid velocity through the membrane. The first term on the right-hand side of (14-10) applies to the laminar flow region and the second to the turbulent region. For a spherical particle:

$$a_v = \pi D_P^2/(\pi D_P^3/6) \text{ or}$$ (14-11)

$$D_P = 6/a_v$$

Substitution of (14-11) into (14-10), for just the laminar-flow region, and rearrangement into the bulk-flow flux form gives

$$N = \frac{\rho\epsilon^3(P_0 - P_L)}{(150/36)(1 - \epsilon)^2 a_v^2\mu l_M}$$ (14-12)

Comparing (14-12) to (14-7), we see that the term (150/36) in (14-12) corresponds to the term (2t) in (14-7). Thus, τ appears to have a value of 2.08, which seems reasonable. Accordingly, (14-12) can be used as a first approximation to the pressure drop for flow through a porous membrane when the pores are not straight cylinders. For gas flow, the density may be taken as the arithmetic average of the densities at the upstream and downstream faces of the membrane.

EXAMPLE 14.2

It is desired to pass water at 70°F through a supported polypropylene membrane, with a skin of 0.003 cm thickness and 35% porosity, at the rate of 200 m³/m²-day. The pores can be considered as straight cylinders of uniform diameter equal to 0.2 micron. If the pressure on the downstream side of the membrane is 150 kPa, estimate the required pressure on the upstream side of the membrane. The pressure drop through the support is negligible.

SOLUTION

Equation (14-4) applies, where in SI units:

$N/\rho = 200/(24)(3600) = 0.0023 \text{ m}^3/\text{m}^2-\text{s}$ $\epsilon = 0.35$
$D_P = 0.2 \times 10^{-6} \text{ m}$ $l_M = 0.00003 \text{ m}$
$P_L = 150 \text{ kPa} = 150,000 \text{ Pa}$ $\mu - 0.001 \text{ Pa-s}$

From (14-4), $P_0 = P_L + \dfrac{32\mu l_M(N\rho)}{\epsilon D_P^2}$

$$= 150,000 + \frac{(32)(0.001)(0.00003)(0.00232)}{(0.35)(0.2 \times 10^{-6})^2} = 309,000 \text{ Pa or 309 kPa}$$

■

Liquid Diffusion

Consider the diffusion of species through the pores of a membrane from a fluid feed to a sweep fluid when identical total pressures but different component concentrations exist on both sides of the membrane. In that case, bulk flow through the membrane due to a pressure difference does not occur and if species diffuse at different rates, a separation can be achieved. If the feed mixture is a liquid of solvent and solutes i, the solute transmembrane flux is given by a modified form of Fick's law:

$$N_i = \frac{D_{e_i}}{l_M}(c_{i_0} - c_{i_L}) \tag{14-13}$$

where D_{e_i} is the effective diffusivity, and c_i is the concentration of i in the liquid in the pores at the two faces of the membrane. In general the effective diffusivity is given by

$$D_{e_i} = \frac{\epsilon D_i}{\tau} K_{r_i} \tag{14-14}$$

where D_i is the ordinary molecular diffusion coefficient (diffusivity) of the solute i in the solution, ϵ is the volume fraction of pores in the membrane, τ is the tortuosity, and K_r is a restrictive factor that accounts for the effect of pore diameter, d_p, when the ratio of molecular diameter, d_m, to pore diameter exceeds about 0.01. The restrictive factor is approximated by Beck and Schultz [17] with:

$$K_r = \left[1 - \frac{d_m}{d_p}\right]^4, (d_m/d_p) \leq 1 \tag{14-15}$$

From (14-15), when $(d_m/d_p) = 0.01$, $K_r = 0.96$, but when $(d_m/d_p) = 0.3$, $K_r = 0.24$. When $d_m > d_p$, $K_r = 0$, and the solute cannot diffuse through the pore. This is the sieving effect illustrated in Figure 14.6c. In general, as illustrated in the following example, transmembrane fluxes for liquids through microporous membranes are very small because effective diffusivities are very low.

EXAMPLE 14.3

Beck and Schultz [18] measured effective diffusivities of urea and several different sugars, in aqueous solutions, through microporous membranes of mica, which were especially prepared to give almost straight, elliptical pores of almost uniform size. Based on the following data for a membrane and two solutes, estimate the transmembrane fluxes for the two solutes in g/cm²-s at 25°C. Assume that the aqueous solutions on either side of the membrane are sufficiently dilute that no multicomponent diffusional effects are present.

Membrane:

Material	Microporous mica
Thickness, μm	4.24
Average pore diameter, Angstroms	88.8
Tortuosity, τ	1.1
Porosity, ϵ	0.0233

Solutes (in aqueous solution at 25°C):

Solute	$D_i \times 10^6$ cm²/s	molecular diameter, d_m, Å	c_{i_0} (g/cm³)	c_{i_L} (g/cm³)
1 Urea	13.8	5.28	0.0005	0.0001
2 β-Dextrin	3.22	17.96	0.0003	0.00001

SOLUTION

Calculate the restrictive factor and effective diffusivity from (14-15) and (14-14), respectively. For urea (1):

$$K_{r_1} = \left[1 - \left(\frac{5.28}{88.8}\right)\right]^4 = 0.783; \quad D_{e_1} = \frac{(0.0233)(13.8 \times 10^{-6})(0.783)}{1.1} = 2.29 \times 10^{-7} \text{ cm}^2/\text{s}$$

For β-dextrin (2):

$$K_{r_2} = \left[1 - \left(\frac{17.96}{88.8} \right) \right]^4 = 0.405; D_{e_2} = \frac{(0.0233)(3.22 \times 10^{-6})(0.405)}{1.1} = 2.78 \times 10^{-8} \, \text{cm}^2/\text{s}$$

Because of the large differences in molecular size, the two effective diffusivities differ by almost an order of magnitude.

Calculate transmembrane fluxes from (14.13) noting the given concentrations are at the two faces of the membranes. Concentrations in the bulk solutions on either side of the membrane will differ from the concentrations at the faces depending upon the magnitudes of the mass-transfer resistances in boundary layers or films adjacent to the two faces of the membrane.

$$\text{For urea: } N_1 = \frac{(2.29 \times 10^{-7})(0.0005 - 0.0001)}{4.24 \times 10^{-4}} = 2.16 \times 10^{-7} \, \text{g/cm}^2\text{-s}$$

$$\text{For β-dextrin: } N_2 = \frac{(2.768 \times 10^{-8})(0.0003 - 0.00001)}{(4.24 \times 10^{-4})} = 1.90 \times 10^{-8} \, \text{g/cm}^2\text{-s}$$

Note that these fluxes are extremely low.   ∎

Gas Diffusion

When the mixture on either side of a microporous membrane is a gas, the rate of species diffusion can again be expressed in terms of Fick's law. If pressure and temperature on either side of the membrane are equal and the ideal gas law holds, (14-13) can be written in terms of a partial pressure driving force:

$$N_i = \frac{D_{e_i} c_M}{P l_M} (p_{i_0} - p_{i_L}) \tag{14-16}$$

where c_M is the total concentration of the gas mixture given as P/RT by the ideal gas law. Thus, (14-16) can be written alternatively as

$$N_i = \frac{D_{e_i}}{RT l_M} (p_{i_0} - p_{i_L}) \tag{14-17}$$

For a gas, diffusion may occur by ordinary diffusion, as with a liquid, and/or in series with Knudsen diffusion when pore diameter is very small and/or total pressure is low. In the Knudsen flow regime, collisions occur primarily between gas molecules and the pore wall, rather than between gas molecules. Thus, in the absence of a bulk flow effect, (14-14) is modified for gas flow:

$$D_{e_i} = \frac{\epsilon}{\tau} \left[\frac{1}{(1/D_i) + (1/D_{K_i})} \right] \tag{14-18}$$

where D_{K_i} is the Knudsen diffusivity, which from the kinetic theory of gases applied to a straight cylindrical pore, is given by

$$D_{K_i} = \frac{d_p \bar{v}_i}{3} \tag{14-19}$$

where $\bar{v}_i$ is the average molecule velocity given by:

$$\bar{v}_i = (8RT/\pi M_i)^{1/2} \tag{14-20}$$

where M is the molecular weight. Combining (14-19) and (14-20):

$$D_{K_i} = 4,850 d_p (T/M_i)^{1/2} \tag{14-21}$$

where D_K is cm^2/s, d_p is cm, and T is K. When Knudsen flow predominates, as it often does for the micropores in membranes, the permeability ratio for species A and B is given from a combination of (14-1), (14-17), (14-18), and (14-21):

$$\frac{P_{M_A}}{P_{M_B}} = \left(\frac{M_B}{M_A}\right)^{1/2} \qquad (14\text{-}22)$$

Except for gaseous species of widely differing molecular weight, the permeability ratio from (14-22) is not large, and the separation of gases by microporous membranes at low to moderate pressures that are equal on both sides of the membrane to minimize bulk flow is almost always impractical, as illustrated in the following example. However, it is important to note that the separation of the two isotopes of UF$_6$ by the United States government was accomplished by Knudsen diffusion, with a permeability ratio of only 1.0043, on a large scale at Oak Ridge using thousands of stages and many acres of membrane surface.

EXAMPLE 14.4

A gas mixture of hydrogen (H) and ethane (E) is to be partially separated with a composite membrane having a 1-μm-thick porous skin with an average pore size of 20 Å and a porosity of 30%. The tortuosity can be assumed to be 1.5. The pressure on either side of the membrane is 10 atm and the temperature is 100°C. Estimate the permeabilities of the two components in barrers.

SOLUTION

From (14-1), (14-17), and (14-18), the permeability can be expressed in gmol-cm/cm^2-s-atm:

$$P_{M_i} = \frac{\epsilon}{RT\tau}\left[\frac{1}{(1/D_i) + (1/D_{K_i})}\right],$$

where $\epsilon = 0.30$, $R = 82.06$ cm^3-atm/mol-K, $T = 373$ K, and $\tau = 15$

At 100°C, the ordinary diffusivity is given by $D_H = D_E = D_{H,E} = 0.86/P$ in cm^2/s with total pressure P in atm. Thus, at 10 atm, $D_H = D_E = 0.086$ cm^2/s. Knudsen diffusivities are given by (14-21) with pore diameter, d_p, equal to 20×10^{-8} cm.

$$D_{K_H} = 4,850(20 \times 10^{-8})(373/2.016)^{1/2} = 0.0132 \text{ cm}^2/\text{s}$$

$$D_{K_E} = 4,850(20 \times 10^{-8})(373/30.07)^{1/2} = 0.00342 \text{ cm}^2/\text{s}$$

For both components, diffusion is controlled mainly by Knudsen diffusion. For hydrogen: $\dfrac{1}{(1/D_H) + (1/D_{K_H})} = 0.0114$ cm^2/s. For ethane: $\dfrac{1}{(1/D_E) + (1/D_{K_E})} = 0.00329$ cm^2/s.

$$P_{M_H} = \frac{0.30(0.0114)}{(82.06)(373)(1.5)} = 7.45 \times 10^{-8} \frac{\text{mol-cm}}{\text{cm}^2\text{-s-atm}}$$

$$P_{M_E} = \frac{0.30(0.00329)}{(82.06)(373)(1.5)} = 2.15 \times 10^{-8} \frac{\text{mol-cm}}{\text{cm}^2\text{-s-atm}}$$

To convert to barrer as defined in Example 14.1, note that

$$76 \text{ cmHg} = 1 \text{ atm and } 22,400 \text{ cm}^3(\text{STP}) = 1 \text{ mol}$$

$$P_{M_H} = \frac{7.45 \times 10^{-8}(22,400)}{(10^{-10})(76)} = 220,000 \text{ barrer}$$

$$P_{M_E} = \frac{2.15 \times 10^{-8}(22,400)}{(10^{-10})(76)} = 63,400 \text{ barrer}$$

■

Nonporous Membranes

The transport of components through nonporous (dense) solid membranes is the predominant mechanism of membrane separators for reverse osmosis (liquid), gas permeation (gas), and pervaporation (liquid and vapor). As indicated in Figure 14.6d, gas or liquid components absorb into the membrane at the upstream face, diffuse through the solid membrane, and desorb at the downstream face.

Liquid diffusivities are several orders of magnitude less than gas diffusivities, and diffusivities of solutes in solids are a few orders of magnitude less than diffusivities in liquids. Thus, differences between diffusivities in gases and solids are enormous. For example, at 1 atm and 25C, diffusivities in cm^2/s for water are as follows:

Water vapor in air	0.25
Water in ethanol liquid	1.2×10^{-5}
Dissolved water in cellulose acetate solid	1×10^{-8}

As might be expected, small molecules fare better than large molecules for diffusivities in solids. For example, from the *Polymer Handbook* [19], diffusivities in cm^2/s for several components in low-density polyethylene at 25°C are

Helium	6.8×10^{-6}
Hydrogen	0.474×10^{-6}
Nitrogen	0.320×10^{-6}
Propane	0.0322×10^{-6}

Regardless of whether a nonporous membrane is used to separate a gas or liquid mixture, the *solution-diffusion model* of Lonsdale, Merten, and Riley [20] is most often applied to analyze experimental permeability data and design membrane separators. This model is based on Fick's law for diffusion through solid nonporous membranes based on the driving force, $c_{i_0} - c_{i_L}$, where the concentrations are those for the solute dissolved in the membrane. The concentrations in the membrane are related to the concentrations or partial pressures in the fluid adjacent to the membrane faces by assuming thermodynamic equilibrium at the fluid–membrane interfaces. This assumption has been validated experimentally by Motanedian et al. [21] for the case of permeation of light gases through dense cellulose acetate membranes at up to 90 atm.

Solution-Diffusion for Liquid Mixtures

Figures 14.7a and b show typical solute concentration profiles for liquid mixtures with porous and nonporous membranes, respectively. Included in these diagrams is the drop in concentration across the membrane and, also, possible drops due to resistances in the fluid boundary layers or films on either side of the membrane. For porous membranes, of the type considered in the previous section, the concentration profile is continuous from the bulk feed liquid to the bulk permeate liquid because liquid is present continuously from one side to the other. The concentration c_{i_0} is the same in the liquid feed just adjacent to the membrane surface and in the liquid just within the entrance of the pore. This is not the case for the nonporous membrane in Figure 14.7b. Solute concentration c'_{i_0} is that in the feed liquid just adjacent to the upstream membrane surface, whereas c_{i_0} is that in the membrane just adjacent to the upstream membrane surface. In general, c_{i_0} is considerably smaller than c'_{i_0}, but the two are related by a thermodynamic equilibrium partition coefficient K_i, defined by

$$K_{i_0} = c_{i_0}/c'_{i_0} \tag{14-23}$$

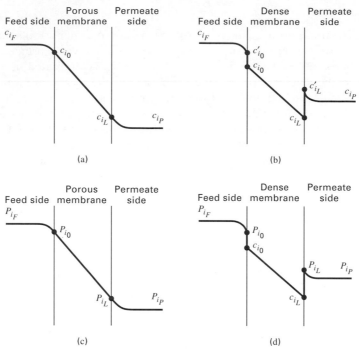

Figure 14.7 Concentration and partial pressure profiles for solute transport through membranes. Liquid mixture with (a) a porous and (b) a nonporous membrane; gas mixture with (c) a porous and (d) a nonporous membrane.

Similarly, at the other face:

$$K_{i_L} = c_{i_L}/c'_{i_L} \tag{14-24}$$

Fick's law applied to the nonporous membrane of Figure 14.7b is:

$$N_i = \frac{D_i}{l_M}(c_{i_0} - c_{i_L}) \tag{14-25}$$

where D_i is the diffusivity of the solute in the membrane. If (14-23) and (14-24) are combined with (14-25), and the partition coefficient is assumed to be independent of concentration, such that $K_{i_0} = K_{i_L} = K_i$, we obtain for the flux

$$N_i = \frac{K_i D_i}{l_M}(c'_{i_0} - c'_{i_L}) \tag{14-26}$$

If the mass-transfer resistances in the two fluid boundary layers or films are negligible:

$$N_i = \frac{K_i D_i}{l_M}(c_{i_F} - c_{i_P}) \tag{14-27}$$

In (14-26) and (14-27), $K_i D_i$ is the permeability, P_{M_i}, for the solution-diffusion model, where K_i accounts for the solubility of the solute in the membrane and D_i accounts for diffusion through the membrane. Because D_i is generally very small, it is important that the membrane material offers a large value for K_i and/or a small membrane thickness.

  Both D_i and K_i, and therefore P_{M_i}, depend on the solute and the membrane. When solutes dissolve in a polymer membrane, it will swell, causing both D_i and K_i to increase.

Table 14.5 Factors That Influence Permeability
of Solutes in Dense Polymers

Factor	Value Favoring High Permeability
Polymer density	low
Degree of crystallinity	low
Degree of cross-linking	low
Degree of vulcanization	low
Amount of plasticizers	high
Amount of fillers	low
Chemical affinity of solute for polymer (solubility)	high

Other polymer membrane factors that can influence D_i, K_i, and P_{M_i} are listed in Table 14.5. However, the largest single factor is the chemical structure of the polymer. Because of the many factors involved, it is important to obtain experimental permeability data on the membrane and feed mixture of interest.

Solution-Diffusion for Gas Mixtures

Figures 14.7c and d show typical solute profiles for gas mixtures with porous and nonporous membranes, respectively, including the effect of fluid boundary layer or film mass-transfer resistances. For the porous membrane, a continuous partial pressure profile is shown. For the nonporous membrane, a concentration profile is shown within the membrane where the solute is dissolved in the membrane. Fick's law, given by (14-25), holds for transport through the membrane. Assuming that thermodynamic equilibrium exists at the two fluid–membrane interfaces, the concentrations in Fick's law can be related to the partial pressures adjacent to the membrane faces by Henry's law, which is a linear relation that is most conveniently written for membrane applications as

$$H_{i_0} = c_{i_0}/p_{i_0} \tag{14-28}$$

$$\text{and } H_{i_L} = c_{i_L}/p_{i_L} \tag{14-29}$$

If we assume that H_i is independent of total pressure and that the temperature is the same at both membrane faces:

$$H_{i_0} = H_{i_L} = H_i \tag{14-30}$$

Combining (14-25), (14-28), (14-29), and (14-30), the flux is

$$N_i = \frac{H_i D_i}{l_M}(p_{i_0} - p_{i_L}) \tag{14-31}$$

If the external mass-transfer resistances are neglected, $p_{i_F} = p_{i_0}$ and $p_{i_L} = p_{i_P}$:

$$N_i = \frac{H_i D_i}{l_M}(p_{i_F} - p_{i_P}) = \frac{P_{M_i}}{l_M}(p_{i_F} - p_{i_P}) \tag{14-32}$$

where

$$P_{M_i} = H_i D_i \tag{14-33}$$

Thus, the permeability depends on both the solubility of the gas component in the membrane and the diffusivity of that component in the membrane. An acceptable rate of transport through the membrane can be achieved only by using a very thin membrane and a high pressure on the feed side. The permeability of a gaseous component in a polymer membrane is subject to the factors listed in Table 14.5. Light gases do not interact with the polymer or cause it to swell. Thus, a light-gas–permeant-polymer combination is readily characterized experimentally. Often both solubility and diffusivity are measured. An extensive tabulation is given in the *Polymer Handbook* [19]. Representative data at 25°C are given in Table 14.6. In general, diffusivity decreases and solubility increases with increasing molecular weight of the gas species. The effect of temperature over a modest range of about 50°C can be represented for both solubility and diffusivity by Arrhenius equations. For example,

$$D = D_0 e^{-E_D/RT} \qquad (14\text{-}34)$$

In general, the modest effect of temperature on solubility may act in either direction. However, an increase in temperature can cause a substantial increase in diffusivity and, therefore, a corresponding increase in permeability. Typical activation energies of diffusion in polymers, E_D, range from 15 to 60 kJ/mol.

The application of Henry's law for rubbery polymers is well accepted, particularly for low-molecular-weight penetrants, but is less accurate for glassy polymers, for which alternative theories have been proposed. Foremost is the dual-mode model first proposed by Barrer and co-workers [22–24] as the result of a comprehensive study of sorption and diffusion in ethyl cellulose. In this model, sorption of the penetrant occurs by ordinary dissolution in the polymer chains, as described by Henry's law, and by Langmuir sorption into holes or sites between chains of glassy polymers. When the downstream pressure is negligible compared to the upstream pressure, the permeability for Fick's law is given by

$$P_{M_i} = H_i D_i + \frac{D_{L_i} a b}{1 + bP} \qquad (14\text{-}35)$$

Table 14.6 Coefficients for Gas Permeation in Polymers

	Gas Species					
	H_2	O_2	N_2	CO	CO_2	CH_4
Low-Density Polyethylene:						
$D \times 10^6$	0.474	0.46	0.32	0.332	0.372	0.193
$H \times 10^6$	1.58	0.472	0.228	0.336	2.54	1.13
$P_M \times 10^{13}$	7.4	2.2	0.73	1.1	9.5	2.2
Polyethylmethacrylate:						
$D \times 10^6$	—	0.106	0.0301	—	0.0336	—
$H \times 10^6$	—	0.839	0.565	—	11.3	—
$P_M \times 10^{13}$	—	0.889	0.170	—	3.79	—
Polyvinylchloride:						
$D \times 10^6$	0.5	0.012	0.0038	—	0.0025	0.0013
$H \times 10^6$	0.26	0.29	0.23	—	4.7	1.7
$P_M \times 10^{13}$	1.3	0.034	0.0089	—	0.12	0.021
Butyl Rubber:						
$D \times 10^6$	1.52	0.081	0.045	—	0.0578	—
$H \times 10^6$	0.355	1.20	0.543	—	6.71	—
$P_M \times 10^{13}$	5.43	0.977	0.243	—	3.89	—

Note. Units: D in cm^2/s; H in cm^3 (STP)/cm^3-Pa; P_M in cm^3 (STP)-cm/cm^2-s-Pa.

where the second term refers to Langmuir sorption with D_{L_i} = diffusivity of Langmuir sorbed species, P = penetrant pressure, and a,b = Langmuir constants for sorption site capacity and site affinity, respectively.

Koros and Paul [25] found that the dual-mode theory accurately represents data for the sorption of CO_2 in polyethylene terephthalate below its glass-transition temperature of about 85°C. Above that temperature, the rubbery polymer obeys just Henry's law. Mechanisms of diffusion for the Langmuir mode have been suggested by Barrer [26].

The ideal dense polymer membrane has a high permeance, P_{M_i}/l_M, for the penetrant molecules and a high separation factor (selectivity) between the components to be separated. The separation factor is defined similarly to relative volatility in distillation:

$$\alpha_{A,B} = \frac{(y_A/x_A)}{(y_B/x_B)} \tag{14-36}$$

where y_i is the mole fraction in the permeate leaving the membrane, corresponding to the partial pressure p_{i_P} in Figure 14.7d, while x_i is the mole fraction in the retentate on the feed side of the membrane, corresponding to the partial pressure p_{i_F} in Figure 14.7d. Unlike the case of distillation, y_i and x_i are not in equilibrium.

For the separation of a binary gas mixture of species A and B in the absence of boundary layer or film mass-transfer resistances, the transport fluxes are given by (14-32):

$$N_A = \frac{H_A D_A}{l_M}(x_A P_F - y_A P_P) \tag{14-37}$$

$$N_B = \frac{H_B D_B}{l_M}(x_B P_F - y_B P_P) \tag{14-38}$$

When no sweep gas is used, the ratio of N_A to N_B is simply the ratio of y_A to y_B in the permeate gas. Thus,

$$\frac{N_A}{N_B} = \frac{y_A}{y_B} = \frac{H_A D_A(x_A P_F - y_A P_P)}{H_B D_B(x_B P_F - y_B P_P)} \tag{14-39}$$

If the downstream (permeate) pressure, P_P, is negligible compared to the upstream pressure, P_F, such that $y_A P_P \ll x_A P_F$ and $y_B P_P \ll x_B P_F$, (14-39) can be rearranged and combined with (14-36) to give an *ideal separation factor:*

$$\alpha_{A,B}^* = \frac{H_A D_A}{H_B D_B} = \frac{P_{M_A}}{P_{M_B}} \tag{14-40}$$

Thus, a high separation factor can be achieved from a high solubility ratio, a high diffusivity ratio, or both. The separation factor depends on both transport phenomena and thermodynamic equilibria.

When the downstream pressure is not negligible, (14-39) can be rearranged to obtain an expression for $\alpha_{A,B}$ in terms of the pressure ratio, $r = P_P/P_F$, and the mole fraction of A on the feed or retentate side of the membrane. Combining (14-36), (14-40), and the definition of r with (14-39):

$$\alpha_{A,B} = \alpha_{A,B}^* \left[\frac{(x_B/y_B) - r\alpha_{A,B}}{(x_B/y_B) - r} \right] \tag{14-41}$$

Because $y_A + y_B = 1$, we can substitute into (14-41) for x_B, the identity:

$$x_B = x_B y_A + x_B y_B$$

Table 14.7 Membrane Separation Factors of Binary Pairs for Two Membrane Materials

	PDMS, Silicon Rubbery Polymer Membrane	PC, Polycarbonate Glassy Polymer Membrane
$P_{M_{He}}$, barrer	561	14
α_{He,CH_4}	0.41	50
α_{He,C_2H_4}	0.15	33.7
$P_{M_{CO_2}}$, barrer	4,550	6.5
α_{CO_2,CH_4}	3.37	23.2
α_{CO_2,C_2H_4}	1.19	14.6
$P_{M_{O_2}}$, barrer	933	1.48
α_{O_2,N_2}	2.12	5.12

to give

$$\alpha_{A,B} = \alpha_{A,B}^* \left[\frac{x_B \left(\frac{y_A}{y_B} + 1 \right) - r\alpha_{A,B}}{x_B \left(\frac{y_A}{y_B} + 1 \right) - r} \right] \qquad (14\text{-}42)$$

If we combine (14-36) and (14-42) and replace x_B with $1 - x_A$, we obtain for the separation factor:

$$\alpha_{A,B} = \alpha_{A,B}^* \left[\frac{x_A(\alpha_{A,B} - 1) + 1 - r\alpha_{A,B}}{x_A(\alpha_{A,B} - 1) + 1 - r} \right] \qquad (14\text{-}43)$$

Equation (14-43) is an implicit equation for $\alpha_{A,B}$, in terms of the pressure ratio, r, and x_A, that is readily solved for $\alpha_{A,B}$ by the formula for a quadratic equation. In the limit when $r = 0$, (14-43) reduces to $\alpha_{A,B} = \alpha_{A,B}^* = (P_{M_A}/P_{M_B})$. Many experimental investigators report values of $\alpha_{A,B}^*$. For example, Table 14.7, taken from the *Membrane Handbook* [6], gives data at 35°C for various binary pairs with polydimethyl siloxane (PDMS), a rubbery polymer, and bisphenol-A-polycarbonate (PC), a glassy polymer. For the rubbery polymer, permeabilities are high, but separation factors are low. The opposite is true for the glassy polymer. For a given feed composition, the separation factor places a definite limit on the degree of separation that can be achieved.

EXAMPLE 14.5

Air can be separated into nitrogen-enriched and oxygen-enriched streams by gas permeation with a number of different dense polymer membranes. In all cases, the membrane is more permeable to oxygen. A total of 20,000 scfm of air is compressed, cooled, and treated to remove moisture and compressor oil prior to being sent to a membrane separator at 150 psia and 78°F. Assume the composition of the air is 79 mol% N_2 and 21 mol% O_2. A low-density polyethylene membrane in the form of a thin-film composite is being considered with solubilities and diffusivities given in Table 14.6. If the membrane skin is 0.2 μm thick, calculate the material balance and area in ft^2 for the membrane as a function of the cut (fraction of feed permeated). Assume a pressure of 15 psia on the permeate side with perfect mixing on both sides of the membrane, such that compositions on both sides are uniform and equal to exit compositions. Neglect pressure drop and mass transfer resistances external to the membrane. Comment on the practicality of the membrane for making a reasonable separation.

SOLUTION

Assume standard conditions are 0°C and 1 atm (359 ft³/lbmol)

$$n_F = \text{Feed flow rate} = \frac{20{,}000}{359}(60) = 3{,}343 \text{ lbmol/h}$$

For the low-density polyethylene membrane, from Table 14.6, and applying (14-33), letting $A = O_2$ and $B = N_2$:

$$P_{M_B} = H_B D_B = (0.228 \times 10^{-6})(0.32 \times 10^{-6})$$

$$= 0.073 \times 10^{-12} \text{ cm}^3 \text{ (STP)-cm/cm}^2\text{-s-Pa}$$

or, in American engineering units,

$$P_{M_B} = \frac{(0.073 \times 10^{-12})(2.54 \times 12)(3600)(101{,}300)}{(22{,}400)(454)(14.7)} = 5.43 \times 10^{-12} \frac{\text{lbmol-ft}}{\text{ft}^2\text{-h-psia}}$$

Similarly, for oxygen: $P_M = 16.2 \times 10^{-12} \dfrac{\text{lbmol-ft}}{\text{ft}^2\text{-h-psia}}$

Permeance values are based on a 0.2-μm-thick membrane skin (0.66×10^{-6} ft)

From (14-1), $\quad \overline{P}_{M_i} = P_{M_i}/l_M$

$$\overline{P}_{M_B} = 5.43 \times 10^{-12}/0.66 \times 10^{-6} = 3.58 \times 10^{-6} \text{ lbmol/ft}^2\text{-h-psia}$$

$$\overline{P}_{M_A} = 16.2 \times 10^{-12}/0.66 \times 10^{-6} = 24.55 \times 10^{-6} \text{ lbmol/ft}^2\text{-h-psia}$$

Material balance equations:

For N_2

$$x_{F_B} n_F = y_{P_B} n_P + x_{R_B} n_R \tag{1}$$

where n = flow rate in lbmol/h and subscripts F, P, and R refer, respectively, to the feed, permeate, and retentate. Let

$$\theta = \text{cut} = n_P/n_F, \text{ then } (1 - \theta) = n_R/n_F$$

Substituting the definition of θ in (1) gives

$$x_{R_B} = \frac{x_{F_B} - y_{P_B}\theta}{1 - \theta} = \frac{0.79 - y_{P_B}\theta}{1 - \theta} \tag{2}$$

Similarly, for O_2,

$$x_{R_A} = \frac{0.21 - y_{P_A}\theta}{1 - \theta} \tag{3}$$

Separation factor:

From the definition of the separation factor, (14-36), since both fluid sides are well mixed:

$$\alpha_{A,B} = \frac{y_{P_A}/x_{R_A}}{(1 - y_{P_A})/(1 - x_{R_A})} \tag{4}$$

Transport equations:

The transport of A and B through the membrane of area A_M, with partial pressures at exit conditions because of perfect mixing, can be written as:

$$y_{P_B} n_P = A_M \overline{P}_{M_B}(x_{R_B} P_R - y_{P_B} P_P) \tag{5}$$

$$y_{P_A} n_P = A_M \overline{P}_{M_A}(x_{R_A} P_R - y_{P_A} P_P) \tag{6}$$

where A_M is the membrane area normal to flow, n_P, through the mambrane. The ratio of (6) to (5) is y_{P_A}/y_{P_B}, and subsequent manipulations lead to (14-43), where

$$r = P_P/P_R = 15/150 = 0.1$$

and $\alpha^*_{A,B} = \alpha_{O_2,N_2} = \overline{P}_{M_{O_2}}/\overline{P}_{M_{N_2}} = (10.7 \times 10^{-6})/(3.5 \times 10^{-6}) = 2.99$

From (14-43):

$$\alpha_{A,B} = \alpha = 2.99 \left[\frac{x_{R_A}(\alpha - 1) + 1 - 0.299}{x_{R_A}(\alpha - 1) + 1 - 0.1} \right] \tag{7}$$

Equations (3), (4), and (7) contain four unknowns: x_{R_A}, y_{P_A}, θ, and $\alpha_{A,B} = \alpha$. The variable θ is bounded between 0 and 1, so values of θ are selected in that range. The other three variables are computed in the following manner. Combine (3), (4), and (7) to eliminate α and x_{R_A}. Solve the resulting nonlinear equation for y_{P_A}. Then solve (3) for x_{R_A} and (4) for α. Solve (6) for the membrane area, A_M. Alternatively, the three equations can be solved simultaneously with a computer program such as Mathcad. The following results are obtained:

θ	x_{R_A}	y_{P_A}	$\alpha_{A,B}$	A_M, ft^2
0.01	0.208	0.406	2.602	22,000
0.2	0.174	0.353	2.587	462,000
0.4	0.146	0.306	2.574	961,000
0.6	0.124	0.267	2.563	1,488,000
0.8	0.108	0.236	2.555	2,035,000
0.99	0.095	0.211	2.548	2,567,000

Note that the separation factor remains almost constant, varying by only 2% with a value of about 86% of the ideal value. The maximum oxygen content of the permeate (40.6 mol%) occurs with the smallest amount of permeate ($\theta = 0.01$). The maximum nitrogen content of the retentate (90.5 mol%) occurs with the largest amount of permeate ($\theta = 0.99$). With a retentate equal to 60 mol% of the feed ($\theta = 0.4$), the nitrogen content of the retentate has been increased only from 79 to 85.4 mol%. Furthermore, the membrane area requirements are very large. The low-density polyethylene membrane is not very practical. To achieve a more reasonable separation, say with $\theta = 0.6$ and a retentate of 95 mol% N_2, it is advisable to use a membrane with an ideal separation factor of 5 in a membrane module that approximates crossflow or countercurrent flow of permeate and retentate with no mixing and a much higher permeance for oxygen. For higher purities, a membrane cascade of two or more stages should be considered. These alternatives are developed in the next two subsections.   ∎

Module Flow Patterns

In Example 14.5, perfect mixing was assumed on both sides of the membrane. Three other idealized flow patterns, shown in Figure 14.8, have received considerable attention; all assume no mixing and are comparable to the idealized flow patterns used to design heat exchangers. These patterns are (b) countercurrent flow; (c) cocurrent flow; and (d) cross-flow. For a given cut, θ, the flow pattern can significantly affect the degree of separation and the membrane area. For flow patterns (b) to (d), fluid on the feed or retentate side of the membrane flows along and parallel to the upstream surface of the membrane. For countercurrent and cocurrent flow, permeate fluid at a given location on the downstream side of the membrane consists of fluid that has just passed through the membrane at that location plus the permeate fluid flowing to that location. For the crossflow case, there is no flow of permeate fluid along the membrane surface. The permeate fluid that has just passed through the membrane at a given location is the only fluid there. For a given

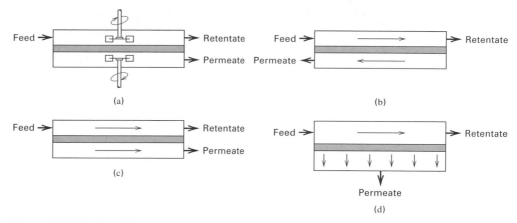

(a)

(b)

(c)

(d)

Permeate

Figure 14.8 Idealized flow patterns in membrane modules: (a) perfect mixing; (b) countercurrent flow; (d) cocurrent flow; (d) crossflow.

module geometry, it is not always obvious which idealized flow pattern to assume. This is particularly true for the spiral-wound module of Figure 14.5b. If the permeation rate is high, the fluid issuing from the downstream side of the membrane may continue to flow perpendicularly to the membrane surface until it finally mixes with the bulk permeate fluid flowing past the surface. In that case, the idealized crossflow pattern might be appropriate. Hollow-fiber modules may be designed to approximate idealized countercurrent, cocurrent, or crossflow patterns. The module shown in Figure 14.5d is approximated by a countercurrent flow pattern.

Walawender and Stern [27] present solution methods for all four flow patterns of Figure 14.8 under the assumptions of a binary feed with constant pressure ratio, r, and constant ideal separation factor, $\alpha^*_{A,B}$. Exact analytical solutions are possible for the perfect mixing case (as shown in Example 14.5) and for the crossflow case, but numerical solutions are necessary for the countercurrent and cocurrent flow cases. A reasonably simple, but approximate, analytical solution for the crossflow case, derived by Naylor and Backer [28], is presented here.

Consider a membrane module with the crossflow pattern shown in Figure 14.9. The feed passes across the upstream membrane surface in plug flow with no longitudinal mixing. The pressure ratio, $r = P_P/P_F$, and the ideal separation factor, $\alpha^*_{A,B}$, are assumed constant.

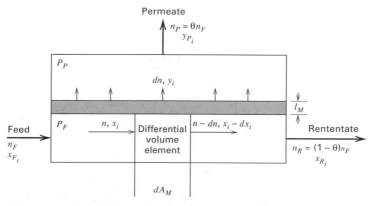

Figure 14.9 Crossflow model for membrane module.

Boundary layer or film mass-transfer resistances external to the membrane are assumed negligible. At the differential element, the local mole fractions in the retentate and permeate, respectively, are x_i and y_i, and the penetrant molar flux is dn/dA_M. Also, the local separation factor is given by (14-43) in terms of the local x_A, r, and $\alpha_{A,B}^*$. An alternative expression for the local permeate composition in terms of y_A, x_A, and r, is obtained by combining (14-36) and (14-41).

$$\frac{y_A}{1 - y_A} = \alpha_{A,B}^* \left[\frac{x_A - r y_A}{(1 - x_A) - r(1 - y_A)} \right] \tag{14-44}$$

A material balance for A around the differential volume element gives:

$$y_A \, dn = d(n x_A) = x_A \, dn + n \, dx_A \quad \text{or}$$

$$\frac{dn}{n} = \frac{dx_A}{y_A - x_A} \tag{14-45}$$

If (14-36) is combined with (14-45) to eliminate y_A, we obtain

$$\frac{dn}{n} = \left[\frac{1 + (\alpha - 1)x_A}{x_A(\alpha - 1)(1 - x_A)} \right] dx_A \tag{14-46}$$

where $\alpha = \alpha_{A,B}$.

In the solution to Example 14.5, it was noted that $\alpha = \alpha_{A,B}$ is relatively constant over the entire range of cut, θ. Such is generally the case when the pressure ratio, r, is small. If the assumption of constant $\alpha = \alpha_{A,B}$ is made in (14-46) and integration is carried out from the intermediate location of the differential element to the final retentate, that is, from n to n_R and from x_A to x_{R_A}, the result is

$$n = n_R(1 - \theta) \left[\left(\frac{x_A}{x_{R_A}} \right)^{\left(\frac{1}{\alpha-1}\right)} \left(\frac{1 - x_{R_A}}{1 - x_A} \right)^{\left(\frac{\alpha}{\alpha-1}\right)} \right] \tag{14-47}$$

The mole fraction of A in the final permeate and the total membrane surface area are obtained by integrating the values obtained from solving (14-44) to (14-46):

$$y_{P_A} = \int_{x_{F_A}}^{x_{R_A}} y_A \, dn/\theta n_F \tag{14-48}$$

By combining (14-48) with (14-46), (14-47), and the definition of α, the integral in n can be transformed to an integral in x_A, which when integrated gives

$$y_{P_A} = x_{R_A}^{\left(\frac{1}{1-\alpha}\right)} \left(\frac{1 - \theta}{\theta} \right) \left[(1 - x_{R_A})^{\left(\frac{\alpha}{\alpha-1}\right)} \left(\frac{x_{F_A}}{1 - x_{F_A}} \right)^{\left(\frac{\alpha}{\alpha-1}\right)} - x_{R_A}^{\left(\frac{\alpha}{\alpha-1}\right)} \right] \tag{14-49}$$

where $\alpha = \alpha_{A,B}$ can be estimated from (14-43) by using $x_A = x_{F_A}$.

The differential rate of mass transfer of A across the membrane is given by

$$y_A \, dn = \frac{P_{M_A} dA_M}{l_M} [x_A P_F - y_A P_P] \tag{14-50}$$

from which the total membrane surface area can be obtained by integration:

$$A_M = \int_{x_{R_A}}^{x_{F_A}} \frac{l_M y_A \, dn}{P_{M_A}(x_A P_F - y_A P_P)} \tag{14-51}$$

EXAMPLE 14.6

For the conditions of Example 14.5, compute exit compositions for a spiral-wound module that approximates crossflow.

SOLUTION

From Example 14.5: $\alpha^*_{A,B} = 2.99$; $r = 0.1$; $x_{F_A} = 0.21$

From (14-43), using $x_A = x_{F_A}$; $\alpha_{A,B} = 2.603$

An overall module material balance for O_2 (A) gives

$$x_{F_A} n_F = x_{R_A}(1 - \theta)n_F + y_{P_A}\theta n_F \quad \text{or} \quad x_{R_A} = \frac{(x_{F_A} - y_{P_A}\theta)}{1 - \theta} \tag{1}$$

Solving (1) and (14-49) simultaneously with a program such as Mathcad gives the following results:

θ	x_{R_A}	y_{P_A}	Stage α_S
0.01	0.208	0.407	2.61
0.2	0.168	0.378	3.01
0.4	0.122	0.342	3.74
0.6	0.0733	0.301	5.44
0.8	0.0274	0.256	12.2
0.99	0.000241	0.212	1,120.

Comparing these results to those of Example 14.5, we see that for crossflow, the permeate is richer in oxygen and the retentate is richer in nitrogen. Thus, for a given cut, θ, crossflow is more efficient than perfect mixing.

Also included in the preceding table is the calculated degree of separation for the stage, α_S, defined on the basis of the mole fractions in the permeate and retentate exiting the stage by

$$(\alpha_{A,B})_S = \alpha_S = \frac{(y_{P_A}/x_{R_A})}{(1 - y_{P_A})/(1 - x_{R_A})} \tag{2}$$

Recall that the ideal separation factor, $\alpha^*_{A,B}$, for this example is 2.99. Also, if (2) is applied to the perfect mixing case of Example 14.5, it is found that α_S is 2.603 for $\theta = 0.01$ and decreases slowly with increasing θ until at $\theta = 0.99$, $\alpha_S = 2.548$. Thus, for perfect mixing, $\alpha_S < \alpha^*$ for all θ. Such is not the case for crossflow. In the table, $\alpha_S < \alpha^*$ for $\theta > 0.2$, and α_S increases with increasing θ. For $\theta = 0.6$, α_S is almost twice α^*. ∎

Calculations of the degree of separation of a binary mixture in a membrane module utilizing cocurrent or countercurrent flow patterns involve the numerical solution of ordinary differential equations. Derivation of these equations and FORTRAN computer codes for their solution are given by Walawender and Stern [27]. A representative solution is shown in Figure 14.10 for the separation of air (20.9 mol% O_2) for conditions of $\alpha^* = 5$, $r = 0.2$. For a given cut, θ, it is seen that the best separation is achieved with countercurrent flow. The curve for cocurrent flow lies between those for crossflow and perfect mixing. The perfect mixing case for binary mixtures is extended to multicomponent mixtures by Stern et al. [29], who present an iterative procedure. As with crossflow, countercurrent flow also offers the possibility of a separation factor for the stage, α_S, defined by (2) earlier, that can be considerably greater than α^*.

Cascades

A single membrane module or a number of such modules arranged in parallel or in series without recycle constitutes a single-stage membrane separation process. The extent to which a feed mixture can be separated in a single stage is limited and, as shown in the previous subsection, is determined by the separation factor, α. This factor depends, in

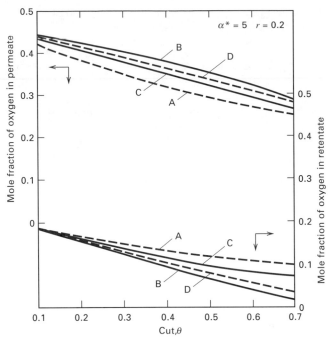

Figure 14.10 Effect of membrane module flow pattern on degree of separation of air. A, perfect mixing; B, countercurrent flow; C, cocurrent flow; D, crossflow.

turn, on the module flow pattern, the permeability ratio (ideal separation factor), the cut, θ, and the driving force for mass transfer through the membrane. To achieve a higher degree of separation than possible with a single stage, a countercurrent cascade of stages, such as are used in distillation, absorption, stripping, and liquid–liquid extraction, or a hybrid process that couples a membrane separator with another separation operation, such as distillation or adsorption, can be applied.

A countercurrent recycle cascade of membrane separators, similar to a distillation column, is shown in Figure 14.11. The feed enters at stage F, somewhere near the middle of the column. Permeate is enriched in components of high permeability in an enriching section, while the retentate is enriched in components of low permeability in a stripping section. The final permeate is withdrawn from stage 1, while the final retentate is withdrawn

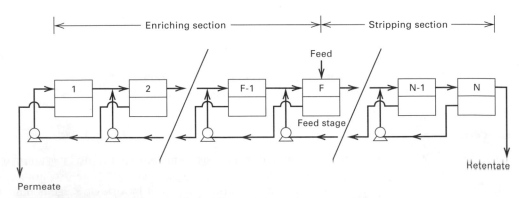

Figure 14.11 Countercurrent recycle cascade of membrane separators.

from stage N. For a cascade, additional factors that affect the degree of separation of the feed are the number of stages and the recycle ratio (permeate recycle rate/permeate product rate). As discussed by Hwang and Kammermeyer [30], it is best to manipulate the cut and reflux rate at each stage so as to force the compositions of the two streams entering each stage to be identical. For example, the composition of retentate leaving Stage 1 would be identical to the composition of permeate leaving Stage 3. This corresponds to the least amount of entropy production for the cascade and, thus, the highest second-law efficiency. Such a cascade is referred to as ideal. Calculation methods for cascades are discussed by Hwang and Kammermeyer [30] and utilize the single-stage methods that depend upon the module flow pattern, as discussed in the previous section. The calculations are best carried out with a computer program, but results for a binary mixture can be conveniently displayed on a McCabe–Thiele type diagram in terms of the mole fraction in the permeate leaving each stage, y_i, versus the mole fraction in the retentate leaving each stage, x_i. For a membrane cascade, the equilibrium curve becomes the selectivity curve in terms of the separation factor for the stage, α_S.

In Figure 14.11, it is assumed the pressure drop on the upstream side of the membrane is negligible. Thus, only the permeate must be pumped, if a liquid, or compressed, if a gas, to be sent to the next stage. In the case of gas permeation, compression costs can be high. Thus, membrane cascades for gas permeation are often limited to just two or three stages, with the most common configurations shown in Figure 14.12. Compared to a single stage, the two-stage stripping cascade is designed to obtain a more pure retentate, whereas a more purer permeate is the goal of the two-stage enriching cascade. The addition of the premembrane stage, shown in Figure 14.12c, may be attractive when the feed concentration is low in the component to be passed preferentially through the membrane, the desired permeate purity is high, the separation factor is low, and/or a high recovery of the more permeable component is desired. An example of the application of the enrichment cascades is given by Spillman [31] for the removal of carbon dioxide from natural gas (assumed to be methane) using cellulose acetate membranes in spiral-wound modules that approximate

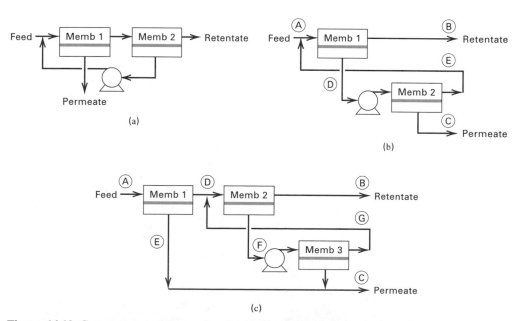

Figure 14.12 Common recycle cascades. (a) Two-stage stripping cascade. (b) Two-stage enriching cascade. (c) Two-stage enriching cascade with additional premembrane stage.

Table 14.8 Separation of N_2 and CH_4 with Membrane Cascades

Case 1: Single Membrane Stage:

	Stream		
	A Feed	B Retentate	C Permeate
Composition (mole%)			
CH_4	93.0	98.0	63.4
CO_2	7.0	2.0	36.6
Flow rate (MM/SCFD)	20.0	17.11	2.89
Pressure (psig)	850	835	10

Case 2: Two-Stage Enriching Cascade (Figure 14.12b):

	Stream				
	A	B	C	D	E
Composition (mole%)					
CH_4	93.0	98.0	18.9	63.4	93.0
CO_2	7.0	2.0	81.1	36.6	7.0
Flow rate (MM/SCFD)	20.00	18.74	1.26	3.16	1.90
Pressure (psig)	850	835	10	10	850

Case 3: Two-Stage Enriching Cascade with Premembrane Stage (Figure 14.12c):

	Stream						
	A	B	C	D	E	F	G
Composition (mole%)							
CH_4	93.0	98.0	49.2	96.1	56.1	72.1	93.0
CO_2	7.0	2.0	50.8	3.9	43.9	27.9	7.0
Flow rate (MM/SCFD)	20.00	17.95	2.05	19.39	1.62	1.44	1.01
Pressure (psig)	850	835	10	840	10	10	850

Note. MM, million.

crossflow. The ideal separation factor, $\alpha^*_{CO_2,CH_4}$ is 21. Results of computer calculations are given in Table 14.8 for a single stage, a two-stage enriching cascade (Figure 14.12b), and a two-stage enriching cascade with an additional premembrane stage. Carbon dioxide flows through the membrane faster than methane. In all three cases, the feed is 20 million (MM) scfd of 7 mol% CO_2 in methane at 850 psig (about 865 psia) and the retentate is 98 mol% in methane. For each stage, the downstream membrane pressure is 10 psig (about 25 psia). In Table 14.8, for all three cases, stream A is the feed, stream B is the final retentate, and stream C is the final permeate. Case 1 achieves a 90.2% recovery of methane. Case 2 increases that recovery to 98.7%. Case 3 achieves an intermediate recovery of 94.6%. The following degrees of separation are computed from data given in Table 14.8:

	α_S for Membrane Stage		
Case	1	2	3
1	28	—	—
2	28	57	—
3	20	19	44

We can also compute overall degrees of separation for the cascades, α_C, of cases 2 and 3, giving values of 210 and 51, respectively.

Concentration Polarization

Thus far, all of the resistance to mass transfer has been assumed to be associated with the membrane. Thus, concentrations in the fluid at the upstream and downstream faces of the membrane have been assumed equal to the respective bulk fluid concentrations on either side of the membrane. When mass-transfer resistances external to the membrane are not negligible, concentration or partial pressure gradients exist in the boundary layers or films adjacent to the membrane surfaces, as was illustrated for four cases in Figure 14.7. For given bulk fluid concentrations, the presence of these resistances reduces the driving force for mass transfer across the membrane and, therefore, the flux of penetrant. The buildup or depletion of species in the boundary layer or film due to mass-transfer resistance is referred to as *concentration polarization*.

In general, the solution-diffusion mechanism for gas permeation is quite slow compared to diffusion in the gas boundary layers or film adjacent to the membrane, so external mass-transfer resistances are negligible and $P_{i_F} = P_{i_0}$ and $P_{i_P} = P_{i_L}$ in Figure 14.7d. Thus, concentration polarization is commonly neglected for gas permeation. Because diffusion in liquid boundary layers and films can be slow, concentration polarization cannot be neglected in membrane processes that involve liquids, such as dialysis, reverse osmosis, and pervaporation. The need to consider the effect of concentration polarization is of particular importance in reverse osmosis, where the effect can reduce the water flux and increase the salt flux.

Consider a membrane process of the type in Figure 14.7a, involving liquids. At steady state, rates of mass transfer of a penetrating species, i, through the three resistances are as follows:

$$N_i = k_{i_F}(c_{i_F} - c_{i_0}) = \frac{P_{M_i}}{l_M}(c_{i_0} - c_{i_L}) = k_{i_P}(c_{i_L} - c_{i_P}) \tag{14-52}$$

If these three equations are combined to eliminate the intermediate concentrations, c_{i_0} and c_{i_L}, we obtain

$$N_i = \frac{c_{i_F} - c_{i_P}}{\dfrac{1}{k_{i_F}} + \dfrac{l_M}{P_{M_i}} + \dfrac{1}{k_{i_P}}} \tag{14-53}$$

where k_{i_F} and k_{i_P} are mass-transfer coefficients for the feed-side and permeate-side boundary layers or films. The three terms in the denominator of the right-hand side are the resistances to the mass flux. In general, the mass-transfer coefficients depend on fluid properties, flow channel geometry, and flow regime (turbulent or laminar). In the laminar-flow regime, a long entry region may exist where the mass-transfer coefficient changes with the distance, L, from the entry of the membrane channel. Estimation of the coefficients is complicated by fluid velocities that change because of mass exchange between the two fluids.

Typical mass-transfer coefficients for channel flow are obtained from the general empirical film-model correlation [32]:

$$N_{\text{Sh}} = k_i d_H / D_i = a N_{\text{Re}}^b N_{\text{Sc}}^{0.33} (d_H / L)^d \tag{14-54}$$

where
$$N_{Re} = d_H v \rho / \mu$$
$$N_{Sc} = \mu / \rho D_i$$
$$d_H = \text{hydraulic diameter}$$
$$v = \text{velocity}$$

The constants a, b, and d are as follows:

Flow Regime	Flow Channel Geometry	d_H	a	b	d
Turbulent,	Circular tube	D	0.023	0.8	0
$(N_{Re} > 10{,}000)$	Rectangular channel	$2hw/(h+w)$	0.023	0.8	0
Laminar,	Circular tube	D	1.86	0.33	0.33
$(N_{Re} < 2{,}100)$	Rectangular channel	$2hw/(h+w)$	1.62	0.33	0.33

where

$w = $ width of channel

$h = $ height of channel

$L = $ length of channel

EXAMPLE 14.7

A dilute solution of solute A in solvent B is passed through a tubular membrane separator, with the feed flowing through the tubes. At a certain location, the solute concentrations are 5.0×10^{-2} kmol/m^3 and 1.5×10^{-2} kmol/m^3, respectively, on the feed and permeate sides. The permeance of the membrane for solute A is 7.3×10^{-5} m/s. If the tube-side Reynolds number is 15,000, the feed-side solute Schmidt number is 500, the diffusivity of the feed-side solute is 6.5×10^{-5} cm^2/s, and the inside diameter of the tube is 0.5 cm, estimate the flux of the solute through the membrane if the mass-transfer resistance on the permeate side of the membrane is negligible.

SOLUTION

The flux of the solute is given by a modification of (14-53):

$$N_A = \frac{c_{A_F} - c_{A_P}}{\dfrac{1}{k_{A_F}} + \dfrac{1}{\overline{P}_{M_A}} + 0}$$

$$c_{A_F} - c_{A_P} = 5 \times 10^{-2} - 1.5 \times 10^{-2} = 3.5 \times 10^{-2} \text{ kmol/m}^3 \qquad \textbf{(1)}$$

$$\overline{P}_{M_A} = 7.3 \times 10^{-5} \text{ m/s}$$

From (14-54) for turbulent flow in a tube, since $N_{Re} > 10{,}000$:

$$k_{A_F} = 0.023 \frac{D_A}{D} N_{Re}^{0.8} N_{Sc}^{0.33} = 0.023 \left(\frac{6.5 \times 10^{-5}}{0.5} \right) (15{,}000)^{0.8} (500)^{0.33}$$

$$= 0.051 \text{ cm/s or } 5.1 \times 10^{-4} \text{ m/s}$$

From (1), $N_A = \dfrac{3.5 \times 10^{-2}}{\dfrac{1}{5.1 \times 10^{-4}} + \dfrac{1}{7.3 \times 10^{-5}}} = 2.24 \times 10^{-6}$ kmol/s $-$ m^2

The fraction of the total resistance due to the membrane is

$$\frac{\dfrac{1}{7.3 \times 10^{-5}}}{\dfrac{1}{5.1 \times 10^{-4}} + \dfrac{1}{7.3 \times 10^{-5}}} = 0.875 \text{ or } 87.5\%$$

∎

14.4 DIALYSIS AND ELECTRODIALYSIS

In a dialysis membrane separation process, shown in Figure 14.13, the feed is a liquid, at pressure P_1, containing solvent, solutes of type A, and solutes of type B and/or insoluble, but dispersed, colloidal matter. A sweep liquid or wash of the same solvent is fed at pressure P_2 to the other side of the membrane. The membrane is thin with micropores of a size such that solutes of type A can pass through by a concentration driving force. Solutes of type B are larger in molecular size than those of type A and pass through the membrane only with difficulty or not at all. This transport of solutes A and B through the membrane is called dialysis. Colloids do not pass through the membrane. With pressure $P_1 = P_2$, the solvent may also pass through the membrane, but by a concentration driving force acting in the opposite direction. The transport of the solvent is called osmosis. By elevating P_1 above P_2, solvent osmosis can be reduced or eliminated. The products of a dialysis unit (dialyzer) are a liquid *diffusate* (permeate) containing solvent, solutes of type A, and smaller amounts of solutes of type B; and a *dialysate* (retentate) of the solvent and remaining solutes of types A and B, and colloidal matter. Ideally, the dialysis unit would enable a perfect separation between solutes of type A and solutes of type B and any colloidal matter. However, at best only a fraction of the solutes of type A are recovered in the diffusate, even when solutes of type B do not pass through the membrane.

For example, when dialysis is used to recover sulfuric acid from an aqueous stream containing sulfate salts, the following results are obtained, as reported by Chamberlin and Vromen [33]:

	Streams in		Streams out	
	Feed	Wash	Dialysate	Diffusate
Flow rate, gph	400	400	420	380
H_2SO_4, g/L	350	0	125	235
$CuSO_4$, g/L as Cu	30	0	26	2
$NiSO_4$, g/L as Ni	45	0	43	0

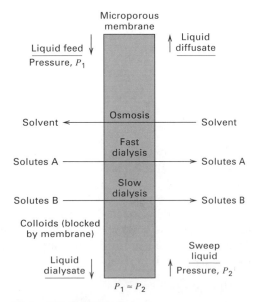

Figure 14.13 Dialysis.

Thus, about 64% of the H_2SO_4 is recovered in the diffusate, accompanied by only about 6% of the $CuSO_4$, and essentially no $NiSO_4$.

Dialysis is closely related to other membrane processes that use other driving forces for separating liquid mixtures, including (1) reverse osmosis, which depends upon a transmembrane pressure difference for solute and/or solvent transport; (2) electrodialysis and electro-osmosis, which depend upon a transmembrane electrical potential difference for solute and solvent transport, respectively; and (3) thermal osmosis, which depends upon a transmembrane temperature difference for solute and solvent transport.

Dialysis is attractive when the concentration differences for the main diffusing solutes are large and the permeability differences between those solutes and the other solute(s) and/or colloids is large. Although dialysis has been known since the work of Graham in 1861 [34], commercial applications of dialysis do not rival reverse osmosis and gas permeation. Nevertheless, dialysis has been applied to a number of separations, including (1) recovery of sodium hydroxide from a 17–20 wt% caustic viscose liquor contaminated with hemicellulose to produce a pure diffusate of 9–10 wt% caustic; (2) recovery of chromic, hydrochloric, and hydrofluoric acids from contaminating metal ions; (3) recovery of sulfuric acid from aqueous solutions containing nickel sulfate; (4) removal of alcohol from beer to produce a reduced-alcohol beer; (5) recovery of nitric and hydrofluoric acids from spent stainless-steel pickle liquor; (6) removal of mineral acids from organic compounds; (7) removal of low-molecular-weight contaminants from polymers; and (8) purification of pharmaceuticals. Also of great importance is hemodialysis, in which urea, creatine, uric acid, phosphates, and chlorides are removed from blood without removing essential higher-molecular-weight compounds and blood cells. This dialysis device is called an artificial kidney.

Typical microporous membrane materials used in dialysis are hydrophilic, including cellulose, cellulose acetate, various acid-resistant polyvinyl copolymers, polysulfones, and polymethylmethacrylate, typically less than 50 μm thick and with pore diameters of 15 to 100 Å. The most common membrane modules are plate-and-frame and hollow-fiber. Compact hollow-fiber hemodialyzers, which are widely used, typically contain several thousand 200-μm-diameter fibers with a wall thickness of 20–30 μm and a length of 10–30 cm. Dialysis membranes can be thin because pressures on either side of the membrane are essentially equal.

At a differential location in a dialyzer, the rate of mass transfer of solute across the dialysis membrane is given by

$$dn_i = K_i(c_{i_F} - c_{i_P})\, dA_M \tag{14-55}$$

where K_i is the overall mass-transfer coefficient, which is given in terms of the individual coefficients from (14-53):

$$\frac{1}{K_i} = \frac{1}{k_{i_F}} + \frac{l_M}{P_{M_i}} + \frac{1}{k_{i_P}} \tag{14-56}$$

The determination of the membrane area is made by integrating (14-55) taking into account the module flow patterns, the bulk concentration gradients, and the individual mass-transfer coefficients in (14-56).

One of the oldest membrane materials for use with aqueous solutions is porous cellophane, for which solute permeability is given by (14-14) with $P_{M_i} = D_{e_i}$ and $\overline{P}_{M_i} \cdot l_M$. In the presence of a solution, cellophane will swell to about twice its dry thickness. The wet thickness should be used for l_M. Typical values of parameters given in (14-13) to (14-15) for commercial cellophane are as follows:

Wet thickness = l_M = 0.004 to 0.008 cm;   porosity = ϵ = 0.45 to 0.60

Tortuosity = τ = 3 to 5;   pore diameter = D = 30 to 50 Å

If solute does not interact with the membrane material, the diffusivity, D_{e_i}, in (14-14) is the ordinary molecular diffusion coefficient, which depends only on solute and solvent properties. However, the membrane may have a profound effect on the solute diffusivity if any of a number of membrane–solute interactions occur, including covalent, ionic, and hydrogen bonding; physical adsorption and chemisorption; and membrane polymer flexibility. Thus, it is preferred to measure $\overline{P}_{M_i}$ experimentally using process fluids.

Although the transport of solvents, such as water, which usually occurs in a direction opposite to the solute, could be formulated in terms of Fick's law, it is more common to measure the solvent flux and report the so-called *water-transport number*, which is the ratio of the water flux to the solute flux, with a negative value indicating transport of solvent in the same direction as the solute. The membrane can also interact with the solvent and even curtail solvent transport. Ideally, the water transport number should be a small value less than +1.0. The ideal experimental dialyzer is a batch cell with a variable-speed stirring mechanism on both sides of the membrane so that external mass transfer resistances, $1/k_{i_F}$ and $1/k_{i_P}$ in (14-56), are made negligible. Stirrer speeds greater than 2,000 rpm may be required.

A common dialyzer is the plate-and-frame type of Figure 14.5a. However, for dialysis applications, the frames are arranged vertically. A typical unit might contain 100 square frames, each 0.75×0.75 m on 0.6-cm spacing, equivalent to 56 m^2 of membrane surface. The dialysis rate for sulfuric acid might be 5 lb/day-ft^2. More recently developed dialysis units utilize hollow fibers of 200-μm inside diameter, 16-μm wall thickness, and 28-cm length packed into a heat-exchanger-type module to give 22.5 m^2 of membrane area in a volume that might be one-tenth of the volume of an equivalent plate-and-frame unit.

In a plate-and-frame dialyzer, the flow pattern is nearly countercurrent. Because total flow rates change little and solute concentrations are typically small, it is common to estimate the solute transport rate by assuming a constant overall mass-transfer coefficient with a log-mean concentration driving force. Thus, from (14-55):

$$n_i = K_i A_M (\Delta c_i)_{\text{LM}} \tag{14-57}$$

where K_i is given by (14-56).

EXAMPLE 14.8

A countercurrent-flow, plate-and-frame dialyzer is to be sized to process 0.78 m^3/h of an aqueous solution of 300 kg/m^3 of H_2SO_4 and smaller amounts of copper and nickel sulfates. A wash water rate of 1.0 m^3/h is to be used, and it is desired to recover 30% of the acid at 25°C. From batch laboratory experiments with an acid-resistant vinyl membrane, in the absence of external mass-transfer resistances, a permeance of 0.025 cm/min for the acid and a water transport number of +1.5 are measured. Membrane transport of copper and nickel sulfates is negligible. For these flow rates, experience with plate-and-frame dialyzers indicates that flow will be laminar and the combined external liquid-film mass-transfer coefficients will be 0.020 cm/min. Determine the membrane area required in m^2.

SOLUTION

$$m_{H_2SO_4} \text{ in feed} = 0.78(300) = 234 \text{ kg/h}$$

$$m_{H_2SO_4} \text{ transferred} = 0.3(234) = 70 \text{ kg/h}$$

$$m_{H_2O} \text{ transferred to dialysate} = 1.5(70) = 105 \text{ kg/h}$$

$$m_{H_2O} \text{ in entering wash} = 1.0(1,000) = 1,000 \text{ kg/h}$$

$$m_P \text{ leaving} = 1,000 - 105 + 70 = 965 \text{ kg/h}$$

For mixture densities, assume aqueous sulfuric acid solutions and use the appropriate table in *Perry's Chemical Engineers' Handbook*:

$$\rho_F = 1,175 \text{ kg/m}^3 \quad \rho_R = 1,114 \text{ kg/m}^3 \quad \rho_P = 1,045 \text{ kg/m}^3$$

$$m_F = 0.78(1,175) = 917 \text{ kg/h} \quad m_R \text{ leaving} = 917 + 105 - 70 = 952 \text{ kg/h}$$

Sulfuric acid concentrations:

$$c_F = 300 \text{ kg/m}^3 \qquad\qquad c_\text{wash} = 0 \text{ kg/m}^3$$

$$c_R = \frac{(234 - 70)}{952}(1{,}114) = 192 \text{ kg/m}^3 \quad c_P = \frac{70}{965}(1{,}045) = 76 \text{ kg/m}^3$$

The log mean driving force for H_2SO_4 with countercurrent flow of feed and wash:

$$(\Delta c)_\text{LM} = \frac{(c_F - c_P) - (c_R - c_\text{wash})}{\ln\left(\dfrac{c_F - c_P}{c_R - c_\text{wash}}\right)} = \frac{(300 - 76) - (192 - 0)}{\ln\left(\dfrac{300 - 76}{192 - 0}\right)} = 208 \text{ kg/m}^3$$

The driving force is almost constant in the membrane module, varying only from 224 to 192 kg/m^3.

From (14-56),

$$K_{H_2SO_4} = \frac{1}{\dfrac{1}{P_M} + \left(\dfrac{1}{k}\right)_\text{combined}} = \frac{1}{\dfrac{1}{0.025} + \dfrac{1}{0.020}} = 0.0111 \text{ cm/min or } 0.0067 \text{ m/h}$$

From (14-57), using mass units instead of molar units:

$$A_M = \frac{m_{H_2SO_4}}{K_{H_2SO_4}(\Delta c_{H_2SO_4})_\text{LM}} = \frac{70}{0.0067(208)} = 50 \text{ m}^2 \qquad\qquad \blacksquare$$

Electrodialysis

Electrodialysis began in the early 1900s as a modification to dialysis by the addition of electrodes and direct current to increase the rate of dialysis in electrolyte solutions. However, since the 1940s, electrodialysis has developed into a membrane separation process that differs from dialysis in many ways. Today, electrodialysis refers to an electrolytic process for separating an aqueous electrolyte feed solution into a concentrate or brine and a dilute or desalted water (diluate) by means of an electric field and ion-selective membranes. A typical electrodialysis process is shown in Figure 14.14, where the four ion-selective membranes shown are of two types arranged in an alternating-series pattern. The cation-selective membranes (C) carry a negative charge, and thus attract and pass positively charged ions (cations), while retarding negative ions. The anion-selective membranes (A) carry a positive charge that attracts and permits passage of negative ions (anions). Both types of membranes are impervious to water. The net result is that both anions and cations are concentrated in compartments 2 and 4, from which concentrate is withdrawn, and ions are depleted in compartment 3, from which the diluate is withdrawn. Compartment pressures are essentially equal. Compartments 1 and 5 are bounded on the far sides by the anode and cathode, respectively. A direct-current voltage is applied (e.g. with a battery or direct-current generator) across the anode and cathode, causing current to flow by metallic conduction of electrons through wiring from the anode to the cathode and then through the cell by ionic conduction from the cathode back to the anode. Both electrodes are chemically neutral metals, with the anode being typically stainless steel and the cathode typically platinum-coated tantalum, niobium, or titanium. Thus, the electrodes are neither oxidized nor reduced.

But half reactions must occur at the two electrodes. Typically, the most easily oxidized species is oxidized at the anode and the most easily reduced species is reduced at the cathode. With inert electrodes, the result at the cathode is the reduction of water by the half reaction

$$2H_2O + 2e^- \rightarrow 2OH^- + H_{2(g)}, \quad E^0 = -0.828 \text{ V}$$

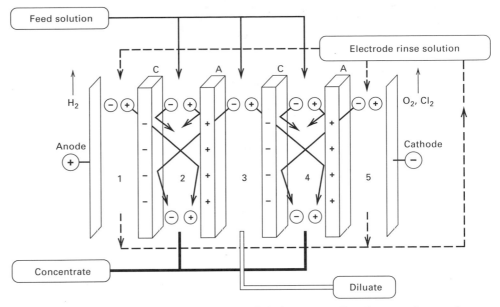

Figure 14.14 Schematic diagram of the electrodialysis process. C, cation transfer membrane; A, anion transfer membrane. [Adapted from W. S. W. Ho and K. K. Sirkar, editors, "Membrane Handbook," Van Nostrand Reinhold, New York (1992).]

The oxidation half reaction at the anode is

$$H_2O \rightarrow 2e^- + \tfrac{1}{2}O_{2_{(g)}} + 2H^+, E^0 = -1.23 \text{ V}$$

or, if chloride ions are present:

$$2Cl^- \rightarrow 2e^- + Cl_{2_{(g)}}, E^0 = -1.360 \text{ V}$$

where the electrode potentials are the standard values at 25°C for one molar solutions of ions and partial pressures of one atmosphere for the gaseous products. Values of E^0 can be corrected for nonstandard conditions by the Nernst equation.

The corresponding overall cell reactions are:

$$3H_2O \rightarrow H_{2_{(g)}} + \tfrac{1}{2}O_{2_{(g)}} + 2H^+ + 2OH^-$$

or

$$2H_2O + 2Cl^- \rightarrow 2OH^- + H_{2_{(g)}} + Cl_{2_{(g)}}, E^0_{\text{cell}} = -2.058 \text{ V}$$

The net reaction for the first case is

$$H_2O \rightarrow H_{2_{(g)}} + \tfrac{1}{2}O_{2_{(g)}}, E^0_{\text{cell}} = -2.188 \text{ V}$$

The electrode rinse solution that circulates through compartments 1 and 5 is typically acidic to neutralize the OH ions formed in compartment 1 and prevent precipitation of compounds such as $CaCO_3$ and $Mg(OH)_2$.

The most widely used ion-exchange membranes for electrodialysis, first reported by Juda and McRae [35] in 1950, are (1) cation-selective membranes containing negatively charged groups fixed to a polymer matrix, and (2) anion-selective membranes containing positively charged groups fixed to a polymer matrix. The former, shown schematically in Figure 14.15, includes fixed anions, mobile cations (called counterions), and mobile anions

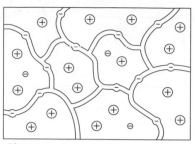

≋ Matrix with fixed charges
⊕ Counter ion
⊖ Co-ion

Figure 14.15 Cation-exchange membrane. [From H. Strath-mann, *Sep. and Purif. Methods,* **14**(1), 41–66 (1985) with per-mission.]

(called co-ions). The latter are almost completely excluded from the polymer matrix by electrical repulsion, called the Donnan effect. For perfect exclusion, only cations are transferred through the membrane. In practice, the exclusion is better than 90%.

A typical cation-selective membrane is made of polystrene cross-linked with divinylben-zene and sulfonated to produce fixed sulfonate, $-SO_3^-$, anion groups. A typical anion-selective membrane of the same polymer contains quaternary ammonium groups such as $-NH_3^+$. Membranes are 0.2–0.5 mm in thickness and reinforced with a screen to provide mechanical stability. The membranes, which are made in flat sheets, contain 30 to 50% water and have a network of pores too small to permit water transport.

A cell pair or unit cell consists of one cation-selective membrane and one anion-selective membrane. Although Figure 14.14 shows an electrodialysis system with two cell pairs, a commercial electrodialysis system is a large stack of membranes patterned after a plate and frame configuration that, according to Applegate [2] and the *Membrane Handbook* [6], may contain 100 to 600 cell pairs. In a stack, membranes of from 0.4 to 1.5 m² surface area each are separated by from 0.5 to 2 mm with spacer gaskets. The total voltage or electrical potential applied across the cell includes (1) the electrode potentials discussed earlier, (2) overvoltages due to gas formation at the two electrodes, (3) the voltage required to overcome the ohmic resistance of the electrolyte in each compartment, (4) the voltage required to overcome the resistance in each membrane, and (5) the voltage required to overcome concentration polarization effects caused by mass-transfer resistances in the electrolyte solutions adjacent to the membrane surface. For large stacks, the latter three voltage increments predominate and depend upon the current density (amps flowing through the stack per unit surface area of membranes). A typical voltage drop across a cell pair is 0.5–1.5 V. Current densities are in the range of 5–50 mA/cm². Thus, a stack of 400 membranes (200 unit cells) of 1 m² surface area each might require 200 V at 100 A. Typically 50 to 90% of brackish water is converted to potable water, depending on concentrate recycle.

As the current density is increased for a given membrane surface area, the concentration-polarization effect increases. A schematic diagram of this effect for a single cation-selective membrane is shown in Figure 14.16, where c_m refers to cation concentrations in the membrane, c_b refers to bulk electrolyte cation concentrations, and superscripts c and d refer to concentrate side and dilute side, respectively. The maximum or limiting current density is reached when c_m^d reaches zero. Typically, an electrodialysis cell is operated at 80% of the limiting current density, which is determined by experiment. The corresponding cell voltage or resistance is also determined experimentally.

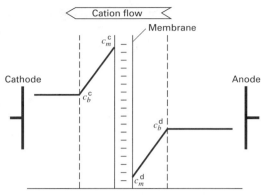

Figure 14.16 Concentration-polarization effects for a cation-exchange membrane. [From H. Strathmann, *Sep. and Purif. Methods,* **14**(1), 41–66 (1985) with permission.]

The amounts of gases formed at the electrodes at the two ends of the stack are governed by *Faraday's law of electrolysis.* During electrolysis, one Faraday (96,520 coulombs) of electricity reduces at the cathode and oxidizes at the anode an equivalent of oxidizing and reducing agent corresponding to the transfer of 6.023×10^{23} (Avogadro's number) electrons through wiring from the anode to the cathode. In general, it takes a very large quantity of electricity to form appreciable quantities of gases in an electrodialysis process.

Of more importance in the design or operation of an electrodialysis process are the membrane area and electrical energy requirements as discussed by Applegate [2] and Strathmann [36]. The membrane area is estimated from the current density, rather than from a permeability, and mass-transfer resistances by applying Faraday's law:

$$A_M = \frac{zFQ\Delta c}{i\xi} \tag{14-58}$$

where

A_M = total area of all cell pairs, m²

z = electrochemical valence of the ions being transported through the membranes

F = Faraday's constant (96,520 amp-s/equivalent)

Q = volumetric flow rate of the diluate (potable water), m³/s

Δc = difference between feed and diluate ion concentration in equivalents/m³

i = current density, amps/m² of a cell pair, usually about 80% of i_{max}

ξ = current efficiency < 1.00

The last variable accounts for the fact that not all of the current is effective in transporting the selected ions through the membranes. Inefficiencies are caused by a Donnan exclusion of less than 100%, some transfer of water through the membranes, current leakage through manifolds, etc.

Power consumption is given by

$$P = IE \tag{14-59}$$

where

P = power, W, I = electric current flow through the stack, and E = voltage across the stack.

The electrical current flow is given by a rearrangement of (14-58):

$$I = \frac{zFQ\Delta c}{n\xi} \qquad \textbf{(14-60)}$$

where n is the number of cell pairs.

The main application of electrodialysis is to the desalinization of brackish water in the salt concentration range of 500 to 5,000 ppm (mg/L). Below this range, ion exchange is more economical, whereas above this range, to 50,000 ppm, reverse osmosis is preferred. However, electrodialysis cannot produce water with a very low dissolved solids content because of the high electrical resistance of dilute solutions. Other applications include recovery of nickel and copper from electroplating rinse water; deionization of cheese whey, fruit juices, wine, milk, and sugar molasses; separation of salts, acids, and bases from organic compounds; and recovery of organic compounds from their salts. Bipolar membranes, prepared by laminating a cation-selective membrane and an anion-selective membrane back-to-back, can be used to produce sulfuric acid and sodium hydroxide from a sodium sulfate solution.

EXAMPLE 14.8

Estimate membrane area and electrical energy requirements for an electrodialysis process to reduce the salt (NaCl) of 24,000 m^3/day of brackish water from 1,500 mg/L to 300 mg/L with a 50% conversion. Assume each membrane has a surface area of 0.5 m^2 and each stack contains 300 cell pairs. A reasonable current density is 5 mA/cm^2 and the current efficiency is 0.8 (80%).

Solution

Use (14-58) to estimate membrane area, with $z = 1$.

$$F = 96,520 \text{ A/equiv} \quad Q = (24,000)(0.5)/(24)(3,600) = 0.139 \text{ m}^3/\text{s}$$

$$\text{MW}_{\text{NaCl}} = 58.5 \quad i = 5 \text{ mA/cm}^2 = 50 \text{ A/m}^2$$

$$\Delta c = (1,500 - 300)/58.5 = 20.5 \text{ mmol/L or } 20.5 \text{ mol/m}^3 = 20.5 \text{ equiv/m}^3$$

$$A_M = \frac{(1)(96,520)(0.139)(20.5)}{(50)(0.8)} = 6,876 \text{ m}^2$$

Each stack contains 300 cell pairs with a total area of $0.5(300) = 150 \text{ m}^2$. Therefore, number of stacks = 6,876/150 = 46 in parallel

From (14-60), electrical current flow is given by

$$I = \frac{(1)(96,500)(0.139)(20.5)}{(300)(0.8)} = 1,146 \text{ A or } I/\text{stack} = 1,146/46 = 25 \text{ A}.$$

To obtain the electrical power, we need to know the average voltage drop across each cell pair. Assume a value of 1 V. From (14-59):

$$P = (1,146)(1)(300) = 344,000 \text{ W} = 344 \text{ kW}$$

Additional energy is required to pump feed, recycle concentrate, and electrode rinse.

It is also instructive to estimate the amount of feed that would be electrolyzed (say, as water to hydrogen and oxygen gases) at the electrodes. From the half-cell reactions presented earlier, half a molecule of H_2O is electrolyzed for each electron or, 0.5 mol H_2O is electrolyzed for each faraday of electricity.

1,146 amps = 1,146 coulombs/s or $(1,146)(3,600(24) = 99,010,000$ coulombs/day or 99,010,000/ 96,520 = 1,026 faradays/day. This electrolyzes $(0.5)(1,026) = 513$ mol/day of water. The feed rate is 12,000 m^3/day, or

$$\frac{(12,000)(10^6)}{18} = 6.7 \times 10^8 \text{ mol/day}$$

Therefore, the amount of water electrolyzed is negligible. ∎

14.5 REVERSE OSMOSIS

Osmosis, from the Greek word for "push," refers to the passage of a solvent, such as water, through a dense membrane that is permeable to the solvent (A), but not the solute(s) (B) (e.g., inorganic ions). The first recorded account of osmosis was given in 1748 by Nollet, whose experiments were conducted with water, an alcohol, and an animal-bladder membrane. The important aspects of osmosis are illustrated by example in Figure 14.17, where all solutions are at 25°C. In the initial condition (a), seawater of approximately 3.5 wt% dissolved salts and at 101.3 kPa is on the left side of the membrane, while pure water at the same pressure is on the right side. The dense membrane is permeable to water, but not to the dissolved salts. By osmosis, water passes from the right side to the seawater on the left side, causing dilution with respect to dissolved salts. At equilibrium, the condition of Figure 14.17b is reached, wherein some pure water still resides on the right side and seawater, less concentrated in salt, resides on the left side. The pressure, P_1, on the left side is now greater than the pressure, P_2, on the right side, with the difference, π, referred to as the *osmotic pressure*.

The process of osmosis is not useful as a separation process because the solvent is transferred in the wrong direction, resulting in mixing rather than separation. However, the direction of transfer of solvent through the membrane can be reversed, as shown in Figure 14.17c by applying a pressure, P_1, on the left side of the membrane, that is higher than the sum of the osmotic pressure and the pressure, P_2, on the right side: that is, $P_1 - P_2 > \pi$. Now water in the seawater is transferred to the pure water, and the seawater becomes more concentrated in dissolved salts. This phenomenon, called *reverse osmosis*, can be used to partially remove a solvent from a solute-solvent mixture. As discussed later, an important factor in developing a reverse osmosis separation process is the osmotic pressure, π, of the feed mixture. In general, as discussed in more detail later, π is proportional to the solute concentration.

In a reverse osmosis (RO) membrane separation process, as shown in Figure 14.18, the feed is a liquid at high pressure, P_1, containing solvent (e.g., water) and solubles (e.g., inorganic salts and, perhaps, colloidal matter). No sweep liquid is used, but the other side of the membrane is maintained at a much lower pressure, P_2. A dense membrane, such as an acetate or aromatic polyamide, is used that is permselective for the solvent. To withstand the large pressure differential, the membrane must be thick. Accordingly, asymmetric or thin-wall composite membranes, having a thin, dense skin or layer on a thick, porous support, are used. The products of reverse osmosis are a permeate of almost pure solvent and a retentate of solvent-depleted feed. However, a perfect separation between the solvent and solute is not achieved, since only a fraction of the solvent in the feed is transferred to the permeate.

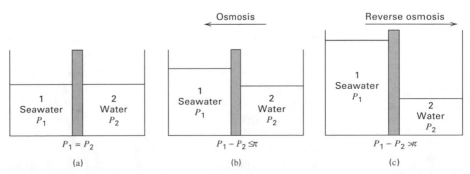

Figure 14.17 Osmosis and reverse osmosis phenomena. (a) Initial condition. (b) At equilibrium after osmosis. (c) Reverse osmosis.

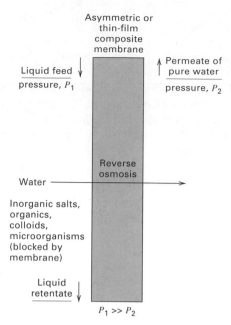

Figure 14.18 Reverse osmosis.

Reverse osmosis is applied to the desalinization and purification of seawater, brackish water, and wastewater. Prior to 1980, multistage flash distillation was the main process for the desalinization of water. By 1990, this situation was dramatically reversed, making RO the dominant process for new construction. The dramatic shift from a thermally driven process to a more economical pressure-driven process was made possible through the development by Loeb and Sourirajan [7] of an asymmetric membrane that allows pressurized water to pass through at a high rate, while almost preventing transmembrane flows of dissolved salts, organic compounds, colloids, and microorganisms. Today more than 1,000 RO desalting plants are producing more than 750,000,000 gallons per day of potable water worldwide.

According to Baker et al. [5], the use of RO to desalinize water is accomplished mainly with spiral-wound and hollow-fiber membrane modules utilizing cellulose triacetate, cellulose diacetate, and aromatic polyamide membrane materials. Cellulose acetates are susceptible to biological attack, and acidic or basic hydrolysis back to cellulose, making it necessary to chlorinate the feed water and control the pH within the range of 4.5 to 7.5. Polymamides are not susceptible to biological attack and resist hydrolysis in the pH range of 4 to 11. However, polyamides are attacked by chlorine.

The preferred membrane for the desalinization of seawater, which contains about 3.5 wt% dissolved salts and has an osmotic pressure of 350 psia, is a spiral-wound, multileaf module of polyamide thin-film composite operating at a feed pressure of 800 to 1,000 psia. With a transmembrane water flux of 9 gal/ft²-day (0.365 m³/m²-day), this module can recover 45% of the water at a purity of about 99.95 wt%. A typical cylindrical module is 8 inches in diameter by 40 inches long, containing 365 ft² (33.9 m²) of membrane surface. Such modules resist fouling by colloidal and particulate matter, but the seawater must be treated with sodium bisulfate to remove oxygen and/or chlorine.

For the desalinization of brackish water containing less than 0.5 wt% dissolved salts, hollow-fiber modules of high packing density, and containing fibers of cellulose acetates or aromatic polyamides, are used if fouling is not serious. Because the osmotic pressure is much lower (<50 psi), feed pressures can be less than 250 psia. Transmembrane fluxes may be as high as 20 gal/ft²-day.

Other uses of reverse osmosis, usually on a smaller scale than the desalinization of water to produce potable water, include (1) the treatment of industrial wastewater to remove heavy metal ions, nonbiodegradable substances, and other components of commercial value; (2) the treatment of rinse water from electroplating processes to obtain a metal ion concentrate and a permeate that can be reused as a rinse; (3) the separation of sulfites and bisulfites from effluents in pulp and paper processes; (4) the treatment of wastewater in dyeing processes; (5) the recovery of constituents having food value from wastewaters in food processing plants (e.g. lactose, lactic acid, sugars, and starches); (6) the treatment of municipal water to remove inorganic salts, low-molecular-weight organic compounds, viruses, and bacteria; (7) the dewatering of certain food products such as coffee, soups, tea, milk, orange juice, and tomato juice; and (8) the concentration of amino acids and alkaloids. In such applications, membranes must have chemical, mechanical, and thermal stability to be competitive with other processes.

As with all membrane processes where the fluid feed being separated is a liquid, three resistances to mass transfer must be considered: the membrane resistance and the two fluid-film or boundary-layer resistances on either side of the membrane. If the permeate is pure solvent, then there is no film resistance on that side of the membrane.

Although the driving force for the transport of water through the dense membrane is the concentration or activity difference in and across the membrane, common practice is to use a driving force based on osmotic pressure. Consider the reverse osmosis process of Figure 14.17c. At equilibrium, solvent chemical potentials or fugacities on the two sides of the membrane must be equal. Thus,

$$f_A^{(1)} = f_A^{(2)} \tag{14-61}$$

From definitions in Table 2.2, rewrite (14-61) in terms of activities:

$$a_A^{(1)} f_A^0\{T, P_1\} = a_A^{(2)} f_A^0\{T, P_2\} \tag{14-62}$$

For pure solvent, A, $a_A^{(2)} = 1$. For seawater, $a_A^{(1)} = x_A^{(1)} \gamma_A^{(1)}$. Substitution into (14-62) gives

$$f_A^0\{T, P_2\} = x_A^{(1)} \gamma_A^{(1)} f_A^0\{T, P_1\} \tag{14-63}$$

Standard-state, pure-component fugacities f^0 increase with increasing pressure. Thus, if $x_A^{(1)} \gamma_A^{(1)} < 1$, then from (14-63), $P_1 > P_2$. The pressure difference $P_1 - P_2$ is shown as a hydrostatic-head difference in Figure 14.17b. This difference, which can be observed experimentally, is the osmotic pressure, π.

To relate π to solvent or solute concentration, we apply the Poynting correction of (2-28), which for an incompressible liquid of specific volume, v_A, gives

$$f_A^0(T, P_2) = f_A^0(T, P_1) \exp\left[\frac{v_{A_L}(P_1 - P_2)}{RT} \right] \tag{14-64}$$

Substitution of (14-63) into (14-64) gives

$$\pi = P_1 - P_2 = -\frac{RT}{v_{A_L}} \ln(x_A^{(1)} \gamma_A^{(1)}) \tag{14-65}$$

Thus, osmotic pressure is a thermodynamic quantity that replaces activity.

For a mixture, on the feed or retentate side of the membrane, that is dilute in the solute, $\gamma_A^{(1)} = 1$. Also, $x_A^{(1)} = 1 - x_B^{(1)}$ and $\ln(1 - x_B^{(1)}) \approx -x_B^{(1)}$. Substitution into (14-65) gives

$$\pi = P_1 - P_2 = RT x_B^{(1)} / v_{A_L} \tag{14-66}$$

Finally, since $x_B^{(1)} \approx n_B/n_A$, $n_A v_{A_L} = V$, and $n_B/V = c_B$, (14-66) becomes

$$\pi \approx RTc_B \tag{14-67}$$

which was cited in Exercise 1.8. For applications to the reverse osmosis of seawater, Applegate [2] suggests the approximate expression

$$\pi = 1.12T\sum \overline{m}_i \tag{14-68}$$

where π is in psia, T is in K, and $\sum \overline{m}_i$ is the summation of molalities of all dissolved ions and nonionic species in the solution in mol/L. More exact expressions for π are developed by Stoughton and Lietzke [38].

In the general case, when reverse osmosis takes place with solute on each side of the membrane, then at equilibrium, $(P_1 - \pi_1) = (P_2 - \pi_2)$. Accordingly, as discussed by Merten [37], the driving force for the transport of solvent through the membrane is $\Delta P - \Delta \pi$, and the rate of mass transport is

$$N_{H_2O} = \frac{P_{M_{H_2O}}}{l_M}(\Delta P - \Delta \pi) \tag{14-69}$$

where

ΔP = hydraulic pressure difference across the membrane

$\quad = P_{feed} - P_{permeate}$

$\Delta \pi$ = osmotic pressure difference across the membrane

$\quad \pi_{feed} - \pi_{permeate}$

Often, $\pi_{permeate} \approx 0$ because the permeate is almost pure solvent.

The flux of solute (e.g., salt) is given by (14-26) in terms of membrane concentrations, and thus is independent of the ΔP across the membrane. Accordingly, the higher the ΔP, the purer the permeate water. Alternatively, the flux of salt may be expressed for ease of application in terms of *salt passage*, *SP*, defined by

$$SP = (c_{salt})_{permeate}/(c_{salt})_{feed} \tag{14-70}$$

Values of *SP* decrease with increasing ΔP. *Salt rejection* is given by $SR = 1 - SP$.

For brackish water of 1,500 mg/L as NaCl, at 25°C, (14-68) predicts $\pi = 17.1$ psia. For seawater of 35,000 mg/L as NaCl, at 25C, (14-68) predicts $\pi = 385$ psia, while Stoughton and Lietzke [38] give 368 psia. From (14-69), ΔP must be greater than $\Delta \pi$ for reverse osmosis to occur. For the desalinization of brackish water by RO, ΔP is typically 400–600 psi, while for seawater, it is 800–1,000 psi. To prevent membrane fouling and scaling, feed water pretreatment, consisting of prefiltration, flocculation, and chemical treatment, is required.

Concentration polarization is particularly important on the feed side of the reverse-osmosis membrane. This effect is illustrated in Figure 14.19, where typical concentrations are shown for water, c_w, and salt, c_s. Because of the high pressure, the activity of water on the feed side is somewhat higher than that of near-pure water on the permeate side, thus providing the necessary driving force for water transport through the membrane. The flux of water to the membrane carries with it salt by bulk flow. However, because the salt cannot readily penetrate the membrane, the concentration of the salt in the liquid adjacent to the surface of the membrane, c_{s_I}, is greater than that in bulk of the feed, c_{s_F}. This difference causes mass transfer of salt by diffusion from the membrane surface back to the bulk feed. The back rate of salt diffusion depends on the mass-transfer coefficient for the film or boundary layer on the feed side. The lower the mass-transfer coefficient, the higher the value of c_{s_I}. The value of c_{s_I} is important because it fixes the osmotic pressure, and thus influences the driving force for water transport according to (14-69).

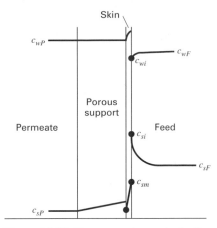

Figure 14.19 Concentration-polarization effects in reverse osmosis.

Consider steady-state transport of water with back-diffusion of salt. A salt balance at the upstream membrane surface gives

$$N_{H_2O}c_{s_F}(SP) = k_s(c_{s_I} - c_{s_F})$$ (14-71)

Solving for c_{s_I} gives

$$c_{s_I} = c_{s_F}\left(1 + \frac{N_{H_2O}(SP)}{k_s}\right)$$ (14-72)

Values of k_s are estimated from (14-54). The concentration-polarization effect is seen to be most significant for high water fluxes and low mass-transfer coefficients.

Pressure drop on the feed side of the membrane is also important because, by (14-69), it causes a reduction in the driving force for water transport. Because of the complex geometries used for both spiral-wound and hollow-fiber modules, it is best to estimate pressure drops from experimental data. Feedside pressure drops for spiral-wound modules and hollow-fiber modules range from 43 to 85 and 1.4 to 4.3 psi, respectively [6].

A schematic diagram of a typical reverse osmosis process for the desalinization of water is shown in Figure 14.20. The source of feed water may be a well or surface water, which is pumped through a series of pretreatment steps to ensure a long membrane life. Of particular importance is pH adjustment. The pretreated water is then fed by a high-pressure-discharge pump to an appropriate parallel-and-series network of reverse osmosis modules of the spiral-wound or hollow-fiber type. The concentrate, which leaves the membrane system at a high pressure that is 10–15% lower than the inlet pressure, is then routed through a power-recovery turbine, which reduces the net power consumption of the process by 25 to 40% while reducing the pressure of the concentrate to an appropriate low level. The permeate, which may be 99.95 wt% pure water and about 50% of the feed water, is sent to a series of posttreatment steps before it is ready to drink.

EXAMPLE 14.9

At a certain location in a spiral-wound membrane, the bulk conditions on the feed side are 1.8 wt% NaCl, 25°C, and 1,000 psia, while bulk conditions on the permeate side are 0.05 wt% NaCl, 25°C, and 50 psia. For the particular membrane being used, the permeance values are 1.1×10^{-5} g/cm²-s-atm for H_2O and 16×10^{-6} cm/s for the salt. If mass-transfer resistances are negligible on each side of the membrane, calculate the flux of water in gal/ft²-day and the flux of salt in g/ft²-day.

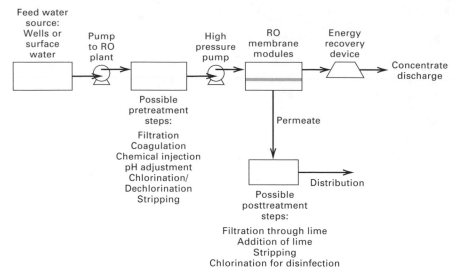

Figure 14.20 Reverse osmosis process.

SOLUTION

Bulk salt concentrations are approximately

$$\frac{1.8(1,000)}{58.5(98.2)} = 0.313 \text{ mol/L on feed side}$$

$$\frac{0.05(1,000)}{58.5(99.95)} = 0.00855 \text{ mol/L on permeate side}$$

For water transport, using (14-68) for osmotic pressure, noting that dissolved NaCl gives 2 ions per molecule:

$$\Delta P - (1,000 - 50)/14.7 = 64.6 \text{ atm}$$

$$\pi_{\text{feedside}} = 1.12(298)(2)(0.313) = 209 \text{ psia} = 14.2 \text{ atm}$$

$$\pi_{\text{permeate side}} = 1.12(298)(2)(0.00855) = 5.7 \text{ psia} = 0.4 \text{ atm}$$

$$\Delta P - \Delta \pi = 64.6 - (14.2 - 0.4) = 50.8 \text{ atm}$$

$$P_{M_{\text{H}_2\text{O}}}/l_M = 1.1 \times 10^{-5} \text{ g/cm}^2\text{-s-atm}$$

From (14-69),

$$N_{\text{H}_2\text{O}} = (1.1 \times 10^{-5})(50.8) = 0.000559 \text{ g/cm}^2\text{-s or}$$

$$\frac{(0.000559)(3,600)(24)}{(454)(8.33)(1.076 \times 10^{-3})} = 11.9 \text{ gal/ft}^2\text{-day}$$

For salt transport:

$$\Delta c = 0.313 - 0.00855 = 0.304 \text{ mol/L or } 0.000304 \text{ mol/cm}^3$$

$$P_{M_{\text{NaCl}}}/l_M = 16 \times 10^{-6} \text{ cm/s}$$

From (14-26):

$$N_{\text{NaCl}} = 16 \times 10^{-6}(0.000304) - 4.86 \times 10^{-9} \text{ mol/cm}^2 \text{ s}$$

or

$$\frac{(4.86 \times 10^{-9})(3,600)(24)(58.5)}{1.076 \times 10^{-3}} = 0.95 \text{ g/ft}^2\text{-day}$$

We see that the flux of salt is very much smaller than the flux of water. ∎

14.6 GAS PERMEATION

In gas permeation (GP), shown in Figure 14.21, the feed gas, at high pressure P_1, contains some low-molecular-weight species (MW < 50) to be separated from small amounts of higher-molecular-weight species. Usually a sweep gas is not used, but the other side of the membrane is maintained at a much lower pressure, P_2, often near ambient pressure. The membrane, often dense but sometimes microporous, is permselective for certain of the low-molecular-weight species in the feed gas, shown in Figure 14.21 as the A species. If the membrane is dense, these species are absorbed at the surface and then transported through the membrane by one or more mechanisms. Thus, permselectivity depends on both membrane absorption and the membrane transport rate. Usually all mechanisms are formulated in terms of a partial pressure or fugacity driving force using the solution-diffusion model of (14-32). The products are a permeate that is enriched in the A species and a retentate that is enriched in B. A near-perfect separation is generally not achievable. If the membrane is microporous, as for example in high-temperature applications, pore size is extremely important because it is usually necessary to block the passage of species B. Otherwise, unless molecular weights of A and B differ appreciably, only a very modest separation is achievable, as was discussed in connection with Knudsen diffusion, (14-22).

Since the early 1980s, applications of GP with dense polymeric membranes have increased dramatically. Applications include (1) separation of hydrogen from methane; (2) adjustment of H_2-to-CO ratio in synthesis gas; (3) O_2 enrichment of air; (4) N_2 enrichment of air; (5) removal of CO_2; (6) drying of natural gas and air; (7) removal of helium; and (8) removal of organic solvents from air.

Gas permeation must compete with distillation at cryogenic conditions, absorption, and pressure-swing adsorption. Some of the advantages of gas permeation, as cited by Spillman and Sherwin [39], are low capital investment, ease of installation, ease of operation, absence of rotating parts, high process flexibility, low weight and space requirements, and low environmental impact. In addition, if the feed gas is already at so high a pressure that a gas compressor is not needed, then no utilities are required.

Since 1986, the most rapidly developing application for GP has been air separation, for which available membranes have separation factors for O_2 with respect to N_2 of 3 to 7.

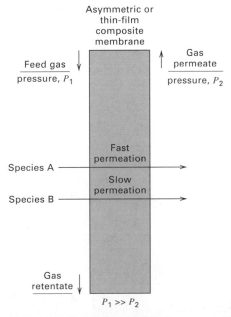

Figure 14.21 Gas permeation.

However, product purities are economically limited to a retentate of 95–99% N_2 and a permeate of 30–45% O_2. Thus, the largest application of GP for air separation is the production of nitrogen rather than oxygen.

Gas permeation also competes very favorably with other separation processes for hydrogen recovery because of the high separation factors achieved. For example, the rate of permeation of hydrogen through a typical dense polymer membrane is more than 30 times that for nitrogen. A typical GP process might achieve a 95% recovery of 90% pure hydrogen from a feed gas containing 60% hydrogen.

Early applications of GP used dense (nonporous) membranes of cellulose acetates and polysulfones, which are still predominant, although polyimides, polyamides, polycarbonates, polyetherimides, sulfonated polysulfones, Teflon, polystyrene, and silicone rubber are also finding applications for temperatures to at least 70°C. Although plate-and-frame and tubular modules can be used for gas permeation, almost all large-scale applications use spiral-wound or hollow-fiber modules because of their higher packing density. Commercial membrane modules for gas permeation are available from more than 20 suppliers. Feed-side pressure is typically 300 to 500 psia, but is as high as 1,650 psia. Typical refinery applications involve feed-gas flow rates of 20 million scfd, but flow rates as large as 300 million scfd have been reported [40]. When the feed gas contains condensables, it may be necessary to preheat the gas prior to entry into the membrane system to prevent condensation on the membrane as the retentate becomes richer in the high-molecular-weight species. For high-temperature applications where polymers cannot be used, membranes of glass, carbon, and inorganic oxides are available, but are limited in their selectivity.

For dense membranes, external mass-transfer resistances or concentration polarization effects are generally negligible, and (14-32) with a partial-pressure driving force can be used to compute the rate of species transport through the membrane. As discussed earlier in the subsection on module flow patterns, the appropriate partial-pressure driving force depends on the flow pattern. Cascades of the type discussed earlier are used to increase the degree of separation.

Progress is being made in the development of a method for the prediction of permeability of gases in glassy and rubbery homopolymers, random copolymers, and block copolymers. Teplyakov and Mcarcs [41] present correlations at 25°C for the diffusion coefficient, D, and solubility, S, applied to 23 different gases for 30 different polymers. Predicted values for glassy polyvinyltrimethylsilane (PVTMS) and rubbery polyisoprene are listed in Table 14.9. Typically, D and S agree with experimental data to within $\pm 20\%$ and $\pm 30\%$, respectively.

Gas permeation separators are claimed to be relatively insensitive to changes in feed flow rate, feed composition, and loss of membrane surface area [42]. This claim is tested in the following example.

EXAMPLE 14.10

The feed to a membrane separator consists of 500 lbmol/h of a mixture of 90% H_2 (H) and 10% CH_4 (M) at 500 psia. Permeance values based on a partial-pressure driving force are

$$\overline{P}_{M_H} = 3.43 \times 10^{-4} \text{ lbmol/h-ft}^2\text{-psi and } \overline{P}_{M_M} = 5.55 \times 10^{-5} \text{ lbmol/h-ft}^2\text{-psi}$$

The flow patterns in the separator are such that the permeate side is well mixed and the feed side is in plug flow. The pressure on the permeate side is constant at 20 psia and there is no pressure drop on the feed side.

(a) Compute the membrane area and permeate purity if 90% of the hydrogen is transferred to the permeate.
(b) For the membrane area determined in part (a), calculate the permeate purity and hydrogen recovery if
    (1) the feed rate is increased by 10%.
    (2) the feed composition is reduced to 85% H_2.
    (3) 25% of the membrane area becomes inoperative

Table 14.9 Predicted Values of Diffusivity and Solubility of Light Gases in a Glassy and a Rubbery Polymer

Permeant	$D \times 10^{11}$, m^2/s	$S \times 10^4$, gmol/m^3-Pa	P_M, barrer
Polyvinyltrimethylsilane (Glassy Polymer)			
He	470	0.18	250
Ne	87	0.26	66
Ar	5.1	1.95	30
Kr	1.5	6.22	29
Xe	0.29	20.6	18
Rn	0.07	69.6	15
H_2	160	0.54	250
O_2	7.6	1.58	37
N_2	3.8	0.84	9
CO_2	4.0	13.6	160
CO	3.7	1.28	14
CH_4	1.9	3.93	22
C_2H_6	0.12	30.2	10
C_3H_8	0.01	98.1	2.8
C_4H_{10}	0.001	347	1.2
C_2H_4	0.23	17.8	12
C_3H_6	0.038	77.6	9
C_4H_8 (1)	0.0052	293	4.5
C_2H_2	0.58	16.8	32
C_3H_4 (m)	0.17	138.1	70
C_4H_6 (e)	0.053	318.5	50
C_3H_4 (a)	0.15	186.5	83
C_4H_6 (b)	0.03	226.1	20
Polyisoprene (Rubber-like Polymer)			
He	213	0.06	35
Ne	77.4	0.08	18
Ar	14.6	0.58	25
Kr	7.2	1.78	25
Xe	2.7	5.68	45
Rn	1.2	18.7	64
H_2	109	0.17	54
O_2	18.4	0.47	26
N_2	12.2	0.26	10
CO_2	12.6	3.80	140
CO	12.1	0.38	14
CH_4	8.0	1.14	27
C_2H_6	3.3	8.13	79
C_3H_8	1.6	25.4	123
C_4H_{10}	1.5	86.4	390
C_2H_4	4.3	4.84	62
C_3H_6	2.7	20.3	163
C_4H_8 (1)	1.5	73.3	333
C_2H_2	5.7	4.64	80
C_3H_4 (m)	4.1	35.3	433
C_4H_6 (e)	2.9	79.6	690
C_3H_4 (a)	4.5	47.4	640
C_4H_6 (b)	3.4	40.0	410

Note. m, methylacetylene; e, ethylacetylene; a, allene; b, butadiene.

SOLUTION

The following independent equations apply to all parts of this example. Component material balances:

$$n_{i_F} = n_{i_R} + n_{i_P}, \quad i = \text{H, M} \tag{1,2}$$

Dalton's law of partial pressures: $P_k = p_{H_k} + p_{M_k}, \quad k = F, R, P$ (3,4,5)

Partial pressure–mole relations: $p_{H_k} = P_k n_{H_k}/(n_{H_k} + n_{M_k}), \quad k = F, R, P$ (6,7,8)

Solution-diffusion transport rates are obtained using (14-32), assuming a log-mean partial-pressure driving force based on the exiting permeate partial pressures on the downstream side of the membrane because of the assumption of perfect mixing on that side:

$$n_{i_P} = \overline{P}_{M_i} A_M \left[\frac{p_{i_F} - p_{i_R}}{\ln\left(\dfrac{p_{i_F} - p_{i_P}}{p_{i_R} - p_{i_P}}\right)} \right], \quad i = \text{H, M} \tag{9,10}$$

Thus, we have a system of 10 equations in the following 18 variables:

$$A_M \quad n_{H_F} \quad n_{M_F} \quad P_F \quad P_R \quad P_P$$

$$\overline{P}_{M_H} \quad n_{H_R} \quad n_{M_R} \quad p_{H_F} \quad p_{H_R} \quad p_{H_P}$$

$$\overline{P}_{M_M} \quad n_{H_P} \quad n_{M_P} \quad p_{M_F} \quad p_{M_R} \quad p_{M_P}$$

Thus, eight variables must be fixed. For all parts of this example, the following five variables are fixed:

$$\overline{P}_{M_H} \text{ and } \overline{P}_{M_M} \text{ given above}$$

$$P_F = 500 \text{ psia} \quad P_R = 500 \text{ psia} \quad P_P = 20 \text{ psia}$$

For each part, three additional variables must be fixed.
(a)

$$n_{H_F} = 0.9(500) = 450 \text{ lbmol/h}$$

$$n_{M_F} = 0.1(500) = 50 \text{ lgmol/h}$$

$$n_{H_P} = 0.9(450) = 405 \text{ lbmol/h}$$

Solving Equations (1)–(10) above, using a PC program such as MathCad, we obtain

$$A_M = 3{,}370 \text{ ft}^2$$

$$n_{M_P} = 20.0 \text{ lbmol/h} \quad n_{H_R} = 45.0 \text{ lbmol/h} \quad n_{M_R} = 30.0 \text{ lbmol/h}$$

$$p_{H_F} = 450 \text{ psia} \quad p_{M_F} = 50 \text{ psia} \quad p_{H_R} = 300 \text{ psia}$$

$$p_{M_R} = 200 \text{ psia} \quad p_{H_P} = 19.06 \text{ psia} \quad p_{M_P} = 0.94 \text{ psia}$$

(b) Calculations are made in a similar manner using Equations (1)–(10). Results for parts (1), (2), and (3) are:

	Part		
	(1)	(2)	(3)
Fixed:			
n_{H_F}, lbmol/H	495	425	450
n_{M_F}, lbmol/h	55	75	50
A_M, ft^2	3,370	3,370	2,528
Calculated, in lbmol/h:			
n_{H_P}	424.2	369.6	338.4
n_{M_P}	18.2	25.9	11.5
n_{H_R}	70.8	55.4	111.6
n_{M_R}	36.8	49.1	38.5
Calculated, in psia:			
p_{H_F}	450	425	450
p_{M_F}	50	75	50
p_{H_R}	329	265	372
p_{M_R}	171	235	128
p_{H_P}	19.18	18.69	19.34
p_{M_P}	0.82	1.31	0.66

From the above results, the following are computed:

	Part			
	(a)	(b1)	(b2)	(b3)
Mol% H$_2$ in permeate	95.3	95.9	93.5	96.7
% H$_2$ recovery in permeate	90	85.7	87.0	75.2

From these results, we see that when the feed rate is increased by 10% (part b1), the hydrogen recovery drops about 5%, but the permeate purity is maintained. When the feed composition is reduced from 90% to 85% hydrogen (part b2), the hydrogen recovery decreases by about 3% and the permeate purity decreases by about 2%. With 25% of the membrane area inoperative (part b3), the hydrogen recovery decreases by about 17%, but the permeate purity is about 1% higher. Overall, percentage changes in hydrogen recovery and purity are less than the percentage changes in feed flow rate, feed composition, and membrane area, thus tending to confirm the insensitivity of gas permeation separators to changes in operating conditions. ∎

14.7 PERVAPORATION

As shown in Figure 14.22, pervaporation (PV) differs from dialysis, reverse osmosis, and gas permeation in that the phase state on one side of the membrane is different from that on the other side. The feed to the membrane module is a liquid mixture (e.g., an alcohol–water azeotrope) at a pressure, P_1, that is usually ambient or elevated high enough to maintain a liquid phase as the feed is depleted of species A and B to produce the product retentate. A composite membrane is used that is selective for species A, but species B usually has some finite permeability. The dense, thin membrane film is in contact with the liquid side. The retentate is enriched in species B. Generally, a sweep fluid is not used on the other side of the membrane, but a pressure, P_2, is maintained at or below the dew point of the permeate, making it vapor. Often, P_2 is a vacuum. Vaporization may occur near the downstream face of the membrane, such that the membrane can be considered to operate with two zones, a liquid-phase zone and a vapor-phase zone, as shown in Figure 14.22.

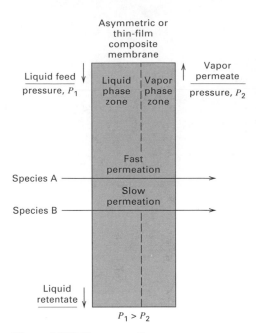

Figure 14.22 Pervaporation.

Alternatively, the vapor phase may only exist on the permeate side of the membrane. The vapor permeate is enriched in species A. Overall permeabilities of species A and B depend upon their solubilities in and diffusion rates through the membrane. Generally, the solubilities cause the membrane to swell.

The term pervaporation is a combination of the two words, *per*mselective and e*vapora-tion*. It was first reported in 1917 by Kober [43], who studied several experimental techniques for removing water from albumin/toluene solutions. Although the economic potential of PV was shown by Binning et al. [44] in 1961, commercial applications were delayed until the mid-1970s, when adequate membrane materials first became available. Major commercial applications now include (1) dehydration of ethanol; (2) dehydration of other organic alcohols, ketones, and esters; and (3) removal of organics from water. The separa-tion of organic mixtures is receiving much attention.

Pervaporation is best applied when the feed solution is dilute in the main permeant because sensible heat of the feed mixture provides the enthalpy of vaporization of the permeant. If the feed is rich in the main permeant, a number of membrane stages may be needed, with a small amount of permeant produced per stage and reheating of the retentate between stages. Even when only one membrane stage is sufficient, the feed may be heated before entering the membrane module.

Many pervaporation separation schemes have been proposed [6], with three of the more important ones shown in Figure 14.23. A hybrid process for integrating distillation with pervaporation to produce 99.5 wt% ethanol from a feed of 60 wt% ethanol is shown in Figure 14.23a. The feed is sent to a distillation column operating at near-ambient pressure, where a bottoms product of nearly pure water and an ethanol-rich distillate of 95 wt% is produced. The distillate purity is limited because of the 95.6 wt% ethanol in water azeotrope. The distillate is sent to a pervaporation step where a permeate of 25 wt% alcohol and a retentate of 99.5 wt% ethanol is produced. The permeate vapor is condensed under vacuum and recycled to the distillation column. The vacuum is sustained with a vacuum pump. The dramatic difference in separability of the pervaporation membrane as compared to vapor–liquid equilibrium for distillation is shown in Figure 14.24, taken from Wesslein et

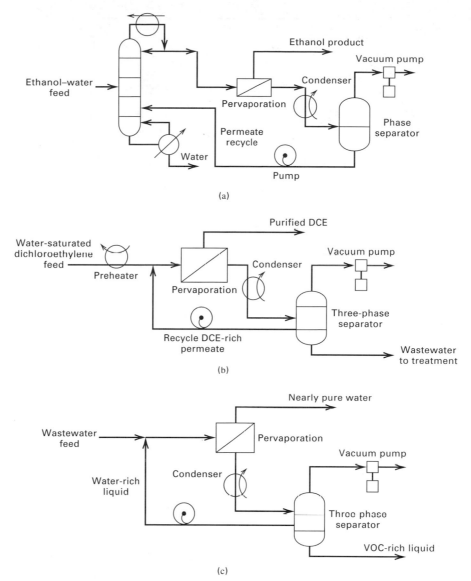

Figure 14.23 Pervaporation processes. (a) Hybrid process for removal of water from ethanol. (b) Dehydration of dichloroethylene. (c) Removal of volatile organic compounds (VOCs) from wastewater.

al. [45]. For pervaporation, the compositions refer to a liquid feed (abscissa) and a vapor permeate (ordinate) at 60°C for a polyvinylalcohol (PVA) membrane and a vacuum of 15 torr. For this membrane, there is no limitation on ethanol purity and the separation index is very high for feeds containing more than 90 wt% ethanol.

A pervaporation process for dehydrating dichloroethylene (DCE) is shown in Figure 14.23b. The liquid feed, which is DCE saturated with water (0.2 wt%), is preheated to 90°C at 0.7 atm and sent to a PVA membrane system, which produces a retentate of almost pure DCE (<10 ppm H_2O) and a permeate vapor of 50 wt% DCE under vacuum. Following condensation, the two resulting liquid phases are separated, with the DCE-rich phase recycled back to the membrane system and the water-rich phase sent to an air stripper, steam stripper, adsorption unit, or hydrophobic pervaporation membrane system for residual DCE removal.

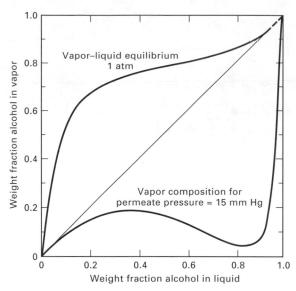

Figure 14.24 Comparison of ethanol–water separabilities. [From M. Wesslein et al., *J. Membrane Sci.,* **51,** 169 (1990).]

Pervaporation can be used for the removal of VOCs (e.g., toluene and trichloroethylene) from wastewater by pervaporation with hollow-fiber modules of silicone rubber, as shown in Figure 14.23c. The retentate is almost pure water (<5 ppb of VOCs) and the permeate, after condensation, is (1) a water-rich phase that is recycled to the membrane system and (2) a nearly pure VOC phase.

A pervaporation module typically operates adiabatically with the enthalpy of vaporization supplied by sensible enthalpy of the feed. Consider the pervaporation of a binary liquid mixture of components A and B. Assume constant pure-component liquid specific heats, and ignore heat of mixing. For an enthalpy datum temperature of T_0, an enthalpy balance, in terms of mass flow rates, m, liquid sensible heats, and heats of vaporization, gives

$$
\begin{aligned}
(m_{A_F}C_{P_A} &+ m_{B_F}C_{P_B})(T_F - T_0) \\
&= [(m_{A_F} - m_{A_P})C_{P_A} + (m_{B_F} - m_{B_P})C_{P_B}](T_R - T_0) \\
&\quad + (m_{A_P}C_{P_A} + m_{B_P}C_{P_B})(T_P - T_0) + m_{A_P}\Delta H_A^{\text{vap}} \\
&\quad + m_{B_P}\Delta H_B^{\text{vap}}
\end{aligned}
$$
(14-73)

where enthalpies of vaporization are evaluated at T_P. After collection of terms, (14-73) reduces to

$$
\begin{aligned}
(m_{A_F}C_{P_A} + m_{B_F}C_{P_B})(T_F - T_R) &= (m_{A_P}C_{P_A} + m_{B_P}C_{P_B})(T_P - T_R) \\
&\quad + (m_{A_P}\Delta H_A^{\text{vap}} + m_{B_P}\Delta H_B^{\text{vap}})
\end{aligned}
$$
(14-74)

The temperature of the permeate, T_P, is the permeate dew point at the permeate vacuum upstream of the condenser. The retentate temperature is computed from (14-74).

Membrane selection is critical in the commercial application of PV, when used in the presence of organic compounds. For water permeation, hydrophilic membrane materials are preferred. For example, a three-layer membrane is often used for the dehydration of ethanol, with water being the main permeating species. The support layer is porous polyester, which is cast on a microporous polyacrylonitrile or polysulfone membrane. The final layer, which provides the separation, is dense PVA of 0.1 μm in thickness. This composite combines chemical and thermal stability with adequate permeability. Hydrophobic membranes, such as silicone rubber and Teflon, are preferred when organics are the permeating species.

Commercial membrane modules for PV are almost exclusively of the plate-and-frame type because of the ease of using gasketing materials that are resistant to organic solvents and the ease of providing heat exchange for evaporation and high-temperature operation. However, considerable interest is evident in the use of hollow-fiber modules for the removal of VOCs from wastewater. Because feeds are generally clean and operation is at low pressure, membrane fouling and damage can be minimal, resulting in useful membrane lives of 2–4 years.

Various models for the transport of a permeant through a membrane by pervaporation have been proposed, based on the solution-diffusion model. They all assume equilibrium between the upstream liquid and the upstream membrane surface, and between the downstream vapor and the other side of the membrane. Transport through the membrane follows Fick's law with a concentration gradient of the permeant in the membrane as the driving force. However, because of the phase change and nonideal-solution effects in the liquid feed, simple equations like (14-55) for dialysis and (14-32) for gas permeation do not apply to pervaporation.

A particularly convenient PV model is that of Wijmans and Baker [46]. They express the driving force for permeation in terms of a partial vapor pressure difference. Because pressures on the both sides of the membrane are low, the gas phase follows the ideal gas law. Therefore, at the upstream membrane surface (1), permeant activity for component i is expressed as

$$a_i^{(1)} = f_i^{(1)}/f_i^{(0)} = p_i^{(1)}/P_i^{s(1)} \tag{14-75}$$

where P_i^s is the vapor pressure at the feed temperature. The liquid on the upstream side of the membrane is generally nonideal. Thus, from Table 2.2:

$$a_i^{(1)} = \gamma_i^{(1)} x_i^{(1)} \tag{14-76}$$

Combining (14-75) and (14-76):

$$p_i^{(1)} = \gamma_i^{(1)} x_i^{(1)} P_i^{s(1)} \tag{14-77}$$

On the downstream vapor side of the membrane (2), the partial pressure is

$$p_i^{(2)} = y_i^{(2)} P_P^{(2)} \tag{14-78}$$

Thus, the driving force can be expressed as $(\gamma_i^{(1)} x_i^{(1)} P_i^{s(1)} - y_i^{(2)} P_P^{(2)})$

The corresponding permeant flux, after dropping unnecessary superscripts, is

$$N_i = \frac{P_{M_i}}{l_M}(\gamma_i x_i P_i^s - y_i P_P) \tag{14-79}$$

or

$$N_i = \overline{P}_{M_i}(\gamma_i x_i P_i^s - y_i P_P) \tag{14-80}$$

where γ_i and x_i refer to the feed-side liquid, P_i^s is the vapor pressure at the feed-side temperature, y_i is the mole fraction in the permeant vapor, and P_P is the total permeant pressure.

Unlike gas permeation where P_{M_i} depends mainly on the permeant, the polymer, and temperature, the permeability for pervaporation depends additionally on the concentrations of permeants in the polymer, which can be large enough to cause polymer swelling and cross-diffusion effects. For a binary system it is best to back-calculate and correlate the permeant flux with feed composition at a given feed temperature and permeate pressure. Because of these nonideal effects, the selectivity can be a strong function of feed concentration and permeate pressure, causing inversion of selectivity in some cases, as illustrated in the following example.

EXAMPLE 14.11

Wesslein et al. [45] present the following experimental data for the pervaporation of liquid mixtures of ethanol (1) and water (2) at a feed temperature of 60°C for a permeate pressure of 76 mmHg, using a commercial polyvinylalcohol membrane:

| wt% ethanol | | Total Permeation Flux |
Feed	Permeate	kg/m²-h
8.8	10.0	2.48
17.0	16.5	2.43
26.8	21.5	2.18
36.4	23.0	1.73
49.0	22.5	1.46
60.2	17.5	0.92
68.8	13.0	0.58
75.8	9.0	0.40

At 60°C, vapor pressures are 352 and 149 mmHg for ethanol and water, respectively.

Liquid-phase activity coefficients at 60°C for the ethanol (1)–water (2) system are given by the van Laar equations:

$$\ln \gamma_1 = 1.6276 \left[\frac{0.9232 x_2}{1.6276 x_1 + 0.9232 x_2} \right]^2$$

$$\ln \gamma_2 = 0.9232 \left[\frac{1.6276 x_1}{1.6276 x_1 + 0.9232 x_2} \right]^2$$

Calculate values of permeance for water and ethanol from (14-80).

SOLUTION

For the first row of data, the mole fractions in the feed mixture (x_i) and the permeate (y_i), using molecular weights of 46.07 and 18.02 for ethanol and water, respectively, are

$$x_1 = \frac{0.088/46.07}{\dfrac{0.088}{46.07} + \dfrac{(1.0 - 0.088)}{18.02}} = 0.0364$$

$$x_2 = 1.0 - 0.0364 - 0.9636$$

$$y_1 = \frac{0.10/46.07}{\dfrac{0.10}{46.07} + \dfrac{0.90}{18.02}} = 0.0416$$

$$y_2 = 1.0 - 0.0416 = 0.9584$$

The activity coefficients for the feed mixture are

$$\gamma_1 = \exp \left\{ 1.6276 \left[\frac{0.9232(0.9636)}{1.6276(0.0364) + 0.9232(0.9636)} \right]^2 \right\} = 4.182$$

$$\gamma_2 = \exp \left\{ 0.9232 \left[\frac{1.6276(0.0364)}{1.6276(0.0364) + 0.9232(0.9636)} \right]^2 \right\} = 1.004$$

From the given total mass flux, the component molar fluxes are

$$N_1 = \frac{(2.48)(0.10)}{46.07} = 0.00538 \frac{\text{kmol}}{\text{h} - \text{m}^2}$$

$$N_2 = \frac{(2.48)(0.90)}{18.02} = 0.1239 \frac{\text{kmol}}{\text{h} - \text{m}^2}$$

From (14-80), the permeance values are

$$\overline{P}_{M_1} = \frac{0.00538}{(4.182)(0.0364)(352) - (0.0416)(76)} = 0.000107 \frac{\text{kmol}}{\text{h} - \text{m}^2 - \text{mmHg}}$$

$$\overline{P}_{M_2} = \frac{0.1239}{(2.004)(1.0 - 0.0364)(149) - (1.0 - 0.0416)(76)} = 0.001739 \frac{\text{kmol}}{\text{h} - \text{m}^2 - \text{mmHg}}$$

Results for the other feed conditions are computed in a similar manner:

wt% Ethanol		Activity Coefficient in Feed		Permeance, kmol/h-m²-mmHg	
Feed	Permeate	Ethanol	Water	Ethanol	Water
8.8	10.0	4.182	1.004	1.07×10^{-4}	1.74×10^{-3}
17.0	16.5	3.489	1.014	1.02×10^{-4}	1.62×10^{-3}
26.8	21.5	2.823	1.038	8.69×10^{-5}	1.43×10^{-3}
36.4	23.0	2.309	1.077	6.14×10^{-5}	1.17×10^{-3}
49.0	22.5	1.802	1.158	4.31×10^{-5}	1.10×10^{-3}
60.2	17.5	1.477	1.272	1.87×10^{-5}	8.61×10^{-4}
68.8	13.0	1.292	1.399	7.93×10^{-6}	6.98×10^{-4}
75.8	9.0	1.177	1.539	3.47×10^{-6}	6.75×10^{-4}

The PVA membrane is hydrophilic. Thus, as the concentration of ethanol in the feed liquid increases, the sorption of feed liquid by the membrane decreases, resulting in a reduction of polymer swelling. The preceding results show that as swelling is reduced, the permeance of ethanol decreases more rapidly than that of water, thus increasing the selectivity for water. For example, the selectivity for water can be defined as

$$\alpha_{2,1} = \frac{(100 - w_1)_P/(w_1)_P}{(100 - w_1)_F/(w_1)_F}$$

where w_1 = weight fraction of ethanol. For the cases of 8.8 and 75.8 wt% ethanol in the feed, the selectivities for water are, respectively, 0.868 (more selective for ethanol) and 31.7 (more selective for water).   ■

SUMMARY

1. The separation of liquid and gas mixtures with membranes is an emerging separation operation. Applications greatly accelerated in the 1980s. The products of separation are the retentate and the permeate.
2. The key to an efficient and economical membrane separation process is the membrane. It must have good permeability, high selectivity, stability, freedom from fouling, and a long life (2 or more years).
3. Commercialized membrane separation processes include dialysis, electrodialysis, reverse osmosis, gas permeation, and pervaporation.

4. Most membranes for commercial separation processes are natural or synthetic, glassy or rubbery polymers. However, for high-temperature ($>200°C$) or operations with chemically reactive mixtures, ceramics, metals, and carbon find applications.

5. To achieve high permeability and selectivity, dense, nonporous membranes are preferred. For mechanical integrity, membranes of 0.1 to 1.0 mm in thickness are incorporated as a surface layer or film onto or as part of a much thicker asymmetric or composite membrane.

6. To achieve a high surface area per unit volume, membranes are fabricated into spiral-wound or hollow-fiber modules. Less surface is available in plate-and-frame, tubular, and monolithic modules.

7. Permeation through a membrane can occur by a variety of mechanisms. For a microporous membrane, the mechanisms include bulk flow (with no selectivity), liquid diffusion, gas diffusion, Knudsen diffusion, restrictive diffusion (including sieving), and surface diffusion. For a nonporous membrane, a solution-diffusion mechanism, involving absorption, diffusion, and desorption, is commonly assumed.

8. Flow patterns in membrane modules have a profound effect on overall permeation rates. Idealized flow patterns for which theory has been developed include perfect mixing, countercurrent flow, cocurrent flow, and crossflow.

9. To overcome the limit of separation in a single membrane-module stage, modules can be arranged in series and/or parallel cascades.

10. In gas permeation, boundary-layer or film mass-transfer resistances on either side of the membrane are usually negligible compared to the membrane resistance. For the membrane separation of liquid mixtures, however, the external mass-transfer effects, referred to as concentration polarization, can be significant.

11. For most membrane separators, the component mass-transfer fluxes through the membrane can be formulated as the product of two terms: concentration, partial pressure, fugacity, or activity driving force; and a permeance $\overline{P}_{M_i}$, which is the ratio of the permeability, P_{M_i}, to the membrane thickness, l_M.

12. In the dialysis of a liquid mixture, small solutes of type A are separated from the solvent and larger solutes of type B with a microporous membrane. The driving force is the concentration difference across the membrane. The transport of solvent can be minimized by adjusting the pressure difference across the membrane to equal the osmotic pressure.

13. In electrodialysis, a series of alternating cation- and anion-selective membranes are used with a direct-current voltage across an outer anode and an outer cathode to concentrate an electrolyte.

14. In reverse osmosis, the solvent of a liquid mixture is selectively transported through a dense membrane. By this means, seawater can be desalinized. The driving force for transport of the solvent through the membrane is the fugacity difference, which is commonly expressed in terms of $\Delta P - \Delta \pi$, where π is the osmotic pressure.

15. In gas permeation, mixtures of gases are separated by differences in permeation rates through dense membranes. The driving force for each component is its partial pressure difference, Δp_i, across the membrane. Both the permeance and permeability depend on the absorptivity of the membrane for the particular gas species (usually as a Henry's law constant) and the diffusivity of the species through the membrane. Thus, $P_{M_i} = H_i D_i$.

16. In pervaporation, a liquid mixture is separated with a dense membrane by pulling a vacuum on the permeate side of the membrane so as to evaporate the permeate. The driving force may be approximated as a fugacity difference expressed by $(\gamma_i x_i P_i^s \quad y_i P_P)$. The permeability can vary greatly with concentration because of membrane swelling.

REFERENCES

1. Lonsdale, H.K., *J. Membrane Sci.,* **10,** 81 (1982).

2. Applegate, L.E., *Chem. Eng.,* **91**(12), 64–89 (1984).

3. Havens, G.G., and D.B. Guy, *Chem. Eng. Progress Symp. Series,* **64**(90), 299 (1968).

4. Bollinger, W.A., D.L. MacLean, and R.S. Narayan, *Chem. Eng. Progress,* **78**(10), 27–32 (1982).

5. Baker, R.W., E.L. Cussler, W. Eykamp, W.J. Koros, R.L. Riley, and H. Strathmann, *Membrane Separation Systems—A Research and Development Needs Assessment,* Report DE 90-011770, Department of Commerce, NTIS, Springfield, VA (1990).

6. Ho, W.S.W., and K.K. Sirkar, Eds., *Membrane Handbook,* Van Nostrand Reinhold, New York (1992).

7. Loeb, S., and S. Sourirajan, Advances in Chemistry Series, Vol. 38, *Saline Water Conversion II* (1963).

8. Henis, J.M.S., and M.K. Tripodi, U.S. Patent 4,230,463 (1980).

9. Wrasidlo, W. J., U.S. Patent 3,951,815 (1977).

10. Barrer, R.M., *J. Chem. Soc.,* 378–386 (1934).

11. Barrer, R. M., *Diffusion in and through Solids, Cambridge Press, London (1951).*

12. Mahon, H. I., U.S. Patent 3,228,876 (1966).

13. Mahon, H. I., U.S. Patent 3,228,877 (1966).

14. Hsieh, H.P., R.R. Bhave, and H.L. Fleming, *J. Membrane Sci.,* **39,** 221–241 (1988).

15. Bird, R.B., W.E. Stewart, and E.N. Lightfoot, *Transport Phenomena, John Wiley and Sons, New York, pp 42–47 (1960).*

16. Ergun, S., *Chem. Eng. Progress,* **48,** 89–94 (1952).

17. Beck, R.E., and J.S. Schultz, *Science,* **170,** 1302–1305 (1970).

18. Beck, R.E., and J.S. Schultz, *Biochim. Biophys. Acta,* **255,** 273 (1972).

19. Brandrup, J., and E.H. Immergut, Eds., *Polymer Handbook,* 3rd ed., John Wiley and Sons, New York (1989).

20. Lonsdale, H.K., U. Merten, and R.L. Riley, *J. Applied Polym. Sci.,* **9,** 1341–1362 (1965).

21. Motamedian, S., W. Pusch, G. Sendelbach, T.-M. Tak, and T. Tanioka, *Proceedings of the 1990 International Congress on Membranes and Membrane Processes,* Chicago, Vol. II, pp. 841–843.

22. Barrer, R.M., J.A. Barrie, and J. Slater, *J. Polym. Sci.,* **23,** 315–329 (1957).

23. Barrer, R.M., and J.A. Barrie, *J. Polym. Sci.,* **23,** 331–344 (1957).

24. Barrer, R.M., J.A. Barrie, and J. Slater, *J. Polym. Sci.,* **27,** 177–197 (1958).

25. Koros, W.J., and D.R. Paul, *J. Polym. Sci., Polym. Physics Edition,* **16,** 1947–1963 (1978).

26. Barrer, R.M., *J. Membrane Sci.,* **18,** 25–35 (1984).

27. Walawender, W.P., and S.A. Stern, *Separation Sci.,* **7,** 553–584 (1972).

28. Naylor, R.W., and P.O. Backer, *AIChE J.,* **1,** 95–99 (1955).

29. Stern, S.A., T.F. Sinclair, P.J. Gareis, N.P. Vahldieck, and P.H. Mohr, *Ind. Eng. Chem.,* **57**(2), 49–60 (1965).

30. Hwang, S.-T., and K.L. Kammermeyer, *Membranes in Separations,* Wiley-Interscience, New York, pp. 324–338 (1975).

31. Spillman, R.W., *Chem. Eng. Progress,* **85**(1), 41–62 (1989).

32. Strathmann, H., "Membrane and Membrane Separation Processes," in Vol. A16, *Ullmann's Encyclopedia of Industrial Chemistry,* VCH, FRG, p. 237, (1990).

33. Chamberlin, N.S., and B.H. Vromen, *Chem. Engr.* **66**(9), 117–122 (1959).

34. Graham, T., *Phil. Trans. Roy. Soc. London,* **151,** 183–224 (1861).

35. Juda, W., and W.A. McRae, *J. Amer. Chem. Soc.,* **72,** 1044 (1950).

36. Strathmann, H., *Sep. and Purif. Methods,* **14**(1), 41–66 (1985).

37. Merten, U., *Ind. Eng. Chem. Fundamentals,* **2,** 229–232 (1963).

38. Stoughton, R.W., and M.H. Lietzke, *J. Chem. Eng. Data,* **10,** 254–260 (1965).

39. Spillman, R.W., and M.B. Sherwin, *Chemtech,* 378–384 (June 1990).

40. Schell, W.J., and C.D. Houston, *Chem. Eng. Progress,* **78**(10), 33–37 (1982).

41. Teplyakov, V., and P. Meares, *Gas Sep. and Purif.,* **4,** 66–74 (1990).

42. Rosenzweig, M.D., *Chem. Eng.,* **88**(24), 62–66 (1981).

43. Kober, P.A., *J. Am. Chem. Soc.,* **39,** 944–948 (1917).

44. Binning, R.C., R.J. Lee, J.F. Jennings, and E.C. Martin, *Ind. Eng. Chem.,* **53,** 45–50 (1961).

45. Wesslein, M., A. Heintz, and R.N. Lichtenthaler, *J. Membrane Sci.,* **51,** 169 (1990).

46. Wijmans, J.G., and R.W. Baker, *J. Membrane Sci.,* **79,** 101–113 (1993).

47. Rautenbach, R., and R. Albrecht, *Membrane Processes,* John Wiley and Sons, New York (1989).

48. Rao, M.B., and S. Sircar, *J. Membrane Sci.,* **85,** 253–264 (1993).

EXERCISES

Section 14.1

14.1 Explain, as completely as you can, how membrane separations differ from:

(a) Absorption and stripping

(b) Distillation

(c) Liquid–liquid extraction

(d) Extractive distillation

14.2 For the commercial application of membrane separators discussed at the beginning of this chapter, calculate the permeabilities of hydrogen and methane in barrer.

14.3 A new asymmetric polyimide polymer membrane has been developed for the separation of N_2 from CH_4. At 30°C, permeance values are 50,000 and 10,000 barrer/cm for N_2 and CH_4, respectively. If this new membrane is used to perform the separation in Figure 14.25, determine the mem-

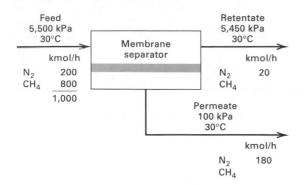

Figure 14.25 Data for Exercise 14.3.

brane surface area required in m², and the kmol/h of CH_4 in the permeate. Base the driving force for diffusion through the membrane on the arithmetic average of the partial pressures of the entering feed and the exiting retentate, with the permeate-side partial pressures at the exit condition.

Section 14.2

14.4 A hollow-fiber module has 4,000 ft² of membrane surface area based on the inside diameter of the fibers, which are 42 μm i.d. × 85 μm o.d. × 1.2 m long each. Determine:

(a) The number of hollow fibers in the module.

(b) The diameter of the module, assuming the fibers are on a square spacing of 120 μm center-to-center.

(c) The membrane surface area per unit volume of module (packing density) m²/m³. Compare your result with that in Table 14.4.

14.5 A typical spiral-wound module made from a flat sheet of membrane material is 0.3 m in diameter and 3 m long. If the packing density (membrane surface area/unit module volume) is 500 m²/m³, determine the center-to-center spacing of the membrane in the spiral, assuming a collection tube 1 cm in diameter.

14.6 A monolithic membrane element, of the type shown in Figure 14.4d, contains 19 flow channels that are 0.5 cm in inside diameter by 0.85 m long. If nine of these elements are placed into a cylindrical module of the type shown in Figure 14.5, determine reasonable values for:

(a) Module volume in m³.

(b) Packing density in m²/m³. Compare your value with values for other membrane modules given in Table 14.4.

Section 14.3

14.7 Water at 70°C is to be passed through a porous polyethylene membrane of 25% porosity with an average pore diameter of 0.3 micron and an average tortuosity of 1.3. The

pressures on the downstream and upstream sides of the membrane are 125 and 500 kPa, respectively. Estimate the flow rate of water through the membrane in m³/m²-day.

14.8 A porous glass membrane, with an average pore diameter of 40 Å, is to be used to separate light gases at 25°C under conditions where Knudsen flow may be dominant. The downstream pressure is 15 psia, while the upstream pressure is not greater than 120 psia. The membrane has been calibrated with pure helium gas, giving a constant permeability of 117,000 barrer over the operating pressure range. Experiments with pure CO_2 over the pressure range give a permeability of 68,000 barrer.

Assuming that helium is in Knudsen flow, predict the permeability of CO_2. Is the value in agreement with the experimental value? If not, suggest an explanation. Reference: Kammermeyer, K., and L.O. Rutz, *C.E.P. Symp. Ser.,* **55** (24), 163–169 (1959).

14.9 Two mechanisms for the transport of gas components through a porous membrane that are not discussed in Section 14.3 or illustrated in Figure 14.6 are (1) partial condensation in the pores by some components of the gas mixture to the exclusion of other components and subsequent transport of the condensed molecules through the pore, and (2) selective adsorption on pore surfaces of certain components of the gas mixture and subsequent surface diffusion across the pores. In particular, Rao and Sircar [48] have found that the latter mechanism provides a potentially attractive means for separating hydrocarbons from hydrogen for low-pressure gas streams. In porous carbon membranes with continuous pores 4–15 Å in diameter, little pore void space is available for the Knudsen diffusion of hydrogen when the hydrocarbons are selectively adsorbed.

Typically, the membranes are not more than 5 μm in thickness. Measurements at 295.1 K of permeabilities for five pure components and a mixture of the five components are as follows:

| | Permeability, barrer | | |
| | As a | In the | mol% in the |
Component	Pure Gas	Mixture	Mixture
H_2	130	1.2	41.0
CH_4	660	1.3	20.2
C_2H_6	850	7.7	9.5
C_3H_8	290	25.4	9.4
nC_4H_{10}	155	112.3	19.9
			100.0

A refinery waste gas mixture of the preceding composition is to be processed through such a porous carbon membrane. If the pressure of the gas is 1.2 atm and an inert sweep gas is used on the permeate side such that partial pressures of feed gas components on that side are close to zero, determine the permeate composition on a sweep-gas-free basis when the composition on the upstream-pressure side of the membrane is that of the feed gas. Explain why the component permeabilities differ so drastically between experiments with the pure gas and the gas mixture.

14.10 A mixture of 60 mol% propylene and 40 mol% propane at a flow rate of 100 lbmol/h and at 25°C and 300 psia is to be separated with a polyvinyltrimethylsilane polymer (see Table 14.9 for permeabilities). The membrane skin is 0.1 μm thick, and spiral-wound modules are used with a pressure of 15 psia on the permeate side. Calculate the material balance and membrane area in m^2 as a function of the cut (fraction of feed permeated) for:

(a) Perfect mixing flow pattern.

(b) Crossflow pattern.

14.11 Repeat part (a) of Excrcisc 14.10 for a two-stage stripping cascade and a two-stage enriching cascade, as shown in Figure 14.12. However, select just one set of reasonable cuts for the two stages of each case so as to produce 40 lbmol/h of final retentate.

14.12 Repeat Example 14.7 with the following changes:

Tube-side Rcynolds number = 25,000
Tube inside diameter = 0.4 cm
Permeate-side mass transfer coefficient = 0.06 cm/s

How important is concentration polarization?

Section 14.4

14.13 An aqueous process stream ot 100 gal/h at 20°C contains 8 wt% Na$_2$SO$_4$ and 6 wt% of a high-molecular-weight substance (A). This stream is processed in a continuous countercurrent flow dialyzer using a pure water sweep of the same flow rate. The membrane is a microporous cellophane with pore volume = 50%, wet thickness = 0.0051 cm, tortuosity = 4.1, and pore diameter = 31Å. The molecules to be separated have the following properties:

	Na$_2$SO$_4$	**A**
Molecular weight	142	1,000
Molecular diameter, Å	5.5	15.0
Diffusivity, cm^2/s × 10^5	0.77	0.25

Calculate the membrane area in m^2 for only a 10% transfer of A through the membrane, assuming no transfer of water. What is the percent recovery of the Na$_2$SO$_4$ in the diffusate? Use log-mean concentration driving forces and assume that the mass-transfer resistances on each side of the membrane are each 25% of the total mass-transfer resistances for Na$_2$SO$_4$ and A.

14.14 A dialyzer is to be used to separate 300 L/h of an aqueous solution containing 0.1 M NaCl and 0.1 M HCl. Laboratory experiments with the microporous membrane to be used give the following values for the overall mass transfer coefficient K_i in (14-57), for a log-mean concentration driving force:

	K_i, cm/min
Water	0.0025
NaCl	0.021
HCl	0.055

Determine the membrane area in m^2 for 90, 95, and 98% transfer of HCl to the diffusate. For each of the three cases, determine the complete material balance in kmol/h. A sweep of 300 L/h can be assumed.

14.15 A total of 86,000 gal/day of an aqueous solution of 3,000 ppm of NaCl is to be desalinized to 400 ppm by elec-trodialysis, with a 40% conversion. The process will be conducted in four stages, with three stacks of 150 cell pairs in each stage. The fractional desalinization will be the same in each stage and the expected current efficiency is 90%. The applied voltage for the first stage is 220 V. Each cell pair has an area of 1,160 cm^2. Calculate the current density in mA/cm^2, the current in A, and the power requirement in kW for the first stage. Reference: Mason, E.A., and T.Λ. Kirkham, *C.E.P. Symp. Ser.,* **55**(24), 173–189 (1959).

Section 14.5

14.16 A reverse osmosis plant is being used to treat 30,000,000 gal/day of seawater at 20°C containing 3.5 wt% dissolved solids to produce 10,000,000 gal/day of potable watcr with 500 ppm of dissolved solids, and the balance as brine containing 5.25 wt% dissolved solids. The feed-side pressure is 2,000 psia, while the permeate pressure is 50 psia. A single stage of spiral-wound membranes is used that approximates crossflow. If the total membrane area is 2,000,000 ft^2, estimate the permeance for water and the salt passage.

14.17 A reverse osmosis process is to be designed to handle a feed flow rate of 100 gal/min. Three designs have been proposed, differing in the % recovery of potable water from the feed:

Design 1: A single stage consisting of four units in parallel to obtain a 50% recovery

Design 2: Two stages in series with respect to the retentate (four units in parallel followed by two units in parallel)

Design 3: Three stages in series with respect to the retentate (four units in parallel followed by two units in parallel followed by a single unit)

Draw the three designs and determine the percent recovery of potable water for designs 2 and 3.

14.18 The production of paper involves a pulping step to break down wood chips into cellulose and lignin. In the Kraft process, an aqueous pulping feed solution, known as white

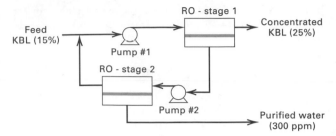

Figure 14.26 Data for Exercise 14.18.

liquor, is used that consists of dissolved inorganic chemicals such as sodium sulfide and sodium hydroxide. Following removal of the pulp (primarily cellulose), a solution known as weak (Kraft) black liquor (KBL) is left, which is regenerated to recover white liquor for recycle. In the conventional process, a typical 15 wt% (dissolved solids) KBL is concentrated to 45 to 70 wt% by multieffect evaporation. It has been suggested that reverse osmosis might be used to perform an initial concentration to perhaps 25 wt%. Higher concentrations may not be feasible because of the very high osmotic pressure, which at 180°F and 25 wt% solids is estimated to be 1,700 psia. The osmotic pressure for other conditions can be scaled with (14–68) using wt% instead of molality.

A two-stage RO process, shown in Figure 14.26, has been proposed to carry out this initial concentration for a feed rate of 1,000 lb/h at 180°F. A feed pressure of 1,756 psia is used for the first stage to yield a permeate of 0.4 wt% solids. The feed pressure to the second stage is 518 psia to produce water of 300 ppm dissolved solids and a retentate of 2.6 wt% solids. Permeate-side pressure for both stages is 15 psia. Equation (14-69) can be used to estimate membrane area, where the permeance for water can be taken as 0.0134 lb/ft²-hr-psi in conjunction with an arithmetic-mean osmotic pressure for plug flow on the feed side. Complete the material balance for the process and estimate the required membrane areas for each stage. Reference: Gottschlich, D.E., and D.L. Roberts. Final Report DE91004710, SRI International, Menlo Park, CA, Sept. 28, 1990.

Section 14.6

14.19 Gas permeation can be used to recover VOCs from air at low pressures using a membrane material that is highly selective for the VOCs. In a typical application, 1,500 scfm (0°C, 1 atm) of air containing 0.5 mol% acetone (A) is fed to a spiral-wound membrane module system at 40°C and 1.2 atm. A liquid-ring vacuum pump on the permeate side establishes a pressure of 4 cmHg. A silicone rubber, thin-composite membrane with a 2 μm-thick skin gives permeabilities of 4 barrer for air and 20,000 barrer for acetone.

If the retentate is to contain 0.05 mol% acetone and the permeate is to contain 5 mol% acetone, determine the membrane area required in m², assuming crossflow. References: (1) Peinemann, K.-V., J.M. Mohr, and R.W. Baker, *C.E.P. Symp. Series*, **82**(250), 19–26 (1986); (2) Baker, R.W., N.

Yoshioka, J.M. Mohr, and A.J. Khan, *J. Membrane Sci.*, **31**, 259–271 (1987).

14.20 The separation of air into nitrogen and oxygen is widely practiced. Cryogenic distillation is most economical for processing 100 to 5,000 tons of air per day, while pressure-swing-adsorption is favorable for 20 to 50 tons/day. For small-volume users requiring less than 10 tons/day, gas permeation finds applications where for a single stage, either an oxygen-enriched air (40 mol% oxygen) or 98 mol% nitrogen can be produced. It is desired to produce 5 ton/day (2,000 lb/ton) of 40 mol% oxygen and nitrogen, ideally of 90 mol% purity, by gas permeation. Assume pressures of 500 psia (feed side) and 20 psia (permeate). Two companies, who can supply the membrane modules, have provided the following data:

	Company A	Company B
Module type	Hollow-fiber	Spiral-wound
$\bar{P}_M$ for O_2, barrer/μm	15	35
$\bar{P}_{M_{O_2}}/\bar{P}_{M_{N_2}}$	3.5	1.9

Determine the required membrane area in m² for each company. Assume that both module types approximate cross-flow.

14.21 A joint venture has been underway for several years to develop a membrane process to separate CO_2 and H_2S from high-pressure sour natural gas. Typical feed and product conditions are:

	Feed Gas	Pipeline Gas
Pressure, psia	1,000	980
Composition, mol%:		
CH_4	70	97.96
H_2S	10	0.04
CO_2	20	2.00

To meet these conditions, the following hollow-fiber membrane material targets have been established:

	Selectivity
CO_2–CH_4	50
H_2S–CH_4	50

where selectivity is the ratio of permeabilities.

$$P_{M_{CO_2}} = 13.3 \text{ barrer},$$

and membrane skin thickness is expected to be 0.5 μm.

Make calculations to show whether the targets can realistically meet the pipeline gas conditions in a single stage with a reasonable membrane area. Assume a feed gas flow rate of 10×10^3 scfm (0°C, 1 atm) with crossflow. Reference: Stam, H., in *Future Industrial Prospects of*

Membrane Processes, L. Cecille and J.-C. Toussaint, Eds., Elsevier Applied Science, London, pp. 135–152 (1989).

Section 14.7

14.22 Pervaporation is to be used to separate ethyl acetate (EA) from water. The feed rate is 100,000 gal/day of water containing 2.0 wt% EA at 30°C and 20 psia. The membrane is dense polydimethylsiloxane with a 1 μm-thick skin in a spiral-wound module that approximates crossflow. The permeate pressure is 3 cmHg. The total measured membrane flux at these conditions is 1.0 L/m²-h with a separation factor given by (14-36) of 100 for EA with respect to water. A retentate of 0.2 wt% EA is desired for a permeate of 45.7 wt% EA. Determine the required membrane area in m² and estimate the temperature drop of the feed. Reference: Blume, I., J.G. Wijans, and R.W. Baker, *J. Membrane Sci.,* **49,** 253–286 (1990).

14.23 For a temperature of 60°C and a permeate pressure of 15.2 mmHg, Wesslein et al. [45] measured a total permeation flux of 1.6 kg/m²-h for a 17.0 wt% ethanol in water feed, giving a permeate of 12 wt% ethanol. Otherwise, conditions were those of Example 14.11. Calculate the permeances of ethyl alcohol and water for these conditions. Also, calculate the selectivity for water.

14.24 The separation of benzene (B) from cyclohexane (C) by distillation at 1 atm is impossible because of a minimum-boiling-point azeotrope at 54.5 mol% benzene. However, extractive distillation with furfural is feasible. For an equimolar feed, cyclohexane and benzene products of 98 and 99 mol%, respectively, can be produced. Alternatively, the use of a three-stage pervaporation process, with selectivity for benzene using a polyethylene membrane, has received attention, as discussed by Rautenbach and Albrecht [47]. Consider the second stage of this process where the feed is 9,905 kg/h of 57.5 wt% B at 75°C. The retentate is 16.4 wt% benzene at 67.5°C and the permeate is 88.2 wt% benzene at 27.5°C. The total permeate mass flux is 1.43 kg/m²-h and the selectivity for benzene is 8. Calculate the flow rates of retentate and permeate in kg/h and the required membrane surface area in m².

Chapter 15

Adsorption, Ion Exchange, and Chromatography

Adsorption, ion exchange, and chromatography are *sorption* operations, in which certain components of a fluid phase, called solutes, are selectively transferred to insoluble rigid particles suspended in a vessel or packed in a column. Sorption, which is a general term introduced by J.W. McBain [*Phil. Mag.,* **18,** 916–935 (1909)], includes selective transfer to the surface and/or into the bulk of a solid or liquid. Thus, absorption of gas species into a liquid and penetration of fluid species into a nonporous membrane are also sorption operations. In a general sorption process, the sorbed solutes are referred to as *sorbate,* and the sorbing agent is the *sorbent.*

In an adsorption process, molecules, as shown in Figure 15.1a, or atoms or ions in a gas or liquid diffuse to the surface of a solid, where they bond with the solid surface or are held there by weak intermolecular forces. The adsorbed solutes are referred to as *adsorbate,* whereas the solid material is the *adsorbent.* To achieve a very large surface area for adsorption per unit volume, highly porous solid particles with small-diameter interconnected pores are used.

In an ion-exchange process, as shown in Figure 15.1b, ions of positive charge (*cations*) or negative charge (*anions*) in a liquid solution, usually aqueous, replace dissimilar and displaceable ions of the same charge contained in a solid *ion exchanger,* which also contains immobile, insoluble, and permanently bound co-ions of the opposite charge. Thus, ion exchange can be cation or anion exchange. Water softening by ion exchange involves a cation exchanger, in which the following reaction occurs to remove calcium ions.

$$Ca_{(aq)}^{2+} + 2NaR_{(s)} \leftrightarrow CaR_{2_{(s)}} + 2Na_{(aq)}^+$$

where R is the residual material of the ion exchanger. The exchange of ions is reversible and does not cause any permanent change to the structure of the solid ion exchanger. Thus, it can be used and reused unless fouled by organic compounds in the liquid feeds that attach to exchange sites on and within the ion exchanger. The ion-exchange concept can be extended to the removal of essentially all inorganic salts from water by a two-step process called *demineralization* or *deionization.* In the first step, a cation resin exchanges hydrogen ions for cations such as calcium, magnesium, and sodium. In the second step, an anion resin exchanges hydroxyl ions for strongly and weakly ionized anions such as sulfate, nitrate, chloride, and bicarbonate. The hydrogen

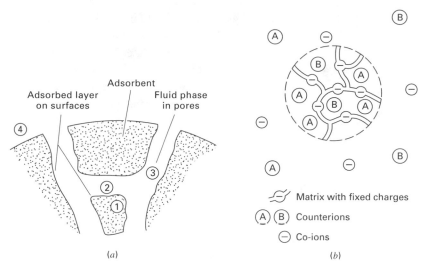

Figure 15.1 Sorption operations with solid-particle sorbents. (a) Adsorption. (b) Ion exchange.

and hydroxyl ions that enter the water combine to form water. Regeneration of the cation and anion resins is usually accomplished with sulfuric acid and sodium hydroxide, respectively.

In a chromatographic process, the sorbent may be a solid adsorbent, an insoluble, nonvolatile liquid absorbent contained in the pores of a granular solid support, or an ion exchanger. In either case, the solutes to be separated move through the chromatographic separator, with an inert eluting fluid, at different rates because of repeated sorption, desorption cycles.

During adsorption and ion exchange, the solid separating agent becomes saturated or nearly saturated with the molecules, atoms, or ions transferred from the fluid phase. To recover the sorbed substances and allow the adsorbent to be reused, it is regenerated by desorbing the sorbed substances. Accordingly, these two separation operations are carried out in a cyclic manner. In chromatography, regeneration occurs continuously, but at changing locations in the separator.

Adsorption processes may be classified as *purification* or *bulk separation,* depending on the concentration in the feed fluid of the components to be adsorbed. Although there is no sharp dividing concentration, Keller [1] has suggested 10 wt%. Early applications of adsorption involved only purification. For example, adsorption with charred wood to improve the taste of water has been known for centuries. The decolorization of liquid solutions by adsorption with bone char and other materials has been practiced for at least five centuries. Adsorption of gases by a solid (charcoal) was first described by C.W. Scheele in 1773. Commercial applications of bulk separation by gas adsorption began in the early 1920s, but did not escalate until the 1960s, following the inventions by Milton [2] of synthetic molecular-sieve zeolites, which provide high adsorptive selectivity, and by Skarstrom [3] of the *pressure-swing cycle,* which made possible the efficient operation of a fixed-bed cyclic gas-adsorption process. The commercial-scale bulk separation of liquid mixtures also began in the 1960s, following the invention by Broughton and Gerhold [4] of the simulated moving bed for adsorptive separation.

The use of ion exchange dates back to at least the time of Moses, who, while leading his followers out of Egypt into the wilderness, sweetened the bitter waters of Marah with a tree [Exodus 15:23–26]. In ancient Greece, Aristotle observed that the salt content of water is reduced when it percolates through certain sands. Systematic studies of ion exchange were published in 1850 by both Thompson and Way, who experimented with cation exchange in soils before the discovery of the existence of ions.

The first major application of ion exchange, which occurred almost 100 years ago, was for water treatment to remove the ions responsible for water hardness, such as calcium. Initially, the ion exchanger was a porous, natural mineral zeolite containing silica. In 1935, synthetic, insoluble, polymeric resin ion exchangers were introduced. Today they are dominant for water-softening and deionizing applications, but natural and synthetic zeolites still find some use.

Since the invention of chromatography by M.S. Tswett [5], a Russian botanist, in 1903, it has found widespread use as an analytical and preparative laboratory technique. Tswett separated a mixture of structurally similar yellow and green chloroplast pigments in leaf extracts by dissolving the extracts in carbon disulfide and passing the solution through a column packed with chalk particles. The pigments were separated by color. Hence, the name chromatography was coined by Tswett in 1906 from the Greek words *chroma*, meaning color, and *graphe*, meaning writing. Chromatography has revolutionized the laboratory chemical analysis of liquid and, particularly, gas mixtures. The large-scale, commercial applications described by Bonmati et al. [6] and Bernard et al. [7], however, did not begin until the 1980s.

The pressure-swing gas adsorption process is primarily used for the dehydration of air and for the separation of air into nitrogen and oxygen. A small unit for the dehydration of compressed air is described by White and Barkley [8] and shown in Figure 15.2. The unit consists of two fixed-bed adsorbers, each 12.06 cm in diameter and packed with 11.15 kg of 3.3-mm-diameter Alcoa F-200 activated alumina beads to a height of

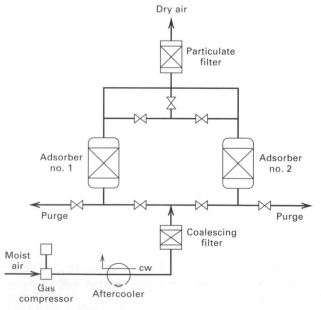

Figure 15.2 Pressure-swing adsorption for the dehydration of air.

1.27 m. The external porosity (void fraction) of the bed is 0.442 and the alumina-bead bulk density is 769 kg/m³.

The unit operates on a 10-min cycle, with 5 min for adsorption of water vapor from the air and 5 min for regeneration, which consists of depressurization, purging of the water vapor, and a 30-s repressurization. While one bed is adsorbing, the other bed is being regenerated. The adsorption (drying) step takes place with air entering at 21°C and 653.3 kPa (6.45 atm) with a flow rate of 1.327 kg/min, passing up through the bed with a pressure drop of 2.386 kPa. The dew-point temperature of the air at system pressure is reduced from 11.2 to −61°C by the adsorption process. During the 270-second period of purging, about a third of the dry air leaving one bed is directed to the other bed as a downward-flowing purge to regenerate the adsorbent. The purge is exhausted at a pressure of 141.3 kPa. By conducting the purge flow countercurrent to the entering air flow, the highest degree of water vapor desorption is achieved.

Other equipment, shown in Figure 15.2, includes an air compressor, an aftercooler; piping and valving to switch the beds from one step in the cycle to the other; a coalescing filter to remove aerosols from the entering air; and a particulate filter for the exiting dry air to remove adsorbent fines. If the dry air is needed only at low-to-moderate pressures, an air turbine can be installed to recover energy while reducing the air pressure.

During the 5-min adsorption period of the cycle, the capacity of the adsorbent for water must not be exceeded. In this example, the water content of the air is reduced from 1.27×10^{-3} kg H_2O/kg air to the very low value of 9.95×10^{-7} kg H_2O/kg air. To achieve this exiting water vapor content, only a small fraction of the adsorbent capacity is utilized during the adsorption step, with most of the adsorption occurring in the first 0.2 m of the 1.27-m bed height.

The bulk separation of gas and liquid mixtures by adsorption is an emerging separation operation. Important progress is being made in the development of new and more selective adsorbents and in more efficient operation cycles. In addition, attention is being directed to hybrid separation systems that include membrane and other types of separation steps. Already, the three sorption operations addressed in this chapter have found numerous applications, some of which are listed in Table 15.1, compiled from listings in Rousseau [9]. The applications cover a very wide range of species molecular weight.

This chapter discusses sorbents, including their equilibrium, sieving, transport, and kinetic properties with respect to solutes being removed from solutions; techniques for conducting cyclic operations; and equipment configuration and design. Both equilibrium-stage and rate-based models are developed. Although emphasis is on adsorption, the basic principles of ion exchange and chromatography are also presented. Further descriptions of the three sorption operations are given by Rousseau [9] and Ruthven [10].

15.1 SORBENTS

To be suitable for commercial applications, a sorbent should have (1) high selectivity to enable sharp separations, (2) high capacity to minimize the amount of sorbent needed, (3) favorable kinetic and transport properties for rapid sorption, (4) chemical and thermal stability, including extremely low solubility in the contacting fluid, to preserve the amount of sorbent and its properties, (5) hardness and mechanical strength to prevent crushing and erosion, (6) a free-flowing tendency for ease of filling or emptying vessels, (7) high

Table 15.1 Industrial Applications of Sorption Operations

1. Adsorption
   Gas purifications:
     Removal of organics from vent streams
     Removal of SO_2 from vent streams
     Removal of sulfur compounds from gas streams
     Removal of water vapor from air and other gas streams
     Removal of solvents and odors from air
     Removal of NO_x from N_2
     Removal of CO_2 from natural gas
   Gas bulk separations:
     N_2/O_2
     H_2O/ethanol
     Acetone/vent streams
     C_2H_4/vent streams
     Normal paraffins/isoparaffins, aromatics
     CO, CH_4, CO_2, N_2, A, NH_3/H_2
   Liquid purifications:
     Removal of H_2O from organic solutions
     Removal of organics from H_2O
     Removal of sulfur compounds from organic solutions
     Decolorization of solutions
   Liquid bulk separations:
     Normal paraffins/isoparaffins
     Normal paraffins/olefins
     p-xylene/other C_8 aromatics
     p- or m-cymene/other cymene isomers
     p- or m-cresol/other cresol isomers
     Fructose/dextrose, polysaccharides
2. Ion Exchange
   Water softening
   Water demineralization
   Water dealkalization
   Decolorization of sugar solutions
   Recovery of uranium from acid leach solutions
   Recovery of antibiotics from fermentation broths
   Recovery of vitamins from fermentation broths
3. Chromatography
   Separation of sugars
   Separation of perfume ingredients
   Separation of C_4–C_{10} normal and isoparaffins

resistance to fouling for long life, (8) no tendency to promote undesirable chemical reactions, (9) the capability of being regenerated when used with commercial feedstocks that contain trace quantities of high-molecular-weight species that are strongly sorbed and difficult to desorb, and (10) relatively low cost.

Adsorbents

Most solids are able to adsorb species from gases and liquids. However, only a few have a sufficient selectivity and capacity to make them serious candidates for commercial adsorbents. Of considerable importance is a large specific surface area (area per unit volume), which is achieved by adsorbent manufacturing techniques that result in solids with a microporous structure. By the definition of the International Union of Pure and Applied Chemistry (IUPAC), a micropore is <20 Å, a mesopore is 20–500 Å, and a macropore is >500 Å (50 nm). Typical commercial adsorbents, which may be granules, spheres, cylindrical pellets, flakes, and/or powders of size ranging from 50 μm to 1.2 cm, have specific surface areas from 300 to 1,200 m^2/g. Thus, just a few grams of adsorbent can have a surface area equal to that of a football field (120 $\times$ 53.3 yards or 5,350 m^2)! Such a large area is made possible by a particle porosity from 30 to 85 vol% with average pore diameters from 10 to 200 Å. Consider a cylindrical pore of diameter d_p and length L. The surface area-to-volume ratio is

$$S/V = \pi d_p L / (\pi d_p^2 L / 4) = 4/d_p \qquad \textbf{(15-1)}$$

If the fractional particle porosity is ϵ_p and the particle density is ρ_p, the specific surface area, S_g, in area per unit mass of adsorbent is

$$S_g = 4\epsilon_p / \rho_p d_p \qquad \textbf{(15-2)}$$

Thus, if ϵ_p is 0.5, ρ_p is 1 g/cm^3 = 1 $\times$ 10^6 g/m^3, and d_p is 20 Å (20 $\times$ 10^{-10} m) substitution into (15-2) gives S_g = 1,000 m^2/g, a desirable value.

Depending upon the type of forces between the fluid molecules and the molecules of the solid, adsorption may be classified as *physical adsorption* (van der Waals adsorption) or *chemisorption* (activated adsorption). Physical adsorption from a gas occurs when the intermolecular attractive forces between molecules of a solid and the gas are greater than those between molecules of the gas itself. In effect, the resulting adsorption is like condensation, which is exothermic and thus is accompanied by a release of heat. The magnitude of the heat of adsorption can be less than or greater than the heat of vaporization and changes with the extent of adsorption. Physical adsorption, which may be a monomolecular (unimolecular) layer, or may be two, three or more layers thick (multimolecular), occurs rapidly. If unimolecular, it is reversible; if multimolecular, such that capillary pores are filled, hysteresis may occur. The density of the adsorbate is of the order of magnitude of the liquid rather than the vapor state. As physical adsorption takes place, it begins as a monolayer, becomes multilayered, and then, if the pores are close to the size of the molecules, capillary condensation occurs, and the pores fill with adsorbate. Accordingly, the maximum capacity of a porous adsorbent can be more related to the pore volume than to the surface area. However, for gases at temperatures above their critical temperature, adsorption is confined to a monolayer.

In contrast, chemisorption involves the formation of chemical bonds between the adsorbent and adsorbate in a monolayer, often with a release of heat much larger than the heat of vaporization. Chemisorption from a gas generally takes place only at temperatures greater than 200°C and may be slow and irreversible. Commercial adsorbents rely on physical adsorption; catalysis relies on chemisorption.

Adsorption from a liquid is a more difficult phenomenon to measure experimentally or describe. When the fluid is a gas, experiments are conducted with pure gases or with mixtures. The amount of gas adsorbed is determined from the measured decrease in total pressure. When the fluid is a liquid, no simple procedure for determining the extent of adsorption from a pure liquid exists; consequently, experiments are only conducted using liquid mixtures, including dilute solutions. When porous particles of adsorbent are immersed in a liquid mixture, the pores, if sufficiently larger in diameter than the molecules in the liquid, fill with liquid. At equilibrium, because of differences in the extent of physical adsorption among the different molecules of the liquid mixture, the composition of the liquid in the pores differs from that of the bulk liquid surrounding the adsorbent particles. The observed exothermic heat effect is referred to as the *heat of wetting,* which is much smaller than the heat of adsorption from the gas phase. As with gases, the extent of equilibrium adsorption of a given solute increases with concentration and decreases with temperature. Chemisorption can also occur with liquids.

Listed in Table 15.2 are six major types of solid adsorbents in use. Included are the nature of the adsorbent and representative values of the mean pore diameter, d_p, particle porosity (internal void fraction), ϵ_p, particle density, ρ_p, and specific surface area, S_g. In addition, for some adsorbents, the capacity for adsorbing water vapor at a partial pressure of 4.6 mmHg in air at 25°C is listed, as taken from Rousseau [9]. Not included is the specific pore volume, V_p, which, can be computed from the other properties by

$$V_p = \epsilon_p / \rho_p \qquad \text{(15-3)}$$

Also not included in Table 15.2, but of interest when the adsorbent is used in fixed beds, are the bulk density, ρ_b, and the bed porosity (external porosity), ϵ_b, which are related:

$$\epsilon_b = 1 - \frac{\rho_b}{\rho_p} \qquad \text{(15-4)}$$

In addition, the true solid density (also called the crystalline density), ρ_s, can be computed from a similar expression:

$$\epsilon_p = 1 - \frac{\rho_p}{\rho_s} \qquad \text{(15-5)}$$

The specfic surface area of an adsorbent, S_g, is measured by adsorbing gaseous nitrogen, using the well-accepted BET method (Brunauer, Emmett, and Teller [11]). Typically, the BET apparatus operates at the normal boiling point of N_2 (−195.8°C) by measuring the equilibrium volume of pure N_2 physically adsorbed on several grams of the adsorbent at a number of different values of the total pressure in the vacuum range of 5 to at least 250 mmHg. Brunauer, Emmett, and Teller derived a theoretical equation to model the adsorp-

Table 15.2 Representative Properties of Commercial Porous Adsorbents

Adsorbent	Nature	Pore Diameter d_p, Å	Particle Porosity ε_p	Particle Density ρ_p, g/cm^3	Surface Area S_g, m^2/g	Capacity for H$_2$O Vapor at 25°C and 4.6 mmHg, wt% (Dry Basis)
Activated alumina	Hydrophilic, amorphous	10–75	0.50	1.25	320	7
Silica gel:	Hydrophilic/ hydrophobic,					
Small pore	amorphous	22–26	0.47	1.09	750–850	11
Large pore		100–150	0.71	0.62	300–350	—
Activated carbon:	Hydrophobic, amorphous					
Small pore		10–25	0.4–0.6	0.5–0.9	400–1200	1
Large pore		>30	—	0.6–0.8	200–600	—
Molecular-sieve carbon	Hydrophobic	2–10	—	0.98	400	—
Molecular-sieve zeolites	Polar-hydrophilic, crystalline	3–10	0.2–0.5	—	600–700	20–25
Polymeric adsorbents	—	40–25	0.4–0.55	—	80–700	—

tion by allowing for the formation of multimolecular layers. Furthermore, they assumed that the heat of adsorption during monolayer formation (ΔH_{ads}) is constant and that the heat effect associated with subsequent layers is equal to the heat of condensation (ΔH_{cond}). The BET equation is

$$\frac{P}{v(P_0 - P)} = \frac{1}{v_m c} + \frac{(c - 1)}{v_m c}\left(\frac{P}{P_0}\right) \tag{15-6}$$

where:

P = total pressure
P_0 = vapor pressure of adsorbate at test temperature
v = volume of gas adsorbed at STP (0°C, 760 mmHg)
v_m = volume of monomolecular layer of gas adsorbed at STP
c = constant related to the heat of adsorption $\approx \exp[(\Delta H_{cond} - \Delta H_{ads})/RT]$

Data for v as a function of P are plotted, according to (15-6) as $P/[v(P_0 - P)]$ versus P/P_0, from which v_m and c are determined from the slope and intercept of the best straight-line fit of the data. The value of S_g is then computed from

$$S_g = \frac{\alpha v_m N_A}{V} \tag{15-7}$$

where:

N_A = Avogadro's number = 6.023×10^{23} molecules/mol
V = Volume of gas per mole at STP conditions (0°C, 1 atm) = 22,400 cm^3/mol

The quantity α is the surface area covered per adsorbed molecule. If we assume spherical molecules arranged in close two-dimensional packing, the projected surface area is:

$$\alpha = 1.091\left(\frac{M}{N_A \rho_L}\right)^{2/3} \tag{15-8}$$

where:

M = molecular weight of the adsorbate
ρ_L = density of the adsorbate in g/cm^3, taken as the liquid at the test temperature

Although the BET surface area may not always represent the surface area available for adsorption of a particular molecule, the BET test is reproducible and widely used in the characterization of adsorbents.

The specific pore volume, typically cm^3 of pore volume/g of adsorbent, is determined for a small mass of adsorbent, m_p, by measuring the volumes of helium, V_{He}, and mercury, V_{Hg}, displaced by the adsorbent. The helium is not adsorbed, but fills the pores. At ambient pressure, the mercury cannot enter the pores because of unfavorable interfacial tension and contact angle. The specific pore volume, V_p, is then determined from

$$V_p = (V_{Hg} - V_{He})/m_p \tag{15-9}$$

The particle density is obtained from

$$\rho_p = \frac{m_p}{V_{Hg}} \tag{15-10}$$

The true solid density is obtained from

$$\rho_s = \frac{m_p}{V_{He}} \tag{15-11}$$

The particle porosity is then obtained from (15-3) or (15-5).

The distribution of pore volume over the range of pore size, which is of great importance in adsorption, is measured by mercury porosimetry for large-diameter pores (>100 Å); by gaseous nitrogen desorption for pores of 15–250 Å in diameter; and by molecular sieving, using molecules of different diameter, for pores <15 Å in diameter. In mercury porosimetry, the extent of mercury penetration into the pores is measured as a function of applied hydrostatic pressure. A force balance along the axis of a straight pore of circular cross-section for the pressure and the interfacial tension between the mercury and the adsorbent surface gives

$$d_p = -\frac{4\sigma_I \cos\theta}{P} \tag{15-12}$$

where for mercury: σ_I = interfacial tension = 0.48 N/m and θ = contact angle = $140°$. With these values, (15-12) becomes

$$d_p\,(\text{Å}) = \frac{21.6 \times 10^5}{P\,(\text{psia})} \tag{15-13}$$

Thus, forcing mercury into a 100-Å-diameter pore requires a pressure of 21,600 psia.

The nitrogen desorption method for determining pore-size distribution in the more important 15–250 Å diameter range is an extension of the BET method described earlier for measuring specific surface area. By increasing the nitrogen pressure above 600 mmHg, the multilayer adsorbed films reach the point where they bridge the pore, resulting in capillary condensation. At $P/P_0 = 1$, the entire pore volume is filled with nitrogen. Then, by reducing the pressure in steps, nitrogen is desorbed selectively, starting with the larger pores. This selectivity occurs because of the effect of pore diameter on the vapor pressure of the condensed phase in the pore, as given by the Kelvin equation:

$$P_p^s = P^s \exp\left(-\frac{4\sigma v_L \cos\theta}{RTd_p}\right) \tag{15-14}$$

where:

P_p^s = vapor pressure of liquid in pore
P^s = the normal vapor pressure of liquid on a flat surface
σ = surface tension of liquid in pore
v_L = molar volume of liquid in pore

Thus, the vapor pressure of the condensed phase in the pore is less than its normal vapor pressure for a flat surface. The effect of d_p on P_p^s can be significant. For example, for liquid nitrogen at $-195.8°$C, $P^s = 760$ torr, $\sigma = 0.00827$ N/m, $\theta = 0$, and $v_L = 34.7$ cm^3/mol. Equation (15-14) then becomes

$$d_p(\text{Å}) = 17.9/\ln(P^s/P_p^s) \tag{15-15}$$

From (15-15), for $d_p = 30$ Å, $P_p^s = 418$ torr, a reduction in vapor pressure of almost 50%. At 200 Å, the reduction is only about 10%. At 418 torr pressure, only pores less than 30 Å in diameter remain filled with liquid nitrogen. For greater accuracy in applying the Kelvin equation, a correction is needed for the thickness of the adsorbed layer. The use of this correction is discussed in detail by Satterfield [12]. For a monolayer, this thickness for nitrogen is about 0.354 nm, corresponding to a P/P_0 of between 0.05 and 0.10. At $P/P_0 = 0.60$ and 0.90, the adsorbed thicknesses are 0.75 and 1.22 nm, respectively. The correction is applied by subtracting twice the adsorbed thickness from d_p in (15-14) and (15-15).

EXAMPLE 15.1

Using data from Table 15.2, determine the volume fraction of pores in silica gel (small-pore type) filled with adsorbed water vapor when its partial pressure is 4.6 mmHg and the temperature is 25°C. At these conditions, the partial pressure is considerably below the vapor pressure of 23.75 mmHg. In addition, determine whether the amount of water adsorbed is equivalent to more than a monolayer, if the area of an adsorbed water molecule is given by (15-8) and the specific surface area of the silica gel is 830 m²/g.

SOLUTION

Take 1 g of silica gel particles as a basis. From (15-3) and data in Table 15.2, $V_p = 0.47/1.09 = 0.431$ cm³/g. Thus, for 1 g , pore volume is 0.431 cm³. From the capacity value in Table 15.2, amount of adsorbed water $= 0.11/(1 + 0.11) = 0.0991$ g. Assume density of adsorbed water is 1 g/cm³, volume of adsorbed water $= 0.0991$ cm³, fraction of pores filled with water $= 0.0991/0.431 = 0.230$, and surface area of 1 g $= 830$ m². From (15-8):

$$\alpha = 1.091 \left[\frac{18.02}{(6.023 \times 10^{23})(1.0)} \right]^{2/3} = 10.51 \times 10^{-16} \, \text{cm}^2/\text{molecule}$$

$$\text{Number of water molecules adsorbed} = \frac{(0.0991)(6.023 \times 10^{23})}{18.02} = 3.31 \times 10^{21} \, \text{molecules}$$

$$\text{Number of water molecules in a monolayer for 830 m}^2 = \frac{830(100)^2}{10.51 \times 10^{-16}} = 7.90 \times 10^{21}$$

Therefore, only 3.31/7.90 or 42% of one monolayer is adsorbed. ∎

The four most widely used adsorbents in decreasing order of commercial usage are carbon (activated and molecular-sieve), molecular-sieve zeolites, silica gel, and activated alumina. In Table 15.2, activated alumina, Al_2O_3, which includes activated bauxite, is made by removing water from hydrated colloidal alumina. Activated alumina has a moderately high specific surface area, with a capacity for adsorption of water sufficient to dry gases to less than 1 ppm moisture content. Because of its great affinity for water, activated alumina is widely used for the removal of water from gases and liquids.

Silica gel, SiO_2, which is made from colloidal silica, has a high surface area and high affinity for water and other polar compounds. Related silicate adsorbents include magnesium silicate, calcium silicate, various clays, Fuller's earth, and diatomaceous earth. Silica gel is also highly desirable for water removal. Both small-pore and large-pore types are available.

Activated carbon is made by processes that involve the partial oxidation of a number of materials, including coconut shells, fruit nuts, wood, coal, lignite, peat, petroleum residues, and bones. Because activated carbon is hydrophobic and has a high specific surface area, it is particularly useful for processes involving nonpolar and weakly polar organic molecules. Macropores within the carbon particles help transfer molecules to the micropores. Two commercial grades are produced, one with large pores for liquid applications and one with small pores for gas adsorption. As shown in Table 15.2, activated carbon is relatively hydrophobic and has a large surface area. Accordingly, it has found wide application for the purification and separation of gas and liquid mixtures containing nonpolar and weakly polar organic compounds, which adsorb much more strongly than water. In addition, the bonding strength of adsorption on activated carbon is low, resulting in a low heat of adsorption and ease of regeneration of the adsorbent.

Unlike activated carbon, which typically has pore diameters starting from 10 Å, molecular-sieve carbon (MSC) has much smaller pores ranging from 2 to 10 Å, making it possible to separate N_2 from air. The small pores, in one process, are made by depositing coke in the pore mouths of activated carbon.

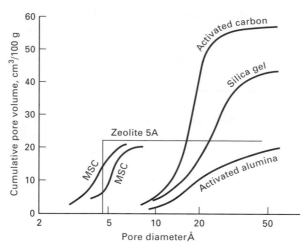

Figure 15.3 Representative cumulative pore-size distributions of adsorbents.

Most commercial adsorbents have a range of pore sizes, as shown in Figure 15.3, where the cumulative pore volume is plotted against pore diameter. Exceptions are the molecular-sieve zeolites, which are crystalline inorganic polymers of aluminosilicates and alkali or alkali-earth cation elements, such as Na, K, Mg, and Ca, with the general stoichiometric unit cell formula

$$M_{x/m}[(AlO_2)_x(SiO_2)_y]z\ H_2O$$

where M is the cation with valence m, z is the number of water molecules in each unit cell, and x and y are integers such that $y/x \geq 1$. The cations balance the charge of the AlO_2 groups, each having a net charge of -1. To activate the zeolite, the water molecules are removed by raising the temperature or pulling a vacuum. This leaves the remaining atoms spacially intact in interconnected cagelike structures with six identical window apertures each, the size of which ranges from 3.8 to about 10 Å, depending on the cation and the crystal structure. These aperatures act as sieves, which permit small molecules to enter the crystal cage, but exclude large molecules. Thus, compared to the other types of adsorbents, molecular-sieve zeolites are highly selective because all apertures have the same size. The properties and applications of five of the most commonly used molecular-sieve zeolites are given in Table 15.3, taken from Ruthven [13]. The zeolites separate not only by molecular size and shape, but also by polarity. Thus, they can also separate

Table 15.3 Properties and Applications of Some Molecular-Sieve Zeolites

Designation	Cation	Unit Cell Formula	Aperture Size, Å	Typical Applications
3A	K^+	$K_{12}[(AlO_2)_{12}(SiO_2)_{12}]$	2.9	Drying of reactive gases
4A	Na^+	$Na_{12}[(AlO_2)_{12}(SiO_2)_{12}]$	3.8	H_2O, CO_2 removal; air separation
5A	Ca^{2+}	$Ca_5Na_2[(AlO_2)_{12}(SiO_2)_{12}]$	4.4	Separation of air; separation of linear paraffins
10X	Ca^{2+}	$Ca_{43}[(AlO_2)_{86}(SiO_2)_{106}]$	8.0 ⎫	Separation of air;
13X	Na^+	$Na_{86}[(AlO_2)_{86}(SiO_2)_{86}]$	8.4 ⎬	removal of mercaptans

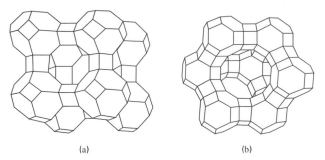

(a)   (b)

Figure 15.4 Line structures of molecular-sieve zeolites:
(a) Type A unit cell. (b) Type X unit cell.

molecules of similar size. Some zeolites have circular aperatures, whereas others have elliptical apertures. Adsorption in zeolites is actually a selective and reversible filling of crystal cages, so surface area is not a pertinent factor. Although naturally occurring zeolite minerals have been known for more than 200 years, molecular-sieve zeolites were first synthesized by Milton [2], who used very reactive materials at temperatures of 25–100°C.

The structure of the unit cell of a type A zeolite is shown in Figure 15.4a as a three-dimensional structure of silica and alumina tetrahedra, each formed by four oxygen atoms surrounding a silicon or aluminum atom. Oxygen and silicon atoms have two negative and four positive charges, respectively, causing the tetrahedra to build uniformly in four directions. Aluminum, with a valence of 3, causes the alumina tetrahedron to be negatively charged. The added cation provides the balance. In Figure 15.4a, an octahedron of tetrahedra is evident with six faces, with one near-circular window aperture at each face. A type X zeolite is shown in Figure 15.4b. This unit cell structure results in a larger window aperature. Monographs by Barrer [14] and Breck [15] cover many important aspects of zeolites.

Of lesser commercial importance are polymeric adsorbents. Typically, they are spherical beads, 0.5 mm in diameter, made from microspheres about 10^{-4} mm in diameter. They are produced by polymerizing styrene and divinylbenzene for adsorbing nonpolar organics from aqueous solutions, and by polymerizing acrylic esters for adsorbing polar solutes. They are regenerated by leaching with organic solvents.

Ion Exchangers

The first ion exchangers were naturally occurring inorganic aluminosilicates (zeolites), used in experiments in the 1850s to exchange between ammonium ions in fertilizers and calcium ions in soils. Industrial water softeners using zeolites were introduced about 1910. However, the zeolites were unstable in the presence of mineral acids. The instability problem was solved by Adams and Holmes [16] in 1935, when they synthesized the first organic polymer ion-exchange resins by the polycondensation of phenols and aldehydes. Depending upon the nature of the phenolic group, the resin contains either sulfonic ($-SO_3^-$) or amine ($-NH_3^+$) groups, used for the reversible exchange of cations or anions. Today, the most widely used ion exchangers are synthetic organic polymer resins based on styrene- or acrylic-acid-type monomers, as described by D'Alelio in U.S. Patent 2,366,007 (Dec. 26, 1944).

Ion-exchange resins are generally solid gels in spherical or granular form, which consist of (1) a three-dimensional polymeric network, (2) ionic functional groups attached to the network, (3) counterions, and (4) a solvent. Strong-acid, cation-exchange resins and strong-base, anion-exchange resins that are fully ionized over the entire pH range are based on

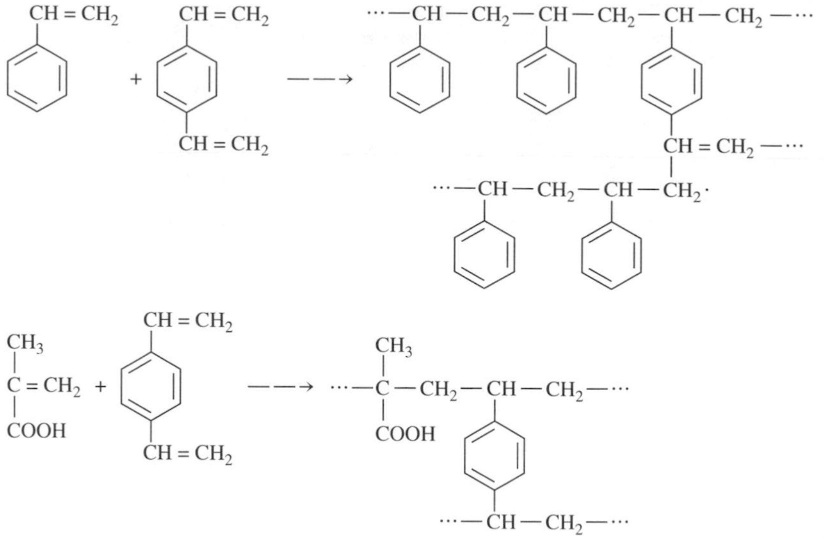

Figure 15.5 Ion-exchange resins: (a) Resin from styrene and divinylbenzene; (b) Resin from acrylic and methacrylic acid.

the copolymerization of styrene and a cross-linking agent, divinylbenzene, to produce the three-dimensional cross-linked structure as shown in Figure 15.5a. The degree of cross-linking is governed by the ratio of divinylbenzene to styrene. Weakly acid cation exchangers are sometimes based on the copolymerization of acrylic acid and methacrylic acid, as shown in Figure 15.5b. These two cross-linked copolymers swell in the presence of organic solvents and have no ion-exchange properties.

To convert the copolymers to water-swellable gels with ion-exchange properties, ionic functional groups are added to the polymeric network by reacting the copolymers with various chemicals. For example, if the styrene–divinylbenzene copolymer is sulfonated, as shown in Figure 15.6a, a cation-exchange resin, as shown in Figure 15.6b, is obtained with (-SO_3^-) groups permanently attached or fixed to the polymeric network to give a negatively charged matrix and exchangeable, mobile positive hydrogen ions (cations). The hydrogen ion can be exchanged on an equivalent basis with other cations, such as Na^+, Ca^{2+}, K^+, or Mg^{2+}, to maintain neutrality of the polymer. For example, two H^+ ions are exchanged for one Ca^{2+} ion. The exchangeable ions are called *counterions*. The liquid whose ions are being exchanged also contains other ions of unlike charge, such as Cl^- for a solution of NaCl, where Na^+ is exchanged. These other ions are called *co-ions*. Often the liquid treated is water, which dissolves to some extent in the resin and causes it to swell. Other solvents, such as methanol, are also soluble in the resin. If the styrene–divinylbenzene copolymer is chloromethylated and then aminated, a strong-base, anion-exchange resin is formed, as shown in Figure 15.6c, which can exchange Cl^- ions for other anions, such as OH^-, HCO_3^-, SO_4^{2-}, and NO_3^-.

Commercial ion exchangers in the hydrogen, sodium, and chloride form are available under the trade names of Amberlite, Duolite, Dowex, Ionac, and Purolite. Typically, they are in the form of spherical beads from about 40 μm to 1.2 mm in diameter. When saturated with water, the beads have typical moisture contents from 40 to 65 wt%. When water-swollen, they have a particle density of 1.1–1.5 g/cm³. When packed into a bed, they have bulk densities from 0.56 to 0.96 g/cm³ with fractional bed porosities of 0.35–0.40.

When water is demineralized by ion exchange, potential organic foulants must be removed from the water feed. As discussed by McWilliam [17], this can be accomplished by coagulation, clarification, prechlorination, and the use of ion-exchanger traps.

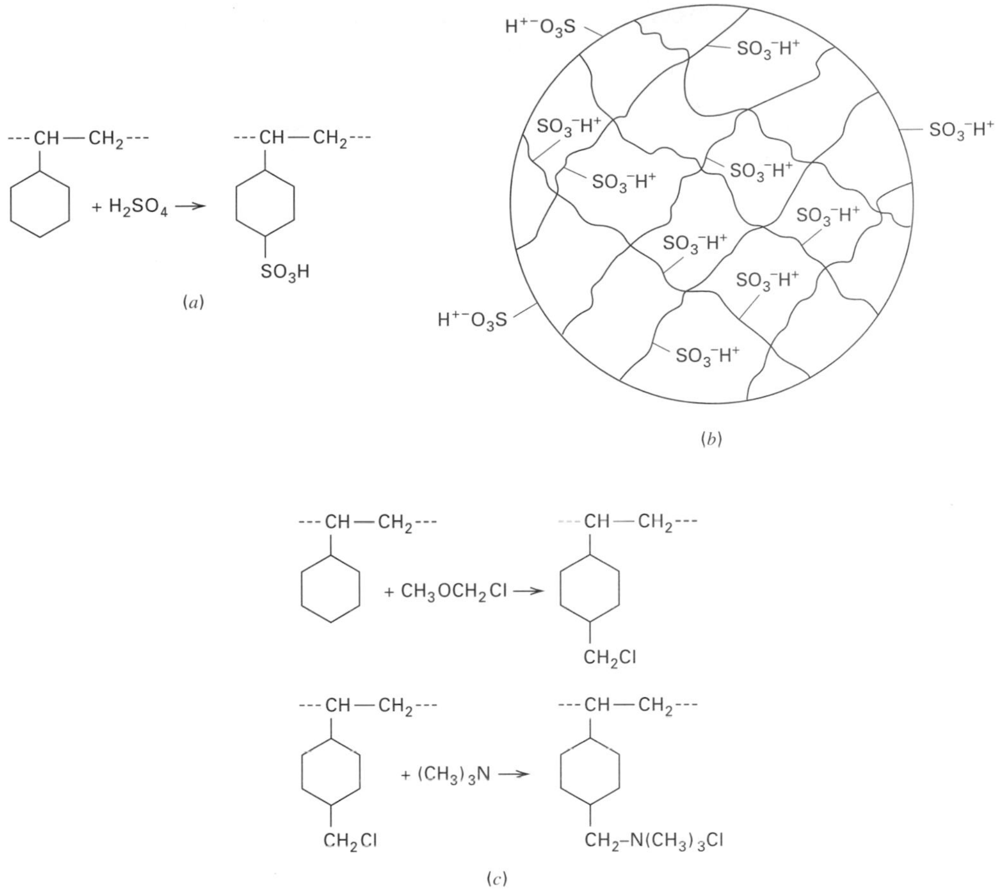

Figure 15.6 Introducing ionic functional groups into resins. (a) Sulfonation to a cation exchanger. (b) Fixed and mobile ions in a cation exchanger. (c) Chloromethylation and amination to an anion exchanger.

The maximum ion-exchange capacity of a strong-acid cation or strong-base anion exchanger is stoichiometric, based on the number of equivalents of mobile charge in the resin. Thus, 1 mol H^+ is one equivalent, whereas 1 mol Ca^{2+} is two equivalents. The exchanger capacity is usually quoted as eq/kg of dry resin or eq/L of wet resin. The wet capacity depends on the water content and degree of swelling of a given resin, whereas the dry capacity is fixed. For the copolymer of styrene and divinylbenzene, the maximum capacity is determined on the assumption that each benzene ring in the resin contains one sulfonic-acid group.

EXAMPLE 15.2

A commercial ion-exchange resin is made from 88 wt% styrene and 12 wt% divinylbenzene. Estimate the maximum ion-exchange capacity in eq/kg resin (same as meq/g resin).

SOLUTION

Basis: 100 g of resin before sulfonation.

	M	g	gmol
Styrene	104.14	88	0.845
Divinylbenzene	130.18	12	0.092
		100	0.937

Therefore, sulfonation at one location on each benzene ring requires 0.937 mol of H_2SO_4 to attach the sulfonic acid group ($M = 81.07$) and split out one water molecule. This is 0.937 equivalents with the addition in weight of $0.937(81.07) = 76$ g. Total dry weight of sulfonated resin = $100 + 76 = 176$ g maximum ion-exchange capacity, or

$$\frac{0.937}{(176/1,000)} = 5.3 \text{ eq/kg(dry)}$$

Depending on the extent of cross-linking, resins from copolymers of styrene and divinylbenzene are listed as having actual capacities of from 3.9 (high degree of cross-linking) to 5.5 (low degree of cross-linking). Although a low degree of cross-linking favors dry capacity, almost every other ion-exchanger property, including wet capacity and selectivity, is improved by cross-linking, as discussed by Dorfner [18]. ■

Sorbents for Chromatography

Sorbents (called *stationary phases*) for chromatographic separations come in a wide variety of forms and chemical compositions because of the many ways in which chromatography is applied. Figure 15.7 shows a classification of analytical chromatographic systems, taken from Sewell and Clarke [19]. The mixture to be separated, after injection into the carrier fluid to form the *mobile phase,* may be a liquid (*liquid chromatography*) or a gas (*gas chromatography*). Often, the mixture is initially a liquid, but is vaporized without decomposition by the carrier gas, giving a gas mixture for the mobile phase. Gas carriers are inert and do not interact with the sorbent or components of the feed. Liquid carriers (solvents) can interact and must be selected carefully.

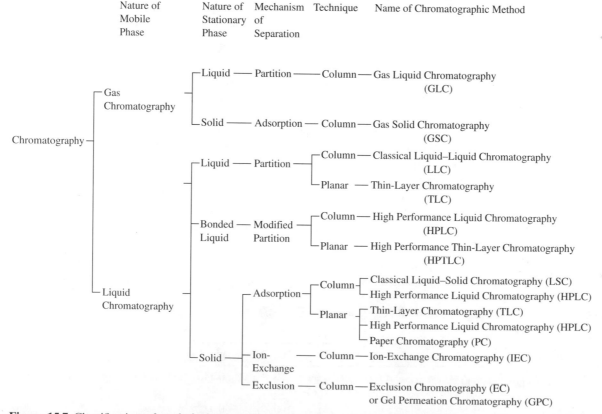

Figure 15.7 Classification of analytical chromatographic systems. [From P.A. Sewell and B. Clarke, *Chromatographic Separations,* John Wiley and Sons, New York (1987), with permission.]

The stationary sorbent phase is a solid, a liquid supported on or bonded to a solid, or a gel. With a porous solid adsorbent, the mechanism or mode of separation is adsorption. If an ion-exchange mechanism is desired, a synthetic polymer resin ion exchanger is used. With a polymer gel or a microporous solid, a separation based on sieving, called *exclusion,* can be applied. Unique to chromatography are the liquid-supported or liquid-bonded solids, where the mechanism is absorption into the liquid, also referred to as a *partition mode* of separation or *partition chromatography*. With mobile liquid phases, there is a tendency for the stationary liquid phase to be stripped or dissolved. Therefore, methods of chemically bonding the stationary liquid phase to the solid bonding support have been developed.

All sorbents can be used in columns. In packed columns >1 mm inside diameter, the sorbents are in the form of particles. In capillary columns <0.5 mm inside diameter, the sorbent is the inside wall or is a coating on the wall. If the inside wall of the capillary is not coated, the capillary column is referred to as a wall-coated open tubular (WCOT) column. If the coating is a layer of fine particulate support material to which a liquid absorbent is added, the column is called a support-coated open tubular (SCOT) column. If the wall is coated with a porous adsorbent only, the column is referred to as a porous-layer open tubular (PLOT) column.

Each type of sorbent can be applied to sheets of glass, plastic, or aluminum for use in *thin-layer* (or planar) *chromatography* or to a sheet of cellulose material for use in *paper chromatography*. If a pump, rather than gravity, is used to pass a liquid mobile phase through a packed column, the name *high-performance liquid chromatography* (HPLC) is used.

The two most common adsorbents used in chromatography are porous alumina and porous silica gel. Of lesser importance are carbon, magnesium oxide, and various carbonates. Alumina is a polar adsorbent and is preferred for the separation of components that are weakly or moderately polar, with the more polar compounds retained more selectively by the adsorbent and, therefore, eluted from the column last. In addition, alumina is a basic adsorbent, preferentially retaining acidic compounds. Silica gel is less polar than alumina and is an acidic adsorbent, preferentially retaining basic compounds, such as amines. Carbon is a nonpolar (apolar) stationary phase with the highest attraction for larger nonpolar molecules.

Adsorbent-type sorbents are better suited for the separation of a mixture on the basis of chemical type (e.g., olefins, esters, acids, aldehydes, alcohols) than for separation of individual members of a homologous series. For the latter, partition chromatography is preferred, wherein an inert solid support, often silica gel, is coated with a liquid phase. For application to gas chromatography, the liquid must be nonvolatile. For liquid chromatography, the stationary liquid phase must be insoluble in the mobile phase. Since this is difficult to achieve, the stationary liquid phase is usually bonded to the solid support. An example of a bonded phase is the result of reacting silica with a chlorosilane. Both monofunctional and bifunctional silanes are used, as shown in Figure 15.8, where R is a

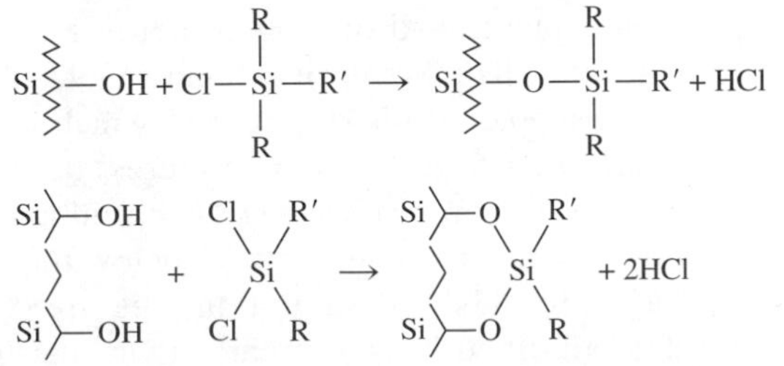

Figure 15.8 Bonded phases from the reaction of surface silanol groups with (a) Monofunctional and (b) Bifunctional chlorosilanes.

methyl (CH_3) group and R' is a hydrocarbon chain (C_6, C_8, or C_{18}) where the terminal CH_3 group is replaced with a polar group, such as -CN or -NH_2. If the resulting stationary phase is more polar than the mobile phase, the technique is referred to as *normal-phase chromatography*. Otherwise, the name *reverse-phase chromatography* is used.

In liquid chromatography, the order of elution from the column of the solutes in the mobile phase can also be influenced by the solvent carrier of the mobile phase by matching the solvent polarity with the solutes and using more polar adsorbents for less polar solutes and less polar adsorbents for more polar solutes.

EXAMPLE 15.3

For the separation of each of the following mixtures, select an appropriate mode of chromatography from Figure 15.7: (a) gas mixture of O_2, CO, CO_2, and SO_2, (b) vaporized mixture of anthracene, phenanthrene, pyrene, and chrysene, and (c) aqueous solution containing Ca^{2+} and Ba^{2+}.

SOLUTION

(a) Use gas–solid chromatography, that is, with a gas mobile phase and a solid adsorbent stationary phase.
(b) Use partition or gas–liquid chromatography, that is, with a gas mobile phase and a bonded liquid coating on a solid for the stationary phase.
(c) Use ion-exchange chromatography, that is, with a liquid as the mobile phase and polymer resin beads as the stationary phase. ∎

15.2 EQUILIBRIUM CONSIDERATIONS

In adsorption, a dynamic phase equilibrium is established for the distribution of the solute between the fluid and the solid surface. This equilibrium is usually expressed in terms of (1) concentration (if the fluid is a liquid) or partial pressure (if the fluid is a gas) of the adsorbate in the fluid and (2) solute *loading* on the adsorbent, expressed as mass, moles, or volume of adsorbate per unit mass or per unit BET surface area of the adsorbent. Unlike vapor–liquid and liquid–liquid equilibria, where theory is often applied to estimate phase distributions, particularly in the form of K-values for the former type of equilibrium, no acceptable theory has been developed to estimate fluid–solid adsorption equilibria. Thus, it is necessary to obtain experimental equilibrium data for a particular solute, or mixture of solutes and/or solvent, and a sample of the actual solid adsorbent material of interest. If the data are taken over a range of fluid concentrations at a constant temperature, a plot of solute loading on the adsorbent versus concentration or partial pressure in the fluid, called an adsorption isotherm, is made. This equilibrium isotherm places a limit on the extent to which a solute is adsorbed from a given fluid mixture on an adsorbent of given chemical composition and geometry for a given set of conditions. The rate at which the solute is adsorbed is also an important consideration and is discussed in the next section.

Pure Gas Adsorption

For pure gases, experimental physical adsorption isotherms have shapes, that are classified into five types by Brunauer et al. [20], as shown in Figure 15.9 and discussed in considerable detail by Brunauer [21]. The simplest isotherm is Type I, which corresponds to unimolecular adsorption. This type applies often to gases at temperatures above their critical temperature. The more complex Type II isotherm is associated with multimolecular adsorption of the BET type and is observed for gases at temperatures below their critical temperature and for pressures below, but approaching the saturation pressure (vapor pressure). The heat of adsorption for the first adsorbed layer is greater than that for the succeeding layers, each of which is assumed to have a heat of adsorption equal to the heat of condensation (vaporization). Both Types I and II are desirable isotherms, exhibiting strong adsorption.

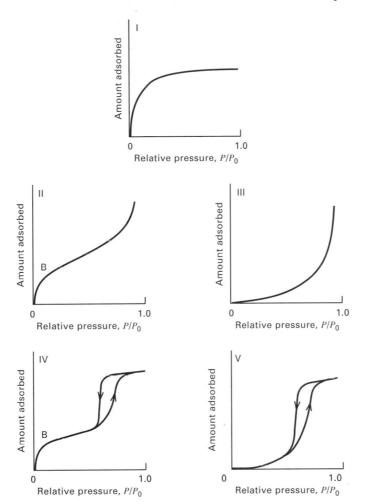

Figure 15.9 Brunauer's five types of adsorption isotherms.

The Type III isotherm in Figure 15.9, with its convex nature, is undesirable because the extent of adsorption is low except at high pressures. According to the BET theory, it corresponds to multimolecular adsorption where the heat of adsorption of the first layer is less than that of succeeding layers. Fortunately, this type of isotherm is rarely observed, an example being the adsorption of iodine vapor on silica gel. In the limit, as the heat of adsorption of the first layer approaches zero, adsorption is delayed until the saturation pressure is approached.

The derivation of the BET equation (15-6) assumes that an infinite number of molecular layers can be adsorbed. Thus, the equation precludes the possibility of capillary condensation. In a development by Brunauer et al. [20], subsequent to the BET equation, the number of layers is restricted by pore size, and capillary condensation is assumed to occur at a reduced vapor pressure in accordance with the Kelvin equation (15-14). The resulting equation is quite complex, but predicts adsorption isotherms of Types IV and V in Figure 15.9, where we see that the maximum extent of adsorption occurs before the saturation pressure is reached. Type IV is the capillary condensation version of Type II; Type V is the capillary condensation version of Type III.

As shown in Figure 15.9, a hysteresis phenomenon can occur in multimolecular adsorption regions for isotherms of types IV and V. The upward adsorption branch of the hysteresis

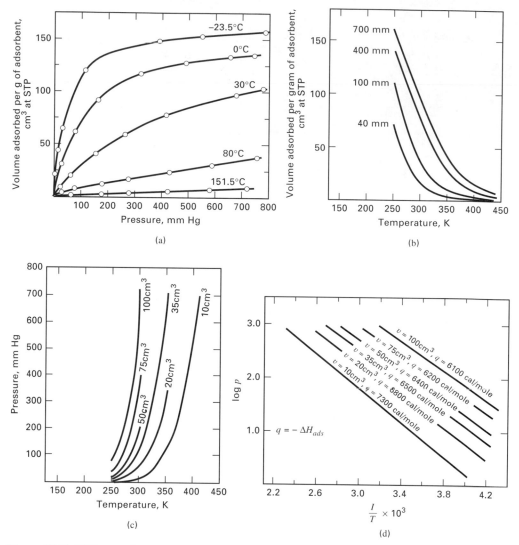

Figure 15.10 Different displays of adsorption equilibrium data for NH_3 on charcoal. (a) Adsorption isotherms. (b) Adsorption isobars. (c) Adsorption isosteres. (d) Isosteric heats of adsorption. [From S. Brunauer, *The Adsorption of Gases and Vapors,* Vol. I, Princeton University Press (1943), with permission.]

loop is due to simultaneous multimolecular adsorption and capillary condensation. Only capillary condensation occurs during the downward desorption branch of the loop. Hysteresis can also occur throughout any isotherm when strongly adsorbed impurities are present. Thus, measurements of pure gas adsorption require adsorbents with clean pore surfaces, normally achieved by preevacuation.

Physical adsorption data of Titoff [22] for ammonia gas on charcoal, as discussed by Brunauer [21], are shown in Figure 15.10. The five adsorption isotherms of Figure 15.10a cover pressures from vacuum to almost 800 mmHg and temperatures from −23.5 to 151.5°C. For ammonia, the normal boiling point is −33.3°C and the critical temperature is 132.4°C. For the lowest-temperature isotherm, up to 160 cm^3 (STP) of ammonia per gram of charcoal is adsorbed, which is equivalent to 0.12 g NH_3/g charcoal. All five isotherms are of Type I. When the amount adsorbed is low (<25 cm^3/g), the isotherms are almost linear and the following form of Henry's law, called the *linear isotherm,* is obeyed:

$$q = kp \qquad (15\text{-}16)$$

where q is equilibrium loading or amount adsorbed/unit mass of adsorbent (specific amount adsorbed), k is an empirical, temperature-dependent constant, and p is the partial pressure of the gas. As the temperature increases, the amount adsorbed decreases because of Le Chatelier's principle for an exothermic process. This is shown more clearly for the same data in the crossplot of *adsorption isobars* in Figure 15.10b, where absolute temperature is employed. A third method of displaying the experimental data is in the form of *adsorption isosteres,* also obtained by crossplotting, as shown in Figure 15.10c. These curves, representing constant amounts adsorbed, resemble vapor pressure plots, for which the adsorption form of the Clausius–Clapeyron equation,

$$\frac{d \ln p}{dT} = \frac{-\Delta H_{\text{ads}}}{RT^2} \qquad (15\text{-}17)$$

or
$$\frac{d \log p}{d(1/T)} = \frac{-\Delta H_{\text{ads}}}{2.303RT}$$

is applied to determine the heat of adsorption; which is negative because the effect is exothermic. The result is shown in Figure 15.10d, where it is seen that $-\Delta H_{\text{ads}}$ is initially 7,300 cal/mol, but decreases as the amount adsorbed increases, reaching 6,100 cal/mol at 100 cm^3/g. These values can be compared to the heat of vaporization of NH_3, which at 30°C is 4,600 cal/mol.

Experimental adsorption isotherm data for 18 different pure gases and a variety of solid adsorbents are summarized and analyzed by Valenzuela and Myers [23]. The data show that adsorption isotherms for a given pure gas at a fixed temperature vary considerably with the adsorbent. This is shown for pure propane vapor in the narrow temperature range of 25–30°C in Figure 15.11 for pressures up to about 101.3 kPa. The highest specific adsorption is with Columbia G-grade activated carbon, while the lowest is with Norton Z-900H, a zeolite molecular sieve. Columbia G-grade activated carbon has about twice the adsorbate capacity of Cabot Black Pearls activated carbon.

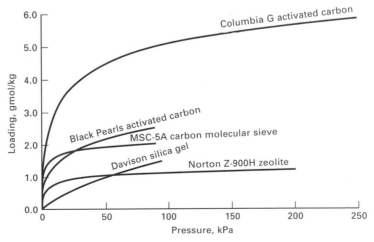

Figure 15.11 Adsorption isotherms for pure propane vapor at 298–303 K.

Table 15.4 Comparison of Equilibrium Adsorption of Pure Gases on 20–40 mesh Columbia L Activated Carbon Particles ($S_g = 1,152$ m^2/g) at 38°C and ~1 atm

Pure gas	q, mol/kg	T_b, °F	T_c, °F
H_2	0.0241	−423.0	−399.8
N_2	0.292	−320.4	−232.4
CO	0.374	−313.6	−220.0
CH_4	0.870	−258.7	−116.6
CO_2	1.64	−109.3	87.9
C_2H_2	2.67	−119	95.3
C_2H_4	2.88	−154.6	48.6
C_2H_6	3.41	−127.5	90.1
C_3H_6	4.54	−53.9	196.9
C_3H_8	4.34	−43.7	216.0

The literature data compiled by Valenzuela and Myers [23] also show that for a given adsorbent, the loading depends strongly on the gas. This is illustrated in Table 15.4 for a temperature of 38°C and a narrow pressure range of 97.9 to 100 kPa from the data of Ray and Box [24] for Columbia L activated carbon. Included in the table are normal boiling points and critical temperatures. As might be expected, the species are adsorbed in approximately the inverse order of volatility.

The correlation of experimental adsorption isotherms for pure gases is the subject of a number of published articles and books. As summarized by Yang [25], approaches have ranged from empirical to theoretical. For practical applications, the classical equations of Freundlich and Langmuir are still dominant because of their simplicity and ability to correlate isotherms of Type I in Figure 15.9.

Freundlich Isotherm

The equation attributed to Freundlich [26], but which was actually devised earlier by Boedecker and van Bemmelen, according to Mantell [27], is empirical and nonlinear in pressure:

$$q = kp^{1/n} \tag{15-19}$$

where k and n are temperature-dependent constants. Generally, n lies in the range of 1 to 5. With $n = 1$, (15-19) reduces to the Henry's law equation (15-16). Experimental q–p isothermal data can be fitted to (15-19) by a nonlinear curve-fitting computer program or by converting (15-19) to a linear form as follows, and using a graphical method or a linear regression program:

$$\log q = \log k + (1/n) \log p \tag{15-20}$$

If the graphical method is employed, the data are plotted as $\log q$ versus $\log p$. The best straight line through the data has a slope of $(1/n)$ and an intercept of $\log k$. In general, k decreases with increasing temperature, while n increases with increasing temperature and approaches a value of 1 at high temperatures. Equation (15-19) is derived by assuming a heterogeneous surface with a nonuniform distribution of the heat of adsorption over the surface, as discussed by Brunauer [21].

Langmuir Isotherm

The Langmuir equation [28], which is restricted to Type I isotherms, is derived from simple mass-action kinetics, assuming chemisorption. Assume that the surface of the pores of the adsorbent is homogeneous (ΔH_{ads} = constant) and that the forces of interaction between adsorbed molecules are negligible. Let θ be the fraction of the surface covered by adsorbed molecules. Therefore, $(1 - \theta)$ is the fraction of the bare surface. Then, the net rate of adsorption is the difference between the rates of adsorption and desorption:

$$dq/dt = k_a p(1 - \theta) - k_d \theta \qquad (15\text{-}21)$$

At equilibrium, $dq/dt = 0$ and (15-21) reduces to

$$\theta = \frac{Kp}{1 + Kp} \qquad (15\text{-}22)$$

where K is the adsorption equilibrium constant ($= k_a/k_d$). Here,

$$\theta = q/q_m \qquad (15\text{-}23)$$

where q_m is the maximum loading corresponding to complete coverage of the surface by the gas. Thus, the Langmuir adsorption isotherm is restricted to a monomolecular layer. Combining (15-23) with (15-22), we obtain

$$q = \frac{K q_m p}{1 + Kp} \qquad (15\text{-}24)$$

At low pressures, if $Kp \ll 1$, (15-24) reduces to the linear Henry's law form (15-16), while at high pressures where $Kp \gg 1$, $q = q_m$. At intermediate pressures, (15-24) is nonlinear in pressure. Although originally devised by Langmuir for chemisorption, (15-24) has been widely applied to physical adsorption data.

The quantities K and q_m in (15-24) are treated as empirical constants, obtained by fitting the nonlinear equation directly to experimental data or by employing the following linearized form, numerically or graphically:

$$\frac{p}{q} = \frac{1}{q_m K} + \frac{p}{q_m} \qquad (15\text{-}25)$$

Using (15-25), the best straight line is drawn through a plot of points p/q versus p, giving a slope of $(1/q_m)$ and an intercept of $1/(q_m K)$. If the theory is reasonable, K should change rapidly with temperature, but q_m should not because it is related through v_m by (15-7) to the specific surface area of the adsorbent, S_g. It should be noted that the Langmuir isotherm predicts an asymptotic limit for q at high pressure, whereas the Freundlich isotherm does not.

Other Adsorption Isotherms

Valenzuela and Myers [23] fit pure gas adsorption-isotherm data to the more complex three-parameter isotherms of (1) Toth:

$$q = \frac{mp}{(b + p^t)^{1/t}} \qquad (15\text{-}26)$$

where m, b, and t are constants for a given adsorbate–adsorbent system and temperature, and (2) Honig and Reyerson (called the UNILAN equation):

$$q = \frac{n}{2s} \ln \left[\frac{c + pe^s}{c + pe^{-s}} \right] \qquad (15\text{-}27)$$

where n, s, and c are constants for a given adsorbate–adsorbent system and temperature. The Toth and UNILAN isotherms reduce to the Langmuir isotherm for $t = 1$ and $s = 0$, respectively.

EXAMPLE 15.4

The following experimental data for the equilibrium adsorption of pure methane gas on activated carbon (PCB from Calgon Corp.) at 296 K were obtained by Ritter and Yang [*Ind. Eng. Chem. Res.*, **26**, 1679–1686 (1987)]:

q, cm³ (STP) of CH₄/g carbon	45.5	91.5	113	121	125	126	126
$P = p$, psia	40	165	350	545	760	910	970

Fit the data to: (a) the Freundlich isotherm, and (b) the Langmuir isotherm. Which isotherm provides a better fit to the data?

SOLUTION

By using the linearized forms of the isotherm equations, a spreadsheet or other computer program can be used to do a linear regression to obtain the constants.

(a) Using (15-20), we obtain $\log k = 1.213$, $k = 16.34$, $1/n = 0.3101$, and $n = 3.225$.

Thus, the Freundlich equation is $q = 16.34\, p^{0.3101}$.

(b) Using (15-25), we obtain $1/q_m = 0.007301$, $q_m = 137.0$, $1/(q_m K) = 0.5682$, and $K = 0.01285$.

Thus, the Langmuir equation is $q = \dfrac{1.760\,p}{1 + 0.01285\,p}$

The predicted values of q from the two isotherms are as follows:

	q, cm³ (STP) of CH₄/g carbon		
p, psia	Experimental	Freundlich	Langmuir
40	45.5	51.3	46.5
165	91.5	79.6	93.1
350	113	101	112
545	121	115	120
760	125	128	124
910	126	135	126
970	126	138	127

For this example, the Langmuir isotherm fits the data significantly better than the Freundlich isotherm. Average percent deviations, in q, are computed to be 1.01% and 8.64%, respectively. One reason for the better fit of the Langmuir isotherm is the trend of the data to an asymptotic value for q at the highest pressures. ∎

Gas Mixtures and Extended Isotherms

Commercial applications of physical adsorption involve mixtures rather than pure gases. If the adsorption of all components in the gas except one (A) is negligible, then the adsorption of A is estimated from its pure gas adsorption isotherm using the partial pressure of A. If the adsorption of two or more components in the mixture is significant, the situation is quite complicated. Experimental data show that one component can increase, decrease, or have no influence on the adsorption of the other, depending on interactions of adsorbed molecules. A simple theoretical treatment is the extension of the Langmuir equation by

Markham and Benton [29], who neglect interactions and assume that the only effect is the reduction of the vacant surface area for the adsorption of A because of the adsorption of other components.

Consider a binary gas mixture of A and B. Let θ_A = fraction of the surface covered by A and θ_B = fraction of the surface covered by B. Then, $(1 - \theta_A - \theta_B)$ = fraction of vacant surface. At equilibrium:

$$(k_A)_a p_A (1 - \theta_A - \theta_B) = (k_A)_d \theta_A \tag{15-28}$$

$$(k_B)_a p_B (1 - \theta_A - \theta_B) = (k_B)_d \theta_A \tag{15-29}$$

Solving these equations simultaneously, and combining the results with (15-23) for each component, gives

$$q_A = \frac{(q_A)_m K_A p_A}{1 + K_A p_A + K_B p_B} \tag{15-30}$$

$$q_B = \frac{(q_B)_m K_B p_B}{1 + K_A p_A + K_B p_B} \tag{15-31}$$

where $(q_i)_m$ is the maximum amount of adsorption of species i for coverage of the entire surface. Equations (15-30) and (15-31) are readily extended to a multicomponent mixture:

$$q_i = \frac{(q_i)_m K_i p_i}{1 + \sum_j K_j p_j} \tag{15-32}$$

In a similar fashion, as shown by Yon and Turnock [30], the Freundlich equation can be combined with the Langmuir equation to give the following relation for gas mixtures:

$$q_i = \frac{(q_i)_0 k_i p_i^{1/n_i}}{1 + \sum_j k_j p_j^{1/n_i}} \tag{15-33}$$

where $(q_i)_0$ is the maximum loading, which may differ from $(q_i)_m$ for a monolayer. Equation (15-33) represents data for nonpolar multicomponent mixtures in molecular sieves reasonably well. Unfortunately, Broughton [31] has shown that the extended Langmuir equation lacks thermodynamic consistency; such is also the case for the extended Langmuir–Freundlich equation. Nevertheless, for practical application, their simplicity makes them the isotherms of choice.

As with multicomponent (three or more components) vapor–liquid and liquid–liquid phase equilibria, experimental data for binary and multicomponent gas–solid adsorbent equilibrium are scarce and less accurate than corresponding pure gas data. Valenzuela and Myers [23] include experimental data on adsorption of gas mixtures from nine published studies on 29 binary systems, for which pure gas adsorption isotherms were also obtained. They also describe procedures for applying the Toth and UNILAN equations to multicomponent mixtures based on the ideal-adsorbed-solution (IAS) theory of Myers and Prausnitz [32]. Unlike the extended Langmuir equation (15-32), which is explicit in the amount adsorbed, the IAS theory, though more accurate, is not explicit and requires an iterative solution procedure. Additional experimental data for higher-order (ternary and/or higher) gas mixtures are given by Miller, Knaebel, and Ikels [33] for 5A molecular sieves and by Ritter and Yang [34] for activated carbon. Yang [25] presents a discussion of existing theories on adsorption of gas mixtures, together with comparisons of these theories with mixture data for activated carbon and zeolites. The data on zeolites are the most difficult to correlate, with the simplified statistical thermodynamic model (SSTM) of Ruthven and Wong [35] giving the best results.

EXAMPLE 15.5	The experimental work of Ritter and Yang, cited in Example 15.4, also includes adsorption isotherms for pure CO and CH_4, and a binary mixture of $CH_4(A)$ and CO(B). For the pure gases, Ritter and Yang give relations over a temperature range of 296–480 K, for the two Langmuir constants. At 294 K, these constants are as follows:	

	q_m, cm^3(STP)/g	K, psi^{-1}
CH_4	133.4	0.01370
CO	126.1	0.00624

With these constants, use the extended Langmuir equation to predict the specific adsorption volumes (STP) of CH_4 and CO for a vapor mixture of 69.6 mol% CH_4 and 30.4 mol% CO at 294 K and a total pressure of 364.3 psia. Compare the results with the following experimental data of Ritter and Yang:

Total volume adsorbed, cm^3/STP)/g	114.1
Mole fractions in adsorbate:	
CH_4	0.867
CO	0.133

SOLUTION

$$p_A = y_A P = 0.696(364.3) = 253.5 \text{ psia}$$
$$p_B = y_B P = 0.304(364.3) = 110.8 \text{ psia}$$

From (15-30):

$$q_A = \frac{133.4(0.0137)(253.5)}{1 + (0.0137(253.5) + (0.00624)(110.8)} = 89.7 \text{ cm}^3(\text{STP})/\text{g}$$

$$q_B = \frac{126.1(0.00624)(110.8)}{1 + (0.0137(253.5) + (0.00624)(110.8)} = 16.9 \text{ cm}^3(\text{STP})/\text{g}$$

The total amount adsorbed = $q = q_A + q_B = 89.7 + 16.9 = 106.6$ cm^3/(STP)/g, which is 6.6% lower than the experimental value. Estimated mole fractions in the adsorbate are $x_A = q_A/q = 89.7/106.6 = 0.841$ and $x_B = 1 - 0.841 = 0.159$. These adsorbate mole fractions deviate from the experimental values by a mole fraction of 0.026. For this example, the extended Langmuir isotherm gives reasonable results. ∎

Liquid Adsorption

When porous adsorbent particles are immersed in a pure gas, the pores fill with the gas, and the amount of adsorbed gas is determined by the decrease in total pressure. With a liquid, the pressure does not change, and no simple experimental procedure has been devised for determining the extent of adsorption of a pure liquid. If the liquid is a homogeneous binary mixture, it is customary to designate one component the solute (1) and the other the solvent (2). The assumption is then made that the change in composition of the bulk liquid in contact with the porous solid is due entirely to adsorption of the solute. That is, adsorption of the solvent is tacitly assumed not to occur. If the liquid mixture is dilute in the solute, the consequences are not serious. If, however, experimental data are obtained over the entire concentration range, the distinction between solute and solvent is arbitrary and the resulting adsorption isotherms, as discussed by Kipling [36], can exhibit curious shapes that are unlike those obtained for pure gases or gas mixtures.

Let:

n^0 = total moles of binary liquid brought into contact with adsorbent

m = mass of adsorbent

x_1^0 = mole fraction of solute in the mixture before contact with adsorbent

x_1 = mole fraction of solute in the bulk solution after adsorption equilibrium is achieved

q_1^e = apparent moles of solute adsorbed per unit mass of adsorbent

A solute material balance, assuming no adsorption of solvent and a negligible change in the total moles of liquid mixture, gives

$$q_1^e = \frac{n^0(x_1^0 - x_1)}{m} \tag{15-34}$$

If data are obtained at constant temperature over the entire concentration range and then processed with (15-34) and plotted as adsorption isotherms, the resulting curves are not of the type shown in Figure 15.12a. Instead, curves of the type shown in Figure 15.12b and c are obtained, where negative adsorption appears to occur in Figure 15.12c. Such isotherms are probably best referred to as *composite isotherms* or *isotherms of concentration change,* as suggested by Kipling [36]. Likewise, the adsorption loading, q_1^e, of (15-34) is more correctly referred to as the *surface excess.*

Under what conditions are composite isotherms of the form shown in Figs. 15.12b and c obtained? This is shown by several examples in Figure 15.13, where various combinations of hypothetical adsorption isotherms for solute (A) and solvent (B) are shown together with the resulting composite isotherms. Thus, when the solvent is not adsorbed, as seen in Figure 15.13a, a composite curve without negative adsorption is obtained. In all other cases of Figure 15.13, negative values of the surface excess are obtained.

Valenzuela and Myers [23] tabulate literature values for the equilibrium adsorption of 25 different binary liquid mixtures. With one exception, all 25 mixtures give composite isotherms of the forms shown in Figure 15.12b and c. The one exception is a mixture of cyclohexane and n-heptane with silica gel, for which the surface excess is almost negligible (0 ± 0.05 mmol/g) over the composition range of $x_1 = 0.041$ to 0.911. They also include literature references to 354 sets of binary-mixture data, 25 sets of ternary-mixture data, and 3 sets of data for higher-order mixtures.

When data for the binary mixture are only available in the dilute region, the amount of adsorption, if any, of the solvent may be constant and all changes in the total amount adsorbed are due to just the solute. In that case, the adsorption isotherms are of the form of Figure 15.12a, which resembles the form obtained with pure gases. It is then common

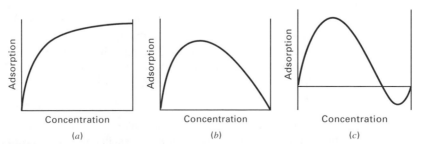

Figure 15.12 Representative isotherms of concentration change for liquid adsorption.

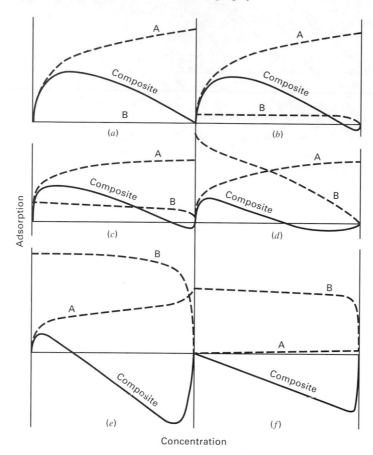

Figure 15.13 Origin of various types of composite isotherms for binary liquid adsorption. [From J.J. Kipling, *Adsorption from Solutions of Non-electrolytes,* Academic Press, London (1965), with permission.]

to fit the data with concentration forms of the Freundlich equation (15-19) or the Langmuir equation (15-24):

$$q = kc^{1/n} \tag{15-35}$$

$$q = \frac{Kq_m c}{1 + Kc} \tag{15-36}$$

Candidate systems for this case are small amounts of organic compounds dissolved in water and small amounts of water dissolved in hydrocarbons. For liquid mixtures that are dilute in two or more solutes, the multicomponent adsorption may be estimated from a concentration form of the extended Langmuir equation (15-32) based on the constants, q_m and K, obtained from experiments on the single solutes. However, when solute–solute interactions are suspected, it may be necessary to determine the constants from multicomponent data.

EXAMPLE 15.6

Small amounts of VOCs in water can be removed by adsorption. Generally, two or more VOCs are present. An aqueous stream continuing small amounts of acetone (1) and propionitrile (2) is to be treated with activated carbon. Single-solute equilibrium data available from Radke and

Prausnitz [37] have been fitted to the Freundlich and Langmuir isotherms, (15-35) and (15-36), with the average deviations indicated, for solute concentrations up to 50 mmol/L:

Acetone in Water (25°C):		Absolute Average Deviation of q, %
$q_1 = 0.141c_1^{0.597}$	(1)	14.2
$q_1 = \dfrac{0.190c_1}{1 + 0.146c_1}$	(2)	27.3
Propionitrile in water (25°C):		
$q_2 = 0.138c_2^{0.658}$	(3)	10.2
$q_2 = \dfrac{0.173c_2}{1 + 0.0961c_2}$	(4)	26.2

where:

$\quad q_i$ = amount of solute adsorbed, mmol/g
$\quad c_i$ = solute concentration in aqueous solution, mmol/L

Use these single-solute results with an extended Langmuir-type isotherm to predict the equilibrium adsorption in a binary-solute aqueous system containing 40 and 34.4 mmol/L, respectively, of acetone and propionitrile at 25C with the same adsorbent. Compare the results with the following experimental values from Radke and Prausnitz [37]:

$$q_1 = 0.715 \text{ mmol/g}, \qquad q_2 = 0.822 \text{ mmol/g}, \qquad \text{and } q_{\text{total}} = 1.537 \text{ mmol/g}$$

SOLUTION

From (15-32), the extended Langmuir isotherm for the liquid phase is

$$q_i = \frac{(q_i)_m K_i c_i}{1 + \sum_j K_j c_j} \tag{5}$$

From (2), $(q_1)_m = 0.190/0.146 = 1.301$ mmol/g.
From (4), $(q_2)_m = 0.173/0.0961 = 1.800$ mmol/g.
From (5):

$$q_1 = \frac{1.301(0.146)(40)}{1 + (0.146)(40) + (0.0961)(34.4)} = 0.749 \text{ mmol/g}$$

$$q_2 = \frac{1.800(0.0961)(34.4)}{1 + (0.146)(40) + (0.0961)(34.4)} = 0.587 \text{mmol/g}$$

$$q_{\text{total}} = 1.336 \text{ mmol/g}$$

Compared to experimental data, the percent deviations for q_1, q_2, and q_{total}, respectively, are 4.8%, -28.6%, and -13.1%. Better agreement is obtained by Radke and Prausnitz using an IAS theory. It is expected that a concentration form of (15-33) would also give better agreement, but that requires that the single-solute data be refitted for each solute to a Langmuir–Freundlich isotherm of the form

$$q = \frac{q_0 k c^{1/n}}{1 + k c^{1/n}} \tag{6}$$

■

Ion Exchange Equilibria

Ion exchange differs from adsorption in that one sorbate (a counterion) is exchanged for a solute ion, and the exchange is governed by a reversible, stoichiometric chemical-reaction equation. Thus, the selectivity of the ion exchanger for one counterion over another may be just as important as the capacity of the ion exchanger. Accordingly, for ion exchange, we apply the law of mass action to obtain an equilibrium ratio rather than fit data to a sorption isotherm such as the Langmuir or Freundlich equation.

As discussed by Anderson [38], two cases are important. In the first case, the counterion initially in the ion exchanger is exchanged with a counterion from an acid or base. For example:

$$Na^+_{(aq)} + OH^-_{(aq)} + HR_{(s)} \leftrightarrow NaR_{(s)} + H_2O_{(l)}$$

Note that the hydrogen ion leaving the ion exchanger immediately reacts with the hydroxide ion in the aqueous solution to form water, leaving no counterion on the right-hand side of the reaction. Accordingly, the ion exchange will continue until the aqueous solution is depleted of sodium ions or the ion exchanger is depleted of hydrogen ions.

In the second case, which is more common than the first, the counterion being transferred from the ion exchanger to the fluid phase remains as an ion. For example, the exchange of counterions A and B is expressed by the reaction

$$A^{n\pm}_{(l)} + nBR_{(s)} \leftrightarrow AR_{n_{(s)}} + nB^+_{(l)} \qquad (15\text{-}37)$$

where A and B must both be either cations (positive charge) or anions (negative charge). For this case, at equilibrium, we can define a conventional chemical equilibrium constant according to the law of mass action:

$$K_{A,B} = \frac{q_{AR_n} c^n_{B^\pm}}{q^n_{BR} c_{A^{n\pm}}} \qquad (15\text{-}38)$$

where molar concentrations c_i, and q_i refer to the liquid and ion exchanger phases, respectively. The constant, $K_{A,B}$ is not a rigorous thermodynamic equilibrium constant because (15-38) is written in terms of concentrations instead of activities. Although (15-38) could be corrected by including activity coefficients, it is usually applied in the form shown, with $K_{A,B}$ referred to as a *molar selectivity coefficient* for A entering the ion exchange resin and displacing B. For the resin phase, concentrations are in equivalents per unit mass or unit bed volume of ion exchanger. For the liquid solution, concentrations are in equivalents per unit volume of solution. For dilute liquid solutions, $K_{A,B}$ is reasonably constant for a given pair of counterions and a particular resin of a given degree of cross-linking.

When exchange is between two counterions of equal charge, (15-38) can be reduced to a simple equation in terms of just the equilibrium concentrations of A in the liquid solution and in the ion exchange resin. Because of (15-37), the total concentrations C and Q in equivalents of counterions in the liquid solution and the resin, respectively, remain constant during the exchange process. Accordingly:

$$c_i = C x_i / z_i \qquad (15\text{-}39)$$

$$q_i = Q y_i / z_i \qquad (15\text{-}40)$$

where x_i and y_i are equivalent fractions, rather than mole fractions, of A and B, such that

$$x_A + x_B = 1 \qquad (15\text{-}41)$$

$$y_A + y_B = 1 \qquad (15\text{-}42)$$

and z_i = valence of counterion i. Combining (15-38) to (15-42) results in

$$K_{A,B} = \frac{y_A(1 - x_A)}{x_A(1 - y_A)} \tag{15-43}$$

Thus, at equilibrium, x_A and y_A are independent of the total equivalent concentrations C and Q. Such is not the case when the two counterions are of unequal charge, as in the exchange of Ca^{2+} and Na^+. A derivation for this general case gives

$$K_{A,B} = \left(\frac{C}{Q}\right)^{n-1} \frac{y_A(1 - x_A)^n}{x_A(1 - y_A)^n} \tag{15-44}$$

For unequal counterion charges, we see that $K_{A,B}$ depends on the ratio C/Q and on the ratio of charges, n.

When experimental data for $K_{A,B}$ for a particular binary system of counterions with a particular ion exchanger are not available, the method of Bonner and Smith [39], as modified by Anderson [38], is used for screening purposes or preliminary calculations. In this method, the molar selectivity coefficient is

$$K_{ij} = K_i/K_j \tag{15-45}$$

where values for relative molar selectivities K_i and K_j are given in Table 15.5 for cations with an 8% cross-linked strong-acid resin and in Table 15.6 for anions with strong-base resins. For values of K in these tables, the units of C and Q are, respectively, eq/L of solution and eq/L of bulk bed volume of water-swelled resin.

A typical cation-exchange resin of the sulfonated styrene–divinylbenzene type, such as Dowex 50, as described by Bauman and Eichhorn [40] and Bauman, Skidmore, and Osmun [41], has an exchangeable ion capacity of 5 ± 0.1 mcq/g of dry resin. As shipped, the water-wet resin might contain 41.4 wt% water. Thus, the wet capacity is $5(58.6/100) = 2.9$ meq/g of wet resin. If the bulk density of a drained bed of wet resin is 0.83, the bed capacity is 2.4 eq/L of resin bed.

As with other separation processes, a separation factor, $S_{A,B}$, which ignores the valence of the exchanging ions, can be defined for an equilibrium stage. For the binary case, in terms of equivalent ionic fractions:

$$S_{A,B} = \frac{y_A(1 - x_A)}{x_A(1 - y_A)} \tag{15-46}$$

Table 15.5 Relative Molar Selectivities for Cations with 8% Cross-linked Strong-Acid Resin

Li^+	1.0	Zn^{2+}	3.5
H^+	1.3	Co^{2+}	3.7
Na^+	2.0	Cu^{2+}	3.8
NH_4^+	2.6	Cd^{2+}	3.9
K^+	2.9	Be^{2+}	4.0
Rb^+	3.2	Mn^{2+}	4.1
Cs^+	3.3	Ni^+	3.9
Ag^+	8.5	Ca^{2+}	5.2
UO_2^{2+}	2.5	Sr^{2+}	6.5
Mg^{2+}	3.3	Pb^{2+}	9.9
		Ba^{2+}	11.5

Table 15.6 Approximate Relative
Molar Selectivities for Anions with
Strong-Base Resins

I⁻	8	OH⁻ (Type II)	0.65
NO_3^-	4	HCO_3^-	0.4
Br^-	3	CH_3COO^-	0.2
HSO_4^-	1.6	F^-	0.1
NO_2^-	1.3	OH⁻ (Type I)	0.05–0.07
CN^-	1.3	SO_4^{2-}	0.15
Cl^-	1.0	CO_3^{2-}	0.03
BrO_3^-	1.0	HPO_4^{2-}	0.01

which is identical to (15-43). Experimental data for an exchange between Cu^{2+} (A) and Na^+ (B) (counterions of unequal charge) with Dowex 50 cation resin over a wide range of total solution normality at ambient temperature are shown in terms of y_A and x_A in Figure 15.14, from Subba Rao and David [42]. At low total solution concentration, the resin is highly selective for copper ion, whereas at high total solution concentration, the selectivity is reversed to favor sodium ion slightly. A similar trend was observed by Selke and Bliss [43,44] for exchange between Ca^{2+} and H^+ using a similar resin, Amberlite IR-120. This sensitivity of the selectivity is shown dramatically in Figure 15.15, from Myers and Byington [45], where the natural logarithm of the separation factor, S_{Cu^{2+},Na^+}, as computed from the data of Figure 15.14 with (15-46), is plotted as a function of equivalent ionic fraction, $x_{Cu^{2+}}$. For dilute solutions of Cu^{2+}, S_{Cu^{2+},Na^+} ranges from about 0.5 at a total concentration of 4 N to 60 at 0.01 N. In terms of K_{Cu^{2+},Na^+} of (15-44), with $n = 2$, the corresponding variation is computed to be only from about 0.6 to 2.2.

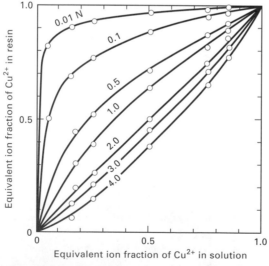

Figure 15.14 Isotherms for ion exchange of Cu^{2+} and Na^+ on Dowex 50-X8 as a function of total normality in the bulk solution. [From A.L. Myers and S. Byington, *Ion Exchange Science and Technology*, M. Nijhoff, Boston (1986), with permission.]

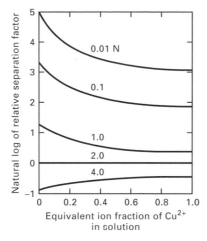

Figure 15.15 Relative separation factor of Cu^{2+} and Na^+ for ion exchange on Dowex 50-X8 as a function of total normality in the bulk solution. [From A.L. Myers and S. Byington, *Ion Exchange Science and Technology*, M. Nijhoff, Boston (1986), with permission.]

EXAMPLE 15.6

An Amberlite IR-120 ion-exchange resin similar to that of Example 5.2, but with a maximum ion-exchange capacity of 4.90 meq/g of dry resin, is used to remove cupric ion from a waste stream containing 0.00975 M $CuSO_4$ (19.5 meq Cu^{2+}/L solution). The spherical resin particles range in diameter from 0.2 to just over 1.2 mm. The equilibrium ion-exchange reaction is of the divalent–monovalent type:

$$Cu^{2+}_{(aq)} + 2HR_{(c)} \leftrightarrow CuR_{2_{(s)}} + 2H^+_{(aq)}$$

As ion exchange takes place, the milliequivalents of cations in the aqueous solution and in the resin remain constant.

Experimental measurements by Selke and Bliss [43,44] show an equilibrium curve of the type of Figure 15.14 at ambient temperature that is markedly dependent on the total equivalent concentration of the aqueous solution, with the following equilibrium data for the cupric ion with a 19.5 meq/liter solution:

c, meq Cu^{2+}/L Solution	0.022	0.786	4.49	10.3
q, meq Cu^{2+}/g Resin	0.66	3.26	4.55	4.65

These data follow a highly nonlinear isotherm.

(a) From the data, compute the molar selectivity coefficient, K, at each value of c for Cu^{2+} and compare it to the value estimated from (15-45) using Table 15.5.
(b) Predict the milliequivalents of Cu^{2+} exchanged at equilibrium from 10 L of 20 meq Cu^{2+}/L using 50 g of dry resin with 4.9 meq of H^+/g.

SOLUTION

(a) Selke and Bliss do not give a value for the resin capacity, Q, in eq/L of bed volume. Assume a value of 2.3. From (15-44):

$$K_{Cu^{2+},H^+} = \left(\frac{C}{Q}\right)\frac{y_{Cu^{2+}}(1 - x_{Cu^{2+}})^2}{x_{Cu^{2+}}(1 - y_{Cu^{2+}})^2}$$

where $(C/Q) = 0.0195/2.3 = 0.0085$

$$x_{Cu^{2+}} = c_{Cu^{2+}}/19.5 \text{ and } y_{Cu^{2+}} = q_{Cu^{2+}}/4.9$$

Using the above values of c and q from Selke and Bliss:

q, meq Cu^{2+}/g	$x_{Cu^{2+}}$	$y_{Cu^{2+}}$	K_{Cu^{2+},H^+}
0.66	0.00113	0.135	1.35
3.26	0.0403	0.665	1.15
4.55	0.230	0.929	4.04
4.65	0.528	0.949	1.30

The average value of K is 2.0. The values in Table 15.5 when substituted into (15-45) predict

$$K_{Cu^{2+},H^+} = 3.8/1.3 = 2.9$$

which is somewhat higher.

(b) Assume a value of 2.0 for K_{Cu^{2+},H^+} with $Q = 2.3$ eq/L. The total solution concentration, C, is 0.02 eq/L. Equation (15-44) becomes

$$2.0 = \left(\frac{0.02}{2.3}\right) \frac{y_{Cu^{2+}}(1 - x_{Cu^{2+}})^2}{x_{Cu^{2+}}(1 - y_{Cu^{2+}})^2} \tag{1}$$

Initially, the solution contains $(0.02)(10) = 0.2$ equivalents of cupric ion with $x_{Cu^{2+}} = 1.0$. Let a = equivalents of Cu exchanged. Then, at equilibrium, by material balance:

$$x_{Cu^{2+}} = \frac{0.02 - (a/10)}{0.02} \tag{2}$$

$$y_{Cu^{2+}} = \frac{(a/50)}{0.0049} \tag{3}$$

Substitution of (2) and (3) into (1) gives

$$2.0 = 0.0087 \frac{\left[\frac{(a/50)}{0.0049}\right]\left[1 - \frac{0.02 - (a/10)}{0.02}\right]^2}{\left[\frac{0.02 - (a/10)}{0.02}\right]\left[1 - \frac{(a/50)}{0.0049}\right]^2} \tag{4}$$

Solving (4), a nonlinear equation, for a gives 0.1887 equivalents of Cu exchanged. Thus, $0.1887/[(0.020)(10)] = 0.944$ or 94.4% of the cupric ion is exchanged. ∎

Equilibria in Chromatography

As discussed in Section 15.1, separation by chromatography involves sorption mechanisms of many types, including adsorption on porous solids, absorption or extraction (partitioning) in liquid-supported or bonded solids, and ion exchange in synthetic resins. Thus, at equilibrium, depending upon the sorption mechanism, equations such as (15-19), (15-24), (15-32), and (15-33) for gas adsorption; (15-35) and (15-36) for liquid adsorption; (6-17) to (6-20) for gas absorption; (8-1) for liquid extraction; and (15-38) (15-43), and (15-44) for ion exchange apply.

When the equilibrium (distribution or partition) constant is defined as

$$K_i = q_i/c_i \tag{15-47}$$

where q is concentration in the stationary phase and c is concentration in the mobile phase, solutes with the highest equilibrium constants will elute from the chromatographic column at a slower rate than solutes with the smallest equilibrium constants.

15.3   KINETIC AND TRANSPORT CONSIDERATIONS

For the adsorption of a solute onto the porous surface of an adsorbent, the following steps are required:

1. External (interphase) mass transfer of the solute from the bulk fluid by convection, through a thin film or boundary layer, to the outer solid surface of the adsorbent
2. Internal (intraphase) mass transfer of the solute by pore diffusion from the outer surface of the adsorbent to the inner surface of the internal porous structure
3. Surface diffusion along the porous surface
4. Adsorption of the solute onto the porous surface

For chemisorption, which involves bond formation, the rate of the fourth kinetic step may be slow and even controlling; for physical adsorption, however, step 4 is almost instantaneous because it depends only on the collision frequency and orientation of the molecules with the porous surface. Thus, only the first three steps need be considered here.

During regeneration of the adsorbent, the reverse of the four steps occurs, where the rate of physical desorption is instantaneous. Adsorption and desorption are accompanied by heat transfer because of the exothermic heat of adsorption and the endothermic heat of desorption. However, although external mass transfer is limited to a convective mechanism, external heat transfer from the particle outer surface occurs not only by convection through the film or boundary layer surrounding each solid particle in the bed, but also by thermal radiation between particles when the fluid is a gas, and by conduction at points of contact by adjacent particles. Conduction and radiation mechanisms for heat transfer can also exist within the particle, in addition to convective heat transfer by the fluid within the pores.

In a fixed bed of adsorbent particles, solute concentration and temperature change continuously with time and location. For a given particle at a particular time, profiles of temperature and solute concentration in the fluid are as shown in Figure 15.16a and b for

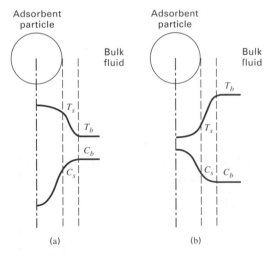

Figure 15.16 Solute concentration and temperature profiles for a porous adsorbent particle surrounded by a fluid: (a) Adsorption. (b) Desorption.

adsorption and desorption, respectively, where subscripts b and s refer to bulk fluid and particle outer surface, respectively. The fluid concentration gradient is usually steepest within the particle, whereas the temperature gradient is usually steepest in the fluid film or boundary layer surrounding the particle. Thus, although the major resistance to heat transfer is usually external to the adsorbent particle, the major resistance to mass transfer usually resides in the adsorbent particle. All four gradients in Figure 15.16 approach asymptotic values at the end points.

External Transport

Rates of convective mass and heat transfer between the outer surface of a particle and the surrounding bulk fluid during an adsorption process are given, respectively, by

$$n_i = \frac{d\mathcal{N}_i}{dt} = k_c A(c_{b_i} - c_{s_i}) \tag{15-48}$$

$$q = \frac{dQ}{dt} = hA(T_s - T_b) \tag{15-49}$$

For a spherical particle surrounded by an infinite, quiescent fluid, the mass- and heat-transfer coefficients are at their minimum values. Assume an insoluble solid spherical particle of radius R_p and diameter $D_p = 2R_p$, suspended in an infinite fluid medium. The particle is heated so that, at steady state, its surface temperature is constant at T_s. The fluid medium is absolutely quiescent (no free convection) and radiation is ignored so that heat transfer through the fluid is by conduction only. The thermal conductivity k of the fluid is constant, and the temperature far from the particle is T_b. Fourier's second law of heat conduction in the fluid, for spherical coordinates, is

$$\frac{d}{dr}\left(kr^2\frac{dT}{dr}\right) = 0 \tag{15-50}$$

for $r \geq R_p$, where r is the radial distance from the center of the particle. The boundary conditions are

$$T\{r = R_p\} = T_s \tag{15-51}$$
$$T\{r = \infty\} = T_b \tag{15-52}$$

If (15-50) is integrated twice with respect to r, we obtain:

$$T = -\frac{C_1}{r} + C_2 \tag{15-53}$$

Substitution of the boundary conditions, (15-51) and (15-52), results in an expression for the temperature profile in the fluid:

$$\frac{T - T_b}{T_s - T_b} = \frac{R_p}{r}, r \geq R_p \tag{15-54}$$

The heat flux at the outer surface of the particle is given by Fourier's first law of heat conduction applied to the fluid adjacent to the particle:

$$\left.\frac{q}{A}\right|_{r=R_p} = -k\left.\frac{dT}{dr}\right|_{r=R_p} \tag{15-55}$$

From (15-54):

$$\left.\frac{dT}{dr}\right|_{r=R_p} = -\frac{(T_s - T_b)}{R_p} \tag{15-56}$$

We can also apply Newton's law of cooling for the heat flux:

$$\left.\frac{q}{A}\right|_{r=R_p} = h(T_s - T_b) \tag{15-57}$$

Combining (15-55) to (15-57):

$$h = k/R_p \tag{15-58}$$

which rearranges into a Nusselt number form:

$$N_{\text{Nu}} = hD_p/k = 2 \tag{15-59}$$

A similar development for convective mass transfer using Fick's laws of diffusion gives

$$N_{\text{Sh}_i} = k_c D_p/D_i = 2 \tag{15-60}$$

where D_i is the diffusivity of component i in the mixture.

When the fluid flows past a single particle, convection increases the convective mass- and heat-transfer coefficients above the values computed from (15-59) and (15-60). Furthermore, the transport coefficients now vary around the periphery of the particle, with the largest value occurring where the fluid flow first impinges on the particle. Correlations of experimental transport data are usually developed for coefficients averaged over the surface of the particle. Typical correlations are those of Ranz and Marshall [46,47] for Nusselt numbers as high as 30 and Sherwood numbers to 160:

$$N_{\text{Nu}} = 2 + 0.60\, N_{\text{Re}}^{1/2} N_{\text{Pr}}^{1/3} \tag{15-61}$$

$$N_{\text{Sh}_i} = 2 + 0.60\, N_{\text{Sc}_i}^{1/3} N_{\text{Re}}^{1/2} \tag{15-62}$$

where:

N_{Pr} = Prandtl number – $C_P\mu/k$
N_{Sc_i} = Schmidt number = $\mu/\rho D_i$
N_{Re} = Reynolds number = $D_p G/\mu$

All fluid properties are evaluated at the average temperature of the film or boundary layer. Equations (15-61) and (15-62) reduce to (15-59) and (15-60), respectively, when the fluid mass velocity, G, is zero.

When particles are packed in a bed, the fluid flow patterns are restricted, and the single-particle correlations of (15-61) and (15-62) cannot be used to estimate the average external transport coefficients for the particles in the bed. However, Ranz [48] showed that equations of the same form as (15-61) and (15-62) correlate external transport data for beds packed with spherical particles. Nevertheless, most early investigators, starting with Gamson, Thodos, and Hougen [49], developed correlations in the form of the Chilton and Colburn [50] *j*-factors:

$$j_D = (N_{\text{St}_M})(N_{\text{Sc}})^{2/3} = f\{N_{\text{Re}}\} \tag{15-63}$$

$$j_H = (N_{\text{St}})(N_{\text{Pr}})^{2/3} = f\{N_{\text{Re}}\} \tag{15-64}$$

with $N_{\text{St}_M} = k_c\rho/G$ and $N_{\text{St}} = h/C_P G$

where different Reynolds-number functions apply to different regions. Various forms of the Reynolds number have been used, including $D_p G/\mu$ and $D_p G/\epsilon_b\mu$, in attempts to account for bed void fraction, ϵ_b, where G is the superficial mass velocity based on the empty-bed cross-sectional area, and G/ϵ_b (a larger value) is the effective mass velocity

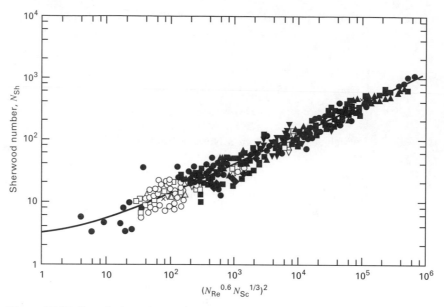

Figure 15.17 Correlation of experimental data for Sherwood number of external mass transfer in a packed bed. [From N. Wakao and T. Funazkri, *Chem. Eng. Sci.*, **33**, 1375 (1978), with permission.]

through the bed. Notable among correlations of this type are those of Sen Gupta and Thodos [51], Petrovic and Thodos [52], and Dwivedi and Upadhay [53].

A more recent study by Wakao and Funazkri [54] reanalyzed 37 sets of previously published mass-transfer data, with Sherwood number corrections for axial dispersion. The resulting correlation, which represents a return to the form of (15-62), is

$$N_{Sh_i} = \frac{k_c D_p}{D_i} = 2 + 1.1 \left(\frac{D_p G}{\mu}\right)^{0.6} \left(\frac{\mu}{\rho D_t}\right)^{1/3} \tag{15-65}$$

The correlation is compared to 12 sets of gas-phase and 11 sets of liquid-phase data in Figure 15.17. The data cover a Schmidt number range of 0.6 to 70,600, a Reynolds number range of 3 to 10,000, and a particle diameter from 0.6 to 17.1 mm. Particle shapes include spheres, short cylinders, flakes, and granules. By analogy, the corresponding equation for fluid-particle convective heat transfer in packed beds is

$$N_{Nu} = \frac{h D_p}{k} = 2 + 1.1 \left(\frac{D_p G}{\mu}\right)^{0.6} \left(\frac{C_P \mu}{k}\right)^{1/3} \tag{15-66}$$

When (15-65) and (15-66) are used with beds packed with nonspherical particles, D_p is the equivalent diameter of a spherical particle. The following suggestions have been proposed for computing the equivalent diameter from geometric properties of the particle. These suggestions may be compared by considering a short cylinder with diameter, D, equal to the length, L.

1. D_p = diameter of a sphere with the same external surface area:

$$\pi D_p^2 = \pi D L + \pi D^2/2 \text{ and } D_p = (DL + D^2/2)^{0.5} = 1.225D$$

2. D_p = diameter of a sphere with the same volume:

$$\pi D_p^3/6 = \pi D^2 L/4 \text{ and } D_p = (3D^2 L/2)^{1/3} = 1.145D$$

3. D_p = 4 times the hydraulic radius, r_H, where for a packed bed,

$$4r_H = 6/a_v$$

a_v = external particle surface area/volume of particle

Thus,

$$a_v = \frac{\pi DL + \pi D^2/2}{\pi D^2 L/4} = \frac{6}{D} \text{ and } D_p = 4r_H = \frac{6D}{6} = 1.0D$$

The use of the hydraulic radius concept is equivalent to replacing D_p in the Reynolds number by $\Psi D_p'$, where Ψ is the sphericity and D_p' is given by Suggestion 2: $D_p' = $ (6/particle volume/6)$^{1/3}$.

Thus, for a cylinder of $D = L$,

$$\Psi = \frac{1.0D}{\left[\frac{6}{\pi}\left(\frac{\pi D^3}{4}\right)\right]^{1/3}} = 0.874$$

This is equivalent to defining the sphericity as

$$\Psi = \frac{\text{Surface area of a sphere of same volume as particle}}{\text{Surface area of particle}}$$

Suggestions 2 and 3 are widely used. Suggestion 3 is conveniently applied to crushed particles of irregular surface, but with no obvious longer or shorter dimension, that is, isotropic in shape. In that case, D_p' is taken as the size of the particle and the sphericity is approximately 0.65, as discussed by Kunii and Levenspiel [55].

EXAMPLE 15.7

Acetone vapor in a nitrogen stream is removed by adsorption in a fixed bed of activated carbon. At a location in the bed where the pressure is 136 kPa, the bulk gas temperature is 297 K, and the bulk mole fraction of acetone is 0.05, estimate the external gas-to-particle mass-transfer coefficient for acetone and the external particle-to-gas heat-transfer coefficient. Additional data are as follows:

Average particle diameter = 0.0040 m and Gas molar velocity = 0.00352 kmol/m²-s

SOLUTION

Because the temperature and composition are known only for the bulk gas and not at the particle external surface, use gas properties at bulk gas conditions. From the ChemCAD process simulation program, relevant properties for use in (15-65) and (15-66) are as follows:

Viscosity $\mu = 0.0000165$ Pa-s (kg/m-s); Density $\rho = 1.627$ kg/m³
Thermal conductivity $k = 0.0240$ W/m-K $= 0.024 \times 10^{-3}$ kJ/m-K-s
Heat capacity at constant pressure = 31.45 kJ/kmol-K
Molecular weight = $M = 29.52$
Thus, specific heat $C_P = 31.45/29.52 = 1.065$ kJ/kg-K

Other parameters are:

Gas mass velocity $G = 0.00352 (29.52) = 0.1039$ kg/m²-s
Assume a sphericity, ψ, of 0.65; therefore, $D_p = 0.65 (0.004) = 0.0026$ m

The diffusivity, D_i, of acetone in nitrogen at 297 K and 136 kPa is independent of the composition and is approximately 0.085×10^{-4} m²/s.

$$N_{Re} = D_p G/\mu = 0.0026 (0.1039)/(0.0000165) = 16.4$$

$$N_{Sc} = \mu/\rho D_i = 0.0000165/[(1.627)(0.0000085)] = 1.19$$

$$N_{Pr} = C_P \mu/k = (1.065)(0.0000165)/(0.000024) = 0.73$$

From (15-65): $N_{Sh} = 2 + 1.1 \ (16.4)^{0.6} \ (1.19)^{1/3} = 8.24$

which from Figure 15.17 is well within the data range of the correlation. Thus, the mass-transfer coefficient for acetone is

$$k_c = N_{Sh}(D_i/D_p) = 8.24(0.0000085/0.0026) = 0.027 \ \text{m/s} = 0.088 \ \text{ft/s}$$

From (15-66): $N_{Nu} = 2 + 1.1(16.4)^{0.6}(0.73)^{1/3} = 7.31$

$$h = N_{Nu}(k/D_p) = 7.31 \ (0.0240/0.0026) = 67.5 \ \text{W/m}^2\text{-K or } 11.9 \ \text{Btu/h-ft}^2\text{-}^\circ\text{F} \blacksquare$$

Internal Transport

Porous adsorbent particles have a sufficiently high effective thermal conductivity that temperature gradients within the particle are usually negligible. However, internal (intraphase) mass transfer in the particle must be considered. Mechanisms for mass transfer in the pores are those described for porous membranes in Section 14.3. However, in membranes, the transport is through the membrane. In sorption applications, transport is only into the interior of the particle during sorption and from the interior of the particle in desorption.

The mathematical model of internal transport in porous particles during adsorption or desorption is very similar to that for catalytic chemical reactions in porous catalyst pellets. The first pore model was that of Thiele [56], who considered a first-order irreversible reaction taking place isothermally on the surface of a single, straight, cylindrical pore, closed at one end. Thiele's treatment was extended to a porous spherical pellet by Wheeler [57], who utilized an effective diffusivity applied to the case of sorption.

Consider the porous spherical pellet in Figure 15.18, where the fluid concentration, c, refers to the solute. A material balance in moles or mass per unit time over the spherical-

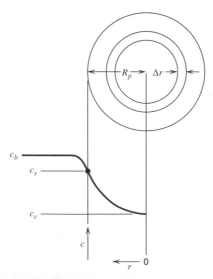

Figure 15.18 Typical solute concentration profile in adsorbent particle.

shell volume of thickness Δr involves diffusion of the solute into the shell at radius $r +$ Δr, adsorption within the shell, and diffusion out of the shell at radius r. Using Fick's first law:

$$4\pi(r + \Delta r)^2 D_e \left.\frac{\partial c}{\partial r}\right|_{r+\Delta r} = 4\pi r^2 \,\Delta r \frac{\partial q}{\partial t} + 4\pi r^2 \left.\frac{\partial c}{\partial r}\right|_r \tag{15-67}$$

Dividing by $4\pi\Delta r$, taking the limit as $\Delta r \to 0$, and collecting terms gives

$$D_e \left(\frac{\partial^2 c}{\partial r^2} + \frac{2}{r}\frac{\partial c}{\partial r}\right) = \frac{\partial q}{\partial t} \tag{15-68}$$

The variable q is the amount adsorbed per unit volume of porous pellet. The effective diffusivity, D_e, applies to the entire surface area of the spherical shell, even though only about 50% of it is available as pores for diffusion. For liquid–phase diffusion in the pores, the effective diffusivity is given by (14-14), which involves the volume fraction of pores in the pellet, the solute molecular diffusivity in the fluid within the pore, the pore tortuosity, and a possible restrictive factor for relatively large solute molecules. For gas-phase diffusion in the pores, the effective diffusivity is given by (14-18), which accounts for the possibility of Knudsen diffusion, with diffusivity D_K for very small pore diameters and/or low total pressures. Although (14-14) and (14-18) strictly apply only to equimolar counterdiffusion, they can be used as an approximation for unimolecular diffusion for fluids dilute in the solute molecules. A diffusion mechanism not accounted for directly in (14-18) is that of surface diffusion along the pore wall due to the concentration gradient of the adsorbate (adsorbed solute) along the wall.

Fick's first law for molecular diffusion through a fluid in a pore can be written

$$n_i = -D_i A(dc_i/dx) \tag{15-69}$$

where n is the molar rate of ordinary diffusion of i through the fluid in the x-direction, perpendicular to the cross-sectional area, A, for diffusivity, D_i, and concentration, c_i, in moles/unit volume of fluid. A modified Fick's first law applies to surface diffusion, as suggested by Schneider and Smith [58]. Thus,

$$(n_i)_s - -(D_i)_s b d(c_i)_s/dx \tag{15-70}$$

where b is the perimeter of the surface, $(c_i)_s$ is the surface concentration of adsorbate in moles/unit surface area, and $(D_i)_s$ is the surface diffusivity as defined by (15-70).

For convenience, (15-70) is converted, as follows, to the flux form of (15-69) so that the two mechanisms of diffusion can be combined in a single transport rate equation. The flux form of (15-69) is:

$$N_i = n_i/A = -D_i(dc_i/dx) \tag{15-71}$$

The corresponding flux form of (15-70) is obtained by dividing both sides by the cross-sectional area of the pore and converting the surface concentration, $(c_i)_s$, in moles/unit surface area to the concentration, q, in mol/g of adsorbent, by using the product of the pore surface/pore volume times the reciprocal of the adsorbent particle density times the particle porosity. The result is:

$$(N_i)_s = -(D_i)_s \frac{\rho_p}{\epsilon_p}\left(\frac{dq_i}{dx}\right) \tag{15-72}$$

Assuming linear adsorption according to Henry's law:

$$q_i - K_i c_i \tag{15-73}$$

Substituting (15-73) into (15-72) and adding the result to (15-71), the total flux is

$$N_i = -\left[D_i + (D_i)_s \frac{\rho_p K_i}{\epsilon_p} \right] \frac{dc_i}{dx} \tag{15-74}$$

In terms of the effective diffusity employed in (15-68):

$$D_e = \frac{\epsilon_p}{\tau} \left\{ \left[\frac{1}{(1/D_i) + (1/D_K)} \right] + (D_i)_s \frac{\rho_p K_i}{\epsilon_p} \right\} \tag{15-75}$$

Equation (15-75) needs to be used with caution, because, as discussed by Riekert [59], the tortuosity, τ, for pore volume diffusion is not necessarily the same as that for surface diffusion.

    Based on the study by Sladek, Gilliland, and Baddour [60], values of the surface diffusivity of light gases for physical adsorption are typically in the range of 5×10^{-3} to 10^{-6} cm^2/s with the larger values applying to cases of a low differential heat of adsorption. For nonpolar adsorbates, the surface diffusivity in cm^2/s may be estimated from the correlation [60],

$$D_s = 1.6 \times 10^{-2} \exp[-0.45(-\Delta H_{\text{ads}})/mRT] \tag{15-76}$$

where $m = 2$ for conducting adsorbents such as carbon and $m = 1$ for insulating adsorbents.

EXAMPLE 15.8

Porous silica gel of 1.0 mm particle diameter, with a particle density of 1.13 g/cm^3, a porosity of 0.486, an average pore radius of 11 Å, and a tortuosity of 3.35 is to be used to adsorb propane from helium. At 100°C, diffusion in the pores is controlled by Knudsen and surface diffusion. Estimate the effective diffusivity. The differential heat of adsorption is −5,900 cal/mol. At 100°C, the adsorption constant (for a linear isotherm) is 19 cm^3/g.

SOLUTION

From (14-21), the Knudsen diffusivity for propane is $D_K = 4,850(22 \times 10^{-8})(373/44.06)^{1/2} = 3.7 \times 10^{-3}$ cm^2/s. From (15-76), using $m = 1$, $D_s = 1.6 \times 10^{-2} \exp[(-0.45)(5,900)]/[(1)(1.987)(373)] = 4.45 \times 10^{-4}$ cm^2/s.

Equation (15-75) reduces to $D_e = \dfrac{\epsilon_p}{\tau} D_K + \dfrac{\rho_p K}{\tau} D_s = \dfrac{0.486}{3.35}(3.17 \times 10^{-3}) + \dfrac{(1.13)(19)}{3.35}(4.45 \times 10^{-4}) = 0.46 \times 10^{-3} + 2.85 \times 10^{-3} = 3.31 \times 10^{-3}$ cm^2/s

    Experiments by Schneider and Smith [58] give a value of 1.22×10^{-3} cm^2/s for D_e with a value of 0.88×10^{-3} for the contribution of surface diffusion. Thus, the estimated contribution from surface diffusion is high by a factor of about 3. In either case, the fractional contribution due to surface diffusion is large. A detailed review of surface diffusion is given by Kapoor, Yang, and Wong [61].   ∎

Mass Transfer in Ion Exchange and Chromatography

As discussed by Helfferich [62], two major mass-transfer resistances occur in ion exchange. The first is the external mass-transfer resistance due to the film or boundary layer surrounding the ion-exchange bead. The second is the internal diffusional resistance due to the resin bead. Either or both resistances can be rate-controlling; in either case, the diameter of the resin bead is an important factor. In general, external mass-transfer film diffusion is rate-controlling at very low exchange-ion concentrations, say below 0.01 N, whereas internal mass transfer (particle diffusion) controls at high concentrations (say above 1.0 N). It has also been observed that a large separation factor, as defined by (15-46), favors external mass-transfer control, and that divalent ions diffuse appreciably more slowly than monovalent ions through the resin bead. Usually, the rate-determining step is not the chemical reaction between the exchanging ions and the resin.

The external mass-transfer coefficient for flow of fluid through a fixed bed of ion-exchange resin beads is estimated from the same relation, (15-65), that is used for applications to adsorption in fixed beds. For internal mass transfer, it is customary to assume that the ion-exchange resin bead is a single quasi-homogeneous phase and that the diffusivity of the diffusing ion is constant at a given temperature. Under these conditions, (15-68) can be applied, where D_e is a diffusivity determined by experiment. In general, such diffusivities depend upon (1) the size and charge of the ion, with the smaller, monovalent ions diffusing the fastest; (2) the degree of cross-linking and resin swelling, with larger diffusivities favored by swelling and a small degree of cross-linking; and (3) temperature.

The most fundamental measurements of diffusivity in ion-exchange resins have been made with isotopes of the ions to obtain self-diffusion coefficients that are independent of ion concentration. Typical data are those of Soldano [63], shown in Figure 15.19 for Na^+, Zn^{2+}, and Y^{3+} in a sulfonated styrene–divinylbenzene cation exchanger at temperatures of 0.2 and 25°C. Recall that typical order-of-magnitude diffusivities for small molecules are as follows:

0.1 cm^2/s in the gas phase
1×10^{-5} cm^2/s in the liquid phase
1×10^{-7} cm^2/s in polymers

From Figure 15.19, it is seen that diffusivities depend strongly on the degree of cross-linking and the charge on the ion, with values much less than those found in liquids, especially for the divalent and trivalent ions, which have diffusivities even smaller than those observed for small molecules in polymers.

No new fundamental principles are required for formulating mass-transfer relations for chromatography. When packed beds are used, (15-65) and (15-66) are applied to determine external transport coefficients. If a coated flat plate or a tube with a coated inner wall is used, correlations of the type discussed in Chapter 3 are applicable. In some cases, an entry region of finite length exists, particularly for laminar flow, such that the transport coefficients vary with axial location, decrease in value with length, and eventually approach

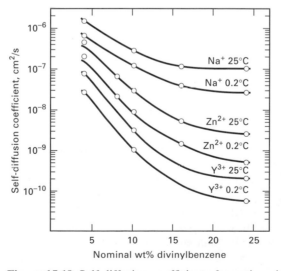

Figure 15.19 Self-diffusion coefficients for cations in a resin as a function of cross-linking with divinylbenzene. [From B.A. Soldano, *Ann. NY Acad. Sci.,* **57,** 116 (1953), with permission.]

an asymptotic value. For internal diffusion in the sorbent, Fick's second law is applied where the effective diffusivity depends on factors discussed earlier in this section.

15.4 SORPTION SYSTEMS

A variety of equipment configurations and operating procedures are employed for commercial sorption separation operations. This variety is due mainly to the wide range of sorbent particle sizes used and the need, in most applications, to regenerate the solid sorbent.

Adsorption

For adsorption, the most widely used equipment configurations and operating procedures are listed in Table 15.7. For analysis purposes, the listed devices may be classified into the three modes of operation, shown schematically in Figure 15.20. An agitated vessel, shown in Figure 15.20a, is used with a batch of liquid to which is added a powdered adsorbent such as activated carbon, of particle diameter typically less than 1 mm, to form a slurry. With good agitation and small particles, the external resistance to mass transfer from the bulk liquid to the external surface of the adsorbent particles is small. For small adsorbent particles, the internal resistance to mass transfer within the pores of the particles is also small. Accordingly, the rate of adsorption is rapid. The required residence time of the slurry in a well-mixed agitated vessel is determined by how fast equilibrium is approached. The main application of this mode of operation is the removal of very small amounts of dissolved, and relatively large molecules, such as coloring agents, from water. Generally, the spent adsorbent, which is removed from the slurry by sedimentation or filtration, is discarded because of the difficulty of desorbing large molecules. The slurry adsorption system, also called *contact filtration*, is also operated continuously.

The cyclic-batch operating mode using a fixed bed, shown schematically in Figure 15.20b, is widely used with both liquid and gas feeds. Adsorbent particle size ranges from 0.05 to 1.2 cm. Bed pressure drop decreases with increasing particle size, but the solute transport rate increases with decreasing particle size. The optimal particle size is determined mainly from these two factors. To avoid jiggling or fluidizing the bed during adsorption, the flow of the liquid or gas feed is often downward. For removal of small amounts of dissolved hydrocarbons from water, the spent adsorbent is removed from the vessel and reactivated thermally at high temperature or it is discarded. Applications of fixed-bed adsorption, also called *percolation,* include the removal of dissolved organic compounds from water. For

Table 15.7 Common Commercial Methods for Adsorption Separations

Phase Condition of Feed	Contacting Device	Adsorbent Regeneration Method	Main Application
Liquid	Slurry in an agitated vessel	Adsorbent discarded	Purification
Liquid	Fixed bed	Thermal reactivation	Purification
Liquid	Simulated moving bed	Displacement purge	Bulk separation
Gas	Fixed bed	Thermal swing (TSA)	Purification
Gas	Combined fluidized bed-moving bed	Thermal swing (TSA)	Purification
Gas	Fixed bed	Inert-purge swing	Purification
Gas	Fixed bed	Pressure swing (PSA)	Bulk separation
Gas	Fixed bed	Vacuum swing (VSA)	Bulk separation
Gas	Fixed bed	Displacement purge	Bulk separation

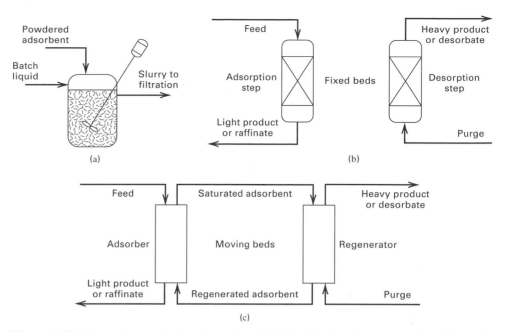

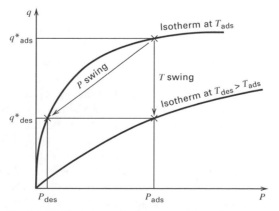

Figure 15.20 Contacting modes for adsorption. (a) Stirred-tank, slurry operation. (b) Cyclic fixed-bed, batch operation. (c) Continuous countercurrent operation.

purification or bulk separation of gases, the adsorbent is almost always regenerated in place by one of the four methods listed in Table 15.7.

In the *thermal (temperature)-swing adsorption* (TSA) method, the adsorbent is regenerated by desorption at a temperature higher than that used during the adsorption step of the cycle, as shown in Figure 15.21. The temperature of the bed is increased by (1) heat transfer from heating coils located in the bed followed by pulling a moderate vacuum or (2) more commonly, by heat transfer from an inert, nonadsorbing, hot purge gas, such as steam. Following desorption, the bed is cooled before the adsorption step of the cycle is resumed. Because heating and cooling of the bed requires hours, a typical cycle time for TSA is hours to days. Therefore, if the quantity of adsorbent in the bed is to be reasonable, TSA is practical only for purification involving small rates of adsorption. Instead of using

Figure 15.21 Schematic representation of pressure-swing and thermal-swing adsorption.

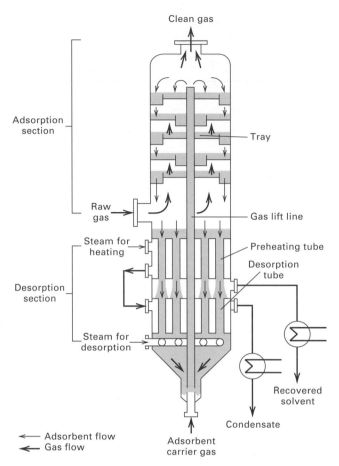

Clean gas

Adsorption section

Tray

Raw gas

Gas lift line

Steam for heating

Preheating tube

Desorption tube

Desorption section

Steam for desorption

Recovered solvent

Condensate

⟵ Adsorbent flow
⟵ Gas flow

Adsorbent carrier gas

Figure 15.22 Purasiv process with a fluidized bed for adsorption and moving bed for desorption. [From G.E. Keller, "Separations: New Directions for an Old Field," *AIChE Monograph Series,* **83** (17) (1987), with permission.]

a fixed bed, a fluidized bed can be used for adsorption and a moving bed for desorption, as shown in Figure 15.22, provided that the adsorbent particles are attrition-resistant. In the adsorption section, sieve trays are used with the raw gas passing up through the perforations and fluidizing the adsorbent particles. The fluidized solids flow like a liquid across the tray, into the downcomer, and onto the tray below. From the adsorption section, the solids pass to the desorption section, where, as moving beds, they first flow down through preheating tubes and then through desorption tubes. Steam is used for indirect heating in both sets of tubes and for stripping in the desorption tubes. Moving beds, rather than fluidized beds on trays, are used in the desorption section because the stripping-steam flow rate is insufficient for fluidizing the solids. At the bottom of the unit, the regenerated solids are picked up by a carrier gas, which flows up through a gas-lift line to the top, where the solids settle out onto the top tray to repeat the adsorption part of the cycle. According to Keller [64], this configuration, which was announced in 1977, is used in more than 50 units worldwide to remove small amounts of solvents from air. Other applications of TSA include the removal of moisture, CO_2, and pollutants from gas streams.

In the *inert-purge swing* method of regeneration, desorption is at the same temperature and pressure as the adsorption step, because the gas used for purging is nonadsorbing (inert) or only weakly adsorbing. This method is used only when the solute is weakly adsorbed, easily desorbed, and of little or no value. The purge gas must be inexpensive so that it does not have to be purified before recycle.

In the *pressure-swing adsorption* (PSA) cycle, adsorption takes place at an elevated pressure, whereas desorption occurs at near-ambient pressure, as is shown in Figure 15.21. PSA is used for bulk separations because the bed can be depressurized and repressurized rapidly, making it possible to operate at cycle times of seconds to minutes. Because of these short times, the beds need not be large even when a substantial fraction of the feed gas is adsorbed. If adsorption takes place at near-ambient pressure and desorption under vacuum, the cycle is referred to as *vacuum-swing adsorption* (VSA). PSA and VSA are widely used for the bulk separation of air. If a zeolite adsorbent is used, equilibrium is rapidly established and nitrogen is preferentially adsorbed. The nonadsorbed, high-pressure product gas is a mixture of oxygen and argon with a small amount of nitrogen. If a carbon molecular-sieve adsorbent is used, the particle diffusivity of oxygen is observed to be about 25 times that of nitrogen. As a result, the selectivity of adsorption is controlled by mass transfer, and oxygen is preferentially adsorbed. The resulting high-pressure product gas is nearly pure nitrogen. In both cases, the adsorbed gas, which is desorbed at low pressure, is quite impure. For the separation of air, large plants use VSA because it is more energy-efficient than PSA. Small plants often use PSA because that cycle is simpler.

In the *displacement-purge (displacement desorption) cycle,* a strongly adsorbed purge gas is used in the desorption step to displace the adsorbed species. Another step is required to recover the purge gas. The displacement-purge cycle is considered only where TSA, PSA, and VSA cannot be used because of pressure or temperature limitations. One application is the separation of medium-molecular-weight linear paraffins (C_{10}-C_{18}) from mixtures of branched-chain and cyclic hydrocarbons by adsorption on 5A zeolite. Ammonia, which is easily separated from the paraffins by flash vaporization, is used as the purge.

Most commercial applications of adsorption involve fixed beds that cycle between adsorption and desorption. Thus, compositions, temperature, and/or pressure at a given location in the bed vary with time. Alternatively, a continuous, countercurrent operation, where such variations do not occur, can be envisaged, as shown in Figure 15.20c and discussed in detail by Ruthven and Ching [65]. The main difficulty with such a scheme is the need to circulate the solid adsorbent, as a moving bed, to achieve a steady-state operation. The first commercial application of countercurrent adsorption and desorption was the moving-bed Hypersorption process for the recovery, by adsorption on activated carbon, of light hydrocarbons from various gas streams in petroleum refineries, as discussed by Berg [66]. Only a few units were installed because of problems with attrition of the adsorbent, difficulties in regenerating the adsorbent when heavier hydrocarbons were present in the feed gas, and unfavorable economics compared to those of distillation. Newer adsorbents with a much higher resistance to attrition and possible applications to more difficult separations are reviving interest in moving-bed units.

A successful alternative countercurrent system for commercial application to the bulk separation of liquid mixtures is the simulated countercurrent system, shown in Figure 15.23 and known generally as the UOP Sorbex process. As described by Broughton [67], the bed is held stationary in one column, which is equipped with a number (perhaps 12) of liquid feed entry and discharge locations. By shifting, with a rotary valve (RV), the locations of feed entry, desorbent entry, extract (adsorbed) removal, and raffinate (non-adsorbed) removal, a countercurrent movement of solids is simulated by a downward movement of liquid. For the valve positions shown in Figure 15.23, locations 2 (entering desorbent), 5 (exiting extract), 9 (entering feed), and 12 (exiting raffinate) are operational, with all other numbered lines closed. However, liquid is also circulated down through and back up

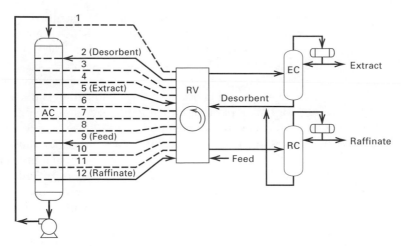

Figure 15.23 Sorbex simulated moving-bed process for bulk separation. AC, adsorbent chamber; RV, rotary valve; EC, extract column; RC, raffinate column. [From D.B. Broughton, *Chem. Eng., Progress,* **64** (8), 60–65 (1968), with permission.]

(external to the column) to the top of the column by a pump. Ideally, an infinite number of entry and exit locations on the column would exist and the valve would continuously change the four operational locations. Since this is impractical, a finite number of locations are used and valve changes are made periodically. In Figure 15.23, when the valve is moved to the next position, Lines 3, 6, 10, and 1 become operational. Thus, raffinate removal is relocated from the bottom of the bed to the top of the bed. The result is that the bed has no top or bottom. As discussed by Gembicki et al. [68], 78 Sorbex-type commercial units were installed during 1962–1989 for the bulk separation of *p*-xylene from C_8 aromatics; *n*-paraffins from branched and cyclic hydrocarbons; olefins from paraffins; *p*- or *m*-cymene (or cresol) from cymene (or cresol) isomers; and fructose from dextrose and polysaccharides.

Ion Exchange

Ion exchange employs the same modes of operation as shown for adsorption in Figure 15.20. Although the use of fixed beds in a cyclic operation is most common, stirred tanks are used for batch contacting, with an attached strainer or filter to separate the resin beads from the solution after equilibrium conditions are approached. Agitation is mild to avoid resin attrition, but sufficient to achieve complete suspension of the resin. To increase resin utilization and achieve high ion-exchange reaction efficiency, much effort has been expended in the development of continuous countercurrent contactors, two of which are shown in Figure 15.24. The Higgins contactor [69] operates as a moving, packed bed by using intermittent hydraulic pulses to move incremental portions of the bed from the contacting section, where ion exchange takes place, up, around, and down to the backwash region, down to the regenerating section, and back up through the rinse section to the contacting section to repeat the cycle. Liquid moves countercurrently to the resin. The Himsley contactor [70] has a series of trays, on each of which the resin beads are fluidized by the upward flow of liquid. Periodically, the flow is reversed to move incremental amounts of resin from one stage to the stage below. The batch of resin at the bottom is lifted to

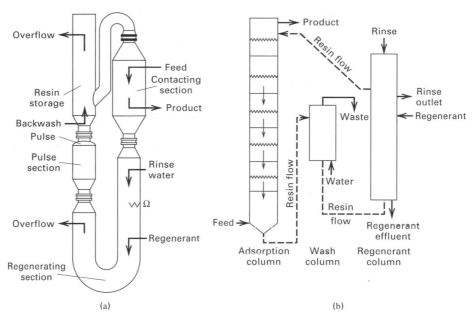

Figure 15.24 Continuous countercurrent ion-exchange contractors. (a) Higgins moving packed-bed process. (b) Himsley fluidized-bed process.

the wash column, then to the regeneration column, and then back to the top of the ion-exchange column for reuse.

Chromatography

Operation modes for large-scale, commercial application of chromatography are of two major types, as discussed in a book edited by Ganetsos and Barker [71]. The first, and the most common, is a transient mode that is a scaled-up version of an analytical chromatograph, referred to as large-scale, batch (or elution) chromatography. Packed columns of diameter up to 4.6 m and packed heights to 12 m have been reported. As shown in Figure 15.25 and discussed by Wankat in Chapter 14 of the *Handbook* edited by Rousseau [9], a recycled solvent or carrier gas is fed continuously into the sorbent-packed column. The feed mixture and recycle is pulsed into the column by an injector. A timer or detector (not shown) splits the effluent from the column, sending it to different separators (condensers, evaporators, distillation columns, etc.). Each separator is designed to remove a particular feed component from the carrier fluid. An additional cleanup step is required to purify the carrier fluid before it is recycled to the column. Separator one produces no product because it handles an effluent pulse that contains the carrier fluid and two or more of the feed components, which are recovered and recycled to the column. Thus, if properly designed and operated, the batch chromatograph operates somewhat like a batch distillation column, producing a nearly pure cut for each component in the feed and slop cuts for recycle. The system shown in Figure 15.25 is designed to separate a binary system. If, say, three more separators are added, the system can separate a five-component feed into five nearly pure products.

The second major type of large-scale chromatograph is the countercurrent flow or simulated countercurrent flow mode already discussed for adsorption. This mode is more

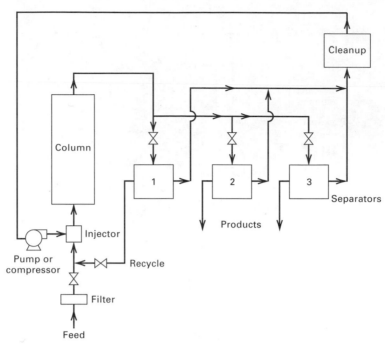

Figure 15.25 Large-scale, batch elution chromatography process.

efficient, but is more complicated and can only separate a mixture into two products. A third mode, not yet commercialized on a large scale, is the continuous cross-current (or rotating) chromatograph, first conceived by Martin [72] and shown schematically in Figure 15.26. The packed annular bed rotates slowly about its axis, past the feed inlet point. *Eluant* (solvent or carrier gas) enters the top of the bed uniformly over the entire cross-sectional area. Both feed and eluant are fed continuously and are carried downward and around by the rotation of the bed. Because of the different selectivities of the feed compo-

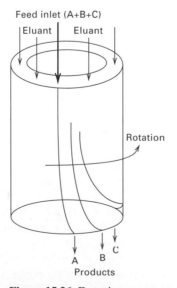

Figure 15.26 Rotating, cross-current, annular chromatograph.

nents for the sorbent, each component traces a different helical path since each spends a different amount of time in contact with the sorbent. Thus, each component is eluted from the bottom of the packed annulus at a different location. In principle, a multicomponent feed can be separated continuously into nearly pure components following separation of the carrier fluid from each eluted fraction.

Slurry Adsorption (Contact Filtration)

Three modes of adsorption from a liquid in an agitated vessel are of interest. The first is the *batch mode* in which a batch of liquid is contacted with a batch of adsorbent for a period of time, followed by discharge of the slurry from the vessel, and a filtration step to separate the solids from the liquid. The second is the *continuous mode*, in which liquid and adsorbent are continuously added to and removed from the agitated vessel. In the third mode, called the *semibatch* or *semicontinuous mode*, the liquid is continuously fed to and removed from the agitated vessel, where it is contacted with the adsorbent, which is retained in a contacting zone of the vessel until it is nearly spent. Models for each of these three modes are developed next, followed by examples of their application. In all models, the slurry is assumed to be perfectly mixed by agitation in the turbulent regime to produce a fluidized bed of sorbent. Perfect mixing is approached by using a liquid depth of from one to two vessel diameters, four vertical wall baffles, and one or two marine propellers or pitched-blade turbines on a vertical shaft. With a proper impeller rotation rate, the axial flow achieves complete suspension. For semicontinuous operation, a clear liquid region is maintained above the suspension region for liquid withdrawal.

Because small particles are used in slurry adsorption and because the relative velocity between the particles and the liquid in an agitated slurry is low (small particles tend to move with the liquid), the rate of adsorption is assumed to be controlled by external, rather than internal, mass transfer.

Batch Mode

The rate of adsorption of solute, as controlled by external mass transfer, is

$$-\frac{dc}{dt} = k_L a(c - c^*) \tag{15-77}$$

where c is the concentration of solute in the bulk liquid; c^* is the concentration in equilibrium with the loading on the adsorbent, q; k_L is the external liquid-phase mass-transfer coefficient; and a is the external surface area of the adsorbent per unit volume of liquid. Starting from feed concentration, c_F, the instantaneous bulk concentration, c, at time t, is related to the instantaneous adsorbent loading, q, by material balance:

$$c_F Q = cQ + qS \tag{15-78}$$

where the adsorbent is assumed to be initially free of adsorbate, Q is the liquid volume (assumed to remain constant for dilute feeds), and S is the mass of adsorbent. The equilibrium concentration, c^*, is given by an appropriate adsorption isotherm: a linear isotherm, the Langmuir isotherm (15-36), or the Freundlich isotherm (15-35). For example, a rearrangement of the latter gives

$$c^* = (q/k)^n \tag{15-79}$$

To solve the system of equations for c and q as a function of time, starting from c_F at $t = 0$, (15-78) is combined with the equilibrium isotherm, for example, (15-79), to eliminate q. The resulting equation is combined with (15-77) to eliminate c^* to give an ODE for c

in t, which is integrated analytically or numerically. Corresponding values of q are then obtained from (15-78).

If the equilibrium is represented by a linear isotherm,

$$c^* = q/k \qquad (15\text{-}80)$$

an analytical integration gives

$$c = \frac{c_F}{\beta}[\exp(-k_L a\beta t) + \alpha] \qquad (15\text{-}81)$$

where

$$\beta = 1 + \frac{Q}{Sk} \qquad (15\text{-}82)$$

$$\alpha = \frac{Q}{Sk} \qquad (15\text{-}83)$$

As the contact time approaches infinity, adsorption equilibrium is approached and for the linear isotherm, from (15-81) or combining (15-78), with $c = c^*$, and (15-80):

$$c\{t = \infty\} = c_F\alpha/\beta \qquad (15\text{-}84)$$

Continuous Mode

When both liquid and solids flow continuously through a perfectly mixed vessel, (15-77) is converted to an algebraic equation because, as in a perfectly mixed reaction vessel (CSTR), the concentration, c, throughout the vessel, is equal to the exit (outlet) concentration, c_{out}. Thus, in terms of the residence time in the vessel, t_{res}:

$$\frac{c_F - c_{\text{out}}}{t_{\text{res}}} = k_L a(c_{\text{out}} - c^*) \qquad (15\text{-}85)$$

or, rearranging:

$$c_{\text{out}} = \frac{c_F + k_L a t_{\text{res}} c^*}{1 + k_L a t_{\text{res}}} \qquad (15\text{-}86)$$

Equation (15-78) becomes

$$c_F Q = c_{\text{out}}Q + q_{\text{out}}S \qquad (15\text{-}87)$$

where Q and S are now flow rates. An appropriate adsorption isotherm relates c^* to q_{out}.

For a linear isotherm, (15-80) becomes $c^* = q_{\text{out}}/k$, which when combined with (15-87) and (15-86) to eliminate c^* and q_{out}, gives

$$c_{\text{out}} = c_F\left(\frac{1 + \gamma\alpha}{1 + \gamma + \gamma\alpha}\right) \qquad (15\text{-}88)$$

where α is given by (15-83) and

$$\gamma = k_L a t_{\text{res}} \qquad (15\text{-}89)$$

The corresponding q_{out} is given by a rearrangement of (15-87):

$$q_{\text{out}} = \frac{Q(c_F - c_{\text{out}})}{S} \qquad (15\text{-}90)$$

For a nonlinear adsorption isotherm, such as (15-35) or (15-36), (15-85) and (15-87) are combined with the isotherm equation, but it may not be possible to express the result explicitly in q_{out}. In that event, a numerical solution is required, as illustrated later in Example 15.9.

Semicontinuous Mode

The most difficult mode to model is the semicontinuous mode, where the adsorbent is retained in the vessel, but the feed liquid enters and exits the vessel at a fixed, continuous flow rate. Both concentration, c, and loading, q, vary with time. With perfect mixing, the outlet concentration is given by (15-86), where t_{res} is the residence time of the liquid in the suspension, and c^* is related to q in the suspension by an appropriate adsorption isotherm. The variation of q in the batch of solids is given by (15-77), rewritten in terms of the change in q, rather than c:

$$S\frac{dq}{dt} = k_L a(c_{out} - c^*)t_{res}Q \tag{15-91}$$

where, for this mode, S is the batch mass of adsorbent in the suspension and Q is the steady volumetric liquid flow rate.

Both (15-91) and (15-86) involve c^*, which can be replaced by a function of instantaneous q by selecting an appropriate isotherm. The resulting two equations are then combined to eliminate c_{out}. The resulting ODE is then integrated analytically or numerically to obtain q as a function of time, from which c_{out} as a function of time can be determined from (15-86) and the isotherm. The time-average value of c_{out} is then obtained by integration of c_{out} with respect to time. For a linear isotherm, the derivation is left as an exercise.

EXAMPLE 15.9

An aqueous solution containing 0.010 mol phenol/L is to be treated at 20°C with activated carbon to reduce the concentration of phenol to 0.00057 mol/L. From Example 4.13, the adsorption equilibrium data are well fitted to the Freundlich equation:

$$q = 2.16c^{1/4.35} \tag{1}$$

or

$$c^* = (q/2.16)^{4.35} \tag{2}$$

where q and c are in mmol/g and mmol/L, respectively. In terms of kmol/kg and kmol/m³, (2) becomes

$$c^* = (q/0.01057)^{4.35} \tag{3}$$

All three modes of slurry adsorption are to be considered. From Example 4.13, the minimum amount of adsorbent is 5 g/L of solution. Laboratory experiments with adsorbent particles 1.5 mm in diameter in a well-agitated vessel have confirmed that the rate of adsorption is controlled by external mass transfer with $k_L = k_c = 5 \times 10^{-5}$ m/s. Particle surface area is 5 m²/kg of particles.

(a) Using twice the minimum amount of adsorbent in an agitated vessel operated in the batch mode, determine the time in minutes to reduce the phenol content to the desired value.
(b) For operation in the continuous mode with twice the minimum amount of adsorbent, determine the required residence time in minutes in the agitated vessel. How does this compare to the batch time of part (a)?
(c) For operation in the semicontinuous mode with 1,000 kg of activated carbon, a liquid feed rate of 10 m³/h, and a liquid residence time equal to 1.5 times the value computed in part (b), determine the run time to obtain a composite liquid product with the desired phenol concentration. Are the results reasonable, or should changes be made to the specifications?

SOLUTION

(a) Batch mode:

$$S/Q = 2(5) = 10 \, g/L = 10 \, kg/m^3$$

$$k_L a = 5 \times 10^{-5} (5)(10) = 2.5 \times 10^{-3} s^{-1}; \quad c_F = 0.010 \, mol/L = 0.010 \, kmol/m^3$$

From (15-78),
$$q = \frac{c_F - c}{S/Q} = \frac{0.010 - c}{10} \tag{4}$$

Substituting (4) into (3),
$$c^* = \left(\frac{0.10 - c}{0.1057}\right)^{4.35} \tag{5}$$

Substituting (5) into (15-77),
$$-\frac{dc}{dt} = 2.5 \times 10^{-3}\left[c - \left(\frac{0.010 - c}{0.1057}\right)^{4.35}\right] \tag{6}$$

where, $c = c_F = 0.010 \, kmol/m^3$ at $t = 0$ and we want t for $c = 0.00057 \, kmol/m^3$. By numerical integration of (6), $t = 1,140 \, s = 19 \, min$.

(b) Continuous mode:

Equation (15-85) applies, where all quantities are the same as those determined in part (a) and $c_{out} = 0.00057 \, kmol/m^3$. Thus,

$$t_{res} = \frac{c_F - c_{out}}{k_L a(c_{out} - c^*)}$$

where c^* is given by (3) with $q = q_{out}$, and q_{out} is obtained from (15-87). Thus,

$$t_{res} = \frac{0.010 - 0.00057}{2.5 \times 10^{-3}\left[0.00057 - \left(\dfrac{0.010 - 0.00057}{0.1057}\right)^{4.35}\right]} = 6,950 \, s \text{ or } 1.93 \, h$$

This residence time is appreciably longer than the batch time of 1,140 s. In the batch mode, the concentration driving force for external mass transfer is initially $(c - c^*) = c_F = 0.010$ $kmol/m^3$ and gradually declines to a much smaller final value, at 1,140 s, of

$$(c - c^*) = c_{final} - \left(\frac{0.010 - c_{final}}{0.1057}\right)^{4.35}$$

$$= 0.000543 \, kmol/m^3$$

For the continuous mode with perfect mixing in the vessel, the concentration driving force for external mass transfer is always at the final batch value of 0.000543 $kmol/m^3$.

(c) Semicontinuous mode:

Equation (15-91) applies with

$$S = 1,000 \, kg, \quad c_F = 0.010 \, kmol/m^3$$

$$Q = 10 \, m^3/h, \quad t_{res} = 10,425 \, s, \quad k_L a = 2.5 \times 10^{-3} \, m/s$$

c^* is given in terms of q by (3) and c_{out} is given by (15-86).

Combining (15-91), (3), and (15-86) to eliminate c^* and c_{out} gives, after simplification,

$$\frac{dq}{dt} = \left(\frac{\gamma}{1 + \gamma}\right)\frac{Q}{S}\left[c_F - \left(\frac{q}{0.01057}\right)^{4.35}\right] \tag{7}$$

where γ is given by (15-89) and the time, t, is the time that the adsorbent remains in the vessel. For values of γ, Q/S, and c_F equal, respectively, to 26.06, 0.01 m^3/h-kg, and 0.010 $kmol/m^3$, (7) reduces to

$$\frac{dq}{dt} = 0.00963\left[0.010 - \left(\frac{q}{0.01057}\right)^{4.35}\right] \tag{8}$$

Table 15.8 Results for Part (c), Semicontinuous Mode, of Example 15.9

| Time t, h | q, kmol/kg | kmol/m^3 | |
		c_{out}	c_{cum}
0.0	0.0	0.000370	0.000370
5.0	0.000481	0.000371	0.000370
10.0	0.000962	0.000398	0.000375
15.0	0.001440	0.000535	0.000401
15.7	0.001506	0.000570	0.000407
20.0	0.001905	0.000928	0.000476
21.0	0.001995	0.001052	0.000501
22.0	0.002084	0.001195	0.000529
23.0	0.002172	0.001356	0.000561
23.2	0.002189	0.001390	0.000568
23.3	0.002197	0.001407	0.000572

where t is in hours and q is in kmol. By numerical integration of (8), starting from $q = 0$ at $t = 0$, we obtain q as a function of t as given in Table 15.8. Included are corresponding values of c_{out} computed from (15-86) combined with (3) to eliminate c^*:

$$c_{out} = \frac{c_F + \gamma(q/0.01057)^{4.35}}{1 + \gamma} = \frac{0.010 + 26.06(q/0.01057)^{4.35}}{27.06}$$

Also included in Table 15.8 are the cumulative values of c, c_{out}, for all of the liquid effluent that exits the vessel during the period from $t = 0$ to $t = t$, as obtained by integrating c_{out} with respect to time: $c_{cum} = \int_0^t c_{out} \, dt/t$.

From the results in Table 15.8, it is seen that the loading, q, increases almost linearly during the first 10 h, while the instantaneous phenol concentration c_{out} in the exiting liquid remains almost constant. At 15.7 h, c_{out} has increased to the specified value of 0.00057 kmol/m^3, but c_{cum} is only 0.000407 kmol/m^3. Therefore, the operation can continue. Finally, at between 23.2 and 23.3 h, c_{cum} reaches 0.00057 kmol/m^3 and the operation must be terminated. During operation, the vessel contains 1,000 kg or 2 m^3 of adsorbent particles. With a liquid residence time of almost 3 h, the vessel must contain 10(3) = 30 m^3. Thus, the vol% solids in the agitated vessel is 6.7. This is reasonable. If the mass of adsorbent in the vessel is increased to 2,000 kg, giving almost 12 vol% solids, the time of operation is doubled to 46.5 h. ∎

Fixed-Bed Adsorption (Percolation)

In the continuous and semicontinuous modes of operation in slurry adsorption, the liquid exiting the vessel always contains unadsorbed solute. If a fixed bed is used, it is possible to obtain a nearly solute-free liquid or gas effluent until the adsorbent in the bed approaches saturation. A fixed bed is frequently used for gas purification and bulk separation.

Consider the flow, down through a fixed bed of adsorbent, of a fluid containing an adsorbable component (the solute). If (1) external and internal mass-transfer resistances are very small; (2) plug flow is achieved; (3) axial dispersion is negligible; (4) the adsorbent is initially free of adsorbate; and (5) the adsorption isotherm begins at the origin, then equilibrium between the fluid and the adsorbent is achieved instantaneously, resulting, as shown in Figure 15.27, in a shock like wave, called a *stoichiometric front*, that moves as a sharp concentration front through the bed. This is *ideal fixed-bed adsorption*. Upstream of the front, the adsorbent is saturated with adsorbate and the concentration of solute in the fluid is that of the feed, c_F. The loading of adsorbate on the adsorbent is the q_F in equilibrium with c_F. The length (height) and weight of the bed section upstream of the

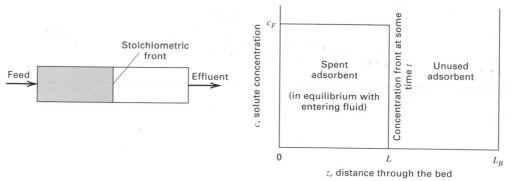

Figure 15.27 Stoichiometric (equilibrium) concentration front for ideal fixed-bed adsorption.

front are LES and WES, respectively, where ES refers to the equilibrium section, called the *equilibrium zone*.

In the upstream region, the adsorbent is spent. Downstream of the stoichiometric front and in the exit fluid, the concentration of the solute in the fluid is zero, and the adsorbent is still adsorbate-free. In this section of the bed, the length and weight are LUB and WUB, respectively, where UB refers to unused bed.

After a period of time, called the *stoichiometric time*, the stoichiometric wave front reaches the bottom of the bed, the concentration of the solute in the fluid abruptly rises to the inlet value, c_F, no further adsorption is possible, and the adsorption step is terminated. This point is referred to as the *breakpoint* and the stoichiometric wave front becomes the ideal *breakthrough* curve.

For ideal fixed-bed adsorption, the location of the concentration wave front L, in Figure 15.27, as a function of time, is obtained solely by material balance and adsorption equilibrium considerations. Thus, at equilibrium, the loading in equilibrium with the feed is designated by $q_F = f\{c_F\}$, where $f\{c_F\}$ is given by an appropriate adsorption isotherm. By material balance on the adsorbate before breakthrough occurs: Solute in entering feed = adsorbate. Accordingly:

$$Q_F c_F t_{\text{ideal}} = q_F S L_{\text{ideal}}/L_B \qquad (15\text{-}92)$$

where Q_F is the volumetric flow rate of feed, c_F is the concentration of the solute in the feed, t_{ideal} is the time for an ideal front to reach $L_{\text{ideal}} < L_B$, q_F is the loading per unit mass of adsorbent that is in equilibrium with the feed concentration, S is the total mass of adsorbent in the bed, and L_B is the total bed length.

$$L_{\text{ideal}} = \text{LES} = \left(\frac{Q_F c_F t_{\text{ideal}}}{q_F S}\right) L_B \qquad (15\text{-}93)$$

$$\text{LUB} = L_B - \text{LES} \qquad (15\text{-}94)$$

$$\text{WES} = S\left(\frac{\text{LES}}{L_B}\right) \qquad (15\text{-}95)$$

$$\text{WUB} = S - \text{WES} \qquad (15\text{-}96)$$

In a real fixed-bed adsorber, the assumptions leading to (15-92) are not valid. Internal transport resistance and, in some cases, external transport resistance are finite. Axial dispersion can also be significant, particularly at low flow rates in shallow beds. These

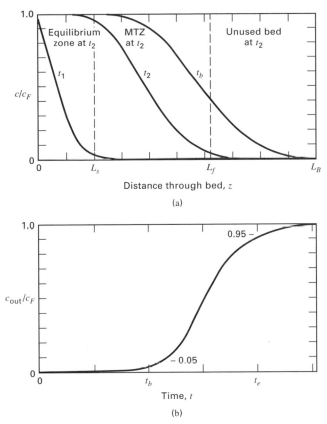

Figure 15.28 Solute wave fronts in a fixed-bed adsorber with mass-transfer effects. (a) Concentration–distance profiles. (b) Break-through curve.

factors contribute to the development of broader concentration fronts like those in Figure 15.28. In Figure 15.28a, typical solute concentration profiles for the fluid are shown as a function of distance through the bed at increasing times t_1, t_2, and t_b from the start of flow through the bed. At t_1, no part of the bed is saturated. At t_2, the bed is almost saturated for a distance L_s. At L_f, the bed is almost clean. Beyond L_f, little mass transfer occurs at t_2 and the adsorbent is still unused. The region between L_s and L_f is the mass-transfer zone, MTZ, where adsorption takes place. Because it is difficult to determine where the MTZ zone begins and ends, L_f can be taken where $c/c_F = 0.05$, with L_s at $c/c_F = 0.95$. From time t_2 to time t_b, the S-shaped front moves through the bed. At t_b, the leading point of the MTZ just reaches the end of the bed. This is the breakthrough point. Rather than using $c/c_F = 0.05$, the breakthrough concentration can be taken as the minimum detectable or maximum allowable solute concentration in the effluent fluid.

Figure 15.28b is a typical plot of the ratio of the outlet-to-inlet solute concentration in the fluid as a function of time from the start of flow. The S-shaped curve is called the breakthrough curve. Prior to t_b, the outlet solute concentration is less than some maximum permissible value, say, $c_{out}/c_F = 0.05$. At t_b, this value is reached, the adsorption step is discontinued, and the regeneration part of the cycle is initiated or the spent adsorbent is discarded. If the adsorption step were to be continued for $t > t_b$, the outlet solute concentration would be observed to rise rapidly, eventually approaching the inlet concentration as the outlet end of the bed became saturated. The time to reach $c_{out}/c_F = 0.95$ is designated t_e.

The steepness of the breakthrough curve determines the extent to which the capacity of an adsorbent bed can be utilized. Thus, the shape of the curve is very important in determining the length of an adsorption bed. For the ideal case, with a stoichiometric wave front, (15-92) applies and all of the bed is utilized before breakthrough occurs. As the width of the breakthrough curve and the corresponding width of the MTZ for the concentration profiles increase, less and less of the bed capacity can be utilized. The situation is further complicated by the fact that the steepness of the concentration profiles shown in Figure 15.28a increases or decreases with time, depending on the shape of the adsorption isotherm, as shown by DeVault [73], in the following manner.

Assume: (1) plug flow of the fluid through the bed at a constant actual (interstitial) velocity, u; (2) instantaneous equilibrium of the solute in the bulk fluid with the adsorbate; (3) no axial dispersion; and (4) isothermal conditions. The bed is not initially free of adsorbate and/or the feed to the bed starting at time $t = 0$ is not at constant composition. The superficial fluid velocity is $\epsilon_b\, u$. A mass balance on the solute for the flow of fluid through a differential adsorption-bed length, dz, over a differential time duration, dt, gives

$$\epsilon_b u A_b c|_z = \epsilon_b u A_b c|_{z+\Delta z} + \epsilon_b A_b \Delta z \frac{\partial c}{\partial t} + (1 - \epsilon_b) A_b \Delta z \frac{\partial q}{\partial t} \tag{15-97}$$

Dividing by Δz and taking the limit as $\Delta z \to 0$ gives

$$\frac{\partial c}{\partial t} + u \frac{\partial c}{\partial z} + \frac{(1 - \epsilon_b)}{\epsilon_b} \frac{\partial q}{\partial t} = 0 \tag{15-98}$$

where q is the adsorption loading/unit volume of adsorbent particles, given by an appropriate adsorption isotherm. By the chain rule:

$$\frac{\partial q}{\partial t} = \frac{\partial q}{\partial c} \frac{\partial c}{\partial t} \tag{15-99}$$

This hyperbolic PDE (15-98) gives $c = f\{z, t\}$. Therefore, by the rules of implicit partial differentiation:

$$u_c = \left(\frac{\partial z}{\partial t}\right)_c = -\frac{\left(\dfrac{\partial c}{\partial t}\right)}{\left(\dfrac{\partial c}{\partial z}\right)} \tag{15-100}$$

where u_c is the velocity of the concentration wave front, $\partial z/\partial t$ at constant c. Combining (15-98) to (15-100):

$$u_c = \frac{u}{1 + \left(\dfrac{1 - \epsilon_b}{\epsilon_b}\right)\dfrac{dq}{dc}} \tag{15-101}$$

This equation gives the velocity of the concentration wave front for the solute in terms of the interstitial fluid velocity and the slope, dq/dc, of the adsorption isotherm. If dq/dc is constant, the wave front moves at a constant value.

In general, the concentration wave front moves through the bed at a velocity, u_c, that is much less than the interstitial fluid velocity. For example, suppose that $\epsilon_b = 0.5$ and the equilibrium adsorption isotherm is given by $q = 5{,}000c$. Then $dq/dc = 5{,}000$. Then, from (15-101), $u_c/u = 0.0002$. If the interstitial velocity is 3 ft/s, the velocity of the concentration wave front is only 0.0006 ft/s. If the bed were 6 ft in height, it would take 2.78 h for the concentration wave front to pass through the bed. If the adsorption isotherm is curved, regions of the wave front at a higher concentration move at a velocity different from

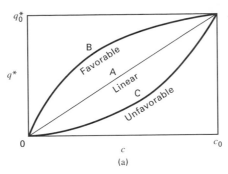

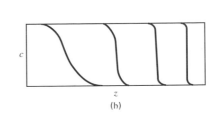

Figure 15.29 Effect of shape of isotherm on sharpness of concentration wavefront. (a) Isotherm shapes. (b) Self-sharpening wavefront caused by a favorable adsorption isotherm.

regions at a lower concentration. Thus, for a linear isotherm (curve A in Figure 15.29a), the width of the MTZ and the wave pattern remain constant. For a favorable isotherm of the Freundlich or Langmuir type (curve B in Figure 15.29a), high-concentration regions move faster than low-concentration regions, and the wavefront steepens with time until a constant pattern front (CPF) is developed, as shown in Figure 15.29b. For the much less common unfavorable type of isotherm (Curve C in Figure 15.29a), low-concentration regions travel faster and the wavefront broadens with time. If the bed is initially clean of adsorbate and a feed of constant solute concentration enters the bed starting at $t = 0$, then the wave front is sharp as shown by the vertical line in Figure 15.27, independent of the type of adsorption isotherm.

For the general case where external and internal mass-transfer resistances are finite and/or axial dispersion is not negligible, methods for predicting concentration profiles and breakthrough curves have been the subject of much study. As will be shown, when mass transfer resistances are a factor, the concentration fronts develop quite differently from the equilibrium fronts just described. Solutions for a number of simplified cases are discussed in detail by Ruthven [10]. The PDE for the governing dynamic behavior is a modification of (15-98):

$$-D_L \frac{\partial^2 c}{\partial z^2} + \frac{\partial (uc)}{\partial z} + \frac{\partial c}{\partial t} + \frac{(1 - \epsilon_b)}{\epsilon_b} \frac{\partial \overline{q}}{\partial t} = 0 \qquad \textbf{(15-102)}$$

where the first term accounts for axial dispersion with eddy diffusivity D_L, the second term permits an axial variation in fluid velocity, and the fourth term is now based on $\overline{q}$, the volume-average adsorbate loading per unit mass. Thus, the latter term accounts for the variation of q throughout the adsorbent particle, due to internal mass-transfer resistance, by averaging the rate of adsorption over the adsorbent particle. The volume-average adsorbate loading is given by

$$\overline{q} = \left(\frac{3}{R_p^3} \right) \int_0^{R_p} r^2 q \, dr \qquad \textbf{(15-103)}$$

where R_p is the radius of the adsorbent particle.

Equation (15-102) gives the concentration of solute in the bulk fluid as a function of time and location in the bed. Equation (15-68) gives the concentration of the solute in the fluid within the pores of an adsorbent particle. These two equations are coupled together by the continuity condition at the particle surface:

$$D_e \left(\frac{\partial c}{\partial r} \right)_{R_P} = k_c(c - c_{R_p}) \qquad \textbf{(15-104)}$$

where k_c is the external mass-transfer coefficient and D_e is the effective diffusivity in the particle, as discussed in Section 15.3. The simultaneous solution of (15-102), (15-103), (15-68), and (15-104) is a formidable task, which can be avoided by using the linear driving force (LDF) model formulated by Glueckauf [74,75] and discussed in detail by Yang [25] and Ruthven [10]. This model, which is widely used to simulate and design fixed-bed adsorption, is based on the following relation, which replaces (15-68) and (15-104):

$$\frac{\partial \overline{q}}{\partial t} = k(q^* - \overline{q}) = kK(c - c^*) \tag{15-105}$$

where q^* is the adsorbate loading in equilibrium with the solute concentration, c, in the bulk fluid; c^* is the concentration in equilibrium with average loading $\overline{q}$; k is the overall mass-transfer coefficient, which includes both external and internal transport resistances; and K is the adsorption equilibrium constant for a linear adsorption isotherm of the form $q = Kc$.

A suitable relationship for the factor kK is

$$\frac{1}{kK} = \frac{R_p}{3k_c} + \frac{R_p^2}{15D_e} \tag{15-106}$$

where the first term on the RHS represents the external mass-transfer resistance, $k_c a_v$, since for a sphere, the surface area/unit volume, a_v, is given by

$$4\pi R_p^2 / [(4/3)\pi R_p^3] = 3/R_p$$

The second term in (15-106) represents the internal resistance, which was first developed by Glueckauf [75], but can also be derived by assuming a parabolic adsorbate loading profile, in the particle, as shown by Liaw et al. [76]. Thus, let

$$q = a_0 + a_1 r + a_2 r^2 \tag{15-107}$$

where the constants a_i depend on time and location in the bed, but are independent of r. Because $\partial q / \partial r = 0$ at $r = 0$ (symmetry condition), $a_1 = 0$. Equating Fick's first law for diffusion into the particle at the particle surface, to the rate of accumulation of adsorbate within the particle, assuming that effective diffusivity is independent of concentration, we obtain

$$\frac{\partial \overline{q}}{\partial t} = D_e a_v \left. \frac{\partial q}{\partial r} \right|_{r=R_p} = \frac{3D_e}{R_p} \left. \frac{\partial q}{\partial r} \right|_{r=R_p} \tag{15-108}$$

At the particle surface, from (15-107):

$$q_{R_p} = a_0 + a_2 R_p^2 \tag{15-109}$$

Substituting (15-107) with $a_1 = 0$ into (15-103) and integrating gives

$$\overline{q} = a_0 + \frac{3}{5} a_2 R_p^2 \tag{15-110}$$

Combining (15-109) and (15-110) to eliminate a_0 gives

$$a_2 = \frac{5}{2R_p^2} (q_{R_p} - \overline{q}) \tag{15-111}$$

From (15-107):

$$\left. \frac{\partial q}{\partial t} \right|_{r=R_p} = 2a_2 R_p \tag{15-112}$$

Combining (15-110), (15-111), and (15-108):

$$\frac{\partial \overline{q}}{\partial t} = \frac{15 D_e}{R_p^2} \left(q_{R_p} - \overline{q} \right) \tag{15-113}$$

Comparing (15-105) with (15-113), we see that the internal resistance is given by the second term in (15-106).

The analytical solution of a simplified form of (15-102), which assumes negligible axial dispersion, constant fluid velocity, u, and the LDF mass-transfer model, is summarized by Ruthven [10] and discussed in detail by Klinkenberg [77]. The solution was first obtained in terms of Bessel functions by Anzelius [78] for the analogous problem of heating or cooling a packed bed of depth z with a fluid. A useful approximate solution is that of Klinkenberg [79]:

$$\frac{c}{c_F} \approx \frac{1}{2} \left[1 + \mathrm{erf} \left(\sqrt{\tau} - \sqrt{\xi} + \frac{1}{8\sqrt{\tau}} + \frac{1}{8\sqrt{\xi}} \right) \right] \tag{15-114}$$

where

$$\xi = \frac{kKz}{u} \left(\frac{1 - \epsilon_b}{\epsilon_b} \right) = \text{Dimensionless distance coordinate} \tag{15-115}$$

$$\tau = k \left(t - \frac{z}{u} \right) = \text{Dimensionless time coordinate corrected for displacement} \tag{15-116}$$

$$\mathrm{erf}(-x) = -\mathrm{erf}(x) \tag{15-117}$$

$$\mathrm{erf}(x) = \frac{2}{\sqrt{\pi}} \int_0^x e^{-\eta^2} \, d\eta \tag{15-118}$$

where ξ and τ are coordinate transformations for z and t, which convert the equations to a much simpler form. The approximation (15-114) is accurate to <0.6% error for $\xi > 2.0$. The erf{{x}, which is included as a function in most spreadsheet programs, is 0.0 at $x = 0$ and asymptotically approaches a value of 1.0 for $x > 2.0$, where x is a dummy variable.

Klinkenberg [79] also includes the following approximate solution for profiles of solute concentration in equilibrium with the average sorbent loading:

$$\frac{c^*}{c_F} = \frac{\overline{q}}{q_F^*} \approx \frac{1}{2} \left[1 + \mathrm{erf} \left(\sqrt{\tau} - \sqrt{\xi} - \frac{1}{8\sqrt{\tau}} - \frac{1}{8\sqrt{\xi}} \right) \right] \tag{15-119}$$

where $c^* = \overline{q}/K$ and $c^*/c_F = \overline{q}/q_F^*$, where q_F^* is the loading in equilibrium with c_F.

EXAMPLE 15.10

Air at 70°F and 1 atm, containing 0.9 mol% benzene, enters a fixed-bed adsorption tower at a flow rate of 23.6 lb/min. The tower is 2 ft in inside diameter and is packed to a height of 6 ft with 735 lb of 4 × 6 mesh silica gel (SG) particles having an effective diameter of 0.26 cm and an external void fraction of 0.5. The adsorption isotherm for benzene has been experimentally determined for the conditions of interest and found to be linear over the concentration range of interest, as given by

$$q = Kc^* = 5,120 \, c^* \tag{1}$$

where:

q = lb benzene adsorbed per ft³ of silica gel particles

c^* = equilibrium concentration of benzene in the gas, in lb benzene per ft³ of gas

Mass-transfer experiments, simulating the conditions of the 2-foot-diameter bed, have been carried out and fitted to a linear driving force LDF model:

$$\frac{\partial \bar{q}}{\partial t} = 0.206 K(c - c^*) \qquad (2)$$

where time is in minutes. The constant $k = 0.206$ min^{-1} includes resistances both in the gas film and in the adsorbent pores, with the latter resistance dominant.

Using the approximate concentration-profile equations of Klinkenberg [77], compute a set of breakthrough curves and determine the time when the concentration of benzene in the exiting air rises to 5% of the inlet concentration. Assume isothermal and isobaric operation. Compare the breakthrough time with the time predicted by the equilibrium model.

SOLUTION

For the equilibrium model, the bed becomes completely saturated with benzene at the inlet concentration.

MW of entering gas $= 0.009(78) + 0.991(29) = 29.44$
Density of entering gas $= (1)(29.44)/(0.730)(530) = 0.076$/lb/ft^3
Gas flow rate $= 23.6/0.0761 = 310$ ft^3/min

Benzene flow rate in entering gas $= \dfrac{(23.6)}{29.44}(0.009)(78) = 0.562$ lb/min

or

$$c_F = \frac{0.562}{310} = 0.00181 \text{ lb benzene/ft}^3 \text{ of gas}$$

From (1),

$$q = 5{,}120(0.00181) = 9.27 \frac{\text{lb benzene}}{\text{ft}^3 \text{ SG}}$$

The total adsorption of benzene at equilibrium $= \dfrac{9.27(3.14)(2)^2(6)(0.5)}{4} = 87.3$ lb

$$\text{Time of operation} = \frac{87.3}{0.562} = 155 \text{ min}$$

For the actual operation, taking into account external and internal mass-transfer resistances, from (15-115), and (15-116),

$$\xi = \frac{(0.206)(5{,}120)z}{u}\left(\frac{1-0.5}{0.5}\right) = 1{,}055 \, z/u$$

$$u = \text{interstitial velocity} = \frac{310}{0.5\left(\dfrac{3.14 \times 2^2}{4}\right)} = 197 \text{ ft/min} \qquad (3)$$

$$\xi = \frac{1{,}055}{197}z = 5.36 \, z$$

where z is in feet. When $z =$ bed height $= 6$ ft, $\xi = 32.2$ and $\tau = 0.206\left(t - \dfrac{z}{197}\right)$ \qquad (4)

where t is in minutes. For $t = 155$ min (the ideal time), and $z = 6$ ft (the bed height), using (4), $\tau = 32$.

Thus, breakthrough curves should be computed from (15-114) for values of τ and ξ no greater than about 32. For example when $\xi = 32.2$ (exit end of the bed), and $\tau = 30$, which corresponds to a time $t = 145.7$ minutes, the concentration of benzene in the exiting gas, from (15-114), is

$$\frac{c}{c_F} = \frac{1}{2}\left[1 + \text{erf}\left(30^{0.5} - 32.2^{0.5} + \frac{1}{8(30)^{0.5}} + \frac{1}{8(32.2)^{0.5}}\right)\right]$$

$$= \frac{1}{2}[1 + \text{erf}(-0.1524)] = \frac{1}{2}\text{erfc}(0.1524)$$

$$= 0.4147 \text{ or } 41.47\%$$

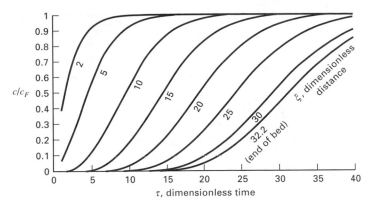

Figure 15.30 Gas concentration breakthrough curves for Example 15.10.

This far exceeds the specification of $c/c_F = 0.05$ or 5% at the exit. Thus, the time of operation of the bed is considerably less than the ideal time of 155 min. Figure 15.30 shows breakthrough curves computed from (15-114) over a range of the dimensionless time, τ, for values of the dimensionless distance, ξ, of 2, 5, 10, 15, 20, 25, 30, and 32.2, where the latter corresponds to the exit end of the bed. For $c/c_F = 0.05$ and $\xi = 32.2$, τ is seen to be about 20 (19.9 by calculation). From (4), with $z = 6$ ft, the time to breakthrough is $t = \dfrac{20}{0.206} + \dfrac{6}{197} = 97.1$ min which is 62.3% of the ideal time.

Figure 15.30 or (15-114) can be used to compute the bulk concentration of benzene at various locations in the bed for $\tau = 20$. The results are as follows:

ξ	z, ft	c/c_F
2	0.373	1.00000
5	0.932	0.99948
10	1.863	0.97428
15	2.795	0.82446
20	3.727	0.53151
25	4.658	0.25091
30	5.590	0.08857
32.2	6.000	0.05158

We can also compute, at $\tau = 20$, the adsorbent loading, at various positions in the bed, from (15-119), using $q = 5,120c$. The maximum loading corresponds to c_F. Thus, $q_{max} = 9.28$ lb benzene/ft³ of SG. Breakthrough curves for the solid loading are plotted in Figure 15.31. As expected, those curves are displaced to the right from the curves of Figure 15.30. At $\tau = 20$:

ξ	z, ft	$\dfrac{c^*}{c_F} = \dfrac{\bar{q}}{q_F^*}$	$\bar{q}, \dfrac{\text{lb benzene}}{\text{ft}^3 \text{ SG}}$
2	0.373	0.99998	9.28
5	0.932	0.99883	9.27
10	1.863	0.96054	8.91
15	2.795	0.77702	7.21
20	3.727	0.46849	4.35
25	4.658	0.20571	1.909
30	5.590	0.06769	0.628
32.2	6.000	0.03827	0.355

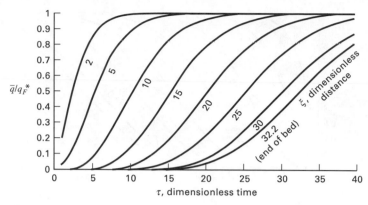

Figure 15.31 Adsorbent loading breakthrough curves for Example 15.10.

The values of $\bar{q}$ in the preceding table are plotted in Figure 15.32 and integrated over the 6-foot bed length to obtain the average bed loading:

$$\bar{q}_{\text{avg}} = \int_0^6 \bar{q}\,dz/6$$

The result is 5.72 lb benzene/ft^3 of SG, which is 61.6% of the maximum loading based on the inlet benzene concentration.

If the bed were increased in height by a factor of 5, to 30 ft, $\xi = 161$. The ideal time of operation would be 780 min or 13 h. With mass-transfer effects taken into account, as before, the dimensionless operating time to breakthrough is computed to be $\tau = 132$, or breakthrough time from (4) is

$$t = \frac{132}{0.206} + \frac{30}{197} = 641 \text{ min}$$

which is 82.2% of the ideal time. This represents a substantial increase in bed utilization. ■

Scale-up for Constant Pattern Front

In Example 15.10, the wave front (of the type shown in Figure 15.28a), broadens as it moves through the bed. This is shown in Figure 15.33, where MTZ, the width of the mass-

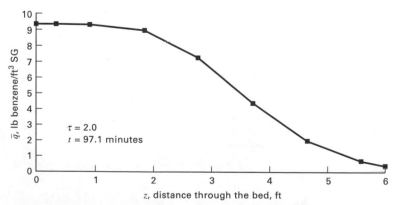

Figure 15.32 Adsorbent loading profile for Example 15.10.

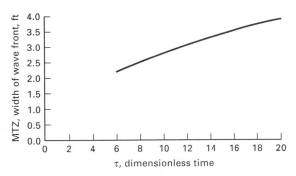

Figure 15.33 Broadening of wave front in Example 15.10.

transfer zone, is plotted against the dimensionless time, τ, up to the value of 20 where the front breaks through the 6-foot-long bed. The MTZ in Figure 15.33 is based on a range of c/c_F from 0.95 to 0.05. As seen, MTZ increases from about 2 feet at $\tau = 6$ to about 4 feet at $\tau = 20$. As shown, with increasing τ, the rate of broadening slows. However, for a deeper bed, it is found that even at $\tau = 100$, the wavefront is still slowly broadening.

The continual broadening of the wavefront determined in Example 15.10 is typical of that obtained with a linear adsorption isotherm (curve A in Figure 15.29a). The wavefront also continues to broaden with an unfavorable isotherm (curve C in Figure 15.29a). But, when the isotherm is of the favorable Langmuir or Freundlich type (curve B in Figure 15.29a), wave-front broadening rapidly diminishes and an asymptotic or *constant pattern front* (CPF) is developed. For such a front, MTZ is constant and curves of c/c_F and $\overline{q}/q^*$ become coincident.

The bed depth at which the CPF is approached depends upon the nonlinearity of the adsorption isotherm and the importance of adsorption kinetics. The mathematical proof of the existence of an asymptotic wavefront solution is given by Cooney and Lightfoot [80], including the case of axial dispersion. Initially, the wavefront broadens because of mass-transfer resistance and/or axial dispersion. Eventually, the opposite influence of a favorable isotherm, as shown in Figure 15.29b, comes into play and an asymptotic wavefront pattern is approached. For a constant pattern front, Sircar and Kumar [81] present some analytical solutions and Cooney [82] presents a rapid approximate method, illustrated with the Freundlich and Langmuir isotherms, to estimate concentration profiles and breakthrough curves when mass-transfer and equilibrium parameters are available.

When the constant pattern front assumption is valid, it can be used to determine the length of a full-scale adsorbent bed from breakthrough curves obtained in small-scale laboratory experiments. This widely used technique is described by Collins [83] for purification applications. The adsorbent bed is considered to be the sum of two sections, analogous to those mentioned for ideal fixed-bed adsorption. Thus, the total bed length is estimated to be the sum of the length, LES, of the ideal fixed-bed adsorber plus an additional length, called the LUB, that depends on the observed width of the MTZ and the shape of the c/c_F profile within that zone. The total required bed length is

$$L_B = \text{LES} + \text{LUB} \tag{15-120}$$

For the ideal fixed-bed adsorber, with MTZ = 0, LUB is not necessary, but if $L_B > $ LES, then LUB is the length of unused bed. However, when an MTZ is present, then an LUB is necessary and is referred to as the equivalent length of unused bed. To determine LUB from an experimental breakthrough curve, for the same feed composition and superficial velocity to be used in the commercial adsorber, and for a CPF, the front is located such

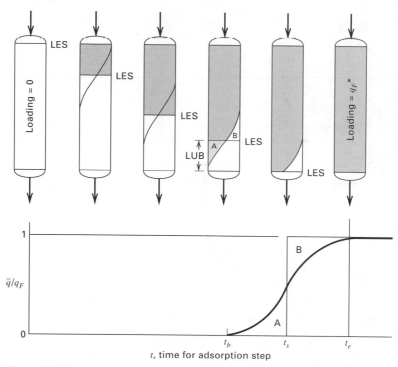

Figure 15.34 Determination of bed length from laboratory measurements.

that in Figure 15.34, area A is equal to area B. Then:

$$\text{LUB} = \frac{(t_s - t_b)}{t_s} L_e \qquad \textbf{(15-121)}$$

where L_e is the length of the experimental bed. For the ideal case, a solute mass balance for a cylindrical bed of diameter D gives

$$c_F Q_F t = q_F \rho_b \pi \frac{D^2}{4} (\text{LES}) \qquad \textbf{(15-122)}$$

where t is the time to breakthrough, from which the LES can be determined.

Instead of positioning the stoichiometric front for equal areas in Figure 15.34, the LUB can be determined from the experimental breakthrough curve data by computing t_s from

$$t_s = \int_0^{t_e} \left(1 - \frac{c}{c_F}\right) dt \qquad \textbf{(15-123)}$$

EXAMPLE 15.11

Collins [83] presents the following experimental data for the adsorption of water vapor from nitrogen in a fixed bed of 4A molecular sieves:

Bed depth = 0.88 ft, temperature = 83°F (negligible temperature change), pressure = 86 psia (negligible pressure drop), G = entering gas flow rate = 29.6 lbmol/h-ft^2, entering water content = 1,440 ppm (by volume), initial adsorbent loading = 1 lb/100 lb sieves, and bulk density of bed = 44.5 lb/ft^3.

c_{exit}, ppm (by volume)	Time, h	c_{exit}, ppm (by volume)	Time, h
<1	0–9.0	650	10.8
1	9.0	808	11.0
4	9.2	980	11.25
9	9.4	1,115	11.5
33	9.6	1,235	11.75
80	9.8	1,330	12.0
142	10.0	1,410	12.5
238	10.2	1,440	12.8
365	10.4	1,440	13.0
498	10.6		

Determine the bed height required for a commercial unit to be operated at the same temperature, pressure, and entering gas mass velocity and water content to obtain an exiting gas with no more than 9 ppm (by volume) of water vapor with a time to breakthrough of 20 h.

SOLUTION

$$c_F = \frac{1,440(18)}{106} = 0.02592 \text{ lb } H_2O/\text{lbmol } N_2$$

$$\frac{Q_F}{\pi D^2/4} = 29.6 \text{ lbmol } N_2/\text{h-ft}^2 \text{ of bed cross-section}$$

$q_F = 29.6(0.02592)(9)/[(0.88)(44.5)] + 1/100 = 0.186 \text{ lb } H_2O/\text{lb solid}$
Initial moisture content of bed = 0.01 lb H_2O/lb solid

From (15-122),
$$\text{LES} = \frac{(0.02592)(29.6)(20)}{(0.186 - 0.01)(44.5)} = 1.96 \text{ ft}$$

Use integration method to obtain LUB. From the data:

$$t_e = 12.8 \text{ h } (1,440 \text{ ppm) and } t_b = 9.4 \text{ h } (9 \text{ ppm})$$

By numerical integration of the breakthrough curve data, using (15-123): $t_s = 10.93$ h

From (15-121),
$$\text{LUB} = \left(\frac{10.93 - 9.40}{10.93}\right)(0.88) = 0.12 \text{ ft}$$

From (15-120): $L_B = 1.96 + 0.12 = 2.08$ ft or a bed utilization of $\frac{1.96}{2.08} \times 100\% = 94.2\%$ ∎

Thermal-Swing Adsorption

Thermal (temperature)-swing adsorption (TSA), in its simplest configuration, is carried out with two fixed beds in parallel, operating cyclically, as in Figure 15.20b. While one bed is adsorbing solute at near-ambient temperature, $T_1 = T_{ads}$, the other bed is regenerated by desorbing adsorbate at a higher temperature, $T_2 = T_{des}$, at which the equilibrium adsorbate loading is much less for a given concentration of solute in the fluid, as illustrated in Figure 15.21. Although the desorption step might be accomplished in the absence of a purge fluid by simply vaporizing the adsorbate, readsorption of some solute vapor would occur upon cooling the bed. Thus, it is best to remove the desorbed adsorbate with a purge. The desorption temperature is high, but not so high as to cause deterioration of the adsorbent. TSA is best applied to the removal of contaminants present at low concentrations in the feed fluid. In that case, nearly isothermal adsorption and desorption is achieved. An ideal cycle involves four steps: (1) adsorption at T_1 to breakthrough, (2) heating of the bed to T_2, (3) desorption at T_2 to a low adsorbate loading, and (4) cooling of the bed to T_1. Practical cycles do not operate with isothermal steps. Instead, Steps 2 and 3 are combined for the regeneration part of the cycle, with the bed being simultaneously heated and desorbed with preheated purge gas until the temperature of the effluent approaches that of the inlet purge. Steps 1 and 4 may also be combined because, as discussed in detail by Ruthven [10], the thermal wave precedes the MTZ front. Thus, adsorption takes place at essentially the feed-fluid temperature.

The heating and cooling steps cannot be accomplished instantaneously because of the relatively low bed thermal conductivity. Although heat transfer can be done indirectly from jackets surrounding the beds or from coils located within the beds, bed temperature changes are more readily achieved by preheating or precooling a purge fluid, as shown in Figure 15.35. The purge fluid can be a portion of the feed or effluent, or some other fluid. The purge fluid can also be used in the desorption step. When the adsorbate is valuable and easily condensed, the purge fluid might be a noncondensable gas. When the adsorbate is valuable, but not easily condensed, and is essentially insoluble in water, steam may be used as the purge fluid, followed by condensation of the steam to separate it from the desorbed adsorbate. When the adsorbate is not valuable, fuel and/or air can be used as the purge fluid, followed by incineration. Often the amount of purge used in the regeneration step is much less than the amount of feed sent to the bed in the adsorption step. In Figure 15.35, the feed fluid is a gas. The spent bed is heated and regenerated with preheated feed gas, which is then cooled to condense the desorbed adsorbate.

Because of the time to heat and cool a fixed bed, cycle times for TSA are long, usually extending over periods of hours or days. The longer the cycle time, the longer the required bed length, and the greater the percent utilization of the bed during adsorption. However, for a given cycle time, when the width of the MTZ is an appreciable fraction of the bed height, such that the capacity of the bed is poorly utilized, consideration should be given to a *lead-trim bed* arrangement of two beds in series for the adsorption step. When the lead bed is spent, it is switched to regeneration. At this point in time, the trim bed has an MTZ occupying a considerable portion of the bed, and that bed becomes the lead bed,

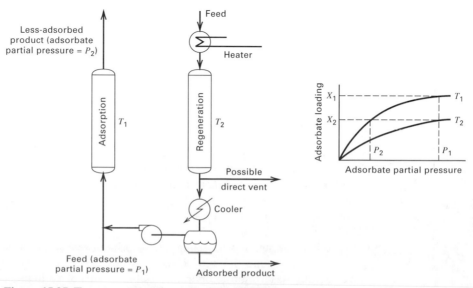

Figure 15.35 Temperature-swing adsorption cycle.

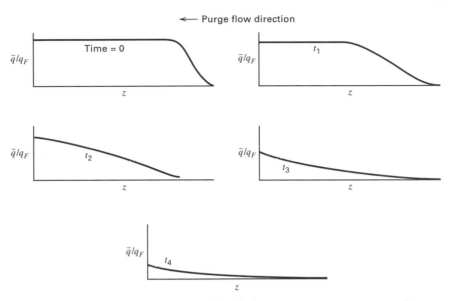

Figure 15.36 Sequence of loading profiles during countercurrent regeneration.

with a regenerated bed becoming the trim bed. In this manner only a fully spent bed is switched to regeneration. Thus, a total of three beds are used. If the flow rate of the feed stream is very high, beds in parallel may be required for both adsorption and desorption.

The adsorption step is usually conducted with the feed fluid flowing downward through the bed. The flow direction for desorption can be either downward or upward, but the upward, countercurrent direction is usually preferred because it is more efficient. Consider the sequence of loading fronts shown in Figure 15.36, for regeneration countercurrent to adsorption. Although the bed is shown in a horizontal position, it must be positioned vertically. The feed fluid flows downward, entering at the left and leaving at the right. At time $t = 0$, breakthrough has occurred, with a loading profile as shown at the top, where the MTZ is seen to be about 25% of the bed. If the purge fluid for regeneration also flows downward (entering at the left), all of the adsorbate will have to move through the unused portion of the bed. Thus, some of the desorbed adsorbate will be readsorbed in the unused section and then desorbed a second time. If countercurrent regeneration is used, the unused portion of the bed is never contacted with desorbed adsorbate. During a countercurrent regeneration step, the loading profile changes progressively, as shown for a series of times in Figure 15.36. The right-side end of the bed, where the purge enters, is desorbed first. At the end of regeneration, the residual loading may be uniformly zero or, more likely, finite and nonuniform as shown at the bottom of Figure 15.36. If the latter, then the useful cyclic capacity, called the *delta loading*, is as shown in Figure 15.37.

Calculations of the concentration and loading profiles during desorption are only approximated by (15-114) and (15-119) because the loading is not uniform at the beginning of desorption. A numerical solution for the desorption step can be obtained in the following fashion using a procedure discussed by Wong and Niedzwiecki [84]. Although their method was developed for the adsorption step, it is readily applied to desorption. In the absence of axial dispersion and for constant fluid velocity, (15-102) and (15-105) are rewritten as

$$u\frac{\partial\phi}{\partial z} + \frac{\partial\phi}{\partial t} + \left(\frac{1-\epsilon_b}{\epsilon_b}\right)kK(\phi - \psi) = 0 \qquad \textbf{(15-123)}$$

$$\frac{\partial\psi}{\partial t} = k(\phi - \psi) \qquad \textbf{(15-124)}$$

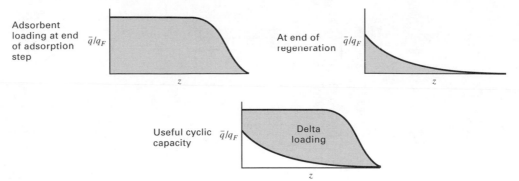

Figure 15.37 Delta loading for regeneration step.

where

$$\phi = c/c_F \tag{15-125}$$

$$\psi = \overline{q}/q_F^* \tag{15-126}$$

and c_F and q_F^* are the values at the beginning of the adsorption step. The boundary conditions are as follows:

At $t = 0$: $\phi = \phi\{z\}$ at the end of the adsorption step
                       $\psi = \psi\{z\}$ at the end of the adsorption step

where, for countercurrent desorption, it is best to let z start from the bottom of the bed (called z') and increase in the direction of purge-gas flow. Thus, u in (15-123) is positive.

At $z' = 0$: $\phi = 0$ (no solute in the entering purge gas)
                     $\psi = 0$

Partial differential equations (15-123) and (15-124) in independent variables z and t can be converted to a set of ordinary differential equations (ODEs) in independent variable t by the method of lines (MOL), which was first applied to parabolic PDEs by Rothe in 1930, as discussed by Liskovets [85], and subsequently to elliptic and hyperbolic PDEs. The MOL is developed in detail by Schiesser [86]. The lines refer to the z'-locations of the ODEs. To obtain the set of ODEs, the z'-coordinate is divided into N increments or $N + 1$ grid points that are usually evenly spaced. For many problems, 20 increments are sufficient. Letting i be the index for each grid point in z', starting from the end where the purge gas enters, and discretizing $\partial\phi/\partial z'$, (15-123) and (15-124) become

$$\frac{d\phi_i}{dt} = -u\left(\frac{\Delta\phi}{\Delta z'}\right)_i - \left(\frac{1 - \epsilon_b}{\epsilon_b}\right)kK(\phi_i - \psi_i), i = 1, N + 1 \tag{15-127}$$

$$\frac{d\psi_i}{dt} = k(\phi_i - \psi_i), i = 1, N + 1 \tag{15-128}$$

where the initial conditions ($t = 0$) for ϕ_i and ψ_i are as given above. Before we can integrate (15-127) and (15-128), we must provide a suitable approximation for $(\Delta\phi/\Delta z')_i$. In general, for a moving-front problem of the hyperbolic type here for adsorption and desorption, the simple central difference

$$\left(\frac{\Delta\phi}{\Delta z'}\right)_i \approx \frac{\phi_{i+1} - \phi_{i-1}}{2\,\Delta z'}$$

is not adequate. Instead, Wong and Niedzwiecki [84] found that a five-point, biased, upwind, finite-difference approximation, discussed by Schiesser [87], is very effective. This approximation, which is derived from a Taylor's series analysis, places emphasis on conditions upwind of the moving front. At an interior grid point:

$$\left(\frac{\Delta\phi}{\Delta z'}\right)_i \approx \frac{1}{12\,\Delta z'}\left[-\phi_{i-3} + 6\phi_{i-2} - 18\phi_{i-1} + 10\phi_i + 3\phi_{i+1}\right] \qquad \textbf{(15-129)}$$

Note that the coefficients of the ϕ-factors, inside the square brackets, sum to 1. At the last grid point, $N + 1$, where the purge gas exits, (15-129) is replaced by

$$\left(\frac{\Delta\phi}{\Delta z'}\right)_{N+1} \approx \frac{1}{12\,\Delta z'}\left[3\phi_{N-3} - 16\phi_{N-2} + 36\phi_{N-1} - 48\phi_N + 25\phi_{N+1}\right] \qquad \textbf{(15-130)}$$

For the first three node points, the following approximations replace (15-129):

$$\left(\frac{\Delta\phi}{\Delta z'}\right)_1 \approx \frac{1}{12\,\Delta z'}\left[-25\phi_1 + 48\phi_2 - 36\phi_3 + 16\phi_4 - 3\phi_5\right] \qquad \textbf{(15-131)}$$

$$\left(\frac{\Delta\phi}{\Delta z'}\right)_2 \approx \frac{1}{12\,\Delta z'}\left[-3\phi_1 - 10\phi_2 + 18\phi_3 - 6\phi_4 + \phi_5\right] \qquad \textbf{(15-132)}$$

$$\left(\frac{\Delta\phi}{\Delta z'}\right)_3 \approx \frac{1}{12\,\Delta z'}\left[\phi_1 - 8\phi_2 + 0\phi_3 + 8\phi_4 - \phi_5\right] \qquad \textbf{(15-133)}$$

However, because values of ϕ_1 (at $z' = 1$) are given as a boundary condition, (15-131) is not needed.

Equations (15-127) to (15-133) with boundary conditions for ϕ_1 and ψ_1 constitute a set of $2N$ ODEs as an initial-value problem, with time as the independent variable. However, the values of ϕ_i and ψ_i at the different axial locations can change with time at vastly different rates. For example, in Figure 15.36 for desorption fronts, if we divide the bed length, L, into 20 equal-width increments, starting from the right-hand side where the purge gas enters, we see that initially ψ_{21}, where the purge gas exits, is not changing at all, while ψ_5 is changing rapidly. Near the end of the desorption step, ψ_{21} is changing rapidly, while ψ_5 is not. The same observations hold for ϕ_i. This type of response is referred to as *stiffness*, as described by Schiesser [87] and in *Numerical Recipes* by Press et al. [88]. If we attempt to integrate the set of ODEs with simple Euler or Runge–Kutta methods, not only do we encounter truncation error, but, with time, the computed values of ϕ_i and ψ_i go through enormous instability characterized by wild swings between large and impossible positive and negative values. Even if the length is divided into many more than 20 increments and very small time steps are used, instability is still often encountered.

The integration of a stiff set of ODEs is most efficiently carried out by variable-order/variable-step-size implicit methods of the type first developed by Gear [89]. These methods are included in a widely available software package called ODEPACK, described by Byrne and Hindmarsh [90]. The subject of stiffness is also discussed in Chapter 13.

EXAMPLE 15.12

In Example 15.10, benzene is adsorbed from air at 70°F and 1 atm onto silica gel in a fixed-bed adsorber, 6 ft in length. Breakthrough occurs at close to 97.1 min for $\phi = 0.05$. At that time, values of $\phi = c/c_F$ and $\psi = \overline{q}/q_F^*$ in the bed are distributed as follows, where z' is measured backwards from the exit of the bed for the adsorption step. These results were obtained by the numerical method just described, as applied to the adsorption step, and are in close agreement with the approximate, analytical Klinkenberg solution given in Example 15.10.

z', ft	$\phi = c/c_F$	$\psi = \bar{q}/q_F^*$
0	0.05227	0.03891
0.3	0.07785	0.05913
0.6	0.11314	0.08776
0.9	0.16008	0.12690
1.2	0.22017	0.17850
1.5	0.29394	0.24387
1.8	0.38042	0.32310
2.1	0.47678	0.41459
2.4	0.57825	0.51469
2.7	0.67861	0.61786
3.0	0.77108	0.71728
3.3	0.84969	0.80603
3.6	0.91057	0.87858
3.9	0.95281	0.93207
4.2	0.97848	0.96690
4.5	0.99172	0.98636
4.8	0.99731	0.99531
5.1	0.99921	0.99857
5.4	0.99987	0.99960
5.7	1.00000	1.00000
6.0	1.00000	1.00000

If the bed is regenerated isothermally with pure air at 1 atm and 145°F, and the desorption of benzene during the heat-up period is neglected, determine the loading, $\bar{q}$, profile at times of 15, 30, and 60 min for pure stripping air interstitial velocities of: (a) 197 ft/min, and (b) 98.5 ft/min.

$$\text{At 145°F and 1 atm, the adsorption isotherm is } q = 1,000\, c^* \tag{1}$$

giving an equilibrium loading of about 20% of that at 70°F. Assume that k is unchanged from the value of 0.206 in Example 15.10.

SOLUTION

This problem is solved by the MOL with 20 increments in z', using the subroutine LSODE in ODEPACK to integrate the set of ODEs. The user supplies the FORTRAN MAIN program and the subroutine FEX, shown in Table 15.9, for the derivative functions given by (15-127) to (15-130) and (15-132) to (15-133). The program LSODE includes detailed instructions for writing these two routines. Note that the program in Table 15.9 actually includes both the adsorption and desorption steps for desorption conditions of 30 min at 197 ft/min.

The computed loading profiles for all conditions are plotted in Figure 15.38a and b, for desorption interstitial velocities of 197 and 98.5 ft/min, respectively, where z is the distance from the feed gas inlet end of the bed for the adsorption step. The curves are similar to those shown in Figure 15.30. For the 197 ft/min case, desorption is almost complete at 60 min with less than 1% of the bed still loaded with benzene. If this velocity were used, this would allow $97.1 - 60 = 3.1$ min for heating and cooling the bed before and after desorption. For the 98.5 ft/min case at 60 min, about 5% of the bed is still loaded with benzene. This may be acceptable, but the resulting adsorption step would take a little longer because initially the bed would not be clean. Several cycles are required to establish a cyclic steady state, whose development is considered in the next section on pressure-swing adsorption. ∎

Pressure-Swing Adsorption

Pressure-swing adsorption (PSA) and vacuum-swing adsorption (VSA), in their simplest configurations, are carried out with two fixed beds in parallel, operating in a cycle, as in

Table 15.9 FORTRAN Computer Program for Example 15.12

```
PROGRAM tsa
IMPLICIT DOUBLE PRECISION(A-H,O-Z)
EXTERNAL FEX
DIMENSION C(40),ATOL(60),RWORK(4162),IWORK(90),CH(40),DL(40)
COMMON CF,VEL,AK,A(20)
open (unit=3,file='n1.out')
write(3,*)'desorption velocity=197ft/min, desorption time=30min'
NEQ=60
CF0=0.00181
CF1=0.0
TCA=97.1
NUMCYCLE=1
MXSTEP=2000
DO 55 I=1,20
55      C(I)=0.0

DO 56 I=21,40
C(I)=0.0
56      CONTINUE

T=0.D0
TOUT=0.0
ITOL=2
RTOL=1.D-6
DO 57 I=1,60
ATOL(I)=1.0 D-12
57      CONTINUE

ITASK=1
ISTATE=1
IOPT=1
IWORK(6)=2000
LRW=4162
LIW=90
MF=22

CF=CF0
C0=CF
AK=5120.
Q0=AK*C0

CALL LSODE(FEX,NEQ,C,T,TOUT,ITOL,RTOL,ATOL,ITASK,ISTATE,
1       IOPT,RWORK,LRW,IWORK,LIW,JEX,MF)
WRITE(3,*)'CONDITIONS AT THE BEGINNING '
WRITE(3,*)'TIME(SEC)=',TOUT
WRITE(3,*)'CONC. GAS PHASE'
WRITE(3,*)C0, (C(I),I=1,20)
WRITE(3,*)'LODING gm/gm'
WRITE(3,*)Q0,(C(I),I=21,40)
write(3,128)
128     format(////)
C0DL=1.0
Q0DL=1.0
DO 989 I=1,20
DL(I)=C(I)/C0
DL(I+20)=C(I+20)/Q0
```

Table 15.9 FORTRAN Computer Program for Example 15.12 (*continued*)

```
989       CONTINUE
          WRITE(3,*)'DIMENSIONLESS CONDITIONS AT THE BEGINNING'
          WRITE(3,*)'TIME(SEC)=',TOUT
          WRITE(3,*)'DIMENSIONLESS GAS CONCENTRATION C/CF'
          WRITE(3,*)C0DL, (DL(I),I=1,20)
          WRITE(3,*)'DIMENSIONLESS LOADING Q/Q0'
          WRITE(3,*)Q0DL, (DL(I),I=21,40)
          write(3,129)
129       format(////////)
          DO 1000 KK=1,NUMCYCLE
C-----------------------------------------------------------------------
C----------ADSORPTION STEP----------------------------------------------
C-----------------------------------------------------------------------
          T=0.0

          CF=CF0
          C0=CF
          AK=5120.
          Q0=AK*C0

          VEL=197.0
          ISTATE=1
          TOUT= 97.1

          CALL LSODE(FEX,NEQ,C,T,TOUT,ITOL,RTOL,ATOL,ITASK,ISTATE,
     1    IOPT,RWORK,LRW,IWORK,LIW,JEX,MF)
          IF(KK.EQ.1)GOTO18
          IF((KK/25)*25.NE.KK)GOTO81
18         WRITE(3,*)'CONDITIONS AT THE END OF ADSORPTION STEP'
          WRITE(3,*)'STEP TIME(SEC)=',TOUT
          WRITE(3,*)'CONC. OF GAS PHASE'
          WRITE(3,*)C0,(C(I),I=1,20)
          WRITE(3,*)'LOADING gm/gm'
          WRITE(3,*)Q0,(C(I),I=21,40)
          WRITE(3,741)
741       FORMAT(///)
          C0DL=1.0
          Q0DL=1.0
          DO 990 I=1,20
          DL(I)=C(I)/C0
          DL(I+20)=C(I+20)/Q0
990       CONTINUE
          WRITE(3,*)'DIMENSIONLESS CONDITIONS AT THE END OF ADSORPTION'
          WRITE(3,*)'STEP TIME(SEC)=',TOUT
          WRITE(3,*)'DIMENSIONLESS GAS CONCENTRATION C/CF'
          WRITE(3,*)C0DL,(DL(I),I=1,20)
          WRITE(3,*'DIMENSIONLESS LOADING Q/Q0'
          WRITE(3,*)Q0DL,(DL(I),I=21,40)
          WRITE(3,238)
238       FORMAT(////////)
C-----------------------------------------------------------------------
C------DESORPTION BY TEMPERATURE SWING----------------------------------
C-----------------------------------------------------------------------
81        T=0.0
          VEL=197.0
```

Table 15.9 FORTRAN Computer Program for Example 15.12 (*continued*)

```
        TOUT=30.0
        ISTATE=1

        CF=CF1
        C0=CF
        AK=1000.0
        Q0=AK*C0

        DO 91 I=1,40
        CH(I)=C(I)
 91     CONTINUE

        DO 92 I=1,19
        C(I)=CH(20-I)
 92     CONTINUE
        C(20)=CF0

        DO 95 I=1,19
 95     C(20+I)=CH(40-I)
        C(40)=CF0*5120.
        CALL LSODE(FEX,NEQ,C,T,TOUT,ITOL,RTOL,ATOL,ITASK,
     1  ISTATE,IOPT,RWORK,LRW,IWORK,LIW,JEX,MF)

        DO 93 I=1,40
        CH(I)=C(I)
 93     CONTINUE

        DO 94 I=1,19
        C(I)=CH(20-I)
        C(20+I)=CH(40-I)
 94     CONTINUE
        C0=C(20)
        Q0=C(40)
        C(20)=CF1
        C(40)=1000.*CF1
        IF(KK.EQ.1)GOTO38
        IF((KK/25)*25.NE.KK)GOTO1000
 38     WRITE(3,*)'CONDITIONS AT THE END OF DESORPTION '
        WRITE(3,*)'STEP TIME(SEC)=',TOUT
        WRITE(3,*)'CONC. OF GAS PHASE '
        WRITE(3,*)C0,(C(I),I=1,20)
        WRITE(3,*)'LOADING gm/gm'
        WRITE(3,*)Q0,(C(I),I=21,40)
        WRITE(3,264)
 264    FORMAT (///)
         C0DL=C0/CF0
         Q0DL=Q0/(CF0*5210.)
         DO 991 I=1,20
         DL(I)=C(I)/CF0
         DL(I+20)=C(I+20)/(CF0*5120.)
 991     CONTINUE
         WRITE(3,*)'DIMENSIONLESS CONDITIONS AT THE END OF DESORPTION'
         WRITE(3,*)'STEP TIME(SEC)=',TOUT
         WRITE(3,*)'DIMENSIONLESS GAS CONCENTRATION C/CF'
         WRITE(3,*)C0DL,(DL(I),I=1,20)
```

Table 15.9 FORTRAN Computer Program for Example 15.12 (*continued*)

```
              WRITE(3,*)'DIMENSIONLESS LOADING Q/Q0'
              WRITE(3,*)Q0DL,(DL(I),I=21,40)
              WRITE(3,365)
365           FORMAT(////////)
1000          CONTINUE

C--------------------------------------------------------------------------
C--------------------------------------------------------------------------
C--------------------------------------------------------------------------
              WRITE(3,60)IWORK(11),IWORK(12),IWORK(13)
 60           FORMAT(/12H NO. STEPS =,I4,11H NO. F-S =,I4,11H NO. J-S =,I4)
              STOP
 80           WRITE(3,90)ISTATE
 90           FORMAT(///22H ERROR HALT.. ISTATE =,I3)
              close(unit=3)
              STOP
              END
C--------------------------------------------------------------------------
              SUBROUTINE FEX (NEQ,T,C,CDOT)
              IMPLICIT DOUBLE PRECISION(A-H,O-Z)
              DIMENSION C(40), CDOT(40)
              COMMON CF,VEL,AK,A(20)
              E=0.5
              C0=CF
              Q0=AK*C0
              DZ=6.0/20.0       !FT
              AA=-VEL
              BB=-(1.0-E)/E

              R4FDX=1./(12.*DZ)
              A(1)=R4FDX*
     1        (-3.*C0-10.*C(1)+18.*C(2)-6.*C(3)+1.*C(4))

              A(2)=R4FDX*
     1        (1.*C0-8.*C(1)+0.*C(2)+8.*C(3)-1.*C(4))

              A(3)=R4FDX*
     1        (-1.*C0+6.*C(1)-18.*C(2)+10.*C(3)+3.0*C(4))

              DO 455 I=4,19
              A(I)=R4FDX*
     1        (-1.*C(I-3)+6.*C(I-2)-18.*C(I-1)+10.*C(I)+3.*C(I+1))
455           CONTINUE
              A(20)=R4FDX*
     1        (3.*C(16)-16.*C(17)+36.*C(18)-48.*C(19)+25.*C(20))

              DO 676 I=1,20
              CDOT(20+I)=0.206*(AK*C(I)-C(20+I))
              CDOT(I)=AA*A(I)+BB*CDOT(20+I)
676           CONTINUE

              RETURN
              END
```

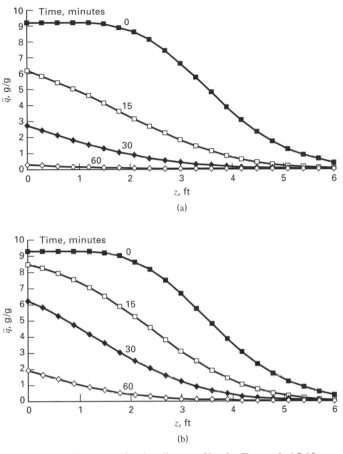

Figure 15.38 Regeneration loading profiles for Example 15.12. (a) Regeneration air interstitial velocity = 197 ft/min. (b) Regeneration air interstitial velocity = 98.5 ft/min.

Figure 15.39. Unlike TSA, where thermal means is used to effect the separation, PSA and VSA use mechanical work to increase the pressure or create a vacuum. While one bed is adsorbing at one pressure, the other bed is desorbing at a lower pressure, as was illustrated in Figure 15.21. Unlike TSA, which can be used to purify gases or liquids, PSA and VSA are used only with gases, because a change in pressure has little or no effect on the equilibrium loading for liquid adsorption. PSA was originally used only for purification, as in the removal of moisture from air by the "heatless drier," which was invented by C.W. Skarstrom in 1960 to compete with TSA. However, by the early 1970s, PSA was being applied to bulk separations such as the partial separation of air to produce either nitrogen or oxygen and to the removal of impurities and pollutants from other gas streams. PSA can also be used for vapor recovery, as discussed and illustrated by Ritter and Yang [91].

A typical sequence of steps in the Skarstrom cycle, operating with two beds, is shown in Figure 15.40. Each bed operates alternately in two half-cycles of equal duration: (1) pressurization followed by adsorption, and (2) depressurization (blowdown) followed by a purge. The feed gas is used for pressurization, while a portion of the effluent product gas is used for purge. Thus, in Figure 15.40, while adsorption is taking place in bed 1, part of the gas leaving bed 1 is routed to bed 2 to purge that bed in a direction countercurrent to the direction of flow of the feed gas during the adsorption step. When moisture is to

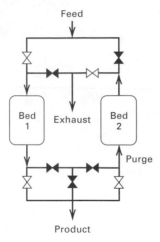

Figure 15.39 Pressure-swing adsorption cycle.

be removed from air, the dry air product is produced during the adsorption step in each of the two beds. In Figure 15.40, the adsorption and purge steps represent less than 50% of the total cycle time. In many commercial applications of PSA, these two steps consume a much greater fraction of the cycle time because pressurization and blowdown can be completed rapidly. Therefore, cycle times for PSA and VSA are short, typically seconds to minutes. Thus, small beds have relatively large throughputs. With the valving shown in Figure 15.39, the entire cyclic sequence can be programmed to operate automatically. With some valves open and others closed, as in Figure 15.39, adsorption takes place in bed 1 and purge takes place in bed 2. During the second half of the cycle, the valve openings and beds are switched.

Since the introduction of the Skarstrom cycle, numerous improvements have been made to increase product purity, product recovery, adsorbent productivity, and energy efficiency, as discussed by Yang [25] and by Ruthven, Farooq, and Knaebel [92]. Among these modifications are the use of (1) three, four, or more beds; (2) a pressure equalization step in which both beds are equalized in pressure following purge of one bed and adsorption

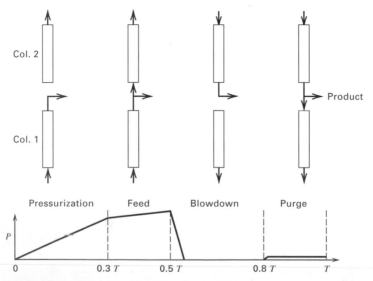

Figure 15.40 Sequence of cycle steps in PSA.

in the other; (3) pretreatment or guard beds to remove strongly adsorbed components that might interfere with the separation of other components; (4) purge with a strongly adsorbing gas; and (5) the use of an extremely short cycle time to approach isothermal operation, if a longer cycle causes an undesirable increase in temperature during adsorption and an undesirable decrease in temperature during desorption.

Separations by PSA and VSA are controlled by adsorption equilibrium or adsorption kinetics, where the latter refers to mass transfer external and/or internal to the adsorbent particle. Both types of control are important commercially. For the separation of air with zeolites, adsorption equilibrium is the controlling factor. Nitrogen is more strongly adsorbed than oxygen and argon. For air with 21% oxygen and 1% argon, oxygen of about 96% purity can be produced. When carbon molecular sieves are used, oxygen and nitrogen have almost the same adsorption isotherms, but the effective diffusivity of oxygen is much larger than that of nitrogen. Consequently, a nitrogen product of very high purity (>99%) can be produced.

PSA and VSA cycles have been modeled successfully for both equilibrium and kinetic-controlled cases. The models and computational procedures are similar to those used for TSA. The models are particularly useful for optimizing cycles. Of particular importance in PSA and TSA is the determination of the cyclic steady state. In TSA, following the desorption step, the regenerated bed is usually clean. Thus, a cyclic steady state is closely approached in one cycle. In PSA and VSA, this is often not the case, and complete regeneration is seldom achieved or necessary. It is only required to attain a cyclic steady state whereby the product obtained during the adsorption step has the desired purity and at cyclic steady state, the difference between the loading profiles after adsorption and desorption is equal to the solute entering in the feed. Starting with a clean bed, the attainment of a cycle steady state for a fixed cycle time may require tens or hundreds of cycles. Consider an example from a study by Mutasim and Bowen [93] on the removal of ethane and carbon dioxide from nitrogen with 5A zeolite, at ambient temperature with adsorption and desorption for 3 min each at 4 bar and 1 bar, respectively, in beds 0.25 m in length. Figure 15.41a and b shows the computed development of the loading and gas concentration profiles at the end of each adsorption step for ethane, starting from a clean bed. At the end of the first cycle, the bed is still clean beyond about 0.11 m. By the end of the 10th cycle, a cyclic steady state has almost been attained, with a clean bed existing

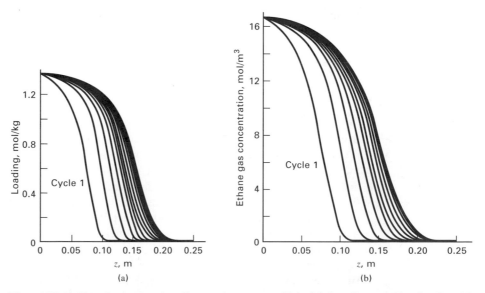

Figure 15.41 Development of cyclic steady-state profiles. (a) Loading profiles for first 11 cycles. (b) Ethane gas concentration profiles for first 16 cycles.

only near the very end of the bed. Experimental data points for ethane loading at the end of 10 cycles agree reasonably well with the computed profile from a mathematical model.

Modeling of PSA and VSA cycles is carried out with the same equations as for TSA. However, the assumptions of negligible axial diffusion and isothermal operation may be relaxed. For each cycle, the pressurization and blowdown steps are often ignored and the initial conditions for adsorption and desorption are the final conditions for the desorption and adsorption steps, respectively, of the previous cycle. This is illustrated in the following example.

EXAMPLE 15.13

Ritter and Yang [91] conducted an experimental and theoretical study of the use of PSA to recover dimethyl methylphosphonate (DMMP) vapor from air. For the following data and operating conditions, starting with a clean bed, use the method of lines with a stiff integrator to estimate the concentration and loading profiles for the beds, the percent of the feed gas recovered as essentially pure air, and the average mole fraction of DMMP in the effluent gas leaving the desorption step during the third cycle.

Feed Gas Conditions:
  236 ppm by volume of DMMP in dry air at 294 K and 3.06 atm
Adsorbent:
  BPL activated carbon, 5.25 g in each bed, 0.07 cm average particle diameter, and 0.43 bed porosity.
Bed dimensions: 1.1 cm i.d. by 12.8 cm each
Langmuir adsorption isotherm: $q = \dfrac{48,360 p_{DMMP}}{1 + 98,700 p_{DMMP}}$,

where q is in g/g and p is in atm.
Overall mass-transfer coefficient: $k = 5 \times 10^{-3}$ s^{-1}
Cycle conditions (all at 294 K):
  1. Pressurization with pure air from p_L to p_H in negligible time.
  2. Adsorption at $p_H = 3.06$ atm with feed gas for 20 mins
u = interstitital velocity = 10.465 cm/s.
  3. Blowdown from p_H to p_L with no loss of DMMP from the adsorbent or gas in the voids of the bed in negligible time.
  4. Desorption at $p_L = 1.07$ atm with product gas (pure air) for 20 min. Interstitial velocity, u, corresponding to use of 41.6% of the product gas leaving the adsorption step.

SOLUTION

This example can be solved using the same equations and numerical techniques employed in Example 15.12, but noting that the units of q are different and a Langmuir isotherm replaces Henry's law. If the bed is not clean following the first desorption step, the results for the second and third cycles will differ from the first. The results are not presented here, but the calculations are required in Exercise 15.30. ∎

Continuous Countercurrent Adsorption Systems

For the bulk separation of liquid mixtures, continuous countercurrent systems are preferred over batch systems because they maximize the mass-transfer driving force. This advantage is particularly important for difficult separations where selectivity is not high and/or where mass-transfer rates are low. Ideally, a continuous countercurrent system involves a bed of adsorbent moving downward in plug flow and the liquid mixture flowing upward in plug flow through the bed void space. Unfortunately, such a system has not been successfully developed because of problems of adsorbent attrition, liquid channeling, and nonuniform flow of adsorbent particles. Successful commercial systems are based on a simulated countercurrent system using a stationary bed, as described earlier and shown in Figure 15.23. Regardless of whether the continuous countercurrent system involves a moving bed or a simulated moving bed, the mathematical model is the same. Either a rate-based model or an equilibrium-stage model, of the type developed for gas absorption and stripping, is employed. Unlike the previously discussed adsorption-system models, the continuous countercurrent system is treated as a steady-state operation. The models can be applied to the purification of a fluid or the bulk separation of a liquid mixture. Only the former, which is the simpler case, is considered here. The rate-based case is developed by Ruthven [10].

McCabe–Thiele and Kremser Methods for Purification

Consider a binary mixture, dilute in a solute that is to be removed by adsorption in a simulated, continuous, countercurrent system of the type shown in Figure 15.42a. Only the solute is adsorbed. Feed F, with solute concentration c_F, enters the adsorption section at plane P_1, from which adsorbent S leaves with a solute loading q_F. Purified feed called the raffinate, with solute concentration c_R, leaves the adsorption section at plane P_2, countercurrent to adsorbent of loading q_R, which enters at the top of the bed. At plane P_3, a purge called the desorbent, D, with solute concentration c_D, enters at the bottom of the desorption section, from which the adsorbent leaves to enter the adsorption section. We assume that the desorbent does not adsorb and exits from the desorption section as extract E, with solute concentration c_E, at plane 4, where recycled adsorbent enters the desorption bed to complete the cycle.

If the system is dilute in the solute, if the solute adsorption isotherms for the feed solvent and the purge fluid are identical, and the system operates at constant temperature and pressure, the McCabe–Thiele diagram for the solute resembles that shown in Figure 15.42b, where the operating and equilibrium lines are straight because of the dilute condition. Note that the proper directions for mass transfer require that the adsorption and desorption operating lines lie below and above, respectively, the equilibrium line. These three lines are as follows:

Adsorption:
$$q = \frac{F}{S}(c - c_F) + q_F \qquad \text{(15-134)}$$

Desorption:
$$q = \frac{D}{S}(c - c_D) + q_R \qquad \text{(15-135)}$$

Equilibrium:
$$q = Kc \qquad \text{(15-136)}$$

where F, S, and D are solute-free mass flow rates.

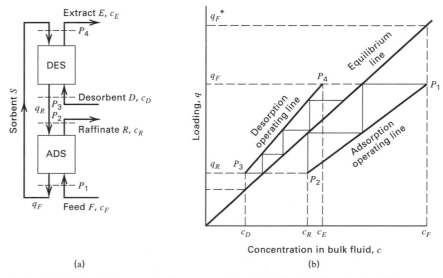

Figure 15.42 (a)

Figure 15.42 (b)

Figure 15.42 Continuous, countercurrent adsorption–desorption system. (a) System sections and flow conditions. (b) McCabe–Thiele diagram.

Comparing (15-134) to (15-136) with Figure 15.42b shows that, of necessity, the slopes of the lines must be such that

$$\frac{F}{S} < K < \frac{D}{S}$$

Thus, because more purge than feed is required, this system is only economical when the purge fluid is inexpensive. From the equilibrium and operating lines in Figure 15.42b, 2 and 3.3 equilibrium stages are determined for the adsorption and desorption sections, respectively, by stepping off stages in the McCabe–Thiele diagram. When the equilibrium and operating lines are straight, as in Figure 15.42b, the algebraic Kremser method, rather than the graphical McCabe–Thiele method, can be employed. The Kremser equation, discussed in Section 6.4, is written in the following end-point form for the adsorption or desorption section:

$$N_t = \frac{\ln\left[\dfrac{c_1 - q_1/K}{c_2 - q_2/K}\right]}{\ln\left[\dfrac{c_1 - c_2}{q_1/K - q_2/K}\right]} \tag{15-137}$$

where 1 and 2 refer to opposite ends of the section, such as planes 1 and 2 in Figure 15.42a, which are chosen so that $q_1 > q_2$.

If the operating conditions for the two sections can be altered so as to place the equilibrium line for desorption below that for adsorption, it becomes possible to use a portion of the raffinate for desorption. This situation, shown in Figure 15.43, is achieved by desorbing at elevated temperature or, in the case of gas adsorption, at reduced pressure. Now, as shown in Figure 15.43, F/S can be greater than D/S. With a portion of the raffinate used in bed 2, the net raffinate product is $F - D$. Note that in this case, the two operating lines must intersect at the point (q_R, c_R). By adjusting the ratio D/F, this point can be moved closer and closer to the origin so as to achieve any raffinate purity, c_R, desired, but at the expense of more theoretical stages and, therefore, deeper beds. For a computed number of theoretical stages N_t, the bed height L can be determined from

$$L = N_t(\text{HETP}) \tag{15-138}$$

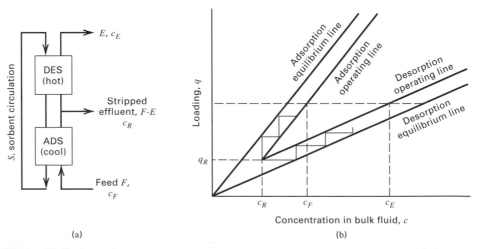

(a)   (b)

Figure 15.43 Continuous, countercurrent system with a temperature swing. (a) System sections and flow conditions. (b) McCabe–Thiele diagram.

of theoretical stages N_t, the bed height L can be determined from

$$L = N_t(\text{HETP}) \tag{15-138}$$

Values of HETP, which depend on mass transfer resistances and axial dispersion, must be determined from experimental measurements.

McCabe–Thiele Method for Bulk Separation

Since the invention in 1961 of the Sorbex simulated moving-bed process, as described by Broughton [67] and depicted in Figure 15.23, the bulk separation of mixtures by liquid adsorption has been widely practiced. In that process, a feed mixture is separated into a purified raffinate and concentrated extract. For a binary mixture of A and B, the cyclic fixed-bed process of Figure 15.23 is closely equivalent to the moving-bed process in Figure 15.44, where A is the more strongly adsorbed component. The extract is rich in A, while the raffinate is rich in B. The Sorbex process resembles a binary distillation operation, with the feed mixture entering at the middle, but two sections are used above and below the feed. The net flow of A is downwards through bed II, following adsorption in bed I. Component B is desorbed in bed II and A is desorbed in bed III to produce the A-rich extract product. The net flow of B is upwards through bed I, following desorption in bed II. Raffinate, rich in B, is withdrawn between beds IV and I.

The Sorbex process requires the addition of a third component, C, called the desorbent or eluant, to facilitate the desorption of A in bed III and the adsorption of B in bed IV. In bed III, C is adsorbed while A is desorbed. In bed IV, C is partially desorbed, while

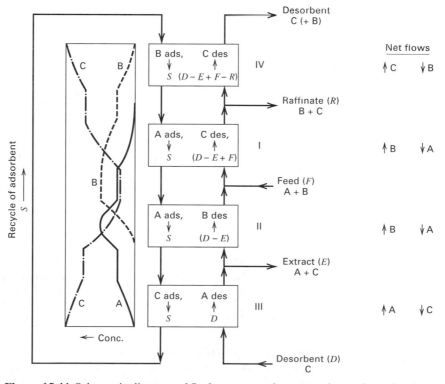

Figure 15.44 Schematic diagram of Sorbex process for separating A from B using eluant C. [From D.B. Broughton, *Chem. Eng. Progress*, **64** (8), 60–65 (1968), with permission.]

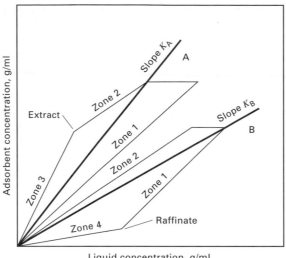

Figure 15.45 McCabe diagram for Sorbex process. [From D.B. Broughton, *Chem. Eng. Progress,* **64** (8), 60–65 (1968), with permission.]

more, the adsorption properties of C must be such that it can displace A from the adsorbent or be displaced from the adsorbent by B, depending upon the relative flow rates of solids and liquid. Adsorbent rich in C and leaving bed III is recycled to bed IV.

Typical concentration profiles of A, B, and C are included in Figure 15.44. Under cyclic steady-state conditions, the following flow ratio constraints are necessary for the limiting case of a perfect separation between A and B, where the flows are indicated in Figure 15.44, and the adsorption isotherms are linear and independent of each other and the concentration of C:

$$\frac{S}{D+F-E-R} > \alpha_{C,B}, \quad \frac{S}{D+F-E} > \alpha_{B,A}, \quad \frac{S}{D-E} < \alpha_{A,B}, \quad \text{and} \frac{S}{D} < \alpha_{C,A}$$

where $\alpha_{i,j} = K_i/K_j$ = relative selectivity for linear adsorption isotherms. With these constraints, a typical composite McCabe–Thiele diagram is given in Figure 15.45. The heavy lines are the equilibrium lines for A and B. The light lines are the corresponding operating lines for the different zones. The McCabe–Thiele construction for nonlinear adsorption isotherms is similar to that in Figure 15.45.

EXAMPLE 15.14

One hundred pounds per minute (dry basis) of air at 80°F and 1 atm with 65% relative humidity is dehumidified isothermally and isobarically to 10% relative humidity in a continuous, counter-current moving-bed adsorption unit. The adsorbent is dry silica gel (SG) having a particle-diameter range of 1.42 to 2.0 mm. Over the water partial pressure range of interest, the adsorption isotherm is given by measurements of Eagleton and Bliss [94] as

$$q_{H_2O} = 29c_{H_2O} \tag{1}$$

with concentration in lb H_2O/lb dry air and loading in lb H_2O/lb dry SG. If 1.5 times the minimum flow rate of silica gel is used, determine the number of equilibrium stages required.

SOLUTION

For relative humidities of 65% and 10%, the corresponding moisture contents are, from a humidity chart, 0.0143 and 0.0022 lb H_2O/lb dry air, respectively.

In this case, Figure 15.42b applies for just the adsorption section. Using the nomenclature in that figure:

$F = 100$ lb/min, $c_F = 0.0143$ lb H_2O/lb dry air, $c_R = 0.0022$ lb H_2O/lb dry air, and $q_R = 0$.

The value of q_F depends on the adsorbent flow rate, S, which is 1.5 times the minimum value. At minimum adsorbent rate, exiting adsorbent is in equilibrium with the entering gas. Therefore, from (1): $q_F^* = 29(0.0143) = 0.415$ lb H_2O/lb dry SG. The amount of water vapor adsorbed is $F(c_F - c_R) = 100(0.0143 - 0.0022) = 1.21$ lb/min. Therefore: $S_{min} = \dfrac{1.21}{0.415} = 2.92$ lb dry SG/min. If 1.5 times the minimum amount of silica gel is used: $S = 1.5\, S_{min} = 1.5(2.92) = 4.38$ lb dry SG/min. By material balance: $q_F = \dfrac{1.21}{4.38} = 0.276$ lb H_2O/lb dry SG. From (15-137), with $K = 29$ from (1) and letting F be at plane 1 and R at plane 2:

$$N_t = \frac{\ln\left[\dfrac{0.0143 - 0.276/29}{0.0022 - 0}\right]}{\ln\left[\dfrac{0.0143 - 0.0022}{0.276/29 - 0}\right]} = 3.2 \text{ stages}$$
■

Ion-Exchange Cycle

Although ion exchange has a wide range of applications, water softening with gel resins continues to be the major one. Usually a fixed bed is used, which is operated in a cycle of four steps: (1) loading, (2) displacement, (3) regeneration, and (4) washing. The solute ions removed from the water in the loading step are mainly Ca^{2+} and Mg^{2+}, which are absorbed by the resin while an equivalent amount of Na^+ is transferred from the resin to the water as the feed solution flows down through the bed. If mass transfer is rapid, the solution and resin are at equilibrium at all points in the bed. With a divalent ion (e.g., Ca^{2+}) replacing a monovalent ion (e.g., Na^+), the equilibrium expression is given by (15-44), where A is the divalent ion. If $(Q/C)^{n-1}K_{A,B} \gg 1$, equilibrium for the divalent ion is very favorable (see Figure 15.29a) and a self-sharpening front of the type shown in Figure 15.29b develops. In that case, which is common, the ion exchange is well approximated using the simple stoichiometric or shock-wave front theory for adsorption, assuming plug flow. As the front moves down through the bed, the resin behind or upstream of the front is in equilibrium with the feed composition. Ahead or downstream of the front, the water is essentially free of the divalent ion(s). Breakthrough occurs when the front reaches the end of the bed.

Suppose that the only cations in the feed are Na^+ and Ca^{2+}. Then, from (15-44):

$$K_{Ca^{2+},Na^+}\left(\frac{Q}{C}\right) = \frac{y_{Ca^{2+}}(1 - x_{Ca^{2+}})}{x_{Ca^{2+}}(1 - y_{Ca^{2+}})} \tag{15-139}$$

where Q is the total concentration of the two cations in the resin, in eq/L of bed of wet resin, and C is the total concentration of the two ions in the solution, in eq/L of solution. One mole of Na^+ is 1 equivalent, while 1 mole of Ca^{2+} is 2 equivalents. The quantities y_i and x_i are equivalent (rather than mole) fractions. From Table 15.5, using (15-45), the molar selectivity factor is

$$K_{Ca^{2+},Na^+} = 5.2/2.0 = 2.6$$

For a given loading step during water softening, values of Q and C remain constant. Thus, for a given equivalent fraction, $x_{Ca^{2+}}$ in the feed, (15-139) is solved for the equilibrium $y_{Ca^{2+}}$. By material balance, for a given bed volume, the time t_L for the loading step is computed. The loading wave-front velocity u_L is $u_L = L/t_L$ where L is the height of the

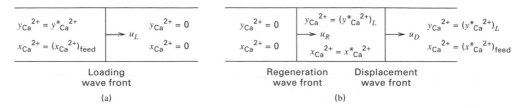

Figure 15.46 Ion exchange in a cyclic operation with a fixed bed. (a) Loading step. (b) Displacement and regeneration steps.

bed. Equivalent fractions ahead of and behind the loading front are shown in Figure 15.46a. Typically, feed-solution superficial mass velocities are about 15 gal/h-ft^2, but can be much higher at the expense of larger pressure drops.

At the end of the loading step, the bed voids are filled with feed solution, which must be displaced from the bed. This is best done with a regeneration solution, which is usually a concentrated salt solution that flows upwards through the bed. Thus, the displacement and regeneration steps are combined. Following displacement, mass transfer of Ca^{2+} from the resin beads to the regenerating solution takes place while an equivalent amount of Na^+ is transferred from the solution to the resin. In order for equilibrium to be favorable for regeneration with Na^+, it is necessary for $(Q/C)K_{Ca^{2+},Na^+} \ll 1$. In that case, which is just the opposite for loading, the wave front during regeneration sharpens quickly into a shocklike wave. This criterion can be satisfied by using a saturated salt solution to give a large value for C.

During displacement and regeneration, two concentration waves move through the bed. The first is the displacement front; the second, the regeneration front. For plug flow and negligible mass-transfer resistance, the resin and solution are in equilibrium at all locations in the bed. Again (15-139) is used to solve for the equilibrium equivalent fractions, which are shown for the displacement and regeneration steps in Figure 15.46b. The displacement time, t_D, is determined from the interstitial velocity, u_D, of the fluid during displacement:

$$t_D = L/u_D$$

The regeneration time, t_R, is determined by material balance, from which the regeneration wave-front velocity is $u_R = L/t_R$. In general, the mass velocity of the regeneration solution is less than that of the feed solution. The cycle is completed by displacing, with water, the salt solution in the bed voids. The cycle calculations are illustrated by the following example.

EXAMPLE 15.15

Hard water, containing 500 ppm (by weight) of magnesium carbonate and 50 ppm of NaCl, is to be softened at 25°C in an existing fixed bed of gel resin of a cation capacity of 2.3 eq/L of bed volume. The bed is 8.5 ft in diameter and packed to a height of 10 ft, with a wetted-resin void fraction of 0.38. During the loading step, the recommended throughput is 15 gal/min-ft^2. During displacement, regeneration, and washing, the flow rate is reduced to 1.5 gal/min-ft^2. The displacement and regeneration solutions are water saturated with NaCl (26 wt%). Determine: (a) flow rate of feed solution, L/min, (b) loading time to breakthrough, h, (c) loading wave-front velocity, cm/min, (d) flow rate of regeneration solution, L/min, (e) displacement time, h, (f) additional time for regeneration, h, (g) regeneration wave-front velocity, cm/min, (h) amount of regeneration solution for one cycle, L, and (i) Washing time, h.

SOLUTION

M of $MgCO_3 = 83.43$

Concentration of $MgCO_3$ in feed $= \dfrac{500\,(1,000)}{83.43\,(1,000,000)} = 0.006$ mol/L or 0.012 eq/L

M of NaCl $= 58.45$

Concentration of NaCl in feed $= \dfrac{50\,(1,000)}{58.45\,(1,000,000)} = 0.000855$ mol/L or eq/L

(a) Bed cross-section area = $3.14(8.5)^2/4$ = 56.7 ft².

Feed solution flow rate = 15(56.7) = 851 gpm or 3,219 L/min

(b) Behind the loading wave front: $x_{Ca^{2+}} = \dfrac{0.012}{0.012 + 0.000855} = 0.9335$

Since no NaCl in the feed is exchanged: C_L = 0.012 eq/L and Q = 2.3 eq/L

From Table 15.5, K_{Mg^{2+},Na^+} = 3.3/2 = 1.65

From (15-139), for Mg²⁺ instead of Ca²⁺ as the exchanging ion, with $x_{Mg^{2+}}$ = that of the feed from Figure 15.46a:

$$1.65\left(\frac{2.3}{0.012}\right) = \frac{(y^*_{Mg^{2+}})(1 - 0.9335)}{(0.9335)(1 - y^*_{Mg^{2+}})}$$

Solving: $y^*_{Mg^{2+}}$ = 0.9998. Thus, sodium ion is displaced from the resin almost completely.

Bed volume = (56.7)(10) = 567 ft³ or 16,060 L
Total bed capacity = 2.3 (16,060) = 36,940 eq
Mg²⁺ absorbed by resin = 0.9998(36,940) = 36,930 eq
Mg²⁺ entering bed in feed solution = 0.012(3,219) = 38.63 eq/min
$t_L = \dfrac{36,930}{38.63}$ = 956 min or 15.9 h

(c) $u_L = L/t_L$ = 10/956 = 0.01046 ft/min or 0.319 cm/min.
(d) Flow rate of regeneration solution = (1.5/15) (3,219) = 321.9 L/min.
(e) Displacement time = time for 321.9 L/min to displace liquid in the voids.

Void volume = 0.38 (16,060) = 6,103 L and $t_D = \dfrac{6103}{321.9}$ = 19 min

(f) For a 26 wt% NaCl solution at 25°C, density from *Perry's Chemical Engineers' Handbook* − 1.19443 g/cm³.

Flow rate of Na⁺ in regeneration solution = $\dfrac{321.9(1,000)(1.19443)(0.26)}{58.45}$ = 1,710 eq/min

NaCl concentration in regenerating solution = $\dfrac{1,710}{321.9}$ = 5.31 eq/L = c_R

From (15-139), noting conditions in Figure 15.46b: $\dfrac{Q}{c_R} K_{Mg^{2+},Na^+} = 1.65\left(\dfrac{2.3}{5.31}\right)$ = 0.715

This is less than 1, but not much less than 1. Therefore, the regeneration wave front may not sharpen rapidly. Assume a shock-wave-like front anyway.

From (15-139), $0.715 = \dfrac{(0.09998)(1 - x^*_{Mg^{2+}})}{x^*_{Mg^{2+}}(1 - 0.9998)}$

Solving: $x^*_{Mg^{2+}}$ = 0.9998
So downstream of the regeneration wave front, but upstream of the displacement wave front, the liquid contains very few sodium ions. ∎

Chromatographic Separations

The separation of multicomponent mixtures into more than two products usually requires more than one separation device. For example, if a four-component mixture (A, B, C, D) is to be separated by distillation into pure products, a sequence of three trayed columns is almost always used. If the order of decreasing volatility is A, B, C, and D, the first column might produce a distillate of nearly pure A; the second column a distillate of nearly pure B; and the third column a distillate of nearly pure C and a bottoms of nearly pure D. Four other sequences are possible, depending upon the selection of the split for each column.

Chromatography is one of the few separation techniques that can separate a multicomponent mixture into nearly pure components in a single device, generally a column packed with a suitable sorbent. The degree of separation depends upon the length of the column and the differences in component affinities for the sorbent.

As an example, consider a mixture of three components, A, B, and C, in order of decreasing affinity for the sorbent, S. If the separation is achieved by adsorption, then A is the most strongly adsorbed. A feed mixture, insufficient to load the sorbent, is introduced as a pulse into one end (feed end) of the packed chromatographic column. The resulting initial concentrations for the three components are shown in Figure 15.47a, where most of the bed remains clean of adsorbates. An *elutant*, such as a carrier gas or solvent that has little or no affinity for the sorbent, is now introduced continuously into the feed end of the bed, causing the three components to desorb, with C desorbing most readily. However, as the desorbed components are carried down the bed by the elutant into cleaner regions of the bed, the components are successively readsorbed and then redesorbed to produce three waves, as shown in Figure 15.47b. Because of the differences in affinities for the sorbent, the three waves, which initially overlap considerably, gradually overlap less (Figure 15.47c), and finally, if the column is long enough, become completely separated, as in Figure 15.47d and e. In that case, the components are eluted from the column, one at a time. In Figure 15.47e, all components but A have been eluted. As the separated waves elute, the area under each component wave is proportional to the mass of the component moving through the packed column.

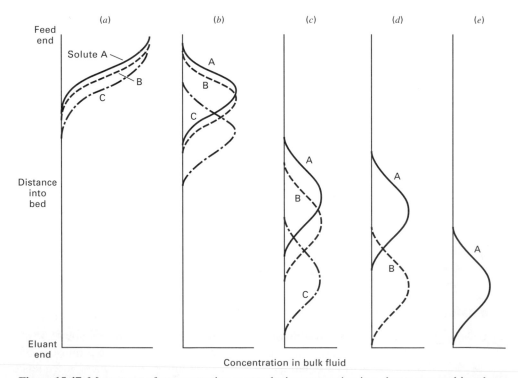

Figure 15.47 Movement of concentration waves during separation in a chromatographic column.

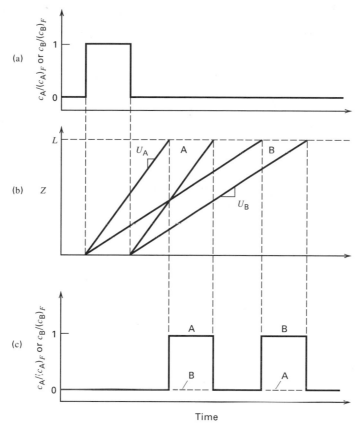

Figure 15.48 Ideal solute wave pulses in a chromatographic column.

Equilibrium Wave Pulse Theory for Linear Isotherm

A simple and useful wave theory for chromatography is based on isothermal plug flow, negligible axial dispersion, and local equilibrium everywhere. This theory, when developed for adsorption, results in the stoichiometric wave front that was shown in Figure 15.27. For chromatography, where solutes are pulsed into the column, a wave pulse rather than a wave front results, as shown in Figure 15.48a. For a stoichiometric (equilibrium) wave, the pulse is a square wave rather than a Gaussian-distribution-like wave of the type shown in Figure 15.47c. The latter type of wave results when axial dispersion occurs, mass-transfer resistances are important, radial variations of the fluid velocity occur, and/or the pulse of solutes is not a square wave.

If the sorbent is nonporous, such as a gel, and the adsorption isotherm is linear for each solute i ($q = Kc$), then (15-101) applies and each solute wave velocity, u_i, in terms of the interstitial fluid velocity is given by

$$u_i = \frac{u}{1 + \dfrac{1 - \epsilon_b}{\epsilon_b} K_i} \qquad \textbf{(15-140)}$$

This equation applies to both the leading and trailing edges of the feed pulse to produce solute movement diagrams, used extensively by Wankat [95]. When the elutant is dilute in the solutes, such that the sorption equilibrium constants do not depend on composition, but only on temperature, (15-140) applies independently to each solute in a multicomponent mixture. For a strongly sorbed solute, such as B in Figure 15.48b, K is large, and the second term in the denominator dominates, so that solute moves much slower through the bed than does the elutant. For a weakly sorbed solute, such as A in Figure 15.48b, K is small and the denominator is not much greater than 1. For such a solute, its velocity may not be significantly smaller than the elutant velocity.

In Figure 15.48b, the wave velocities of both the leading and trailing edges of the wave pulses are constant, with $u_A > u_B$. For each solute, the pulse time to move through a column of length L is $t_i = L/u_i$. The wave pulse of A reaches the end of the column in less time than the B wave pulse. In Figure 15.48c, the product concentration ratios, c_A/c_{A_F}, and c_B/c_{B_F}, are shown at the end of the bed as a function of time. The widths of these product waves are identical to the widths of the feed pulses, as illustrated in the following example. This simple equilibrium wave pulse theory for linear isotherms can be used to obtain an approximate estimate of the separation achievable in a chromatographic column. Unfortunately, the estimate is not conservative when computing the necessary column length, because the pulses broaden, as shown after the following example.

EXAMPLE 15.16

An aqueous solution of 3 g/cm^3 each of glucose (G), sucrose (S), and fructose (F) is to be separated in a chromatographic column, packed with an ion-exchange resin of the calcium form. In the range of expected solute concentrations, the sorption isotherms are linear and independent, with $q_i = K_i c_i$, where q_i is in grams sorbent per 100 cm^3 resin and c_i is in grams solute per 100 cm^3 solution. From experiment:

Solute	K
Glucose	0.26
Sucrose	0.40
Fructose	0.66

The superficial solution velocity is 0.031 cm/s and bed void fraction is 0.39. If a 500-second pulse, t_P, of feed is followed by elution with pure water, what length of column packing is needed to separate the three solutes if sorption equilibrium is assumed? How soon after the first pulse begins can a second 500-second pulse begin?

SOLUTION

Interstitial solution velocity = 0.031/0.39 = 0.0795 cm/s.
Wave velocity for glucose from (15-140):

$$u_G = \frac{0.0795}{1 + \left(\dfrac{1 - 0.39}{0.39}\right)(0.26)} = 0.0565 \text{ cm/s}$$

Similarly, $\qquad u_S = 0.0489$ cm/s and $u_F = 0.0391$ cm/s

The smallest difference in wave velocities is between glucose and sucrose. Therefore, the separation between these two waves determine the column length. The minimum column length, assuming equilibrium, corresponds to the time at which the trailing edge of the glucose wave pulse, together with the leading edge of the sucrose wave pulse, leaves the column. Thus, if t_P is the duration of the first pulse and L is the length of the packing:

$$t_P + \frac{L}{u_G} = \frac{L}{u_S} \tag{1}$$

Thus, $\qquad\qquad\qquad 500 + \dfrac{L}{0.0565} = \dfrac{L}{0.0489}$

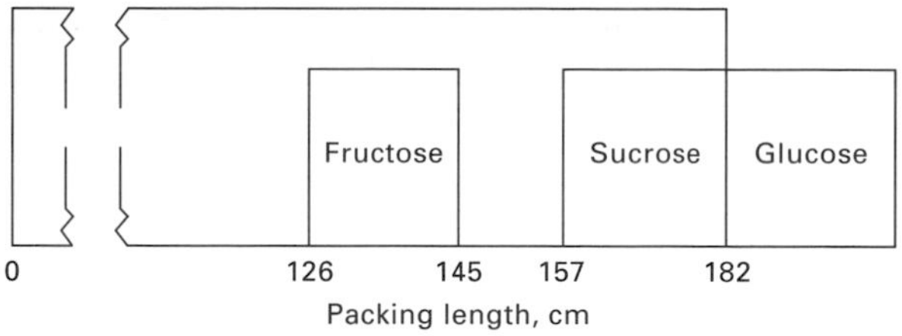

Figure 15.49 Locations of solute waves of first pulse for Example 15.16 at 3,718 s.

Solving, length of packing, $L = 182$ cm. The glucose just leaves the column at

$$500 + \frac{182}{0.0565} = 3,718 \text{ s}$$

The locations of the three wave fronts in the column at 3,718 s are shown in Figure 15.49.

The time at which the second pulse begins is determined so that the trailing edge of the first fructose wave pulse just leaves the column as the second pulse of glucose begins to leave the column. This time, based on the fructose, is $500 + \dfrac{182}{0.0391} = 5,155$ s. It takes the leading edge of a glucose wave $\dfrac{182}{0.0565} = 3,220$ s to pass though the column. Therefore, the second pulse can begin at $5,155 - 3,220 = 1,935$ s. This establishes the following ideal cycle: pulse: 500 s, elute: 1,435 s, pulse: 500 s, elute: 1,435 s, etc. In the real case, where we account for mass-transfer resistance, as shown in the next example, the column will have to be longer. ■

Analytical Solution for Rate-Based Chromatography with Linear Isotherm

A plot of solute concentrations in the elutant as a function of time is a chromatogram. When mass-transfer resistances, axial dispersion, and other non-ideal phenomena are not negligible, the solute concentrations in a chromatogram will not appear as square waves, but will exhibit the wave shapes in Figure 15.47. Carta [96] developed analytical solutions for chromatographic response to periodic injections of rectangular feed pulses, taking into account mass-transfer resistances for solute mixtures having linear, independent adsorption isotherms. Carta's solution for the LDF approximation is readily applied to the determination of the necessary length of packing and frequency of feed pulses for the chromatographic separation of a feed mixture.

For each solute in the feed, (15-102) is simplified by neglecting axial dispersion and assuming a constant interstitial fluid velocity, u, through the packing:

$$\frac{\partial c_i}{\partial t} + u \frac{\partial c_i}{\partial z} + \frac{(1 - \epsilon_b)}{\epsilon_b} \frac{\partial \overline{q}_i}{\partial t} = 0 \qquad \textbf{(15-142)}$$

The linear driving force approximation (15-105), the linear isotherm, and (15-106) for the overall mass-transfer resistance are assumed to apply for each solute. For periodic, rectangular feed pulses, the boundary conditions for feed pulses of duration, t_F, each followed by an elution period of duration, t_E, are for each solute concentration, $c_i\{z, t\}$:

Initial condition:
At $t = 0$, $c_i\{z, 0\} = 0$ $\qquad$ **(15-143)**

Feed pulse:
At $z = 0$, $c_i\{0,t\} = (c_i)_F$ for $(j - 1)(t_F + t_E) < t < j(t_F + t_E) - t_E$ $\qquad$ **(15-144)**

Elution period:
At $z = 0$, $c_i\{0, t\} = 0$ for $j(t_F + t_E) - t_E < t < j(t_F + t_E)$ $\qquad$ **(15-145)**

where $j = 1, 2, 3, \ldots$ is an index that accounts for the periodic nature of the feed and elution pulses. Thus, with $j = 1$, the feed pulse takes place from $t = 0$ to $t = t_F$ and the elution pulse is from t_F to $t_F + t_E$.

Carta solved the linear system of (15-142), (15-105), and (15-106) for conditions (15-143) to (15-145) by the Laplace transform method to obtain the following series solution, in terms of dimensionless parameters, which is applied to each solute in the feed pulse:

$$X = \frac{r_F}{2r} + \frac{2}{\pi} \sum_{m=1}^{\infty} \left[\frac{1}{m} \exp\left(-\frac{m^2 n_f}{m^2 + r^2} \right) \sin\left(\frac{m\pi r_F}{2r} \right) \cos\left(\frac{m\theta_f}{r} - \frac{m\pi r_F}{2r} - \frac{m\beta n_f}{r} - \frac{mrn_f}{m^2 + r^2} \right) \right]$$

(15-146)

where,
$$X = c/c_F \tag{15-147}$$

$$r = \frac{k}{2\pi K}(t_F + t_E) \tag{15-148}$$

$$r_F = \frac{k}{\pi K} t_F \tag{15-149}$$

$$n_f = \frac{(1 - \epsilon_b)kz}{\epsilon_b u} \tag{15-150}$$

$$\theta_f = \frac{kt}{K} \tag{15-151}$$

$$\beta = \frac{\epsilon_b}{(1 - \epsilon_b)K} \tag{15-152}$$

$$K = q/c \tag{15-153}$$

and where

$$k = \frac{1}{\dfrac{R_p}{3k_c} + \dfrac{R_p^2}{15D_e}} \tag{15-154}$$

When nonlinear adsorption isotherms such as the Freundlich equation (15-35), the Langmuir equation (15-36), or extensions thereof to multicomponent mixtures [e.g., concentration forms of (15-32) or (15-33)] are necessary, the analytical solution of Carta is not applicable. However, the method of lines, using five-point biased upwind finite-difference approximations, as described earlier in the section on thermal-swing adsorption, can be applied to obtain a numerical solution.

EXAMPLE 15.17

Use Carta's equation with the following properties to compute the chromatogram for the conditions of Example 15.16 with a packing length of 182 cm. Does a significant overlap of peaks result?

Property	Glucose	Sucrose	Fructose
K	0.26	0.40	0.66
D_e, cm^2/s	1.1×10^{-8}	1.8×10^{-8}	2.8×10^{-8}
k_c, cm/s	5.0×10^{-3}	5.0×10^{-3}	5.0×10^{-3}

$\epsilon_b = 0.39$, $R_p = 0.0025$ cm, $u = 0.0795$ cm/s, $z = 182$ cm, $t_E = 2,000$ s, and $t_F = 500$ s

SOLUTION

Values of k and the computed dimensionless parameters from (15-148) to (15-152) are as follows:

	Glucose	Sucrose	Fructose
r	40.22	42.66	40.06
r_F	16.09	17.07	16.03
n_f	94.13	153.6	238.0
θ_f	0.1011 t	0.1072 t	0.1007 t
β	2.459	1.598	0.9687
k, s^{-1}	0.0263	0.0248	0.0665

where t is in seconds.

Values of $X = c/c_F$ are computed with these parameters using (15-146) for values of time, t, in the neighborhood of times for the equilibrium-based waves. The resulting chromatogram for glucose is shown in Figure 15.50a, compared to the equilibrium rectangular wave (shown as a dashed line) determined in Example 15.16 using (15-140). The areas under the two curves should be identical. The equilibrium-based wave appears to be centered in time within the mass-transfer-based wave.

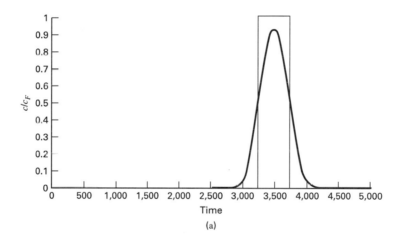

(a)

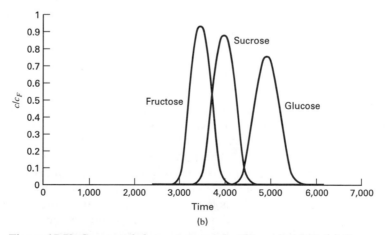

(b)

Figure 15.50 Computed chromatograms for Example 15.17. (a) Comparison of ideal to nonideal wave for fructose. (b) Computed chromatogram for nonideal eluant.

In Figure 15.50b, the complete computed chromatogram is plotted for the three carbohydrates. It is seen that the effect of mass transfer is to cause the peaks to overlap significantly. To obtain a sharp separation, it is necessary to lengthen the column or reduce the feed pulse time, t_F.

      ■

SUMMARY

1. Sorption is a generic term for the selective transfer of a solute from the bulk of a liquid or gas to the surface and/or into the bulk of a solid or liquid. Thus, sorption includes adsorption and absorption. The sorbed solute is commonly called the sorbate.

2. For commercial applications, a sorbent should have high selectivity, high capacity, rapid solute transport rates, stability, strength, and ability to be regenerated. An adsorbent should have small pores so as to give a large surface area per unit volume.

3. Physical adsorption of pure gases and gas mixtures is easily measured. Adsorption of pure liquids and liquid mixtures is not easily measured.

4. The most widely used commercial adsorbents are carbon (activated and molecular-sieve), molecular-sieve zeolites, silica gel, and activated alumina.

5. The most widely used ion exchangers are water-swellable solid gel resins based on the copolymerization of styrene and a cross-linking agent, such as divinylbenzene. They can be cation or anion exchangers. Ions are exchanged stoichiometrically on an equivalent basis. Thus, Ca^{2+} is exchanged for 2 Na^+.

6. Sorbents for chromatographic separations are typically solid adsorbents, liquid absorbents supported on or bonded to an inert solid, or a gel.

7. The most commonly used adsorption isotherms for gases and liquids are Henry's law (linear isotherm), the Freundlich isotherm, and the Langmuir isotherm. The last asymptotically approaches the linear isotherm at low concentrations and an asymptotic value, representing maximum surface coverage, at high concentrations. For mixtures, extended versions of the Freundlich and Langmuir isotherms are often used.

8. Ion-exchange equilibrium is most commonly represented by an equilibrium constant based on the law of mass action. Because of the dilute conditions common in chromatography, a linear equilibrium isotherm is commonly employed.

9. For physical adsorption, the rate of adsorption is almost instantaneous after the solute reaches the sorbing surface. Thus, only external and internal mass-transfer resistances need be considered. External mass-transfer coefficients are generally obtained from empirical correlations of the Chilton–Colburn j-factor type. Internal mass transfer is generally based on a modified Fick's first law using an effective diffusivity that depends on particle porosity, pore tortuosity, bulk molecular diffusivity, and surface diffusivity. Diffusivities in ion-exchange resin gels depend strongly on the degree of cross-linking.

10. A wide variety of sorption systems are used, including slurry adsorption in various modes of operation, fixed-bed adsorption, and simulated continuous countercurrent adsorption. When sorbent regeneration is necessary, the system must be operated on a cycle. For fixed beds, the most common cycles are temperature-swing adsorption (TSA) and pressure-swing adsorption (PSA). Ion exchange almost always includes a regeneration step, using a displacement fluid. In a chromatographic separation, adsorption and regeneration take place in the same column.

11. For the design and operation of all sorption systems, the adsorption isotherm is of great importance because it relates, at equilibrium, the concentration of the solute in the fluid to its loading as a sorbate in and/or on the sorbent. Most commonly,

the overall rate of adsorption is expressed in the form of a linear driving force (LDF) model, where the driving force is the difference between the bulk concentration and the concentration in equilibrium with the loading. The coefficient in the LDF equation is a combined overall mass-transfer coefficient and area for sorption.

12. In ideal fixed-bed operation, solute–sorbate equilibrium between the flowing fluid and the static bed is assumed everywhere. For plug flow and negligible axial dispersion, the result is a sharp concentration front that moves like a shock wave (stoichiometric front) through the bed. Upstream of the front, the sorbent is spent and in equilibrium with the feed mixture. Downstream of the front, the sorbent is clean of sorbate. Typically, the stoichiometric front travels through the bed at a much slower velocity than the interstitial velocity of the fluid feed. The time for the concentration front to reach the end of the bed is the breakthrough time.

13. When mass-transfer effects are taken into account, the concentration front broadens into an S-shaped curve such that at breakthrough only a portion of the sorbent is fully loaded. When mass-transfer coefficients and sorption isotherms are known, these curves can be readily computed with the Klinkenberg equations. Alternatively, when the shapes of experimental concentration fronts appear to exhibit a constant pattern, because of favorable adsorption equilibrium, commercial-size adsorption beds can be scaled-up directly from experimental breakthrough data by the method of Collins.

14. Thermal-swing adsorption (TSA) can be used to remove small concentrations of solutes from gas and liquid mixtures. Typically, adsorption is carried out at ambient temperature and desorption at an elevated temperature. Because heating and cooling of the bed between the adsorption and desorption steps is not instantaneous, TSA cycles are long, typically hours or days. The desorption step, starting with a partially loaded bed, can be computed numerically by the method of lines, using a stiff integrator.

15. Pressure-swing adsorption (PSA) is used to separate air and enrich hydrogen-containing streams. Adsorption is carried out at an elevated or ambient pressure, whereas desorption occurs at ambient pressure or in a vacuum; the latter is often referred to as vacuum-swing adsorption (VSA). Because pressure swings can be made rapidly, PSA cycles are short, typically seconds or minutes. Usually, it is not necessary to regenerate the bed completely. When that circumstance exists and the cycle is fixed, a number of cycles may be needed to approach a cyclic steady-state operation.

16. Although continuous, countercurrent adsorption with a moving bed is difficult to achieve in practice, a simulated system is used for bulk liquid separations. Scale-up design calculations for such a system are conveniently made with the McCabe–Thiele graphical method.

17. Design calculations for ion-exchange operations are based on the equilibrium assumption for both the loading and regeneration steps.

18. In chromatography, the feed is periodically pulsed into a column packed with sorbent. Between feed pulses, an elutant is passed through the column, causing the less strongly sorbed solutes to move through the column more rapidly than other solutes. If the column is long enough, a multicomponent feed can be completely separated, with solutes eluted one by one from the column. In the absence of mass-transfer resistances, a rectangular feed pulse is separated into individual solute rectangular pulses, whose position–time curves are readily established. When mass-transfer effects are important, the rectangular pulses take on a Gaussian distribution that can be predicted by the analytical solution of Carta, provided that a linear adsorption isotherm applies and axial dispersion is negligible.

REFERENCES

1. Keller, G.E. II, in *Industrial Gas Separations*, T.E., Whyte, Jr., C.M. Yon, and E.H. Wagner, eds., ACS Symposium Series No. 223, American Chemical Society, Washington, D.C., p. 145 (1983).
2. Milton, R.M., U.S. Patents 2,882,243 and 2,882,244 (1959).
3. Skarstrom, C.W., U.S. Patent 2,944,627 (1960).
4. Broughton, D.B., and C.G. Gerhold, U.S. Patent 2,985,589 (May 23, 1961).
5. Ettre, L.S., and A. Zlatkis, Eds., *75 Years of Chromatography—A Historical Dialog*, Elsevier, Amsterdam (1979).
6. Bonmati, R.G., G. Chapelet-Letourneux, and J.R. Margulis, *Chem. Engr.*, **87** (6), 70–72 (1980).
7. Bernard, J.R., J.P. Gourlia, and M.J. Guttierrez, *Chem. Engr.*, **88** (10), 92–95 (1981).
8. White, D.H., Jr., and P.G. Barkley, *Chem. Eng. Progress*, **85** (1) 25–33 (1989).
9. Rousseau, R.W., Ed., *Handbook of Separation Process Technology*, Wiley-Interscience, New York (1987).
10. Ruthven, D.M., *Principles of Adsorption and Adsorption Processes*, John Wiley and Sons, New York (1984).
11. Brunauer, S., P.H. Emmett, and E. Teller, *J. Am. Chem. Soc.*, **60**, 309 (1938).
12. Satterfield, C.N., *Heterogeneous Catalysis in Practice*, McGraw-Hill, New York (1980).
13. Ruthven, D.M., in *Kirk-Othmer Encyclopedia of Chemical Technology*, Vol. 1, 4th ed., Wiley-Interscience, New York (1991).
14. Barrer, R.M., *Zeolites and Clay Minerals as Sorbents and Molecular Sieves*, Academic Press, New York (1978).
15. Breck, D.W., *Zeolite Molecular Sieves*, John Wiley and Sons, New York (1974).
16. Adams, B.A., and E.L. Holmes, *J. Soc. Chem. Ind.*, **54**, 1–6T (1935).
17. McWilliams, J.D., *Chem. Engr.*, **85** (12), 80–84 (1978).
18. Dorfner, K., *Ion Exchangers, Properties and Applications*, 3rd ed., Ann Arbor Science, Ann Arbor, MI (1971).
19. Sewell, P.A., and B. Clarke, *Chromatographic Separations*, John Wiley and Sons, New York (1987).
20. Brunauer, S., L.S. Deming, W.E. Deming, and E. Teller, *J. Am. Chem. Soc.*, **62**, 1723–1732 (1940).
21. Brunauer, S., *The Adsorption of Gases and Vapors*, Vol. I, *Physical Adsorption*, Princeton University Press (1943).
22. Titoff, A., *Z. Phys. Chem.*, **74**, 641 (1910).
23. Valenzuela, D.P., and A.L. Myers, *Adsorption Equilibrium Data Handbook*, Prentice-Hall, Englewood Cliffs, NJ (1989).
24. Ray, G.C., and E.O. Box, Jr., *Ind. Eng. Chem.*, **42**, 1315–1318 (1950).
25. Yang, R.T., *Gas Separation by Adsorption Processes*, Butterworths, Boston (1987).
26. Freundlich, H., *Z. Phys. Chem.*, **73**, 385–423 (1910).
27. Mantell, C.L., *Adsorption*, 2nd ed., McGraw-Hill, New York, p. 25 (1951).
28. Langmuir, J., *J. Am. Chem. Soc.*, **37**, 1139–1167 (1915).
29. Markham, E.C., and A.F. Benton, *J. Am. Chem. Soc.*, **53**, 497–507 (1931).
30. Yon, C.M., and P.H. Turnock, *AIChE Symp. Series*, **67** (117), 75–83 (1971).
31. Broughton, D.B., *Ind. Eng. Chem.*, **40**, 1506–1508 (1948).
32. Myers, A.L., and J.M. Prausnitz, *AIChE J.*, **11**, 121–127 (1965).
33. Miller, G.W., K.S. Knaebel, and K.G. Ikels, *AIChE J.*, **33**, 194–201 (1987).

34. Ritter, J.A., and R.T. Yang, *Ind. Eng. Chem. Res.*, **26**, 1679–1686 (1987).
35. Ruthven, D.M., and F. Wong, *Ind. Eng. Chem. Fundam.*, **24**, 27-32 (1985).
36. Kipling, J.J., *Adsorption from Solutions of Nonelectrolytes*, Academic Press, London (1965).
37. Radke, C.J., and J.M. Prausnitz, *AIChE J.*, **18**, 761–768 (1972).
38. Anderson, R.E., "Ion-Exchange Separations," in *Handbook of Separation Techniques for Chemical Engineers*, 2nd ed., P.A. Schweitzer, Ed., McGraw-Hill, New York (1988).
39. Bonner, O.D., and L.L. Smith, *J. Phys. Chem.* **61**, 326–329 (1957).
40. Bauman, W.C., and J. Eichhorn, *J. Am. Chem. Soc.*, **69**, 2830–2836 (1947).
41. Bauman, W.C., J.R. Skidmore, and R.H. Osmun, *Ind. Eng. Chem.*, **40**, 1350–1355 (1948).
42. Subba Rao, H.C., and M.M. David, *AIChE Journal*, **3**, 187–190 (1957).
43. Selke, W.A., and H. Bliss, *Chem. Eng. Prog.*, **46**, 509–516 (1950).
44. Selke, W.A., and H. Bliss, *Chem. Eng. Prog.*, **47**, 529–533 (1951).
45. Myers, A.L., and S. Byington, in *Ion Exchange: Science and Technology*, A.E. Rodrigues, Ed., Martinus Nijhoff, Boston, pp. 119–145.
46. Ranz, W.E., and W.R. Marshall, Jr., *Chem. Eng. Prog.*, **48**, 141–146 (1952).
47. Ranz, W.E., and W.R. Marshall, Jr., *Chem. Eng. Prog.*, **48**, 173–180 (1952).
48. Ranz, W.E., *Chem. Eng. Prog.*, **48**, 247–253 (1952).
49. Gamson, B.W., G. Thodos, and O.A. Hougen, *Trans. AIChE*, **39**, 1–35 (1943).
50. Chilton, T.H., and A.P. Colburn, *Ind. Eng. Chem.*, **26**, 1183–1187 (1934).
51. Sen Gupta, A., and G. Thodos, *AIChE J.*, **9**, 751–754 (1963).
52. Petrovic, L.J., and G. Thodos, *Ind. Eng. Chem. Fundamentals*, **7**, 274–280 (1968).
53. Dwivedi, P.N., and S.N. Upadhay, *Ind. Eng. Chem. Process Des. Dev.*, **16**, 157–165 (1977).
54. Wakao, N., and T. Funazkri, *Chem. Eng. Sci.*, **33**, 1375–1384 (1978).
55. Kunii, D., and O. Levenspiel, *Fluidization Engineering*, 2nd ed., Butterworth-Heinemann, Boston, Chap. 3 (1991).
56. Thiele, E.W., *Ind. Eng. Chem.*, **31**, 916–920 (1939).
57. Wheeler, A., *Advances in Catalysis*, Vol. 3, Academic Press, New York, pp. 249–327 (1951).
58. Schneider, P., and J.M. Smith, *AIChE J.*, **14**, 886–895 (1968).
59. Riekert, L., *AIChE J.*, **31**, 863–864 (1985).
60. Sladek, K.J., E.R. Gilliland, and R.F. Baddour, *Ind. Eng. Chem. Fundam.*, **13**, 100–105 (1974).
61. Kapoor, A., R.T. Yang, and C. Wong, "Surface Diffusion," *Catalyst Reviews*, **31**, 129–214 (1989).
62. Helfferich, F., *Ion Exchange*, McGraw-Hill, New York (1962).
63. Soldano, B.A., *Ann. NY Acad. Sci.*, **57**, 116–124 (1953).
64. Keller, G.E., "Separations: New Directions for an Old Field," *AIChE Monograph Series*, **83** (17) (1987).
65. Ruthven, D.M., and C.B. Ching, *Chem. Eng. Sci.*, **44**, 1011–1038 (1989).
66. Berg, C., *Trans. AIChE*, **42**, 665–680 (1946).
67. Broughton, D.B., *Chem. Eng. Progress*, **64** (8), 60–65 (1968).

68. Gembicki, S.A., A.R. Oroskar, and J.A. Johnson, in *Encyclopedia of Chemical Technology,* 4th edition, Vol. 1, John Wiley, New York, pp. 573–600 (1991).

69. Higgins, I.R., and J.T. Roberts, *Chem. Engr. Prog. Symp. Ser.,* **50** (14), 87–92 (1954).

70. Himsley, A., Canadian Patent 980,467 (Dec. 23, 1975).

71. Ganetsos, G., and P.E. Barker, Ed., *Preparative and Production Scale Chromatography,* Marcel Dekker, New York (1993).

72. Martin, A.J.P., *Disc. Faraday Soc.,* **7**, 332 (1949).

73. DeVault, D., *J. Am. Chem Soc.,* **65**, 532–540 (1943).

74. Glueckauf, E., *Trans. Faraday Soc.,* **51**, 1540–1551 (1955).

75. Glueckauf, F., and J.E. Coates, *J. Chem. Soc.,* 1315–1321 (1947).

76. Liaw, C.H., J.S.P. Wang, R.A. Greenkorn, and K.C. Chao, *AIChE J.,* **25**, 376–381 (1979).

77. Klinkenberg, A., *Ind. Eng. Chem.,* **46**, 2285–2289 (1954).

78. Anzelius, A., Z., *Angew. Math u. Mech.,* **6**, 291–294 (1926).

79. Klinkenberg, A., *Ind. Eng. Chem.,* **40**, 1992–1994 (1948).

80. Cooney, D.O., and E.N. Lightfoot, *IEC Fundamentals,* **4**, 233–236 (1965).

81. Sircar, S., and K. Kumar, *Ind. Eng. Chem. Process Des. Dev.,* **22**, 271–280 (1983).

82. Cooney, D.O., *Chem. Eng. Comm.,* **91**, 1–9 (1990).

83. Collins, J.J., *Chem. Eng. Prog. Symp. Ser.* **63** (74), 31–35 (1967).

84. Wong, Y.W., and J.L. Niedzwiecki, *AIChE Symposium Series,* **78** (219), 120–127 (1982).

85. Liskovets, O.A., *Differential Equations* (a translation of *Differentsial'nye Uravneniya*) **1**, 1308–1323 (1965).

86. Schiesser, W.E., *The Numerical Method of Lines Integration of Partial Differential Equations,* Academic Press, San Diego (1991).

87. Schiesser, W.E., *Computational Mathematics in Engineering and Applied Science,* CRC Press, Boca Raton, FL (1994).

88. Press, W.H., S.A. Teukolsky, W.T. Vetterling, and B.P. Flannery, *Numerical Recipes in FORTRAN,* 2nd ed., Cambridge University Press, Cambridge (1992).

89. Gear, C.W., *Numerical Initial Value Problems in Ordinary Differential Equations,* Prentice-Hall, Englewood Cliffs, NJ (1971).

90. Byrne, G.D., and A.C. Hindmarsh, *J. Comput. Phys.,* **70**, 1–62 (1987).

91. Ritter, J.A., and R.T. Yang, *Ind. Eng. Chem. Res.,* **30**, 1023–1032 (1991).

92. Ruthven, D.M., S. Farooq, and K.S. Knaebel, *Pressure-Swing Adsorption,* VCH, New York (1994).

93. Mutasim, Z.Z., and J.H. Bowen, *Trans. I. Chem. E.,* **69**, Part A, 108–118 (March 1991).

94. Eagleton, L.C., and H. Bliss, *Chem. Eng. Progress,* **49**, 543–548 (1953).

95. Wankat, P.C., *Rate-Controlled Separations,* Elsevier Applied Science, New York (1990).

96. Carta, G., *Chem. Eng. Sci.,* **43**, 2877–2883 (1988).

EXERCISES

Section 15.1

15.1 Porous particles of activated alumina have a BET surface area of 310 m^2/g, a particle porosity of 0.48, and a particle density of 1.30 g/cm^3. Determine: (a) specific pore volume in cm^3/g, (b) true solid density, g/cm^3, and (c) approximate pore diameter in angstroms from (15-2).

15.2 Carbon molecular sieves are available in two forms from a Japanese manufacturer:

	Form A	Form B
Pore volume, cm^3/g	0.18	0.38
Average pore diameter	5 Å	2.0 μm

Estimate the surface area of each form.

15.3 Representative properties of small-pore silica gel are as follows: pore diameter = 24 Å; particle porosity = 0.47; particle density = 1.09 g/cm^3; and specific surface area = 800 m^2/g

(a) Are these values reasonably consistent? (b) If the adsorption capacity for water vapor at 25°C and 6 mmHg partial pressure is 18% by weight, what fraction of a monolayer is adsorbed?

15.4 The following data were obtained in a BET apparatus for adsorption equilibrium of nitrogen on silica gel (SG) at

−195.8°C. Estimate the specific surface area in m^2/g of silica gel. How does your value compare with that in Table 15.2?

N$_2$ Partial Pressure, torr	Volume of N$_2$ Adsorbed in cm^3 (0°C, 1 atm) per Gram SG
6.0	6.1
24.8	12.7
140.3	17.0
230.3	19.7
285.1	21.5
320.3	23.0
430	27.7
505	33.5

15.5 Estimate the maximum ion-exchange capacity in meq/g resin for an ion-exchange resin made from 8 wt% divinylbenzene and 92 wt% styrene.

Section 15.2

15.6 Shen and Smith [*Ind. Eng. Chem. Fundam.,* **7**, 100–105 (1968)] measured equilibrium adsorption isotherms at four different temperatures for pure benzene vapor on silica gel, having the following properties: surface area = 832 m^2/g, pore volume = 0.43 cm^3/g, particle density = 1.13 g/cm^3, and average pore diameter = 22 Å.

The adsorption data are as follows:

Partial Pressure of Benzene, atm	Moles Adsorbed/g Gel × 10^5			
	70°C	90°C	110°C	130°C
5.0×10^{-4}	14.0	6.7	2.6	1.13
1.0×10^{-3}	22.0	11.2	4.5	2.0
2.0×10^{-3}	34.0	18.0	7.8	3.9
5.0×10^{-3}	68.0	33.0	17.0	8.6
1.0×10^{-2}	88.0	51.0	27.0	16.0
2.0×10^{-2}	—	78.0	42.0	26.0

(a) For each temperature, obtain a best fit of the data to (1) linear, (2) Freundich, and (3) Langmuir isotherms. Which isotherm(s), if any, fit the data reasonable well?
(b) Do the data represent less than a monolayer of adsorption?
(c) From the data, estimate the heat of adsorption. How does this value compare to the heat of vaporization (condensation) of benzene?

15.7 The separation of propane and propylene is accomplished by distillation, but at the expense of more than 100 trays and a reflux ratio of greater than 10. Consequently, the use of adsorption has been investigated in a number of studies. Jarvelin and Fair [*Ind. Eng. Chem. Research,* **32**, 2201–2207 (1993)] measured adsorption equilibrium data at 25°C for three different zeolite molecular sieves (ZMSs) and activated carbon. The data were fitted to the Langmuir isotherm with the following results:

Adsorbent	Sorbate	q_m	K
ZMS 4A	C_3	0.226	9.770
	$C_3^=$	2.092	95.096
ZMS 5A	C_3	1.919	100.223
	$C_3^=$	2.436	147.260
ZMS 13X	C_3	2.130	55.412
	$C_3^=$	2.680	100.000
Activated carbon	C_3	4.239	58.458
	$C_3^=$	4.889	34.915

where q and q_m are in mmol/g and p is in bar.
(a) Which component is most strongly adsorbed by each of the adsorbents? (b) Which adsorbent has the greatest adsorption capacity? (c) Which adsorbent has the greatest selectivity? (d) Based on equilibrium considerations, which adsorbent is best for the separation?

15.8 Ruthven and Kaul [*Ind. Eng. Chem. Res.,* **32**, 2047–2052 (1993)] measured adsorption isotherms for a series of gaseous aromatic hydrocarbons on well-defined crystals of NaX zeolite over ranges of temperature and pressure. For 1,2,3,5-tetramethylbenzene at 547 K, the following equilibrium data were obtained with a vacuum microbalance:

q, wt%	7.0	9.1	10.3	10.8	11.1	11.5
p, torr	0.012	0.027	0.043	0.070	0.094	0.147

Obtain a best fit of the data to the linear, Freundlich, and Langmuir isotherms, with q in mol/g and pressure in atm. Which isotherm gives the best fit?

15.9 Lewis, Gilliland, Chertow, and Hoffman [*J. Am. Chem. Soc.,* **72**, 1153–1157 (1950)] measured adsorption equilibria for pure propane, pure propylene, and binary mixtures thereof, on activated carbon and silica gel. Adsorbate capacity was high on carbon, but selectivity was poor. Selectivity was high on silica gel, but capacity was low. For silica gel (751 m^2/g), the following pure component data were obtained at 25°C:

Propane		Propylene	
P, torr	q, mmol/g	P, torr	q, mmol/g
11.1	0.0564	34.2	0.3738
25.0	0.1252	71.4	0.7227
43.5	0.1980	91.6	0.7472
71.4	0.2986	194.3	1.129
100.0	0.3850	198.3	1.168
158.9	0.5441	271.5	1.401
227.5	0.7020	353.2	1.562
304.2	0.843	550.7	1.918
387.0	1.010	555.2	1.928
468.0	1.138	760.6	2.184
569.0	1.288		
677.8	1.434		
775.0	1.562		

The following mixture data were measured at 25°C, over a pressure range of 752–773 torr:

Total Pressure, torr	Millimoles of Mixture Adsorbed/g	y_{C_3}, Mole Fraction in Gas Phase	x_{C_3}, Mole Fraction in Adsorbate
769.2	2.197	0.2445	0.1078
760.9	2.013	0.299	0.2576
767.8	2.052	0.4040	0.2956
761.0	2.041	0.530	0.2816
753.6	1.963	0.5333	0.3655
766.3	1.967	0.5356	0.3120
754.0	1.974	0.6140	0.3591
753.6	1.851	0.6220	0.5550
754.0	1.701	0.6252	0.7007
760.0	1.686	0.7480	0.723
—	2.180	0.671	0.096
760.0	1.993	0.8964	0.253
760.0	1.426	0.921	0.401

(a) Fit the pure component data to Freundlich and Langmuir isotherms. Which gives the best fit? Which component is most strongly adsorbed?

(b) Use the results of the Langmuir fits in part (a) to predict the binary mixture adsorption using the extended Langmuir equation, (15-32). Are the predictions adequate?

(c) Ignoring the pure component data, fit the binary mixture data to the extended Langmuir equation, (15-32). Is the fit better than that obtained in part (b)?

(d) Ignoring the pure component data, fit the binary mixture data to the extended Langmuir–Freundlich equation, (15-33). Is the fit adequate? Is the fit better than that in part (c)?

(e) For the binary mixture data, compute the relative selectivity,

$$\alpha_{C_3, C_3^=} = y_{C_3}(1 - x_{C_3})/[x_{C_3}(1 - y_{C_3})]$$

for each condition. Does α vary widely or is the assumption of constant α reasonable?

15.10 In Example 15.6, pure-component liquid-phase adsorption data are used with the extended Langmuir isotherm to predict a binary-solute data point. Use the following mixture data to obtain the best fit to an extended Langmuir–Freundlich isotherm of the form

$$q_i = \frac{(q_0)_i k_i c_i^{1/n_i}}{1 + \sum_j k_j c_j^{1/n_j}} \qquad \text{(1)}$$

Data for binary-mixture adsorption on activated carbon (1000 m²/g) at 25°C for acetone (1) and propionitrile (2) are as follows:

Solution Concentration, mol/L		Loading, mmol/g	
c_1	c_2	q_1	q_2
5.52E − 5	7.46E − 5	0.0192	0.0199
6.14E − 5	7.71E − 5	0.0191	0.0198
1.06E − 4	1.35E − 4	0.0308	0.0320
1.12E − 4	1.46E − 4	0.0307	0.0319
3.03E − 4	2.32E − 3	0.0378	0.263
3.17E − 4	2.34E − 3	0.0378	0.264
3.25E − 4	3.89E − 4	0.0644	0.0672
1.42E − 3	1.58E − 3	0.161	0.169
1.42E − 3	1.61E − 3	0.161	0.169
1.43E − 3	1.60E − 3	0.161	0.169
2.09E − 3	3.84E − 4	0.250	0.0390
2.17E − 3	3.85E − 4	0.251	0.0392
4.99E − 3	5.24E − 3	0.291	0.307
5.06E − 3	5.31E − 3	0.288	0.305
7.41E − 3	2.42E − 2	0.237	0.900
7.52E − 3	2.47E − 2	0.236	0.896
2.79E − 2	7.59E − 3	0.802	0.251
4.00E − 2	3.44E − 2	0.715	0.822
4.02E − 2	3.42E − 2	0.717	0.834

15.11 Sircar and Myers [*J. Phys. Chem.,* **74**, 2828–2835 (1970)] measured liquid-phase adsorption at 30°C for a binary mixture of cyclohexane (1) and ethyl alcohol (2) on activated carbon. Assuming no adsorption of ethyl alcohol, they used (15-34) to obtain the following results:

x_1	q_1^e, mmol/g	x_1	q_1^e, mmol/g
0.042	0.295	0.440	0.065
0.051	0.485	0.470	0.000
0.072	0.517	0.521	−0.129
0.148	0.586	0.537	−0.362
0.160	0.669	0.610	−0.643
0.213	0.661	0.756	−1.230
0.216	0.583	0.848	−1.310
0.249	0.595	0.893	−1.180
0.286	0.532	0.920	−1.230
0.341	0.383	0.953	−0.996
0.391	0.192	0.974	−0.470

(a) Plot the data as q_1^e against x_1. Explain the shape of the curve.

(b) In what regions of concentration could the Freundlich isotherm be fitted to the data? Make the fits.

15.12 Both the adsorptive removal of small amounts of toluene from water and small amounts of water from toluene are important in the process industries. Activated carbon is particularly effective for removing soluble organic compounds (SOCs) from water. Activated alumina is effective for removing soluble water from toluene. Fit each of the following two sets of equilibrium data for 25°C to both the Langmuir and Freundlich isotherms. For each case, which isotherm provides the better fit? Could a linear isotherm be used?

Toluene (in Water) Activated Carbon		Water (in Toluene) Activated Alumina	
c, mg/L	q, mg/g	c, ppm (by Weight)	q, g/100g
0.01	12.5	25	1.9
0.02	17.1	50	3.1
0.05	23.5	75	4.2
0.1	30.3	100	5.1
0.2	39.2	150	6.5
0.5	54.5	200	8.2
1	90.1	250	9.5
2	70.2	300	10.9
5	125.5	350	12.1
10	165	400	13.3

15.13 Derive (15-44). Use this equation to solve the following problem. Sulfate ion is to be removed from 60 L of water by exchanging it with chloride ion on 1 L of a strong-base resin with relative molar selectives as listed in Table 15.6 and an ion-exchange capacity of 1.2 eq/L of resin. The water to be treated has a sulfate ion concentration of 0.018 eq/L and a chloride ion concentration of 0.002 eq/L. Following the attainment of equilibrium ion exchange, the treated water will be removed and the resin will be regenerated with 30 L of 10 wt% aqueous NaCl.

(a) Write the ion-exchange reaction.

(b) Determine the value of $K_{SO_4^{2-},Cl^-}$.

(c) Calculate equilibrium concentrations $c_{SO_4^{2-}}$, c_{Cl^-}, $q_{SO_4^{2-}}$, and q_{Cl^-} in eq/L for the initial ion-exchange step.

(d) Calculate the concentration of Cl^- in eq/L for the regenerating solution.

(e) Calculate $c_{SO_4^{2-}}$, c_{Cl^-}, $q_{SO_4^{2-}}$, and q_{Cl^-} upon reaching equilibrium in the regeneration step.

(f) Are the separations sufficiently selective?

15.14 Silver ion in methanol was exchanged with sodium ion using Dowex 50 cross-linked with 8% divinylbenzene by Gable and Stroebel [*J. Phys. Chem.,* **60**, 513–517 (1956)]. The molar selectivity coefficient was found to vary somewhat with the equivalent fraction of Na^+ in the resin as follows:

x_{Na^+}	0.1	0.3	0.5	0.7	0.9
K_{Ag^+,Na^+}	11.2	11.9	12.3	14.1	17.0

If the wet capacity of the resin is 2.5 eq/L and the resin is initially saturated with Na^+, calculate the equilibrium equivalent fractions if 50 L of 0.05 M Ag^+ in methanol is treated with 1 L of wet resin.

15.15 Ion exclusion is a process that uses ion-exchange resins to separate nonionic organic compounds from ionic species contained in a polar solvent, usually water. The resin is presaturated with the same ions as in the solution, thus eliminating ion exchange. However, in the presence of the polar solvent, resins undergo considerable swelling by absorbing the solvent. Experiments have shown that a nonionic solute will distribute between the solution outside the resin and the solution within the resin, while the ions can only exchange.

A feed solution of 1,000 kg contains 6 wt% NaCl, 35 wt% glycerol, and 47 wt% water. This solution is to be treated with Dowex 50 ion-exchange resin in the sodium form, after prewetting with water, to recover 75% of the glycerol. The following data for the glycerol distribution coefficient,

$$K_d = \frac{\text{mass fraction in solution inside resin}}{\text{mass fraction in solution outside resin}}$$

were reported by Asher and Simpson [*J. Phys. Chem.,* **60**, 518-521 (1956)]:

Mass Fraction Glycerol in Solution Outside Resin	K_d	
	6 wt% NaCl	12 wt% NaCl
0.10	0.75	0.91
0.20	0.80	0.93
0.30	0.83	0.95
0.40	0.85	0.97

If the prewetted resin contains 40 wt% water, determine the kilograms of resin (dry basis) required.

Section 15.3

15.16 Benzene vapor in an air stream is adsorbed in a fixed bed of 4×6 mesh silica gel packed to an external void fraction of 0.5. The bed is 2 feet in inside diameter and the air flow rate is 25 lb/min (benzene-free basis). At a location in the bed where the pressure is 1 atm, the temperature is 70°F, and the bulk mole fraction of benzene is 0.005, estimate the external gas-to-particle mass-transfer and heat-transfer coefficients.

15.17 Water vapor in an air stream is to be adsorbed in a 12.06-cm inside diameter column packed with 3.3-mm-diameter Alcoa Γ-200 activated alumina beads with an external porosity of 0.442. At a location in the bed where the pressure is 653.3 kPa, the temperature is 21°C, the gas flow rate is 1.327 kg/min, and the dew-point temperature is 11.2°C, estimate the external gas-to-particle mass-transfer and heat-transfer coefficients.

15.18 For the conditions of Example 15.7, estimate the effective diffusivity of acetone vapor in the pores of activated carbon with the following properties: particle density = 0.85 g/cm³, particle porosity = 0.48, average pore diameter = 25 Å, and tortuosity = 2.75
Consider both bulk and Knudsen diffusion, but ignore surface diffusion.

15.19 For the conditions of Exercise 15.16, estimate the effective diffusivity of benzene vapor in the pores of silica gel with the following properties: particle density = 1.15 g/cm³, particle porosity = 0.48, average pore diameter = 30 Å, and tortuosity = 3.2
Consider all mechanisms of diffusion. The adsorption equilibrium constant is given in Example 15.10, and the differential heat of adsorption is −11,000 cal/mol.

15.20 For the conditions of Exercise 15.17, estimate the effective diffusivity of water vapor in the pores of activated alumina with the following properties: particle density = 1.38 g/cm³, particle porosity = 0.52, average pore diameter = 60 Å, and tortuosity = 2.3
Consider all mechanisms of diffusion except surface diffusion.

Section 15.4

15.21 Adsorption with activated carbon, made from bituminous coal, of soluble organic compounds (SOCs) to purify surface and ground water is a proven technology, as discussed by Stenzel [*Chem. Eng. Prog.*, **89** (4), 36–43 (1993)]. The less soluble organic compounds, such as chlorinated organic solvents and aromatic solvents, are the more strongly adsorbed. Water containing 3.3 mg/L of trichloroethylene (TCE) is to be treated with activated carbon to obtain an effluent with only 0.01 mg TCE/L. At 25°C, adsorption equilibrium data for TCE on activated carbon are correlated with the following Freundlich equation:

$$q = 67 \ c^{0.564} \qquad (1)$$

where:

q = mg TCE/g carbon and c = mg TCE/L solution

The TCE is to be removed by slurry adsorption using a powdered form of the activated carbon, with an average particle diameter of 1.5 mm. In the absence of any laboratory data on mass-transfer rates, assume that the rate of adsorption for the small particles is controlled by external mass transfer with a Sherwood number of 30. Particle surface area is 5 m²/kg. The molecular diffusivity of TCE in low concentrations in water at 25°C may be determined from the Wilke–Chang equation.
(a) Determine the minimum amount of adsorbent needed.
(b) For operation in the batch mode with twice the minimum amount of adsorbent, determine the time to reduce the TCE content to the desired value.
(c) For operation in the continuous mode using twice the minimum amount of adsorbent, determine the required residence time.
(d) For operation in the semicontinuous mode at a feed rate of 50 gpm and for a liquid residence time equal to 1.5 times that computed in part (c), determine the amount of activated carbon to give a reasonable vol% solids in the tank and a run time of not less than 10 times the liquid residence time.

15.22 Repeat Exercise 15.21 for water containing 0.324 mg/L of benzene (B) and 0.630 mg/L of m-xylene (X)
Adsorption isotherms at 25°C for these low concentrations are essentially independent and are given by

$$q_B = 32 \ c_B^{0.428} \qquad (1)$$

$$q_X = 125 \ c_X^{0.333} \qquad (2)$$

The feed concentrations of the SOCs in the feed are to be reduced to 0.002 mg/L each.

15.23 Repeat Exercise 15.21 for water containing 0.223 mg/L chloroform, whose concentration is to be reduced to 0.010 mg/L. The adsorption isotherm at 25°C is given by

$$q = 10 \ c^{0.564} \qquad (1)$$

15.24 Three fixed-bed adsorbers containing 10,000 lb of granules of activated carbon (ρ_b = 30 lb/ft³) each are to be used to treat 250 gpm of water containing 4.6 mg/L of 1,2-dichloroethane (D) to reduce the concentration to less than 0.001 mg/L. Each carbon bed has a height equal to twice the diameter. Two beds are to be placed in series so that when bed 1 (the lead bed) becomes saturated with D at the feed concentration, that bed is removed. Bed 2 (the trailing bed), which is partially saturated at this point, depending upon the width of the MTZ, becomes the lead bed, and previously idle bed 3 takes the place of bed 2. While bed 1 is off-line, its spent carbon is removed and replaced with fresh carbon. The spent carbon is incinerated. The equilibrium adsorption isotherm for D is given by $q = 8 \ c^{0.57}$, where q is in mg/g and c is in mg/L. Once the cycle is established, how often must the carbon in a bed be replaced? What is the maximum width of the *MTZ* that will allow saturated loading of the lead bed?

15.25 The fixed-bed adsorber series arrangement of Exercise 15.24 is to be used to treat 250 gpm of water containing 0.185 mg/L of benzene (B) and 0.583 mg/L of m-xylene (X). However, because the two solutes may have considerably different breakthrough times, more than two operating beds in series may be needed. The adsorption isotherms are given in Exercise 15.22, where q is in mg/g and c is in mg/L. From laboratory measurements, the widths of the mass-transfer zones are estimated to be MTZ_B = 2.5 ft and MTZ_X = 4.8 ft. Once the cycle is established, how often must the carbon in the bed be replaced?

15.26 Air at 80°F, 1 atm, 80% relative humidity, and a superficial velocity of 100 ft/min passes through a 5-ft-high bed of 2.8-mm-diameter spherical particles of silica gel (ρ_b = 39 lb/ft³). The adsorption equilibrium isotherm at 80°F is given by

$$q_{H_2O} = 15.9 \ p_{H_2O} \qquad (1)$$

where q is in lb H_2O/lb gel and p is in atm. The overall mass-transfer coefficient can be estimated from (15-106), using an effective diffusivity of 0.05 cm²/s and with k_c estimated from (15-65). Using the approximate concentration-profile equations of Klinkenberg, compute a set of breakthrough curves and determine the time when the humidity of the exiting air reaches 0.0009 lb H_2O/lb dry air. Assume isothermal and isobaric operation. Compare the time to breakthrough with the time for the equilibrium model. At breakthrough, what is the approximate width of the mass-transfer zone. What is the average loading of the bed at breakthrough?

15.27 A train of four 55-gallon cannisters of activated carbon is to be used to reduce the nitroglycerine (NG) content of 400 gph of wastewater from 2,000 ppm by weight to less than 1 ppm. Each cannister has a diameter of 2 ft and holds 200 lb activated carbon (ρ_b = 32 lb/ft³). Each cannister is equipped with a liquid-flow distributor to promote plug flow through the bed of carbon. The effluent from the first cannister is monitored so that when a 1 ppm threshold of NG is reached, that cannister is removed from the train and a fresh cannister is added to the end of the train. The spent carbon is mixed with coal for use as a fuel in a coal-fired power plant at the process site. Using the following pilot-plant data, estimate how many cannisters are needed each month and the monthly cannister cost at $700 per cannister.

Pilot-plant data:

Tests with the same 55-gallon cannister to be used in the commercial process; water flow rate = 10 gpm; NG content in feed = 1,020 ppm by weight.

Breakthrough correlation:

$t_B = 3.90 \, L - 2.05$, where t_B = time, h, at breakthrough of the 1 ppm threshold and L = bed depth of carbon in feet.

15.28 Air at a flow rate of 12,000 scfm (60°F, 1 atm) and containing 0.5 mol% ethyl acetate (EA) and no water vapor is to be treated with activated carbon (C) ($\rho_b = 30 \, \text{lb/ft}^3$) with an equivalent particle diameter of 0.011 ft in a fixed-bed adsorber to remove the ethyl acetate, which will be subsequently stripped from the carbon by steam at 230°F. Based on the following data, determine the diameter and height of the carbon bed, assuming adsorption at 100°F and 1 atm and a time-to-breakthrough of 8 h with a superficial gas velocity of 60 ft/min. If the bed height-to-diameter is unreasonable, what change in design basis would you suggest?

Adsorption isotherm data (100°F) for EA:

p^{EA}, atm	q, lb EA/lb C	p^{EA}, atm	q, lb EA/lb C
0.0002	0.125	0.0020	0.227
0.0005	0.164	0.0050	0.270
0.0010	0.195	0.0100	0.304

Breakthrough data at 100°F and 1 atm for EA in air at a gas superficial velocity of 60 ft/min in a 2-ft dry bed:

Mole Fraction EA in Effluent	Time, Min	Mole Fraction EA in Effluent	Time, Min
0.00005	60	0.00100	95
0.00010	66	0.00250	120
0.00025	75	0.00475	160
0.00050	84		

15.29 In Examples 15.10 and 15.12, benzene is adsorbed from air at 70°F in a 6-ft-high bed of silica gel and then stripped with air at 145°F. If the bed height is changed to 30 ft, the following data are obtained for breakthrough at 641 minutes for the adsorption step:

z, ft	$\phi = c/c_F$	$\psi = \overline{q}/q_F^*$	z, ft	$\phi = c/c_F$	$\psi = \overline{q}/q_F^*$
0–12	1.000	1.000	22	0.825	0.808
13	1.000	1.000	23	0.722	0.701
14	1.000	1.000	24	0.599	0.575
15	1.000	1.000	25	0.468	0.444
16	0.999	0.999	26	0.343	0.321
17	0.997	0.997	27	0.235	0.217
18	0.992	0.990	28	0.150	0.137
19	0.978	0.975	29	0.090	0.081
20	0.951	0.944	30	0.050	0.044
21	0.901	0.890			

If the bed is regenerated isothermally with pure air at 1 atm and 145°F, and the desorption of benzene during the heat-up period is neglected, determine the loading, $\overline{q}$, profile at a time sufficient to remove 90% of the benzene from the bed if an interstitial pure air velocity of 98.5 ft/min is used. Values of k and K at 145°F are given in Example 15.12.

15.30 Use the method of lines with a five-point biased upwind finite-difference approximation and a stiff integrator to perform PSA cycle calculations that approach the cyclic steady state for the data and design basis in Example 15.13, starting from: (a) a clean bed, and (b) a bed saturated with the feed. Are the two cyclic steady states essentially the same?

15.31 Solve Example 15.13 for $P_L = 0.12$ atm and an interstitial velocity during desorption that corresponds to the use of 44.5% of the product gas from the adsorption step.

15.32 For the separation of air by PSA, adsorption of both O_2 and N_2 must be considered. Develop a model for this case taking into account two species mass balances, overall mass balance, two species mass-transfer rates, and two extended Langmuir isotherms. Each of the two main steps can be isothermal and isobaric. Can your PDE equations still be solved by the method of lines with a stiff integrator? If so, outline a procedure for doing it.

15.33 Two adsorption-based separation processes not considered in this chapter because of lack of significant commercial application are (1) parametric pumping, first conceived by R.H. Wilhelm in the early 1960s, and (2) cycling zone adsorption, invented by R.L. Pigford and co-workers in the late 1960s. The status of and future for these two processes was assessed by Sweed in 1984 [AIChE Symp. Series, **80** (233), 44–53 (1984)]. Describe in detail each of these processes. Can either be used for both gas-phase and liquid-phase adsorption?

15.34 A gas mixture containing 55 mol% propane and 45 mol% propylene is to be separated into products containing 10 and 90 mol% propane by adsorption in a continuous, countercurrent adsorption system operating at 25°C and 1 atm. The adsorbent is silica gel, for which equilibrium data are given in Exercise 15.9. Determine by the McCabe–Thiele method: (a) the adsorbent flow rate per 1,000 m³ of feed gas at 25°C and 1 atm if 1.2 times the minimum rate is used, and (b) the number of theoretical stages required.

15.35 Repeat Example 15.15, except for a feed containing 400 ppm (by weight) of $CaCl_2$ and 50 ppm of NaCl.

15.36 An aqueous solution, buffered to a pH of 3.4 by sodium citrate and containing 20 mol/m^3 each of glutamic acid, glycine, and valine, is separated in a chromatographic column, packed with Dowex 50W-X8 in the sodium form to a depth of 470 mm. The resin is 0.07 mm in diameter and packs to a bed void fraction of 0.374. Equilibrium data follow Henry's law , as in Example 15.16, with the following dimensionless constants, determined by Takahashi and Goto [*J. Chem. Eng. Japan,* **24,** 121–123 (1991)]:

Solute	K
Glutamic acid	1.18
Glycine	1.74
Valine	2.64

The superficial solution velocity is 0.025 cm/s. Using equilibrium theory, what pulse duration can be used to achieve complete separation? How long must the elution step be before the second pulse can begin?

15.37 Repeat Exercise 15.36, but using Carta's equation to account for mass transfer with the following effective diffusivities:

Solute	D_e, cm^2/s
Glutamic acid	1.94×10^{-7}
Glycine	4.07×10^{-7}
Valine	3.58×10^{-7}

Assume $k_c = 1.5 \times 10^{-3}$ cm/s. Establish a cycle of feed pulses and elution periods that will give the desired separation.

Index